# Eighth International Conference on Environmental Science and Technology (ICEST 2016)

Houston, Texas, USA
6-10 June 2016

Volume 1 of 2

**Editors:**

**George A. Sorial** **Jihua Hong**

ISBN: 978-1-5108-3972-4

**Printed from e-media with permission by:**

Curran Associates, Inc.
57 Morehouse Lane
Red Hook, NY 12571

**Some format issues inherent in the e-media version may also appear in this print version.**



Printed by Curran Associates, Inc. (2017)

For permission requests, please contact American Academy of Sciences
at the address below.

American Academy of Sciences
9720 Town Park Drive, Suite 18
Houston TX 77036
USA

Phone: 504-486-6332

info@AASci.org

**Additional copies of this publication are available from:**

Curran Associates, Inc.
57 Morehouse Lane
Red Hook, NY 12571 USA
Phone: 845-758-0400
Fax: 845-758-2633
Email: curran@proceedings.com
Web: www.proceedings.com

# TABLE OF CONTENTS

***Groundwater*** 



***Drinking Water***



***Water Quality Assessment/Management***

119



### *Nitrogen-Phosphorus Wastewater Treatment / Sludge Treatment*

134



142

### *Sludge Treatment*



### *Municipal Wastewater Biotreatment*

***Reactions and Degradation of Wastewater Contaminants***



***Nanotechnology Applications***

**AIR POLLUTION AND AIR QUALITY CONTROL**

***Aerosol***



**Air Quality Assessment**

### *Bio-response and Ecotoxicology*



### *Bioavailability and Bio-accumulation*

***Microbiology and Microbial Degradation***

# INTRODUCTION

The Eighth International Conference on Environmental Science and Technology 2016 was held in Houston, Texas, USA, June 6-10, 2016. The Program included 16 sections, containing 60 sessions with approximately 600 platform and poster presentations. This conference series strives to provide a platform for an extremely diverse group of environmental topics for engineers and scientists from around the world.

Authors of the presentations accepted for the program were invited to submit their papers to the Conference Organizing Committee. More than 120 papers were received and then reviewed by the editors, session chairs, and the members of the Scientific/Technical Committee of the conference. Those papers and abstracts accepted for publication were assembled into two volumes.

Sections are arranged basically according to their order listed in the original program except the sessions entitled Land (Soil, Waste Solid) Pollution and Remediation. This exception was made to balance the length of the two volumes. The conference also consisted of having a plenary session with four speakers from different universities at the United States and other countries.

Environmental Science and Technology 2016 (Volome 1) contains the following sections:

- Plenary Presentation
- Water Pollution and Water Quality Control
- Air Pollution and Air Quality Control
- Bio-Assessment and Toxicology

Sections included in Environmental Science and Technology 2016 (Volume 2) are as follows:

- Land (Soil, Waste Solid) Pollution and Remediation
- Ecosystem Assessment and Restoration
- Bio-Assessment and Toxicology
- Wetlands and Sediments
- Global Warming and Climate Change
- Metals (Distribution, Removal, Remediation, Speciation, and Phytoremediation)
- Chlorinated and Other Persistent Organic Compounds
- Modeling
- GIS for Environmental Assessment, Database, and Remote Sensing Applications
- Environmental Analysis and Measurements
- Society and the Environment
- Environmental Planning and Management

• Renewable Energy Development

We would like to especially thank the session chairs who were instrumental in the success of the conference. The Conference was sponsored and organized by the American Academy of Sciences, with financial contributions from the co-sponsors and supporting organizations.

The papers in these proceedings represent the authors' results and opinions. No sponsors, cosponsors, participating organizations or editors should be construed as endorsing any specific contents or conclusions in the proceedings.

George A. Sorial, Ph.D.

University of Cincinnati

Jihua Hong, Ph. D.

American Academy of Sciences

# PLENARY PRESENTATION

# GEOLOGICAL AND ANTHROPOGENIC IMPRINTS ON GROUND WATER QUALITY: BEWARE OF THE ANECDOTAL MODELS

***Barry J. Hibbs***
(Department of Geosciences and Environment, California State University, Los Angeles, 5151 State University Drive, Los Angeles, CA 90032, USA. bhibbs@exchange.calstatela.edu)

Geological and mineralogical associations have been used to determine baseline water quality variations in aquifers for well over a century. Geological factors can be misidentified and misunderstood in establishing water quality relationships in aquifers however, especially in groundwater basins undergoing rapid development. The problem is particularly acute when time series and spatial changes in aquifer water quality occur. Time series and spatial changes in groundwater quality may be related to physical changes in aquifer dynamics that are either natural or anthropogenic.

The effectiveness of a generic, but widely accepted model in explaining water quality in one aquifer may be a weak analog for explaining water quality variations in another, apparently similar aquifer. Too often interpreters grasp at the most simple or apparent explanation for a process based on anecdotal evidence; e.g., "agriculture creates the elevated salinity," "industrial activity is the source of the elevated trace elements in the aquifer." A possible misinterpretation might arise merely on account of the fact that there is substantial agricultural or industrial activity in the groundwater basin. Reasons for misidentifying processes related to water quality is often due to a poor understanding of the geologic history of the groundwater basin, mischaracterization and misidentification of flow processes/hydraulics, and incomplete knowledge of the primary and secondary mineralogical assemblages in the aquifer. Insufficient training in evaluating the geological, hydraulic, and water/rock/soil interactions is another factor that comes into play.

The purpose of this presentation is to summarize our workgroup studies where earlier anecdotal models of water quality did not account for the actual water quality processes in the aquifers. These studies will serve as examples of flagrant risks associated with accepting anecdotal models to explain water quality variations in aquifers.

# WHO IS RESPONSIBLE FOR THE ENVIRONMENT?

***Erica Schoenberger***
(Department of Geography and Environmental Engineering, The Johns Hopkins University, 3400 N. Charles Street, Baltimore, MD 21218, USA. ericas@jhu.edu)

We are all in some sense responsible for the environmental damages humans have inflicted on the earth. At the same time, none of is exactly responsible for those harms. I contribute to those damages by living my ordinary life, and if I ceased living tomorrow, the harms would continue to accumulate. What is the nature of responsibility in such a situation? What kinds of choices are open to us and how may we tilt the system in the direction of clean and away from dirty?

# THE FUTURE OF HUMAN ENVIRONMENT AND HEALTH: SUSTAINABILITY THROUGH INTEGRATED APPROACH

***Momoh A. Yakubu***
(Department of Environmental and Interdisciplinary Sciences, College of Science, Engineering and Technology, Texas Southern University, 3100 Cleburne Avenue, Houston, TX 77004, USA. Yakubu_ma@tsu.edu)

Environmental health concern brings together experts, researchers, scholars and students from diverse backgrounds. These experts are working so hard in their respective fields to protect the human environment from the devastation and calamity that is waiting to befall it. Despite these efforts and colossal resources expended over these years, we are still faced with bad news of the impending doom for the human environment and health. Recently, our young generations are increasingly diagnosed with devastating diseases like dementias, Parkinson, infertility, attention deficit disorder, Alzheimer's, diabetes, etc. at very tender age. Some of these diseases have been linked to the quality of our environment and the increasing concentrations of disrupting chemical in our water. The fear of climate change is real, the recorded increase in global temperature with the attending droughts, heat waves, intense rainfalls resulting in floods, loss of vegetation and deforestations. The oceans are warming up and glaciers are melting with increasing sea levels. All of these attest to this fear. Despite advances in our understanding of our environment and the dangers locking around it – we are still on a crazing speed to self-destruction. There is therefore, the need to harness and integrate the various efforts of environmental-natural scientists along with others and stake holders – activists, government and global agencies, community leaderships, regulators, developed and developing nations to work seamlessly together to protect and sustain the future of human environment. NOW, we have the need to create a new generation of scientists to integrate a rigorous discipline depth with the ability to interact with other disciplines and work in an increasingly interdisciplinary, collaborative team to serve our global and human environment. This is more urgent as knowledge creation and innovation frequently occur at the interface of diverse disciplines.

# MANUFACTURED NANOMATERIALS: NEW EMERGING CONTAMINANTS AND THEIR POTENTIAL IMPACT TO DRINKING WATER RESOURCES

***Endalkachew Sahle-Demessie***

(U.S. EPA, Office of Research and Development, National Risk Management Laboratory, 26 W. Martin Luther King Drive, Cincinnati, OH 45268, USA. Sahle-demessie.endalkachew@epa.gov)

As nanomaterials become increasingly part of everyday consumer products, it is imperative to measure their potential release during production, use, and disposal, and assess their impact on the health of human and the ecosystem. This compels the research to better understand how the properties of manufactured nanomaterials (MNMs) lead to their accumulation and redistribution in the environment whether they could become emerging pollutants or can affect the mobility and bioavailability of other toxins. If nanoparticles are released from their matrix, the high surface to volume ratio and reactivity of MNMs makes them highly dynamic in environmental systems. The resulting transformations of the MNMs affect their fate, transport, and toxic properties. Photodegradaton of nanocomposites in the environment during the use or end-of-life could release of MNM. This study aims in better understanding of the human and environmental risks of using consumer products, such as polymer nano-composites.

This presentation focuses on the characterization, occurrence in the environment, detection and analysis, toxicity, fate and transport of MNM and the efficacy of water treatment methods in removing them from drinking water supplies.

# WATER POLLUTION

# AND

# WATER QUALITY CONTROL

# SUSTAINABLE RIVER BANK EROSION MANAGEMENT BY SOFT ENGINEERING APPROACH: A CASE STUDY OF THE BRAHMAPUTRA

***Prashanta Kumar Bordoloi***
(Department of Civil Engineering, The Assam Kaziranga University, NH37, Karaikhowa, Jorhat-785006, Assam, India)

The Brahmaputra River basin (catchments area 924,000 km$^2$), the least studied one in Asia, is currently experiencing severe river bank erosion, of its reach 640 km resulting from climatic and ecological changes in the basin' The Brahmaputra valley receives in average 2000mm of rainfall annually, 65% of which is received within four summer months from June to September leading to flooding and erosion of the river banks. Since the formation Task Force Against Erosion in 1964, 0.4 million square kilometer land is eroded by the river causing agony to the inhabitants. Government has taken up hard engineering measure like flood proofing with embankment and deflecting flow from eroded reach by structural measures like spars with mixed result. Mostly erosion is caused by the auxillary channels. In 2009, soft engineering measures were taken at vulnerable areas of Jhanjimukh for containing erosion to minimum with people's participation with great success. The measure consists of initiation of siltation in the auxiliary channel launching porcupine like RCC fencing post, and planting riverine plants with elasticity property to retard the flow velocity on the eroded reach. Silt deposition has been a continuous process on the eroded reach. Locally available elastic planting materials are planted on the fresh sedimentation on the eroded reach, which is protected by the plants in a sustainable way. This paper discusses the potential of containing river bank erosion with unique soft engineering technique in a sustainable way and its prospect of wider adaptability in the alluvial world.

# FRESHWATER MUSSELS OF THE UPPER SAN ANTONIO RIVER WATERSHED AND LOWER CIBOLO CREEK IN BEXAR, GUADALUPE, WILSON, AND KARNES COUNTIES: ABUNDANCE AND DENSITIES

*Larry Larralde*
(San Antonio River Authority, 600 E. Euclid Street, San Antonio, Texas 78283, USA)

Three mussel species under review for federal listing under the United States Endangered Species Act (ESA) and legally threatened in the State of Texas have historically been found in the San Antonio River (SAR) basin. One species, golden orb Quadrula aurea has been recently found in numerous locations in the San Antonio River and its tributaries (Lower Cibolo Creek). Previous efforts to survey the lower Cibolo Creek watershed began in the early 1990's and no records have been located regarding the West Side Creeks. Therefore, the San Antonio River Authority (SARA) proactively decided that basin wide monitoring efforts be implemented as soon as possible.

Sample data included baseline demographics such as density, richness, diversity, and age-class structure for the native mussel community. Data will be collected and distributed to regulatory agencies to assist in the decision making process for listing or delisting candidate species of concern. Sample efforts for 2014-2015 included the West Side Creeks of the Upper San Antonio Watershed from the Texas Commission on Environmental Quality (TCEQ) segment 1911, and the Lower Cibolo Creek, TCEQ segment 1902. Quantitative sample methods (Systematic Sampling) were implemented with a 0.25m$^2$ quadrat and catchment along with qualitative timed visual and tactile searches conducted at 46 collection sites across five lower Cibolo Creek sample reaches.

Qualitative-catch-per-unit-efforts (CPUE no. /h) under visual timed searches was the preferred and only method utilized for the West Side Creeks. Qualitative results from all four creeks yielded one native mussel species, paper pondshell Utterbackia imbecillis, from Martinez Creek. The CPUE for Martinez creek was 0.83/h., and overall, five live specimens were collected with a CPUE estimate of 0.26/h mussels for all four creeks. The total area searched and total search time were 2,762m$^2$ over 19 search hours. Quantitative results from lower Cibolo Creek yielded a total of eighty – one observed individuals and a population(s) estimate (n=78) representative of five native species (species richness, n=5), yellow sandshell Lampsilis teres (n=38), golden orb Quadrula aurea, (n=32), paper pondshell Utterbackia imbecillis (n=3), pistolgrip Quadrula verrucosa (n=7), and Louisiana fatmucket Lampsilis hydiana (n=1). Non-native asian clam, corbicula spp. were common across all reaches. Species richness per site ranged from 0 - 2 species at 0.83 ± 0.12 (mean ± SE), with the highest (2 spp.) occurring at 10 sites and the lowest (no catch) at 22 sites. Species richness in quadrat samples and species richness in timed searches were found to be significantly correlated (Spearman's rho, r=0.36, P = 0.01). The range of quadrats per sample site (n = 13-17) was a contributing factor for the total number of quadrats (n=720) aimed at providing for good spatial coverage and estimating for precise densities. Only one fourth (26%) of our overall quantitative sample had measurable data indicative of low mussel densities at most sites that ranged from 0.0m$^2$ – 0.024m$^2$. Quantitative mean mussel densities for all sites was 0.034m$^2$ ± 0.01 and seemed to be the standard. Our species of conservation concern, golden orb, was collected at 35% of all our sample sites in low densities and represented by thirty - two specimens. The mean CPUE for golden orb for all sites was 0.13/h ± 0.05. Qualitative timed searches were more successful and exhibited mussel presence 52% of the time whereas quantitative efforts were at 26%.

A Paired 2 Sample t-test was ran to test the Null hypothesis comparing the effectiveness of quantitative (quadrat) and qualitative (timed searches) mussel surveys and results were significant at the <.05 level (P = 0.008 ) thus rejecting the null hypothesis. Qualitative results produced an overall CPUE range of 0.0 – 2.33 /h and a mean CPUE for the community from all sites of 0.40/h ± 0.08. Shell lengths were measured to provide insight on community size class structure, however, class estimates could be biased due to a small sample size. Louisiana fatmucket and pistolgrip lacked representative juveniles, and conversely all other species except paper pondshell were represented by medium to large sized adults. Golden orb was represented by only a few juveniles and more representative in the 31-40mm class. The

---

largest class was represented by pistolgrip, yellow sandshell, and golden orb in moderation, however, there are also some recent recruitment issues from several species. Habitat utilization was dominated by golden orb and yellow sandshell located in glide and run combinations. There also seemed to be disjointed mussel community where no individuals were located in the farthest upstream reach and just below the middle reach where very few were found. The lower Cibolo creek mussel community has high species richness however densities are relatively low and evenness among species is lacking. The Golden orb is a state threatened mussel found in relatively low abundance throughout the San Antonio River and its tributaries, therefore, this mussel and the watershed that it is found should obtain focused conservation efforts to protect it.

# BIOACCUMULATION OF ARSENIC, CADMIUM, MERCURY, AND LEAD IN FISHES OF THE DALIAO RIVER ESTUARY, CHINA

Chunye Lin, Kai Lei, Deqi Jiao, Mengchang He
(State Key Joint Laboratory of Environmental Simulation and Pollution Control, School of Environment, Beijing Normal University, Beijing 100875, China)

The contamination of toxic metals in the coastal aquatic ecosystem is extensively concerned owing to their high bioaccumulation in marine products and potential risks to human health through seafood consumption. As an important industrial base with oil, chemical, and steal factories in the northeast part of China, the Liao River Watershed has been subjected to heavy anthropogenic influences as a result of rapid economic developments in the past ca. 50 years. This has resulted in significant stress to the Daliao River Estuary in the Liaodong Bay of Bohai Sea (China). In this research, sediment, water, and fishes were sampled in the Daliao River Estuary. The concentrations of As, Cd, Hg, and Pb in the sediment, water, fish were determined. The bioaccumulation factor and hazard quotient via fish consumption were calculated. The results indicated that the concentrations of As, Cd, Hg, and Pb in fishes were 0.57~8.18, 0.01~3.33, 0.01~8.29, 0.07~1.71 mg/kg, respectively. The concentrations of As, Cd, Hg, and Pb in sediment were 4.61~11.41, 0.05~0.67, 0.01~0.10, and 16.57~39.18 mg/kg, respectively; while their concentrations in water were 3.84~9.12, 0.05~0.64, 0.01~0.05, and 0.69~4.37 □g/L. The bioaccumulation factor revealed that heavy metals were more accumulated in the liver and intestine of fishes. Hg and Cd showed the high bioaccumulation level in the fishes. Hazard quotient showed that fish consumption may pose a health risk to local residents, primarily due to the high concentration of As in the fishes.

# NITRIFICATION AND DENITRIFICATION BY ALGAE-ATTACHED AND FREE-LIVING MICROORGANISMS DURING CYANOBACTERIAL BLOOM IN LAKE TAIHU

***Xiaofeng Chen***, Haiyang Jiang (Yangzhou University, Yangzhou, China)
Liuyan Yang (Nanjing University, Nanjing, China)

Cyanobacterial bloom may stimulate the epiphytic and pelagic microorganisms to take great roles in nitrification and denitrification process. In order to validate this hypothesis, a 4-weeks floating mesocosms experiment, involved cyanobacterial decay-growth-decay period, was performed at Lake Taihu. In addition to conventional detection of the physical and chemical properties, quantitative real-time PCR was used for the determination of nitrification and denitrification genes (AOA, AOB, nirS and nirK). Results showed that total nitrogen removed from the treatment with cyanobacteria added was about 3.62 mg $L^{-1}$, 40% of which was organic nitrogen, indicating a nitrogen transformation and removal mechanism occurred in the system. Variations of the biogeochemical properties suggested that remineralization and coupling nitrification and denitrification was the primary pathway that nitrogen removed. This study reveals that algal bloom can accelerate nitrogen removal efficiency in lake ecosystem.

# TAILINGS STORAGE FACILITIES AND ENVIRONMENTALLY SUSTAINABLE MINING

***Erica Schoenberger***
(The Johns Hopkins University, Baltimore, MD USA)

Tailings dam failures account for about three-fourths of major mining-related disasters. Tailings dams are not built like water retention dams. They are constructed from mine wastes and built up as mining proceeds. Nevertheless, water management is the critical problem. Tailings storage facilities (TSF) must be maintained in perpetuity. The technical challenges of constructing a TSF that can withstand extremes of climate, seismic events and often difficult terrain are substantial. I argue, however, that the principal cause of TSF failures is political and economic rather than technical.

When mining companies are held to the highest standards, they can and do meet them. Whether or not they are held to those standards depends in part on the regulatory environment. How exigent are the regulations, how comprehensive are they, and how well are they enforced? The answers to these questions have in part to do with the influence of the industry in particular jurisdictions compared with other land-intensive uses, especially as this bears on regulatory capacity and competence. Second, the social composition of the surrounding population also matters. Local populations with political and financial resources will have a much greater chance of escaping environmental disasters than those without such resources. Third, building the design of the TSF into the design of the mine and submitting the design for peer review provides important safeguards.

In this paper, I analyze the performance of three mines. Two of them suffered major TSF dam collapses with widespread and continuing environmental damage: the Ok Tedi mine in Papua New Guinea (PNG), and the Mount Polley mine in British Columbia. The third mine – the McLaughlin mine in Northern California – is a rare success story in which all of the environmental dislocations necessarily associated with mining were confined on site and, to a significant degree, remediated after active mining ceased.

The environmental damages of mining are closely linked to social harms (through impacts on livelihoods, exposure to environmental toxins and the like), so it is particularly worthwhile getting at the causes of both success and failure in an effort to determine whether mining can increasingly be done in a way that contains and remediates environmental harms.

# STATE APPROACHES TO WATERSHED-BASED PERMITTING

***Danielle Stephan*** (US Environmental Protection Agency, 1200 Pennsylvania Ave, NW, Washington, DC 20460)
Greg Currey (Tetra Tech, Inc., 10306 Eaton Place, Suite 340, Fairfax, VA 22030)

The United States Environmental Protection Agency (USEPA) has been an advocate for State Permitting Authorities to consider using a Watershed-based Permitting approach where pollution is a result from a variety of sources within a defined watershed. The primary difference between this approach and the traditional approach to permitting is the consideration of watershed goals and the impact of multiple pollutant sources and stressors, including nonpoint source contributions rather than a focus on a single point source. Watershed-based permitting may encompass a variety of activities ranging from synchronizing permits within a basin to developing water quality-based effluent limits using a multiple-discharger modeling analysis. The types of permitting activities will vary from watershed to watershed, depending on the unique circumstances in the watershed and the sources affecting watershed conditions. The ultimate goal of watershed-based NPDES permitting, however, is to develop and issue NPDES permits that consider the conditions of the entire watershed and address the diverse sources within the watershed.

Several states have emerged as champions of the watershed-based Permitting approach, and EPA has developed a variety of case studies to demonstrate real world examples of how the watershed-based permitting approach can be used to address pollution problems in a given watershed. This presentation will discuss the concepts of a watershed permitting analytical approach and an NPDES watershed framework as they relate to state approaches to watershed planning and permitting. Specifically, the presentation will describe the drivers, processes and outcomes of a number of state watershed permitting approaches. Examples of cases studies that will be presented may include the following:

1) *Tualatin River Watershed Permit for Clean Water Services in Portland, Oregon*: This case study focuses on the components of the watershed-based permit issued to Clean Water Services (CWS), a public utility in Portland, Oregon. The permit incorporates the NPDES requirements for all four of CWS's advanced wastewater treatment facilities, its two industrial storm water permits, and its municipal separate storm sewer permit into a single permit and includes provisions for water quality credit trading involving temperature (thermal load), biochemical oxygen demand (BOD), and ammonia.

2) *Neuse River Watershed Permit, North Carolina* - This case study focuses on the components of the watershed-based permit issued to the Neuse River Compliance Association and the group compliance mechanisms used by the co-permittees to meet the terms of the permit.

3) *General Permit for Nitrogen and Phosphorus Dischargers to the Chesapeake Bay Watershed in the Commonwealth of Virginia* - This case study focuses the general permit issued to approximately 180 significant and new and expanding municipal dischargers of nitrogen and phosphorus in the Commonwealth of Virginia discharging to the Chesapeake Bay watershed.

# RAIN HARVESTING FOR SUSTAINABLE WATERSHED MANAGEMENT – LOCAL WATER SUPPLY AND REGIONAL WATER QUALITY IMPROVEMENT. NEAL SHAPIRO

Neal Shapiro
(Santa Monica City, Office of Sustainability & the Environment, Santa Monica, CA, USA)

Throughout the United States, water quality and adequate water supplies are worsening. Around the world, rising population demand is using freshwater supplies faster than can be replenished by nature or society, resulting in short-term deficits and long-term unreliability; millions of people die or suffer health impacts from unsafe or inadequate water supplies; climatic oscillations in temperature gradients and precipitation from what civilization has considered "normal" for the last 50 years will make long-term, reliable storage management more challenging; water quality of surface water bodies are degrading from more intense stormwater runoff pollution; and energy costs to move surface water hundreds of miles through aging infrastructure from distant watersheds to consumers are unsustainable; such infrastructure is exposed to natural disasters which would disrupt water flow to urbanized centers. The clean water supply pie shrinks, yet water demand grows. The solution is to grow the supply pie in a sustainable way to keep pace with demand while also managing water demand to be as efficient as technically possible.

With the participation of water consumers, water management professionals must implement a new sustainable water demand and supply side management approach, including well-tested scientific and technical products focused increasingly on non-traditional, local non-potable water resources, such as rainwater. Such solutions exist, and are proven and working today to improve water supply and quality.

Why rainwater? Rain harvesting has many benefits: it is a local, independent supply, e.g. water security and resiliency; it eliminates stormwater, and stormwater pollution; it reduces combined sanitary sewer overflows, helping to comply with Clean Water Act NPDES regulations; and it lowers the use of municipal water, such that more freshwater can be left in stressed surface waters and aquifers. Onsite rain harvesting effectively delivers a water supply to an end use at an acceptable water quality and minimal energy cost. By keeping rainwater onsite, less pollution enters surface water bodies, where water-quality violations can occur.

Rain harvesting technology has been around for decades, but has improved dramatically over the last 10 years. These systems are safe and reliable, and modular and scalable. Such systems use the appropriate harvesting and treatment technologies to meet appropriate end uses of non-potable and even potable.

The national leader for rain harvesting is the American Rainwater Catchment Systems Association. ARCSA's mission is to promote sustainable rain and stormwater harvesting practices to help solve potable, non-potable, stormwater, water quality and energy challenges throughout the world. ARCSA is in the forefront in seeking, producing and implementing solutions, nationally and internationally, fostering integration and transfer of the latest scientific research and engineering applications. With partners, ARCSA produced plumbing design standards documents and recently a comprehensive rainwater harvesting manual.

Innovative rain harvesting case studies demonstrate the effectiveness of this sustainable, low impact development and green infrastructure strategy. From using rainwater and stormwater for irrigation to indoor flushing, these examples show how the technology works, how economies grow, how pollution is kept out of the aquatic environment and how local water supply is increased.

# IMPACT OF WATERSHED MANAGEMENT ACTIVITIES IN LAND USE PATTERN AND PEOPLE'S LIVELIHOOD IN NEPAL

Ajay Karki
(Department of Soil Conservation and Watershed Management, Po Box 4719, Babarmahal, Kathmandu, Nepal)

This paper examines the impact of watershed management activities on landuse pattern and livelihood upliftment, implemented by Department of Soil Conservation and Watershed Management in Kaski District of western Nepal. An JICA funded development project was implemented in Kaski district for five years. Main objective of the project was to improve the watershed condition by making (bio)/engineering structures and to uplift the people's livelihood by implementing the income generation activities. Baseline data and report of the project were used mainly to prepare this paper. Following are the main result of the project:

1. Irrigated area increased significantly
2. Green coverage area increased
3. River bank cut off decreased and rehabilated area increased
4. Landslide decreased, soil erosion decreased
5. Income of people increased

# NATURAL ATTENUATION OF INDICATOR BACTERIA IN NATURAL WATER SYSTEMS

***Amin Kiaghadi*** and Hanadi S. Rifai (University of Houston, Houston, TX, USA)

**ABSTRACT:** The presence of pathogens in surface water bodies is considered to be the main cause of several different types of diseases worldwide. Monitoring for Fecal Indicator Bacteria (FIB) in water bodies has conventionally relied on *E. coli* and Enterococci for fresh and marine waters, respectively. High concentrations of FIB in rivers, bays and estuaries, including in Galveston Bay near Houston, Texas where this study was undertaken, have been extensively documented in the general literature. While FIB concentrations are relatively high in Galveston Bay (Texas, USA), they are much lower when compared to FIB concentrations in the Houston Metropolitan area. This result could be explained by considering the Houston Ship Channel (HSC) as a treatment reactor that naturally attenuated the bacterial loads emanating from the Houston Metropolitan watersheds. Higher salinity, lower velocities, larger flows in the HSC, and tidal influences from Galveston Bay support this assumption. There are no studies to date which investigate the concept of natural attenuation of pathogenic pollutants in surface waters. The main purpose of this study is to address this deficiency and introduce the term "natural attenuation" for pathogenic pollutants in natural water systems. The fate and transport of the bacterial load transportation within the HSC can be categorized into three mechanisms: 1) reactive- dilution mixing of fresh and marine water, 2) advective-reactive transport along the HSC, and 3) dilution mixing with Galveston Bay water. This study investigates whether each of the aforementioned attenuating mechanisms is responsible for reducing FIB concentrations and loads into Galveston Bay. Several lines of evidence are presented that demonstrate natural attenuation of FIBs within the HSC: 1) lower Enterococci concentrations in tributary tidal waters compared to those in fresh waters for the tributaries discharging into the HSC, 2) 70% reduction in bacterial loads within the Channel itself, 3) a general decreasing trend in Enterococci geometric means from upstream to downstream within the HSC, and 4) lower Enterococci concentrations in Galveston Bay at the confluence with the HSC.

## INTRODUCTION

The presence of pathogens in surface water bodies is considered to be the main cause of several different types of diseases worldwide. The annual cost of marine-born microbial contamination in the U.S. alone exceeds $900 million (Dwight et al. 2005, Ralston et al. 2011). Elevated Fecal Indicator Bacteria FIB levels in natural water systems are typically associated with urbanization, storm water discharges, releases from wastewater plants, illegal dumping, and domestic and wildlife animals. Ingestion or inhalation of water during recreational activities, consumption of raw seafood, and contaminated water contact with open wounds are the main causes of diseases. Monitoring for FIB in water has been widely used to reduce the cost of marine-born microbial contamination (Desai et al. 2010, Goodwin et al. 2012). For this purpose, fecal coliform, and more recently *E. coli* for fresh waters and Enterococci for marine waters have been suggested as measures of pathogenic contamination (Ostrolenk et al. 1947, Cabelli et al. 1979). Different environmental factors can affect FIB concentrations including physico-chemical conditions such as salinity, pH, Dissolved Oxygen (DO), and water temperature, and biological-environmental conditions such as microbial predation and the effect of sunlight (Anderson et al. 2005, Mill et al. 2006).

In 2010, Desai et al. reported elevated levels of *E. coli* in almost all of the Houston Metropolitan area bayous (Desai et al. 2010). Waters originating from Houston watersheds discharge to the Houston Ship Channel (HSC), a 55 mi navigational man-made tidal channel, and eventually to Galveston Bay and its recreational beaches. Elevated values of FIBs in Galveston Bay have been reported by several public agencies such as the Texas Commission on Environmental Quality (TCEQ) and the Texas Department of State Health Services (TDSHS). To date, the sources of FIB contributing to Upper Galveston Bay have not been elucidated. However, a rational hypothesis would associate bacterial loads from Houston area watersheds to the presence of FIB in Galveston Bay. However, FIB concentrations in Galveston Bay are significantly lower when compared to the FIB concentrations in Houston area watersheds. This change could be explained by considering the HSC as a treatment reactor that naturally attenuates the bacterial loads emanating from Houston watersheds. Higher salinity, lower velocities, larger flows in the HSC, and tidal influences from Galveston Bay support this observation. This concept is novel for FIB, and to the best of the knowledge of the authors has yet to be addressed in this context.

Natural attenuation is defined as "observed reduction in contaminant concentrations as contaminants migrate from the source in environmental media due to a number of fate and transport processes" (Wiedemeier et al. 1999). Most of the research related to natural attenuation has been mostly completed in ground water systems and groundwater-surface water interactions (Kao et al. 2010, Ouvrard et al. 2013, Freitas et al. 2015). Very little research has been done on the concept of natural attenuation for surface water; most has focused on Uranium and pharmaceuticals and in wetlands for nutrients (Casey and Klaine 2001, Manaka et al. 2007, Acuna et al. 2015). There are no studies to date to investigate the natural attenuation of pathogenic pollutants in surface waters. The main purpose of this study is to fill this gap and analyze the "natural attenuation" for pathogenic pollutants in natural water systems. This research relies on statistical methods to study the role of the tidally influenced HSC in naturally attenuating bacterial loads from the Houston Metropolitan area.

## MATERIALS AND METHODS

**Study Area Data.** The HSC is tidally influenced and begins at the Port of Houston, near downtown Houston, Texas, extends through Galveston Bay and enters the Gulf of Mexico. The average depth and width of the channel are 13.7 and 162 m respectively. In this study, a total length of 46 km upstream of Morgan's Point (the inlet to Galveston Bay) was studied (**Error! Reference source not found.**). The Texas Commission on Environmental Quality (TCEQ) Surface Water Quality Web Reporting Tool was used to obtain the water quality data for the HSC. Thirteen water quality monitoring (WQM) stations in the HSC, one in Upper Galveston Bay, and nineteen additional stations in bayous were used in the study (Figure 1). The period of record covers the years 2000-2013 since there was no available data for Enterococci prior to 2000. As noted previously, *E.coli* and Enterococci are the FIB for fresh and marine waters respectively. Thus, stations above the tide line (see Figure 1) have *E.coli* data while the tidal stations contain Enterococci data. Flow and rainfall data were compiled from the USGS and the Harris County Flood Control District database (HCFCD, 2014), respectively. Hourly tide data at Morgan's Point were obtained from the National Oceanic and Atmospheric Administration (NOAA) Center for Operational Oceanographic Products and Services (CO-OPS, 2014) tides and currents database.

**Conceptual Model of Natural Attenuation in the Study Area**. Galveston Bay is an estuarine system with relatively low tide height and the HSC is considered a pathway for bacteria loads that emanate from Houston area watersheds to Galveston Bay (**Error! Reference source not found.**). The fate and transport of FIB loads within the HSC can be categorized into three mechanisms: 1) reactive-dilution mixing of bayou fresh water with HSC marine water, 2) advective-reactive transport along the Channel, and 3) dilution mixing with Galveston Bay water (**Error! Reference source not found.**). The HSC has long travel and residence times due to the very low advective flows within the channel. Consequently, the HSC can be considered as a series of mixing reactors with natural attenuation capacities that reduce the bacterial contamination. However, bacteria loads from Houston area watersheds and other sources such as leakage in large barge sewage systems and bird feces can counteract the aforementioned capacity and increase the bacteria concentration.

To demonstrate the effect of the first mechanism (dilution-mixing with HSC water), in a given tributary, one should be able to observe 1) a noteworthy change in FIB concentrations between two WQM stations located above and below the tide line, respectively, and 2) a significant reduction in the HSC bacteria loads discharged to Galveston Bay when compared to the cumulative input loads from Houston area watersheds. The first observation was tested by converting the freshwater *E. coli* data at WQMs above the tide line into saline Enterococci-equivalents. The conversion ratio was obtained by applying a linear regression to the data from 8 WQM stations near the tidal boundaries where both indicator data were available. For the second observation, an existing Environmental Fluid Dynamics Code (EFDC) simulation of the HSC developed by Howell (2012) was used to obtain the discharge flows from the HSC at Morgan's

Point (station 11252, see Figure 1). The FIB load for the HSC was calculated by multiplying the geometric mean concentration of Enterococci at station 11252 with the extracted discharge rate from EFDC. Input loads from Houston area watersheds were calculated using flow data from USGS gages and the converted Enterococci-equivalents data from non-tidal WQM stations (except for Goose and Vince bayous where the closest WQM stations to the tidal boundaries were used). The calculated HSC load was then compared to the FIB input loads from Houston.

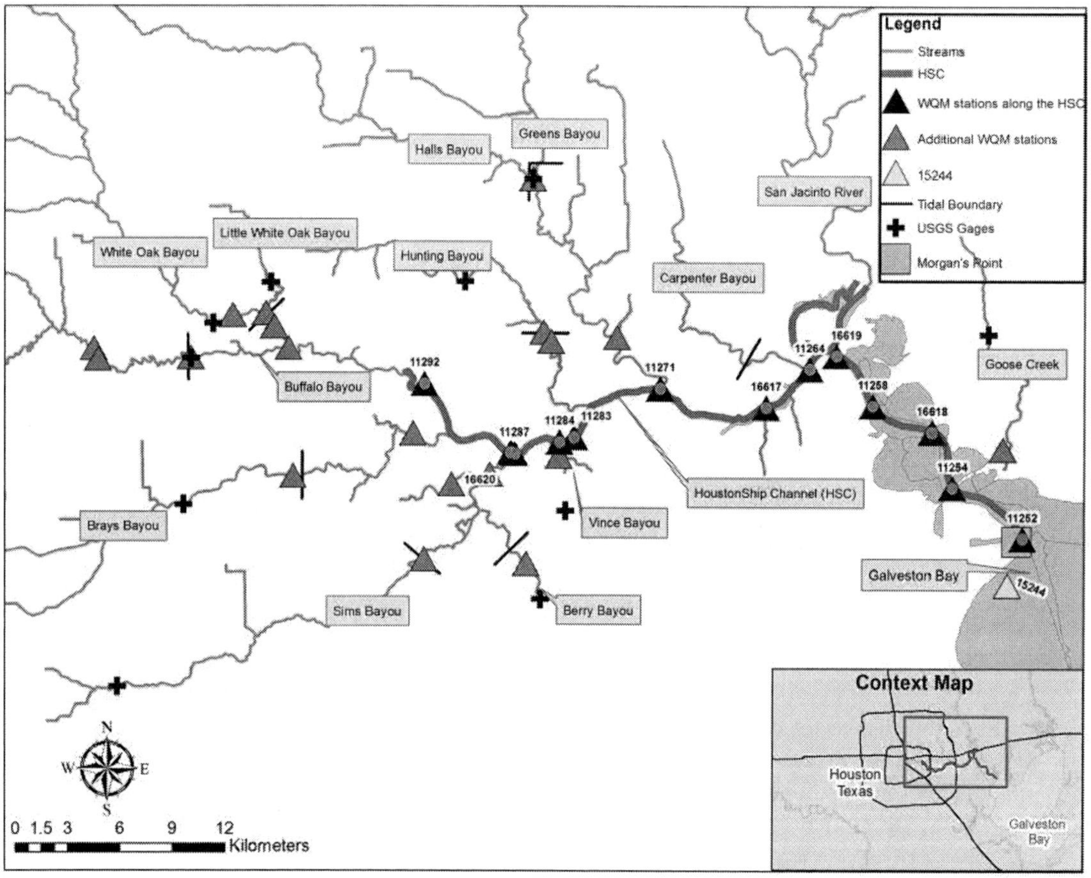


Figure 1- Houston Ship Channel (HSC), USGS gages and Water Quality Monitoring Stations

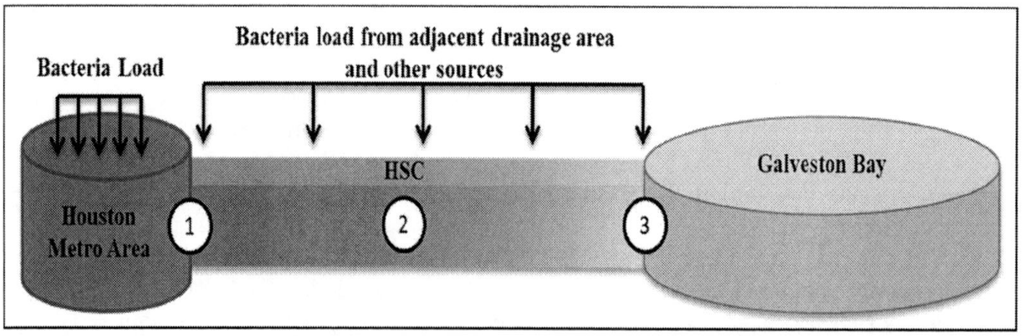


Figure 2- Ba□teria load pathway from Houston Metropolitan area through the HSC leading to Galveston Bay

Different components of natural attenuation can occur during the advective-reactive transport along the channel (second mechanism) that can be attributed to variation in the volume of water (dilution), tide effects, and ambient environmental factors. Thus, a spatial analysis for FIB concentrations within the HSC was undertaken in addition to a correlation analysis between FIB concentrations and tides and ambient environmental factors.

Lastly, and to assess attenuation of FIB due to dilution-mixing with Galveston Bay waters (third mechanism), FIB concentrations in the WQM station located at the mouth of the HSC (station 11252) were compared to those at the WQM station located in Upper Galveston Bay (station 15244).

**Statistical Analyses.** Normality and log-normality tests were conducted using SPSS software via Kolmogorov-Smirnov and Shapiro-Wilk methods. The results showed that none of the water quality data sets used in this study were normal or log-normal (P-Value<<0.05) for any station. Therefore, in this study, geometric mean and median instead of mean; Mann-Whitney U test instead of t-test; and Spearman correlation test instead of Pearson Correlation test were used. To find any possible correlation between concentrations/values of FIB and other environmental variables and also with distance from the channel outlet (Morgan's Point), visual analysis, regression analysis and Spearman's correlation test were used.

## RESULTS

The average log-scale conversion ratio between Enterococci and *E. coli* was found to be 0.78, which is comparable to the 0.73 value obtained by taking the ratio of the EPA standards for the two FIBs (log35/log126 MPN/dL).

### Reactive-Dilution Mixing of Bayou Fresh Water with HSC Marine Water

***Relationship between FIB in tributaries and the receiving HSC.*** As shown in Figure 3, a rapid drop in Enterococci geometric mean concentrations is observed when comparing the fresh water (above the tide line) segments of tributaries with their saline water (tidal) counterparts for all tributaries. This observation has also been confirmed using the Mann-Whitney U test which showed a significant difference in Enterococci bacteria concentrations between fresh and saline waters in all tributaries except for Greens Bayou (P-value=0.09). The Greens Bayou anomaly can be explained by the fact that the distance between the tidal and above tide line WQM stations is relatively longer in comparison to the other tributaries. Moreover, the above tide line WQM station is located at the confluence of Greens and Halls Bayous causing some uncertainties due to rapid mixing.

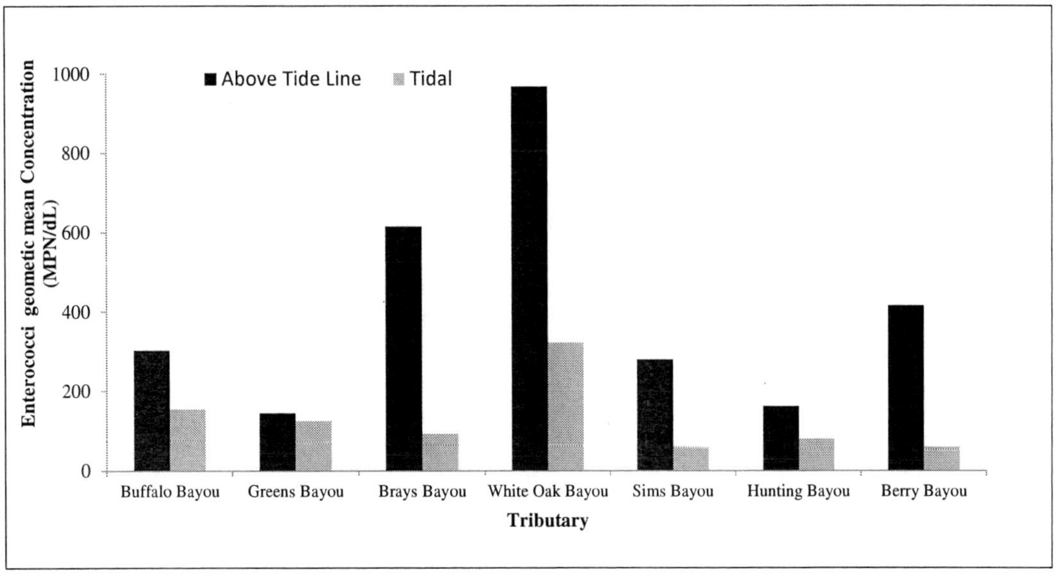


Figure 3- Enterococci geometric mean concentration before and after tide line in different tributaries

***Input and output bacterial loads.*** Although White Oak, Brays, and Little White Oak bayous had the highest Enterococci concentrations, the variation in Enterococci geometric mean concentrations did not exceed an order of magnitude in the different tributaries. For the flow, however, the variation was up to three orders of magnitude: Greens, Buffalo, and Brays had the highest flows whereas the lowest flows were found in Goose, Vince and Berry bayous. The highest load contributions were calculated for Greens, Buffalo, and Brays bayous with median loads of 11, 7.4, and 2 trillion MPN per day, respectively.

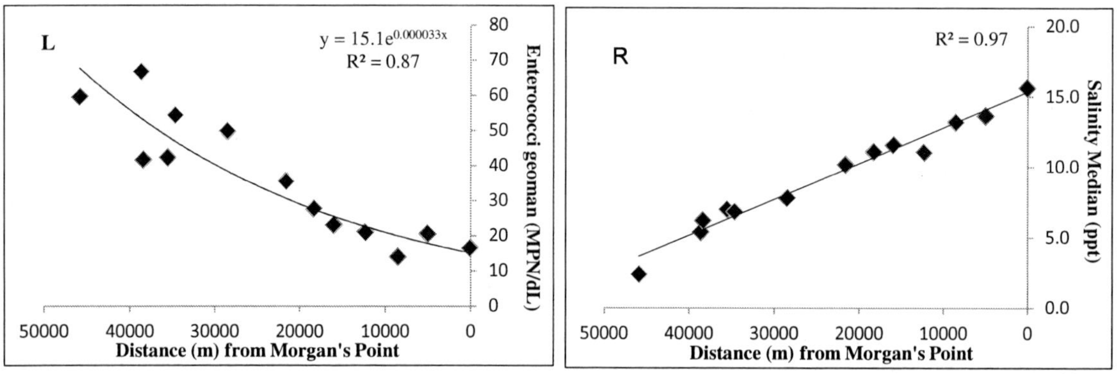


Figure 4- Enterococci geometric mean concentrations (L) and Salinity median values (R) as a function of distance from outfall of Channel into Galveston Bay

The sum of all bacteria loads discharging to the HSC was estimated to be 21.6 trillion MPN per day. EFDC model results at Morgan's Point (**Error! Reference source not found.**) show a median advective velocity of 0.108 m/s. A flow rate of 444 $m^3/s$ was estimated by multiplying the aforementioned velocity with the HSC cross sectional area obtained using ArcGIS (Version 10.1, ESRI, Redlands, California), Digital Elevation Model (DEM) data and the EFDC model grid. Finally, by considering the Enterococci geometric mean concentration of 17 MPN/dL at Morgan's Point (station 11252, see **Error! Reference source not found.**) the total output bacteria load from the HSC was estimated to be 6.5 trillion MPN per day, a 70% reduction when compared to the input load.

**Advective-Reactive Transport along the Channel**

Along the channel (in 13 WQM stations), Enterococci concentrations ranged from non-detect ($\leq 10$ MPN/dL) to more than 24,000 MPN/dL. As demonstrated in **Error! Reference source not found.**L, a general decreasing trend can be observed in Enterococci geometric mean concentrations from upstream to downstream in the HSC. This trend has also been confirmed using Spearman's correlation test (P-value<0.05). A curvilinear regression analysis showed that an exponential trend (P-value<0.05, $R^2$=0.87) achieved the best fit for the data. The observed decreasing trend supports the occurrence of natural attenuation of FIB within the HSC

The median for salinity increases as the flow moves towards the mouth of the HSC as would be expected due to the mixing of fresh and marine water from Galveston Bay (see **Error! Reference source not found.**R). Significant negative correlation between salinity and Enterococci concentration (P-value<0.05) was confirmed using Spearman's correlation test. A stronger correlation was obtained by analyzing median concentrations of salinity and geometric means of Enterococci at WQM stations. The Mann-Whitney U test was applied to evaluate the effect of tide conditions on Enterococci concentrations. The results showed Enterococci concentrations were significantly lower (P-Value<0.05) in samples collected in ebb (receding) tides compared to ones collected in higher when the tide is rising.

**Dilution Mixing with Galveston Bay Water**

The results of the Mann-Whitney U test for Enterococci data from the two WQM stations located at the entrance of the HSC into Galveston Bay (station 11252 within the HSC and 15244 in Galveston Bay, see Figure 1) indicated a significant difference between them. Additionally, for the samples collected in flood tides. This is in agreement with what would be expected as salinity is Galveston Bay (WQM 15244),

81% of the samples had a value below the detection limit. This number was only 35% in station 11252 that is located only 3.5 km upstream of station 15244. Lastly, the Enterococci geometric mean concentration for station 15244 was found to be 7 MPN/dL, a value that is less than half of the 17 MPN/dL reported for station 11252. It can be concluded that dilution-mixing is occurring as HSC water mixes into Galveston Bay.

## DISCUSSION

In natural water systems, coastal estuaries can be characterized as a transitional phase for bacteria loads from fresh waters to their destination saline receptors. Such tidal systems have normally higher salinities compared to the fresh water streams that discharge bacteria loads but lower salinities than the terminal receptor, in this case, Galveston Bay. The reduction in FIB concentrations in estuary systems compared to fresh waters that have been reported widely in the literature (McLaughlin et al. 2007, Lewis et al. 2013) can be attributed to the natural attenuation capacities of these sentry water bodies such as the HSC.

Enterococci concentrations along the HSC showed a high degree of spatial variability in keeping with similar trends reported by others (Cui et al. 2013, De Luca-Abbott et al. 2000). Mixing marine waters from Galveston Bay with fresh waters emanating from the greater Houston Metropolitan area can significantly change the physico-chemical characteristics of water. Salinity and tide oscillations were the most important variables affecting the reduction of FIBs within the HSC; confirming similar findings by Mill et al. 2006, Johnston et al. 2015.

In fresh water streams, Desai and Rifai (2013) showed a Bacteria Diurnal Sag (BDS) in a non-tidal urban watershed in Houston caused by changes in water temperature and solar radiation intensities. They concluded that time and numbers of samples were the most important factors in water-quality monitoring. In an estuary system, tidal mechanisms can cause significant variation in the salinity and consequently affect FIB concentrations at different times of the day suggesting that FIB measurements in WQM stations are highly sensitive to the time of sampling. Thus, for a tidal system such as the HSC, the time of sampling should also be adjusted to the tide mechanisms that could vary from month to month different in different months.

## CONCLUSION

Several lines of evidence were presented that demonstrate natural attenuation of FIBs within the HSC: 1) lower Enterococci concentrations in tributary tidal waters compared to those in fresh waters for the tributaries discharging into the HSC, 2) a 70% reduction in bacteria loads within the Channel itself, 3) a general decreasing trend in Enterococci geometric means from upstream to downstream within the HSC, and 4) lower Enterococci concentrations in Galveston Bay at the confluence with the HSC.

Water bodies similar to the HSC with natural attenuation capacities can act as sentries for coastal pathogenic contamination by reducing overall bacteria loads to receiving bays and estuaries thus reducing the risk of diseases. It would be important to maintain relatively high water quality in such systems in order to sustainably use their natural attenuation capacities for bay protection. In addition, bacteria concentrations should be monitored relatively frequently to capture the inherent variabilities in FIB levels caused by sunlight, temperature, and tide effects.

## AKNOWLEGMENTS

The research was funded by the Texas Commission on Environmental Quality and the U. S. EPA; their support is gratefully acknowledged.

## REFERENCES

Acuna, V., von Schiller, D., Garcia-Galan, M.J., Rodriguez-Mozaz, S., Corominas, L., Petrovic, M., Poch, M., Barcelo, D. and Sabater, S. (2015) Occurrence and in-stream attenuation of wastewater-derived pharmaceuticals in Iberian rivers. *Science of the Total Environment* 503, 133-141.

Anderson, M.L., Whitlock, J.E. and Harwood, V.J. (2005) Persistence and differential survival of fecal indicator bacteria in subtropical waters and sediments. *Applied and Environmental Microbiology* 71(6), 3041-3048.

Cabelli, V.J., Dufour, A.P., Levin, M.A., McCabe, L.J. Haberman, P.W. (1979)Relationship of microbial indicators to health effects at marine bathing beaches. *American Journal of Public Health* 69(7), 690-696.

Casey, R.E. and Klaine, S.J. (2001) Nutrient attenuation by a riparian wetland during natural and artificial runoff events. *Journal of Environmental Quality* 30(5), 1720-1731.

Cui, H.L., Yang, K., Pagaling, E. and Yan, T. (2013) Spatial and Temporal Variation in Enterococcal Abundance and Its Relationship to the Microbial Community in Hawaii Beach Sand and Water. *Applied and Environmental Microbiology* 79(12), 3601-3609.

De Luca-Abbott, S., Lewis, G.D. and Creese, R.G. (2000) Temporal and spatial distribution of enterococcus in sediment, shellfish tissue, and water in a New Zealand Harbour. *Journal of Shellfish Research* 19(1), 423-429.

Desai, A.M. and Rifai, H.S. (2013) Escherichia coli Concentrations in Urban Watersheds Exhibit Diurnal Sag: Implications for Water-Quality Monitoring and Assessment. *Journal of the American Water Resources Association* 49(4), 766-779.

Desai, A.M., Rifai, H., Helfer, E., Moreno, N. and Stein, R. (2010) Statistical Investigations into Indicator Bacteria Concentrations in Houston Metropolitan Watersheds. *Water Environment Research* 82(4), 302-318.

Dwight, R.H., Fernandez, L.M., Baker, D.B., Semenza, J.C. and Olson, B.H. (2005) Estimating the economic burden from illnesses associated with recreational coastal water pollution - a case study in Orange County, California. *Journal of Environmental Management* 76(2), 95-103.

Freitas, E.T.F., Montoro, L.A., Gasparon, M. and Ciminelli, V.S.T. (2015) Natural attenuation of arsenic in the environment by immobilization in nanostructured hematite. *Chemosphere* 138, 340-347.

Goodwin, K.D., McNay, M., Cao, Y.P., Ebentier, D., Madison, M. and Griffith, J.F. (2012) A multi-beach study of Staphylococcus aureus, MRSA, and enterococci in seawater and beach sand. *Water Research* 46(13), 4195-4207.

Howell, N.L. (2012) Bed and suspended sediments as source and transport mechanisms for polychlorinated biphenyls in HSC estuary system, University of Houston, Houston, TX.

Johnston, K.K., Dorsey, J.H. and Saez, J.A. (2015) Stratification and loading of fecal indicator bacteria (FIB) in a tidally muted urban salt marsh. *Environmental Monitoring and Assessment* 187(3).

Kao, C.M., Chien, H.Y., Surampalli, R.Y., Chien, C.C. and Chen, C.Y. (2010) Assessing of Natural Attenuation and Intrinsic Bioremediation Rates at a Petroleum-Hydrocarbon Spill Site: Laboratory and Field Studies. *Journal of Environmental Engineering-Asce* 136(1), 54-67.

Lewis, D.J., Atwill, E.R., Pereira, M.D.C. and Bond, R. (2013) Spatial and Temporal Dynamics of Fecal Coliform and Escherichia coli Associated with Suspended Solids and Water within Five Northern California Estuaries. *Journal of Environmental Quality* 42(1), 229-238.

Manaka, M., Seki, Y., Okuzawa, K., Kamioka, H. and Watanabe, Y. (2007) Natural attenuation of dissolved uranium within a small stream of central Japan. *Limnology* 8(2), 143-153.

McLaughlin, K., Ahn, J.H., Litton, R.M. and Grant, S.B. (2007) Use of salinity mixing models to estimate the contribution of creek water fecal indicator bacteria to an estuarine environment: Newport Bay, California. *Water Research* 41(16), 3595-3604.

Mill, A., Schlacher, T. and Katouli, M. (2006) Tidal and longitudinal variation of faecal indicator bacteria in estuarine creek in south-east Queensland, Australia. *Marine Pollution Bulletin* 52(8), 881-891.

Ostrolenk, M., Kramer, N. and Cleverdon, R.C. (1947) Comparative Studies of Enterococci and Escherichia coli as Indices of Pollution. *Journal of Bacteriology* 53(2), 197-203.

Ouvrard, S., Chenot, E.D., Masfaraud, J.F. and Schwartz, C. (2013) Long-term assessment of natural attenuation: statistical approach on soils with aged PAH contamination. *Biodegradation* 24(4), 539-548.

Ralston, E.P., Kite-Powell, H. and Beet, A. (2011) An estimate of the cost of acute health effects from food and water-borne marine pathogens in the USA. *Journal of Water and Health* 9(4), 680-694.

Wiedemeier, T.H., Rifai, H.S., Newell, C.J. and Wilson, J.T. (1999) *Natural Attenuation of Fuels and Chlorinated Solvents in the Subsurface.*

#546

# SURVEY OF PHYSICAL-CHEMICAL QUALITY OF SUPERFICIAL WATERS AT THE SURROUNDING AREA OF THE SEBKHA BASIN (MOKNINE, TUNISIA)

TAOUFIK NAAMOUN (University of Sfax, Faculty of Sciences, Department of Geology, Route de Soukra Km 3,5; P.B. 1171, Sfax, 3000, Tunisia)

The present investigation was conducted to control the physical-chemical parameters of superficial waters in the sebkha basin to evaluate their degree of pollution. Also, we aim to study the environmental impacts of the anthropogenic inputs on the different compartments of the sebkha. Twenty two water samples were collected and analyzed. The first results have shown that the obtained temperatures reflect those of the air the days corresponding to the field campaigns. The pH is oscillating between near neutral to slightly basic. The electro conductivity reveals that all samples are excessively mineralized. The salinity ranges from saline to hyper saline. Moreover, most samples contain potential reserves of hazardous metals. However, most of the measured concentrations exceed the limits for cadmium and lead for the sewage. Their concentrations reached 299ppb and 2.58ppm respectively.

# ASSESSMENT OF WATER BALANCE IN THE SOUTH-WEST COASTAL REGION OF BANGLADESH

***Nepal C Dey***[1*], Mahmood Parvez[1] and Nur A. Khondaker[2]
([1]Research and Evaluation Division, BRAC, 75 Mohakhali, Dhaka 1212; [*]Department of Environmental and Occupational Health, George Washington University, USA; [2]FAO Representation in Bangladesh, House # 32, Road # 8, Dhanmondi R/A)

Coastal environment and morphology are relatively less suitable for traditional crop cultivation, mainly because of soil and water salinity and inefficient water management system. As availability of fresh water is a limited resource mainly in the lean period, it is very important to assess how efficiently water can be managed considering development of more conservation sources for rain water harvesting to use in the lean period. This paper has been extracted from the water balance study conducted in South-West Coastal Region of Bangladesh in 2014. The aim of this paper is to make an assessment of utilizable water resources by looking at the inflow-outflow process in the South-West coastal region of Bangladesh. This paper focuses specially on irrigation adoption based on surface water availability in order to propose an effective planning, and on efficient management of the water resources. Assessment of monthly water balance for Khulna, Patuakhali and Barguna districts has been made based on the various inflow and outflow components in the catchment. Groundwater fluctuation method has been applied to estimate the monthly aquifer recharge. Availability of groundwater and surface water has been assessed based on demand and supply of water for the catchment. Temporal and spatial variation of soil and water salinity has been studied thematically. Study shows over 37 percent of total rainfall results into surface-runoff in this region. As distribution of rainfall is not uniform across the study area, situation of water balance also varies in two major districts, Khulna and Patuakhali of the South-West region. Most of the amount of run-off occurs during monsoon season (June to September) when there is relatively low scarcity of irrigation water. On the other hand, there is a lack of availability of irrigation water during dry season. Two different conditions of groundwater resources have been found in the coastal area, declining trend over the last three decades in Khulna area, and increasing trend in Patuakhali-Barguna area. Groundwater table fluctuates within 1-4m from the surface round the year in the coastal area. The salinity level of groundwater remains above the critical (3 ds/m) in most of the months in Khulna region and within the standard level with 1.6-2.0 dS/m throughout the year in Patuakhali region. Water level in major rivers in this region shows an increasing trend over the last three decades. So, sufficient river water is available in this area, but due to high salinity during most of the time surface water irrigation is not very suitable. From water availability analysis it is evident that a significant part of future irrigation water demand can be met by properly conserving the runoff resulting from precipitation using surface structures. The difference in annual irrigation water volume and the surface storable amount (runoff) is the total annual deficit that can be met from other sources, viz. subsurface, improving water productivity, improving water management practice, such as AWD, conjunctive use of saline water and other sources. The area has abundance of water resources to meet irrigation requirement even with increased cropping intensity in future, but requires proper planning to reduce the pressure on groundwater use.

# APPLICATION OF WATER QUALITY INDEX FOR GROUNDWATER QUALITY ASSESSMENT OF NANDED TEHSIL, (M.S.) INDIA

***Vasant Wagh,*** Dipak Panaskar, Ranjitsinh Pawar, Aniket Muley and Shrikant Mukate
(Swami Ramanand Teerth Marathwada University, Dnyanteerth, Vishnupuri, Nanded (MS), India)

**ABSTRACT:** Groundwater is an essential natural resource constituent of our life support system, where, it is used for drinking and irrigation purpose. An attempt has been made to recognize the hydro chemical characteristics of groundwater to develop water quality index model (WQI) in Nanded Tehsil. A total of 50 representative groundwater samples were collected from dug/bore wells during post-monsoon 2012 and analyzed for major cations and anions. The groundwater quality were assessed by different physicochemical parameters such as pH, electrical conductivity (EC), total dissolved solids (TDS), total hardness (TH), calcium ($Ca^{2+}$), magnesium ($Mg^{2+}$), sodium ($Na^{+}$), potassium ($K^{+}$), carbonate ($CO^{-}_{3}$), bicarbonate ($HCO^{-}_{3}$), chloride ($Cl^{-}$), nitrate ($NO_3^{-}$), sulphate ($SO_4^{-}$) and phosphate ($PO_4^{-}$). The Hydrochemical results were compared with Indian standards (BIS) which illustrate that TDS (16%), TH (22%), Ca (2%), Cl (2%) and Na (12%) samples exceeds the permissible limits. Water quality index (WQI) used to classify water quality as excellent, good, poor, and unsuitable categories. WQI shows that 34% samples are excellent, 60% samples are good and 6% unsuitable for drinking purpose. Through clutching hydro chemical investigation, GIS based IDW technique was used to signify the spatial variation WQI of study area.

**Keywords:** Groundwater, Water Quality Index, GIS, Nanded.
Email: wagh.vasant@gmail.com

## INTRODUCTION

Groundwater is an important natural resource for drinking water particularly in rural areas of many counties like India. Poor drinking water quality leads to extensive acute and chronic diseases and, in many countries is a major cause of death (USEPA 2007). In the world the quality of water resources is being gradually more degraded as a consequence of its intensified anthropogenic exploitation (Causape et al. 2004). Generally, overexploitation of groundwater leads to quality and quantity is distorted due to human induced activities, agricultural practices, domestic sewage etc. Just the once, the groundwater is polluted, the restoration of such water is still remained a major challenge, therefore, it is crucial to develop the water quality monitoring system for fitness of water for different purposes. Groundwater quality is mainly defined in terms of its physical, chemical and biological characteristics for drinking. Naturally, groundwater quality depends on geological and geochemical composition of rocks, residence time of water, dissolution and precipitation of minerals and oxidation reduction reactions.

The water quality index (WQI) is most important tool to assess the water quality which is extensively helpful to researcher and policy makers for the sustainable management of groundwater. The WQI was first time suggested by Horton (1965). The WQI provides a single numerical digit by integrating water quality parameters with respective regulatory standards, which understand the list of constituents and their concentration present in water (Satmbuk-Giljanovic 1999, Stigter et al 2006). The objectives of the present work are to perform the groundwater quality assessment based on WQI method and generation of WQI map to represent its spatial distribution by using IDW technique for Nanded Tehsil, Maharashtra.

## STUDY AREA

Nanded Tehsil is administrative block of Nanded District located in South Eastern Part of Deccan Volcanic Province of Marathwada region of Maharashtra state. The study area is located between latitudes $19^{0}$ 3' N to $19^{0}$ 17'N and longitude $77^{0}$ 10' E to $77^{0}$ 25' E (Fig. 1). The study area comprises with semi arid region and tropical climate. The temperature of study area is ranging from 13 to 46 $^{0}$C. The average rainfall of the study area is 900 mm, where, 88 % rain receives under the influence of South West Monsoonal winds. Godavari River flows in South West to South East direction having alluvial plain along its coast in the study area. The thickness of alluvium varies at different places and maximum 20 m encountered in flood plain areas of rivers. The study area is principally irrigated by Godavari and its sub-tributary Asna River and also left bank canal taking off water from Siddheshwar dam of the Purna project, of Parbhani district. The North East part of the study area is the convergence point of Asna and Godavari River. The study area underlain by geological formation mainly Deccan Basalt flows of vesicular, amygdaloidal, weathered, fractured basalt. Black cotton soil and loamy to sandy soil are main soil types covering the study area. The soil has high productivity value for the crops like Cotton, Jawar, Soyabean etc due to rich in plant nutrients.

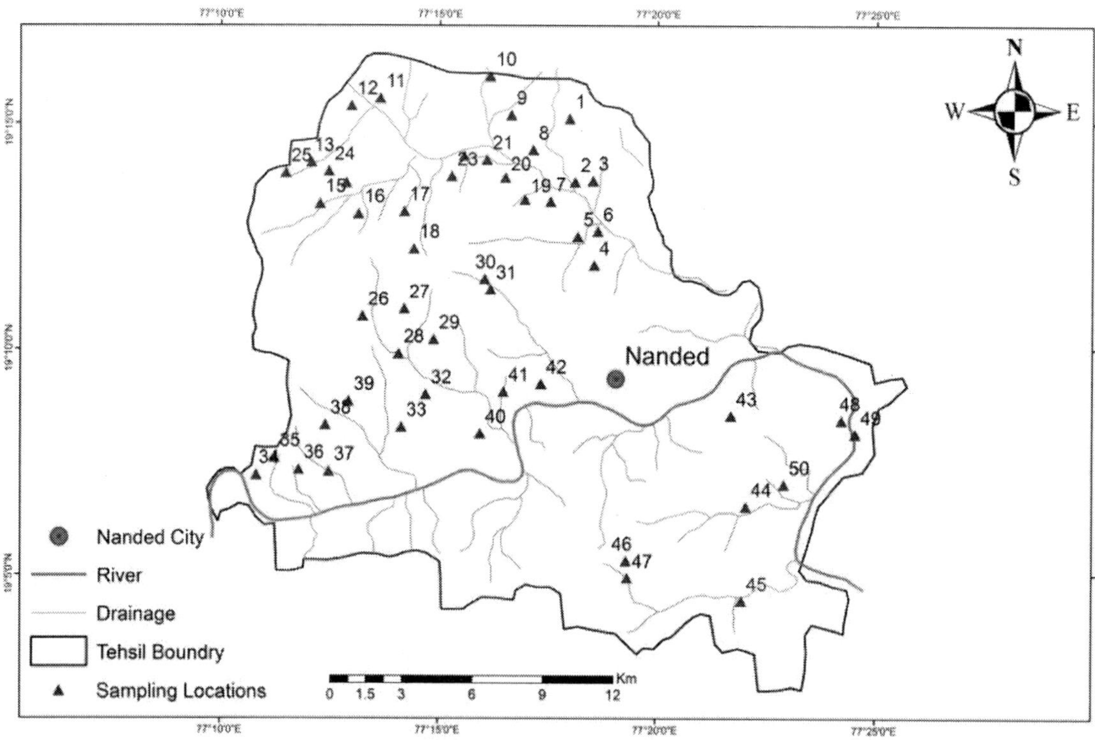


**FIGURE 1**. Study Area Map with sample locations

## MATERIAL AND METHODOLOGY

A fifty representative groundwater samples were collected from dug wells and bore wells from different locations of Nanded Tehsil for the period of post-monsoon season 2012. Survey of India, Toposheet no's 56 E/5, 56 E/8, 1:50,000 scales were used for the preparation of water quality index map of the study area. The groundwater samples were collected in pre-washed one-litre polyethylene cans after pumping the wells for 5-10 minutes; afterwards these cans were properly sealed, labelled and brought to laboratory for further physicochemical analysis. Location coordinates were recorded by using GPS device. pH and Electrical conductivity measured in the field itself by using portable digital meter. Calcium, magnesium carbonate, bicarbonate and chloride were analyzed in the laboratory by gravimetric analysis following the standard methods of American Public Health Association (APHA 2005). The sodium and potassium ion was determined by using flame photometer (Elico CL 361). The sulphate, phosphate, and nitrate were determined by using spectrophotometer (Shimadzu UV-1800). The Arc GIS 9.3 software was used for the preparation of water quality index maps.

**Water Quality Index Method.** Water Quality Index (WQI) is a technique of rating that provides the composite influence of individual water quality parameters on the overall quality of water for human consumption (Mitra et al 2006). It is computed in three steps (Horton 1965, Yidana 2010). In first step the weight is assigned to each parameter according their relevance to drinking quality of water. The most significant parameter weighted with 5 and least significant with 1. Second step is calculation of relative weight (RWi), computed with the help of eq. No. 1. The third step is quality rating scale (qi) calculation based on dividing the parameter concentration in each water sample by its respective standard (BIS 1998) multiplied by 100 and its computed with the help of eq. No. 2. The Wi and qi are used to calculate the Sub index SIi of each parameter following eq. No. 3. Finally, the water quality index (WQI) is calculated from eq. No. 4.

$$RWi = Awi / \sum AWi \tag{1}$$

Where, RWi = Relative weight,
AWi = Assigned Weight
n = Number of parameters

The quality rating scale for each parameter is calculated with the following equation

$$qi = (ci/si)*100 \tag{2}$$

Where, qi = Quality rating for $i^{th}$ parameter
Ci = Concentration of $i^{th}$ chemical parameter of water sample (mg/l)
Si = permissible standard for $i^{th}$ parameter set by BIS

Sub index (SIi) for each parameter is determined by equation 3 and final WQI is calculated by following equation 4

$$(SIi = RWi * qi) \tag{3}$$

Where, SIi = sub index of $i^{th}$ parameter
qi = rating based on concentration of $i^{th}$ parameter

$$WQI = \sum SIi \tag{4}$$

## RESULTS AND DISCUSSION

**Groundwater Quality Assessment.** The descriptive statistics including minimum, maximum, average and standard deviation of the groundwater samples collected during post-monsoon 2012 are tabulated in Table 1. A total 14 physicochemical parameters were assessed to evaluate the groundwater suitability for drinking purpose. Analytical results for all the physicochemical parameters were compared with BIS (1998) standards, shows that the average pH value in study area is 8.3 which is slightly alkaline in nature where, 6 samples i.e 12% are above the permissible limit. In the study area 16% samples surpass the permissible limit (2000 mg/l) of TDS. Such, high concentration in groundwater is due to leaching of salts from soil and anthropogenic activities, which leads to corrosion of metals. The total hardness ranges from 88.0 to 946.0 mg/l with an average of 233.3 mg/l. It was found that TH value in all groundwater samples are within permissible limit except sample no. 44 (946.0 mg/l). The sample no. 44 is located close to Tuppa industrial area and quality affected due to the percolation untreated effluent or leaching of minerals. Only 2% samples exceed the permissible limit of Calcium which may be due to erosion of parent rock material such as limestone and dolomite. Sodium in groundwater ranges from 17.2 to 630.2 mg/l, where, 12% samples exceed the permissible limit. Generally, the sodium salts occur in groundwater due to rock weathering or dissolution of soil salts stored by the influence of evaporation and it also, indicates its high solubility behavior (Stallard and Edmond 1983). The high chloride concentration was found in sample no. 44 which is far beyond the permissible limit. Consumption of such high chloride content water affects on taste, indigestion, corrosion and palatability, which also leads to heart and kidney related diseases (CPCB 2008). It is observed that the values of Magnesium, Potassium, Carbonate, Bicarbonate, Sulphate and Phosphate are within permissible limit.

**TABLE 1.** Descriptive Statistics of groundwater quality parameters (n = 50)

| | **pH** | **EC** | **TDS** | **TH** | **Ca** | **Mg** | **Na** | **K** | **$CO_3$** | **$HCO_3$** | **Cl** | **$NO_3$** | **$SO_4$** | **$PO_4$** | **WQI** |
|---|---|---|---|---|---|---|---|---|---|---|---|---|---|---|---|
| **Min.** | 7.8 | 816 | 522.2 | 88 | 5.6 | 8.3 | 17.2 | 2 | 0 | 80 | 63.9 | 0.4 | 3.9 | 0 | 38.1 |
| **Max.** | 8.8 | 64400 | 41216 | 946 | 352.7 | 81.4 | 630.2 | 8.6 | 80 | 460 | 1043.7 | 7.2 | 45.6 | 0.3 | 398.7 |
| **Avg.** | 8.3 | 4792.4 | 3067.1 | 233.3 | 36 | 34.6 | 107.4 | 2.7 | 25 | 181.4 | 175.5 | 3.3 | 15.7 | 0.1 | 74.7 |
| **S.D.** | 0.2 | 13120.4 | 8397 | 135.1 | 50.5 | 16.8 | 112.8 | 1.3 | 16.3 | 71.5 | 144.4 | 1.7 | 10.2 | 0.1 | 71 |

**TABLE 2.** Relative weight of each water parameter

| Parameters | pH | TDS | TH | Ca | Mg | Na | K | $CO_3$ | $HCO_3$ | Cl | $NO_3$ | $SO_4$ | $PO_4$ | Total |
|---|---|---|---|---|---|---|---|---|---|---|---|---|---|---|
| Weight (Wi) | 3 | 5 | 3 | 2 | 2 | 3 | 2 | 2 | 2 | 5 | 5 | 3 | 2 | 39 |
| Relative weight (RWi) | 0.077 | 0.128 | 0.077 | 0.051 | 0.051 | 0.077 | 0.051 | 0.051 | 0.051 | 0.128 | 0.128 | 0.077 | 0.051 | 1 |
| BIS Standards (1998) | 8.5 | 2000 | 300 | 75 | 30 | 100 | 10 | -- | 200 | 250 | 45 | 200 | -- | |

**TABLE 3.** Classification of groundwater samples based on Water Quality Index

| Category | Water Type | Number of Samples | % of Samples |
|---|---|---|---|
| 0-50.0 | Excellent | 17 | 34.0 |
| 50.01-100.0 | Good | 30 | 60.0 |
| 100.01-200.0 | Poor | -- | -- |
| >200.01 | Unsuitable | 03 | 6.0 |
| Total | | 50 | 100 |

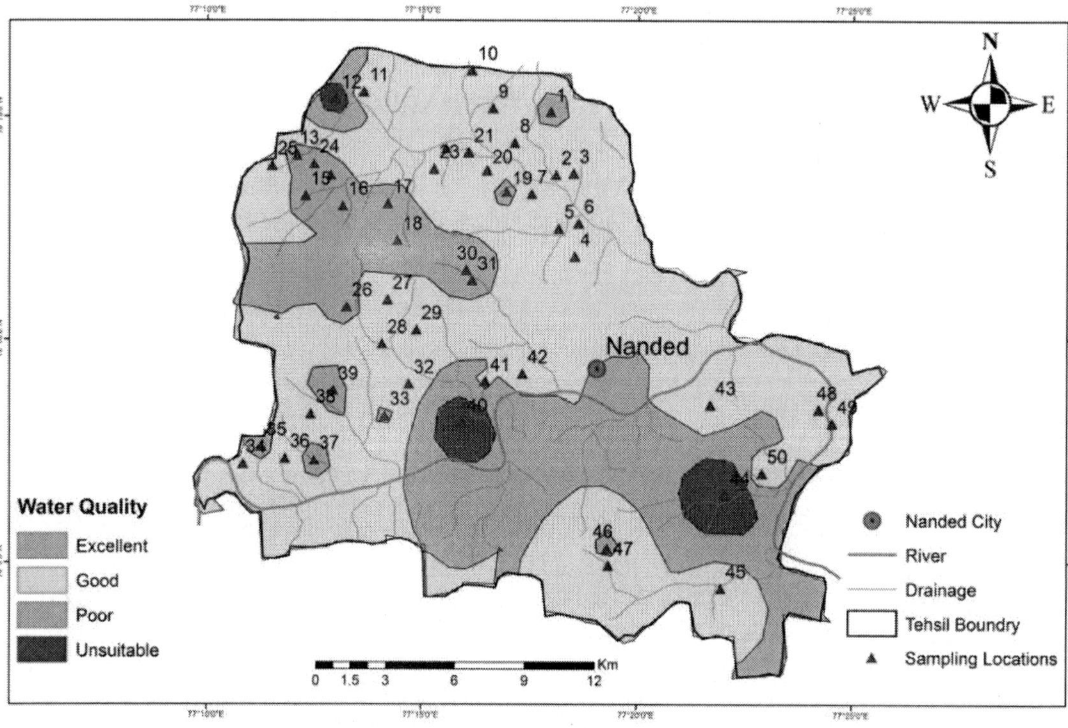


**FIGURE 2.** Spatial distribution of Water Quality Index

**Spatial Distribution of Water Quality Index.** The groundwater quality of Nanded tehsil for drinking purpose was evaluated through water quality index method. The water quality index values were interpolated in Arc GIS using IDW technique for preparation of water quality index map of the study area. The different categories in the map was based on the classification shown in Table 3. The water quality index value varies from 38.9 to 398.7 hence it shows large variation of water quality among the groundwater samples. The water quality index were classified into four water types namely Excellent (0-50.0), Good (50.01-100.0), Poor (100.01-200.0) and unsuitable (above 200.01). It is observed that 34% groundwater samples of in the North Western and few patches in N-NE and S-SW part of the study area comes under excellent water type. The good water type represents 60% groundwater samples identified in the study area as shown in Figure 2, hence; 94% samples are fit for drinking purpose. Only 3 samples, located in central part, S-SE near urban area and one patch in North part of the study area are unsuitable for drinking purpose.

## CONCLUSIONS

The present study was carried out to investigate the groundwater quality for drinking purpose in Nanded Tehsil. WQI method has been applied to categorize the different water class and GIS was used to represent its spatial extent in the study area. Analytical results suggests that groundwater in the Tehsil slightly alkaline and moderately hard to very hard in nature. TDS is found higher in 16% of samples, due to the leaching of salts. The concentration of sodium over calcium suggests that cation exchange process in the groundwater at many places. The water quality index of the study area revealed that majority of the samples belongs to excellent to good water type, hence, fit for drinking. Moreover, only 3 samples are unsuitable for drinking. These samples are mainly affected due to high concentration of TDS, TH and Sodium. This study suggests that the possible contamination sources of groundwater such as industrial/domestic waste and fertilizers must be controlled and monitored to protect this precious resource.

## ACKNOWLEDGEMENTS

The author expresses thanks to University Grant Commission for financial assistance to carry out this research work and Director, School of Earth Sciences for providing necessary facilities.

## REFERENCES

Apha, A. (2005). WEF, 2005. *Standard methods for the examination of water and wastewater*, *21*, 258-259.

Bureau of Indian Standard (BIS), (1998) "Drinking Water—Specification (First Revision) IS-10500." *BIS, New Delhi, India.*

Causapé, J., Quílez, D., & Aragüés, R. (2004). Assessment of irrigation and environmental quality at the hydrological basin level: II. Salt and nitrate loads in irrigation return flows. *Agricultural Water Management*, *70*(3), 211-228.

CPCB (2008) Guidelines for water quality management. Central Pollution Control Board, Parivesh Bhavan, East Arjun Nagear, New Delhi

Horton, R. K. (1965). An index number system for rating water quality.*Journal of Water Pollution Control Federation*, *37*(3), 300-306.

Mitra, B. K., Choichi Sasaki, and Enari Keijirou (2006) "Spatial and temporal variation of ground water quality in sand dune area of aomori prefecture in Japan." In *2006 ASAE Annual Meeting*, p. 1. American Society of Agricultural and Biological Engineers

Scheidt, D. J., & Kalla, P. I. (2007). Everglades ecosystem assessment: water management and quality, eutrophication, mercury contamination, soils and habitat: monitoring for adaptive management: a R-EMAP status report. USEPA Region 4, Athens, GA. *USEPA Region*, *4*, 98.

Stallard, R. F., & Edmond, J. M. (1983). Geochemistry of the Amazon: 2. The influence of geology and weathering environment on the dissolved load. *Journal of Geophysical Research: Oceans*, *88*(C14), 9671-9688.

Štambuk-Giljanović, N. (1999). Water quality evaluation by index in Dalmatia. *Water Research*, *33*(16), 3423-3440.

Stigter, T. Y., Ribeiro, L., & Dill, A. C. (2006). Application of a groundwater quality index as an assessment and communication tool in agro-environmental policies–Two Portuguese case studies. *Journal of Hydrology*, *327*(3), 578-591.

U.S. Environmental Protection Agency (USEPA) (2007) Dallas, TX (2000-05). Chapter 3: Exposure scenario selection. Retrieved Feb 200. RCRA Delisting Technical Support Document. P8

Yidana, S. M., & Yidana, A. (2010). Assessing water quality using water quality index and multivariate analysis. *Environmental Earth Sciences*, *59*(7), 1461-1473.

# HYDROCHEMICAL CHARACTERIZATION OF GROUNDWATER IN BHOKAR TALUKA OF NANDED DISTRICT, MAHARASHTRA, INDIA

***Dipak B. Panaskar,*** Vasant M. Wagh and Ranjitsinh S. Pawar
(School of Earth Sciences, Swami Ramanand Teerth Marathwada University, Nanded, Maharashtra, India)[17]

**ABSTRACT:** This study is carried out to analyse the Hydrochemical characterisation of Groundwater in Bhokar Taluka, where, fluoride contamination has been recognized as one of the major serious problems. In the study area groundwater is the main source of drinking water, with this contest; 89 representative groundwater samples were collected and analysed physicochemical parameters such as pH, EC, TDS, TH, Ca, Mg, Na, K, $CO_3$, $HCO_3$, TA, Cl, $SO_4$ and F for pre and post monsoon seasons 2011. It is observed that, pH has remained alkaline in nature in both seasons. Fluoride concentration was recorded in the range 0 to 6.75 mg/l and 0.02 to 6.8 mg/l with an average 0.82 and 0.65 mg/l in pre and post monsoon season; where, the desirable limit is 0.8 to 1.5 mg/l (ISI, 1983; ICMR, 1987). According to Wilcox (1955) classification, EC of the majority of the groundwater samples fall in permissible water class. TDS classification proved that majority of the groundwater samples fall in non-saline and slightly saline class. TH classification shows that majority of the groundwater samples of the study area fall in very hard water class. The Ca, Na, Cl, and $SO_4$ content of the majority of groundwater samples under the highest desirable limit given by WHO.

**Keywords:** Groundwater, Hydrochemical, Watershed, Fluoride, Bhokar
Email: dbpanaskar@gmail.com

## INTRODUCTION

Groundwater is primary source of water for domestic, agricultural and industrial uses in many countries like India. It is estimated that around one third of world population uses groundwater for drinking purpose (UNEP 1999). The groundwater demand has increased multiple over the years due to population growth, urbanisation, industrialization and agricultural intensification. The groundwater quality is being deteoriated mostly due to overexploitation, increase use of pesticides and fertilizers, improper management of domestic and industrial waste in rural and urban area of many counties. In the developing world, 80 % of all diseases are directly related to poor drinking water and unsanitary conditions (Olajire and Imeokparia, 2001). The chemical composition of groundwater is controlled by many factors that include composition of groundwater from precipitation, geological structure and mineralogy of groundwater from the watersheds and aquifers, and geological processes within the aquifer (Andre et al., 2005).

Bhokar Taluka, the study area is under investigation, where, fluoride contamination in groundwater is reported which remains unexplored so far. Rural dwellers of the study area directly depend on groundwater for drinking and irrigation purpose. Geologically study area is covered by basalt and is underlined by granite from 4 to 20 meter depth and granite is one of the important natural sources of fluoride contamination.

India is one of the 23 nations, which is reported to be endemic to fluorosis disease (UNICEF 1999). In India, nearly 60-65 million people drink fluoride contaminated groundwater leads to 2.5 to 3 million people affected by fluorosis in states like Andhra Pradesh, Bihar, Gujrat, Madhya Pradesh, Punjab, Rajasstan, Assam etc.(Athavale and Das1999; Susheela1999; Rao 2008; Sharma et al.2011). In India the desirable limit of fluoride in drinking water is 0.8 to 1.5 mg/l (ISI, 1983; ICMR, 1987). A higher concentration of fluoride causes dental and skeletal fluorosis and concentration less than 0.8 mg/l cause dental carries.

The objective of the study is Hydrochemical characterization and appraises their suitability for drinking purpose, in addition particularly focusing on fluoride contamination in groundwater of Bhokar Taluka of Nanded District, Maharashtra.

## STUDY AREA

For the present study the Sita Nadi Watershed area of Bhokar Taluka, Nanded District has been selected (Fig. 1). The study area is bounded by longitude $77^0$ 35'and $78^0$ E and latitude $19^{0}$'3' and $19^0$ 17'N. Geologically, study area is covered by basalt and is underlined by granite from 4 to 20 meter depth and granite is one of the important natural sources of fluoride contamination. Quartz vein is also intruded between Divshi and Kolgaon villages. The width of quartz vein is about 4 meter, and quartz is one of the important sources for fluoride contamination in water.

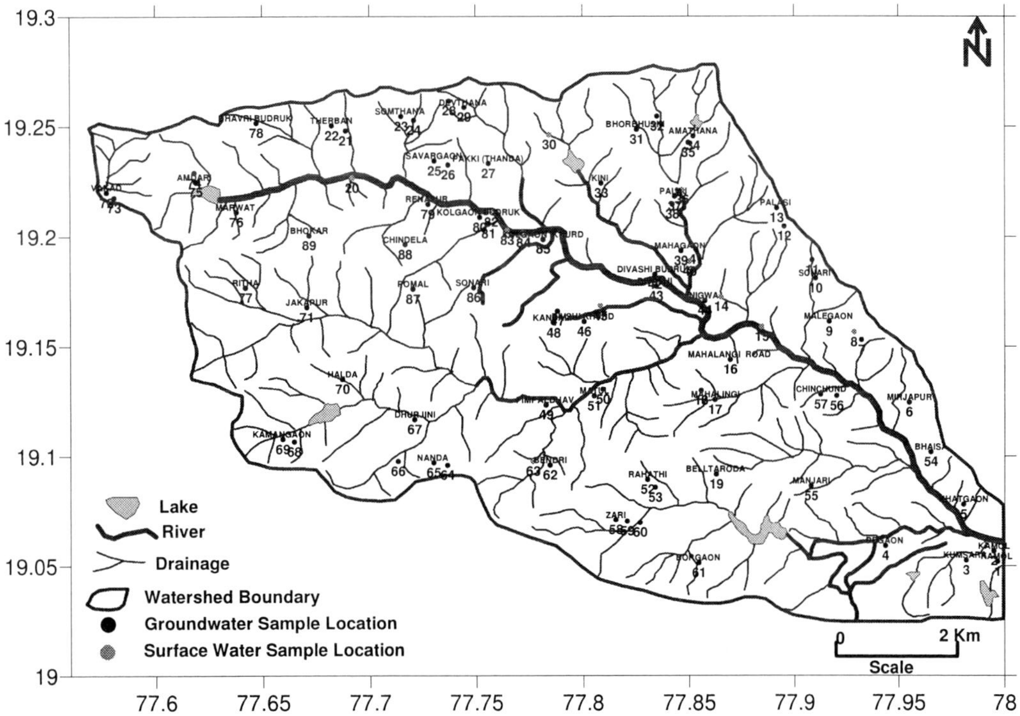


**FIGURE 1.** Study Area Map with sample locations

## MATERIAL AND METHODOLOGY

Total 89 representative groundwater samples have been collected for pre monsoon and post monsoon season 2011. pH and Electrical conductivity (EC) were measured in the field itself by using portable digital meter. The left over physicochemical parameters such as Ca, Mg, $CO_3$, $HCO_3$ and Cl were analyzed in the laboratory by following the standard methods of American Public Health Association (APHA 2005). The sodium (Na) and potassium (K) ions was determined using flame photometer. The $SO_4$, $PO_4$, $NO_3$ and F were determined by using spectrophotometer. Surfer 9.0 software was used for the preparation of spatial distribution map of fluoride.

**Table 1.** Descriptive statistics of Physicochemical Parameters of Pre Monsoon Season 2011

| Sr. No. | pH | EC | TDS | TH | Ca | Mg | Na | K | Cl | $HCO_3$ | $CO_3$ | TA | $SO_4$ | F |
|---|---|---|---|---|---|---|---|---|---|---|---|---|---|---|
| **Min.** | 7.5 | 574 | 367.36 | 46 | 3.21 | 2.92 | 8.1 | 5.5 | 18.46 | 35 | 0 | 45 | 29 | 0 |
| **Max.** | 8.9 | 5740 | 3673.6 | 712 | 193.99 | 172.5 | 406.4 | 266.4 | 710 | 440 | 40 | 455 | 132 | 6.75 |
| **Avg.** | 8.31 | 1630.5 | 1043.5 | 310.29 | 28.94 | 56.45 | 90.72 | 28.94 | 147.73 | 173.71 | 17.02 | 190.73 | 65.67 | 0.82 |
| **StD.** | 0.28 | 936.17 | 599.1 | 161.38 | 27.25 | 42.11 | 80.33 | 54.93 | 118.79 | 65.75 | 8.07 | 67.25 | 25.08 | 1.01 |

## RESULTS AND DISCUSSION

The quality of groundwater is very important because it is the main factor which decides its suitability for domestic, industrial and agricultural purpose. Physicochemical parameters of groundwater samples from Bhokar watershed were determined and assessed in order to understand the variation of the various parameters and their interrelationship of various locations. Data obtained during the course of both field and laboratory analyses of groundwater samples have been discussed as minimum, maximum, average and standard deviation of the parameters, with locations during the pre and post monsoon seasons 2011 have been presented in Table 1 and 2.

**Table 2.** Descriptive statistics of Physicochemical Parameters of Post Monsoon Season 2011

| Sr. No. | pH | EC | TDS | TH | Ca | Mg | Na | K | Cl | $HCO_3$ | $CO_3$ | TA | $SO_4$ | F |
|---|---|---|---|---|---|---|---|---|---|---|---|---|---|---|
| Min. | 7.5 | 523 | 334.72 | 68 | 9.62 | 14.5 | 9.5 | 5.3 | 50.6 | 100 | 0 | 100 | 20 | 0.02 |
| Max. | 8.9 | 4256 | 2723.84 | 642 | 137.9 | 329.66 | 242.1 | 170.5 | 412.5 | 550 | 20 | 560 | 120 | 6.8 |
| Avg. | 8.22 | 1494.3 | 956.4 | 307.4 | 49.0 | 122.26 | 80.72 | 20.55 | 192.92 | 317.87 | 7.81 | 325.7 | 56.73 | 0.65 |
| StDev. | 0.26 | 744.9 | 476.7 | 133.2 | 24.62 | 70.13 | 64.45 | 33.38 | 81.21 | 92.11 | 5.69 | 91.83 | 23.35 | 0.94 |

**Table 3.** Classification of EC (Wilcox, 1955), TDS and TH.

| Range | Class | Sample Number (%) | |
|---|---|---|---|
| | | Pre 11 | Post 11 |
| **EC (µS/cm)** | | | |
| < 250 | Excellent | 00 | 00 |
| 250-750 | Good | 06(6.8) | 13(14.6) |
| 750-2250 | Permissible | 64(71.9) | 64(71.9) |
| 2250-5000 | Doubtful | 18(20.2) | 12(13.5) |
| > 5000 | Unsuitable | 01(1.1) | 00 |
| **TDS (mg/l)** (Rabinove et al. 1958) | | | |
| < 1000 | Non Saline | 57 (64.04) | 52 (46.28) |
| 1000-3000 | Slightly Saline | 32 (35.95) | 36(32.04) |
| 3000-10000 | Moderately Saline | 00 | 1 (0.89) |
| > 10000 | Very Saline | 00 | 00 |
| **TH (mg/l)** (Sawyer and McCarty, 1967) | | | |
| < 75 | Soft | 02 (3.4) | 02 (2.2) |
| 75 - 150 | Moderate | 16 (18) | 06 (6.7) |
| 150 - 300 | Hard | 27 (30.3) | 35 (39.3) |
| > 300 | Very Hard | 43 (48.3) | 46 (51.7) |

pH of groundwater varies from 7.5 to 8.9 with an average 8.31 for pre and post monsoon season 2011. It can be seen that pH has remained alkaline in both the seasons. pH values of 100 % groundwater samples for both the seasons are above the highest desirable limit, but below maximum permissible limit set by WHO (1998).

Electrical Conductivity (EC) of groundwater samples varies from 574 to 5740 µS/cm and 523 to 4256 µS/cm with an average 1630.45 µS/cm and 1494.34 µS/cm for pre and post monsoon season 2011 respectively. In pre monsoon season high EC is recorded due to ion exchange and solublization in the aquifer. According to Wilcox (1955) classification of EC (Table 3) the majority of the groundwater samples 71.9 % in the Pre and post monsoon seasons 2011 of the study area falls in permissible water class. Total Dissolved Salts (TDS) of groundwater samples varies wide ranges 367.36 to 3673.6 mg/l and 334.72 to 2723.84 mg/l with an average TDS is 1043.52 mg/l and 956.38 mg/l for pre and post monsoon season 2011 respectively. Generally, the higher concentration of TDS in the groundwater is seen due to the contribution of salts from the soil, weathered parent rock materials and also more residence time in an aquifer body (Darr et al. 2011) According to the classification of TDS (Table 3) the majority of the groundwater samples 64.04 %, 46.28 % and 35.95 %, 32. 04 % falls in non saline and slightly saline class in pre and post monsoon season respectively, however only one sample (sample no. 41) in post monsoon season falls in moderately saline class.

Total Hardness (TH) of groundwater samples varies from wide range of 46 to 712 mg/l and 68 to 642 mg/l with an average 310.25 mg/l and 307.37 mg/l in pre and post monsoon 2011 respectively. Majority of the groundwater samples 48.3 % and 51.7 % in the pre and post monsoon season of 2011 respectively show high TH values than maximum permissible limit given by WHO (1998). According to the classification of TH (Table 3) the majority of the groundwater samples 48.3 % and 51.7 % fall in very hard water class in the pre and post monsoon season. Hardness was slightly increased during post monsoon season; which can be attributed to the leaching content like calcium and magnesium from soil, which is responsible for increase in hardness due to precipitation. Calcium (Ca) of groundwater samples widely varies from 3.21 to 193.99 mg/l and 9.62 to 137.9 mg/l with an average 28.94 mg/l and 49.0 mg/l in the pre and post monsoon season 2011respectively. Majority of the groundwater samples 94.4 % and 86.5 % in the pre and post monsoon season of 2011 show less Ca values than desirable limit given by WHO (1998) indicating groundwater is good for drinking purposes. Magnesium (Mg) of groundwater samples show wide range from 2.92 to 172.5 mg/l and 14.5 to 329.66 mg/l with an average 56.46 mg/l and 122.26 mg/l in the pre and post monsoon season of 2011 respectively. Majority of the groundwater samples 66.3 % in the pre monsoon season 2011 shows less Mg values than maximum permissible limit and 54 % in the post monsoon season of 2011 show High Mg values than maximum permissible limit given by WHO (1998).

Sodium (Na) of groundwater samples ranges from 8.1 to 406.4 mg/l and 9.5 to 242.1 mg/l with an average 90.72 mg/l and 80.72 mg/l in pre and post monsoon season of 2011 respectively. Majority of the groundwater samples 86.5 % and 91 % in the pre and post monsoon season of 2011 show Less Na content than the highest desirable limit. Potassium (K) of groundwater samples varies from 5.5 to 266.4 mg/l and 5.3 to 170.5 mg/l with an average 28.94 mg/l and 20.55 mg/l in the pre and post monsoon season of 2011 respectively. Chloride (Cl) of groundwater samples varies from 18.46 to 710 mg/l and 50.6 to 412.5 mg/l with an average 147.73 mg/l and 192.92 mg/l in the pre and post monsoon season of 2011 respectively. High chloride content in groundwater may result from natural and anthropogenic sources such as agricultural run-off, use of fertilizers, animal feeds and landfill leachate etc (Taiwo et al. 2011)

Carbonate ($CO_3$) of groundwater samples varies from 0 to 40 mg/l and 0 to 20 mg/l with an average 17.02 mg/l and 7.81 mg/l in the pre and post monsoon season of 2011 respectively. Bicarbonate ($HCO_3$) of groundwater samples varies from 35 to 440 mg/l and 100 to 550 mg/l with an average 173.71 mg/l and 317.87 mg/l in the pre and post monsoon season of 2011 respectively. The $CO_3$ and $HCO_3$ are possibly derived from weathering of silicate rocks, dissolution of $CO_3$ precipitates *and* atmospheric and soil $CO_2$ gas (Jeong 2001). Alkalinity (TA) of groundwater samples varies from 45 to 455 mg/l and 100 to 560 mg/l with an average 190.73 mg/l and 325.67 mg/l in the pre and post monsoon season of 2011 respectively. Sulphate (SO4) of groundwater samples varies from 29 to 132 mg/l and 20 to 120 mg/l with an average 65.67 mg/l and 56.73 mg/l in pre and post monsoon season of 2011 respectively.

Fluoride (F) of groundwater samples in the pre monsoon season of 2011 ranges from 0 to 6.75 mg/l. The maximum Fluoride was recorded from sample No. 11 (1.71 mg/L), 34 (1.91), 37 (1.71), 39 (6.75), 43 (4.02), 48 (1.90), 50 (1.81), 66 (1.81), 75 (1.95), 80 (2.12), 85 (3.98) and 89 (1.83). The spatial distribution of fluoride is as shown in figure 2 (a). The average content of Fluoride in groundwater is 0.82 mg/l which is considerably high, consumption of such water leads to dental and skeletal fluorosis (ICMR 1987). In the post monsoon season 2011, the Fluoride ranges from 0.02 to 6.8 mg/l. The maximum Fluoride was found at sample No. 37 (4.02), 39 (6.8), 42 (2.46), 43 (1.7), 50 (1.72), 85 (3.02). The spatial distribution is as shown in figure 2 (b). The average Fluoride value of the groundwater is 0.65 mg/l, this concentration of fluoride in groundwater may causes dental caries (ICMR 1987). Majority of the groundwater samples 77.5 %, and 87.6 % in the pre and post monsoon season illustrate that less F content than the highest desirable limit.

The desirable limit of fluoride in drinking water is 0.8 to 1.5 mg/l (ISI, 1983; ICMR, 1987). A higher concentration of fluoride causes dental and skeletal fluorosis and concentration less than 0.8 mg/l cause dental carries. The source of fluoride contamination in surface water and groundwater is either natural or anthropogenic or a mixture of both. The higher values of fluoride in groundwater can be attributed to leaching of fluoride from natural sources like granite, mica and other fluoride bearing minerals present in study area and anthropogenic activities like application of phosphate fertilizers because study area has intensive agricultural belt and farmers are using lot of phosphate fertilizers.

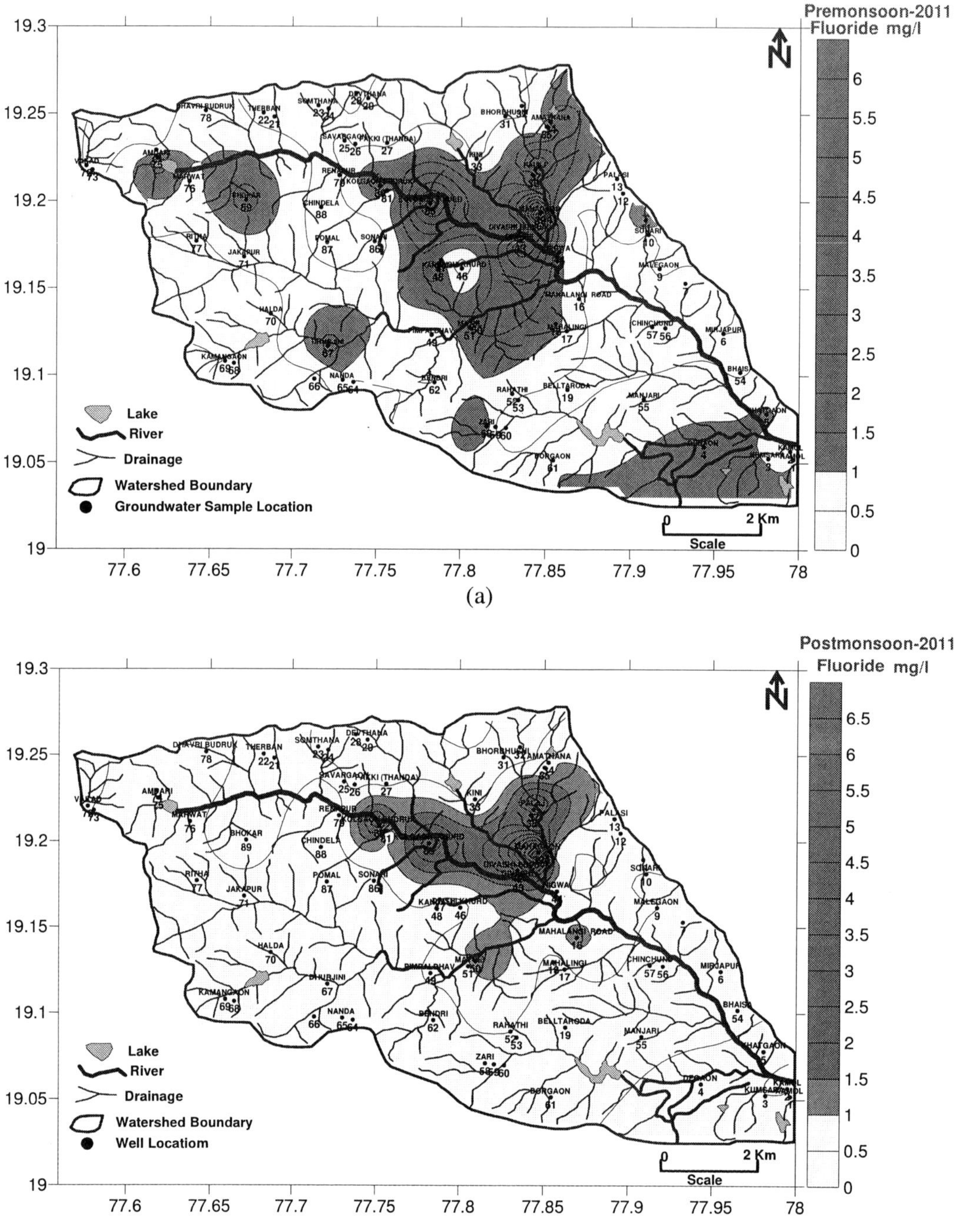


(b)

**FIGURE2.** Fluoride Concentration in Study Area (a) Pre-monsoon 2011, (b) post-monsoon 2011

Majority of study area is covered by basalt and is underlined by granite from 4 to 20 meter depth and granite is one of the important natural sources of fluoride contamination. Quartz vein is also intruded between Divshi and Kolgaon villages. The width of quartz vein is about 4 meter, and quartz is one of the important sources for fluoride contamination in water.

The major trend of the lineaments NW- SE and NE- SW. The major lineaments trend in the NW- SE direction while the minor lineaments trend in the NE- SW direction. The one major lineament i.e. Sita lineament trending in NW- SE direction which follows the Sita River course and passes from Renapur, Kolgaon Budruk, Kolgaon Khurd, Divshi Budruk and Divshi Khurd villages.

## CONCLUSIONS

Hydrochemical characterisation of groundwater of Bhokar Taluka was assessed to determine drinking suitability in pre and post monsoon season of 2011. Physicochemical interpretation confirms that groundwater is alkaline and moderately hard to very hard in nature in both the season of the study area. High EC is recorded in pre monsoon season due to ion exchange and solublization in the aquifer. The higher concentration of TDS in the groundwater is found due to contribution of salts from the soil and weathered rock materials. Hardness was slightly increased in post monsoon season; it can be attributed to the leaching of calcium and magnesium contents from soil. High chloride content in groundwater resulted from natural and anthropogenic sources. It is confirm that $CO_3$, $HCO_3$ and $SO_4$ contents are within permissible limit in both the seasons.

It is interesting to note that the groundwater samples (39, 42 and 43) which are in the vicinity of Sita lineament show higher concentration of fluoride. This can be attributed to the granitic rock as a source of fluoride which is underlined by basalt in this area. Since, the lineament is a weak zone; the fluoride from the granitic rock is getting leached out and contaminates the groundwater in this region. In the pre-monsoon season the fluoride concentration is more and extensively distributed as compared to post-monsoon season in the study area.

For sustainable management of groundwater in study area needs regular monitoring, epidemiological surveillance for dental and skeletal fluorosis check up, installation of fluoride removal plant and conduct fluorosis awareness program though public participation .

## ACKNOWLEDGEMENTS

One of the author, express thanks to Department of Science and Technology (DST), New Delhi, for financial assistance under major research project.

## REFERENCES

Andre, L., Franceschi, M., Pouchan, P., & Atteia, O. (2005). Using geochemical data and modelling to enhance the understanding of groundwater flow in a regional deep aquifer, Aquitaine Basin, south-west of France. *Journal of Hydrology*, *305*(1), 40-62.

APHA, A. (2005). WEF, 2005. *Standard Methods for the Examination of Water and Wastewater*, *21*, 258-259.

Athavale, R. N., & Das, R. K. (1999). Beware! Fluorosis is zeroing in on you. *Down to Earth*, *8*, 24-25.

Dar, I. A., Sankar, K., Shafi, T., & Dar, M. A. (2011). Investigation of groundwater quality in hardrock terrain using Geoinformation System. *Environmental Monitoring and Assessment*, *176*(1-4), 575-595

ICMR (1987), Indian Council of Medical Research; Manual of Standards of Quality for drinking water Supplies.

International Nutrition Foundation, & Micronutrient Initiative. (1999).*Preventing Iron Deficiency in Women and Children: Background and Consensus on Key Technical Issues and Resources for Advocacy, Planning, and Implementing National Programmes: UNICEF*. Micronutrient Initiative.

ISI (1983), (Indian Standard Institute) Indian Standard Specifications for drinking water, IS 10500.

Jeong, C. H. (2001). Effect of land-use and urbanization on hydrochemistry and contamination of groundwater from Taejon area, Korea. *Journal of Hydrology*, *253,*194–210

Olajire, A. A., & Imeokparia, F. E. (2001). Water quality assessment of Osun River: Studies on inorganic nutrients. *Environmental Monitoring and Assessment*, *69*(1), 17-28.

Rabinove CJ, Long Ford RH, Brook Hart JW (1958) Saline water sources of north Dakota. US Geological Survey Water Supply Paper 1428, 72.

Rao, N. S. (2009). Fluoride in groundwater, Varaha River Basin, Visakhapatnam District, Andhra Pradesh, India. *Environmental Monitoring and Assessment*, *152*(1-4), 47-60.

Sawyer, C. N., & McCarty, P. L. (1967). Chemistry for sanitary engineers. In *Chemistry for sanitary engineers*. McGraw-Hill

Sharma, P., Sarma, H. P., & Mahanta, C. (2012). Evaluation of groundwater quality with emphasis on fluoride concentration in Nalbari district, Assam, Northeast India. *Environmental Earth Sciences*, *65*(7), 2147-2159.

Susheela, A. K. (1999). Fluorosis management programme in India. *Current Science- Bangalore* , *77*, 1250-1256.

Taiwo, A. M., Adeogun, A. O., Olatunde, K. A., & Adegbite, K. I. (2011). Analysis of groundwater quality of hand-dug wells in peri-urban area of Obantoko, Abeokuta, Nigeria for selected physico-chemical parameters.*Pacific J. Sci. & Technol*, *12*(1), 527-534.

UNEP, 1999. *Global Environment Outlook 2000. Earthscan*, London.

UNICEF (1999) State of the report on the extent of fluoride in drinking water and the resulting endemicity in India. Report by fluorosis and Rural Development Foundation for UNICEF, New Delhi

Wilcox, L. (1955). Classification and Use Of Irrigation Waters.

World Health Organization. (1998). *Guidelines for Drinking-Water Quality*. Vol. 2, Health criteria and other supporting information: addendum.

# MODELLING OF ARSENIC REMOVAL FROM GROUNDWATER BY ELECTRODIALYSIS

***Rose Marie O. Mendoza*** (Far Eastern University-NRMF, Institute of Medicine, Quezon City, Philippines)
Meng-Wei Wan and Chi-Chuan Kan (Chia Nan University of Pharmacy and Science, Department of Environmental Engineering and Science, Tainan, Taiwan)
Maria Lourdes P. Dalida (National Graduate School of Engineering, Department of Chemical Engineering, University of the Philippines, Diliman, Quezon City, Philippines)

**ABSTRACT:** A model for Arsenic removal from ground water was evaluated. At varying experimental conditions, the second order homogeneous ordinary differential equation model proposed by Yu and Admassu on ion removal (from pulp and paper mill process stream) was able to predict the concentration, current utilization and removal of As ions from ground water samples efficiently. Two model parameters specifically for arsenic systems was derived: A is a parameter used to describe the characteristics of the solution, while $\lambda^2/B$ is a parameter that is related to the stack electrical field strength. Post hoc analysis of the model and $R^2$ values of 0.9899 and above was obtained for all parameters considered which indicate that the model can adequately describe the removal of arsenic from contaminated waters. The model was used to determine the effects of experimental parameters on Arsenic removal from ground water samples in terms of applied potential, feed flow velocity and initial feed concentration. A very satisfactory agreement between model predicted and experimental values of concentration, current utilization and removal efficiencies were obtained.

## INTRODUCTION

Arsenic-related incident covers almost majority of Asia. Bangladesh, Cambodia, China, India, Myanmar, Nepal, Pakistan, Taiwan and Vietnam are countries recently has been identified to use arsenic (As) contaminated ground water for irrigation. This poses serious health hazard to people eating food from the crops irrigated with arsenic contaminated water, in addition to the fact As accumulating in irrigated soils poses a serious threat to sustainable agriculture in affected areas (Brammer and Ravensoft, 2009). Bangladesh's As-contaminated groundwater has been extensively used to irrigate paddy rice, during dry season of 75% of the total area given to rice cultivation with arsenic levels of 4-8 μg/g and reaches to 83 μg/g (Rahman and Hasegawa, 2011). Because of the similar agro-ecological and hydrogeological conditions of these South East Asian regions the threat of arsenic contamination can easily reach neighboring countries. So far, arsenic contamination is widely detected to the alluvial plains of South and South East Asia where rice (paddy) is the main crop irrigated with ground water. According to Ravensoft (2007a,b), the GIS-based-geoogical-geochemical-hydrological models predict widespread pollution of ground waters in Indonesia, Malaysia and the Philippines – where water supplies were not yet tested for Arsenic.

Mendoza et al. (2014) in their study mentioned the viability of Electrodialysis (ED) as a process to remove inorganic As from contaminated waters. The study also stressed that experiments on arsenic systems should be carefully modelled and optimized to yield higher removal rates, while generating lesser waste by-products which can be more toxic than its source. Galvanin et al. (2015) opined that model-based design of experiment is a satisfactory approach to minimize the number of trials required for actual experimental runs.

To date, there is very less attention focused on model parameters for natural aqueous systems, particularly on arsenic removal and separation using electrodialysis (ED). The Nernst-Plank equation is one of the most utilized and widely-used model for a time-dependent transport and conversion on mass through the motion of charged chemical species in fluid medium (Kirby et al., 2010). Galvanin et al. (2015) utilized the Nernst-Plank approach to model the electrodialytic recovery of multi-valent electrolytes to accomplish desalination using NaCl while Zourmand et al. (2015) combined the Nernst-Plank model with that of Navier-Stokes model to develop a mass transfer model for the prediction of ion transport through electrodialysis (ED) cell using also NaCl solution. Mohammadi et al. (2005) as well as Sadrzadeh and Mohammadi (2007) provided a statistical model of metal ion removal from textile and semiconductor effluents using electrodialysis and electrodyliasis reversal by a Taguchi L9 OA design. It was only Mendoza et al. (2014) who used the electrodialysis (ED) process in removing arsenic from as-spiked water and come up with an optimum experimental conditions based on maximized As removal or separation. Thus, the burden of treating arsenic contaminated waters that will produce streams of higher arsenic concentration is one of the drawback of the study. The need to initially model arsenic systems to predict its removal from contaminated water will be of great contribution to the environmental sustainability of the said process. Through a model that will mimic the removal of arsenic from aqueous systems and with several related studies, lesser concentrated wastes will be generated which can gear the ED process towards zero-waste technology.

This paper investigates the removal of As from actual groundwater samples through the determination of the concentration and limiting current profile using the Yu and Admassu model (2000) for metal ions removal (from pulp and paper industry) and to come up with an model-generated predictions on arsenic removal at varied experimental conditions. The study also includes the model determination of the effect of experimental parameters such applied potential, feed flow velocity, initial feed concentration and operating time which is then later compared to the actual experimental data.

## MATERIALS AND METHODS

**Materials.** Groundwater samples used in this experiment were obtained from raw water source, well no. 9 and 10 of Taiwan Water Corporation's Water Treatment Plant in Beigang Township, Yunlin, Taiwan. Characteristics of the groundwater used for this study was presented in Table 1. The samples were filtered immediately through a 0.45μm pore size cellulose acetate syringe filter to minimize arsenic III oxidation with air. To further preserve arsenic species in the filtered sample, ethylenediaminetetraacetic acid (EDTA) was added immediately to the filtrate. This preserves the distribution of arsenic species by chelation to metal ions, prevents drastic changes in sample pH and minimizes microbial activity. Filtered samples were stored in PET containers at reduced head space and refrigerated at 4°C (Bednar et al., 2004).

**TABLE 1 Characteristics of the Ground Water Used in This Study**

| Cations | Concentration mg/L | Anions | Concentration mg/L |
|---|---|---|---|
| $Na^+$ | 145.50 | $Cl^-$ | 130.559 |
| $K^+$ | 34.515 | $NO_3^-$ | <0.1 |
| $Mg^{2+}$ | 18.97 | $SO_4^{2-}$ | 2.285 |
| $Ca^{2+}$ | 20.75 | $HCO_3^-$ | 119.70 |
| $Mn^{2+}$ | 0.093 | $PO_4^{3-}$ | 6.299 |
| $Fe^{tot}$ | <0.01 | TOC | 1.19 mg. C |
| $As^{tot}$ | 0.702 | | |

*Others: $E_h$ = -162.2 mV; pH = 7.68; T = 23°C; EC = 1388 μS/cm*

*(Obtained from: Taiwan Water Corporation's Water Treatment Plan, Well no. 9 and no. 10, Beigang County, Yunlin, Taiwan)*

The ORP and pH of the solutions was monitored using Suntex pH-ORP meter model SP-300. Current and Voltage measurements were performed using a Hao Ling HL 833 digital ammeter and voltmeter, respectively. The total As, As III and As V concentrations and concentrations of cations (Na, Mg, Ca, K, Mn and Fe) of the feed, diluate stream and concentrate stream was determined using a Perkin Elmer 2000 Optima DV Inductively Coupled Plasma - Optical Emission spectrophotometer equipped with a hydride generator (HG-ICP-OES). Anion concentration in the form of nitrates, nitrites, phosphates, sulphates and chlorides, were analyzed using a Dionex DX-120 ion chromatograph (IC) equipped with a Reagent Free Controller (RFC), carbonates were obtained by titrimetric methods and TOC by direct method for drinking water and wastewaters (Method 10129) using Hach DR 500 UV-VIS Spectrometer. Conductivities of the solution were determined using Suntex EC-410 digital conductivity meter attached to the control panel.

**Methods.** The tailor-made electrodialysis (ED) provided by Shell Kwong Sir Enterprise Co., was used to conduct As removal experiments in the recirculating batch mode. The ED stack is composed of a single hydraulic and single electrical stack with 20 cell pairs situated 0.9 mm apart. Operating temperature for the whole system was set at 25°C ± 1°C. Maximum parameters for the ED stack as recommended by the manufacturer were indicated in the study of Mendoza et al. (2014). Optimum experimental settings such as applied voltage of 17V, initial feed concentration (Co) of 700 ppb, feed flow rate of 0.033 l/s and a minimum operating time of 92 minutes used by Mendoza et al. (2014) in their study on As-spiked water was considered for model experimental runs. Groundwater samples were pre-treated in the pre-oxidation tank to transform $As^{3+}$ to $As^{5+}$, then fed into the feed tank. From the feed tank, contaminated groundwater sample was introduced in the ED stack at an optimum feed flow velocity, and the sample was desalted in 92 minutes using 17V as the applied voltage. As the feed was introduced, the tap water as the electrode flush solution was circulated into the anodic and cathodic compartments. The concentrate stream was dumped into the electrode flush tank and eventually became part of the electrode flush solution.

The samples from the diluate stream were immediately digested with 5% potassium iodide (KI), 5% vitamin C and concentrated HCl. The mixture was set aside for 25 minutes for stabilization and diluted to its marked volume using Ultrapure 18.2 mΩ deionized water. A solid phase extraction membrane, Supelco LC-SAX strong anion resin was used to immediately separate As III from As V at an elution time of 1 minute (Jang et al., 2011). The total As, As III and As V of the sample was analyzed using a Perkin Elmer Inductively Coupled Plasma-Optical Emission Spectrophotometer (HG-ICP-OES) Optima DV2000 at 228.812 nm, using Hydride Generation. A reducing agent in the form of 5% sodium borohydride ($NaBH_4$) in 0.5% sodium hydroxide (NaOH) was used counter current with the sample to prevent further sample oxidation or metal complexation phenomena. Calibration standards of 1, 5, 10, 50, 100 and 700 ppb are freshly prepared during sample analysis. Two 2 sample replicates, check samples and standards are included to countercheck the accuracy and precision of the determination of the concentrations using HG-ICP-OES. $R^2$ for each analysis was established at 0.999 or higher. Cations and anion content as well as TOC and bicarbonate were also evaluated.

## RESULTS AND DISCUSSION

In a published paper in 2000 by Z.Yu and W. Admassu, a model for the concentration and current profile was evaluated and used in removing metal ion from pulp and paper mill process stream. The model describes the transport of metal ions in the boundary layers at membrane surfaces and the limiting current density as a function of low velocity and removal efficiency. The general transport equation for the molar flux (J) driven by diffusion, migration and convection in the ED channels is governed by the Nernst-Planck equation (Yu and Admassu, 2000):

$$J_{\pm} = -z_{\pm}v_{\pm}FC_{\pm}\nabla\emptyset - D\nabla C_{\pm} + C_{\pm}u \tag{1}$$

The model was anchored on the following assumptions: (1) the potential is a function of location due to a constant current supply, and the external applied potential is considered constant thus potential gradient will also be considered constant. Thus, equation (1) becomes:

$$\frac{\partial C_{\pm}}{\partial t} + u\nabla C_{\pm} = z_{\pm}v_{\pm}F\nabla\cdot(C_{\pm} + \nabla\emptyset) + D_{\pm}\nabla^2 C_{\pm} \tag{2}$$

The motion of the charged particles will bring current into the system. The current intensity for the solution is defined as i=FΣzJ which gives the equation for the aqueous solution as:

$$\mathrm{i}_{\pm} = -F^2\nabla\emptyset\Sigma z_{\pm}^2 v_{\pm}C_{\pm} - F\Sigma z_{\pm}\mathrm{D}_{\pm}\nabla C_{\pm} + Fu\Sigma z_{\pm}C_{\pm} \tag{3}$$

To simplify the model equations the following assumptions were made:(1) for a quasi-steady state the time variation of concentration and the diffusion term along the flow direction is negligible compared to convention; and (2) neutrality of charge is satisfied in the whole channel which makes $z_+C_+ + z_-C_- = 0$

and (3) the influence of the flow profile of fluid is negligible which allowed the system to be considered as a plug flow fluid. The dilute nature of the solution makes its viscosity as that of water which will now flow the schematic flow in Figure 1 (Yu and Admassu, 2000). This follow that:

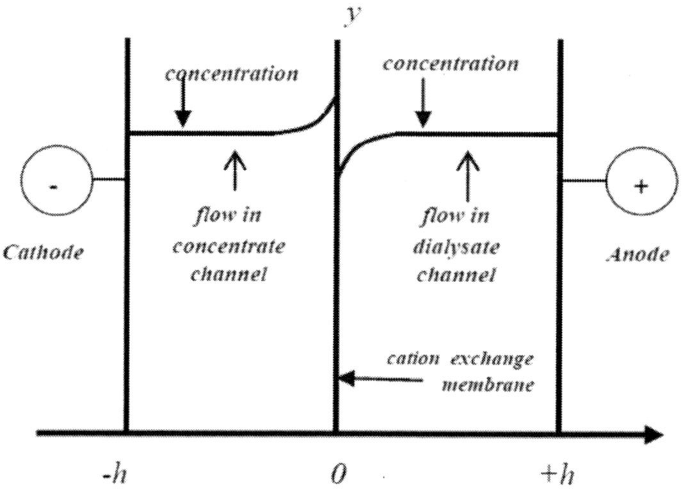


**FIGURE 1.** Flow Scheme of the Solution in the ED Stack

Simplifying equation (3) and inserting it to equation (2) results in:

$$u\frac{\partial C_+}{\partial y} = D_+\frac{\partial^2 C_+}{\partial x^2} + z_+ v_+ F\frac{d\emptyset}{d\emptyset}\frac{\partial C_+}{\partial x} \tag{4}$$

For the current density and simplification of equation (3) is based on the assumption previously expressed considering a constant potential gradient within the stack. This makes the effect of convection in equation $uC_\pm$ negligible. The current in the anionic membrane $i_- = 0$ since the current density in the cation exchange membrane is the only one being considered in the model. Thus, the current in the cationic membrane is:

$$i_+ = -D_+ F z_+\frac{dC_+}{dx} - \frac{F^2 z_+^2 D_+ C_+}{RT}\frac{d\emptyset}{dx} \tag{5}$$

and by principle of electroneutrality and non-ideality on membranes,

$$\frac{dC_+}{dx} = \frac{F z_+ C_+}{RT}\frac{d\emptyset}{dx} \tag{6}$$

Equation (2) is simplified by setting $A = u/D_+$ and another parameter as

$$B = {}^{z_+ v_+ F}/_{D_+} \cdot {}^{d\emptyset}/_{dx} \tag{7}$$

from a deduced dimensionless reduced concentration $C^* = C_w/C_0$, equation (2) becomes

$$A\frac{\partial c^*}{\partial y} = \frac{\partial^2 c^*}{\partial x^2} + B\frac{\partial c^*}{\partial x} \tag{8}$$

where $C_0$ is the initial concentration and $C_w$ is defned as the wall concentration in the boundary layers near the interface as shown in Figure 3 (Yu and Admassu, 2000). Analyzing Figure 2, the boundary conditions for equation (16) are set as: $x = 0$, $C^* = C_s$ $(C^*|_{x=0})$, $x \rightarrow h$ $(x>x_a)$, $C^* = C_h(C^*|_{x=h})$ and $y = 0$, $C^* = 1$ for $x > x_a$.

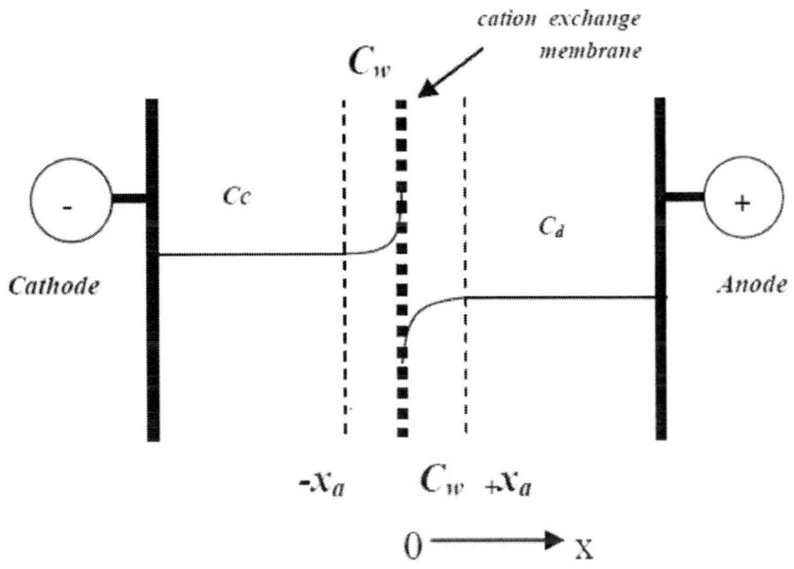


**FIGURE 2.** Boundary layer used in this study for the ED process

The transport of ions in this model assumes that it is occurs and only significant along the boundary layers, which is in the region where x is greater than 0 but less than $x_a$. From Figure 3 we can say that $x_a \ll h$ when $x > x_a$ and x → h, then x → ∞. This analysis of the boundary layers generates a general solution for equation (8) as:

$$C^* (x,y) = f(y)g(x) \tag{9}$$

which can be written as:

$$\frac{A}{f}\frac{df}{dy} = \frac{1}{g}\left(\frac{d^2g}{dx^2} + B\frac{dg}{dx}\right) \tag{10}$$

In the dialysate channel, the cation transport decreases along the flow direction (y). Introducing a constant $-\lambda^2$ to represent separation constant in the dialysate channel based on the boundary conditions set, equation (10) becomes:

$$\frac{df}{dy} + \frac{\lambda^2}{A}f = 0 \tag{11}$$

which is a second-order homogeneous ordinary differential equation with the form;

$$\frac{d^2g}{dx^2} + B\frac{dg}{dx} + \lambda^2 g = 0 \tag{12}$$

And an auxillary equation (Kreyzig, 2006):

$$K^2 + BK + \lambda^2 = 0 \tag{13}$$

Equation (13) has two roots:

$$K_{1,2} = -\frac{B}{2}\left\{1 \mp (1 - \frac{4\lambda^2}{B^2})^{1/2}\right\} \tag{14}$$

The general solution for Equation (12) (Kreyzig, 2006) is:

$$g(x) = N_1exp(K_1x) + N_2exp(K_2x) \tag{15}$$

Combining equation (11) and equation (13), the reduced concentration in the dialysate channel is:

$$C^*(x,y) = f(y)g(x) = \{m_1\exp(K_1x) + m_2\exp(K_2x)\}\exp(-\frac{\lambda^2}{A}y) \tag{16}$$

The integration constants $m_1$ and $m_2$ was determined using boundary conditions. For the purpose of simplification of equation (16) the following assumptions were made: $4\lambda^2/ B^2 \ll 1$ and $\lambda^2/ B \ll B$ so that $K_2 \approx -B$; at $y = 0$, $C^* = 1$ for $x > x_a$, x becomes large, exp (-Bx) → 0 faster than exp $(-\lambda^2/ Bx)$ because $\lambda^2/ B \ll B$. This yields,

$$m_1 = \exp(\frac{\lambda^2}{B} h) \tag{17}$$

and

$$m_2 = \frac{\exp\left(\frac{\lambda^2}{B}h\right)\left\{\frac{\left(\frac{d\phi}{dx}\right)_{m,d}}{\frac{d\phi}{dx}} + \left(\frac{RT}{z_+F}\right)\left(\frac{\lambda^2}{B}\right)/\left(\frac{d\phi}{dx}\right)\right\}}{1+\left(\frac{d\phi}{dx}\right)_{m,d}/\left(\frac{d\phi}{dx}\right)} \tag{18}$$

and gives the concentration profile at any location in the ED dialysate and concentrate channel, respectively as :

$$\frac{C_W}{C_0} = \left|1 - \frac{\left(\frac{d\phi}{dx}\right)_{m,d}/\frac{d\phi}{dx}}{1+\left(\frac{d\phi}{dx}\right)_{m,d}/\frac{d\phi}{dx}}\exp(-Bx)\right|\exp\left(-\frac{\lambda^2}{A}y\right) \tag{19}$$

For the current density profile equations (5), (6) and (19) yields:

$$i_{w.d} = -2D_+Fz_+C_0\left[B\frac{\left(\frac{d\phi}{dx}\right)_{m,d}/\frac{d\phi}{dx}}{1+\left(\frac{d\phi}{dx}\right)_{m,.d}/\frac{d\phi}{dx}}\right]\exp(-Bx)\exp\left(-\frac{\lambda^2}{A}y\right) \tag{20}$$

Equation (19) and (20) is then simplified by integrating both sides with the boundary condition $C_s = C_w|_{x=0}$ which results to:

$$C_w - C_s = -\frac{i_+}{2D_+Fz_+}x_a \tag{21}$$

where $i_+$ is the current flowing in the opposite direction of $x$, F is the Faraday's constant (96,480 C/mol), $D_+$ diffusion coefficient (cm²/s) of cation in the solution, z is the charge of ion and $x_a$ (mm) average thickness of the boundary layer. $C_w - C_s$ will be greater than 0 which is considered as the concentration of in the diluate compartment when boundary conditions $C_s = C_w|_{x=0}$ is considered. When $C_s \to 0$, $i_+ = i_{lim}$ and :

$$i_{lim} = \frac{2D_+Fz_+}{x_a}C_w|_{x\to 0} \tag{22}$$

From Kocherginsky (2010), equation (22) has a form:

$$i_{lim} = C_oe^{\frac{-u^*}{u}} \tag{23}$$

where $u^* = D_+ \lambda^2 y$ . Equation (23) indicates the limiting current density is a function of both initial concentration $C_o$ and flow velocity $u$. The magnitude of the parameter $\lambda^2$ is a function of initial concentration ($C_o$) removal efficiency (R%) and physical ED stack characteristics.

For equations (19), in the boundary layers, when $x \approx 0$, the concentration becomes independent of x as x → h. This means that the significant removal or separation of ions mainly occurs at the membrane walls/surfaces. As a function of y, the decrease in the concentrate channel results to an increase in concentration of the concentrate channel. Due to the transport of ions in the boundary layer, the system will tend to polarize when y = L as x →0. This makes the field strengths at membrane surface different in the cationic and anionic membrane surfaces. In this model, the field strengths are estimated using the continuity condition of electrical current at specified locations.

From equation (20), the value of current at entrance (y=0) is at maximum and decreases continuously as it approaches y = L. At the membrane surface (x=0), the current density also reaches the maximum value. In the boundary layer, the current density decreases as $x \rightarrow x_a$, beyond the boundary layer, $x \rightarrow h$ and $i \rightarrow 0$. This strongly suggests that concentration or concentration gradient is one of the factors that can influence current and current density.

From Ohms Law, the intensity of current is directly proportional to the electric field. As such there is a direct variation that can be observed between current density and the surface electrical field (dø/dx)m,d which in turn show a direct relationship between ion transport. This resulted in the direct proportionally between current density and initial concentration.

For parameter estimation of the model equations for concentration profile and current density at any location in the ED channel, R% is defined based on as a function of λ is obtained as:

$$S\% = R\% = \frac{C_o - C}{C_o} \, x\, 100 = \left(1 - exp\left(-\frac{\lambda^2}{A} L\right)\right) x\, 100 \qquad (24)$$

Equation (20) is true only when 0 < (R%) < 100. An evaluation of the concentrate channel, though not so significant in this study must also be performed to facilitate transport equation facilitation. For the concentrate channel, the bulk concentration increases along the flow direction (Yu and Admassu, 2000). To compare the separation of concentrate to that of the dialysate channel, another separation constant in the form of ξ is defined related to the gain efficiency in the concentrate channel which leads to:

$$R\% = \frac{C_L - C_o}{C_o} \, x\, 100 = \left[\exp\left(\frac{\xi^2}{A} y\right) - 1\right] x\, 100 \qquad (25)$$

and yields,

$$\xi^2 = \frac{A}{L} \, ln\left(1 + \frac{R'\%}{100}\right) \qquad (26)$$

**TABLE 2 Typical Values Used For Model Calibration And Model Parameters for the ED Process**

| Geometric Sizes | Symbol | Value | Source |
|---|---|---|---|
| Flow distance | L (mm) | 360 mm | This study |
| Flow width | W(mm) | 170 mm | This study |
| Half-ch,annel width | h (mm) | 5 mm | This study |
| Average film thickness | xa | 0.45 mm | This study |
| Number of cell pairs | | 20 | This study |
| *Characteristics of Feed Solution* | | | |
| Acerage flow speed | $u$ (c,m/s) | 0.0033 | This study |
| Valence (as H2AsO4) | z | 1 | Clifford and Ghurye, 2005 |

| | | | |
|---|---|---|---|
| Initial Concentration | Co (ppb) | 700 | This study |
| Diffusion Coefficient | D (cm2/s) | $0.809 \times 10^{-5}$ | Leaist (2007) |
| Density | ρ (g/cm3) | 1 | |
| Viscosity of water | μ( kg/ m s) | $900 \times 10^{-6}$ | Leaist (2007) |
| *Applied Conditions* | | | |
| Temperature | T (K) | 298 | This study |
| Constant electrical field | dθ/dx (vol/m) | 100 kg-m/$s^2$C | Yu and Admassu (2000) |
| Removal Effieiciensy | E% | 99.56 | This study |
| *Physical Constant* | | | |
| Faraday Constant | F (C/mol) | 96480 | |
| Gas Constant | R (J/K mol) | 8.413 | |

Non-ideality of membrane assumption ensures that there is no retention of cations at the membrane surfaces and in the membrane and indicate that the amount cations removed from the dialysate channel (R%) is equal to the amount of ions gained in the concentrate channel (R'%) which makes the magnitude of $\lambda^2 > 0$.

To estimate the magnitude of $\lambda^2/B$, model parameters based on typical values in ED process for KCl is used by the paper (Yu and Admassu, 2000), which gives an initial value for for Co and an average R% of 50 for batch ED process (Mc Neil and McCoy, 1989). This follows that:

$$\frac{\lambda^2}{B} = \frac{1}{L}\left|ln\left(1 - \frac{R\%}{100}\right)\right|\frac{uRT}{z_+D_+F(d\emptyset/dx)} = 0.018/\text{mm}. \quad (27)$$

$$B = \frac{z_+v_+F}{D_+}\frac{d\emptyset}{dx} = \frac{z_+F}{RT}\frac{d\emptyset}{dx} = 3.823/\text{mm} \quad (28)$$

$$\frac{\xi^2}{B} = 0.011/\text{mm}. \quad (29)$$

For the model calibration, Table 2 presents the nown physical parameters of the ED stack as well as the physical and chemical properties of the solution and the membranes while to facilitate model parameter estimation, typical values for ED process in the table below is used.

The very good agreement between the experimental and model predicted values was dictated by the $R^2$ values indicated in Table 3 which was found to be greater than 0.98 on all parameters set for the study. Two model parameters were considered for this study; A is a parameter used to describe the characteristics of the solution, while $\lambda^2/B$ is a parameter that is related to the stack electrical field strength. The increasing values of $\lambda^2/B$ with respect to increase in initial feed concentration indicate efficient current utilization in the stack and efficient collision of ions during desalination within the stack channel while a constant value of A as the initial concentration increases indicates that ion transport is not limited to solution characteristics, rather it is more dictated by current and potential. Figure 3a shows that the removal of As ions from the feed for both experimental and predicted values are more efficient at applied potentials above 10V.

**Effect of experimental parameters on Arsenic removal from groundwater using ED.** Model predicted values was compared with that of the experimental values in evaluating the effect of experimental parameters on arsenic removal. From Figure 3a model predicted values agrees satisfactorily with that of experimental values, confirming a direct relationship between applied potential %removal at variable concentration. The %removal was found to increase as the applied potential is increased at high or low initial feed concentrations. The same satisfactory agreement was found between % removal and feed flow velocity at high and low concentrations. However, an inverse relationship was found between %removal and feed flow velocity at varied concentrations as illustrated in Figure 3b. This can be due to the fact that the initial feed flow velocity was complemented by the applied potential resulting in a very fast transit of ions to the boundary layers, which indicates that in an ED system, very low or very high feed flow velocity are both non ideal conditions that may lead to neither increased energy consumption or polarization. Higher removal rate was found to at feed flow velocity of 0.033 l/s.

**TABLE 3 Parameters obtained for Yu and Admassu Model**

| Varied Parameters | Correlation Coefficient, $R^2$ | $\lambda^2/B$ $mm^{-1}$ | A $cm^{-1}$ |
|---|---|---|---|
| **Initial Feed Concentration[a]** | | | |
| 200 ppb | 0.993 | 0.63690 | 42.0272 |
| 500 ppb | 0.966 | 0.67654 | 42. 0272 |
| 700 ppb | 0.981 | 0.73093 | 42.0272 |
| **Feed Flow Velocity[b]** | | | |
| 0.0025 $m^3/s$ | 0.994 | 1.30843 | 30.9024 |
| 0.0033 $m^3/s$ | 0.999 | 1.30843 | 42.0272 |
| 0.0042 $m^3/s$ | 0.999 | 1.30843 | 51.9160 |
| **Applied Potential[c]** | | | |
| 5V | 0.980 | 0.03133 | 42.0272 |
| 10V | 0.993 | 0.00084 | 42.0272 |
| 17V | 0.993 | 0.00004 | 42.0272 |

[a] *Apllied Potential (E)=3-20V; feed flow rate = 0.0033 $m^3/s$; Time = 92 min.*
[b] *Apllied Potential (E)=17V; Initial feed concentration = 700 ppb; Time = 92 min.*
[c] *Feed flow rate = 0.0033 $m^3/s$; Initial feed concentration = 700 ppb; Time = 92 min.*

As observed from Figure 4, an increase in removal rate corresponds to decreases in current utilization. An increase in the electrical field strength parameter as applied potential is increased is a result of stack current utilization. The current in the stack is utilized to transport ions and co-ions across the membrane surfaces. As the ions are transported across, the ions in the diluate stream also decrease which results in the lesser current requirement. As shown in Figure 4, as the removal rates approaches the 90% and above region as current utilization approaches to minimum, and as expected electrical field strength is reduced. The reduction in the electrical field strength in this manner does not necessarily indicate polarization since it can also mark as the end of the desalination process. Both model generated vales and experiment a values accounts for such trend.

## CONCLUSIONS

The Yu and Admassu model which was initially used to model the concentration and current density profile for the removal of metal ion from pulp and paper mill process stream was evaluated and found to satisfactorily predict removal of arsenic from contaminated water streams and determine the effect of experimental parameters on arsenic removal. Two model parameters was derived: A is a parameter used to describe the characteristics of the solution, while $\lambda^2/B$ is a parameter that is related to the stack electrical field strength. The increasing values of $\lambda^2/B$ with respect to increase in initial feed concentration indicate efficient current utilization in the stack and efficient collision of ions during desalination within the stack channel while a constant value of A as the initial concentration increases indicates that ion transport is not limited to solution characteristics, rather it is more dictated by current and potential. Results of this study has been found to support several literature on ED process indicating that removal rate increases as applied voltage and current density increases. In terms of initial feed concentration, the remarkably high removal rates from 500 – 700 ppb is brought about by the increase in ionic particles that intensifies the current in the stack. Medium feed flow velocity was known to be ideal for arsenic systems. The experimental data generated yielded 2 parameters than can be utilized for arsenic systems.

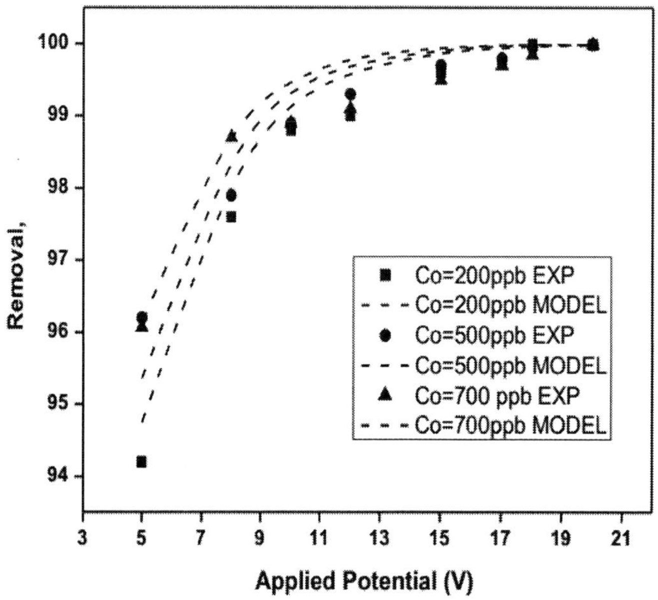


(a)

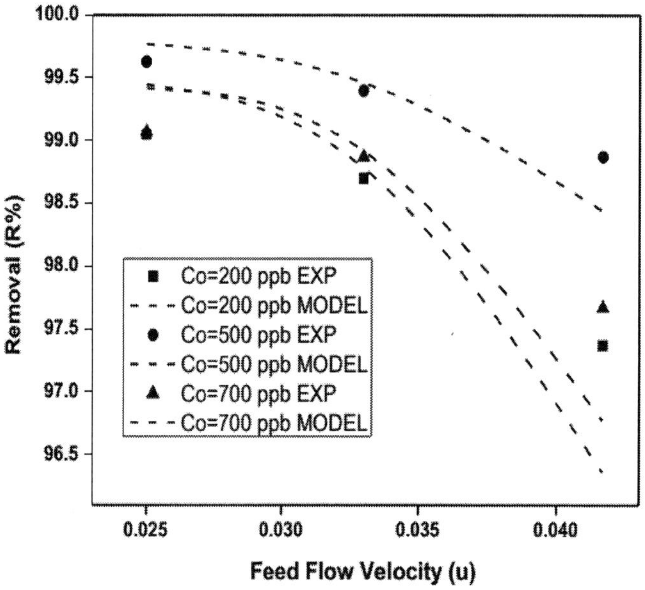


(b)

**FIGURE 3. (a)** Experimental and model predicted values on the effect of applied potential on As removal efficiency at variable concentrations; (b). The comparison between experimental values on the effect of feed flow velocity (u) to removal efficiency at varied initial feed concentration and constant applied potential of 15V and feed flow velocity of 0.033 l/s (2.0 l/min) and retention time of 92 minutes.

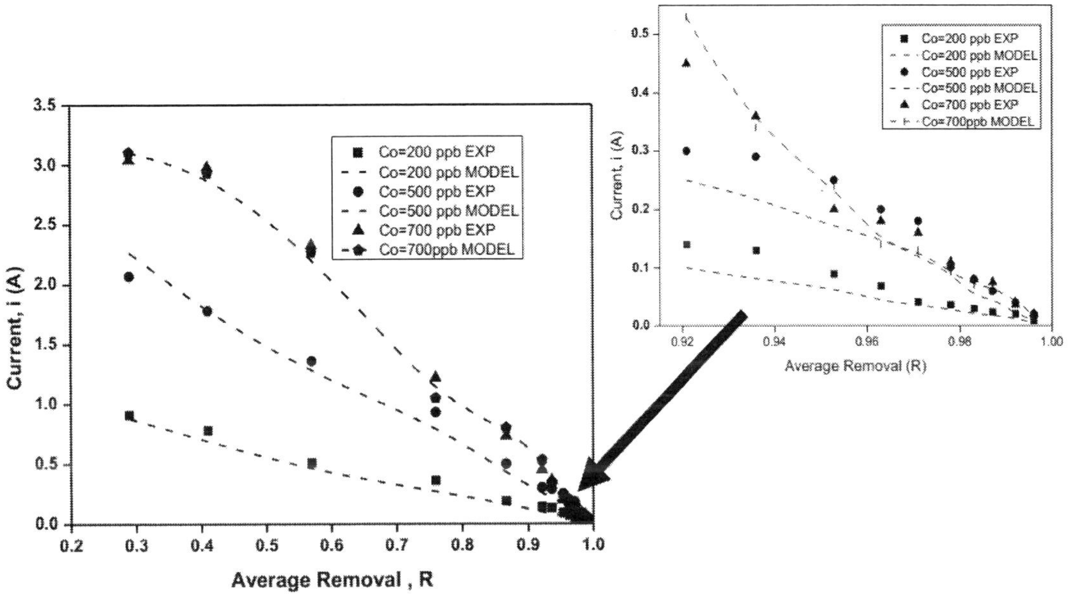


**FIGURE 4.** The experimental and model predicted values of current utilization and removal efficiency at a constant applied potential of 17V and feed flow velocity of 0.033 l/s (2.0 l/min) at variable concentrations.


## ACKNOWLEDGEMENT

This paper acknowledges the Taiwan National Science Council for financially supporting this research under Contract No. NSC 101-2221-E-041-010-MY3 in collaboration with the Commission on Higher Education-Higher Education Development Program (CHED-HEDP), Republic of the Philippines.


## REFERENCES


Bednar, A.J., Gabarino, J.R.,Burkhadt, M.R., Ranville, J.F., Wildeman, T.R. (2004). "Field and laboratory arsenic speciation methods and their application to natural-water analysis". *Water Research* **38,** (2). 355-364.

Brammer, H. Ravenscroft, P. (2009). "Arsenic in groundwater: A threat to sustainable agriculture in South and South East-Asia." *Environmental International*.**35**, 647-654.

Chekioua, A., Delimi, R. (2015). "Purification of $H_2SO_4$ by Pickling Bath by Fe (II) ions using Electrodialysis process." *Energy Procedia*. **74**, 1418-1433.

Galvanin, F., Marchesini, R., Barolo, M., Bezzo, F., Fidaleo, M. (2015). "Optimal design of experiments for parameter identification in electrodialysis models." *Chem. Eng. Res. Design.* **In press, accepted manuscript**, Available on line 12, November 2015.

Ghurye, G., Clifford, D.,(2004). "As(III) Oxidation using chemical and solid-phase oxidants." *J.AWWA*. **96**(1)84-96.

Kirby, B.J. (2010). "Micro- and Nanoscale Fluid Mechanics: Transport in Microfluidic Devices: Chapter 11:Species and Charge Transport", http://www.Kirbyresearch.com/index.cfm/wrap/textbook/microfluidicsnanofludicsch11.html, retrieved on line December 22, 2015.

Mendoza, RM. O., Kan, C-C., Dalida, ML. P., Wan, M-W. (2014). "Feasibility studies on arsenic removal from aqueous solutions by Electrodialysis." *J. Environmental Science and Health, Part A*. **49**, 545-554.

Mohammadi T., Moheb A., Sadrzadeh M., Rasmi A. (2005)."Modelling of metal ion removal from waste water by electrodialysis." *Separation Purification Technology,* **41**, 73-82.

Rahman, M.A., Hasegawa, H. (2011). "High levels of inorganic arsenic in rice in areas where arsenic-contaminated water is used for irrigation and cooking." *Science of Total Environment,***409** (22), 4645-4655.

Sadrzadeh, M., Rami A., Mohammadi, T., (2007). "Separation of different ions from Waste water at various operating conditions using electrodialysis." *Desalination*. **54**,147-156.

Zourmand, Z., Faridirad, F., Kasiri, N., Mohammadi, T. (2015). "Mass transfer modelling of desalination through and electrodiaysis cell." *Desalination.* **359**, 41-51.

# PHOSPHATE REMOVAL FROM SYNTHETIC GROUNDWATER BY MIXED ALGAE CULTURE

***Eda Tuna Öztürk,*** Güneş Özden and Serdar Göncü
(Anadolu University, Eskişehir, TURKEY)

**ABSTRACT:** Extremely high levels of phosphate in groundwater can cause health problems. Chemical and physicochemical treatment methods can be applied to reduce phosphate levels in groundwater as well as biological methods. Mixed algae culture was collected from Porsuk Dam, Eskisehir, Turkey and was used in the experiment without any pre-treatment. The growth kinetic of algae culture was investigated in BG-11 medium. The growth kinetic was evaluated by using turbidity and chlorophyll-a measurements. Maximum growth rate was calculated as 1,6 $day^{-1}$ by using Monod kinetic equation. During calculations, phosphate was chosen as the substrate. For phosphate removal studies from synthetic groundwater, algae culture in exponential phase was used. Change in the phosphate concentrations by time were measured by SM 4500-P E Ascorbic Acid Method. The initial phosphate concentration was 8,2 mg $PO_4^{-3}$/L and was reduced to 0,04 mg $PO_4^{-3}$/L in 9 days. It is shown that mixed algae culture can be used for phosphate removal from groundwater as ex-situ bioremediation processes.

## INTRODUCTION

Natural and anthropogenic sources are the main reasons for phosphate contamination in groundwater. Atmospheric deposition, weathering of soluble inorganic materials, runoff, sedimentation, natural decomposition of rocks and minerals and decaying biomass are the natural sources of phosphate. In addition, fertilizers, animal wastes, detergents, drinking water treatment, synthetic material development surface, wastewater and septic effluent, industrial discharge, phosphate mining and forest fires are the anthropogenic sources for phosphate (Fadiran et al., 2008).

Phosphorus can be removed from water sources by physical, chemical and biological treatment techniques. Physical methods such as electrodialysis and reversed osmosis are expensive and inefficient to remove phosphorus. Chemical techniques, the addition of metal salts, is one of the common method for phosphorus removal. Although, chemical treatment technique is highly effective, remnants of this technique brings additional problems (Clark et al., 1997; Morse et al., 1998). On the other hand, biological treatment methods outstand among the other treatment techniques with its low operational costs, less residue problems and wide range of operational conditions. In-situ and ex-situ methods can be listed as two different approaches for biological groundwater treatment. Recently these two methods are becoming popular for obtaining drinking water from groundwater (Bedient et al., 1997).

Groundwater has become an important water sources. Generally, phosphorus levels are found to be lower in groundwater due to the adsorption processes in the aquifer and vadose zone. Phosphorus contamination in surface water is mostly caused by particulate phosphorus (PP) while dissolved phosphorus (DP) in groundwater. In the vicinity of Eskisehir city, groundwater can be found down to 2 m bgs (below ground surface), causing an increased risk of phosphorus contamination due to the lack of adsorption in aquifer. Moreover, Porsuk Stream, passing through the city, contributes the phosphorus contamination of groundwater with its high phosphorus level up to 1 mg $PO_4^{-3}$/L (Uğurluoğlu, 2013).

## MATERIALS and METHODS

Mixed algae culture was collected from Porsuk Dam, Eskisehir, Turkey (Figure 1). Porsuk Dam, which is located on Porsuk Stream, has been received intensive amount of pollutants like nitrogen and phosphorus. Because of having high levels of nutrients, it is classified as an eutrophic lake.

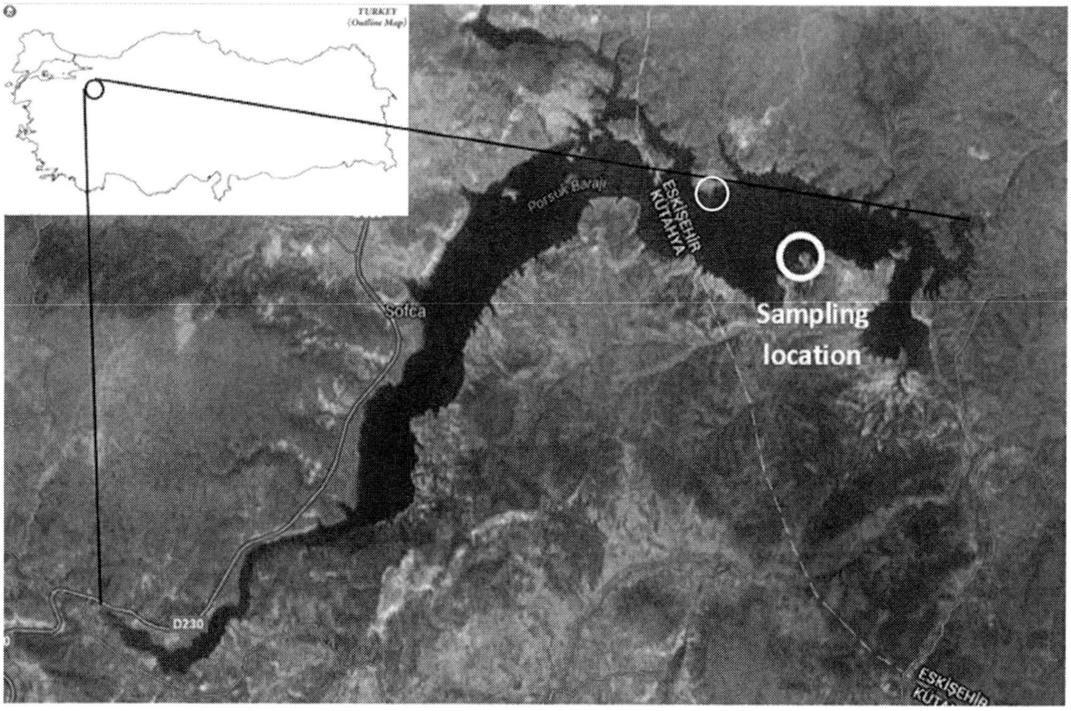


**FIGURE 1.** Porsuk Dam, is located northwest of Eskişehir, Turkey.

Algae culture was used in the experiment without any pre-treatment and isolation techniques. Mixed algae culture has been examined morphologically under the microscope and found that *Chlorella* and *Scenedesmus* are the most dominated species in the mixed culture (Figure 2).

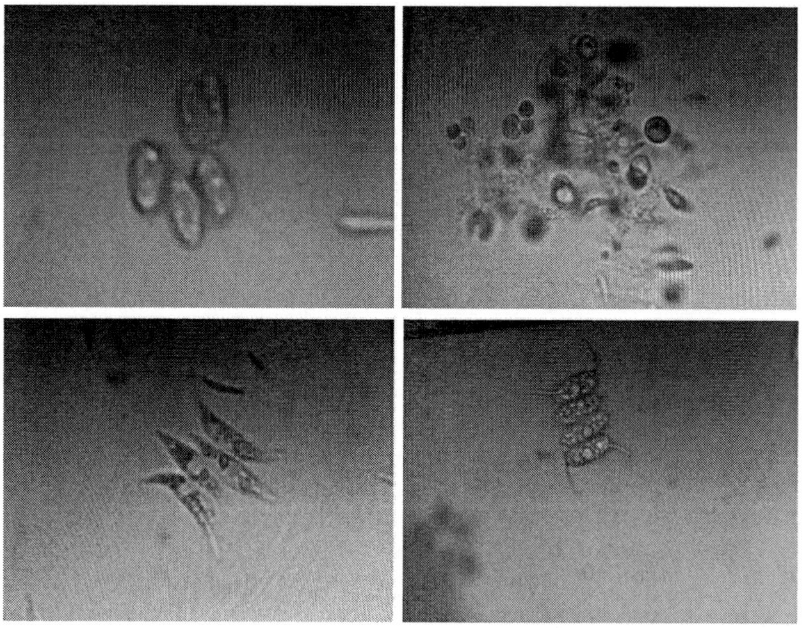

**FIGURE 2.** Mixed algae culture species under microscope.

The growth kinetic of algae culture was investigated in BG-11 medium. Algae : medium ratio was 1:10 in growth kinetic experiment and total volume was chosen 1L. The optimum conditions for algae culture were chosen as 3000 lux light intensity, room temperature, pH range between 8 – 8,2. Also natural atmospheric $CO_2$ concentration has been used for aeration of reaction vessel (Cassidy, 2011). The experimental setup is seen in Figure 3. The growth kinetic was investigated by using turbidity and chlorophyll-a measurements. Turbidity was measured by Micro TPW Turbidimeter and chlorophyll-a was measured according to Standard Methods 10200 H. by HACH DR5000 spectrophotometer.

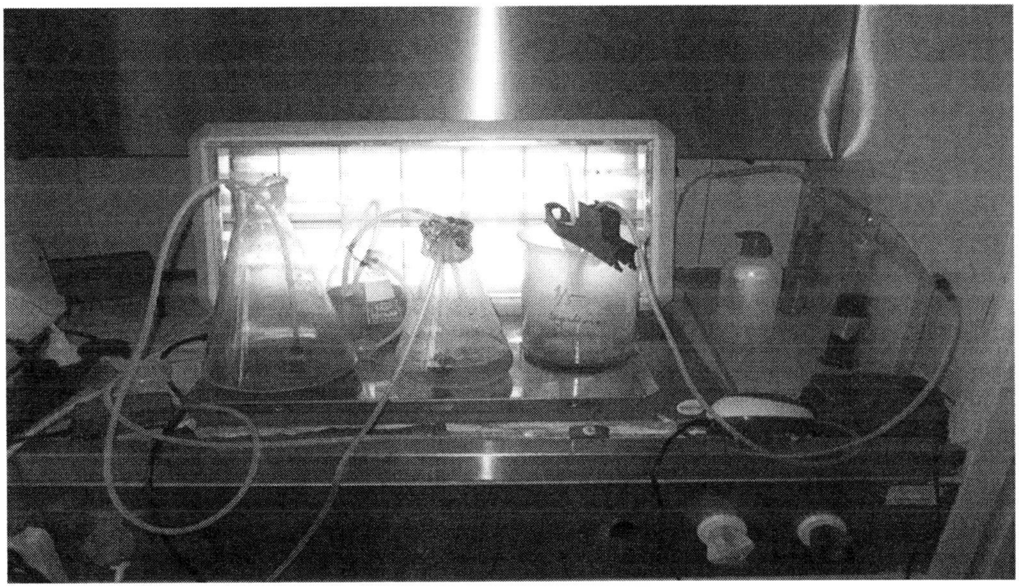

**FIGURE 3.** Experimental setup of mixed algae culture.

For phosphate removal studies from synthetic groundwater, mixed algae culture was used in exponential phase. Change in the phosphate concentrations were determined by SM 4500-P E Ascorbic Acid Method in daily time steps.

## RESULTS AND DISCUSSIONS

As seen in Figure 4, during lag phase which is the first day, algae culture adapted themselves to growth conditions. $1^{st} – 3^{rd}$ days are the exponential days, which is a period characterized by cell doubling. After $3^{rd}$ day, due to lack of essential nutrient, growth rate and death rate of the algae concentration were equal.

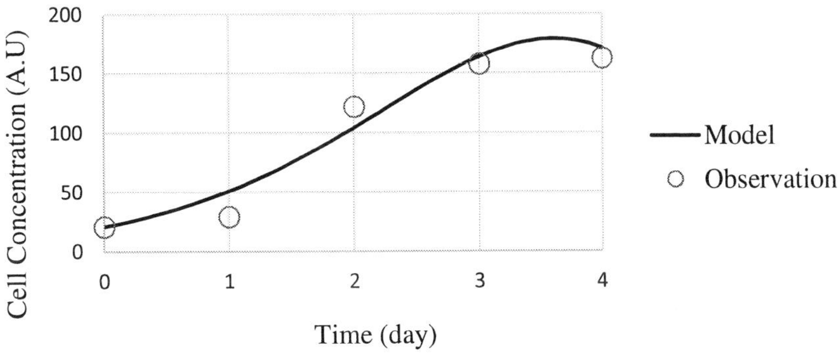


**FIGURE 4.** Mixed algae growth curve.

Monod kinetics parameters have been determined by using Excel solver module with using observation values. It is seen that model results corresponded to observation results (Figure 4 and 5).

Maximum growth rate was calculated as 1,6 $day^{-1}$ by using Monod kinetic equation. During calculations, phosphate was chosen as the substrate. Half- velocity constant (Ks) was found 5,5 mg/L (Figure 5).

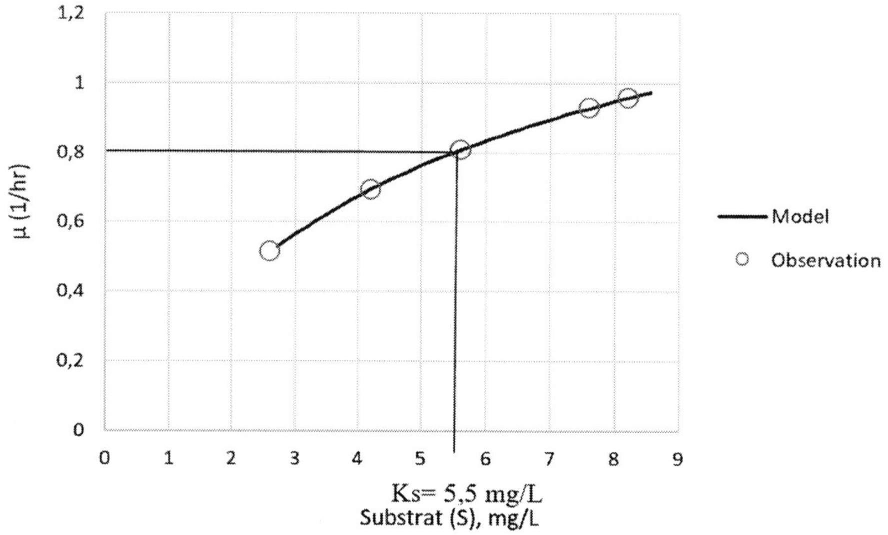


**FIGURE 5.** Growth kinetics of mixed algae culture.

As shown in the Figure 6, phosphate concentration decreased from 8,2 mg $PO_4^{-3}$/L to 0,04 mg $PO_4^{-3}$/L in 9 days. The efficiency of phosphate removal was calculated 99.5%. Algae population has increased with using phosphate as a nutrient. Meanwhile, turbidity and pH levels have increased with a reason of algae growth.

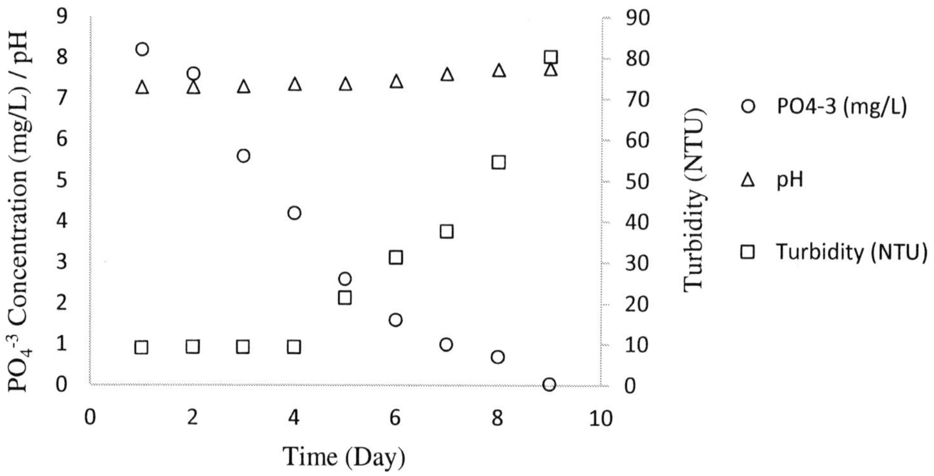


**FIGURE 6.** $PO_4^{-3}$, pH and Turbidity Change in Time.

## CONCLUSION

This study demonstrates that it is possible to calculate specific growth rate and half-specific velocity of mixed algae culture by Monod Equation. Also, phosphate can be removed from synthetic groundwater by mixed algae culture. All the experiments have been done with synthetic groundwater in the laboratory conditions. In reality, groundwater temperature is around 12 °C which is not an optimum temperature for many algae species. To apply this study for real conditions, either different algae culture could be tried or different temperature conditions could be examined.

**ACKNOWLEDGEMENTS**

The authors thank to Anadolu University, Scientific Research Project Funding (AU BAP) for their financial support [Project number: 1601F023].

**REFERENCES**

Bedient, P.B., Rifai, H.S., and Newell C.J. 1997. *Ground Water Contamination* Transport and *Remediation.* Prentice Hall PTR, Upper Saddle River, NJ.

Clark, T., Stephenson, T., and Pearce, P.A. 1997. "Phosphorus Removal by Chemical Precipitation in a Biological Aerated Filter." *Water Research* 31(10):2557-2563.

Fadiran, A.O., Dlamini, S.C., Mavuso, A. 2008. "A Comparative Study of the Phosphate Levels in Some Surface and Ground Water Bodies of Swaziland." *Bull. Chem. Soc. Ethiop."* 22(2):197-206.

Cassidy, K.O. 2011 *Evaluating Algal Growth at Different Temperatures.* M.S. Thesis, University of Kentucky UKnowledge, Lexington, Kentucky.

Morse, G.K., Brett, S.W., Guy, J.A., Lester, and J.N. 1998. "Review: Phosphorus Removal and Recovery Technologies". *The Science of the Total Environment* 212(1):69-81.

Uğurluoğlu, A. 2013. *Seydisuyu Havzası Yeraltı Su Kalitesinin ve Kirlilik Düzeyinin Belirlenmesi.* M.S. Thesis, Anadolu University, Eskişehir, Turkey.

# GROUNDWATER POLLUTANT LOADING IN URBAN WATERSHEDS ALONG LINED AND UNLINED CHANNELS

***Barry Hibbs***
(California State University, Los Angeles, USA)

Lined and unlined urban channels and lined storm drains are used worldwide to convey storm runoff away from urban areas. They also convey dry weather flows such as car wash runoff, lawn runoff, and other "nuisance flows" in urban centers. Areas in southern California and elsewhere have shallow water tables where dry weather flows in concrete lined storm drains consist primarily of perennial groundwater baseflows. In these areas, groundwater leaks through cracks, joints, weepholes, and intentional dewatering structures into storm drains, which eventually flows into natural channels, coastal wetlands, and estuaries. The "lined" sections of these channels are hardly impervious; in fact, the lining of the channels often affords excellent opportunities to collect groundwater samples at discrete points that are usually not present along unlined channels where well control is sparse or where technical or legal issues prevent installation of piezometers. Frequently, as much as 95% of the nutrient and trace element loading to urban catchments during dry weather is from seepage of groundwater that is laden with pollutants. Unlined channels mitigate several contaminants more effectively than lined channels due to biologically active hyporheic zone and riparian vegetation. This project presents several examples of our experiences in conducting stream/aquifer studies in lined and unlined channels in Southern California. Sub-projects include evolution of groundwater quality along flowpaths, detection of unwanted recharge mounds in shallow groundwater systems due to leaky water main pipes, and geologic factors in water quantity and water quality fluxes in urban catchments. Tools that are utilized and described include isotopic tracers (O-H-S-N stable isotopes); trace element speciation analysis; and innovative sampling methods.

# ARSENIC CONTENT IN DRINKING WATER SOURCES AND OCCURENCE OF CANCER IN NALBARI AND KARUP DISTRICT

***Dipali Das Deka*** and Jogen Chandra Kalita
(Tihu College and Gauhati University, ASSAM, INDIA)

**ABSTRACT**: At the present time people are aware of about the drinking water contaminants and health risk. Since 1970's, when sampling for contaminants increased, hundreds of manufactured chemicals have been identified in various ground water and other drinking water supplies. Source water contaminants of concern include arsenic, asbestos, radon, agricultural chemicals and hazardous water. Of these, the strongest evidence for cancer risk involves arsenic, which is linked to cancer of the liver, lung, bladder and kidneys. A survey was conducted during 2011to2014 in some selected areas of Nalbari and Kamrup district of Assam to know the occurrence of cancer and different types of diets consumed with the help of a prepared questionnaire. Water samples were collected in different regions accordingly during pre and post monsoon period. Spectophotometric method was used to find out the arsenic in drinking water sources. Our study confirmed a higher level of arsenic content in drinking water sources in all the selected spots.Prevelrnce of cancer in the study area was also reported high.

## INTRODUCTION

Assam is situated in far North East corner of India. The total geographical area of the state is 78, 438 sq kms, which accounts for about 2.4 percent of the country's total geographical area. It has been reported that there is a very high incidence of cancers of all sites in general and tobacco and pesticides related cancers in particular in North East India. Not only Assam but the entire North East region of India has different customs, food habits, life style, and diverse ethnic group. The people of North East India use different types and pattern of tobacco, which is not seen in other parts of the India. There is extensive use of pesticides in tea- gardens in North east, which can lead to widespread occupational and environmental exposures. Epidemiological studies reveal that Assam states have very high incidence of esophageal cancer . The present study area Kamrup and Nalbari district is part of the state of Assam. It has been also reported Kamrup district has highest number of cancer cases (1988)of different types followed byNagaon Barpeta ,Sonitpur ,Dhubri and Nalbari ( Annual Report: 2013-2014 Dr B Borooah Cancer Institute). The following data will give an idea about the current trends of cancer in Assam(Table1).

Information on ground water quality of North Eastern India is scanty. Available literature shows that ground water of Assam valleys are highly ferruginous (Aowel1981). Recent research reveal that the concentration of arsenic in groundwater exceeds the permissible level(50 µg/lbased on the water consumption of 2litre per day WHO in parts of Assam(20 disricts out of 24 districts.(Sing2004) In the State, of the total 56, 180 groundwater samples tested in a three-tier system during the past few years, under a joint plan of action by the State PHED, UNICEF and the IIT Guwahati, 7.95 per cent of the samples were found to be containing over 50 parts per billion (ppb) arsenic and 29.75 per cent were found to be containing over 10 ppb arsenic. The 50 ppb is the Government of India norm, while 10 ppb is the World Health Organisation (WHO) norm for determining safe drinking water in the case of arsenic(Assam Tribune2014). of 2litre per day WHO in parts of Assam(20 disricts out of 24 districts.(Sing2004) reveal concentration of arsenic in groundwater exceeds the permissible level(50 µg/lbased on the water consumption of 2-litre per day WHO in parts of Assam(20 disricts out of 24 districts (Sing 2004).

## OBJECTIVES

The main objective of this work was to determine the level of arsenic in drinking water sources in Nalbari and Kamrup district of Assam, India and prevalence of different types of cancer.

**TABLE 1. District -Wise Break –Up of Patients in the State of Assam during 2011- 2012, 2012- 2013, 2013-2014**

| SL No. | District | 2011-12 | 2012-13 | 2013-14 |
|---|---|---|---|---|
| 1. | Kokrajhar | 73 | 61 | 134 |
| 2. | Dhubri | 239 | 212 | 442 |
| 3. | Goalpara | 145 | 171 | 334 |
| 4. | Barpeta | 333 | 321 | 646 |
| 5. | Morigaon | 181 | 166 | 363 |
| 6. | Nagaon | 483 | 504 | 1002 |
| 7. | Sonitpur | 267 | 280 | 643 |
| 8. | Lakhimpur | 100 | 144 | 259 |
| 9. | Dhemaji | 26 | 36 | 90 |
| 10. | Tinsukia | 65 | 68 | 155 |
| 11. | Dibrugarh | 83 | 84 | 159 |
| 12. | Sivsagar | 110 | 98 | 205 |
| 13. | Jorhat | 191 | 191 | 337 |
| 14. | Golaghat | 125 | 140 | 273 |
| 15. | Krbi Anglong | 75 | 71 | 131 |
| 16. | Dima Hasao | 09 | 11 | 17 |
| 17. | Cachar | 29 | 19 | 47 |
| 18. | Karimganj | 24 | 20 | 29 |
| 19. | Hailakandi | 09 | 07 | 15 |
| 20. | Bongaingaon | 129 | 131 | 281 |
| 21. | Chirang | 55 | 50 | 117 |

SOURCE: Annual report Dr. B.Barooah Cancer Institute (2013-13,2013-2014).

## METHODOLOGY

A survey was conducted during 2011to2014 in some selected areas of Nalbari and Kamrup district of Assam to know the occurrence of cancer and different types of diets consumed with the help of a prepared questionnaire. During the survey, information was also taken for different sources of drinking water. Water samples collected from some selected areas. Spectroscopic method has been applied to find out the concentration of Arsenic in the ground water.

## RESULTS

We have studied 20 different spots of Nalbari and Kamrup (10 from each) where occurrence of different types of cancer also reported. It was observed that in all the selected spots arsenic was present in the drinking water sources. The result is shown in table 2 and table 3.

## CONCLUSIONS

Our study confirmed a higher level of arsenic content in drinking water sources in all the selected spots. Prevalence of cancer in the study area was also reported high. The result indicated that exposure of arsenic in drinking water may be located with occurrence of cancer. However further large scale investigation would certainly throw light behind this higher occurrence of cancer in these area.

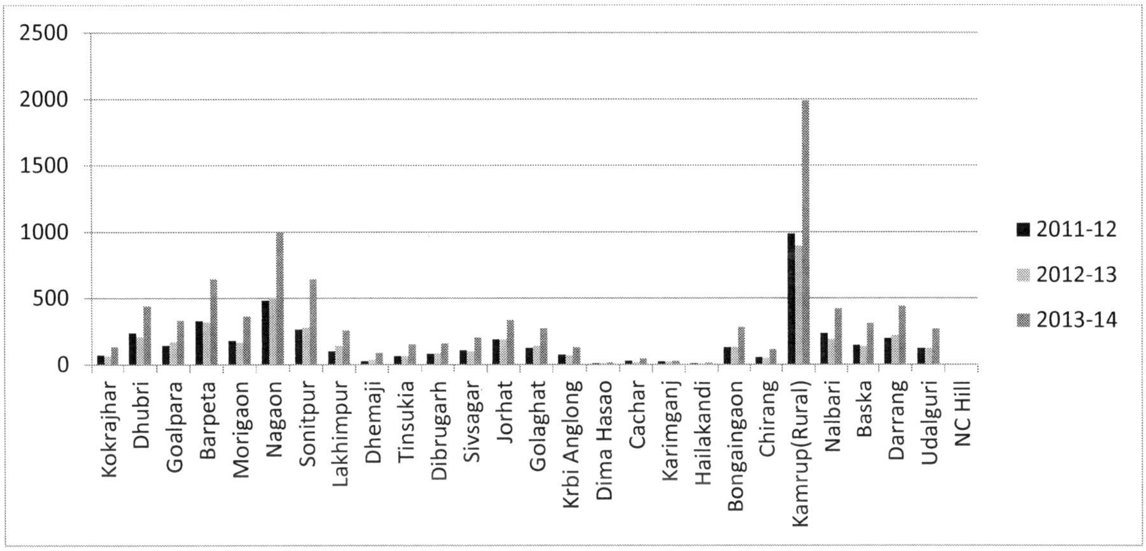


**FIGURE 1.** Graphical Representation of the Table-1

**TABLE 2.** Arsenic content in drinking water sources in different places of Nalbari District

| Sl No | locality | Total population | Cancer patient found | Source of drinking water | Depth (in feet) | Arsenic (µg/L) |
|---|---|---|---|---|---|---|
| 1 | Haribhanga | 4786 | 07 | Tube well | 170 | 510 |
| 2 | Mugkuchi | 2225 | 10 | Tubewell | 185 | 550 |
| 3 | Bhogpur | 1250 | 11 | Tubewell | 200 | 570 |
| 4 | Bala | 1236 | 12 | Tubewell | 185 | 580 |
| 5 | Kaljar | 2616 | 12 | Tubewell | 200 | 590 |
| 6 | Nizzuliki | 2844 | 12 | Tubewell | 200 | 610 |
| 7 | Mathurapur | 3656 | 15 | Tubewell | 185 | 640 |
| 8 | Ward no3 | 1335 | 15 | Tubewell | 210 | 710 |
| 9 | Parmankhoa | 504 | 15 | Tubewell | 200 | 720 |
| 10 | Nannatari | 1833 | 17 | Tubewell | 200 | 770 |

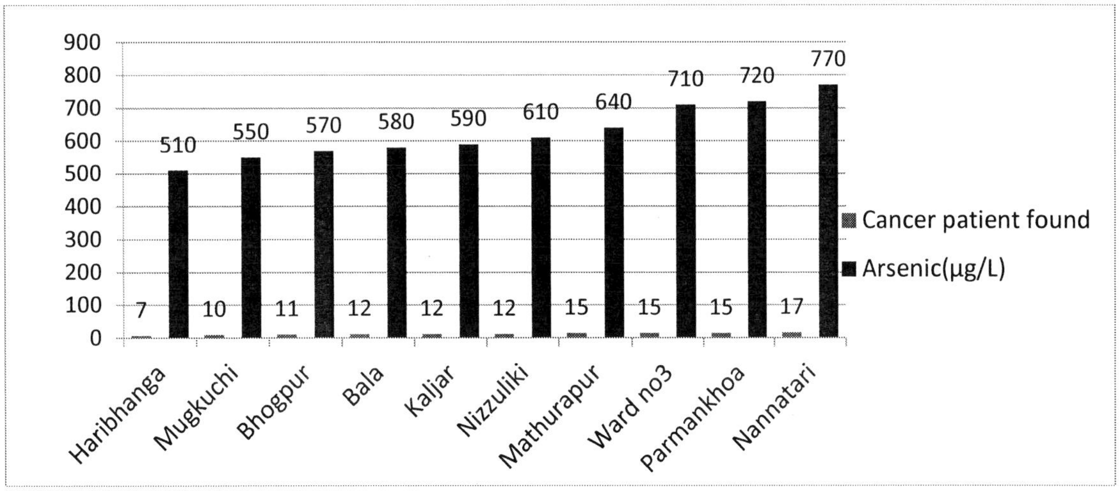


**FIGURE 2.** Graphical Representation of the Table 2

**TABLE 3.** Arsenic content in drinking water sources in different places of Kamrup district

| SlNo | Locality | Total Population | Cancer patient Found | Source of drinking water | Depth (in feet) | Arsenic (μg/L) |
|---|---|---|---|---|---|---|
| 1 | Belguri(Sonapur) | 4917 | 02 | Ring well | 85 | 560 |
| 2 | Dadara | 2000 | 05 | Tubewell | 195 | 650 |
| 3 | Chaudhurypara | 5030 | 10 | Tubewell | 180 | 670 |
| 4 | Kinnan | 100 | 02 | Tubewell | 185 | 680 |
| 5 | Pamohi | 1200 | 11 | Ringwell | 80 | 680 |
| 6 | DobokNo2 | 1570 | 07 | Tubewell | 200 | 680 |
| 7 | RampurNo1 | 7211 | 15 | Ringwell | 180 | 700 |
| 8 | Boko(Kenduguri) | 5000 | 11 | Ring well | 90 | 720 |
| 9 | DimuNo1 | 1045 | 11 | Tubewell | 180 | 740 |
| 10 | Silkijhar | 4086 | 15 | Tubewell | 200 | 800 |

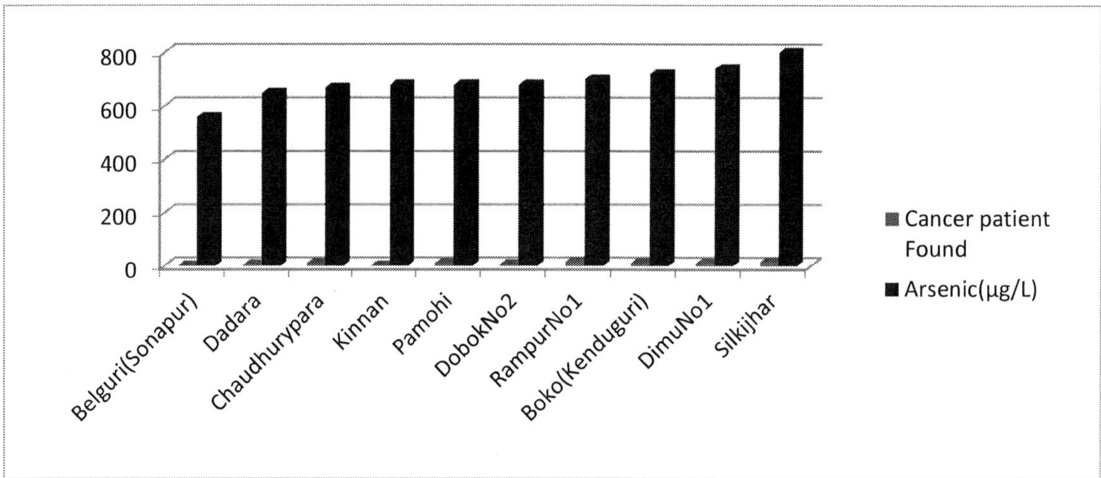


**FIGURE 3.** Graphical Representation of the Table-3

## ACKNOWLEDGEMENTS

All the family member of the patients including the patients themselves and all the physicians and staff of DR .B. BAROOAH CANCER INSTITUTE is gratefully acknowledged.

## REFERENCES

Aowal, A.F.S.A. 1981. "Design of an Iron eliminator for hand tube wells." *Journal of IWWA*, .XIII(1):65.

Chen C.J.andChuangY.C., Lin T.M.andWuHY .1985. "Malignant neoplasm among residents of black foot disease-endemic area in Tiwan: High-Arsenic artesian well water and cancer". *Cancer Res*, 45(11Pt2):5895-5899.

International Agency for Research on Cancer 1987. IARCmonographs on the evaluation of carcinogenic risks to humans: overall evaluation of carcinogenicity: an updating of IARCmonographs: 1-42, IARC publ. Suppl, 7,100.

Majumder D.N. 2010. ".Effect of chronic intake of arsenic-contaminated water on liver". *Toxicol. Appl. Pharmacol*, 206:169-175.

Patowary Ajit.2014. "Arsenic Content in state drinking water high".*The AssamTribune*.Sept 6:1

Sing AK 2004. "Arsenic Contamination inGroundwater of North Eastern India" In: *Proc of Nat . Seminar on Hydrology with focal theme on Water Quality.*

WHO 2001. Env Health criteria 224: *Arsenic & Arsenic Compounds.* WHO, Generia

WHO, 2010. *Arsenic in Drinking Water.* Fact Sheet.

Zhang A.H., HuangX.X. JiangX.Y., Luo P.,Guo Y.C.and Xue S.Z. 2000. "The progress of study on endemicarsenism due to burning arsenic containing coal in Guizho province". In: *Metal Ions in Biology and Medicin.* Conteno JA.Collery P, Vernet G, Finkleman RO, Gibb H, EtenneJ-Ceds, vol 6, France: John Libbey Eurotext LTD, 53-55.

# SIMULTANEOUS NITRATE AND PHOSPHATE REMOVAL FROM DRINKING WATER USING IMMOBILIZED POLYETHYLENEIME

***Enrico T. Nadres*** and Debora F. Rodrigues
(University of Houston, Houston, Texas 77008, USA)

A new biosorbent containing polyethylemeine was synthesized for the removal of nitrate and phosphate. The adsorption capacity and optimum conditions for the removal of these contaminants was initially investigated with batch experiments and then by packed bed columns. The batch results demonstrated that the optimum pH for nitrate adsorption was around 4-5 and the adsorption capacity was 100 mg/g and 30 mg/g for nitrate-N and phosphate, respectively. The column experiments aimed to investigate the effect of flow rate, influent concentration and bed height in the removal of the contaminants by the new adsorbent. The adsorption capacity of the new adsorbent was determined to be 120 mg $NO_3$-N/g with the Thomas model. The volumetric capacity was determined to be 0.35 eq/L by fitting the results in Adams-Bohart model. Regeneration investigation of the adsorbent demonstrated that the adsorbent regenerates more that 90% of its capacity even after 3 adsorption-desorption cycles with 0.1 M HCl. When real environmental water sample was passed through the column, 97 % of nitrate-N and 70 % of phosphate.

# SUSPENDED BACTERIA COMMUNITY OF PROCESSING UNITS IN THE TYPICAL DRINKING WATER TREATMENT PLANT OF CHINA

Weiying Li, ***Feng Wang****, Junpeng Zhang and Wanqi Qi
(Tongji University, Shanghai, China)

For reducing the amount of organic pollutants in drinking water, advanced water treatment process (AWTP) were adapted on the basis of conventional water treatment process (CWTP) in China, but little attention was paid to suspended bacteria. Focused on a drinking water treatment plant employing AWTP and CWTP simultaneously, this paper aimed at revealing suspended bacteria community structure of different water treatment units by 454-pyrosequencing technique and seeking possible bacteria pivotal for the effects of AWTP. Results showed that conventional water treatment process played little impact on microbial diversity comparing to AWTP. Although complex community composition was beneficial for organics removal, the dominant bacteria varied slightly. Proteobacteria phylum (54.7%~99%) and Betaproteobacteria class (28.6%~86.3%) were dominant during the procedure of drinking water production, while the primary genus was *Limnohabitans* in both the effluent of CWTP (17.4%) and AWTP (25.6%). Furthermore, redundancy analysis implied that the TOC level was positively correlated with the relative abundance of *Limnohabitans*, which was inferred to be one of the crucial contributors for the biodegradable function of activated carbon filtration tank. The paper provided reference for better developing the microbial effects during water treatment procedure in the future.

# FULL-SCALE STUDIES OF FACTORS RELATED TO BIOLOGICAL STABILITY OF WATER DISTRIBUTION SYSTEM IN SOUTH CHINA

Weiying Li, ***Junpeng Zhang***, Feng Wang and Wanqi Qi
(Tongji University, Shanghai, China)

Bacterial regrowth in drinking water, especially opportunistic pathogens growth, leads to the deterioration of water quality and gives risks to the public health. AOC (assimilable organic carbon), BDOC (biodegradable dissolved organic carbon), BRP (bacterial regrowth potential) and HPC (heterotrophic plate count) as biological evaluation methods were examined in a full-scale drinking water distribution system (DWDS) and bench tests. According to the statistical analysis of big data, HPC was significantly correlated with AOC, BRP, BDOC, $COD_{Mn}$, ammonia, free chlorine and temperature. Among these factors, temperature had great effects on bacterial regrowth. When temperature ranged from 25°C to 30°C, HPC numbers were 35.67 times higher than those with water temperature ranging from 5 to 10°C. Organic carbon in DWDS facilitated microbial growth, $AOC_{P17}$ was more likely to be utilized by microbes at high temperature and $AOC_{NOX}$ increased as temperature rose. Disinfectant affected interaction between nutrients and bacteria. Chlorine imposed greater effects on bacterial regrowth than on chloramine. The bench test using flow cytometry showed that when chlorine was higher than 0.5mg/L or chloramine was more than 0.75mg/L, the HPC level was low and AOC concentration almost unchanged. Conversely, the HPC level increased quickly and declined slightly, with chlorine lower than 0.15mg/L or chloramine less than 0.25mg/L. The biologically stable values with or without different disinfectants were also different after being incubated for 30 days.

# OPTIONS FOR THE REMOVAL OF ARSENIC FROM DRINKING WATER IN RURAL AREAS OF BIHAR, INDIA

Astha Kumari[1], Sudhir Nigam[2] and ***Nityanand Singh Maurya***[1]
([1]Department of Civil Engineering, National Institute of Technology Patna, India; [2]Lakshmi Narain College of Technology and Science, Bhopal 462021, India)

Arsenic is a ubiquitous toxic metalloid of natural origin and ranked as 20$^{th}$, 14$^{th}$, & 12$^{th}$ most abundance element in earth crust, seawater and human body respectively. Arsenic is the major constituent of at least 245 different minerals. Though it is a natural origin element, but several anthropogenic activities such as manufacturing of transistors, lasers and semiconductors, as well as in the processing of glass, pigments, textiles, paper, metal adhesives, use of chemical weapons, fertilizer and pesticides, cosmetic industries, combustion of fossil fuels, wood preservation, urban waste management, mining etc are responsible for spreading out it in the environment. It is an essential constituent of human body, but its continuous ingestion may cause several health problems such as palmplantar hyperkeratosis, hepatic damage, damage to the central neural system, hair loss, chronic headache, skin cancer and cancer of internal organs such lungs, liver, kidney and bladder; colour change on the skin, and hard patches on the palms and soles of the feet. In most cases its adverse effect is not easily detectable because many of its symptoms are also indicative of other illnesses.

Elevated arsenic concentration in groundwater is now well recognized environmental problem and global threat to human health. Approximately 70 countries are currently experiencing high level concentration of arsenic in drinking water. The recent literatures indicate at-least 150 million persons are under high risk of arsenicosis mostly from south-eastern Asian countries such as Bangladesh, India and China. Bihar, the 3$^{rd}$ largest populated states of India is also experiencing groundwater contamination due to arsenic. As early as in 2002, arsenic was first detected in village Semaria Ojha Patti of Shahpur block, Bhojpur district. According to Government of Bihar report presently at-least 13 out of 38 districts of Bihar are found to be severely affected by arsenic contamination. Since groundwater is only source of drinking water, thus entire population of approximately 1590 habitations is compulsorily consuming contaminated water. Currently several agencies have installed mini water supply system based on adsorption/ion-exchange process that requires imported/patented media and constant power supply. Majority of the affected population lives in rural areas deprived off from basic infrastructure such as electric power supply and road connectivity. Thus the operation and maintenance of such treatment plants is a big challenge. To make the treatment process sustainable it is necessary to search suitable alternatives which can work in rural areas of Bihar.

Currently various treatment methods such as precipitation/coagulation and filtration, membrane filtration, ion-exchange and adsorption etc are available for the treatment of arsenic contaminated water. But their suitability depends on various characteristics of the water to be treated and socio-economic condition of the consumers. Therefore a treatment method suitable for one place may not solve the problems at other place. These warrant evaluating the available treatment options to search suitable alternatives for rural areas.

The objective of the present study is to evaluate the existing treatment technologies employed for the removal arsenic in rural areas of Bihar. Further various conventional treatment options and recent advancements in arsenic removal processes will be studied and a comparative chart will be prepared on the basis of their removal efficiency, necessity of pre-treatment, electricity, skilled manpower, land and installation as well as operational cost. The outcomes of the study will be helpful in the selection of suitable method for a specific area.

To achieve the above objectives, survey will be conducted to evaluate the existing treatments installed in rural areas of Bihar. Further review of published literature including textbooks, papers presented in reputed conference will be conducted.

On the basis of above studies a comparative chart will be prepared, which will provide ranking of various treatment options for the removal of arsenic from water.

# POND-WETLAND COMPLEXES AS PROCESSOR OF DRINKING WATER SOURCE

***Weidong Wang*** and Chengqing Yin.
(Research Center for Eco-Environmental Sciences, Chinese Academy of Sciences, Beijing, China)

Shijiuyang Constructed Wetland (110 ha) is a drinking water source treatment wetland with primary structural units of ponds and plant-bed/ditch systems. The wetland can process about 250,000 tonnes of source water in the Xincheng River every day and supplies purified raw water for Shijiuyang Drinking Water Plant. Daily data for 28 months indicated that the major water quality indexes of source water had been improved by one grade. The percentage increase for dissolved oxygen and the removal rates of ammonia nitrogen, iron and manganese were 73.63%, 38.86%, 35.64%, and 22.14% respectively. The treatment performance weight of ponds and plant-bed/ditch systems was roughly equal but they treated different pollutants preferentially. Most water quality indexes had better treatment efficacy with increasing temperature and inlet concentrations. These results revealed that the pond-wetland complexes exhibited strong buffering capacity for source water quality improvement. The treatment cost of Shijiuyang Drinking Water Plant was reduced by about 30.3%. Regional rainfall significantly determined the external river water levels and adversely deteriorated the inlet water quality, thus suggesting that the "hidden" diffuse pollution in the multitudinous stream branches as well as their catchments should be the controlling emphases for river source water protection in the future. The combination of pond and plant-bed/ditch systems provides a successful paradigm for drinking water source pretreatment. Two other larger drinking water source treatment wetlands with ponds and plant-bed/ditch systems are in operation or one in construction in the stream networks of the Yangtze River Delta and more people will be benefited.

# INTERACTION BETWEEN HYDROLYSIS AND FLOCCULATION, AND ITS INFLUENCE ON GENERATION OF DISINFECTION BY-PRODUCTS WITH ALUMINUM

***Hong Shen*** , Xin Chen •, Hongbin Chen
(Collaborative Innovation Center for Clean Water, College of Environmental Science and Engineering, Tongji University, Shanghai 200092, P.R. China)

**Introduction.** Aluminum (Al) is commonly found in natural waters, and its salts are often used as coagulants in drinking water treatment. Hence, the effects of Al ions ($Al^{3+}$) on disinfection and the security of drinking water should not be ignored. In this study, using tannic acid as the model compound of natural organic matters (NOMs), the effects on disinfection by-products (DBPs) and carcinogenic factors (CF) after the addition of $Al^{3+}$ with $Al^{3+}$/ initial tannic acid molar ratios (MRs) of 0, 1.7, 3.4 and 5.1 (referred to as $MR_0$, $MR_{1.7}$, $MR_{3.4}$ and $MR_{5.1}$, respectively) during chlorination of synthetic water was investigated.

**Experimental procedures.** The chlorination experiment systems contain a model compound composed of NOMs (tannic acid adjusted to 2 ± 0.1 mg-C/L DOC), different concentrations of metal ions ($Al^{3+}$), potassium dihydrogen phosphate (1 m mol/L) and sodium hypochlorite (20 mg-Cl/L). The chlorination experiment was conducted under a headspace-free condition in 60-mL amber glass screw-cap vials with PTFE-lined septa, at a stable temperature (20 ± 1°C) for 7 d of contact time. The aluminum sulfate was added in concentrations of 0, 0.1, 0.2 and 0.3 mg-$Al^{3+}$/L (hereafter referred to with the abbreviations $MR_0$, $MR_{1.7}$, $MR_{3.4}$ and $MR_{5.1}$, respectively) according to the Al/ initial tannic acid molar ratios. The pH was maintained at 7 ± 0.02 by the addition of 1 M $H_2SO_4$ or NaOH. During the 7 d chlorination period, samples were quenched using ammonium chloride before being extracted for DBPs analysis.

**Results and discussion.** The results demonstrated that the presence of $Al^{3+}$ in chlorination led to an increase in both DBPs (except for $MR_{5.1}$) and carcinogenic factors in finished water. These effects arose through a transformation of the distribution ratios of trihalomethanes (THMs) and haloacetic acids (HAAs) in the DBPs after chlorine disinfection. Also, the addition of $Al^{3+}$ resulted in inhibition efficiency for THMs, but promotion efficiency for HAAs. With the increase of $Al^{3+}$, the inhibition efficiency for THMs and the promotion efficiency for HAAs were both reduced. According to analyses of the results by ultraviolet spectrophotometer (UV) and size exclusion chromatography (SEC), an equilibrium system can be assumed to exist between hydrolysis and flocculation in tannic acid solution. It could also be inferred that the addition of $Al^{3+}$ affected the equilibrium system, which had a direct or indirect influence on the chemical properties of tannic acid, and finally resulted in changing the generation and distribution ratios of the THMs and HAAs of DBPs during chlorination.

# DEFLUORIDATION POTENTIAL OF VERMICULITE MODIFIED WITH HEXADECYLTRIMETHYLAMMONIUM

***Tayo Oladipo Ologundudu,*** John Ogony Odiyo, Georges-Ivo Ekosse
(University of Venda, Thohoyandou, Limpopo Province, South Africa)

Vermiculite modified with hexadecyltrimethylammonium, HDTMA, was investigated for its defluoridation potential, as fluorosis is a major health challenge especially in the rural areas across Africa and Asia. 1 M sodium chloride was used to activate the clay mineral before its agitation with 0.2 M HDTMA at 40 °C for 24 hrs. Fluoride removal experiments were carried out using batch adsorption technique coupled with optimisation of experimental parameters such as pH, mass of adsorbent, contact time and temperature. Using 8 mg/L fluoride solution, the results showed a fluoride removal of 51% at pH 4, 2 g, 70 min and 25 °C. The $pH_{pzc}$, 6.4, of modified vermiculite enhanced the fluoride removal via electrostatic attraction, while the FTIR analysis revealed that the $CH_3$ and $CH_2$ bands introduced to vermiculite by the organic cation, was responsible for defluoridation. The antimicrobial properties of HDTMA, coupled with the defluoridation potential of this adsorbent, showed that vermiculite modified with HDTMA could be developed by further research into a domestic water defluoridation unit. This approach would be effective and economical for practical use especially in non-metropolitan environments.

# EVALUATION OF WATER QUALITY OF THE ASI (ORONTES) RIVER BETWEEN 1997 AND 2008

Hasan Göksel Özdilek[1] , Ziya S. Cetiner[2] and Mustafa Kemal Sangün[3]

[1] Çanakkale Onsekiz Mart University, Faculty of Engineering, Terzioglu Campus 17020 Canakkale, Turkey. E-mail:hgozdilek@comu.edu.tr

[2] Çanakkale Onsekiz Mart University, Faculty of Engineering, Geological Engineering Department, Terzioglu Campus 17020 Canakkale, Turkey

[3] Mustafa Kemal University, Faculty of Science and Letters, Chemistry Department, Tayfur Sokmen Campus 31100 Antakya, Hatay, Turkey

**ABSTRACT:** This paper assesses the changes in water quality of the Orontes (Asi) River, an internationally shared river in the Middle East starting from Lebanon, crossing through Syria and ending in southern Turkey. For this purpose, historical water quality data covering the years between 1997 and 2008 were examined at two water quality monitoring stations, one of which is situated just after the Syrian-Turkish border and the other is located within a large city, Antakya. Results from two comprehensive water quality surveys are also evaluated during October 2005 and February 2006. Since available historical data includes thirteen water quality parameters, which were sampled once a month, our assessment provides comprehensive record of long-term changes in water quality. Water quality index (WQI) developed by the Canadian Minister of the Environment was also used to classify water quality in the river for two additional comprehensive water surveys conducted in October 2006 and February 2007. It was found that wet-weather poses a risk in terms of water quality compared to dry weather conditions. In the lower Orontes River catchment, it was found that a noticeable trend of increasing sodium load, especially in wet years, between the two water quality monitoring stations operated between 1997 and 2008.

**Key Words:** Asi river, water quality index, heavy metals, boron, transboundary waters

## INTRODUCTION

In general, water quality is defined by the physical, chemical and biological parameters affecting the well-being of a water body. Water quality is crucially important because of its close linkage with public health and ecological quality. Contamination of water bodies from various sources deprives current and future generations of a birthright and jeopardizes ecological integrity. While point sources of water pollution resulting from independent effluents can be measured directly at the site of concern, nonpoint sources of pollution is a chiefly wet weather driven phenomenon that causes change in pollutant concentrations in a river course.

Changes in water quality can occur for many reasons, such as irrigation schemes, precipitation amounts, extensive use of fertilisers and other types of agricultural aids, mechanization of agricultural activities, exhaust gases, and urban sprawl, all of which can negatively affect natural water quality and sediment transport in river ecosystems (Wei et al. 2009; Tong et al. 2009). Rivers are important since they contribute to pollution control through transport and removal of contaminants and excess nutrients. The factors just aforementioned are all nonpoint sources that may introduce natural nutrients, bacteria, minerals, heavy metals, and toxins into the water bodies.

Recently, many studies related to surface water quality have been conducted in different parts of the globe (Srapper et al. 2000; Woitke et al. 2003; Jarvi et al. 2006; Neal et al. 2010; Bayram et al. 2011; Kumarasamy et al. 2014, Tavakoli et al. 2014, McClain and Maher, 2016), but determination of trends in water quality is a relatively newer issue that requires analysis of reliable and long-term data.

For over a half-century, the Orontes River and many of its tributaries have been used for the discharge of industrial and domestic wastewaters at various locations. Refineries, olive oil production plants, and municipal waste discharges have all contributed to the pollution of the Asi River. This is an international dispute between Syria and Turkey. Similar experiences have been experienced in different parts of the world, because water is an essential part of life (Begon et al. 2006). However, despite the fact that its catchment has the second highest human population density (after the Smaller Meander River situated in western Turkey), the Orontes River is the second richest river (following the Sakarya River in north-western Turkey) in terms of the number of native fish species present (Akbulut et al. 2009). The length of its main branch is 380 km, the total length of the river and its tributaries is nearly 520 km, and its total catchment area is nearly 24,000 km$^2$. There are some dams on the river and its tributaries, namely Quattena, Ar-Rastan, Mahardeh, Apamia and Zeyzoun Dams in Syria, and Tahtaköprü (on Karasu branch) and Yarseli Dam (on Beyazcay branch) in Turkey.

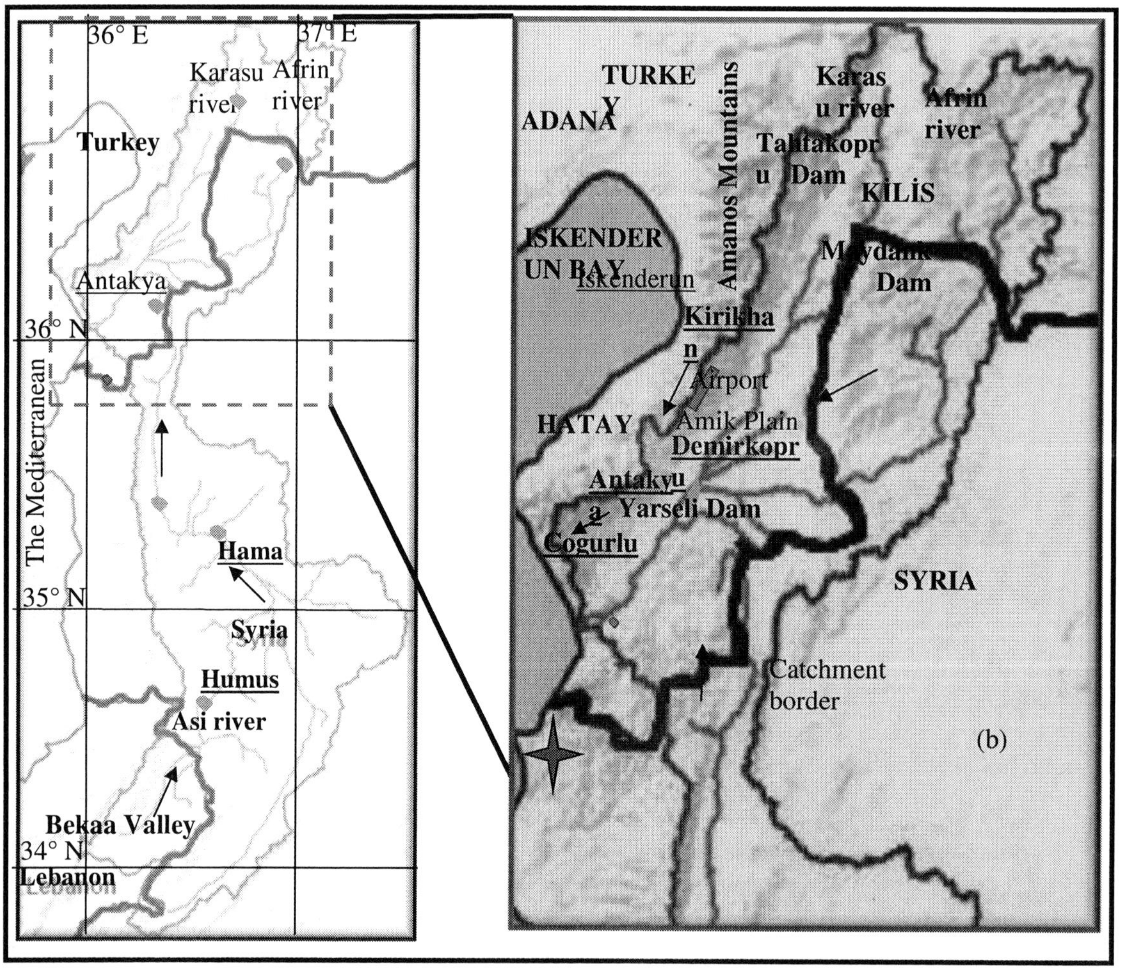


Figure 1. The Asi basin (a) as a whole and (b) its Turkish part in detail

The aim of this paper is to investigate changes in water quality focusing on water quality between 1997 and 2008 in the downstream portion of a transboundary Orontes River (Asi river) situated in a region, where water is crucially scarce. This river is located in the north-eastern Mediterranean. It extends relatively parallel to the eastern Mediterranean coast with a total length of approximately 450 km (Arisoy and Turkoglu, 1998).

The Orontes River originates from a stream located in the Bekaa Valley in the north-eastern tip of the Lebanon and passes through important cities, such as Baalbeck and Hermel (in Lebanon), Humus, Hama, Muhradah, Jisr Ash-Shughur (in Syria), and Antakya (in Turkey), (listed from south to north). The river, defines the border between Syria and Turkey and is fed by Karasu and Afrin (flowing from north to south) just before Antakya (provincial centre of Hatay), where it flows into the Mediterranean Sea in the district of Samandag.

Kirikhan, located within the middle of the Karasu river, had a population of 70,543 and Reyhanli, close to the Afrin river, had a population of 60,073 in 2000. There are no wastewater treatment plants serving these two areas. Some of the municipalities contributing to effluent discharges that do not have domestic wastewater treatment facilities include Kucukdalyan (population equivalent of 8,095), Narlica (population equivalent of 12,750), Ekinci (population equivalent of 6,459), Guzelburc (population equivalent of 6,281), Kuzeytepe (population equivalent of 6,051), Odabasi (population equivalent of 10,192), Serinyol (population equivalent of 12,517), and Ovakent (population equivalent of 6,288). However, the wastewater discharge of Antakya did not affect the second sampling station situated in the middle of the city. Antakya, the province center of Hatay, has been operating its trickling filter type domestic wastewater treatment plant since 2007 although prior to this year wastewater treatment plant was inactive. The population of Antakya was 186,213 in 2007 and the most recent population is 347,974 (TUIK, 2013).

## MATERIALS AND METHODS

First, historical monthly water quality data sampled between 1985 and 2008 was assessed with the parameters taken into account in the model as water acidity (pH), electrical conductivity (EC), sulphate, chlorine, total hardness, and water temperature. The Canadian Water Quality Index was used to detect the trend at Demirkopru water quality monitoring station just after Syrian-Turkish border.

For boron and stream flow, the length of monitoring period and available number of samples are given in Table 2. These samples were taken on monthly basis to evaluate water quality of the river. The aim for this research was to determine the following important points.

(1) Background level of the stream flow water quality of the Orontes River at Demirkopru (just after the Syrian-Turkish border) and

(2) Increase or decrease in water quality in the case of the examined parameters at Antakya city centre

The river samples were analysed for major, minor and trace elements by filtering to 0.45 μm in situ. Boron and sodium samples were determined by inductively coupled plasma mass spectrometer (ICP-MS) on acidified samples. Calibration was conducted using prepared standards, which were cross-checked against internationally certified standards to quality check and correct for potential instrumentation drift. Analysis included three replicates and for every six samples a set of our standards was run. If the standards analyzed differed from their "real values" by more than 5%, the analysis was redone. The lowest detection value was 10 $\mu g\ L^{-1}$ for boron and 20 $\mu g\ L^{-1}$ for sodium.

Historical stream flow records were received from the Turkish Electrical Power Sources. Corrections have not been made to allow for exact representation of stream flows at water gauges. It is generally assumed that the additional volumetric input of effluents into the Karasu, Afrin (stream flow recorded since 1968), and Asi river between two water gauging stations is relatively insignificant.

In general, two tributaries (Afrin and Karasu) contribute only 14% of the total flow measured downstream of the river, approximately 1 km away from where it flows into the Mediterranean Sea. The only exceptions to this occurred in May, likely due to snowmelt in upper parts in addition to higher surface runoff of the Karasu catchment, and in December when notable precipitation occurs in northern part of the Karasu catchment. According to the official records on meteorological parameters, Kilis, which is the origination point of the Afrin, gets 17% of its total annual precipitation in December.

Finally, two comprehensive water quality surveys were taken into account to examine whether there is a trend in water quality in Turkish part of the river. One survey was performed in the dry period (October 2006), and the other was completed in the wet period (February 2007).

**CWQI 1.0 Model.** The CWQI model presents a mathematical framework for evaluating ambient water quality situations related to water quality objectives set up for different purposes, such as irrigation, domestic water supply, and aquatic life. The model has three measures of difference from important water quality objectives: scope ($F_1$) defined as the number of variables violating water quality limits, frequency ($F_2$), the number of samplings the objectives are not met, and amplitude ($F_3$) the amount by which water quality limits are not met. The index yields a numerical score between 0 (worst water quality) and 100 (the best attainable water quality). These numbers are divided into five descriptive classes to make the presentation simpler and water quality index (WQI) is computed as

$$F_1 = 100 \text{ x} \frac{\text{Number of failed parameters}}{\text{Total number of parameters}} \quad (4)$$

$$F_2 = 100 \text{ x} \frac{\text{Number of failed tests}}{\text{Total number of tests}} \quad (5)$$

and $F_3$ (amplitude) represents the amount by which failed test values violate their objectives. $F_3$ is computed according to the following three steps.

1st Step. The number of times by which an individual level is more than or less than (when the objective is a minimum level) the objective, is expressed an "excursion" and stated as follows. When the test value must not exceed the objective criteria:

$$\text{excursion}_i = \frac{\text{Failed test value i}}{\text{Objective j}} - 1 \quad (6)$$

When the objective is at a minimum level and the parameter must be larger than this minimum.

$$\text{excursion}_i = \frac{\text{Objective j}}{\text{Failed test value i}} - 1 \quad (7)$$

where,
i and j = 1, 2, 3, ..., n

2nd Step. Normalized sum of excursion (NSE), which is defined by the summation the excursion of individual tests from their objective criteria divided by the total number of tests (both those within objective limits and those violating objective limits), is computed as,

$$NSE = \frac{\sum_{i=1}^{n} excursion\ i}{number\ of\ tests} \quad (8)$$

3rd Step. $F_3$ is computed by an asymptotic arithmetical function that scales the NSE to output a score between 0 and 100:

$$F_3 = \frac{\text{NSE}}{0.01\text{NSE}+0.01} \quad (9)$$

Finally, the WQI is computed as

$$\text{WQI} = 100 - \frac{\sqrt{(F_1^2+F_2^2+F_3^2)}}{1.732} \quad (10)$$

If WQI is between 95 and 100, the water is categorized as excellent; whereas, if it falls below 44, the water is classified as poor. This ranking is based on aquatic and irrigation uses of water resources.

## RESULTS AND DISCUSSION

Based on historical data obtained from the General Directorate of Electrical Power Resources Survey and Development Administration of Turkey, water quality index at Demirkopru gauging station (between 1984 and 2007) first dropped to "marginal" status in 1999, while prior to this it was assessed to be in "fair" status. A status ranging from "good" to "marginal" indicates that water quality, is protected generally but is occasionally impaired based on the six available water quality parameters including pH, water temperature, electrical conductivity (EC), sulphate, chlorine, and total hardness constituents (Figure 2). Instead of 1250 $\mu S\ cm^{-1}$ limit for EC, a value of 1000 $\mu S\ cm^{-1}$ was adopted because the Mann-Kendall trend analysis showed an increasing trend in EC at Demirkopru water quality monitoring station over the past few years. The WQI computed for Demirkopru based on years showed a worsening trend. In 2002, the water quality was evaluated to be at "marginal" status second time in the time series examined (covers the years between 1985 and 2007). One more step was taken to determine whether WQI at Demirkopru water quality monitoring station is increasing or diminishing based on the six parameters listed above. The Mann-Kendall trend analysis was performed for WQI scores computed for yearly data between 1985 and 2007. The outcome of this test showed that the WQI has decreased significantly ($S=-105$; $SE(S)=37.86$; $u=-2.75 < t=-2.07$). It is clear that the water quality at the Demirkopru water quality monitoring station, based only on the six parameters mentioned above, decreased dramatically between 1985 and 2007. This finding indicates that the upstream segments (before Turkish catchment) of the river provide a great deal of contamination. The only marginal state of the WQI was observed in 2002. However, the WQI is close to

64, which is within marginal water quality, after the year 2002. 2002 is the same year that the Zeyzoun Dam collapsed and damaged some parts in the watershed as well as river bed. This might also be related with the sediment contamination. However, further studies are needed to potentially link sediment quality with water quality.

**Dry Weather and Wet Weather Conditions.** Water quality parameters for October 2005 and February 2006 samples were determined at seven different stations. They were also evaluated by the WQI. The water quality parameters examined included: EC (the limit value was chosen to be 1000 μS $cm^{-1}$), water temperature (should be less than 22 °C), pH (should be between 6.5 and 8.5 for potable waters), surface active material (should not be greater than 0.2 mg $L^{-1}$), chemical oxygen demand (COD should be less than 30 mg $L^{-1}$), total suspended solids (TSS should be less than 25 mg $L^{-1}$), nitrate ($NO_3$-N, which should not exceed 0.5 mg $L^{-1}$), total phosphorus (TP should be below 0.65 mg $L^{-1}$), zinc (should be no more than 3 μg $L^{-1}$), iron (should be less than 100 μg $L^{-1}$), nickel (should be less than 20 μg $L^{-1}$), potassium (should be less than 650 μg $L^{-1}$), dissolved oxygen (should exceed 6.5 mg $L^{-1}$), selenium (should be less than 4.6 μg $L^{-1}$), boron (should be less than 1 mg $L^{-1}$), and manganese (should be less than 0.05 mg $L^{-1}$). The measured values were compared to the toxicity limits of constituents which mainly depend on the water hardness.

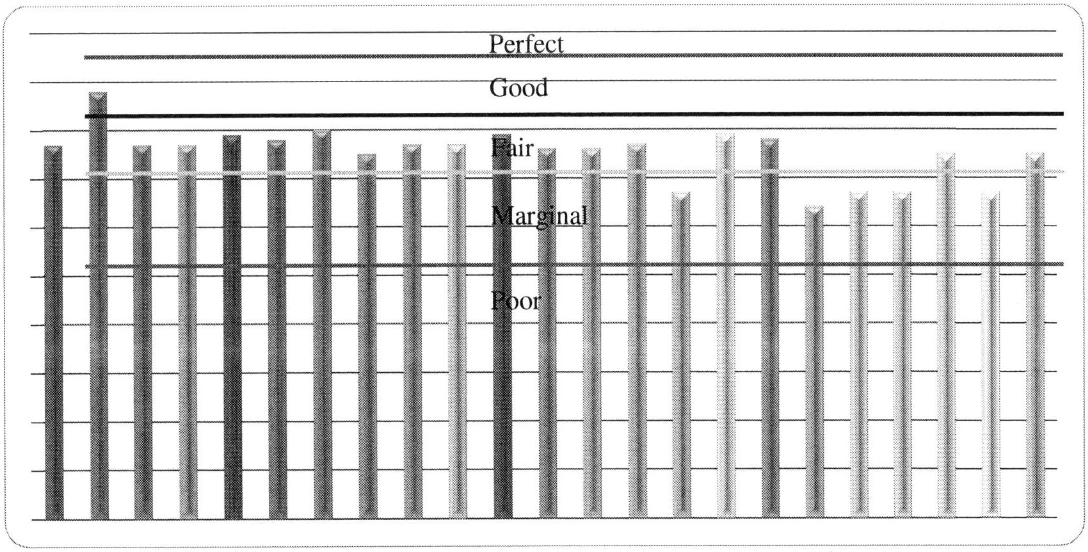


Figure 2. Water quality index with respect to years at Demirkopru water quality monitoring station

Water hardness was determined to range between 417-442 mg $L^{-1}$ $CaCO_3$ at seven subsequently located sampling stations in October 2005. At this time, the streamflow was 13.2 $m^3$ $s^{-1}$ at Demirkopru, and 22.9 $m^3$ $s^{-1}$ at Antakya. In February 2006, the streamflow was 43.5 $m^3$ $s^{-1}$ at Demirkopru and 177.4 $m^3$ $s^{-1}$ at Antakya (measured by the staff of the General Directorate of Electrical Power Resources Survey and Development Administration of Turkey). With the increase of streamflow, the water hardness of the seven sampling stations ranged from 187 to 195 mg $L^{-1}$ $CaCO_3$ in February 2006.

Copper levels were compared with the fish and aquatic life criteria (Pennsylvania Code, 2009) and were computed to be 3.5 μg $L^{-1}$ for October 2005 (calculations were based on calcium carbonate hardness as mg $L^{-1}$ and all measured values as triplicates from seven stations were determined to be on the safe side). On the other hand, all samples from all sampling stations exceeded the limit determined for copper (2.8 μg $L^{-1}$) in February 2006. Especially, after the city centre (Antakya), copper concentrations were determined to be more than 7 μg $L^{-1}$. For momentary sampling results, neither October 2005 nor February 2006 sampling results exceeded limit values, which were computed as 52 μg $L^{-1}$ for October 2005 and 24 μg $L^{-1}$ for February 2006.

For lead, neither October 2005 nor February 2006 samples were found to exceed their long-term limit values, which were 12 μg $L^{-1}$ for October and 5 μg $L^{-1}$ for February 2006. Furthermore, the momentarily sampling limit values were determined as nearly 300 μg $L^{-1}$ for October 2005 and 130 μg $L^{-1}$ for February 2006.

For selenium, only the first sampling station (just after the Syrian-Turkish border) and the last two stations (close to the river delta) were found to exceed the limit for selenium concentration of 4.6 μg $L^{-1}$ in October 2005. On the other hand, the first two stations were determined to be problematic due to higher selenium compared to the limit value (4.6 μg $L^{-1}$) in February 2006. For nickel all sampling stations and the sampling times were found to violate the limit value (approximately 5 μg $L^{-1}$) based on water hardness. The limit value of selenium exists only for continuous concentration for fish and aquatic life.

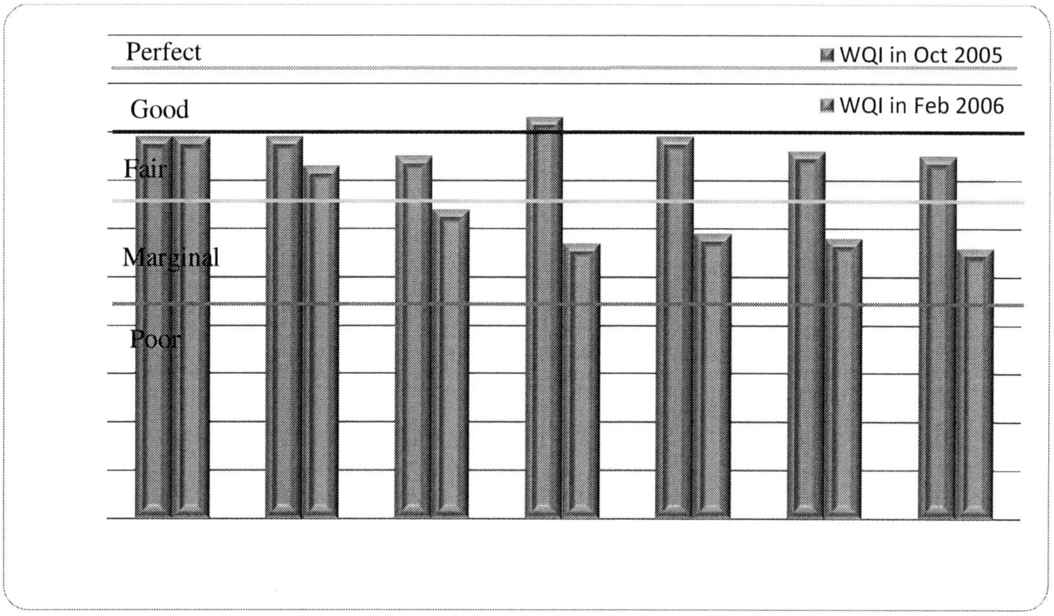


Figure 3. Water quality index result of two different surveys performed on seven stations on the generally within Hatay Province, Turkey

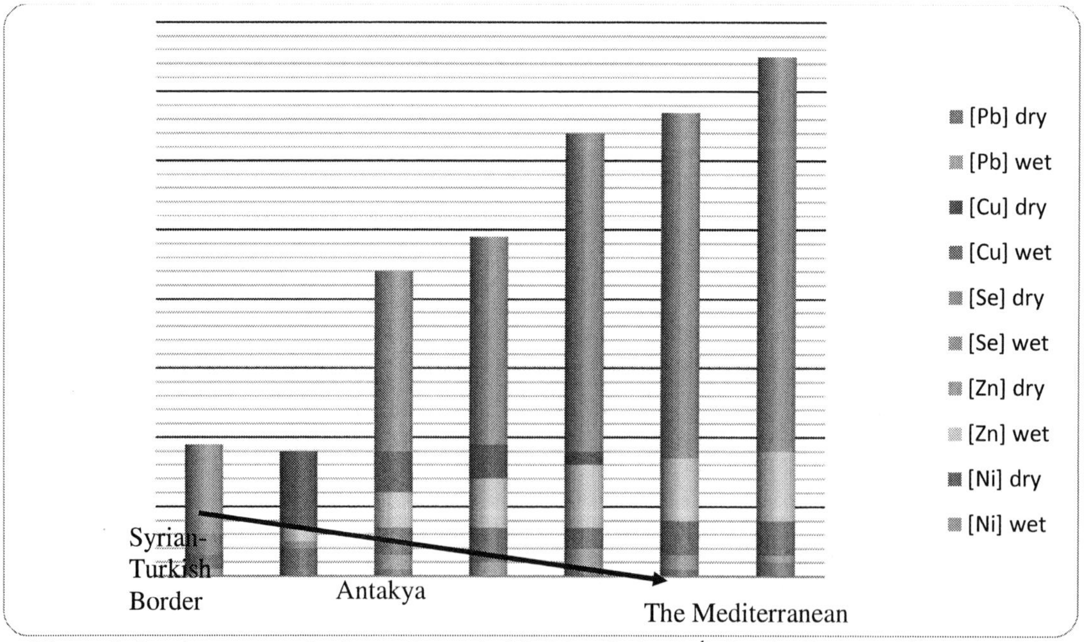


Figure 5. Lead, copper, selenium, zinc and nickel levels (μg $L^{-1}$) for the Orontes River's Turkish part along seven sampling stations in October 2005 (dry period) and February 2006 (wet period)

For zinc, only the first sampling station was found to have a higher concentration compared to the limit value (6 μg $L^{-1}$) in October 2005. Surprisingly, all sampling stations, with the exception of the first one located just after the Syrian-Turkish border, had an almost three-fold higher zinc concentration than the limit value (5.3 μg $L^{-1}$) in February 2006. Figure 4 represents the measured concentrations of the above mentioned water quality constituents along the seven sampling sites.

Although not illustrated in Figure 4, aluminium concentrations indicated a risk in February 2006 due to the fact that the limit value for momentary sampling is approximately 750 μg $L^{-1}$; whereas aluminium concentrations, in October 2005, at sampling stations 3 through 7 were two to three times above the limit value (Pennsylvania Code, 2009).

For chromium, all sampling stations were found to have lower concentrations than the limit value (determined as 1 mg $L^{-1}$) in February 2006, and the limit value (determined as 1.9 mg $L^{-1}$) in October 2005. For vanadium, both samplings were found to have lower concentrations than the limit value (510 μg $L^{-1}$) determined for momentary sampling. Similarly, both samplings were not found to exceed the limit value of barium (21 mg $L^{-1}$) and of cobalt (95 μg $L^{-1}$).

On one hand, anthropogenic pressures on natural landscapes regulates stream flow more homogenously, but on the other hand, water quality deteriorates due to floods affected by changing basin structure.. In a relatively unoccupied river basin, the River Murray, sodium concentration increased more than thirty fold between Jingellic (2322 km away from the delta) and Tailem Bend (87 km away from the delta) (Mackay et al, 1988).

## CONCLUSION

There has been a noticeable trend of increasing sodium load, especially in wet years, between the two water quality monitoring stations located in the lower Orontes River catchment. Water quality has been improved for a number of reasons, such as wastewater treatment facilities, better waste management plans, and stricter use of chemical aids in agriculture. The increase in some water quality constituents along the river especially during the wet period signifies sediment contamination as well as wash-off of these constituents. However, a comprehensive sediment quality or background analyses need to be completed in the study area.

## REFERENCES


Akbulut (Emir), N., Bayarı, S., Akbulut, A., Yalçın, Ş., (2009). Rivers of Turkey, in: Tockner, K., Robinson, C. T., and Uehlinger, U. (Eds.) Rivers of Europe. Academic Press, Inc., London, pp. 643-672.

Arisoy, A.I. and Turkoglu, Y., (1998). Türkiye'nin Sınıraşan ve Sınır Oluşturan Sularının Siyasi Coğrafya Açısından Değerlendirilmesi (in Turkish). Seminar, Gazi University, Ankara.

Bayram, A., Onsoy, H., Akinci, G, Bulut, V. N., (2011). Variation of total organic carbon content along the Stream Harşit, Eastern Black Sea Basin, Turkey. *Environ. Monit. Assess.* 182 (1-4), 85-95.

Begon, M., Townsend, C. R., Harper, J.L., (2006). Ecology: From individuals to ecosystems, fourth ed. Blackwell Publishing, Malden, MA.

Kumarasamy, P., Dahms, H.U., Jeon, H.J., Rajendran, A., James, R.A., (2014). Irrigation water quality assessment-an example from the Tamiraparani river, Southern India. Arabian J of Geoscience. 7 (12): 5209-5220.

Jarvi, H. P., Neal, C., Withers, P. J. A., (2006). Sewage-effluent phosphorus: A greater risk to river eutrophication than agricultural phosphorus? *Sci. Total Environ.* 360 (1-3): 246-253.

Mackay, N., Hillman, T., Rolls, J., (1988). Water Quality of the River Murray: a review of monitoring 1978-1986. Murray-Darling Basin Commission, Canberra, A. C. T., Water Qual. Rep. No. 1.

McClain and Maher (2016).

Neal, C., (2000). The fractionation of the elements in river waters with respect to the continental crust: a UK perspective based on a river enrichment factor approach. *Hydrol. Earth Syst. Sci.* 4: 499-509.

Neal, C., Williams, R. J., Bowes, M. J., Harrass, M. C., Neal, M., Rowland, P., Wickham, H., Thacker, S., Harman, S., Vincent, C., Jarvie, H. P., (2011). Decreasing boron concentrations in UK rivers: Insights into reductions in detergent formulations since the 1990s and within-cathment storage issues. *Sci. Total Environ.* 408: 1374-1385.

Pennsylvania Code, (2009). §93.8c. Human Health and aquatic life criteria for toxic substances. Accessed from the internet on December 22, 2011 from: http://www.pacode.com/secure/data/025/chapter93/s93.8c.html

Scheumann, W., Sagsen, I., Tereci, E., (2011). Orontes River Basin: Downstream Challenges and Prospects for Cooperation in Turkey's Water Policy, in: Kibaroglu, A., Scheumann, W., Kramer, A. (Eds.) Springer-Verlag, Berlin Heidelberg, pp. 301-312.

Tavakoli, A., Kerachian, R., Nikoo, M. R., Soltani, M., & Estalaki, S. M. (2014). Water and waste load allocation in rivers with emphasis on agricultural return flows: application of fractional factorial analysis. Environmental Monitoring and Assessment, 186(9), 5935–5949.

The General Directorate of Electrical Power Resources Survey and Development Administration of Turkey (2010).

Water Quality Monitoring Stations 1997 and 1998 reports. Unpublished data.

Tong, S.T.Y., Liu, A. J., Goodrich, J.A., (2009). Assessing the water quality impacts of future land-use changes in an urbanising watershed. *Civil Eng. Environ. Syst*. 26 (1): 3-18.

TUIK, (2013). Address Based Population Data (in Turkish) 2013. Accessed from the internet on January 20, 2015 from: http://rapor.tuik.gov.tr

Wei, G. L., Yang, Z. F., Cui, B. S., Li, B., Chen, H., Bai, J. H., Dong, S. K., (2009). Impact of Dam construction on Water Quality and Self Purification capacity of the Lancing River, China. *Wat. Res. Man*. 23: 1763-1780.

Woitke, P., Wellmitz, J., Helm, D., Kube, P., Lepom, P., Litheraty, P., (2003). Analysis and assessment of heavy metal pollution in suspended solids and sediments of the river Danube. *Chemosphere*. 51 (8): 633-642.

# WATER QUALITY ASSESSMENT FOR MELEN WATERSHED IN THE MARMARA REGİON, TURKEY

***Ersin Orak***[a], Atilla Akkoyunlu[b]

a) Çanakkale Onsekiz Mart University Environmental Eng. Department Çanakkale, Turkey
E-Mail : ersinorak@comu.edu.tr

b) Boğaziçi University Department of Civil Engineering 34342 Bebek/İstanbul Turkey
E-mail: akkoyun@boun.edu.tr

**ABSTRACT:** Water quality assessment is necessary for management of water resources. As known environmental processes are very complex . Reflection of this complexity is also seen in water quality models.Therefore finding efficient and sound models in many cases are very difficult. Water quality class assessment for inland water bodies in Turkey, is evaluated for each parameters according to the Water Quality Standards for Inland Waterbodies. This standard is only for a single pollution parameter. In this study authors tried to express the class of water body with the coverage of overall water qualities measured. For this aim, fuzzy syntetic evaluation (FSE), and self-organizing map (SOM) are applied for Melen River and compared with traditional method. FSE and SOM are easy to understand and apply in water quality assessments for watershed water quality management. The Melen River is in the Marmara region of Turkey located in north. This river is used for many application like irrigation of the agricultural lands, drinking water supply and several different industries' process water, It supplies potable water to Istanbul which is most densely populated.mega city in Turkey.

## INTRODUCTİON

Water management is now considered among the most important issue for natural resources in any country. Assessment of water quality is very important because it refers to the physical chemical and biological characteristics of water. Many countries have their own management strategies for their water sources protection (Chang, Chen, & Ning, 2001). Water quality assessment is very important for the management applications. Modelling of environmental processes have high complexity, dynamism, and non-linearity. Therefore, finding efficient models in many cases are very difficult (Kalteh, Hjorth, & Berndtsson, 2008). Some new methods are invented to overcome these problems. In this study, two of them are used, namely These are Fuzzy Logic (FL), and Self-Organazing Map (SOM).

Fuzzy logic was introduced by Zadeh (1965) and has become one of the most common approaches in the field of Artificial Intelligence (AI). FL is a multi-valued logic that is an alternative to Aristotelian Logic (Seguí et al., 2013). FL includes fuzzification, application of the fuzzy rule base to data, and finallty defuzzification (Icaga, 2007). Fuzzy synthetic evaluation (FSE) is a model, which based on FL, using to assess multiple criteria decision-making (Xu et al., 2010).

Artificial Neural Networks (ANNs) based on supervised and unsupervised learning, are alternative tools for ecological data processing, which become in extensive use over the last decades. Supervised learning is carried out for data estimation (e.g., prediction, revealing the environment–community causality relationships) based on a priori knowledge. Unsupervised learning is useful in extracting information from the data (e.g., ordination, classification) without previous knowledge (T. S. Chon & Park, 2008).

Self-Organizing Maps (SOM) are competitive and unsupervised forms of ANNs pioneered by the Finnish professor, Professor Teuvo Kohonen (1981) (Kohonen, 2001). One of the most widely used ANN algorithms is SOM, which has been used extensively for information extraction without prior knowledge and the efficiency of visualization (T.-S. Chon, 2011).

Water quality class assessment in Turkey, is evaluated for each parameters according to the Water Quality Standards for Inland Waterbodies which are in "Water Pollution Control Regulations of Turkey" (WPCR, 31.12.2004).

## STUDY AREA

The Melen River is in the northern part of Turkey. Its watershed is located in the western Black Sea Region of Turkey (Fig. 1) with a total area of 2,437 km$^2$. It has a transitional climate between Western Black Sea, Marmara and Mediterranean regions. It is used mainly for irrigation agriculture, drinking water, and several different industries' process water. The used water is directly discharged into the Melen River. The other important use of Melen River is that it supplies water to Istanbul City, which is highly populated. The water supplied from Melen River satisfies more than 50% of Istanbul's water demand due to its high water potential and relatively less degraded water quality (Akiner & Akkoyunlu, 2012), (Koklu, Sengorur, & Topal, 2010), (Erturk et al., 2010), and (Karakaya & Evrendilek, 2010).

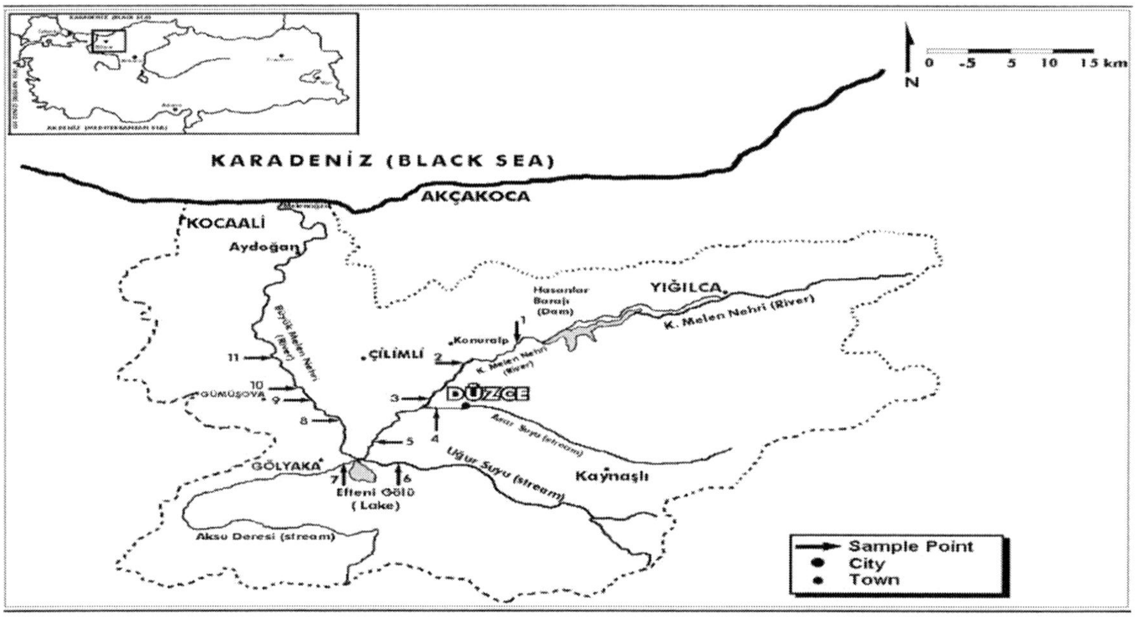


Figure 1 Location map of Melen catchment

In this study, there are twelve observation station, which these are Aksu Creek, Aksu Hasanlar Dam In, Karadere Hasanlar Dam In (quarry), Asar River Before K.Melen River, B.Melen River Bridge Before Pakmaya, B.Melen River Ugurlu Village, K.Melen River Hasanlar Dam In, K.Melen River Hasanlar Dam Bottom Sluice Out, Lahna Stream Before B.Melen, K:Melen River Pasa Konagi, Ugur Creek, and six water quality parameters, namely, dissolved oxygen (DO), ammonium nitrogen ($NH_4$-N), nitrite ($NO_2$-N), nitrate ($NO_3$-N), biological oxygen demand (BOD), and total phosphorous (TP), are measured from 1995 to 2009. Data for this research is provided from the State Hydraulic Works of Turkey (DSI, 2009).

## METHODOLOGY

In the study, water quality classes are given according to the "Water Pollution Control Regulations of Turkey" and two different methodologies, such FL, and SOM are employed for the data treatment and modeling.

**Fuzzy Logic.** The applications of the fuzzy logic are very wide are including process control, management and decision making, operations research etc.

Models based on fuzzy rules are enough to demonstrate uncertainty and inaccuracies in knowledge and data. These models can reflect qualitative sides of knowledge and human outcome processes without a precise quantitative analysis. Thus, they are less certain than traditional numerical models. Nevertheless, the gains are in simplicity, computational speed and flexibility that result from the use of these models with compensation loss in precision (Lermontov, Yokoyama, Lermontov, & Machado, 2009).

In the beginning of 1990s, classification of river water quality by fuzzy theory has started. Kung et al. (1992) published the application of fuzzy clustering analysis to water quality classification.Fuzzy synthetic evaluation was used by Lu et al. (1999) to investigate and compare the water quality of Fei- Tsui Reservoir in Taiwan. The conclusion of this search showed that most of the water quality records were evaluated as the same trophic state by both the Carlson index and FSE (Lu & Lo, 2002).

Fuzzy evaluation method is applied in three steps. First, the membership functions are identified, which is followed by fuzzy relations matrix (R) and the standard inference mechanisms (Şen, 2010). Finally, the water quality values are obtained through a fuzzy operator.

***Determination of membership function.*** In the determination of water quality classes of the different points along the Melen River "Water Quality Standards for Inland Water Bodies" have been employed. This study defines four membership functions (μ), in crisp sets an element may belong the set or not but in fuzzy sets membership degree value between 0 to 1 (Şen & Altunkaynak, 2009). Water quality classes according to the Water Quality Standards for Inland Waterbodies are given below. The standard parameters values related to this study are presented in Table 1 (WPCR, 31.12.2004). A water quality classification in the regulation is as follows:

- Class I: High quality water
- Class II: Slightly polluted water
- Class III: Polluted water
- Class IV: Extremly polluted water

Table 1. The Water Quality of Inland Surface Waters in River, Lakes, and Dam Reservoirs

| Parameters | 1 | 2 | 3 | 4 |
|---|---|---|---|---|
| Dissolved Oxygen (mg O2/L) | 8 | 6 | 3 | < 3 |
| Ammonium Nitrogen (mg NH4+-N/L) | 0.2 | 1 | 2 | > 2 |
| Nitrite Nitrogen (mg $NO2^{-}$-N/L) | 0.002 | 0.01 | 0.05 | > 0.05 |
| Nitrate Nitrogen (mg $NO3^{-}$-N/L) | 5 | 10 | 20 | > 20 |
| Total Phosphorus (mg P/L) | 0.02 | 0.16 | 0.65 | > 0.65 |
| Biochemical Oxygen Demand (BOD) (mg/L) | 4 | 8 | 20 | > 20 |

The calculation of membership functions are as follows:

$$\mu_1 = \begin{cases} 1 & if\ 0 \le x \le D_1 \\ \dfrac{D_2 - x}{D_2 - D_1} & if\ D_1 \le x \le D_2 \\ 0 & if\ x \ge D_2 \end{cases} \qquad \mu_2 = \begin{cases} \dfrac{x - D_1}{D_2 - D_1} & if\ D_1 < x < D_2 \\ 1 & if\ x = D_2 \\ \dfrac{D_2 - x}{D_2 - D_1} & if\ D_2 < x < D_3 \end{cases}$$

$$\mu_3 = \begin{cases} 0 \quad if\ x \le D_1\ or\ x \ge D_3 \\ \dfrac{x - D_1}{D_2 - D_1} & if\ D_1 < x < D_2 \\ 1 & if\ x = D_2 \\ \dfrac{D_2 - x}{D_2 - D_1} & if\ D_2 < x < D_3 \end{cases} \qquad \mu_4 = \begin{cases} 0 & if\ x \le D_3 \\ \dfrac{x - D_3}{D_4 - D_3} & if\ D_3 \le x \le D_4 \\ 1 & if\ x \ge D_4 \end{cases} \qquad (1)$$

where Di (i=1, 2, 3, 4) is the value of the water quality class in, x is the sample water quality value, and μ is membership degree value.

Evaluation matrix (R) is set by using the membership values. R is given below, where, i represents DO, NH4-N, NO2-N, NO3-N, TP, BOD; and j is for class I, class II, class III, and class IV water quality.

$$R = \begin{bmatrix} \mu_{11} & \mu_{12} & \mu_{13} & \mu_{14} \\ \mu_{21} & \mu_{22} & \mu_{23} & \mu_{24} \\ \mu_{31} & \mu_{32} & \mu_{33} & \mu_{34} \\ \mu_{41} & \mu_{42} & \mu_{43} & \mu_{44} \\ \mu_{51} & \mu_{52} & \mu_{53} & \mu_{54} \\ \mu_{61} & \mu_{62} & \mu_{63} & \mu_{64} \end{bmatrix} \quad (2)$$

***Determination of weights.*** For each pollutant parameter it is necessary to determine a weight, which impacts one another, because each pollutant has its own characteristics. In this study, weights are distributed equally for the main pollutants, which are DO, BOD, TP, and components of nitrogen. The weight matrix is given in the following expression.

$W = \{w_{DO}(0.25), w_{BOD}(0.25), w_{TP}(0.25), w_{NH4\text{-}N}(0.25/3), w_{NO2\text{-}N}(0.25/3), w_{NO3\text{-}N}(0.25/3)$

***Fuzzy Synthetic Evaluation.*** FSE, which has been described earlier, can be calculated as follows:

$$FSE = W.R = \{b1, b2, b3, b4\} \quad (3)$$

The quality class of water body is evaluated depending on the maximum value of the four values (b1, b2, b3, b4).

**Self Organizing Map (SOM).** Ecological data processing is achieved by Artificial Neural Networks (ANNs), which depend on supervised and unsupervised learnings. Although supervised learning is carried out for data estimation (e.g., prediction, showing the environment–community causality relationships) based on a priori knowledge (i.e., templates), unsupervised learning is useful in extracting information from the data (e.g., sorting, classification) without previous knowledge. Self-organizing maps (SOMs), which is a unsupervised learning neural networks based on the Kohonen network, as in Figure 2, are extensively used in the extraction of information from ecological data (Chon and Park).

The term ''self-organizing'' refers to the ability to learn and organize information without being given the associated-dependent output values for the input pattern (Astel, Tsakovski, Barbieri, & Simeonov, 2007). The self-organizing map describes a mapping from a higher dimensional input space to a lower dimensional map space. The SOM compresses information while preserving the most important topological and/or metric relationships of the primary data elements on the display. It may also be thought to produce some kind of abstractions. These two aspects, visualization and abstraction, can be utilized in a number of ways in complex tasks such as process analysis, machine perception, control, and communication (Kohonen, 2001) .

Ecological data are thought to be difficult to analyze due to numerous biological and environmental factors involved in a complex manner in environment organism relationships. The SOM has advantages for information extraction without prior to knowledge and the efficiency of visualization(T.-S. Chon, 2011). The SOM is a fascinating neural network method that has found increasing interest in water resources applications (Kalteh et al., 2008).

The main advantages of the SOM algorithm are its non-linearity and an ability to preserve the topological structure of the data. The typical structure of an SOM consists of two layers; an input layer which contains one neuron for each variable in the data set and a Kohonen or output layer which is connected to every neuron in the input layer through adjustable weights or network parameters(Kalteh et al., 2008).

Since Chon et al. (1996) first applied the SOM to pattern benthic communities in streams, SOMs have been implemented in various aspects of water research, e.g., classifying biological and environmental data and patterning long-term data. The SOM has also proven to be a practical tool in assessing water quality and managing ecosystem (Céréghino & Park, 2009).

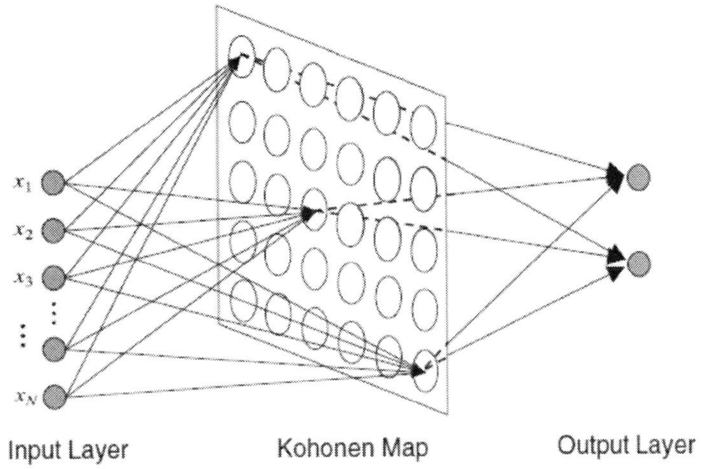


Figure 2. Schematic diagram of SOM.

There are three procedures for applying of SOM as explained in the following three sub-sections: Data gathering and normalization, Training and Extracting information.

***Data gathering and normalization.*** Normalizing the pollution parameters is very important, because it transforms the parameters to the range of 0-1, with all parameters having equal importance (Kalteh et al., 2008). In our study, the input data, DO, $NH_4$-N, $NO_2$-N, $NO_3$-N, TP, and BOD, are the normalized.

***Training.*** It is a iterative procedure and is recommended that the iterations number should be at least 500 times the number of neurons in the output layer (Kalteh et al., 2008). The training strategy is "winner takes all" (Lu & Lo, 2002).

***Extracting information.*** After the training of SOM, the results can be post-processed based on visualization, clustering, or local modelling purposes. Figure 3 shows an example of SOM clustering (Kalteh et al., 2008).

.

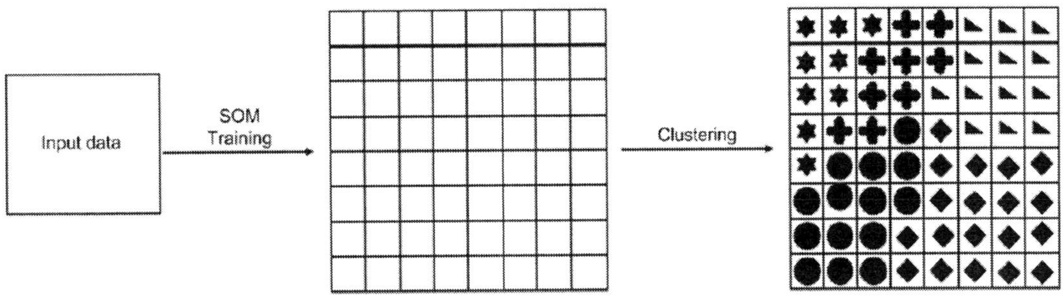


Figure 3. The diagram of a two-level clustering of the SOM. Different symbols represent different clusters

**RESULTS**

In this study, there are twelve observation station and dissolved oxygen (DO), ammonium nitrogen ($NH_4$-N), nitrite ($NO_2$-N), nitrate ($NO_3$-N), biological oxygen demand (BOD), and total phosphorous (TP) were measured from 1995 to 2009. The name of observation stations, measurement date, and matching of FSE results and SOM results (%) are given in Table3.

Table 3: The observation stations on the Melen River

| Observation Station | Measurement Date | Matching of FSE Results and SOM Results (%) |
|---|---|---|
| Aksu Creek | jan.1995 - Oct.2009 | 98.5 |
| Aksu Hasanlar Dam In | Dec.2006 - Dec.2008 | 94 |
| Karadere Hasanlar Dam In (quarry) | Dec.2006 - Dec.2008 | 94 |
| Asar River Before K.Melen River | Dec.2006 - Oct.2009 | 60 |
| B.Melen River Bridge Before Pakmaya | jan.1995 - Nov.2006 | 83 |
| B.Melen River after Pakmaya | Dec.2006 - Oct.2009 | 33.3 |
| B.Melen River Ugurlu Village | jan.1995 - Oct.2009 | 78 |
| K.Melen River Hasanlar Dam In | Dec.2006 - Dec.2008 | 94 |
| K.Melen River Hasanlar Dam Bottom Sluice Out | Dec.2006 - Nov.2009 | 89.3 |
| Lahna Stream Before B. Melen | Dec.2006 - Oct.2009 | 82.6 |
| K:Melen River PasaKonagi | jan.1995 - Oct.2009 | 69 |
| Ugur Creek | jan.1995 - Nov.2009 | 94.7 |

## CONCLUSİON

The Melen River Basin takes place in the north of the Turkey which supports a variety of sectors as irrigation systems in agricultural lands, drinking water and several different industries. In the basin, pollution sources are municipal and industrial wastewater and agricultural run-off. Due to the importance of Melen River for water supply to Istanbul City its water quality assessment is very important.

Ecologocical data analysis is considered to be difficult because many biological and environmental factors are involved in a complex manner in ecological relationships. In this study, Water Quality Standards for Inland Waterbodies in Turkey are used for showing to water classes for each parameters. Besides, Fuzzy Synthetic Evaluation, and Self-Organizing Map give an overall evaluation which may speed up the management activities in a watershed.. The other advantages of FL and SOM are easy to understand and apply to water quality assessments for watershed quality management.

The results of FSE and SOM are highly similar. According to the results SOM values are getting higher in the case of high pollution. Maybe more polluted regions SOM gives more safely results.

## REFERENCES


Akiner, M. E., & Akkoyunlu, A. (2012). Modeling and forecasting river flow rate from the Melen Watershed, Turkey. *Journal of Hydrology, 456–457*(0), 121-129. doi:http://dx.doi.org/10.1016/j.jhydrol.2012.06.031

Astel, A., Tsakovski, S., Barbieri, P., & Simeonov, V. (2007). Comparison of self-organizing maps classification approach with cluster and principal components analysis for large environmental data sets. *Water Research, 41*(19), 4566-4578. doi:http://dx.doi.org/10.1016/j.watres.2007.06.030

Céréghino, R., & Park, Y. S. (2009). Review of the Self-Organizing Map (SOM) approach in water resources: Commentary. *Environmental Modelling & Software, 24*(8), 945-947. doi:http://dx.doi.org/10.1016/j.envsoft.2009.01.008

Chang, N.-B., Chen, H. W., & Ning, S. K. (2001). Identification of river water quality using the Fuzzy Synthetic Evaluation approach. *Journal of Environmental Management, 63*(3), 293-305. doi:http://dx.doi.org/10.1006/jema.2001.0483

Chon, T.-S. (2011). Self-Organizing Maps applied to ecological sciences. *Ecological Informatics, 6*(1), 50-61. doi:http://dx.doi.org/10.1016/j.ecoinf.2010.11.002

Chon, T. S., & Park, Y. S. (2008). Self-Organizing Map. In S. E. Jørgensen & B. D. Fath (Eds.), *Encyclopedia of Ecology* (pp. 3203-3210). Oxford: Academic Press.

Erturk, A., Gurel, M., Ekdal, A., Tavsan, C., Ugurluoglu, A., Seker, D. Z., . . . Ozturk, I. (2010). Water quality assessment and meta model development in Melen watershed – Turkey. *Journal of Environmental Management, 91*(7), 1526-1545. doi:http://dx.doi.org/10.1016/j.jenvman.2010.02.021

Icaga, Y. (2007). Fuzzy evaluation of water quality classification. *Ecological Indicators, 7*(3), 710-718. doi:http://dx.doi.org/10.1016/j.ecolind.2006.08.002

Kalteh, A. M., Hjorth, P., & Berndtsson, R. (2008). Review of the self-organizing map (SOM) approach in water resources: Analysis, modelling and application. *Environmental Modelling & Software, 23*(7), 835-845. doi:http://dx.doi.org/10.1016/j.envsoft.2007.10.001

Karakaya, N., & Evrendilek, F. (2010). Water quality time series for Big Melen stream (Turkey): its decomposition analysis and comparison to upstream. *Environmental Monitoring and Assessment, 165*(1-4), 125-136. doi:10.1007/s10661-009-0932-7

Kohonen, T. (2001). *Self-Organizing Maps*: Springer-Verlag New York, Inc.

Koklu, R., Sengorur, B., & Topal, B. (2010). Water Quality Assessment Using Multivariate Statistical Methods—A Case Study: Melen River System (Turkey). *Water Resources Management, 24*(5), 959-978. doi:10.1007/s11269-009-9481-7

Lermontov, A., Yokoyama, L., Lermontov, M., & Machado, M. A. S. (2009). River quality analysis using fuzzy water quality index: Ribeira do Iguape river watershed, Brazil. *Ecological Indicators, 9*(6), 1188-1197. doi:http://dx.doi.org/10.1016/j.ecolind.2009.02.006

Lu, R.-S., & Lo, S.-L. (2002). Diagnosing reservoir water quality using self-organizing maps and fuzzy theory. *Water Research, 36*(9), 2265-2274. doi:http://dx.doi.org/10.1016/S0043-1354(01)00449-3

Seguí, X., Pujolasus, E., Betrò, S., Àgueda, A., Casal, J., Ocampo-Duque, W., . . . Darbra, R. M. (2013). Fuzzy model for risk assessment of persistent organic pollutants in aquatic ecosystems. *Environmental Pollution, 178*(0), 23-32. doi:http://dx.doi.org/10.1016/j.envpol.2013.02.014

Şen, Z., & Altunkaynak, A. (2009). Fuzzy system modelling of drinking water consumption prediction. *Expert Systems with Applications, 36*(9), 11745-11752. doi:http://dx.doi.org/10.1016/j.eswa.2009.04.028

WPCR. (31.12.2004). Water Pollution Control Regulation. *Resmi Gazete 25687, 25687.*

Xu, Y., Yeung, J. F. Y., Chan, A. P. C., Chan, D. W. M., Wang, S. Q., & Ke, Y. (2010). Developing a risk assessment model for PPP projects in China — A fuzzy synthetic evaluation approach. *Automation in Construction, 19*(7), 929-943. doi:http://dx.doi.org/10.1016/j.autcon.2010.06.006

# ANALYSIS OF ELEVATED CONCENTRATIONS IN TRACE ELEMENTS AND NUTRIENTS IN URBAN CREEK

***Michael Harrison*** and Barry Hibbs
(California State University, Los Angeles, USA)

**ABSTRACT:** Dry weather runoff (imported and recycled water) contributes to perennial flows in the Malibu Creek Watershed located in southern California. Our study investigates an urban creek for selenium, nutrients, and stable water isotopes to further understand inputs and sources. We have found concentrations of 9.9 to 32 ug/L Se and 0.37 to 6.6 mg/L $NO_3^-N$ from groundwater seeps that flow within a storm drain. From the input of a storm drain where dry weather runoff provides perennial flow, concentrations range from 4.1 to 11 ug/L Se and 1.5 to 4.9 mg/L $NO_3^-N$. Concentrations decrease from 4.1 to 0.74 ug/L Se and 0.12 to 0.59 mg/L $NO_3^-N$ downstream. Moderately high concentrations of nitrate-nitrogen correlating with elevated concentrations of selenium suggest that enriched selenium is oxidized by elevated concentrations of nitrate found in local shallow groundwater and urban runoff. Our results coupled with seasonal analysis of hydrogen and oxygen isotopes further suggest perennial input and mixing of local groundwater throughout the riverine system.

## INTRODUCTION

The Malibu Creek Watershed is located within the Santa Monica Mountains in Los Angeles and Ventura County, California. The upper portion of the watershed exposes the Monterey-Modelo Formation, a Miocene marine deposit. The Monterey-Modelo Formation has been noted for its contribution of high selenium, metals, phosphate, nitrogen, and salinity, content in the Malibu Creek Watershed (LVMWD, 2010).

**FIGURE 1.** Map of residential community north of El Camino Real Creek.

The majority of cities within the upper Malibu Creek Watershed and use recycled water primarily for irrigation. Recent studies in the Malibu Creek Watershed have suggested high concentrations of phosphorus and nitrate may be more of a result of dry weather runoff from urban landscapes (Hibbs, 2014).

To further understand the geologic and anthropogenic controls on nutrient loading we examine El Camino Real Creek. El Camino Real Creek lies amongst Monterey-Modelo Formation strata and fed perennially through a storm drain from a residential community (Figure 1). Studying a creek within the Monterey-Modelo Formation serves as a controlled area that removes the variable that come with larger watershed studies.

## MATERIALS AND METHODS

Several stations along El Camino Real Creek were sampled repeatedly to understand the spatial and hydrochemical differences throughout the creek (Figure 2). Samples were collected approximately every 6 weeks for a year period starting October 2014. Each station were measured for temperature, conductivity, and dissolved oxygen were conducted with an YSI 63 YSI 550A meter. Samples were collected filtered and analyzed for nutrients (nitrate and orthophosphate), ions, selenium, and water isotopes.

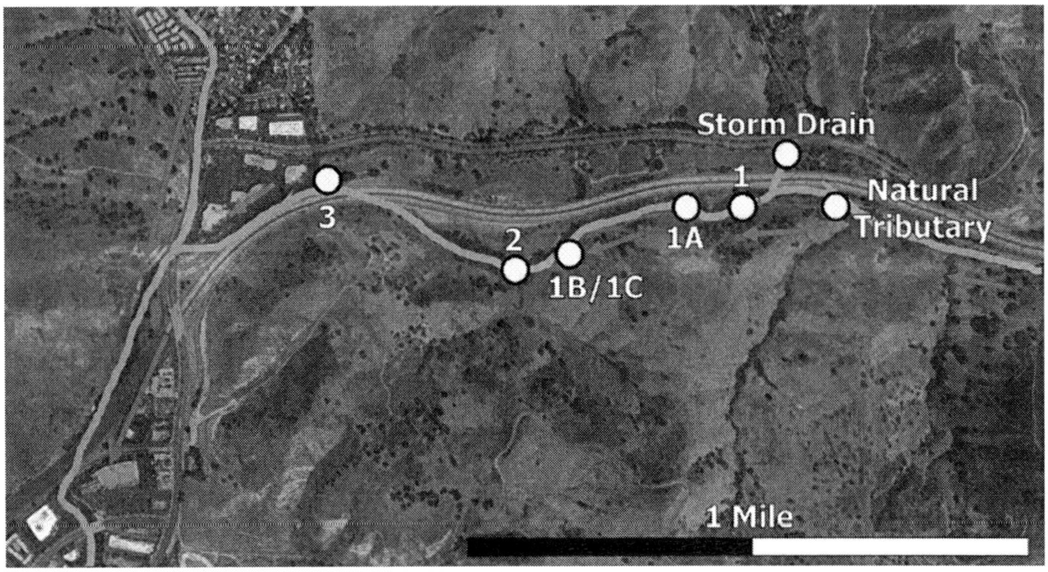


**FIGURE 2.** Map of sampling stations throughout El Camino Real Creek.

## RESULTS AND DISCUSSION

Continuous data show concentrations of nitrate and phosphate decreasing downstream from the storm drain (Harrison, 2015). From an initial input at the storm drain where dry weather flows consistently water contains an average of 2.57 mg/L $NO_3^-N$ and an average of 1.60 mg/L $PO_4$. Nutrient values decrease to concentrations averaging 0.22 mg/L $NO_3^-N$ and 0.62 mg/L $PO_4$ where El Camino Real Creek converges with Las Virgenes Creek at its most downstream station. The natural tributary source flow developed during the winter season has significantly low concentrations of $NO_3^-N$ that average at 0.31 mg/L and 0.10 $PO_4$.

Downstream from the storm drain, El Camino Real Creek has a trend of decreasing $NO_3^-N$ and an initial decrease before stabilization of $PO_4$ values (Figure 3). Decreases in nitrate and phosphorus are due to transformative process such as denitrification and vegetation uptake, along with some mixing and dilution with groundwater baseflow that has ambient concentrations of nutrients. Stabilization of phosphorus downstream may be a result of groundwater leaching from phosphatic bearing strata in equilibrium with vegetation uptake or from accumulation of phosphorus from recycled water sorbing onto soils within the creek.

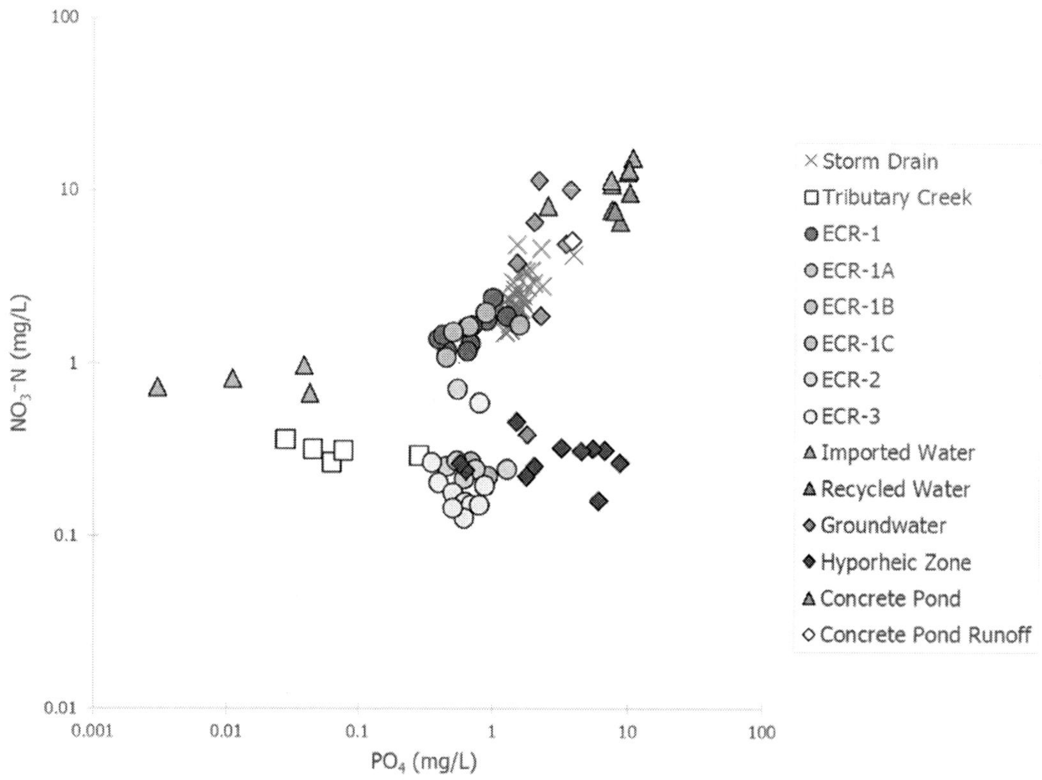


**FIGURE 3.** Concentrations of $NO_3^-N$ and $PO_4$ throughout El Camino Real Creek and surrounding area.

Local samples of recycled and imported water were established as endmembers for stable water isotope mixing analysis. Imported water includes tap water and recycled wastewater. Both are used for urban water applications in the entire study area. Recycled water and tap water have relatively light isotope values compared to local water derived by precipitation recharge.

The average isotope values for the recycled and imported endmembers are -9.3‰ $\delta^{18}O$ and -74‰ $\delta D$. The average isotope values for the samples collected at the storm drain are -7.3‰ $\delta^{18}O$ and -63‰ $\delta D$. Samples from the natural tributary have stable isotope values of -6‰ $\delta^{18}O$ and -41‰ $\delta D$. The average isotope values for the samples collected along El Camino Real are -6.4‰ $\delta^{18}O$ and -52‰ $\delta D$. Samples downstream from the storm drain (Stations 1 through 3) along El Camino Real Creek fit along a mixing line of the plotted stable isotope data with samples collected closest to the storm drain containing the most imported water (Figure 4).

Stable isotopes of hydrogen and oxygen show mixing of recycled water with local groundwater downstream, which demonstrates that nutrients in this creek are not strictly dominated by geologic sources. Isotope data also shows a strong correlation between high concentrations of $NO_3^-N$ and $PO_4$ with recycled water fractions. Stations that have a percentage of imported and recycled water generally have higher values of $NO_3^-N$ and $PO_4$. This correlation strengthens the point that high nutrients from the storm drain are mixing with the natural tributary and groundwater downstream.

We have found Se concentrations of from groundwater seeps that flow within an urban storm drain. From the input of a storm drain concentrations of Se range from 4.1 to 11 ug/L. Concentrations of Se decrease from 4.1 to 0.74 ug/L downstream, where El Camino Real Creek converges with Las Virgenes Creek. Within the hyporheic zone at Station 1A Se concentrations range from 0.48 to 1.3 ug/L.

Groundwater samples range from 0.37 to 6.6 mg/L $NO_3^-N$ while surface water samples range from 1.5 to 4.9 mg/L $NO_3^-N$ upstream near the storm drain. Further downstream, after the creek flows move through a riparian area, nitrate concentrations in surface water decrease to 0.12 to 0.59 mg/L $NO_3^-N$. Hyporheic samples range from 0.16 to 0.38 mg/L $NO_3^-N$.

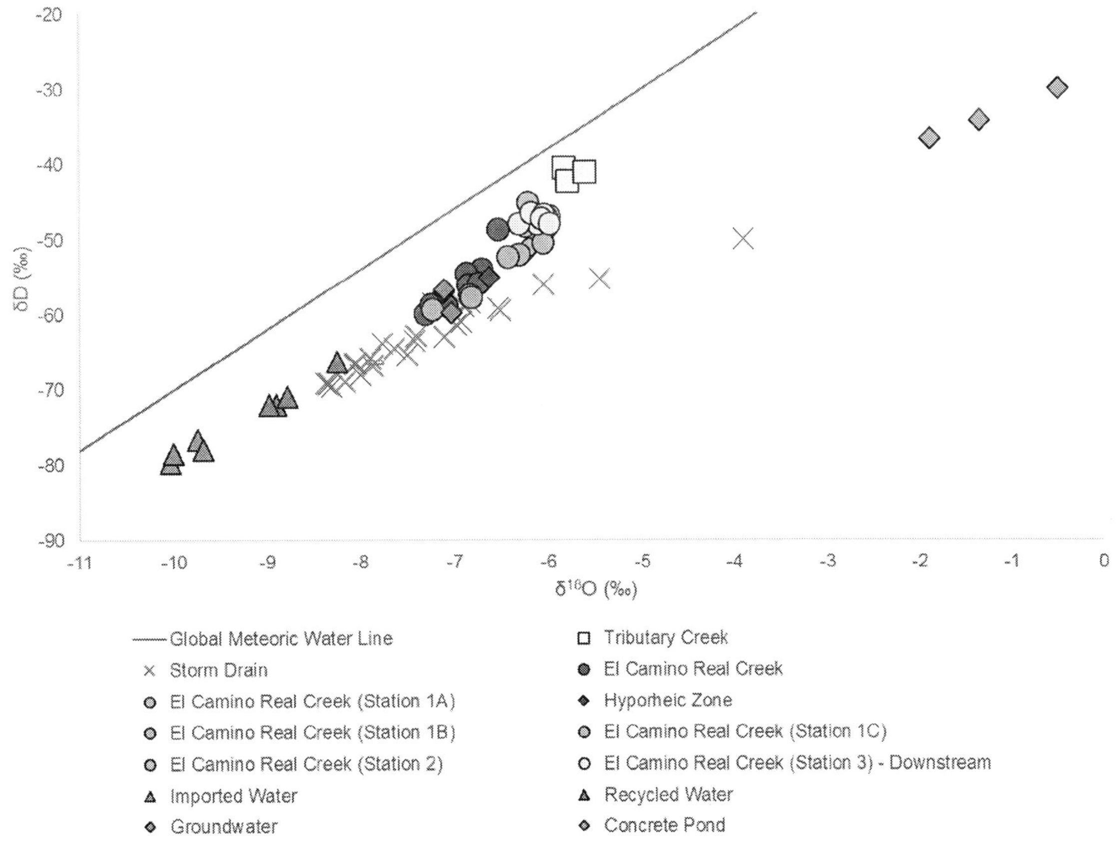


**FIGURE 4.** Stable water isotope data from samples collected at El Camino Real Creek along with endmembers.

Moderately high concentrations of nitrate-nitrogen are correlated with elevated concentrations of selenium in surface water and groundwater near the creek. Elevated concentrations of nitrate in surface water are associated with recycled water used for urban irrigation, and/or legacy nitrate sources from groundwater baseflow. Seasonal analysis of hydrogen and oxygen isotopes suggest perennial input and mixing of local groundwater throughout the riverine system.

## CONCLUSIONS

This investigation suggest that enriched selenium found in local groundwater is oxidized by elevated concentrations of nitrate found in groundwater and dry weather urban runoff (imported and recycled water) that penetrates the hyporheic zone. Being able to identify source flows and the seasonal hydrochemical changes has helped understanding of the water rock interaction of recycled water with Monterey-Modelo Formation.

Many streams within the Malibu Creek Watershed contain pollutants that have not met water quality standards by regulatory agencies. By understanding the geologic and anthropogenic controls of nutrients, trace metals, and source flows further regulatory action can be made in urbanized watersheds.

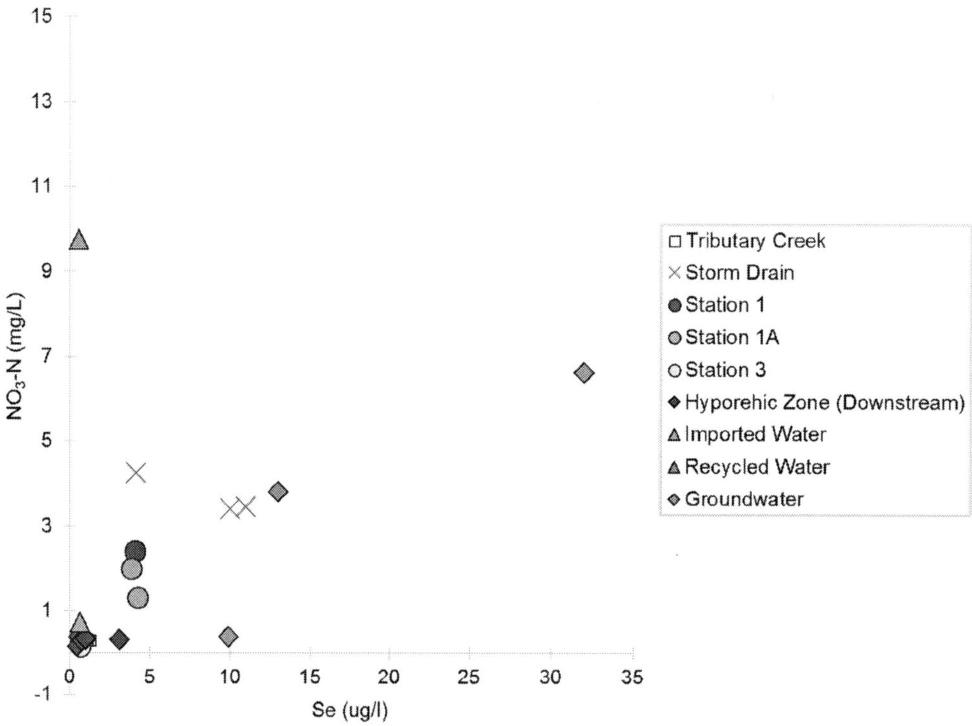


**FIGURE 5.** Selenium verses nitrate-nitrogen graph of samples collected at El Camino Real Creek along with endmembers.

## ACKNOWLEDGEMENTS

Funding for this project is provided by the National Science Foundation under Grant # HRD-1363399.

## REFERENCES

Harrison, M., Hibbs, B., 2015, Isotopic and hydrochemical identification of source waters and pollutants in Malibu Creek Watershed, CA: in Water is Not for Gambling, Utilizing Science to Reduce Uncertainty: *Proceedings of 2015 Joint Conference of University Council on Water Resources (UCOWR), and The National Institute for Water Research (NIWR)*: 123 – 128.

Hibbs, B. 2014. *New models for origin of salinity in upper Malibu Creek watershed and nearby watersheds.* Presented at California State University, Los Angeles, CA. August 19, 2014.

Las Virgenes Municipal Water District. n.d. Section 3: *Natural Source Assessment - The Monterey / Modelo Formation.* Available at http://www.lvmwd.com/Home/ShowDocument?id=2273. Accessed April 12, 2016.

# INFLUENCE OF POLLUTANT PROCESS UNCERTAINTY ON THE PREDICTION OF STORMWATER QUALITY IN URBAN CATCHMENTS

***Buddhi Wijesiri***, Prasanna Egodawatta, James McGree, Ashantha Goonetilleke
(Queensland University of Technology (QUT), Brisbane, Queensland, Australia)

**ABSTRACT:** Stormwater quality in urban catchments is influenced by the variability in pollutant build-up and wash-off processes. The process variability results in uncertainty which influences stormwater quality modelling outcomes. This highlights the need for quantitative assessment of process uncertainty prior to utilising modelling outcomes for decision making. However, poor characterisation of process variability limits the quantitative assessment of process uncertainty. It was hypothesised that the temporal variations in the particle load <150μm and >150μm during pollutant build-up and wash-off are the primary factors that characterise process variability. Accordingly, the incorporation of the characteristics of build-up and wash-off process variability into mathematical models enables accounting of process uncertainty as part of the modelling outcomes. Compared to the case where process uncertainty is not considered, it was found that the upper limit of uncertainty associated with stormwater quality predictions changes when process uncertainty is accounted for in the uncertainty assessment. Furthermore, the uncertainty assessment revealed that the impact of build-up process uncertainty on predicted stormwater quality is greater than that of wash-off process uncertainty. This confirmed that build-up process variability contributes significantly to the uncertainty associated with the prediction of stormwater quality.

## INTRODUCTION

Stormwater pollution is a major concern in the context of urban water management, as stormwater runoff transports a range of pollutants such as particulate solids and particle-bound heavy metals into urban receiving waters (Brown and Peake, 2006; Zhao and Li, 2013). Therefore, it is necessary to improve stormwater quality through effective pollution mitigation. The designing of pollution mitigation strategies requires information on catchment stormwater quality, which are commonly generated using stormwater quality models. However, two types of uncertainty: uncertainty in modelling of pollutant processes (build-up and wash-off) and uncertainty inherent to these processes primarily influence the interpretation of modelling outcomes (Loucks et al., 2005).

Process modelling uncertainty arises from several sources such as model structure, model parameters and input and calibration data (Dotto et al., 2012). On the other hand, uncertainty inherent to pollutant build-up and wash-off stems from the variability associated with these processes (Zoppou, 2001). The behaviour of particles that transport pollutants such as heavy metals and hydrocarbons during build-up and wash-off generates process variability. The temporal variations in load and composition of particle-bound pollutants also contribute to variability in build-up and wash-off processes. In this context, particles <150μm and >150μm play a significant role due to differences in particle behaviour and pollutant affinity for the different particle size fractions (Hvitved-Jacobsen et al., 2010; Wijesiri et al., 2015a, b). Although modelling uncertainty is generally understood, only limited research studies have investigated process uncertainty (Sun et al., 2012; Wijesiri et al., 2016). Accurate accounting of process uncertainty is necessary for informed decision making for effective stormwater pollution mitigation.

The research study was undertaken to investigate the impact of process uncertainty on the prediction of stormwater quality in urban catchments. It was hypothesised that the temporal variations in the particle load <150μm and >150μm during build-up and wash-off primarily influence process variability. Accordingly, prediction of stormwater quality and assessment of associated uncertainty were based on the mathematical models of pollutant build-up and wash-off, in which process variability is accurately accounted. The methodology described by Wijesiri et al. (2015a) was adopted for incorporating process variability into these mathematical models.

## MATERIALS AND METHODS

A 105ha urban residential catchment, namely, Highland Park (Figure 1), located in Gold Coast, Australia was selected for the investigation. Build-up and wash-off sampling were undertaken on asphalt paved road sites within the catchment. Particulate build-up samples were collected for different antecedent dry periods using a portable wet vacuum system. A rainfall simulator was used to simulate storm events with different intensities and durations for wash-off sample collection. Further details of the sampling procedures can be found in Egodawatta et al. (2007) and Herngren et al. (2006). The samples were analysed for particle size distribution and particulate solids load (after wet sieving into particle size fractions <150μm and >150μm) using standard test methods.

**FIGURE 1. Aerial view of catchment and road sites (Map Data: Google, DigitalGlobe.**

For the selected catchment, a runoff model was developed using the Mike URBAN – SWMM module, which is a physically based hydrologic modelling software (MikeUrban, 2014). The catchment-specific hydrologic parameters were determined using field data. As such, model calibration was not required for this investigation. Boundary conditions for the runoff model were set using rainfall data selected from a ten year period (2004-2014) of rainfall records for the study area. Runoff was then simulated for sixty-five selected storm events which occurred in two representative years, 2005 and 2008.

Prediction of catchment stormwater quality and quantification of uncertainty were undertaken based on two scenarios: (1) using the primary build-up and wash-off models (Equations 1 and 2) which were proposed by Ball et al. (1998) and Sartor and Boyd (1972), respectively, where process variability is poorly characterised. The detailed discussions on the application of these models in relation to stormwater quality modelling can be found in Obropta and Kardos (2007) and Zoppou (2001); and (2) using modified build-up and wash-off models (Equations 3 and 4), proposed by Wijesiri et al. (2015a) which were derived incorporating process variability.

$$B_{primary} = \alpha t^{\beta} \quad (1)$$

$$W_{primary} = W_0 \left(1 - \exp(-kIt)\right) \quad (2)$$

$$B_{revised} = [B_{(<150)} = \alpha_{(<150)}\, t^{-\beta(<150)}] + [B_{(>150)} = \alpha_{(>150)}\, t^{\beta(>150)}] \quad (3)$$

$$W_{revised} = B_{(<150)} \left(1 - \exp(-k_{(<150)}It)\right) + B_{(>150)} \left(1 - \exp(-k_{(>150)}It)\right) \quad (4)$$

Where: B – build-up load; W – amount of material washed-off; Wo – amount of material available at the beginning of storm event; t – antecedent dry period/storm event duration; I – storm event intensity; α, β, k – build-up and wash-off coefficients (subscripts indicate corresponding particle size fraction).

Given the catchment area, particulate build-up and wash-off could be predicted using the primary and revised models. The model parameters were estimated employing the 'non-linear least squares regression' technique using MATLAB in-built function 'nlinfit' (MathWorks, 2013). The proportional error model defined as $y = f+\theta f\varepsilon$ was assumed as the form of the error term using the variable $\varepsilon \sim N(0, 1)$ with the function value $f$ and the error parameter $\theta$ (initial estimate of $\theta$ set to be the default value =1). In order to quantify the uncertainty associated with model predictions, a large number of mathematical simulations (10,000) of build-up and wash-off were performed. These simulations were performed by primarily accounting for residual errors and parameter uncertainty. This enabled the determination of specific upper and lower uncertainty limits for each prediction. In this investigation, upper and lower uncertainty limits were set to account for 95% uncertainty interval.

## RESULTS AND DISCUSSION

**Accounting for Build-Up and Wash-Off Process Uncertainty.** The uncertainty associated with catchment stormwater quality predicted as Event Mean Concentration (EMC) was compared between primary and revised models. As evident from Figure 2, the uncertainty bandwidth (difference between upper and lower uncertainty limits) corresponding to the EMCs predicted with the revised models shows an increase. However, it was found that the predicted EMCs of primary and revised models were slightly different. Therefore, the relative uncertainty bandwidth (RUB), which is the ratio between uncertainty bandwidth and predicted EMC, was also analysed. Figure 3 confirms the increase in uncertainty associated with EMCs predicted using the revised model. This increase in uncertainty can be related to the revised mathematical form of the primary models, which in fact result from the incorporation of process variability. Moreover, it is important to note that this change in uncertainty should not be considered as a consequence of common reasons such as adoption of complex model structures which results in the increase in uncertainty.

Accordingly, it is possible to conclude that accurate characterisation of process variability enables the accounting of process uncertainty as a part of the uncertainty associated with model predictions. This means that typical uncertainty assessment techniques, which primarily focus on modelling uncertainty, can also be employed together with the approach derived in this investigation in order to assess the overall uncertainty that influences the interpretation of stormwater quality predictions. However, it is also important to note that the current uncertainty assessment techniques need to be improved as they have drawbacks such as relying on subjective criteria, effect of prior knowledge and accounting only for a few uncertainty sources (Dotto et al., 2012). For example, commonly adopted GLUE method is influenced by user defined likelihood measures (Freni et al., 2009). Further, uncertainty analyses, which adopt the classical Bayesian approach, are prevented from being objective due to the influence of prior knowledge on Bayesian updating (Freni and Mannina, 2010).

Moreover, when applying the approach derived for assessing process uncertainty, practical approaches would be required to assist stormwater management personnel for understanding and quantifying process uncertainty. This is due to inadequate knowledge of how the assessment of process uncertainty can be utilised for informed decision making in the context of stormwater pollution mitigation. For example, the knowledge created in the current research study can be effectively utilised by the academic researchers, industry practitioners, and technology developers for creating stormwater quality modelling software which are embedded with tools to assess both process and modelling uncertainty. Such modelling tools will help stormwater water management personnel to accurately interpret the information relating to stormwater quality, leading to enhanced stormwater pollution mitigation.

**Characterisation of the Influence of Process Uncertainty.** It is evident from Figure 2 that the change in uncertainty is predominantly influenced by the increase in the upper uncertainty limit. This change in upper uncertainty limit is consistently greater than that of the lower uncertainty limit. This shows the changes in the range in which stormwater quality predictions vary, and on the other hand reflects the influence of process variability on pollutant loads entrained in stormwater runoff. As such, in a case where process uncertainty is not accounted for, the predicted catchment stormwater quality may not be appropriately interpreted, leading to ineffective decision making in relation to the designing of stormwater pollution mitigation strategies.

Moreover, in relation to the revised models, the variation in RUBs over different build-up events is greater than the variation in RUBs over different wash-off events (Fig. 3). This implies that compared to the pollutant wash-off process variability, the pollutant build-up process variability plays a more important role in influencing catchment stormwater quality. Therefore, stormwater pollution mitigation strategies need to specifically account for temporal variations in particle-bound pollutant load and composition during build-up.

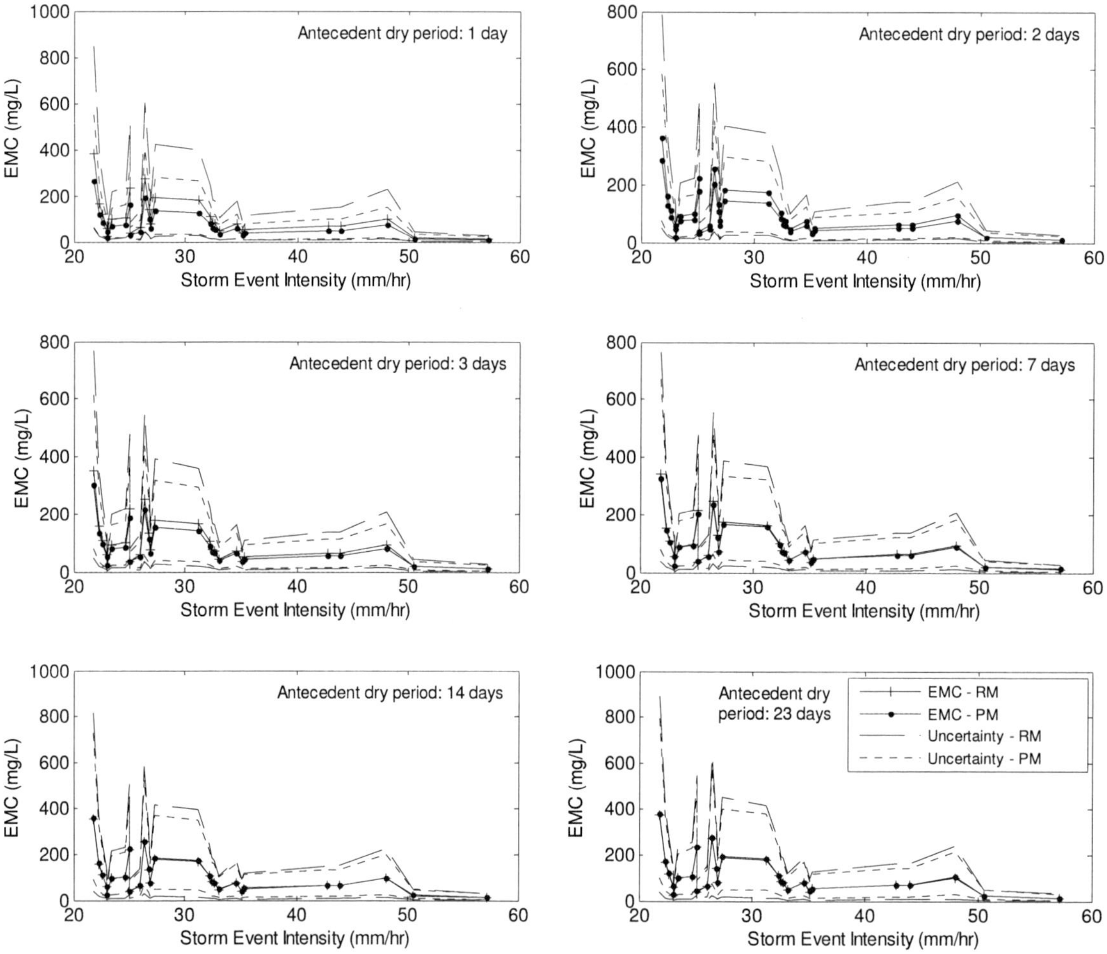


**FIGURE 2. Comparison of uncertainty limits between primary and revised models – year 2005 storm events; note: PM – primary models, and RM – revised models.**

## CONCLUSIONS

The investigation discussed in this paper derived an approach to quantitatively assess the uncertainty inherent to pollutant build-up and wash-off processes. The approach enables the accounting of process uncertainty as an integral part of the uncertainty associated with stormwater quality predictions. Accounting of process uncertainty significantly changes the upper limit of the uncertainty associated with predicted EMCs of particulate solids. The change in uncertainty limits relates to the limits within which stormwater quality predictions vary. This highlights the influence of process variability on the variations in pollutant loads released into stormwater runoff. Further, the influence of build-up process uncertainty on predicted stormwater quality was found to be significant compared to that of wash-off process uncertainty.

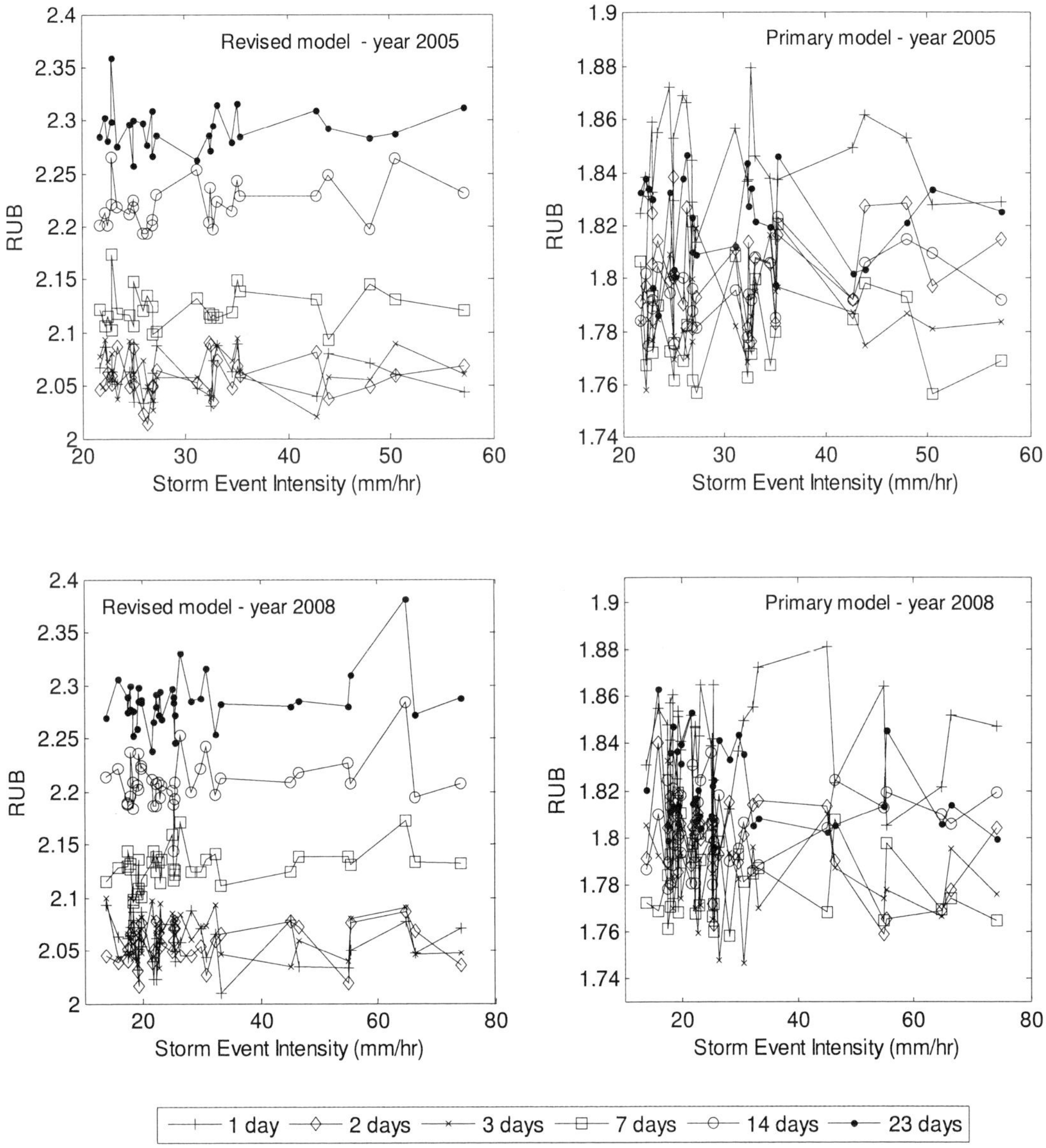


**FIGURE 3. Comparison of relative uncertainty bandwidth (RUB) between primary and revised models.**


## ACKNOWLEDGEMENTS

The authors would like to acknowledge Queensland University of Technology (QUT) for the scholarship provided to the first author for undertaking doctoral studies.


## REFERENCES


Ball, J. E., Jenks, R., Aubourg, D. (1998). An Assessment of the Availability of Pollutant Constituents on Road Surfaces. *Science of the Total Environment, 209*(2–3), 243-254.

Brown, J. N., Peake, B. M. (2006). Sources of Heavy Metals and Polycyclic Aromatic Hydrocarbons in Urban Stormwater Runoff *Science of the Total Environment, 359*, 145-155.

Dotto, C. B. S., Mannina, G., Kleidorfer, M., Vezzaro, L., Henrichs, M., McCarthy, D. T., Freni, G., Rauch, W., Deletic, A. (2012). Comparison of Different Uncertainty Techniques in Urban Stormwater Quantity and Quality Modelling. *Water Research, 46*(8), 2545-2558.

Egodawatta, P., Thomas, E., Goonetilleke, A. (2007). Mathematical Interpretation of Pollutant Wash-off from Urban Road Surfaces Using Simulated Rainfall. *Water Research, 41*(13), 3025-3031.

Freni, G., Mannina, G. (2010). Uncertainty in Water Quality Modelling: The Applicability of Variance Decomposition Approach. *Journal of hydrology, 394*(3–4), 324-333.

Freni, G., Mannina, G., Viviani, G. (2009). Uncertainty in Urban Stormwater Quality Modelling: The Influence of Likelihood Measure Formulation in the GLUE Methodology. *Science of the Total Environment, 408*(1), 138-145.

Herngren, L., Goonetilleke, A., Ayoko, G. A. (2006). Analysis of Heavy Metals in Road-deposited Sediments. *Analytica Chimica Acta, 571*(2), 270-278.

Hvitved-Jacobsen, T., Vollertsen, J., Nielsen, A. H. (2010). *Urban and Highway Stormwater Pollution: Concepts and Engineering*: CRC Press, Taylor and Francis Group.

Loucks, D. P., Van Beek, E., Stedinger, J. R., Dijkman, J. P., Villars, M. T. (2005). *Water Resources Systems Planning and Management: An Introduction to Methods, Models and Applications*: Paris: UNESCO.

MathWorks. (2013). MATLAB & Simulink (Version R2013a): MathWorks Inc. Massachusetts, USA.

MikeUrban. (2014). *Mike Urban Collection System - User Guide*. Danish Hydraulic Institute.

Obropta, C. C., Kardos, J. S. (2007). Review of Urban Stormwater Quality Models: Deterministic, Stochastic, and Hybrid Approaches. *Journal of the American Water Resources Association, 43*(6), 1508-1523.

Sartor, J. D., Boyd, G. B. (1972). *Water Pollution Aspects of street Surface Contaminants* EPA-R2-72-081. U.S. Environmental Protection Agency, Washington, D.C.

Sun, S., Fu, G., Djordjević, S., Khu, S.-T. (2012). Separating Aleatory and Epistemic Uncertainties: Probabilistic Sewer Flooding Evaluation Using Probability Box. *Journal of hydrology, 420–421*(0), 360-372.

Wijesiri, B., Egodawatta, P., McGree, J., Goonetilleke, A. (2015a). Incorporating Process Variability into Stormwater Quality Modelling. *Science of the Total Environment, 533*(0), 454-461.

Wijesiri, B., Egodawatta, P., McGree, J., Goonetilleke, A. (2015b). Influence of Pollutant Build-up on Variability in Wash-off from Urban Road Surfaces. *Science of the Total Environment, 527-528*(0), 344-350.

Wijesiri, B., Egodawatta, P., McGree, J., Goonetilleke, A. (2015c). Process Variability of Pollutant Build-up on Urban Road Surfaces. *Science of the Total Environment, 518–519*(0), 434-440.

Wijesiri, B., Egodawatta, P., McGree, J., Goonetilleke, A. (2016). Assessing Uncertainty in Pollutant Build-up and Wash-off Processes. *Environmental Pollution, 212*, 48-56.

Zhao, H., Li, X. (2013). Risk Assessment of Metals in Road-deposited Sediment along an Urban–rural Gradient. *Environmental Pollution, 174*(0), 297-304.

Zoppou, C. (2001). Review of Urban Storm Water Models. *Environmental Modelling & Software, 16*(3), 195-231.

# CONTAMINANT REDUCTION BY PERVIOUS CONCRETE PAVEMENT AND BAMBOO BIORETENTION BASIN

***Vincent Hwang*** (Southeastern Educational Society, Mayagüez, Puerto Rico 00680)
Amber Masters (University of Puerto Rico, Mayagüez, PR, 00680)
Evelyn Montalvo (Southeastern Educational Society, Mayagüez, Puerto Rico 00680)
Sangchul Hwang* (University of Puerto Rico, Mayagüez, PR, 00680)

Pervious concrete pavement (PCP) and bamboo bioretention basin (BBB) were combined for urban stormwater runoff volume control and water quality enhancement. A nanogeopolymer pervious concrete pavement (GPCP) was produced with the optimized mix design of water-to-binder at 36.0%, coal fly ash-to-binder at 29.5% and nano-iron oxide-to-binder at 0.78%. A regular pervious concrete pavement (RPCP) was produced an optimum mix design of water-to-binder at 36.7% and coal fly ash-to-binder at 21.4%. For both GPCP and RPCP, gravels in size of 4.75-12.5 mm were used at a mass ratio of gravel-to-binder at 4:1. A miniature bamboo species (*Dracaena sanderiana*) was used to accommodate the lab-scale experiment. GPCP or RPCP was served as a pre-treatment for urban stormwater runoff capture and infiltration as well as reduction of phosphate and fecal coliform, whereas BBB was a post-treatment for further pollution abatement especially for nitrate reduction. For comparison purposes, a typical impervious concrete pavement (IMCP) (i.e., control) was also run in parallel. Results showed that both PCPs excelled phosphate and fecal coliform reduction by >90% and >80%, respectively, but were poor in nitrate reduction. However, BBB reduced nitrate by >60%. Either GPCP+BBB or RPCP+BBB outperformed IMCP+BBB in stormwater runoff control and water quality enhancement. An enhanced growth of *D. sanderiana*, judged by the Chlorophyll a on the leaf and the body mass, was also observed during the experiment. Therefore, it is concluded that a treatment train of PCP and BBB would effectively control urban storm water in terms of runoff volume and water quality enhancement.

# TOXICOLOGY OF AZO DYES WITH RESPECT TO THEIR METABOLICALLY PRODUCED AND CHEMICALLY RELATED AROMATIC AMINES

**King-Thom Chung**
(The University of Memphis, Memphis, TN, USA)

Azo dyes are synthetic organic dyes and are widely used in various industries. The public health concern of azo dyes can be traced to the discovery of the medical use of Prontosil, which is the first chemotherapeutical drug studied by Gerhardt Domagk, *et al.* in the 1930s. The active component of Prontosil a reductively cleaved (azo reduction) product, *p*-aminophenyl- sulfonamide that contains sulfur. Thus, the significance of azo reduction is revealed in the production of the biological active compounds, mainly aromatic amines. Many aromatic amines can be metabolically produced by human intestinal microflora and to a lesser extent by liver azo-reductase.

Some azo dyes such as *4*-aminoazobenzene, *o*-aminoazotoluene, Methyl Yellow, Methyl, Yellow derivatives, Sudan azo dyes, Para Red, Trypan Blue, and others with structures containing free amine groups can be carcinogenic when the azo linkage is not being cleaved. However, the carcinogenicity of many other azo dyes is due to their cleaved component aromatic amines such as benzidine, benzidine congeners, *p*-phenylenediamine. etc.. Other structurally related compounds such as *4*-aminobiphenyl, *o*-toluidine, aniline, *p*-nitroaniline, and some monocyclic aromatic amines such as *2, 4*-diaminotoluene, *2*-nitro-*p*-phenylenediamine, *2*-amino-*4*-nitrophenol, *4*-nitro-*o*-phenylenediamine, *m*-phenylenediamine, and *o*-phenylenediamine are also reported to be carcinogenic.

*p*-Phenylenediamine (*p*-PDA) is usually mixed with $H_2O_2$ for hair dyeing. Oxidized *p*-PDA becomes a diaminophenazine or Bandrowski base, which is extremely mutagenic. Commercial hair dye *p*-PDA was reported to be contaminated with carcinogenic *4*-aminobiphenyl. *p*-PDA is also a contact allergen that can induce throat irritation in the pharynx and larynx, and cause dermatitis and bronchial asthma. Allergenic aromatic amines also include *o*-phenylenediamine, *m*-phenylene-diamine, *p*-toluidine, *p*-sulfanilic acid, *p*-aminobenzoic acid, *2*-amino-*p*-cresol, *p*-nitroaniline, *2-1*-amino-naphthol-sulfonic acid, etc.

Azo dyes such as Disperse Blue 106, Disperse Blue 124, Methyl Yellow, *4*-aminoazobenzene and *o*-aminoazotoluene were reported to be allergenic. Azo dye Trypan Blue has also been shown to be carcinogenic and teratogenic. Food certified azo dyes Tartrazine and Sunset Yellow were also reported to be allergenic to humans. Sulfonamide drugs, some of which may not be azo dyes cleaved products, have also been reported to induced human maladies such as allergy, urinary tract disorders, haempoietic disorders, porphyria, hypersensitive reactions, etc. Various human disorders caused by many other aromatic amines were also reported. Azo dyes and their metabolites, primarily aromatic amines, are therefore significant chemicals affecting human health and damaging our environments.

# SIGNIFICANCE AND APPLICATION OF THE GLOBAL DRINKING WATER QUALITY INDEX: LEBANON CASE STUDY

***Mey Jurdi****, Sami Ramia (American University of Beirut, Beirut Lebanon)
Samira Korfali (Lebanese American University, Beirut, Lebanon)
Nabil Amacha (Lebanese University)
Joumana Nasr, Rola Ajib and Sara Chehab (American University of Beirut)

The management of the quality of potable piped water supplies remains a major challenge in developing countries in the absence of sufficient technical and financial resources. Proper water quality monitoring is an essential component of risk assessment based on which proper quality management could be recommended. As such, it is important to enhance the quality of monitoring activities under the prevailing deficient and bureaucratic systems.

Hence, the objectives of the research work were to examine the significance and application of the Global Drinking Water Quality Index in monitoring the quality and the safety of piped water supplies in Lebanon. As such, a drinking water quality index was adapted to reflect on acceptability, health and safety of potable water supplies. The quality indicators were chosen to reflect on the major characteristics of the water supply in Lebanon. The water quality assessment included parameters relating to water desirability (turbidity, color, hardness etc.), acceptability (pH, ammonia, chlorides, sulfates etc.), safety (nitrates, copper, cadmium, fecal coliform, etc.) and conditions of distribution network (iron, copper, zinc etc.). Fluoride and mercury levels were not included in the computation of the index as levels ae relatively minimal in water supplies, as documented by earlier studies. On the other hand, the total hardness was included as mostly ground water sources feed distribution networks. Moreover, the sensitivity of the developed index to reflect on water quality changes and to identify and flag problematic distribution areas was also examined.

Results showed that the development and use of such a quality index can help identify changes in the quality of the water supply in a way that summarizes complex scientific data into a simpler form of risk assessment and reporting and risk communication. Moreover, it can be adopted and used as a screening tool for further evaluation; this is important given the deficient water quality monitoring systems that require more extensive resources.

# COLIFORM COUNTS AND PLANKTON SPECIES IN LAWAYE RIVER, BATANGAS AS INDICATORS OF POLLUTION

***Natividad F. Lacdan***, Jose Rafael L.Lopez, and Renz Michael F.Pasilan
(University of the Philippines Manila, Manila, Philippines)

Lawaye River in San Juan, Batangas, Philippines is an important source of water for local residents. It finally drains into the waters of the Verde Island Passage Corridor. However, there has been insufficient researches regarding the water quality of the river. This study was conducted to determine the water quality of Lawaye River, using coliform counts and plankton species composition. Four sampling sites were set up along the river: boundary between Rosario and San Juan towns, upstream, midstream and downstream. Fecal coliform count was quantified using the multiple tube fermentation method while detection of the coliform, *Escherichia coli* involved the Presumptive, Confirmatory and Completed tests along with the IMVic profiles. Plankton analysis involved the identification of various plankton genera found in the river. Nygaard's and Palmer's indices were then used to determine the river's trophic state and pollution status. The coliform concentration in all sites was found to be well beyond the normal range. Plankton analysis showed that sampling areas in Lawaye River were found to have different levels of organic pollution. The Palmer's and Nygaard's Indices demonstrated that the downstream and midstream sites registered the highest probability of having organic pollution, while the sites found on the upstream and boundary were comparatively normal and oligotrophic. Factors, such as improper waste disposal, effluents from drainage and agricultural runoffs could have contributed to the observed results.

# EFFECTS OF SUPER-SATURATION CONTROL STRATEGIES ON HYDROXYAPATITE (HAP) CRYSTALLIZATION PROCESS FOR PHOSPHORUS RECOVERY FROM WASTEWATER

Hongliang Dai[a,b], Xiwu Lu[a,b*]
[a]School of Energy and Environment, Southeast University, No. 2 Sipailou Road, Nanjing 210096, China.
E-mail: daihongliang103@163.com;
[b]ERC Taihu Lake Water Environment (Wuxi), No. 99 Linghu Road, Wuxi 214135, China.
E-mail: 230149342@seu.edu.cn; xiwulu@seu.edu.cn

**ABSTRACT:** The HAP crystallization process for phosphorus removal from wastewater contributes to an environmental friendly production due to the fact that it helps reduce or eliminate the water eutrophication as well as increases the recovery of mineral resources. However, the generated microcrystalline with poor settle-ability in high levels of super-saturation solution has a negative effect on the phosphorus recovery efficiency. In order to overcome the drawback, multiple reagent feed ports (four feed ports) and different recirculation ratio (1.0, 1.5, 2.0, 2.5, 3.0) were investigated to control the levels of super-saturation in an air-agitated reactor with calcite as seeds. The results showed that the approach of multiple reagent feed ports could improve the conversion ratio of orthophosphate, but it had a limited effect (~3% improvement) on phosphorus recovery efficiency (deposition on the seeds). With the increase in the recirculation ratio, the recovery efficiency was increased gradually and reached an optimal value of 85.63% under the recirculation ratio of 2.5 and four feed ports. This is because of that the adopted strategies could reduce the level of super-saturation by diluting the concentration of the reagents and inhibit large numbers of microcrystalline coinstantaneous occurrence. Meanwhile, the crystallized products were detected and analyzed by scanning electron microscope (SEM) with X-ray spectroscopy (EDS) and X-ray diffraction (XRD), which were proved to be HAP with a lofty purity. Collectively, these results demonstrated that super-saturation control using conventional approaches had a limited improvement in the phosphorus recovery efficiency in the form of HAP and the new control strategies for super-saturation dispersion should be developed in the further study.

**Key Words**: Phosphorus recovery; Super-saturation control; Hydroxyapatite; Induced crystallization; Air-agitated reactor.

## INTRODUCTION

Phosphorus is a limited and non-renewable mineral resource, making a significant contribution to the development of agriculture and industry [1-2]. Word phosphorus reserves will run out in between 50 to 100 years [3]. Meanwhile, excessive phosphorus content in wastewater will result in a serious eutrophication [4]. For these reasons, phosphorus recovery from wastewater can prevent the loss of nutrient resources and alleviate the environmental impacts of excess phosphorus on receiving waters. Municipal wastewater as hidden treasure of phosphorus resource satisfies the world's phosphorus demand of 15~20% by its recovery and recycling [5-6]. Many wastewater treatment plants built on the enhanced biological phosphorus removal (EBPR) process provide a phosphate-rich stream (anaerobic supernatant, ~20 mg/L) to make phosphorus recovery and reuse feasible [5, 18].

Induced crystallization for phosphorus recovery in the form of hydroxyapatite (HAP) from the wastewater is an economical and efficient approach, possessing the characteristics of effective phosphorus recovery and high-quality crystallization products [7-9]. The mechanism of HAP crystallization follows three chemical stages: the formation of super-saturation, nucleation (crystal birth) and crystal growth [10]. According to the classical nucleation theory, primary homogenous nucleation will form small particles (microcrystalline) in the high-level super-saturation solution [11]. The generated tiny particles called microcrystalline with poor settleability have a negative effect on the solid-liquid separation and subsequent recovery of crystallized products [12-13]. Karapinar et al. [10] showed that the increase of super-saturation levels in solution would lead to a decrease in phosphorus efficiency (precipitation on seeds), though the total precipitation efficiency was increased. Generally, the levels of super-saturation can be reduced by using the lower reagent concentrations. However, high levels of super-saturation (SI>0) are produced even at low reagent concentration during HAP crystallization process due to the smaller solubility products ($2.35\times10^{-59}$ mol/L, 25℃) [14].

Conventional approaches to control the levels of super-saturation are the multiple reagents feed points and reaction liquid reflux for studying the sparingly soluble compounds' recovery *via* crystallization reaction [12-13]. Seckler et al. [15] studied the calcium phosphate precipitation in FBR and showed that microcrystalline aggregation process could be improved by distributing the super-saturation more evenly throughout the reactor with a recirculation stream. Taty Costodes and Lewis [13] studied nickel hydroxy-carbonate precipitation using the approach of multiple reagents feed points, and the results showed that the amount of fines produced inside the reactor was significantly reduced using multiple reagent feed points. This was attributed to effective control of the high levels of super-saturation produced near the reagent inlet point. Mokone et al. [12] studied the effect of control strategies used to manage super-saturation during the precipitation reactions of copper and zinc sulphide in FBR, and the results showed that extremely high super-saturation prevalent during metal sulphide precipitation was difficult to control using conventional approaches and suggested that the seeded FBR was unsuitable for this application. This is because FBR can't achieve perfect mixing with an even distribution of super-saturation [13, 19]. Meanwhile, the reaction of HAP crystallization merely occurs in alkaline conditions and needs a completely mixed state. Increasing the pH of wastewater by adding alkali (sodium hydroxide) requires complicated and expensive equipment to add chemical with a pH sensor. The air-agitated reactor is more popular due to stripping of $CO_2$ to increase the pH by aeration and providing a perfectly mixed system [16-17].

In the present study, conventional super-saturation control strategies (multiple reagent feed points and recirculation flow) were taken to investigate the effects on process efficiency and particle characteristics for HAP crystallization process in an air-agitated reactor. The performance of the phosphorus recovery was investigated by measuring the orthophosphate content and the saturation index (SI) of solution under different operation conditions. Meanwhile, the precipitate's characteristics were analyzed by scanning electron microscopy (SEM) with energy dispersive X-ray analysis (EDS), and X-ray diffraction (XRD). The results are expected to solve the problem of low phosphorus recovery efficiency due to the homogeneous nucleation in the HAP crystallization process

## MATERIALS AND METHODS

**Experimental Apparatus and Procedures.** The experimental setup (**Fig. 1**) consisted of a continuously air-agitated crystallization column of 2.5 L working volume with an overall height of 1300 mm. The reactor consists of precipitation area, buffer area and reaction area, with four sampling points and four feed ports. The stock solutions for HAP formation were continuously pumped into the reactor using peristaltic pumps. Experiments were carried out at laboratory temperature (25±2 ℃). In order to prevent the loss of seeds, the outlets of reactor were coated with a 120 ASTM mesh sieve (0.12 mm). As the removal efficiency depends on the active specific surface area of the seeds, the size of calcite seeds was 0.15~0.21 mm, and the seeds dosage were 30 g/L. The seeds inside the reactor can promote heterogeneous primary nucleation and/or secondary nucleation.

**Table 1. Process conditions in the air-agitated crystallization column**

| Parameter | Value |
|---|---|
| Inlet P concentration | ~23 mg/L |
| Inlet Ca/P ratio | 2.5 |
| Inlet P flow rate | 80 ml/min |
| Inlet Ca flow rate | 8 ml/min |
| Re-circulation flow ratio | 1.0, 1.5, 2.0, 2.5, 3.0 |
| Reagent feed ports | 4 |
| Working volume | 2.5 L |

Other conditions given in Table 1 were carried out at constant Ca/P ratio (2.5) and aeration rate rates (300 L/h). Precipitation is induced by addition of a precipitating reagent into the reactor. When investigating the performance of the system, the reactor must be run continuously for at least 2 days with the "saturated" solution of HAP so as to allow a first layer of HAP to cover the seeds.

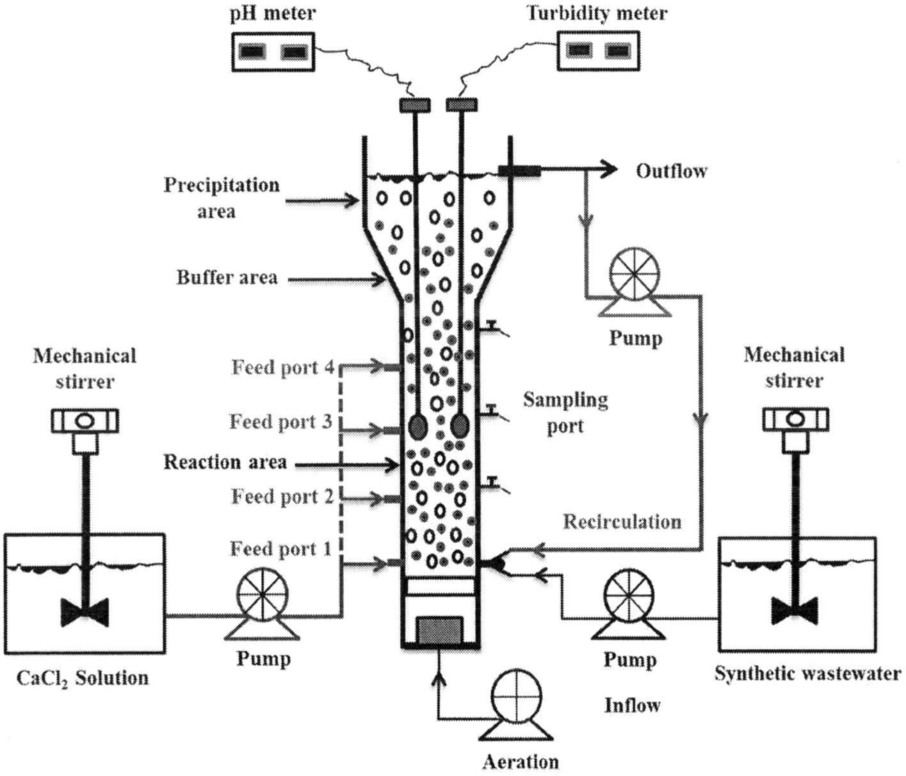


**Fig.1. Schematic diagram of a continuously air-agitated crystallization column for phosphorus recovery**

**Solution Preparation.** The anaerobic supernatant was taken from an EBPR process (anaerobic-anoxic/nitrifying two sludge process) established by our group [18], and the concentrations of constituents were given in table 2. An air stripping for alkalinity reduction and pH increase was used to pretreat the anaerobic supernatant. In order to investigate the proposed process on recovering phosphorus in form of HAP, two stock solutions ($CaCl_2$ and anaerobic supernatant) were continuously pumped into the reactor with the volume ratio equal to 0.1, and the final molar ratio of $Ca^{2+}$ and $PO_4^{3-}$ was 2.5 by adjusting the concentration of $CaCl_2$ solution. All the reagents used throughout this investigation were analytical grade and originated from Aladdin (China).

**Table 2. Characteristics of anaerobic supernatant of EBPR process used in experiments.**

| Parameters | $PO_4^{3-}$-P (mg/L) | $NH_4^+$-N (mg/L) | COD (mg/L) | $Mg^{2+}$ (mg/L) | pH | Turbidity (NTU) | Alkalinity (mg/L) |
|---|---|---|---|---|---|---|---|
| Anaerobic supernatant | 23.1±2.5 | 36.4±4.3 | 75.7±8.6 | 2.5±1.1 | 7.6±0.3 | 30.5±2.1 | 302.7±22.6 |

**Calculation Methods of Several Evaluation Parameters**

***Process parameters.*** The generated microcrystalline of homogeneous nucleation in the liquid phase and abrasion of the grains in the air-agitated reactor lead to small particle products, which leave the reactor with the effluent, together with the remaining phosphorus in solution, the fraction of the phosphorus is not possible to recover.

For the interpretation of the crystallization process, the following parameters were taken into account: the amount of the recovered phosphorus ($X_R$: Recovery efficiency), the amount of orthophosphate transformation ($X_T$: Transformation ratio), and the amount of microcrystalline with poor settleability in solution ($X_M$: Microcrystalline ratio). Three evaluation parameters $X_R$, $X_M$, $X_T$ were calculated according to equations (1), (2), (3) and (4), respectively.

$$X_R = \left(C_{in} - C_{out}\right)/C_{in} \times 100\,\% \tag{1}$$

$$X_M = \left(C_{out} - C_{out\text{-}d}\right)/C_{in} \times 100\,\% \tag{2}$$

$$X_T = \left(C_{in} - C_{out\text{-}d}\right)/C_{in} \times 100\,\% \tag{3}$$

$$X_T = X_R + X_M \tag{4}$$

Where $C_{in}$ is the phosphorus concentration of influent (mg/L); $C_{out}$ is the total phosphorus concentration of effluent from crystallization reactor (mg/L); $C_{out\text{-}d}$ is the phosphorus concentration of effluent which is rapidly filtrated by 0.45 μm membranes (Millipore HAWP01300, USA) after sampling.

***Saturation index calculations.*** Hydroxyapatite (HAP, $Ca_5(PO_4)_3OH$) is a sparingly soluble salt ($Ksp=2.35\times10^{-59}$ mol/L, 25 °C) [14]. Theoretically, the principal chemical reaction of HAP formation is based on the reaction shown in equations (5).

$$5Ca^{2+} + OH^- + 3PO_4^{3-} = Ca_5(PO_4)_3OH \downarrow \tag{5}$$

Thermodynamic properties of HAP crystallization can be well reflected by the saturation index (SI). In this study, SI calculations for HAP in the effluent were performed with the aid of Visual MINTEQ (Version 3.0). The calculation of SI was defined as follows:

$$SI = \log\left(IAP/Ksp\right) \tag{6}$$

Where, *IAP* is the ion activity product, and *Ksp* represents the thermodynamic solubility product of the precipitate phase. The solution is deemed to be unsaturated when *SI* is below zero, and no precipitation occurs. The solution is supersaturated when *SI* is above zero, and precipitation occurs spontaneously.

**Analysis and Characterization.** Water samples of 10 ml were taken from the effluent of the reactor at different times, then acidified with 2 M hydrochloric acid and stored in PE plastic containers at 4 ℃ until analysis. Water samples were filtrated rapidly with 0.45 μm membranes (Millipore HAWP01300, USA) before acidification for investigating the concentration of dissolvable phosphorus. The chemical oxygen demand (COD), total phosphorus, ammonia and alkalinity were carried out in accordance with Standard Methods [20]. Calcium and magnesium were analyzed by an atomic absorption spectrometer (PerkinElmer AAS, USA). Turbidity and pH measurements were conducted with a turbidity meter (HACH SOLITAX-1720E) and pH meter (YSI Pro Plus, USA), respectively.

The settling rate of the fine particles (microcrystalline) which remained in suspension was measured after operating the process 30 d (without the seed material) into an Imhoff settling cone and then measuring the change in the height of the suspension as a function of time for 2 h. The settling results are represented as the ratio between the suspension height (H) at a specific time and the initial suspension height ($H_o$) [12].

The pellets removed from the bottom of the reactor were dried in an oven at 60 ˚C for 24 h and then used for microstructural analysis. Crystallized products used for morphological and micro-structural analysis were obtained from the reactor at the end of experiment (i.e. after 30 d). Morphology of phosphorus-coated seeds and microcrystalline were determined using scanning electron micrography (SEM) with energy dispersive spectrometry (EDS) techniques (Hitach S-4800, Japan). X-ray diffraction analysis (XRD) for the products structure was carried out utilizing a D8 ADVANCE diffractometer (Bruker AXS, Japan).

## RESULTS AND DISCUSSION

**The performance of Air-Agitated Crystallization Column on Phosphorus Recovery.** The generated small particles (i.e. microcrystalline) with poor settleability during the crystallization process have a negative effect on the phosphorus recovery from wastewater. The completely air-agitated crystallization reactor coupled with two-sludge process could successfully recover P from the anaerobic supernatant in by our previous study [18]. In this section, an experiment was performed to evaluate the phosphorus recovery efficiencies from anaerobic supernatant in a continuously single air-agitated reactor, and the effects of different sampling ports for conversion efficiency ($X_T$) and microcrystalline ratio ($X_M$) were also investigated in the process.

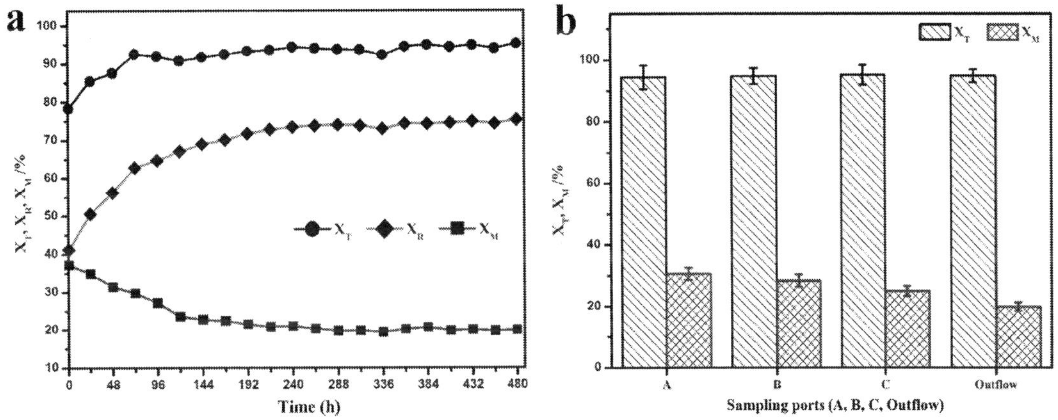


**Fig. 2. The performance of air-agitated crystallization column on phosphorus recovery (a: the variation of $X_T$, $X_R$, and $X_M$ as a function of time; b: the variation of $X_T$ and $X_M$ in different sampling ports).**

Fig. 2 showed how the performance of air-agitated crystallization column on phosphorus recovery during 480 h operations. At the beginning of the process operation, the white precipitations were produced and the turbidity of the solution was increased. This demonstrated that homogeneous nucleation happened in the high super-saturation solution and a large number of microcrystalline were produced [10]. As illustrated in Fig. 2a, $X_R$ was approximately 42 %, but the $X_T$ and $X_M$ were 78.34 % and 37.13 %, respectively, in the beginning of the process operation. With the operation of the system, the $X_R$ was continually improved. After 120 h, the $X_R$ was reached to 70.08 % and $X_M$ was down to 22.39 %, then the stable recovery efficiency was maintained throughout the following operational period: the average was 72.38 % of $X_R$ and the range was 19.47~23.78 % of $X_M$, indicating that phosphorus had been successfully transferred from the solid phase to the liquid phase and induced crystallization with seed crystal had a starting period before products completely covered [7, 21]. At beginning of the experiment, different surface structure and constituent between newly formed crystals and seed crystals hampered the microcrystalline from depositing upon the surface of the seed. As the seed crystals were gradually covered by the crystallization products, the crystals could better growth on the surface of the seed crystals and effectively improve the phosphorus recovery in the process [21-22]. Aldaco et al. [19] showed that the similar variation tendency of $X_R$ when studying the calcium fluoride recovery from fluoride wastewater in a fluidized bed reactor.

The results of $X_T$ and $X_M$ in different sampling ports (A, B, C and Outflow) are presented in Fig. 2b. The $X_M$ was decreased with the increase of the sampling ports' height, whereas the $X_T$ was almost constant. In the air-agitated reactor for HAP crystallization, the rising velocity slowed down due to the presence of precipitation area and buffer area, which lead to settling behavior of partial microcrystalline. Therefore, the value of $X_M$ in the outflow port was lower than the sampling ports of the reaction area, but the $X_T$ had not been affected by the position of the sampling ports. The higher microcrystalline ratio in the reaction system indicated that the solution was in a high super-saturation state and occurring homogeneous crystallization immediately. Because of the fine size and poor settleability of microcrystalline, it is difficultly to adhere to the surface of the seed crystal and easy to flow out from the rector, resulting in the low phosphorus recovery efficiency [10, 15]. Therefore, super-saturation control and microcrystalline capture are necessary to avoid the homogeneous nucleation and consequently to achieve a high phosphorus recovery in further investigations.

**Effects of the Number of Feed Points (FP) on Process Efficiency.** When the two solutions ($CaCl_2$ and phosphorus-containing wastewater) were continuously fed into the reactor with a certain proportion and the incoming mixed stream creates super-saturation, homogeneous nucleation occurs immediately. An increase in recovery efficiency is obtained if the level of super-saturation produced near the inlet point is controlled below that required for homogeneous nucleation [12-13]. Multiple reagents feed points as a conventional approach was utilized to control the levels of super-saturation when studying the sparingly soluble compounds recovery *via* crystallization reaction [12-13]. In order to primarily understand the effects of the number of feed points on process efficiency, different numbers (1, 2, 3, and 4) of feed points were used in the study. All data presented as a function of the number of feed points was obtained after operating the process of 30 d. Fig. 3 shows that the approach of multiple reagent feed ports could improve the conversion ratio (~5% improvement) of orthophosphate, but it had a limited effect (~3 % improvement) on the phosphorus recovery efficiency (deposition on the seeds). The SI and pH of the solution, on the other hand, was slightly lower than the one feed port was employed. The amount of microcrystalline produced was substantial (~21%) when the one feed port configuration was used and slightly increased using four FP (~23%). In spite of multiple reagents feed ports could reduce super-saturation at the reagent inlet points, the SI was always above zero due to the smaller solubility products of HAP ($2.35\times10^{-59}$ mol/L, 25℃) [14]. Therefore, increasing the number of feed ports could not decrease the microcrystalline ratio and improve the phosphorus recovery efficiency during the HAP crystallization process.

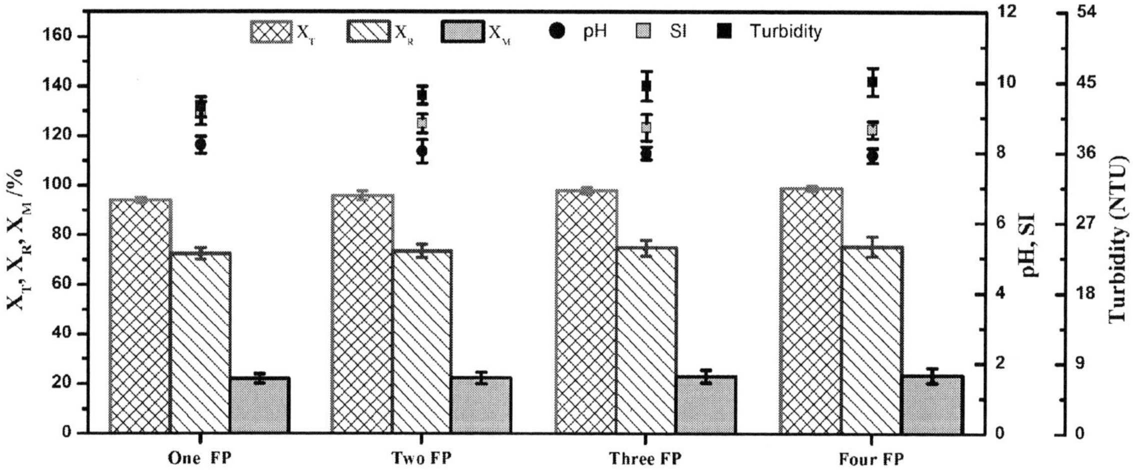


**Fig. 3. The variation of process efficiency, pH, SI and turbidity as a function of the number of feed points (without recirculation flow).**

**Effects of Recirculation Flow Rate on Process Efficiency.** Process efficiency, pH, SI and turbidity as a function of the different recirculation flow ratios were shown in Fig. 4. The conversion ratio did not be affected by the change of the recirculation flow ratio. However, the amount of microcrystalline recovered from suspension and the SI, pH and turbidity of the solution appeared to be affected. With the increase of the recirculation ratio, the recovery efficiency was increased gradually and reached an optimal value of 85.63 % under the recirculation ratio of 2.5. Because it can reduce the level of super-saturation by diluting the concentration of the reagents and inhibiting the large number of microcrystalline occurred simultaneously and also provide a perfectly mixed state of the reaction system and amounts of microcrystalline as seeds for heterogeneous nucleation. The results agreed with the obtained by Seckler et al. [15] studying the phosphorus recovery in the form of HAP in FBR.

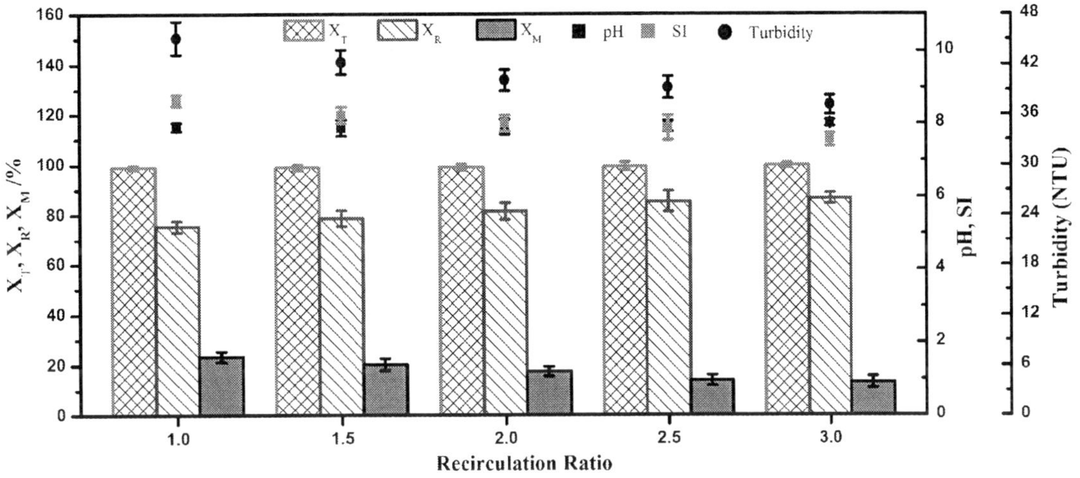


**Fig. 4. The variation of process efficiency, pH, SI and turbidity as a function of the different recirculation flow ratios (four feed ports).**

**Long-term Performance of the Air-Agitated Crystallization Column for Phosphorus Recovery.** In order to evaluate the phosphorus recovery efficiencies and the stability of the air-agitated crystallization column operated with four feed ports and recirculation ratio of 2.5, a long term continuous flow (30 d) experiment was carried out. As shown in Fig. 5, a stable and high phosphorus recovery efficiency and low microcrystalline ratio were maintained throughout the operational period: the average was 86.82 % and 13.14 %, varied in the ranges of 83.45~87.39 % and 12.48~14.83 %, respectively, which were increased by 19.95 % and reduced by 39.14 % compared with the initial operation condition. The results indicated that conventional approaches using multiple reagents feed points and recirculation flow had a limited effect on the inhibition of homogeneous nucleation and the improvement of phosphorus recovery efficiency.

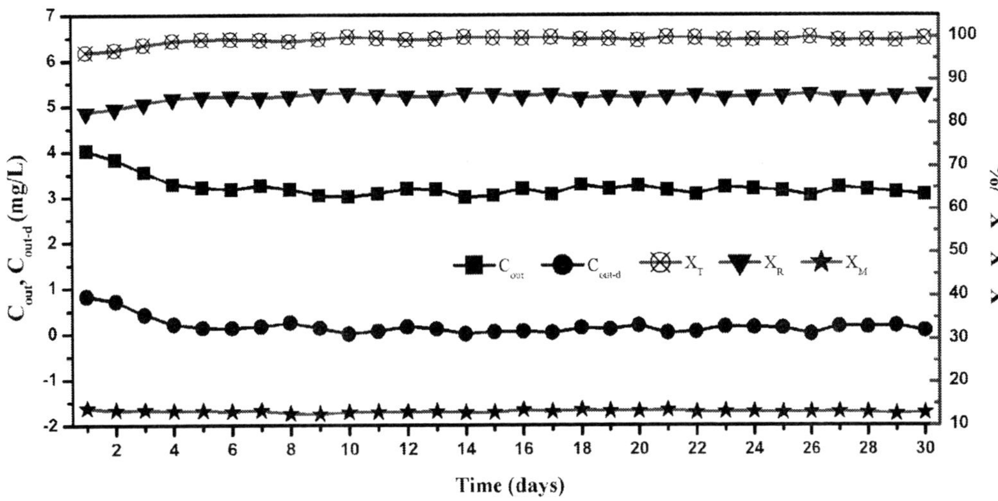


**Fig. 5. The performance of air-agitated crystallization column as a function of time (30 d, four feed ports and recirculation flow ratio of 2.5).**

**Characteristics of Crystallization Products.** Settling characteristics of microcrystalline particles was studied by a static settling experiment, and the charge on the surface of the microcrystalline using zeta potential measurements. Morphologies and structures of crystallized products were characterized by means of scanning electron micrograph (SEM) with energy dispersive spectrometry (EDS), and X-ray diffraction (XRD).

***Setting characteristics and zeta potential of microcrystalline.*** In order to overcome the drawbacks of microcrystalline with poor settleability on the phosphorus recovery efficiency, the settling characteristics of microcrystalline was measured over a period of 2 h. Fig. 6 shows the results for the settling behavior of the microcrystalline particles which could not be recovered from suspension. The results showed that the microcrystalline particles of HAP had relatively long time (90 min) to completely precipitate under quiescent conditions. The electrostatic forces of microcrystalline particles had a negative effect on their aggregation and precipitation. Therefore, the charge on the surface of the microcrystalline using zeta potential measurement was investigated. The results show that the zeta potential of the microcrystalline was ~20 mV (data not shown). Thus, the poor settling behavior of the HAP microcrystalline can be ascribed to the highly negative surface charge.

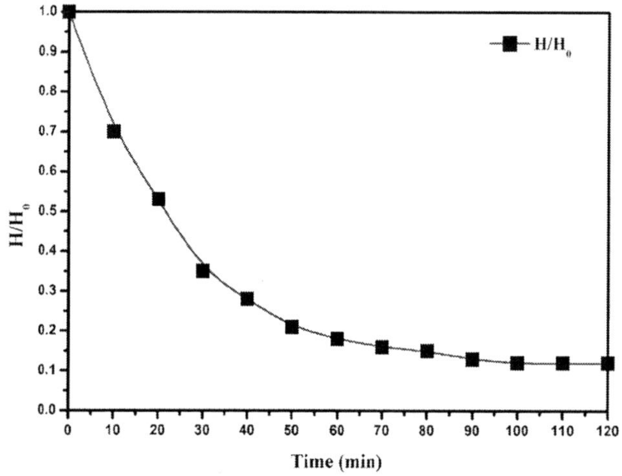


**Fig. 6. Settling characteristics of microcrystalline particles obtained from the air-agitated crystallization column after operating 30 d.**

***Morphologies and structures of crystallized products.*** Morphology features of the seeds and microcrystalline were examined by SEM. Fig. 7 (S1-S2) showed SEM micrographs of seed particles before and after HAP crystallization in the air-agitated crystallization column. Comparison of these micrographs showed that crystals formation caused a rougher surface of the seed crystal. The micrograph in Fig. 7 (S2) revealed that the amorphously distributed crystalline substance covered all the calcite seeds, which were assumed to be HAP [7, 21, 22]. The micrographs of microcrystalline (Fig.7 (S3)) showed a looser crystal structure compared with coated seeds.

Fig.7 (E1-E3) presented the quantitatively analyzed elemental composition of the seeds and crystal products by EDS. The prominent elements in the seed surface and microcrystalline were calcium and phosphorus, with traces of magnesium and chlorine. As shown in Fig.7 (E2-E3), the mole ratio (Ca/P) of the coated seeds was 2.48 and microcrystalline was 1.69. Compared with the theoretical Ca/P of HAP (1.67) and its precursors (1.33~1.59) [21, 23], the higher Ca/P of crystal products on the seed surface due to the main component of seed crystal was calcium carbonate. The ratio of the microcrystalline (1.69) was similar to the theoretical ratio of HAP (1.67) [25]. The presence of magnesium (Fig.7 (E2-E3)) indicated that the co-precipitation of calcium phosphate with magnesium ammonium phosphate had happened [24].

XRD analysis was carried out to investigate the structure of crystallized products in the air-agitated crystallization column. The XRD spectrograms of the seed crystal and microcrystalline were presented in Fig. 7 (X1-X3). The results illustrated that all the products had matched well with the standard peaks of HAP from ICCD (International Centre for Diffraction Data, PDF 73-0293). HAP formation was noted as being poorly crystalline from XRD, which can be ascribed to its small crystallite size and generated amorphous phase of calcium phosphate (ACP, OCP, and DCPD) [26]. These similar phenomena were found in the experiments for production of HAP powder performed by several researchers [6, 7, 18, 22]. Meanwhile, the presence of XRD spectrograms of calcite revealed that the co-precipitation of phosphorus with carbonate [27]. There were no the XRD spectrograms of MAP and dolomite due to the trace amounts of magnesium in wastewater.

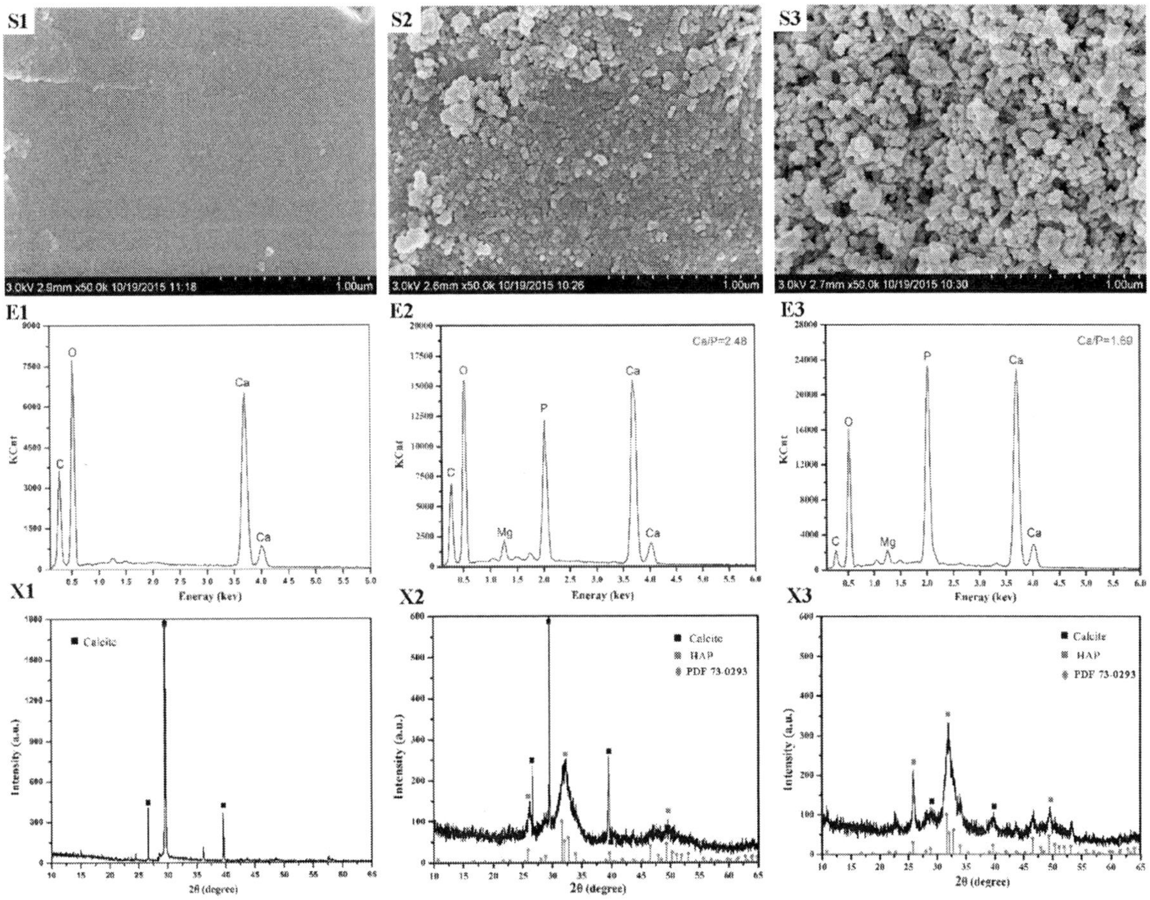


**Fig. 7. SEM micrographs, EDS analysis graph and X-Ray diffraction diagrams of seeds and microcrystalline in the air-agitated crystallization column (after 30 d, SEM micrographs (S1-S3), EDS analysis graph (E1-E3) and X-Ray diffraction diagrams (X1-X3) of initial seeds, coated seeds and microcrystalline, respectively).**

## CONCLUSIONS

After 30 d operation under a condition of multiple reagent feed points (four FP) and recirculation ratio (2.5), the phosphorus recovery efficiency ($X_R$) and microcrystalline ratio ($X_M$) were maintained at the average of 86.82 % and 13.14 %, respectively, which were only increased by 19.95 % and reduced by 39.14 % compared with the initial operation condition. The microcrystalline particles had a relatively long time (90 min) to completely precipitate under quiescent conditions due to the highly negative surface charge (~20 mV) of microcrystalline particles inhibited their aggregation and precipitation. SEM micrographs showed that the surfaces of the calcite seeds were coated with a large amount of amorphously distributed crystalline substance, which was speculated to be HAP and its precursors (ACP, OCP and DOCP). The results of EDS and XRD analysis were all verify the speculation. EDS mapping also showed the Ca/P (1.69) of microcrystalline was similar to the theoretical ratio of HAP (1.67). The results presented in this study showed a limited improvement on the phosphorus recovery in the form of HAP, associated with the conventional techniques of taking the measures of multiple reagent feed points and recirculation stream. New control strategies for super-saturation dispersion should be developed in the further study.

## ACKNOWLEDGEMENTS

This research has been supported by the National Twelfth Five-year Major Projects (2012ZX07101_005), the National Natural Science Foundation of China (51078074) and the Fundamental

Research Funds for the Central Universities: the Research Innovation Program for College Graduates of Jiangsu Province (KYLX15_0073). We thank the anonymous reviewers for their constructive comments that improved the manuscript.

## REFERENCES

[1] Guan, W., Ji, F.Y., Chen, Q.K., Yan, P., Zhang, Q., 2013. Preparation and phosphorus recovery performance of porous calcium-silicate-hydrate, Ceram. Int. 39(2), 1385-1391.

[2] Rahman, M.M., Salleh, M.A.M., Rashid, U., Ahsan, A., Hossain, M.M., Ra, C.S., 2014. Production of slow release crystal fertilizer from wastewaters through struvite crystallization- A review, Arab. J. Chem. 7(1), 139-155.

[3] Cordell, D., Drangert, J. O., White, S., 2009. The story of phosphorus: global food security and food for thought. Global Environ. Chang. 19(2), 292-305.

[4] Le Corre, K.S., Valsami-Jones, E., Hobbs, P., Parsons, S.A., 2009. Phosphorus Recovery from Wastewater by Struvite Crystallization: A Review, Crit. Rev. Env. Sci. Tec. 39(6), 433-477.

[5] Yuan, Z., Pratt, S., Batstone, D.J., 2012. Phosphorus recovery from wastewater through microbial processes. Curr. Opin. Biotech. 23 (6), 878-883.

[6] Qiu, G., Law, Y.M., Das, S., TING, Y.P., 2015. Direct and complete phosphorus recovery from municipal wastewater using a hybrid microfiltration-forward osmosis membrane bioreactor process with seawater brine as draw solution. Environ. Sci. Technol. 49(10), 6156-6163.

[7] Song, Y.H., Weidler, P.G., Berg, U., Nüesch, R., Donnert, D., 2006. Calcite-seeded crystallization of calcium phosphate for phosphorus recovery. Chemosphere, 63 (2), 236-243.

[8] Hosni, K., Moussa, S.B., Chachi, A., Amor, M.B., 2008. The removal of PO43– by calcium hydroxide from synthetic wastewater: optimization of the operating conditions. Desalination 223 (1), 337-343.

[9] Okano, K., Uemoto, M., Kagami, J., Miura, K., Aketo, T., Toda, M., Honda K., Ohtake, H., 2013. Novel technique for phosphorus recovery from aqueous solutions using amorphous calcium silicate hydrates (A-CSHs). Water Res. 47 (7), 2251-2259.

[10] Karapinar, N., Hoffmann, E., Hahn, H.H., 2006. P-recovery by secondary nucleation and growth of calcium phosphates on magnetite mineral. Water Res. 40 (6), 1210-1216.

[11] Nývlt, J., Sohnel, O., Matuchova, M., Broul, M., 1985. In: The Kinetics of Industrial Crystallization, Academia, Prague.

[12] Mokone, T.P., Van Hille, R.P., Lewis, A.E., 2012. Metal sulphides from wastewater: Assessing the impact of super-saturation control strategies. Water Res. 46 (7), 2088-2100.

[13] Costodes, V.T., Lewis, A.E., 2006. Reactive crystallization of nickel hydroxy-carbonate in fluidized-bed reactor: fines production and column design. Chem. Eng. Sci. 61 (5), 1377-1385.

[14] Kumar, R., Prakash, K.H., Cheang, P., Khor, K.A., 2004. Temperature driven morphological changes of chemically precipitated hydroxyapatite nanoparticles. Langmuir 20 (13), 5196-5200.

[15] Seckler, M.M., Bruinsma, O.S.L., Van Rosmalen, G.M., 1996. Phosphate removal in a fluidized bed-I. Identification of physical processes. Water Res. 30 (7), 1585-1588.

[16] Battistoni, P., Fava, G., Pavan, P., Musacco, A., Cecchi, F., 1997. Phosphate removal in anaerobic liquors by struvite crystallization without addition of chemicals: preliminary results. Water Res. 31 (11), 2925-2929.

[17] Suzuki, K., Tanaka, Y., Osada, T., Waki, M., 2002. Removal of phosphate, magnesium and calcium from swine wastewater through crystallization enhanced by aeration. Water Res. 36 (12), 2991-2998.

[18] Shi, J., Lu, X.W., Yu, R., Zhu, W.T., 2012. Nutrient removal and phosphorus recovery performances of a novel anaerobic-anoxic/nitrifying/induced crystallization process. Bioresour. Technol. 121(2), 183-189.

[19] Aldaco, R., Garea, A., Irabien, A., 2007. Calcium fluoride recovery from fluoride wastewater in a fluidized bed reactor. Water Res. 41(4), 810-818.

[20] APHA, 2005. Standard Methods for the Examination of Water & Wastewater, twenty-first ed. American Public Health Association, American Water Works Association and Water Environment Federation, Washington, DC, U.S.A.

[21] Kim, E.H., Lee, D.W., Hwang, H.K., Yim, S., 2006. Recovery of phosphates from wastewater using converter slag: Kinetics analysis of a completely mixed phosphorus crystallization process. Chemosphere 63 (2), 192-201.

[22] Chen, X., Kong, H., Wu, D., Wang, X., Lin, Y., 2009. Phosphate removal and recovery through crystallization of hydroxyapatite using xonotlite as seed viron. Sci. 21 (5), 575-80.

[23] Wang, L., Nancollas, G.H., 2008. Calcium orthophosphates: crystallization and dissolution. Chem. Rev. 108 (11), 4628-4669.

[24] Ohlinger, K.N., Young, T.M., Schroeder, E.D., 1998. Predicting struvite formation in digestion. Water Res. 32 (12), 3607-3614.

[25] Tervahauta, T., van der Weijden, R.D., Flemming, R.L., Leal, L.H., Zeeman, G., Buisman, C.J., 2014. Calcium phosphate granulation in anaerobic treatment of black water: a new approach to phosphorus recovery. Water Res. 48(1), 632-642.

[26] Combes, C., Rey, C., 2010. Amorphous calcium phosphates: synthesis, properties and uses in biomaterials. Acta Biomater. 6 (9), 3362-3378.

[27] Song, Y.H., Hahn, H.H., Hoffmann, E., 2002. Effects of solution conditions on the precipitation of phosphate for recovery: A thermodynamic evaluation. Chemosphere 48 (10), 1029-1034.

# PHYTOREMEDIATION OF NUTRIENT-RICH WASTEWATERS USING DUCKWEED

***Summer Lentini*** and ***Emily Smith*** (Wilkes University, Wilkes-Barre, PA, USA)
David Petrik (T&M Associates, Bethlehem, PA, USA)
Sara Arana (IES Engineers, Blue Bell, PA, USA)

Release of treated municipal wastewaters after biological treatment and containing nutrients such as nitrates and phosphates and their effects on surface waters has been well documented for many decades. Traditional treatment methods of reducing nutrient loads on receiving water bodies involve biological and/or chemical methods many of which are energy-intense and relatively expensive in the long run. Preliminary studies have shown that phytoremediation, which is a technology that uses plants to remove contaminants, can be used for effective removal of nitrates and phosphates. This study focuses on the potential of an aquatic macrophyte, *Lemna minor* (duckweed), to remove phosphates and nitrates from wastewater. Tests conducted with varying duckweed areal densities, water depths, and initial nutrient concentrations indicate nitrates and phosphates removal consistently at 95% and above levels.

# EFFECTS OF ANAEROBIC DIGESTION ON THE COMBUSTION OF SEWAGE SLUDGE

***Emrehan Berkay Çelebi***, Ayşegül Aksoy and F. Dilek Sanin
(Middle East Technical University, Ankara, Turkey)

**ABSTRACT:** Around the world, especially in the developing countries, sludge management is a topic gaining increasing importance, as the number of treatment plants and the people connected to these plants are on the rise. Old approaches seeing sludge as a waste, including landfilling, leave its place for new applications where sludge acts as an energy source or a resource as fertilizer. Combustion or co-combustion of sludge exploits its energy potential while providing stabilization and reduction in volume. Besides combustion, anaerobic digestion has been in use for a long time, serving to both stabilize the sludge and provide energy in terms of biogas. Anaerobic digestion is known to reduce the organic content of sludge, in turn, reducing its calorific value. The effects of anaerobic digestion enhanced by ultrasound pretreatment, on the combustion properties of sludge are demonstrated experimentally. 2.5 L anaerobic batch reactors were used to quantify how the calorific value of the sludge changes with digestion time. On the average, over 71 days, methane production from the sonicated reactors was 8.6 L, 32% more than the control. Observed removals were 49% for VS and 38% for COD. Calorific values of the samples were reduced by 30% on the average. The ash content remained relatively the same.

## INTRODUCTION

Sludge is produced in abundance all around the world as the byproduct of wastewater treatment. Its quantity is expected to increase along with increased wastewater production and new treatment plants (Kelessidis and Stasinakis, 2012). In many countries, landfilling the sludge after dewatering had been the most popular method of disposal. However, due to stricter regulations (Landfill Directive in EU), this will no longer be applicable. It is likely that revenue generating disposal methods like land application and co-combustion will gain more popularity (Kelessidis and Stasinakis 2012).

The three major benefits of co-combustion include firstly the utilization of sludge's heating value, instead of direct disposal. Thus, sludge is transformed into an energy source. Secondly, sludge replaces some part of the original fuel, leading to the conservation of that fuel source. Finally, remainder of the sludge after co-combustion, ash, is significantly easier to dispose of than sludge itself (Onaka 2000). Also as studied by Wzorek (2012), sludge itself can be used as an energy source during cement production furthermore, its ash can be incorporated into cement without having adverse effects.

The important parameters for combustion of sludge include water content, lower heating value, and the amount of ash produced (Werther and Ogada 1999). The water in the sludge that would be used as a fuel results in negative effects on the overall combustion process such as lowering of the combustion temperature. These effects need to be mitigated by drying the sludge up to or over 90% dry solids (Werther and Ogada 1999). Calorific value or the lower heating value indicates the amount of energy that will be released during combustion. Sewage sludge that has not been subjected to any stabilization procedure is expected to have a calorific value around 17 kJ/g, whereas, digestion processes will reduce this value to about 10 kJ/g (Werther and Ogada 1999). When compared to the calorific values of high quality coal (33 kJ/g) and lignite (6 kJ/g), it is seen that sludge can provide as much energy as medium quality coal.

Biogas resulting from anaerobic digestion provides another opportunity to use the energy potential of sludge. While conventional digesters are used all around the world for decades, new practices aim to increase organic loading rates, biogas yield and its production rate (Pérez-Elvira et al. 2006; Braguglia et al. 2012a). For the purpose of the latter, pretreatment studies are commonly conducted. Xu et al. (2011) observed an increase in VS reduction and biogas production when ultrasound pretreatment was applied. Braguglia et al. (2012b) compared ultrasound and ozone pretreatment methods and demonstrated 20% increase in biogas production. Another study showed that during anaerobic digestion in the form of a BMP assay, sonication provided a higher increase in biogas yield at lower food to microorganism (F/M) ratios (Braguglia et al. 2012a).

Although both combustion and anaerobic digestion of sewage sludge are investigated separately in the literature, effects of digestion on fuel properties of sludge have not been studied extensively. Within the scope of this study, sludge samples taken from a municipal wastewater treatment plant were subjected to ultrasound pretreatment and then anaerobic digestion in the form of large batch reactors. Calorific values were measured during digestion to quantify the change in the total amount of energy that can be harvested.

## MATERIALS AND METHODS

**Sludge Samples.** Ankara Central Wastewater Treatment Plant was the source for both the waste activated sludge (WAS) that was digested, and the anaerobic seed sludge. The plant is designed to treat 760,000 $m^3$ of wastewater daily, using a conventional activated sludge treatment system for the wastewater line. The plant includes anaerobic digesters with a solids retention time (SRT) of 14 days, where mixed primary and secondary sludge are digested together (Official Website of Ankara Metropolitan Municipality, General Directorate of Water and Sewage Administration, 2014). The WAS samples were taken from the return activated sludge line and the seed samples were taken from the digesters. Samples were left to settle for a day at 4°C following sampling to obtain concentrated sludge.

**Ultrasound Pretreatment and Setup.** WAS samples were sonicated for different time periods after settling using a laboratory scale sonicator, Sartorius Labsonic P. Sonication frequency was set as 24 kHz and the amplitude was 100%. Probe diameter was selected as 22 mm. The frequency of the device was 1, which resulted in the maximum power of 255 W. The volume of the samples sonicated were 350 mL, yielding a constant sonication density of 0.73 W/mL. Sonication duration was set to be 10 minutes while the F/M ratio was 1 (g VS/ g VSS) following the prior studies of Apul and Sanin (2010) and Köksoy and Sanin (2010). Temperature control was applied during sonication. Beakers holding 350 mL sludge sample each were placed in an ice bath for the whole duration to keep the samples at 25 to 30 °C. The seed sludge taken from the digester (ADS) was not sonicated to preserve the microorganisms required for anaerobic digestion.

**Analytical Methods.** For the calorific value analysis, samples were dried at 105°C to constant weights. Then they were sieved through a 60 mesh size sieve prior to combustion in the calorimeter (LECO AC500). Elemental compositions of the samples were also determined using an analyzer (LECO TruSpec CHN) after drying. For TS and VS measurements, Standard Methods 2540B and 2540E; for VS and VSS measurements, Standard Methods 2540D and 2540E were used, respectively (APHA, AWWA, and WEF, 2005). Ash content was determined according to ASTM Standard Method D3174-12 (ASTM, 2012). COD analyses were conducted using Hach COD kits and Hach DR 3900 spectrophotometer in accordance with the USEPA approved dichromate method (Jirka and Carter, 1975). The amount of biogas produced were measured using a gas collection column filled with a brine solution consisting of 10% NaCl (w/v) and 2% $H_2SO_4$ (v/v) to prevent the dissolution of the produced biogas. Composition of the biogas was determined by a gas chromatograph (GC) equipped with a thermal conductivity detector (TCD) (Agilent Technologies, 6890N). The carrier gas in the system had a velocity of 29 cm/s and the HP-Plot Q capillary column had the dimensions of 30.0 m × 530 µm × 40.0 µm. Temperature program set the temperature at 45 °C for the first minute, and then increased to 65 °C at a rate of 10 °C/min.

## RESULTS

Reactors were operated for 71 days until the biogas production was negligible. Monitored parameters include TS, VS, COD, calorific value, ash content, elemental composition and biogas volume and composition.

Shown in Figure 1 are the methane productions from the reactors. First of the non-sonicated reactors (U0A) showed a cumulative production of 6.71 L, while the second of the non-sonicated reactors (U0B) had 6.31 L. Sonicated reactors had more methane production, from first of the sonicated reactors (U10A) 8.81 L was observed, while the second sonicated reactor (U10B) produced 8.43 L. On the whole, a 32.1% increase was documented. This increase is higher than the increase in biogas production which was calculated to be 15.2%, indicating that more of the produced biogas is methane when sonicated sludge is digested.

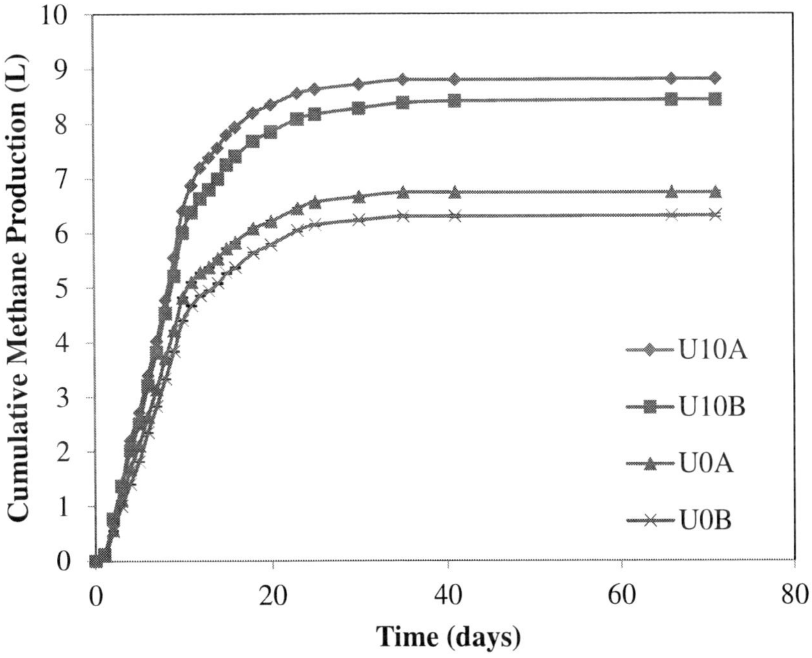


**FIGURE 1.** Cumulative methane productions of sonicated and non-sonicated reactors

When VS removal is considered, all four reactors showed a similar pattern, with sonicated reactors being one or two days ahead of the non-pretreated reactors, in terms of the progression of digestion, especially during the first two weeks (Figure 2). U10A had a VS concentration of 21583 mg/L which is reduced to 10817 mg/L, showing a 49.9% removal. U10B had 21267 mg/L initially and 10717 mg/L after the digestion, totaling to 49.6% removal in VS. U0A started with a VS concentration of 21717 mg/L and it was reduced to 11150 mg/L, displaying a removal rate of 48.7%. Finally, U0B had its VS concentration reduced from 21633 mg/L to 11100 mg/L, a 48.7% reduction overall. COD concentrations of the samples were also measured during digestion. U10A had its COD reduced from 42025 mg/L to 25950 mg/L, with a 38.3% reduction. The other sonicated reactor, U10B, showed a 39.0% decrease from 41800 mg/L to 25000 mg/L. As expected, sonicated samples exhibited more COD removal. Control groups of U0A and U0B showed 37.9% (from 43275 mg/L to 26875 mg/L) and 37.5% (from 43650 mg/L to 27300 mg/L) reductions, respectively.

As displayed in Figure 3, calorific values of sludge from each of the reactors were measured during anaerobic digestion to show how it changes with respect to time. Prior to digestion, the first of the sonicated reactors, U10A, had 14730 J/g, while the second one, U10B, had 14680 J/g. Control reactor U0A had 15552 J/g and the second control reactor U0B had 15550 J/g. On the very first day, the initial calorific values were reduced by about 10%. From this point forward, the decrease showed a declining pattern up to day 30. During this time, sonicated reactors displayed a lower calorific value, as more of the carbon was converted to biogas (and methane). After the 30-day mark, calorific values were stabilized and final values were 9845 J/g for U10A (33.1% decrease), 9950 J/g for U10B (32.2% decrease), 10183 J/g for U0A (30.0% decrease) and finally 10108 J/g for U0B (30.4% decrease). Following the same pattern for biogas and methane productions, and VS and COD removal, sonication reduced the calorific value further compared to the case with no pretreatment. It should be noted that while the reduction in calorific value is significant by itself, it gets magnified when the reduction in TS concentration is also considered.

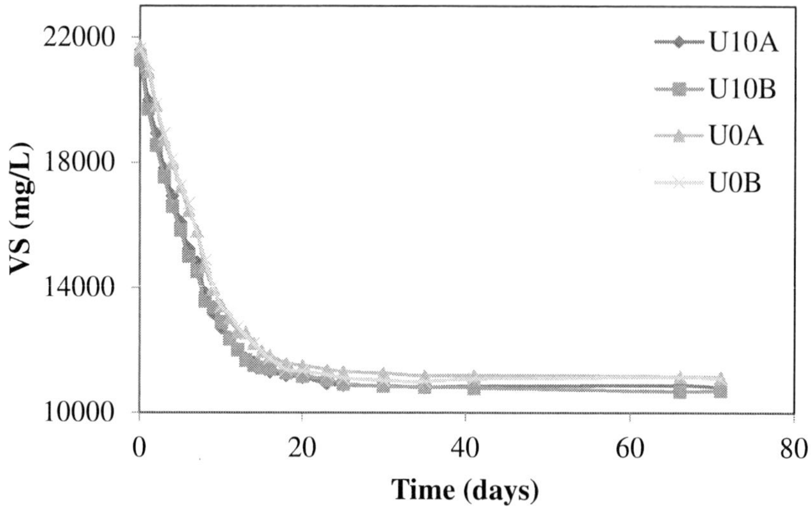


**FIGURE 2.** Changes in the VS concentrations with respect to time

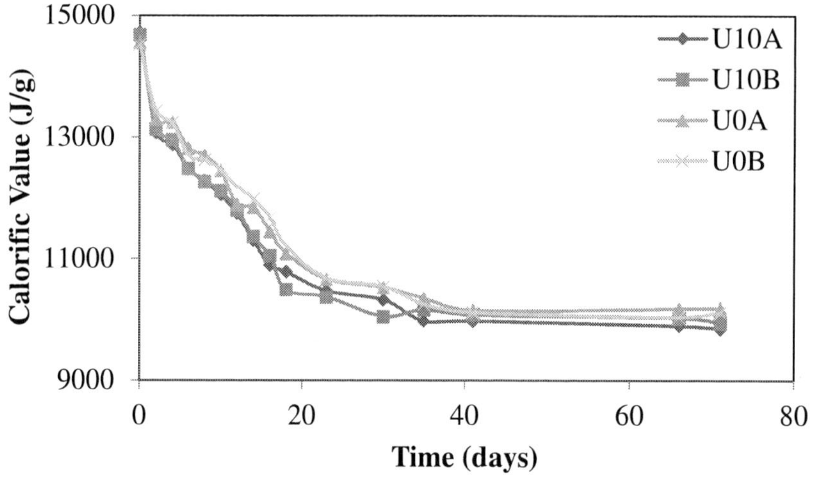


**FIGURE 3.** Changes in the calorific values with respect to time

Ash concentrations of the reactors were tracked in both mg/L and percent of dry solids. The concentrations fluctuate between 13000 mg/L and 14000 mg/L for the sonicated reactors and 12500 mg/L and 13000 mg/L for the non-sonicated reactors. On the other hand, the relative amount of ash within the sludge was found to increase as the total solids amount decreased. The initial ash contents which were 36.5% of U10A, 37.6% of U10B, 34.2% of U0A and 34.3% of U0B were found to increase to 52.3% for U10A, 52.2% for U10B, 49.0% for U0A and 49.0% for U0B. In other words, at the end of digestion, about half the sludge consisted of incombustible and inert ash for both sonicated and non-pretreated sludge.

Elemental analysis was done throughout the digestion duration to quantify the changes in carbon, hydrogen and nitrogen contents of the sludge. On the average, sludge from all four reactors started with 31.9% carbon and the non-sonicated reactors ended up both with 24.7%. Sonication reduced this further to 22.7% for U10A and 23.5% for U10B. Reason for this additional loss can be attributed to the increase in biogas production due to sonication. Hydrogen levels decreased as digestion progressed, as it took part in methane yielding reactions during digestion. Initially, samples contained 4.7% hydrogen, and this value decreased while showing some fluctuations, towards 3.4% finally. Nitrogen content remained between 1.8% and 2.7% throughout digestion, without showing any significant changes.

**CONCLUSION**

The overview of the results indicates that ultrasound pretreatment has indeed improved VS and COD removals and methane yield. Overall methane productions were 8.6 L for sonicated reactors and 6.5 L for control reactors. VS removals were 49% on the average while the COD removal rates were 38%. Ash content of the samples started with about 35% and they increased to 51% on the average. As for the calorific value of the sludge, overall, 30% reduction was observed throughout the samples, magnified further by the reduction of the amount of combustible solids.

While ultrasound pretreatment is shown to enhance anaerobic digestion significantly, energy calculations should be done to include the complete system before judging ultrasound pretreatment coupled with anaerobic digestion as worthy or not. Also, while calorific values were indeed reduced by anaerobic digestion, overall situation may demand the sludge to be digested before combustion to negate some of the associated costs of transportation and handling. To sum up, results should be evaluated on a case-by-case basis in order to enhance the energy yield from the whole sludge treatment scheme.

**REFERENCES**

APHA, AWWA, & WEF. (2005). *Standard Methods for the Examination of Water and Wastewater* (21st ed.). Washington, DC.: American Public Health Association, American Water Works Association, Water Environment Federation.

Apul, O. G., & Sanin, F. D. (2010). Ultrasonic pretreatment and subsequent anaerobic digestion under different operational conditions. *Bioresource Technology*, *101*(23), 8984–8992.

ASTM. (2012). *D3174-12, Standard Test Method for Ash in the Analysis Sample of Coal and Coke from Coal*. West Conshohocken, PA: ASTM International.

Braguglia, C. M., Gagliano, M. C., & Rossetti, S. (2012a). High frequency ultrasound pretreatment for sludge anaerobic digestion: Effect on floc structure and microbial population. *Bioresource Technology*, *110*, 43–49.

Braguglia, C. M., Gianico, A., & Mininni, G. (2012b). Comparison between ozone and ultrasound disintegration on sludge anaerobic digestion. *Journal of Environmental Management*, *95*(SUPPL.).

Jirka, A. M., & Carter, M. J. (1975). Micro semi-automated analysis of surface and wastewaters for chemical oxygen demand. *Analytical Chemistry*, *47*(8), 1397–1402.

Kelessidis, A., & Stasinakis, A. S. (2012). Comparative study of the methods used for treatment and final disposal of sewage sludge in European countries. *Waste Management*, *32*(6), 1186–1195.

Köksoy, G. T., & Sanin, F. D. (2010). Effect of digester F/M ratio on gas production and sludge minimization of ultrasonically treated sludge. *Water Science and Technology*, *62*(7), 1510–1517.

Official Website of Ankara Metropolitan Municipality, General Directorate of Water and Sewage Administration. (2014). Retrieved March 23, 2015, from https://www.aski.gov.tr/en/

Onaka, T. (2000). Sewage can make Portland cement: a new technology for ultimate reuse of sewage sludge. *Water Science and Technology*, *41*(8), 93–98.

Pérez-Elvira, S. I., Nieto Diez, P., & Fdz-Polanco, F. (2006). Sludge minimisation technologies. *Reviews in Environmental Science and Biotechnology*, *5*(4), 375–398.

Werther, J., & Ogada, T. (1999). Sewage sludge combustion. *Progress in Energy and Combustion Science*, *25*(1), 55–116.

Wzorek, M. (2012). Characterisation of the properties of alternative fuels containing sewage sludge. *Fuel Processing Technology*, *104*, 80–89.

Xu, H., He, P., Yu, G., & Shao, L. (2011). Effect of ultrasonic pretreatment on anaerobic digestion and its sludge dewaterability. *Journal of Environmental Sciences*, *23*(9), 1472–1478.

# CHANGE OF pH AND ORP WITH APPLIED DC ON LEACHATE ACTIVATED SLUDGE

***Gülizar Kurtoğlu Akkaya****, Elif Sekman, Selin Top, Senem Yazici Guvenc, Ece Sagir, Mahir Ince and Mehmet Sinan Bilgili (Yildiz Technical University, Istanbul, TURKEY)

Leachate from landfill is a very high polluted wastewater having a complex structure. Few reasearchers was made for leachate treatment by SMEBR (submerged electro membrane bioreactor) system. First of all, leachate activated sludge taken from the aeration tank of a leachate treatment facility (Istanbul Odayeri Sanitary Landfill) is consistently ventilated until the run is performed. DC (direct current) is given to 500 ml activated sludge samples at different currents and exposure durations. After each run, samples taken from near of anode and cathode, pH and ORP are transiently measured. Afterwards, obtained results are commented according to current and exposure durations.

# COMBUSTION CHARACTERISTICS AND KINETICS OF THE PULPING EFFLUENT SLUDGE USING THERMOGRAVIMETRIC ANALYSIS

***Kanhaiya Lal*** and Anurag Garg
(Indian Institute of Technology Bombay, Mumbai, INDIA)

Examining the performance of carbon products of various activation levels sometimes play important role, especially to determine the operating conditions of large scale activated carbon manufacturing plants. This study compares the removal performance of phenol from water using biochar and two physically (steam) activated carbons (having two different activation levels) produced from the same biomass, Albizia, an invasive plant. First, biochar was produced from Albizia by pyrolysis process at 850 °F. This biochar was then physically activated at 1500 °F with two different activation times (residence times of 20 minutes and 35 minutes). The iodine numbers (IN) of each of these three products, biochar and two activated carbons were determined to be 132 mg/g, 835 mg/g and 1103 mg/g respectively. Batch kinetic experiments and isotherm studies were performed to compare the effect of activation levels of activated carbons (IN of 1103 mg/g and 835 mg/g) with biochar (used as baseline with IN of 132 mg/g) on phenol adsorption. Batch kinetic studies (with 1 mg/L of phenol concentration and 2 h mixing time) showed that phenol removal can be increased more than 1.5 times with a target dosage range 200 mg/L to 400 mg/L if activated carbons are used rather than biochar. The equilibrium data were found well represented by Freundlich isotherm model.

# REDUCED SLUDGE GROWTH AT HIGH BULK LIQUOR DO INDUCED BY INCREASED MAINTENANCE

***Anwar Khursheed*** (AMU, Aligarh-202002, India)
A. A. Kazmi & Meena Sharma (Indian Institute of Technology Roorkee, Roorkee-247667, India)
Rubia Z. Gaur (Agriculture Research OrganisationVolcani Centre,Israel 50250)
Abid Ali Khan (Jamia Millia Islamia, New Delhi-110055, India)
Vinay Kumar Tyagi (TERI University, New Delhi 110070, India)

Sludge handling is a matter of great concern being largest in volume out of the constituents removed during wastewater treatment. Sludge reduction by physico-chemical methods results in buildup of chemicals, which may require further treatment. Owing these reasons various biologically sustainable methods of sludge reduction including application of high oxygenation have been successfully tested.

Experiments on actual sewage in two lab scale SBRs were conducted under normal (1.5-2.5 mgDO/L and high DO (HDO: 3-6.5 mgDO/L) regimes. it was observed that microorganism allocated substrate between maintenance and growth in the form of maintenance coefficient. This could be induced by endogenous respiration owing to high SRT, predation on bacteria, chemical toxicity, adverse environment, viral attack on bacteria. Every treatment plant may experience one or more maintenance causing factors nevertheless high SRT and prevailing environmental conditions are imminent and therefore considered as primary maintenance ($m_p$), while remaining is classified as secondary ($m_s$).

Average yield coefficient reduction at HDO was 32.7% and 28.2% compared to stoichiometric and at normal DO respectively (Figure). The observed primary and secondary maintenance was 0.11gCOD/gVSS.d (±0.01) at an SRT of 25.2d (±2.0) and 0.096gCOD/gVSS.d (±0.045) at an SRT of 24.2d (±3.6d) respectively.

# EFFECT OF THERMAL-ALKALINE PRETREATMENT ON MICROBIAL COMMUNITIES IN AN ANAEROBIC DIGESTION OF HIGH-SOLID SLUDGE

***Wang Tao***, Gao Peng and Dai Xiaohu
(Tongji University, Shanghai, China)

In order to obtain more methane, most studies in literatures were centered on enhancing the hydrolysis of sludge anaerobic digestion which was proved as un-efficient. In this paper, a thermal-alkaline pretreatment method, i.e., pretreating sludge at 100°C for 60 min and then adjusting the initial pH 9.0 was proposed to significantly enhance methane production from an anaerobic digestion of high-solid sludge. During the fermentation time of 60-80d, the optimal condition increased the biogas production to 1.20-fold of the control. In this study, the underlying mechanisms for the improved methane production under the optimal conditions were investigated by using 454 pyrosequencing and fluorescent in situ hybridization (FISH) to analyze the microbial communities shift in the anaerobic digestion reactors of high-solid sludge. It was found that thermal-alkaline pretreatment increased the abundances of *Bacteroidetes sp.* and *Firmicutes sp.*, which was able to enhance the degradation of macromolecular organic material, indicated that more volatile fatty acids were produced after thermal-alkaline pretreatment. Meanwhile, the abundance of methane-producing archaea (such as *Methanosaeta sp.* and *Methanosarcina sp.*) in the reactor of pretreated high-solid sludge was also higher than that of the control, which benefited the methanogenesis of volatile fatty acids. Therefore, thermal-alkaline pretreatment of high-solid sludge increased the abundances of bacteria and archaea involved in an anaerobic digestion system, which resulted in the efficient production of methane.

# EFFICIENT ANAEROBIC PRODUCTION OF VOLATILE FATTY ACIDS FROM SEWAGE SLUDGE BY ALKYL POLYGLUCOSE

***Jingyang Luo***, Leiyu Feng, Yinglong Su,Yinguang, Chen
(Tongji University, Shanghai, P.R.China)[44]

Volatile fatty acids (VFAs), the preferred carbon sources for biological nutrients removal, can be produced from sewage sludge (SS) during anaerobic digestion, by which both sludge reduction and recovery of carbon resources are accomplished. The addition of alkyl polyglucose (APG) remarkably stimulated the VFAs production, especially the propionic acid, in the anaerobic digestion system via simultaneously accelerating solubilization and hydrolysis, enhancing acidification, inhibiting methanogenesis and balancing carbon to nitrogen (C/N) ratio of SS. Also the feasible operation time for maximum VFAs accumulation was shortened in the presence of APG. The VFAs accumulation reached 2988 ± 60 mg COD/L at 0.3 g APG/g total suspended solids (TSS) within 4 d, while it was only 578.1 mg COD/L in the control within 6 d. The corresponding yield of propionic acid was 1312 ± 25 mg COD/L, 7.9-fold of the control. Mechanism investigations have found that both the solubilization and hydrolysis of SS were accelerated by reducing the surface tension and increasing the zeta potential of fermentation liquid, and the acidification was enhanced, while the methanogenesis was inhibited in the presence of APG. Further exploration revealed that the activities of key enzymes responsible for organic compounds hydrolysis and VFAs production were all greatly promoted due to the adjustment of C/N ratio of substrates by APG. Moreover, the microbial community in the APG reactors was shifted to be beneficial for the VFA production via 454 high-throughput pyrosequencing analysis.

# PRETREATMENT PERFORMANCE OF A NOVEL AWFR FOR DECENTRALIZED DOMESTIC WASTEWATER

***Juanhong Li*** and Xiwu Lu
(Southeast University, Nanjing, Jiangsu, China)

**ABSTRACT:** Anaerobic reactor is of great interest for domestic wastewater treatment due to the reduced-carbon footprint concept. In this work, a novel wool-felt filter reactor (AWFR) was designed and applied to treat real decentralized domestic wastewater. A lab-scale AWFR (with HRT of 3 day) was operated for 90 days in winter (temperature 8-15°C). The pollutant removal of lab-scale AWFR and the archaeal communities were investigated. Results demonstrated that the removal efficiency of COD, TN and TP were 49.5±5.7%, 13.8±4.0% and 18.6±6.7%, respectively. The average methane yield was 115L/kg·$COD_{removal}$. In addition, the acetate and propionate were slightly accumulated in effluent due to lower temperature in winter. Based on the phylogenetic analysis of High-throughput MiSeq sequencing, *Methanobacterium* and *Methanolinea* were the predominant genera in the AWFR performance, indicating that hydrogenotrophic methanogenesis pathway played a critical role for methane production of the AWFR system in winter. Overall, the AWFR with a stable performance is a promising cost-efficient pretreatment in decentralized domestic wastewater treatment.

## INTRODUCTION

Recent years, more attention has been focused on anaerobic techniques in treating decentralized domestic wastewater, particularly in the rural areas of China. Anaerobic treatment has proven to be an eco-friendly technology due to less sludge production, higher efficiency and higher energy methane recovery (Bandara et al., 2012; Gao et al., 2014a; Lew et al., 2011; Smith et al., 2013). A number of full-scale anaerobic treatment processes have recently been built and applied in tropical regions (Aiyuk et al., 2006). Nevertheless, without of heating systems, low operation temperature has been cite as a barrier to affecting the performance of anaerobic process and shift of microbial communities. Therefore, various anaerobic reactors have been properly designed to investigate the temperature effect on the performance of anaerobic treatment process (Bandara et al., 2012; Feng et al., 2008; Lew et al., 2011; Smith et al., 2013; Sumino et al., 2012). As a modern high-rate reactor, the anaerobic filter (AF) is widespread application in wastewater treatment. Carriers in the AF play a crucial role in anaerobic performance due to the large surface for microorganism attach on. In fact, many types of carriers are developed for application in wastewater treatment, such as PVC, glass, elastic fiber, nonwoven, and so on(Loupasaki & Diamadopoulos, 2013; Van Haandel et al., 2006; Wu et al., 2013). Wool felt is a natural porous and low-cost carrier. Nonetheless, to our knowledge, wool-felt carrier has not been used in anaerobic wastewater treatment.

In order to enhance performance of anaerobic process in treating decentralized domestic wastewater at low temperature, a novel anaerobic wool-felt filter reactor was designed as an alternative biotechnology for pre-treatment. Archaeal communities play an important role in organic matter removal. Molecular technologies including FISH, PCR-DGGE, and T-RFLP are effective to evaluate the major archaeal communities, but are inadequate to gain in-depth knowledge of archaeal communities (Ariesyady., 2007;Bandara., 2012;Lew., 2009). Unlike these molecular technologies, high-throughput MiSeq sequencing deeply reveals the structure of archaeal communities and the archaeal mechanism during the anaerobic operation.

The main objective of this study was to evaluate the performance of the anaerobic wool-felt filter reactor (AWFR) treating decentralized domestic wastewater in winter. Moreover, high-throughput MiSeq sequencing as a next generation sequencing tool was first conducted to investigate the archaeal communities in the AWFR system in winter.

---

## MATERIALS AND METHODS

**The AWFR System Configuration.** The AWFR consisted of Polyvinyl chloride polymer column with 50 L effective working volume packed with vertical wool felt carriers. The reactor was continuously fed with decentralized domestic wastewater via a peristaltic pump. The AWFR was operated in winter at an HRT of 3 days for 91 days. The temperature of the reactor ranged from 8 °C to 15 °C. The characteristic of the decentralized domestic wastewater was as follows: COD 229±35 mg/L, TP 2.6±0.7 mg/L, TN 26.8±4.4 mg/L, pH 6.9±0.2. Seed anaerobic sludge was obtained from a municipal wastewater treatment plant in Wuxi, China. The concentration of total suspended solids (TSS) and volatile suspended solids (VSS) were 29.5 g/L and 16.8g/L, respectively.

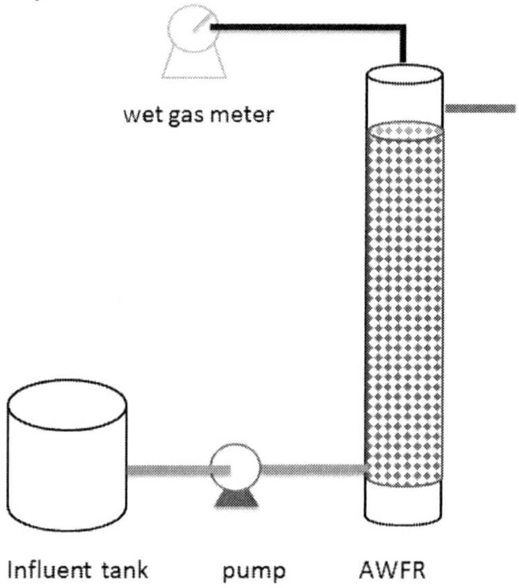


**FIGURE 1.** Schematic diagram of the experimental AWFR

**Chemical Analysis and Microbial Analysis.** Chemical oxygen demand (COD), total nitrogen (TN), total phosphorus (TP), total suspended solids (TSS) and volatile suspended solids (VSS) were measured according to the Standard Methods (APHA, 1998). Biogas production was measured using a wet gas meter. The volatile fatty acids were analyzed using gas chromatography GC3900 (Teng Hai, Shandong, China) equipped with a flame ionization detector. The pH was measured with a portable YSI-pH100 meter. Attached biofilm samples were collected from AWFR on day 10 and day 90. The V3-V4 region of archaeal 16S rRNA genes was amplified using primers 349F and 806R. The PCR products were cloned and sequenced by Illumina MiSeq System.

## RESULTS AND DISCUSSION

**Performance of the AWFR System.** The removal efficiency of COD, TN and TP are presented in Fig2a, b and c. Generally, the results indicated that the AWFR had a stable performance on organic matter degradation in winter with temperature ranged from 8 to 15 °C. Despite that the influent COD concentration ranged from 176 mg/L to 312 mg/L, the average effluent COD concentration was 113±13mg/L, with a removal efficiency of 49.5±5.7% in winter. Compared to the previous study(Wu et al., 2013), the COD removal efficiency was greater, probably due to the higher activity of microorganism and the configuration of carrier in AWFR. These results were subject to confirmation by sequencing analysis. In addition, the influent concentration of TN and TP fluctuated from 20.4 mg/L to 35.7 mg/L, and 1.3 mg/L to 4.0 mg/L, respectively. The effluent TN concentration was slightly lower than influent, the average TN removal efficiency was 13.8±4.0%. The TN removal might be mainly attributed to microbial assimilation. Meanwhile, average TP removal efficiency was 18.6-6.7% throughout the whole operation. These findings were similar to previous study (Feng et al., 2008). Different with TN removal pathway, the mechanism of

TP removal might be related to biofilm adsorption and microbial assimilation. Obviously, the concentration of TN and TP effluent could not comply with discharge standards. Hence, future studies should focus on the post-treatment performance of the AWFR.

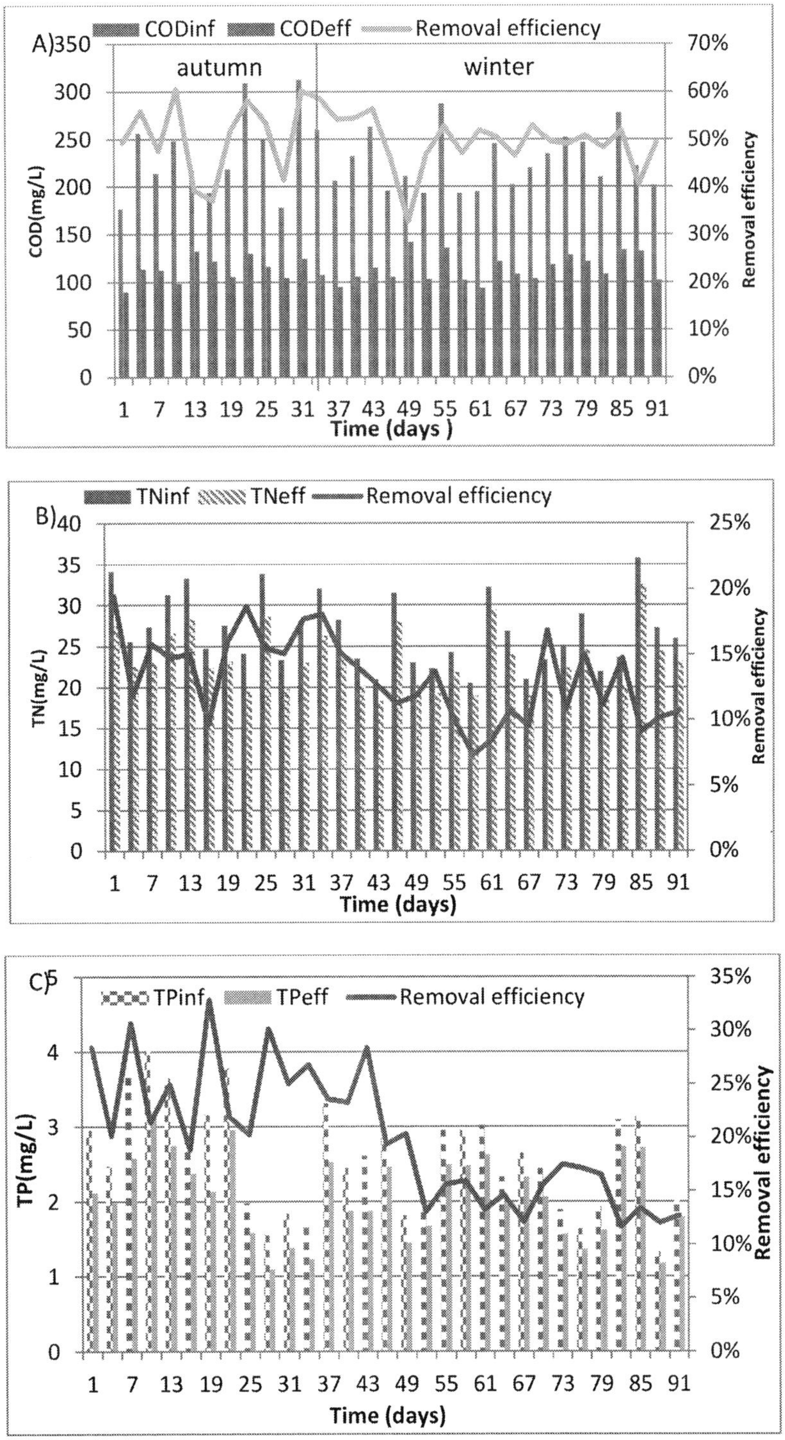


**FIGURE 2.** Variation of (A) COD, (B) TN, (C) TP and removal efficiencies

**VFAs and Biogas Production.** To evaluate the performance of AWFR in winter, VFAs and daily biogas production were to measure. As shown in Fig. 3a, acetate and propionate were the main components of the effluent VFAs. It was observed that acetate and propionate increased from autumn to winter in response to a decreased in COD removal efficiency. Acetate and propionate accumulation results were also observed in previous study (Gao et al., 2014a). Meanwhile, the daily $CH_4$ production decreased from autumn to winter. This result was consistent with an increase in VFAs effluent. However, the average $CH_4$ production in winter was still $876 \pm 418$mL/d, whereas the average methane yield was 115L/kg·CODremoval. Considering these findings, we postulated that low temperature might be responsible for the microbial activity and the production of biogas.

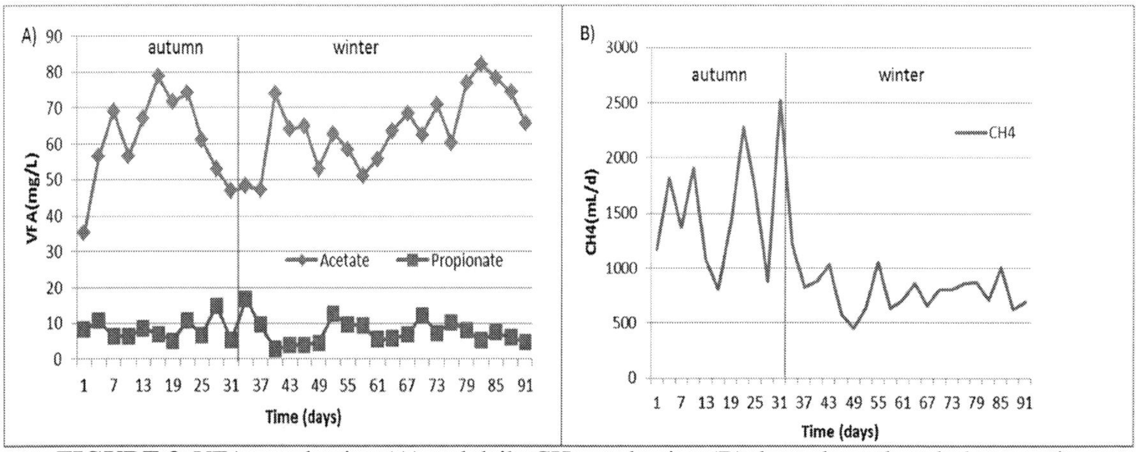


**FIGURE 3.** VFAs production (A) and daily $CH_4$ production (B) throughout the whole operation

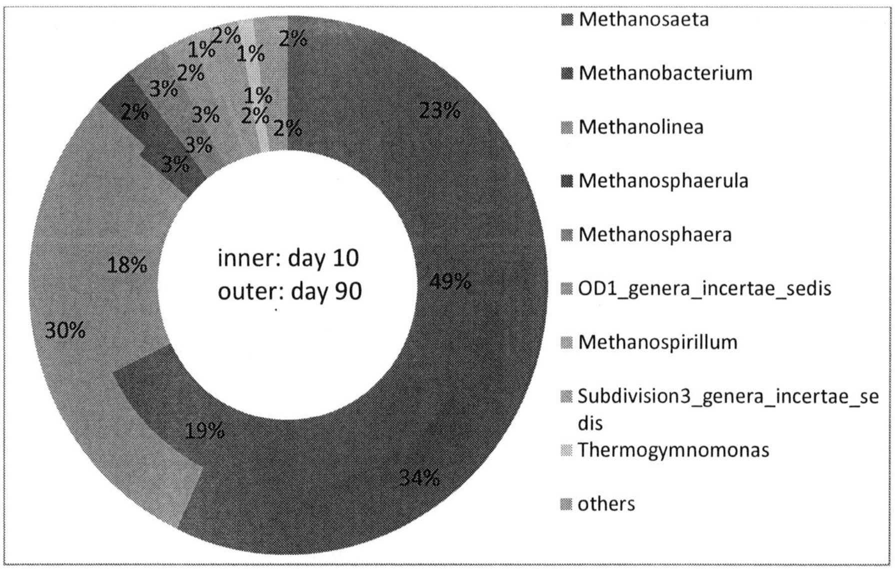


**FIGURE 4**. Archaeal community structures and compositions at genus level

**Diversity of Microbial Community in the AWFR System.** Microbial communities in the system are crucial for organic matter degradation. In order to get deep insight into the information of archaeal community structure, 16S rRNA gene PCR products were used for sequencing by Illumina MiSeq System (Fig. 4). The taxonomic classification of the samples was compared at order and genus levels. According to the RDP Classifier, the sequencing results showed that archaeal community had a lower diversity affiliated with three major orders (*Methanosarcinales*, *Methanobacteriales* and *Methanomicrobiales*). These findings are in agreement with previous studies (Bandara et al., 2012; Gomec et al., 2008). At genus level, it was remarkably observed that the genus *Methanosaeta* (49%) played a crucial role in autumn. It was well-known that *Methanosaeta* affiliated with *Methanosarcinales* was a type of acetoclasic methanogene (Fey & Conrad, 2000). Unlike in autumn, the hydrogenotrophic *Methanobacterium* was the most dominated genus in winter (day 90), whereas *Methanolinea* affiliated with the order *Methanomicrobiales*, as a type of hydrogenotrophic methanogen, was the subdominant genus. These genera were also detected by previous studies (Bandara et al., 2012; Gomec et al., 2008). Difference with traditional PCR-DGGE technique, the genera *Methanosphaerula*, *Methanosphaera* and *Methanospirillum* were detected at low relative abundance. These findings indicated that hydrogenotrophic methanogenesis pathway play a key role for methane production in winter.

## CONCLUSIONS

The AWFR was achieved good performance in treating decentralized domestic wastewater with an HRT of 3 days. The results showed that removal efficiencies of COD, TN and TP 49.5±5.7%, 13.8±4.0% and 18.6±6.7% were achieved, respectively. And the daily $CH_4$ production was stable in winter. The hydrogenotrophic genera *Methanobacterium*, and *Methanolinea* had crucial functions in the conversion of organic matter to methane in winter. In addition, post-treatment is required to remove nutrients (N, P). Based on the study, AWFR has proved to be a sustainable option for pre-treatment decentralized domestic wastewater.

## ACKNOWLEDGEMENTS

This research was supported by China National Water Pollution Control and Management Technology Major Projects (2012ZX07101-005).

## REFERENCES

Aiyuk, S., Forrez, I., van Haandel, A., Verstraete, W. 2006. "Anaerobic and complementary treatment of domestic sewage in regions with hot climates—A review". *Bioresour. Technol.* **97**(17), 2225-2241.

APHA. 1998. Standard methods for the examination of water and wastewater. *Washington, DC.*

Ariesyady, H.D., Ito, T., Okabe, S. 2007. "Functional bacterial and archaeal community structures of major trophic groups in a full-scale anaerobic sludge digester". *Water. Res.* **41**(7), 1554-1568.

Bandara, W.M., Kindaichi, T., Satoh, H., Sasakawa, M., Nakahara, Y., Takahashi, M., Okabe, S. 2012. "Anaerobic treatment of municipal wastewater at ambient temperature: Analysis of archaeal community structure and recovery of dissolved methane". *Water. Res.* **46**(17), 5756-5764.

Feng, H., Hu, L., Mahmood, Q., Qiu, C., Fang, C., Shen, D. 2008. "Anaerobic domestic wastewater treatment with bamboo carrier anaerobic baffled reactor". *Int. Biodeter. Biodegr.* **62**(3), 232-238.

Fey, A., Conrad, R. 2000. "Effect of temperature on carbon and electron flow and on the archaeal community in methanogenic rice field soil". *Appl. Environ. Microbiol.* **66**(11), 4790-4797.

Gao, D.-W., Hu, Q., Yao, C., Ren, N.-Q. 2014a. "ATreatment of domestic wastewater by an integrated anaerobic fluidized-bed membrane bioreactor under moderate to low temperature conditions". *Bioresour. Technol.* **159**, 193-198.

Gomec, C.Y., Letsiou, I., Ozturk, I., Eroglu, V., Wilderer, P.A. 2008. "Identification of Archaeal population in the granular sludge of an UASB reactor treating sewage at low temperatures". *J. Environ. Sci. Health Part A.* **43**(13), 1504-1510.

Lew, B., Lustig, I., Beliavski, M., Tarre, S., Green, M. 2011. " An integrated UASB-sludge digester system for raw domestic wastewater treatment in temperate climates". *Bioresour. Technol.* **102**(7), 4921-4924.

Lew, B., Tarre, S., Beliavski, M., Green, M. 2009. " Anaerobic degradation pathway and kinetics of domestic wastewater at low temperatures". *Bioresour. Technol.* **100**(24), 6155-6162.

Loupasaki, E., Diamadopoulos, E. 2013. "Attached growth systems for wastewater treatment in small and rural communities: a review". *J. Chem. Technol. Bio.* **88**(2), 190-204.

Smith, A.L., Skerlos, S.J., Raskin, L. 2013. "Psychrophilic anaerobic membrane bioreactor treatment of domestic wastewater". *Water. Res.* **47**(4), 1655-1665.

Sumino, H., Murota, R., Miyashita, A., Imachi, H., Ohashi, A., Harada, H., Syutsubo, K. 2012. "Treatment of low-strength wastewater in an anaerobic down-flow hanging sponge (AnDHS) reactor at low temperature". *J. Environ. Sci. Health Part A.* **47**(12), 1803-1808.

Van Haandel, A., Kato, M.T., Cavalcanti, P.F., Florencio, L. 2006. "Anaerobic reactor design concepts for the treatment of domestic wastewater". *Rev. Environ. Sci. Biotechnol.***5**(1), 21-38.

Wu, Y., Zhu, W., Lu, X. 2013."Identifying key parameters in a novel multistep bio-ecological wastewater treatment process for rural areas". *Ecol. Eng.* **61**, 166-173.

# A COMPARATIVE STUDY ON LIGHT GREY WATER FROM BUILDINGS WITH DIFFERENT FUNCTIONS

Giresunlu, E., ***Beler-Baykal, B.*** and Afacan, E.
(Istanbul Technical University, Istanbul, TURKEY)

**ABSTRACT:** Grey water is one of the streams generated upon segregation of domestic wastewater, originating mainly from various washing functions in the household and excludes toilet wastewater. It constitutes about three fourths of conventional domestic wastewater by volume, and may be returned to almost any point in the water cycle upon proper treatment. The need for characterization of this stream will become more obvious/critical, as grey water reuse becomes widespread. This work aims to contribute to grey water reuse by characterizing grey water collected from buildings with different functions but from similar sources/points of origin, i.e. wash basins, showers/bath tubs. Raw grey water characteristics in all three buildings were comparable to weak domestic wastewater or had better quality for physicochemical parameters. There were however differences in terms of quality when the three sources were compared. The highest pollution level was observed in the student residence hall (SRH) followed by the five-star hotel (FSH) and the university building (UB). The results of microbiological analyses revealed that grey water contains appreciable amounts of microbiological indicators. SRH had the highest pollution potential in terms of microbiological indicators, with comparable quality to weak domestic wastewater for total coliforms and to strong domestic wastewater for fecal coliforms.

## INTRODUCTION

Several domestic wastewater management approaches are being proposed for reusing domestic wastewater as a source rather than discarding it as a waste. One of those approaches is stream segregation, which is based on the separation of different domestic wastewater streams at their sources of generation. Grey water is one of the streams generated upon such segregation, originating mainly from various washing functions in the household. Actually it is the entire domestic wastewater with the exception of that which comes from toilets. It constitutes about three fourths of wastewater volume, with a lower pollution potential as compared to conventional domestic wastewater (Beler-Baykal, 2015). Grey water may further be categorized as light and dark grey water (Birks and Hills, 2007) based on the pollution potential of each sub-stream. Of those, light grey water is that part which comes from wash basins, showers and bath tubs.

Characterization of grey water is an important issue for its management and it will become more significant as grey water reuse becomes widespread. Depending on the function of the building from which grey water is collected and the specific source in the household it originates from, its characteristics may change considerably. Table 1 lists several pieces of work from literature focusing upon characteristics of light grey water originating from different types of buildings. A survey of the table reveals that the quality of grey water is variable even if it originates from exactly the same origin in the same type of buildings. As grey water contains about 40% of organic matter in domestic wastewater (Beler-Baykal, 2015), it is the pollutant studied the most in the literature. Moreover, due to their relatively low concentrations, nitrogen and phosphorus are not considered as problematic pollutants.

A survey of Table 1 reveals that more than half of the articles did not report microbiological indicators in untreated grey water. This is probably because the infection risk is generally attributed to fecal contamination, and grey water is segregated at the source without coming in contact with feces (Winblad et al., 2004). However, concentrations of the microbiological indicators reported in Table 1 show that grey water may contain those indicators at high levels, therefore they need to be controlled carefully when grey water reuse is intended.

As a considerable amount of organic matter is contained in grey water, the first consideration in treatment is to remove organics. Therefore, treatment systems with a biological treatment unit are generally preferred. Although after proper treatment grey water may be returned to almost any point in the water cycle, reclaimed grey water is most frequently used for toilet flushing and irrigation, in

**TABLE 1.** Characteristics of light grey water.

| Source | Grey water fraction | Parameter | Concentration | Unit | Reference |
|---|---|---|---|---|---|
| Hotel | Bath/shower and wash basin | COD | 171 | mg/L | March et al. (2005) |
| | | TSS | 44 | mg/L | |
| | | TN | 11.4 | mg/L | |
| | | Turbidity | 20 | NTU | |
| Hotel | Bath/shower and wash basin | COD | 72.7±50 | mg/L | Gual et al. (2008) |
| | | TSS | 32.2±7 | mg/L | |
| | | TN | 4.1±2 | mg/L | |
| | | Turbidity | 38.8±20 | NTU | |
| Student residence hall | Bath/shower and wash basin | BOD | 46.4±26.6 | mg/L | Birks & Hills (2007) |
| | | sBOD | 31.2 | mg/L | |
| | | COD | 96.3±52.6 | mg/L | |
| | | TSS | 36.8±29.4 | mg/L | |
| | | TP | 0.86±0.82 | mg/L | |
| | | TKN | 4.6±2.8 | mg/L | |
| | | Turbidity | 26.5±21 | NTU | |
| | | Total coliform | 7.34 | $log_{10}$ cfu/100 mL | |
| | | E. coli | 5.59 | $log_{10}$ cfu/100 mL | |
| | | Enterococci | 3.40 | $log_{10}$ cfu/100 mL | |
| Student residence hall | Bath/shower and wash basin | BOD | 95 | mg/L | Friedler et al. (2008) |
| | | COD | 270 | mg/L | |
| | | TSS | 147 | mg/L | |
| | | TKN | 11 | mg/L | |
| Student residence hall | Bath/shower and wash basin | BOD | 20 | mg/L | Winward et al. (2008) |
| | | COD | 86 | mg/L | |
| | | TSS | 29 | mg/L | |
| | | Total coliform | 5.4±0.8 | $log_{10}$ cfu/100 mL | |
| | | E. coli | 2.8±0.8 | $log_{10}$ cfu/100 mL | |
| | | Enterococci | 2.8±0.9 | $log_{10}$ cfu/100 mL | |
| Student residence hall | Bath/shower and wash basin | BOD | 43.9±10.6 | mg/L | Kadewa et al (2010) |
| | | COD | 151±88 | mg/L | |
| | | Turbidity | 57.6±14.7 | NTU | |
| University building | Wash basins | COD | 248 | mg/L | Sarioglu (2011) |
| | | TSS | <10 | mg/L | |
| | | $NH_4$-N | 0.45 | mg/L | |
| | | Turbidity | 12 | NTU | |

addition to uses such as car washing, groundwater recharge, fire extinguishing, etc. where the water does not need to be at the quality of drinking water.

This work aims to contribute to grey water reuse by characterizing grey water collected from buildings with different functions but from similar sources/points of origin, i.e light grey water. Within the context of this work, a five-star hotel (FSH) and a student residence hall (SRH) both collecting grey water from bathtubs, showers and wash-basins, and a university building (UB) collecting from wash-basins of restrooms were investigated. Characterization was made taking both physicochemical parameters (Chemical Oxygen Demand (COD), soluble COD, Biochemical Oxygen Demand (BOD), Total Suspended Solids (TSS), Volatile Suspended Solids (VSS), Ammonical-N ($NH_3$-N), Total Kjeldahl Nitrogen (TKN), Orthophosphate ($PO_4$-P) and Total Phosphorous (TP)) and microbiological indicators (Total Coliforms (TC), Fecal Coliforms (FC), E. coli (EC) and enterococci (ENT)) into account.

## MATERIALS AND METHODS

The five-star hotel and the student (girls) residence hall investigated collect their grey water from washbasins and showers/bathtubs. After proper treatment, treated grey water is reused for toilet flushing and irrigation in the hotel, while it is reused for toilet flushing only in the student residence hall. The main treatment unit employed in both buildings is membrane bioreactors (MBR). Grey water samples were taken from the equalization tank, which are located before the MBRs, to characterize raw grey water quality.

On the other hand, in the university building, the grey water collection system is for research purposes only and the grey water is not reclaimed. Water generated at the lavatories' washbasins are collected in a tank, which also serves the purpose of equalization.

Sampling in the five-star hotel was done every-week throughout one year, while the student residence hall and the university building grey water were sampled every month, covering a complete academic year. Samples were taken as grab samples.

In terms of physicochemical parameters, samples were analyzed for organic matter through COD, sCOD and BOD and solid matter concentrations were determined using TSS and VSS. $NH_3$-N, TKN, $PO_4$-P and TP were the parameters analyzed for assessing the occurrence of nutrients. All physicochemical analyses were carried out according to APHA Standard Methods (Rice et al., 2012) except for COD and sCOD, as ISO 6060 method was used for these analyses.

Microbiological quality was analyzed in terms of total coliforms and fecal coliforms, as these indicator groups have been used conventionally over the past years. Additionally, E. coli and enterococci were also monitored since they were identified as indicators which had a better correlation to occurrence of diseases (USEPA, 1986: WHO, 1999). ISO 8199 was the method used for all microbiological analyses. All analyses were carried out with three different dilutions and in triplicate.

## RESULTS AND DISCUSSION

The results obtained in this work for physicochemical parameters are summarized in Figure 1 and Table 2, and for microbiological indicators in Figure 2.

The average concentrations of organic matter and solid matter are shown in Figure 1-a, together with their counterparts for weak conventional domestic wastewater quality as given by Metcalf & Eddy (2003). Average BOD was less than 110 mg/l in all cases indicating that all three buildings showed lower pollution potential in terms of organic matter as compared to weak conventional domestic wastewater, with only one value exceeding this concentration. The lowest pollution potential in terms of solid matter was observed for the university building and it was followed by the five-star hotel, both lower as compared to the typical values for weak conventional domestic wastewater. However, the student residence hall showed somewhat lower quality in terms of solids as compared to weak conventional domestic wastewater.

Figure 1-b, showing the average nutrient concentrations in grey water samples in terms of TKN, $NH_3$-N, TP and $PO_4$-P, reveals that nutrients in all of the grey water samples were at a level much lower than weak conventional domestic wastewater. As observed from both Table 2 and Figure 1-b, the highest TKN and TP average concentrations were calculated for the hotel grey water, as 6.99 and 2.29 mg/L.

All in all, analysis of the results of physicochemical parameters has revealed that raw grey water characteristics in all three buildings were comparable to weak domestic wastewater or had better quality as can be observed from Figure 1. Moreover, priority components in this group were organic matter and solids. Specifically, solid concentration for grey water treatment is important as one of the most used systems for grey water reclamation employs membranes, as in the case of two of the buildings studied, which tend to be clogged very fast if the suspended solid concentrations are high.

There were however differences in terms of quality when the three sources were compared, as summarized in Table 2. The highest concentrations in terms of the parameters representing the priority components organic matter and solids was observed in the student residence hall followed by the five-star hotel and the university building. Organic matter concentrations were the highest in the student residence hall with averages as 217 mg/l COD, 76 mg/l BOD, followed by the five-star hotel with 120 mg/l COD, 32 mg/l BOD, which was about half of that of student residence hall, although both buildings collected from the same sources. The university building had lower values, 89 mg/l COD, 54 mg/l BOD, which was expected, because the source of grey water was wash-basins only.

The analyses showed that the highest TSS and VSS concentrations were observed in the student residence hall with 145 and 96 mg/L, respectively, similar to the organic matter.

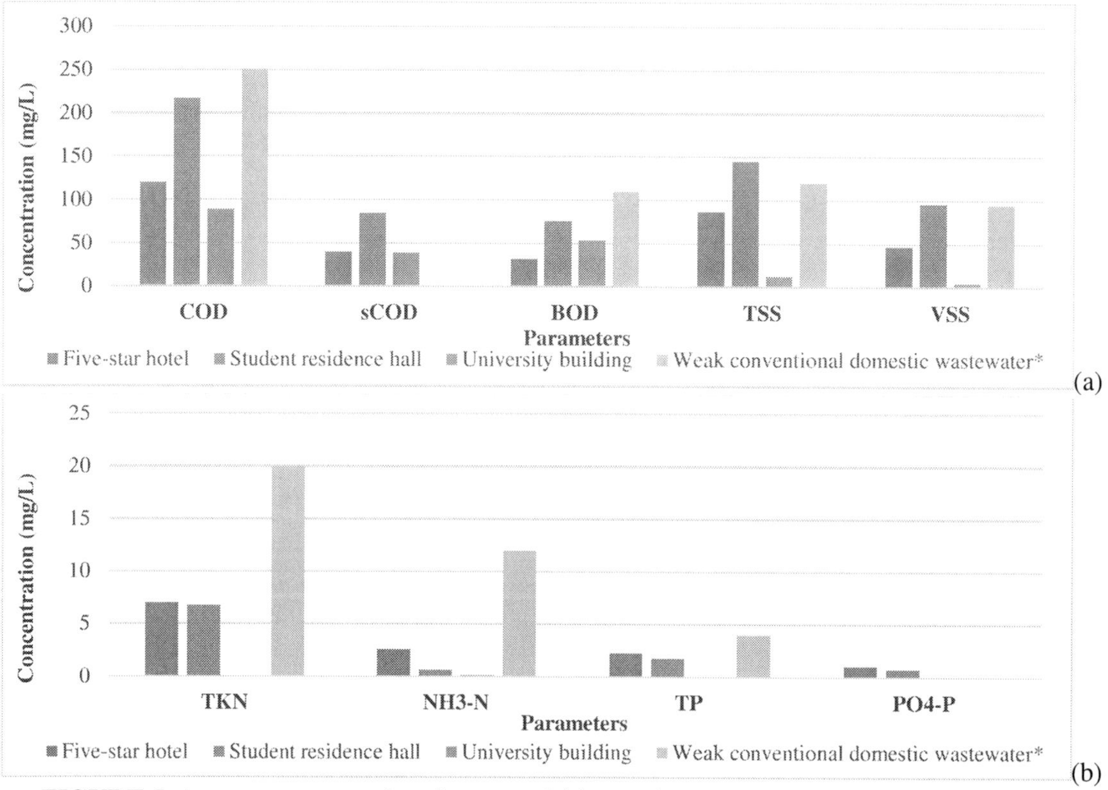


**FIGURE 5.** Average concentrations in terms of (a) organic matter and solid concentrations and (b) nutrients, as compared to typical concentrations of weak conventional domestic wastewater. *Metcalf & Eddy, 2003

concentrations. The five-star hotel was the second most polluted grey water source in terms of TSS and VSS concentrations as 87 and 47 mg/L, respectively. Solid concentrations in university building were very low, as shown in Table 2.

The highest pollution potential regarding organic matter and solids were thus observed in the student residence hall. This could be attributed to occasional food washes and dish-washing practiced in wash-basins together with concentrated personal care products use/activities in the girls' residence hall.

A survey of the BOD/COD ratios showing the tendency towards biodegradation and sCOD/COD ratios as an indicator of particulate fraction reveals that the organic matter in grey water was mostly non-biodegradable or slowly biodegradable and in particulate form in the five-star hotel (35%, 39%) and in the student residence hall (26%, 33%). Based on these results, the option of membrane filtration could also be evaluated in favor of leaving out the biological step, with low values of BOD and particulate organics predominating solid matter, especially in the five-star hotel.

Data regarding microbiological quality were limited in comparison to physicochemical parameters, nevertheless they are plotted in Figure 2. Microbiological analyses were carried out using three samples for five-star hotel, four samples for student residence hall and three samples for university building, all worked in sets of triplicates. The results however were supported by another project undertaken in the same hotel (Beler-Baykal, 2014).

The results of the microbiological analyses showed that grey water collected from the student residence hall had the highest pollution potential with mean concentrations of 7.18 $\log_{10}$ cfu/100 mL of total coliform, 6.95 $\log_{10}$ cfu/100 mL of fecal coliform, 6.92 $\log_{10}$ cfu/100 mL of E. coli and 4.16 $\log_{10}$ cfu/100 mL of enterococci. Five-star hotel grey water had lower pollution potential as compared to student residence hall. Grey water from the university building showed the lowest pollution potential in terms of all microbiological indicators. The only source free of enterococci was the five-star hotel. The characteristic microbiological indicator concentrations for conventional domestic wastewater are given in Table 2. A comparison with these values reveals that the average concentrations of total coliforms in the five-star hotel and the university building were comparable to weak domestic wastewater characteristics. On the other hand, university building samples showed lower concentrations as compared to weak domestic wastewater. In terms of fecal coliforms, the five-star hotel and the student residence hall grey water were comparable to strong conventional domestic wastewater. In this case, the university building grey water was comparable to weak domestic wastewater.

**TABLE 2.** Results of physicochemical analyses (n: sample number)

| Location | Five-star hotel | | | | Student residence hall | | | | University building | | | |
|---|---|---|---|---|---|---|---|---|---|---|---|---|
| Parameters (mg/L) | min | max | avg | n | min | max | avg | n | min | max | avg | n |
| COD | 30 | 710 | **120** | 41 | 99 | 598 | **217** | 8 | 38 | 121 | **89** | 7 |
| sCOD | 30 | 125 | **40** | 37 | 59 | 94 | **85** | 5 | 30 | 57 | **39** | 7 |
| BOD | 8 | 82 | **32** | 35 | 54 | 124 | **76** | 7 | 19 | 78 | **54** | 5 |
| TSS | 16 | 838 | **87** | 41 | 44 | 328 | **145** | 8 | 4 | 18 | **12** | 7 |
| VSS | 9 | 376 | **47** | 40 | 4 | 216 | **96** | 8 | 0 | 7 | **4** | 7 |
| TKN | 2.11 | 35.65 | **6.99** | 36 | 4.05 | 16.89 | **6.76** | 8 | - | - | **-** | 0 |
| $NH_3$-N | 0.58 | 5.55 | **2.58** | 36 | 0 | 0.86 | **0.62** | 8 | 0.04 | 0.21 | **0.14** | 7 |
| TP | 0.73 | 8.35 | **2.29** | 36 | 1.35 | 2.17 | **1.78** | 8 | - | - | **-** | 0 |
| $PO_4$-P | 0 | 4.01 | **1.05** | 36 | 0.27 | 1.19 | **0.75** | 8 | 0 | 0.01 | **0** | 7 |

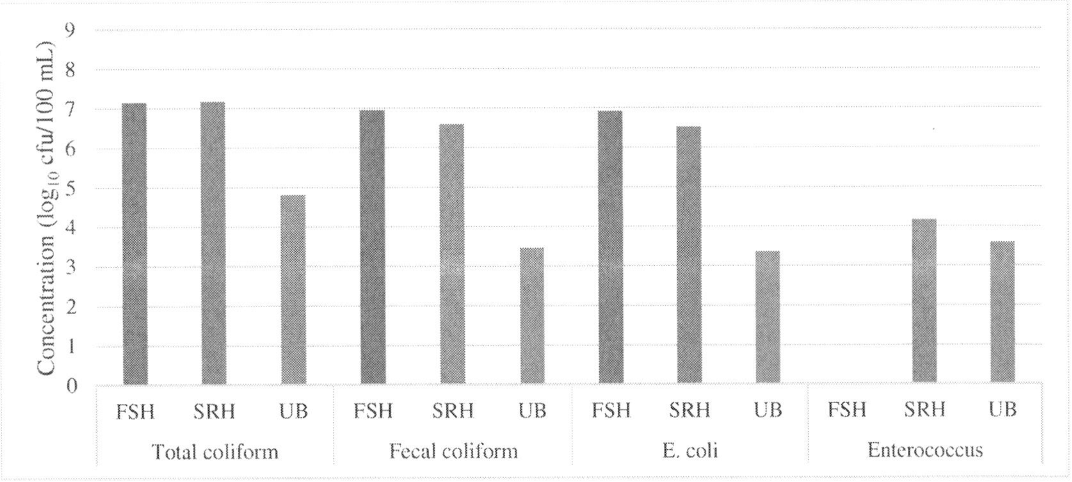


**FIGURE 6.** Average concentrations of microbiological indicators.

A comparison with the literature values listed in Table 1 reveals that, COD and TN values in the hotel grey water in this work falls within the range of values reported, whereas the average concentration of TSS in this study is almost twice as high as the ones listed. Student residence hall grey water falls within the range reported in the literature in terms of organics and nitrogen, while TSS concentration is comparable and TP concentration is higher in this study as compared to those reported. University building grey water showed lower or comparable concentrations as compared to its counterpart given in the table. Microbiological indicators were investigated in only two pieces of work reported in Table 1, and both of them concentrated on student residence hall grey water. Values obtained in this work for the student residence hall were higher as compared to the literature data. Moreover, hotel grey water showed lower quality in terms of all indicators tested. The university building grey water showed higher concentrations in terms of total coliforms, while showing better quality in terms of E. coli and enterococcus. These differences show that even if the function of the buildings where grey water is collected from and the source of grey water are the same with those in the literature, the quality of grey water may change significantly.

The data obtained in this work has revealed that light grey water quality tends to show variations according to the function of the building where it is collected, even if the sources of collection are the same. Characterization of the specific grey water to be reclaimed may play a major role in the selection of grey water reclamation schemes.

**TABLE 2.** Microbiological indicator concentrations in domestic wastewater (Metcalf & Eddy, 2003)

| | Weak | Intermediate | Strong |
|---|---|---|---|
| Total coliform ($log_{10}$cfu/100 mL) | 6-8 | 7-9 | 8-10 |
| Fecal coliform ($log_{10}$cfu/100 mL) | 3-5 | 4-6 | 5-8 |

## CONCLUSION

Raw grey water in all three buildings showed characteristics comparable to weak domestic wastewater or had better quality in terms of physicochemical parameters. Organic matter concentrations were lower than typical values for weak domestic wastewater, while solids were comparable or better. All sources showed similar characteristics in terms of nutrients, which were considerably lower as compared to weak domestic wastewater.

On the other hand, although all are light grey water, the three sources showed different quality when compared to each other. Student residence hall has shown the highest pollution potential in terms of organic matter and solids and it was followed by the five-star hotel and the university building respectively.

The results of the analyses showed that the larger fraction of organic matter was in particulate form, and the average BOD concentrations were low, which raises the question whether a biological unit should be employed or not.

Even though the physicochemical parameters showed low pollution potential, the amount of microbiological indicators were appreciable. Total coliform concentrations were comparable to the typical values of weak domestic wastewater, while the fecal coliform concentrations could go as high as strong domestic wastewater. Therefore, microbiological indicators should not be overlooked while reusing grey water, to sustain public health.

In summary, the results of this work have revealed that even if the origins of the grey water are the same in terms of the source where they are collected, the quality of grey water may change significantly depending upon the function of the buildings. Characterization of the specific grey water to be processed is crucial in determining the reclamation scheme.

## REFERENCES


Beler-Baykal, B. 2014. Atiksularin akimlarina ayrilmasi ve gri su geri kazanimi (Stream segregation and grey water reclamation) in Turkish, online at http://www.csb.gov.tr/db/tay/webmenu/webmenu 13378.pdf

Beler-Baykal, B. 2015. "Stream segregation in household use: A review of grey water as an alternative source of water and yellow water as an alternative source of fertilizers", *Water Quality Exposure and Health,* 7(1), 27-37.

Birks, R. and Hills, S. 2007. "Characterization of indicator organisms and pathogens in domestic greywater for recycling." *Environmental Monitoring and Assessment*, 126, 61-69.

Friedler, E. and Alfiya, Y. 2010. "Physicochemical treatment of office and public buildings greywater." *Water Science and Technology*, 62(10), 2357-2363.

Gual, M., Moia, A. and March, J.G. 2008. "Monitoring of an indoor plant for osmosis rejection and greywater reuse to flush toilets in a hotel." *Desalination*, 219(2008), 81-88.

ISO 6060:1989 "Water quality – Determination of the chemical oxygen demand"

ISO 8199:005 "Water quality -- General guidance on numeration of microorganisms by culture with membrane filtration"

Kadewa, W.W., Le Corre, K., Pidou, M., Jeffrey, P.J. and Jefferson, B. 2010. "Comparison of grey water treatment performance by a cascading sand filter and a constructed wetland." *Water Science and Technology*, 62(7), 1471-1478.

March, J.G., Gual, M. and Ramonell, J. 2005. "A kinetic model for chlorine composition in grey water." *Desalination*, 181(2005), 267-273.

Metcalf & Eddy. 2003. *Wastewater engineering treatment and reuse* 4th ed., McGraw Hill, New York, NY, USA.

Rice, E. W., Bridgewater, L., Water Environment Federation, American Water Works Association, and American Public Health Association. 2012. *Standard methods for the examination of water and wastewater* (22nd 2012 ed.). Washington, D.C: American Public Health Association.

Sarioglu, S. 2011. *Low load greywater and municipal wastewater treatment by membrane bioreactor.* M.S. Thesis, Institute of Environmental Sciences, Bogazici University.

USEPA. 1986. *Ambient Water Quality Criteria for Bacteria.*

Winblad, U., Simpson-Hérbert, M., Calvert, P. & Stockholm Environment Institute. 2004. *Ecological sanitation*, 2., rev. and enlarg ed, Stockholm Environment Institute Stockholm, Stockholm.

WHO. 1999. *Health-based monitoring of recreational water: The feasibility of a new approach* (The Annapolis Protocol). Geneva, Switzerland.

Winward, G.P., Avery, L.M., Frazer-Williams, R., Pidou, M., Jeffrey, P., Stephenson, T. and Jefferson, B. 2008. "A study of the microbial quality of grey water and an evaluation of treatment technologies for reuse." *Ecological Engineering*, 32, 187-197.

# A PILOT-SCALE STUDY OF A MODIFYED WASTEWATER TREATMENT WITH SLUDGE REDUCTION EFFECT

***Chenyi Shi***, Yushan Wang, Xiange Wei, Wuzhen Guo, Meishan Lin, Jinxin Tan, and Lianpeng Sun.
(Sun Yat-Sen University, Guangzhou 510275, Guangdong, China)

**ABSTRACT:** The MOSA (Modified Oxic-Settling-Anaerobic) process generates less excess sludge during the process of wastewater treatment, but many studies on the MOSA process are undertaken at the laboratory scale or based on synthetic wastewater, which are not sufficient for practical application or to achieve ideal sludge reduction. A 10-$m^3$/d pilot-scale system was established to treat municipal sewage. The A/O process and the A/O-MOSA process were performed in sequence to study the sludge reduction efficiency of the MOSA process by comparing the effluent quality and the sludge yield of the two processes. The study indicated that, compared with the separate A/O process, the A/O-MOSA process achieved a 35% sludge reduction without any influence on the effluence properties, but resulted in a small increase in the total phosphorus (TP) concentration. After adding $Al_2(SO4)_3$ into the aerobic basin, the TP concentration in the effluent can be kept under 1.0mg/L while simultaneously achieving a 33% sludge reduction. The results indicate the good stability and sludge reduction efficiency of the MOSA process.

## INTRODUCTION

The Oxic-Settling-Anaerobic (OSA) process inserts an anaerobic reactor between the aeration basin and secondary settling basin, exhibits excellent performance in sludge reduction. Subsequently, many researchers have further studied the OSA process and have obtained good results in reducing excess sludge (Chen et al., 2003; Novak et al., 2007; Wu et al., 2013; Sun et al., 2015), highlighting the representative mechanisms of excess sludge reduction, such as energy uncoupling and sludge decay(Chen et al., 2003), extracellular polymeric substances and metal complex degradation (Novak et al., 2007). Tan and Sun established a modified Oxic-Settling-Anaerobic (MOSA) process by changing the sludge interchange rate and mode (Tan, 2014). Many studies on MOSA have been undertaken at the laboratory scale and not sufficient for practical application or to achieve ideal sludge reduction. Therefore, this study was undertaken at the pilot scale using municipal sewage and aimed at evaluating the performance of sludge reduction and the effluent properties of the MOSA process to lay a foundation for applying the MOSA process at full scale.

## MATERIALS AND METHODS

Based on the A/O process in the municipal wastewater treatment plant (WWTP) in Zhenan, Foshan, China, which treats 100000 $m^3$ of wastewater every day, this A/O-MOSA pilot-scale system was operated in the same manner as the A/O process in the plant. For this system, the hydraulic retention time (HRT) of the anaerobic and aerobic basins was 1.35h and 5.05h, respectively, while the sludge concentration was in the range of 2000 to 3500 mg/L and the dissolved oxygen (DO) was between 2.0-4.0 mg/L. Figure 1 schematizes the 10-$m^3$/d pilot system based on the MOSA process, with a sludge retention time (SRT) of 5d and an interchange

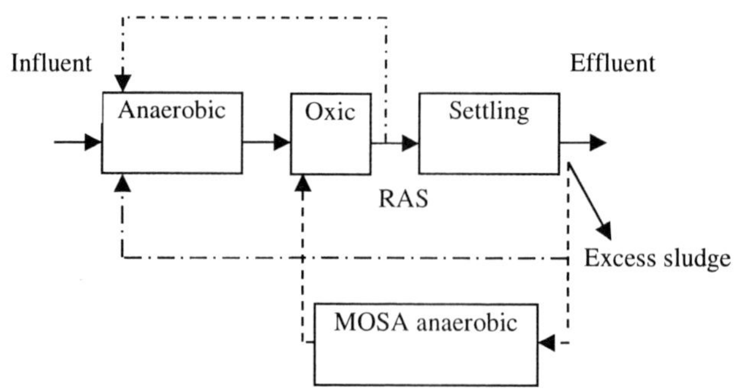


**FIGURE 1**. Schematic diagram of the A/O-MOSA process

rate (IR) of 10%. The system was operated at 4 cycles. The sludge was pumped into the MOSA reactor from the secondary settling tank of the A/O process and then flowed into an oxic tank. The amount of wasted sludge was calculated according to the growth of the sludge in the oxic tank. This system has been steadily operated for 212 days (A/O process was applied for the first 60 days, the A/O-MOSA process was applied for the next 130 days, and the last 22 days were used for the chemical phosphorus removal experiment under the A/O-MOSA system.) After a 2-week sludge cultivation under anaerobic conditions, the MOSA reactor was started and inserted into the A/O process, which formed the complete A/O-MOSA system. The data were collected in periods from the 61th to 90th, 156th to 190th, and 191th to 212th day, which were denoted as A/O-MOSA I, A/O-MOSA II, and A/O-MOSA III, respectively. For the 91th ~115th day, the influent properties fluctuated because of equipment failure in the pilot-scale system and maintenance of the WWTP. The temperature for the A/O and A/O-MOSA I period was 26-33°C, while that for the A/O and A/O-MOSA I periods was 14-20°C. All of the quality parameters were measured according to the Standard Methods (State Environmental Protection Administration of China, 2002).

## RESULTS AND DISCUSSION

Before and after starting the MOSA reactor, the removal effects of COD and SS are shown in Figure 2 and Figure 3. We can observe that the system had a high efficiency for the removal of COD and SS. During the operation of the A/O process, the average removal rates of COD was 79.1%. And 82.6% in A/O-MOSA I, 77.1% in A/O-MOSA II. There was no significant difference in the removal rate of organic matter in various stages. Switching in the MOSA anaerobic tank had no adverse effects on the A/O process; the MOSA anaerobic still retained a higher removal rate for organic matter.

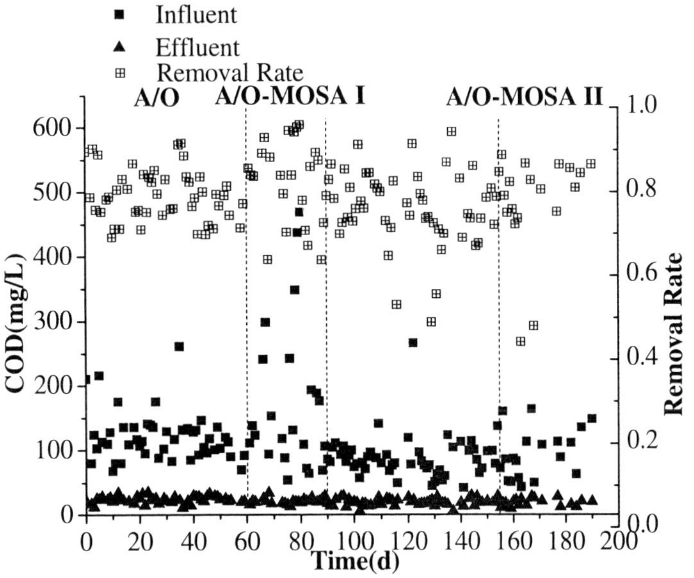


**FIGURE2.** COD removal rate over the entire operational period

From Figure 3, the removal rate of SS was found to be elementarily unanimous, the SS of the effluent remained at 13~17 mg/L. The average removal rate of SS was 75.1%; and 78.3% and 82.8% in the A/O-MOSA I and A/O-MOSA II system, respectively. In this research, the insertion of the MOSA anaerobic reactor improved the treatment efficiency of the effluent SS.

The removal of $NH_4$-N and TN in the system is shown in Figure 4 and Figure 5.. From Figure 4, starting the MOSA reactor did not negatively impact the removal of $NH_4$-N. In the entire experimental stage, the concentration of $NH_4$-N in the effluent was under 0.6 mg/L, and the removal rate reached 97.8%. The system had a good removal efficiency for $NH_4$-N.

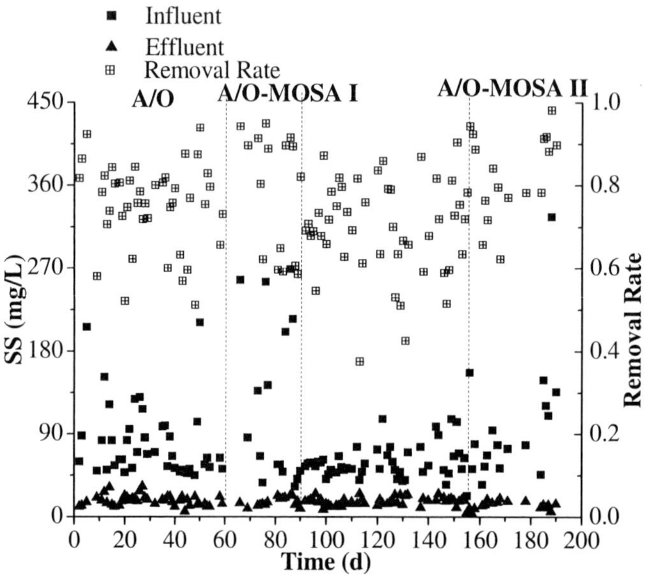


**FIGURE 3**. SS removal rate over the entire operation period

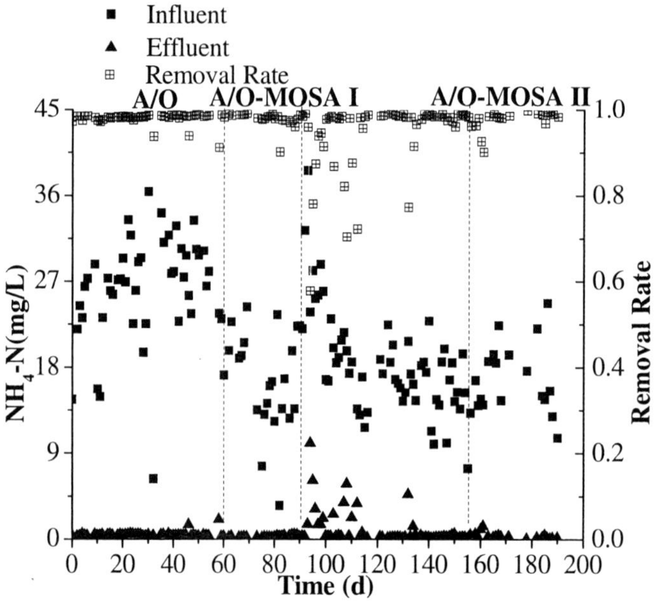


**FIGURE 4**. $NH_4$-N removal rate over the entire operational period

From Figure 5, we can see that the average removal rates of TN in A/O, A/O-MOSA I and A/O-MOSA II were 33.4%, 38.2% and 30.4%, respectively. Compared to A/O, after inserting the MOSA reactor, the removal efficiency of TN exhibited a trend from increasing to decreasing. The main reasons for this behavior were as follows: (1) In the A/O process, C/N is required to be as high as 8.0 (Rasool et al., 2014). In this study, C/N in A/O, A/O-MOSA I, and A/O-MOSA II was 3.0, 7.5, and 4.0. In A/O MOSA I, the greatest carbon source was found, contributing to the best denitrification effect. (2) The optimum temperature of denitrification is 20~35°C (Ma, et al., 2008).. During the period of operation of the A/O and A/O-MOSA I systems, the temperature was 26~33°C, which is in the range of the most suitable temperature for denitrification. In the A/O-MOSA II operation period, the temperature was 14~22°C, which affected the denitrification rate, resulting in the decrease of the TN removal efficiency. The removal of phosphorus in this system is shown in Figure 6. During the effective operation of the system, the effluent TP removal rate was only 26.7%. Both the A/O and A/O-MOSA processes had an undesirable TP removal effect, and inserting the MOSA anaerobic reactor resulted in a further decrease of the TP removal efficiency. The primary reasons for the observed behavior were as follows: (1) The SS concentration in the effluent was high and the average concentration was 17mg/L, directly leading to the poor dephosphorization effect. (2)Lack of carbon source. Wentzel et al. (1991) believed that to remove 1.0 mg of phosphorus requires 7.5~10.7 mg of COD of VFAs. During the operational period, the average concentrations of influent of COD were 130mg/L (3)In the course of system operation, the average concentration of $NO_3^-$ in the returned sludge reached 7.4mg/L, the PAOs could not become the dominant bacteria; as a result, the biological phosphorus removal process was restricted. (4) The sludge that entered into the MOSA anaerobic reactor released plenty of phosphorus in the anaerobic condition, and the soluble TP in the exchanged sludge was as high as 70 mg/L, which increased the TP load in the process and impacted the process. In addition, the A/O-MOSA system had a lower sludge yield, which reduced the discharge of excess sludge, indirectly leading to the extension of SRT. Consequently, the phosphorus in the system could not be discharged in time.

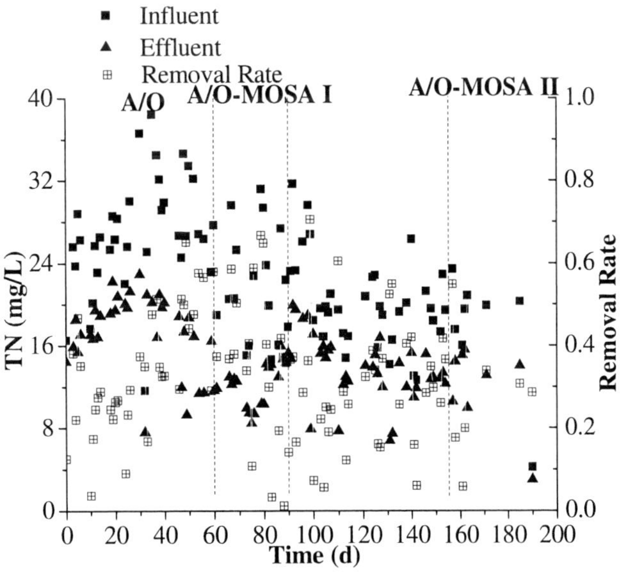


**FIGURE 5**. TN removal rate over the entire operational period

To improve the TP removal effect in the effluent for the A/O-MOSA system, this experiment added chemical agents to intensify phosphorus removal. From the 36th day in the A/O-MOSA II operation (A/O-MOSA III stage), we added an $Al_2(SO_4)_3$ solution to the tail end of the aerobic tank and the continuous run for 22 days.. The TP changes in A/O-MOSA III operational period is shown in Figure 7. Figure 7 shows that in the A/O-MOSA III stage, the average concentration of the influent TP was 1.90mg/L, while the effluent TP was 0.57mg/L, with an average removal rate of 61.0%.

The apparent sludge yield is calculated by the type 1 equation(Tan, 2014):

$$Y_{obs} = \frac{(MLSS_2 - MLSS_1) \times V_a + W}{\sum_{1}^{n} (COD_1 - C_2) \times Q} \quad (1)$$

The $Y_{obs}$ of A/O, A/O-MOSA I, A/O-MOSA II and A/O-MOSA III were 0.244, 0.152, 0.167 and 0.165 kg-MLSS/kg-COD, respectively. Therefore, the sludge reduction effect of A/O-MOSA I, A/O-MOSA II and A/O-MOSA III was 38%, 32% and 33%, respectively. The average sludge reduction effect was 35%. The sludge reduction effect of A/O-MOSA II was relatively lower, and the primary cause was the impact from the temperature reduction. Zhang et al.(2006) found that improving the temperature was beneficial to the sludge reduction effect in the OSA process. Before and after using the chemical phosphorus removal method, the sludge reduction effects exhibited no significant differences. Chemical phosphorus removal had no adverse effects on the sludge reduction effect in the A/O-MOSA process.

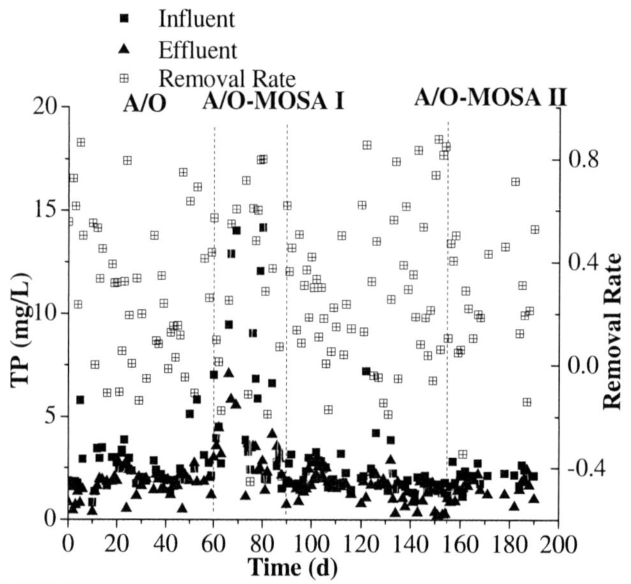


**FIGURE6**. TP removal rate over the entire operational period

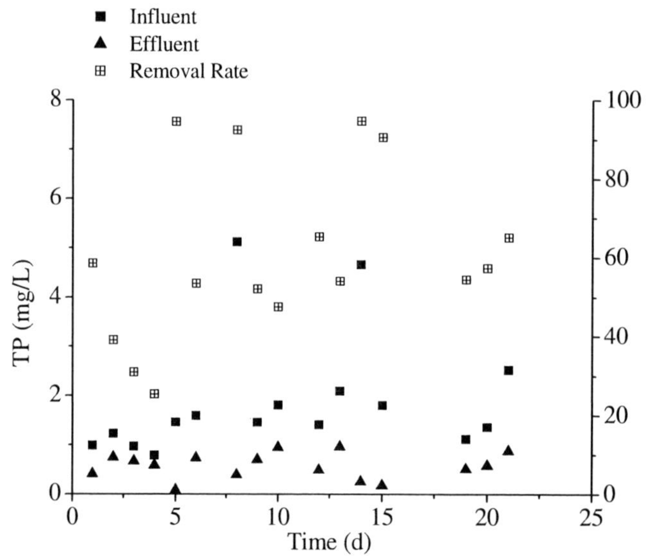


**FIGURE7**. TP removal rate using the chemical phosphorus removal method

## CONCLUSIONS

This research is based on a practical system and established a 10-m³/d A/O-MOSA pilot-scale model using actual sewage and involves a multi-stage comparison study. The experimental conclusions were as follows:

(1) In the experimental process, no significant changes for each indicator in the effluent were found, except for TP. The insertion of the MOSA anaerobic reactor did not cause an obviously negative impact on the original technology.

(2) Chemical phosphorus removal through the addition of $Al_2(SO_4)_3$ solved the roblem of the MOSA process: a poor removal efficiency for TP, which could reduce the level of TP in the effluent to

below 1.0mg/L. Compared with the A/O process, the A/O-MOSA process provided a favorable sludge reduction effect, with a sludge reduction rate of 35%. Meanwhile, adding a chemical phosphorus removal step did not negatively influence the sludge reduction, with the sludge reduction rate remaining high at 33%.

**ACKNOWLEDGEMENTS**

This work is supported by the Science & Research Program of Guangdong (Contract No. 2012A032300005), the Science & Research Development Program of Foshan (2012AA100091)

**REFERENCES**

Chen, G. H., An, K. J., Saby, S., et al., 2003. Possible cause of excess sludge reduction in an oxic-settling-anaerobic activated sludge process (OSA process). *Water Res.* 37(16), 3855-3866

Ma, J., Peng, Y. Z., Wang, L., Wang, et al., 2008. Effect of temperature on denitrificaiton and profiles of pH during the process. *China Environ. Sci.* 28 (11), 1004-1008. (in Chinese).

Novak, J. T., Chon, D. H., Curtis, B. A., et al., 2007. Biological solids reduction using the Cannibal process. *Water Environ. Res.* 79(12), 2380-2386.

State Environmental Protection Administration of China. 2002. Standard Methods for the Examination of Water and Waste Water (Fourth Version), *Environmental Science publication* in China, Beijing. (in Chinese).

Sun, L. P., Chen, J. F., Guo, W. Z., et al., 2015. Study of the sludge reduction in an oxic-settling-anaerobic activated sludge process based on UNITANK. *Water Sci. Technol.* 71 (1), 111-116.

Tan, J. X. *A pilot-scale study of sludge reduction in a modified Oxic-Settling-Anaerobic (MOSA) process based on A/O*. 2014. Master Dissertation, Sun Yat-sen University, Guangzhou, China. (in Chinese).

Wang, L. K., Zeng, G. M., Yang, Z. H., et al., 2014. Operation of partial nitrification to nitrite of landfill leachate and its performance with respect to different oxygen conditions. *Biochem. Eng. J.* 87,62-68.

Wentzel, M. C., Lotter, L. H., Ekama, G. A., et al., 1991. Evaluation of biochemical models for biological excess phosphorus removal. *Water Sci. Technol.* 23(4-6), 567-576.

Wu, K., Li, S. Y., Jiang, F., Wang, G. J., et al., 2013. A new oxic-settling-anaerobic (NOSA®) activated sludge process for minimizing excess sludge in secondary biological treatment plants: a pilot-scale evaluation of the absorption–biodegradation process. *Water Sci. Technol.* 68(3), 530-536.

Zhang, B., Ji, F. Y., Guan C., et al., 2006. Research on municipal wastewater sludge reduction in OSA process and its influencing factors. *Journal of Chongqing Jianzhu University.* (01), 92-95. (in Chinese).

# DISSOLVED METHANE RECOVERY FROM ANAEROBIC ATTACHED GROWTH REACTOR TREATING SYNTHETIC DOMESTIC WASTEWATER

***Brian Crone*** and George Sorial (University of Cincinnati, Cincinnati, OH, USA)
Jay Garland (United States Environmental Protection Agency, Cincinnati, OH, USA)

Anaerobic treatment processes have the potential to be energy neutral or energy producing when treating Domestic Wastewater. To date the largest barrier to this accomplishment has been the loss of between 10-100% of the potential energy in the form of dissolved methane in the effluent. The magnitude of potential energy loss is dependent on process temperature and configuration. For this study a novel reactor design that combines anaerobic attached growth processes and non-porous hollow fiber degassing membranes for treatment synthetic domestic waste water and simultaneous recovery of dissolved methane will be evaluated. Biomass will be cultured directly on membrane fibers which have been functionalized to improve bacterial and archaeal adhesion and which will be evacuated using a vacuum pump. Direct culturing on the degassing membrane will allow for removal of biogas constituents into the lumen of the hollow fibers as constituents are produced and before they can diffuse into the bulk solution. Removal of dissolved carbon dioxide will decrease carbonic acid concentrations and the need for additional alkalinity. This will improve process stability and reduce operating costs. Biogas recovery will also provide thermodynamically favorable conditions for Chemical Oxygen Demand (COD) removal by maintaining low hydrogen and methane partial pressures. It is postulated that the combination of the anaerobic biological treatment and physical degassing into a single reactor will reduce capital and operating costs while achieving efficient treatment efficiencies and high potential energy recovery.

# A NOVEL AND COST-EFFECTIVE SEWAGE TREATMENT PROCESS WITH ENERGY RECOVERY AND AUTOTROPHIC NITROGEN REMOVAL

**Dawen Gao**
(Northeast Forestry University, Harbin 150040, China)

A system combined an upflow anaerobic fixed bed (UAFB) and an expanded granular sludge bed (EGSB) was designed and verified as a success for treating real sewage with simultaneous energy recovery and autotrophic nitrogen removal. The effects of the temperature and HRT on the combined system were investigated. The impact of temperature (stepwise decreased from 30°C to 20°C and 10°C) was a primary focus, aiming to reveal the response of the anaerobic digestion (AD) and anammox efficiency to the temperature variation. As the temperature decreases, the sCOD removal rate was 90.6%, 90.0% and 84.7% respectively; TN removal was 69.4%, 48.8 %, 38.4% respectively; $NH_4^+$-N removal was 91.3%, 74.9%, 65.1% respectively. Methanogenic activity of UAFB was significantly influenced by low temperatures, while the unavoidable growth of heterotrophic organisms in EGSB also contributed to the sCOD removal by utilizing a certain amount of VFAs, even at 10°C. Lower working temperature (10/20°C) limited the growth and activity of AOB and anammox bacteria, but improved the NOB activity. Shorting HRT reduced the COD removal, but the effluent COD was still less than 50 mg/L. When decreasing HRT from 4 h to 2 h, the ammonia removal changed from 88.7% to 66.2%, but the reactor recovered and run stably with increasing HRT to 4 h. Compared with $A^2/O$ wastewater treatment process, the integrated sewage treatment process could decrease the construction investment cost by 52.1%, cut down the floor space by more than 90% and save the running expense by 14.9% annually. At the same time, the economic value created by using methane into electric power generation and sewage recycling was about 20.5% running cost.

# ENRICHMENT OF DENITRIFYING METHANOTROPHIC BACTERIA OF THE NC10 PHYLUM FROM ACTIVATED SLUDGE

***Shubham Singh*** and Jih-Gaw Lin
(National Chiao Tung University, Taiwan)

Anaerobic treatment processes in wastewater treatment industries discharge methane in gaseous as well as in aqueous form. Methane is considered a powerful greenhouse gas that has a global warming potential 25 times greater than carbon dioxide over a 100-year period. While methane in the gaseous phase can be easily collected and purified, dissolved methane is difficult to recover. Therefore, a post-treatment process is needed in anaerobic systems to oxidize dissolved methane, thereby reducing greenhouse gas emissions and making anaerobic wastewater treatment a more eco-friendly technology. Nitrite dependent denitrifying anaerobic methane oxidation (N-Damo) is a novel process for simultaneous carbon and nitrogen removal. This process is assigned to a methanotroph that has been identified as a member of the phylum NC 10 and named "*Candidatus* Methylomirabilis oxyfera (*M. oxyfera*)". This bacterium utilizes methane as electron donor as well as carbon source and nitrite as electron acceptor to oxidize methane into carbon dioxide. In this study, we are enriching the denitrifying methanotrophic bacteria of the NC10 Phylum (*M. oxyfera*) by using activated sludge from Xin-Feng wastewater treatment plant, Taiwan as inoculum. A 3 L bioreactor (working volume 2.5 L) was setup and has been in operation in sequencing batch mode for better biomass retention with a hydraulic retention time (HRT) of 10 days. For the first 34 days, the reactor was operated at ambient temperature (19-26°C) and after day 35 temperature was controlled at 30°C. The pH is also controlled between 7-7.5 manually by adding 1 M HCl. Temperature, pH, alkalinity, oxidation-reduction potential (ORP) and dissolved oxygen (DO) are continuously monitored. Nitrite loading rate (NiLR), nitrate loading rate (NaLR) and total nitrogen loading rate are 2.1 mg $NO_2^-$ -N $L^{-1}$ $d^{-1}$, 2 mg $NO_3^-$ -N $L^{-1}$ $d^{-1}$, and 4.1 mg N $L^{-1}$ $d^{-1}$ respectively. From day 7 onwards, both nitrite and nitrate are completely removed in the effluent. Since day 34, reactor temperature was changed to 30°C and agitation was reduced to 50 RPM from 100 RPM. There was a sudden drop in ORP after adjusting temperature and agitation speed, and nitrate removal efficiency was also decreased from 100% to 75%. It took another 5 days to reach nitrate removal efficiency to its original value. After day 39, total nitrogen removal was 4.1 mg N $L^{-1}$ $d^{-1}$. From day 65, improved medium was provided, which also caused some turbulence in removal efficiency but recovered to stable state by day 76. So far by our knowledge, this is the fastest cultivated and acclimated n-damo of the NC10 Phylum. In the future we plan to increase the total nitrogen rate exponentially and will monitor the subsequent removal efficiency.

# SURFACE ENGINEERED GREEN POLYMER FOR ENHANCED WATER DECONTAMINATION

Gurpreet Kaur Khaira and ***Moushumi Ghosh****
(Department of Biotechnology, Thapar University, Patiala-147004, Punjab, India)
*Email: mghosh@thapar.edu

**ABSTRACT:** We report the microbicidal activity of an exopolymer produced by *Klebsiella terrigena* and attributes for disinfection of water. The exopolymer possessed an intrinsic ability to rapidly bind waterborne pathogens due to surface structure through quarternization. The latter could significantly ($p<0.05$) bind and inactivate key water borne pathogens-*Salmonella typhimurium, Shigella flexneri* 2a, *Aeromonas hydrophila, Yersinia enterocolitica, Listeria monocytogenes* and *Escherichia coli* O157:H7 ATCC to an order of 4 log at ambient temperature over a period of 30 minutes. Physical and chemical analysis revealed that the quarternized biopolymer possessed total sugar and total protein content of 66.8% and 2.45% (w/w), uronic acid (2.83%) and pyruvic acid (7.4%) attributable to acidic polysaccharide. The 'surface engineered' or quarternized biopolymer or green polymer exhibited high polydispersity index, stability over wide pH (5-9) and temperature range (25-45$^0$C) as well as low dosage requirement and safety demonstrable from animal studies. These results suggest a potential of this biomaterial for affording water hygiene through water disinfection and a future scope for development of the biopolymer based sensors for determining the extent of decontamination achieved.

***Keywords:*** bioflocculants, cell membrane, water borne pathogens, *K. terrigena*, quaternization, water borne pathogens

## 1. INTRODUCTION

The quality of water has been strongly recognized as an indicator of socio-economic development. This is because the quality of water, whether used for drinking or domestic purposes, food production or recreational purposes has an important impact on health. Water of poor quality can cause disease outbreaks and can contribute to background rates of disease manifesting themselves on different time scales. Waterborne pathogens present a formidable threat to water; these pathogens include various types of bacteria, protozoan parasites, viruses and other microorganisms (Cabral 2010). Surface and groundwater, both are contaminated by bacteria and viruses, whereas parasitic protozoa appear predominantly in surface water. Their entry into drinking water occurs upon contamination of the water source by sewage and animal waste, or when wells are not properly sealed or constructed.

Waterborne pathogenic bacteria are extremely adept in acquiring resistance to conventionally used synthetic disinfectants for rendering water potable. Disinfectants derived from natural molecules are currently envisioned as promising futuristic options for water treatment since they do not induce resistance and are completely non-toxic to human. Surface modification of non-toxic, biodegradable, and environment friendly microbial extracellular polymers containing amino groups can lead to an effective and safe substitute to chemical biocides as the cationic biopolymers containing quaternary ammonium groups have gained wide attention in the recent years (Kurane et al., 1986; Siedenbiedel and Tiller, 2012, Tan et al., 2013). The role and subsequent importance of quaternary ammonium compounds (QACs) as membrane-active agents with a target site, predominantly at the cytoplasmic (inner) membrane in bacteria or the plasma membrane (Tan et al., 2013) is well established.

Chitosan have been modified through various techniques, the intensive methylation of chitosan generates the N,N,N-trimethyl derivative, characterized by possessing positive charges in chains as a consequence of the quaternization of the amino groups in the C-2 position of the chitosan structure (Sieval et al., 1998). Quaternized chitosan is a water-soluble polyelectrolyte with enhanced antibacterial activity used in various biomedical applications. The activity in these polymers is described as the result of ionic interaction between chitosan positive charges and the negatively charged cell surface of bacteria. Polysaccharide microbial polymers with similar structures to chitosan can be chemically modified in order to enhance or impart specific properties. The biopolymeric derivatives can be synthesized by either covalent addition of a substituent containing a quaternary ammonium group, or by quaternization of the amino groups of the parent polymer (Britto et al., 2011). It is the latter method which has received most attention with alkylation agents such as alkyl halide which is usually employed.

The biopolymer produced by *Klebsiella terrigena* has been reported in previous studies to be effective in removal of *Cryptosporidium* oocysts and *Salmonella* (Ghosh et al., 2009a,b) by flocculation. In an attempt to endow antibacterial functions we chemically altered the surface properties of a biopolymeric flocculant produced by *K. terrigena* and chitosan for the sake of comparison. Alteration were brought about in the native biopolymers through quaternization, and the trimethyl biopolymeric derivatives (N,N,N trimethyl biopolymer and N,N,N trimethyl chit*osan,* abbreviated as TMB andTMC, respectively) were investigated for their antibacterial activity against the major water borne pathogens. Fourier transform infrared (FT-IR) absorbance spectroscopy (in the mid-IR range, usually defined as 4000-400cm$^{-1}$ or 2500-25000nm) has sufficient resolution to distinguish intact microbial cells at the strain level. The resulting spectra for microbial cells represent the "total" biochemical composition of those cells. FTIR spectroscopy measures vibrations of functional groups and highly polar bonds. Therefore spectral features of many important biochemical constituents create a "fingerprint" for a bacterial strain. The importance of FT-IR spectroscopy as a physicochemical method to determine the global chemical features of cells suggests its adequacy to study the molecular changes after treatment with antibacterial agents (Ashtiani et al., 2009). The present study reports the development of biopolymeric flocculant with antibacterial activity with a view to enhance the scope of applicability of the bioflocculant as a natural antibacterial agent for water treatment.

## MATERIALS AND METHODS

All chemicals were purchased from Sigma chemicals company (Sigma, Mo, USA). All other reagents were of the highest grade available commercially.

### Biopolymer Producing Strain and Indicator Strains

***Biopolymer producing strain.*** An industrial wastewater isolate, *Klebsiella terrigena* (Accession number EU082029), was used in this study (Ghosh et al., 2009a,b). The amino rich biopolymer producing strain was preserved in glycerol stock solutions at -80°C.The strain was revived in LB agar plate and incubated overnight at 37°C and a single colony was picked and transferred in FIB broth and was grown to mid log phase at 37°C.

***Indicator strains.*** For testing the antibacterial activity of antibacterial polymer(s), selected bacterial strains, *Salmonella typhimurium* ATCC 25315, *Shigella flexneri* Type 2a, *Aeromonas hydrophila* ATCC 35654, *Yersinia enterocolitica* ATCC 9610, *Listeria monocytogenes* ATCC19111 and *Escherichia Coli* O157:H7 ATCC 32150 were maintained in Tryptone Soy Broth (TSB) with 2% glycerol and stored at 4°C. The cultures were revived in Brain Heart Infusion (BHI) agar plate and incubated overnight at 37°C.Prior to the experiments, single colonies were picked and transferred in BHI broth which was incubated for 18-24h at 37°C.

**Biopolymer Production.** Cells were removed from the culture medium by centrifugation and the biopolymer was separated by the addition of two volumes of ethanol (99.5%) to 500mL of concentrated supernatant, and allowed to precipitate at 40°C for 24hr. The precipitated polymer was collected by filtration (Whatman GF Filter) and dialyzed extensively using dialysis tubing (Cellulose, MWCO 12000 Da), against deionized water. Crude biopolymer was purified by addition of a 10% solution of cetylpyridinium chloride. The precipitated polymer complex was collected by centrifugation at 10,000rpm for 20min at 4°C and redissolved in 10% NaCl solution. Three volumes of ethanol were added to recover the purified biopolymer, which was further dialyzed and lyophilized to powder and stored until further use (Ghosh et al., 2009a,b).

**Synthesis and Characterization of N, N, N Trimethyl Biopolymeric Derivative (TMB).** Purified biopolymer (150mg) was dissolved in dimethylsulfate (2.4mL) and deionized water (0.6mL). The solution was then filtered to eliminate the impurities. Sodium hydroxide (0.18mg) and sodium chloride (0.132mg) were added to the resulting suspension, followed by stirring the solution at ambient temperature for six hours. The product was precipitated using acetone, filtered and vaccum dried. White precipitates obtained were redissolved in deionized water (20mL) and subjected to dialysis using a dialysis membrane for one day. Dialyzed solution was then lyophilized to get white fluffy powder (4.5mg) (Belalia et al., 2008).

NMR analysis is as follows that $^{1}H$ NMR spectra of biopolymer and N, N, N-trimethyl derivative (TMB) was recorded using Bruker Avance II (400 MHz) spectrometer. For this analysis, samples were dissolved in $D_2O$.

**Evaluation of Flocculating Activity.** TMB was evaluated for its ability to flocculate suspended solids of varied size; the native biopolymer was used as a comparison. A standard solution of suspended solids of 2000 NTU was made by suspending active carbon, silica, magnesium hydroxide, cellulose and yeast in 100mL water. A bacterial suspension of 2000 NTU was also used for this study. A suspension of desired turbidity was prepared by diluting the suspension form stock to desired turbidity with a turbidimeter (Cyber Scan TBDIR1000 Meter, Eutech, Netherlands). An assay was performed by adding 10mL $CaCl_2$ (5mM), 0.5mL biopolymer (2mg/l) and 9.5mL of distilled water to 80mL of these solutions. The pH was adjusted to 7±0.2 and the solutions were allowed to stand at room temperature for 5min. 20mL of aliquots were withdrawn from the upper phase and its turbidity was measured. Its flocculating activity was calculated by recording its optical density with a spectrophotometer at 550nm. Flocculating activity of purified biopolymer and TMB in water was measured against distilled water as control according to method of Kurane et al., (1986). Activity (%) was defined and calculated as (B–A)/ B × 100. The activity was expressed as the mean value from duplicate determinations.

**Determination of Antibacterial Activity.** The antibacterial activity of biopolymeric derivative was determined against the six selected pathogenic cultures. Prior to experiments, the respective cultures were revived in BHI broth by incubating at 37°C for 18hr. For a preliminary screening uniform sized discs prepared from whatman filter paper were soaked in sterile solution of 100µg/ml of each native biopolymer and TMB, placed on the assay plate containing 100 µl of culture on BHI broth. All plates were incubated at 37°C for 12 hr, after which the diameters of inhibition zones were observed. The experiments were performed in triplicates.

*Killing kinetics:* Antibacterial activity of TMB was assayed by microdilution method, using a sterile 96 well-microtiter plate (Raafat et al., 2008).Briefly, serial twofold dilutions of TMB solutions were prepared in the appropriate culture medium in sterile 96-well round bottom polystyrene microtiter plates (Bioscreen C, Thermolab systems, Helsinki, Finland).Final TMB concentrations used were 1-100µg/ml. Minimal inhibitory concentration (MIC) was determined by microtiter broth dilution method (Andrews 2001). The indicator bacterial pathogenic cultures were grown in the respective broth at 37°C to an optical density of 1 at 600nm and subsequently diluted in the same medium toabout $10^7$CFU/ml. The solution of biopolymer and TMB were added, and their inhibitory effect was observed. Each well of the microtiter plate then received 100µL ofthe inoculated medium, and the plates were incubated at 37°C for up to 24hr. MIC was defined as the lowest concentration of compound required to completely inhibit microbial growth after incubation. The MIC determinations were carried out in triplicates, with two independent experiments performed. The surviving $\log_{10}$CFU/ml was recorded against time for each of the different quaternized biopolymer concentrations.

**Leakage Assays**

***Preparation of washed cell suspensions.*** Pregrown bacterial cells were separated by centrifugation at 11,000×g for 10min from the culture medium (BHI). Cell pellets were washed in 10mM PBS (pH 7.0) and diluted with the same buffer to approximately $10^7$CFU/ml at an absorbance (OD 600nm) of 1.

***Leakage of glucose, lactate dehydrogenase (LDH) and protein from treated bacterial cells.*** To examine the effect of quaternized derivative on cell leakage and the viability of the indicator strain, inoculum of the test organism (1ml) was inoculated into sterile deionized water (10ml) with or without quaternized derivative in a culture tube. The mixture, containing water-soluble biopolymeric derivative (TMB) and indicator cultures were incubated at 37°C with shaking (120rpm) for 12h. At regular interval of 1h, aliquots were withdrawn for determination of protein and glucose contents and lactate dehydrogenase (LDH) activity. The cell suspension was centrifuged at 8,000×g for 15min, and the supernatant was measured for LDH activity, protein and glucose contents. The glucose content was analyzed by a glucose assay kit (DiaSys Diagnostic Systems GmbH, Holzheim, Germany). A sample or glucose standard (0.0–3.0mg/dl, 200μl) was added to reagent (1ml) containing glucose dehydrogenase. After incubation at 25°C for 15min, the absorbance at 334 nm was recorded. The LDH activity was analyzed by an LDH assay kit (Clontech, USA). A 200μl sample was added to 1ml reagent containing NADH and incubated at 25°C. LDH activity was then determined by measuring the rate of decrease of the NADH concentration which was monitored by recording the change of absorbance at 334nm. The protein concentration was measured by absorbance at 280nm.

**Extraction and FT-IR Analysis of Lipopolysaccharides (LPS) of Target Pathogen.** Lipopolysaccharide from the pathogen treated with quaternized derivatives and control cells was extracted by hot phenol-water method as described by Westphal and Jann, (1965) with some modifications. In brief, bacterial suspensions ($10^8$ CFU/ml) were centrifuged at 10,000 × g for 5min. The pellets were washed twice in PBS (pH=7.2) (0.15M) containing 0.15mM $CaCl_2$ and 0.5mM $MgCl_2$. Pellets were then re-suspended in 10ml PBS and sonicated for 10min on ice. In order to eliminate contaminating protein and nucleic acids, treatment with proteinase K, DNase and RNase was performed prior to extraction step. For this purpose, proteinase K (100μg/ml) (Roche, Mannheim, Germany) was added to the cell mixture and the tubes were kept at 65°C for an additional hour. Mixture was subsequently treated with RNase (40μg/ml) (Roche, Mannheim, Germany) and DNase (20μg/ml) (Roche, Mannheim, Germany) in the presence of 1μl/ml 20% MgSO4 and 4μl/ml chloroform and incubation was continued at 37°C overnight. At the next step, an equal volume of hot (65–70°C) 90% phenol was added to the mixtures followed by vigorous shaking at 65-70°C for 15min. Suspensions were then cooled on ice, transferred to 1.5ml polypropylene tubes and centrifuged at 8500 × g for 15min. Supernatants were transferred to 15ml conical centrifuge tubes and phenol phases were re-extracted by 300μl distilled water. Sodium acetate at 0.5M final concentration and 10 volumes of 95% ethanol were added to the extracts and samples were stored at –20°C overnight in order to precipitate LPS. Tubes were then centrifuged at 2000×g, 4°C for 10min and the pellets were re-suspended in 1ml distilled water. Extensive dialysis against double distilled water at 4°C was carried out at the next step until the residual phenol in the aqueous phases was totally eliminated. Final purified LPS product was lyophilized and stored at 4°C.

LPS were analyzed by Nicolet magma, FTIR spectrometer equipped with a Maillard-I-Alanine doped deuterated triglycine sulphate (DLa TGS) detector using KBr pellets. To prevent absorption from ambient, the system was purged with dry nitrogen. For each spectrum, 64 scans were co-added at spectral resolution of $4cm^{-1}$. The spectra covered the wave number ranging from 4000-400$cm^{-1}$ frequencies for all sharp bands were determined accurately from the original baseline corrected spectra of corresponding group using a software OPUS (version 6.5). The different FT-IR spectra graphs were drawn by origin software (version 6.0).

**Electron Microscopy.** Indicator pathogenic cells treated with TMB as described above were used to visualize structural damage, following 20, 40 and 60 minutes of incubation; the cells were washed twice with phosphate buffer saline (PBS) and fixed in a fixative (2% glutaraldehyde, 2% paraformaldehyde, and 0.5% $CaCl_2$ in 0.1M cacodylate buffer). Samples were post fixed in 1.33% osmium tetroxide, dehydrated in graded ethanol and infiltrated with a 1:1 mixture ofpropylene oxide. For SEM, cells were sputtered with a 300Å thick layer of gold using an ion coater and examined in a scanning electron microscope (JSM-7100F, JEOL, Japan).

## RESULTS AND DISCUSSION

**Synthesis and Characterization of Quaternized Biopolymeric Derivative(s).** Amino group present in the polysaccharide backbone of the bioflocculant were transformed into quaternary ammonium compounds synthetically so that the polymer that could effectively bind, would also be able to inactivate the water borne pathogen simultaneously. Methylation is an effective approach in quaternization of polymers. Synthesis of trimethyl biopolymeric derivatives leads to methylation of the amino groups in the C-2 position of biopolymer to form quaternary amino groups with fixed positive charges on the repeating units of the quaternized polymer chain. Also the advantage ofintroducing methyl groups at different levels on the polymeric backbone can be controlled externally with ease. Quaternization process using either dimethyl sulphate have been well documented for obtaining quaternary ammonium compounds of chitosan (de Britto et al., 2007; Belalia et al., 2008; Tan et al., 2013). Thus, the suitability and applicability of the same was attempted in case of the selected biopolymeric flocculant in the present study.

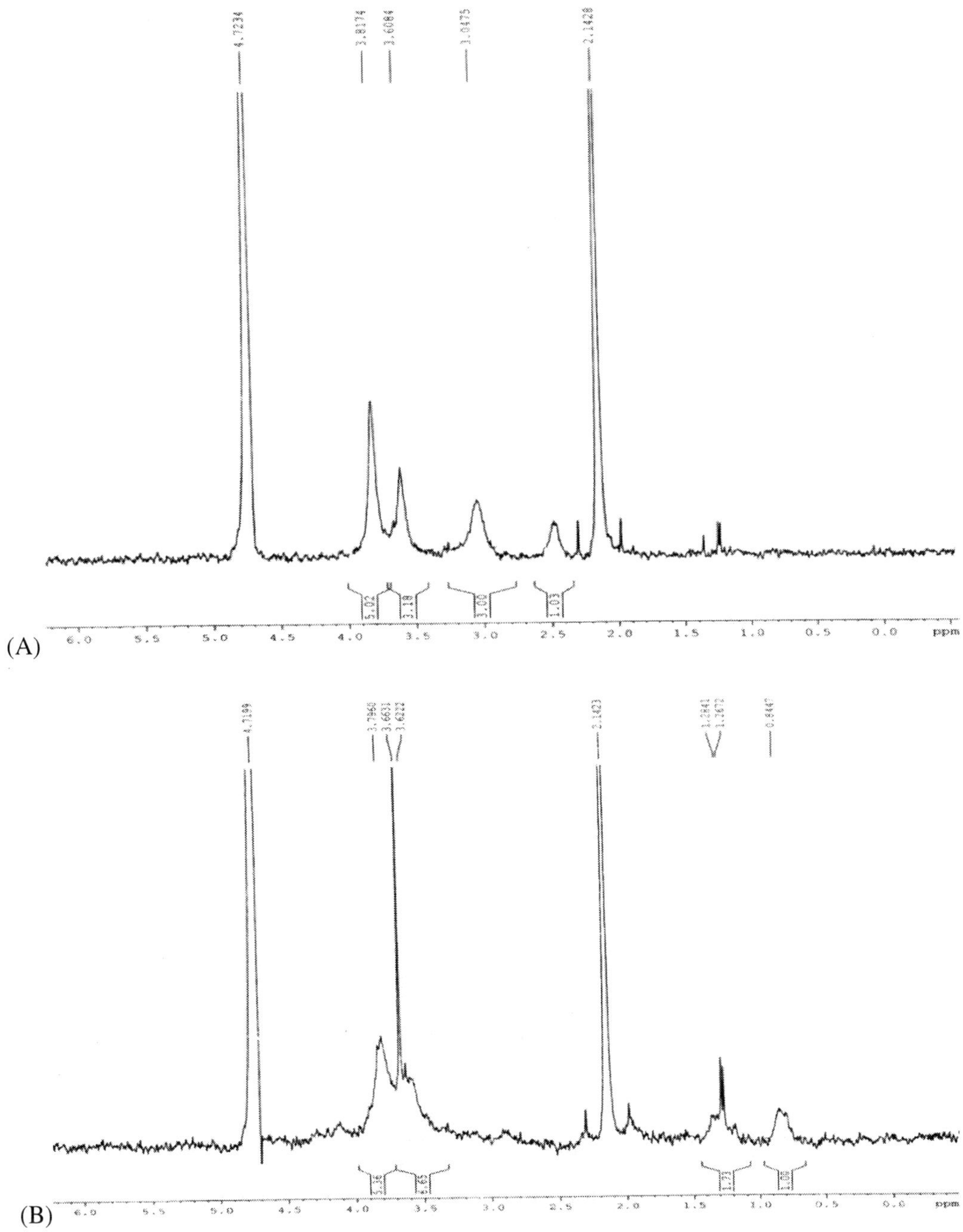


**Fig. 1.** $^1$H NMR spectrum of biopolymer (A) and TMB (B) dissolved in $D_2O$.

Fig. 1 depicts the $^1$H-NMR spectra of the biopolymer and quaternized biopolymer (TMB), respectively. The spectra revealed an intense signal at 3.66 ppm corresponding to the trimethylammonium group. Sieval et al. (1998) indicated the peak at 3.6 ppm is assigned to the trimethyl amino group, the peak at 3.1 ppm is assigned to dimethyl amino groups and the peaks between 4.7 and 5.7 ppm are assigned to $^1$H protons. As evident from the enlargement of the peak at 3.6 ppm, a longer reaction time resulted in an increase in the degree of quaternization of the polymer.

**Flocculating Efficiency of the Quaternized Biopolymer.** Results revealed that TMB aggregated a wide range of colloidal particles, over concentrations of 2-10ppm. As illustrated in Fig. 2, the most effective flocculation was achieved in suspensions comprising very low and low sized particles at even low concentration of the biopolymer. Whereas the efficacy of flocculation of suspension of medium sized particles ranged from 30-40% that too at relatively higher concentration of 8-10ppm of the biopolymer. The suspensions of large sized particles showed 50-65% flocculant activity at the same concentration range that flocculated in the range of 70-90% in case of low and very low sized particle suspension. This might be due to non-uniform distribution of pore size on the biopolymer surface; also the number of sites available for binding low and very low sized particles might be more than that of binding medium and large sized particles. However, large variations in the degree of flocculation were observed, where the kaolin particles (size~2µ) showed highest flocculating activity (90%). The quaternized biopolymer was found to have good flocculating activity against bacteria (~0.5-0.8µm) similar to its native counterpart.

This phenomenon might be due to the increase of polymer and particle complexes that start to form and the maximum removal of turbidity is at the dosage value when almost all polymer sites are occupied by the particles. Beyond that optimum dosage value, the turbidity removal decreases probably due to lack of polymer sites for interaction with the colloidal particles. The flocculant dosage leading to the maximum flocculation efficiency value is considered as optimal dosage value. The mechanism of the flocculation with TMB may be explained by a combination of charge neutralization and polymer bridging.

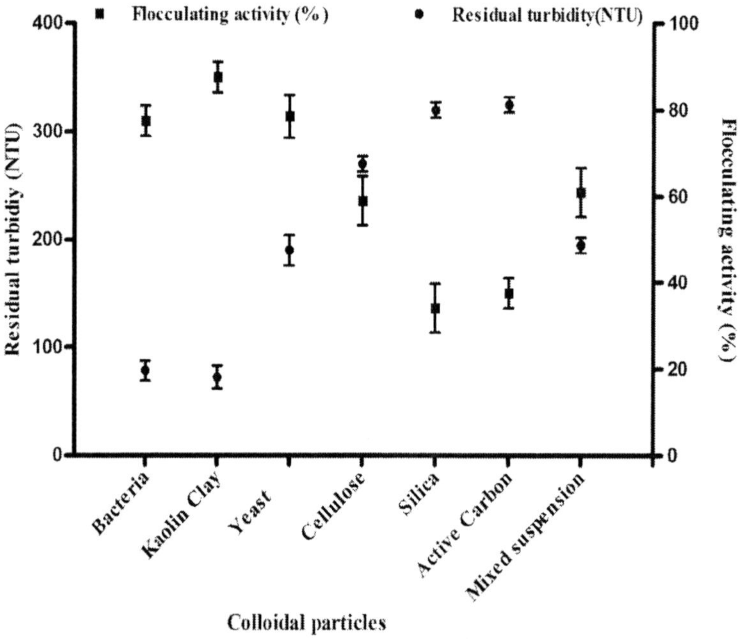


**Fig. 2.** Efficiency of removal of varied size colloidal particles in solution applied with different doses of TMB and native biopolymer ranging from 2ppm-10ppm. The corresponding size of the particles are following [~0.5-0.8µm (bacterial cell suspension), ~4-7µ (Kaolin, yeast cell suspension, cellulose), ~15-20µ (silica), ~100µ (Active carbon)]

**Biocidal Activity of Quaternized Biopolymeric Flocculant Against Pathogens.** The antibacterial activity of biopolymer and TMB was assessed thereafter against *the* respective pathogens by observing the zone of inhibitions and determining MICs. Results obtained by the agar disc diffusion assay of the quaternized biopolymer indicated a strong inhibition by TMB, the native biopolymer failed to inhibit (data not shown). The orders of zone of inhibition with TMB were found to be *Salmonella> Shigella> Aeromonas> Yersinia> E.coli 0157:H7> Listeria.* The gram- positive *Listeria* was inhibited the least by both the quaternized derivatives.TMB significantly ($p<0.05$) inhibited the growth of the pathogens in 60 to 90 minutes at concentrations ranging from 50 to 100μg/ml.The effect of TMB on the pathogens is shown in Table 1(a) where a reduction of 4 log CFU/ml was evident after 60-70min of treatment and the respective MIC values are depicted in Table 1(b). The higher antibacterial activity of quaternized biopolymeric derivative could be due to the permanent positive charges on the quaternized biopolymeric flocculant chain, as a consequence of the quaternization of the amino groups. The activity resulted from the interaction between the positively charged amino groups of the biopolymer and negatively charged cell surface of gram negative bacteria and also due to the flocculant activity of the biopolymer.

**Table. 1(a)** Killing kinetics for BF (Bioflocculant), N,N,N- trimethyl biopolymer, (TMB), chitosan and N,N,N-trimethyl chitosan, (TMC) against indicator waterborne pathogens. Each value is expressed as mean ± SD (n = 3)

| | BF | | Chitosan | | TMB | | | TMC | | |
|---|---|---|---|---|---|---|---|---|---|---|
| | 25 | 100 | 25 | 100 | 25 | 75 | 100 | 25 | 75 | 100 |
| ***Salmonella typhimurium* ATCC 23564** | NZ | 14±0.6 | NZ | 10±0.4 | 14±0.5 | 22±0.4 | 25±0.5 | NZ | 12±0.4 | 15±0.7 |
| ***Shigella flexneri 2a*** | NZ | 12±0.2 | NZ | 8±0.6 | 15±0.3 | 24±0.6 | 25±0.9 | NZ | 10±0.3 | 12±0.6 |
| ***Aeromonas hydrophila* ATCC 35654** | NZ | 13±0.7 | NZ | 11±0.3 | 13±0.1 | 20±0.3 | 23±0.5 | 9±0.3 | 14±0.7 | 18±0.4 |
| ***Yersinia enterocolitica* ATCC 9610** | NZ | 13±0.5 | NZ | 12±0.2 | 10±0.6 | 19±0.6 | 21±0.3 | 10±0.4 | 17±0.9 | 19±0.5 |
| ***Escherichia coli O157:H7* ATCC 32150** | NZ | 14±0.2 | NZ | 11±0.8 | 10±0.7 | 16±0.8 | 16±0.5 | 9±0.1 | 16±0.4 | 18±0.6 |
| ***Listeria monocytogenes* ATCC 19111** | NZ | 13±0.1 | NZ | 9±0.7 | NZ | 14±0.5 | 16±0.2 | 8±0.2 | 11±0.4 | 11±0.5 |

ND: Not determined; NZ: No zone observed; Diameters includes the diameter of disc (6 mm)

Table 1(b): Minimum inhibitory concentration (MIC, μg/mL) of Chitosan and quaternized bioflocculant derivatives- TMC and TMB against six bacterial pathogens (in BHI media)

| **Bacteria** | **TMC** | **TMB** |
|---|---|---|
| ***S. typhimurium*** | 74 | 60 |
| ***A. hydrophilla*** | 71 | 62 |
| ***Y. enterocolitica*** | 67 | 65 |
| ***E. coli O157:H7*** | 70 | 68 |
| ***L. monocytogenes*** | 80 | 72 |
| ***S. flexneri 2a*** | 78 | 62 |

**Biochemical Responses of Water Borne Pathogens.** Insights on mode of action of antibacterial agents are crucial both for structure refinement/design modification of antibacterial agents and fate of target pathogens. In order to elucidate the possible mechanism of the inhibitory action of TMB against target pathogen was further assessed for its membrane potential, permeability intactness through measurement of glucose, protein leakage and lactate dehydrogenase activity. Cellular damage and death was also studied by electron microscopy and FT-IR spectroscopy.

Table 2 summarizes the effect of TMB and TMC on leakage of protein from target pathogen in the extracellular media. Treatment of pathogens with a concentration of 76μg/ml of TMB led to rapid leakage of protein from the bacterial cells within a period of 3h. The amount of protein detected in post treatment increased gradually over time for all organisms. In order to elucidate the possible mechanism of the bactericidal action of quaternized biopolymers, leakage of glucose and Lactate dehydrogenase (LDH) enzyme from treated pathogens were studied. In the present study, treatment of pathogen with a concentration of 76μg/ml of TMB led to rapid leakage of glucose and LDH within a period of 3h (Table 2), though a minor increase in glucose and LDH levels in the extracellular media was observed after 6h, no further increase in the levels of glucose, protein or LDH occurred thereafter.

**Table 2.** Effect of biopolymeric derivative (TMB) (75 μgml$^{-1}$) on leakage of glucose concentration, lactate dehydrogenase and protein level in the extracellular media of bacterial pathogen cultured at 37 °C for up to 3 h. Control is bacterial cells in distilled water without TMB treatment. Each value is expressed as mean ± SD (n = 3).

| **Bacterial strain** | **LDH activity (U/L)** | | **Glucose (mg/dl)** | | **Protein absorbance (280 nm)** | |
|---|---|---|---|---|---|---|
| | **0 h** | **3 h** | **0 h** | **3 h** | **0 h** | **3 h** |
| | 0 | 17.44 ± 0.08 | 0 | 0.42 ± 0.06 | 0 | 0.616 ± 0.04 |
| **Control** | 0 | 0.02±0.001 | 0 | 0.03±0.001 | 0 | 0.018±0.004 |

Several authors have proposed that the antimicrobial action of chitosan could be explained by a more direct disturbance of membrane functions. The reactive amino groups in chitosan could interact with a multitude of anionic groups on the surface of the cell to alter its permeability. This causes the leakage of intracellular components such as glucose and LDH, resulting in a destabilized cell membrane beyond repair and subsequent cell death.

**Analysis of Lipopolysaccharides (LPS) Isolated from Treated Pathogens.** The LPS of inactivated pathogenic strains following biopolymer, chitosan, TMB and TMC treatment was examined by mid-infrared spectroscopy (4000-400cm$^{-1}$) (Fig. 4). Treatment with biopolymer, chitosan and TMC resulted in no changes on the LPS region of the gram negative pathogens, whereas the structure of amide, polysaccharide and fingerprint region was completely altered on treatment with TMB. Hence, results reveal significant effect of TMB on the structure of LPS region of target pathogens. The results obtained in this study are similar to those reported in most instances Rafaat etal(2005)suggesting membrane damage by TMB to be the principal mechanism for inactivation.

In order to correlate spectroscopic data with ultra-structural changes caused by biopolymer treatment, scanning electron micrograph data was collected for treated bacterial cells at various time intervals. The intracellular changes were observed in TMB treated pathogens when compared to the non-treated cells including cytoplasmic membrane disruption and blebbing. Besides, remarkable modifications of cell membrane and disruption of cell membranes occurred after only a short period of exposure (Fig. 5).

Overall, the observations of our study indicated the efficacy of TMB in inactivating water borne pathogens by directly damaging their membranes; whether flocculation of the target pathogens is followed by inactivation, remains to be examined. Based on the results of this study, a potential application of the TMB for enhanced water safety is implicated.

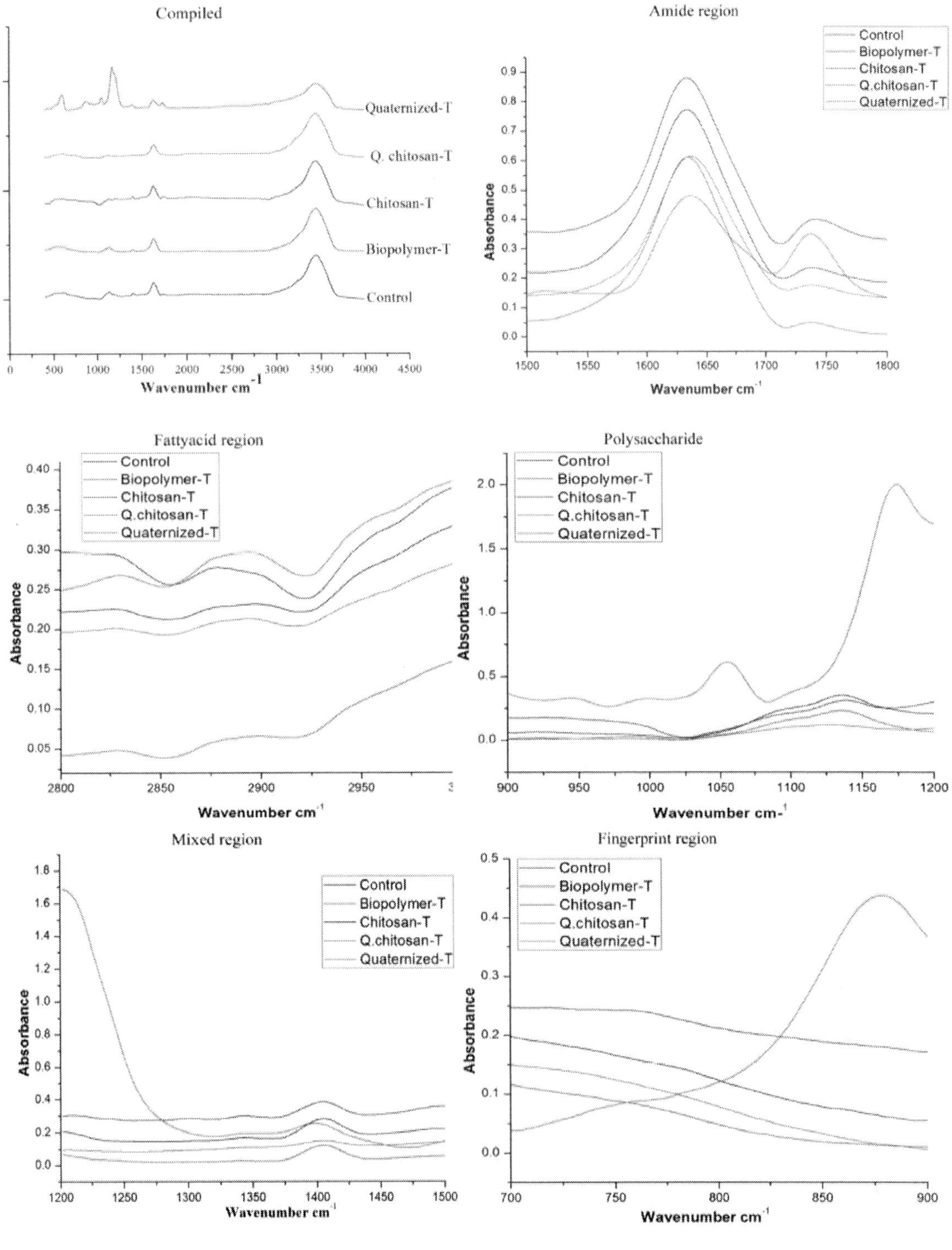


**Fig. 4.** Representative FT-IR spectra of LPS isolated from the pathogens (control), biopolymer treated bacterial cells (Biopolymer-T) and quaternized biopolymer treated bacterial cells (Quaternized-T).

## CONCLUSION

This study reported the biocidal effects of a polymeric flocculant produced by *K. terrigena*. Quarternization was effective in rendering biocidal properties to the native biopolymer, the quaternized biopolymeric flocculant led to rapid inactivation of water borne bacterial pathogenic strains. Cell death occurred presumably by a rapid release of cytoplasmic enzymes, changes in electrical conductivity and

FTIR spectra. Gross morphological changes on cell membranes of bacteria were observed from electron micrographs of exposed cells. The newly developed biocidal biopolymeric flocculant may have significant applications in both removing and inactivating bacterial pathogens transmitted through water.

| Bacteria | 0 min | 30min | 60min |
|---|---|---|---|
| Representative pathogen | 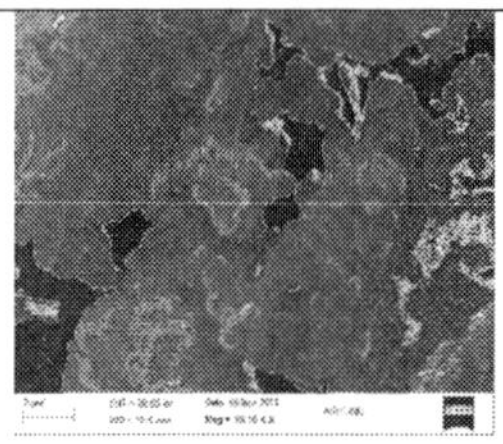 | 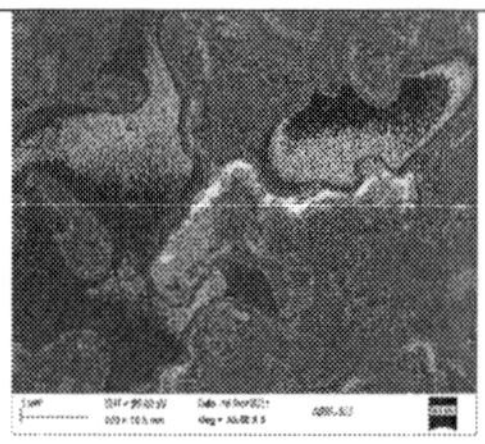 | 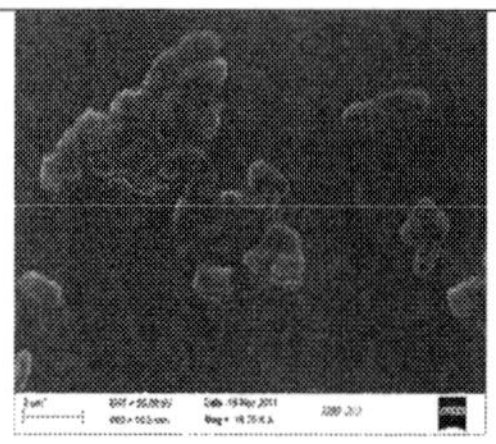 |

**Fig. 5.** Scanning electron micrographs of indicator bacteria treated with TMB in mid-logarithmic growth at $10^7$CFU/ml at various time intervals.


## ACKNOWLEDGEMENT

The authors would like to acknowledge DST UKEIRI for financial assistance in carrying out this work.


## REFERENCES


Ahmed, A.M., Shimamoto, T., 2014. Isolation and molecular characterization of *Salmonella enterica*, *Escherichia coli* O157: H7 and *Shigella* spp. from meat and dairy products in Egypt. Int. J. Food Microbiol. 168, 57-62.

Ashtiani, M.T. H., Monajemzadeh, M., Kashi, L. 2009. Trends in antimicrobial resistance of fecal *Shigella* and *Salmonella* isolates in Tehran, Iran. Indian J. Pathol. Microbiol. 52, 52-55

Belalia, R., Grelier, S., Benaissa, M., Coma, V. 2008. New bioactive biomaterials based on quaternized chitosan. J. Agric. Food Chem. 56, 1582–1588.

Britto, D., Assis, B.G. 2007. A novel method for obtaining a quaternary salt of chitosan. Carbohydr. Polym. 69, 305–310.

CDC 2009 Health, United States, 2009 with Special Feature on Medical Technology. Centers for Disease control and Prevention, United States.

de Britto, D., Celi Goy, R., Campana Filho, S.P., Assis, O.B. 2011. Quaternary salts of chitosan: history, antimicrobial features, and prospects. Int. J. Carbohydr. Chem.2011, 1-12.

Ghosh, M., Pathak, S., Ganguli, A. 2009a. Effective removal of C*cryptosporidium* by a novel bioflocculant. Water Environ. Res. 81, 160–164.

Ghosh, M., Pathak, S., Ganguli, A. 2009b. Application of a novel biopolymer for removal of *Salmonella* from poultry wastewater. Environ. Technol. 30, 337–344.

Kotloff, K.L., Winickoff, J.P., Ivanoff, B., Clemens, J.D., Swerdlow, D.L., Sansonetti, P.J., Adak, G.K., Levine, M.M., 1999. Global burden of *Shigella* infections: implications for vaccine development and implementation of control strategies. Bull. World Health Organ. 77, 651-666.

Kurane, R., Takeda, K., Suzuki, T. 1986. Screening for and characteristics of microbial flocculants. Agric. Biol. Chem, 50, 2301-2307.

Raafat, D., Bargen, K., Haas, A., Sahl, H. 2008. Insights into the mode of action of chitosan as an antibacterial compound. Appl. Environ. Microbiol. 74, 3764–3773.

Rolfo, F., Marin, G. H., Silberman, M., Pattin, J., Giugnio, S., Gatti, B., Bettiol, M., Rigoni, A., 2011. Epidemiological study of shigellosis in an urban area of Argentina. J. Infect. Dev. Ctries. 6, 324-328.

Siedenbiedel, F., Tiller, J. C. 2012. Antimicrobial polymers in solution and on surfaces: overview and functional principles. Polymers. 4, 46-71.

Sieval, A.B., Thanou, M., Kotze, A.F., Verhoef, J.C., Brussee, J., Junginger, H.E. 1998. Preparation and NMR characterization of highly substituted N-trimethyl chitosan chloride. Carbohydr. Polym. 36, 157-165.

Tan, H., Ma, R., Lin, C., Liu, Z. Tang, T. 2013. Quaternized chitosan as an antimicrobial agent: antimicrobial activity, mechanism of action and biomedical applications in orthopaedics. Int. J. Mol. Sci. 14, 1854–1869.

Westphal, O. 1965. Bacterial lipopolysaccharide-extraction with phenol water and further application of procedure. Methods carbohydr. chem. 1, 83-91.

WHO 2005 Guidelines for the control of shigellosis, including epidemics due to *Shigella dysenteriae* type 1. World Health Organization, Geneva.

Woodward, D.L., Rodgers, F.G., 2000. Surveillance of antimicrobial resistance in *Salmonella*, *Shigella* and *Vibrio cholerae* in Latin America and the Caribbean: A collaborative project. Can. J. Infect. Dis. 11, 181.

# VOLATILE FATTY ACID PRODUCTION DURING ACIDIFICATION OF OLIVE MILL WASTEWATER

Havva BAG[1], Secil ERDEM[1], ***Canan CAN YARIMTEPE***[1]*, Orhan Ince[2] and Nilgün AYMAN OZ[1]
[1]Canakkale Onsekiz Mart University, Canakkale, Türkey
[2] Istanbul Technical University, İstanbul, Türkey

**ABSTRACT**: This study aims to investigate acidification potential of oil mill wastewater (OMW) via acidification reactors. During the study, tVFA production was determined; pH, sCOD, TSS, Total Volatile Fatty Acid concentration and composition were monitored. Acidification reactors were performed for 60 days and according to datas for first 60 days; tVFA concentration ( in terms of Acetic Acid), gradually increased form 450 mg/l to 5278 mg/l. During the process, changes in Volatile Fatty Acid composition have been observed. While each VFA concentration were increased, also other VFA compounds, not detected in influent, were determined in the effluent. During acidification process, changes in sCOD concentration were monitored daily. No major changes were observed in organic matter concentration during process. Therefore, loss of organic matter was not mentioned. Although there is no change in organic matter concentration but there is increase in tVFA concentration. Therefore it is indicated that organic matter change form thus acidification process was performed successfully.

## INTRODUCTION

Marmara and Aegean region including Canakkale is main part for olive oil production in Turkiye. Olive mill wastewater (OMW), has high amount of organic matter and also toxic and inhibitor materials such as lipids and phenolic compounds. Therefore it is not easy to treat such strong wastewater with conventional treatment processes. In literature several treatment methods have been tested for OMW such as physico chemical treatment, advance treatment and biological treatment. As physico chemical treatment ;centrifuge (Turano et al., 2002), physicochemical treatment (Ginos et al., 2006), membrane filtration (Boari ve ark., 1984; Bradley 1980) have been examined. In addition as advance treatment; ozone (Amat et al., 2003), fenton oxidation (Rivas et al., 2001), combination of ozone and fentone (Filipakopoulou ve ark., 1999) have been tested. However due to high operation cost and sludge production of chemical treatment technologies, new alternative treatment process are required. In literature there are also some studies about biological treatment of OMW such as aerobic treatment (Bertin et al., 2001) and anerobic treatment (El-Gohary et al., 2009; Azbar et al., (2008). However, anaerobic treatment of OMW is limited due to high polyphenol and lipid concentrations.

Because of its complex composition, a sufficient treatment solution have not been addressed for OMW management, so far. However, it is possible to produce clean energy such as methane and hydrogen and recover resource metarials such VFAs (Volatile Fatty Acids) and phenolic compounds from olive mill wastewater. With this study, environmental pollution originating from olive mill wastewater can be reduced in terms of color and turbidity, at the same time energy production and resource recovery in a form of VFA can be provided.

In literature there is limited studies about VFA production from OMW. Scoma et al. (2013) have studied the effects of HRT on acidification from pretreated OMW (according to phenolic compounds). They have tested 1, 3, 5 and 7 day of HRT and have selected optimum HRT as 1 day. In optimum conditions, it is reported that while initial VFA concentrations are at very low levels, final VFA concentrations are measured as 10 gr/l. In addition acetic acid, butyric acid and propionic acid were found the dominant VFAs.

Recently, acidification phase receives considerable attention and relevant studies covered a wide range of operational parameters. Since acidogenic reactor plays a crucial role in the conversion of complex organic substrates to VFA (intermediate fermentation product), the initial aim of this study is to optimise the acidogenic reactor to provide feed with high amount of VFA to second stage treatment.

Therefore, the aim of this study is investigation of acidification potential of OMW via anaerobic sequencing batch reactors.

## MATERIALS AND METHODS

The raw OMW was collected from an Olive Oil Production Factory in Canakkale, Turkiye in December, 2014. OMW composition is given in Tablo 1.

**Table 1. OMW Composition**

| Parameters | Unit | OMW |
|---|---|---|
| pH | | 4,991 |
| Conductivity | µs/cm2 | 13950 |
| Turbidity | NTU | 16100 |
| Color | ptCO | 38200 |
| TS | mg/L | 48544 |
| TVS | mg/L | 41272 |
| TSS | mg/L | 20900 |
| COD | mg/L | 87716 |
| TOC | mg/l | 23883 |
| SCOD | mg/L | 40306 |
| Total Phenol | mg/L | 6560 |

Acidification process were perfomed with 1 liter anaerobic reactors consisting separate gas and sample ports which were submerged in a water bath with a temperature control and magnetic stirrer units. Seed sludge which is used in acidification reactors was obtained from a yeast factory's acidification reactors, in Izmir. Table 2. shows sludge sludge properties.

**Table 2. Seed sludge properties**

| Parameter | Unit | Value |
|---|---|---|
| Total Solid Matter | mg/l | 163382 |
| Total Volatile Solid | mg/l | 60612 |
| Total Suspended Solid | mg/l | 88305 |
| Total Volatile Suspended Solid | mg/l | 35092 |

With aim of determination optimal conditions for acidification; organic loading rate was increased gradually as 5000 mg COD/l, 10000 mg COD/l and 20000 mg COD/l while microorganism concentration was stable at 5000 mg TVS/l in reactors. In this case; study was started with 1 gr COD/ gr TVS as S/X ratio and increased gradually to 2 gr COD/ gr TVS and 4 gr COD/ gr TVS for testing. Acidification reactors were performed for 2 days hydraulic retention time at pH 5-5,5. During the study; pH, sCOD and TSS were determined according to standard methods (APHA, 1998). After acidification process; in olive mill wastewater samples, volatile fatty acid concentration especially acetic acid concentration was monitored with GC. Acidification efficiencies were determined as acidification rate and calculated by following equation:

$$\text{Acidification rate, \%} = VFA_{COD}/\text{Total COD}$$

## RESULTS AND DISCUSSIONS

With the aim of acidification potential investment of OMW, anaerobic sequencing batch reactors were performed as acidification reactor. The study was started with 1 g COD/ g TVS as S/X ratio and increased gradually to 2 g COD/ g TVS, 4 g COD/ g TVS and 8 g COD/ g TVS for testing at pH 5-5,5 Acidification performed for 2 days hydraulic retention time. The best acidification rate was achieved at 4 g COD/ g VSS as S/X ratio, 2 days as HRT and 5,5±0,1. Therefore, acidification reactors have been performed in optimum conditions with raw OMW. Acidification reactors were performed for 60 days and according to data for first 60 days; tVFA concentration ( in terms of Acetic Acid), was gradually increased form 450 mg/l to 5278 mg/l. Total VFA concentrations in acidification reactor operated for 60 days at optimal conditions are shown in Figure 1.

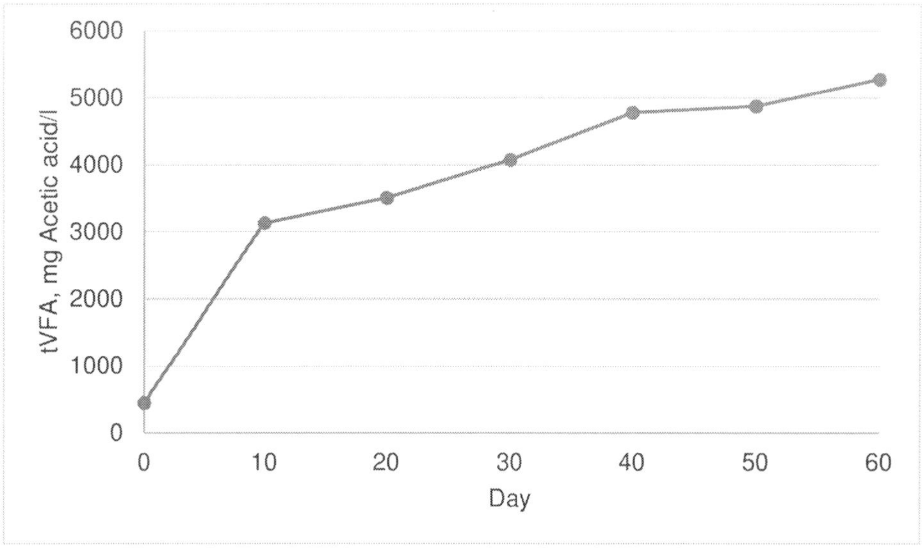


**Figure 1. Total VFA concentrations in acidification reactor**

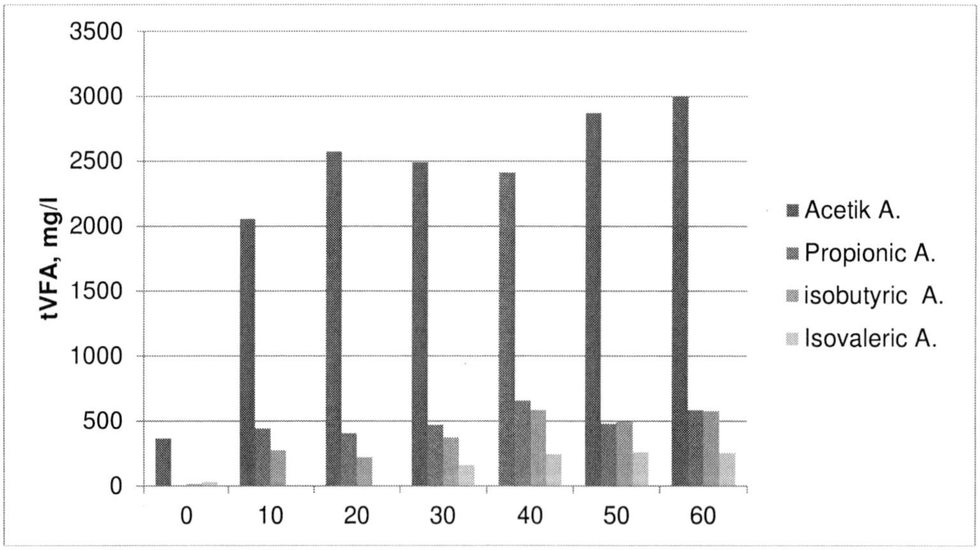


**Figure 2. Total volatile fatty acid composition in acidification reactor**

During the process, Volatile Fatty Acid composition was also monitored. After acidification process, changes in Volatile Fatty Acid composition have observed. Figure 2 shows influent and effluent Total volatile fatty acid composition in acidification reactor operated for 60 days.

While each VFA concentration were increased, also other VFA compounds, not detected in influent, were determined in effluent. During acidification process, changes in sCOD concentration were monitored daily. Figure 3. shows the changes in sCOD concentration during acidification.

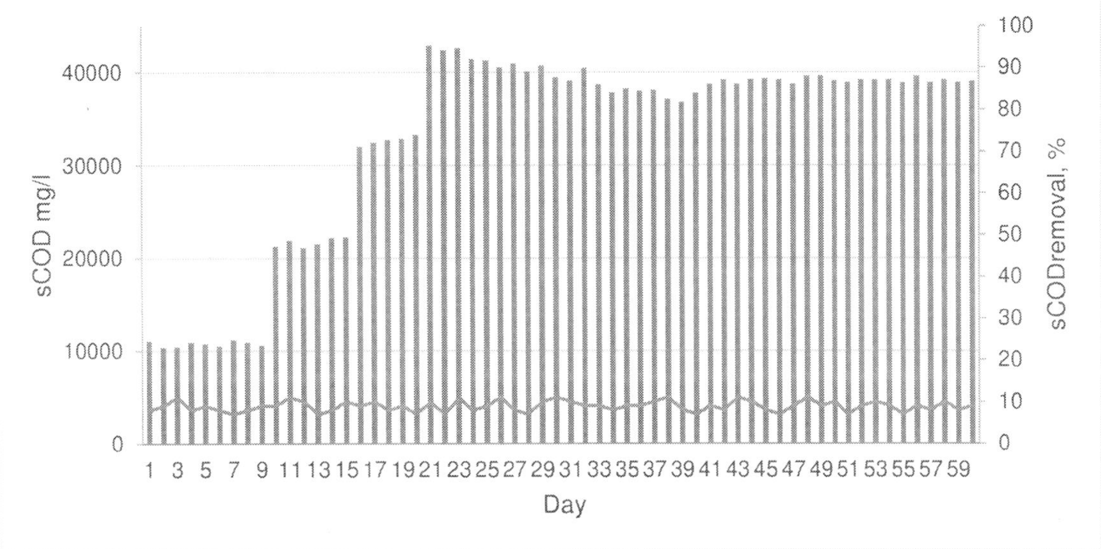


**Figure 3. changes in sCOD concentration during acidification**

No major changes were observed in organic matter concentration during process. Therefore loss of organic matter was not mentioned. Although there is no change in organic matter concentration but there is an increase in tVFA concentration. Therefore it indicates that acidification process has been performed successfully.

## CONCLUSIONS

In biological systems; efficiency is directly related to the amount of organic matter which can be converted into VFA. Therefore, it is important to convert organic matter in wastewater to VFA by a process such as acidification prior to biological processes. The results obtained from acidification can be enhanced by combination of the advance processes. At this point, acidogenic fermentation can be used to produce mainly organic acids (e.g. VFA) with several advantages. It is strategically important to test alternative technologies in the area of wastewater treatment especially for complex wastewater such as OMW. The operation of acidification reactors may offer two main benefits: sustainable clean energy production and the potential for pollution reduction. With the tVFA concentration increase, more efficient results can obtained from next treatment processes for energy and resource recovery.

## ACKNOWLEDGEMENT

Financial support by The Scientific and Technological Research Council of Turkey (TUBITAK) was gratefully acknowledged (Project No.: 114Y179).

## REFERENCES

AMAT A.M., Arques A., Beneyto H., Garcı´a A., Miranda M.A., Seguı´ S., Ozonisation coupled with biological degradation for treatment of phenolic pollutants: a mechanistically based study, *Chemosphere* , 53, 79– 86, (2003).

APHA, AWWA, WPCF, *Standard Methods for the Examination of Water and Wastewater, 20th.*, American Public Health Association, Washington, DC, (1998).

AZBAR, N. Ursillo, P. and Speece, R., Effect of process configuration and substrate complexity on the performance of anaerobic processes, *Water Research*, 35 (3), 817-829, (2000).

BERTİN, L., Majone, M., Di Gioi, D., Fav, F., An aerobic fixed-phase biofilm reactor system for the degradation of the low-molecular weight aromatic compounds occurring in the effluents of anaerobic digestors treating olive mill wastewaters, *Journal of Biotechnology* , 87, 161–177, (2001).

BOARI, G., Brunetti, A., Passino, R., and Rozzi, A. Anaerobic digestion of olive oil mill wastewaters. *Agricultural wastes,* 10(3), 161-175, (1984).

BRADLEY R.M. and Baruchello L. *Primary wastes in the olive oil industry Effluent Water Treatment Journal*, 20, 176–177, (1980).

GINOS, A., Manios, T., Mantzavinos, D., Treatment of olive mill effluents by coagulation–flocculation-hydrogen peroxide oxidation and effect on phytotoxicty, *Journal of Hazardous Materials*, 133, 135–142, (2006).

RIVAS F.J., Beltra´n F.J., Gimeno O., Frades J., Treatment of olive oil mill wastewater by Fenton's reagent. *J Agric Food Chem* 49, 1873– 80, (2001).

TURANO, E., Curcio, S., De Paola, M. G., Calabrò, V., & Iorio, G., An integrated centrifugation–ultrafiltration system in the treatment of olive mill wastewater. *Journal of Membrane Science*, 209(2), 519-531, (2002).

# DECOLORIZATION OF ANAEROBICALLY DIGESTED MOLASSES SPENTWASH BY FUNGAL STRAIN: ISOLATION AND SCREENING OF STRAINS

***Mrityunjay Singh Chauhan*** (Maulana Azad National Institute of Technology, Bhopal, INDIA)
Anil K. Dikshit (Indian Institute of Technology Bombay, Mumbai, INDIA)

India is among the top five sugar producing countries in the world with an annual production of 17.6 million tons of sugar. Molasses is the mother liquor left-over after crystallization of sugar from concentrated cane juice. Almost all of the Indian distilleries use molasses as raw material for alcohol production. For production of one litre of alcohol, about 4-10 kg of molasses is required and about 15 litres of wastewater (commonly known as spent wash) is generated. Spent wash is characterized by its high organic and inorganic loading coupled with dark brown or black colour. Looking to its pollution potential, distillery industry is categorized among top 15 polluting industries in India.

This study was carried out with the objective of isolating fungal species suitable for decolorizing anaerobically digested molasses spentwash (ADMS). A total of eight fungal species were isolated from the sludge of anaerobic digester or the lagoon. The ADMS was pretreated with 15 g/L poly aluminum chloride. Fungal isolates were grown in 10% ADMS amended with nutrients over 15 days, where V-1, V-7 and V-8 showed significant growth. The secondary screening was carried out for the above three strains in 50% ADMS enriched with nutrients, only V-7 and V-8 showed growth with V8 showing profuse growth. The growths of V-7 and V-8 along with decolorization were monitored over 15 days. About 70% decolorization of ADMS could be achieved by V-8 isolate with subsequent reduction in COD within the three days of inoculation. Fungal Isolate was named as V-8(IITB) and identified as *Aspergillus niger*.

Results of isolation, screening, growth pattern, identification and feasibility of treating ADMS are discussed in this paper.

# IMPROVEMENT OF BACTERIAL BIODEMULSIFIER BIOSYNTHESIS IN PERMEABILIZING AGENT-ENHANCED UTILIZATION OF RAPE OIL

***Yuyan Zhang***, Kaiming Peng, Yansong Wei, and Xiangfeng Huang
(Tongji University, Shanghai, China)

Biodemulsifiers have obtained growing interest in the applications for separation of the oil and water phases from emulsions in petroleum oil recovery and oily wastewater treatment, owing to their superiorities of wide applicability and environmental compatibility over traditional chemical demulsifiers. Biodemulsifier-producing strains prefer hydrophobic organic compound (HOC) as a carbon source, in which transmembrane transport of HOC is generally the rate-limiting step. Permeabilizing agent-enhanced HOC bioavailability is a promising strategy to increase the yields of biodemulsifiers.

In this study, twenty-four kinds of permeabilizing agents with different concentrations, as medium additives, were separately used for improving rape oil utilization to enhance the biosynthesis of bacterial demulsifier *Alcaligenes* sp. S-XJ-1, including surfactants (Tweens, Spans, Triton X-100, CHAPS, CTAB, SDS, rhamnolipid, and saponin), organic solvents (glycerol, DMSO, toluene, alcohol, hexane, and 1-dodecanol), and others (EDTA, biotin, lysozyme, glycine, and proline). The demulsification capability was evaluated by bottle test and Turbiscan Lab Expert. Cell membrane permeability was characterized by transmission electron microscopy (TEM) and fluorescein diacetate (FDA) assay.

Results showed that saponin, Span60, CHAPS, 1-dodecanol, DMSO, biotin, glycine, or proline could simultaneously promote the yields and demulsification capability of the S-XJ-1. Among them, 0.05% of saponin was optimal to enhance the yields, demulsification ratio, and demulsification speed, which increased by 2.4 times, 18%, and 4.0 times, respectively. For cells cultivated with 0.05% saponin, the TEM images showed that their cell membranes became incomplete and the average thickness of their cell walls decreased about 4 nm, and FDA assay indicated that their fluorescence intensities increased by 5.9 times. We also found that oil solubilization and stimulated cellular metabolism contributed to the increased yields, and that the demulsification boost was related to responses of cell-surface properties and composition. Thus, saponin addition overcame the permeability barrier of the cell envelope and accelerated the transport of rape oil into cells, providing benefits for demulsifier biosynthesis. This study provides a simple, yet effective strategy to enhance biodemulsifier production.

# SURFACTANT REMOVAL BY METAL OXIDES AND ADSOLUBILIZATION OF ORGANICS IN EXHAUSTED MATERIAL

Suman Koner (Jalpaiguri Govt. Engineering College, Jalpaiguri, India)
Anjali Pal (Indian Institute of Technology, Kharagpur, India)
***Asok Adak*** (Indian Institute of Engineering Science and Technology, Shibpur, Howrah, India. Email: asok@civil.iiests.ac.in)

**ABSTRACT:** The present study deals with the removal of ionic surfactants using oppositely charged metal oxides and use of exhausted metal oxides for removal of other organic pollutants like crystal violet (CV) dye and 2,4-dichlorophenoxyacetic acid (2,4-D) through adsolubilization process. Sodium dodecyl sulfate (SDS), a representative of anionic surfactant was removed from its highly concentrated wastewater by neutral alumina and exhausted adsorbent called surfactant modified alumina (SMA) was used to treat CV bearing wastewater. In similar fashion cetyltrimethylammonium bromide (CTAB), a representative of cationic surfactant was removed by silica gel waste (SGW) collected from a local factory at Kolkata, India. The surfactant modified silica gel waste (SMSGW) was utilized as an adsorbing media for removal of 2,4-D from agricultural wastewater.

**Keywords:** Adsorption, Adsolubilization, Surfactant, Dye, Herbicide.

## INTRODUCTION

The agricultural, industrial and domestic activities result in the generation of large amounts of wastewater containing a number of 'pollutants' which are potentially hazardous to human health, terrestrial and aquatic ecosystem. Among them, surfactants, dyes and pesticides are widely used. Surfactants are harmful to human beings, fishes and vegetation. They are responsible for causing foams in rivers and effluent treatment plants to reduce the quality of water. Surfactants cause short term as well as long-term changes in ecosystem. The main environmental concern with dyes is their absorption and reflection of sunlight entering the water and thus causing reduction in photosynthesis and DO level in river (Donia et al., 2009). Most of the pesticides including 2,4-D are potentially carcinogenic and poor biodegradability of these compounds make these environmentally stable (Carter 2000; Chao et al., 2008).

The conventional methods for surfactant removal from the water environment involve processes such as chemical oxidation, reverse osmosis, chemical precipitation, adsorption, various biological methods etc. Use of metal oxides for treatment of surfactant bearing wastewater has drawn much attention in recent years. The ionic surfactants are adsorbed on the surface of metal oxide due to (1) electrical attraction between charged surface and ionic head groups of surfactant and (2) interaction between the long hydrocarbon chains of surfactant molecules through formation of monolayer, partial bilayer or bilayer structure depending on surfactant concentration (Somasundaran and Fuerstenau, 1966). These surfactant layers formed on solid surface have the ability to solubilize different organics within their structure. This phenomenon is known as adsolubilization (Kitiyanan et al., 1996). Thus, the metal oxides serve dual purposes – firstly they remove the ionic surfactant and then remove different other organics commonly present in the wastewater.

Thus the present study aims at removal of two ionic surfactant viz. SDS and CTAB by using alumina and SGW respectively from highly concentrated solution and then at reuse of surface modified metal oxides such as SMA and SMSGW for the removal of CV dye and 2,4-D herbicide respectively from water environment.

## MATERIALS AND METHODS

---

**Reagents.** SDS (BDH, India), CTAB (Hi-Media, India), CV dye (BDH, India) and 2,4-D (Merck, India) were used as received. All other chemicals used in this study were of high purity and used without further purification. All the chemicals were of Analytical Reagent Grade.

**Adsorbent.** Alumina was supplied by SRL, India and used as such without further grinding and sieving. The granulation of neutral alumina was 70 – 290 mesh ASTM, molecular weight was 101.96 and zero point charge ($Z_{pc}$) was 9.15.

SGW was prepared from solid waste generated from a local silica gel factory located at Kolkata, India. After collection, the material was thoroughly washed with tap water followed by distilled water, dried at 100°C and then sieved. About 70% of the total material lies in the particle size between 150 - 300 microns and that was used for the study. The zero point charge of SGW was found to be 2.0.

For preparation of SMA, alumina (200 g) was shaken for 24 hours with 2 L of SDS solution having 20000 mg/L concentration. The NaCl dose was 2500 mg/L and the pH 4.4±0.1. After shaking, the supernatant was discarded and the alumina was washed thoroughly initially with tap water and finally with distilled water. Then the material was dried at 60°C for 24 hours. The alumina thus obtained is SMA and was used as an adsorbent for CV dye removal from water environment.

In similar fashion, 180 g SGW (at a dose of 30 g/L) was shaken with 6 L of CTAB solution having concentration of 7500 mg/L for 2 h. After shaking, the supernatant was discarded and the SGW was washed thoroughly initially with tap water and finally with distilled water and then dried at 60 °C for 24 h. The SGW was thus converted to SMSGW and was used as an adsorbent for 2,4-D removal from wastewater.

**Analytical Method.** Spectrophotometric method was used for the determination of SDS (Adak et al., 2005), CS (Few and Ottewill, 1956) and 2,4-D (Kundu et al., 2005). The CV dye was also measured by spectrophotometric method at its $\lambda_{max}$ of 591 nm.

**Experimental Studies.** The batch experiments were carried out using synthetic samples of SDS and CTAB separately in distilled water at 25±2°C in a mechanical shaker at an agitation speed of 150 rpm. The sorption equilibrium time was studied using initial surfactant concentrations of 2000 mg/L and adsorbent dose of 100 g/L for alumina and 30 g/L for SGW. The pH for both the solution was 6.8±0.1 and the shaking time was varied from 0 to 120 minutes. The experiments were also carried out to check the effectiveness of both metal oxides (alumina & SGW) for treatment of real wastewater. The real wastewater containing AS was collected from small scale laundry and the same containing CS was collected from a textile industry. The major water chemistry parameter for both the wastewater is reported in Table No 1.

The efficiency of the exhausted metal oxides such as SMA and SMSGW were tested. A sample similar to real wastewater was prepared following the composition given by Sapari (Sapari, 1996) containing CV concentration 200 mg/L in the laboratory and was used to check the efficacy of SMA by conducting batch kinetic study and optimum dose calculation. For kinetic study the SMA dose was fixed at 6 g/L and the shaking time was varied from 0 to 105 minutes where for optimum dose calculation SMA quantity was varied from 0 to 8 g/L with a contact period of 60 minutes.

Efficacy of SMSGW was tested in similar fashion described above with the real wastewater containing 2,4-D concentration 32.90 mg/L (2,4-D sodium salt). The 2,4-D sodium salt (Commercially known as Salix salt) are applied to control broad leaf wed in the tea garden. After application of salix salt there was a heavy storm and the wastewater was collected from the surface runoff of a local tea garden situated in Jalpaiguri district of the state West Bengal, India.

Table 1. Quality Parameter for Laundry and Textile Wastewater

| Parameter | Laundry Wastewater | Textile Wastewater |
|---|---|---|
| pH | 9.1 | 9.2 |
| TDS | 2154 mg/L | 720 mg/L |
| Surfactant | 8068 mg/L as AS | 362 mg/L as CS |

## RESULTS AND DISCUSSION

**Surfactant Removal by Metal Oxide.** Kinetic study was carried out in order to find out the equilibrium contact time for SDS or CTAB uptake by alumina and SGW respectively. In case of SDS removal the equilibrium time was found to be 90 minutes (Figure 1a). However, for CTAB adsorption it was only 30 minutes (Figure 1a). The adsorption capacity at equilibrium was 12.4 mg/g for SDS. This value was quite high in case of CTAB adsorption by SGW and it was 42.72 mg/g. In general the lower value of equilibrium contact time compared to activated carbon adsorption would be advantageous, since it would facilitate to reduce the reactor volume in case of continuous mode of operation in the field.

In order to investigate the mechanism of solute adsorption onto the adsorbent, four kinetic models viz., first order reaction model (Benefield Randall, 1980), pseudo-first order equation of Lagergren (1898), second order reaction model and pseudo-second order reaction model of Ho and MckKay (1999) were analyzed and it was found that the reaction followed the pseudo-second order reaction model best (Figure 1b) for both the cases. This indicated that the rate limiting step for both the cases was chemical adsorption or chemisorption (Ho and McKay, 1999).

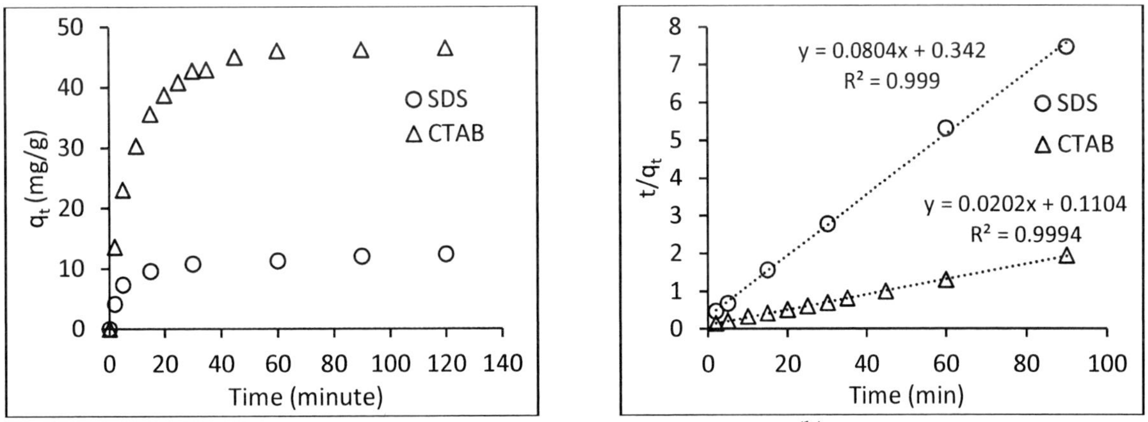


(a) (b)

**Figure 1**. Kinetic study of SDS and CTAB removal by metal oxide - (a) adsorption capacity variation with time and (b) linerized plot of pseudo-second order reaction model.

The experiments were carried out to check the effectiveness of both the metal oxides for treatment of real wastewater. The source and quality of the wastewater has been reported above. A series of 100 mL samples of AS bearing wastewater were shaken for 45 minutes with adsorbents at varying doses (0 to 180 g/L). It was found that the removal of AS increases with the increase of alumina dose and the removal remained almost constant (Figure 2a) beyond the adsorbent dose of 120 g/L. The removal of AS at this stage was found to be 94%. Hence 120 g/L was selected as the optimum adsorbent dose. However, the removal of SDS under similar experimental condition using distilled water spiked sample was only 55%.This might be due to presence of high amount of total dissolved solids (anions and cations) in wastewater which increased the removal efficiency. The presence of various anions and cations generally decease the lateral repulsive force between the head groups of the ionic surfactants causing increased adsorption capacity (Luciani et al., 2001). The kinetics study was conducted with optimum adsorbent dose of 120 g/l at varying time and shown in Figure 2b. The rate of adsorption was very rapid initially. It was noticed that ~ 87% removal of AS took place within half an hour and then gradually increased upto 94% and 96% in 45 and 60 minutes respectively. The equilibrium time for wastewater was found to be less than SDS-spiked in distilled water. The presence of salts in wastewater made the reaction faster than in the case of distilled water.

Experiments were also carried out to find out the effectiveness of SGW for removal of CS from actual wastewater samples under the experimental conditions used for this study. A series of 100 mL portions of CS bearing wastewater were shaken for 30 minutes with varying SGW doses (0 to 20 g/l). It was noticed that the removal was found to be almost constant after a dose of 10 g/l (Figure 2a). At this stage 87% removal was achieved. The adsorption capacity of SGW was found to vary from 13.32 to 73.85 mg of

CS /g of SGW. It was very interesting to observe that the removal efficiency in distilled water spiked sample was 76%. In this case the adsorption capacity of SGW was found to vary from 12.44 to 58.91 mg of CS /g of SGW. Like SDS removal, presence of high amount of total dissolved solids (TDS) in wastewater and also higher pH value which increased the adsorption capacity. Moreover, the adsolubilization of other organics present in the real wastewater e.g. dyestuff was found to take place along with the removal of CS from wastewater. The kinetic study revealed that a contact time of 30 minutes would be sufficient to attain the equilibrium (Figure 2b).

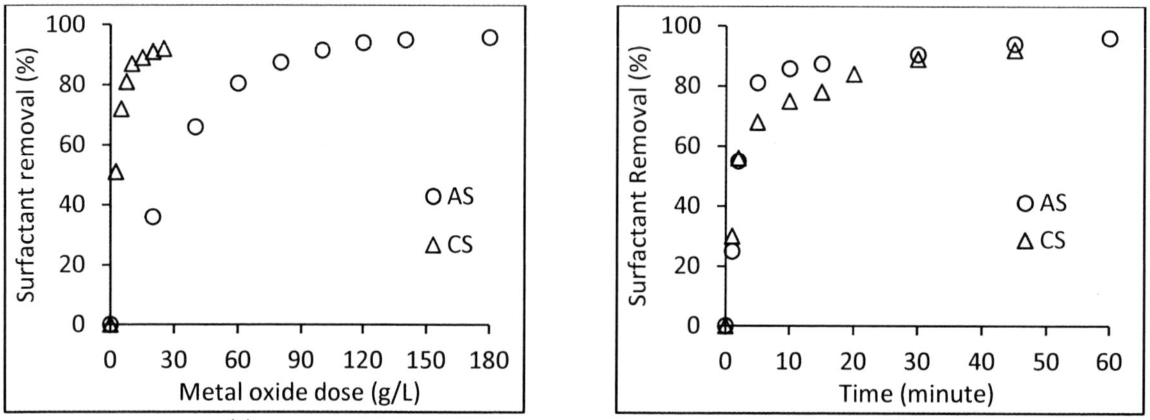


(a) (b)

**Figure 2.** Removal of AS and CS from real wastewater by alumina and SGW respectively – (a) adsorbent dose study and (b) kinetic study

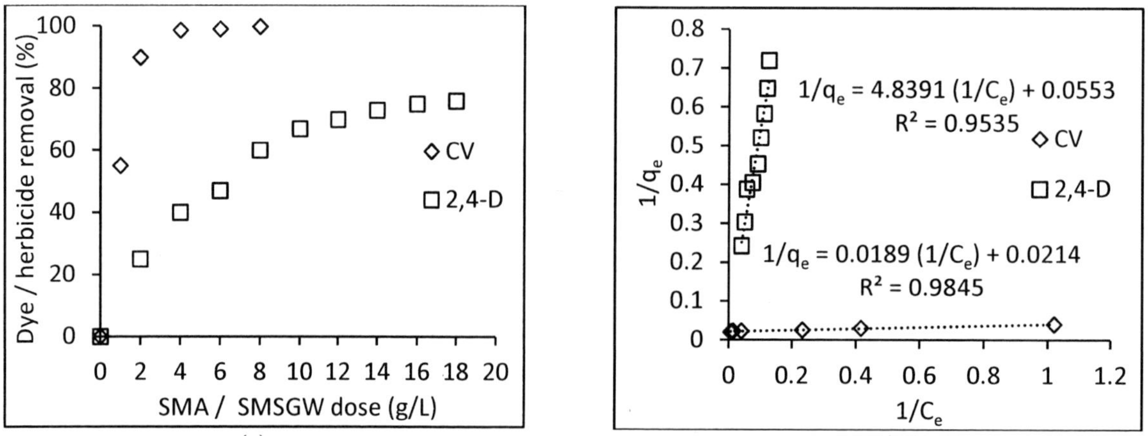


(a) (b)

**Figure 3.** Removal of dye and herbicide from real wastewater by SMA and SMSGW respectively – (a) adsorbent dose study and (b) linearized plot of Langmuir isotherm model

**CV Dye and 2,-4 D Removal By Exhausted Sludge.** Experiments were carried out to find out the effectiveness of SMA for removal of CV from prepared synthetic wastewater following the composition given by Sapari (Sapari, 1996) and the same for SMSGW for 2,4-D removal from a real wastewater collected from a tea garden as mentioned earlier. The crystal violet concentration in wastewater was 200 mg/L. The pH was 10.9±0.1, total dissolved solid (TDS) was 3030 mg/L and the turbidity was 120 NTU. Since the pH of the raw wastewater was 10.9±0.1, it was brought down to 7.0±0.1 by adding acid before being treated with SMA. The effect of adsorbent dose was studied using 0 to 8 g/l of adsorbent. The shaking time was 60 minutes and the temperature was 25±2°C. The optimum adsorbent dose was found to be 6 g/L and corresponding removal of CV was 99 % (Figure 3a). Kinetics studies were carried out with adsorbent dose of 6 g/l. The shaking time was varied from 0 to 105 minutes. The equilibrium time was found to be 30 min only. The adsorption followed Langmuir isotherm model and the maximum adsorption capacity of CV by SMA was found to be 46.72 mg/g at 30 °C (Figure 3b). The adsorption capacity for distilled water spiked CV sample was found to be similar as above.

The performance of SMSGW was also evaluated by applying it for the treatment of real wastewater containing 2,4-D sodium salt concentration of 32.90 mg/L as mentioned earlier. After collection, the wastewater was filtered to remove suspended solids. Figure 3a shows the effect of adsorbent dose. The dose of SMSGW was varied from 0 to 18 g/l and the shaking time was 20 min. It was noticed from the figure that the optimum adsorbent dose was 16 g/l and the capacity as found from Langmuir isotherm was 18.03 mg/g at 30 °C(Figure 3b). However, the adsorption capacity for distilled water spiked 2,4-D sample was found to be 33.9 mg/g. Such lower value of adsorption capacity was due to the presence of dissolved salts in the wastewater. Moreover, higher solubility of 2,4-D sodium salt (45000 mg/l) than 2,4-D acid (900 mg/l) produced lower affinity towards adsorption.

## CONCLUSION

The ionic surfactants could effectively be removed using oppositely charged metal oxides as adsorbent. The equilibrium contact time was very short compared to time requirement for activated carbon adsorbent. The reaction for both cases followed pseudo-second order reaction model. The removal of AS or CS by alumina or SGW respectively was found to be more in actual wastewater compared to distilled water spiked samples due to presence of high TDS in wastewater. The real wastewaters contained 8068 mg/L of AS and 362 mg/L of CS. Dose study revealed that 10 g/L of SGW dose could remove ~87% of CS within 30 minutes. The regeneration of exhausted metal oxide could be done by acid or alkali. Furthermore, the waste adsorbent having surfactant coating on its surface could be used for removal of different organic pollutants from wastewater through adsolubilization process. SMA and SMSGW were used for removal of CV dye and 2,4-D herbicide from real wastewater. The uptake capacity of CV and 2,4-D were found to be 46.72 and 18.03 mg/g respectively. Thus, alumina and SGW were evaluated as adsorbents for efficient removal of surfactants and other organic pollutants like dye and herbicide from wastewater.

## REFERENCES

Adak, A., Bandyopadhyay, M. and Pal, A. 2005. "Spectrophotometric determination of anionic surfactants in wastewater using acridine orange". *Indian Journal of Chemical Technology*, 12(2): 145-148.

Benefield, L.D. and Randall, C.W. 1980. "Biological process design for wastewater treatment". 1st Ed., Prentice-Hall, Inc., USA.

Carter, A.D. 2000. "Herbicide movement in soils: principles, pathways and processes". *Weed Research*. 40: 13–122.

Chao,Y.-F., Chen, P.-C., Wang, S.-L. 2008. "Adsorption of 2,4-D on Mg/Al-$NO_3$ layered double hydroxides with varying layer charge density". *Applied Clay Science*, 40: 193–200.

Donia, M.M., Atia, A.A, Al-amrani, W.A. and El-Nahas A.M. 2009. "Effect of structural properties of acid dyes on their adsorption behaviour from aqueous solutions by amine modified silica". *Journal of Hazardous Material*, 161: 1544-1550.

Few, A.V. and Ottewill, R.H. 1956. "A spectrophotemetric method for the determination of cationic detergents", *Journal of Colloid Science*. 11: 34-38.

Ho, Y.S. and Mckay, G. 1999. "Pseudo second order model for sorption process". *Process Biochemistry*, 34: 451-465.

Kitiyanan, B., O'Haver, J. H., Harwell, J. H, and Osuwan, S. 1996. "Adsolubilization of styrene and isoprene in cetyltrimethylammoniumbromideadmicelle on precipitated silica". *Langmuir*, 12: 2162-2168.

Kundu, S., Pal, A. and Dikshit, A.K. 2005. "UV induced degradation of 2,4-D: kinetics, mechanism and effect of various conditions on the degradation", *Separation and Purification Technology*. 44: 121-129.

Lagergren, S. 1898. "Zurtheorie der sogenannten adsorption gelösterstoffe", Bil. K. Svenska

Luciani, L.,Denoyel, R. and Rouquerol, J. 2001. "Poly (ethoxy) anionic surfactants: micellization and adsorption at the solid/liquid interface". *Colloids and Surfaces A: Physicochemical and Engineering Aspects*, 178: 297-312.

Sapari, N. 1996. "Treatment and reuse of textile wastewater by overland flow". *Desalination*, 106: 179-182.

Somasundaran, P. and Fuerstenau,D.W. 1966. "Mechanism of alkyl sulfonate adsorption at the alumina-water interface". *J. Phys. Chem.*, 70(1): 90-96.
Venteskapsakad, Handl. 24; as cited by Wasay et al. 1996. *Water Research*, 30(5): 1143-1148.

# A COMPARATIVE STUDY ON PRODUCTION OF ACTIVATED CARBON FROM HARDWOOD CHIPS AND PELLETS

***Hafiz Ahmad**** and Brandon Madden (Florida State University, Panama City, FL, USA)
Jordan Mayers (AECOM, Tallahassee, FL, USA)

**ABSTRACT:** Selection of feed-stock from same biomass but different forms sometimes plays an important role in decision-making, especially for building large scale manufacturing plants. Hardwoods in the form of chips and pellets are available in the market and they can be used for commercial production of activated carbon. But the production yields, adsorptive capacity/quality from each form are not known. In this study, first, biochar was produced from hardwood chips and pellets by the pyrolysis process at 850 °F. The biochar was then physically activated (by steam activation) at 1500 °F with various activation (residence) times and different categories of activated carbons were produced. Systematic laboratory tests were performed to determine the yield and iodine number (an indicator of adsorptive capacity, adsorbent quality and surface area) of each product. Models were developed to relate adsorptive capacity (iodine number) and production yield for the activated carbon produced from each type of feed-stock. The study confirmed that chips would be significantly easier to steam activate than pellets and it will give higher yield result than pellets to achieve iodine number of 700 mg/g or more. About 13% more dry wood (by weight) would be required to produce activated carbon with a desirable iodine number of 800 mg/g if hardwood pellets are used rather than chips.

## INTRODUCTION

Activated carbon is the highly porous form of solid carbon. They are commonly produced from carbonaceous raw materials using physical and chemical activation methods (Wang, et al., 2013). These carbon products have highly developed internal surface area and porosity which make them eligible for versatile applications. Activated carbon is used in water treatment, gas purification, metal extraction, air filters, medical purposes, energy storage, soil remediation etc. (Kumar, et al., 2006). This pore structure and pore size distribution of activated carbon significantly depends on the nature of the starting material (Gergova et al., 1993, 1994).

In this study, the feedstocks of interest are hardwood chips and pellets. Hardwood chips are medium-sized solid material made by cutting or chipping larger pieces of wood and are commercially done with wood-chipper. On the other hand, hardwood pellets are compressed biomass made by the hammer milling of dried sawdust. In recent years, the rapid increase in pellet production and its easy and large scale commercial availability has made it a good candidate for activated carbon production like hardwood chips. This leads to an opportunity, which was given to us to evaluate these two feed stocks (hardwood chips and pellets) to produce various grade activated carbons and compare their qualities and yields.

In physical activation, biochar is produced during the first step, by pyrolysis of the precursor. This step removes the non-carbon species and produces char with high percentage of carbon (Dalai and Azargohar, 2007). This char has undeveloped porous structure, which makes its usage limited for many practical applications and requires enhancing its porous structures by the activation process (Marsh et al., 1997, Derbyshire, et al., 1995). The second step of physical activation is high temperature gasification (activation) using oxidizing agents such as steam or carbon dioxide, which produces activated carbon with high porosity. This pore development is due to the penetration of the oxidizing agent into the internal structure of the biochar, opening up the blockage of pores by removing the tar and widening the pores (Dalai and Azargohar, 2007). For a certain temperature, the porosity of the activated carbon usually increases with the increase of activation time. However, the yield of the product decreases with the increase of activation time due to its burn-off. Therefore, there is an immense need for finding a balance between porosity, yield and activation time to evaluate a feed stock for its commercial application to produce activated carbon. In this study, two feed stocks (hardwood chips and pellets) are evaluated and compared for production of activated carbon by finding the relationship between production yield and its porosity derived from different activation times.

## MATERIALS AND METHOD

Biochar and activated carbon from a biomass was prepared by the procedure described in detail elsewhere (Mayers, 2014). A total of 6 tests were performed. Out of these six tests, three were performed for hardwood chips and the other three were for hardwood pellets. Table 1 shows the details of these tests. Initially, the hardwood chips and pellets were cleaned with brushes, all foreign materials were removed and then oven dried. For each test, the dried material was then pyrolysed at 850 °F using a pyrolysis apparatus schematic diagram of which is shown in Figure 1.

**TABLE1: Test conditions for various grade activated carbon production**

| Tests | Hardwood Sample | Preparation Conditions | | |
|---|---|---|---|---|
| | | Pyrolysis Temperatur e (°F) | Activation Temperature (°F) | Activation / residence time (minute) |
| Test-1 | | | | 6 |
| Test-2 | chips | 850 | 1500 | 12 |
| Test-3 | | | | 20 |
| Test-4 | | | | 6 |
| Test-5 | | | | 12 |
| Test-6 | pellets | 850 | 1500 | 20 |

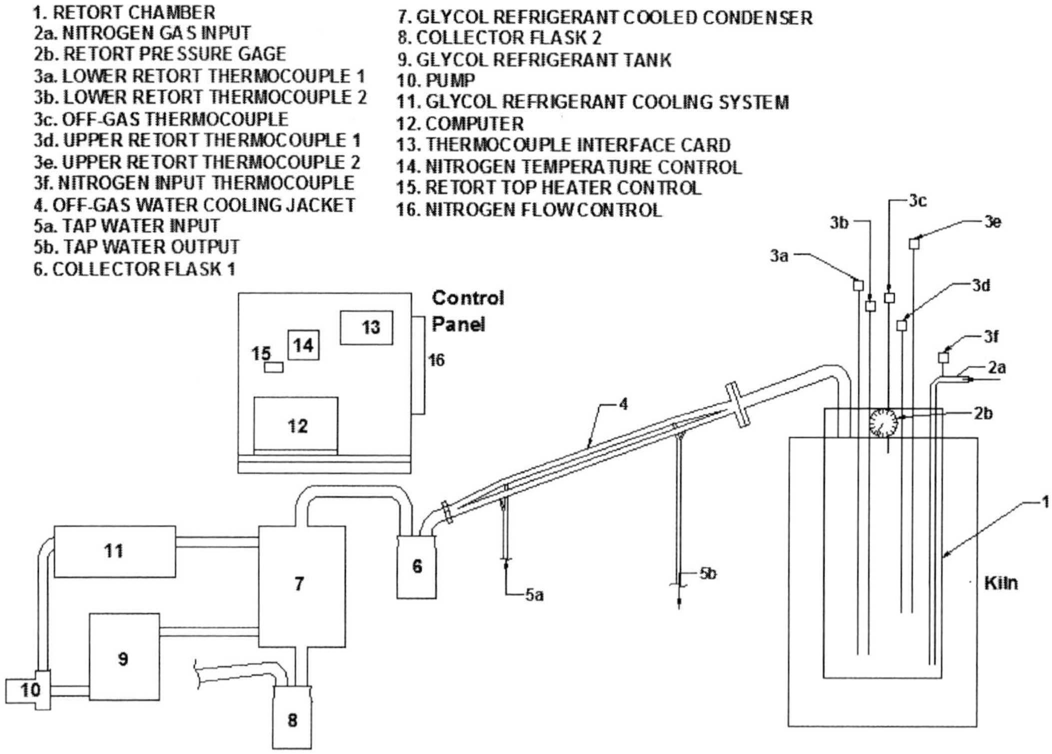


**FIGURE 1.** Schematic diagram of pyrolysis apparatus

The biochar produced from hardwood chips and pellets after pyrolysis (in retort chamber (#1)) was physically activated (steam activation) at 1500 °F for three activation (residence) times (20 minutes, 12 minutes and 6 minutes) and three different grades of activated carbons were produced from each feed stock.

Similar/identical experimental conditions were applied for carbonization and activation process of the two feed stocks. The yield of activated carbon was calculated as follows (Eq. (1)):

$$Y\% = M/M_o \times 100 \qquad (1)$$

Where, M is the mass of activated carbon and $M_o$ is the mass of air dried feed stock (hardwood chips and pellets).

The iodine numbers (mg of iodine adsorbed/g of activated carbon) of each of these six activated carbon products were determined according to ASTM 4607-86 (1994). This number (i.e. mg of iodine adsorbed per g of activated carbon) is referred as iodine numbers throughout this paper. Besides iodine number, another parameter calculated was modified iodine number (mg of iodine adsorbed/g of wood) and it is expressed as follows:

Modified iodine number = mg I / g of wood
= (mg I/g of activated carbon) x (g of activated carbon/g of wood)
= iodine number x yield

## RESULTS AND ANALYSIS

**Comparison of Yields and Iodine Numbers.** Figure 2(a) shows activated carbon yields at various activation/residence time. As seen in this Figure, yield decreases with the increase of activation time. This is due to the fact that burn-off of carbon increases with the increase of activation time. Activated carbon yield from hardwood chips at 6 minute residence time was 17.18%, and it is 15.29% when residence time is 12 minute and it went down to 12.77% when residence time was increased to 20 minutes. Similar pattern was also observed for hardwood pellets. Activated carbon yields from pellets were 17.17%, 14.58%, 10.67% for 6, 12 and 20 minute residence times respectively.

Comparing the yields of hardwood chips and pellets at each activation time, it was found that yield of activated carbon from hardwood chips is higher than that of pellets in all cases (Figure 2(a)). As for example, activated carbon yield from chips for the 20 minute residence time is 12.77% while it is 10.67% for pellets for the same residence time. This indicates, activated carbon production from hardwood chips could be more favorable that pellets.

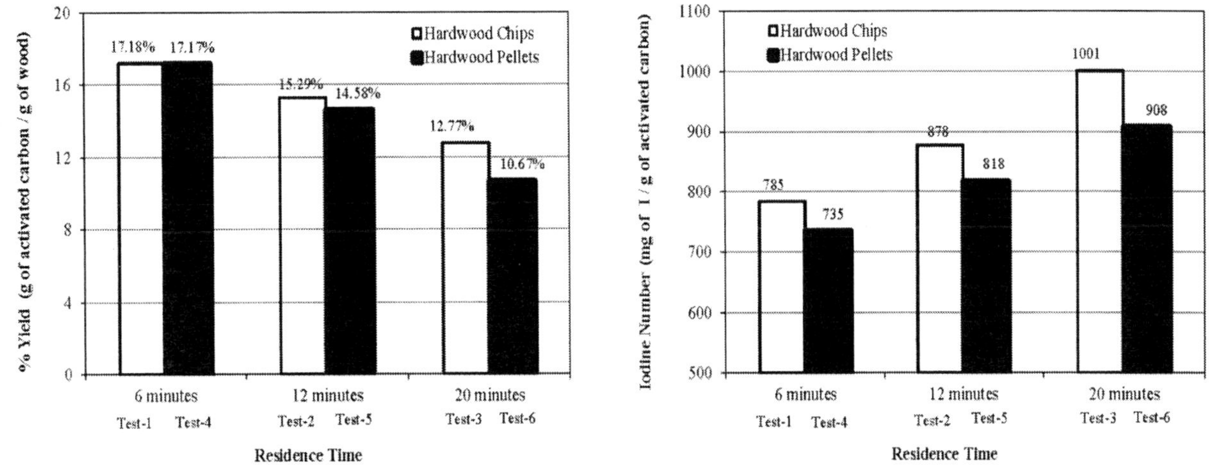


**FIGURE 2.** (a) Activated carbon yield at various activation times, (b) Iodine numbers of activated carbon produced at various activation times.

Though increase of residence time reduces the yield of activated carbons, the opposite phenomena occur for the quality of the carbon product. In other words, with the increase of temperature, the iodine number increases (Figure 2(b)) indicating more porous, more surface area and more adsorptive capacity of the products. Iodine number for the activated carbon produced from hardwood chips was 785 mg/g for the 6 minute residence time. This iodine number increases to 878 mg/g and 1001 mg/g for the residence time of 12 minutes and 20 minutes respectively. Similar pattern observed for hardwood pellets. Iodine numbers determined for hardwood pellets were 735 mg/g, 818 mg/g, 908 mg/g for the residence time of 6 minutes, 12 minutes and 20 minutes respectively.

Comparing the iodine numbers of hardwood chips and pellets (Figure 2(b)), it is seen that the iodine number of activated carbon from hardwood chips is higher than that of pellets in all cases. Iodine number of activated carbon from hardwood chips is 785 mg/g while it is 735 mg/g for pellets for the same residence time (6 minutes). For 12 minutes residence time, iodine numbers of chips and pellet are 878 and 818 mg/g respectively. For the 20 minute residence time the iodine number for chips is 1001 mg/g while it is 908 mg/g for pellet for the same residence time. This again indicates that activated carbon production from hardwood chips could provide better quality product than that of pellets.

**Model Development.** As mentioned above and seen in Figure 2(b), the increase in residence time increases the iodine number (i.e. quality of the product). However, this increase in iodine number was achieved at the expense of the yield of activated carbon (Figure 2(a)). So, it is imperative to find a relationship between iodine number and yield of activated carbon derived from each feed stock (hardwood chips and hardwood pellets) and compare them. Iodine numbers vs. yield of activated carbon from feed stocks are plotted and presented in Figure 3. These plots within the range tested exhibit a linear correlation (the correlation coefficients of plots, $R^2$, are greater than 0.99 in both cases). Considering linear relationship (within the test range) of iodine number and product yield for hardwood chips and pellets, following relationships were determined (Figure 3).

Hardwood chips: $$Y = -4896.9X + 1625.9 \qquad (2)$$

Hardwood pellets: $$Y = -2631.2X + 1192.3 \qquad (3)$$

Where, Y and X are iodine number (mg I/g of activated carbon) and yield (g of activated carbon/g of wood) respectively.

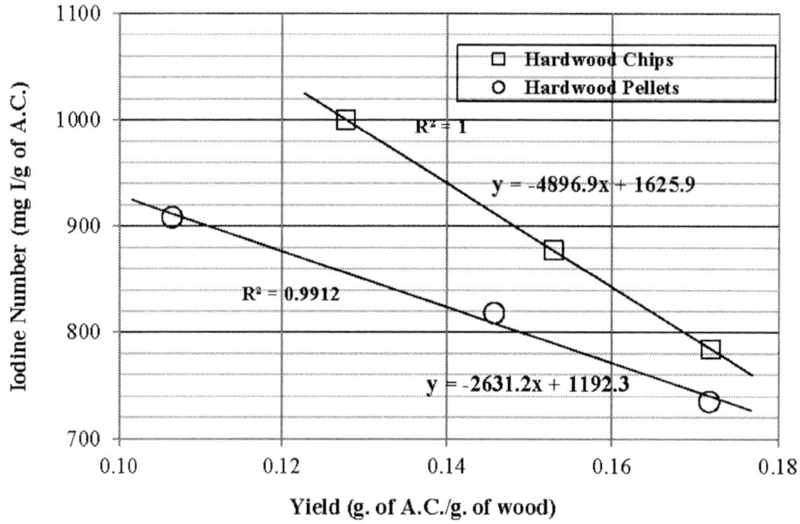


**FIGURE 3.** Iodine number vs. activated carbon yield from woods

Comparing the hardwood pellets with chips, Figure 3 shows that the yield of activated carbon from hardwood chips is significantly higher that pellets at the higher iodine numbers (700 mg/g). In other words, to achieve a certain yield, higher iodine number can be produced from hardwood chips than that from hardwood pellets. As for example, from Figure 3, it is seen that to achieve a 14% yield (wood basis), approximate iodine number from hardwood pellet is 820 mg/g. On the other hand, from same 14% yield (wood basis), approximate iodine number from hardwood chips is 940 mg/g.

In industry, it is more conventional to express the porosity characteristics per unit mass of raw material instead of the product (Dalai and Azargohar, 2007). So, to further investigate, plots of modified iodine number (mg. I/g of dry wood) vs. iodine number (mg I/g of activated carbon) are plotted for hardwood chips and pellets and shown in Figure 4. These curves are generated by using the models ((equation (2) for hardwood chips and equation (3) for hardwood pellets). For different iodine numbers the corresponding yield was calculated from the equation. Then, the modified iodine number was determined by taking the product of iodine number and corresponding yield (which was detailed earlier).

It can be seen that from Figure 4 that dry wood basis porosity (i.e. modified iodine number) exhibits a maximum value when it reaches to a certain porosity of activated carbon (i.e. iodine number). To achieve the iodine number 800 mg/g, modified iodine numbers were found 120 mg/g for hardwood pellet and 138 mg/g hardwood chips (Figure 4). These infers that, 1 mg of iodine can be adsorbed in 8.33 mg of wood pellets while same 1 mg of iodine can be adsorbed in 7.25 mg of wood chips indicating about 13% more wood is needed to achieve the same porosity if hardwood pellets are used rather than chips.

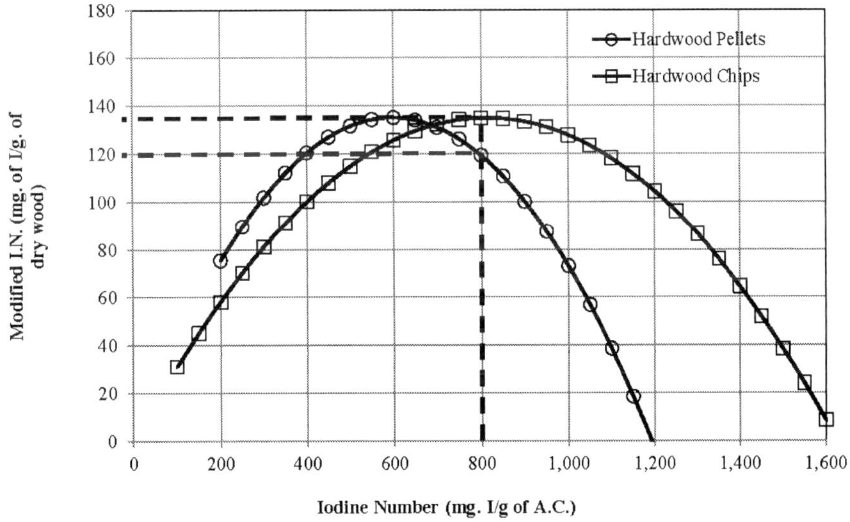


**FIGURE 4.** Modified iodine number vs. iodine number

## CONCLUSIONS

Various grades of activated carbons were produced from hardwood chips and hardwood pellets by the method of physical activation. The effect of activation times (for 1500°F temperature) on adsorption capacity, production yields was studied. The study demonstrates that by increasing temperature, the adsorptive capacity of the activated carbon can be increased but this will result in decrease of the product yield. Models were developed to relate adsorptive capacity (iodine number) and production yield for the products produced from each type of feed-stock (i.e. hardwood chips and pellets). The study confirmed that chips would give higher yield result than pellets to achieve iodine number of 700 mg/g or more. Additionally, 13% more dry wood would be required to produce activated carbon with a desirable iodine number of 800 mg/g if hardwood pellets are used rather than chips.

## ACKNOWLEDGEMENT

The authors gratefully acknowledge the financial support provided by the Pyrolysis Tech. LLC., Gladwyne, Pennsylvania.

## REFERENCES

Dalai, A.K. and R. Azargohar. 2007. "Production of Activated Carbon from Biochar Using Chemical and Physical Activation: Mechanism and Modeling", *In Materials, Chemical, and Energy from Forest Biomass; Argyropoulos, D.*; ACS Symposium Series; American Chemical Society: Washington D.C.

Derbyshire, F., M. Jagtoyen and M. Thwaites, "Porosity in Carbons: Characterization and Applications" In W. PatrickJr. (Ed.), Edward Arnold, London, 1995, Chapter 9.

Gergova, K., N. Petrov and S. Eser. 1994. "Adsorption Properties and Microstructure of Activated Carbons Produced from Agricultural Byproducts by Steam Pyrolysis" *Carbon* 32, 693–702.

Gergova, K., N. Petrov, V. Minkova. 1993. "A Comparison of Adsorption Characteristics of Various Activated Carbon" *J. Chem. Tech. Biotechnol.* 56, 77–82.

Kumar, B.G.P., K. Shivakamy, L.R. Miranda and M. Velan. 2006. "Preparation of Steam Activated Carbon from Rubberwood Sawdust (Hevea brasiliensis) and its Adsorption Kinetics" *J. Hazard Mater*, 136: 922–929.

Marsh, H., E.A. Heintz and F. Rodriguez-Reinoso. 1997. "Introduction to Carbon Technologies", First ed., Universidad de Alicante, Secretariado de Publicaciones, Chapter 2.

Mayers, J. 2014. *Pyrolysis and Activation of an Invasive Species.* M.S. Thesis, Florida State University, Tallahassee, FL.

Wang, X., D. Li, W. Li, J. Peng, H. Xia, L., Zhang, S. Guo and G. Chen. 2013. "Optimization of Mesoporous Activated Aarbon from Coconut Shells by Chemical Activation with Phosphoric Acid" *BioResources*, 8(4): 6184-6195.

# MOLECULAR SIMULATION AND VALIDATION STUDIES OF RESORCINOL ADSORPTION ON ORDERED MESOPOROUS CARBON

Bing Chao, ***Zaki Uddin Ahmad*** and Daniel Dianchen Gang*
(University of Louisiana at Lafayette, Lafayette, LA70504, USA)

**ABSTRACT:** Ordered Mesoporous Carbons (OMCs) with well-controlled pore structure and narrow pore size distribution demonstrated great potential as highly functional adsorbents. The pore size and surface chemistry of OMCs were considered as two most important factors that affect the adsorption capacity of organic compounds. The objective of this study was to optimize the structure of OMCs for resorcinol adsorption by changing the pore size and oxygen content using computational approach. New rhombic OMC models with varied pore size and oxygen content were constructed using Materials Visualizer module. The specific surface area, total pore volume and resorcinol adsorption capacity results were calculated by Forcite and sorption module in Materials Studio package. The simulation results were validated by the experimental data. Experimentally, the OMCs were synthesized using sucrose as carbon precursor by hard-template method. The tunable pore size (4 nm to 15 nm) and oxygen content of the OMCs are obtained by adjusting the amount of boric acid as a pore-expanding reagent. The experimental results such as BET surface area, pore size distribution and adsorption capacity of resorcinol were compared with the simulation results.

## INTRODUCTION

Ordered Mesoporous Carbons (OMCs) with well-controlled pore structure and narrow pore-size distribution demonstrate great promise for their adsorption of inorganic and organic contaminants. Several methods (Yuan et al., 2007; Kim et al., 2005; Peng et al., 2008) have been proposed and studied experimentally to prepare OMCs and investigate the relationship between OMCs' characterizations and adsorption capacity. Based on their studies, the pore size and surface chemistry of OMC are identified as two important factors that affected the adsorption capacity for organic compounds. Even though extensive experimental studies have been carried out on adsorption process over OMCs, the interactions between OMCs and adsorbates are not fully understood. As a complementary technique, molecular simulation method has been introduced. Zhou et al. (2008) constructed a hexagonal model of MCM-41 and studied $CO_2$ adsorption on MCM-41 by the Grand Canonical ensemble Monte Carlo (GCMC) simulation. Chen et al. (2009) constructed a MCM-41 model using Materials Visualizer module and investigated the toluene adsorption. In all these works, it was generally assumed that rectangular structure is the repetitive unit. However, OMC prepared by SBA-15 as the hard template possessed two-dimensional hexagonal (p6mm) pore structure (Fig.1). The rectangular model subjected to several limitations such as difficulties on controlling the pore size and forming the hexagonal structure. So in this study, a new rhombic template was proposed as the repetitive unit of OMC models.

Herein, we present a study of OMCs with different pore sizes and oxygen contents using Materials Studio to investigate the structural information and adsorption behaviour and find out the optimal pore size and oxygen content for resorcinol adsorption. In the experimental phase, the OMCs with varied pore sizes and oxygen contents were synthesized using boric acid as the pore-expanding reagent following the work of Dai et al. (2010).

## MATERIALS AND METHODS

**Construction of OMC models.** In this study, a new OMC model using rhombus shaped repetitive cell and ZSM-5 (Fig. 2(a)) as the unit cell was proposed. ZSM-5 is an aluminosilicate zeolite abundant in micropores. The unit cell was modified by changing all the other atoms to carbon atoms. The obtained structures were optimized by Forcite module. The lattice parameter of angle was set to $\alpha = 90°$, $\beta = 90°$ and $\gamma = 120°$ in order to build a hexagonal array structure.

The OMC models with different pore sizes (from 4 nm to 12 nm) can be obtained by changing the radius of the carving cylindrical channel. OMCs model with varied oxygen content was modelled by adding O atoms on the unsaturated C atom sites. The final models generated were geometrically optimized by Forcite module in Materials Studio packages via COMPASS force field.

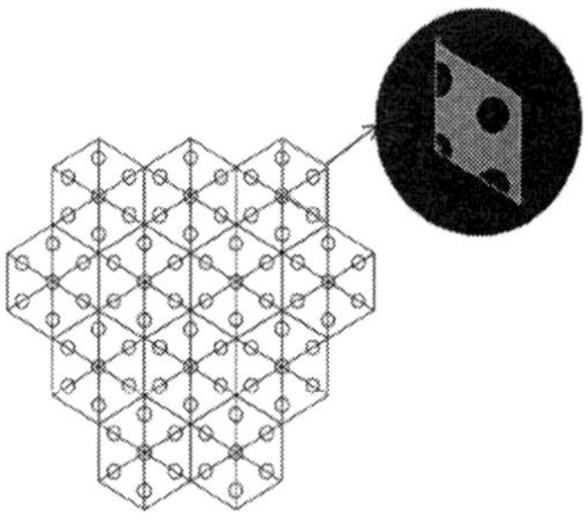

**FIGURE 1.** Repetitive unit for OMC model

**Fabrication of OMCs.** The synthesis of SBA-15 followed the method reported by Zhao et al. (1998) with modification on the reaction temperature and period. The mixture of silica source (Tetraethyl Orthosilicate, TEOS) and surfactant Pluronic P123 (PEO-PPO-PEO, BASF) was placed in a water bath for 4h at 40°C, followed by aging for 24h at 90°C. The SBA-15 was obtained after calcination in air at 550°C for 8h. Different amount of boric acid were added to obtain different pore size expansion. The carbon replicas were denoted as OMC-x. The x value indicated the molar ratio of boric acid to carbon precursor, which is varied from 0 to 12. In a typical synthesis of OMC-1, 0.27 g of boric acid (99.8 wt%), 1.5 g of sucrose and 4 drops of concentrated $H_2SO_4$ were dissolved in 5.0 ml distilled water. After adding 2 g of SBA-15, the mixture was heated at 100°C for 6 h and subsequently at 160°C for another 6 h. The resulted composite was impregnated again with an aqueous solution of 0.09 g of boric acid, 0.51 g of sucrose and 15 drops of concentrated $H_2SO_4$. The dried composite was carbonized at 700°C for 8 h under the nitrogen atmosphere. Finally, the OMC-1 was obtained by the removal of SBA-15 using 48% HF solution at room temperature.

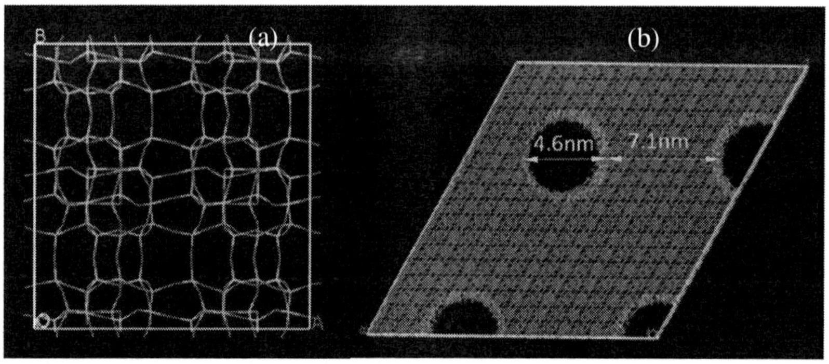


**FIGURE 2.** The model of (a) ZSM-5 repetitive unit (b) OMC

## RESULT AND DISCUSSION

**Verification of the OMC model.** A reliable model of OMC materials is the basic criteria for simulating the adsorption process. The optimized OMC with the same unit cell parameter of OMC-0 is shown in Fig. 2(b). The radius of the pore is 4.6 nm and wall thickness is 7.1 nm, respectively. The optimized structure parameter and element contents of the OMC model and OMC-0 sample made by sucrose were listed in Table 1. The decrease in surface area and pore volume was caused by the ideally ordered structure assumed for the OMC model. Similar results were obtained in the MCM-41 model built by Chen et al. (2009). However, such difference has no significant effect on describing pore structure of the experimentally produced OMCs, while other parameters match the real experimental results.

The adsorption process was simulated by Soption module which was based on GCMC theory. The loading of resorcinol in equilibrium was 157 adsorbed resorcinol per cell. After calculating, the simulated adsorption capacity of OMC model was 41 mg/g, which approached the experimental data of 36.5 mg/g. Therefore, the OMC model is a reliable model for adsorption simulation.

**TABLE 1.** Lattice parameters and element contents of OMC model and OMC-0 sample

| Name | Pore size (nm) | Wall thickness (nm) | Oxygen content (%) | Carbon content (%) | Density (g/cm$^3$) | Pore volume V(cm$^3$/g) | SSA (m$^2$/g) |
|---|---|---|---|---|---|---|---|
| OMC model | 4.6 | 7.1 | 11.2 | 85.3 | 1.13 | 1.36 | 969.3 |
| OMC-0 sample | 4.6 | 7.1 | 13.3 | 83.7 | 1.12 | 1.58 | 1404.2 |

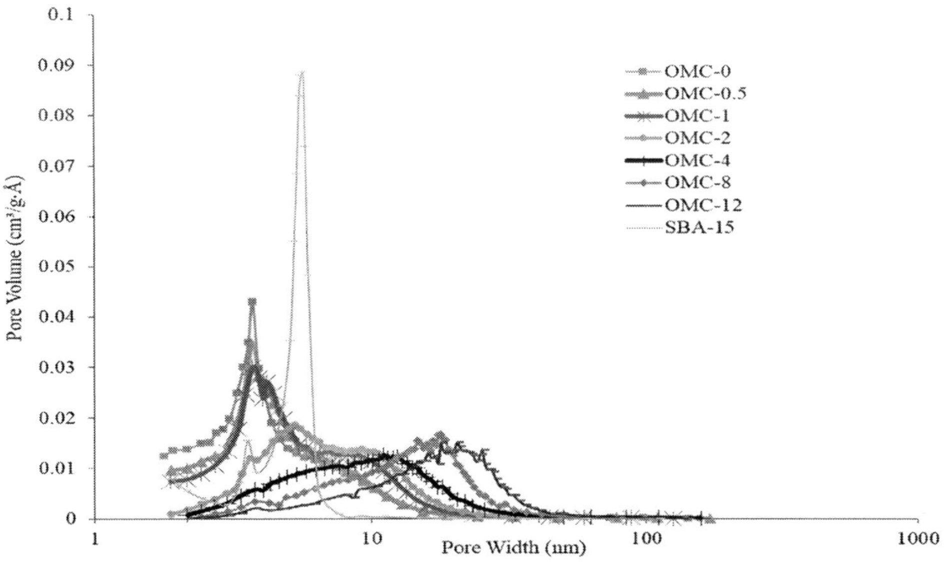


**FIGURE 3**. Pore size distributions for SBA-15 and OMCs

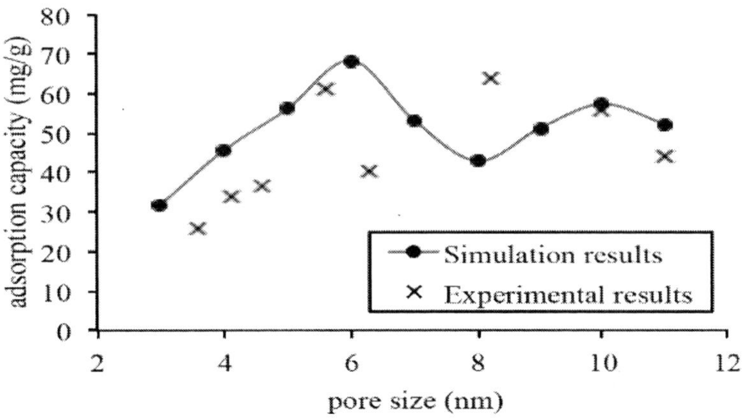


**FIGURE 4.** Comparison of the simulation results and experimental results: adsorption capacity vs. pore size

**Resorcinol**

**Adsorption on OMC.**

***Effect of pore size.*** The increase in the pore size with the increase in boron content was confirmed by pore size distributions (Fig. 3). The pore size distributions were determined by the desorption branches of the isotherms. The OMC-4, OMC-8 and OMC-12 give flatter peaks, which indicated wider ranges of pore size and less ordered structure. The largest pore size (11.3 nm) was obtained by OMC-12, which was extremely close to the unit cell length. For these three samples, the wall thickness was too thin to support the rigid carbon structure and therefore, less ordered carbon structure was obtained. The same observation was obtained by Ryoo et al. (2001). In general, the OMCs made with different ratio of boric acid to sucrose have tailored pore sizes ranging from 4.6 nm to 11.3 nm. In order to investigate the effect of pore size on the adsorption performance, a series of OMC models with different pore sizes (from 3 nm to 11 nm) were built using Materials Studio. The unit cell parameter was maintained at 11.7 nm, which was the same as the experimental OMCs. The simulation results are shown in Table 2. The relationships between the pore sizes with adsorption capacities were plotted and compared with the experimental data as shown in Fig. 4. The simulation results and experimental data showed a similar trend with average relative error of 0.1938. That means the simulation results well fitted the experimental results. The highest adsorption capacity could be obtained at the diameter of 6 nm, which also yielded a larger surface area. The larger pore size did not promise a better adsorption capacity. From simulation results, it was observed that the specific surface area was the main factor affecting the adsorption capacity, which satisfied the conclusion drawn by You et al. (2009).

**TABLE 2.** Structural parameter and adsorption capacity of the OMC model

| Pore size (nm) | Density ($g/cm^3$) | Specific surface area ($m^2/g$) | Wall thickness (nm) | Adsorption capacity (mg/g) |
|---|---|---|---|---|
| 3 | 1.06 | 808.53 | 8.7 | 31.5 |
| 4 | 0.955 | 973.2 | 7.7 | 45.6 |
| 5 | 0.857 | 1132.2 | 6.7 | 56.1 |
| 6 | 0.77 | 1347.6 | 5.7 | 68.3 |
| 7 | 0.73 | 1148.5 | 4.7 | 53.3 |
| 8 | 0.666 | 954.3 | 3.7 | 42.8 |
| 9 | 0.613 | 963.2 | 2.7 | 51.2 |
| 10 | 0.579 | 986.7 | 1.7 | 57.3 |
| 11 | 0.543 | 990.3 | 0.7 | 52.2 |

***Effect of oxygen content.*** The effects of oxygen content on the surface of the OMC model were evaluated by constructing the OMC model doped with different amount of oxygen atoms and pore size of 4.6 nm as the basic structure. Oxygen content was the only variable in this iteration of simulation. The oxygen atoms existed in the form of hydroxyl groups and carboxyl groups and distributed on the surface of the pore structure. The simulation results are shown in Table 3.

**TABLE 3.** Adsorption simulation results for OMC model with different oxygen contents

| Pore size (nm) | Oxygen content (%) | Density ($g/cm^3$) | SSA ($m^2/g$) | Adsorption capacity (mg/g) |
|---|---|---|---|---|
| 4.6 | 0 | 0.87 | 1028.3 | 41.3 |
| 4.6 | 2.9 | 0.92 | 1030.5 | 43.6 |
| 4.6 | 4.4 | 1.03 | 1031.4 | 45.1 |
| 4.6 | 6.2 | 1.28 | 1033.5 | 45.8 |
| 4.6 | 9.6 | 1.49 | 1035.7 | 46.7 |
| 4.6 | 12.3 | 1.69 | 1036.8 | 47.2 |

It was observed from Fig. 5 that the adsorption capacity of resorcinol increased with additional oxygen atoms distributed on the pore surface. Comparing with the influence of specific surface area with the influence of oxygen-containing functional groups, the specific surface area increased the adsorption capacity up to approximate 116% while the influence of oxygen-containing functional group for the adsorption of resorcinol was nearly negligible confirming that the adsorption process was more of a physical process rather than a chemical process.

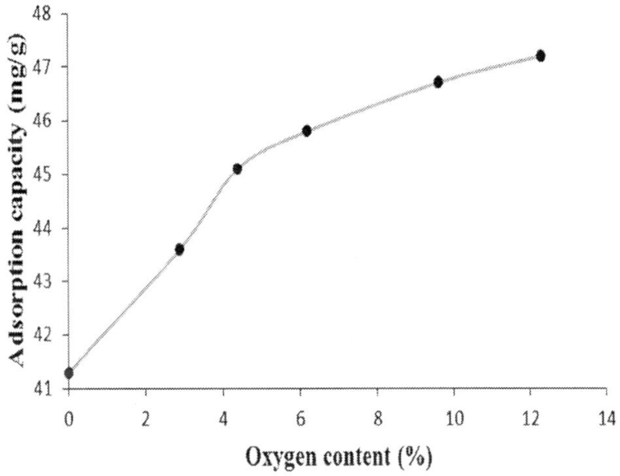


**Figure 5**. The plot of adsorption capacity vs. oxygen content

**CONCLUSION**

OMC materials were synthesized using sucrose as the carbon precursor and boric acid as the pore expanding reagent via hard template method. The experimentally obtained results were used for validating the results of computational simulation on the adsorption of resorcinol. In the simulation, OMC models with different pore sizes and oxygen contents were constructed to evaluate the adsorption performance of OMC materials for resorcinol removal. The GCMC simulation on resorcinol adsorption suggested that (a) the optimal pore sizes were found to be around 6 nm and the corresponding adsorption capacity of 68.3 mg/g; (b) with the increase in oxygen content, the adsorption capacity increased with growth rate of 14%. The improvement of adsorption capacity was not significant comparing with the influence of specific surface area, since the adsorption process was more of a physical process rather than chemical process.

**ACKNOWLEDGEMENTS**

This work was supported by the Louisiana Board of Regents under BORSF (2010-2015)-LaSpace and by NASA under award NNX10AI40H.

**REFERENCES**

Chen, H., H. Xi, X. Cai and Y. Qian. 2009. "Experimental and molecular simulation studies of a ZSM-5-MCM-41 micro-mesoporous molecular sieve." *Microporous and Mesoporous Materials*. 118(1-3):396-402.

Dai, W., M. Zheng, Y. Zhao, S. Liao, G. Ji and J. Cao. 2010. "Template synthesis of three-dimensional cubic ordered mesoporous carbon with tunable pore sizes." *Nanoscale Res. Lett.*. 5(1): 103-107.

Guo, R., J. Guo, F. Yu, D. D. Gang. 2013. "Synthesis and surface functional group modifications of ordered mesoporous carbons for resorcinol removal." *Microporous and Mesoporous Materials*. 175: 141-146.

Kim, D. J., H. I. Lei, J. E. Yie, S. Kim and J. M. Kim. 2005. "Ordered mesoporous carbons: implication of surface chemistry, pore structure and adsorption of methyl mercaptan." *Carbon*. 43(9): 1868-1873.

Lee, H. I., J. H. Kim, D. J. You, J. E. Lee, J. M. Kim, W. S. Ahn, C. Pak, S. H. Joo, H. Chang and D. Seung. 2008. "Rational synthesis pathway for ordered mesoporous carbon with controllable 30 to 100-angstrom pores." *Advanced Material*. 20(4): 757-762.

Peng, X. D., D. Cao and W. Wang. 2008. "Heterogeneity characterization of ordered mesoporous carbon adsorbent CMK-1 for methane and hydrogen storage: GCMC simulation and comparison with experiment." *Journal of Physical Chemistry*. 112(33): 13024-13036.

Ren, H., W. Shou, C. Ren and D. Gang. 2016. "Preparation and post-treatments of ordered mesoporous carbons (OMC) for resorcinol removal." *International Journal of Environmental Science and Technology*. 13(6): 1505-1514.

Ryoo, R., S. H. Joo, S. Jun, T. Tsubakiyama and O. Terasaki. 2001. "Ordered mesoporous carbon molecular, sieves by template synthesis: the structural varieties." *Studies in Surface Science and Catalysis*. 135: 150.

Shou, W., B. Chao, Z. U. Ahmad and D. Gang. 2016. "Ordered mesoporous carbon preparation by the in situ radical polymerization of acrylamide and its application for resorcinol removal." *Journal of Applied Polymer Science*. 133(19): 43426.

You, C., Y. Xuewu, Y. Wang, S. Zhang, J, Kong, D. Zhao and B. Liu. 2009. "Electrocatalytic oxidation of NADH based on bicontinuous gyroidal mesoporous carbon with low overpotential." *Electrochemistry Communication*. 11(1): 227-230.

Yuan, X., S. Zhou, W. Xing, H. Cui, X. Dai, X. Liu and Z. Yan. 2007. "Aqueous dye adsorption on ordered mesoporous carbons." *Journal of Colloid and Interface Science*. 310:83-89.

Zhao, D., S. Sun, Q. Li, J. Feng and G. D. Stucky. 2000. "Morphological control of highly ordered mesoporous silica SBA-15." *Chemistry of Materials*. 12(2): 275-279.

Zhao, J. 2014. *Simulation on the self-assembly behavior of ordered mesoporous carbons and their performance in adsorption*. M.S. Thesis, South China University of Technology, China.

Zhou, S., Y. Huang, J. Hu, H. Liu and J. Jiang. 2008. "Computer simulation of adsorption of $CO_2$, $N_2$ and flue gas in a mimetic MCM-41." *Journal of Physical Chemistry C*. 112:11295-11300.

# PHENOL REMOVAL FROM WATER USING BIOCHAR AND ACTIVATED CARBON FROM ALBIZIA: AN INVASIVE PLANT EVALUATION

***Hafiz Ahmad***, Korhan Adalier, Brandon Madden and
Douglas Brown (Florida State University, Panama City, FL, USA)

**ABSTRACT:** Examining the performance of carbon products of various activation levels sometimes play important role, especially to determine the operating conditions of large scale activated carbon manufacturing plants. This study compares the removal performance of phenol from water using biochar and two physically (steam) activated carbons (having two different activation levels) produced from the same biomass, Albizia, an invasive plant. First, biochar was produced from Albizia by pyrolysis process at 850 °F. This biochar was then physically activated at 1500 °F with two different activation times (20 minutes and 35 minutes). The iodine number (an indicator of adsorbent quality, surface area and porosity) of each of these three products, biochar and two activated carbons were determined to be 132 mg/g, 835 mg/g and 1103 mg/g respectively. Batch kinetic experiments and isotherm studies were performed to compare the effect of activation levels of activated carbons (iodine number of 1103 mg/g and 835 mg/g) with biochar (used as base-line with iodine number of 132 mg/g) on phenol adsorption. Batch kinetic studies showed that phenol removal can be increased more than 1.5 times with a target dosage range 200 mg/L to 400 mg/L if activated carbons are used rather than biochar. The equilibrium data represented by Freundlich isotherm model also supported the above results showing the enhanced performance of activated carbon compared to biochar on phenol removal.

## INTRODUCTION

Phenols are toxic organic pollutants, common in industrial effluent such as oil refineries, coal tar, disinfectants, pharmaceutical and steel industries (Lin and Juang, 2009). Besides these industrial sources, phenolic derivatives exist in pesticides, insecticides, domestic wastewater. The US Environmental Protection Agency (EPA) regulations call for lowering phenol content in the wastewater to less than 1 mg/L (Banat, et al., 2000). Removal of the phenolic pollutant from wastewater is very difficult by conventional biological treatment process as they are resistant to microbial degradation. Besides biodegradation, other common phenol removal methods that are in practice include stripping, solvent extraction, oxidation, ion exchange and adsorption methods. Among them, the carbon adsorption method has become a popular method in recent years because of its large adsorption capacities, fast adsorption kinetics and relative ease of regeneration (Subramanyam and Das, 2009).

The adsorption of phenol depends on various factors such as surface areas, micro-pore volumes, adsorption capacities of the carbon. These characteristics are controlled by the operating conditions (e.g. temperature, heating rate, residence time etc.) during their production/activation. This operating condition is controlled to achieve the desired activation level (i.e. quality) of the product for its end-use. As for example, for the targeted application of carbon to remove phenol from aqueous solution, various qualities of carbon products can be produced by controlling two major operating conditions (temperature and residence time) during pyrolysis and activation process. However, the prediction of optimum activation level for most efficient removal of phenol from aqueous solution is quite difficult without laboratory tests and/or model study. Therefore, it is essential to test the effectiveness especially for the products which are derived from low-cost, less-known invasive plants like Albizia.

In our recent research (Mayers, 2014), biochar and various grades of activated carbons were produced from Albizia, an invasive fast-growing plant of tropical regions. The biochar was produced by pyrolysis method and this biochar was further physically activated to produce activated carbon of two activation levels by controlling the activation residence times. Activation level and porosity of these carbon products from this low-cost feedstock were assessed by iodine numbers, a popular parameter used to estimate relative surface area, adsorptive capacity (Wang et al., 2013). To evaluate the performance of phenol removal from water, in this paper, bio-char (iodine number 132 mg/g) is used as a base-line. Then, two different activated carbons (iodine numbers 835 mg/g and 1103 mg/g) were used to compare their effectiveness with the base-line (i.e. biochar).

## MATERIALS AND METHOD

**Adsorbent Preparation.** Biochar and activated carbon from Albizia plant was prepared by the procedure described in detail elsewhere (Mayers, 2014). Initially, the Albizia chips were cleaned with brushes, all foreign materials were removed and then oven dried. The dried material was then pyrolysed at 850 °F using a pyrolysis apparatus schematic diagram of which is shown in Figure 1.

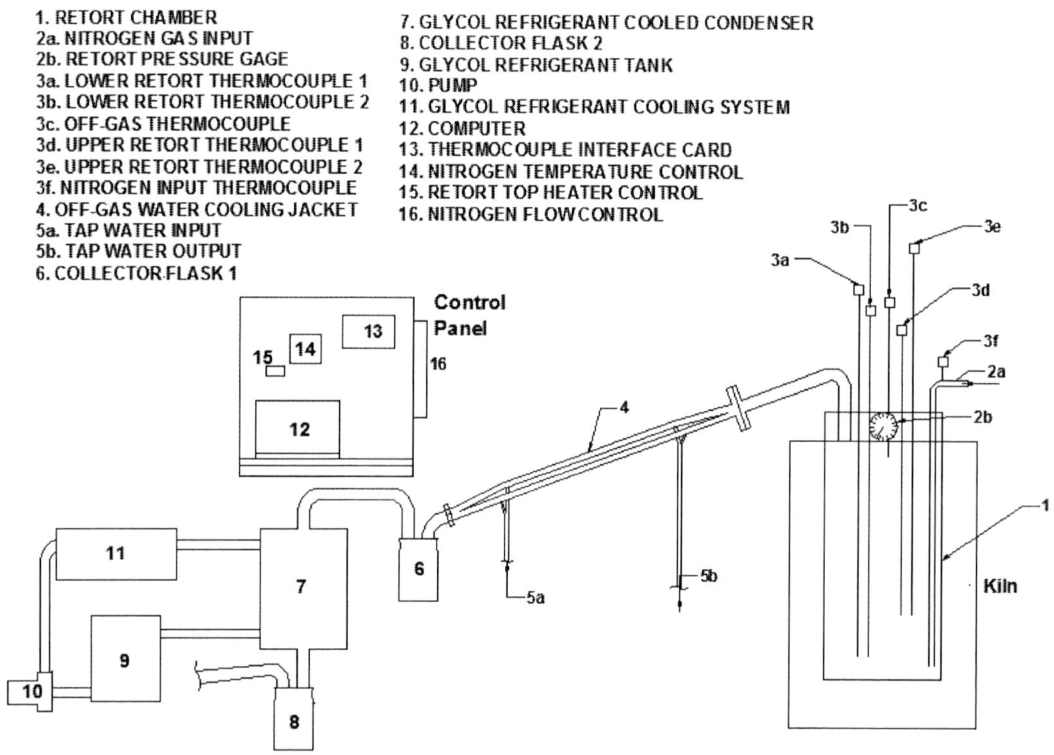


**FIGURE 1.** Schematic diagram of pyrolysis apparatus

**TABLE 1: Iodine number, surface area and major preparation conditions of adsorbents**

| Absorbent | Preparation Conditions | | | Adsorbent Quality | |
|---|---|---|---|---|---|
| | Pyrolysis Temperature (°F) | Activation Temperature (°F) | Activation time (minute) | Iodine Number (mg/g) | Surface Area ($m^2/g$) |
| Biochar (referred BC#132 afterwards) | 850 | - | - | 132 | 130 |
| Activated Carbon (referred AC#835 afterwards) | 850 | 1500 | 20 | 835 | 823 |
| Activated Carbon (referred AC#1103 afterwards) | 850 | 1500 | 35 | 1103 | 1009 |

The biochar produced after pyrolysis (in retort chamber #1) is then physically activated (steam activation) at 1500 °F for two residence times (20 minutes and 35 minutes) and two different grades of activated carbons were produced. The iodine numbers (IN) of each of these three products were determined. The surface areas of the same products were estimated using the model proposed by Mianowski et al. 2007. The values of iodine number (IN), surface area of each product and their production operating conditions are shown in Table 1.

**Adsorption Procedure.** All the three carbon products (BC#132, AC#835, AC#1103) were first ground using a standard mortar and pestle then passed through a #250 sieve. The samples were then dried at 320 °F for a period of 24 hours. Different amount of adsorbents (dose ranging from 20 mg/L to 400 mg/L) were then added to each 1000 mL flask containing 1.0 mg/L (i.e. 1000 µg/L) phenol solutions. All these 1000 mL flasks were sealed and installed in a water shaker (120 rpm) with a constant temperature of 77 °F and the adsorption experiments were run at a batch mode. After equilibrium (2 hour), the mixing was stopped, and the mixed solution was filtered with filter paper (Whatman No. 42) and analyzed for the residual phenol concentration, which was determined by measuring its absorbance at 510 nm using a Perkin Elmer Lamda XLS spectrophotometer (Method 420.1, EPA). Phenol removal percentage (R%) was calculated as follows (Eq. (1)):

$$R\% = [(C_0 - C_e)/C_i] \text{ x } 100 \qquad (1)$$

The amount of adsorption at equilibrium, $q_e$ (mg/g), was calculated by (Eq. (2)):

$$q_e = [(C_0 - C_e)V]/M \qquad (4)$$

Where, $C_0$ and $C_e$ (mg/L) are the liquid-phase concentrations of phenol at initial and equilibrium, respectively. V (L) is the volume of the solution, and M (g) is the mass of dry adsorbent used.

## RESULTS AND ANALYSIS

**Comparison of Phenol Removal.** Figure 2 shows the phenol removal using various doses of BC#132, AC#835 and AC#1103 from the solution of pH 6.2. Carbon dosage was varied from 24 mg/L to 400 mg/L. The results show that the activated carbons (AC#835 and AC#1103) are more effective than biochar (BS#132) for the quantitative removal of 1 mg/L (i.e. 1000 µg/L) of phenol in 1000 ml. The phenol removal can be increased more than 1.5 times with a target dosage range 200 mg/L to 400 mg/L if activated carbons

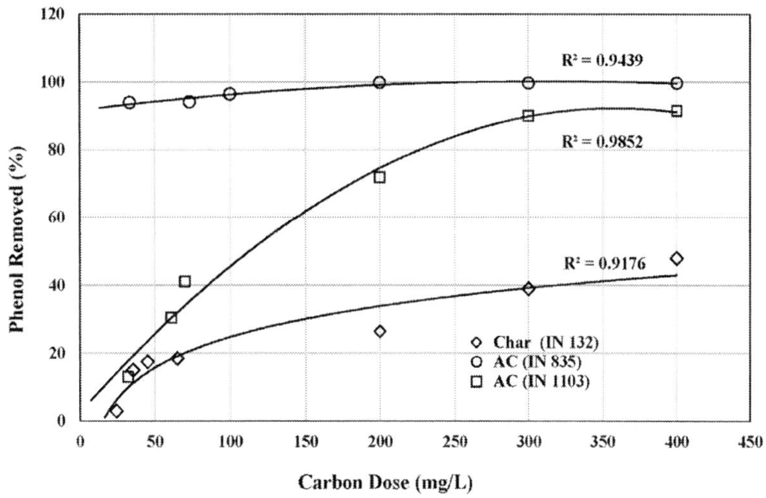


**FIGURE 2.** Effect of carbon dose on removal of phenol

are used rather than biochar. For example, for biochar (base-line), 39% phenol removal is achieved with a dose of 300 mg/L but at that same dose, AC#1103 achieves a removal of 90% while AC#835 achieves 99.7%.

The removal of phenol using BS#132 and AC#1103 significantly improves up to the doses of 200 mg/L. Beyond this dose (200 mg/L), the change of removal efficiency is relatively small for these two products. However, for AC#835, a low carbon dosage of 100 mg/L can achieve significant (96%) removal of phenol. Beyond this dosage, the change of removal efficiency of AC#835 is minor. It is evident from the results that among these three products, the best phenol removal can be achieved by the activated carbon with IN of 832 mg/g (AC#835).

**Phenol Adsorption on the Adsorbents.** Phenol adsorption onto three different adsorbents were calculated and shown in Figure 3 (a). Comparing the capacity of phenol adsorption at 2 hour time, it is clear that activated carbons (AC#835 and AC#1103) are much better performer than biochar (BC#132, base-line) for phenol removal. Maximum phenol adsorption achieved was 5 mg/g using AC#835 while lowest adsorption was 1.2 mg/g on BC#132 (Figure 3(a)) with a 200 mg/L of dose. High phenol adsorption was also achieved by using AC#1103 (3.6 mg/g) with the same dose. This reveals that activation is needed to enhance quality of biochar. However, optimum activation condition to capture maximum phenol is approximately 1500 °F temperature and 20 minute residence time (that resulted AC#835).

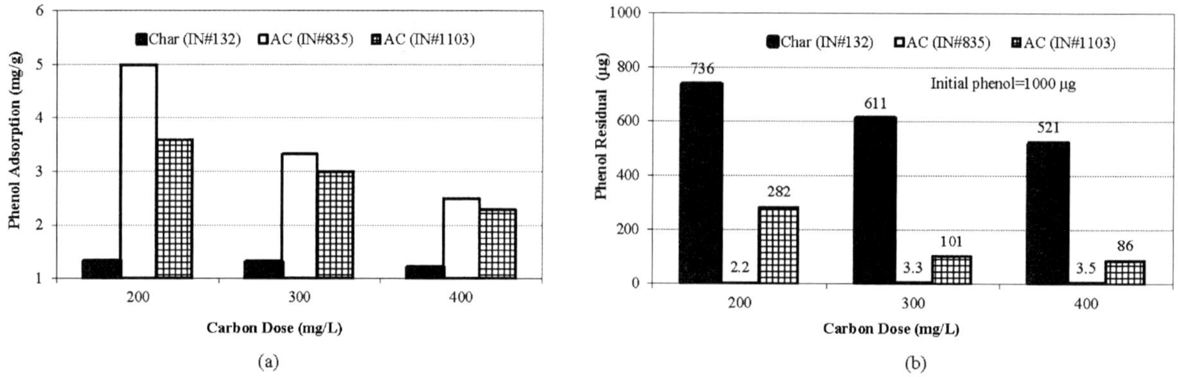


**FIGURE 3.** (a) Phenol Adsorption by carbon products, (b) Phenol residual at equilibrium.

For AC#835, phenol removal due to addition of this product beyond the dose of 200 mg/L is relatively small (Figure 2). At this dose, 99.8% phenol is removed using this AC#835. Also, Figure 3 (a) shows that the dose of 200 mg/L exhibits better phenol adsorption rate (5 mg/g) for this product compared to the other doses (i.e. 300 mg/L, 400 mg/L). This suggests that the 200 mg/L of AC#835 dose could be used for high performance phenol treatment process.

**Phenol Residual in Water.** Figure 3(b) shows the results of residual phenol in the solution at equilibrium (2 hour). As seen in this Figure, the maximum phenol residual in the solution was 736 μg when BC#132 is used while residual was lowest (2.2 μg) when AC#835 is used. In other words, maximum phenol was removed when AC#835 was used. Low residual of phenol was also found when AC#1103 was used. This again shows that biochar activation significantly improves the adsorption capacity of the adsorbent and thereby improves the phenol removal efficiency. This also implies that by enhancing the surface area and porosity to an optimum level, the biochar can be converted to a high-performing product (like AC#835).

**Adsorption Isotherm.** Adsorption isotherm plays a vital role for the analysis and design of adsorption systems as well as model prediction (Mukherjee et al., 2007). It describes how solutes interact with adsorbents, and is critical in optimizing the use of adsorbents (Hameed and Rahman, 2008). The Freundlich isotherm models were commonly used to describe the relationship between the amount of phenol adsorbed and its equilibrium concentration in solutions. The Freundlich isotherm is given by (Eq. (3)):

$$q_e = K_F C_e^{1/n} \qquad (3)$$

where $q_e$ (mg/g) amount of adsorption at equilibrium, $C_e$ (mg/L) is the liquid-phase concentrations of phenol at equilibrium. $K_F$ and n are Freundlich constants. The linear form of the above Equation is (Eq. (4)):

$$\log q_e = \log K_F + (1/n) \log C_e \qquad (4)$$

which will have a straight line with a slope of 1/n and an intercept of log($K_F$) when log($q_e$) is plotted against log($C_e$). $K_F$ ((mg/g)(L/mg)$^{1/n}$) is an indicator of the adsorption capacity. The larger $K_F$ value indicates that the adsorbent has a larger adsorption capacity and a favorable adsorbent (Mukherjee et. al., 2007). On the other hand, 1/n value ranging between 0 and 1 is a measure of adsorption intensity. The value of 1/n less than 1 represents a favorable adsorption condition (McKay, et al., 1982).

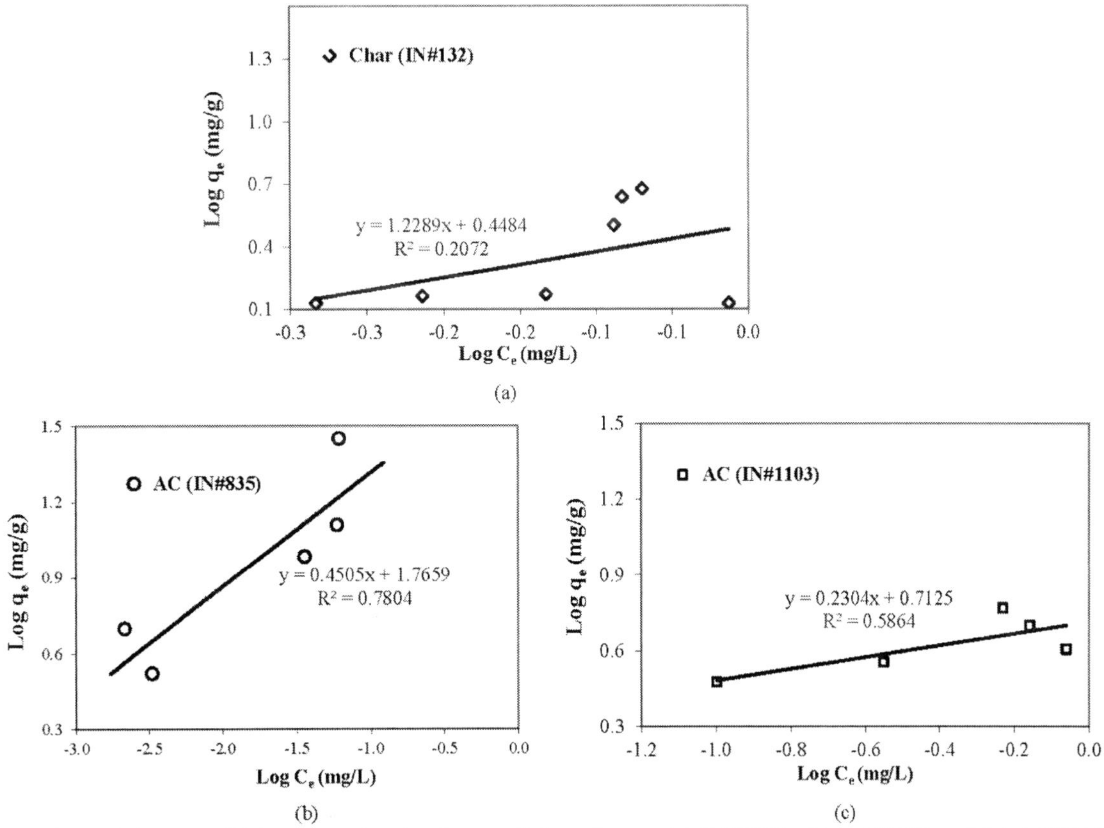


**FIGURE 4.** Freundlich adsorption isotherm of phenol onto (a) BC#132, (b) AC#837 and (c) AC#1103.

The data obtained in this study were used to plot log $q_e$ vs. log $C_e$ and presented in Figure 4. The values of constants $K_F$ and 1/n are also calculated and shown in Table 2. A comparison of the values of $K_F$ reveals that AC#835 has the highest $K_F$ value (58.33) indicating it has the largest adsorption capacity and most favorable adsorbent among the three products. The lowest $K_F$ value of 2.81 obtained for BC#132 suggests that it has the lowest adsorption capacity. Furthermore, comparing the value of 1/n for the three products, it is found that 1/n for AC#835 and AC#1103 are less than 1 while the 1/n for BC#132 is more than 1. These values again indicate the easy adsorption by AC#835 and AC#1103 compared to BC#132. For AC#835 adsorbent ($R^2$=0.7804), the Freundlich isotherm model gave a slightly better fit than the AC#1103 ($R^2$=0.0.5864). However, BC#132 results show poor fit of the Freundlich isotherm model. This model study also suggests that biochar activation to an optimum level enhances the performance of the adsorbent.

**TABLE 2. Freundlich isotherm coefficients**

| Absorbent | $K_F$ ((mg/g)(L/mg)$^{1/n}$) | 1/n | $R^2$ |
|---|---|---|---|
| Biochar (BC#132) | 2.81 | 1.2289 | 0.2072 |
| Activated Carbon (AC#835) | 58.33 | 0.4505 | 0.7804 |
| Activated Carbon (AC#1103) | 5.16 | 0.2304 | 0.5864 |

## CONCLUSIONS

The phenol adsorption performance by various grades carbon products from Albizia was investigated using batch tests. The study demonstrates that it is possible to remove phenol from aqueous solution in a safe manner using biochar and activated carbon produced from Albizia, an invasive plant. However, preparation method of these carbon products has significant effect on phenol adsorption capacity. Comparing the adsorption performance, the study suggests that phenol removal can be enhanced significantly when biochar is activated and converted to activated carbon. Activated carbon produced with a 20 minute residence time (i.e. AC#835) at temperature 1500 °F proved to be most effective for phenol removal. The removal efficiency of this carbon product is 1.5 times more than that of biochar with a target dosage range 200 mg/L to 400 mg/L. Even a small dose of 100 mg/L of AC#835 can achieve more than 96% removal efficiency.

## REFERENCES

Banat, F.A., B. Al-Bashir, S. Al-Asheh and O. Hayajneh. 2000. "Adsorption of phenol by bentonite", *Environ. Pollut.* 107: 391-398.

Hameed, B.H., A.A. Rahman, 2008 "Removal of phenol from aqueous solutions by adsorption onto activated carbon prepared from biomass material", *J. Hazard. Mater.* 160: 576–581.

Lin, S-H. and R-S. Juang. 2009. "Adsorption of phenol and its derivatives from water using synthetic resins and low-cost natural adsorbents: a review". *J Environ. Manag.* 90(3): 1336–1349.

Mayers, J. 2014. *Pyrolysis and Activation of an Invasive Species.* M.S. Thesis, Florida State University, Tallahassee, FL.

McKay, G., H.S. Blair, J.R. Gardner. 1982. "Adsorption of dyes on chitin-1: equilibrium studies", *J. Appl. Polym. Sci.* (27): 3043–3057.

Mianowski, A., M. Owczarek, and A. Marecka. 2007. "Surface Area of Activated Carbon Determined by the Iodine Adsorption Number." Energy Sources, Part A, 29.9, pp 839-850. Taylor & Francis, LLC, Abingdon, UK.

Mukherjee, S., S. Kumar, A. K. Misra, M. Fan. 2007. "Removal of phenol from water environment by activated carbon, bagasse ash and wood charcoal", *Chem. Eng. J.* 129: 133-142.

Subramanyam, B. and A. Das. 2009. "Study of the adsorption of phenol by two soils based on kinetic and isotherm modeling analyses", *Desalination* 249(3): 914–921.

Wang X., D. Li, W. Li, J. Peng, H. Xia, L., Zhang, S. Guo and G. Chen. 2013. "Optimization of mesoporous activated carbon from coconut shells by chemical activation with phosphoric acid" *BioResources*, 8(4): 6184-6195.

# EXPERIMENTAL STUDY OF CONGO RED AND DIRECT RED 80 ADSORPTION FROM WATER ONTO CARBON NANOTUBES

***Madhu Agarwal**** & Priti Kumari (Department of Chemical Engineering, Malaviya National Institute of Technology Jaipur, India)

**ABSTRACT**: This study investigated the potential use of carbon nanotubes as an adsorbent for the removal of Congo red & Direct red 80 dyes from aqueous solution. Batch adsorption study was carried out for removal of both the dyes from aqueous solution in the presence of carbon nanotubes. The effect of various operating conditions such as adsorbent dose, initial dye concentration, and pH, temperature & contact time on dye adsorption on carbon nanotubes was studied. It was found that for both the dyes maximum removal was achieved at pH 2. The removal of dyes increases with adsorbent dosage but decreases with initial dye concentration. Langmuir isotherm gave a better fit to the adsorption compare to the Freundlich for both dyes. Adsorption kinetics followed the pseudo second order kinetic. The adsorption efficiency of carbon nano-tubes was estimated to 125 mg/g for Congo red & 79.325mg/g for direct red 80 at adsorbent dose of 0.010g/50 ml of 10ppm dye solution at equilibrium time of 80 min. and 105 min. respectively and at optimum dose of 0.015 the capacity becomes 100 mg/g for Congo red & 66 mg/g for direct red 80.

## INTRODUCTION

Dyes are widely used in industries such as textiles, paper, paints, leather, rubber, plastics, cosmetics, food and drug etc. to color their products. Textiles industries consumed more than 60 % dyes of the total world production. It is estimated that 10-20% of dye was lost during the dying process and release as effluent (Zhiqiao et al.,2007). Due to their chemical structures, dyes are resistant to fading on exposure to light, water and many chemicals and, therefore, are difficult to be decolorized once released into the aquatic environment (Sharma and. Janveja,2008). Many of the organic dyes are hazardous and may affect aquatic life and even the food chain. Therefore, it is very important to develop new systems that can be used for removing dyes from waters. Research is being carried out for removal dyes using physicochemical, chemical and biological treatment technologies, such as chemical coagulation–flocculation (Pavan et al., 2008) different type of oxidation processes (Neelmegani et al., 2004), membrane filtration and adsorption (Purkait et al., 2007) for the removal of coloured dye from wastewater. Among these advanced treatments, at this moment, adsorption is considered more effective and less expensive than other technologies. Activated carbon is the most used material owing to its adsorption capacity, but it is very expensive and its regeneration is difficult (Sharma and. Janveja,2008). Since adsorption was best suitable and reasonable method for removal of dyes hence many authors were used different adsorbents for dye removal. Nagarethinam and Mariappan, 2002, used the bamboo dust carbon for adsorption of CR dye and they achieved a maximum adsorption capacity of 101.9mg/g for removal of CR dye. Recently cylindrical graphene- CNT hybrid was developed by Lunhong and Jing, 2012 and they used it for adsorption of Methylele blue (MB) dye and achieved a maximum adsorption capacity of 81.97 mg/g for the removal of MB from aqueous solution. The removal efficiency reached 97% for low concentration (10mg/l initial) of MB. In recent times carbon Nano tube has got greater attention in field of scientific development. It has shown a great prospect in developing the future Nano structure with novel properties. The high strength and stiffness property of CNTs have made it a potential reinforcing material. In the present research multiwall carbon nanotube (MWCNT) in the range of 8-15nm subsequently applied for the dyes removal. The effect of factor such as initial pH, adsorbent dose (mg/ml) and contact time (t) on Congo red (CR) and direct red 80(DR 80) removal percentage was investigated and optimized. Isotherm of adsorption was examined by fitting the experimental data to various isotherm equations, and it was found that Langmuir is the best model with high correlation coefficient for both dyes. The kinetics of adsorption using pseudo first order and second order were tested with experimental data for their validation.

## MATERIAL AND METHODS

**Batch Experimental Studies.** Batch experiments were conducted to investigated the effect of parameters like pH, contact time And adsorbent dose, on the adsorptive removal of CR & DR 80. For each experimental run, 50ml of 10mg/L of CR & DR 80 solution at various pH in the range of 1-8 was mixed with 10mg/50ml to 25mg/50ml of adsorbent. This mixture was agitated in incubator shaker with constant speed of 200 rpm till equilibrium reached. Thereafter, all the samples were centrifuged at 4000rpm for 20min to settle down suspended particles. After centrifugation clear supernatant samples were obtained and the supernatant solution was estimated accurately by UV spectrophotometer for remaining dye in the solution.The kinetics of adsorption was determined by analyzing adsorptive uptake of the dye from the aqueous solution at different time interval. For adsorption isotherms, different concentration of dye solution was agitated with known amount of adsorbent till the equilibrium.

The amount of dye adsorbed was calculated from the following equation:

$$q = \frac{(C_0 - C_e)V}{m} \quad \textbf{(1)}$$

Where $q$ is the amount of dye adsorbed per unit weight of adsorbent (mg/g); $C_0$ the initial concentration (mg/l); $C_e$ the concentration of CR or DR 80 in solution at equilibrium time (mg/l); $V$ the solution volume (l); $m$ is the adsorbent dosage (mg/ml).

## RESULT AND DISCUSSION

**SEM analysis of commercial CNT.** The CNTS is characterized by SEM to observe very fine structures of samples. Fig 1 shows a SEM image of CNTs .The noodle shapes in the figure are CNTs having average diameter 12 nm. CNTs formed in cluster and very thin, many roots of carbon not properly converted into carbon nanotubes due to presence of some impurities.

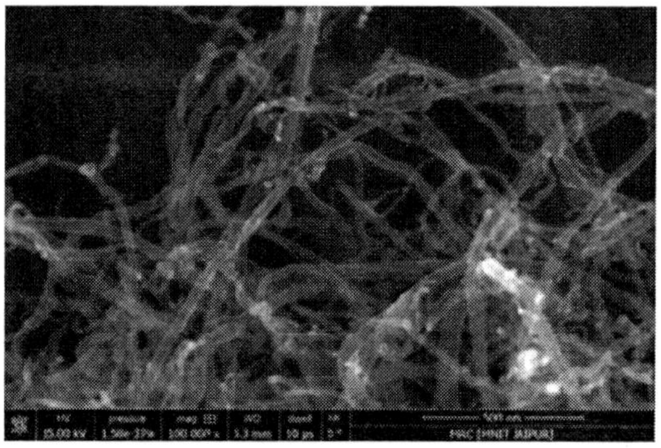

**FIGURE 1:** SEM Images of CNT

**Effect of pH.** The medium pH is a critical parameter in adsorption process as it strongly affects the structural stability and color intensity of CR and DR 80 dye. After adjusting the pH value (1-8), solution was kept in incubator. After equilibrium time (150 min) the dye solution was withdrawn and centrifuged. The supernatant was checked for presence of dye in UV spectrophotometer. Fig 2 Shows that at pH 2-5 more than 90% dye has been removed thereafter removal of dye decreases as pH increases. The maximum removal (99.1%) was obtained at pH 2.Thats why further experiment is conducted at pH 2.

**Effect of Adsorbent Dosage and Initial Concentration.** The effect of adsorbent dosage on CR adsorption was conducted by varying adsorbent dosage from 10mg/50ml to 25mg/50ml at dye concentration of 10 ppm. With increase in adsorbent dosages (CNT) the percent removal of dye also increases for both CR and DR 80, due to the increase in adsorbent surface area hence, increase in total number of exchange sites. As depicted in fig 3(a) and 3(b) when adsorbent dosage change from 15 mg/ml to 20 mg/ml, there is slight change in percent removal, therefore, it was consider that 15 mg can be taken as optimum dose for both the dyes.

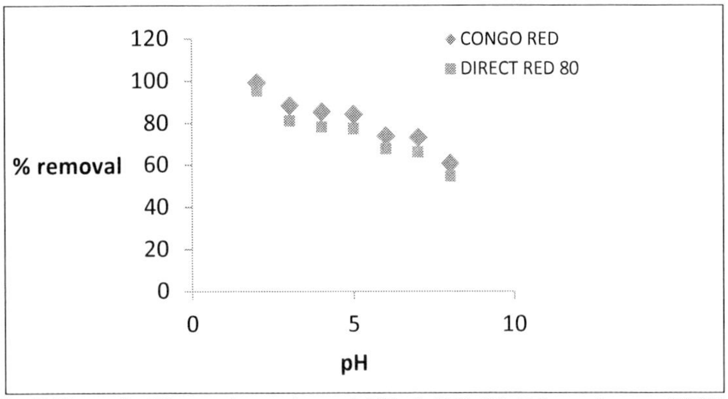


**FIGURE 2.** Effect of pH on adsorption of CR & DR 80 (10ppm) onto CNT (15mg/ml) at 303K, agitation speed 200rpm for the equilibrium time

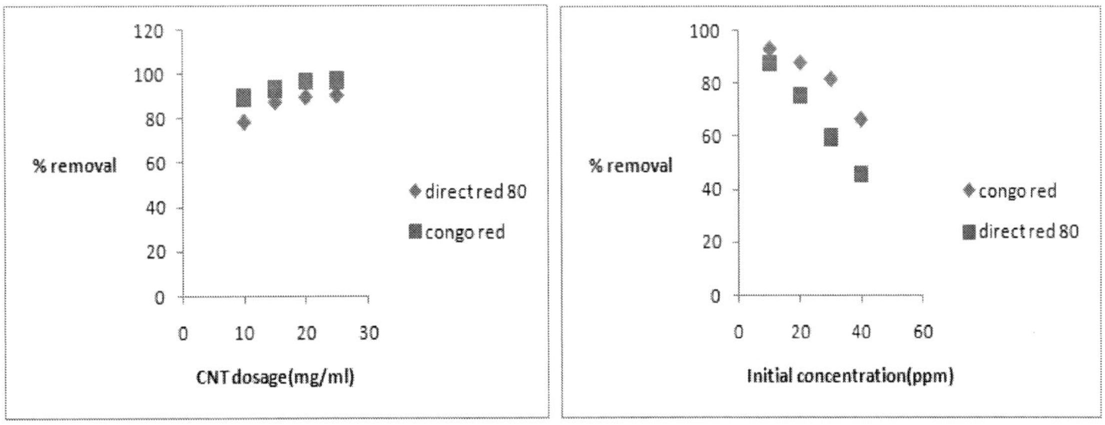


(a) (b)

**FIGURE 3.** Effect of (a) Adsorbent Dosage (10, 15, 20, 25 Mg) at Dye Concentration 10ppm (b) Initial Concentration (10, 20, 30, 40ppm) at Adsorbent Dosage 15mg, Contact Time 80 min for CR and 105 min for DR 80, Temperature (303K) & Agitation Speed 200 Rpm for Congo Red & Direct Red 80

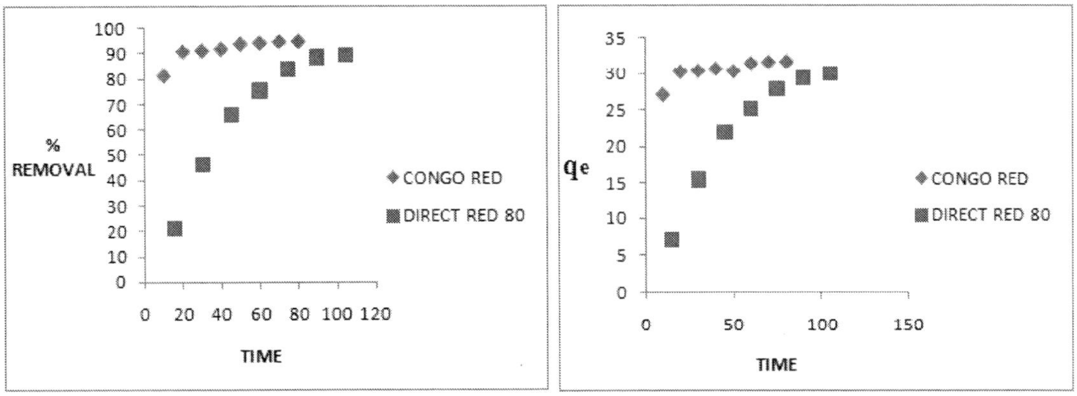


(a) (b)

**FIGURE 4.** Effect of Contact Time on (a) Percent Removal (b) Amount Adsorbed per unit mass of adsorbent at 10 ppm dye concentrations, 303 K temp., and adsorbent dose 15mg /50ml dye solution

The effect of initial dye concentration on adsorption of the CR and DR 80 was investigated in the concentration range 10-40 ppm. The results are shown in fig 3(a, b). When concentration was 10ppm percent removal was maximum as compare to 20, 30, 40 ppm, hence further study is carried out at 10 ppm.

The decrease in percent removal with increase in initial dye concentration was more for DR 80 than CR; this may be due to less affinity of DR 80 with CNT compared to CR.

**Effect of Contact Time and Concentration of Dye on Adsorption.** The Fig 4(a, b) shows percent removal and the amount of dye adsorbed per unit mass of adsorbent with respect to time. Both percent removal and qe (mg/g) increases with increase in contact time. The percent removal of dye increases sharply within initial 15min. after that it increases slowly and at 80min it become constant for CR. For DR 80 the increase in percent removal was slow from beginning and become constant at 105 min. More than 90% removal was achieved for both the dyes at equilibrium time.The amount of dye adsorbed per unit mass of adsorbent was very high for CR at beginning but same was very low for DR 80. Close to equilibrium time both dyes adsorbed efficiently on adsorbent under consideration (fig 4(b).

**Adsorption Kinetics.** In order to investigate the adsorption process of CR and DR 80 on the adsorbent (CNT), pseudo first-order and pseudo second-order kinetics models were fitted to experimental data. The pseudo-first-order equation used is given below;

$$\log(q_e - q_t) = \log q_e - \frac{k_1}{2.303}t, \qquad \textbf{(2)}$$

Where $q_e$ and $q_t$ are the amount of CR and DR 80 adsorbed at equilibrium and at a time t, $k_1$ is the first order rate constant (min). The slope and intercept of plot Log ($q_e$-$q_t$) v/s t were used to determine the pseudo first order rate constant $k_1$ and $q_e$. Pseudo second order model is;

$$\frac{t}{q_t} = \frac{1}{k_2 q_e^2} + \frac{1}{q_e}t, \qquad \textbf{(3)}$$

Where $k_2$ (g/mg min) is the rate constant, slope and intercept of plots t/qt v/s t are used to calculate $q_e$ and $K_2$. The results obtained for constants were summarized in Table 1. The fig. 5 (a,b) shows that pseudo second order kinetics model better fits to the experimental data obtained for both dyes.

**TABLE 1.** Kinetic Parameters for CR and DR 80 Removal using 15 mg/ml Adsorbent Dose over Dye Concentration of 10 ppm

| Type Of Dye | Pseudo First Order Model | | | Pseudo Second Order Model | | |
|---|---|---|---|---|---|---|
| | $K_1$ | $q_{ecal}$ | $R^2$ | $K_2$ *10 | $q_{ecal}$ | $R^2$ |
| Congo Red | 0.02533 | 1.7576 | 0.648 | 0.155 | 33.33 | 0.999 |
| Direct Red 80 | 0.0368 | 4.75882 | 0.939 | 0.00162 | 62.5 | 0.944 |

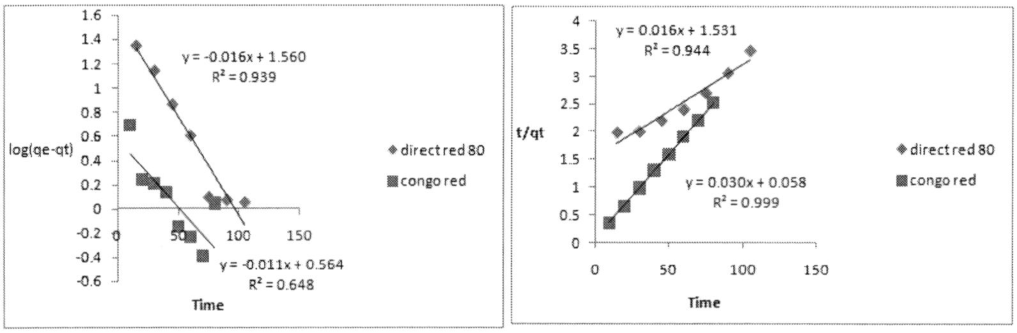


**(a)** **(b)**

**FIGURE 5.** Plots for Different Kinetics Model (a) Pseudo 1st Order and (b) Pseudo Second Order for Removal of CR And DR 80 At 10ppm Concentration, 303 K, Adsorbent Dose 15g/50ml & Equilibrium Time 80 min

Correlation coefficient for pseudo second order is closer to unity compare to pseudo first order kinetics. Therefore sorption can be approximated more appropriately by the pseudo-second-order kinetic model than the first-order kinetic model for the adsorption of CR and DR 80 as depicted in table 1.

**Adsorption Isotherm.** The dyes adsorption equilibrium data were fitted for Langmuir; Freundlich and Temkin Isotherms (fig. 6). The optimization of an adsorption process requires an understanding of the driving forces that govern the interaction between adsorbent and adsorbate. The $R^2$ value in case of Langmuir isotherm for CR is 0.996 and for DR 80 is 0.9997. The isotherm results indicate that the adsorption of CR and DR 80 onto adsorbent (CNT) is consistent with Langmuir isotherm model as shown in fig. 6(a). This value is greater than those $R^2$ values calculated for the Freundlich & Temkin isotherm, listed in table 2. Langmuir isotherm represented by,

$$\frac{C_e}{q_e} = \frac{C_e}{q_{max}} + \frac{1}{q_{max}k_L} \quad \textbf{(4)} \qquad R_L = 1/(1+K_a C_o) \quad \textbf{(5)}$$

Where $R_L$ is equilibrium parameter express as essential characteristics of Langmuir isotherm; Ce is the concentration of dye (ppm) at equilibrium. The constant $Q_m$ is the adsorption capacity (mg/g) and $K_L$ is related to energy of adsorption (l/mg). Value of $Q_m$ and $K_L$ calculated from the slope and intercept of the linear equation of isotherm.

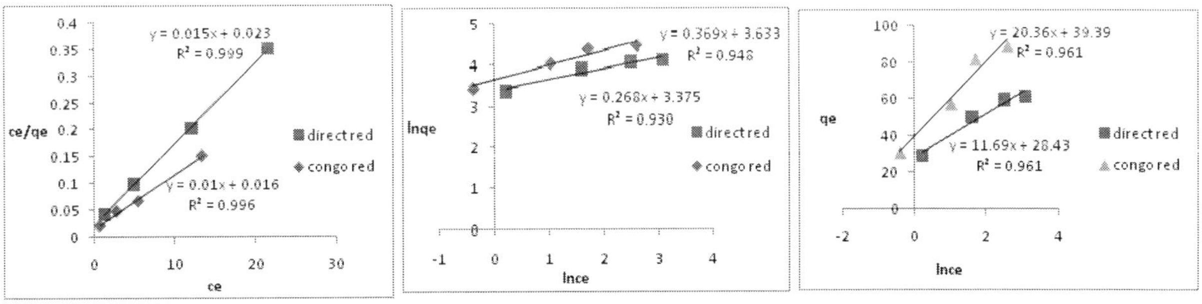


**(a)** **(b)** **(c)**

**FIGURE 6.** Isotherm of Adsorption for Removal of Congo Red And Direct Red 80 Dyes by Commercial CNTS (a)Langmuir (b) Freundlich (c) Temkin

**TABLE 2.** Isotherm Constants of CR and DR 80 onto Multiwall CNT

| Type of Dye | | | | Temkin | | | Langmuir | | | |
|---|---|---|---|---|---|---|---|---|---|---|
| | 1/n | $K_f$ | $R^2$ | $B_1$ | $K_T$ | $R^2$ | $Q_m$ | Ka | $R^2$ | $R_L$ |
| Congo Red | .37 | 37.84 | 0.947 | 20.35 | 6.92 | 0.96 | 100 | 0.616 | 0.996 | 0.0392-.14 |
| Direct Red 80 | 0.26 | 29.25 | 0.93 | 11.7 | 11.36 | 0.96 | 65.789 | 0.661 | 0.9997 | .0692-.17 |

**TABLE 3.** Comparison of adsorption capacity of various adsorbents for adsorption of CR and DR 80

| Adsorbent | Dye | Adsorption Capacity(mg/g) | References |
|---|---|---|---|
| CNT | Direct Red 80 | 66.00 | Present Work |
| Mixture Almond Shells | Direct Red 80 | 22.42 | Doulati et al., 2008 |
| Orange Peel | Direct Red 80 | 21.05 | Arami et al., 2005 |
| CNT | Congo Red | 100.0 | Present Work |
| Magnetically Modified Fodder Yeast Cell | Congo Red | 49.7 | Safarik et al., 2007 |
| Palm Kernel Seed Coat | Congo Red | 66.23 | Yuzhu and Viraghavan 2002 |
| Fungus Aspergillus Niger | Congo Red | 14.72 | Oladoja and Akinladi 2009 |
| Carbonized Coir Pith | Congo Red | 6.7 | Yuzhu and Viraghavan 2002 |
| Banana Peel | Congo Red | 18.2 | Annadurai et al., 2002] |

## CONCLUSION

The study presented revealed that MWCNT can be used as a good and low cost adsorbent with high efficiency as reported in table 3. With high adsorption capacity of 100 mg/g for CR in equilibrium time 80 min & for DR 80 adsorption capacity of 66.0 mg/g in equilibrium time 105 min. At pH 2 and at adsorbent dose of 0.015g percent removal was more than 95% for both the dyes. Langmuir isotherm gave a better fit to the adsorption isotherm compared to the Freundlich and Temkin adsorption isotherm for both dyes. A theoretical adsorption capacity of 125 mg/g for Congo red & 79.325mg/g for direct red 80 was obtained at lower adsorption dose (0.010g). Adsorption Kinetics followed the second order kinetics model.

## REFERENCES


Annadurai, G., R.S. Juang, and D.J. Lee. 2002. "Use of cellulose-based wastes for adsorption of dyes from aqueous solutions". *J. Hazard. Mater.* B92: 263–274.

Arami, M., N. Yousefi Limaee, N.M. Mahmoodi, and N. Salman Tabrizi. 2005. "Removal of dyes from colored textile wastewater by orange peel adsorbent: equilibrium and kinetic studies". *J. Colloid Interf. Sci.* 288: 371–376.

Doulati Ardejani, F., Kh. Badii , N. Yousefi Limaee, S.Z. Shafaei, and A.R. Mirhabibi. 2008. "Adsorption of direct red 80 dye from aqueous solution onto almond shells: effect of pH,initial concentration and shell type". *Journal of Hazardous Materials.* 151: 730–737.

Lunhong, Ai and J. Jing. 2012. "Removal of methylene blue from aqueous solution with self-assembled cylindrical graphene–carbon nanotube hybrid".*Chemical Engineerin Journal.*192:156-163.

Nagarethinam, K., and M. Mariappan,2002. "Adsorption of congo red on various activated carbons, a comparative study".*Water Air Soil Pollut.*138: 289–305.

Neelmegani, R., V. Baskaran, R. Dhancekar, and T.Viruthgiri. 2004. "Decolorisation of synthetic dyes using rise straw attached pleurotus oslereths". *Indian J. Chem. Techo.* 11: 622-625.

Oladoja and Akinlabi. 2009. "Congo red biosorption on palm kernel seed coat". *Ind. Eng. Chem. Res.*48: 6188–6196.

Pavan, F. A., S. L.P. Dias, E. C. Lima, and E.V. Benvenutti. 2008. "Removal of Congo red from aqueous solution by anilinepropylsilica xerogel". *Dyes and Pigments.* 76: 64-69.

Purkait, M.K., A Maiti, Das Gupta, and S. De.2007". "Removal of congo red using activated carbon and its regeneration"*J. Hazard. Mater.* 145: 287–295.

Safarik,I.,L.F.T.Rego,M.Borovska, E.Mosiniewicz-Szablewska, F. Weyda, andM. Safarikova.2007."New magnetically responsive yeast-based biosorbent for the efficient removal of water-soluble dyes". *Enzyme Microb. Technol.* 40:1551–1556.

Sharma, J., and B. Janveja. 2008. "A study on removal of congo red dye from the effluents of textile industry using rice husk carbon activated by steam". *Rasayan J. Chem.* 1(4): 936-942.

Yuzhu, F.,and T.Viraraghavan. 2002. "Removal of Congo red from an aqueous solution by fungus Aspergillus Niger". *Adv. Environ. Res.* 7: 239–247.

Zhiqiao, H., S. Shuang, Z. Huamin, Y. Haiping, and C. Jianmeng. 2007. "C.I. Reactive Black 5 decolorization by combined sonolysis and ozonation". *Ultrasonics Sonachem.* 14 (3):298-304.

# COPPER(II) BIOSORPTION CHARACTERISTICS OF A NOVEL BIOSORBENT COMBINATION: FUNGUS AND AGRICULTURAL WASTE

Jingyao Wang and Chongwei Cui* (Harbin Institute of Technology, Harbin, China)

**ABSTRACT**: All previous attempts have examined fungus and agricultural waste separately. The present work is therefore novel as it uses a combination of fungus and agricultural waste for removal of copper(II) from aqueous solutions. The biosorbent was corn stalk fermented by *Aspergillus niger* (fermented corn stalk). Results show that the content of O in fermented corn stalk, compared with that of in raw corn stalk, declined as the C, N, P, and K increased, and that the zeta potential was lower. Fermented corn stalk exhibited a short biosorption equilibrium time (60 min), high biosorption capacity (11.2 mg/g), and high removal efficiency at pH 3. The fact that the biosorption capacity of fermented corn stalk was better than that of raw corn stalk under identical conditions can be attributed to the sprawl of *Aspergillus niger* (biosorption capacity was better than raw corn stalk)and the enzyme system secreted by *Aspergillus niger* which changed the surface properties of raw corn stalk. The biosorption mechanism was involved in chelation with functional groups such as N-H, C=O, C-O, and C=C, and the biosorption data obtained were well described by the pseudo-second order and the Freundlich isotherm model.

## INTRODUCTION

Recently, low cost adsorbent materials from the forestry, fishing, and agriculture industries have attracted the attention of several researchers thanks to their high volume, low value, and underutilized biomaterials, and because they contain high levels of cellulose, hemicellulose, and lignin (Abdolali et al., 2014). Corn is one of the most widely produced crops in the world and it can be used as an alternative raw material for the preparation of various functional polymers. However, due to the intermolecular hydrogen bonds involving cellulose hydroxyl groups in corn stalk, its biosorption capacity is generally low. Zheng used acrylonitrile to modify corn stalk to biosorb Cd(II) so that its biosorption capacity was triple that of raw corn stalk (Zheng et al., 2010a). The corn stalk was also treated to be corn stalk graft copolymers as a kind of biosorbent to biosorb heavy metal (Zheng et al., 2010b). Although these chemical modifications can improve biosorption capacity, we must still consider other aspects, such as secondary pollution of chemical modification and cost. Therefore, microbiological modification may be a processing technology. *Aspergillus nige*r, as one kind of mycelium fungus, was commonly used in the fermentation industry, because it has abundant enzymes (including amylase, pectinase, and cellulose) and is non-toxic and inexpensive (Odetallah et al., 2005). For the past two decades, researchers have been conducting biosorption studies with *Aspergillus nige*r, since it is easy to acquire and to separate from the liquid medium (Amini et al., 2009). The Cr(III) from tanning effluents was removed by *Aspergillus niger* and isolated from the tanning environment (Sepehr et al., 2012). Malihe Amini studied the biosorption of nickel(II) from aqueous solutions by *Aspergillus niger* with response surface methodology methods (Amini et al., 2009). Baik separated the components of *Aspergillus niger* mycelium and used them to biosorb heavy metal and found that chitosan contributed most to the biosorptive capacity (Baik et al., 2002). To the best of our knowledge, there are no reports on the application of a combination of fungus and agricultural waste in the removal of heavy metals. However, this combination of biosorbents has two advantages: first one was some enzymes secreted by the fungus can modify agricultural waste and change the properties of cell surfaces, and second was the propagation of fungi, which have a higher biosorption capacity than agricultural waste, can lead to the improvement of biosorption capacity. Hence, we decided to use *Aspergillus niger* to ferment corn stalk and called the combination of biosorbents "fermented corn stalk." The aim of this study was to testify to the effectiveness of this combination of biosorbents by comparing the biosorption capacities of fermented corn stalk and raw corn stalk under various dosages, deciding on the best pH for fermented corn stalk, and characterizing the biosorbents with SEM-EDX and Fourier infrared (FTIR) spectroscopy. Subsequently, the fermented corn stalk was systematically studied using a combination of kinetics and isothermal modeling.

## MATERIALS AND METHODS

***Microorganism.*** *Aspergillus niger* 98003, purchased from CMCC; Spore suspension: washed the aerial mycelium with water and adjusted the concentration to $10^{6.}$ The spore concentration of the final suspension was determined by counting in a hemocytometer.

***The preparation of corn stalk biosorbent.*** Corn stalk washed with distilled water, crushed with grinder (FW100, Tianjin Theis brand), sieved with a standard sieve to obtain powder with a particle diameter of 0.3 mm. The powder was dried in an electrically heated blast dry box to a constant weight.

***The preparation of fermented corn stalk.*** Spore suspension was added to the corn stalk powder at a ratio of 1:1. Seven days later, when the aerial mycelia of *Aspergillus niger* covered the entire medium, they were collected and placed in a dry box at 60° C to ensure constant weight.

***Comparison of raw corn stalk and fermented rice straw.*** Specific dosages (0.1, 0.3, 0.5, 0.7, 1.0 g) of raw corn stalk and fermented corn stalk were each placed in 100 mL of $CuSO_4.5H_2O$ solution, with an initial $Cu^{2+}$ concentration of 30 mg/L (since the concentration of most industrial wastewater is under 50mg/L) and shaken at 30° C and 100 rpm for two hours.

***The effect of different initial acidity on fermented corn stalk.*** Fermented corn stalks were placed in 100 mL of $CuSO_4.5H_2O$ solution with an initial $Cu^{2+}$ concentration of 30 mg/L and initial pH values of 2.0, 3.0, 4.0, 5.0, and 5.3 (the natural pH) achieved by addition of $H_2SO_4$. The flasks were shaken at 30° C and 100 rpm for two hours to ensure the best pH value.

***Experiment of fermented corn stalk on biosorption kinetics.*** Ten 0.1 g samples of fermented corn stalk were placed in 100 mL of $CuSO_4$ solution with an initial $Cu^{2+}$ concentration of 30 mg/L and were shaken at 30° C and 100 rpm for 1, 3, 5, 10, 15, 30, 60, 90, 120, and 150 minutes.

***Experiment of fermented corn stalk on biosorption equilibrium.*** Nine 0.1 g samples of fermented corn stalk were placed in 100 mL of $CuSO_4$ solution with initial $Cu^{2+}$ concentrations of 10, 30, 60, 90, 120, 150, 200, 250, and 300 mg/L and were shaken at 30° C and 100 rpm for two hours.

***Data processing.*** The $Cu^+$ solution was filtered through a membrane with a pore diameter of 0.45 um and analyzed by Inductively Coupled Plasma Mass Spectrometry to determine the concentrations of the $Cu^{2+}$ before and after biosorption (Göksungur et al., 2005).

***FTIR and SEM-EDX analysis of fermented rice straw and raw rice straw.*** The raw rice straw and fermented rice straw samples were screened before and after biosorption by Fourier Transform Infrared Spectroscopy (FTIR 1730 model, Perkin-Elmer Corporation) using the KBr tablet method (Lodeiro et al., 2006). SEM in combination with EDX can provide information about the morphology and elemental component of the two kinds of biosorbents (Chen et al., 2014).

## RESULTS AND DISCUSSION

***Comparison of raw corn stalk and fermented corn stalk with different dosages.*** The high sorption capacity was the key factor of a biosorbent (Won et al., 2014). Fig. 1 shows that the biosorption value for fermented corn stalk was 3.4 mg/g. By contrast, the biosorption value for raw corn stalk was only 1.8 mg/g when the dosage was 0.1g. Furtherrmore, the biosorption values for fermented corn stalk were higher than those for raw corn stalk for all other dosages. Which can be ascribed to two points: the sprawl aerial mycelia of *Aspergillus niger* accounted for part of the biosorbents, thus affecting the proportion of corn stalk, and cellulase and protease may have enzymolysized the components of corn stalk during the fermentation process, thus causing the biosorption properties to change.

***Effect of pH and zeta potential analysis.*** Another parameter which can influence heavy metal removal is pH value. It can affect the characterization and availability of metal ions and change the surface functional group chemistry of biosorbents (Abdolali et al., 2014). Since the pH of the 30mg/L $Cu^{2+}$ solution was 5.3, we decided the range of pH should be from 1 to 5.3. Fig. 2 shows that biosorption of $Cu^{2+}$ by fermented corn stalk increased sharply as the pH increased (from 1 to 3) and attained maximum biosorption at pH 3. After that, the biosorption efficiency decreased as the pH increased. Which ascribed to the $Cu^{2+}$ ions had to compete with $H_3O^+$ ions for biosorption sites and functional groups of fermented corn stalk were more protonated when pH was lower. As the pH increased to 3, the $Cu^{2+}$ biosorption capacity also increased to its peak. This can be attributed to the increase in negative charge on the biosorbent surface and the reduction of $H^+$ ions (Hu et al., 2015). The Zeta potential stands for the charge of the sample surface. The Zeta potential of raw corn stalk and fermented corn stalk was -19.6 (biosorption capacity = 1.8 mg/g) and -23.67 (biosorption capacity = 3.4 mg/g), respectively, which indicates that the biosorption process was related to electrostatic attraction. In addition, fermented corn stalk was more negative than raw corn stalk, which means the former can attract more positive ions( better biosorption capacity).

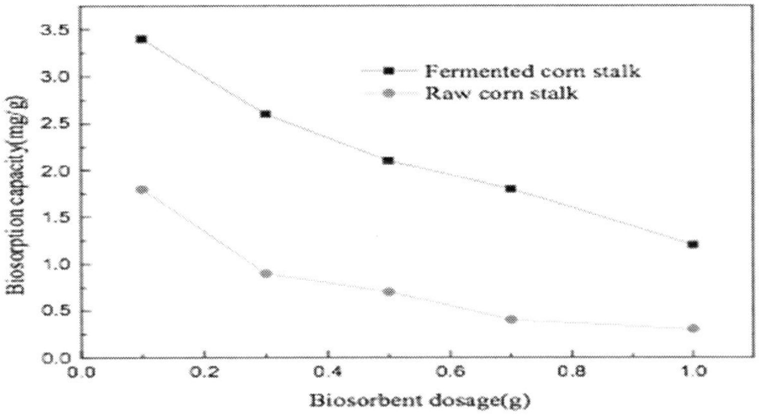


FIRURE 1. Biosorption Value

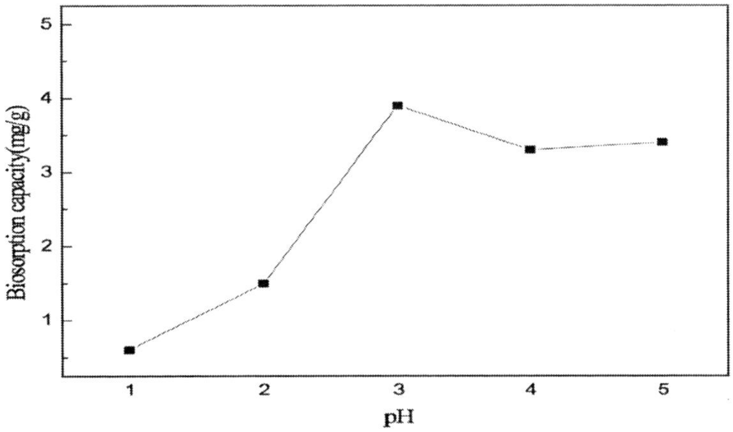


FIRURE 2. Biosorption of $Cu^{2+}$

***FTIR of raw corn stalk and fermented corn stalk.*** Fig.3 shows that the bands observed at 1722–1740cm$^{-1}$ were assigned to a C-O stretching of carboxylic acid or pectin ester (Jacques et al., 2007). The bands observed at 2920 cm$^{-1}$ indicated the presence of C–H group (Chen et al., 2011a). The peak at 1420 cm$^{-1}$ corresponds to the C=C stretching in the aromatic ring. In addition, the $CH_3$ deformation and –OH bending could have occurred at 1376cm (Petrović et al., 2016). The bands at 1605 cm$^{-1}$ corresponded to the N–H amine stretching band of the amino groups (Chen et al., 2011b) and some bonds, such as 834 and 665 in the fingerprint region, stand for out-of-plane vibration of N-H belonging to amine. Two additional bonds after fermentation. The strong peak at 1635 cm$^{-1}$ is C=O stretching vibration of carboxylic acid that exists in the intermolecular hydrogen bond (Jacques et al., 2007) The main components of *Aspergillus niger* were chitosan and lipids, the reproduction of which contributed to the increase of C=O. Another peak at 1106 cm$^{-1}$ is assigned to C-O stretching vibration (Subbaiah et al., 2011), which may be caused by cellulose hydrolyzation, because when cellulase was secreted glucose was generated.After biosorption, the wavenumber of peaks such as at 1605, 1425, 1731, and 665 cm$^{-1}$ shifted, indicating the involvement of N–H, C=O, C=C, and C-O in adsorption. Although all of the above-mentioned groups were involved in the $Cu^{2+}$ biosorption on the corn stalk, carbonyl was the main functional group that involved in this biosorption. The FTIR analysis confirmed that the complexation process could occur in the biosorption of $Cu^{2+}$ on the corn stalk.

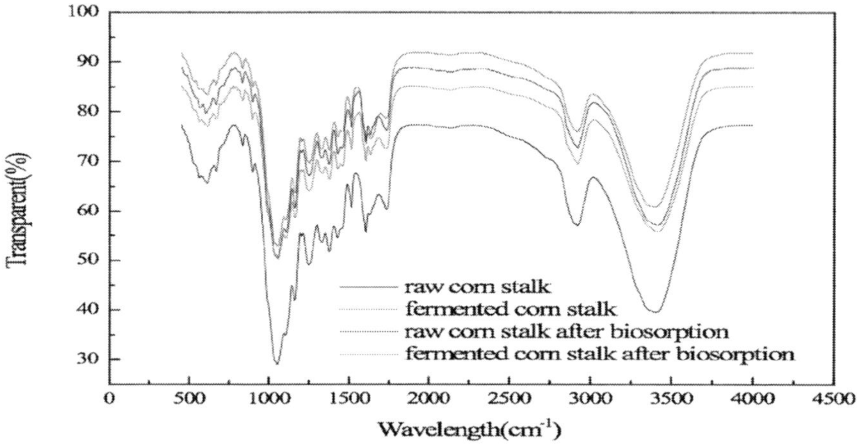


FIGURE.3. FTIR of raw corn stalk and fermented corn stalk before and after biosorption

***SEM-EDX of raw corn stalk and fermented corn stalk.*** Fig. 4a shows that raw corn stalk has irregular fragments and pores on the otherwise smooth surface, which can be caused by cellulose and lignin which have rigid structures, whereas corn cobs have loose structures. After fermentation using *Aspergillus niger*, the structure became more rough and some *Aspergillus niger* spores emerged (Fig. 4b.), because the enzymatic hydrolysis of cellulose on the surface of the fiber modified the surface characteristics either by removing or masking the functional groups or by exposing more binding sites (Pehlivan et al., 2012). That give rise to the change in corn stalk surface properties, together with the appearance of *Aspergillus niger*, and led to the improvement of biosorption capacity. In addition, EDX results shown some differences (at%) between raw corn stalk and fermented corn stalk. After fermentation, the content of O for fermented corn stalk declined when compared to that for raw corn stalk as the C, N, P, and K increased. The reasons can be concluded as follows: The main component of corn stalk is cellulose. When it was degraded by the *Aspergillus niger*, polysaccharide glucose molecules were released and the glucose was primarily composed of C and O, which led to a decrease of C and O on the fermented corn stalk. However, during this time, in the later stage of aerial mycelia growth, a type of reproductive organ (spores) can be produced. The spores primarily consisted of C (67.63%) according to my previous analysis of XPS. Thus, compared with raw corn stalk, the content of C increased after fermentation. In addition, with the extent of the *Aspergillus niger* mycelium, chitin, and protein, the main content of *Aspergillus niger* cells increased, which led to an increase in N, P, and K.

FIGURE.4a. SEM image of raw corn stalk

FIGURE 4b. SEM image of fermented corn stalk

***Biosorption kinetics of $Cu^{2+}$ by fermented corn stalk.*** The sorption kinetics reflect the relationship between sorption rate and sorption time (Ding et al., 2012). Lagergren's first-order kinetic equation(Fig.5a) and a pseudo-second-order kinetic equation(Fig.5b) were adopted to simulate the relationship between $Cu^{2+}$ biosorption capacity and biosorption time. The pseudo-second order equation was better ($R^2 = 0.995$) than Lagergren's first-order equation ($R^2 = 0.958$) to simulate the dynamic $Cu^{2+}$ sorption process by fermented corn stalk. The data which fit in the pseudo-second-order equation (Fig. 5) meant that a chemical process such as ion exchange took place, little physical adsorption was involved in the chemical reaction as the data fit in Lagergren's first-order models as well ($R^2$>0.90). This result was similar to the study by Zheng, whose biosorbents were corn stalk xanthates (XMCS) with alkali treatment, and the process fit in the pseudo-second-order model (Zheng & Meng, 2016).

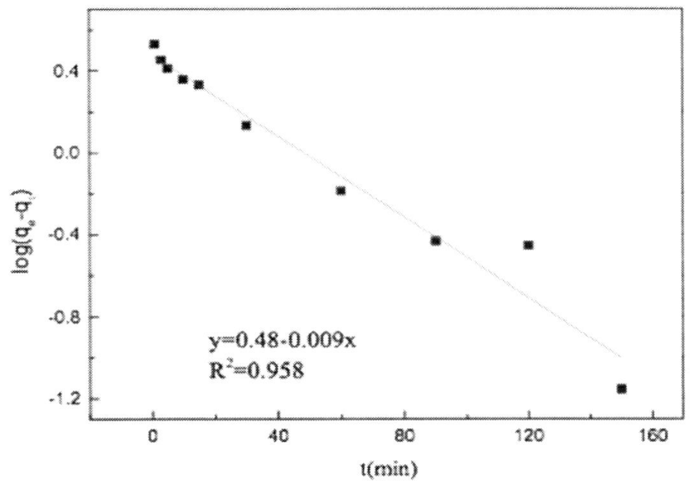


FIGURE 5a. Biosorption kinetics of $Cu^{2+}$ (Lagergren's first-order kinetic)

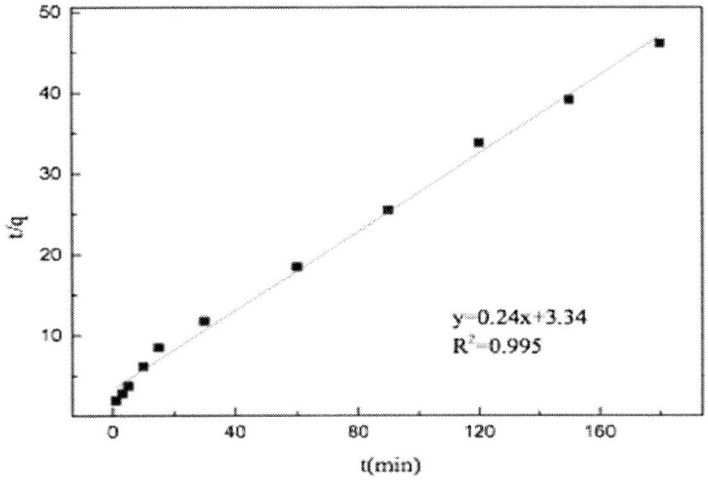


FIGURE 5b. Biosorption kinetics of $Cu^{2+}$ (pseudo-second-order kinetic)

***Isotherm biosorption of fermented corn stalk.*** The sorption isotherm model describes the relationship between sorption capacity and equilibrium concentration (Fig.6). The theoretical maximum uptake from the Langmuir isotherm(Fig. 6a) was 28.49 mg/g and the efficiency of sorption was 0.843. The results were clearly better in the Freundlich model ($R^2$=0.997)(Fig.6b). The obtained values of 1/n (0.1<1/n<1) indicated a higher adsorbent ability of $Cu^{2+}$ in this study (Chen et al., 2011b). Since the Freundlich model assumes that the surface is heterogeneous and the adsorption process is multilayer in nature, we conclude that there was a heterogeneous distribution of the available active sites onto the fermented corn stalk surface (Yu & Luo, 2014). The Freundlich relationship is an empirical equation and stands for a heterogeneous process of biosorption. Similar results were acquired by Buasri, who modified corn cobs with phosphoric acid (PA) to remove $Zn^{2+}$ from aqueous solution (Buasri et al., 2012). However, unmodified rice straw acted in accordance with the Langmuir model in Ding's study, and his degree of fitting was 100% (Ding et al., 2012). Thus, this phenomenon indicates that the microbial treatment affected the biosorbent materials, changed the functional groups and binding sites on the surface, and led to the properties of biosorption. This result was similar to the study by Ebrahimi, whose biosorbents were Alhagi maurorum seeds, and the process fit very well in the Freundlich model (Ebrahimi et al., 2015).

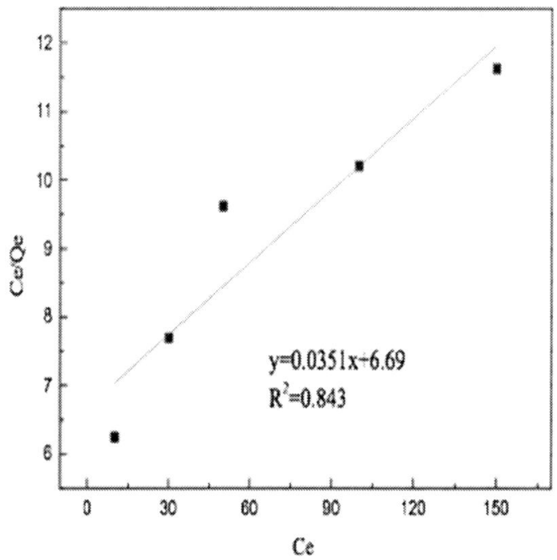


FIGURE 6a. Isotherm biosorption of fermented corn stalk (Langmuir isotherm)

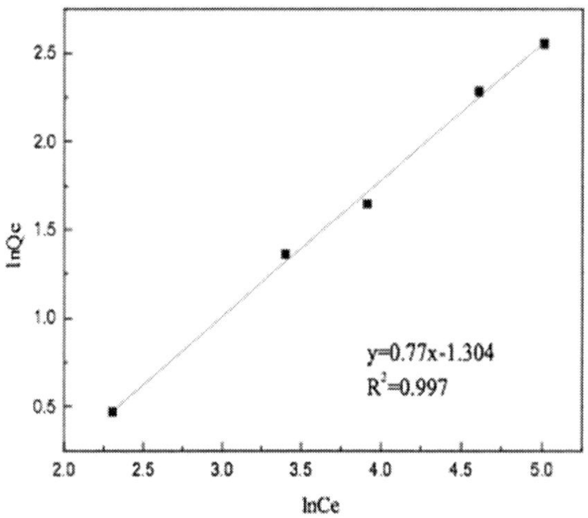


FIGURE 6c. Isotherm biosorption of fermented corn stalk (Freundlich model)

## CONCLUSIONS

(1) Fermented corn stalk exhibited a short biosorption equilibrium time (60 min), high biosorption capacity (11.2 mg/g), and high removal efficiency at pH 3. The content of O in fermented corn stalk, compared with that of in raw corn stalk, declined as the C, N, P, and K increased, and that the zeta potential was lower. The biosorption data obtained were well described by the pseudo-second order and the Freundlich isotherm model.

(2) The biosorption mechanism was involved in electrostatic attraction and chelation with functional groups such as N-H, C=O, C-O, and C=C, and The fact that the biosorption capacity of fermented corn stalk was better than that of raw corn stalk under identical conditions can be attributed to the sprawl of *Aspergillus niger* (biosorption capacity was better than raw corn stalk)and the enzyme system secreted by *Aspergillus niger* which changed the surface properties of raw corn stalk.
(3) Although the process of biosorption is environmentally-friendly, and biosorption capacity of fermented corn stalk was higher than that of raw corn stalk, the capacity of the combination always be the focus of our attention. In further study, the fungi can be mutant by UV or modified by genetic engineering, to improve the performance of this combination biosorbent.

**ACKNOWLEDGEMENTS**

This study was financially supported by heavy metal wastewater treatment technology received from Heilongjiang Academy of Sciences (GC13C3042) and the heavy metal pollution emergency control of the National Environmental Protection Public Welfare Projects (201209048).

**REFERENCES**

Abdolali, A., Guo, W.S., Ngo, H.H., Chen, S.S., Nguyen, N.C., Tung, K.L. 2014. Typical lignocellulosic wastes and by-products for biosorption process in water and wastewater treatment: A critical review. *Bioresource Technology*, **160**, 57-66.

Amini, M., Younesi, H., Bahramifar, N. 2009. Biosorption of nickel(II) from aqueous solution by Aspergillus niger: Response surface methodology and isotherm study. *Chemosphere*, **75**(11), 1483-1491.

Baik, W.Y., Bae, J.H., Cho, K.M., Hartmeier, W. 2002. Biosorption of heavy metals using whole mold mycelia and parts thereof. *Bioresource Technology*, **81**(3), 167-170.

Buasri, A., Chaiyut, N., Tapang, K., Jaroensin, S., Panphrom, S. 2012. Equilibrium and Kinetic Studies of Biosorption of Zn(II) Ions from Wastewater Using Modified Corn Cob. *APCBEE Procedia*, **3**, 60-64.

Chen, C., Wen, D., Wang, J. 2014. Cellular surface characteristics of Saccharomyces cerevisiae before and after Ag(I) biosorption. *Bioresource Technology*, **156**, 380-383.

Chen, S., Yue, Q., Gao, B., Li, Q., Xu, X. 2011a. Preparation and characteristics of anion exchanger from corn stalks. *Desalination*, **274**(1–3), 113-119.

Chen, S., Yue, Q., Gao, B., Li, Q., Xu, X. 2011b. Removal of Cr(VI) from aqueous solution using modified corn stalks: Characteristic, equilibrium, kinetic and thermodynamic study. *Chemical Engineering Journal*, **168**(2), 909-917.

Ding, Y., Jing, D., Gong, H., Zhou, L., Yang, X. 2012. Biosorption of aquatic cadmium(II) by unmodified rice straw. *Bioresource Technology*, **114**, 20-25.

Ebrahimi, A., Ehteshami, M., Dahrazma, B. 2015. Isotherm and kinetic studies for the biosorption of cadmium from aqueous solution by Alhaji maurorum seed. *Process Safety and Environmental Protection*, **98**, 374-382.

Göksungur, Y., Üren, S., Güvenç, U. 2005. Biosorption of cadmium and lead ions by ethanol treated waste baker's yeast biomass. *Bioresource Technology*, **96**(1), 103-109.

Hu, H., Zhang, J., Lu, K., Tian, Y. 2015. Characterization of Acidosasa edulis shoot shell and its biosorption of copper ions from aqueous solution. *Journal of Environmental Chemical Engineering*, **3**(1), 357-364.

Jacques, R.A., Lima, E.C., Dias, S.L.P., Mazzocato, A.C., Pavan, F.A. 2007. Yellow passion-fruit shell as biosorbent to remove Cr(III) and Pb(II) from aqueous solution. *Separation and Purification Technology*, **57**(1), 193-198.

Lodeiro, P., Barriada, J.L., Herrero, R., Sastre de Vicente, M.E. 2006. The marine macroalga Cystoseira baccata as biosorbent for cadmium(II) and lead(II) removal: Kinetic and equilibrium studies. *Environmental Pollution*, **142**(2), 264-273.

Monser, L., Adhoum, N. 2002. Modified activated carbon for the removal of copper, zinc, chromium and cyanide from wastewater. *Separation and Purification Technology*, **26**(2–3), 137-146.

Odetallah, N.H., Wang, J.J., Garlich, J.D., Shih, J.C.H. 2005. Versazyme supplementation of broiler diets improves market growth performance. *Poultry Science*, **84**(6), 858-864.

Pehlivan, E., Altun, T., Parlayici, Ş. 2012. Modified barley straw as a potential biosorbent for removal of copper ions from aqueous solution. *Food Chemistry*, **135**(4), 2229-2234.

Petrović, M., Šoštarić, T., Stojanović, M., Milojković, J., Mihajlović, M., Stanojević, M., Stanković, S. 2016. Removal of Pb2+ ions by raw corn silk (Zea mays L.) as a novel biosorbent. *Journal of the Taiwan Institute of Chemical Engineers*, **58**, 407-416.

Sepehr, M.N., Nasseri, S., Zarrabi, M., Samarghandi, M.R., Amrane, A. 2012. Removal of Cr (III) from tanning effluent by Aspergillus niger in airlift bioreactor. *Separation and Purification Technology*, **96**, 256-262.

Subbaiah, M.V., Vijaya, Y., Reddy, A.S., Yuvaraja, G., Krishnaiah, A. 2011. Equilibrium, kinetic and thermodynamic studies on the biosorption of Cu(II) onto Trametes versicolor biomass. *Desalination*, **276**(1–3), 310-316.

Won, S.W., Kotte, P., Wei, W., Lim, A., Yun, Y.-S. 2014. Biosorbents for recovery of precious metals. *Bioresource Technology*, **160**, 203-212.

Yu, L., Luo, Y.-m. 2014. The adsorption mechanism of anionic and cationic dyes by Jerusalem artichoke stalk-based mesoporous activated carbon. *Journal of Environmental Chemical Engineering*, **2**(1), 220-229.

Zheng, L., Dang, Z., Yi, X., Zhang, H. 2010a. Equilibrium and kinetic studies of adsorption of Cd(II) from aqueous solution using modified corn stalk. *Journal of Hazardous Materials*, **176**(1–3), 650-656.

Zheng, L., Dang, Z., Zhu, C., Yi, X., Zhang, H., Liu, C. 2010b. Removal of cadmium(II) from aqueous solution by corn stalk graft copolymers. *Bioresource Technology*, **101**(15), 5820-5826.

Zheng, L., Meng, P. 2016. Preparation, characterization of corn stalk xanthates and its feasibility for Cd (II) removal from aqueous solution. *Journal of the Taiwan Institute of Chemical Engineers*, **58**, 391-400.

# SUN-CORAL POWDER AS ADSORBENT FOR PHOSPHORUS REMOVAL IN WASTEWATER

**M.T.G Vianna** and Marcia Marques
(Rio de Janeiro State University, Rio de Janeiro, Brazil)

**ABSTRACT:** Searching for low-cost and efficient strategies for phosphorus (P) removal is justified, once its discharge throughout untreated wastewater is related to eutrophication of water bodies. In Brazil, two exotic species known as sun-coral introduced during the 80's are currently spread along 2,000 km of the Brazilian coast representing a serious threat to the Brazilian marine biodiversity. Since they have exoskeletons with high concentration of calcium carbonate, the present investigation focused on the removal of phosphorus from wastewater, using exoskeleton's powder in three forms: raw (RSC); physically modified (SCA) and chemically modified (SCC). A Central Composite Design was applied to the assays along to studies of kinetics, thermodynamics and equilibrium, besides sorbent's characterization by XRD and SEM-EDX techniques. The maximum P removal capacities of RSC, SCA and SCC were 6826, 7062 and 9597 mg P $kg^{-1}$, respectively. In all cases, the adsorption followed the pseudo-second-order kinetic model being the process thermodynamically appropriate. For Real wastewater samples, RSC and SCA adsorbents removed around 80% and the SCC material removed 100% of the P contained in the wastewater. Based on these results, exoskeleton powder proved to be an attractive option for wastewater treatment and may contribute to the control of sun-coral population.

## INTRODUCTION

The excessive release of phosphorus (P), nitrogen (N) and the organic load, by discharge of domestic and industrial effluents associated to other suitable environmental conditions causes eutrophication of water bodies. This well-known phenomenon is responsible for aquatic ecosystems degradation; damage to reservoirs; offensive odour and bad taste in treated water (Su et al., 2013). Thus, removal of phosphorus from effluents is an important strategy aiming the preservation of aquatic systems (Karageorgiou et al., 2007). The form in which orthophosphate is found in water ($PO_4^{3-}$, $HPO_4^{2-}$, $H_2PO_4^-$ or $H_3PO_4$) depends on the pH. In a typical wastewater, the predominant form is $HPO_4^{2-}$ (Mangwandi et al., 2014). Orthophosphate is considered to be the only type of P directly assimilated by most plants and algae (Loganathan et al., 2014).

The method most commonly used for phosphorus removal from wastewater is chemical precipitation, although is relatively expensive and demands proper sludge disposal. Alternative methods for phosphorus removal include biological treatment, electro-dialysis, reverse osmosis, ultrafiltration and adsorption. Among them, adsorption seems to be an effective treatment option because of simplicity of operation, design and economic point of view (Karageorgiou et al., 2007).

Natural materials such as apatite, bauxite, limestone, sand, shells, among others, have been used as adsorbents for removal of phosphates from water (Vohla et al., 2011). In common, these materials have high concentration of calcium (Ca), iron (Fe) and aluminium (Al) in their structures with high yield for phosphorus removal, through adsorption processes.

The introduction of exotic species in aquatic environments has long been considered an environmental problem that affects the integrity of natural communities, mechanisms such as competition, predation and parasitism, and alters food webs and nutrient cycles (Bax et al., 2003). Some of these invasive species have high concentrations of calcium in their exoskeleton composition, potentially useful in water treatment, particularly for phosphorus removal.

**The sun coral problem in the Brazilian coast.** Sun coral (Fig. 1) refers to two species (*Tubastraea coccinea* and *Tubastraea tagusensis*, Scleractinia order, Dendrophylliidae family) native from Asia and very abundant in the Indian and Pacific oceans. However, the sun corals are invasive species that rapidly colonizes large areas, in recent decades, its range distribution has increased, and today, the species is considered pantropical, being observed in Caribbean Sea (Puerto Rico and Curaçao), Gulf of Mexico (including the Flower Garden Banks Marine Sanctuary and Florida Keys) and Western Atlantic (Brazilian Coast) (Sammarco et al., 2013). In Brazilian Coast, Sun coral was first observed the 1980s, in an oil and gas platforms in Northern of Rio de Janeiro state. Currently, the sun coral's occupation extends over 2000 km of the Brazilian coast (Fig. 1). These corals are opportunistic species with high production of oocytes, early childbearing age, short time incubation embryo and hermaphroditism (Creed and De Paula, 2007). Sun coral produces substances with anti-fouling and anti-predatory properties and releases allelopathic substances that cause necrosis to Brazilian endemic species posing serious threat to the Brazilian marine biodiversity and affect species economically important.

**FIGURE 7.** Sun coral (Photo taken by Joel Creed/Projeto Coral-Sol/BrBio) and its occupation along the Brazilian coastline (modified map from Google Earth).

During recent years, the Public Prosecutor's Office has moved civil actions in different Brazilian states, requesting an injunction to protect the marine environment and control sun coral expansion. Currently, attempts to control the expansion of coral sun infestation includes manual and mechanical removal and inoculation by specific bacteria and viruses. One strategy to promote the removal of these species would be to assign use to the extracted material, making the extraction economically attractive. This study aims to investigate the capacity of exoskeleton (in powder form) of sun coral in removing P from synthetic and real wastewaters.

## MATERIALS AND METHODS

**Materials.** All chemicals were purchased from Sigma-Aldrich and all solution were prepared using deionized Milli-Q water (Millipore TM – resistivity of $\geq 18.2$ MΩ $cm^{-2}$). All P solutions were prepared by dissolving the primary standard potassium dihydrogen phosphate ($KH_2PO_4$) in Milli-Q water. For pH adjustment, solutions of 0.1N sodium hydroxide (NaOH) and 0.1N hydrochloric acid (HCl) and a multiparameter 5 Star Orion (Method: 4500 H + B) were used (Eaton et al., 2005).

**Preparing Sun Coral Powder.** The sun coral was obtained from Ilha Grande Bay, Rio de Janeiro, Brazil. The material collected was dehydrated in an oven at 60 °C during 72 h. The sun coral exoskeleton was then, washed at room temperature to remove the remaining impurities. Subsequently, the exoskeleton was submerged in vessels containing sodium hypochlorite 10% (NaClO-10% w/v). The exoskeleton was washed with hot Milli-Q water (90 ° C) and then, crushed and sieved with a 200-mesh sieve (75 μm). The washed exoskeleton was divided into three portions: The first portion was used with no further modification (RSC);

the second portion was kept in a muffle furnace at 700 °C for 2 h (SCA) before use; the third portion was kept in a muffle furnace at 900 °C for 2 h (SCC).

**Characterization of Sun Coral Powders.** The XRD patterns were recorded on a Brucker-AXS D4 Endeavour powder X-ray diffractometer using $Co_{K\alpha}$ radiation (40 kV, 40 mA). The measurements were performed in a θ scanning range of 4–80° with a 0.020 step size and 91s of measurement time for each step at ambient temperature and pressure. Scanning electron microscopy with energy dispersive X-ray analysis (SEM-EDX), model FEI Quanta 400 FEG, was used to examine the morphology and chemical composition of the sun coral powder before and after adsorption experiments.

**Phosphorus Analysis.** The determination of total phosphorus in water followed the 4500-P E ascorbic acid Method described by APHA in 4500-P E method (Eaton et al., 2005). The amount of phosphorous removed by adsorption in relation to the mass of adsorbent was determined according to Equation 1:

$$\frac{P_{removed}(mg)}{ads_{mass}(Kg)} = \frac{[P_i - P_f]\cdot V_{Psolution}}{ads_{mass}} \qquad (1)$$

Where $P_i$ and $P_f$, are the initial and final concentration of P, respectively; $V_{Psolution}$ (L) is the volume of phosphorus solution utilized in the experiment; $ads_{mass}$ (Kg) is the amount of exoskeleton powder utilized (RSC, SCA or SCC) in each batch. The result is expressed in milligrams of phosphorus removed per kilograms of adsorbent.

**Experimental Setup: Factorial Design.** Three independent variables were selected as the most important ones for the process of phosphorus removal by the adsorbents RSC, SCA and SCC: (i) pH, (ii) adsorbate/adsorbent ratio and (iii) temperature. In order to assess the best interactions between variables with the purpose of optimizing the phosphorus removal, a central composite design (CCD) was used (Rodrigues and Iemma, 2014). The variables and levels (Table 1) were chosen based on the ranges observed in studies with domestic wastewater available in literature.

**TABLE 3.** Coded levels (5) for independent variables (3) selected to be used in study for assays of phosphorus removal from water by RSC, SCA and SCC powder as sorbent.

| Variables | Levels | | | | |
|---|---|---|---|---|---|
| | **-1.68** | **-1** | **0** | **+1** | **+1.68** |
| pH | 5.32 | 6.0 | 7.0 | 8.0 | 8.68 |
| Adsorbate/adsorbent ratio | 0.4 | 1.25 | 2.5 | 3.75 | 4.6 |
| Temperature °C | 23.6 | 27 | 32 | 37 | 40.4 |

The experiments were performed in batch mode as predicted by the experimental design. The study was conducted on an orbital shaker (150 rpm) in 250 ml Erlenmeyer flasks containing 100 ml of water with dissolved phosphorous in amounts to fit to a mass of 400 mg adsorbent respecting the adsorbate/adsorbent ratio predicted by experimental design.

In total, 18 runs (including the central points of CCD) were performed in random sequence. This planning allows obtaining mathematical models with linear and quadratic parameters (multiple regressions) of the studied variables. By combining the different levels presented in the experimental design, it was possible to examine the independent effects of each variable in response to removal, as well as the effect due to the interaction between them.

**Kinetic, Equilibrium and Themodynamic Studies.** The assay with better P removal is repeated. During the batch experiment, aliquots of 2 ml were removed from the solution to measure the P concentration. Aliquots were taken at intervals of 2 min 30 sec and after 30 min the experiment was stopped. The results were applied to the following kinetic models: pseudo-first-order and pseudo-second-order; For equilibrium studies, all variables were fixed, excepting for the initial P concentration. The concentration range varied from 5 to 100 mg $l^{-1}$ by placing the P solutions in contact with 400 mg of each material used as adsorbent. After the equilibrium period, the P concentration was analysed to obtain the constants of Langmuir and

Freudlich models; Thermodynamics was evaluated using five temperatures (25, 30, 35, 40 and 45 °C), in the equilibrium time. Free energy, enthalpy and entropy of process, were obtained using Van't Hoff and Free Gibbs Energy equations.

**P Removal from Real Wastewater.** Experiments were carried out with a real effluent obtained after biological treatment in a municipal wastewater treatment plant located in Niteroi city, Rio de Janeiro, Brazil. The P concentration of the real effluent was measured before and after adsorption treatment. Batch studies were conducted in triplicates, using an orbital shaker and 250 ml Erlenmeyer flasks containing 100 ml of real wastewater. The temperature and the pH of the effluent were not changed due to practical reasons and to avoid additional costs in a future full-scale plant. The results obtained in the experiment was compared with the predicted by the theoretical model.

## RESULTS AND DISCUSSION

**Phosphorus removal studies.** The range of P removal, in the applied range of parameters, using RSC varied from 287.6 to 3685.5 mg P $kg^{-1}$; for CSA varied from 315.0 to 3712.5 mg P $kg^{-1}$; and for SCC varied from 355.0 to 4567.5 mg P $kg^{-1}$. In all materials, the maximum removal occurred in run 12 with pH 7.0, adsorbate/adsorbent ratio of 4.6 (P concentration of 18.4 mg $l^{-1}$) and temperature of 32 °C. Using the experimental results, the regression model for P removal in each material were obtained:

$$P\ removal\left(\frac{mg\ P}{Kg\ RSC}\right) = 1892.50 + 37.49\,(pH) + 1005.9\,(Adb/ads\ ratio) + 38.18\,(Adb/ads\ ratio)^2 - 12.03\,(Temp) \quad R^2=0{,}99 \qquad (2)$$

$$P\ removal\left(\frac{mg\ P}{Kg\ SCA}\right) = 1926.76 + 38.42\,(pH) + 1005.35\,(Adb/ads\ ratio) + 33.75\,(Adb/ads\ ratio)^2 - 8.78\,(Temp) \quad R^2=0{,}99 \qquad (3)$$

$$P\ removal\left(\frac{mg\ P}{Kg\ CSQ}\right) = 2455.06 + 1250.07\,(Adb/ads\ ratio) \quad R^2=0{,}99 \qquad (4)$$

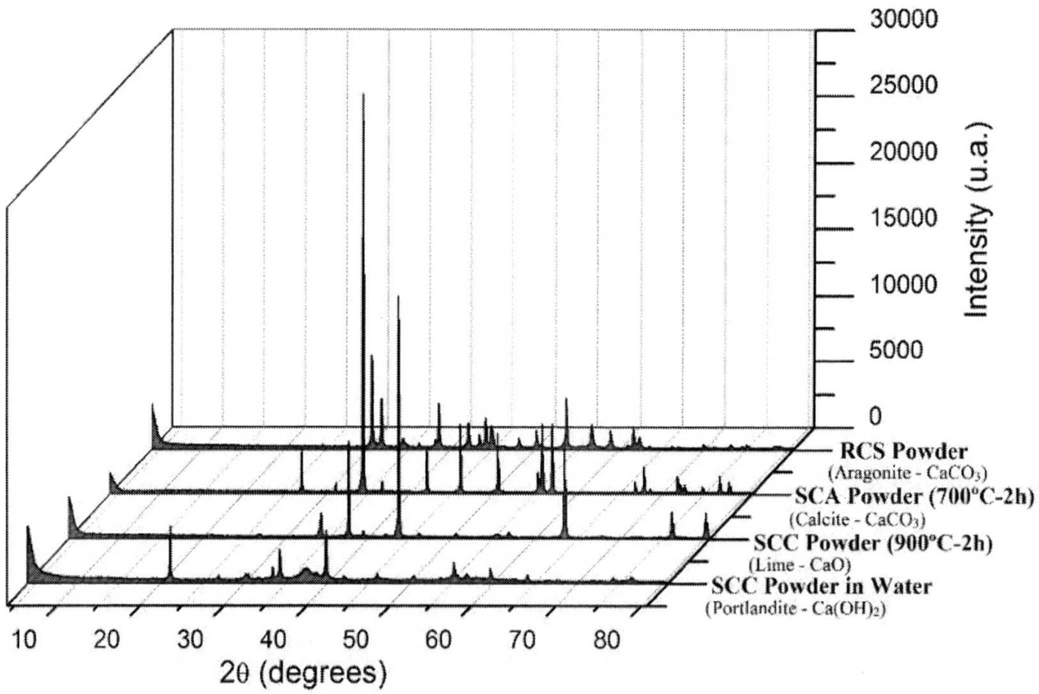


**FIGURE 2.** XRD patterns – Processes overview

**Characterization of sun coral powders.** The XRD patterns (Fig. 2) reveals that the RSC is composed by orthorhombic calcium carbonate, namely Aragonite; the crystalline structure of RSC changes when exposed to heat at 700 °C, forming a rhombohedral crystal of calcium carbonate, called calcite (SCA). When exposed to 900 °C, the material changed chemically by decomposition of calcium carbonate forming calcium oxide, the calcium oxide obtained exhibits a cubic crystalline form called lime (SCC). After adsorption processes the crystalline structure of RSC and SCA materials remained unchanged, however, the SCC reacted with water producing calcium hydroxide with cubic crystalline structure, called Portlandite. The Portlandite material remained unchanged after adsorption processes. Figure 3 shows the SEM images of SCC after phosphorus adsorption processes. The EDX analysis was used to determine the composition and content on the material surface. The EDX spectra confirmed the existence of P adsorbed to the surface pf the sun coral powder. The EDX technique applied together with SEM image revealed where the P uptake occurred in the surface material.

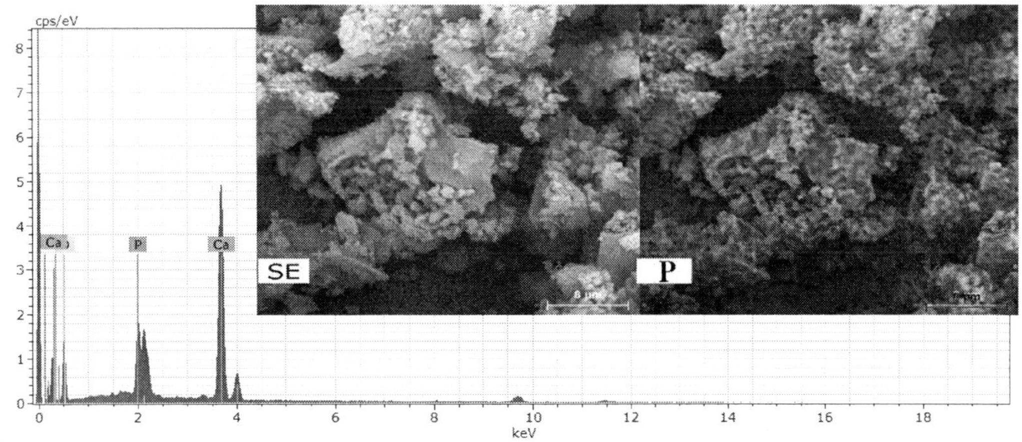


**FIGURE 3.** SEM-EDX images for SCC material, before and after adsorption.

**TABLE 4.** Best results obtained in Kinetic, Equilibrium and Termodynamic Studies

| | Kinetic Model | $R^2$ | $k_2$ (g mg $P^{-1}$ $min^{-1}$) | |
|---|---|---|---|---|
| RSC | Pseudo-second-order | 0.9944 | 0.1952 | |
| SCA | Pseudo-second-order | 0.9981 | 0.3431 | |
| CSS | Pseudo-second-order | 0.9993 | 0.1843 | |
| | **Equilibrium Model** | $R^2$ | $Q_{max}$ (mg $g^{-1}$) | $K_L$ (l mg-1) |
| RSC | Langmuir | 0.9979 | 6.826 | 0.7157 |
| SCA | Langmuir | 0.9989 | 7.062 | 10.351 |
| CSS | Langmuir | 0.9979 | 9.597 | 41.513 |
| | **Thermodynamic** | ΔS (J $K^{-1}$ $mol^{-1}$) | ΔH (KJ $mol^{-1}$) | ΔG* (KJ $mol^{-1}$) |
| RSC | Spontaneous and Exothermic | 10.59 | -22.20 | -25.35 |
| SCA | Spontaneous and Exothermic | 13.71 | -22.19 | -26.27 |
| CSS | Spontaneous and Exothermic | 44.99 | -16.24 | -29.65 |

* Temperature of 25° C

**Kinetic, Equilibrium and Thermodynamic studies.** According to kinetic results, the best-fit model was the pseudo-second order model ($R^2$ close to 1.0) and all processes reached the equilibrium around 20 min; For Equilibrium, the results revealed that Langmuir model is the one that best describes the P adsorption by sun coral-based adsorbents, which indicates that the adsorption sites are distributed homogeneously on the surface of the adsorbent. The highest capacity for P removal from water achieved by sun coral-based adsorbents RSC, SCA, SCC under the studied conditions were 6826, 7062 and 9597 mg P $kg^{-1}$, respectively; The negative value of change in enthalpy (ΔH) shows that the adsorption is exothermic. Negative values of free energy (ΔG) indicated the feasibility and spontaneity of adsorption.

**P removal from real wastewater.** The levels applied for different independent variables were encoded and included in the P removal model obtained by CCD (Table 3). The mass of adsorbent used in all experiments were 37 mg to fit the +1.68 coded value of adsorbate/adsorbent ratio. To minimize the influence of P consumption by microbiological routes, taking into account the kinetic parameters, the assays were performed within 20 min. Table 3 shows high levels of P removal. The initial concentration of P in wastewater (1.71 mg $l^{-1}$) reduced to 0.37, 0.36 and 0.02 mg P $l^{-1}$ when the wastewater was treated with RSC, SCA and SCC, representing reductions of 78.17, 79.14 and 98.6% respectively. These results show that the powder obtained with exoskeleton of sun coral is a promising material for using in filters aiming P removal. The model obtained with spiked water showed good fit when used in real wastewater; in all cases, the difference between predicted and real values was below 2%.

**TABLE 3.** P removal by sun coral powders in real wastewater.

| | Variables | CCD uncoded | CCD code | P removal predicted | Amount of P removed | SD | P removal (%) |
|---|---|---|---|---|---|---|---|
| RSC | pH<br>Temperature<br>P total | 6.58<br>27.4 °C<br>1.71 mg $l^{-1}$ | -0.42<br>-0.92<br>1.68 | 0.136 mg P in 37 $mg^{-1}$ | 0.1337 mg P in 37 $mg^{-1}$ | 0.0004 | 78.17 |
| CSA | pH<br>Temperature<br>P total | 6.58<br>27.4 °C<br>1.71 mg $l^{-1}$ | -0.42<br>-0.92<br>1.68 | 0.137 mg P in 37 $mg^{-1}$ | 0.1353 mg P in 37 $mg^{-1}$ | 0.0044 | 79.14 |
| CSQ | pH<br>Temperature<br>P total | 6.58<br>27.4 °C<br>1.71 mg $l^{-1}$ | -0.42<br>-0.92<br>1.68 | 0.169 mg P in 37 $mg^{-1}$ | 0.1687 mg P in 37 $mg^{-1}$ | 0.0044 | 98.64 |

## CONCLUSIONS

The highest capacity for P removal from water achieved by sun coral-based adsorbents RSC, SCA, SCC under the studied conditions were 6826, 7062 and 9597 mg P $kg^{-1}$ respectively. In all cases, the adsorption process followed the Langmuir model, the pseudo-second-order kinetic model and was thermodynamically appropriate. According to the results, sun coral-based adsorbents proved to be a good option for wastewater treatment, particularly when associated to the mandatory removal of these invasive species from the coastal areas. Next steps of this research involves the use of the obtained parameters to project lab-scale and real-scale filters aiming P removal. The real-scale filter will be installed in an Engineered Ecosystem located in Ilha Grande, Rio de Janeiro, Brazil.

## ACKNOWLEDGMENTS

The authors acknowledge the financial support to the project provided by the Research Support Foundation of Rio de Janeiro (FAPERJ) and the scholarship provided to the first author by the Coordination for the Improvement of Higher Education Personnel (CAPES).

## REFERENCES

Bax, N., Williamson, A., Aguero, M., Gonzalez, E., Geeves, W., 2003. Marine invasive alien species: a threat to global biodiversity. *Mar. policy* 27, 313–323.

Creed, J.C., De Paula, A.F., 2007. Substratum preference during recruitment of two invasive alien corals onto shallow-subtidal tropical rocky shores. *Mar. Ecol. Prog. Ser.* 330, 101–111.

Eaton, A.D., Clesceri, L.S., Greenberg, A.E., 2005. *Standard methods for the examination of water and wastewater* American Public Health Association. Washington, DC 20001–23710.

Karageorgiou, K., Paschalis, M., Anastassakis, G.N., 2007. Removal of phosphate species from solution by adsorption onto calcite used as natural adsorbent. *J. Hazard. Mater.* 139, 447–52.

Loganathan, P., Vigneswaran, S., Kandasamy, J., Bolan, N.S., 2014. Removal and recovery of phosphate from water using sorption. *Crit. Rev. Environ. Sci. Technol.* 44, 847–907.

Mangwandi, C., Albadarin, A.B., Glocheux, Y., Walker, G.M., 2014. Removal of ortho-phosphate from aqueous solution by adsorption onto dolomite. *J. Environ. Chem. Eng.* 2, 1123–1130.

Rodrigues, M.I., Iemma, A.F., 2014. *Experimental Design and Process Optimization,* Ilustred. ed. CRC press, Boca Raton, FL.

Sammarco, P.W., Porter, S. a, Sinclair, J., Genazzio, M., 2013. Depth distribution of a new invasive coral (Gulf of Mexico)–Tubastraea micranthus, comparisons with T. coccinea, and implications for control. *Manag. Biol. Invasions* 4, 291–303.

Su, Y., Cui, H., Li, Q., Gao, S., Shang, J.K., 2013. Strong adsorption of phosphate by amorphous zirconium oxide nanoparticles. *Water Res.* 47, 5018–26.

Vohla, C., Kõiv, M., Bavor, H.J., Chazarenc, F., Mander, Ü., 2011. Filter materials for phosphorus removal from wastewater in treatment wetlands—A review. *Ecol. Eng.* 37, 70–89.

# COMPARATIVE STUDY OF ADSORPTION OF DYES ONTO ACTIVATED CARBON AND CHITOSAN IMPREGNATED ACTIVATED CARBON

***Pratyusha Reddy Reddy****, George A. Sorial
(University of Cincinnati, Cincinnati, OH, USA)

Dyes are organic compounds that impart color when applied to a substrate. The presence of dyes in wastewater makes in aesthetically unacceptable and are known to pose problems to the aquatic environment. It is important to treat the dye containing wastewater before its discharge into water bodies. Among the wastewater treatment methods available, adsorption is commonly used for the removal of dyes from wastewater. Activated carbon is a popular adsorbent used for treating organic compounds present in water and wastewater. Activated carbon is expensive, hence, other low-cost alternatives are being investigated. One such low-cost option could be the modification of commercial activated carbon with chitosan, a polymer present in the hard shells of crustaceans which could lead to higher adsorption capacities.

In this work, three different types of dyes adsorption behavior are evaluated by two adsorbents-commercial activated carbon (Filtrasorb F400) and commercial activated carbon impregnated with chitosan. The dyes used are acid blue 113 (AB113), basic blue 3 (BB3) and direct blue 71 (DB71). Single solute and multi solute adsorption studies are carried for each of the adsorbents. The single solute adsorption isotherms are modelled against the two well-known isotherm equations Fruendlich and Langmuir isotherm equations. The multi solute are evaluated against predictions by the Ideal Adsorbed Solution Model.

# NATURAL IRON-BASED MATERIAL AS A COST-EFFECTIVE SOLUTION FOR THE TREATMENT OF ARSENIC CONTAMINATED WATERS FROM GOLD MINE

***Małgorzata Szlachta*** and Patryk Wójtowicz
(Wrocław University of Technology, Wrocław, Poland)

Arsenic is a chemical element widely known for its high toxicity to humans and aquatic organisms. It may occur naturally in the aquatic environment or may be released into the water through industrial sources. In some areas, excessive quantities of arsenic found in natural waters often originate from the mining activities. The impact of mining operations on the natural environment depends on the amount of arsenic bearing minerals in the exploited rock. The gold mines are a potential source of arsenic due to arsenopyrite, which is usually present in the sulphide ores associated with sediment-hosted gold deposits. To prevent contamination of rivers, lakes and groundwater, the arsenic species should be captured and effectively removed from gold mine effluents. The adsorption/ion exchange process due to its technical applicability for the treatment of large volumes of water may offer a good solution for the robust and cost-effective treatment of mine process waters and effluents.

The main objective of this work was to evaluate the efficiency of natural iron-based material in the treatment of arsenic contaminated mine waters. The material used in the study was crushed iron ore composed of hematite, maghemite and limonite originating from Romania. To investigate the performance of tested material, the batch adsorption technique was applied. The adsorption experiments were carried out in the laboratory scale and included adsorption kinetics and multi-load tests as well as experiments with various material dose and solution pH. In the first phase of the research, the model solutions spiked with arsenic(III) and (V) were used for the tests. Furthermore, with the aim of practical application, the performance of natural iron-based material was verified with the use of real arsenic contaminated waters from the gold mines in Finland.

It was confirmed that arsenic has a strong affinity for iron rich surfaces, therefore the application of natural material investigated in the study may provide suitable and cost-effective treatment of arsenic-rich mine waters. It was also found that the efficiency of the process depends on operating parameters, including contact time, material dose and solution pH. However, it was observed that arsenite was more difficult to remove from water compared to the arsenate. Therefore, when considering the treatment of mine process waters and effluents by adsorption using studied natural material the peroxidation step to convert trivalent arsenic to pentavalent is recommended.

The problem presented in this work is currently being addressed in the international SUSMIN project "Tools for sustainable gold mining in the EU".

# CONVERSION OF BLACK LIQUOR INTO ACTIVATED CARBON FOR CIPROFLOXACIN REMOVAL FROM WASTEWATER

***Anirudh Gupta*** and Anurag Garg
(Indian Institute of Technology Bombay, Mumbai, INDIA)

Black liquor is a major by-product generated from a pulp and paper mill. It can be considered as a precursor for activated carbon (AC) preparation due to high carbon content. In the present study, the effectiveness of black liquor derived AC was investigated for the treatment of synthetic wastewater contaminated by ciprofloxacin (CPX), a pharmaceutically active compound toxic to aquatic organisms. The black liquor was transformed into an adsorbent using chemical activation process mediated by $ZnCl_2$. The Brunauer-Emmet-Teller (BET) surface area and total pore volume of the adsorbent were found to be 600 $m^2/g$ and 0.324 $cm^3/g$, respectively. The $pH_{PZC}$ (i.e. point of zero charge) of the adsorbent was slightly below the neutral pH (= 6.6). The batch adsorption runs were conducted in an orbital shaker with 50 ml of synthetic wastewater (initial CPX concentration = 20 – 100 mg/l) with an adsorbent dosage of 2 g/l. The CPX concentration was measured at 278 nm using UV-visible spectrophotometer. The equilibrium time for the CPX adsorption was found to be 12 h. The adsorption capacity of the waste derived adsorbent was ranged between 8 – 32 mg/g at an optimum pH of 7.0. The equilibrium data was found fit to Freundlich and Langmuir isotherm models. The CPX adsorption on AC was found to be spontaneous and favourable. The adsorption kinetic data could fit well in pseudo first order equation and intra-particle diffusion model.

# MULTIWALLED CARBON NANOTUBES AS A NOVEL SOLID-PHASE EXTRACTION ADSORBENT FOR TPH DETERMINATION IN CONTAMINATED WATER

***Akinpelu A Adeola***[a], Ilyas Muhammad [a], Ahsan M Shemsi [a]
([a] Center of Environment and Water King Fahd University of Petroleum and Minerals, Dhahran Saudi Arabia)

The application of Multiwalled carbon nanotubes (MWCNTs) as a novel adsorbent of solid-phase extraction (SPE) for the determination of total petroleum hydrocarbon (TPH) was investigated using gas chromatography (GC) coupled with flame ionization detector (FID). Parameters influencing the optimum extraction of TPH such as, elution solvent, solvent volume, elution flow rate and sample loading flow rate, were examined. Elution with hexane and so also Isopropanol followed by Hexane/Dichloromeathane at 3ml/min elution flow rate were observed to be the optimum conditions for efficient extraction of TPH from contaminated water. The calibration curves for 22 TPH standards were linear in the range of 100-5000n/L, with correlation coefficients ($r^2$) between 0.9981 and 0.9991. There was a good precision as seen in (RSD) which is 1.42%-7.04%. It was concluded that MWCNTs packed cartridge coupled with FID is an excellent option for the determination of TPH in contaminated water.

# AMMONIA NITROGEN ADSORPTION BY ZEOLITE AND WETLAND SOILS UNDER DIFFERENT TEMPERATURES

***Wei Huang*** and Weidong Wang
(Research Center for Eco-Environmental Sciences, Chinese Academy of Sciences, Beijing, China)

Adsorption characteristics of ammonia nitrogen on zeolite and soils from natural and constructed wetlands were investigated by batch experiments under different temperatures. The effects of relevant parameters including contact time, temperature, and initial ammonia concentration were examined, respectively. The results showed that although it would need at least one hour to reach exchange equilibrium, ammonia adsorption by zeolite occurred rapidly within the initial fifteen minutes of contact time, and the ammonia adsorption capacity of zeolite increased with temperature (from 4 to 35 Celsius degree). However, the adsorption capacity of soils decreased with the increase of temperature, which was contrary to zeolite. The adsorption capacity of zeolite was 5 to 15 times those of reed-bed rhizospheric soils from constructed wetlands in the Yangtze River delta and from natural wetlands in Baiyangdian Lake of north China plain under the same condition. Ammonia adsorbed by zeolite during the cold seasons were desorbed, and then nitrified in warm seasons, which resulted in a bio-regeneration of zeolite. This could relieve the seasonal fluctuation in ammonia nitrogen removal.

# HETEROGENEOUS CATALYTIC DEGRADATION OF ACETAMINOPHEN AND SIMULTANEOUS OXIDATION/ADSORPTION OF ARSENITE BY Cu-Zn-Fe-LDH

***Hongtao Lu***, Zhiliang Zhu, Jianyao Zhu
(Tongji University, Shanghai, CHINA)

A three metal layered double hydroxide Cu-Zn-Fe-LDH was synthesized and used as a heterogeneous Fenton-like oxidation catalyst and adsorbent to simultaneously remove acetaminophen (Paracetamol) and arsenic. The degradation of acetaminophen and oxidation/adsorption of arsenite were studied in aqueous solution under natural water pH range. Degradation of Paracetamol was accelerated with decrease of pH or increase in $H_2O_2$ concentration and catalyst dosage. When the catalyst dosage was 0.5 g•$L^{-1}$ and $H_2O_2$ dosage of 30 mmol•$L^{-1}$ , the acetaminophen was totally degraded within 24h. The synthesized Cu-Zn-Fe-LDH was also a possible efficient adsorbent for arsenate removal from aqueous solutions, the maximum adsorption capacity was 126.13 mg•$g^{-1}$. In the presence of hydrogen peroxide, the arsenite was gradually oxidized into arsenate and adsorbed at the same time by Cu-Zn-Fe-LDH. The simulate water sample with coexistence of arsenic and Paracetamol had been treated by the Cu-Zn-Fe-LDH material. The residual arsenic concentration was below 10 μg•$L^{-1}$ and the Paracetamol were not detected in the solution by UPLC. This material can be used for simultaneous removal of acetaminophen and arsenic (As(III) and As(V)).

# ELECTROSPRAYING SACCHAROMYCES CEREVISIAE IMMOBILIZED ONTO COMPOSITE NANOFIBROUS MATS FOR HEAVY METALS ADSORPTION

***Hongbing Deng***, Xiaodan Qiu, Zhaoyang Zeng
(Wuhan University, Wuhan, China)

*Saccharomyces cerevisiae* (SCV) immobilized electrospun polymer-layered silicate nanofibrous mats were fabricated via electrospraying method. These organic-inorganic composite nanofibrous mats, including cellulose acetate/rectorite (CA/REC) and polycaprolactone/chitosan/REC (PCL/CS/REC) nanofibrous mats, have three-dimensional nanofibrous structure, ultrafine fiber diameter and big specific surface area. SCV colonies distribute on the surface of polymer-layered silicate nanofibrous mats with fiber diameter range from 52 to 292 nm. With SCV, the composite nanofibrous mats get obvious adsorption capacity, especially for heavy metals such as $Cr^{6+}$, $Pb^{2+}$ and $Cu^{2+}$. The surface areas of composite nanofibrous mats increase with the addition of REC, a kind of layered silicate with super large surface area and interlayer distance. Within a certain range, the more REC is added, the higher adsorption capacities are obtained. And CS, also an adsorption agent, could remarkably improve the adsorption effect of the nanofibrous mats. SCV-PCL/CS/REC composite nanofibrous mats have an adsorption capacity of 215.1±19.8 mg/g toward $Pb^{2+}$. SCV immobilized electrospun polymer-layered silicate nanofibrous mats can serve as a bio-polymer-layered silicate composite thin film for the adsorption of heavy metals in water.

# SYNERGISTIC ADSORPTION AND DEGRADATION OF AQUEOUS CONTAMINANTS USING MAGNETIC MESOPOROUS ADSORBENT

Yaocheng Deng, ***Lin Tang***, Jingjing Wang
(Hunan University, Changsha, China)

Magnetic mesoporous materials, due to their high surface areas, large pore volumes, excellent physical-chemical properties and convenience for recovery, have been widely investigated as promising candidates for the treatment of aqueous contaminants. And meanwhile, the large surface areas make mesoporous materials not only good adsorbent, but also excellent support for some other oxidant or reductant, which will greatly enhance the removal efficiency for environmental contaminants. Considering the excellent characteristics of magnetic mesoporous materials mentioned above, we have conducted a series of studies based on them for environment remediation.

Hexavalent chromium (Cr(VI)) is a highly toxic pollutant, which is able to cause carcinogenesis, mutation to humans and animals. Therefore, developing effective methods to remove Cr(VI) from the environment is of great significance to human health and ecological system. To effectively remove Cr(VI) in aqueous solution, we have prepared some adsorbents based on magnetic mesoporous materials. First of all, we considered mesoporous silica (SBA-15). We synthesized polyaniline grafted magnetic mesoporous silica composite (PANI-Fe/SBA-15) for the adsorption and reduction of Cr(VI). The results showed that the prepared PANI-Fe/SBA-15 presented great removal efficiency for Cr(VI) due to the efficient adsorption and synergistic reduction by PANI, and the maximum adsorption capacity reached to 193.85 mg $g^{-1}$, which presented great potential for realistic application. Mesoporous carbon is also a good adsorbent with better bioavailability, so we also incorporated ordered mesoporous carbon with polyaniline and iron nanoparticles, to prepare a new adsorbent (PANI-Fe/OMC). The adsorption capacity of the functionalized material was 2- and 10-fold compared with pristine SBA-15 and the magnetic mesoporous carbon (Fe/OMC), and the adsorption process of Cr(VI) was demonstrated to be the chemisorption. Besides, carbon nanorods (CMK-3) with hexagonally ordered uniform mesoporous structure have shown good performance as adsorptive media due to its unique physical and chemical proprieties. To improve the adsorption capacity and obtain reusability of the mesoporous material, we doped iron inside CMK-3 for magnetic property (Fe/CMK-3). We found its synergistic effect on adsorption and reduction of Cr(VI). Under the optimal conditions, Fe/CMK-3 showed the maximal removal capacity of 256.85 mg $g^{-1}$, which was larger than that of pure CMK-3 as 182.03 mg $g^{-1}$. The excellent performance of Fe/CMK-3 could be contributed to the following two aspects. On one hand, the large specific surface area presented great adsorption capacity for Cr(VI); on the other hand, due to the existence of $Fe^0$, Cr(VI) could be reduced to Cr(III) with lower toxicity. Besides, the synthesized Fe/CMK-3 can be regenerated and showed high removal efficiency even after seven cycles.

Additionally, we have also prepared some mesoporous materials for the removal of organic pollutant in aqueous solution. For example, we prepared nanoscale zero-valent iron particles immobilized on mesoporous silica for the reductive degradation of aqueous p-nitrophenol, and phosphorus-doped ordered mesoporous carbons embedded with Pd/Fe bimetal nanoparticles for the dechlorination of 2,4-dichlorophenol.

Briefly, magnetic mesoporous adsorbent based composites can efficiently remove heavy metal or organic pollutants with combined adsorption and degradation process, and present great potential for wastewater treatment.

# PERSULFATE ACTIVATION IN THE PRESENCE OF FORMIC ACID FOR CARBON TETRACHLORIDE AND CHROMIUM (VI) REMOVAL

***Xiaogang Gu*** and Shuguang Lu (East China University of Science and Technology, Shanghai, China)

**ABSTRACT:** In this study, carbon tetrachloride (CT) and Cr(VI) removal performance was investigated by means of the thermal activation of persulfate, in the presence of formic acid to turn the oxidizing capacity of the reaction mixture into a reductive one by the generation of carbon dioxide anion free radicals ($CO_2^{-}\bullet$). The Effects of the initial concentrations of persulfate, formic acid, and contaminants on CT and Cr(VI) removal performance were evaluated, respectively. The results indicated that CT and Cr(VI) could be effectively removed in the thermally activated PS system in the presence of FA. CT and Cr(VI) removal performance was found to increase with increasing PS and formic concentrations. And Cr(VI) removal decreased with increasing the initial Cr(VI) concentration, while no obvious influence of the initial CT concentration was observed.

## INTRODUCTION

Carbon tetrachloride (CT, a kind of chlorinated aliphatic hydrocarbon) and hexavalent chromium Cr (VI) have been frequently detected in the groundwater of the National Priority List (NPL) sites in USA (Choi et al., 2009). In-situ chemical oxidation (ISCO) is becoming an effective method for in-situ soil and groundwater remediation, which is capable of breaking down many contaminants in solid or aqueous phases. Among the ISCO oxidants, persulfate (PS) is a promising new oxidant. ISCO using PS has been documented to treat aquifers contaminated groundwater (Tsitonaki et al., 2010).

Thermally activated PS system has been carried out to effectively degrade the toxic organic contaminants (Gu et al., 2011). And the addition of formate to the thermally activated PS system has been successfully applied to degrade trichloroacetic acid (Mora et al., 2009). To our best knowledge, the performance of perchlorinated hydrocarbons and Cr(VI) reduction in the thermally activated PS system in the presence of formic acid (FA) has never been studied, and it is expected that thermally activated PS system in the presence of FA may generate more reductive radicals that can remove CT and Cr(VI) in groundwater efficiently. Therefore, the objective of this study was to investigate the performance of CT and Cr(VI) removal in the thermally activated PS system in the presence of FA.

## EXPERIMENTAL

**Materials.** Sodium persulfate (98.0%), formic acid (99.0%), and potassium iodide (99.0%) were purchased from Shanghai Jingchun Reagent Co. Ltd. (Shanghai, China). Carbon tetrachloride (CT, 99.5%) and n-Hexane (97%) were purchased from Shanghai Lingfeng Chemical Reagent Co. Ltd. Methyl viologen dichloride (MV2+) was purchased from Aladdin Reagent Ltd Co. (Shanghai, China). Ultrapure water from a Milli-Q water process (Classic DI, ELGA, Marlow, U.K.) was used for preparing aqueous solutions.

**Experimental Procedures.** All reactions were conducted in a series of 24 mL borosilicate vials capped with polytetrafluoroethylene (PTFE) lined septa under controlled temperature (20~50 ℃). CT, Cr(VI), and PS solutions were prepared by dissolving reagents at room temperature into Milli-Q water to make stock solutions. Each stock solution and desired volume of FA were added to the volumetric flask at desired concentration, then a series of reaction vials were fully filled. The reaction vials were sacrificed at different intervals for immediate analyses. The initial pH in all experiments was unadjusted except in the test for investigating the influence solution pH.

**Analytical Methods.** The concentration of CT was quantified after extraction with hexane by a gas chromatograph (GC, Agilent 7890A, Palo Alto, CA) equipped with an autosampler (Agilent 7693). The

Cr(VI) concentration was determined using the 1,5-diphenylcarbazide method on a 722N visible spectrophotometer (Hach DR 6000, Loveland, CO).

## RESULTS AND DISCUSSION

**Enhanced CT and Cr(VI) removal with the presence of FA in PS system.** FIGURE 1 shows the CT and Cr(VI) removal performance in the thermally activated PS system in the presence of FA and in different control groups. 2.3% and 3.4% CT lost in the CT alone group and in CT solution with the presence of FA at 50 ℃ within 180 min reaction time, respectively. The loss of the component was likely due to the thermolysis and volatilization of CT in solution. However, the degradation of CT was significantly increased when FA was added to CT solution. Complete CT degradation was achieved for reactions carried out at PS concentration of 20 mM and FA concentration of 30 mM over 180 min. According to Cr(VI), the results of control tests showed that there was no observable Cr(VI) removal when PS or FA was applied solely in the thermal systems. However, Cr (VI) removal did occur when PS and FA were applied together at 50 ℃ (FIGURE 1b).

In the presence of FA, $SO_4\bullet^-$ and HO• generated in the chain reaction were scavenged to yield $HCO_2\bullet$ via Eqs. 1-2. Carbon dioxide radical anion ($CO_2\bullet^-$) formed from the propagation reactions as shown in Eq. 3 (Lin et al., 2008). $CO_2\bullet^-$ is a strongly reducing species, with a redox potential of -2.0 V vs. NHE (normal hydrogen electrode). The addition of FA to the solution can turn the oxidizing capacity of the reaction mixture into a reductive one, since $CO_2\bullet^-$ can transfer an electron rapidly to quinones, nitro and nitroso compounds, pyridinium and viologen ions, porphyrins, oxygen, and many other organic and inorganic compounds (Neta et al, 1998). Therefore, it is deduced that the removal of CT and Cr(VI) in the thermally activated PS system in the presence of FA was due to the role of $CO_2\bullet^-$.

$$SO_4\bullet^- + HCOOH \rightarrow SO_4^{2-} + HCO_2\bullet + H^+ \quad (1)$$

$$HO\bullet + HCOOH \rightarrow H_2O + HCO_2\bullet \quad (2)$$

$$HCO_2\bullet \leftrightarrow CO_2\bullet^- + H^+ \quad (3)$$

**Effects of initial PS, FA and CT concentrations on CT degradation.** To study the effect of PS concentration on CT degradation in the thermally activated PS system in the presence of FA, different PS concentrations were used in the reaction (FIGURE 2). When the PS concentration was more than 20 mM, complete CT degradation was achieved in 180 min, and in contrast CT degraded 40.0% and 9.8% with the PS concentrations of 10 and 5 mM, respectively (FIGURE 2a). The results showed that the rate of CT degradation increased as the initial PS concentration increased. With the increase of PS concentration, the amount of $SO_4\bullet^-$ and HO• were elevated and more $CO_2\bullet^-$ generated through Eqs. 1-3. Liang et al. found that when PS concentration was more than 6.75 mM in the degradation of trichloroethylene in the ferrous activated PS system, increase in PS concentration only resulted in smaller increases in trichloroethylene removal (Liang et al., 2004). Similarly, excess amount of $CO_2\bullet^-$ would lead to radical-radical recombination as shown in Eq. 4, which might stop the increase of CT degradation rate (Cooper et al., 2009).

$$CO_2\bullet^- + CO_2\bullet^- \rightarrow {}^-OOC\text{-}COO^- \quad (4)$$

Different concentrations of FA were applied to evaluate the effect of FA concentration on CT degradation performance and the results are displayed in FIGURE 2(b). It could be seen that CT removal efficiency increased with the increase of FA concentration from 5 to 100 mM. Complete CT degradation could be achieved in 180 min when the FA concentration was over 30 mM, while CT degraded 80.6%, 56.7% and 46.3% with the FA concentrations of 20, 10 and 5 mM, respectively. The increased concentration of FA induced the reactions as shown in Eqs. 1-2 reactions, therefore more $CO_2\bullet^-$ were generated in the system.

The variation of CT degradation efficiency in the thermally activated system in the presence of FA under different initial CT concentrations, ranging from 0.01 mM to 1 mM, was also investigated and the results are shown in FIGURE 2(c). CT degradation rate followed zero order rate kinetics as shown in FIGURE 2(c). When the initial CT concentration increased, the slopes of these lines were almost the same and the CT degradation rates were nearly constant.

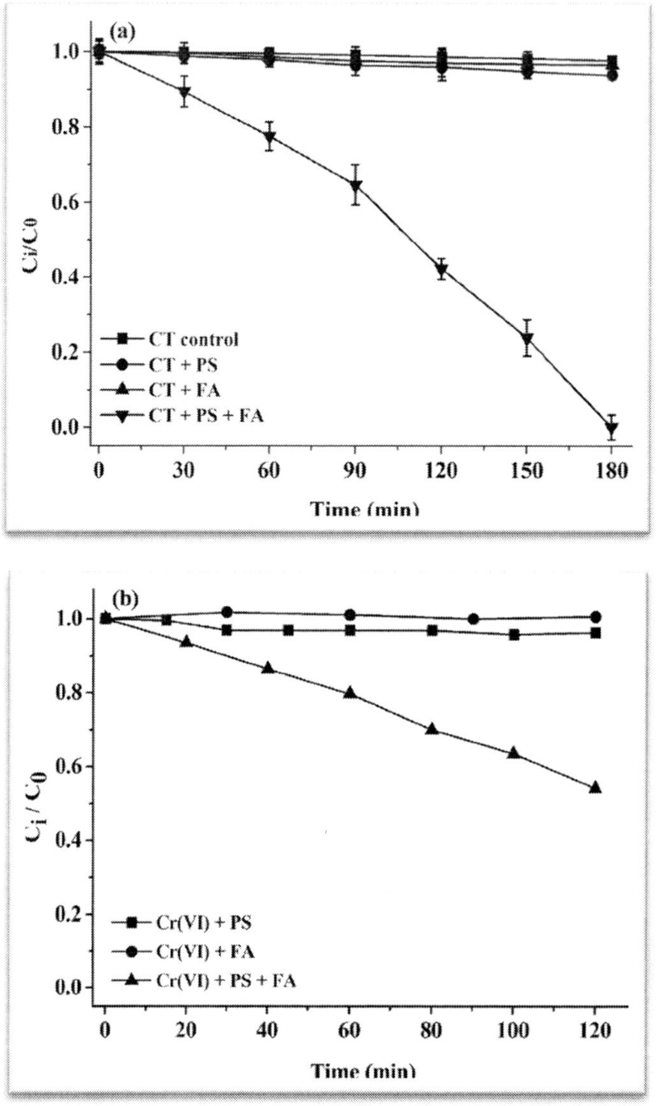


**FIGURE 1. CT (a) and Cr(VI) (b) removal in the thermally activated PS system in the presence of FA (50°C, $[CT]_0$ = 10 μM, $[Cr(VI)_0]$= 0.2 mM, [PS] = 20 mM, [FA] = 30 mM)**

**Effects of initial PS, FA and Cr(VI) concentrations on Cr(VI) removal.** The effect of initial PS concentration at 50 ℃ was investigated by varying the initial PS concentration in the range of 20~100 mM, while keeping the initial FA concentration constant at 100 mM (FIGURE 3a). It can be seen that, with the increase of the initial PS concentration, the removal efficiency increased. When the initial concentrations of persulfate, formic acid, and Cr(VI) were 100 mM, 100 mM, and 0.2 mM, respectively, Cr(VI) could be almost completely removed at 2 h.

Moreover, the influence of initial FA concentrations of 20 mM to 200 mM was investigated at a temperature of 50 ℃, and the results are shown in FIGURE 3b. Complete removal of Cr(VI) was achieved in 2 h at the experimental conditions, and the removal efficiency increased with the increase of FA concentration. The results of FIGURE 3a and 3b strongly suggested that the removal of Cr(VI) proceeded more rapidly with increasing PS and FA concentrations, and PS played a much more important role than the initial FA concentration. The influence of PS and FA concentrations on Cr(VI) removal were consistent with the results of CT degradation.

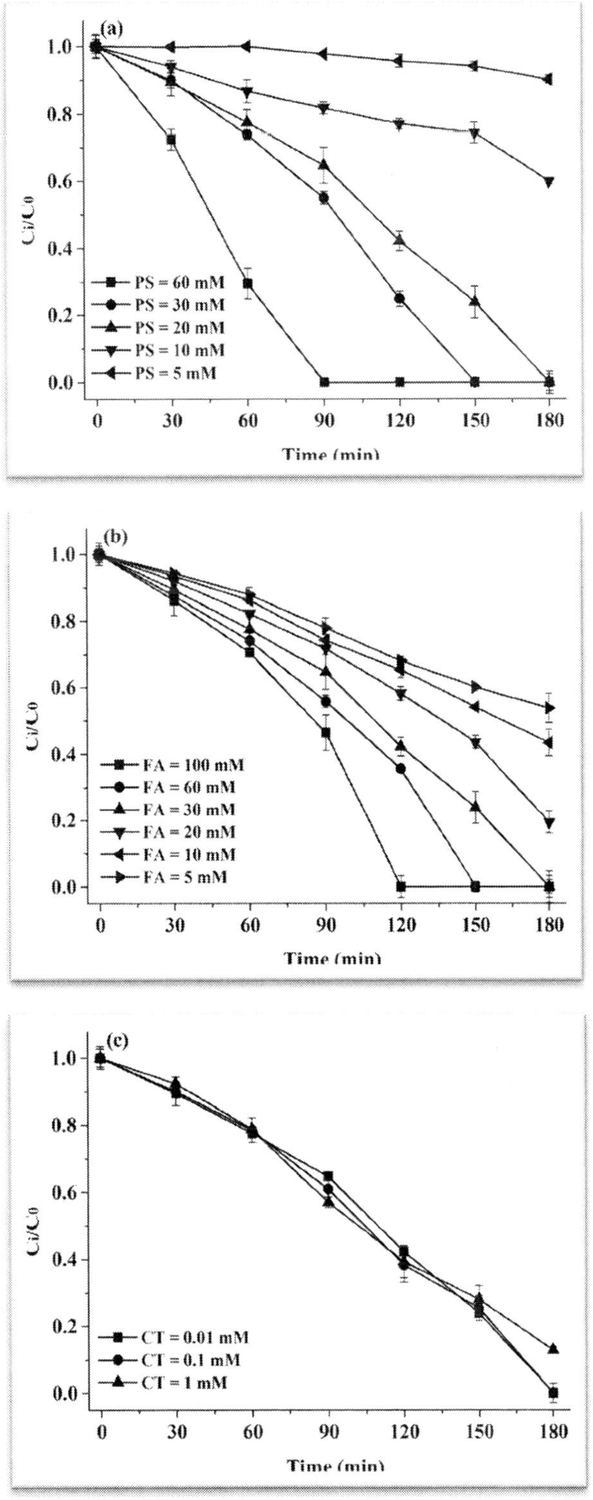


**FIGURE 2. Effect of initial concentration on CT degradation. (a) PS (50 ℃, $[CT]_0$ = 10 μM, [FA] = 30 mM); (b) FA (50 °C, $[CT]_0$ = 10 μM, [PS] = 20 mM); (c) CT (50 ℃, [PS] = 20 mM, [FA] = 30 mM).**

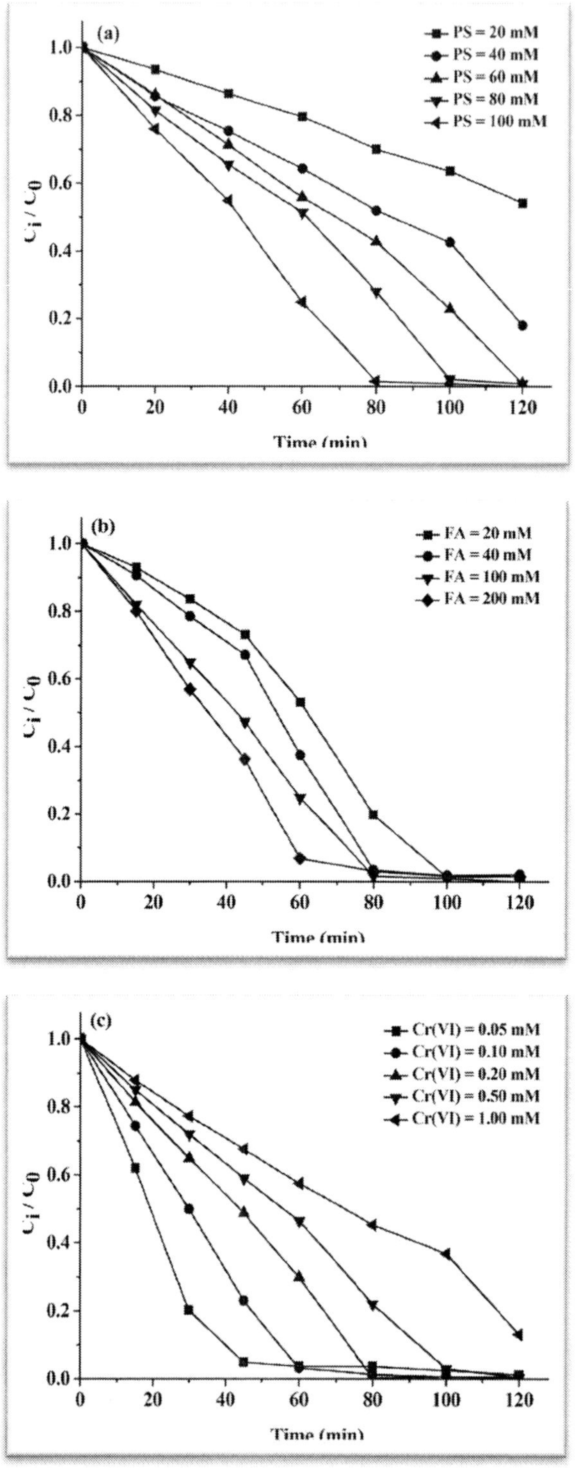


**FIGURE 3. Effect of initial concentration on Cr(VI) removal. (a) PS (50 °C, [Cr(VI)$_0$]= 0.2 mM, [FA] = 100 mM); (b) FA (50 ℃, [Cr(VI)$_0$]= 0.2 mM, [PS] = 100 mM); (c) CT (50 ℃, [PS] = 100 mM, [FA] = 100 mM).**

As shown in FIGURE 3c, concentration, various Cr(VI) concentrations (0.05 mM~1.00 mM) were conducted to investigated the effect on Cr(VI) removal. Unlike CT tests, the removal efficiency of Cr(VI) decreased with the increase of initial Cr(VI) concentration, and the slopes of Cr(VI) removal were different. However, the removal rates of Cr(VI) were nearly constant.

## CONCLUSIONS

This study confirmed that CT and Cr (VI) could be effectively removed in the thermally activated PS system in the presence of FA at 50 ℃, and contaminants removal followed a model. CT degradation rate increased with increasing PS or FA dosage, and no obvious effect of the initial CT on CT degradation rate was observed. Cr (VI) removal was found to increase with increasing PS and formic concentrations, and decrease with increasing the initial Cr(VI) concentration. In conclusion, this study demonstrated that thermally activated PS system in the presence of FA is a promising technique in ISCO remediation for CT and Cr(VI)-contaminated sites.

## ACKNOWLEDGEMENT

This study was financially supported by the grant from the National Natural Science Foundation of China (No.41373094) and China Postdoctoral Science Foundation (2015M570341).

## REFERENCES

Choi, J. Y., K. H. Choi, W. J. Lee. 2009. "Effects of transition metal and sulfide on the reductive dechlorination of carbon tetrachloride and 1,1,1-trichloroethane by FeS". *J. Hazard. Mater.* 162: 1151-1158.

Cooper W. J., C. J. Cramer, N. H. Martin, S. P. Mezyk, K. E. O'Shea, C. Sonntag. 2009. "Free radical mechanisms for the treatment of methyl tert-butyl ether (MTBE) via advanced oxidation/reductive processes in aqueous solutions." *Chem. Rev.* 109: 1302-1345.

Gu, X. G., S. G. Lu, L. Li, Z. F. Qiu, Q. Sui, K. F. Lin, Q. S. Luo. 2010. "Oxidation of 1,1,1-trichloroethane stimulated by thermally activated persulfate". *Ind. Eng. Chem. Res.* 50(19): 11029-11036.

Lin, M. Z., Y. Katsumura, Y. Muroya, H. He, T. Miyazaki, D. Hiroishi. 2008. "Pluse radiolysis of sodium formate aqueous solution up to 400℃: absorption spectra, kinetics and yield of carboxyl radical $CO_2^{\cdot -}$". *Radiat. Phys. Chem.* 77: 1208-1212.

Liang, C. J., C. J. Bruell, M. C. Marley. 2004. "Persulfate oxidation for in situ remediation of TCE. I. activated by ferrous ion with and without a persulfate-thiosulfate redox couple." *Chemosphere* 55: 1213-1223.

Mora, V. C., J. A. Rosso, G. C. Roux, D. O. Martire, M. C. Gonzalez. 2009. "Thermally activated peroxydisulfate in the presence of additives: A clean method for the degradation of pollutants". *Chemosphere, 75*: 1405-1409.

Neta P., R. E. Huie, A. B. Ross. 1998. "Rate constants for reactions of inorganic radicals in aqueous solution." *J. Phys. Chem. Ref. Data.* 17: 1027-1284.

Tsitonaki, A., B. Petri, M. Crimi , H. Mosbaek, R. L. Siegrist, P. L. Bjerg. 2010. "In situ chemical oxidation of contaminated soil and groundwater using persulfate: A review". *Crit. Rev. Env. Sci. Tec.* 40(1): 55-91.

# ENHANCED SETTLING OF SUSPENDED SOLIDS IN OLIVE MILL WASTEWATER BY APPLICATION OF HIGH FREQUENCY ULTRASOUND

Alev Cagla Uzun, ***Canan Can Yarımtepe***, Nilgün Ayman Oz
(Canakkale Onsekiz Mart University, Canakkale, Türkey)

**ABSTRACT**: The aim of the study is to investigate effect of high frequency ultrasound on biogas and methane production from olive mill wastewater (OMW) in anaerobic batch reactors. OMW was sonicated at 582 kHz for different durations. Optimum time was selected as 10 minutes based on total chemical oxygen demand (COD) and suspended solids. Then, anaerobic batch reactors were fed with ultrasound pretreated and raw OMW to determine impact of the process on biogas and methane production. Sonication of OMW before anaerobic batch tests increased both biogas production and methane yield in anaerobic reactors by 8% and 37%, respectively. The results showed that high frequency ultrasound before anaerobic processes has a positive effect especially on methane production rate. Another outcome of the study, it has been observed that high frequency ultrasound caused enhanced settling of suspended solids of OMW samples. After 60 minutes of high frequency ultrasound, OMW sample was left for two hours and suspended solids settled quickly. After settling, 54% decrease in SS concentration at supernatant was recorded.

## INTRODUCTION

Anaerobic treatment is a sustainable option for olive mill effluents, which contain very high organic matter concentration. Olive mill wastewaters also contain high particulate matter, which cannot be efficiently removed during settling process. The character of the wastewater limits the performances of anaerobic reactors according to COD removal and biogas production. Low frequency ultrasound before anaerobic treatment has been proposed for facilitating hydrolysis step and increasing biodegradation rate. In recently, there is also an interest in high frequency ultrasound, which is an oxidative process, for wastewaters containing complex compounds. So far, several studies have been reported in literature on application of high frequency ultrasound on different wastewaters and waste activated sludge whereas combination of the process with anaerobic batch process for OMW has not been reported so far. Therefore, the aim of this study is investigation of effect of high frequency ultrasonic pretreatment on settling of suspended solids, biogas and methane production in anaerobic batch processes.

## MATERIALS AND METHODS

The OMW was collected from a factory which produced olive oil with three phase process, in Canakkale, Turkey at November 12, 2012. The composition of the OMW sample is summarized in Table 1.

Complete analysis were carried out according to the Standard Methods (APHA, 1998). Meinhardt Ultrasound multifrequency at 582 kHz was used during ultrasound experiments. Sonication was performed during 60 min on 500 mL of OMW. During sonication process, changes in tCOD, turbidity and suspended solids were monitored. After ultrasound, untreated and ultrasound pretreated samples were fed to anaerobic batch reactors. Glass reactors with 900 ml active volume were used in the anaerobic batch reactors. Anaerobic seed sludge was taken from a brewery factory's EGSB reactor.

Untreated and ultrasound pretreated OMW samples were added to reactors with seed sludge and dilution water and F/M ratio was set to 0,5 in reactors. Biogas production was monitored with milli gas counter. Biogas composition was determined by Gas Chromatografy.

## RESULTS AND DISCUSSIONS

In this study, effect of high frequency ultrasound on biogas and methane production of OMW was investigated. Ultrasound was applied to OMW during 60 min. However, after 10 minutes there was not a change in tCOD concentration. Therefore, optimum time for application of ultrasound was chosen as 10 minutes. Fig. 1. shows variation in total COD concentration and Fig. 2. Shows total COD removal efficiencies during ultrasound at 582 kHz.

Table 1. OMW Composition

| **Parameter** | **Unit** | **Value** |
|---|---|---|
| pH | - | 5,142 |
| Turbidity | NTU | 11688±1060 |
| Total Chemical Oxygen Demand (TCOD) | mg/L | 68543±3337 |
| Soluble Chemical Oxygen Demand (SCOD) | mg/L | 40506±1719 |
| Biochemical Oxygen Demand (BOD) | mg/L | 17119 |
| Conductivity | µs/cm | 9190 |
| Total Nitrogen (TN) | mg/L | 219 |
| Total Organik Carbon (TOC) | mg/L | 14835 |
| Suspended Solids (SS) | mg/L | 12598±1032 |
| Total Solid (TS) | mg/L | 49143 |
| Phenol | mg/L | 5056 |
| Color | pt | 29807±8887 |
| Soluble Color | pt | 11035±4410 |

After 10 min sonication, tCOD concentration was decreased about 16% compared with the raw OMW sample. High frequency ultrasound has been previously reported as effective in breaking down organic matter in sludge samples (Braguglia et. al., 2012). Fig. 3. shows variation in turbidity and suspended solids concentration and Fig. 4. shows turbidity and suspended solids removal efficiencies in during ultrasound at 582 kHz.

22% decrease in SS concentration has been recorded after 10 min sonication. These results showed pretreatment have a significant impact on disrupting organic matter. After 10 minutes ultrasound pretreatment, the OMW sample was fed to anaerobic batch reactor at F/M ratio of 0.5. Another anaerobic batch reactor was fed with raw OMW under the same operating conditions as a control reactor.

During the batch anaerobic tests, biogas production and methane content were monitored for raw and pretreated OMW. Fig. 5. shows cumulative biogas production and Fig. 6. shows methane production in raw and ultrasound pretreated anaerobic batch reactors.

The comparison between batch anaerobic tests carried out with untreated and sonicated OMW showed an increase in both biogas production by 8% and methane production by 37%. In the literature, there is no reported study for high frequency ultrasound assisted anaerobic treatment of OMW. Ultrasound assisted anaerobic treatment can be used for complex wastewaters such as OMW to obtain a better efficiencies in anaerobic processes such as COD removal and methane production. However, the research area needs further investigations.

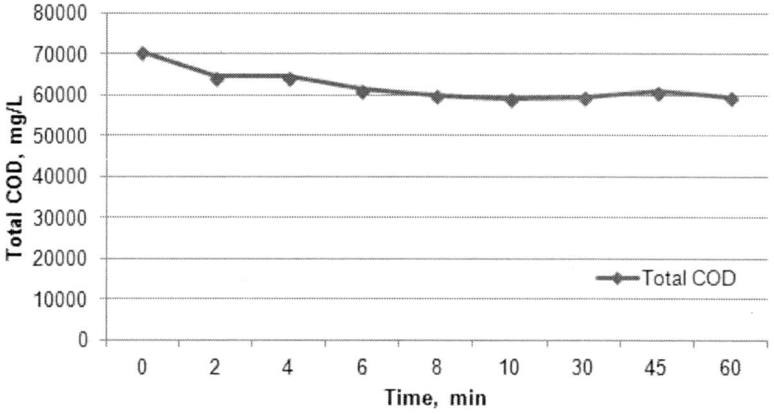


Figure 1. Variation in total COD concentration during ultrasound at 582 kHz

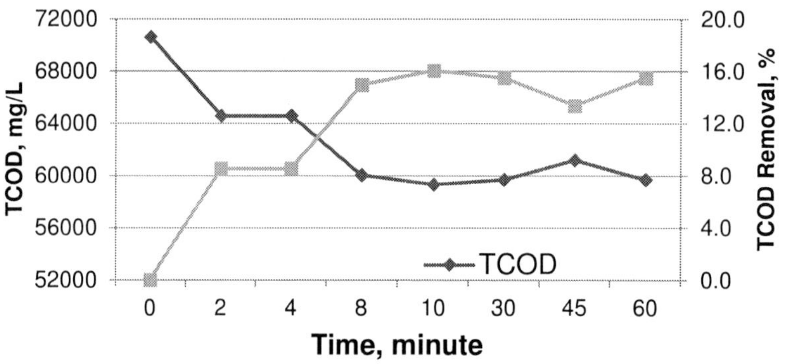


Figure 2. Total COD removal efficiencies during ultrasound at 582 kHz

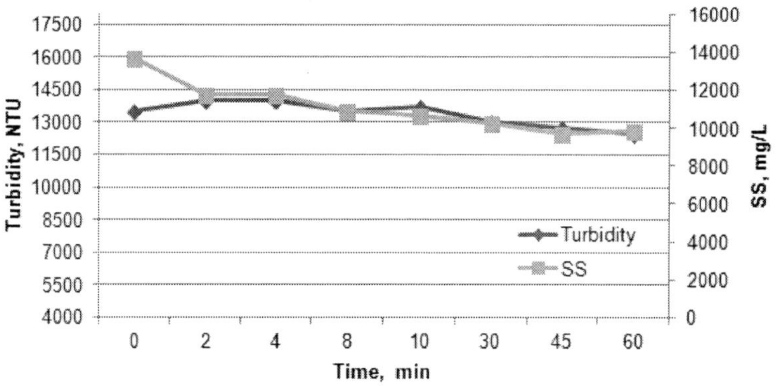


Figure 3. Variation in turbidity and suspended solids during ultrasound at 582 kHz

Another outcome of the study, after high frequency ultrasound was applied to OMW samples, enhanced settling of suspended solids of OMW was recorded. After application of 60 minutes of high frequency ultrasound, OMW sample was left for two hours and then suspended solids settled quickly.

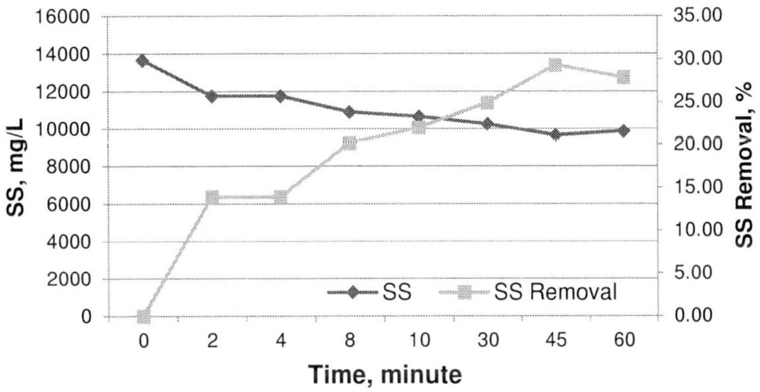


Figure 4. turbidity and suspended solids removal efficiencies in during ultrasound at 582 kHz

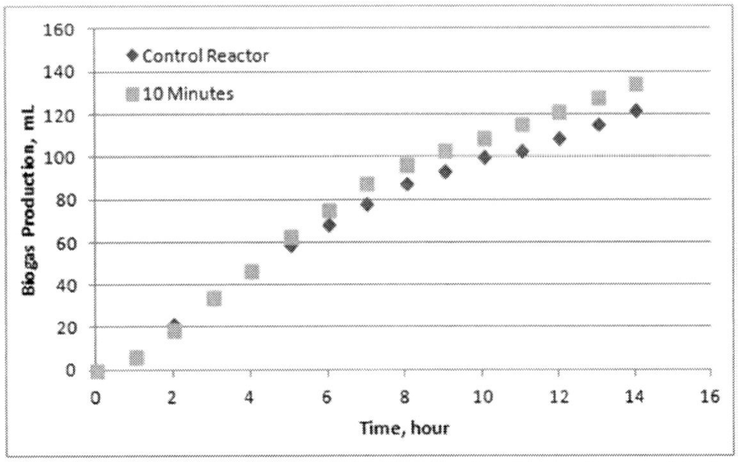


Figure 5. Cumulative biogas production in raw and ultrasound pretreated anaerobic batch reactors

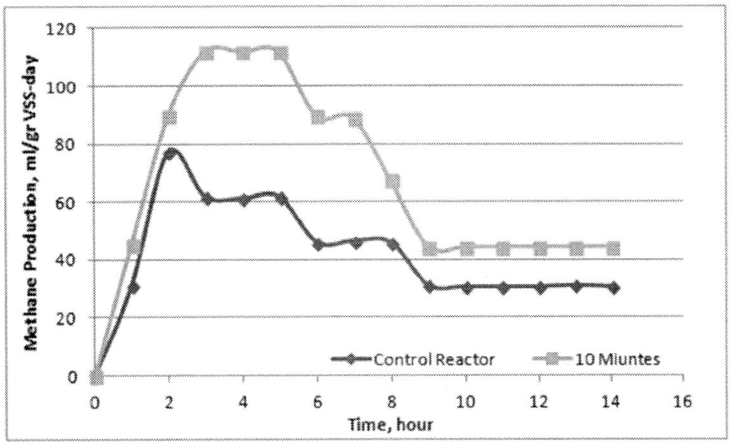


Figure 6. Methane production in raw and ultrasound pretreated anaerobic batch reactors

After settling, 54% decrease in SS concentration at supernatant was recorded. It should also be mentioned that raw OMW could not settle with gravity even after 1 day. High frequency ultrasound has been previously reported as effective in settling of microorganisms in water samples (Bosma et. al., 2003) and sludge (Guangming et. al., 2006). But there is no report on enhancement of settling for wastewater samples.

## CONCLUSIONS

As a result, application of ultrasound to complex industrial wastewaters such as OMW showed promising results. Especially, as a pretreatment prior to the anaerobic system, it can enhance both pollutant removal efficiencies and biogas production. However, there is a need for a further investigation.

## ACKNOWLEDGEMENT

Financial support by The Scientific and Technological Research Council of Turkey (TUBITAK) was gratefully acknowledged (Project No.: 111Y112).

## REFERENCES

APHA, AWWA, WPCF, Standard Methods for the Examination of Water and Wastewater, 20th ed., American Public Health Association, Washington, DC, 1998.

Bosma R., Spronsen, W.A., Tramper J. and Wijffels H. R. (2003) Ultrasound, a new separation technique to harvest microalgae, Journal of Applied Phycology 15: 143–153.

Zhang, Guangming, Panyue Zhang, and Yanming Chen. (2006) Ultrasonic enhancement of industrial sludge settling ability and dewatering ability, Tsinghua Science & Technology 11.3: 374-378.

# STUDY OF A FLUIDIZED BED REACTOR FOR PHENOL PHOTOCATALYSIS

***Guillermo Rincon**** and Enrique La Motta
(University of New Orleans, New Orleans, Louisiana, USA)

A bench-scale tubular reactor with recirculation was built in order to study the efficiency of the photocatalytic oxidation of phenol on fluidized titanium oxide-coated silica gel beads. A UV-C lamp was used as source of photons. A bed of silica gel beads was fluidized by means of fluid recirculation and forced to follow upward helical flow around the lamp. Anatase was successfully synthetized on silica gel particles with average diameters 224, 357 and 461 microns, as confirmed by scanning electron micrographs, through a sol-gel technique using a titanium tetraisopropoxide / hydrochloric acid / ethanol precursor.

Experimental results confirmed that the efficiency of phenol photocatalytic degradation decreases with increasing pollutant concentration. Also, the highest removal was achieved at initial pH 3, and it decreased with increasing pH. When NaCl was added to the solution, COD removal increased with increasing salinity. Additionally, it was found that dissolved oxygen is indispensable for photocatalysis to proceed, and saturation of the treated mixture with oxygen was achieved by keeping the liquid in contact with pure oxygen at 1 atm.

Finally, a statistical model based on the exponential decay fit of phenol-derived COD vs irradiation time data in the experimental reactor was proposed (minimum $R^2$ = 0.9840) to allow for the accurate estimation of the reactor efficiency for any given initial COD.

# DEVELOPMENT OF $BiVO_4$ PHOTOANODE FOR DEGRADING POLLUTANTS

***Oin Shi*** and Hui Wang* (Beijing Forestry University, Beijing, China)
Zhaoyong Bian (Beijing Normal University, Beijing, China)

Chlorinated phenols (CPs) are detrimental environmental pollutants due to their hazardous impact. The efficient photoelectrochemical degradation of CPs has attracted much attention by using various photo-anodes. Bismuth vanadate ($BiVO_4$) has been identified as one of the most promising photoanode materials. However, the typical efficiencies of unmodified $BiVO_4$ photoanodes for degradation of pollutants were not impressive as they suffer from excessive electron–hole recombination and poor charge transport properties. In this work, a $FeOOH/BiP_{0.02}V_{0.98}O_4$ modified graphene composite photoelectrocatalyst was prepared by using three strategies to alleviate these limitations, including P-droping, pairing with oxygen evolution catalysts and construction of composite structures. The composite photoelectrocatalyst was characterized by X-ray diffraction (XRD), scanning electron microscopy (SEM), transmission electron microscopy (TEM), atomic force microscopy (AFM), X-ray photoelectron spectroscopy (XPS). Its photoelectrocatalytic performance was investigated via electrochemical measurements (LAS and EIS) in a three-electrode system in a solution containing 0.1 M $NaSO_4$ and 1 mM 2,4-dichlorophenol, to highlight its potential application in photoelectrocatalytic electrooxidation of CPs. The characterization results showed that $BiVO_4$ and $BiP_{0.02}V_{0.98}O_4$ have a uniform scheelite-monoclinic structure at a size of 100-200 nm. A thin FeOOH layer was photo-deposited on the $BiP_{0.02}V_{0.98}O_4$ surface as an OEC layer successfully. Graphene nanosheets could be found wrapping the $FeOOH/BiP_{0.02}V_{0.98}O_4$ particles in the $FeOOH/BiP_{0.02}V_{0.98}O_4$-graphene composite. The photocurrent densities had an elevation with the increase in this order: $BiVO_4$, $BiP_{0.02}V_{0.98}O_4$, $FeOOH/BiP_{0.02}V_{0.98}O_4$, $FeOOH/BiP_{0.02}V_{0.98}O_4$-graphene. It meant that the $FeOOH/BiP_{0.02}V_{0.98}O_4$-graphene photoelectrocatalyst posed the highest catalytic activity for photoelectrochemical oxidation of 2,4-dichlorophenol. The EIS analyses were performed to understand the phenomenon. The fitted values of $R_t$ are 37964, 7351, 3655, and 1236 Ω for pristine $BiVO_4$, $BiP_{0.02}V_{0.98}O_4$, $FeOOH/BiP_{0.02}V_{0.98}O_4$, $FeOOH/BiP_{0.02}V_{0.98}O_4$-graphene electrodes, respectively. The $FeOOH/BiP_{0.02}V_{0.98}O_4$-graphene electrode showed the fastest interfacial electron transfer which is in accordance with its enhanced photoelectroactivity.

# OPTIMIZATION OF OPERATING PARAMETERS ON REACTIVE BLACK 5 DYE REMOVAL BY ELECTROCOAGULATION PROCESS AND ENERGY CONSUMPTION

Mook Wei Tze and Mohamed Kheireddine Aroua (University of Malaya, Kuala Lumpur, Malaysia);
***Małgorzata Szlachta*** (Wrocław University of Technology, Wrocław, Poland)

The textile industry is one the major sector responsible for severe water pollution as it discards large amount of colour wastewater. Reactive dye comprises 20 to 30% of the total dye market as their vivid colours and easily attach to fibre polymer. However, reactive dye is high stability and resistant to biological degradation. Hence, reactive dye wastewater treatment becomes a greater challenge for researchers. Electrocoagulation (EC) is one of the effective and economical method to treat dye wastewater. The simplicity of operation, low energy consumption, low sludge generation and high efficiency made EC become a favourable treatment technique. In EC process, no chemicals are added to form coagulant. Instead, it is generated in situ through electrolytic oxidation of anode material.

In this work, Response Surface Methodology (RSM) and Central Composite Design (CCD) were proposed to optimize the operating parameters on the energy consumption and removal efficiency of Reactive Black 5 dye (RB5) through electrocoagulation process. The RSM is a valuable tool to build an equation model, reduce number of experiments, observe interaction between parameters and obtain optimum conditions for desirable responses.

The effects of initial pH and current density on RB5 removal efficiency and energy consumption were investigated. It was found that with increasing pH more RB5 were removed. At a lower pH, the $H^+$ reduced to hydrogen gas ($H_2$) at the cathode, which reduced the amount of hydroxide ions generated. However, the removal of dye was observed to decrease at a pH beyond 6.5. This could be because the $Fe(OH)_3$ dissolves to form $Fe(OH)_4^-$ and migrates to the anode surface to undergo oxidization; hence anodic generation of $Fe^{2+}$ was negatively affected.

It was also found that the low current density failed to have a sufficient impact upon the RB5 removal process. According to Faraday's law, higher current density enhances coagulant generated to react with RB5 and increases the removal rate. Moreover, the bubble generation rate is directly proportional to the current density, which is beneficial for pollutant removed by $H_2$ flotation. The highest RB5 removal was found at neutral pH and higher current density.

The current density is directly proportional to the electrical energy consumption (EEC) because current plays a major part in EEC calculation. While pH had a least effect as the EEC trend insignificant changed. In order to achieve high RB5 removal and low EEC, the optimum conditions suggested by RSM which were current density of 2.13 mA/cm$^2$ and pH of 6.6. The predicted result was 80.1% and 30.7 kWh/kg dye, which the percentage error between experimental and predicted results was only 3-5%.

# QUARTZ SAND FILTER MEDIUM FOR OILY WASTEWATER BY HYDROPHOBIC MODIFICATION

***Wei Bigui***
(Lanzhou Jiaotong University, Lanzhou 730070, Gansu, China)

The quartz sand was surface dry process modified using Titanate coupling agent DN101 to improve the lipophilic-hydrophobic property of filter medium. The effect of reaction time, the dosage of Titanate coupling agent DN101 and reaction temperature were studied by single factor experiment, Results show that when the reaction time was 70 min, DN101 dosage was 1.2% and reaction temperature was 60 °C, the effect of DN101 dry process modified quartz sand was the best. Lipophilic Hydrophilic Ratio(LHR) could be increased from 1.25 to 11.1, The adsorption capacity of modified quartz sand was greatly improved from 0.17 $mg \cdot g^{-1}$ to 0.25 $mg \cdot g^{-1}$ for 15.61 $mg \cdot L^{-1}$ oily wastewater; The filter removal rate of modified quartz sand was improved from 72.6% to 97.8% for 17.3 $mg \cdot L^{-1}$ oily wastewater. SEM, XPS and FTIR shown that DN101 combined with quartz sand surface functional groups in the form of chemical bonds, formed a uniform and stable cladding layer on the surface of quartz sand.

# CONSTRUCTION OF Fe-TAML ACTIVATOR SYSTEMS TO DEGRADE SULFUR-CONTAINING CONTAMINANTS IN WATERS

***Qingquan Liu*** and Xiyun Cai
(Dalian University of Technology, Dalian, China)

Sulfur-containing compounds commonly undergo in vivo sulfoxidation (transformation of thioethers to sulfoxides or sulfoxides to sulfones), which in turn alter reactivity, translocation, and thus toxicity of the parents. Similarly, sulfur-containing compounds were degraded in soils and waters to form sulfoxide and sulfone. These metabolites are often recalcitrant, compared to the parent, and often detected in high frequency and at high levels in waters, thereby being similar or more toxic than the parent itself. Iron-tetraamidomacrocyclic ligand (Fe-TAML) activators are small molecule mimics of short-circuited P450 enzymes through formation of Iron (V/IV)-oxo-TAML intermediates under the activation of $H_2O_2$. In this study, attempts to degrade sulfur-containing compounds (i.e., pesticides and pharmaceuticals) in waters were conducted using Fe-TAML/$H_2O_2$. Fe-TAML activators exhibited degradation efficiencies of 10%-99% toward 14 sulfur-containing contaminants in water at a catalyst:substrate ratio of 1:500 within 60 min. The degradation processes of sulfur-containing compounds were investigated with fenamiphos as a model compound. Results show that fenamiphos via successive oxidation at the sulphur atom and oxidative hydrolysis at the P-O group, and ultimately all metabolites can be mineralized. These two processes were pH-dependent, both of which accelerated rapidly with increasing pH. For example, fenamiphos was completely oxidized to fenamiphos sulfoxide (FSO) in approximately stoichiometry within 10 min at pH 12 and at a catalyst:substrate ratio of 1:500, and this intermediate proceeded further sulfoxidation to form fenamiphos sulfone ($FSO_2$) at rather low yields (nearly 5%). These two intermediates underwent rapid oxidative hydrolysis to form phenol intermediates (i.e., FSO-OH and $FSO_2$-OH). All phenol compounds were totally mineralized within 5 min. The presence of typical natural water constituents (e.g., DOC, $Mg^{2+}$ and $Ca^{2+}$) did not diminish the catalysis efficiency of Fe-TAML activators. A strongly basic anion exchange resin was combined to the Fe-TAML catalytic system for eliminating sulfur-containing contaminants in wastewater without adjusting its pH. Therefore, Fe-TAML activators may be a choice for the removal of sulfur-containing contaminants in waters.

# p-AgI ANCHORED ON n-$Bi_2O_2CO_3$ SHEETS BY Co-CRYSTALLIZATION WITH EXCELLENT PHOTOCATALYTIC PERFORMANCES UNDER VISIBLE LIGHT

***Lili Zhang*** and Chun Hu
(Chinese Academy of Sciences, Beijing, China)

**Introduction.** The development of efficient and stable visible-light photocatalysts has become one of the most important topics in water purification and hydrogen generation. It is known that the photocatalytic activity of semiconductors depends strongly on the specific surface area, particle size, the band match and crystal facets. Considering these important factors, we prepared a p-n heterojunction photocatalyst, p-AgI/n-$Bi_2O_2CO_3$, by a one-step co-crystallization method for the first time. We found that a strong interaction existed between AgI and $Bi_2O_2CO_3$. The in-situ combination of AgI and $Bi_2O_2CO_3$ led to the transformation of the morphology, the increase of the BET surface area, the reduction of particle size and the change of exposed lattice facets. Importantly, the visible-light photocatalytic activity of p-AgI/n-$Bi_2O_2CO_3$ significantly increased compared to the single AgI and $Bi_2O_2CO_3$. A possible photocatalytic mechanism was discussed in detail.

**Objectives.** The main objective of this study was to verify the strong interaction of p-AgI and n-$Bi_2O_2CO_3$, the enhanced visible light photocatalytic activity of p-AgI/n-$Bi_2O_2CO_3$ and clarify the photo catalytic mechanism.

**Methods.** A novel visible-light-responsive photocatalyst, monodispersed AgI nanoparticles anchored on the surface of $Bi_2O_2CO_3$ sheets (AgI/BOC), was prepared by a one-step co-crystallization method. The characterization of the catalysts were studied via SEM, TEM, XRD, XPS, FT-IR, UV-Vis DRS, $N_2$ adsorption-desorption and Electrochemical measurements. The analysis methods included HPLC, TOC, ICP-OES and ESR.

**Results and Conclusions.** The single BOC mainly exposed the {001} facets, while BOC in AgI/BOC exposed more (110) facets. In comparision with the pristine BOC and AgI, the positive shift in the binding energy of Bi 4f, O 1s and C 1s for carbonate and the negative shift in the binding energy of Ag 3d and I 3d in AgI/BOC revealed that the alternation of the distribution of the electric charge in the composite. The variation of FTIR spectra indicated the significant change of chemical environments of $CO_3^{2-}$ after in-situ combining BOC with AgI. Therefore, a strong interaction existed between AgI and $Bi_2O_2CO_3$. Moreover, the migration and separation of photogenerated carriers have been improved significantly evidenced by elecrochemical impedance spectroscopy, transient current measurement, and photo-luminescence spectra. The formation of p-n junction was also confirmed by elecrochemical measurements. In this case, AgI/BOC exhibited an exceptional performance in contaminant degradation. The 2-chlorophenol (10 ppm) could be completely degraded within 60 min, and around 60.5% of TOC was degraded within 60 min in the suspension of AgI/BOC under visible light, which is much higher than the 7.5% and 2.7% of 2-chlorophenol degradation in the single AgI and BOC suspension under similar conditions, respectively, indicating the excellent activity of AgI/BOC under visible light. After the reaction, no Ag and Bi were detected in the solutions. The activity of AgI/BOC did not decrease after seven successive cycles of degradation testing, which suggested that AgI/BOC was a stable catalyst for visible-light photocatalytic reactions.

# COST ANALYSIS OF ELECTRO DIALYSIS CELL FOR DESALINATING BRACKISH WATER

Naglaa M. Eid
(Faculty of Engineering, UAE University, Al Ain, United Arab Emirates)

Freshwater resources are becoming increasingly limited due to current uses and growing demands. The use and development of non-conventional water resources such as brackish water could be a possible solution to overcome the imminent shortages of freshwater

In this paper, Electro dialysis, the technique based on the transport of ions through selective membranes under the influence of an electrical field, was utilized to desalinate highly concentrated Al Ain brackish groundwater. The design and optimization of an electro dialysis cell for desalinating brackish groundwater was studied.

A number of laboratory experiments were conducted to test the effect of different design parameters on the configured cell and evaluate different operational schemes as well. The tested parameters were the ED applied voltage, initial salt concentration, the hydraulic flow velocity and the ratios of the volume of the solution in the concentrate compartment to the volume of the solution in the dilute compartment. The operating energy of the tested schemes was also evaluated to help in selecting the optimum operation system achieving the highest desired efficiency with the lowest possible energy. Cost analysis for running and maintaining the obtained optimum operation system was performed and the cost was then compared to the corresponding cost of other desalination membrane techniques.

# MIXED METAL OXIDES DERIVED FROM Ce-DOPED Zn-Al LAYERED DOUBLE HYDROXIDE: SYNTHESIS, CHARACTERIZATION AND PHOTOCATALYTIC ACTIVITY

***Jianyao Zhu,*** Zhiliang Zhu, and Hongtao Lu
(Tongji University, Shanghai, China)

In this work, a series of Zn-Al-Ce mixed metal oxide (ZnAlCe-MMO) photocatalysts with different Ce doping content were prepared by calcination of Ce-doped Zn-Al layered double hydroxide (ZnAlCe-LDH) precursors at various temperatures in air atmosphere. The materials were characterized by X-ray diffraction (XRD), scanning electron microscopy (SEM), transmission electron microscopy (TEM), Fourier transform infrared (FTIR) spectroscopy, UV–vis diffuse reflectance spectroscopy (DRS), Brunauer–Emmett–Teller (BET) surface area measurements, thermogravimetric analysis (TGA) and X-ray photoelectron spectroscopy (XPS) analysis. The photocatalytic activities of the samples were evaluated by the photodegradation of rhodamine B (RhB) dye in aqueous solution under simulated solar light irradiation. The results show that the ZnAlCe-MMO samples exhibit much higher photocatalytic activities than ZnAl-MMO, and the optimal Ce doping content is 5 mol% ($n_{Ce/(Zn+Al+Ce)}$). The enhanced photocatalytic activity of the ZnAlCe-MMO is mainly attributed to the increasing in the separation efficiency of electrons and holes. The effect of the calcination temperature was also investigated and the optimal calcination temperature is 750°C. The enhanced photocatalytic performance of LDH calcined at 750°C can be attributed to the formation of well-crystallized mixed metal oxides during calcination. Under optimal conditions, 97.8% degradation efficiency of RhB was achieved after 240 min reaction. The active species trapping experiments suggest that hole ($h^+$) plays the most important role while superoxide radical ($\bullet O_2^-$) and hydroxyl radical ($\bullet OH$) are the other two active species during the photocatalytic process. Moreover, the photocatalyst was demonstrated to have good stability and reusability.

# A NOVEL PROCESS BY A SMALL MOLECULE ENHANCED EFFECTIVE DEGRADATION FOR METHYL ORANGE DYE IN AQUEOUS SOLUTIONS UNDER UV-VIS IRRADIATION

Jingjing Wang, ***Renbi Bai***
(National University of Singapore, 117576 Singapore)

The development of effective technologies to treat recalcitrant organic dye wastewater has long been of great research and practical interest. In this study, a small molecule, formic acid (FA), was employed as a process enhancer in the degradation of methyl orange (MO) dye in aqueous solutions under the condition of UV-Vis light irradiation with air aeration at the ambient temperature. It was found that decolouration of the dye solution was rapidly achieved, reducing the time from more than 10 h without FA to less than 2 h with the presence of FA. Mineralization rate of MO dye reached as high as about 82% in 1.5 h for MO dye at the initial concentration of 25 $mg \cdot L^{-1}$, in contrast to nearly no mineralization for a similar system without the FA added. The study revealed that the generation of $H_2O_2$ species in the system was enhanced and the produced OH· radicals contributed to the effective degradation of the MO dye. The variation of the initial concentration of MO dye, FA dosage and solution pH was all found to have some impact on the degradation efficiency under the same strength of UV-Vis light irradiation and air aeration. The MO dye degradation performance was found to follow a first-order reaction rate to the MO dye concentration in most cases and there existed a positive correlation between the reaction rate constant and the FA dosage used. Compared to the traditional $H_2O_2$/UV-Vis oxidation system, the use of a small molecule, such as FA, as a process-enhancing species, has the advantages of low cost, easy availability, and safe to use, and so on. The study demonstrates a promising approach to use a readily available small molecule to enhance the degradation of recalcitrant organic pollutants, such as MO dye, especially in their pre-treatment before a biological process may be applied.

# PHYSICO-CHEMICAL QUALITIES OF THE TREATED FINAL EFFLUENT DISCHARGES OF SOME WASTEWATER TREATMENT PLANTS IN BUFFALO CITY MUNICIPALITY, EASTERN CAPE, SOUTH AFRICA

T Kulati, OO Okoh and AI Okoh
(Department of Chemistry, University of Fort Hare, Alice 5700, South Africa)

It is well known that successive stages of treatment at a sewage plant reduce the quantity of suspended solids, the organic matter content, the nutrient constituent, the microbial load and the biochemical oxygen demand of sewage, so that the polluting strength of the final effluent becomes a small fraction of that of the sewage received at the plant. Therefore, the efficiency of four wastewater treatment plants that serve the Buffalo City Municipality area in the Eastern Cape Province of South Africa was investigated with regards to nutrient removal and physicochemical characteristics that might lead to the pollution of receiving waterbodies. The study was conducted from September 2012 to August 2013 and standard physicochemical methods were used. The quality of the effluents were acceptable with respect to the temperature (21.3-21.9 °C), pH (7.2-7.4), electrical conductivity (351-410 µS/cm), total dissolved solids (221-260), dissolved oxygen (6.38-7.71 mg/l), biological oxygen demand (2.93-4.52 mg/l), residual chlorine (0.13-0.26), nitrate (6.89-9.20 mg/l); nitrite (0.34-1.92 mg/l). However, the concentrations of chemical oxygen demand (234-260 mg/l), orthophosphate (2.67-5.95 mg/l) and turbidity (4.96-8.59 NTU) were severally outside the compliance levels of the South African Guidelines for the protection of aquatic ecosystems.

# APPLİCATİON OF ELECTROHYDROLYSİS PROCESS FOR VİNEGAR WASTEWATER AND MUNİCİPİAL WASTEWATER MİXTURE

Faruk Hakan CANBAZ, Canan Can YARIMTEPE, Rabia ERARSLAN, Elif ERDOGAN and Nilgun Ayman OZ

(Çanakkale Onsekiz Mart University, Department of Environmental Engineering, Çanakkale/Turkey)

In this study, mixture of vinegar wastewater and municipial wastewater was subjected to different DC voltages (4-8 V) for hydrogen gas production and COD removal by using two different type of electrodes (aluminium and copper) . Effects of applied DC voltage and type of electrodes on hydrogen gas production and COD, sCOD, turbidity, Total Suspended Solids and Color removals were investigated. Vinegar fermentation wastewater was obtained from a Fermentation Company in Izmir and municipial wastewater was obtained from Kepez WWTP in Turkey. Electrohydrolysis process was carried out in a lab-scale 1 L glass reactor with consists a cover supporting two parallel aluminum or copper electrodes. A direct current with different voltages including 4, 5, 6, 7 and 8 voltages were tested by using a power supply. For electrodes comparison, applied DC voltage was constant at 4 V. The highest cumulative hydrogen production (200 ml) for 75 min and hydrogen content of gas (88%) was obtained with the aluminium electrode.The highest cumulative hydrogen production (470 ml) for 75 min and hydrogen content of gas (90%) were obtained with 8 V DC voltage. Net energy production efficiency (consumed energy/produces energy) also reached the highest level (54%) with 8 V DC voltage application. COD removal from wastewater mix with 8 DC voltage application was the most effective voltage with 36%. In the most effective conditions; sCOD, Turbidity, TSS and Color removal efficiencies was determined as 32%, 82%, 44% and 92%, respectively. The results indicated that Hydrogen gas production from electrohydrolysis of vinegar wastewater/municipial wastewater mixture was found to be a fast and effective method with high energy production and pollutant removal efficiencies in terms of turbidity and color.

Financial support by The Scientific and Technological Research Council of Turkey (TUBITAK) was gratefully acknowledged (Project No.: A-2209 (2014-1)).

# REMOVAL AND DEGRADATION PATHWAY OF SULFAMETHOXAZOLE FROM MUNICIPAL WASTEWATER BY ANAEROBIC MEMBRANE BIOREACTOR

***Chun-Hai Wei,*** Claudia Sanchez Huerta and TorOve Leiknes
(King Abdullah University of Science and Technology, Thuwal 23955-6900, Saudi Arabia)

**ABSTRACT:** A widely used antibiotic sulfonamide, Sulfamethoxazole (SMX) with chemical abstracts service (CAS) number of 723-46-6 has been detected in the influent and effluent of municipal wastewater treatment plant and thus in the discharging water bodies, resulting in the potential harmful effects on public health and aquatic ecosystems. Removal and degradation pathway of SMX from synthetic municipal wastewater treatment via a lab-scale mesophilic anaerobic membrane bioreactor (AnMBR) were systematically investigated in this study. During the whole operation of around 100 d under different feed SMX concentrations (10-1000 μg/L) and bulk organics concentrations (chemical oxygen demand 800-2000 mg/L) and the same hydraulic retention time (1 d), both SMX and bulk organics removal were over 96%. SMX degradation followed first order reaction kinetics with rate constant of 0.147 $h^{-1}$ and half-life of 4.7 h based on removal independent on feed concentrations. 5 degradation products were detected: N-Butylbenzenesulfonamide (CAS 3622-84-2) detected under all feed SMX concentrations; Sulfanilamide (CAS 63-74-1) detected under feed SMX concentrations of 100 and 1000 μg/L; Sulfisomidine (CAS 515-64-0), 4-Aminothiophenol (CAS 1193-02-8) and Aniline (CAS 62-53-3) detected under feed SMX concentration of 1000 μg/L only. Thus, a degradation pathway of SMX from municipal wastewater treatment via AnMBR was firstly proposed.

## INTRODUCTION

Organic micro-pollutants (OMPs) in water environment including pharmaceutically active compounds, personal care products, household chemicals, pesticides and etc., have received increasing attention in recent years due to their potential harmful effects on public health and aquatic ecosystems. Among them, antibiotics are incurred more concerns due to their direct toxicity to aquatic creatures as well as potential development of antibiotic resistance genes and species. As a widely used antibiotic sulfonamide, Sulfamethoxazole (SMX) with chemical abstracts service (CAS) number of 723-46-6 has been frequently detected in raw wastewater and effluent of municipal wastewater treatment plant (WWTP) and even in surface water (Boreen et al., 2004; Abegglen et al., 2009; Dirany et al., 2010; Alidina et al., 2014). Eliminating SMX in WWTP is an important solution to control its discharge that may impact ecosystems and potentially follow the water cycle back to drinking water sources.

As a hybrid process of anaerobic digestion and membrane separation for municipal wastewater treatment, anaerobic membrane bioreactor (AnMBR) can achieve high biomass (thus high methane recovery) and good permeate quality due to the complete rejection of suspended solids by microfiltration (MF) or ultrafiltration (UF) membrane (Wei et al., 2014). Previous AnMBR studies on OMPs removal (Monsalvo et al., 2014; Wijekoon et al., 2015; Wei et al., 2016) also showed good SMX removal (over 95%) mainly through biodegradation due to its negatively charged nature at pH>5.7 and low sorption capacity (Boreen et al., 2004). However, there is no information available on SMX's degradation products and pathway in AnMBR so far. Considering the relatively complex molecular structure of SMX, it would be expected the occurrence of some degradation products rather than complete mineralization in AnMBR. Thus, the objective of this study was to investigate the removal and degradation pathway of SMX from synthetic municipal wastewater treatment by AnMBR.

## MATERIALS AND METHODS

**AnMBR Set-Up.** A lab-scale AnMBR set-up consisted of a completely mixed anaerobic activated sludge bioreactor (volume 2 L, temperature 35°C, pH 7, stirrer speed 200 rpm, hydraulic retention time 1 d) and a side-stream crossflow hollow fibre UF membrane module (polyvinylidene fluoride, nominal pore size 30 nm, filtration area 310 $cm^2$). Synthetic municipal wastewater with chemical oxygen demand (COD) of 800 mg/L spiked SMX at different concentrations (10, 100, 1000 μg/L, in the range of real SMX concentration detected in raw municipal wastewater) using concentrated SMX stock solution in methanol (1 g/L) was used as the feed of AnMBR. The details of AnMBR set-up and synthetic municipal wastewater recipe can be found in our previous publication (Wei et al., 2014). Before starting experiment, the AnMBR had been assimilated to the synthetic municipal wastewater spiked low concentration (10 μg/L) SMX (Wei et al., 2016). The whole experiments were classified as 7 phases depending on feed SMX concentration: Phase 1 (1-20 d) feed SMX 10 μg/L, Phase 2 (21-33 d) feed SMX 100 μg/L, Phase 3 (34-54 d) feed SMX 1000 μg/L, Phase 4 (55-61 d) no SMX spiking to wash out residual SMX in AnMBR, Phase 5 (62-88 d), feed SMX 10 μg/L, Phase 6 (89-98) feed SMX 100 μg/L, and Phase 7 (99-113 d) feed SMX 1000 μg/L. SMX removal was intensively investigated in Phase 1-3, where the detection methods of SMX degradation products were also explored. In Phase 5-7, SMX degradation products were intensively investigated while SMX removal was confirmed again. No sludge was wasted except sampling for biomass measurement during the whole operation.

**Analytical Methods.** For SMX quantification, liquid samples (100-1000 ml) spiked the isotope (SMX-d4) to control SMX recovery were firstly pre-concentrated into 10 ml via Dionex Autotrace 280 solid-phase extraction (SPE) instrument (Thermo Scientific) using Oasis HLB SPE cartridges (Waters) and the elution solvents of methanol followed by 10% methanol in methyl tert-butyl ether (MTBE), and further concentrated into the final volume of 1 ml via evaporation under 40°C and nitrogen stripping. Then the concentrated samples were quantified via Agilent Technology 1260 Infinity Liquid Chromatography (LC) with a Phenomenex Luna 5u C18 column using mobile phase of water and methanol containing 0.1% formic acid followed by Applied Biosystems SCIEX QTRAP 5500 Mass Spectrometry (MS) with positive electrospray ionization mode. For degradation products qualification, liquid samples (1-12 L) were firstly pre-concentrated via the above-mentioned SPE using different solvents (dichloromethane, ethyl acetate, acetone, methanol, MTBE, methanol followed by 10% methanol in MTBE) and the above-mentioned evaporation to final volume of 0.1-0.5 ml. Then the final concentrated samples were measured via Agilent Technology 7890A Gas Chromatography (GC) with Agilent J&W HP-5ms column (helium as carrier gas) followed by Agilent Technology 5975C Mass Spectrometry. The GC oven program started at an initial temperature of 60°C for 3 min, and ramped until 320°C with a heating rate of 20°C/min and remained at 320°C for 5 min. The main peaks from total ion chromatography of samples were finally identified via comparing the mass spectra of samples to standard compounds in the NIST mass spectral library. Only compounds with matching probability over 70% and repetitively detected under different extraction solvents could be identified as SMX degradation products. Commercial COD kits (TNT series, Hach Company) were used to measure the COD of influent and permeate based on the method of rapid digestion (150°C, 2 h) followed by colorimetric measurement. Biogas composition and volume (reported under temperature of 25°C and pressure of 1 atm) were measured according to the gas bag method (Ambler and Logan, 2011) based on gas chromatography (SRI 310C). Biomass concentration in terms of mixed liquor (volatile) suspended solids (ML(V)SS) was measured according to the standard method of glass fiber filtration followed by sequential drying at 105°C and 550°C.

## RESULTS AND DISCUSSION

**Bulk Organics Removal.** During the whole operation, MLVSS was in the range of 2.5-4.3 g/L with MLVSS/MLSS ratio of 0.8-0.85, indicating roughly stable biomass characteristics. Summarized results on bulk COD and SMX removal were shown in TABLE 1. The feed COD increased from 800 to 2000 mg/L with feed SMX concentration from 10 to 1000 μg/L due to methanol contribution (1 ml methanol in 1 L feed equivalent to COD of 1200 mg/L) from concentrated SMX (dissolved in pure methanol) spiking. Despite feed COD increase with feed SMX concentration, permeate COD was mostly below 50 mg/L and thus COD removal was always over 96%, indicating the good biological performance as well as no inhibition effects of SMX on bulk organics removal. There was a relative stable biogas composition with methane (80-95%), carbon dioxide (<10%) and nitrogen (<10%). With feed COD increase from 800 to 2000 mg/L, methane production also increased from 80 to 230 ml/$gCOD_{removed}$. These results were similar to our previous study (Wei et al., 2014).

**TABLE 1 Bulk COD and SMX removal in AnMBR (n=5)**

| Phase | COD | | | SMX | | |
|---|---|---|---|---|---|---|
| | Feed (mg/L) | Permeate (mg/L) | Removal (%) | Feed (μg/L) | Permeate (μg/L) | Removal (%) |
| Phase 1 (1-20 d) | 810 | 25.2±5.0 | 96.9±0.6 | 10 | 0.32±0.09 | 96.8±0.9 |
| Phase 2 (21-33 d) | 920 | 27.3±2.9 | 97.0±0.3 | 100 | 2.9±0.8 | 97.1±0.8 |
| Phase 3 (34-54 d) | 2000 | 39.3±9.8 | 98.0±0.5 | 1000 | 25.4±3.3 | 97.5±0.3 |
| Phase 5 (62-88 d) | 810 | 30.0±5.5 | 96.3±0.7 | 10 | 0.33±0.16 | 96.7±1.6 |
| Phase 6 (89-98 d) | 920 | 31.9±3.8 | 96.5±0.4 | 100 | 3.0±0.2 | 97.0±0.2 |
| Phase 7 (99-113 d) | 2000 | 48.6±6.3 | 97.6±0.3 | 1000 | 27.7±3.5 | 97.2±0.4 |

**SMX Removal.** From TABLE 1, during all phases with a large range of feed concentration (10 to 1000 μg/L), SMX removal were over 96%, indicating the very good degradation capacity of AnMBR. A good reproducibility for SMX removal was also confirmed via comparing phases with the same feed SMX concentration (Phase 1 and 5, Phase 2 and 6, Phase 3 and 7). In addition, SMX removal under different feed concentrations (10, 100, and 1000 μg/L) and the same hydraulic retention time (1 d) was fallen into a narrow range of around 97%, indicating its independence on feed concentration. This could deduce that SMX degradation would follow first order reaction kinetics (shown in Equation 1 and 2, where $C$ was SMX concentration at time $t$, $C_0$ was feed SMX concentration, $k$ was rate constant, $t$ was reaction time, $t_{1/2}$ was half-life.) with rate constant of 0.147 $h^{-1}$ and half-life of 4.7 h.

$$C = C_0 e^{-kt} \quad (1)$$
$$t_{1/2} = \ln(2)/k \quad (2)$$

**SMX Degradation Products.** 5 degradation products (shown in TABLE 2) were identified via GC-MS analysis of permeate samples extracted by different SPE solvents (dichloromethane, ethyl acetate, acetone, methanol, MTBE, methanol followed by 10% methanol in MTBE). Under feed SMX concentration of 10 μg/L (i.e., the maximum value in normal raw municipal wastewater without receiving untreated hospital and pharmaceuticals wastewater containing high-concentration SMX) in Phase 1 and 5, only N-Butylbenzenesulfonamide (CAS number 3622-84-2) was detected under all tested SPE solvents and a large range of SPE-evaporation concentrating factor from 1k to 120k (i.e., 1/12 L permeate sample concentrating to 1/0.1 ml via SPE followed by evaporation). Compared with parent compound of SMX (CAS number 723-46-6), N-Butylbenzenesulfonamide as a degradation product does not contain antibiotic sulfonamide group. This indicated that even AnMBR treatment did not achieve complete mineralization of SMX but still produced permeate free of antibiotic functions. In addition, further degradation of N-Butylbenzenesulfonamide was also overserved in Phase 2, where its peak area (i.e., concentration) in 1k concentrated permeate samples using methanol followed by 10% methanol in MTBE as SPE solvents decreased significantly with operation time.

In Phase 2 and 6 with feed SMX concentration of 100 μg/L, Sulfanilamide (CAS number 63-74-1) being an antibiotic sulfonamide was detected besides N-Butylbenzenesulfonamide. In Phase 3 and 7 with feed SMX concentration of 1000 μg/L, Sulfisomidine (CAS number 515-64-0) being an antibiotic sulfonamide, 4-Aminothiophenol (CAS number 1193-02-8) and Aniline (CAS number 62-53-3) without antibiotic functions were detected besides N-Butylbenzenesulfonamide and Sulfanilamide. The reasons behind the sequential occurrence of N-Butylbenzenesulfonamide, Sulfanilamide, and Sulfisomidine, 4-Aminothiophenol, Aniline in Phase 1-3 and 5-7 with increasing feed SMX of 10, 100 and 1000 μg/L might be as follows: 1) All 5 products could be formed under all tested conditions but at different concentrations, resulting in their sequential detection due to the concentrating factor limitation of SPE-evaporation. However, it was unlikely happened because the concentrating factor had been reached to a very high level of 20k (even 120k in Phase 5) in this study. 2) All 5 products could be formed under all tested conditions but degraded further at different rates, resulting in their sequential detection due to their different degradation rates. N-Butylbenzenesulfonamide, Sulfanilamide, and 4-Aminothiophenol, Aniline might follow this rule based on the decreasing sequence of their molecular structural complexity (i.e., generally reverse sequence of degradability). However, Sulfisomidine could not follow this rule due to its similar structure to SMX. 3) Product formation could depend on SMX loading rate to biomass in terms of μg SMX per g MLVSS per day, which was roughly proportional to feed SMX concentration based on the roughly stable biomass concentration in this study. With the sequential increase of feed SMX concentration from 10 to 100 to 1000 μg/L, N-Butylbenzenesulfonamide, Sulfanilamide and Sulfisomidine might be formed sequentially based on the metabolic response of anaerobic microorganism to increasing SMX loading rate.

**TABLE 2 Degradation Products of SMX in AnMBR Identified by GC-MS**

| No. | Compound name | CAS number | Molecular formula | Molecular weight (Dalton) | Molecular structure | GC elution time (min) | Detection under feed SMX concentrations ( μg/L) |
|---|---|---|---|---|---|---|---|
| 1 | Benzenesulfonamide, N-Butyl- | 3622-84-2 | $C_{10}H_{15}NO_2S$ | 213 | 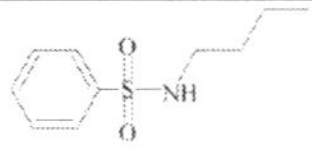 | 11.1-11.2 | 10, 100, 1000 |
| 2 | Sulfanilamide | 63-74-1 | $C_6H_8N_2O_2S$ | 172 | 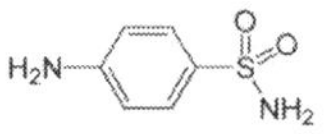 | 12.2 -12.5 | 100, 1000 |
| 3 | Sulfisomidine | 515-64-0 | $C_{12}H_{14}N_4O_2S$ | 278 | 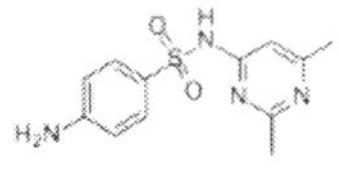 | 14.5 -15.5 | 1000 |
| 4 | 4-Aminothiophenol | 1193-02-8 | $C_6H_7NS$ | 125 |  | 8.1-8.2 | 1000 |
| 5 | Aniline | 62-53-3 | $C_6H_7N$ | 93 | 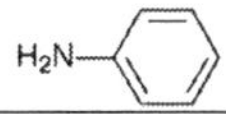 | 5.1-5.5 | 1000 |

To the best of our knowledge, no degradation products of SMX in AnMBR have been reported so far in the literature. Among the detected degradation products of SMX in this study, Sulfanilamide, 4-Aminothiophenol and Aniline have been detected in batch tests using an aerobic cold-adapted bacterium *Pseudomonas psychrophila* HA-4 (Jiang et al., 2014) and using electrochemical degradation at boron-doped diamond electrode (de Amorim et al., 2013) for SMX degradation. N-Butylbenzenesulfonamide and Sulfisomidine are firstly reported as SMX degradation products from this study. Interestingly, 3-Amino-5-methylisoxazole (CAS number 1072-67-9) as a hydrolysis product of SMX has been reported in a few studies for SMX degradation using an aerobic bacterium *Microbacterium sp.* strain BR1 (Ricken et al., 2013), an aerobic cold-adapted bacterium *Pseudomonas psychrophila* HA-4 (Jiang et al., 2014) and an anodic chamber of microbial fuel cell (Wang et al., 2015). However, the 5-methylisoxazole group of SMX was not hydrolyzed but directly accepted protons to form butane (shown in FIGURE 1) in this study. This could be due to the anaerobic bacteria used in this study different with the above-mentioned studies.

**SMX Degradation Pathway.** Based on the identified degradation products, the proposed degradation pathway of SMX via AnMBR treatment was firstly shown in FIGURE 1, which was divided into 3 routes. The first route was that the amine and 5-methylisoxazole group of SMX accepted protons to form N-Butylbenzenesulfonamide, which could be further degraded. The first route was activated under the lowest feed SMX concentration of 10 μg/L in this study. The second route was that the sulfonamide group of SMX accepted a proton to form Sulfanilamide, which was further degraded to Aniline via deamination and desulfurization simultaneously and/or deamination to 4-Aminothiophenol followed by desulfurization. Aniline might be finally mineralized based on its simple molecular structure (Jiang et al., 2014) and the fact of no detection under feed SMX concentration of 100 μg/L, indicating its possible complete degradation. The second route was activated under feed SMX concentration of 100 μg/L in this study. The third route was that the 5-methylisoxazole group of SMX was substituted by 4-Amino-2,6-dimethylpyrimidine to form Sulfisomidine, which might be degraded further based on its similar molecular structure to SMX. The third route might only be activated under feed SMX concentration of 1000 μg/L in this study. In addition, it could be deducted based on the formation pathway of main degradation products that in AnMBR, the amine group of SMX was transformed to ammonia; 5-methylisoxazole group of SMX was transformed to butane (would be easily degraded further by anaerobic microorganism due to its simple chain structure), ammonia and water; sulfonamide group of SMX was transformed to hydrogen sulfide, ammonia and water.

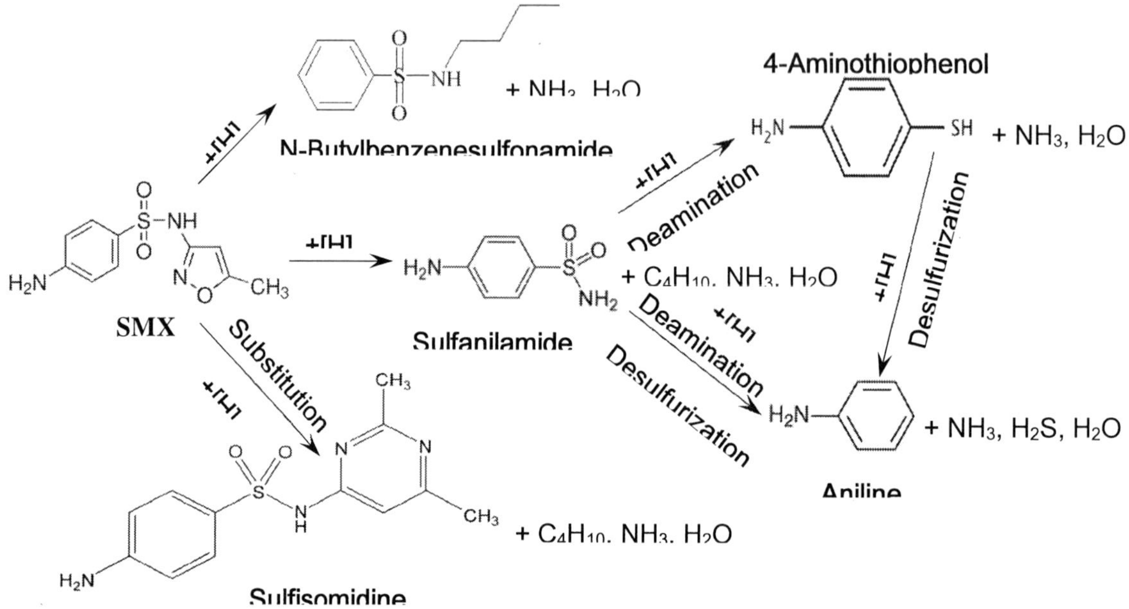


**FIGURE 1 Proposed Degradation Pathway of SMX in AnMBR**

## CONCLUSIONS

In this study, the lab-scale mesophilic AnMBR process for synthetic municipal wastewater (spiked SMX) treatment showed very good removal performance (over 96%) for both bulk organics and SMX in a large range of feed COD (800-2000 mg/L) and SMX (10-1000 µg/L) concentration under the same hydraulic retention time (1 d) and roughly stable biomass characteristics (MLVSS of 2.5-4.3 g/L, MLVSS/MLSS ratio of 0.8-0.85). Methane production increased from 80 to 230 ml/gCOD$_{removed}$ with feed COD increase from 800 to 2000 mg/L. Based on the removal independent on feed concentration, SMX degradation was deduced to follow first order reaction kinetics with rate constant of 0.147 $h^{-1}$ and half-life of 4.7 h. 5 degradation products were identified via GC-MS analysis of concentrated permeate samples via SPE using multiple solvents (dichloromethane, ethyl acetate, acetone, methanol, MTBE, methanol followed by 10% methanol in MTBE) followed by evaporation. N-Butylbenzenesulfonamide (CAS number 3622-84-2) without antibiotic functions was firstly detected as SMX's degradation product under all feed SMX concentrations and especially the sole degradation product under feed SMX concentration of 10 µg/L (i.e., the maximum SMX concentration occurred in normal raw municipal wastewater) in this study. Sulfanilamide (CAS number 63-74-1) still being an antibiotic sulfonamide was detected under feed SMX concentrations of 100 and 1000 µg/L. Sulfisomidine (CAS number 515-64-0) still being an antibiotic sulfonamide, 4-Aminothiophenol (CAS number 1193-02-8) and Aniline (CAS number 62-53-3) without antibiotic functions were detected under feed SMX concentration of 1000 µg/L only. Based on the identified degradation products, a degradation pathway of SMX from municipal wastewater treatment via AnMBR was firstly proposed, which consisted of 3 degradation routes for the formation of N-Butylbenzenesulfonamide, Sulfanilamide (then degraded further to 4-Aminothiophenol and Aniline), and Sulfisomidine, respectively.

## ACKNOWLEDGEMENTS

This work has been financed by KAUST Center Competitive Funding Project on Anaerobic Membrane Bioreactor. Dr. Tao Zhang and Ms. Tong Zhan were greatly appreciated for their kind technical supports on LC-MS and GC-MS.

## REFERENCES

Abegglen, C., A. Joss, C. S. McArdell, G. Fink, M. P. Schlusener, T. A. Ternes and H. Siegrist. 2009. "The Fate of Selected Micropollutants in a Single-house MBR." *Water Res.* 43: 2036-2046.

Alidina, M., C. Hoppe-Jones, M. Yoon, A. F. Hamadeh, D. Li and J. E. Drewes. 2014. "The Occurrence of Emerging Trace Organic Chemicals in Wastewater Effluents in Saudi Arabia." *Sci. Total Environ.* 478: 152-162.

Ambler, J. R. and B. E. Logan. 2011. "Evaluation of Stainless Steel Cathodes and a Bicarbonate Buffer for Hydrogen Production in Microbial Electrolysis Cells using a New Method for Measuring Gas Production." *Int. J. Hydrogen Energ.* 36(1): 160-166.

Boreen, A. L., W. A. Arnold and K. McNeill. 2004. "Photochemical Fate of Sulfa Drugs in the Aquatic Environment: Sulfa Drugs Containing Five-membered Heterocyclic Groups." *Environ. Sci. & Technol.* 38(14): 3933-3940.

de Amorim, K. P., L. L. Romualdo and L. S. Andrade. 2013. "Electrochemical Degradation of Sulfamethoxazole and Trimethoprim at Boron-Doped Diamond Electrode: Performance, Kinetics and Reaction Pathway." *Sep. Purif. Technol.* 120: 319-327.

Dirany, A., I. Sirés, N. Oturan and M. A. Oturan. 2010. "Electrochemical Abatement of the Antibiotic Sulfamethoxazole from Water." *Chemosphere* 81(5): 594-602.

Jiang, B., A. Li, D. Cui, R. Cai, F. Ma and Y. Wang. 2014. "Biodegradation and Metabolic Pathway of Sulfamethoxazole by *Pseudomonas psychrophila* HA-4, a Newly Isolated Cold-Adapted Sulfamethoxazole-Degrading Bacterium." *Appl. Microbiol. Biotechnol.* 98: 4671-4681.

Monsalvo, V. M., J. A. McDonald, S. J. Khan and P. Le-Clech. 2014. "Removal of Trace Organics by Anaerobic Membrane Bioreactors." *Water Res.* 49: 103-112.

Ricken, B., P. F. X. Corvini, D. Cichocka, M. Parisi, M. Lenz, D. Wyss, P. M. Martínez-Lavanchy, J. A. Müller, P. Shahgaldian, L. G. Tulli, H. P. E. Kohler and B. A. Kolvenbach. 2013. "*ipso*-Hydroxylation and Subsequent Fragmentation: a Novel Microbial Strategy to Eliminate Sulfonamide Antibiotics." *Appl. Environ. Microbiol.* 79(18): 5550-5558.

Wang, L., Y. Wu, Y. Zheng, L. Liu and F. Zhao. 2015. "Efficient Degradation of Sulfamethoxazole and the Response of Microbial Communities in Microbial Fuel Cells." *RSC Adv.* 5: 56430-56437.

Wei, C. H., C. Hoppe-Jones, G. Amy and T. Leiknes. 2016. "Organic Micro-Pollutants' Removal via Anaerobic Membrane Bioreactor with Ultrafiltration and Nanofiltration." *J. Water Reuse Desal.* in press, DOI: 10.2166/wrd.2015.138.

Wei, C. H., M. Harb, G. Amy, P. Y. Hong and T. Leiknes. 2014. "Sustainable Organic Loading Rate and Energy Recovery Potential of Mesophilic Anaerobic Membrane Bioreactor for Municipal Wastewater Treatment." *Bioresource Technol.* 166: 326-334.

Wijekoon, K. C., J. A. McDonald, S. J. Khan, F. I. Hai, W. E. Price and L. D. Nghiem. 2015. "Development of a Predictive Framework to Assess the Removal of Trace Organic Chemicals by Anaerobic Membrane Bioreactor." *Bioresource Technol.* 189: 391-398.

# OH RADICAL FORMATIONS IN PHYTO-FENTON PROCESS: DETECTION AND APPLICATION

***Shigeki Nara***, Yoshihiko Inagaki, and ***Yutaka Sakakibara***
(Department of Civil and Environmental Engineering, Waseda University, Tokyo, Japan)

**ABSTRACT:** Phyto-Fenton process utilizing endogenous H2O2 in aquatic plants and iron catalyst has a potential to be a low cost and high performance phytoremediation technology. In this study, OH radical formations and reaction sites in this process were analyzed using an OH radical probe, 3'-(p-aminophenyl) fluorescein (APF) and a confocal laser scanning microscopy. Strong fluorescence was observed at whole parts inside a root, when ferrous sulfate was used as iron catalyst. On the other hand, fluorescence was observed from peripheral parts when using ferrihydrite colloids and/or ferric-ion zeolite. Reaction sites of OH radical formations could be observed in vertical and horizontal sections of aquatic plants. These results indicated Fenton and Fenton like reactions occur simultaneously and the OH radical formation sites could be controlled by the selection of iron compounds. Furthermore, pentachlorophenol (PCP) could be removed by the present Phyto-Fenton process. In addition, tetrachlorohydroquinone (TCHQ) was detected as one of intermediate byproducts in treated water; while no intermediate detected in the absence of iron compounds. A further study will be needed to optimize the OH radical formations as well as decompositions of refractory pollutants.

## INTRODUCTION

Recently, as water consumption in the world is increasing with increasing world population and economic growth, water problem becomes more serious. Especially, in developing countries implementations of modern sewage treatment systems are not always feasible because of high initial and operation costs. We proposed a low cost/high performance water and wastewater treatment process, Phyto-Fenton process (A. R. Reis et al, 2011; Y. Inagaki et al, 2016), where OH radicals can be produced through Fenton and Fenton like reaction utilizing endogenous H2O2 in aquatic plants.

Fenton reaction and Fenton-like reaction are shown in Reactions (1) and (2), respectively. Hydrogen peroxide is converted to OH radicals which can decomposes almost every organic pollutant. This process has a potential to be a low cost and high performance phytoremediation technology. This research was conducted to analyze the mechanism of Phyto-Fenton reaction and reaction sites generating OH radicals in the presence of different iron catalysts.

$$H_2O_2 + Fe^{2+} \rightarrow Fe^{3+} + OH\cdot + OH^- \quad (1)$$

$$H_2O_2 + Fe^{3+} \rightarrow Fe^{2+} + HO_2\cdot + H^+ \quad (2)$$

## MATERIAL AND METHODS

**Observation with Confocal Laser Scanning Microscopy.** In order to identify the OH radical generation and its reaction sites in plants, a fluorescence reagent, 3'-(p-aminophenyl) fluorescein (APF) was used and fluorescence was observed with a confocal laser scanning microscopy (FV-1000D, Olympus). Slight fluorescence was observed in the absence of iron catalyst. However, strong fluorescence could be observed if OH radical reacted with APF (Setsukinai et al., 2003). The mechanism of fluorescence emission from APF is shown in FIGURE 1. A confocal scanning microscopy is enables to observe the emission in three-dimensional structures and thereby to collect a series of images at different depths.

In experiments, plants (0.20 g F.W.) were cultivated for one day in a test tube containing phosphate buffer with 20 mg-Fe/L iron compounds and 5 mmol/L APF. Three types of iron compounds: ferrous sulfate, ferric ion zeolite and ferrihydrite colloids, were used as iron catalyst. After cultivation, fluorescence was observed with the microscopy.

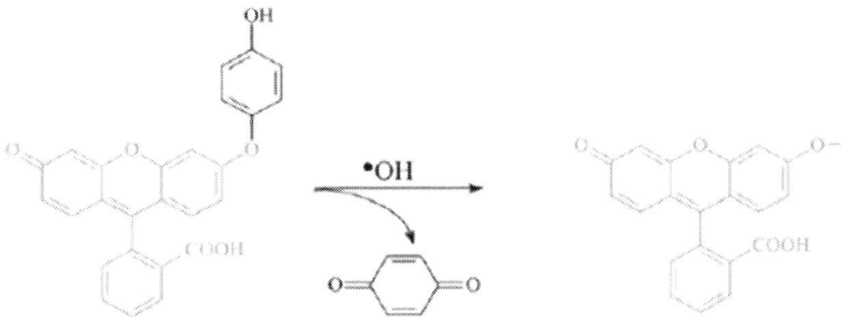


FIGURE 1. illustration of fluorescence probe reaction with OH radicals.

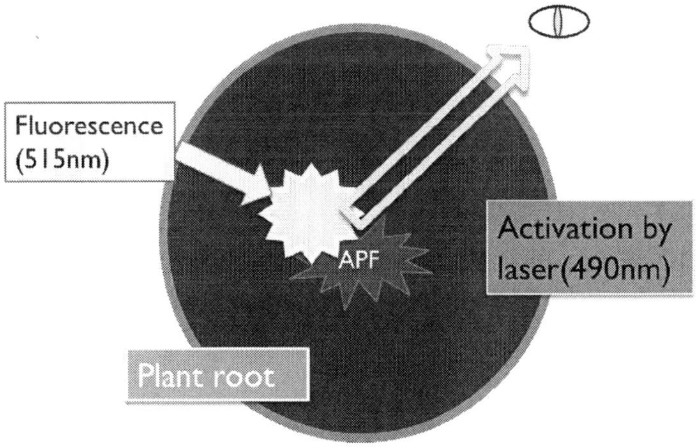


FIGURE 2. Conceptual diagram of confocal laser scanning microscopy in detecting OH radical generation sites in phyto-Fenton process.

### Treatment of Pentachlorophenol(PCP)

***Measurement of PCP concentration.*** Experiments were conducted to confirm removal and decomposition of PCP. Plants were cultivated for one week in test tubes containing PCP with and without iron catalyst. Details of experimental condition are shown in TABLE 1. Procedures in preparation of samples are shown in FIGURE 3. PCP concentration was measured with GC-ECD (6890, Agilent).

TABLE 1. Experimental conditions for Batch treatment of PCP.

| Resctor No. | R1 | R2 | R3 | R4 |
|---|---|---|---|---|
| PCP Conc. (mg/L) | 10.6 | 10.6 | 10.6 | 10.6 |
| Fe Conc. (mg/L) | 0 | 20 | 0 | 20 |
| Plant (g F.W.) | 0 | 0 | 0.20 | 0/2 |
| Type of iron | Ferrous sulfate(FeSO4) | | | |
| pH | Around 4.5 | | | |

***Identification of PCP by-products.*** Batch treatments of PCP were conducted for one week to detect intermediate products from PCP. Plants were cultivated in test tubes, containing PCP with and without iron catalyst. 20mg/L ferrous sulfate was used as iron catalyst. After treatment, solution was analyzed with LCMS (Xevo G2-XS Qtof, Waters).

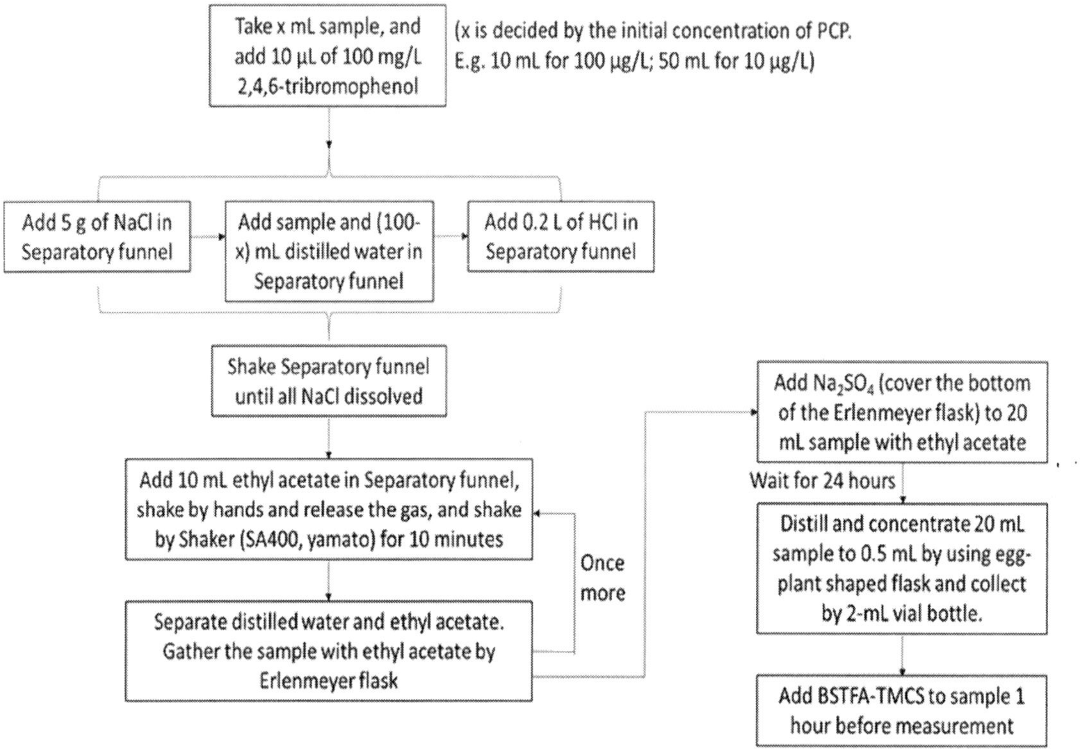


FIGURE 3. Pre-preparation of PCP samples

## RESULTS AND DISCUSSION

**Observation with Confocal Laser Scanning Microscopy.** Fluorescence from plants is shown in FIGUERE 4. In each image, central area indicates X-Y plane, bottom side X-Z plane and right side Y-Z plane, respectively. As shown in (A), strong fluorescence was observed from root at every plane, when ferrous sulfate was used as iron catalyst. On the other hand, when ferric ion zeolite and ferryhidrite colloids were used as iron catalyst, strong fluorescence was observed from the surface of plant's root. It is supposed that the difference of results was attributable to difference in size of ironcatalyst.

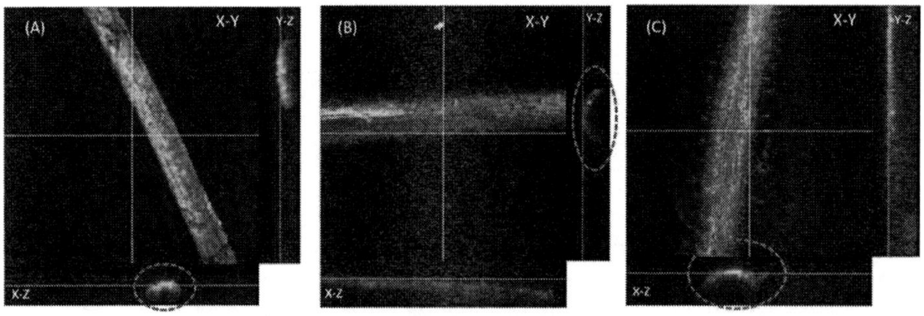


FIGURE 4. Fluorescence from plants in the presence of
(A) ferrous sulfate, (B) ferric-ion zeolite, and (C) ferrihydrite colloids.

**Treatment of Pentachlorophenol(PCP) Measurement of PCP Concentration.** PCP concentrations after treatment are compared in FIGURE 5. From the comparisons, it was found that PCP concentration was the lowest in R4, where iron catalyst and plant were present, indicating that phyto-Fenton reaction is effective to remove PCP.

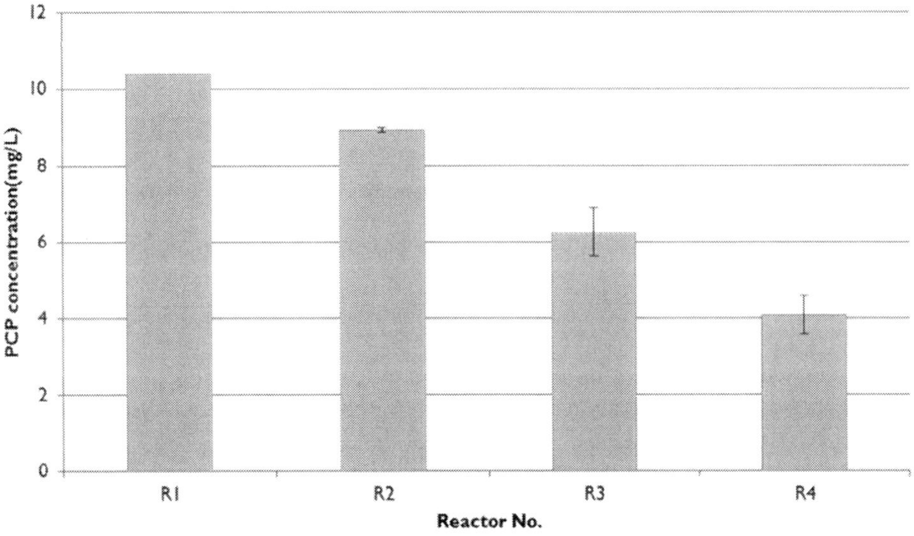


FIGURE 5. PCP concentrations under different conditions.

**Identification of PCP By-Products.** Result of analysis with LCMS is shown in FIGURE 6. A peak of intermediate product was observed when iron catalyst was present. No peak was observed when not iron catalyst was present. The intermediate product was identified as tetrachlorohydroquinone (TCHQ). Zimbron and Reardon (2009) demonstrated that TCHQ was produced through the decomposition of PCP by Fenton Reaction. However, DCMA, more degraded byproduct, was not detected in this study.

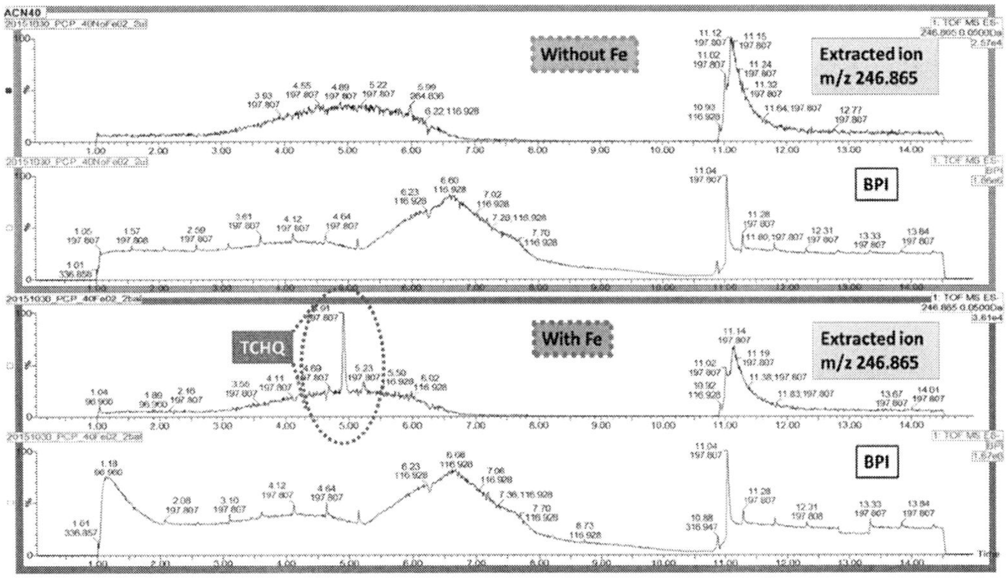


FIGURE 6. UPLC chromatographic profiles.

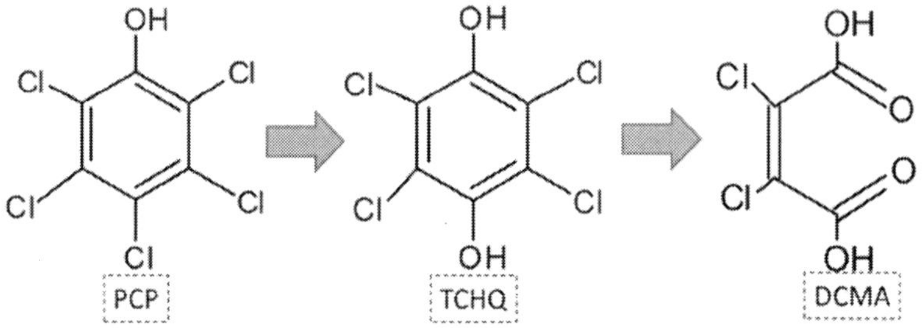


FIGURE 7. PCP degradation pathway by Fenton reaction (Zimbron and Reardon, 2009).

## CONCLUTIONS

We confirmed phyto-Fenton reaction occurs in plants. Moreover, it was found reaction sites of OH radical generations were affected by a type of iron catalyst. In batch treatment, PCP was decomposed to TCHQ by Phyto-Fenton reaction. Furthermore, it was demonstrated a confocal laser microscopy is a useful tool to identify the OH radical generations. A further study will be needed to optimize the process.

## ACKNOWLEDGEMENTS

A part of this work was supported by a Grant-in-Aid for Scientific Research (B) (15H02846), the Japanese Ministry of Education, Culture, Sports, Science and Technology (MEXT).

## REFERENCES

Inagaki Y., Cong V. H., Sakakibara Y. 2016. "Identification and application of Phyto-Fenton Reactions." *Chemosphere*, **144**, 1443-1450

Reis, A.R. and Sakakibara, Y. 2011. "Continuous treatment of endocrine disrupting chemicals by aquatic plants and biological Fenton Reaction.", *JSCE, Ser. G (Environmental Research)*, **67**(7), 725-734.

Setsukunai, K., Urano, Y., Kakinuma, K., Majima, H. J. and Nagano, T. 2003. "Development of Novel Fluorescence Probes That Can Reliably Detect Reactive Oxygen Species and Distinguish Specific Species", *The Journal of biological chemistry*, **278**(5), 3170-3185

Zimbron, J. A. and Reardon, K. F. 2009. "Fenton's oxidation of pentachlorophenol", *Water Research*, **43**(7), 1831-1840

# CONTINUOUS TREATMENT OF WASTEWATERS CONTAINING PPCPS BY AN ELECTROCHEMICAL ADVANCED OXIDATION PROCESS

***Junya Suzuki***, Eiji Kawada, Taiki Maehata and Yutaka Sakakibara (Department of Civil and Environmental Engineering, Waseda University, Tokyo, Japan)

**ABSTRACT:** A novel electrochemical advanced oxidation process (AOP) was proposed and continuous treatments of a synthetic wastewater and tertiary effluent from a municipal WWTP were conducted under different operating conditions. The electrochemical AOP consists of SnO2 anode for O3 production and multiple cathode compartments packed with either granular Pt/Ti electrodes or granular activated carbons (GACs) for OH radical formation through the reduction of produced O3. The synthetic wastewater was prepared by dissolving antibiotics such as tetracycline, erythromycin and sulfamethoxazole. From experiments using the synthetic wastewater, it was found that all PPCPs were decomposed completely. Moreover, decrease in TOC was not observed for Pt/Ti granular cathodes, while significant decrease observed for GAC cathodes. Main intermediate products from PPCPs were identified, which were similar to those in AOPs reported in former studies. Furthermore, experimental results for tertiary effluent demonstrated that superior TOC removals could be achieved for GAC cathodes. These results suggest that PPCPs and other organic pollutants were firstly absorbed on GAC and then decomposed by OH radicals. Energy consumptions were in the range of 0.3-4kWh/m$^3$. From these results, we concluded that the proposed electrochemical AOP would be a feasible alternative to former AOPs because of simpler operation and structure.

## INTRODUCTION

In public water bodies, PPCPs such as antibiotics, antipyretics, and analgesics have been detected and have attracted attention as emerging pollutants. Especially, antibiotics have been deeply concerned, because micro-organisms exposed to antibiotics may acquire antibiotic resistances. As conventional wastewater and water treatment processes are not able to remove antibiotics completely, developments of innovative technology are needed (Yunlong et al., 2014; Bendz et al., 2005).

In advanced oxidation processes (AOPs), sufficient quantity of OH radicals can be produced. As OH radicals are a strong which can decompose almost every organic pollutant, it is considered AOPs are one of the technology applicable to antibiotics removal. There are different types of AOPs such as UV irradiation with injection of ozone and/or H2O2, supersonic treatment, and electrolysis; however most are complicated in operation and reactor configurations (Glaze et al., 1989; Hoare, 1985). In this study, we proposed a novel electrochemical AOP and investigated treatment performances.

Mechanisms of the proposed electrochemical AOP are represented by Reactions (1) to (3).

Anode:

$$3H_2 \rightarrow O_3 + 6H^+ + 6e^- \qquad (1)$$

Cathode:

$$O_3 + e^- \rightarrow O_3 \qquad (2)$$

$$O^-_3 + H_2O \rightarrow \cdot OH + OH^- + O_2 \qquad (3)$$

By an application of electric current, ozone is produced from anode in Reaction (1), and is reduced to OH radicals via ozonide ion in Reactions (2) and (3). Highly concentrated electrolyte is required to increase the yield for Reaction (1). In this respect, we used a solid polymer electrolyte membrane. On the other hand, it was reported that a rate limiting step is the mass transfer of ozone from bulk liquid to cathode surface; therefor granular electrodes were used to enhance overall removal performances (Kishimoto et al., 2005; Seader et al., 1952).

## MATERIALS AND METHODS

**Electrode and Experimental Apparatus.** In this study, $SnO_2$ electrode was used for the generation of ozone. It was prepared referring to former studies (Cui et al., 2009; Parsa et al., 2012), where Ti expanded substrate was coated with solution containing $SnO_2$ precursors, dried and heated at 390 °C. Some electrodes were prepared at different coating times (5-20). Surface area is 2.3 $cm^2$. SPE membrane (Nafion N-117, USA) was sandwiched by $SnO_2$ and stainless mesh.

**FIGURE 1.** shows the schematic diagram of experimental apparatus. Main body was built using stainless and fluorocarbon resin sheet. The anode with SPE membrane was equipped in the reactor. Four cathodic compartments were installed downstream from the anode and granular electrodes (Pt/Ti or activated carbon) were packed inside. Cathode surface area and effective volume were about 1500 $cm^2$ and 35 mL, respectively.

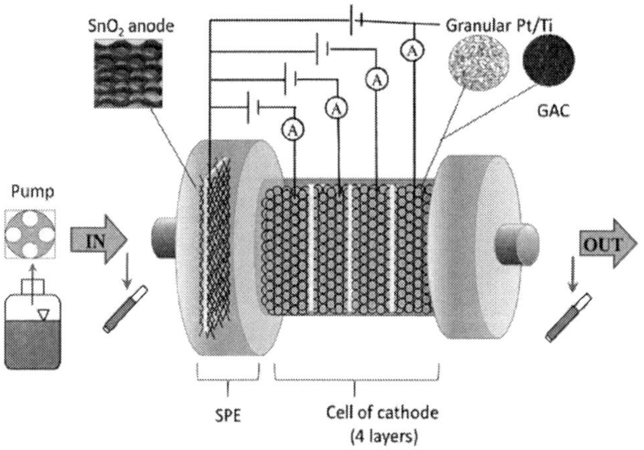


**FIGURE 1.** A main part of experimental apparatus (Electrolytic cell).

**Continuous Antibiotics Removal Experiments.** Electrolytic cell (FIGURE 1.) was used in continuous treatments of antibiotics. Synthetic wastewater was fed continuously and influent and effluent concentrations of antibiotics were measured under different operating conditions (i.e. applied currents, HRT and pH). In this study, electric current passing SPE membrane is defined as "ozone oxidation current", and one passing every cathode as "ozone reduction current". The same electric current was applied to each cathode compartment. Details of experimental procedures were shown in TABLE 1.

Synthetic wastewater containing 1 mg/L each tetracycline (TC), sulfamethoxazole (SMX) or erythromycin with 1mM Na2SO4 was prepared using de-ionized water. Structures of antibiotics were shown in **FIGURE 2**. pH of synthetic wastewater was adjusted using HCl or NaOH solution. Antibiotics and their intermediate products were measured with GC/MS (HP-1100, Agilent Technologies, Germany) and LC/MS/MS (UPLC-QTof, Waters, USA).

**TABLE 1.** Experimental conditions.

| Experiment | Solution | Cathode cell | $O_3$ oxidation current (mA) | $O_3$ reduction current (mA) | HRT (min) | pH |
|---|---|---|---|---|---|---|
| Continuous treatment of PPCPs | 1mg/L antibiotics, 1mM Na2SO4 | Pt/Ti or GAC | 0-75 | 1.5 | 2.5, 5, 10 | 3.5, 5.5, 7.5, 9.5 |

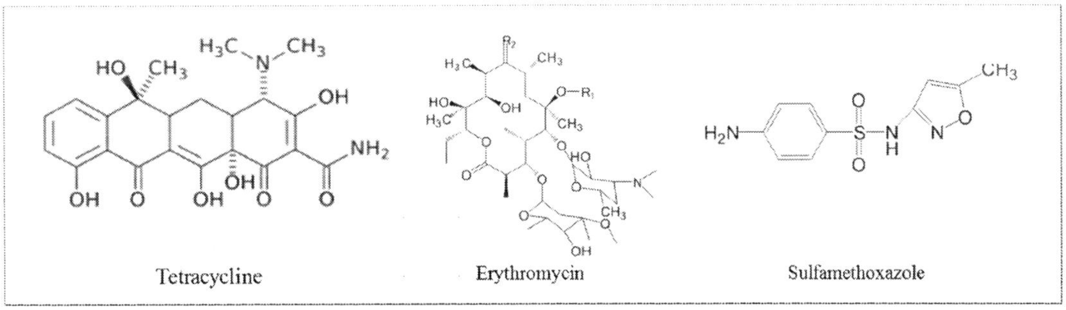


**FIGURE 2.** Antibiotics used in this study.

## RESULTS AND DISCUSSION

**Treatment of PPCPs (Tetracycline).** In continuous experiments, effluent concentrations of antibiotics approached to steady-state values. **FIGURE 2.** shows normalized effluent concentrations of tetracycline against oxidation currents under different HRTs and pH conditions. As shown in Figure, effluent tetracycline concentrations decreased with increasing current. Satisfactory removal was achieved at pH 5.5 and 7.5. Longer the HRT, lower the effluent concentrations. Comparing results for Pt/Ti and GAC electrodes, more tetracycline was removed. This was attributable to adsorption by GAC. In experiments of sulfamethoxazole and erythromycin treatments, similar results were observed (data not shown) and the effectiveness of the proposed process was demonstrated.

**Decrease in Bactericidal Effect.** FIGURE 4. shows the results of bactericidal effect to coliform. The bar indicated by (a) was 10-fold diluted secondary effluent, where number of colonies was about 60 in 100mL sample. The bars (b) and (d) were mixtures of influent 20-fold diluted TC and 35-fold diluted SMX solutions with the 10- fold diluted secondary effluent, respectively. In contrast, the bars (c) and (d) indicate results for effluent samples prepared in the same manner to (b) and (d), respectively. As shown, the numbers of colonies for (b) and (d) were smaller in (b) and (d) than (a), due to bactericidal effects by TC and SMX. However, the numbers of colonies in (a), (c) and (e) were almost identical, demonstrating that bactericidal effects were removed by the present process.

**By-products of Tetracycline.** FIGURE 5. shows intermediate products (marked by red color dotted lines) identified in the effluent in TCA treatments. In comparison, reaction pathways of Tetracycline in AOP in the literature (Joonseon et al., 2010) were shown. Reaction path was comprised of multi-steps reactions. A further study will be needed to quantitative distributions of the intermediates.

**Comparison of Energy Consumption.** FIGURE 5. shows energy consumptions in this study with those for other AOPs under 90% removal efficiency. The energy consumption in this study was about 0.3-

$4kWh/m^3$. Although target pollutants and their concentrations are different, the energy consumptions was similar to or smaller than former AOPs.

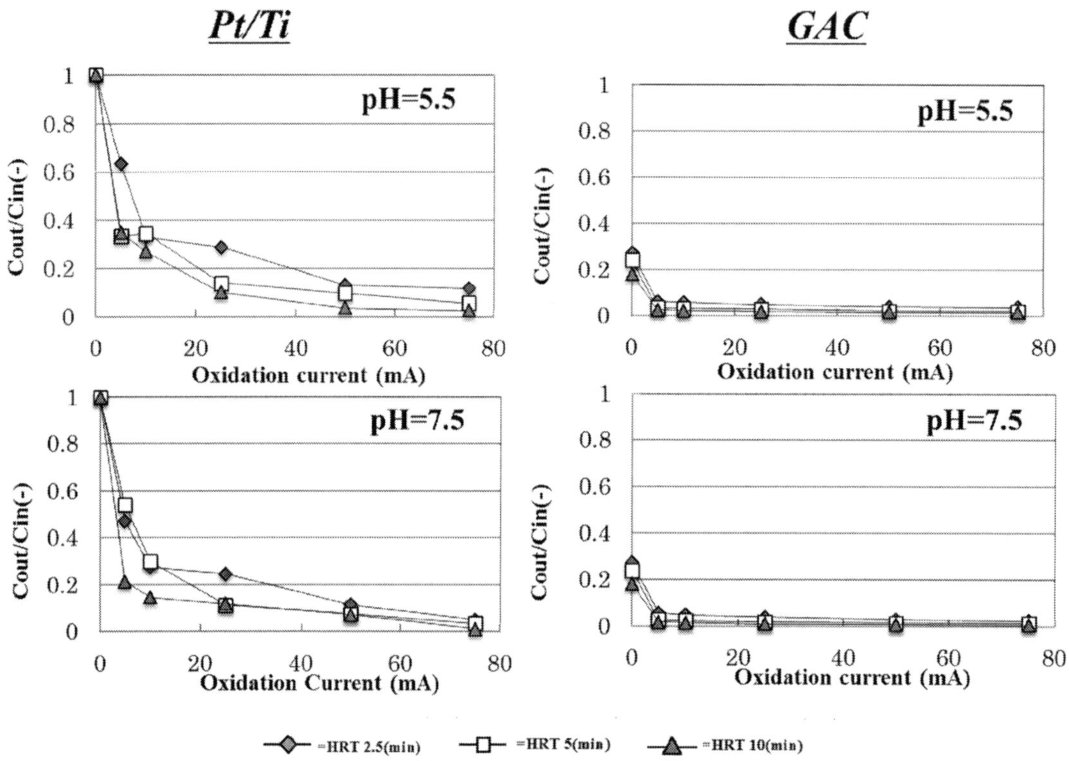


**FIGURE 3.** Continuous treatment experiment (Tetracycline)

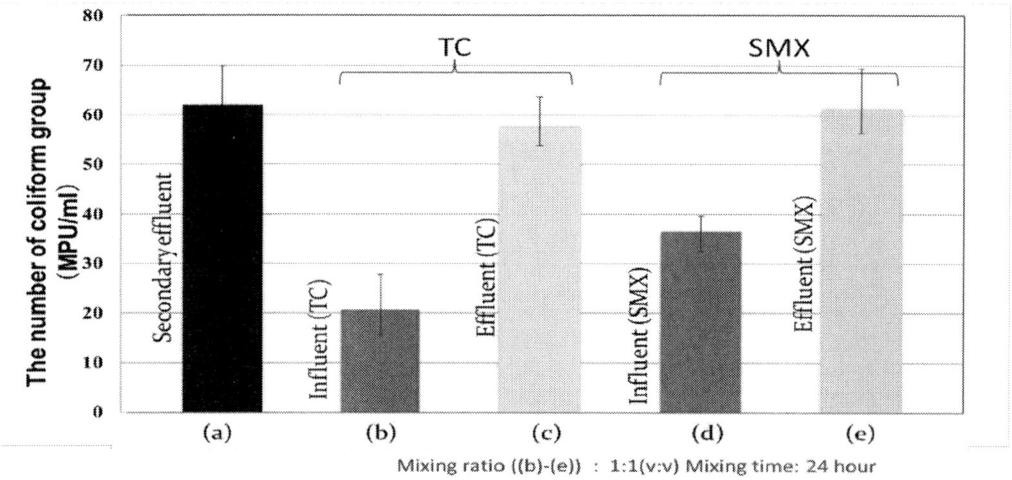


**FIGURE 4.** Decrease in bactericidal effect to coliform

## CONCLUSIONS

From continuous experiments for tetracycline, sulfamethoxazole and erythromycin, the following conclusions were drawn;

- The proposed electrochemical AOP could remove and decomposePPCPs.

- Reduction in bactericidal effect of coliform bacteria was achieved by the process.
- Energy consumption was 0.3-4 $kWh/m^3$, which was similar to or less than former AOPs.
- The present process is very simple in operation and control; and is considered as a feasible alternative to former AOPs.

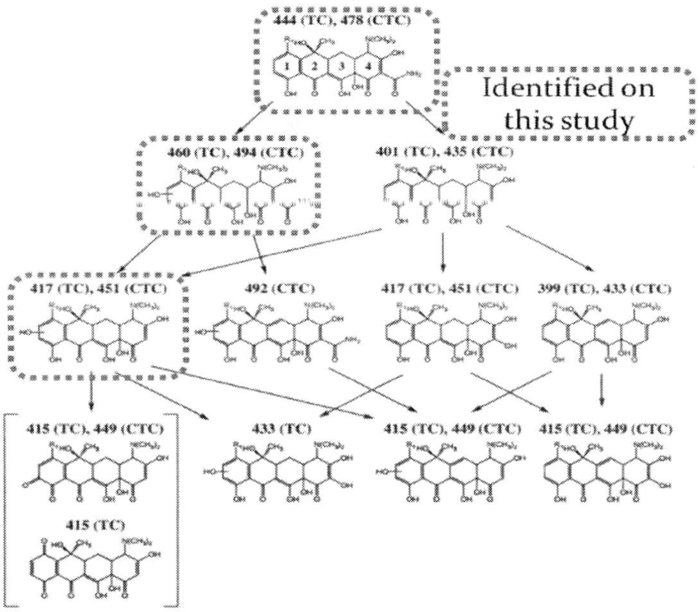


**FIGURE 5.** Reaction pathways of Tetracycline on AOPs (Joonseon et al., 2010) and intermediate products identified in this study.

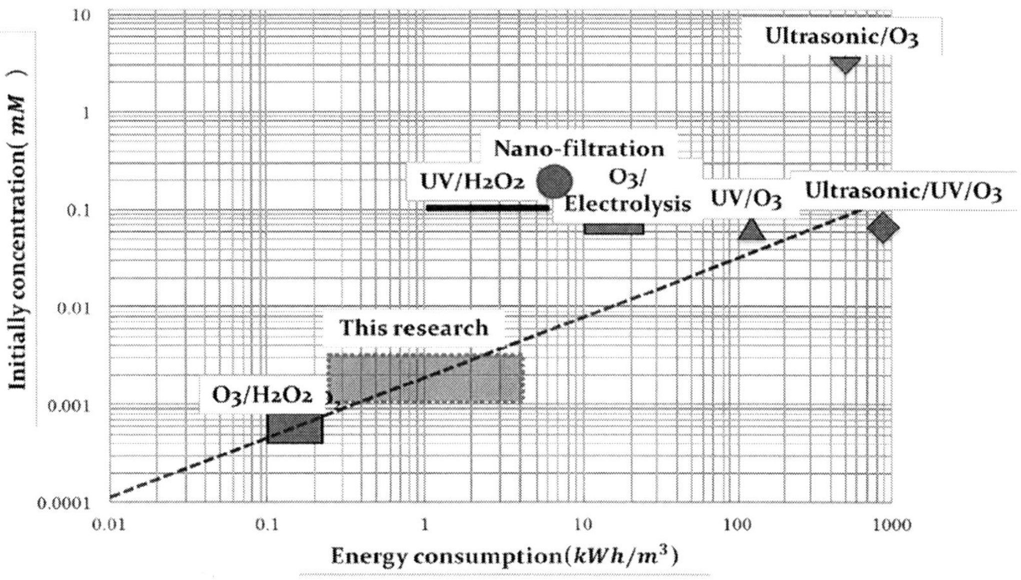


**FIGURE 6.** Comparison of energy consumption under 90% removal efficiency.

## ACKNOWLEDGEMENT

This study was supported in part by Grant-in-Aid for Scientific Research (B) (No. 24360219), the Ministry of Education Culture, Sports, Science and Technology (MEXT) Japan.

## REFERENCES

Bendz D. Paxéus N. A., Ginn T. R. and Loge F. J. 2005. "Occurrence and fate of pharmaceutically active compounds in the environment, a case study: Höje River in Sweden". *Journal of Hazardous Materials*, 122(3): 195-204

Cui Y., Wang Y., Wang B., Zhou H., Chan K. Y., and Li X. Y. 2009. "Electrochemical generation of ozone in a membrane electrode assembly cell with convective flow". *Journal of The Electrochemical Society*, 156(4): E75-E80

Glaze W. H. and Kang J. 1989. "Advanced Oxidation Process. Description of a kinetic model for the oxidation of hazardous materials in aqueous media with ozone and hydrogen peroxide in a semibatch reactor". *Industrial and Engineering Chemistry Research*, 28: 1573-1580

Hoare J. P. 1985. "Standard potentials in aqueous solution". *Marcel Dekker*, New York, 49-66 Joonseon J., Weihua S., William J. C., Jinyoung J., John G. 2010. "Degradation of tetracycline antibiotics: Mechanisms and kinetic studies for advanced oxidation/reduction processes". *Chemosphere*, 78: 533-540

Kishimoto N., Morita Y., Tsuno H., Oomura T. and Mizutani H. 2005. "Advanced Oxidation effect of ozonation combined with electrolysis". *Water Research*, 39:4661-4672

Parsa J. B., Abbasi M. and Cornell A. 2012. "Improvement of current efficiency of the Ti/Sn-Sb-Ni oxide electrode via carbon nanotubes for ozone generation". *Journal of The Electrochemical Society*, 159(5): D265-D269

Seader J.D. and Tobias C. W. 1952. "Ozone by electrolysis of sulfuric acid". *Industrial and Engineering Chemistry*, 44(9): 2207-2211

Yunlong L., Wenshan G. and Huu H. N. 2014. "A review on the occurrence of micropollutants in the aquatic environment and their fate and removal during wastewater treatment". *Science of the Total Environment*, 474: 619-641

# ACUTE TOXICITY OF GRAPHENE NANOPLATELET TO BACTERIAL COMMUNITIES IN ACTIVATED SLUDGE

***Hang Ngoc Nguyen*** and Debora Frigi Rodrigues
(Civil and Environmental Engineering, University of Houston, Houston, TX)

Carbon based nanomaterials are used extensively in many applications in different fields. Graphene is the building block of graphitic materials such as flullerenes, carbon nano-tube and graphite. Graphene consists of monolayer of carbon atoms arranged in a two-dimensional hexagonal lattice. Graphene is stronger than steel, has high electrical conductivity and has antimicrobial properties. Graphene has been increasingly used in research and in industrial applications such as in energy, automotive, aerospace, electronics and biomedical devices. With all these emerging applications, the question we have to ask "Where will all the graphene go after use and disposal?". Wastewater is one of the most common discharges that all the waste including graphene will finally end up. Leaking and bleaching from graphene based devices will introduce graphene in the wastewater. Graphene has been shown to exhibit antimicrobial properties, so its impact to the biological treatment process in wastewater treatment plant needs to be investigated. This study aims to understand how graphene interacts with activated sludge in biological treatment process at the acute toxicity level. We study the acute toxicity of graphene nanoplatelet suspension to the performance of activated sludge in lab scale reactors. The activated sludge was collected directly from a wastewater treatment plant, and then cultivated in the lab for 44 hours with synthetic wastewater before running the reactors with different concentrations of graphene suspension from 0 to 300 mg/L. The performance of the reactors in removal of nutrient, ammonia, phosphate, nitrate and chemical oxygen demand were determined. Genes of interest were quantified using real time PCR with standard curve. Metagenomics of 16S rRNA was also performed to investigate the changes in the population of microbes in activated sludge in the presence of graphene. Initial results show that graphene reduces the nutrient removal in the treatment and has the potential to affect microbial community in the activated sludge.

# ISOLATION OF BACTERIA FROM LAKE WATERS ASSOCIATED WITH WASTEWATER EFFLUENTS CAPABLE OF DEGRADING VARIOUS PHARMACEUTICALS

***Noreen Gallagher*** and Dr. Jeff Lodge
(Thomas H Gosnell School of Life Sciences, Rochester Institute of Technology, Rochester, NY, USA)

Many wastewater treatment plants are not properly equipped for the removal of non-steroidal anti-inflammatory drugs such as ibuprofen and naproxen and analgesics such as acetaminophen. These compounds are continually discharged into surface waters and , although in a very low concentration, their presence is becoming an emerging issue. NSAIDs, analgesics, and their metabolites are biologically active in surface waters and, as a result, possess potentially harmful effects to the environment as well as public health through the consumption of water and food containing residues of these compounds.

One objective of this project is to isolate and characterize heterotrophic bacteria that are capable of degrading NSAIDs (ibuprofen and naproxen) and analgesics (acetaminophen) from several areas in Western NY associated with wastewater effluent outfalls. From the various isolates the best degraders will be chosen to further study their degradation rates, % degradation, and the environmental parameters that are important in enhancing degradation rates. Another objective is to show that these organisms may be important in controlling pharmaceutical levels in surface waters.

Water and sediment samples from three wastewater effluent outfall areas along Lake Ontario in the Rochester area and one outfall area in Onondaga Lake (Syracuse area) were used in various enrichment techniques to isolate potential bacteria capable of degrading ibuprofen, naproxen, and acetaminophen. Potential isolates are than grown on the pharmaceutical of choice as the sole carbon source as a second selection. Organisms shown to grow at high rates on various pharmaceuticals are then further studied to determine degradation rates, % degradation, and environmental parameters important in enhancing degradation rates in lake waters.

Preliminary results from various enrichment experiments have led to the isolation of various heterotrophic bacteria capable of growing on ibuprofen, naproxen, or acetaminophen as their sole carbon source. One isolate from Onondaga Lake (Syracuse) degraded 80% of the acetaminophen in 5 days. One isolate from Charlotte Beach (Rochester) and one from Payne Beach (Rochester) were also able to degrade 80% of acetaminophen in 5 days. Two isolates from Durand Eastman Beach significantly degraded ibuprofen while an isolate from Payne Beach degrade 80% of naproxen. These isolates are being studied further as well as some new recent isolates that were found.

Utilizing various enrichment techniques our lab has been able to isolate various heterotrophic bacteria from Lake Ontario and Onondaga Lake that can degrade ibuprofen, naproxen, and acetaminophen. This data shows that there are endogenous heterotrophs located near wastewater outfalls that can degrade various pharmaceuticals and may be applied to a commercial process to try and reduce the amount of pharmaceuticals entering surface waters.

# CHARACTERIZATION AND QUANTIFICATION OF DOM IN WASTEWATER AND ITS INTERACTION WITH PHARMACEUTICALS

***Sanjeeb Mohapatra****, Neha Sharma and Suparna Mukherji
(Indian Institute of Technology Bombay, Mumbai, India)
Lokesh P. Padhye
(The University of Auckland, Auckland. New Zealand)

Dissolved organic matter (DOM) plays a vital role on contaminant transport in both natural waters and wastewater. It may potentially alter the fate of pharmaceuticals in aqueous systems by forming complexes. The seasonal variation in DOM concentration over three seasons in two major WWTPs in Mumbai, India was determined based on total organic carbon (TOC) measurements. Initial DOM characterization using specific UV absorbance (SUVA) was followed by characterization using fluorescence spectroscopy and parallel factor (PARAFAC) analysis. PARAFAC analysis was performed using the three-dimensional excitation–emission matrix (3-D EEM) representing the three major DOM components. In the samples collected during the summer season, protein-like fluorescence peak representative of tryptophan analogues was found at excitation and emission wavelength of 280 and.340 nm, respectively. Tyrosine analogue peaks were also observed at both the WWTPs for samples collected in the monsoon season. Comparatively higher intensity of tryptophan like fluorescence was observed in the samples collected in the winter season.

In-situ measurements were also carried out to assess interactions between four common pharmaceuticals and two distinct DOM sources, i.e., DOM in untreated wastewater and Suwannee River DOM. The pharmaceuticals selected for this study were atenolol, atorvastatin, sulfamethoxazole, and ranitidine. Pharmaceuticals were spiked in aqueous DOM solution to achieve concentration over the range 100 to 1000 mg/L and quenching of DOM fluorophores due to interaction with the pharmaceuticals was determined. The concentration of pharmaceuticals in solution was quantified using high resolution liquid chromatography tandem mass spectrometry (HR LC-MS/MS). At varying concentrations of pharmaceuticals, and DOM the interaction of pharmaceuticals with different components of DOM was quantified through PARAFAC analysis to reveal the interaction of pharmaceuticals with DOM. The results indicate significant interaction between pharmaceuticals and DOM that may potentially affect the fate and transport of pharmaceuticals in the environment. Such interactions may also be responsible for poor recovery of pharmaceuticals from wastewater.

# DEEP INSIGHTS INTO THE MECHANISM OF 2,4-DI-TERT-BUTYLPHENOL(2,4-D) DEGRADATION BY USING UV/PERSULFATE WITH A NEW MODEL

***Qiongfang Wang***, Yisheng Shao, Naiyun Gao and Xiang Shen (Tongji University, Shanghai, China )
Yisheng Shao (China Academy of Urban Planning & Design, Beijing, China)

In this research, it was the first time to assessed the degradation of 2,4-Di-tert-butylphenol (2,4-D) using a ultraviolet/persulfate（UV/PS） process. The results showed that 2,4-D could be degraded effectively using this process. The process was modeled based on a steady-state assumption that both hydroxyl and sulfate radicals reacted with 2,4-D. The second order rate constants for 2,4-D reacting with $SO_4^{\cdot-}$ and $OH\cdot$ were estimated to be $(3.07 \pm 0.03) \times 10^9$ and $(1.62 \pm 0.13) \times 10^{10}\ M^{-1}s^{-1}$ at pH 7, respectively. The pseudo-first order rate constants ($k_o$) of 2,4-D increased with increasing dosage of PS. The pH did not affect the degradation of 2,4-D directly in acidic condition, while the degradation reduced a little from pH 7 to 10 but was strongly inhibited at pH 11. Dihydrogen phosphate and lower nitrate concentrations had no effect on the degradation of 2,4-D, whereas 50mM nitrate concentration suppressed the degradation. Bicarbonate, hydrogen phosphate and natural organic matter (NOM) inhibited the degradation of 2,4-D dramatically through consuming $SO_4^{\cdot-}$ and $OH\cdot$, which were likely to be the main radical scavengers in natural waters when using UV/PS process to control 2,4-D. The degradation of 2,4-D was much enhanced by adding different dosages of $Cl^-$. The degradation pathways were proposed by Liquid Chromatography-Mass/Mass Spectroscopy (LC–MS/MS) analysis.

# REMOVAL OF ESTRIOL IN WATER USING SEQUENTIALLY COUPLED MEMBRANE FILTRATION/$TiO_2$ PHOTOCATALYTIC PROCESSES

**Irwing M. Ramírez-Sánchez[1], Ashantha Goonetilleke[2], *Erick R. Bandala*[3,4*]**

1. Universidad de las Americas, Puebla. Sta. Catarina Martir, Cholula 72810 Puebla. Mexico.
2. Queensland University of Technology. GPO Box 2434, Brisbane, Queensland 4001, Australia.
3. Desert Research Institute (DRI). 755 E. Flamingo Road. Las Vegas, 89119-7363 Nevada, USA. E-mail: erick.bandala@dri.edu
4. Graduate Program Hydrologic Sciences. University of Nevada, Reno. Reno, NV 89557, USA

**ABSTRACT:** A coupling system of membrane separation and photocatalytic oxidation was evaluated for the removal of estriol (E3) in water. An aqueous solution containing estriol was fed to the membrane unit, which generated a permeate flow with a pollutant concentration that was below the method detection limit and a reject flux with an increased estrogen concentration. Subsequently, the reject stream was treated using a photocatalytic processes. Membrane technology was used in addition to the photocatalytic processes and an NF90 membrane was selected based on the proposed E3 concentration. The results demonstrated that

## INTRODUCTION

Emerging contaminants (ECs) are recognized as significant water pollutants (Czech and Rubinowska, 2013). Many ECs affect human and wildlife endocrine systems once they are released into the environment and are considered endocrine disrupting compounds (EDCs). Natural and synthetic estrogens are among the most important EDCs that occur as pollutants in drinking water and wastewater. The natural estrogens found in water include estrone (E1), 17β-estradiol (E2), and estriol (E3) and the synthetic steroid estrogens found in water include ethinyl estradiol (EE2), mestranol and quinestrol. Removing estrogens from aquatic environments after they have been released is complicated because of their low concentrations, but because of their biological activity, even significantly low amounts of estrogens have been shown to be detrimental (Carvalho et al., 2015).

Several removal methodologies have been proposed for treating estrogens in water. Among them, membrane filtration has been shown to be useful for removing estrogens from drinking water and wastewater (Bodzek and Dudziak, 2006). The capability of nanofiltration (NF), a membrane filtration technique that shares properties with ultrafiltration and reverse osmosis, to remove natural and synthetic estrogens from the water matrix has been studied. Reports show that NF efficiently removes estrogen from water. For example, Schafer et al. (2003) reported 80%-99% E1 retention using diverse membrane types. The authors suggested that retention efficiencies depend on the adsorptive effects, which are driven by hydrogen bonding between E1 and the membranes. Other authors have found a decreasing rejection of E1 after 24 hours of filtration (Hu et al., 2007) and some studies on estradiol desorption have reported negative rejection (McCallum et al., 2008).

The main drawback of using NF to remove estrogens is that it generates reject effluent. Because it is a phase-change process, NF generates a permeate flow that has a lower concentration of the contaminant and a reject flow in which the mass balance of the process increases the contaminant concentration. The concentration of estrogens in the reject solution after NF depends on several factors, including the feed water chemistry, membrane type, and solute properties. Studies have shown that the main mechanisms for estrogens rejection are related with adsorption (Yoon et al., 2004); desorption (McCallum et al., 2008); the octanol/water partition coefficient (Bodzek and Dudziak, 2006); pH, pKa, and the membrane's surface charge (Hu et al., 2007); the presence of natural organic matter; and the initial concentration (Jin et al., 2010).

McCallum et al. (2008) found that when the maximal initial adsorption is reached, steady-state rejection is achieved and size exclusion phenomena govern the rejection. When membrane processes are developed, concentrated rejected effluents could increase health hazards and potential toxicity because of the increased concentration of pollutants. To reduce the effects of concentrated effluent during NF processes, coupled systems to remove and degrade estrogens over sequential stages or coupled hybrid systems have been proposed (Grzechulska-Damszel et al., 2010), such as coagulation–NF or NF advanced oxidation processes (AOPs), to achieve significant (e.g., 95.4%-99.9%) removal efficiencies (Berberidou et al., 2009; Ganiyu et al., 2015). Recently, coupling an NF90 membrane with UV/$TiO_2$ was shown to concentrate and degraded pharmaceuticals (Martínez et al., 2013). Currently, there are very few studies on coupled NF and postconcentrate photocatalytic treatment (Ganiyu et al., 2015) for reducing E3, and therefore more research is needed to expand our understanding of the rates of estrogen concentration and degradation. Coupling AOPs with membrane filtration has been studied as a pretreatment for the membrane feed stream, a posttreatment of nonrejected pollutants in permeate (Pereira et al., 2012), and a treatment for highly concentrated organic compounds in a concentrate stream (Martínez et al., 2013). In all of these cases, separation and degradation were simultaneously achieved in the hybrid processes (Ganiyu et al., 2015). This study analyzed NF using suspended photoactivated $TiO_2$ as a posttreatment process for the concentrate stream before discharge. This report shows that coupling NF with a $TiO_2$-based photocatalytic process effectively concentrates and further degrades estrogen.

## MATERIALS AND METHODS

**Chemicals and Membranes**. All the chemicals used were of analytical grade. Estriol was obtained from Sigma-Aldrich (USA). Estriol pKa is above 10.54. A stock solution of E3 (10 µM) was prepared by adding 0.0028 g of E3 to 1 L of deionized water. The solution was then stored in an amber flask. An NF90 membrane sheet from Dow Chemical was selected and supplied by Sterlitech (USA). The properties of the polyamide layer of NF90 (Zhang et al., 2016) are as follows: MWCo, 200-400 Da; max temperature, 45°C; pH range, 3-10; salt rejection, 85%-95% (NaCl), >97% (CaCl); and top layer composition, thin-film composite polyamide. Aeroxide® $TiO_2$ P25 (previously known as Degussa P25, 50±15 $m^2$ $g^{-1}$ specific surface area, 21 nm average primary particle size, 80:20 anatase/rutile ratio, according to the manufacturer) was obtained from Evonik Industries and used as the photocatalyst.

**First stage: Nanofiltration Processes.** All membrane filtration experiments were carried out in a cross-flow cell as described in Figure 1A. The flow cell was constructed in the lab using Nylamid®, a polyamide engineered plastic. The membrane support was made of sintered stainless steel filter plate (filter rating: 10 µm; maximum pore diameter: 30 µm; permeability: 160 $m^3$ $h^{-1}$ $m^{-2}$ $kpa^{-1}$, according to the manufacturer specification) supplied by Baoji Qixin Titanium Co., Ltd. The water pressure pump used was a DC 24V electric diaphragm pump, which is capable of providing a maximum discharge pressure of 10 bar and a maximum flow rate of 16 L $min^{-1}$. The membrane was set up on two concentric O-rings® to prevent filtration outside the cell. The effective membrane surface area was calculated as 48 $cm^2$ (8 cm x 6 cm). A new membrane was used for each experiment. The operating pressure on the membrane was adjusted using a bypass valve. The working pressure was 6 kg $cm^{-2}$ (588.4 kPa) and a 1 L reservoir was used to store the feed solution and collect the concentrate effluent. The membrane was compacted for 4 hours at 6 kg $cm^{-2}$ using deionized water. The temperature of the feed solution was kept at 22 ± 2 °C using a cylindrical, water-cooled, jacketed, glass vessel with a thermostatic bath (Polystat®, Cole-Palmer). After the membrane was compacted, the concentrate and permeate reservoirs were emptied and filled with a test solution. A test solution of 300 mL was placed in the concentrate reservoir and the filtration process was conducted at 6 kg $cm^{-2}$ using the NF90 membrane. The retinate stream was continuously recirculated to the feed tank. Samples were taken from the concentrate reservoir and filtrate reservoir every 30 minutes during a period of 300 minutes.

**Post-treatment: Photocatalytic Degradation of E3 of Concentrate Solution**. As a post-treatment for the concentrated solution, an aliquot of 100 mL of the NF's concentrate stage was taken for photocatalytic degradation of E3. The lamps were positioned in the center and horizontally above the photoreactor (Figure 1B). The solution was irradiated with two 15 W blacklight lamps (GE, F15T8 BLB) that emitted radiation mainly in the 350-400 nm ($\lambda_{max}$ = 360 nm) wavelength interval. The horizontal and vertical position of the photoreactor was kept constant for the experiment. The temperature was controlled using a thermostatic bath (Polystat, Cole-Palmer). The overall system was placed in a closed box to avoid the effects of sunlight or any other artificial radiation source. Initially a study for photocatalytic degradation of E3 was performed to determine the effect of catalyst loading using 20, 40, 80 and 160 mg $L^{-1}$ of $TiO_2$, from this study the $TiO_2$ photocatalyst load selected was 20 mg $L^{-1}$ to be used in the post-treatment processes. In all the photocatalytic experiments a dark period (no radiation) was allowed for 20 minutes, during which time E3 and the photocatalyst were stirred without any radiation source to ensure adsorption equilibrium in the solution.

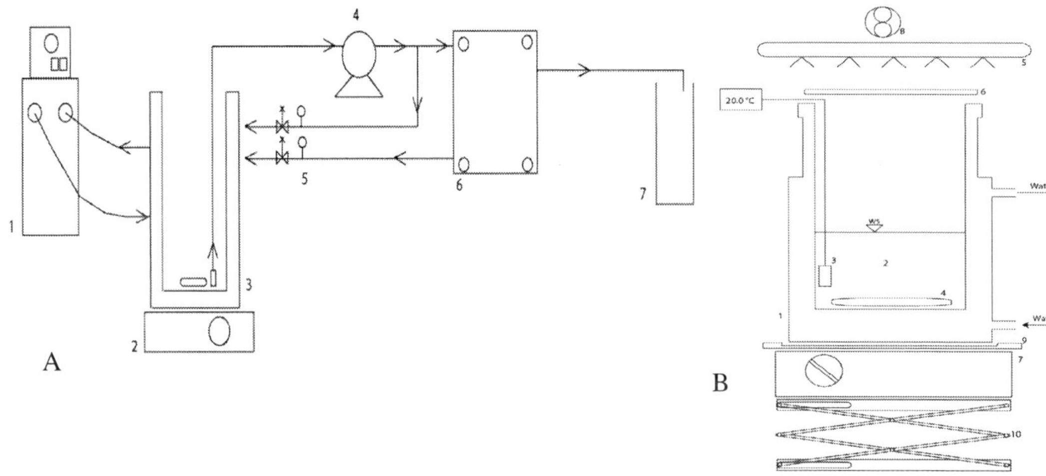


**FIGURE 1 A).** Cross-flow nanofiltration system setup: 1) temperature-controlled water bath, 2) stirring plate, 3) concentrate reservoir, 4) diaphragm pump, 5) pressure gauge, 6) cross-flow filtration cell, and 7) filtrate reservoir. **B).** Photocatalytic reactor setup: 1) glass reactor, 2) testing solution, 3) temperature probe, 4) spin bar, 5) backlight lamp, 6) optical filter (if needed), 7) stirring plate, 8) cooling fan, 9) horizontal position setup, and 10) lab jack lifting platform

**Analytical Methods.** Estriol was monitored using an HPLC system (Waters 1515) equipped with a UV detector (Waters 2787) with an injection volume of 20 μL. Estriol determinations were performed at isocratic analytical mode using a 150 mm × 4.6 mm with Inertsil® ODS-3, 5 μm column thermostated at 25 °C. The UV detection was set at $\lambda$= 280 nm and the mobile phase used was methanol (49%) and ultrapure water (51%) at a flow rate of 1 mL $min^{-1}$. The retention time was 10 minutes and the method detection limit was determined to be 0.1 μM.

## RESULTS AND DISCUSSION

**Concentration of Estriol by Nanofiltration Processes.** The experiment was divided into two phases: the first was nanofiltration, which started with a 1 μM E3 solution, and the second was the photocatalytic degradation of concentrate using $TiO_2$ activated with UV radiation. The nanofiltration process was run for 180 minutes (Figure 2A). The flux throughout the membrane that was obtained during nanofiltration was 17 L $m^{-2}$ $h^{-1}$. Flux evolution was constant and no decline was observed during the experiment. The initial volume of the feed solution was 500 mL and the final volume after nanofiltration was 150 mL. The estriol concentration in permeate was below the method detection limit for the entire experiment. The concentration of E3 in the rejected flow increased continuously, as shown in Figure 2A. The steric hindrance because of the molecular weight of E3 (280 g $mol^{-1}$) and the MWCo of NF90 membranes (200-400 Da) was the main factor that determined the retention capacity. The adsorption/desorption dynamic phenomenon was also an important mechanism that was taken into account for estrogen rejection (Yoon et al., 2004; Hu et al., 2007). Based on this information, it can be assumed that the cross flow tangential velocity on the membrane was sufficiently high to avoid building up a polarization layer with high concentration of E3. This reduced polarization layer kept E3 rejection rate for equal E3 concentration in bulk solution than boundary layer of the membrane. Charge rejection phenomena could be less important because at pH = 7.0, all E3 molecules are not dissociated due to their pKa value (10.54) and the zeta potential of the NF90 membrane (-55 mV) at that pH (Wang et al., 2016).

**3.2 Photocatalytic Degradation of Estriol.** Figure 2B shows the photocatalytic degradation of E3 using different $TiO_2$ catalyst loads. As shown, an increase in the catalyst load generated an increase in the reaction rate, with a load of 160 mg $L^{-1}$ having the highest removal efficiency. These results are in agreement with previous reports that found the same trend. It was also found that further increases in the catalyst load led to a decrease in the efficiency of the process, which was also previously reported (data not shown).

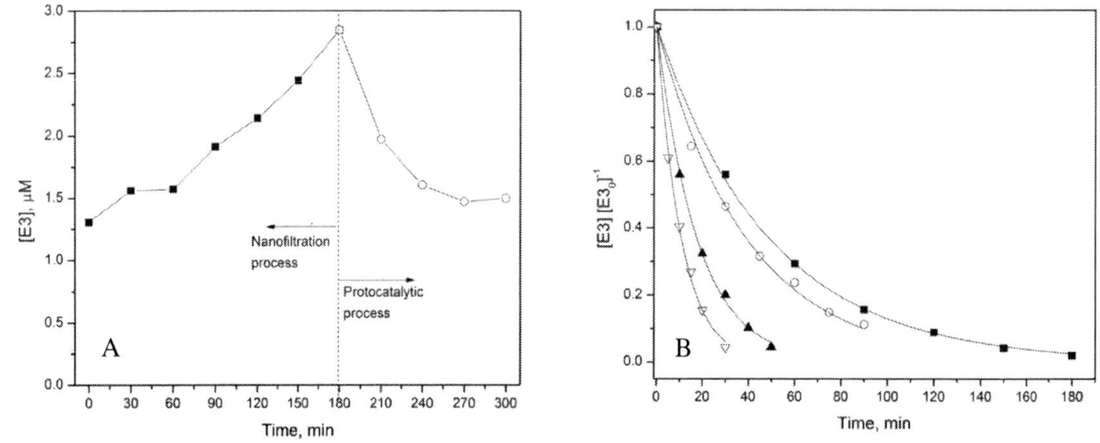


**FIGURE 2 A).** Coupled membrane filtration and $TiO_2$ photocatalytic processes for concentration-degradation of estriol: —■— = concentrate solution; —○— = concentrate solution under photocatalytic process. **B).** Catalyst load effect on degradation of E3 under backlight lamp (GE F15T8 BLB); experimental data: ■ = 20 mg $L^{-1}$, ○ = 40 mg $L^{-1}$, ▲ = 80 mg $L^{-1}$, ▽ = 160 mg $L^{-1}$ and (-) = first-order kinetic model.

The kinetic analysis of the experimental data depicted in Figure 2B fit a pseudo-first-order kinetic model. The kinetic model fitted the experimental data fairly well, which is also shown in Figure 2B (solid lines). For all the catalyst loads tested, approximately the same removal efficiency (i.e., 95%) was achieved and the only difference in the experimental runs was the time of irradiation required. Approximately 180 minutes of photocatalytic treatment were needed to achieve 95% degradation of E3 using 20 mg $L^{-1}$ of the photocatalyst load, whereas the same removal was obtained after only 30 minutes (one sixth of the reaction time) using 160 mg $L^{-1}$ (eight times more load). These results agree with previous reports that have demonstrated that higher concentrations of the photocatalyst overdoses the system, which leads to a decrease in degradation efficiency because of radiation scattering, reduced penetration, and $TiO_2$ agglomeration (Colina-Márquez et al., 2015). Considering that the increase in the catalyst load did not generate the expected improvement in E3 degradation, further experiments were performed using the lowest catalyst load to favor the efficiency of the overall process. After 180 minutes of concentration by the NF processes, an aliquot 100 mL of concentrate solution was taken and submitted to photocatalytic degradation using $TiO_2$. As stated previously, $TiO_2$ was dispersed in a concentrate solution to get 20 mg $L^{-1}$.

The degradation of E3 in the concentrate (Figure 2A) for time periods of 180 minutes and up to 300 minutes showed that it also follows a pseudo-first-order kinetics, which was previously observed for the experiments with variable photocatalyst loads. It is well known that the phenol moiety of E3 is responsible for absorption at 280 nm. For this reason, any change in its structure will produce changes in molar absorption at this wavelength. When the E3 solution was degraded using $TiO_2$ and UV radiation, it was observed that the initial adsorption of the solution decreased as a function of time following a first-order reaction.

## CONCLUSIONS

Further studies of E3 adsorption on the membrane should be conducted to obtain a mass balance. It was confirmed that NF90 membrane can effectively reject estriol and it is feasible concentrating by recalculating the reject stream of the NF process. It has been demonstrated, using the load tested, that UV/$TiO_2$ degradation of estriol can be a feasible coupled process for E3 degradation when E3 found in rejected effluents of a nanofiltration process. Further studies should compare the effects of pH and salt concentration on the rejection and degradation of estriol in water. This work highlights an advanced, coupled NF system and heterogeneous UV/$TiO_2$ photocatalysis to reduce the discharge volume that should be treated when estrogen compounds are removed from the permeate effluents of NF systems.

## ACKNOWLEDGMENTS

The Aeroxide® P25 Evonik catalyst used for this work was provided by Intertrade S.A. de C.V., the supplier of Evonik Industries in Mexico. The research was partially supported by CONACYT under Project CB-2011/168285. The authors thank Dr. M.A. Quiroz Alfaro for his excellent technical help and allowing the use of materials and equipment at the UDLAP's electrochemical lab.

## REFERENCES

Berberidou, C., S. Avlonitis, and I. Poulios. 2009. "Dyestuff Effluent Treatment by Integrated Sequential Photocatalytic Oxidation and Membrane Filtration." *Desalination* 249 (3): 1099–1106.

Bodzek, M., and M. Dudziak. 2006. "Removal of Natural Estrogens and Synthetic Compounds Considered to Be Endocrine Disrupting Substances (EDs) by Coagulation and Nanofiltration." *Pol. J. Environ. Stud.* 15 (1): 35–40.

Carvalho, A. R M, V. V. Cardoso, A. Rodrigues, E. Ferreira, M. J. Benoliel, and E. A. Duarte. 2015. "Occurrence and Analysis of Endocrine-Disrupting Compounds in a Water Supply System." *Environ. Monit. Assess.* 187 (3): 139.

Colina-Márquez, J., F. Machuca-Martínez, and G. Li Puma. 2015. "Modeling the Photocatalytic Mineralization in Water of Commercial Formulation of Estrogens 17- β Estradiol (E2) and Nomegestrol Acetate in Contraceptive Pills in a Solar Powered Compound Parabolic Collector." *Molecules*, 13354–13373.

Czech, B., and K. Rubinowska. 2013. "$TiO_2$-Assisted Photocatalytic Degradation of Diclofenac, Metoprolol, Estrone and Chloramphenicol as Endocrine Disruptors in Water." *Adsorption* 19 (2): 619–630.

Ganiyu, S.O., E. D. Van Hullebusch, M. Cretin, G. Esposito, and M. A. Oturan. 2015. "Coupling of Membrane Filtration and Advanced Oxidation Processes for Removal of Pharmaceutical Residues: A Critical Review." *Sep. Purif. Technol. 156:* 891-914

Grzechulska-Damszel, J., S. Mozia, and A.W. Morawski. 2010. "Integration of Photocatalysis with Membrane Processes for Purification of Water Contaminated with Organic Dyes." *Catal. Today* 156 (3-4): 295–300.

Hu, J. Y., X. Jin, and S. L. Ong. 2007. "Rejection of Estrone by Nanofiltration: Influence of Solution Chemistry." *J. Membr. Sci.* 302 (1-2): 188–96.

Jin, X., J. Hu, and S. L.Ong. 2010. "Removal of Natural Hormone Estrone from Secondary Effluents Using Nanofiltration and Reverse Osmosis." *Water Res.* 44 (2): 638–48.

Martínez, F., M. J. López-Muñoz, J. Aguado, J. A. Melero, J. Arsuaga, A. Sotto, R. Molina, et al. 2013. "Coupling Membrane Separation and Photocatalytic Oxidation Processes for the Degradation of Pharmaceutical Pollutants." *Water Res.* 47 (15): 5647–5658.

McCallum, E. A., H. Hyung, T. Anh Do, C. H. Huang, and J.Hong Kim. 2008. "Adsorption, Desorption, and Steady-State Removal of 17β-Estradiol by Nanofiltration Membranes." *J. Membr. Sci.* 319 (1-2): 38–43.

Pereira, V. J., J. Galinha, M. T. B. Crespo, C. T. Matos, and J. G. Crespo. 2012. "Integration of Nanofiltration, UV Photolysis, and Advanced Oxidation Processes for the Removal of Hormones from Surface Water Sources." *Sep. Purif. Technol.* 95: 89–96.

Schafer, A. I., L. D. Nghiem, and T. D. Waite. 2003. "Removal of the Natural Hormone Estrone from Aqueous Solutions Using Nanofiltration and Reverse Osmosis." *Environ. Sci. Technol.* 37 (1): 182–188.

Wang, J., Z. Wang, Y.Liu, J. Wang, and S. Wang. 2016. "Surface Modification of NF Membrane with Zwitterionic Polymer to Improve Anti-Biofouling Property." *J. Membr. Sci.* 514 (September): 407–417.

Yoon, Y., P.Westerhoff, J. Yoon, and S. A. Snyder. 2004. "Removal of 17β Estradiol and Fluoranthene by Nanofiltration and Ultrafiltration." *J. Environ. Eng.* 130 (12): 1460–1467.

Zhang, R., S.Yu, W. Shi, J.Tian, L. Jin, B. Zhang, L. Li, and Z. Zhang. 2016. "Optimization of a Membrane Cleaning Strategy for Advanced Treatment of Polymer Flooding Produced Water by Nanofiltration." *RSC Adv.* 6 (34): 28844–2853.

# STUDY ON THE INFLUENCE OF ADDITION OF ACIDS ON PARTICLE SIZE AND SURFACE AREA OF TITANIUM DIOXIDE PHOTOCATALYST

Padmini Ellappan[a] and ***Lima Rose Miranda***[a*]
[a] (Carbon Research and Engineering Laboratory, Department of Chemical Engineering, A.C.Tech, Anna University, Chennai, India 600 025. *Email: limamiranda2007@gmail.com)

**ABSTRACT:** An investigation was carried out on the synthesis of titanium dioxide nano powders via the sol gel method, examining the influence of addition of different acids during preparation of the catalyst. The resulting powders were characterized by X-ray diffraction and surface area. The X-ray diffraction revealed that there was formation of a pure crystalline phase, consisting only of anatase for catalyst prepared by the addition of glacial acetic acid. The catalyst prepared using glacial acetic acid yielded a particle size of 7.1 nm and specific surface area of 175.72 $m^2gm^{-1}$. The effect of the particle size on the degradation of Acid Red 88, 3, 4- Dichloroaniline and Nitrobenzene was studied and the studies revealed the importance of particle size in degradation effects.

**Keywords:** solgel; acids; particle size; characterization; acid red 88; nitrobenzene; photocatalysis

## INTRODUCTION

Generally any photocalalytic process considers Titania as the best choice as it fits other desirable criteria, despite the low solar absorbance property. Properties of $TiO_2$ that makes it more advantageous are it is cheap, nontoxic, photolytically and chemically stable, and reusable with a high turnover rate. Modification of the chemical and physical characteristics of $TiO_2$, including absorption range and particle size are rather simple and are considered the most important means of modification of titania's oxidation capabilities.

Different methods have been employed for the synthesis of titanium dioxide, such as chemical precipitation (Pedraza and Vasquez, 1999; Scolan and Sanchez,1998), the sol-gel method (Gartner et al., 2004; Su et al., 2006), hydrothermal (Wu et al., 2002) and solvothermal processes (Kim et al., 2003; Yin et al., 2003), combustion method (Nagaveni et al., 2003), a microemulsion-mediated process (Hong et al., 2003), electrochemical synthesis (Karuppuchamy et al., 2006), fungus-mediated synthesis (Jha et al., 2009) and chemical vapour deposition (CVD) (Jones and Chalker 2003). Venkatachalam et al. (2007) reported the preparation of titanium dioxide nanoparticles via the sol-gel method with different reaction parameters, such as different hydrolyzing agents, molar ratio, aging time and calcinations temperature. They obtained nanocrystals of titanium dioxide with high surface area and different ratios of anatase /rutile.

Sol-gel synthesis has become one of the most promising and important methods employed in nanomaterial production and nanotechnology (Pavasupree et al., 2008). It is a simple route process with moderate conditions such as low temperature and short reaction times producing highly crystalline oxides. This method due to its characteristics mostly used to produce ceramic materials, allowing the control of the particle size, morphology and phase composition (Su et al, 2006; Wang, 2007).

The optimal particle size (and thus surface area) of titania has been studied by many groups over the years and the particle size affects both the amount of pollutants that can adsorb to the surface and the amount of charge carrier recombination that can occur, since recombination is a surface process (Serpone et al 1995, Calza et al 2007, Zhang et al 1998). In general, nanoscale titanium dioxide (1-100 nm) is considered the most active in the degradation of organic compounds, although nano scale catalysts are available and moderately effective (Tahiri et al 1996). The differences in the photocatalytic activity are likely to be related to the differences in the BET surface, impurities, lattice mismatches or density of hydroxyl groups on the catalyst's surface, since these factors could affect the adsorption behavior of a pollutant or intermediate molecule and the life time and recombination rate of electron-hole pairs (Tahiri et al 1996).

The aim of this study was to investigate the effects of addition of different acids during the preparation of the catalyst by solgel method on the $TiO_2$ physical properties such as particle size and surface area.

## MATERIALS AND METHOD

**Chemicals Used.** The chemicals used for the preparation of catalyst were Titanium isopropoxide and glacial Acetic acid supplied by Sigma Aldrich. The other chemicals used were Ethanol, Citric acid, Nitric acid, Sulfuric acid, Ammonia supplied by Central drug house, Mumbai. Water used for preparation of solution was distilled water. . All the chemical reagents were of analytical grade. The model compound chosen for degradation was Acid Red 88, 3, 4- dichloroaniline and Nitrobenzene.

**Preparation of Catalyst.** The following catalysts were prepared (Table 1) using different acids. Titanium dioxide sol was prepared using Titanium isopropoxide as precursor. 8 mL of Titanium isopropoxide was mixed with 42 mL of ethanol and stirred well. To this 100 mL of 50% ethanol was added slowly and continuously stirred for 30 min at a temperature of 80°C for the hydrolysis reaction to take place. To this mixture 10 mL of glacial acetic acid / citric acid/ nitric acid/ sulphuric acid was added.The solution was left for 48 hrs to age. The gel obtained was dried at 110°C for 24 hrs. The dried gel was ground and calcined at 500°C for 4 hours. This catalyst is referred to as $TiO_2$ (sol) [Carp et al 2004].

**Table 1:** Different catalyst prepared

| Acid used | Referred in text as |
|---|---|
| Acetic Acid | Ti-A |
| Citric acid | Ti-C |
| Nitric acid | Ti-N |
| Sulphuric acid | Ti-S |

**Catalyst Characterization**. The crystalline phase was obtained by a Philips Pw 1830 X-ray diffractometer (XRD) operated at 40 kV and 35 mA and a Philips X´ Pert X-ray diffractometer operated at 40 kV and 30 mA, both of which used Cu Kα radiation source at λ of 1.54 Å. XRD data were collected from 2θ of 20-60 degree. Each of the crushed samples was either coated on the inverse side of cello tape and made its surface smooth by pressing with a spatula or spread on a small piece of glass and pressed with another smooth-surfaced glass.

Scherrer's equation was used to calculate the crystallite size as given in equation 1 (Patterson et al., 1939).

$$D = \frac{k\lambda}{\beta \, Cos\, \theta} \tag{1}$$

where d is the crystallite size (nm)
k is the constant whose value is approximately 0.9
λ is the wavelength of the X-ray radiation source (0.154 nm for Cu Kα)
β is the full width at half maximum intensity (FWHM) (radians)
$\theta_B$ is the Bragg angle at the position of the peak maximum

The anatase content was also calculated applying Spurr-Myers equation as given in equation 2 (Spurr et al., 1957).

$$w_A = \frac{1}{1+1.26 I_R/I_A} \tag{2}$$

where $w_A$ is the weight fraction of anatase in the mixture
$I_R$ is the intensity of the diffraction peak of rutile
$I_A$ is the intensity of the diffraction peak of anatase

The specific surface area of the catalysts was determined by nitrogen adsorption at 77 K by BET Quanta chrome instrument. All the samples received a preliminary heat treatment at 100 – 120 °C under $10^-$

[2] Torr vacuum for 4 hours to eliminate the trace amount of water present in the samples. The specific surface area was calculated using the BET method.

The particle size was also calculated using the surface area calculated from BET isotherms determined using BET Quanta chrome instrument as given in equation 3.

$$\text{Particle diameter } D = \frac{6000}{S_{BET} * \rho} \quad (3)$$

where $S_{BET}$ is the surface area ($m^2g^{-1}$) and ρ is the density of the catalyst which is approximately 4.2 $gcm^{-3}$ for Titania based particles.

**Photocatalytic Degradation Studies.** The photocatalytic degradation using the prepared catalyst was studied in an annular photoreactor of capacity of 1000 mL. The visible light of intensity 500 W was chosen for irradiating the solution. The catalyst quantity in all the cases were 0.1 $gL^{-1}$ and irradiation time of 120 min was maintained. Knowing the initial and final concentration of the solution i.e. before and after irradiation, the % degradation (%PCD) was calculated. In the case of AR 88 and NB, initial solution concentration of 50 $mgL^{-1}$ was used while for DCA 10 $mgL^{-1}$ of concentration was used.

## RESULTS AND DISCUSSION

Figure 1 show a series of XRD patterns of the prepared $TiO_2$ samples following the addition of different acids during catalyst preparation. All diffraction peaks show the complete formation of the crystalline anatase phase and were indexed according to the JCPDS card No. 89-4921. The peaks at $2\theta=27°$, 36° and 54.3° indicate the presence of the $TiO_2$ in the rutile phase while peaks at $2\theta=25.5°$, 37.2°, 48°, and 62.6° indicate the presence of the anatase phase. No peaks related to brookite phases were observed in the XRD patterns of Ti-A, Ti-C, Ti-N and Ti-S, whereas rutile phases were observed in all the catalyst prepared except Ti-A. The XRD pattern of the Ti-A was compared with data files and all the sharp peaks observed in the XRD pattern belong to anatase phase of $TiO_2$. The other crystalline forms of $TiO_2$, rutile and brookite, have not been detected. The sol-gel synthesis route using glacial acetic acid presents the advantage of obtaining phase-pure $TiO_2$ nanoparticles.

From the diffractogram, the widths of the bases of the peaks and their intensity characterize the size of the nanocrystals, because small crystals may promote a more intense spreading due to internal reflections that occur in the system. Considering the diffractograms shown in Fig. 1, it can be observed that the samples obtained using acetic acid present ssharp peaks, revealing that the nanocrystals obtained have a smaller size. This fact is evidenced by the calculation of the average crystallite size using the Scherrer equation, as shown in Table 2. The average crystallite size of the nano-$TiO_2$ applying Scherer equation was found to be 7.1 nm.

Comparing the samples synthesized using different acid based on the specific surface area as shown in Table 1, higher specific surface area were observed in Ti-A sample as compared to other catalysts. In addition, in Table 2, it is observed that the average crystallite size calculated using XRD and BET analysis was almost the same. The pore volume and the pore radius of Ti-A was the maximum as compared to other catalyst.

Photocatalysis, using semiconductor particles under sufficient irradiation energy for the simultaneous reduction and oxidation of different redox systems, has been intensively studied for Acid Red 88 (AR88), Nitrobenzene (NB) and 3,4-dichloraniline (DCA) using the prepared nanocrystalline catalyst. Photocatalytic degradation of AR88, DCA and NB in the presence of Visible light are shown in Figures 2, 3 and 4 respectively.

It could be seen for all the three model compounds chosen, the degradation was higher with Ti-A as compared to other catalyst. Ti-A performed better due to small particle size and larger specific area. More active sites are available on the surface of Ti-A hence there is better electron formation leading to better degradation. The differences in the photocatalytic activity of catalysts are likely to be due to differences in the BET-surface, impurities, lattice mismatches or density of hydroxyl groups present on the catalysts surface. Since they will affect the adsorption behaviour of a pollutant or intermediate molecule and the lifetime and recombination rate of electron-hole pairs, there is a difference in the performance of the catalysts prepared.

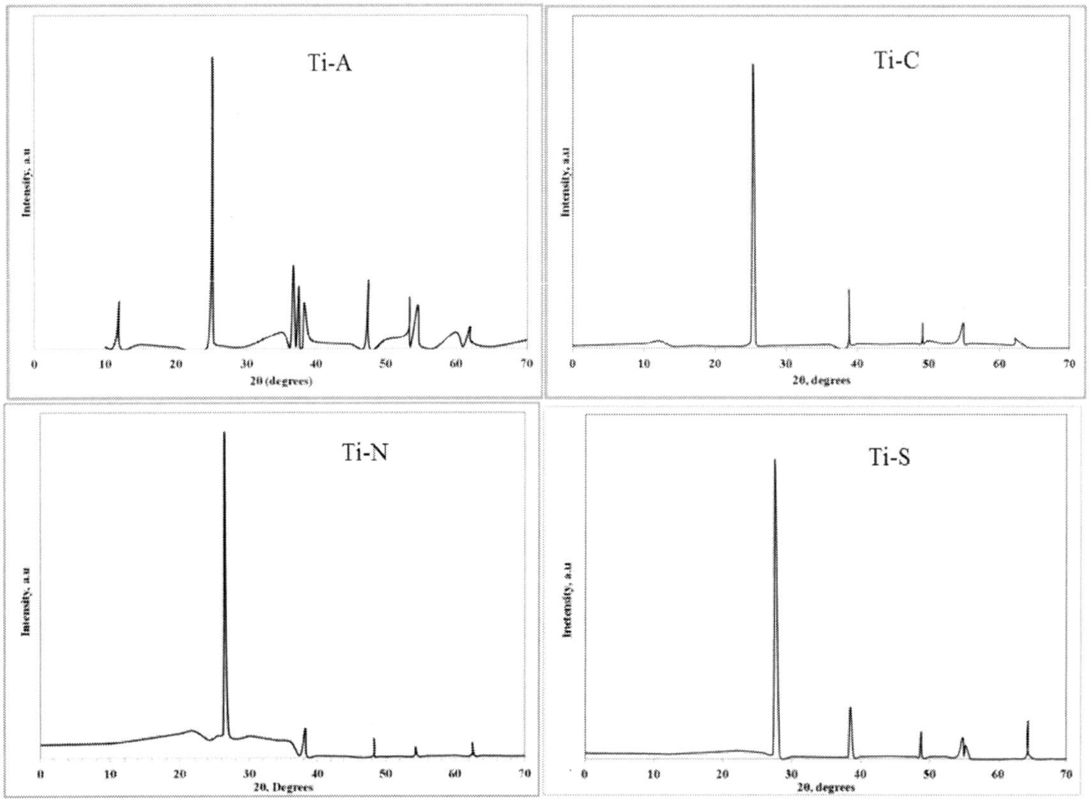


**Fig. 1. XRD analysis of the prepared catalyst**

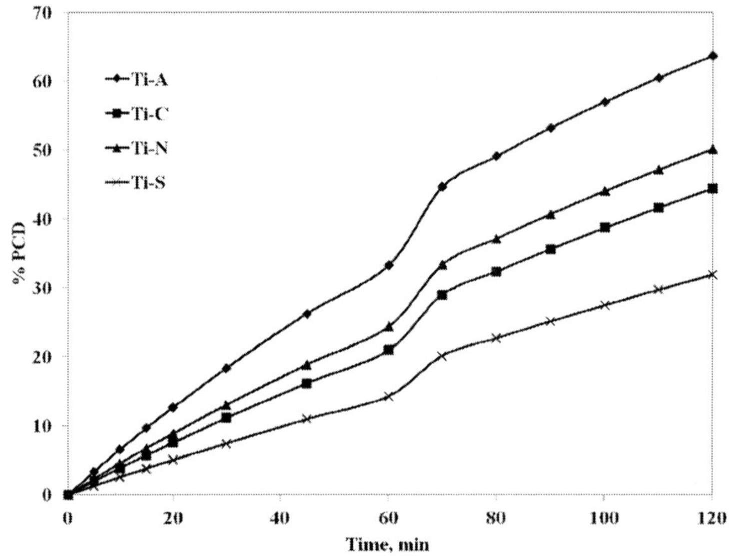


**Fig. 2.** **Photocatalytic activity of different nanocrystalline catalyst on AR88 degradation in the presence of Visible light (AR 88 conc: 50 $mgL^{-1}$, Catalyst dosage: 0.1 $gL^{-1}$, Visible light irradiation: 120min, temp: 30°C)**

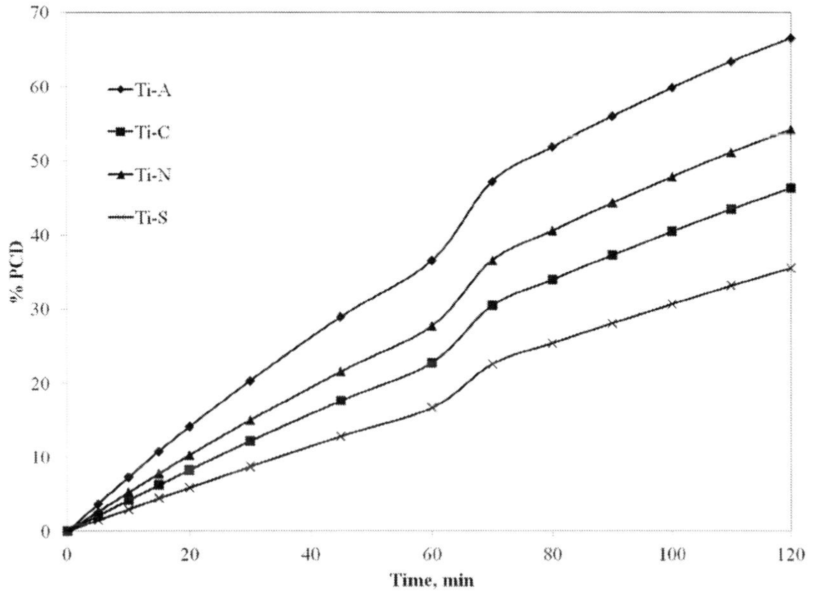


**Fig. 3.** **Photocatalytic activity of different nanocrystalline catalyst on DCA degradation in the presence of Visible light (DCA conc: 10 $mgL^{-1}$, Catalyst dosage: 0.1 $gL^{-1}$, Visible light irradiation: 120min, temp: 30°C)**

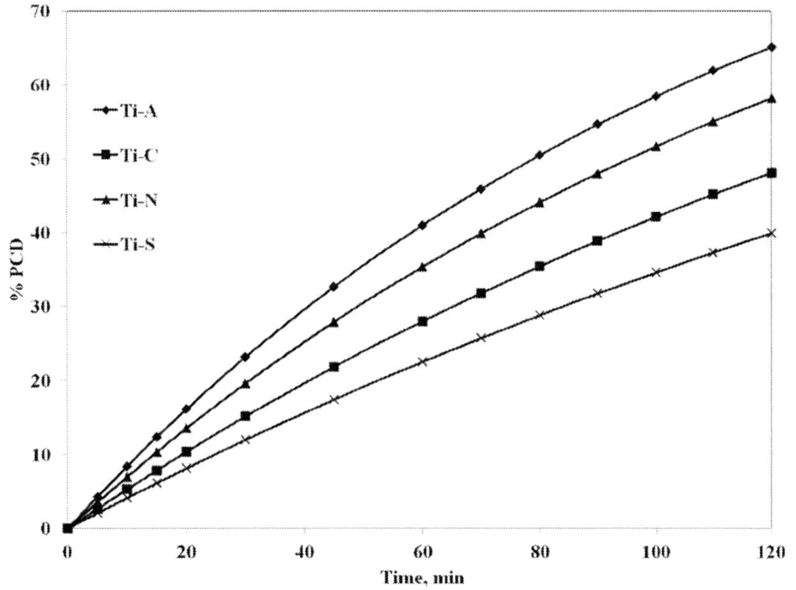


**Fig 4:** **Photocatalytic activity of different nanocrystalline catalyst on NB degradation in the presence of Visible light (NB conc: 50 $mgL^{-1}$, Catalyst dosage: 0.1 $gL^{-1}$, Visible light irradiation: 120min, temp: 30°C)**

## CONCLUSION

The photocatalytic activity of semiconductor catalyst is dependent on surface and structural properties of semiconductor such as crystal composition, surface area, particle size distribution, porosity, band gap and surface hydroxyl density. Average crystal size is of primary importance in heterogeneous catalysis, because it is directly related to the efficiency of a catalyst through the definition of its specific surface area. Hence in this research an effort to prepare smaller particle size and high specific surface area was achieved by the addition of glacial acetic acid. PCD studies also proved that the particle size influences greatly the degradation of the compounds.

**Table 2. Properties of the catalyst**

| Properties / Catalyst | Average crystallite size, nm | Particle size, nm (BET method) | Pore volume, $m^2gm^{-1}$ | Pore radius Å | Specific surface area $m^2gm^{-1}$ | Anatase % |
|---|---|---|---|---|---|---|
| Ti-A | 7.10 | 8.13 | 0.15 | 2.25 | 175.72 | 100.00 |
| Ti-C | 13.17 | 14.25 | 0.09 | 1.26 | 100.25 | 88.19 |
| Ti-N | 11.13 | 12.54 | 0.10 | 1.49 | 113.92 | 92.34 |
| Ti-S | 20.62 | 21.43 | 0.07 | 0.99 | 66.66 | 23.15 |

## REFERENCES


Calza P, Pelizzetti E, Mogyorósi K, Kun R and Dékány I (2007). Size dependent photocatalytic activity of hydrothermally crystallized titania nanoparticles on poorly adsorbing phenol in absence and presence of fluoride ion. *Appl. Catal. B* 72: 314-321. doi:10.1016/j.apcatb.2006.10.019.

Gartner M, Scurtu R, Ghita A, Zaharescu M, Modreanu M, Kokkoris M, Kordas G and Trapalis C. (2004) Spectroellipsometric characterization of sol-gel $TiO_2$- CuO thin coatings. *Thin Solid Films* 455: 417-421 (2004). doi: 10.1016/j.tsf.2004.01.030

Pavasupree S, Jitputti J, Ngamsinlapasathian S. and Yoshikawa S (2008). Hydrothermal synthesis, characterization, photocatalytic activity and dyesensitized solar cell performance of mesoporous anatase $TiO_2$ nanopowders. *Mater. Res. Bull* 43: 149-157. doi: 10.1016/j.materresbull.2007.02.028

Pedraza F and Vasquez A (1999). Obtention of $TiO_2$ rutile at room temperature through direct oxidation of TiCl3. *J. Phys. Chem. Solids*, 60, 445-448 (1999). doi:10.1016/S0022-3697(98)00315-1

Scolan E and Sanchez C (1998) Synthesis and characterization of surface-protected nanocrystalline titania particles. *Chem. Mater* 10: 3217-3223. doi: 10.1021/cm980322q

Serpone N, Lawless D, Khairutdinov R and Pelizzetti E (1995) Subnanosecond Relaxation Dynamics in $TiO_2$ Colloidal Sols (Particle Sizes Rp = 1.0-13.4 nm). Relevance to Heterogeneous Photocatalysis. *J. Phys. Chem.*, 99: 16655-16661. doi: 10.1021/j100045a027.

Su C, Lin KF and Lin YH (2006) Preparation and characterization of high-surface-area titanium dioxide by sol-gel process. *J. Porous Mater* 13: 251-258. doi: 10.1007/s10934-006-8012-7.

Tahiri H, Serpone N and Le van Mao R (1996). Application of concept of relative efficiencies and surface characterization of a new titania photocatalyst designed for environmental remediation. *J. Photochem. Photobiol. A*, 93: 199-203. doi:10.1016/1010-6030(95)04195-8.

Wang G (2007) Hydrothermal synthesis and photocatalytic activity of nanocrystalline $TiO_2$ powders in ethanol–water mixed solutions. *J. Mol. Catal. A: Chem.* 274: 185 -191. doi:10.1016/j.molcata.2007.05.009.

Zhang Z, Wang CC, Zakaria R and Ying JY (1998). Role of Particle Size in Nanocrystalline $TiO_2$-Based Photocatalysts. *J. Phys. Chem. B* 102:10871-10878. doi: 10.1021/jp982948+

# INTERACTION OF ENGINEERED MATERIALS WITH MICROBIAL BIOFILMS AND ITS POTENTIAL APPLICATIONS

***Hengye Jing***, and George A. Sorial (University of Cincinnati, Cincinnati, OH, USA)
Endalkachew Sahle-Demessie* (US Environmental Protection Agency, Cincinnati, OH, USA)

Biofilms are surface-associated and highly-stratified microbial communities, existing in water distriibution systems. Biofilms are mainly composed of extracellular polymeric substance (EPS), which act as a protective layer for bacteria. Pathogenic EPSs are critical in host-microbe interactions, where they aid in adherence and colonization within a host and function in immunomodulation or in water distribution systems. Different strategies have been applied to control waterborne pathogens. For instance, surface modification, or using engineered nanoparticles as anti-bacterial agents to manage their growth associated with biofilm in the water distribution system. Carbon nanotubes (CNTs) have been proven to have effective anti-bacterial and anti-biofilm properties and been widely used for making polymeric nanocomposites. In this study, the potential role of polyethylene (PE) nanocomposite containing 0%, 2 wt% and 4wt% of multi-walled carbon nanotubes (MWNTs) on inhibiting biofilm growth was studied. Two types of bacteria, *Pseudomonas fluorescens* and *Mycobacterium smegmatis* biofilms were investigated. Laser scanning microscopy (LSM) and FilmTracer™ LIVE/DEAD biofilm viability kit were used to trace the biofilm growth between the different surfaces on a time course. PE nanocomposites containing 2 wt% and 4 wt% MWNTs were observed to have less biofilm growth during first four weeks for both two types of biofilms. The nanocomposite containing 4 wt% MWNTs showed a stronger anti-bacterial effect during the first few weeks than the one with 2 wt% MWNTs. Further studies are conducted to explore this reason. Scanning electron microscope (SEM) provided a micro view of surface property of the PE/MWNTs nanocomposite such as surface roughness while atomic force microscopy (AFM) provided topographical mapping of the surface. Contact angle goniometer was also used to determine the surface energy of the different polymeric nanocomposites. SurPASS electrokinetic analyzer determined the streaming surface zeta potential of both PE/MWNTs nanocomposite and biofilms.

# UNDERSTANDING THE FORMATION OF NATURALLY OCCURRING SILVER NANOPARTICLES IN AQUATIC ENVIRONMENT

***Nathaniel F. Adegboyega*-**, William C. Hockaday (Baylor University, Waco, TX, USA )
Virender K. Sharma (Texas A&M, College Station, TX, USA)

Silver nanoparticles (AgNPs) as an emerging contaminant in the environment is of concern to society due to the potential for long-range transport, and the potential for ecological and human toxicity. Aquatic humic substances (HS) can influence AgNP transport, reactivity, removal, dissolution and formation. Therefore, understanding of the nature of HS-AgNP interactions would be essential to environmental behavior and identifying best-practices for AgNP removal from municipal waters. However, knowledge on the detail interactions between AgNPs and silver ion ($Ag^+$) and aquatic humic substances (HS) at the molecular level is scant due to the limited number of analytical tools capable of detecting formation or transformation of AgNPs in the presence of HS.

We applied several analytical techniques including UV-visible spectroscopy, Fourier Transform-Ion Cyclotron Resonance mass spectrometry (FT-ICR MS) and several characterization techniques to investigate the formation of silver nanoparticles in the environment. Our results show that the spontaneous formation of silver nanoparticle in the environment is directly linked to the reduction of silver ions by naturally occurring humic substances. The formation of these silver nanoparticle is particularly controlled by environmental factors such as pH, temperature and light. The presence of $Fe^{2+/3+}$ redox systems were found to influence the amount of silver nanoparticle formation.

An electrospray ionization technique equipped with a Fourier Transform-Ion Cyclotron Resonance mass spectrometry (FT-ICR MS) was also used as an analytical tool to monitor the direct interactions between humic substance and $Ag^+$. Our preliminary characterization of $Ag^+$-HS mixtures suggest a high preference of $Ag^+$ for low molecular species of humic substances.

# OPTIMIZED SYNTHESIS OF POLYMER-BASED GRAPHENE OXIDE NANOCOMPOSITES FOR HEAVY METAL ADSORPTION USING RESPONSE SURFACE METHODOLOGY

***Jem Valerie D. Perez*** and Maria Lourdes P. Dalida
(University of the Philippines, Quezon City, Philippines)
Debora F. Rodrigues
(University of Houston, Houston, TX, USA)

Nanocomposites containing graphene oxide (GO), polyethyleneimine (PEI), and chitosan (CS) were synthesized for chromium (VI) and copper (II) removal from water. Response surface methodology (RSM) was used for the optimization design of the synthesis of the CS-PEI-GO beads formed with different concentrations of PEI (1.0 – 2.0%), GO (500 – 1500 ppm), and glutaraldehyde (GLA) (0.5 – 2.5%) as cross-linking agent. Batch adsorption experiments were performed to obtain responses in terms of percentage removal of both Cr (VI) and Cu (II) ions. A second-order polynomial equation was used to model the relationship between the synthesis conditions and the adsorption responses. Analysis of Variance (ANOVA) showed that GLA concentration has opposite effects on Cr (VI) and Cu (II) adsorption due to different uptake mechanisms by the CS-PEI-GO beads for the two heavy metals. High $R^2$ values of 0.9848 and 0.8327 for Cr (VI) and Cu (II) removal were obtained from the regression analyses, suggesting good correlation between observed and predicted values. Removal of up to 91.10% Cr (VI) and 78.18% Cu (II) were achieved by the beads composed of 2.0% PEI, 1500 ppm GO, and cross-linked with 2.08% GLA. Finally, the structure and surface properties of the CS-PEI-GO beads were characterized using X-ray diffraction (XRD), porosity and BET surface area analysis, Scanning electron microscopy (SEM), Fourier transform infrared spectroscopy (FTIR), and X-ray photoelectron spectroscopy (XPS). Overall, the synthesized CS-PEI-GO beads were proven to be effective in removing both cationic and anionic heavy metal pollutants.

# ORGANIC RADICAL INTERMEDIATE-SPARKED HIGHLY FENTON-CATALYTIC EFFICIENCY OF DANDELION-LIKE TiCuAl-SiO$_2$ NANOSPHERES FROM ORGANIC CONTAMINANT DEGRADATION

***Lai Lyu***, Lili Zhang and Chun Hu
(Chinese Academy of Sciences, Beijing, China)

**INTRODUCTION.** Triggered by growing water pollution issues, research aiming at a more efficient water purification technology has been thriving. Fenton reaction as a typical advanced oxidation process is especially powerful, since it can rapidly and nonselectively degrade various organic pollutants via $^{\bullet}OH$. However, in the conventional Fenton process, the reduction of the oxidation state ($M^{(n+m)+}$) to the reduction state ($M^{n+}$) of the metal by oxidizing $H_2O_2$ is the rate-limiting step, and $H_2O_2$ is decomposed with the formation of $O_2$ as a reaction product in addition to superoxide radical ($O_2^{\bullet-}$), which leads to excessive consumption of $H_2O_2$. Due to these deficiencies, most heterogeneous Fenton catalysts exhibited low activity and utilization efficiency of $H_2O_2$. Recently, we prepared a fibrous Fenton catalyst consisting of dandelion-like titanium-copper-aluminium trimetal-doped silica nanospheres (*d*-TiCuAl-SiO$_2$ Ns) for the first time. We found that a novel electron supply pathway from the generated organic radical intermediates ($^{\bullet}R$) sparked by Ti-Cu system could address the problems for activity and the utilization efficiency of $H_2O_2$. In this system, *d*-TiCuAl-SiO$_2$ Ns exhibited strong Fenton-catalytic activity and supreme utilization efficiency of $H_2O_2$ in the degradation of persistent organic compounds under neutral pH conditions.

**OBJECTIVES.** The main objective of this study was to verify the generation of the organic radical intermediates and clarify the inducement mechanism of the high utilization efficiency of $H_2O_2$.

**METHODS.** The dandelion-like TiCuAl-SiO$_2$ nanospheres (*d*-TiCuAl-SiO$_2$ Ns) was synthesized via an *in situ* hydrothermal process. The characterization of the catalysts were studied via SEM, TEM, XPS, UV-Vis DRS, EPR and cyclic voltammetry. The analysis methods included HPLC, TOC, ICP-OES and ESR. The reaction process was analyzed by ATR-FTIR and *in situ* Raman.

**RESULTS AND CONCLUSIONS.** The XPS revealed that both $Cu^+$ and $Cu^{2+}$ existed on the surface of *d*-TiCuAl-SiO$_2$ Ns and $Ti^{4+}$, $Ti^{3+}$ and $Ti^{2+}$ coexisted on the surface of *d*-TiCuAl-SiO$_2$ Ns with the percentage of 80.7%, 7.4% and 11.9%, respectively. The cyclic voltammetry data suggested that the introduction of Ti into the framework of *d*-TiCuAl-SiO$_2$ Ns reduced the reduction potential of $Cu^{2+}$ and accelerated the electron transfer rate of $Cu^+/Cu^{2+}$. Therefore, the electrons of the produced $^{\bullet}R$ from the contaminant degradation could be successfully transferred to $Cu^{2+}$ due to the lower reduction potential of $Cu^{2+}$ in the Cu-Ti system, which eventually led to the rapid reduction of $Cu^{2+}$ to $Cu^+$ on the surface of the catalyst. In addition, the ATR-FTIR and *in situ* Raman results suggested that Ti greatly promoted the rapid contact between $H_2O_2$ and Cu sites on the surface of *d*-TiCuAl-SiO$_2$ Ns, which was not affected by the pollutants. In this case, *d*-TiCuAl-SiO$_2$ Ns exhibited an exceptional performance in contaminant degradation. The BPA (25 ppm) could be completely degraded within 60 min and around 70.5% of TOC was degraded within 180 min in the suspension of *d*-TiCuAl-SiO$_2$ Ns, which is much higher than the 10-30% of TOC removal for BPA degradation in the reported Fenton processes under similar conditions, indicating the excellent activity of *d*-TiCuAl-SiO$_2$ Ns. After the reaction, no Ti and Al were detected in the solutions and the concentration of total dissolved Cu was only 0.136 mg $L^{-1}$, which was much below the limitation of Cu in EU directives

(<2 mg $L^{-1}$) and USA regulations (<1.3 mg $L^{-1}$). The activity of *d*-TiCuAl-$SiO_2$ Ns did not decrease after eight successive cycles of degradation testing, which suggested that *d*-TiCuAl-$SiO_2$ Ns was a stable catalyst for heterogeneous Fenton reactions. In addition, the utilization efficiency of $H_2O_2$ was maintained at more than 80% prior to the disappearance of BPA and this value did not decrease sharply after that due to the effect of the generated $^{\bullet}R$, indicating the highly catalytic efficiency of *d*-TiCuAl-$SiO_2$ Ns.

# ELECTROSPUN RECTORITE/$TiO_2$/POLYMER NANOFIBROUS MATS FOR ADSORPTION OF HEAVY METALS

***Yingfei Zhan*** and ***Hongbing Deng***
(Wuhan University, Wuhan, China)

$TiO_2$ and rectorite (REC) doped cellulose acetate (CA) based nanofibrous mats were fabricated via electrospinning method, a young nanotechnology. These organic-inorganic composite had enhanced specific surface area, mechanical strength and thermal stability. REC/$TiO_2$/CA composite nanofibrous mats had good three-dimensional structure just like pure polymer nanofibrous mats, and their fiber diameter range from 200 to 835 nm. The tensile strength of nanofibrous mats increased distinctly with the addition of $TiO_2$, which changed from 2.09±0.47 MPa to 4.93±0.72 and 6.75±1.09 MPa. The TGA and DCS curves showed that the thermal stability of nanofibrous mats was enhanced with inorganic doping. And interestingly, the surface areas of $TiO_2$/CA and REC/$TiO_2$/CA (containing 1% wt REC) composite nanofibrous mats were 1.011 and 3.509 $m^2/g$, respectively. Because REC is a kind of layered silicate with super large surface area and interlayer distance. $TiO_2$, the adsorption agent, could remarkably improve the adsorption effect of the nanofibrous mats. And the more REC were added, the higher copper adsorption capacities were obtained. With favorable tensile strength and thermal stability, the REC/$TiO_2$/CA composite nanofibrous mat can serve as a strong thin film for copper adsorption and obstruction in water.

# MULTIFUNCTIONAL GRAPHENE-BASED NANOCOMPOSITE MODIFIED MEMBRANE FILTERS FOR HEAVY METALS AND BACTERIA REMOVAL FROM WATER

**Yvonne Ligaya F. Musico** and Maria Lourdes P. Dalida
(University of the Philippines, Diliman, Quezon City, Philippines)
Debora F. Rodrigues and Catherine M. Santos
(University of Houston, Houston, TX, USA)

Commercially available membrane filters were modified with different polymers, nanomaterials and nanocomposites, namely, poly(*N*-vinylcarbazole (PVK), graphene (G), poly(*N*-vinylcarbazole)-graphene (PVK-G), graphene oxide (GO), and poly(*N*-vinylcarbazole)-graphene oxide (PVK-GO). The presence of graphene-based nanomaterials on the filter surfaces significantly improved the antibacterial property and heavy metal removal capacity of commercial membrane filters. The desorption process of lead (II) from the modified membrane filters using hydrochloric acid indicates that the filters can be reused. The results show that PVK-GO modified membrane filters consistently exhibited superior antimicrobial properties, metal removal capacity, desorption capacity and reusability among all the graphene-based membrane filters investigated. This research demonstrates that PVK-GO modified membrane filters, along with other graphene-based nanocomposite membrane filters, have efficient antibacterial property and high $Pb^{2+}$removal capacity that can be ideal for water and wastewater treatment applications.

# STUDY ON THE PHOTOCATALYTIC DEGRADATION OF MICROCYSTINS BY $TiO_2$ IMMOBILIZED ON FIBERGLASS CLOTH

Deqiang Chen*, Yiqun Chen
(Key Laboratory of Integrated Regulation and Resource Development on Shallow Lakes, Ministry of Education, College of Environment, Hohai University, Nanjing, 210098, China)

The $TiO_2$ immobilized on fiberglass cloth(FGC) was prepared through sol-gel method followed by daubing and coating process to improve the photocatalytic activity and overcome the difficulty of reuse of catalyst and the photodegradation of microcystins-LR(MC-LR) by the prepared catalyst was investigated. The morphology and microstructure of $TiO_2$ loaded on FGC were characterized via XRD and SEM, respectively. The result revealed that the crystalline structure of immobilized $TiO_2$ was nearly unchanged compared with commercial titanium dioxide (P25), and the average size of $TiO_2$ particles was about 15 nm.The process of sol-gel method followed by twice daubing and coating exhibited the relatively high $TiO_2$ loading amount. The $TiO_2$/FGC system can effectively degrade the MC-LR in water under UV light and be effected by pH and initial concentration of MC-LR. The degradation rates of MC-LR reached to 91% and 57% at pH of 4 and 9, respectively. As the initial concentration increased from 50μg/L to 300μg/L, the MC-LR removal decreased from 87% to 60%. Addition of $Fe^{3+}$ can accelerate the decomposition of MC-LR by $TiO_2$/FGC system.

# ELECTROSPUN CARBON NANOFIBERS WITH ZERO VALENT IRON NANOPARTICLES (ZVINPs@ECNFs) FOR HEAVY METALS REMEDIATION IN GROUND AND WASTEWATER

Nikhil. R. Mucha, Lifeng Zhang
(Joint School of Nanoscience and Nanoengineering, North Carolina A&T State University, Greensboro, NC 27401, USA)
***Ramesh Ravella***, Muchha, R. Reddy
(North Carolina A&T State University, Greensboro, NC 27401, USA)

Rapid worldwide industrialization and population growth is leading to extensive environmental pollution. The discharge of wastewater from commercial and industrial wastes, untreated domestic sewage, and chemical contaminants into surface waters leads to a catastrophe. Industrial activities have led to elevated concentrations of a wide range of heavy metal ions in soil and groundwater. Heavy metal ions such as Chromium Cr (VI), Copper and Nickel are highly toxic. Various methods have been attempted to remove heavy metals from water including filtration, chemical precipitation, electrodeposition etc., but these methods suffer from limitations such as disposal of metal residual sludge, membrane clogging, intensive energy consumption, and high cost. Zero Valent Iron nanoparticles (ZVINPs) possess large capacity for remediating Cu (II) in water owing to their large surface area; high reactivity, non-toxicity, and ease of production. In this study, ZVI nanoparticles (ZVINPs) were synthesized using reduction method followed by deposition on electrospun carbon nanofibers (CNFs) and reached a hierarchical nanostructure, i.e. electrospun carbon nanofibers with surface attached ZVI nanoparticles (ZVINPs@CNFs). Zero Valent Iron (ZVI) nanoparticles possess high capacity for remediating heavy metal ions in water owing to their large surface area, high reactivity, non-toxicity, and ease of production. In this study, ZVI nanoparticles (ZVINPs) were synthesized using reduction method followed by deposition on electrospun carbon nanofibers (CNFs) and reached a hierarchical nanostructure, i.e. electrospun carbon nanofibers with surface attached ZVI nanoparticles (ZVINPs@CNFs). This novel nanomaterial has been evaluated for heavy metal removal from a series of solutions with varying concentrations, ZVI loading and pH of the solution. ZVINPs@CNFs outperformed pure ZVINPs in all cases. 0.04 g ZVINPs@CNFs could remove 94.2% Cu (II) ions from 50 ppm Cu (II) solution at pH 5.5 and 0.04 g of standalone ZVI could remove 60% of Cu(II) from 50 ppm solution at pH 5.5. It is envisioned that ZVINPs@CNFs is going to serve as a novel ZVI based nanomaterial for efficient heavy metal remediation in contaminated groundwater as well as in waste water treatment.

# AIR POLLUTION

# AND

# AIR QUALITY CONTROL

# $PM_{2.5}$ CHEMICAL CONSTITUENTS AT A RURAL SITE OF AGRA IN INDO-GANGETIC PLAIN: SOURCES AND TRANSPORT

***Aparna Satsangi***, Nidhi Verma, Anita Lakhani and K. Maharaj Kumari*
*Email: maharajkumari.k@rediffmail.com
(Dayalbagh Educational Institute, Agra, UP, India)

**ABSTRACT:** Fine particulate matter ($PM_{2.5}$ aerosols) has been observed to adversely affect human health, visibility and climate change. $PM_{2.5}$ aerosol samples were collected from January to December 2015 during summer and winter seasons at Sikandarpur, Agra. The samples were analysed for carbonaceous aerosols (Organic and Elemental Carbon- OC and EC) and water soluble ions ($F^-$, $Cl^-$, $NO_3^-$, $SO_4^{2-}$, $Na^+$, $NH_4^+$, $K^+$, $Mg^{2+}$ and $Ca^{2+}$). $PM_{2.5}$ mass concentrations showed higher concentration during winter (109.6±17.8 µg $m^{-3}$) as compared to summer (73.6±20.2 µg $m^{-3}$). Carbonaceous aerosols (Organic Carbon: OC and Elemental Carbon: EC) and most of the water soluble ionic constituents ($NO_3^-$, $SO_4^{2-}$, $NH_4^+$ and $K^+$) also showed enhanced concentrations during winter season while $Ca^+$ high concentrations in summer season. The variation in summer and winter concentrations is due to the difference in meteorological parameters and different sources of emissions in both the seasons. This is further supported by air mass back trajectory analysis. In summer, long range transport from north western region was prominent while in winter season localized contributions dominated which could be due to enhanced biomass burning and other anthropogenic activities that occur in winter season to combat cold.

## INTRODUCTION

Atmospheric aerosols mainly consist of particles derived from natural sources and have a pronounced impact on global climate system (Poschl, 2005). The enhancement of atmospheric particles in present day due to anthropogenic activities substantially alters the aerosol burden of the atmosphere (Andreae, 2001). Such aerosols are laden with fine particles especially those which have aerodynamic diameter less than 2.5. This fine particulate matter is known to adversely affect human health, visibility and climate change. The direct and indirect effects of aerosol may change the radiation process (Liu et al., 2016). In order to limit the adverse impacts and to develop efficient strategies for air quality control, chemical composition of $PM_{2.5}$ needs to be studied. The study of variation of atmospheric composition with transport pathways may also help us to understand the influence of long range transport of pollutants on the atmospheric environment.

Very few studies have been carried out in the Indo-Gangetic plain (IGP) to characterize $PM_{2.5}$ aerosol which covers a vast area of Indian subcontinent and experiences extreme variability in the climate over the annual seasonal cycle (Ram et al., 2010b; Pachauri et al., 2013; Rastogi et al., 2014; Srinivas et al., 2016). It is one of the most populated and polluted regions of India due to large scale urbanization, changes in land use, industrial activities, and regional emission sources. These activities contribute a significant fraction to high aerosol loading over the entire region. As Agra lies in the IGP, it is also influenced by its variable climatic conditions.

## SAMPLING SITE

Agra lies in the Indo-Gangetic plain. It is in a semi-arid zone adjacent to the Thar Desert. It falls in the subtropical climatic belt and is about 1500 km away from the sea. Its borders touch Rajasthan to its west and south, the district of Firozabad to its East and the district of Mathura to its North. The Mathura oil refinery is about 40 km to the north and is in the upwind direction as winds are mostly from west and northwest sector. Industrial activities of Agra include rubber processing, engineering and a few ferrous casting industries based on natural gas. It shows extreme variation in temperature with hot summers and chilly winters. May and June are the hottest months having maximum temperature over 40 °C. Dust storms are most frequent in May and can also occur in April and June which last for a period of 30 minutes to 2 hours. The frequency of dust storms in northern and north western India is high with wind speed ranging from 8 – 12 m $s^{-1}$. The increased wind speed causes uplifting and transport of desert dust. These dust storms cause reduced visibility (800 m – 1 km) and severe air pollution. In winter, temperatures may fall as low as 0°C. The year's coldest months are December and January, when temperature averages around 10–15°C. The season frequently experiences heavy fog which disrupts daily life and hinders visibility.

The site under study is a rural area located in the north eastern region of Agra district in Uttar Pradesh, India. This village, Sikandarpur Mustkil (longitude and latitude: 27.26° N and 78.00° E) is about 15 km away from Agra city and does not have any industrial establishment within a diameter of about 10 kms and the major roads are more than 2 kms away from the site. It is surrounded by agricultural fields where mainly rice, wheat and seasonal vegetables are cultivated. Most of the people in the village cook food on *chulhas* using cowdung cakes and wood. The location of sampling site is shown in Figure 1.

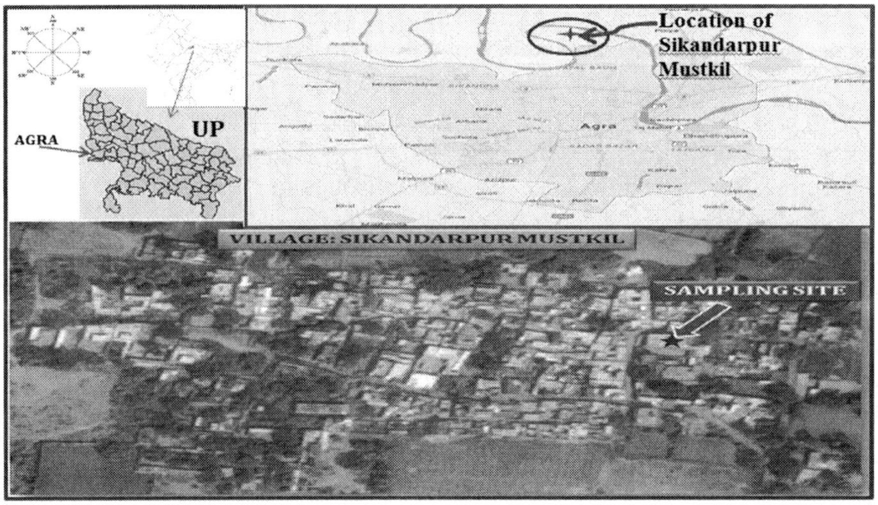


**FIGURE 1. Map showing Sampling site**

## SAMPLING

In 2015, $PM_{2.5}$ aerosol samples were collected during summer (March to June) and winter (January and February) seasons on the roof of a house in Sikandarpur at about 10 m above the ground level. For sample collection, quartz fiber filters (QFF, 47 mm, Pallflex) were weighed and placed in Fine Particulate Sampler (Model: APM 550EL; Make: M/s Envirotech Instruments Private Limited) for 24 hours at a flow rate of 1.0 $m^3$ $hr^{-1}$. The flow meter of the sampler was calibrated with Air Flow Calibrator with an accuracy of 2%. QFFs were pre-annealed for 3 hours at 900°C in a muffle furnace. These filters were wrapped in aluminium foil within sealed polythene plastic bags before and after sampling. The exposed filters were stored in a refrigerator at ~4°C until extraction for chemical analysis to avoid contamination.

## SAMPLE ANALYSIS

**Gravimetric Analysis of $PM_{2.5}$ Mass Concentrations.** $PM_{2.5}$ mass concentration was determined by the gravimetric analysis of exposed filters using electronic microbalance. Each filter was stabilized at constant temperature (25±1°C) and relative humidity (50±1%) for 48 hours before and after sampling, and the net particle loading was obtained as $PM_{2.5}$ mass in μg $m^{-3}$ by the subtraction of pre-sampling weights from post-sampling weights. The typical uncertainty of gravimetric analysis is ±20 μg which represents ±5% of the total aerosol mass of the field samples.

**Carbonaceous Aerosol Analysis.** The OC and EC concentrations were determined using OCEC Laboratory instrument (Model 2000), Sunset Laboratory, Forest Grove, USA using NIOSH-5040 protocol based on thermal-optical transmittance (TOT). A punch of 1.5 $cm^2$ was placed in the sample boat of OCEC Analyzer. Detailed description of the method employed has been given in Satsangi et al. (2012).

**Chemical Analysis of Water Soluble Inorganic Ions (WSIIs).** Atmospheric aerosols were quantified for their water soluble inorganic ions (WSIIs). One-half of each filter paper was cut and extracted in 10 ml deionised water by ultra-sonication for 45 minutes. The extract was then divided into two parts: one part was treated with chloroform to prevent decomposition and was used for the analysis of anions while another part was preserved with 1% nitric acid (v/v) and used for the cation analysis. The concentration of major ions ($F^-$, $Cl^-$, $NO_3^-$, $SO_4^{2-}$, $Na^+$, $NH_4^+$, $K^+$, $Mg^{2+}$ and $Ca^{2+}$) was quantified by ion chromatography using Dionex ICS 1100 Ion Chromatograph system. The extracts were filtered with 0.45 μm PTFE syringe filter (Pall Co. Ltd., USA). Anions were analysed using 6mM NaOH as eluent on the system equipped with guard column (AG11), analytical column (IonPac AS11), and anion self-regenerating suppressor (ASRS 300, 4 mm) while cations were analysed on the system equipped with guard column (CG12A), analytical column (IonPac CS12A), and cation self-regenerating suppressor (CSRS 300, 4 mm) using 20 mM methane sulphonic acid (MSA) as eluent. The recoveries of ions were in the range of 80-90%.

All the reported ion concentrations were corrected for field blanks. Field blanks were the filters brought to the field and installed in the samplers but no air was pumped through the filters. Analytical uncertainties were estimated by calibrating the Dionex ICS 1100 Ion chromatograph with a fresh working standard solution of 1 ppm, prepared from a 1000 ppm stock solution by serial dilution. The variation in peak area was found to be less than 5%. The detection limit for every ion was calculated as three times the standard deviation of the blank. Detection limits (ng $m^{-3}$) of the ionic species were as follows: $F^-$: 5.0, $Cl^-$: 17.2, $NO_3^-$: 9.5, $SO_4^{2-}$: 12.8, $Na^+$: 2.5, $NH_4^+$: 8.7, $K^+$: 16.8, $Mg^{2+}$: 3.2 and $Ca^{2+}$: 7.2.

## TRANSPORT BY AIR MASS TRAJECTORIES

The sources and transport pathways of the chemical components in the atmospheric aerosols may be identified by the simulation of air mass backward trajectory analysis. These air mass back-trajectories were simulated using the National Oceanic and Atmospheric Administration (NOAA) Hybrid Single-Particle Lagrangian Integrated Trajectory (HYSPLIT) model which is based on the GDAS Meteorological Data. Three day back trajectories were calculated at 10:00 hours at a height of 1000 m above the ground level and have been represented in Figure 2. The average mixing height was also calculated every 6 hour for each sampling day by using HYSPLIT model (Draxler and Rolph, 2003).

## METEOROLOGICAL PARAMETERS

Surface meteorological parameters like temperature, relative humidity, wind direction and wind speed were recorded using an automatic weather monitoring system, Envirotech Wind Monitor WM271. The instrument was programmed to collect meteorological data at an interval of 1 minute which was downloaded and computed through software (Metlog, version 1). Average temperature varied from 8.3 to 39.5 °C during the sampling period from January to December 2015. Relative humidity ranged from 16.0 to 94.2% while the wind speed varied from 0.6 to 2.8 m $sec^{-1}$ with an average of 1.7 m/sec. The flow of wind direction was mostly from north or northwest and at other times it was observed to originate from southwest and east directions.

## RESULTS

**$PM_{2.5}$ Mass Concentrations.** $PM_{2.5}$ concentrations varied from 31.5 to 131.5 μg $m^{-3}$ with an average of 94.5 μg $m^{-3}$. Almost 90% of the concentrations were above NAAQS standard value of 60 μg $m^{-3}$. $PM_{2.5}$ mass showed higher concentration during winter (109.6±17.8 μg $m^{-3}$) as compared to summer (73.6±20.2 μg $m^{-3}$).

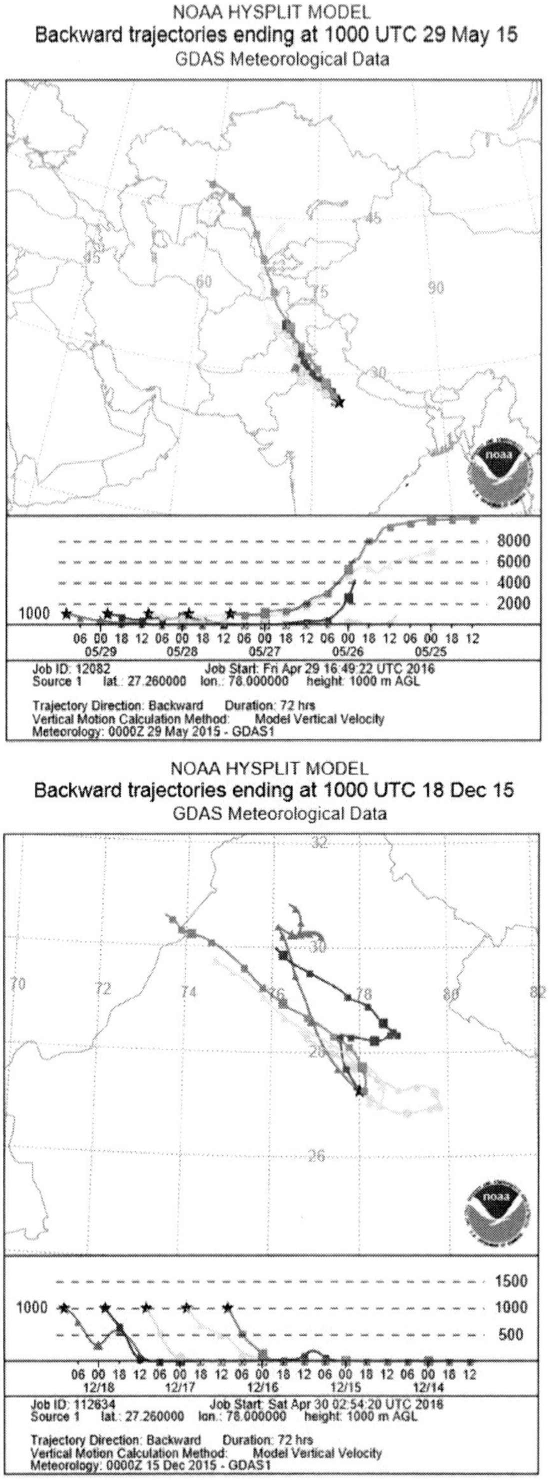


**FIGURE 2. Air mass back trajectories at different heights**

**Concentrations of Carbonaceous Aerosols.** Average OC and EC concentrations were estimated to be 27.9±11.0 μg $m^{-3}$ and 4.6±2.0 μg $m^{-3}$, respectively. Total carbonaceous aerosols (TCA) accounted for about

45% of total $PM_{2.5}$ mass. OC/EC ratios suggest about transformation and sources of carbonaceous aerosols. In the present study, OC/EC ratios were 6.0±1.7.

OC and EC showed distinct seasonal variation in summer and winter seasons. Both OC and EC recorded higher concentrations in winter season as compared to summer. OC and EC concentrations were 18.3 and 3.5 μg $m^{-3}$ in summer and 30.6 and 4.9 μg $m^{-3}$ in winter season, respectively. OC and EC concentrations are comparable to the values reported from an earlier study, Lal Gadi, a rural site in Agra (OC=30.3 μg $m^{-3}$ and EC= 4.0 μg $m^{-3}$; Pachauri et al., 2013). OC/EC ratios were 6.3±1.9 and 5.4±1.2, in winter and summer seasons, respectively. This indicates greater OC and hence, suggests secondary aerosol formation. In addition, burning activities might also be responsible for high ratios.

**Estimation of Secondary Organic Carbon (SOC).** On the basis of EC tracer method, Turpin and Huntzicker (1995) proposed an equation to estimate SOC formation by using the minimum OC/EC ratio.

$$SOC=OC_{total}-EC\times(OC/EC)_{min}$$

where SOC is the secondary organic carbon, $OC_{total}$ is the total OC in $PM_{2.5}$, and $(OC/EC)_{min}$ is the minimum ratio observed in all samples. Castro et al. (1999) stated that for primary emissions, OC/EC ratio should be ~2. OC/EC ratios greater than 2.0 indicate secondary aerosol formation. For the estimation of SOC in the present study, lowest OC/EC ratios in the respective seasons were: [$(OC/EC)_{min}$ in summer=3.3; $(OC/EC)_{min}$ in winter=3.0]. The relative amount of SOC ranged from 20.0 to 47.0% (average=28.0%) in summer and from 32.0 to 65.0% (average=49.0%) of the total organic carbon in winter. This suggests greater SOC formation in winter.

**$K^+$/OC and $K^+$/EC ratios.** $K^+$ is considered as the index of biomass burning due to its release during combustion processes. Hence, $K^+$/OC and $K^+$/EC ratios may be used to characterize whether the emission source of carbonaceous aerosols is biomass burning or fossil fuel combustion. Relatively high $K^+$/EC ratios have been reported for biomass burning (range: 0.21-0.46) and low ratios for fossil-fuel emissions (range: 0.025–0.09). This is because fossil-fuel emission produces very little potassium (Yang, He et al., 2005; Yang, Jian et al., 2005). At Agra, average of $K^+$/EC and $K^+$/OC ratios are 0.56±0.26 and 0.13±0.05 in winter season. These ratios are similar to those observed at Kanpur and Allahabad (Ram and Sarin, 2010a; 2010b) and indicate biomass burning may be the major source of carbonaceous aerosols in winter at Agra. Earlier studies have also reported biomass burning emissions from wood fuel, agriculture-waste and cow dung as a major source of carbonaceous aerosols in India (Venkataraman et al., 2005; Ram and Sarin, 2010a).

**Water Soluble Ionic Constituents.** Average ionic concentrations of major water soluble ions that were analysed have been shown in Table 1. Highest concentrations were of $SO_4^{2-}$ followed by $Cl^-$, $NO_3^-$, $NH_4^+$, $Na^+$, $K^+$ and $Ca^{2+}$. $F^-$ and $Mg^{2+}$ showed low concentrations.

**TABLE 1. Average and Seasonal Variation of Water Soluble ions (μg $m^{-3}$)**

| | $F^-$ | $Cl^-$ | $NO_3^-$ | $SO_4^{2-}$ | $Na^+$ | $NH_4^+$ | $K^+$ | $Mg^{2+}$ | $Ca^{2+}$ |
|---|---|---|---|---|---|---|---|---|---|
| **Winter** | 0.3 | 9.6 | 9.7 | 10.6 | 4.2 | 6.6 | 5 | 0.8 | 3 |
| **Summer** | 0.3 | 6.5 | 4.7 | 5.5 | 4.4 | 2.9 | 3.4 | 0.6 | 3.6 |
| **Average** | 0.3 | 7.4 | 6.3 | 8 | 4.2 | 4.7 | 4.2 | 0.7 | 3.3 |

Most of the ions showed high concentrations in winter season as compared to summer. $NO_3^-$, $SO_4^{2-}$ and $NH_4^+$ were almost double the concentrations recorded in summer, $Cl^-$ and $K^+$ concentrations were also higher as compared to summer concentrations. This also indicates formation of secondary inorganic aerosol in winter. Contrarily, $Ca^{2+}$ was higher in summer season. $Mg^{2+}$ was also higher in winter while $F^-$ did not show any seasonal change. Winter/Summer ratio has been depicted in Figure 3.

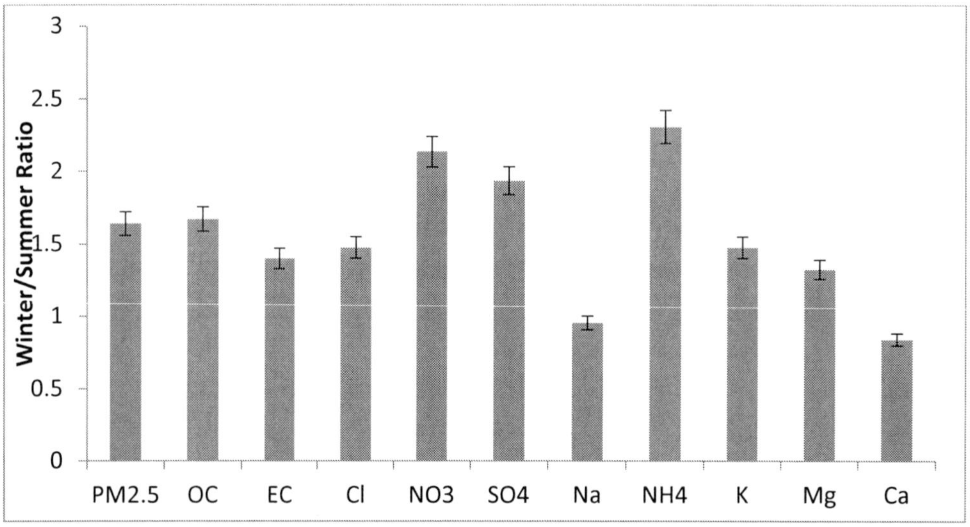


**Figure 3. Winter/Summer ratios of concentrations of selected chemical species.**

WSIS in fine particulate matter is mainly composed of $NH_4^+$, $K^+$, $NO_3^-$ and $SO_4^{2-}$. $NH_4^+$, $NO_3^-$ and $SO_4^{2-}$ are formed from precursor gases through chemical reactions that occur in the atmosphere. Gaseous $NH_3$ first reacts with $H_2SO_4$ to form $(NH_4)_2SO_4$ and the rest combines with $HNO_3$ to form $NH_4NO_3$. Molar $NH_4^+/SO_4^{2-}$ ratios suggest formation of $(NH_4)_2SO_4$ when the ratio is 2.0 or more. In the present study, molar ratios are 2.7 which suggest formation of both $(NH_4)_2SO_4$ and $NH_4NO_3$ salts.

**Source Characterization.** The probable reason for the difference in summer and winter seasons could be due to variation of meteorological parameters and different sources of emissions in both the seasons. This is supported by air mass back trajectory analysis. In summer, long range transport from north western region was noticed. High $Ca^{2+}$ may be the result of dust events that predominate in May and June. When the wind speed exceeds 18km $hr^{-1}$, dust events occur and bring heaps of sand to the present site causing enhanced concentrations of crustal ions especially, $Ca^{2+}$ and $Na^+$. Such an effect was absent in winter season where localized contribution dominated due to enhanced biomass burning and other anthropogenic activities. In order to combat cold, people burn wood, coal and cowdung cakes to keep themselves warm as the temperature becomes very low. $K^+$ is an indicator of biomass burning while $SO_4^{2-}$ and $Cl^-$ indicate coal burning. Concentrations of OC, EC, $K^+$, $SO_4^{2-}$ and $Cl^-$ are enhanced in winter due to the dominance of burning activities. Another reason for increase in winter time concentrations of most of the ions may be stagnant meteorological conditions (low wind speed, low mixing height, low temperature and high relative humidity) as compared to summer. Stable atmospheric conditions in winter prolong the residence time of the atmospheric pollutants and strengthen atmospheric oxidation. In summer, long periods of solar radiation favour the dispersion of pollutants due to convective mixing. This is further enhanced by high temperature and low relative humidity during this period. The average mixing height increases which leads to greater dispersion of primary pollutants.

## CONCLUSIONS

The concentrations of carbonaceous and inorganic aerosol components were reported from Sikandarpur, Agra. Average OC and EC concentrations were estimated to be 27.9±11.0 μg $m^{-3}$ and 4.6±2.0 μg $m^{-3}$, respectively. Both OC and EC recorded higher concentrations in winter season as compared to summer. OC/EC ratios suggest secondary aerosol formation. The relative amount of SOC ranged from 20.0 to 47.0% (average=28.0%) in summer and from 32.0 to 65.0% (average=49.0%) of the total organic carbon in winter. $K^+$/OC and $K^+$/EC ratios suggest that biomass burning may be the major source of carbonaceous aerosols. Water soluble cations had higher concentrations as compared to anions. $SO_4^{2-}$ showed highest concentrations followed by $Cl^-$, $NO_3^-$, $NH_4^+$, $Na^+$, $K^+$ and $Ca^{2+}$ while $F^-$ and $Mg^{2+}$ showed low concentrations.

Molar $NH_4^+/SO_4^{2-}$ ratios of 2.7 suggest formation of both $(NH_4)_2SO_4$ and $NH_4NO_3$ salts. Most of the ions showed high concentrations in winter season as compared to summer. In winter season, $NO_3^-$, $SO_4^{2-}$ and $NH_4^+$ were almost double the concentrations recorded in summer. This also suggests secondary aerosol formation in winter season. Results are further supported by air mass back trajectories.

## ACKNOWLEDGEMENTS

This work is financially supported by Women Scientist Scheme (Department of Science and Technology), India. The authors also thank Director of Dayalbagh Educational Institute and Head, Department of Chemistry for providing necessary facilities.

## REFERENCES

Andreae, M. O. 2001, "The dark side of aerosols", *Nature*, 409: 671–672.

Andreae, M. O.; Merlet, P. 2001. "Emission of trace gases and aerosols from biomass burning". *Glob. Biogeochem. Cycle.* 15: 955– 966.

Draxler R.R. and Rolph G.D. 2003. HYSPLIT (HYbrid Single-Particle Lagrangian Integrated Trajectory) Model access via NOAA ARL READY Website (http://www.arl.noaa.gov/ready/hysplit4.html) NOAA Air Resources Laboratory Silver Spring MD.

Liu, S., M. Chen, and Q. Zhuang. 2016. "Direct radiative effects of tropospheric aerosols on changes of global surface soil moisture." *Climatic Change*: 1-13.

Pachauri T., A. Satsangi, V. Singla, A. Lakhani, K. Maharaj Kumari. 2013. "Characteristics and Sources of Carbonaceous Aerosols in $PM_{2.5}$ during Wintertime in Agra, India". *Aer. Air Qual. Res.*, 13: 977–991.

Pöschl, U. 2005. *Atmospheric Aerosols: Composition, Transformation, Climate and Health Effects*. Angew. Chem. Int. Ed., 44, 7520 – 7540.

Ram K. and Sarin M.M. 2010a. Spatio-temporal variability in atmospheric abundances of EC, OC and WSOC over Northern India. *J. Aer. Sci.* 41(1): 88–98.

Ram K. and Sarin M.M. 2010b. "Day-night variability of EC, OC, WSOC and inorganic ions in urban environment of Indo-Gangetic Plain: Implications to secondary aerosol formation". *Atmos. Environ.* 45(2): 460-468.

Rastogi, N., A. Singh, D. Singh, M.M. Sarin. 2014. "Chemical characteristics of $PM_{2.5}$ at a source region of biomass burning emissions: Evidence for secondary aerosol formation". *Environ. Poll.* 184: 563–569.

Satsangi,A, Tripti Pachauri, Vyoma Singla, Anita Lakhani & K. Maharaj Kumari. 2012. "Organic and Elemental Carbon Aerosols at a Suburban site". *Atmos. Res.* 113: 13–21.

Srinivas, B., N. Rastogi, M.M. Sarin, A. Singh, D. Singh. 2016. "Mass absorption efficiency of light absorbing organic aerosols from source region of paddy-residue burning emissions in the Indo-Gangetic Plain". *Atmos. Environ.* 125: 360–370. South Asian Aerosols and Anthropogenic Emissions: Regional And Global Climate Implications.

Venkataraman C, G. Habib, A. Eiguren-Fernandez, A. H. Miguel and S. K. Friedlander. 2005. "Residential biofuels in South Asia: Carbonaceous aerosol emissions and climate impacts". *Science* 307(5714): 1454–1456.

Yang F., K. He, B. Ye X. Chen, L. Cha, S. H. Cadle et al. 2005. "One-year record of organic and elemental carbon in fine particles in downtown Beijing and Shanghai". *Atmos. Chem. Phys.* 5(1): 1449–1457.

Yang H., Z. Y. Jian, H. Ho, J. Xu, W. S. Wu, H. W. Chun et al. 2005. "The chemical composition of inorganic and carbonaceous materials in $PM_{2.5}$ in Nanjing, China". *Atmos. Environ.* 39(20): 3735–3749.

# AN INNOVATIVE APPROACH TO USE MODIS AOD DATA FOR PM2.5 MONITORING

***Zhiming Yang***, Harris Williams
(North Carolina Central University, Durham, NC 27707, U.SA)

**ABSTRACT**: Current air particular matter (PM) 2.5 monitoring in U.S. is limited in spatial resolution due to high operation and maintenance costs. Satellite-based monitoring could serve as an alternative for PM2.5 monitoring. New aerosol optical depth (AOD) product derived from Moderate Resolution Imaging Spectroradiometer (MODIS) imagery offers a spatial resolution (3km) which is suitable for urban PM2.5 monitoring. The purpose of this study was to develop a new approach to employ MODIS AOD data for PM2.5 monitoring in NC. In this study, Terra MODIS AOD data (Collection 6) collected from 2010 to 2012 were downloaded from the Unites States Geological Survey (USGS) web site. Hourly PM2.5 data for each PM2.5 station was downloaded from the EPA AirNow web site. After the extraction of AOD values, a temporal AOD-PM2.5 correlation was examined for each PM2.5 station annually. Results suggest that the temporal correlation between MODIS AOD value and ground PM2.5 concentration was statistically significant for majority of PM2.5 monitoring stations in NC from 2010 to 2012. This significant AOD-PM2.5 relationship can be used to develop a practical method for PM2.5 monitoring using MODIS AOD. The similar approach could also be applied in other states in U.S.

## INTRODUCTION

Long-term exposure to high concentrations of fine particulate matter (particles with aerodynamic radius less than 2.5 micrometers, or PM2.5) may cause several health issues such as asthma, respiratory infections, lung cancer, and cardiovascular problems (Zanobetti et al. 2009). Current PM) 2.5 monitoring in U.S. is limited in spatial resolution due to high operation and maintenance costs of air monitoring instruments. No PM2.5 monitoring station exists in many urban areas. Also, the PM2.5 data collected from fewer fixed monitoring stations in a city does not fully represent the condition of an entire city, especially in a big city. Satellite-based monitoring could serve as an alternative for PM2.5 monitoring because of its ability to provide more cost-effective observation on spatial pattern and gradient of PM2.5 than traditional air quality monitoring.

New aerosol optical depth (AOD) product derived from Moderate Resolution Imaging Spectroradiometer (MODIS) imagery offers a spatial resolution (3km) which is suitable for PM2.5 monitoring in urban areas. There have been several published studies on the evaluation of the new 3km MODIS AOD product (Munchak et al. 2013, Strandgren et al. 2014, Li et al. 2015). Strandgren et al. (2014) reported that the correlation between PM2.5 and AOD was improved significantly with increasing spatial resolution of the AOD from 10km to 3km while Li et al. (2015) discovered the 3 km AOD had nearly the same correlation with the PM2.5 as the 10 km AOD.

It was noticed that a large area was covered in aforementioned studies in which spatial-temporal AOD-PM2.5 correlation was examined. Seldom has there been any study conducted at a station level, which is essential to test the relationship of AOD-PM2.5 for real applications of MODIS AOD. The purpose of this study is to examine the AOD-PM2.5 correlation in the state of North Carolina at a station level in order to develop an effective method to use MODIS AOD for PM2.5 monitoring in NC.

## MATERIAL AND METHODS

In this study, Terra MODIS AOD data (Atmosphere level 2 Aerosol Product in HDF format, Collection 6) collected in 2010, 2011 and 2012 were ordered and downloaded from the Unites States

Geological Survey (USGS) web site. HDF data processing procedures include the reprojection of MODIS AOD data to a universal traverse Mercator (UTM) projection and conversion of file format to Geo TIFF. Hourly PM2.5 data for each PM2.5 station in North Carolina was downloaded from the EPA AirNow web site. There were seventeen PM2.5 stations providing hourly measurements in NC in during 2011-2012 (Fig 1). The latitude and longitude of each station was used to create a shape file containing all hourly PM2.5 monitoring stations in North Carolina. To match overpass time of Terra satellite, the average of PM2.5 concentration measured at 11 am and 12 pm for each day was selected for the correlation testing.

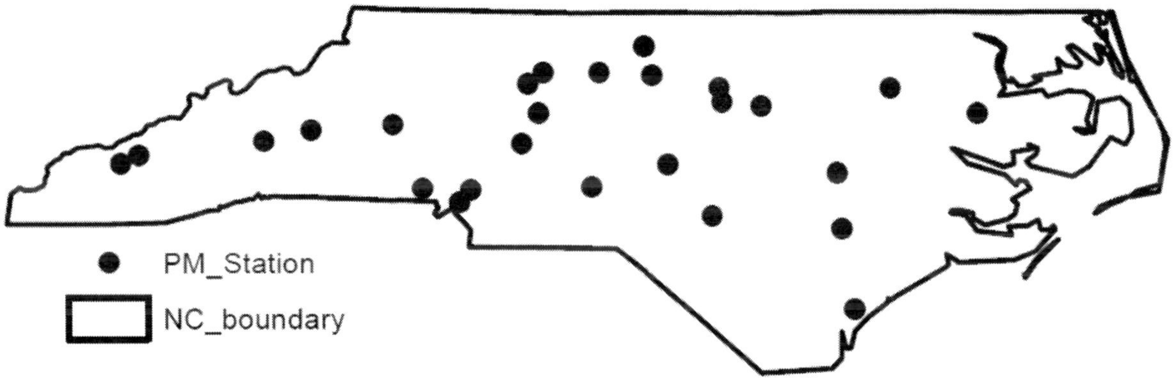


Figure 1. Map of study area and PM2.5 stations

In ArcMap, both the shape file of PM2.5 stations and downloaded MODIS AOD images were overlaid. AOD value at each PM2.5 station was extracted using a function tool entitled as "Extract values to points" in Arctool Box. The attribute table of each newly generated shape file was exported as a txt file and then transferred into Microsoft Excel format. The temporal correlation between MODIS AOD value and ground PM2.5 concentration was examined for each AirNow PM2.5 station. The significance of all correlations was tested at 95% confidence level.

**RESULTS AND DISCUSSION**

Annual temporal AOD-PM2.5 correlation was examined for each AirNow PM2.5 station from 2010 to 2012. The results are shown on Table 1. It can be seen that the temporal correlation between MODIS AOD and ground PM2.5 was statistically significant for 13 out of 17 stations in NC in 2010. In 2011, temporal correlation between MODIS AOD and ground PM2.5 was statistically significant for every AirNow PM2.5 station in NC. In 2012, the temporal correlation between MODIS AOD and ground PM2.5 was statistically significant for majority of AirNow PM2.5 stations in NC. Furthermore, each correlation coefficient for the correlation between MODIS AOD and ground PM2.5 was positive except Millbrook in 2012. If significant, most stations showed medium to strong AOD-PM2.5 correlations in each year. These results suggested that it is promising to use temporal AOD-PM2.5 correlation to develop a method used to retrieve PM2.5 concentration using MODIS AOD data.

Linear equations were developed for stations whose temporal correlations were statically significant for all three years and results are shown Table 2. The slope and intercept of linear equations ranged 0.0079-0.0309 and 1.3689-6.7012 respectively for PM2.5 stations in NC. In the same year, each station has different slope and intercept. Also, no spatial pattern was observed in slope and intercept. These results suggest each PM2.5 station may have its own unique AOD-PM2.5 correlation in NC. This explains why a spatial correlation between AOD and PM2.5 was not significant for almost all days in which both AOD and PM2.5 were sampled. It is likely that mixed AOD values from different stations would lead to failure in setting up significant spatial AOD-PM2.5 correlation in NC.

**TABLE 1. Annual temporal correlation coefficients of AOD-PM2.5 from 2010 to 2012**

| Station Name | 2010 | 2010 | 2012 |
|---|---|---|---|
| BRDED | 0.3127 | 0.3753* | 0.2295 |
| Brryson | 0.6123* | 0.6281* | 0.3445* |
| Clemmons Middle | 0.8190* | 0.6083* | 0.5688* |
| Durham Armory | 0.8205 | 0.8496* | 0.2687 |
| Garinger | 0.6247* | 0.7949* | 0.8887* |
| Grier Middle | 0.7222* | 0.7704* | 0.1657 |
| Hattieaven | 0.7942* | 0.5488* | 0.8591* |
| Hickory | 0.7912* | 0.6137* | 0.8819* |
| Hopedale | 0.9192* | 0.7530* | 0.3760 |
| Leggett | N/A | 0.7855* | 0.6553* |
| Lwattowr | 0.8827* | 0.5809* | 0.2692 |
| Marion | 0.5193* | 0.8692* | 0.7459* |
| Mendnhal | 0.3655 | 0.7333* | 0.5655 |
| Millbrook NCore | 0.8561* | 0.4570* | -0.7785 |
| Montclaire | 0.7967* | 0.8184* | 0.8596* |
| Rockwell | 0.7558* | 0.7694* | 0.6685* |
| W Owen Sch | 0.1524 | 0.4558* | 0.2324 |

**TABLE 2. Linear equations for some AirNow PM2.5 stations in NC**

| Station Name | 2010 | 2011 | 2012 |
|---|---|---|---|
| BRYSON | PM2.5 = 0.0302AOD + 6.2954 | PM2.5 = 0.026 AOD + 5.0558 | PM2.5 = 0.0079AOD +5.4894 |
| CLEMMONS | PM2.5 = 0.019AOD + 5.6951 | PM2.5 = 0.0157AOD + 4.1755 | PM2.5 = 0.0134AOD + 4.3906 |
| GARINGER | PM2.5 = 0.0206AOD + 5.1603 | PM2.5 = 0.0222AOD + 1.9847 | PM2.5 = 0.0235AOD + 2.24 |
| HATTIEAVEN | PM2.5 = 0.0196AOD + 5.3961 | PM2.5 = 0.0172AOD + 1.8613 | PM2.5 = 0.0192AOD + 1.6052 |
| HICKORY | PM2.5 = 0.0117AOD + 6.7012 | PM2.5 = 0.0156AOD + 4.5604 | PM2.5 = 0.0148AOD + 4.3314 |
| MARION | PM2.5 = 0.0128AOD + 6.5031 | PM2.5 = 0.0283AOD + 2.1746 | PM2.5 = 0.0168AOD + 2.2938 |
| MONTCLAIRE | PM2.5 = 0.0309AOD + 2.4352 | PM2.5 = 0.0269AOD + 1.7902 | PM2.5 = 0.0307AOD + 1.3689 |
| ROCKWELL | PM2.5 = 0.0373AOD + 4.1406 | PM2.5 = 0.0213AOD + 3.9573 | PM2.5 = 0.0211AOD + 3.2748 |

The relationship between ground-level PM2.5 and satellite-based AOD is highly associated with optical properties of local aerosols and their vertical distribution (van Donkelaar et al. 2010). In past decade, many researchers have successfully improved correlations between MODIS AOD and ground PM2.5 by incorporating vertical distribution (Boyouk et al. 2010; Chu et al. 2013) or inclusion of meteorological parameters (Koelemeijer et al. 2006; Gupta and Christopher 2009). However, significant amount of money and time are needed to develop a reliable method for the real application of MODIS AOD. The method based on temporal correlations might be more practicable to develop a method for PM2.5 monitoring.

For reliable PM2.5 monitoring by the MODIS satellite in NC, it is necessary to examine the number of days during which both MODIS AOD and PM2.5 were sampled at each station. The results were summarized in Table 4, 5 and 6. In 2010, both AOD and PM2.5 were sampled for less than six days per month and thirty one days per year for every PM2.5 station. In 2011, with the exception of November and December, both AOD and PM2.5 were sampled for less than five days per month and sixty days per year for every PM2.5 station. In 2012, both AOD and PM2.5 were sampled for less than six days per month and thirty seven days per year for every PM2.5 station.

According to results above, there were significant number of days in which both AOD and PM were not sampled for almost all stations in all three years. In most cases above, AOD value did not exist while PM2.5 value was available for each PM2.5 station. This is a well-known "low sampling" problem mentioned in the literature (Gupta and Christopher 2008; Christopher and Gupta 2010). It was caused by the lower number of cloud-free days in the state of North Carolina from 2010 to 2012 since MODIS AOD values were only available during cloud-free days. There are usually a low number of cloud-free days (average 213 sunny days per year) in any given year in NC. Thus it is practicable to use MODIS AOD for PM2.5 monitoring at a weekly level.

**TABLE 3. Number of days in which both AOD and PM2.5 existed in North Carolina in 2010**

| Name | Jan | Feb | Mar | Apr | May | Jun | Jul | Aug | Sep | Oct | Nov | Dec | Total |
|---|---|---|---|---|---|---|---|---|---|---|---|---|---|
| BDED | 0 | 0 | 3 | 5 | 0 | 0 | 3 | 1 | 4 | 3 | 5 | 1 | 25 |
| Bryson | 4 | 0 | 3 | 3 | 0 | 0 | 4 | 4 | 3 | 1 | 3 | 0 | 25 |
| Clemmons | 2 | 2 | 2 | 2 | 2 | 2 | 2 | 2 | 5 | 2 | 6 | 2 | 31 |
| Durham Arm | 0 | 1 | 1 | 0 | 0 | 0 | 0 | 0 | 0 | 0 | 1 | 2 | 5 |
| Grainger | 3 | 1 | 2 | 5 | 1 | 4 | 3 | 1 | 2 | 3 | 4 | 2 | 31 |
| Grier Middle | 3 | 0 | 4 | 3 | 0 | 1 | 2 | 1 | 0 | 1 | 1 | 2 | 18 |
| Hattieaven | 2 | 0 | 2 | 3 | 1 | 0 | 2 | 1 | 2 | 0 | 4 | 1 | 18 |
| Hickory | 1 | 0 | 0 | 1 | 0 | 1 | 1 | 1 | 2 | 2 | 0 | 3 | 12 |
| Hopedale | 2 | 1 | 1 | 1 | 0 | 0 | 1 | 1 | 0 | 4 | 5 | 2 | 18 |
| Lwattowr | 0 | 2 | 1 | 1 | 0 | 0 | 1 | 0 | 1 | 2 | 1 | 1 | 10 |
| Marion | 0 | 1 | 2 | 0 | 0 | 2 | 3 | 2 | 3 | 0 | 4 | 1 | 18 |
| Mendnhal | 1 | 1 | 1 | 2 | 0 | 0 | 2 | 1 | 1 | 1 | 2 | 0 | 12 |
| Millbrook | 0 | 1 | 1 | 1 | 0 | 1 | 1 | 0 | 2 | 0 | 2 | 2 | 11 |
| Montclaire | 3 | 1 | 3 | 1 | 2 | 0 | 3 | 0 | 2 | 1 | 2 | 1 | 19 |
| Rockwell | 3 | 2 | 3 | 3 | 1 | 2 | 2 | 1 | 4 | 3 | 3 | 2 | 29 |
| W Owen sch | 0 | 1 | 1 | 2 | 1 | 0 | 0 | 1 | 1 | 2 | 2 | 2 | 13 |

To be effective in PM2.5 monitoring, the use of station-specific temporal correlation is an alternative approach to develop the application of MODIS AOD for PM2.5 monitoring. The key step is to develop a reliable equation based on temporal AOD-PM2.5 correlation for each PM2.5 station. Also it is feasible to fill spatial PM2.5 gaps based on temporal correlations in NC. For locations near current AirNow PM2.5 stations, PM2.5 concentrations could be retrieved from their corresponding AOD values using the linear equation developed for a nearby AirNow PM2.5 station. For a location which is far from any existing AirNow PM2.5 station, its PM2.5 concentration could be interpolated from PM2.5 concentrations at surrounding locations using various spatial interpolation techniques such as Kriging

**TABLE 4. Number of days in which both AOD and PM2.5 existed in North Carolina in 2011**

| Name | Jan | Feb | Mar | Apr | May | Jun | Jul | Aug | Sep | Oct | Nov | Dec | Total |
|---|---|---|---|---|---|---|---|---|---|---|---|---|---|
| BDED | 2 | 1 | 2 | 3 | 1 | 1 | 2 | 1 | 2 | 3 | 4 | 11 | 33 |
| Bryson | 1 | 1 | 4 | 2 | 2 | 3 | 5 | 7 | 3 | 5 | 4 | 6 | 43 |
| Clemmons | 2 | 2 | 2 | 4 | 5 | 4 | 3 | 4 | 3 | 5 | 8 | 7 | 49 |
| Durham Armory | 2 | 0 | 0 | 1 | 0 | 1 | 0 | 0 | 2 | 0 | 3 | 9 | 19 |
| Garinger | 4 | 1 | 3 | 2 | 1 | 3 | 0 | 3 | 2 | 5 | 7 | 4 | 36 |
| Grier Middle | 2 | 1 | 1 | 1 | 2 | 2 | 2 | 2 | 1 | 4 | 4 | 6 | 28 |
| Hattieaven | 2 | 2 | 1 | 3 | 3 | 3 | 1 | 3 | 3 | 1 | 7 | 8 | 39 |
| Hickory | 0 | 2 | 1 | 0 | 0 | 1 | 0 | 0 | 0 | 1 | 5 | 6 | 18 |
| Hopedale | 3 | 0 | 2 | 1 | 0 | 3 | 0 | 1 | 2 | 2 | 9 | 10 | 35 |
| Leggett | 0 | 0 | 0 | 0 | 0 | 0 | 2 | 0 | 3 | 4 | 7 | 13 | 30 |
| Lwattowr | 0 | 0 | 1 | 0 | 2 | 1 | 0 | 1 | 1 | 1 | 4 | 8 | 19 |
| Marion | 2 | 1 | 3 | 1 | 0 | 4 | 0 | 1 | 2 | 5 | 2 | 9 | 31 |
| Mendnhal | 2 | 1 | 2 | 2 | 2 | 0 | 1 | 3 | 1 | 0 | 4 | 11 | 27 |
| Millbrook | 3 | 0 | 1 | 1 | 0 | 0 | 1 | 1 | 3 | 0 | 5 | 7 | 22 |
| Montclaire | 1 | 0 | 2 | 2 | 0 | 1 | 1 | 3 | 1 | 2 | 5 | 7 | 25 |
| Rockwell | 4 | 1 | 3 | 5 | 2 | 4 | 3 | 4 | 3 | 6 | 9 | 11 | 56 |
| W Owen sch | 3 | 1 | 0 | 0 | 2 | 1 | 0 | 0 | 0 | 3 | 10 | 10 | 25 |

**TABLE 5. Number of days in which both AOD and PM2.5 existed in North Carolina in 2012**

| Name | Jan | Feb | Mar | Apr | May | Jun | Jul | Aug | Sep | Oct | Nov | Dec | Total |
|---|---|---|---|---|---|---|---|---|---|---|---|---|---|
| BDED | 1 | 1 | 1 | 0 | 0 | 1 | 2 | 1 | 1 | 2 | 1 | 0 | 11 |
| Bryson | 0 | 4 | 3 | 1 | 4 | 5 | 2 | 4 | 2 | 5 | 5 | 0 | 35 |
| Clemmons | 3 | 2 | 2 | 4 | 2 | 4 | 4 | 2 | 3 | 5 | 3 | 3 | 37 |
| Durham Armory | 1 | 1 | 0 | 0 | 1 | 0 | 1 | 0 | 1 | 0 | 1 | 0 | 6 |
| Garinger | 2 | 0 | 0 | 0 | 0 | 1 | 3 | 0 | 1 | 3 | 1 | 2 | 13 |
| Grier Middle | 1 | 2 | 1 | 0 | 3 | 0 | 1 | 0 | 3 | 1 | 3 | 2 | 17 |
| Hattieaven | 1 | 2 | 1 | 0 | 1 | 1 | 1 | 1 | 2 | 3 | 2 | 0 | 15 |
| Hickory | 1 | 1 | 2 | 0 | 2 | 0 | 1 | 0 | 0 | 0 | 0 | 1 | 8 |
| Hopedale | 2 | 2 | 3 | 2 | 0 | 0 | 0 | 0 | 0 | 1 | 1 | 1 | 12 |
| Leggett | 5 | 3 | 5 | 3 | 0 | 2 | 5 | 3 | 4 | 2 | 2 | 3 | 37 |
| Lwattowr | 0 | 0 | 1 | 1 | 0 | 0 | 1 | 0 | 1 | 2 | 2 | 2 | 10 |
| Marion | 3 | 2 | 1 | 3 | 2 | 2 | 1 | 1 | 0 | 1 | 1 | 2 | 19 |
| Mendnhal | 1 | 2 | 2 | 1 | 1 | 1 | 1 | 1 | 1 | 2 | 2 | 1 | 16 |
| Millbrook | 0 | 0 | 1 | 1 | 1 | 0 | 0 | 0 | 0 | 0 | 1 | 0 | 4 |
| Montclaire | 1 | 2 | 1 | 2 | 1 | 2 | 1 | 0 | 3 | 0 | 1 | 1 | 15 |
| Rockwell | 3 | 4 | 1 | 3 | 3 | 2 | 3 | 0 | 4 | 2 | 4 | 3 | 32 |
| W Owen sch | 1 | 0 | 1 | 1 | 1 | 1 | 1 | 0 | 1 | 1 | 2 | 0 | 10 |

Despite the limitations of this work, our findings revealed that there is an alternative approach to develop a practical method to use MODIS AOD data for PM2.5 monitoring. This new way would help in better understanding the characteristics of AOD-PM2.5 correlation at city and state level and its applications from an innovative perspective. Further studies are needed to develop a robust retrieval AOD-PM2.5 equation for each AirNow PM2.5 station in North Carolina for a real application of MODIS AOD data.

## CONCLUSIONS

Our results show that no significant spatial correlations between MODIS AOD at daily level were found in the state of North Carolina in 2010, 2011 and 2012. But there was statistically significant temporal correlation between MODIS AOD values and ground PM2.5 concentration for most stations from 2010 to 2012. This significant temporal correlation can be used to develop a practical method for PM2.5 monitoring using MODIS AOD in NC and other states in the United States.

## REFERENCES

Boyouk, N., J. F. Leon, H. Delbarre, T. Podvin and C. Deroo. 2010. Impact of the mixing boundary layer on the relationship between PM2.5 and aerosol optical thickness. *J Atmos. Environ*. 44: 271-277.

Christopher, S.A. and P. Gupta. 2010. Satellite remote sensing of particulate matter air quality: the cloud-cover problem. *J Air Waste Manag Assoc*. 60(5): 596-602.

Chu, D. A., T. Tsa, J. Chen, S. Chang, Y. Jeng, W. Chiang. and N. Lin. 2013. Interpreting aerosol lidar profiles to better estimate surface PM2.5 for columnar AOD measurements.
*Atmospheric Environment* 79: 172-187.

Gupta, P. and S.A. Christopher. 2009. Particulate matter air quality assessment using integrated surface, satellite, and meteorological products: multiple regression approach. *J. Geophys. Res.*114, D14205. http://dx.doi.org/10.1029/2008JD011496.

Koelemeijer RBA, C.D. Homan, and J. Matthijsen. 2006. Comparison of spatial and temporal variations of aerosol optical thickness and particulate matter over Europe. *Atmos Environ* 40:5304; doi:10.1016/j.atmosenv.2006.04.044

Li, R., J. Gong, L. Chen, and Z. Wang. 2015. Estimating Ground-Level PM2.5 using fine-resolution satellite data in the megacity of Beijing, China. *Aerosol and Air Quality Research* 15 1347–1356.

Munchak, L. A., R.C. Levy, S. Mattoo, L.A. Remer, B.N. Holben, J.S. Schafer, C. A. Hostetler, and R.A. Ferrare. 2013. MODIS 3 km aerosol product: applications over land in an urban/suburban region, Atmos. Meas. Tech., 6, 1747–1759, doi:10.5194/amt-6-1747-2013.

Strandgren, J., L. Mei, M. Vountas, J.P. Burrows, A. Lyapustin, and Y. Wang. Study of satellite retrieved aerosol optical depth spatial resolution effect on particulate matter concentration prediction. *Atmos. Chem. Phys. Discuss*., 14, 25869-25899, doi:10.5194/acpd-14-25869-2014.

van Donkelaar A, R.V. Martin, M. Braue, R. Kahn, R. Levy, and C. Verduzco, et al. 2010. Global estimates of ambient fine particulate matter concentrations from satellite-based aerosol optical depth: development and application. *Environ Health Perspect* 118:847–855.

Zanobetti, A., N. Franklin, P. Koutrakis and J. Schwartz. 2009. Fine particulate air pollution and its components in association with cause-specific emergency admissions. *Environmental Health* 8:58–70.

# STUDY ON AEROSOL FORMATION PROPERTIES IN AMMONIA-BASED WFGD PROCESSES

***Huang Rongting,*** Pan Danping, and Yang Linjun
(Southeast University, Nanjing, China)

An ammonia-based wet flue gas desulfurization (WFGD) experimental platform was built to simulate separately the two different aerosol formation processes in the ammonia WFGD system of the heterogeneous reaction mechanism and the entrainment mechanism. The concentrations, the size distributions, the morphology features and the chemical composition of the aerosols formed after WFGD system were tested and investigated with the electrical low pressure impactor (ELPI), the pollutant PM sampler, the field emission scanning electron microscope (FESEM), and the aerosol mass spectrometer (AMS). The comparison was made of the numerical concentration size distributions of aerosols generated under the two formation mechanisms and the typical ammonia WFGD process. The results showed that the heterogeneous reaction mechanism is the dominating one of the two aerosol formation mechanisms in ammonia-based WFGD, which produced the majority of the aerosols emitted from the scrubber, which, if counted in numerical concentration, were mainly submicron particles, with the chemical ingredients consisting mainly of sulfates and ammonium and the size distribution based on numerical concentration being a bimodal one. While the entrainment mechanism had less influences on the aerosol formation in the ammonia WFGD, the aerosols generated under which condition were much larger than those formed under the heterogeneous reaction mechanism. Great differences were found between the aerosols formed under the two mechanisms.

# AIR FLOW SIMULATION AROUND AIRCRAFT

***Jianfeng Liu**** [1], Baoqing Wang[2], Ronghui Chen[3], and Zihui Ren[4]
[1]nkliujf@126.com, [2]wangbaoqing@nankai.edu.cn, [3]361970816@qq.com, [4]renzihuink@163.com
(College of Environmental Science and Engineering, Nankai University, Tianjin 300071, China)

**ABSTRACT:** Ambient environmental quality measurement by using aircraft becomes more and more important. In this paper, the air flow around aircraft is studied and analyzed through CFD numerical simulation method. The results show that the airflow and pressure field around aircraft is obvious different in different position. The suitable position of the instruments on the aircraft can be found by simulation. So, It is suggested that CFD simulation can be useful for obtaining the optimum position of aerosol sampling inlet.

***Keywords:*** Flow analysis; Aircraft; CFD simulation

## INTRODUCTION

An aerosol is defined as a suspension of liquid or solid particles in a gas. In reviewing aerosol measurement it is important to remember the gas. Average particle compositions vary with size, time, and location, and the bulk compositions of individual particles of a given size also vary significantly, reflecting the particles' diverse origins and atmospheric processing[1].Air quality in China is getting worse following the economic expansion, which has continued to increase in the past decades. Around 20-30% of carbonaceous aerosols in the global atmosphere are generated in China[2].These aerosols, together with the soil dusts originated from the deserts in the north/west regions, are occasionally transported to the western United States across the North Pacific Ocean[3]. A great variety of measurement systems, ranging from bulk filters to particle sizing instruments, are used on aircraft to examine ambient aerosols. For accuracy of aircraft measurements, a more fundamental question concerns the location of the instruments on the aircraft[4-5]. In this study, the suitable location will be discussed.

## DESCRIPTION OF THE MODEL

In this study, the sampling instrument is designed to be carried by an aircraft which cruises with a speed of 180~220km/h at 1~3km altitude. A photograph of aircraft shown in Fig. 1. Its length of wing is 17.24 m, the length and height is 14.86 m and 5.58 m. Its maximum flying height is 7000 m, maximum velocity is 292 km/h.

A computational fluid dynamics (CFD) program, FLUENT 14.0 (ANSYS Inc.), was used to air flow around aircraft. As a professional pre-processors software, ICEMCFD can provide a high efficient and reliable analysis model for CAE software. This research adopts the RANS (Reynolds Averaged Navier-Stokes) equation combined with the RNG k–ε model, which is the most appropriate model that characterizes the flow field around aircraft. Numerical simulation of the airflow was performed with CFD assuming (1) the flow is incompressible, isothermal and turbulent, (2) the mean flow is symmetric about the centerline of the system and (3) a steady-state solution exists.

Fig. 2 shows computational domain and boundary conditions of aircraft. For analyzing an external flow, it is necessary to have a computational domain as large as possible, in order to ignore upstream and downstream effects. Considering these effects, 10 times of length of extra space is provided for the aircraft, and 8 times of hight and width of aircraft. Three dimention surface mesh of aircraft shown Fig. 3.

FIG.1. The Photograph of Aircraft

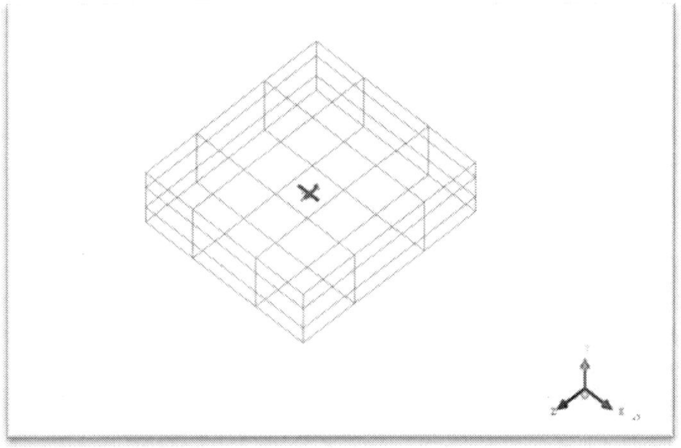

FIG. 2. Computational Domain and Boundary Conditons of Aircraft

FIG. 3. Three Dimention Surface Mesh of Aircraft

Longitudinal profile velocity and pressure distribution of aircraft surface shown in Fig. 4 and Fig.5. From Fig.4 and Fig.5, it is obviously different for flow fields around various shapes at reasonable distances from the boundary and away from regions of high curvature, the flow fields are determined primarily by the overall radius of the cyclindrical portion of the fuselage. It can be seen that the lifting of a finite wing generates line vortices which are shed from the trailing edge of the wing. Due to different curvature around aircraft, the depletion zone and enhancement zone will generated. The steady flow zone located in the bottom of the aircraft. So it is suitable position mounted the sampling instruments.

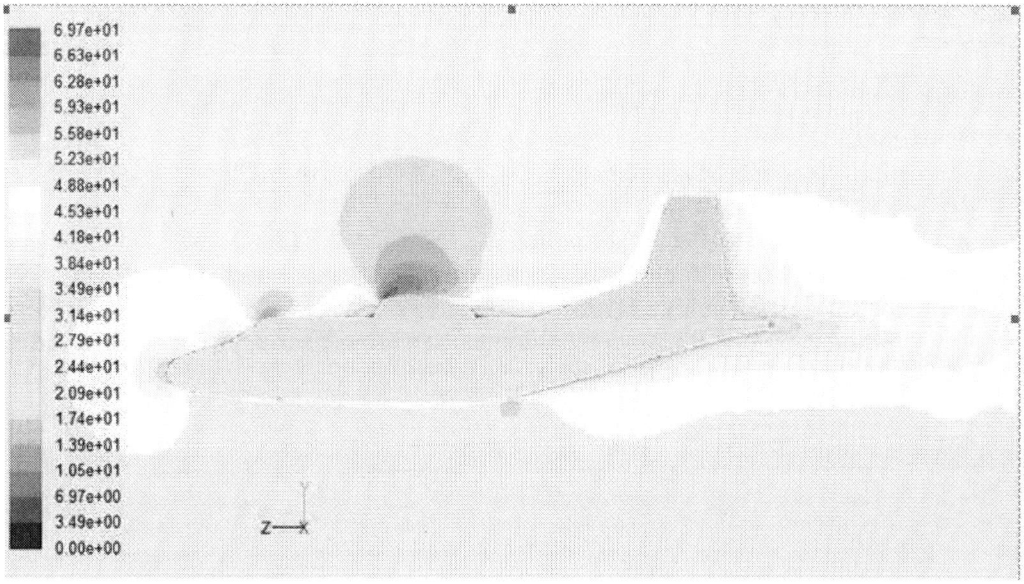


FIG. 4. Longitudinal Profile Velocity Distribution of Aircraft Surface

FIG. 5. Longitudinal Profile Pressure Distribution of Aircraft Surface

## CONCLUSIONS

In this paper, it is shown that the dominant features of the flow around shapes similar to aircraft fuselages can be obtained using CFD simulation applied to flows around aircraft. The flow of the bottom around aircraft is steady flow, So it is suitable position mounted the sampling instruments in order to avoid flow disturbance around aircraft. At the same time, particles tragectory will be simulated by CFD to determined the depletion zone and enhancement zone around aircraft.

**ACKNOWLEDGMENT**

This work was financially supported by National major scientific instrument equipment development special (2011YQ060111).

**REFERENCES**

[1] Peter H. McMurry. "A review of atmospheric aerosol measurements." Atmospheric Environment 34 (2000) 1959-1999.
[2] Gehui Wang et al.. "Aircraft Measurement of Organic Aerosols over China." Environ. Sci. Technol. 41 (2007) 3115-3120.
[3] Wilkening, K. E. et al.. "Trans-Pacific air pollution." Science 290 (2000): 65-67.
[4] W.D. King. "Air Flow and Particle Trajectories around Aircraft Fuselages. I:Theory." Journal of Atmospheric and Ocean Technology 1(1984) 5-13.
[5] B. Dix et al. "Airborne multi-axis DOAS measurements of atmospheric trace gases on CARIBIC long-distance flights." Atmos. Meas. Tech. 2(2009) 639–652.

# BLACK CARBON AND SIZE-RESOLVED PARTICLE NUMBER CONCENTRATIONS DURING OPEN-FIELD BURNING OF CORN STRAW

***Yu-Hsiang Cheng*** and Li-Sing Yang
(Ming Chi University of Technology, New Taipei, Taiwan)

**ABSTRACT:** Information on the effect of open-field burning of agricultural residues on ambient black carbon (BC) mass and particle number concentrations is scarce. In this study, the real-time variation of BC mass and size-resolved particle number concentrations was monitored during a period of corn straw burning at a rural site to understand the effect of such open-field burning on short-term air quality. The measurement results showed that the hourly BC mass, total particle number, and ultrafine particle (UFP) number concentrations during the corn straw burning period could reach 3775 ng $m^{-3}$, $3.99 \times 10^4$ # $cm^{-3}$, and $2.03 \times 10^4$ # $cm^{-3}$, respectively, which were 3.2, 5.3, and 3.8 times the reference values at the site, respectively. Field measurements showed that the hourly absorption Ångström exponent value during the corn straw burning period was 2.0. Furthermore, the measurement results showed a single-mode pattern for the particle number size distribution, and the dominant mode corresponded to 100 nm. Additionally, the measurement results suggested that most of the BC mass concentrations in fresh aerosols during the corn straw burning period were significantly and positively correlated with particle number concentrations in the size range of 56–320 nm.

## INTRODUCTION

A large amount of PM is produced during uncontrolled burning of agricultural residues, immediately degrading the air quality in the surrounding areas and causing acute symptoms in people with chronic respiratory conditions (Ryu et al., 2007; Zhang et al., 2011). Epidemiological studies have consistently reported that short-term exposure to high concentrations of PM can increase cardiovascular morbidity and mortality (Pope et al., 2011; Rice et al., 2013). Therefore, short-term air pollution episodes resulting from open-field burning of agricultural residues could pose a serious public health hazard.

Particularly, toxicological studies have suggested that ultrafine particles (UFPs) may be more harmful to health than large particles are because small particles have a considerably larger adsorbed or condensed surface area than large particles with the same mass do and can be deposited in the alveoli, where they interact with epithelial cells (Oberdörster et al., 2005). Additionally, black carbon (BC) is a major constituent of atmospheric aerosols, and it has an appreciable impact on global climate change, apart from adversely affecting human health (Rich et al., 2005; Jacobson, 2010). Therefore, the risk to human health from exposure to high BC mass and high particle number concentrations within a short period could be more serious than that from exposure to high $PM_{2.5}$ concentrations.

However, information on the effect of open-field burning of agricultural residues on ambient BC mass and particle number concentrations is scarce. In this study, to understand the effect of open-field burning on short-term air quality, real-time variations of the BC mass and size-resolved particle number concentrations were monitored before and during a corn straw burning episode at a rural site. Correlations between the BC mass and size-resolved particle number concentrations during the episode were investigated. Moreover, the particle number size distribution and absorption Ångström exponent were determined for obtaining the characteristics of aerosol emissions from the corn straw open-field burning. The results can be used to address public health concerns and as a reference for managing similar episodes of open-field burning of agricultural residues.

## MATERIALS AND METHODS

In this study, the monitoring location was in the downtown area of Yanshui in southern Taiwan. There are a lot of corn fields in the vicinity of this small town. In this study, the monitoring instruments were set up on the rooftop of a building of Yanshui Junior High School (23°19′17.5″N, 120°15′50.3″E). Two Rack Mount Aethalometer instruments (Model AE31; Magee Scientific Corp., Berkeley, CA, USA) were used to measure the BC mass concentrations in parallel at two different aerosol sampling flow rates. The instruments can be used to evaluate BC mass concentrations at seven wavelengths: 370, 470, 520, 590, 660, 880, and 950 nm. Furthermore, a correction algorithm developed by Cheng and Yang (2015) was applied for minimizing artifacts in determining BC mass concentrations. In the current study, the BC mass concentrations determined at the wavelengths of 370 and 880 nm were treated as ultraviolet BC (UVBC) and BC, respectively, to distinguish particles bound to ultraviolet-absorbing organic compounds (Sandradewi et al., 2008). Additionally, a scanning mobility particle sizer (SMPS) spectrometer (TSI Model 3938; TSI, Inc., Shoreview, MN, USA) was employed to measure particle number size distributions of particles with diameters in the range of 13–750 nm. Local meteorological data were recorded using a Vantage Pro2$^{TM}$ weather station (Davis Instruments Corp., Hayward, CA, USA). The instruments were operated continuously from February 11 to 19, 2014, and the logging interval for all measurements was set at 5 min.

During the whole monitoring period, a remarkable open-field burning of corn straw upwind of the monitoring location was observed between 17:00 and 19:00 (2 h) on February 16, 2014. During this period, the ambient air was filled with smoke. Therefore, the data obtained during this period were considered to correspond to a short-term air pollution episode for agricultural residue burning (episode period). Furthermore, the data monitored at the same time and for the same duration on February 12 (Wednesday), 2014, were considered to represent the reference air quality status (nonepisode period) for this monitoring site because the wind speed and principal wind direction on this day were similar to those on February 16, 2014.

## RESULTS AND DISCUSSION

During the whole monitoring period, the average wind speed, temperature, and relative humidity were 1.7 m $s^{-1}$, 15.9 °C, and 71.4%, respectively. Despite the irregular wind direction patterns on February 17 and 18, 2014, the principal wind direction was NNW–NNE. The daily average wind speed, principal wind direction, temperature, and relative humidity were 1.5 m $s^{-1}$, NNW–NNE, 14.3 °C, and 68.9%, respectively, on February 12, 2014. On February 16, 2014, the daily average wind speed, principal wind direction, temperature, and relative humidity were 1.2 m $s^{-1}$, NW–NE, 17.8 °C, and 73.2%, respectively. During open-field burning of corn straw (episode duration), the 5-min wind speed was between 0.9 and 1.8 m $s^{-1}$ (1.1 m $s^{-1}$ on average), and the principal wind direction was between NW and N.

The average UVBC and BC mass concentrations were 1652 and 1345 ng $m^{-3}$, respectively, and the interquartile ranges were 799–1921 and 681–1641 ng $m^{-3}$ for UVBC and BC mass concentrations, respectively, during the whole monitoring period. Figure 1 shows the diurnal variations of the 5-min and 1-h UVBC and BC concentrations at the monitoring site on February 12 and 16, 2014. The daily average BC mass concentrations on February 12 and 16, 2014, were 1074 and 2013 ng $m^{-3}$, respectively. The daily average BC mass concentration on February 16, 2014, was significantly higher than that on February 12, 2014, because obvious open-field burning was performed on February 16, 2014.

During the episode, the 5-min UVBC concentration can reach 47500 ng $m^{-3}$, and remarkably, it is approximately three times the BC concentration, suggesting that ambient particles in this period were bound to ultraviolet-absorbing organic compounds, which were abundant. The hourly UVBC concentration during the episode increased to 11932 ng $m^{-3}$, which was 8.3 times the value in the nonepisode period. The hourly BC concentration during the episode could increase to 3775 ng $m^{-3}$, which was 3.2 times the value in the nonepisode period.

The absorption Ångström exponent has been used as an index for determining the type of BC emission source (Sandradewi et al., 2008). In this study, the daily average absorption Ångström exponent value for February 12, 2014 was 1.2, which was same as the value on February 16, 2014. The BC concentrations at this rural monitoring site on normal days were lower, and the measurement results indicated that vehicular traffic was the main source of BC aerosols. Nevertheless, the hourly absorption Ångström exponent value during the episode was 2.0, implying that the source of BC aerosols during the episode was related to agricultural residue burning.

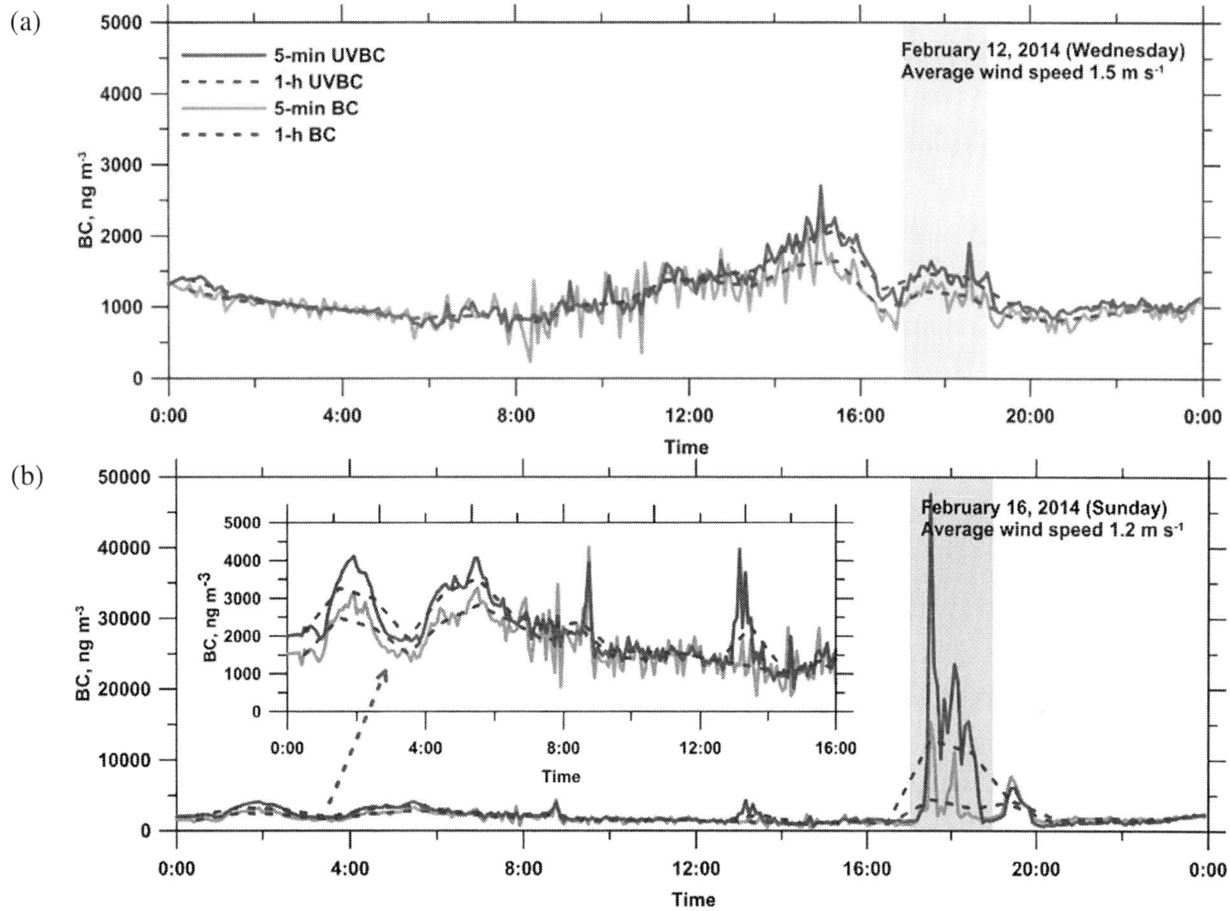


**FIGURE 1.** Diurnal variations of 5-min and 1-h UVBC and BC concentrations at the monitoring site on (a) February 12, 2014 and (b) February 16, 2014.

In this study, the total particle and UFP number concentrations in the size ranges of 13–750 and 13–100 nm were calculated from the raw data of the SMPS, respectively. During the whole monitoring period, the average total particle and UFP number concentrations were $7.06 \times 10^3$ and $3.76 \times 10^3$ # $cm^{-3}$, respectively. The interquartile ranges were $4.18 \times 10^3$–$8.13 \times 10^3$ and $2.22 \times 10^3$–$4.32 \times 10^3$ # $cm^{-3}$ for total particle and UFP number concentrations, respectively. The average ratio of UFP-to-Total particle was 55.1% during the whole monitoring period, and the interquartile range was 47.9%–63.4%. The relative high ratio value of the UFP-to-Total particle usually can be observed during noon to afternoon hours, corresponding to photochemical nucleation processes under higher temperature conditions. Otherwise, the relative low ratio value of the UFP-to-Total particle occurred during pre-dawn hours, which was due to the coagulation of aged aerosols. The low ratio value of the UFP-to-Total particle also could be observed after a couple hours of the agricultural residues burning.

Figure 2 shows the diurnal variation of the 5-min and 1-h total particle and UFP number concentrations at the monitoring site on February 12 and 16, 2014. Time variations similar to those of the BC concentrations were observed for the total particle and UFP number concentrations, indicating that the ambient particles could share a common source with BC aerosols in this rural area. The daily average total particle and UFP number concentrations were 6.13 × $10^3$ and 3.61 × $10^3$ # $cm^{-3}$ for February 12, 2014, respectively, and they were 1.01 × $10^4$ and 4.83 × $10^3$ # $cm^{-3}$ for February 16, 2014, respectively.

On February 16, 2014, an abrupt increase in the total particle and UFP number concentrations was clearly observed during the open-field corn straw burning episode (17:00–19:00). During the episode, the 5-min total particle and UFP number concentrations could reach 1.54 × $10^5$ and 6.72 × $10^4$ # $cm^{-3}$, respectively. These particle number concentrations during the episode were comparable with those at a busy traffic site and could severely affect human health. Additionally, the hourly total particle and UFP number concentrations during the episode increased to 3.99 × $10^4$ and 2.03 × $10^4$ # $cm^{-3}$, respectively, which were 5.3 and 3.8 times the values in the nonepisode period, respectively.

Particles in the size range of 100–180 nm (38.1%) and 56–100 nm (37.6%) were dominant during the episode. By contrast, particles in the size range of 32–56 nm (28.7%) and 56–100 nm (28.3%) dominated during the nonepisode period. According to the measurement results, the particle number size distribution patterns for different periods were considerably different. During open-field corn straw burning, the dominant mode of the particle number size distribution was at 100 nm.

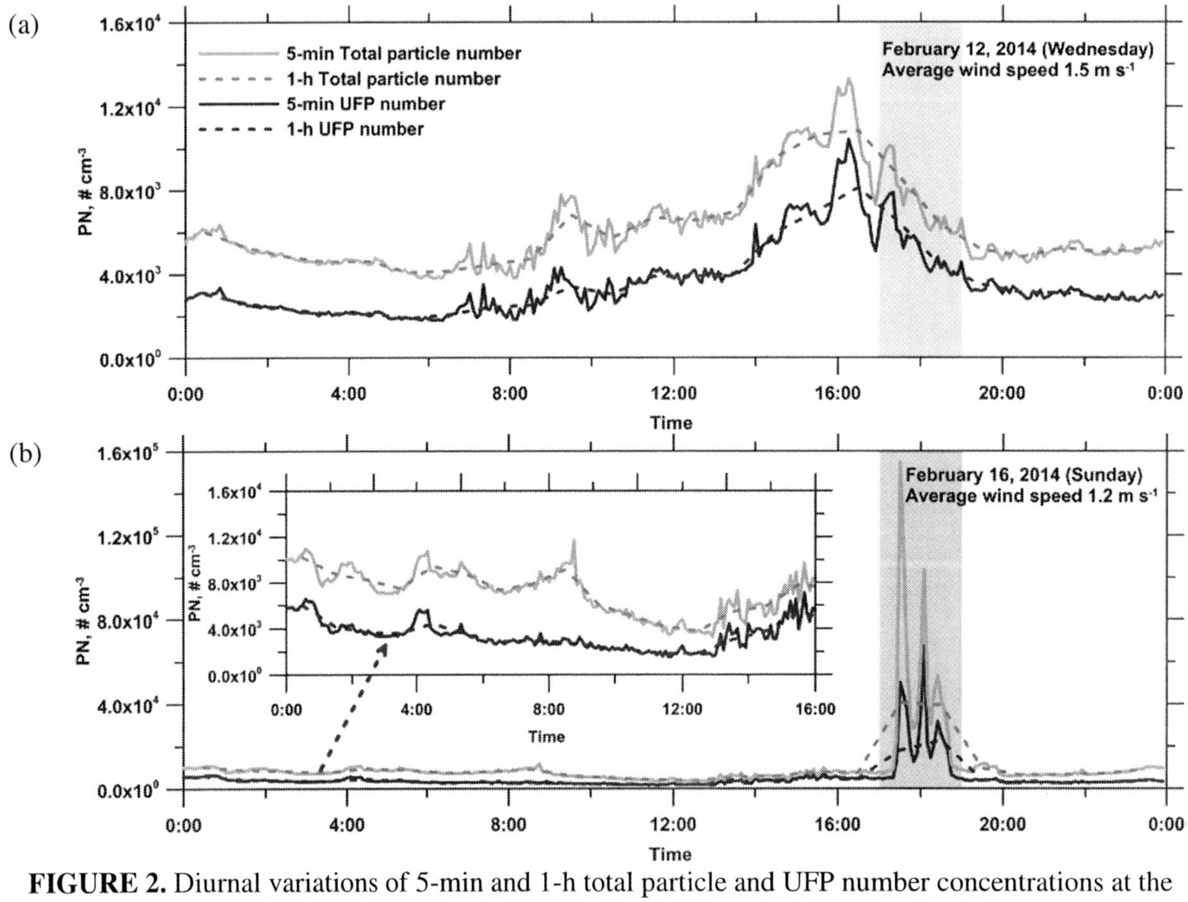


**FIGURE 2.** Diurnal variations of 5-min and 1-h total particle and UFP number concentrations at the monitoring site on (a) February 12, 2014 and (c) February 16, 2014.

## CONCLUSIONS

In this study, the hourly BC concentration during the corn straw burning period increased to 3775 ng $m^{-3}$, which was 3.2 times the value during the same period on the reference day. The 5-min UVBC concentration during this episode reached 47500 ng $m^{-3}$, and it was approximately three times greater than the BC concentration, suggesting that ambient particles were bound to ultraviolet-absorbing organic compounds. According to the field measurement results, the hourly absorption Ångström exponent value was 2.0 during the episode. Additionally, the hourly total particle and UFP number concentrations during the episode increased to $3.99 \times 10^4$ and $2.03 \times 10^4$ # $cm^{-3}$, respectively, which were 5.3 and 3.8 times the reference values at the rural monitoring site, respectively. During the episode, the 5-min total particle and UFP number concentrations reached $1.54 \times 10^5$ # $cm^{-3}$ and $6.72 \times 10^4$ # $cm^{-3}$, respectively. These particle number concentrations were comparable with those at a busy traffic site and could severely affect human health. Furthermore, the particle number concentrations during the episode were dominated by particles in the size range of 56–180 nm. The measurement results showed a single-mode pattern for the particle number size distribution, and the dominant mode was at 100 nm.

## ACKNOWLEDGEMENTS

The authors would like to thank the Ministry of Science and Technology of the Republic of China, Taiwan, for financially supporting this research under Contract No. MOST 102-2221-E-131-002-MY3.

## REFERENCES

Cheng, Y. H., and L. S. Yang, 2015. "Correcting aethalometer black carbon data for measurement artifacts by using inter-comparison methodology based on two different light attenuation increasing rates". *Atmospheric Measurement Techniques Discussions*. 8:2851-2879.

Jacobson, M. Z., 2010. "Short - term effects of controlling fossil - fuel soot, biofuel soot and gases, and methane on climate, Arctic ice, and air pollution health". *Journal of Geophysical Research*. 115:D14209.

Oberdörster, G., E. Oberdörster, and J. Oberdörster, 2005. "Nanotoxicology: An emerging discipline evolving from studies of ultrafine particles". *Environmental Health Perspectives*. 113:823-839.

Pope, C. A., J. C. Hansen, R. Kuprov, M. D. Sanders, M. N. Anderson, and D. J. Eatough, 2011. "Vascular function and short-term exposure to fine particulate air pollution". *Journal of the Air & Waste Management Association*. 61:858-863.

Rice, M. B., P. L. Ljungman, E. H. Wilker, D. R. Gold, J. D. Schwartz, P. Koutrakis, G. R. Washko, G. T. O'Connor, and M. A. Mittleman, 2013. "Short-term exposure to air pollution and lung function in the Framingham Heart Study". *American Journal of Respiratory and Critical Care Medicine*. 188: 1351-1357.

Rich, D. Q., J. Schwartz, M. A. Mittleman, M. Link, H. Luttmann-Gibson, P. J. Catalano, F. E. Speizer, and D. W. Dockery, 2005. "Association of short-term ambient air pollution concentrations and ventricular arrhythmias". *American Journal of Epidemiology*. 161:1123-1132.

Ryu, S. Y., B. G. Kwon, Y. J. Kim, H. H. Kim, and K. J. Chun, 2007. "Characteristics of biomass burning aerosol and its impact on regional air quality in the summer of 2003 at Gwangju, Korea". *Atmospheric Research*. 84:362-373.

Sandradewi, J., A. S. H. Prévôt, E. Weingartner, R. Schmidhauser, M. Gysel, and U. Baltensperger, 2008. "A study of wood burning and traffic aerosols in an Alpine valley using a multi-wavelength Aethalometer". *Atmospheric Environment*. 42:101-112.

Zhang, H., D. Hu, J. Chen, X. Ye, S. X. Wang, J. M. Hao, L. Wang, R. Zhang, and Z. An, 2011. "Particle size distribution and polycyclic aromatic hydrocarbons emissions from agricultural crop residue burning". Environmental Science and Technology. 45:5477-5482.

# INTEGRATED ASSESSMENT OF PARTICULATE MATTER IN KPK AND BALUCHISTAN BY USING GAINS- SOUTH ASIA MODEL

***Sheikh Saeed Ahmad*** and Aisha Khan
(Fatima Jinnah Women University, Rawalpindi, Pakistan)[104]

Pakistan is a developing country existing geographically at a pivoted location between two of the world's largest pollution emitting countries (China and India) which adds to the severity of environmental issues faced by the country. These concerns include air pollution, especially the particulate matters, climate change and extreme weather situations prevailing in Pakistan. Particulate matter is specifically responsible for deteriorating the health and causing significance decrease in life expectancy. The initial step in devising a wide ranging, multifaceted, economically feasible and sustainable solution to deal with the severity of this issue is the quantification of the Particulate Matter. GAINS model is one of the most comprehensive tool, dealing with the air pollutants and greenhouse gases covered by the Kyoto Protocol. This study has utilized this model to analyze the source based anthropogenic emissions of Particulate Matter ($PM_{2.5}$, $PM_{10}$ and $PM_{tsp}$), their impacts and abatement cost, for the duration of 1990-2030, in Khyber Pakhtunkhwa and Baluchistan regions of Pakistan. An overall increasing trend was observed during 1990-2030 for $PM_{2.5}$ (103.21-405.31kT/Y); $PM_{10}$ (132.02-558.52kT/Y) and $PM_{tsp}$ (204.64-775.86kT/Y). The loss in life expectancy caused by the exposure to $PM_{2.5}$ during 1990-2030 was projected to increase from 10.03 months to 20.07 months. Finally, the study calculated the abatement cost for the emissions was projected to reach 21.57 million dollars in 2030. Therefore, this increasing emissions calls for swiftly initiating an efficient, comprehensive and cost effective approach towards the diminution of these Particulate matters, for improving the environment, with the co-benefit of improving the life expectancy of the population of the area under study.

# HEAVY METALS IN ROAD DUST FROM XIANDAO DISTRICT, CHANGSHA CITY, CHINA: HARACTERISTICS, HEALTH RISK ASSESSMENT AND INTEGRATED SOURCE IDENTIFICATION

Jingdong Zhang, ***Fei Li***, Jun Yang
(School of Information and Safety Engineering, Zhongnan University of Economics and Law, Wuhan 430073, China)

The dust properties (dust organic material and pH), the concentrations of metals (Cu, Zn, Pb, Cd, Cr and Fe) in 51 road dust samples from Xiandao District (XDD), China, were investigated. Enrichment factor (EF), multivariate statistics, geostatistics and health risk assessment model were adopted to study the spatial pollution pattern, sources of studied metals and to identify the priority pollutants and regions of concern. The mean EFs of the studied metals revealed the following orders: Cd > Zn≈Pb≈Cu > Cr. For non-carcinogenic effects, the exposure pathway which resulted in the highest levels of exposure risk for children and adults was ingestion, followed by dermal contact and inhalation. Hazard index (HI) values for the studied metals at each site were within the safe level of 1 except maximum $HI_{Cr}$ (1.08) for children. The carcinogenic risk (CR) for Cd and Cr at each site were within the acceptable risk level (1E-06) except $CR_{Cr}$ (1.08E-06) at sampling site 6 for children. Cr was identified as the priority pollutant to human exposure followed by Pb and Cd with simultaneous consideration of the local population distribution. Results based on the integrated source identification indicated Pb, Cr and Zn were probably sourced from traffic-related sources; Cd associated with the dust organic material mainly originated from industrial sources; Cu mainly derived from both sources. Spatially, northwest and northeast of XDD were regarded as the priority regions of concern which were proved to be in close relationship with the land use patterns.

# CHLORINATED PARAFFINS IN CANADIAN HOUSE DUST AND NIST SRM 2585 (ORGANIC CONTAMINANTS IN HOUSE DUST)

***Xinghua Fan***, Hongtao Shang, Cariton Kubwabo, and Pat E. Rasmussen
(Health Canada, Ottawa, ON, Canada)

Chlorinated paraffins (CPs) have been widely used as both flame retardants and plasticizers in rubbers and textiles, in paints and coating, in leather processing, and also as metal working fluids and sealants. They can be roughly classified into three categories according to their carbon chain length: short-chain CPs (C10-C13, SCCPs), medium-chain CPs (C14-C17, MCCPs), and long-chain CPs (C18-C30, LCCPs). The worldwide production of CPs was estimated to be 916,000 tons per year . Animal studies show that CPs (C12, 60% chlorine content) are carcinogenic in rats and mice of both sexes . Although the underlying mechanisms are not clear, the major target organs are the liver, kidney, and thyroid. CPs have been classified as Group 2B - possibly carcinogenic to humans based on animal study results by International Agency for Research on Cancer . Due to their wide use, CPs are ubiquitous in the environment and have been detected in air, soil, sediment, water, and house dust.

House dust is a sink and repository of various contaminants including CPs; it has been recognized as an important source for human exposure, especially for children, to dust-bound contaminants via inhalation, dermal adsorption and unintentional ingestion. The objective of the current study was to develop a method for the determination of SCCPs and MCCPs in indoor dust samples collected under the Canadian House Dust Study (CHDS) . The method was based on those reported by Reth et al. (2005) and Yuan et al. (2012) with some modifications. Dust sample (0.1 g, <80□m) was subjected to sonication extraction, cleanup by solid phase extraction (SPE) (Florisil cartridge topped up with acidified silica gel), and separation and detection by gas chromatography –mass spectrometry (GC/MS) operated in electron capture negative ionization (ECNI) mode. The method demonstrated good sensitivity, with the method detection limits (MDLs) down to 0.21 μg/g for SCCPs and 0.70 μg/g for MCCPs. The overall recoveries of the method were good, with 104±11% and 108±16% for SCCPs and MCCPs, respectively. SCCPs and MCCPs were detected in every sample (n=48), with median concentrations (range) of 6.2 (4.0-58) μg/g for SCCPs and 20 (5.9-900) μg/g for MCCPs. These are the first CP concentration data reported for Canadian house dust, which are relatively lower compared with those reported in other countries. On average, the percentages of SCCP congeners were 16.5%, 21.7%, 33.8%, and 28.0% for C10, C11, C12, and C13, respectively; the percentages of MCCP congeners were 48.6%, 25.8%, 15.8%, and 9.8% for C14, C15, C16, and C17, respectively. The method was also applied to the analysis of SCCPs and MCCPs in NIST standard reference material (SRM 2585, organic contaminants in house dust). To the best knowledge of the authors, this study was the first to report the CP concentrations in SRM 2585, with 7.58±0.43 μg/g for SCCPs and 16.4±2.1μg/g for MCCPs. These results could be useful for the comparison of the accuracy for CP analysis using different analytical methods.

# BEIJING DUST STORM IN SPRING SEASON AND ITS SOURCE APPORTIONMENT

***Liu Yan-Ju*** (Beijing Milu Ecological Research Center, Beijing, China)
Tong-Lin Han, Qing-Yang Liu (Beijing Center for Physical and Chemical Analysis, Beijing, China)

Dust storm of Beijing, usually occurred in Spring, has been paid much attention due to its seriously adverse effect on environment, ecosystem, economy and especially on human health. It has been disturbing the capital city of Beijing in recent years, making its exact source as a hot topic of debate again. This study aims to designate source of the Beijing dust storm by investigating physical and chemical features of both deposit dust and particulate matter during typical dust storm period, together with surface soil dust features of possible source regions, which were classified into four types of soil surface including dry saline lake (also saline soil, etc.), hills (also the mountain, etc.), the agricultural land (also desertified land, grassland, degraded grassland, etc.), sand (also deserts, sand dunes, etc.) based on the specific characteristics of the soil surface, the terrain feature, surface water soluble salt and dust content under diameter of 200 μm, and the topography and landforms.

The result shows that the Beijing storm dust is closely related to dry saline lake due to its higher dust content, yellow-brown color, water delicate slippery feeling, "dirty" particle appearance, higher water soluble salt content, anion and cation content, low water insoluble matter content. The dust is similar to the dry saline lake and the hills in their dry laser particle size and the variation range, and conductivity. Thc proportion and minimum wind speed of dust are far lower than those of sand. MgO, CaO, MnO, $CO_2$ content of dust followed the diluted characteristics of the saline Lake dust; $Fe_2O_3$, FeO, $H_2O$ content has the features of diluted hilly dust; the $TFe_2O_3$ content of storm dust showed the characteristics of diluted hilly and farmland dust; the content of $K_2O$ has the characteristics of diluted agricultural land and sand dust. It implies that the Beijing Dust Storm features were closely related to the dry saline lake, followed by the agricultural land and the hills, and weakly correlated with the sand and desert.

Beijing Dust Storms in March 27-29, and April 27-30, 2012 in Beijing were investigated the Backward trajectory to find that first Storm passed by northwest of Inner Mongolia, Hebei and Shanxi in the north, and the second one passed by Inner Mongolia, northern Hebei, Tianjin, Beijing and surrounding areas. Lead isotope composition in $PM_{10}$ were further compared with those from the soil of the regions which air mass passed by and found that the two storms might be from Dry Salt Lake areas such as Anguli Noel, Ha Natsu Tu Nord, Bai Yin stem Nur etc. Therefore, Inner Mongolia, North Hebei, Tianjin, Beijing and around including dry Saline Lake of Inner Mongolia, certainly contributed to Beijing Dust storm.

# OBSERVABLE CHARACTERISTICS OF CLOUD-TO-GROUND LIGHTNING INDUCED NOX AND $O_3$ OVER THE PRD REGION, CHINA

***Yonglin Liu***, Xinhui Bi, L. Y. Chan, Qinhao Lin, Leilei Fei, Xinming Wang, Ping'an Peng, Guoying Sheng

(Guangzhou Institute of Geochemistry, CAS, Guangzhou, China)

Ground level NOx and O3 from respective air quality monitoring stations along with cloud-to-ground (CG) lightning characteristic parameters deduced from the lightning location system (LLS) for five differing microenvironmental areas relative to mega city, city, municipal town, hilly suburban area and mountainous rural area conditions were examined in Pearl River Delta (PRD) study. We observed NOx enhancement and significant O3 decline on observable lightning stroke days (OLSDs) as compared to non-OLSDs. For observation sites in the upwind of lightning occurred, there were no correlation between lightning parameters and ground level NOx and O3. For stations in downwind, the lightning stroke frequency and average peak current were found to be positively related to ground level NOx after lightning and negatively related to O3. The lightning occurred distance and wind speed were negatively related to ground level NOx. Our LLS data analysis showed that there were high variation of lightning characteristics and phenomenal changes among these areas. Special features and phenomenal changes related to the lightning characteristic parameters, such as OLSDs and observable lightning stroke frequency and density for an OLSD, were also addressed. Observed variation of lightning stroke distribution and thunderstorm activity is used to analyze the corresponding changes in ground level NO, NO2 and O3 over five areas. The ground level NOx and O3 were affected by the transport from the upper or middle troposphere affect by downdraft of the thunderstorm and local microenvironmental meteorology. Microenvironmental variation due to change in topography, degree of urbanization, urban effect, as well as thunderstorm strength was found to affect the spatial distribution of lightning stroke and the severity of lightning activities over the observation areas. This approach increases our understanding of lightning induced NOx and O3 in subtropical China. It also tells us more about the behavior of lightning while the thunderstorm traverses through an observation area. This information is lacking in previous studies.

# ATMOSPHERIC AEROSOL COMPOSITIONS AND SOURCES AT TWO NATIONAL BACKGROUND SITES IN NORTHERN AND SOUTHERN CHINA

***Qiao Zhu***, Ling Yan He and ***Xiao Feng Huang***
(Key Laboratory for Urban Habitat Environmental Science and Technology, Peking University Shenzhen Graduate School, Shenzhen, China.)

Although China's severe air pollution has become a focus in the field of atmospheric chemistry and the mechanisms of urban air pollution there have been researched extensively, few field sampling campaigns have been conducted at remote background sites in China, where air pollution characteristics on a larger scale are highlighted. In this study, an Aerodyne high-resolution time-of-flight aerosol mass spectrometer (HR–ToF–AMS), together with other relevant instruments, was deployed at two of China's national background sites in northern (Lake Hongze site) and southern (Mount Wuzhi site) China in the spring season, in order to characterize submicron aerosol composition and sources. The campaign-average $PM_1$ concentration was 36.2 ± 20.2□g $m^{-3}$ at the northern China background (NCB) site, which was far higher than that at the southern China background (SCB)site (10.9 ± 7.8□g $m^{-3}$). Organic aerosol (OA) (27.1%), nitrate (27.0%), and sulfate (21.9%) contributed the most to the $PM_1$mass at NCB, while OA (43.5%) and sulfate(30.5%) were the most abundant components of the $PM_1$ mass at SCB, where nitrate only constituted a small fraction (4.7%) and might have contained a significant amount of organic nitrates (8–18%).The aerosol size distributions and organic aerosol elemental compositions all indicated very aged aerosol particles at both sites. The OA at SCB was more oxidized with a higher average oxygen to carbon (O/C) ratio (0.98) than that at NCB (0.67), and was found to be the most aged OA in real ambient air ever reported in the literature. Positive matrix factorization analysis was used to classify OA components and only secondary components were identified at both sites, including a semi-volatile oxygenated organic aerosol (SV-OOA) component and a low-volatile oxygenated organic aerosol (LV-OOA) component. LV-OOA accounted for more of the total OA at both sites. Using the total potential source contribution function model, the likely source areas of the major $PM_1$ components at both sites were a large regional scale in East Asia. The responsible sources may not only include emissions from the Chinese mainland but also significantly include emissions from ocean-going cargo ships and neighboring countries.

# DECREASING EMISSION FACTOR OF POLLUTANTS IN ABADAN REFINERY BY RENOVATING THE FURNACE DESIGN

**Abbas Zabihi** (Mazndaran Gas Company, Sari, Iran)
Mohammadreza Raazi Tabari (Azad Univesity of Ahvaz, Ahvaz, Iran)

The significant consumption of petroleum and gas by most industries in the world results in the emission of greenhouse gases into the atmosphere. Abadan refinery has always been the biggest and oldest in the Middle East and has a variety of refined products. After six months of collecting data about pollutant concentration emitted from a stack of old furnaces of units 75 and new unit 200, the emission factor of the pollutant was calculated. The result showed that the emission factor of some hazardous pollutants emitted from old unit 75 was tremendously higher than that of unit 200. This study suggests the installation of a forced fan to provide the excess air and a feed temperature controlling system to control fuel gas consumption. These would make the fuel combustion complete and decrease its consumption. Meanwhile, further results showed that the renovation of unit 75 could lead to a significant annual reduction of some pollutants such as CO, $H_2S$, and $C_xH_x$ to 66 ton, 3 ton, and 800 kg, respectively; this would increase the emission rate of pollutant $SO_2$ up to 150 ton annually. Finally, the new data of pollution coming from unit 75 were compared to that of American refineries units. Results showed that the emission factor of $NO_x$ emitting from renovated furnace of unit 75 was 55% more than that of American units. Moreover, the $SO_2$ emission of unit 75 was significantly higher than that of US refineries. The study suggested that $NO_x$ be used to reduce burners and refine fuel gas, in order to eliminate hydrogen sulfide, decrease nitrogen, and sulfur oxides.

# MODELLING GROUND-LEVEL OZONE CONCENTRATION USING IMPROVED DATA-MINING ALGORITHMS

*S. MOHAN*[1] and P. SARANYA[2]

([1]Environmental and Water Resources Engineering, Department of Civil Engineering, Indian Institute of Technology Madras, Chennai – 600036, India; [2]Environmental and Water Resources Engineering, Department of Civil Engineering, Indian Institute of Technology Madras, Chennai – 600036, India)

Ground level Ozone Concentration is an air-pollutant that is harmful to human health and vegetation. Also, the Environmental risks caused by exposure to ground level ozone have significantly increased during recent years, in many parts of the world as evident from literature. Thus the accurate ozone prediction or forecasting is important for warning the public of potential risks associated with heightened ozone levels. One major producer of ozone is the photochemical reaction between volatile organic components and the anthropogenic nitrogen oxides created by vehicular traffic. Therefore the measurement and monitoring of atmospheric ozone concentration levels is important. In this paper, the use of state-of-the-art machine learning approaches, especially, data mining approach in modelling the concentration of ground level ozone, measured in an industrial environment in the city of Chennai, India, once in 15 minutes over a period of 3 months in a critical season. The prediction is based on concentrations of three gases ($NO_2$, $SO_2$, and VOCs (Benzene, Toluene, p-Xylene)) and three meteorological parameters (ambient temperature, wind speed, wind direction, global radiation, and relative humidity). The analysis of the results indicates that accurate models for the concentration of ground level ozone can be derived with the best performance accuracies indicated by the data mining algorithms. The analysis carried out compares the use of different machine learning classifiers and proved that the Ensemble classifier Bagging performs superior to standard single classifiers, namely Artificial Neural Networks and Support Vector Machines, that are commonly adopted used in this type of analysis. In addition, the performance of the proposed Ensemble classifier technique when different base classifiers are used in optimising the configurations has been compared and found that the ensemble classifier outperforms the other single classifier systems. WEKA software was extensively used for the analysis in the study reported. The research conducted definitely bridges an existing research gap in big-data analytics related to prediction of environmental data, where present research is largely limited to using standard learning algorithms such as Neural Networks and Support Vector Machines often used and also takes the advantages of the standard learning algorithms.

# ADVANCED BUFFER MATERIALS FOR $CO_2$ CONTROL: IMPROVED AIR QUALITY AND ENERGY CONSERVATION IN COMMERCIAL BUILDINGS

Pavithra Ethi Rajan, Glenn Morrison, Fateme Rezaei
(Missouri University of Science and Technology, Rolla, MO, USA)

**ABSTRACT:** The following is novel approach to control the indoor $CO_2$ levels in closed spaces, majorly focusing on commercial buildings. Amine impregnated Silica were identified as suitable candidates and synthesized; the amine loading ratio was varied to create a range of materials. These samples along with other zeolite candidates have been tested at varying levels of $CO_2$ in various set ups to determine the theoretical and real time $CO_2$ adsorption capacity. The research work also focuses on the desorption capacity of the substance. The desorption is allowed to occur gradually over time having concentration gradient as its sole driving force; usually a pressure or temperature swing is applied to desorb the adsorbates out of adsorbent. Results have showed TEPA impregnated silica to have a good adsorption capacity (1.5 mmol/g); and on decreasing the amine content, the silica-TEPA samples showed good desorption rate (31%). At room temperature and pressure conditions, cyclic tests were conducted in a small chamber mimicking real time indoor space. The adsorbent aided in considerable reduction of the $CO_2$ concentration in the confined space.

***Keywords:*** Indoor Air Quality; $CO_2$ control; Amino polymers; TEPA; APS; PEI; Adsorption; Desorption

## INTRODUCTION

The ambient $CO_2$ concentration in outdoor air is about 400 ppm and that of indoor air is typically about 650-700 ppm. American Society of Heating, Refrigerating and Air conditioning Engineers (ASHRAE) recommend safe levels of $CO_2$ to be 1000-1100 ppm [5]. Studies also show that $CO_2$ at increased levels are considered as pollutants and affect cognition and productivity. At slightly higher levels, $CO_2$ induces stiffness and drowsiness; at greater levels the functioning of central nervous system is damaged. This affects and impairs the decision making performance [1].

Ventilation is required to dilute air pollution generated indoors, but the energy used to conditioning this air is responsible for a large fraction of energy consumption in the United States. While removing pollution sources is the best strategy, $CO_2$ that is generated by people cannot be removed. In this research, a novel *passive* $CO_2$ control system is developed that incorporates advanced sorbents into surface coatings. These coatings adsorb $CO_2$ during high-occupancy periods and release the $CO_2$ during lower occupancy periods, thereby reducing peak concentrations of $CO_2$, improving air quality and reducing the ventilation requirements. Research and survey show that out of the total energy consumed in typical buildings, about 40-60% is consumed for maintenance of appropriate ventilation and air circulation [6].

The first leg of this research work was about understanding and studying amine impregnated silica and their adsorption capacity. There are quite a few $CO_2$ sorbents typically preferred for these studies, activated carbon, zeolites, silica supported amino polymers, and several metal-organic frameworks are among those. Silica functionalized with amines are very promising sorbents for $CO_2$ capture. Amines owe their high adsorption capacities to the presence of high density of active sites per unit mass. silica–amines can be categorized into three groups: primary, secondary and tertiary amines [2]. Diamine functional groups and hyperbranched amino-silica were found to have good adsorption capacities[3][4]. The candidates selected suitable for this study were tetraethylenepentamine (TEPA), polyethyleneimine (PEI), (3-aminopropyl)triethoxysilane (APS) and a commercial silica (PD, from PQ-corporation) was used as a mesoporous support. Literature surveys showed TEPA-impregnated silica has been widely accepted as a good candidate in studies involving $CO_2$ capture. The preliminary tests of this research also prove the above statement. Out of the three amine candidates synthesized, PD-TEPA samples have high adsorption capacities and also display good desorption rates. Other than the amine candidates, Mg-MOF-74 was also considered for this study. Although MOFs are typically very good adsorbents, they are expensive and have stability issues. As the main objective of this study is to compare different materials to select a few suitable candidates as buffer material, activated carbon and zeolite 13X were also considered as potential buffer materials.

The next phase was the synthesis and characterization of the selected candidates, which included testing them for their adsorptive performance. The silica - amine samples were synthesized at varying ratios to generate a wide set of materials. Characterization included inspecting all the potential buffer materials by Fourier Transform Infrared Spectroscopy (FTIR). The Thermogravimetric Analysis (TGA) helped find out the $CO_2$ equilibrium adsorption capacity of each sample. For this, the sample is exposed to $CO_2$ gas for a specific period of time. Consecutively, purging nitrogen after $CO_2$ gas, the desorption rate as well was studied. The samples were exposed at both low (500 ppm) and high (3000 ppm) concentration of $CO_2$. The dynamic adsorptive performance of the material was assessed with the help of a fixed bed column by performing $CO_2$ breakthrough measurements. The material should adsorb $CO_2$ rapidly but also desorb readily. More importantly, this happens without the supply of heat or pressure.

In the next step, the sorbents are incorporated into surface coatings such as water-based latex paint and the performance of each sorbent is compared with that of the powdered form. Finally after identifying the best adsorbent candidates, they are applied to walls in a full-sized (8 $m^3$) chamber. The chamber simulates a small room with appropriate ventilation rates, surface area, $CO_2$ source rates, temperature and humidity controls. In this chamber, high and low occupancy periods can be simulated and the dynamic $CO_2$ buffering capability of each coating can be assessed under full-scale, realistic conditions. The primary goal of this paper is to make a comparative study analyzing the adsorption and desorption capacities of promising amino polymers of differing classes. Cyclic runs are studied to study and analyze the change in capacity (if any) of the adsorbent with each cycle.

A small scale chamber test was arranged and run to check the working capacity of the selected adsorbents. This set up consisted of a small confined space (20 L) into which a mixture of $CO_2$ and air was first purged, followed by purging just air. On comparing runs with and without the adsorbent material, it was observed that a considerable amount of $CO_2$ was reduced in presence of adsorbents.

The novelty of this research is in studying the desorption capacities of the adsorbents along with the adsorption capacities; having concentration gradient is the sole driving force here. Also, studies at low concentrations of $CO_2$ are rare. This study generates data on $CO_2$ buffering under realistic conditions that allow for extrapolation to multiple building types and conditions and also generates predictions about the applicability and energy consequences of $CO_2$ buffering for building stock across the U.S.

## EXPERIMENTAL SECTION

**Material Synthesis.** The amine-silica samples and the MOF samples were synthesized in lab. The initial step for all amine sorbents was drying silica overnight in a vacuum oven at about 105 ºC. Similar procedures were followed for the PEI and TEPA samples. 20 mL methanol/g amine was mixed and stirred for an hour. With silica added to this solution, it was left to react overnight. The methanol was then evaporated with the help of rotovap. The sample was recovered by drying in vacuum oven at 60 ºC for 1-2 h. The APS sample had a slightly different procedure where 70 mL toluene/g silica was degassed first. After half an hour, 0.3 ml DI water/g silica was added and let to equilibrate for 2-3 hours. Silica (PD) was added to this solution and later attached to condenser and heat to reflux at 85 ºC. This was left to react overnight; later, to be filtered with 100 mL toluene, 100 mL hexane, 100 mL ethanol. The sample was similarly recovered. The same procedures were followed for all the amine samples with different silica/amine ratio.

Solvothermal synthesis method was applied for synthesizing Mg–MOF–74. About 0.112 g of DHTA was dissolved along with 0.475 g of $Mg(NO_3)_2.6H_2O$ in a solution of 45 ml DMF, 3 ml ethanol, 3 ml DI water. This solution was mixed with the aid of sonication. This was sealed and placed in oven for 21 h at 125 °C. The mother liquor was decanted and replaced with methanol after cooling it down to room temperature. This methanol solvent was decanted 5 times repeatedly over the next 2 days. The product was then dried at 250 °C under vacuum. The sample was left to cool down gradually and then recovered. Commercial zeolite 13X and activated carbon were purchased and used in this study.

**Material Characterization**

***$N_2$ physisorption.*** BET characterization was carried out to determine the physical characteristics such as surface area and pore volume of the adsorbents. Liquid nitrogen BET analysis was performed using 3Flex gas analyzer (Micrometrics). The samples are required to be degassed before analysis and the degassing step was done using Smartvec.

***FTIR.*** The synthesized amine samples were characterized for functional groups using Fourier Transform Infrared Spectroscopy. Nicolet Nexus 470 FTIR spectrometer was used for performing this analysis. EZ-OMNIC software was used to plot and observe the spikes/peaks of the samples. A preparatory step of mixing amines with KBr was carried out before analyzing the amines. This mixing was done at 1:50 ratio of amine: KBr. A spectrum of the background was collected and then the prepared samples of each amine were tested. Each sample was analyzed at 16 scans per test.

**Equilibrium Adsorption Measurements.** TGA was used to determine the equilibrium adsorption measurements. The very first step in TGA is degassing the sample. This pulls out any gas adsorbed on the surface of the adsorbent; completely clearing out the active sites. The TGA follows a series of steps which can be modified as per requirement. For this study, the procedure was such that the material tested, is initially exposed to $CO_2$ gas for 4 hours and then nitrogen is purged for 6 hours. While the first half determines the adsorption capacity of the adsorbent, the second half shows the desorption capacity of the adsorbent. All the potential buffer materials were initially tested at 500 ppm $CO_2$. The amines were tested at 3000 ppm $CO_2$ as well and also with varying procedure. The procedures were varied to study the effect of desorption time, the effect of increasing the flow rate of nitrogen on the desorption capacity.

**Chamber Experiments.** The thermogravimetric analysis gives more of a theoretical adsorptive capacity. To understand how these materials function in real time, a small chamber set up was readied; this followed similar procedure as TGA. The only difference being, TGA uses change in weight over time to determine the capacity, the set-up was in such way that the outlet $CO_2$ concentration could be measured at regular intervals of time. The set-up consists of a small chamber (20 L). A slab of gypsum which acts similar to a wall in a room was placed inside this chamber. For preliminary tests, the adsorbent material was simply dusted on the slab. A mixture of air and $CO_2$ was let in, maintaining the level of $CO_2$ at about 3300-3500ppm. At the same time, the outlet is connected to a box having a $CO_2$ monitor. The sensors in the $CO_2$ monitor continuously pick up the levels of the gas and the data is stored in the computer attached to this monitor. On comparing runs with and without the adsorbent, the effect of the adsorbent can be studied.

## RESULTS AND DISCUSSION

**Physical Properties of Materials.** BET analysis helped in understanding the physical properties of the amine adsorbent. The volume of the pores in the material and the surface area of the adsorbent were calculated. Upon amine functionalization, both surface area and pore volume decreased. On comparing the pore volumes and surface area of all the three amine adsorbents, the TEPA impregnated silica sample showed better characteristics with 0.765 $cm^3/g$ pore volume and about 250 $m^2/g$ of surface area. Table 1 shows the compilation of the $N_2$ physisorption results.

Figure1 shows the peaks obtained from FT-IR tests. It was observed that all the silica – amine samples had similar peaks when compared with each other and against bare silica as well. The major peaks belonged to O-H stretching at about 3000 – 3100 $cm^{-1}$; Si-O-Si at around 1200 $cm^{-1}$; Si-$CH_3$ was recognized by sharp peaks at 1260 $cm^{-1}$ and 750-865 $cm^{-1}$; the N-H of amines were observed at 1600-1650 $cm^{-1}$ ; and Si peaks at 460-480 $cm^{-1}$.

**Table 1**. Physical properties of amine adsorbents

| **Adsorbent** | $S_{BET}$ ($m^2/g$) | $V_{pore}$ ($cm^3/g$) |
|---|---|---|
| PD-Bare | 294.1 | 1.04 |
| PD - TEPA | 247.4 | 0.76 |
| PD - PEI | 20.0 | 0.09 |
| PD - APS | 63.1 | 0.23 |

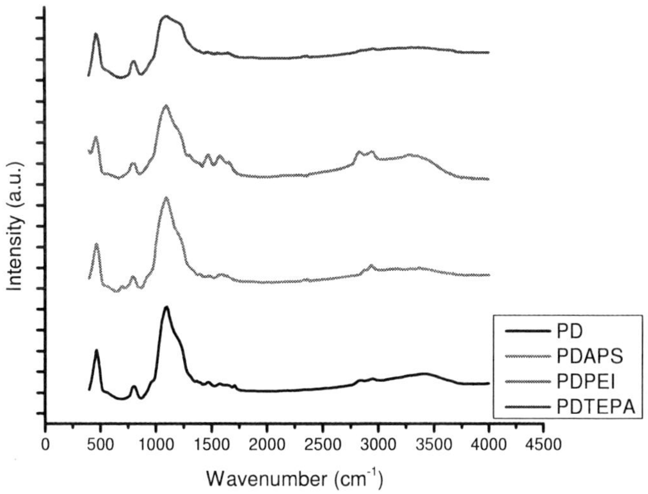


**Figure 1**. FTIR analysis of amine adsorbents

**Table 2.** Equilibrium adsorption measurements at 500 ppm $CO_2$

| **Adsorbent Material** | **Adsorption capacity (mmol/g)** | **Desorption capacity (mmol/g)** |
|---|---|---|
| PD-TEPA | 2.1 | 0.021 (10%) |
| PD - PEI | 1.2 | 0.102 (8.5%) |
| PD - APS | 0.7 | 0.211 (30%) |
| Mg-MOF-74 | 19 | - |

**Equilibrium Adsorption Measurements.** The equilibrium adsorption measurements were initially carried out at 500 ppm $CO_2$ for all the buffer materials. The sample to be tested was first degassed for 30 minutes and then cooled down to room temperature. Once in equilibrium at room temperature, the samples were first subjected to $CO_2$ at a concentration level of 500 ppm for 4 h. At the end of 4 h, the $CO_2$ supply is cut and nitrogen is supplied instead, for 6 h. This pattern helps understand the adsorption and desorption patterns of the samples tested. The same procedure has been followed to test all samples to have a common ground of comparison. It was found that only the amine impregnated silica samples showed desorption along with adsorption (2.1 mmol/g; 10% desorption capacity). The metal organic framework (Mg-MOF-74) though had a very high adsorption capacity (1.9 mmol/g), there was no desorption noticed. Table 1 discusses the adsorption and desorption capacities of various adsorbents considered as potential buffer materials for this study. Figure 2 shows the normalized weight change during adsorption and desorption in the TGA runs.

As observed from the initial results, amine impregnated silica samples are more promising and suitable for this particular study. This is because along with adsorption, the desorption capacity too important. On identifying this category of adsorbents, several samples were synthesized by altering the ratio of amine: silica. Also, from the preliminary tests, it can be observed that the silica-TEPA sample is the most ideal adsorbent as it has a very high adsorption capacity (2.1 mmol/g). Samples of silica-TEPA were synthesized by gradually varying the amine content in each PD-TEPA sample.

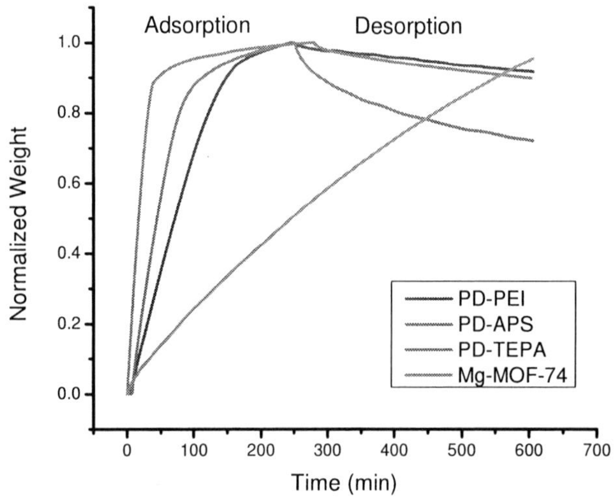


**Figure 2**. TGA plot for adsorption – desorption curves of amine adsorbents at 500pm $CO_2$

**Table 3**. Effect of amine load on adsorption – desorption capacities

| Adsorbent Material | PD:Amine ratio | Adsorption capacity (mmol/g) | Desorption capacity (%) |
|---|---|---|---|
| PD-TEPA_100% | 1:1 | 2.1 | 8 |
| PD-TEPA_50% | 1:0.5 | 1.4 | 31 |
| PD-TEPA_30% | 1:0.3 | 1.08 | 40 |
| PD-TEPA_1% | 1:0.01 | 0.2 | 89 |

The first PD-TEPA sample synthesized contained equal ratio of silica and TEPA (1:1). The consecutive samples were at 1:0.5 (ML1) and 1:0.3 (ML2). Reduction in the amount of amine added to the zeolite reduces the overall adsorption capacity of the adsorbent. But, lower amine content helps increase the desorption capacity. At 1:1 ratio, the sample was observed to have very high adsorption capacity, 2.1 mmol/g, but had a very low desorption capacity, of just about 10%. On reducing the amine content by half, at 1:0.5, the sample now adsorbs a lower quantity of $CO_2$, 1.4mmol/g, but the desorption capacity has gone up to 31%. On further reducing the amine content, to a ratio of 1:0.3, the same pattern is observed. The adsorption capacity is just about 1.08mmol/g and desorption rate has gone up to 40%. The same patter would be observed of the amine content is continued to reduce, a reduction in the overall adsorption capacity but an increase in the overall desorption rate.

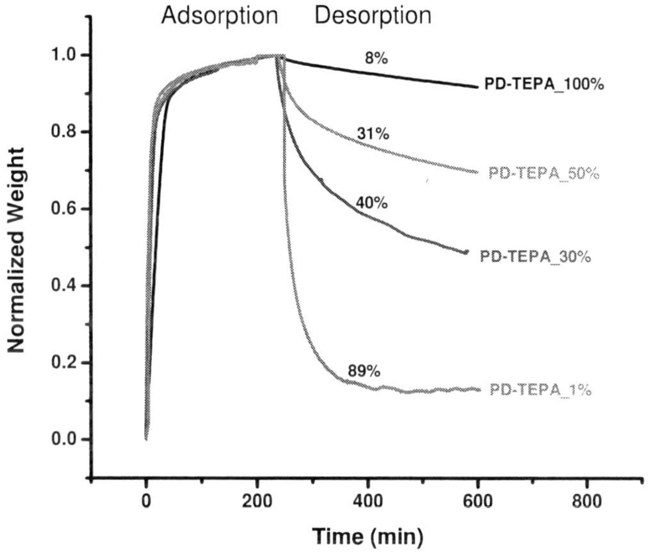


**Figure 3**. TGA plot for adsorption – desorption curves of PD-TEPA samples with varying TEPA loading.

**Table 4**. Effect of varying desorption conditions on desorption capacity

| **PD-TEPA** (desorption condition) | | **Desorption capacity** (%) |
|---|---|---|
| Time (h) | Flow rate (mL/min) | |
| 6 | 40 | 31 |
| 10 | 40 | 37 |
| 15 | 40 | 39 |
| 6 | 80 | 34 |

Among the three PD-TEPA samples, the sample with 1:0.5 ratio of PD:TEPA is of interest as it is more suited for further studies. To understand the properties and functioning of these adsorbents, the chosen sample was used to carry out the same TGA experimental procedure, but with varying conditions. The effect of duration of desorption time and the effect of flow rate of gas on the adsorbent was studied.

As it can be observed from the above plots and table, desorption capacity increases with longer desorption duration and with higher flow rate of gas. The amount of $CO_2$ desorbed increases with increase in desorption time; but the rate at which desorption occurs, decreases. Also, though there is an increase in desorption, on increasing the flow rate of nitrogen, the difference is considerably small. From these results, it can be concluded that the chamber tests can be run at 8-16 cycles (8 hour adsorption; 16 hour desorption).

All the tests so far conducted were based on 1 cycle of adsorption – desorption. Hence it was decided that cyclic runs needed to tested in the TGA to understand how the adsorption and desorption capacity work when run consecutively. And it was observed that the adsorption and desorption of $CO_2$ in the first cycle was similar to the previous results, but on the second and the third runs, the material adsorbed to its initial capacity and desorbed the entire amount adsorbed in that cycle. This too is promising as the overall adsorption and desorption capacities have not varied much

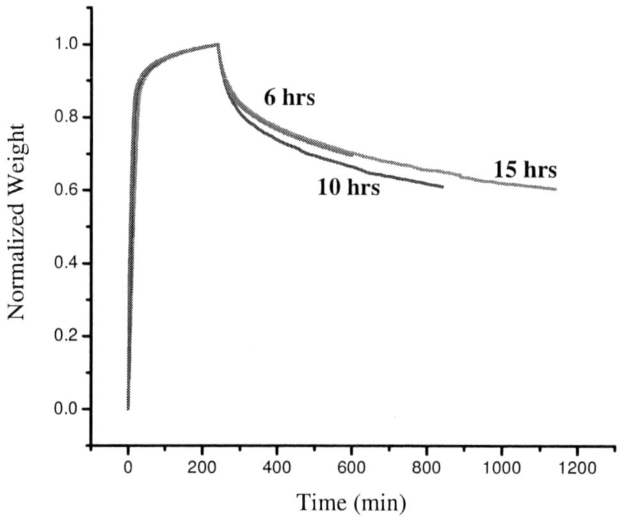


**Figure 4(a)**. Effect of time on desorption capacity

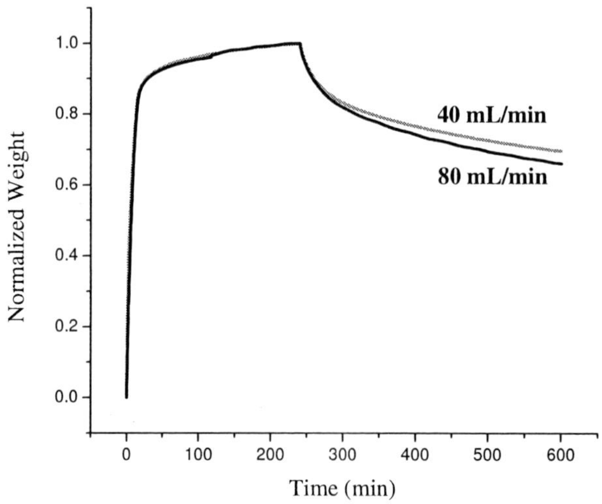


**Figure 4(b)**. Effect of air flow rate on desorption capacity

**Chamber Experiments.** The initial chamber tests showed a very slight reduction in the levels of $CO_2$ with the presence of adsorbents. But this is promising as some reduction in the level of $CO_2$ has been observed; however small the reduction is. This leads to new exploration possibilities. The quantity of adsorbent material used, the flow rate at which the gas enters the chamber are few of the conditions that can be altered to understand the working of the adsorbent.

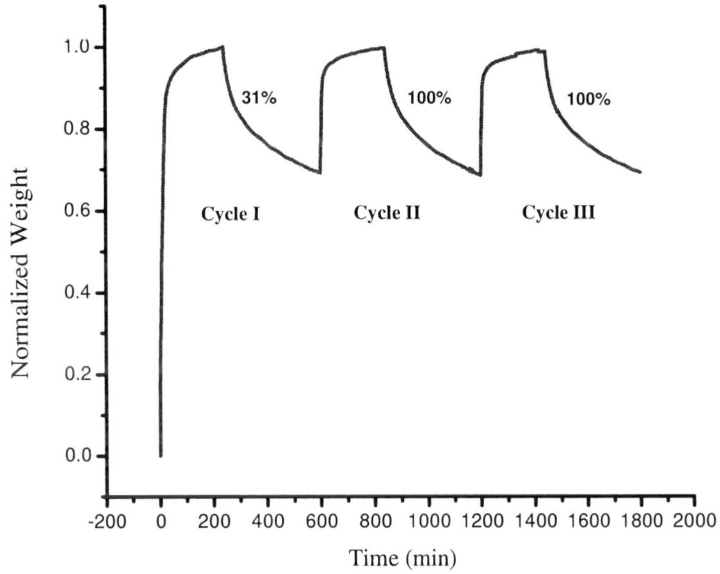


**Figure 5**. TGA plots for cyclic runs of adsorption – desorption

**Table 6**. Effect of varying inlet air conditions on adsorption – desorption

| **Adsorbent quantity** (g) | **Inlet gas flow rate** (mL/min) | **Adsorption capacity** (mmol/g) |
|---|---|---|
| 0.2 | 1000 | 0.68 |
| 1 | 1000 | 1.21 |
| 1 | 100 | 1.91 |
| 1 | 50 | 2.19 |

The first set of experiments was to determine an apt amount of adsorbent material to be used. A no-sample run was conducted for comparative purpose. Each run consisted of 3 cycles and each cycle consisted of 8 hr – adsorption and 16 hr – desorption. Runs were conducted at 0.2g and 1g adsorbent quantities. It is evident from the results that, increasing the amount of adsorbent material increases the adsorption capacity. This is because of the availability of more active sits on increasing the amount of the substance. And, we observe that the $CO_2$ levels do not rapidly fall after turning off the $CO_2$ supply; this is because, the $CO_2$ adsorbed by the material will be released back in to the air and then is flushed out of the chamber. With increase in the amount of material, longer shift occurs as more $CO_2$ is stored. Results indicated that 1.0 g adsorbent runs were more successful. Hence, the next set of runs – to determine the effect of inlet gas rate, was conducted with 1.0 g adsorbent material in the chamber. It was observed that on decreasing the flow rate of inlet gas, the adsorption capacity increased. A similar shift in the adsorption – desorption curves are noticed. Over here, as the over-all gas flow rate is reduced, the pace at which $CO_2$ is flushed out also decreases. Hence at lower flowrates, we find longer shifts. The adsorption capacity of the adsorbent is influenced by the inlet gas flowrate and the amount of adsorbent present.

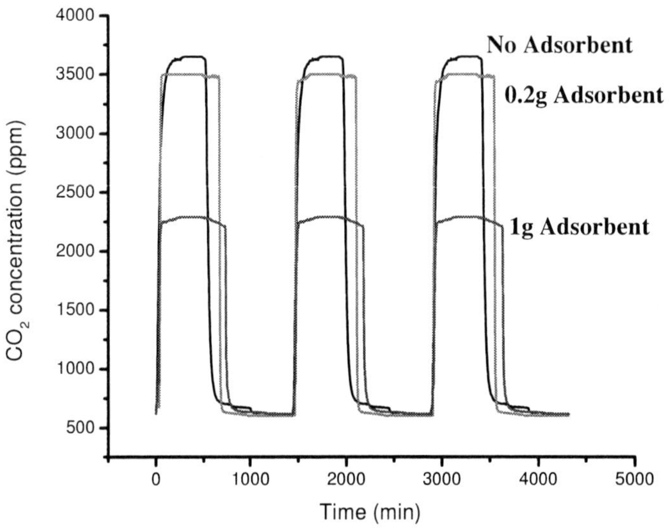


**Figure 6(a)**. Effect of amount of adsorbent on adsorption – desorption

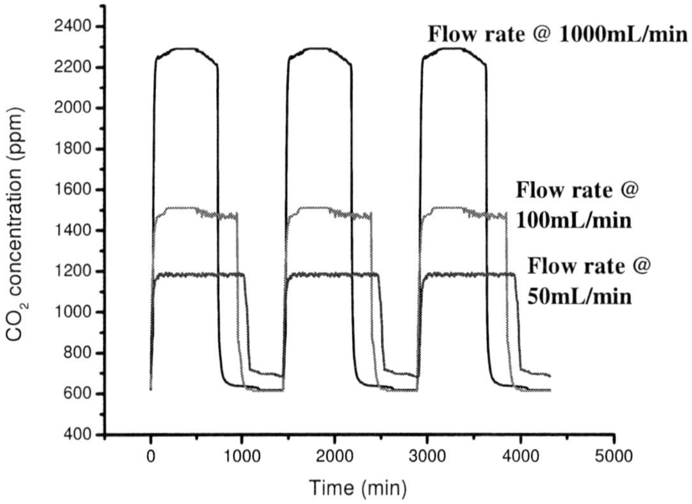


**Figure 6(b)**. Effect of inlet flow rate of air on adsorption – desorption

## CONCLUSION

Amine impregnated silica have always been considered as suitable adsorbents for controlling $CO_2$ levels. It has been predicted and proved in this work that concentration gradient can be a sole driving force for desorption to occur. Also, this study creates a platform on which further studies and experiments related to control of indoor $CO_2$ could be performed. The chamber tests mentioned in the study are small scale testing. A bigger chamber/room should be considered for further studies. Results from bigger chamber/room will help shape the project such that this idea can be implemented shortly.

## ACKNOWLEDGEMENTS

This work was supported by the National Science Foundation (NSF CBET-1549736). We acknowledge the Environmental Research Center (ERC) and the Department of Chemical Engineering of Missouri S&T for the opportunity and the lab facilities.

**REFERENCES**

[a] Department of Civil, Architecture and Environmental Engineering, Missouri University of Science and Technology, 1404, N Pine Street, Rolla, MO 65409, USA

[b] Department of Chemical and Biochemical Engineering, Missouri University of Science and Technology, 1404, N Pine Street, Rolla, MO 65409, USA

[1] Cindy Frei, Harvard Study: Carbon Pollution Impairs Cognitive Function. 2015

[2] Lee, Hsieh, et.al, $CO_2$ Adsorption by Y-type Zeolite Impregnated with Amines in Indoor Air

[3] Wurzbacher, Gebald, et.al, Separation of $CO_2$ from Air by Temperature - Vacuum Swing Adsorption using Diamine Functionalized Silica Gel

[4] Choi, Drese, Application of Amine Tethered Solid Sorbents for Direct $CO_2$ Capture from Ambient Air

[5] ASHRAE Standard 62.1-2013, "Ventilation for Acceptable Indoor Air Quality", plus ASHRAE BOD approved addenda.

[6] Distribution of Energy Consumption, Energy Data Book, US Department of Energy, 2006.

# AMINE IMPREGNATED ACTIVATED CARBON IS A NOVEL MATERIAL FOR $CO_2$ CAPTURE

***Dipa Das,*** Debi Prasad Samal, and B. C. Meikap.
(Indian Institute of Technology Kharagpur, West Bengal, India – 721302)

**ABSTRACT:** The quality of the adsorbents is decided by its textural characteristics and surface chemistry. Prepared activated carbons surfaces are usually nonselective and lack of reactive groups. By suitable modification of their surfaces the quality of adsorption increases. Impregnation is one of the most suitable modifications techniques used for activated carbons. Monoethanol amine (MEA) and diethanol amines (DEA) solutions were selected for impregnation on the surface of the activated carbon with impregnation ratio 0.4.$CO_2$ adsorption test was performed to investigate the effect of the modification on $CO_2$capture at standard temperature and pressure. Characterizations of the prepared samples have been done by physical and chemical analysis method. Amine impregnated sample leads to more $CO_2$uptake at flue gas temperature. This property of amine impregnated activated showed more $CO_2$ uptake with a higher efficiency as compared to simple activated carbon. Among the two activated carbon, the maximum $CO_2$ adsorption capacity ($q_e$) for monoethanolamine impregnated activated carbon(MEA-AC)was found out to be 36.83668 mg/g as compared to diethanol amine impregnated activated sample due to stronger carbamate formation and high stability.

## INTRODUCTION

Nowadays, large concentration of carbon dioxide, is emitted into the atmosphere. 60 percent of the global warming is caused by carbon dioxide (Houghton et al., 2001).Carbon captures and storage (CCS) process involves the separation of the $CO_2$from either stack gas or other intermediate gas streams. Absorption, adsorption, cryogenic distillation and membrane separations are the current methods for $CO_2$ capture from flue gas. Absorption, by using an amine-based solution is frequently used in the industrial processes. The major disadvantages of this process are large amount of energy required for regeneration (Gray et al., 2004; Plaza et al., 2007) and it is very expensive. Adsorption is considered as one of the most important methods for $CO_2$ capture (Reynolds et al., 2005). The solid adsorbent used for adsorption should have high $CO_2$ selectivity and capacity for $CO_2$capture. Among different porous materials, activated carbons are suitable adsorbents for $CO_2$capture (Siriwardane et al.,2005) due to its well developed micro-and mesoporosities. In case of activated carbon, the adsorption capacity decreases when the temperature is high and also water competes with $CO_2$ for adsorption (Wang et al., 2007; Plaza et al., 2010; Plaza et al.,2009; Pevida et al., 2008;Maroto-Valer et al., 2005). For efficient capture of $CO_2$ from flue gas, the adsorbent used should operate at high temperatures and also high adsorption capacity and selectivity (Maroto-Valer et al.,2008). Introduction of amines on the surface of the activated carbon offers more absorption and due to solids it is easy to handle, and no corrosion of the process equipment occurs (Gray et al., 2004;Zhao et al., 2007). Selective adsorption of $CO_2$of amine impregnated activated carbon was much more than the activated carbon adsorbents (Przepiorski et al., 2004; Wei wt al., 2009).The main objective of our work is the development of an amine impregnated activated carbon adsorbent with high adsorption capacity. Impregnation of amine solution on the surface of activated carbon is a suitable modification technique to improve their capture performance for adsorption of $CO_2$.For our current work, green coconut shell based ACs was used as adsorbents for gas adsorption.Monoethanol amine solution and diethanol amine solution were used to impregnate the surface of activated carbon. Carbon dioxide adsorption capacity was studied at standard temperature and pressure for the impregnated activated carbon material.

## MATERIALS AND METHODS

**Preparation of Adsorbents.** Fresh green coconut shells were collected from the nearby local market of IIT Kharagpur and then cut into small pieces followed by washing with tap water for removal of dirty material attached to it. Those washed green coconut shells were kept in the sunlight for 15-20 days till it becomes completely dry. Dried materials were kept inside the oven at 105°C for 48 hours for removal of moisture and other volatile impurities. Then the dried samples were crushed with a locally made crusher and sieved to a size of 512 μm. Chemical activation was done with $ZnCl_2$.20 gm. of dried precursor was well mixed with 100 ml of the concentrated solution of $ZnCl_2$ that contains 20 gm. of $ZnCl_2$ with impregnation ratio (activating agent/precursor) 1:1. The powdered material in slurry form was properly mixed and kept for 24 hours for proper soaking of $ZnCl_2$ on the surface of powdered precursor. The slurry was kept inside the oven at 100 °C for 24 hours. The resulting chemical impregnated samples were placed inside galvanized iron pipe of dimensions of length 8 cm and inner diameter of 1.5 cm and kept inside the furnace; The material inside the furnace was heated (10 °C $min^{-1}$) to the final carbonization temperature of 600 °C under the nitrogen flow rate of 120 $cm^3$ $min^{-1}$at STP. The material was kept inside the furnace for one hour at 600 °C. Then it was cooled under the constant flow of nitrogen gas till the temperature reaches the room temperature. The dried material was washed with 0.5 N HCl for 2-3 times followed by washing with warm distilled water for removal of any kinds of residual organic and mineral matter. Finally the material was washed with cold water till the solution becomes neutral. The sample was dried for 24 hours at 100 °C inside an airoven till it becomes completely dry to form activated carbon. The dried activated carbon was impregnated with diethanol amine solution ($HO\text{-}CH_2\text{-}CH_2\text{-}NH\text{-}CH_2\text{-}CH_2\text{-}OH$) and monoethanol amine solution ($OH\text{-}CH_2\text{-}CH_2\text{-}NH_2$) in the impregnation ratio of 0.4. Two different amine-impregnated activated carbon samples were prepared and then dried in the hot air oven at the temperature of 100 °C for 48 hrs till it became completely dry and kept inside an air tight container for our experimental purpose.

**Characterization of Adsorbents.** Before the use of the adsorbents for specific purpose, physical and chemical characterization has been done by different methods that are discussed in the following section. Density of the amine impregnated activated carbons were found to be 2.85 gm/$cm^3$ and 2.97gm/$cm^3$ for monoethanolamine impregnated activated carbon(MEA-AC) and diethanolamine impregnated activated carbon(DEA-AC) with impregnation ratio 0.4.

## RESULTS AND DISCUSSION

**Proximate and Ultimate Analysis.** Proximate analysis of the sample has been done for determination of the amount of percentage of moisture, volatile matter, fixed carbon and ash content. The result of proximate analysis was shown inTable1.

**TABLE 1. Results of Proximate Analysis**

| Sample | Proximate Analysis | | | |
|---|---|---|---|---|
| | Moisture (%) | Volatile Matter (%) | Ash (%) | Fixed Carbon (%) |
| RAW | 13.834 | 44.969 | 2.564 | 38.633 |
| AC | 12.983 | 7.208 | 1.902 | 77.907 |
| MEA-AC(0.4) | 11.338 | 17.528 | 0.5 | 70.634 |
| DEA-AC(0.4) | 12.439 | 17.12 | 0.45 | 69.991 |

From this analysis it has been observed that the fixed carbon content of AC and amine impregnated AC was very high as compared to raw precursor, which results in better adsorbent for adsorption purpose.

From the Table 1, it has been observed that the raw precursor has high volatile matter and low ash content, so it can be a good starting material for preparation of activated carbon (Singh et al., 2008).During carbonization and activation process, the coconut shell has been decomposed as a result of which the volatile matters were removed from the coconut shell and carbon content increases. $H_2O$ and $H_2$ are removed by $ZnCl_2$ (Yorgunet et al., 2009) instead of removal of hydrocarbons. Elemental analysis of the sample has been done by CHNS analyser and the corresponding values of the different adsorbents were shown in Table 2. It has been seen that due to the amine impregnation the nitrogen content of the samples were more as compared to activated carbon and the raw precursor.

**TABLE 2. Elemental Analysis**

| **Sample** | **Ultimate Analysis** | | | | |
|---|---|---|---|---|---|
| | C (%) | H (%) | N (%) | O (%) | S (%) |
| **RAW** | 46.185 | 9.416 | 0.99 | 33.385 | 10.024 |
| **AC** | 69.04 | 2.963 | 1.16 | 26.791 | 0.046 |
| **MEA-AC(0.4)** | 66.15 | 2.595 | 7.66 | 23.329 | 0.266 |
| **DEA-AC(0.4)** | 66.09 | 2.888 | 5.062 | 25.824 | 0.136 |

**Pore Structure and Surface Area of the Adsorbents.** The textural properties of the prepared samples have been determined by nitrogen adsorption and desorption isotherm at 77K. Before adsorption, the samples were degassed at $300^0$C under vacuum for 5 hr.The total surface area,micropore surface area, and the total pore volume of the prepared samples were found by applying Brunauer-Emmett-Teller(BET) and t-plot method in the relative pressure (P/P°) range of 0.05-0.30 (Brunauer et al.,1938).Micropore volume was found out by DR method. Mesopore volume was found out from deducting micro pore volume from total pore volume. The pore size distributions of the prepared samples have been determined by BJH model (Barrett et al., 1951). The pore structure parameters of the different samples were shown in Table 3.It is observed from the Table 3 that surface area and pore volume of the amine impregnated activated samples, were decreases due to pore blockage by functional groups attached on the surface as widely reported in the literature (Gorgulho et al., 2008).

**TABLE 3. Pore Structure Parameters**

| **Samples** | **$S_{BET}$ ($m^2$/g)** | **$V_T$ ($cm^3$/g)** | **M.P.V ($cm^3$/g)** | **Avg. Pore Radius ($A^O$)** | **Micro Pore Area ($m^2$/g)** |
|---|---|---|---|---|---|
| **RAW** | 59.728 | 0.05031 | 0 | 16.8461 | 0 |
| **AC** | 995.799 | 0.4487 | 0.372 | 9.01198 | 921.71 |
| **MEA-AC(0.4)** | 569.352 | 0.2795 | 0.201 | 9.81965 | 488.716 |
| **DEA-AC(0.4)** | 552.776 | 0.2621 | 0.198 | 9.79635 | 471.324 |

**$CO_2$ Capture Performance of Amine Impregnated Sample.** $CO_2$ batch adsorption analysis was carried out by autosorbiQ manufactured by Quanta chrome instrument. The samples were out-gassed at 523K for 6 hours upto atmospheric pressure. After degassing, under a vacuum, the carbon dioxide adsorption of the sorbents at relative pressure ($P/P_0$) was evaluated at 298K. From the Figure 1, it has been seen that the adsorption of $CO_2$ was carried out till the relative pressure ($P/P_o$) reaches the atmospheric pressure. The maximum volume of $CO_2$adsorbed for MEA-AC (0.4) and DEA-AC (0.4) at atmospheric pressure was found out to be 18.7464 and 13.8062 cc/g. Similarly, the maximum $CO_2$ adsorption capacity ($q_e$)for MEA-AC (0.4), and DEA-AC (0.4) was found out to be 36.83668 and 24.08795 mg/g .From the Figure 1 it can be observed that MEA-AC (0.4) has better adsorption capacity than DEA-AC (0.4). Amines are the important source of basic sites for acidic carbon dioxide gas adsorption. Due to their acid-base properties, amine on the AC surfaces increases the adsorption capacity and selectivity of carbon dioxides. Typically, $CO_2$ is lewis acid and amines are lewis base. There is the presence of a lone pair of electrons on a nitrogen atom on amine functional groups.

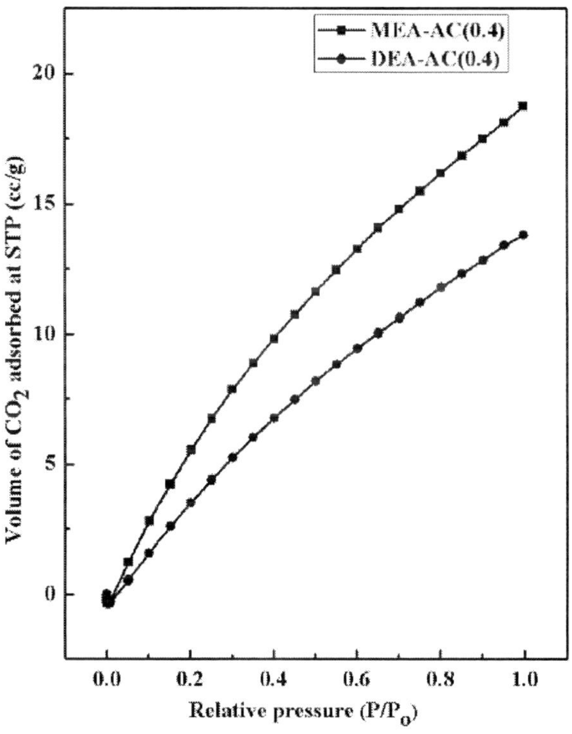


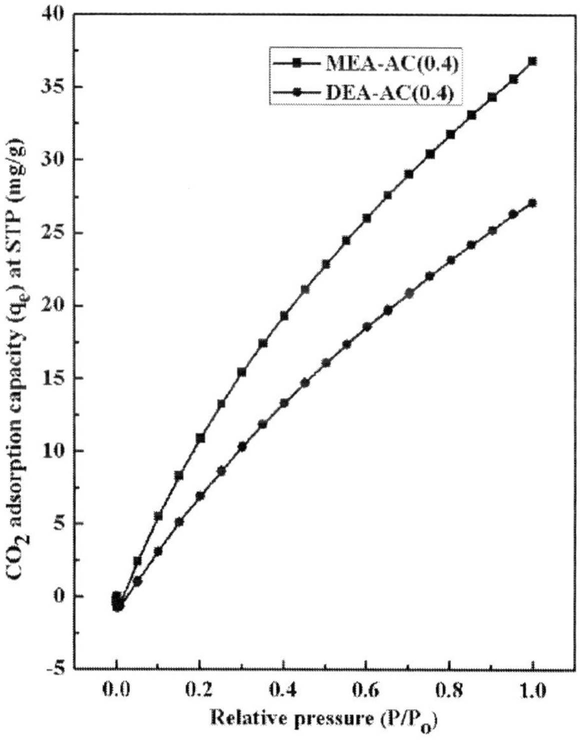


(a) (b)

**FIGURE 1. (a)Volume of $CO_2$ adsorbed vs. Relative pressure and (b)$CO_2$ adsorption capacity vs. Relative pressure for amine impregnated activated carbon adsorbents**

So by which there is nucleophilic attack on the acidic carbon dioxide as a result of which there is formation of carbamate through the carboxyl group (Danckwerts et al.,1979; Caplow et al., 1968).The adsorption between the active site of amine and carbon dioxide is the driving force for capture of carbon dioxide by amine-modified carbon materials. It is observed from the Figure 1 that MEA modified samples have a higher capacity than DEA due to the effect of chemisorptions. $1^{o}$ amine was more stable than $2^{o}$ amine because MEA is of less viscous than DEA that affects the $CO_2$ reaction to a greater extent. The main reason can be described because of the –I effect of two alcohol groups present in $2^{o}$ amine which acts as electron repellent to the existent electron deficient N+ ion present in carbamate. Hence, the stability of $1^{o}$ amine is more compared to $2^{o}$ amine (Danckwerts et al., 1979) and thus MEA containing adsorbents shows greater adsorbent capacity than DEA containing adsorbent. As amine anchoring decreases surface area and increases the amine content. From the graph it can be seen that at same impregnation ratio of the amine impregnated carbon the volume of $CO_2$ adsorption and $CO_2$ adsorption capacity ($q_e$) of MEA-AC was more than that of DEA-AC.From these results, MEA-AC may act as better adsorbent than DEA-AC for $CO_2$ capture.

**Thermo Gravimetric Analysis.** Figure 2 shows the plot of original weight verses temperature for MEA-AC (0.4) and DEA-AC (0.4). When the temperature was raised from room temperature 27 $^{0}$C to a very high temperature 800 $^{0}$C, amine impregnated activated carbons have high thermal stability and three zones of the profile are obtained. 1st stage, accounts for removal of volatile matter, moisture and other impurities. 2nd stage shows the decreasing slope accounts for slow decomposition. In 3rd stage with further rise in temperature the downward slope again increases rapidly for both cases that account to severe erosion of material. Clearly with respect to increasing the temperature from 27-800 °C, the total weight loss for MEA-AC(0.4) and DEA-AC(0.4) obtained are 36.102 % and 42.876 % . By analyzing both graphs we observed that the MEA-AC (0.4) has very high resistant to weight loss compared to the other for the same rise in temperature change. From this result we understand that impregnating activated carbon with monoethanol amine solution results high thermal stability in comparison to diethanol amine solution and it is thermally stable and resistant to surface erosion with rise in temperature and can be used in industrial scale application.

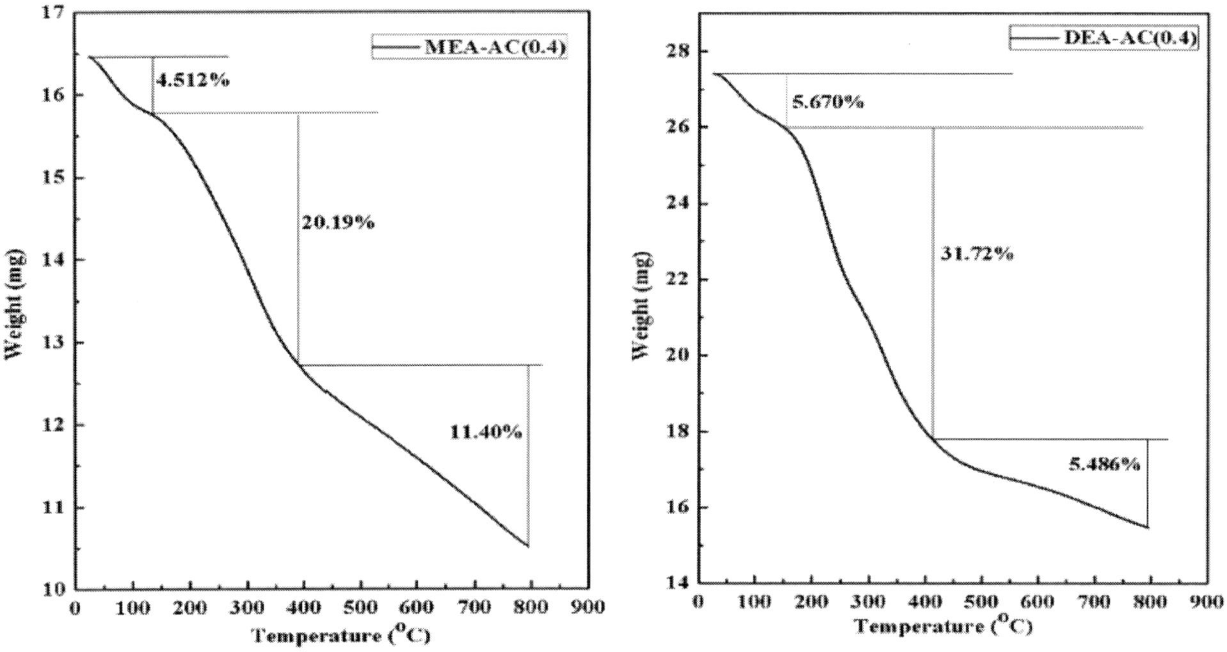


(a) (b)

**FIGURE 2. Thermal gravimetric analysis of (a) MEA-AC (0.4) and (b) DEA-AC (0.4)**

**FTIR Analysis of the Amine Impregnated Activated Sample.** The FTIR spectra of amine impregnated activated carbon of impregnation ratio 0.4 at the absorption band 4000–400 $cm^{-1}$ intervals, has been shown in Figure 3. From the graph, we observe all kinds of functional groups such as carboxylic acid, alcohol, ether, ester, phenols, aliphatic amine, nitro, C-H groups and other aliphatic groups present in amine impregnated activated carbon. The bands at 609.42 and 607.09 $cm^{-1}$ have been assigned to N-H group. At 1320-1000$cm^{-1}$(strong) the functional group presents are alcohol, carboxylic acid, esters and ethers. A band at 1100 $cm^{-1}$ is attributed to C-O stretching. The band at 1054 $cm^{-1}$ may be due to the C-N stretching vibration. The bands at 1735$cm^{-1}$ are due to N-H deformation vibration (Tang et al., 2012)

## CONCLUSIONS

In this work, activated carbons were impregnated by monoethanol amine and diethanol amine solutions with impregnation ratio 0.4.Carbon dioxide adsorptions of the samples were performed. From the ultimate analysis it was found that the nitrogen content of the amine impregnated ACs was higher than that of raw material and activated carbon. Due to the presence of nitrogen group of the amine modified material the carbon dioxide adsorption capacity was more than that of simple activated carbon. From the results of the BET surface area, it has been seen that amine impregnated ACs have less specific surface area, pore volume and pore diameter, due to their blocking of pores by amine functional groups. But at the same time creating many active sites for $CO_2$ adsorption by chemisorptions method. Many active sites have been created on impregnated activated carbon particles and inside the micropores.The MEA impregnated AC shows higher $CO_2$ adsorption capacity. The maximum $CO_2$ adsorption capacity ($q_e$) for monoethanolamine impregnated activated carbon (MEA-AC) was found out to be 36.83668 mg/g than DEA-AC due to stronger carbamate formation. Among the impregnated activated carbon samples, MEA-AC (0.4) is the more suitable adsorbents for $CO_2$ capture.

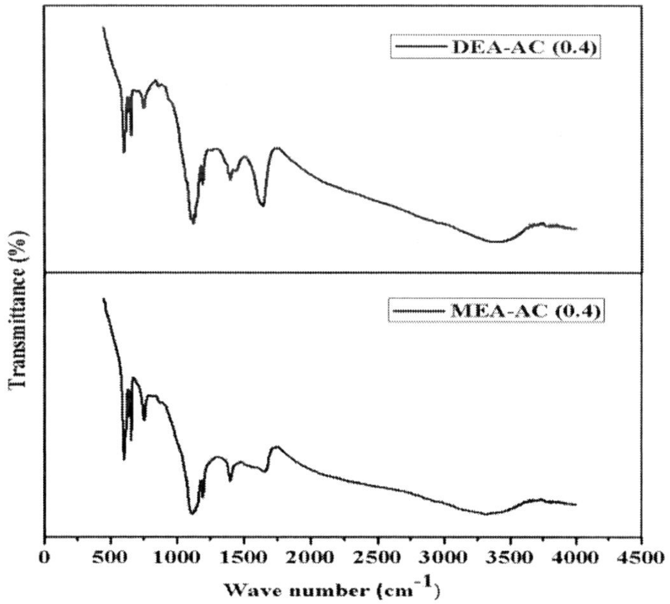


**FIGURE 3. FTIR spectra of MEA-AC (0.4) and DEA-AC(0.4)**

## ACKNOWLEDGEMENTS

This work was supported by the Chemical Engineering Department, Indian Institute of Technology, Kharagpur

## REFERENCES

Barrett, E. P., L. G. Joyner, P. P.Halenda. 1951. "The determination of pore volume and area distributions in porous substances. I. Computations from nitrogen isotherms". *J. Am. Chem. Soc.* 73(1): 373-380.

Brunauer, S., P. H. Emmett, and E. Teller. 1938. "Adsorption of gases in multimolecular layers". *J. Am. Chem. Soc.* 60(2): 309-319.

Caplow, M. 1968. "Kinetics of carbamate formation and breakdown". *J. Am. Chem. Soc.* 90(24): 6795-6803.

Danckwerts, P. V. 1979. "The reaction of $CO_2$ with ethanolamines". *Chem. Eng. Sci.* 34(4): 443-446.

Gorgulho, H. F., J. P.Mesquita, F. Goncalves, M. F. R. Pereira, and J. L.Figueiredo. 2008. "Characterization of the surface chemistry of carbon materials by potentiometric titrationsand temperature-programmed desorption". *Carbon.* 46(12):1544–1555.

Gray, M. L., Y.Soong, K. J.Champagne, J.Baltrus, R. W.Stevens, P.Toochinda, S. S. C.Chuang. 2004. $CO_2$ capture by amine-enriched fly ash carbon sorbents. *Sep. Purif. Technol.*35(1): 31-36.

Houghton, J. T.,Y. D. J. G. Ding, D. J. Griggs, M. Noguer, P. J. van der Linden, X. Dai, and C. A. Johnson. 2001. Climate change 2001: the scientific basis.

Maroto-Valer, M. M., Z. Lu, Y. Zhang, and Z.Tang. 2008. "Sorbents for $CO_2$ capture from high carbon fly ashes". Waste Management. 28(11): 2320-2328.

Maroto-Valer, M. M., Z.Tang, andY. Z. Zhang. 2005. "$CO_2$capture by activated and impregnated anthracites". *Fuel. Proc. Technol.* 86(14): 1487–1502.

Pevida, C., M. G. Plaza, B.Arias, J.Fermoso,F.Rubiera, and J. J.Pis. 2008. "Surface modification of activated carbonsfor $CO_2$ capture".*Appl. Surf. Sci.* 254(22): 7165–7172.

Plaza, M. G., C. Pevida, B.Arias, J. Fermoso, M. D. Casal, C. F. Martín, and J. J. Pis. 2009. "Development of low-cost biomass-based adsorbents for postcombustion $CO_2$ capture". *Fuel,* 88(12): 2442-2447.

Plaza, M. G., C.Pevida, A.Arenillas, F.Rubiera, and J. J.Pis. 2007. "$CO_2$ capture by adsorption with nitrogen enriched carbons". *Fuel.* 86(14):2204-2212.

Plaza, M. G., C.Pevida, C. F. Martín, J. Fermoso, J. J. Pis, and F.Rubiera. 2010. "Developing almond shell-derived activated carbons as $CO_2$ adsorbents". *Sep. Purif. Technol.* 71(1): 102-106.

Przepiórski, J., M. Skrodzewicz, and A. W. Morawski. 2004. High temperature ammonia treatment of activated carbon for enhancement of $CO_2$ adsorption. *Appl. Surf. Sci.* 225(1), 235-242.

Reynolds, S. P., A. D. Ebner, and J. A.Ritter. 2005. "New pressure swing adsorption cycles for carbon dioxide sequestration". *Adsorption,* 11(1): 531-536.

Singh, C. K., J. N.Sahu, K. K.Mahalik, C. R.Mohanty, B. R.Mohan, and B. C.Meikap. 2008. "Studies on the removal of Pb (II) from wastewater by activated carbon developed from Tamarind wood activated with sulphuric acid". *J. Hazard. Mater.* 153(1): 221-228.

Siriwardane, R. V., M. S. Shen, E. P.Fisher, andJ.Losch. 2005."Adsorption of $CO_2$ on zeolites at moderate temperatures". *Energy Fuel.* 19(3): 1153-1159.

Tang, Y. B., Q.Liu, and F. Y.Chen, 2012. "Preparation and characterization of activated carbon from waste ramulusmori". *Chem. Eng. J.* 203, 19-24.

Wang, N., L.Ma, A.Wang, Q.Liu, T.Zhang. 2007."$CO_2$ adsorption on SBA-15 modified by aminosilane". *Chin. J. Catalysis.* 28 (9): 805.

Wei, J., J.Shi, H.Pan,Q. Su, J. Zhu, and Y.Shi. 2009. "Thermal and hydrothermal stability of amino-functionalized SBA-16 and promotion of hydrophobicity by silylation". *Microporous and Mesoporous Materials.* 117(3):596-602.

Yorgun, S., N.Vural, H.Demiral. 2009 Preparation of high-surface area activated carbons from Paulownia wood by $ZnCl_2$ activation. *MicroporousMesoporousMater.* 122(1): 189-194.

Zhao, H.,J. Hu, J.Wang, L. Zhou, and H.Liu. 2007. "$CO_2$ capture by the amine-modified mesoporous materials". *ActaPhysico-ChimicaSinica.* 23(6), 801-806.

# CATALYTIC OXIDATION OF VOCS AND CO OVER OCTAHEDRAL LAYERED BIRNESSITES SYNTHESIZED BY DIFFERENT METHODS

Zhidan Fu, Qing Ye*, Heng Lu, Shuiyuan Cheng, Wang Dao
(Laboratory of Beijing on Regional Air Pollution Control, College of Environmental and Energy Engineering, Beijing University of Technology, Beijing 100124, China)

Catalytic oxidation of volatile organic compounds (VOCs) and CO is one of the most effective technologies. Manganese oxide shows good catalytic performance in a number of oxidation processes. The manganese oxides have long been used as an active catalyst for the oxidation of carbon monoxide, methane, and hydrocarbons. The aim of the present work is to make manganese oxide octahedral layered (OL) birnessites using different methods and compare the catalytic performance of these materials.

The octahedral layered nanofiber-like birnessite (OL-1) was prepared via the oxidation $Mn^{2+}$ by $MnO_4^-$ in the KOH solution. The octahedral layered wool ball-like birnessite (OL-2) catalyst was prepared using the method similar to that for OL-1 fabrication but adding $HNO_3$ instead of KOH to the mentioned. Moreover, the mixed solution was hydrothermally treated in Teflon-lined autoclave for at 170 °C for 4 days. The OL catalysts were prepared via the reduction $MnO_4^-$ by ethanol or glucose aqueous in solution, and the obtained catalysts were denoted as OL-3 and OL-4, respectively. The obtained OL-1, OL-2, OL-3, and OL-4 precursors were filtered, washed, and calcined in air at 300 °C for 2 h. The OL catalysts were characterized by the XRD, BET and $H_2$-TPR techniques. Catalytic activities of the OL samples for the oxidation of CO, benzene or toluene. The reactant mixture was (1% CO + air), (2000 ppm benzene +air) or (2000 ppm toluene +air), the corresponding space velocity was 15,000 mL/(g h), and 40,000 mL/(g h).

The results show that the preparation method significantly had an important influence on the property of the OL catalyst. It is found that all of the samples have birnessite-type octahedral layered structure and an interlayer spacing of ca. 0.72 nm. The surface areas of the OL-1, OL-2, OL-3, and OL-4 samples were 7.6, 13.4, 63.2, and 42.6 $m^2/g$, respectively. Obviously, surface areas and pore volumes of the OL-3 and OL-4 samples were much higher than those of the OL-1 and OL-2 samples. There was co-presence of $Mn^{3+}$, $Mn^{4+}$ and/or $Mn^{2+}$ on the surface of these samples. Based on the manganese ion contents, the average oxidation states of surface Mn species in the birnessite samples were in the range of 3.2-3.5, which was lower than those (3.5-3.9) obtained from the $H_2$-TPR studies. The amounts of oxygen vacancies and lattice oxygen mobility of the OL-1 and OL-4 samples were higher than those of the OL-2 and OL-3 samples. Either in CO oxidation or in benzene or toluene oxidation, the catalytic activity decreased in the order of OL-1 > OL-4 >> OL-3 > OL-2, with the OL-1 sample showing the best performance ($T_{50\%}$ = 115 and $T_{100\%}$ = 150 °C for CO oxidation at 15,000 mL/(g h), $T_{50\%}$ = 200 °C and $T_{95\%}$ = 240 °C for benzene oxidation, and $T_{50\%}$ = 190 °C and $T_{95\%}$ = 230 °C for toluene oxidation at 40,000 mL/(g h)).

# PHOTOCATALYTIC OXIDATION OF FUEL VAPORS: A NOVEL METHOD FOR REDUCING FUEL VAPOR EMISSIONS FROM AUTOMOBILES

**Catherine B. Almquist**
(Department of Chemical and Paper Engineering, Miami University, 650 E High Street, Oxford, Ohio 45056, USA)

Fuel vapor recovery systems on automobiles are required on all models in the US since the year 2000. On each automobile is a carbon canister that recovers fuel vapors during refueling. When the automobile is turned on, air is passed through the carbon filters to desorb the fuel vapors, which are directed to the engine and burned. However, a parked automobile is exposed to temperature swings throughout the day and night, and the fuel vapors from the gasoline tank are continuously challenging the carbon filters. Eventually, the carbon canisters become saturated with fuel vapors while the automobile is not in operation, and the fuel vapors pass through the carbon filter and, thus, are emitted to the environment. This is especially problematic on warm sunny days, when volatile organic carbon emissions contribute to urban smog.

Photocatalytic oxidation of fuel vapors is being investigated for application in fuel vapor recovery systems. The idea is to destroy volatile organic compounds that escape from the carbon filters in an automobile before they are emitted to the environment. With the development of ultraviolet light emitting diodes (UV LEDs), it is possible to incorporate battery-powered UV light sources that are small enough for very niche applications, such as fuel vapor recovery systems. UV LEDs and a suitable photocatalyst, then, can be used to degrade VOCs within the automobile's fuel vapor recovery system. Results of this study include effects of humidity, temperature, and concentration of selected organic compounds on the efficiency and effectiveness of photocatalytically degrading fuel vapors within the fuel vapor recovery system in an automobile.

# SYNTHESIS OF LIAL$_2$-LAYERED DOUBLE HYDROXIDES FOR $CO_2$ CAPTURE OVER A WIDE TEMPERATURE RANGE

**Liang Huang,** Junya Wang, Yanshan Gao, Yaqian Qiao, Qianwen Zheng and Qiang Wang
(Beijing Forestry University, Beijing, Cnina)
Zhanhu Guo (Lamar Universit, TX, USA)
Yufei Zhao (Chinese Academy of Sciences, Beijing, China)
Dermot O'Hare (University of Oxford, Oxford, UK)

Although there have been many reports on layered double hydroxide (LDH) derived $CO_2$ adsorbents, none of them have studied the special case of $LiAl_2$ LDHs. Here we report the first detailed investigation of the performance of $LiAl_2$ LDHs as novel $CO_2$ adsorbents. $LiAl_2$ LDHs were synthesized using both traditional coprecipitation and gibbsite intercalation methods. All the materials were thoroughly characterized using XRD, SEM, TEM, FTIR, BET, and TGA. The $CO_2$ capture performance of these LDHs were investigated as a function of the charge compensating anions, Li/Al ratio in preparation solution, calcination temperature, adsorption temperature, and doping with $K_2CO_3$. The data indicated that $LiAl_2$ LDHs derived compounds can be used as $CO_2$ adsorbents over a wide temperature range (60–400 °C), with a $CO_2$ capture capacity of 0.94 and 0.51 mmol $g^{-1}$ at 60 and 200 °C, respectively. By doping $LiAl_2$-$CO_3$ LDH with 20 wt% $K_2CO_3$, the $CO_2$ adsorption capacity was increased up to 1.27 and 0.83 mmol $g^{-1}$ at 60 and 200 °C, respectively. $CO_2$ adsorption/desorption cycling studies showed that both pure $LiAl_2$ LDH and the $K_2CO_3$ promoted $LiAl_2$ LDH had stable $CO_2$ capture performance even after 22 cycles. Considering its high $CO_2$ capture capacity and good cycling stability, $LiAl_2$ LDH based novel $CO_2$ adsorbents have significant potential for $CO_2$ capture applications.

# NOVEL DeNOx AND DeSOx TECHNOLOGIES FOR FLUE GAS TREATMENT FOR BOILERS AND FURNACES

***Jian Luo***
(KL Chemical Technologies, Inc., Tianjin Binhai EnviroTechnologies, Inc., Tianjin Key Lab of Heavy Metal Containing Waste water Treatment and Resources Recovery)
Qiuhua Zhang
(Tianjin Key Lab of Heavy Metal Containing Waste water Treatment and Resources Recovery, Tianjin, 300350, China)

Luo's Tehnologies for flue gas treatment, including a fabric filter device for particulate (as well as a part of heavy metal elements) removal, a fixed bed reactor for integrated removal of NOx, SOx, and heavy metals have been employed for treating flue gas for industrial and municipal boilers and furnaces. A novel oxidation catalyst for NO oxidation and subsequent removal, a reactant for SOx removal has been included for DeNOx and DeSOx, respectively. Two case studies, the applications of above-mentioned technologies and related devices combination for an industrial boiler with a capacity of 35 t/h in Langfang, Hebei Province and a municipal furnace for home heating in Tianjin (both in China) were shown. In both cases, the emission levels in terms of particuates, NOx, SOx, and heavy metals are well below those stipulated by Chinese nationals standards and are below those for natural gas firing boilers and furnaces. Pictures for devices, monitoring results, and comparison to other technologies will be shown in presentations.

# IMMOBILIZATION OF SELF-STABILIZED PLASMONIC Ag-AgI ON MESOPOROUS $Al_2O_3$ FOR NO ELIMINATION UNDER LED ILLUMINATION

***Chun He,*** Xiuqin Tan, Jiawei Zeng and Yaohua Hou (Sun Yat-sen University, Guangzhou, China)
Dong Shu (South China Normal University, Guangzhou, 510006, China)

Nitrogen oxides ($NO_x$), which result from automobile exhaust gas and industrial combustion of fossil fuels, are a major source of air pollution and cause ozone depletion, acid rain and photochemical smog. This has serious implications on the environment and health of the mankind. Till now, many conventional techniques have been developed to eliminate NO in industrial emission. However, these technologies were not economically feasible for removal of air pollutants at the parts per million (ppm) levels. The ideal process should be developed with further decreasing the economic and environmental costs of potential secondary pollution. Visible light driven (VLD) photocatalysis as a green technology that could use natural sunlight or artificial light to purify air pollutants at ambient condition, provides an attractive alternative to conventional approaches. Currently, the localized surface plasmon resonance (SPR) effect of noble-metal nanoparticles (e.g. Ag, Au, Pt) has become a research focus in the development of VLD photocatalysts. For practical application, the powdery photocatalyst would be simply blown away and time-consuming for nano-photocatalysts recycle. Therefore, to apply photocatalytic techniques in waste gas purification, it is essential to develop a facile process to effectively immobilize the powdery photocatalyst on a proper support

In this work, the plasmonic Ag-AgI with photoinduced self-stability was successfully immobilized on mesoporous $Al_2O_3$ by a wet impregnation-precipitation and in-situ photoreduction method. The immobilized Ag-AgI/$Al_2O_3$ was applied for photocatalytic removal of ppm-leveled NO under illumination of indoor LED tubes, which could obtain 100% NO removal in minutes at optimized conditions. The large dispersion of Ag-AgI on mesoporous $Al_2O_3$ was firmly enough to overwhelm the continuous air flowing, and the enlarged surface area of Ag-AgI could efficiently enhance NO removal efficiency than pure Ag-AgI powder. Most importantly, the plasmonic Ag-AgI/$Al_2O_3$ was stable in activity and can be used repeatedly without deactivation. On the basis of reactive species trapping and reaction intermediate monitoring, the responsible reactive species of photogenerated $^{\bullet}OH$, $\bullet O_2^-$ $H_2O_2$ and $h^+$ were confirmed and the possible reaction mechanism of photocatalytic NO oxidation by Ag-AgI/$Al_2O_3$ was revealed. The present work could provide new perspectives for promoting large-scale environmental applications of supported plasma photocatalysts.

# TARGET-SPECIFIC CAPTURE OF RELEVANT GASEOUS POLLUTANTS USING BIODEGRADABLE POLYMERIC NANOPARTICLES

***Fernanda Delbuque Guerra***, McKenzie L. Campbell, Frank Alexis and Daniel C. Whitehead
(Clemson University, Clemson, SC, USA)

According to the World Health Organization (WHO), as urban populations grow, the quality of urban environments affects the public health with regards to issues ranging from waste disposal, water sanitation and air pollution. Anthropogenic and biogenic sources of chemical compounds can contribute to millions of deaths worldwide annually. Air pollution accounts for 1 in 8 deaths worldwide, corresponding to approximately 7 million deaths in 2012, according to data from the WHO. EPA has set National Ambient Air Quality standards for six principal pollutants; carbon monoxide, lead, nitrogen dioxide, ozone, particle pollution and sulfur dioxide. In regards to ozone levels, volatile organic compounds (VOCs) are of concern along with nitrogen oxides (NOx), once the reaction between these two classes of compounds, in the presence of light, is responsible for emission of ground level ozone. Major sources of NOx and VOCs include industrial facilities and electric utilities, motor vehicle exhaust, gasoline vapors, and chemical solvents. Many VOCs are also hazardous air pollutants and also known for their low, often unpleasant, threshold odor, which can be below parts per billion in some cases . Different nanotechnology approaches have been investigated in an effort to remediate such pollutants from the air. Among the many challenges preventing the global use of many nanomaterials are the high costs of fabrication, on-site recovery, limited scale-up, potential toxicity since many of them rely on metallic nanomaterials, the low off-targetting specificity responsible for decreasing the effectiveness of the material and the limited applicability to few classes of contaminants.

A rendering plant was used as a case study for the determination of the relevant compounds present at an industrial site. The sample analysis revealed different classes of chemical compounds, but aldehydes and carboxylic acids were predominant. Therefore, we developed biodegradable functional nanoparticles comprised of poly(d,l-lactic acid)- poly(ethylene glycol)-poly(ethyleneimine) (PDLLA-PEG-PEI) block co-polymers that are capable of capturing aldehydes and carboxylic acids by simple chemical reaction. Polymeric nanoparticles (NPs) were prepared using nanoprecipitation and surface functionalization with branched PEI was done by conjugation reaction using EDC. Aldehydes were captured by a condensation reaction forming imines, whereas carboxylic acids were captured by acid/base reaction. Our material has shown reductions of aldehyde and carboxylic acid vapors greater than 69% and 76%, respectively, reaching up to 98% in some cases. Further, we successfully demonstrated that a simultaneous capture of mixtures of aldehydes and carboxylic acids is possible, as well as the simultaneous capture of a mixture of two aldehydes. The nanoparticles reacted selectively with target contaminants obviating off-target binding when challenged by other VOCs with orthogonal reactivity. In additions to competition assays we have successfully demonstrated that our material outperforms activated charcoal, a well-known sorbent material widely used for various applications including flue fas treatment processes. The significant advantage of our strategy relies on the ability to tailor the surface functionality of the nanomaterials for a specific target analyte from vapor mixtures.

# WATER SCRUBBING OF $H_2S$ GAS IN DUAL-FLOW SIEVE PLATE COLUMN SCRUBBER

***Swamy Kurella***, Pavan Kishan Bhukya and B. C. Meikap
(IIT Kharagpur, West Bengal, India)

Plate scrubbers can be used for the removal of acid gases from the exhaust gas streams which are being released into the environment from many chemical process industries. The raw syngas which leaves from coal gasification units is having particulate matter and acid gases majorly hydrogen sulfide ($H_2S$) and hydrogen chloride. A three stage lab scale dual-flow sieve plate column scrubber has been designed, fabricated and installed. In this work, the experiments were conducted to observe the removal of $H_2S$ using water as absorbing solution in the dual-flow sieve plate column wet scrubber. The $H_2S$ removal efficiencies were estimated at different liquid to gas ratios for 50 ppm, 100 ppm, 200 ppm and 300 ppm inlet $H_2S$ concentrations. The operated gas flow rates were from $1.93 x 10^{-3}$ $Nm^3/s$ to $2.76 x 10^{-3}$ $Nm^3/s$ and the liquid flow rates were from $27.53 x 10^{-6}$ $m^3/s$ to $48.18 x 10^{-6}$ $m^3/s$. The removal efficiency increased with the liquid flow rate, gas flow rate and inlet $H_2S$ concentration. The maximum percentage removal of $H_2S$ is observed at $2.76 x 10^{-3}$ $Nm^3/s$ gas flow rate and $48.183 x 10^{-6}$ $m^3/s$ liquid flow rate for 300 ppm. A model has also been developed to predict the $H_2S$ gas removal by using the results from the experiments and adding the variables which show impact on the scrubber performance.

# INDUSTRIAL AIR POLLUTION IN SAUDI ARABIA AND THE INFLUENCE OF METEOROLOGICAL VARIABLES

***Jamal A. Radaideh***
(Civil and Environmental Dept., King Faisal University, Saudi Arabia. jalradaideh@kfu.edu.sa)

**ABSTRACT:** This survey study focuses on major air pollutants at the second industrial city in Dammam/Eastern province of Saudi Arabia, which was carried out in two following weeks during April 2015, and aims to estimate the degree of air pollution arise from industry and to determine the influence of meteorological variables on concentrations of air pollutants. The study investigated the distribution of major air pollutants including, carbon monoxide (CO), sulfur dioxide (SO2), nitrogen dioxide (NO2), volatile organic compounds (VOCS), ozone (O3), hydrogen sulfide (H2S), Carbon dioxide (CO2) and respirable particulate matter (PM2.5 and PM10) in ambient air of Dammam industrial estate. Correlation analyses have been used to examine the association of air pollutants with each other and with meteorological variables. It was found that TVOC, PM10, TPS have a positive relationship with relative humidity, but negatively correlated with ambient temperatures. NO2 in contrast, has a reverse correlation with relative humidity and a positive relationship with ambient temperatures. Ozone concentrations vary slightly in dependence of changing meteorological factors (temperatures and relative humidity). Our results suggest that appropriate industrial air pollution control should be applied to coarse particulate (PM10) and gaseous pollutants in the study area.

***Keywords:*** Air pollution monitoring, air pollutants, meteorological variables, 2nd industrial city of Dammam, Saudi Arabia.

## INTRODUCTION

Recently, air pollution has become a major issue in Saudi Arabia and the neighboring GCC (Gulf corporation countries) because of accelerating urban development and growing industrialization.

While Saudi Arabia's economic growth continues to be dominated by oil, the Kingdom has taken steps to diversify its economy. Heavy dependence on petroleum revenue continues, but industry and agriculture now account for a larger share of economic activity. Today, industrial products make up more than 90 percent of the Kingdom's non-oil exports. Saudi Arabia exports petrochemicals, plastics, metal goods, construction materials and electrical appliances to some 90 countries [1]. Petrochemical and other oil-based industries are concentrated at industrial cities in major urban centers. These plants use natural gas and natural gas liquids that were previously flared, as well as refined products from the oil industry to manufacture products that would in turn feed non-oil industries. Concentration on industrial plants in specific areas also facilitates the provision of vital support services, such as water, power and transportation. Since the mid-1970s, attention has centered on industrial development, primarily in the downstream activities of the petroleum sector and on import substitution. In fact, one of the most significant Kingdom's development strategies is the varying selection of industries. Steel plants, fertilizer plants, domestic and export-oriented refineries, and a series of major petrochemical complexes form the basis of the government's attempt to diversify the economy.

One of the problems that occur in Kingdome of Saudi Arabia like other countries in the world is air pollution. Air pollution in Saudi Arabia is now recognized as a significant environmental impact of intensive anthropogenic activities. Heavy traffic in the streets of urban areas increases emissions of O3 as a result of the increased emissions of precursors (NOX and VOCS). Also, the local Saudi climatic conditions (high temperature, intense solar radiation, clear sky) lead to enhanced formation of photochemical pollutants. Hence, the air pollution is the cause of many risks on health and environment, many of the people not realize it, but air pollution impacts our everyday life, causing grave health and environmental damages.

Air pollution is a main environmental risk to health. Through reducing air pollution levels, countries can reduce the trouble of disease from embrace, heart disease, lung cancer, and both chronic and acute respiratory diseases, including asthma. WHO estimates that some 80% of outdoor air pollution-related premature deaths were due to ischemic heart disease and strokes, while 14% of deaths were due to chronic obstructive pulmonary disease or acute lower respiratory infections; and 6% of deaths were due to lung cancer. The 2005 "WHO Air quality guidelines" offer global guidance on thresholds and limits for key air pollutants that pose health risks. The Guidelines indicate that by reducing particulate matter (PM10) pollution from 70 to 20 micrograms per cubic meter (μg/m3), we can cut air pollution-related deaths by around 15% [4].

The multiple sources of air pollution and the heavy load of pollutants are a normal result of accelerated economic growth Saudi Arabia witnessed over the past three decades. Higher rates of air pollution are becoming strongly correlated with economic progress and style of living. Saudi Arabia's energy consumption has climbed dramatically over the past two decades. The energy consumption was 1.7 quadrillion Btu (quads) in 1980, and it was 4.6 quads in 2000. Overall, Saudi Arabia consumes about 1.1% of the world energy consumption. Therefore, Saudi Arabia's carbon emissions from transport sector have jumped in the past 40 years from 3.29 million metric tons in 1971 to 104.42 million metric tons in 2010. In addition the value for CO2 emissions from gaseous fuel consumption (ktonn) in Saudi Arabia was recorded as much as 140,500 ktonn in 2009, while a minimum value of 0.00 ktonn was recorded in 1960 [3].In the Gulf Cooperation Council (GCC) countries total atmospheric emission loads are about 3.85 million tons per year, made of 28% CO, 27% SO2 and 23% particulates. Recent studies have indicated that the Gulf countries emit about 50% of the total of Arab countries' (254 million metric tons of carbon) emissions of CO2 [5]. In terms of per capita carbon emissions, Saudi Arabia is still a regional leader. In 2000, the country's per capita carbon emissions were 3.7 metric tons [4].

The multiple sources of air pollution and the heavy load of pollutants are but a normal consequence of accelerated economic growth Saudi Arabia witnessed over the past three decades. Higher rates of air pollution are becoming strongly correlated with economic progress. Therefore, the Kingdom of Saudi Arabia has paid special attention to monitoring and reducing such emissions through concerted efforts under taken at both national and international levels alike.

Given the widespread concern of ambient air pollution in Saudi Arabia and its broad ranging impacts, the present study is focused on investigation of air pollution in one of the densest industrial areas of the region. The objective of this study is to present the spatial distributions of selected gaseous pollutants ($SO_2$, $NO_2$, CO, $H_2S$, $CO_2$, TVOC and Ozone) and particulate matter ($PM_{2.5}$, $PM_{10}$ and $PM_1$) within the study area and to identify the major impacts of air pollutants. Evaluation of existing air quality findings is occurring using limits recommended by available local and global air quality standards and guidelines, in particular NAAQS of USEPA [6], WHO air quality guidelines [2 and 3], PME standards [7], RCER for Jubail & Yanbu [8] and the European standards [9] will be reference standard for this survey study.

## MATERIALS AND METHODS

The 2nd industrial city in Dammam is one of the highest areas of industrial cities in kingdom of Saudi Arabia, which contain 340 of factories occupied a total area of 25,520,000 m2. It was established in (1978) and has 75,000 of workers distributed over the different factories involving food and drinks, manufactured metal products, machineries and equipment's, chemical products, building materials, textiles, wood products and furniture.

Air sampling using Air quality monitoring mobile station used to conduct the measurements of potential air pollutants.

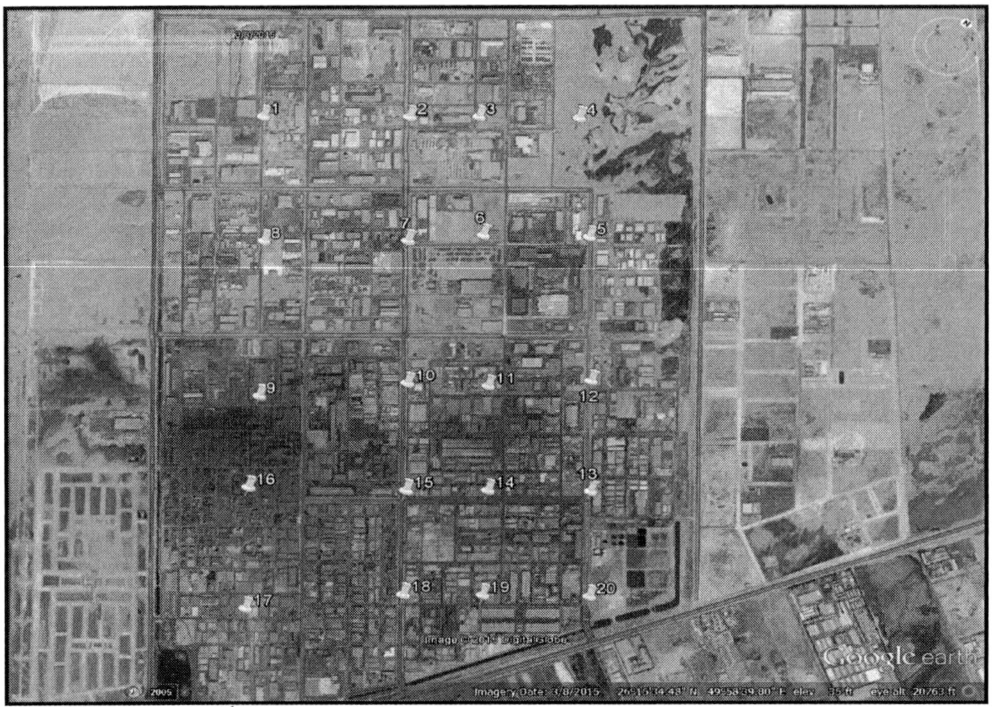


Figure 1. Boundary of $2^{nd}$ industrial city of Dammam as shown by google map showing assigned location points for measurements and sampling (L1 - L20).

The survey study was conducted during two following weeks with noticeable change in ambient outside temperature and varying relative humidity. A total of 200 measurements and samples are collected over the course of study at selected points outside factories and workshops. Data of particulate matter ($PM_{2.5}$, $PM_{10}$ and PM1) and gaseous air pollutants such as nitrogen dioxides (NO2), ozone (O3), carbon monoxide (CO), sulfur dioxide (SO2), carbon dioxide (CO2) and total organic compounds (TVOC) are measured. Meteorological parameters like temperatures and relative humidity are also collected during this survey. Levels of gaseous pollutants are analyzed using Wolf Pack AAQ Instrument. The PM levels are obtained using a particulate matter monitor (AEROCET 531S). The samplers draw air through a filter paper at a known constant rate for 24 hours. The resulting increase in the weight of the filter paper is expressed as the particulate concentration.

## STUDY AREA

The second industrial city of Dammam is located outside the urban boundaries of the city. The area with the coordinates, Latitude: 26°15'31"N Longitude: 49°58'5"E has a length of 15.57 kilometers and houses about 340 different factories.

The sampling station is installed at all points assigned for data collection (L1 – L20). Each point is supposed to represent a specific area from the whole study area. Factories are distributed randomly and equally over the area in a way that any measured air pollutant couldn't be referred to any specific industry.

When instrument is turned on, it starts to record concentrations of gaseous pollutants and particulate matter. After 6-8 minutes the recorded levels of concentrations are displayed on screen for all considered pollutants. The average value of all readings (60-65 readings) for each potential pollutant is presented in Table 1.

## DATA COLLECTION AND ANALYSIS

As soon as the measuring devices are calibrated and turned on to record pollution levels at each assigned location, recorded data pass through rectify units and an average value is displayed. For each potential pollutant about 60 – 65 values are registered during the measuring period ranging from 7-8 minutes. The hourly mean concentrations of the measured pollutants during the two periods of study are presented in Table 1. In addition meteorological parameters such as temperature and relative humidity are recorded continuously and simultaneously.

Table 1: Results of air quality monitoring survey at 2nd industrial city/Dammam

| Location points | $SO_2$ ppm | $NO_2$ ppm | TVOC µg/m$^3$ | $CO_2$ ppm | Ozone ppm | CO ppm | PM1 (µg/m$^3$) | PM2.5 (µg/m$^3$) | PM10 (µg/m$^3$) | TSP (µg/m$^3$) |
|---|---|---|---|---|---|---|---|---|---|---|
| L1 | 0.622 | 0.104 | 76.94 | 541.29 | 0.088 | 1.35 | 17.050 | 64.525 | 385.750 | 527.650 |
| L2 | 0.630 | 0.130 | 54.27 | 561.98 | 0.081 | 1.43 | 14.800 | 56.067 | 159.933 | 170.867 |
| L3 | 0.660 | 0.087 | 74.77 | 548.27 | 0.080 | 1.39 | 15.333 | 65.767 | 438.733 | 612.567 |
| L4 | 0.500 | 0.100 | 58.80 | 541.09 | 0.082 | 1.34 | 15.700 | 60.250 | 264.800 | 309.150 |
| L5 | 0.490 | 0.067 | 27.73 | 520.08 | 0.084 | 1.28 | 14.875 | 56.500 | 198.825 | 221.000 |
| L6 | 0.512 | 0.083 | 33.51 | 559.22 | 0.080 | 1.26 | 14.067 | 61.633 | 491.133 | 710.033 |
| L7 | 0.420 | 0.080 | 37.23 | 554.80 | 0.080 | 1.24 | 13.300 | 57.800 | 384.900 | 497.867 |
| L8 | 0.380 | 0.102 | 89.51 | 602.16 | 0.086 | 1.23 | 14.067 | 62.067 | 405.233 | 539.833 |
| L9 | 0.390 | 0.060 | 8.34 | 567.42 | 0.068 | 1.39 | 13.467 | 62.867 | 421.000 | 578.000 |
| L10 | 0.400 | 0.064 | 42.92 | 560.59 | 0.086 | 1.28 | 15.933 | 64.633 | 434.867 | 611.233 |
| L11 | 0.569 | 0.010 | 164.02 | 531.93 | 0.069 | 1.18 | 26.567 | 209.167 | 1409.767 | 1904.967 |
| L12 | 0.476 | 0.021 | 141.59 | 537.02 | 0.090 | 1.80 | 26.133 | 204.800 | 1525.000 | 2196.700 |
| L13 | 0.461 | 0.030 | 155.40 | 558.55 | 0.096 | 1.37 | 25.800 | 199.167 | 1470.533 | 2129.167 |
| L14 | 0.524 | 0.035 | 135.84 | 538.58 | 0.089 | 1.58 | 27.900 | 218.000 | 1858.350 | 2737.900 |
| L15 | 0.532 | 0.037 | 207.29 | 554.00 | 0.103 | 1.63 | 22.500 | 168.125 | 1229.550 | 1752.350 |
| L16 | 0.692 | 0.027 | 256.40 | 584.13 | 0.092 | 1.56 | 29.033 | 226.133 | 1806.367 | 2767.333 |
| L17 | 0.541 | 0.060 | 158.66 | 611.66 | 0.104 | 1.47 | 31.067 | 239.367 | 1717.633 | 2417.700 |
| L18 | 0.674 | 0.069 | 118.52 | 631.18 | 0.103 | 1.41 | 31.967 | 234.700 | 1495.667 | 2009.000 |
| L19 | 0.817 | 0.053 | 115.73 | 606.75 | 0.093 | 1.63 | 29.700 | 229.400 | 1742.550 | 2634.200 |
| L20 | 0.800 | 0.067 | 67.78 | 546.20 | 0.099 | 1.57 | 30.600 | 228.033 | 1398.233 | 1740.700 |

Conversion factors: CO: 1 ppm ≈ 1.15 mg/m$^3$, $NO_2$: 1 ppm ≈ 378.5 ppb ≈ 1880 µg/m$^3$. $SO_2$: 1 ppb ≈ 2.6 µg/m$^3$ and 1ppm ≈ 2760 µg/m$^3$. $CO_2$: 1 ppm ≈ 1.83 mg/m$^3$. Ozone: 1 ppm ≈ 1966 µg/m$^3$ ≈ 1.143 mg/m$^3$.

## RESULTS AND ANALYSIS OF DATA COLLECTION

**Analysis of TOVC Concentrations.** As Table 1 and the derived Figure 3 show that the concentrations of TVOC are in the range of 27.73 – 89.51 µg/m$^3$ with an average of 53.39 µg/m$^3$ for first measurement batch conducted at locations L1 to L10, while during the second measurement period values are ranging from 115.73 – 256.40 µg/m$^3$ with an average of 152.123 µg/m$^3$ found at locations L11- L20. It is clearly observed that significantly higher TVOC levels are obtained during the second measurement batch at locations L11 – L20 against those recorded at locations L1 – L10 during the first measurement period.

The higher concentration of TVOC at these sampling sites is attributed to decrease in temperatures and relative humidity. The ambient temperatures during the first period were ranging from 35.81°C to 38.60 °C and the relative humidity values were in the range of 9.95% – 15.93%. During the second period the temperatures were in range of 29.56°C to 36.90 °C and the relative humidity values were in the range of 21.24% – 45.08%. The magnitude of air pollution in regard to TVOC couldn't be evaluated because there are no available standards for TOVC to compare with.

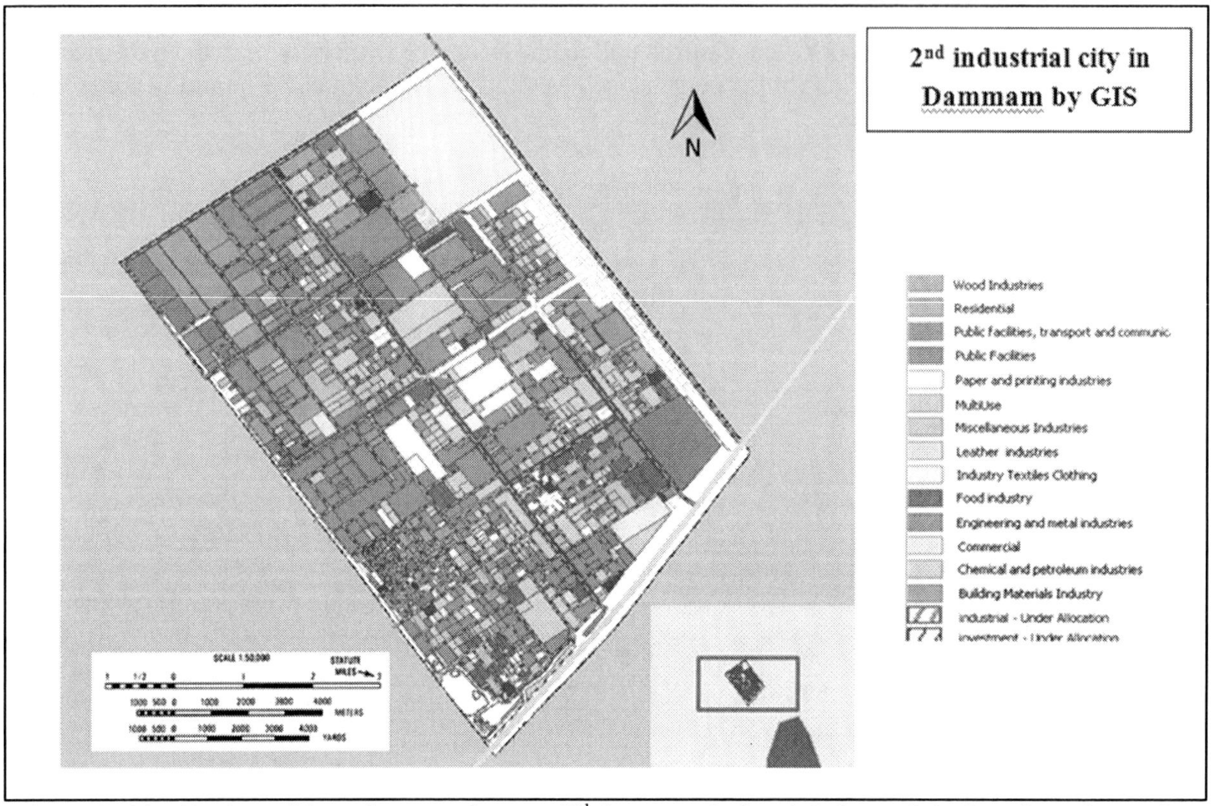


Figure 2. Distribution of factories over the area 2nd dammam industrial city and the location of sampling points (L1-L20).

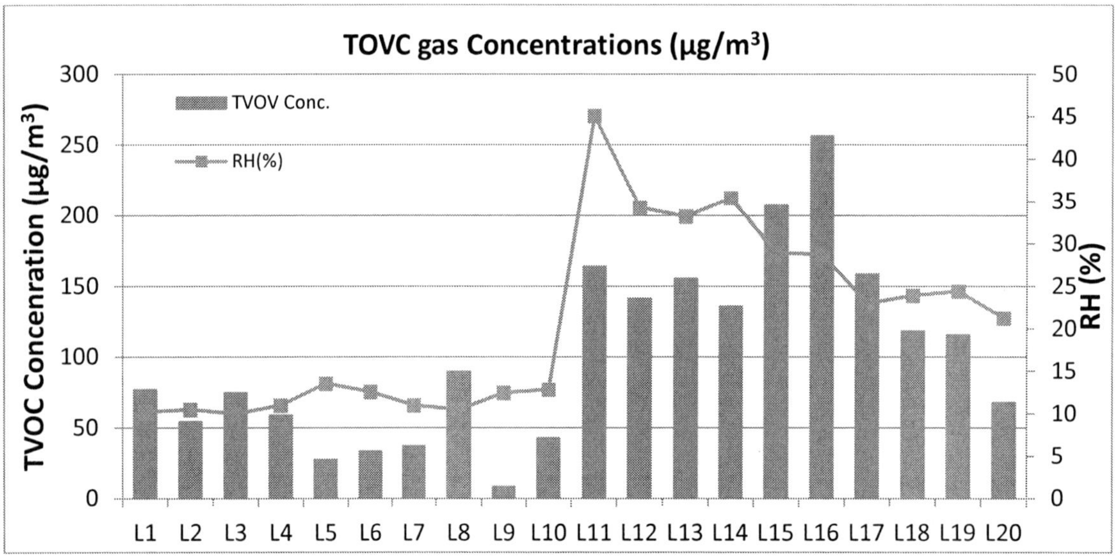


Figure 3. TVOC concentrations monitored at locations L1 – L20.

**Analysis of CO Concentrations.** For all Locations L1- L20, the hourly carbon monoxide concentrations ranged from 1.18 to 1.80 ppm with an overall hourly average of 1.42 ppm as shown in table 1. It is clear from the results also that all the measurements were below 2 ppm. The recorded 1-h measurements were compared with the available national and global standards: this comparison showed that almost all recorded results of the current study were below the 1-h average standards. As shown in table

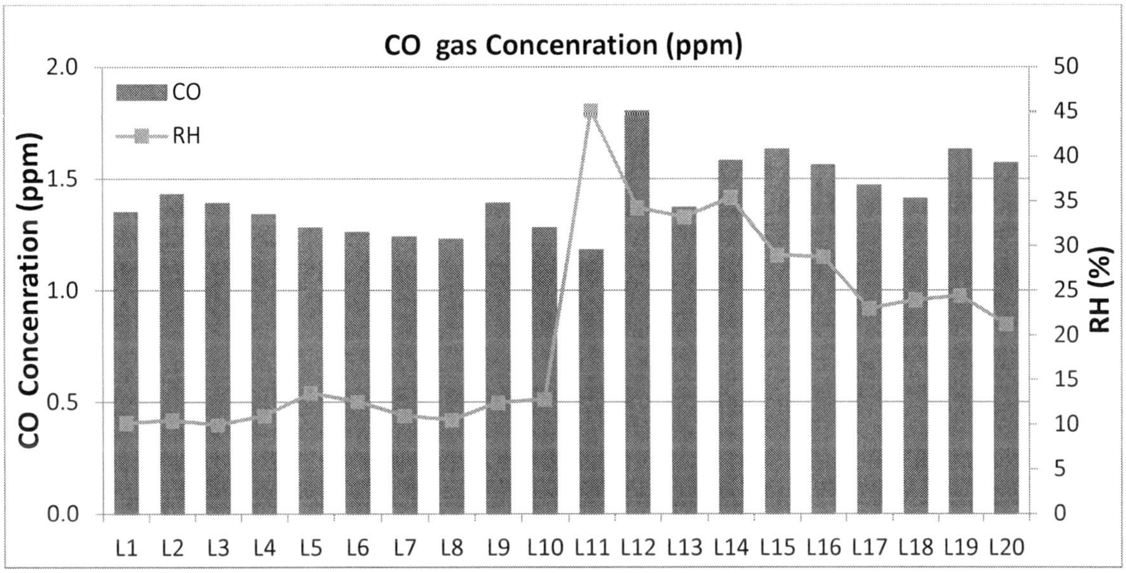


Figure 4. CO concentrations monitored at locations L1 – L20.

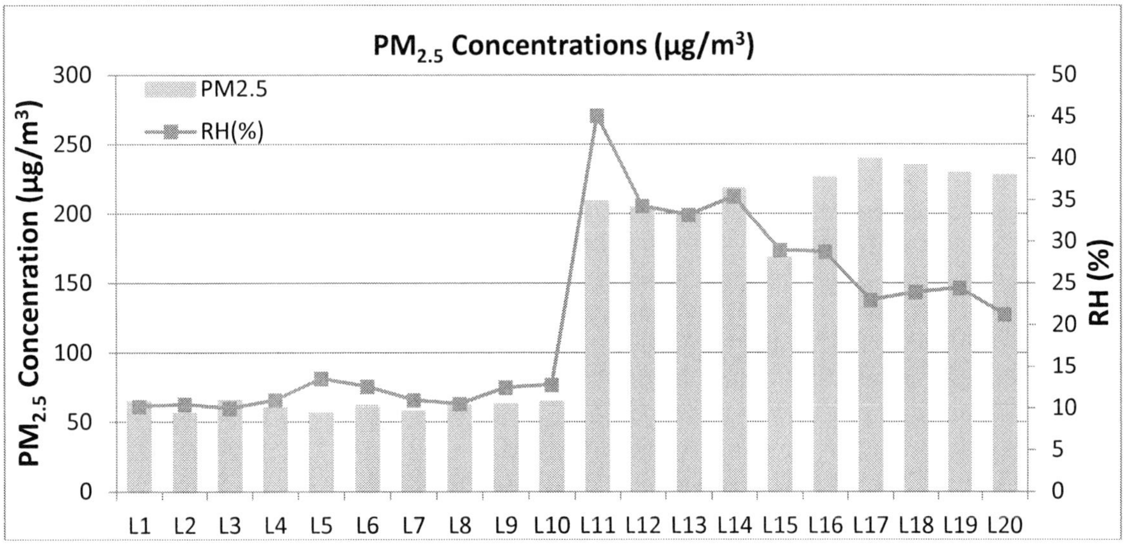


Figure 5. Average values of $PM_{2.5}$ recorded during both survey periods

**Analysis of $PM_{2.5}$ Levels.** As shown in Table 1, the concentration of $PM_{2.5}$ was in the range of 56.5 – 65.77 µg/m$^3$ , in average of 61.21 µg/m$^3$ for the first measurement period and in the range of 168.12- 239.37 µg/m$^3$ with an average of 215.69 µg/m$^3$ for the second period. The variation of particulate concentrations during this period was found to be minimal and below USEPA standards. Significantly higher $PM_{2.5}$ values were obtained at locations L11 to L20 compared to those obtained at locations L1-L10 during second period. The higher concentrations of particulate pollution at these sampling sites may be attributed to re-suspension of dust (road dust, entrainment of dust into the air from nearby areas and vehicular traffic around the study area) and to decrease in temperatures and associated relative humidity. Comparing values of $PM_{2.5}$ with allowable limits, it is clear that the quality of ambient air is below air quality standards defined available standards as shown in table.

**Analysis of $PM_{10}$ Concentrations.** Table 1 presents the average $PM_{10}$ concentrations found at sampling sites during both sampling periods. The average concentrations at locations L1 – L10 were found in the range of 159.93 – 491.13 μg/m$^3$, while the average concentrations at locations L11 – L20 were found significantly higher and ranging from 1229.55 – 1858.35 μg/m$^3$. Comparing the 1-h results at locations L1 – L10 with the available standards showed that they fluctuating around the highest allowable standards value defined by PME [7] standards. However, all of the individual recorded results were far beyond the desired value recommended by national and international standards as Table 1 indicates.

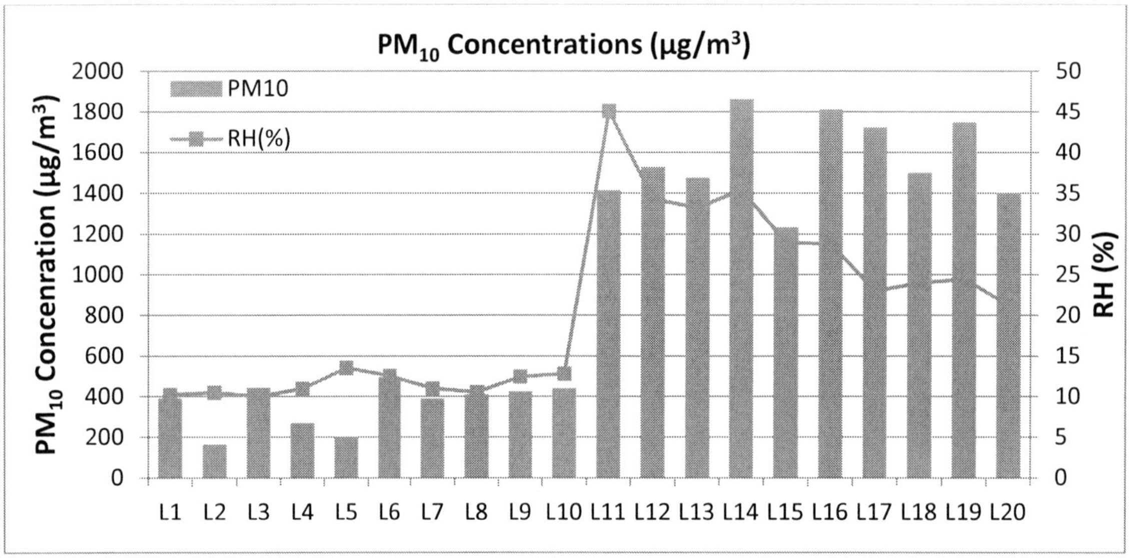


Figure 6. Average values of $PM_{10}$ recorded during both survey periods

**Analysis of Total Suspended Particles (TSP) Concentrations.** (TSP) are tiny airborne particles or aerosols that are ≤ than 100 micrometers are collectively referred to as total suspended particulate matter (TSP). These particles constantly enter the atmosphere from many sources include human activities (combustion products, industrial processes and power generation) and from natural sources (soil, bacteria and viruses, fungi, mold and yeast, pollen, vapors and aerosols).

Table1 presents the average values of TSP concentrations found at sampling sites during both survey periods. The average concentrations at locations L1 – L10 were found in the range of 170.87 – 710.03 μg/m$^3$, while the average concentrations at locations L11 – L20 were found significantly higher and ranging from 1740.70 – 2767.33 μg/m$^3$. The higher levels of suspended particulates at locations L11 – L20 recorded during the second sampling period may be attributed to noticeable decrease in temperatures and relative humidity.

Comparing the 1-h results at locations L1 – L10 with the available standards showed that they fluctuating around the highest allowable standards value defined by PME standards. However, all of the individual recorded results were far beyond the desired value recommended by NAAQS of USA, WHO, and the European standards.

**Analysis of Ozone Concentrations.** The results of ozone monitoring are presented in Table1. As Figure 8 shows, average ozone concentrations are in range of 0.068 to 0.088 ppm for samples taken at locations L1 – L10 during the first sampling period. The overall average for these readings is (ppm). Concentrations taken during the second period are ranging from 0.069 to 0.104 ppm, with an average of (ppm). The ozone concentrations vary slightly among sampling locations and with varying meteorological variables during sampling periods. Comparing obtained levels for ozone with the available standards [6, 7, 8, 9 and 10], it is conclude that measured ozone is matching the reference standards used as evaluation basis for air quality.

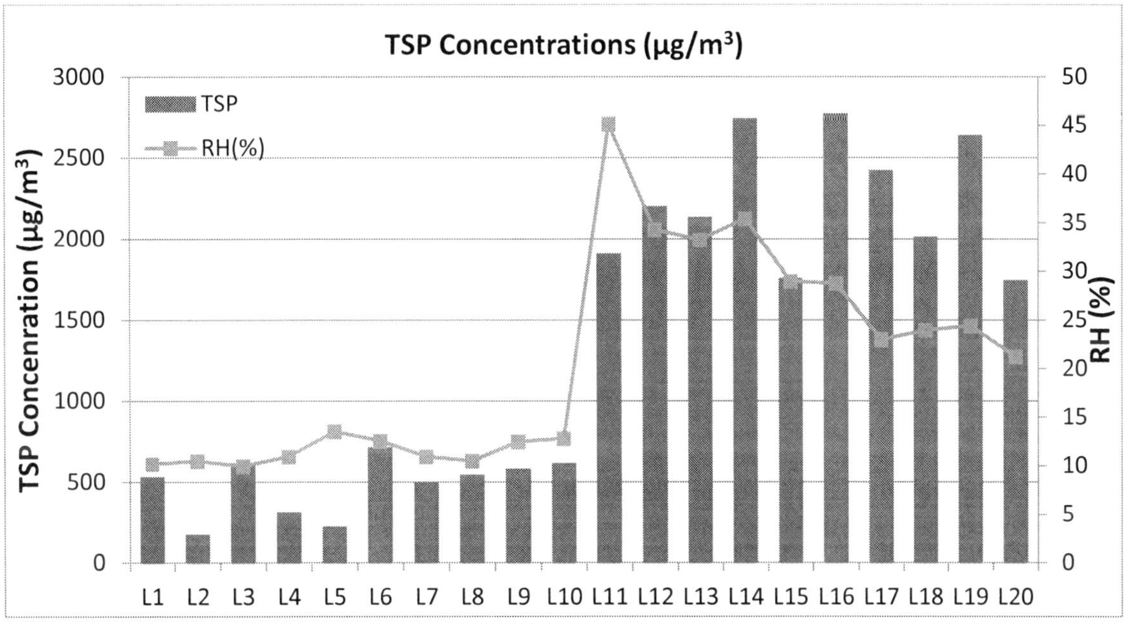


Figure 7. Average values of TSP concentrations recorded during survey periods.

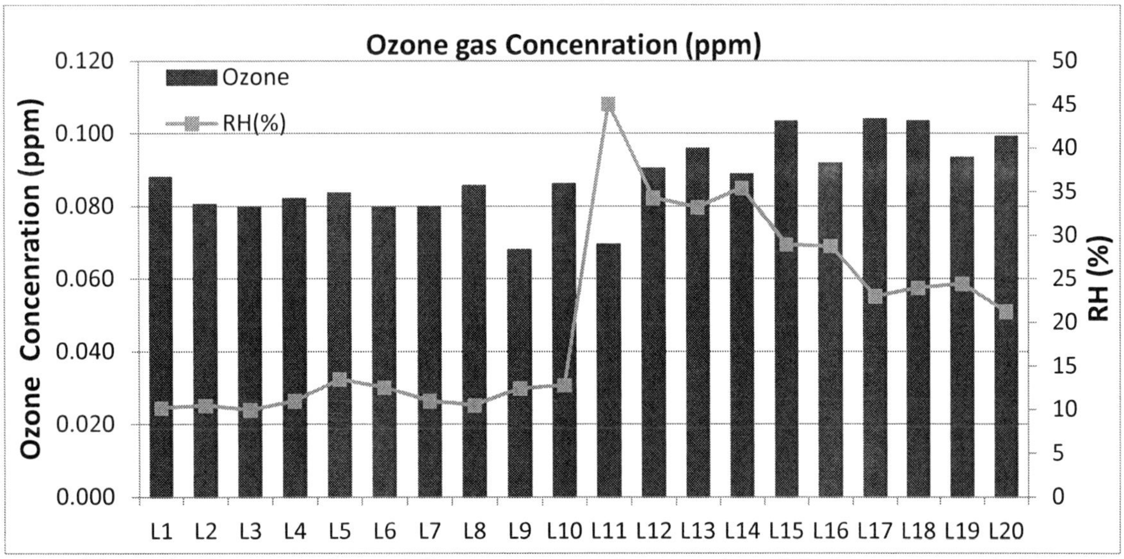


Figure 8. Average values of Ozone concentrations monitored during survey periods.

As indicated in figure 8 the correlation between relative humidity and ozone levels couldn't be clearly demonstrated. However a slight increase in ozone levels at locations L11 – L20 is noticeable and could be referred to the increase in ambient relative humidity during the sampling period.

**Analysis of $NO_2$ Concentrations.** It is apparent from Table 1 and Fgure 9 that the average $NO_2$ concentrations ranged from 0.06 ppm to 0.104 ppm for samples taken at locations L1 – L10 during the first measuring period with an overall average of (0.88 ppm). The $NO_2$ concentrations taken during the second period at Locations L10 – L20 are ranging from 0.01 to 0.069 ppm with an overall average of (0.41 ppm) for this sampling batch. It has been noticed that the results were fluctuating among locations. Comparing existing concentrations taken during both sampling periods with available standards as shown in table, it is obvious that the quality of ambient air is below air quality defined by available standards. In addition,

Figure 9 shows clearly the decrease of $NO_2$ levels taken during the second sampling batch (at L11 – L20) against those monitored during the first sampling period taken at locations L1 –L10). Variations in meteorological parameters (temperatures and relative humidity) could be seen as cause for this concentration drop. Therefore, it is to conclude that $NO_2$ concentrations decrease as relative humidity increase, but decrease as temperature decrease. The negative correlation indicates that probably high temperatures result in higher dispersion and dilution of air pollutants, probably linked with vertical and horizontal turbulence [11].

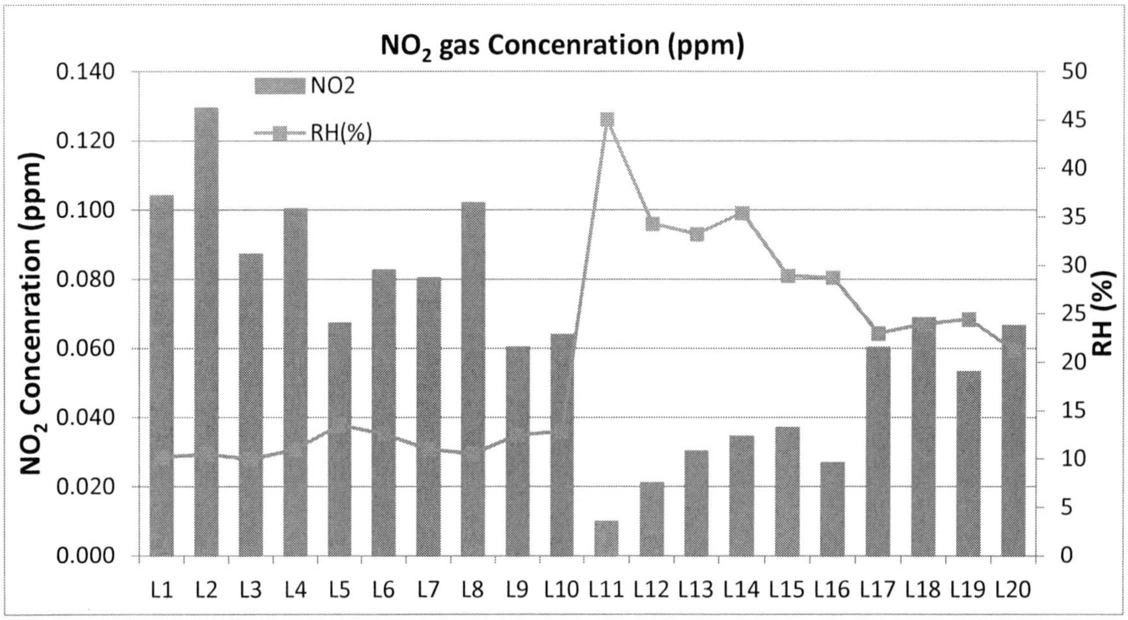


Figure 9. Average values of $NO_2$ concentrations monitored at locations L1 – L20.

**Analysis of $SO_2$ Concentrations**. The concentrations of SO2 in the first measuring period at locations (L1 –L10) are in the range (0.38 – 0.66 ppm) with a daily average equals 0.50 ppm, while in the second period the concentrations ranging from 0.461 – 0.817 ppm with an average 0.53 ppm, slightly higher than during first period, as shown in Table 1 and Figure 10. A clear dependency of this result with the relative humidity couldn't be established. However, the small increase in concentrations at locations L11-L20 compared to levels of SO2 monitored at L1-L10 could be referred to increase in relative humidity.

**Analysis of $CO_2$ Concentrations**. As presented in Table 1, Concentration levels of CO2 are ranging from 541- 602 ppm with an of 555.7 ppm for measurements taken at L1 – L10 while the values of CO2 levels ranged from 531- 631 ppm with an average 570.00 ppm for the points L11-L20. The concentration of CO2 is almost constant at all points throughout the study area and don't experience any change in relation change of meteorological parameters (temperature and relative humidity). Although carbon dioxide is naturally present in the atmosphere as part of the Earth's carbon cycle, monitored levels came as high as it is expected in industrial areas. The main human activity that emits CO2 is the combustion of fossil fuels (coal, natural gas, and oil) for energy and transportation, although certain industrial processes and land-use changes also emit CO2.

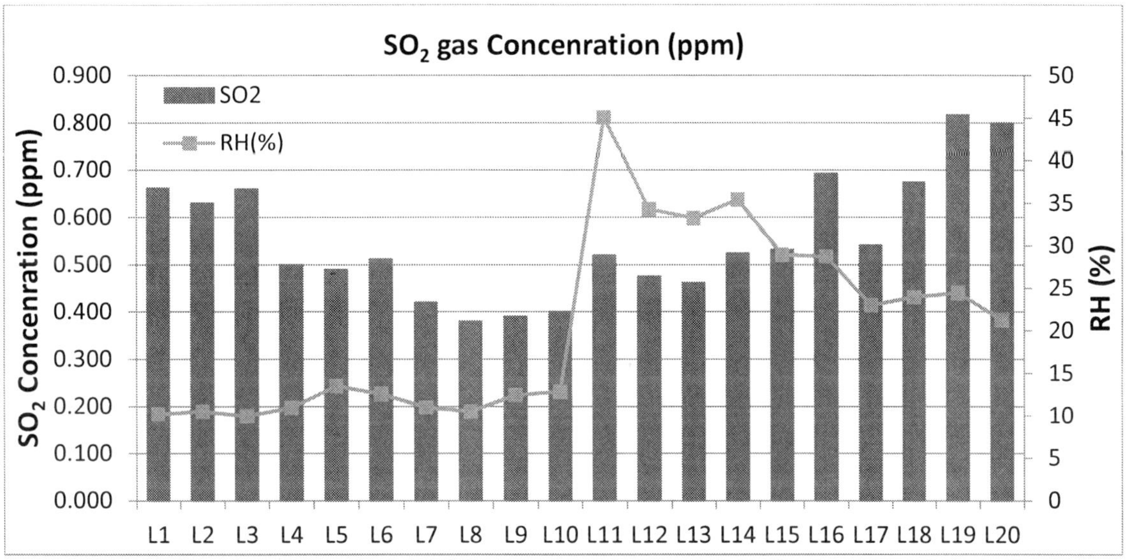

Figure 10. Average values of $SO_2$ concentrations monitored at locations L1 – L20.

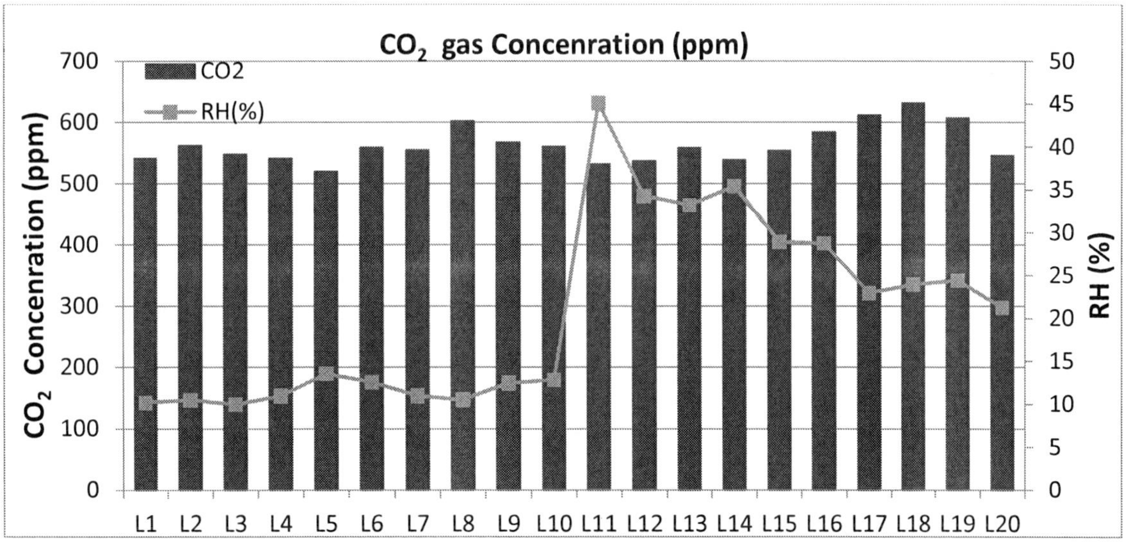

Figure 11. Average values of $CO_2$ concentrations monitored at locations L1 – L20.

## DISCUSSION

Air in the Kingdom of Saudi Arabia is generally not polluted. However, there are certain locations where the concentration of some pollutants could and do reach levels above the acceptable limits. This country is developing rapidly and the volume of imported and domestically produced chemicals and other products are on the increase which will all contribute to various types of air pollutants.

The study thought to examine the relationship between potential air pollutants (gaseous and particulate) among each other and with selected meteorological variables, emitted from factories distributed over the area of 2$^{nd}$ industrial estate of Dammam during 2 weeks in April, 2015. From the results presented for potential air pollutants and with considering Table 2 as basis for evaluation, it is quite evident that the quality of air in and around industrial cities might well be different from that of the rest of the Kingdom.

The analysis and comparison of collected data indicates that particulate matter in form of $PM_{10}$ and $PM_{2.5}$ contributes most to the highest levels of pollution. The recorded levels of these air pollutants reach ranges that exceed limit values adopted by all jurisdictions. For particulate matter, it is a major problem in Saudi Arabia, especially, in eastern province of Saudi Arabia that occasionally experience high particulate and suspended matter concentrations as high as 900 – 1000 µg/m3 [13]. Concentrations as high as monitored (215.69 and 1565.36) µg/m3 are detrimental to infrastructure and are a health hazard to human beings.

Table 2. Comparison of monitored air pollutants at 2nd industrial city based on the available air quality standards. All values are given either in ppm or in $\mu g/m^3$.

| Pollutant | | $PM_{2.5}$ | $PM_{10}$ | $SO_2$ | | $NO_2$ | | CO | $O_3$ |
|---|---|---|---|---|---|---|---|---|---|
| Air quality standards | | µg/m³ | µg/m³ | ppm | µg/m³ | ppm | µg/m³ | ppm | µg/m³ |
| NAAQSs (USA) [6] | | 12*** | 150*** | 0.12* | - | 0.246* | - | 35* | 0.075** |
| CAAQSs (Canada) [12] | | - | - | 0.335* | - | 0.213* | - | 30.6* | 0.082 |
| WHO [10] | | 25*** | 50*** | 0.13* | 20* | 0.07* | 200*** | 26* | 100** |
| EU guideline [9] | | 25 | 50*** | 0.28* | 350* | - | 200 | 0.14* | 120** |
| PME + (RCER) [6] + [7] | | 35*** | 340*** | 0.28* | - | 0.35* | - | 32* | 0.15* |
| Exist. value | L1 - L10 | 61.21* | 358.52* | 0.504* | - | 0.088* | - | 1.319* | 0.081* |
| | L10 - L20 | 215.69* | 1565.36* | 0.604* | - | 0.041* | - | 1.52* | 0.094* |

*: averaged over 1-hr **: averaged over 8-hr ***: averaged over 24-hr Conversion factors: CO: 1 ppm ≈ 1.15 mg/m³, $NO_2$: 1 ppm ≈ 378.5 ppb ≈ 1880 µg/m³. $SO_2$: 1 ppb ≈ 2.6 µg/m³ and 1ppm ≈ 2760 µg/m³. $CO_2$: 1 ppm ≈ 1.83 mg/m³. Ozone: 1 ppm ≈ 1966 µg/m³ ≈ 1.143 mg/m³.

The observed CO level (1.32 and 1.52 ppm) fell below the limit values adopted by NAAQS (USA, Canada), WHO and PME + RCER, but higher than EU allowable limits of 0.14 ppm. Ozone concentration measurements showed that in the study area, moderate ozone concentrations were measured. The recorded average values (0.081 and 0.094) and over 1-hr averaged slightly exceed the selected guideline values, in particular, are higher than the value of (0.075, 8-hr average) adopted by NAAQS of USA.

$SO_2$ maintains higher levels than required indicating a pollution. $NO_2$ maintains levels very close to the WHO guideline value around the world. For particulate matter, it is a major problem in almost all of Asia, exceeding 300 µg/m³ in many cities.

## CONCLUSIONS

The present study analyzed the concentrations of potential air pollutants including $SO_2$, $NO_2$, TVOC, $CO_2$, CO and $O_3$ measured at the 2nd industrial city/Dammam over 2 weeks in April 2015 and shows the daily average of monitored air pollutants are above the permissible limits suggested by available national and international jurisdictions such as PME and RCER, NAAQS of USA and Canada and the WHO guidelines. This result indicates that there are some areas suffering from air pollution and require immediate action plans to deal with the problem.

The interaction between various air pollutants and meteorological parameters is confirmed through the results of study and concluded that levels of observed air pollutants not only depend on the amount of pollutants emitted from various sources but also on meteorological parameters.

Analysis of temperatures and relative humidity clearly demonstrated that air pollutants concentrations increase with increase in relative humidity and decrease of temperatures.

## REFERENCES


[1] "The Petrochemicals and Plastics Sector in the Kingdom of Saudi Arabia" . U.S.-Saudi Arabian Business Council. Retrieved 12 February 2015.

[2] The World Health Organization, (2011) air quality in countries around the globe.

[3] WHO Factsheet N 313, March 2014. WWW.who.int/mediacentre/factsheets/fs303/en/

[4] Abdallah, Kathleen. "Health and Environmental Benefits of Clean Fuels and Vehicles. "Keynote presentation. Cairo, Egypt: UN DESA, May 24, 2006.

[5] Karin B. Yeatts, et.al. Indoor Air Pollutants and Health in the United Arab Emirates. Environmental Health Perspectives • volume 120, issue 5, May, 2012 . Environ Health Perspect; DOI:10.1289/ehp.1104090.

[6] U.S. Environmental Protection Agency, (2011) National Ambient Air Quality Standards (NAAQS).

[7] Ambient Air Quality, Presidency of Meteorology and Environment. Kingdom of Saudi Arabia National Environmental Standard.

[8] Kingdom of Saudi Arabia National Environmental Standard. Ambient Air Quality Standards. Royal Commission Environmental regulations RCER for Jubail & Yanbu, 2004, Vol.I.

[9] Air Quality standards, European Community 2014. New Air quality Directive (The Directive2008/50/EC). http://ec.europa.eu/environment/air/quality/legislation/directive.htm

[10] WHO, Public Health Environment (PHE), Air Quality Guidelines 2005. http://www.who.int/phe/health_topics/outdoorair/outdoorair_aqg/en/

[11] EPA, 2010. US Environmental Protection Agency, air pollution control orientation course. Control emissions technologies. Transport and dispersion of air pollutants. http://www.epa.gov/apti/course 422/ce1

[12] Canadian Air Quality Standards CAAQS of Canada.. http://www.ccme.ca/en/current_priorities /air/caaqs.html.

[13] B.H. Alharbi, M.J.Pasha, and N. Tapper. Assessment of ambient air quality in Riyad City, Saudi Arabia. International Research Journal of environment science. ISSN: 0973-4229

# MONITORING OF HAZARDOUS ATMOSPHERIC RELEASES UNDER EMERGENCY SITUATION BY RADAR TRACER METHOD

***Boris S. Yurchak*** (EPA, Arlington, VA, USA)

**ABSRACT**: In the case of an accident at nuclear or chemical facilities accompanied with emission of hazardous pollutants into the atmosphere, the lack of information about the spatial distribution of pollutants in the vicinity of an enterprise-emitter during approximately an hour after the accident in air and on the ground is experienced, i.e., within the local spatial scale. This makes it difficult to manage the emergency situation on the initial stage of the accident. To get more detailed information at this stage, the Radar Tracer Method (RTM) is suggested. RTM is based on marking aerosol-like pollution with radar light reflectors (chaff) that makes it possible to distinguish the markers and consequently to locate the pollution pattern reliably by radar. The features of this method and its possible fields of application are discussed.

## INTRODUCTION

Bhopal chemical and other similar accidents (e.g., MacKenzie, 1985), Chernobyl and Fukushima nuclear accidents in 1986 and 2011, respectively, highlighted the necessity to have a system for monitoring dispersion and transport of hazardous atmospheric releases (HAR) from potentially dangerous enterprises (PDE) of nuclear and chemical industries in real-time scale. This requirement is due to the health protection issues of enterprise employees and inhabitants in nearby located places as well as to target the logistic measures (e.g., evacuation). Currently, a forecast of the contamination is conducted using the models of pollutant dispersion in the Atmospheric Boundary Layer (ABL). However, the input information for these models is extracted from regular meteorological network and aerological (upper-air) observations with data acquisition frequency of only several times per day. The network of aerological stations is rare. Due to that, the prognostic wind field of scale ~ 30-50 km differs from the actual one, particularly in the transient period of a day, under the stable stratification, and during different inclement weather events. To support and improve systems of decision making under accidences at the PDE, the Radar Tracer Method (RTM) was suggested to be exploited (Zhukov and Yurchak, 1993). The objective of the current paper is to discuss the main features of the method and possible fields of its application.

## PRIMARY PRINCIPLE OF THE RTM

The RTM is aimed for radar 'visualization" of footprint and path of a cloud of hazardous pollutants in the atmosphere. The pollutants can be radiation or chemical in nature. In this method the gas-aerosol impurity produced, for example, during an accident at a hazardous object, is marked with passive radar reflectors (chaff) which are then observed in the atmosphere by means of radar. The chaff was primary used to study the physics of the atmosphere as part of a research designed to monitor the modification of warm clouds and fogs and to study the diffusion of a passive impurity inside and around a cloud (e.g., Skhirtladze and Yurchak, 1979) and in the lower part of the ABL (e.g., Hildebrand, 1977; Zhukov and Yurchak, 1994). A characteristic feature of this study was use of chaff consisting of carbon-graphite materials "Uglen" (Girs et al., 1979) with a sedimentation rate of 10-30 cm/s and "Metka" with sedimentation rate of about 3 cm/s (Bessonov et al., 1987). These materials have a lower inertia than metallized materials containing a glass or polymer base. Although they produce a weaker radar signal, it can be detected on a distance of several tens of kilometers by ordinary weather radars. The advantage of tracking a hazardous impurity by the RTM is its sensitivity to wind shear and entrainment into a cloud. The basic characteristics of the RTM were provided in (Zhukov and Yurchak, 1993). Assuming Gaussian diffusion for time much longer than the Lagrangian time scale $\tau_L$ of turbulence, the following radar equation was obtained for finding the maximum range (in time and distance) for tracking chaff as a function of the diffusion parameters $x_L = \bar{u}\tau_L$ (the spatial scale in the direction of transport, $\bar{u}$ is average wind-direction transport velocity) and $r_L = \tau_L\sqrt{\langle v^2 \rangle}$ (the diffusion scale, $\langle v^2 \rangle$ is the variance of the pulsations of the velocity of a moving particle):

$$\lg \xi = \frac{1}{7}\left[2\lg \sigma_\Sigma + \frac{1}{5}\left(\frac{P_t}{P_r}\right)_{dB} + 2\lg \Pi - 3\lg 4\pi - 6\lg r_L - 4\lg x_L\right] \quad (1)$$

where $\xi = t/\tau_L = \bar{u}t/\bar{u}\tau_L = r/x_L$ is dimensionless range; $\sigma_\Sigma$ is the initial effective scattering surface area of the chaff emission; $\Pi$, $P_t$ and $P_r$ are the radar potential, transmitted pulse power, and radar receiver sensitivity, respectively. This form of the radar equation takes into account the dependence of the dimensionless range on meteorological situation (through the Lagrangian time scale and diffusion scale). The equation is a basis for designing the RTM system for different applications. It should be noted also that chaff can be detected not only by ground-based radars but airborne and satellite Synthetic Aperture Radars (SAR). The SAR application seems very perspective for HAR tracking due to possibility of having the plan position image of chaff clouds. Although the documentary proof cases of chaff tracking by SAR in the atmosphere are currently unknown, there are several papers describing detection of rain clouds over land and ocean by satellite SAR (e.g., Weinmann et al., 2008). It might be expected that radar detection of chaff is more reliable in comparison with rain drops due to anisotropic shape of an elementary scatterer of chaff that is a half-wavelength dipole. Effective chaff detection can be performed with the polarimetric radar (e.g., Martner et al., 1992).

## TRACKING AND MAPPING CLOUDS OF HAR DURING AN ACCIDENT

As depicted in Figure 1, during an accident or planned emission of toxic and/or radioactive gases and aerosols, the reflectors from dispersing devices are mixed by operator command or automatically with the radioactive impurity and are transported together with the impurity into the atmosphere, making it possible to estimate the average transport and transformation of the cloud by wind and turbulent diffusion in the layer of the atmosphere near the ground. At the enterprises, which are not equipped with containers holding the reflectors or with dispersing systems, the radar markers can be introduced into the area of the source of contamination by other methods, for example, dumped from flight vessels or shot in using pyrotechnic or pneumatic systems. In the RTM, the "visualization" of a radioactive cloud is accomplished by observing passive reflectors, which mark the cloud, by means of a radar (e.g., weather radar) located near an enterprise or group of enterprises being monitored. Information about the position and configuration of reflectors is fed into a computer which is coupled with the radar for preliminary processing of the echo signals and data compression. Next, the information is transmitted to the crisis center at the enterprise, where it is displayed on a monitor and serves for in situ calculations of the trajectory and geometric characteristics of the radioactive cloud in real time scale. The data are used to reconstruct the track of the cloud in the atmosphere and on the ground, to give instructions for airborne and ground-based sampling, to predict the spreading of the cloud during the immediate future period in time, and to make recommendations to the public health service (Yurchak, 1998). The working mobile prototype of such system is described in (Gariyants et al., 2006). The mapping of air and ground footprints of chaff releases from 300-metre meteorological tower has been performed during many field simulation experiments.

It should be noted additionally that the RTM can be used for training the personnel of the PDE by means of the natural imitation of HAR. This creates conditions of the simulated emergency situation very close to the real one and increases the capacity of the RTM-based monitoring system installed in the vicinity of the PDE.

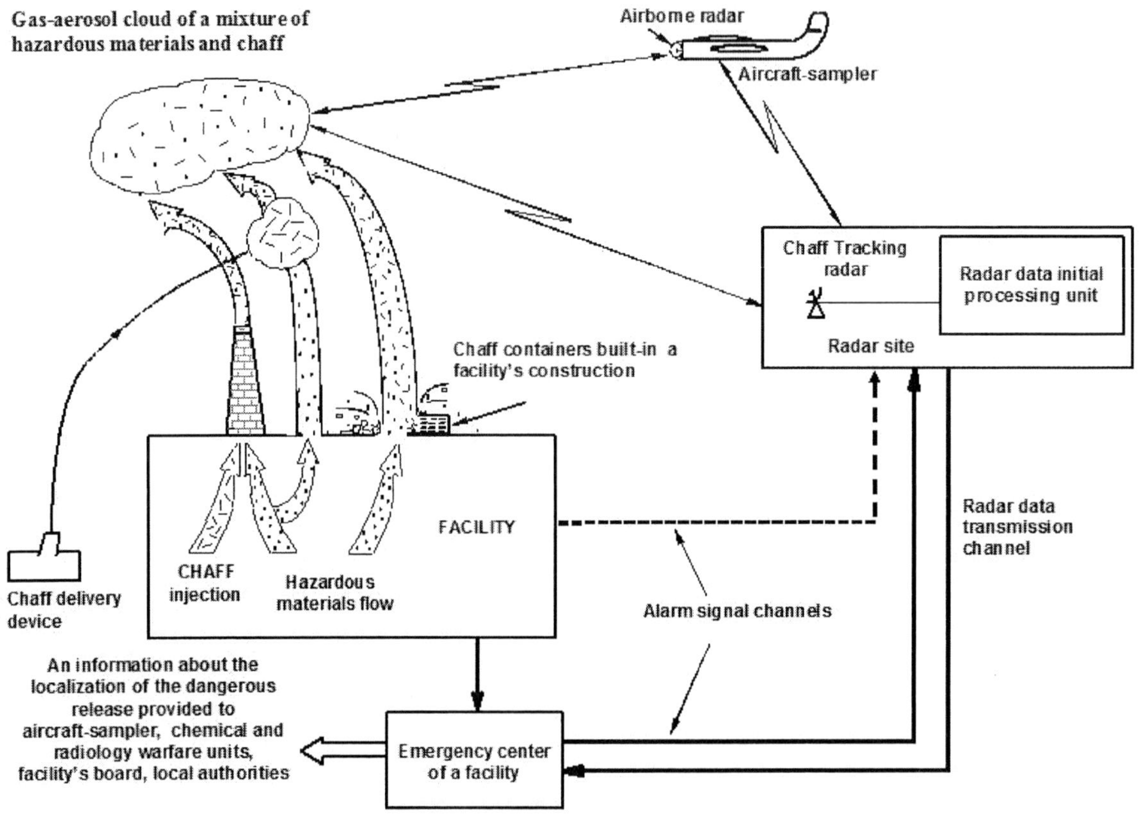


**FIGURE 1. An illustrative scheme of the RTM for tracking and mapping clouds of hazardous atmospheric releases from potentially dangerous facilities/enterprises**

## COMPILATION OF THE RTM WITH THE RADIOLOGICAL MEASUREMENTS

The most efficient known methodology for determining the radioactive contamination of the atmospheric HAR is γ-spectrometry performed by the ground-based spectrometers with their detectors upward. That data are essentially important for the source term assessment. The practical implementation of this method demands knowledge of emission geometry, i.e., the configuration of the cloud (plume) of radioactive substances, the position of the cloud in space relative to the γ-spectrometer, and the distribution of activity within the cloud. Using the RTM, it is shown that carbon chaff are subjected to atmospheric transport and diffusion process to nearly the same degree as the contamination of interest (Bessonov et al., 1987). It results the distribution of the concentration of chaff and HAR can be assumed to be similar, i.e., identical to within a dimensional factor. The isotopic ratio is assumed to be constant for all parts of HAR. Based on these conditions, the spatial distribution of radar reflectivity from chaff $Z(\mathbf{R})$, which marks the HAR cloud, within the radar volume is proportional to the volume activity of an elementary radionuclide source $C(\mathbf{R})$ and differs only by a constant dimension factor $D = C(\mathbf{R})/Z(\mathbf{R})$ at any radius-vector $\mathbf{R}$ of the source relative to the radar. Once the coefficient $D$ has been determined, the desired radioactivity distribution $C(\mathbf{R})$ can be found from the distribution, measured by the radar method. Other radiation characteristics of the radioactive cloud can be calculated from the activity distribution. Practically, the factor $D$ is determined from the formula:

$$D=\left[I_{ph}(E)\right]\left[\frac{1}{\eta(E)}\right]\left[\frac{1}{\varepsilon(E)}\right]\left[\frac{1}{G(E)}\right], \quad (2)$$

where the square brackets represent the following parameters (from left to right): the spectrometer-measured quantity, the physical constant, the instrumental function of the detector and, the dimensional geometrical factor, respectively. All parameters are functions of energy $E$ of unscattered gamma-rays. The last one, $G(E)$, is the subject of determination by the RTM:

$$G(E)=\int_{\tilde{V}} d\mathbf{R}\frac{Z(\mathbf{R})}{4\pi r^2}\exp[-\mu(E)r], \quad (3)$$

where $\mu(E)$ is a liner attenuation factor for photons attenuated by air; $\tilde{V}$ is the volume of HAR (chaff) cloud; $\mathbf{R}$ and $\mathbf{R}_0$ are the radius vector of the source and location of the detector relative to the radar, respectively, $r=\left|\mathbf{R}-\mathbf{R}_0\right|$. Since the distribution $Z(\mathbf{R})$ can be determined by radar, the geometrical factor is determined as well. Analytical expressions for $G(E)$ at different patterns of radar probing and other details of the methodology is provided in (Martinenko and Yurchak, 2001). At the light of contemporary achievements in exploiting of Unmanned Aerial Vehicles (UAV) for environmental monitoring including radiological survey (e.g., Torii and Sanada, 2015), it might be expected that the deployment of described above combined radar-radiometer system on UAV platform after corresponding adaptation will further improve the characteristics of the information support system.

## VALIDATION OF MESOSCALE MODELS

The analysis of results of the ETEX experiments indicates that even the initial material cloud development close to the source decides on the its future behavior (Graziani et al., 1997). The importance of correct determination of the tracer in the mesoscale region is therefore enhanced. This requires that experiments be conducted with increased density of the measurement network close to the source. There are estimates suggesting that the optimal number of samplers, even at 4 km range from the source, must be not less than about 80 (Kummel, 1984). Certainly, in real experiment such demands cannot be met and it is necessary to find other methods for registration of space–temporal characteristics of tracer. The validation procedure compares the prognostic results of theoretical models evaluated with the current meteorological parameters as input data and results of real pattern of impurity dispersion simulated by chaff dispersion under the same meteorological situation, Figure 2. The chaff dispersion is described in terms of the statistical diffusion theory at homogeneous turbulence with allowance for possible wind shifts (Byzova et al., 1991). The governing equation for chaff standard spatial deviations of concentration along axes of Cartesian coordinate system (x, y, z) in this approach is as follow:

$$S_i^2=a_i\left[b_i t-\frac{3}{2}+2\exp(-b_i t)-\frac{1}{2}\exp(-2b_i t)\right]+c_i t^3 , \quad (4)$$

where $i=x, y, z$; $a_i=2K_i\tau_{L,i}$; $b_i=\tau_{L,i}^{-1}$; $c_i=1/6\Gamma_i^2 K_z$; $c_z=0$; $t$ is diffusion time; $\tau_{L,i}$ and $K_i$ are integral time of Lagrangian turbulent scale and the diffusion factor in $i$-th direction, respectively; $\Gamma_i$ is vertical gradient of the $i$-th component of the mean wind velocity; $K_z$ is vertical diffusion coefficient. The first term in equation (4) describes the regime of the impurity cloud expansion with homogeneous turbulence, the second –accounts for the influence of vertical diffusion on longitudinal and lateral ones, given a wind velocity shift. The approximation of field dependences $S_i^2(t)$ by the least square method using (4) permits estimating parameters $a_i$, $b_i$ and $c_i$. These parameters are used to determine the diffusion coefficients and integral time scale of turbulence. The mixing layer height is determined as an upper bound of chaff cloud image at radar Range-Height Indicator. The complete description of measuring diffusion parameters of chaff releases from 300-metre meteorological tower is in (Zhukov and Yurchak, 1994) and field validation experiment - in (Golubenkov et al., 1998). In particular, results of comparison of experimental $S_i$ with calculations by the RIMPUFF model (Mikkelsen et al., 1984) suggest considerable difference, particularly for $S_z$ and $S_y$. Therefore, the applicability of relations for calculating dispersions $S_i$ recommended by that model calls for further studies. The study of applicability of the model based on the Monte-Carlo method (e.g., Hanna, 1984) for calculating diffusion indicates that the features of cloud dispersion derived from this model are in satisfactory agreement with known theoretical results for relative diffusion.

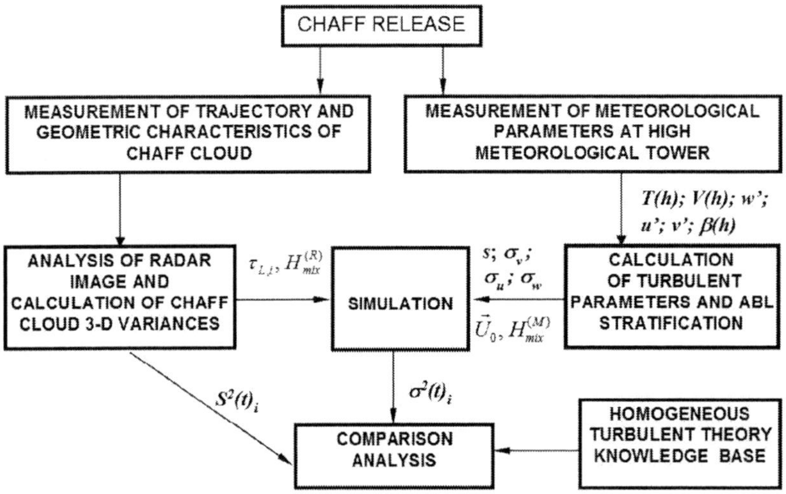


**FIGURE 2. Scheme of validation of meso-scale models using the RTM. Designations are in the text.**

## SUMMARY

The Radar-Tracer Method exploiting the radar chaff was presented. It is based on the well-known principle of marking a difficult-to-detect substance with materials one of whose properties is the same as that of the substance to be detected (for example, aerodynamic properties) while other properties differ so sharply that they make it possible to distinguish the markers reliably by technical means. Several useful attributes of the RTM technique are briefly demonstrated in the cases presented here. These include: (1) Tracking and mapping clouds of harmful atmospheric releases (HAR) during an accident; (2) Determination of the geometrical factor of the radioactive HAR to estimate its irradiation rate; (3) Validation of mesoscale models of dispersion of passive impurity in the atmosphere; (4) Training the personnel of environmental dangerous enterprises. The first two items are the direct actions of the countermeasure procedure in support of decision making person. The last two ones are aimed to improve the preparedness for the possible accident. Feasibility of the RTM was proved during many field experiments which primary feature was the usage of super light carbon chaff.

## REFERENCES

Bessonov, V.A., Yu. P. Grishin, and M.A. Iordansky. 1987. "Product "Metka" for study of structure and dynamics of air fluxes in the near-cloud area" (in Russian). *Meteorol. Hydrol.*, 1:118-120.

Byzova, N. L., E.K. Garger, and V.N. Ivanov. 1991. *Experimental studies of atmospheric diffusion and calculation of dispersion of material* (in Russian). Hydrometeoizdat, Leningrad, USSR.

Gariyants, A.M., V.G. Bulgakov, V.M. Shershakov, and G.P. Zhukov. 2006. "Mobile measurement facilities for the real-time interaction with the RECASS NT system in antiterrorism actions." In S.

Apikyan and D. Diamond (Eds.), *Countering Nuclear and Radiological Terrorism*, pp. 191-198. NATO Security through Science Series. Springer, Netherlands.

Girs, S.P., V.N. Kamentsev, M.N. Serova, G.I. Skhirtladze, and B.S. Yurchak. 1979. "On the possibility of using electrically conducting carbon fibers for investigating air flows in clouds" (in Russian). *Trudy GGO,* 420: 82-88.

Golubrnkov A. V., O.I. Vozzhenikov, G.P. Zhukov, and B.S. Yurchak. 1998. "Validation of meso-scale models using radar chaff data." Unpublished manuscript.

Graziani G., W. Klug, S. Galmarini, G. Grippa. 1997. *Real Time Long Range Transport Model Evaluation.-ETEX second release*. Draft report for modelers meeting. Vienna. Joint Research Center, Ispra, Italy.

Hanna, S.R. 1984. "Applications in air pollution modeling." In F.T Nieuwstadt and H. van Dop (Eds.), *Atmospheric turbulence and air pollution modelling*, pp. 275-307. D. Reidel Publishing Company, Holland.

Hildebrand, P.H. 1977. "A radar study of turbulent diffusion in the lower atmosphere." *J. Appl. Meteorol.*, 21(11): 493-510.

Kummel, M. 1984. "Development of optimal measuring network for environmental monitoring at NPP" (in Russian). In: *The radiation safety support of NPP maintenance*, Book 5, pp. 78-89. Vilnius. Energoatomizdat, Moscow, USSR.

MacKenzie, D. 1985. "After Bhopal, everybody loves factory inspectors." *New Scientist*, Oct. 31, 1985.

Martinenko, V.P., and B.S. Yurchak. 2001. "Radar-spectrometric method for determining thc characteristics of radioactive emission." *Atomic Energy*, 91(2): 676-682.

Martner, B.E., J.D. Marwitz, and R.A. Kropfi. 1992. "Radar observations of transport and diffusion in clouds and precipitation using TRACIR." *J. Appl. Meteorol.*, 9(6): 226-241.

Mikkelsen, T., S.E. Larsen and S. Thykier-Nielsen. 1984. Description of the RISO Puff Diffusion Model. *Nuclear Technology*, 67: 55-65.

Skhirtladze, G.I. and B.S. Yurchak. 1979. "Measurements of coefficient of horizontal turbulent diffusion in cumulus clouds by the radar method." *Izv. of USSR AS, FAO*, 15(2): 146-153.

Torii, T., and Yu. Sanada. 2015. "Radiation measurement by unmanned aircraft after Fukushima Daiichi nuclear power plant accident." *Remotely Piloted Aircraft Systems Symposium*, ICAO Headquarters, Montreal, Canada.

Weinmann, J.A., and F.Z. Marzano. 2008. "An exploratory study to derive precipitation over land from X-band Synthetic Aperture Radar measurements." *J. Appl. Meteorol. Clim.*, 47(2): 562-575.

Yurchak, B.S. 1998. "Radar support for radiation conditions monitoring and survey systems." *Atomic Energy*, 84(5): 335-341.

Zhukov, G.P., and B.S. Yurchak. 1994. "Diffusion of passive material in the atmospheric boundary layer by radar data." *Izv. of USSR AS, FAO,* 30(4): 451-457.

Zhukov, G.P., and B.S. Yurchak. 1993. "An assessment of the possibility of determining the location and configuration of a cloud of harmful atmospheric emissions from industrial enterprises by the radar method" (in Russian). *Meteorol. Hydrol.*, 12: 94-100.

# AN ALTERNATIVE ON-LINE METHOD FOR EVALUATING BROWN CARBON IN THE ATMOSPHERE

***Guohua Zhang***, Xinhui Bi, and Xinming Wang
(Guangzhou Institute of Geochemistry, Chinese Academy of Sciences, Guangzhou, 510640, PR China)

While the dominant light-absorbing aerosol species is thought to be BC, other light absorbing species are also present in atmospheric aerosols, which were recently revealed to be of potentially important in the absorption of near-UV light, and thus the radiative forcing. High contribution of brown carbon (BrC) to light absorption at near-UV wavelength is reported worldwide from satellite and ground observations, and modeling studies. However, On-line estimation of light absorption properties of BrC still presents challenges. In this study, a multi-wavelength (370–950 nm) Aethalometer was applied to obtain the wavelength dependent light absorption coefficient ($\sigma_{abs}$) of aerosols, before and after heated to 250 $^{o}$C on a 5 minute cycle. An alternative method was developed to evaluate the contribution of BrC to light absorption at 370 nm ($\sigma_{abs,BrC}/\sigma_{abs,370nm}$) by subtracting the measured $\sigma_{abs}$ to $\sigma_{abs,BC}$ at 370 nm. The $\sigma_{abs,BC}$ at 370 nm was extrapolated using field measured absorption Ångström exponent (AAE) over the wavelength of 880–950 nm with one-hour resolution. The simultaneous measurements of heated aerosols provided evidence that the influence of BrC on the $\sigma_{abs}$ over 880–950 nm is negligible. Meanwhile, $\sigma_{abs,BrC}/\sigma_{abs,370nm}$ was also estimated by the previously reported methods by assuming AAE to be one (Method I) and based on the light absorption enhancement (Method II). While the estimated $\sigma_{abs,BrC}/\sigma_{abs,370nm}$ based on our developed method and Method I was highly correlated ($r^2$ = 0.78), the difference could be as large as > 20% on average. However, the mean contribution was obtained to be negative with Method II, indicative of net production of BrC when the aerosols were heated. The difference between the obtained $\sigma_{abs,BrC}/\sigma_{abs,370nm}$ was ~40% on average and much higher (> 50%) during noon hours, when there were higher abundance of secondary organic aerosols (SOA) and sulfate. This developed method in this study is beneficial for future studies on the light absorption and climate forcing of BrC. The result indicates a potentially important contribution of light absorption from BrC at UV wavelength in the PRD region, which should not be neglected in the climate model.

# ENVIRONMENTAL IMPACT OF BIOMASS BURNING FOR CIVIL USES ONTO A MOUNTAIN AREA

***F. Petracchini, C. Balducci, M. Perilli, A. Cecinato,* F. Liotta,** *V. Paolini,* ***L. Paciucci.***
(National Research Council of Italy, Monterotondo RM, Italy)

The impact of biofuel combustion for civil uses onto the environment was investigated through determining a list of air toxicants at Leonessa, a town lying in Apennines mountain region, Central Italy, and in its surroundings (Terzone). Attention was focussed on suspended particulate matter ($PM_{10}$), polycyclic aromatic hydrocarbons (PAHs) and regulated pollutants (nitrogen dioxide, ozone and benzene).Two in-field campaigns, each lasting two weeks, were carried out in summer 2012 and winter 2013. Contemporarily, air was monitored in downtown Rome and in other localities of the province. In the summer all pollutants were more abundant in Rome (e.g., PAHs, 0.93 ng/m$^3$, vs. 0.37 ng/m$^3$ at Leonessa and 0.14 ng/m$^3$ at Terzone; benzene, 0.9 μg/m$^3$ in Rome vs. 0.2 μg/m$^3$ at Leonessa), apart from ozone (73 μg/m$^3$ in Rome and >100 μg/m$^3$ at Leonessa). By contrast, in the winter PAHs were more at Leonessa (15.8 ng/m$^3$) than in the capital city (7.0 ng/m$^3$), and benzene was similar (2.3 μg/m$^3$), despite suspended particulates were less (22 vs. 34 μg/m$^3$). Due to lack of other important sources and the scarce impact of transport at mid (inter-regional) scale, biomass burning was identified as the major emitter of PAHs at Leonessa during the winter. Its importance was confirmed by the molecular signatures, pictured by PAH concentration ratios distinct from those of Rome. Other towns too experienced PM levels similar to capital city but higher PAH loads (~9.6 ng/m$^3$), suggesting that uncontrolled biomass burning contributed to pollution across the Rome province. Important differences among PAH diagnostic ratios were observed at the sites also in the summer; nevertheless, in this case the ageing of air masses could play a role. As results of the two in filed compaign can be concluded that biomass burning was the major emitter of PAHs at Leonessa during the winter. In summer period important differences among PAH diagnostic ratios were observed. Despite, the suspended particulates were less than in Rome suggesting the lack of other important sources and the scarce impact of transport at mid scale.

# A PRELIMINARY ASSESSMENT OF MAJOR AIR POLLUTANTS IN THE CITY OF URUMQI, CHINA

F. Petracchini, A. Cecinato, L. Paciucci, ***V. Paolini**** and F. Liotta
(Institute of Atmospheric Pollution Research, National Research Council of Italy,Monterotondo, RM, Italy)

Nitrogen oxides ($NO_2$ and $NO_x$), ozone ($O_3$), sulphur dioxide ($SO_2$), benzene, toluene, ethylbenzene and xylene (BTEX) and ammonia ($NH_3$) in ambient air were preliminarily assessed, by diffusive sampling technique, in the years 2010-2012 through four measurement campaigns in Urumqi, the capital of Xinjiang in NW China. This is a heavily industrialized area where, up to date, very few air pollution studies are available. Impacts of pollutants have been assessed in the various critical areas affected by industrial activity and traffic. Principal component analysis (PCA) was also performed to understand the pollution sources and the correlation between the several pollutants monitored.

Considerably high mean annual pollutants concentrations have been observed, which ranged between, 46 and 158 μg $m^{-3}$ ($SO_2$), 33 and 126 μg $m^{-3}$ ($NO_2$), 82.5 and 140.5 μg $m^{-3}$ ($NO_x$) and 46 and 82 ug $m^{-3}$ (BTEX).

It was found that nitrogen oxides, benzene and xylene were mostly emitted from industrial sources and motor vehicles. Very high consumption rates of fossil fuels for energy generation and domestic heating are mainly responsible for high annual pollution levels. The concentration values of the pollutants measured during different seasonal periods show remarkable differences: in winter period the coal burning in domestic heating facilities was found as the major contributor to $SO_2$ and BTEX pollution. Ammonia pollution was also found in the city area where are located a slaughter facility and a meat treatment facility. Predominant factors influencing pollutions were associated to those elements conditioning air pollutant diffusion: meteorology (wind speed and direction), air stagnation and inversion events, relevant emission sources (especially for BTEX and $NO_x$), emission seasonality and photochemistry.

The analysis conducted can be a consistent pollution control policy framework to reduce emissions and critical pollution episodes.

# METHODOLOGIES FOR A BETTER INTERPRETATION OF THE PRELIMINARY ASSESSMENT: IAPMS PANČEVO

***L. Paciucci****, F. Petracchini, A. Cecinato, ***P. Romagnoli****, V.Paolini, F.Liotta, F. Vichi
(Institute of Atmospheric Pollution Research, National Research Council, Monterotondo, RM, Italy)
M. Biscotto (Politecnico di Torino, Torino, Italy)

Air quality preliminary assessment was carried out in Pančevo, throught eight two-week campaigns, one for each season, were carried out starting from August 2005 up to May 2006 by means of passive samplers for nitrogen dioxide, nitrogen oxides, sulphur dioxide, ozone, ammonia, benzene, toluene, xylenes, and total non methanic hydrocarbons exposed in 35 sites.

The sampling sites were divided into three categories, i.e.: (1) background - B, including both urban and suburban locations; (2) traffic - T, including traffic locations and hot spots; and (3) industrial - I.

By performing statistical analysis was found a high correlation between organic pollutants (benzene, toluene and xylenes; BTX) due to same source origin. In the T category sampling sites, the correlation between benzene and nitrogen dioxide was good as expected from sites close to traffic areas. This result substantiates the categorization of the sites. In the B category the correlation coefficient ($R^2$) between $NO_2$ and BTX was lower than in the other two categories (T and I categories) suggesting that some of the sites may not have background characteristics (they are outliers). As expected, after the removal of the outliers, the correlation coefficients raised up. This correlation is similar to that found in the T category, thus showing an affinity between these two categories (B and T categories). In the I category, unlike in the above mentioned categories, there was a lack of correlation between the organic pollutants and nitrogen dioxide (since traffic is not the major source). Considering the data set for the entire year it was found that sulphur dioxide (regional source) was rather uniform over all categories, nitrogen dioxide was slightly higher for T category, benzene and ammonia showed much higher concentrations in the I category due to the presence of sources in the area.

Regarding the sampling sites, the industrial areas strongly differ from the others because of the prevalence of BTX (Benzene, Toluene and Xylenes) emissions from inner sources, in particular near or inside the refinery. The B and T sites seemed to have the same chemical distribution, thus suggesting a similarity of sources. The overall conclusion was that all sites could be split up in two big clusters: an industrial and an urban one. The I cluster is strongly influenced by the facility emissions, while the urban cluster is less influenced by the industrial emissions and is affected by local emissions, such as traffic and domestic heating. The contribution of traffic sources to BTX and nitrogen oxides concentrations was not negligible, taking into account the differences between T and B values.

# BIOLOGICAL TREATMENT OF CHLOROFORM IN A CONTROLLED AEROBIC TRICKLE BED BIOFILTER

***Keerthisaranya Palanisamy*****, Bineyam Mezgebe and George A. Sorial
(University of Cincinnati, Cincinnati, Ohio, USA)[127]
Endalkachew Sahle-Demessie (United States Environmental Protection Agency, Cincinnati, Ohio, USA)

Chlorination disinfection forms harmful disinfection byproducts (DBPs) when chlorine reacts with the naturally occurring humic and fulvic like substances present in water. DBPs are a group of halogenated organic compounds such as Trihalomethanes (THMs), haloacetic acids (HAAs), chlorophenols, chloral hydrate, and haloacetonitriles (HANs) found in disinfected water that is released into the distribution system. Trihalomethanes are the commonly reported DBPs at environmentally relevant concentrations and chloroform constitutes 68% of the total THMs detected in drinking water. Under the SDWA, the USEPA establishes a MCL of 70 ppb for chloroform stating that people consuming chloroform in excess of this limit could experience adverse effects in liver, kidney, or central nervous system and increased risk of cancer. The level of chloroform detected in public water supplies range from 2 to 88 ppb and reaches as high as 300 ppb in ground water contaminated by landfill leachate. Since chloroform is a volatile organic compound (VOC), it readily evaporates into the atmosphere from disinfected water surface, where it is non-reactive at gaseous state causing long range air pollution. As a result, human exposure through inhalation is reported at an average concentration of 2 to 200 μg/day. Current treatment technologies that counter the problem of DBPs in water treatment facilities such as air stripping, adsorption, ozonation, and advanced oxidation processes are reported to be energy intensive and generate secondary pollutants. Air stripping is a non-destructive mechanism commonly used by water utilities to remove VOCs post disinfection, however it releases concentrated off-gas streams into the atmosphere. Biofiltration is one of the most proven techniques for the removal of gas phase VOCs and is a widely applied off-gas control mechanism in Europe and Japan, but only recently been investigated in the United States.

In this paper, a trickle bed air biofilter (TBAB) is subjected to the biodegradation of chloroform under acidic conditions. TBAB has been shown to efficiently remove both hydrophobic and hydrophilic VOCs such as Methyl Ethyl Ketone, Methyl Isobutyl Ketone, Benzene and n-Hexane and achieves stable removal performance for a wide range of feed concentrations. In this study chloroform is subjected to cometabolitic degradation with ethanol. The aerobic biofilter is operated at acidic conditions to benefit the growth of filamentous fungi as the principle media for biodegradation. The reason behind using fungi is that its aerial mycelia largely improve the surface hydrophobicity resulting in high partition of the VOCs by increasing its mass transfer from the gas phase to microorganisms in the solid phase. The objectives of the research are to: (1) determine the critical loading rate of chloroform that achieves stable removal efficiencies; (2) generate a carbon mass balance closure for the biofilter; (3) evaluate the biodegradation removal rate kinetics and (4) determine the amount of nitrogen consumed for biomass growth. The biofilter is constructed using cylindrical glass sections and packed with pelletized diatomaceous earth to support biological media. Compressed air transfers chloroform and ethanol from calibrated syringe pumps into the biofilter where buffered nutrients are sprayed intermittently to facilitate the activity of the micro-organisms. The biodegradation of chloroform is measured by the decrease in its initial concentration based on Pseudo first order kinetics. The experiment is carried out in five different phases of operation with each phase constituting a different feed ratio of the contaminant and the cometabolite. The biofilter is subjected to treat 5 $ppm_v$ of chloroform mixed with 200, 150, 100, 50 and 25 $ppm_v$ of ethanol. The removal efficiency obtained for chloroform was above 70% for four feed ratios studied. Degradation of chloroform peaked at 82% efficiency for influent chloroform and ethanol concentrations of 5 and 200 $ppm_v$ respectively. This dechlorination process is expected to achieve a sustainable removal of atmospheric chloroform released from aerated disinfected water.

# EFFECT OF METHANOL AND TOLUENE ON REMOVAL OF TRICHLOROETHYLENE IN A FUNGI SEEDED BIOTRICKLING FILTER

***Dhawal Chheda**** and George Sorial
(University of Cincinnati, Cincinnati, OH, USA)

Trichloroethylene (TCE) is a chlorinated solvent which has several applications in the metal degreasing, plastics, dry cleaning, rubber and refrigerant industries. It is listed as a potential mutagenic and carcinogenic compound and when in contact, TCE can affect our nervous, reproductive and immune systems. It is one of the most detected chemicals in the Superfund sites and is therefore ranked at 16$^{th}$ position in the 2005 CERCLA List published by US E.P.A. More than 95% of US rural areas depend on groundwater which is contaminated by TCE and thus, its remediation is of utmost importance. Several technologies like ozonation, oxidation and adsorption have been suggested for removal of TCE from air and water resources. But such methods are either expensive, energy intensive or produce toxic by-products like furans. This research focuses on removal of trichloroethylene in a biotrickling filter (bioremediation) which is seeded with fungi strains and operated at an acidic pH.

TCE is in a highly oxidized state and not a competent carbon source for the microbes. Therefore, methanol is added along with TCE as a primary substrate or a "co-metabolite". Two ratios of methanol-to-TCE are tested: 80% methanol to 20% TCE and 70% methanol to 30% TCE. TCE loading rates ranging from 3.22 g/m$^3$/hr to 12.88 g/m$^3$/hr are tested for each ratio. Thus, methanol loading ranges between 103.71 g/m$^3$/hr and 711.14 g/m$^3$/hr depending on the ratio. This assessment helps in establishing which ratio and what concentration range provides higher elimination capacities for TCE while avoiding the inhibitory effect of methanol at higher concentrations. For a loading rate of 3.22 g/m$^3$/hr, average TCE removal efficiency of 81.4% was achieved in the 80:20 system while it was 79.9% for the 70:30 ratio. For a loading of 6.44 g/m$^3$/hr, the efficiencies attained were 79% and 87.1% for the 80:20 and 70:30 systems respectively. For methanol, removal efficiencies achieved were consistently over 95%. Further analysis of TCE elimination is under examination for different residence times and co-metabolite concentrations.

In 2006, more than 82% of TCE was released from the above mentioned industries. When released from these industries, TCE is often accompanied by methanol and toluene. Therefore, a study is performed to test the biofilter performance for an inlet of a ternary blend of TCE, methanol and toluene. Again, two ratios are tested: 50% toluene to 26% TCE to 24% methanol and 45% toluene to 15% TCE to 40% methanol. Inlet concentration ranges from 10 to 20 ppmv TCE, 23 to 118 ppmv methanol and 17 to 50 ppmv toluene. This study is currently under investigation to check how the dynamics of TCE removal are disrupted in a ternary mixture of organic compounds in a fungi seeded biotrickling filter.

# EVALUATIONS OF THE REMOVAL OF TRIHALOMETHANES BY TWO INDEPENDENT BIO-TRICKLING FILTERS UNDER DIFFERENT OPERATING CONDITIONS

***Bineyam Mezgebe,*** George Sorial, Keerthisaranya Palanisamy
(University of Cincinnati, Cincinnati, OH 45221-0012, USA)
Endalkachew Sahle-Demessie
(US EPA National Risk Management Research Laboratory, Cincinnati, OH 45268, USA)

Disinfection resulted from the reactions between the chlorine and natural organic substances increased the formation of chlorinated by-products including trihalomethanes (THMs). Since chlorination by-products are classified as potential carcinogenic by EPA, they are of importance for human health. THMs have the highest concentration of chlorination by products in drinking water and chloroform is the most abundant of all THMs with a maximum contaminant level (MCL) of 0.070 mg/L. Several physical and chemical removal methods are used to treat chloroform which are expensive and could generate secondary pollutants. Biofiltration is one of the most proven technologies for VOC control as it is environment–friendly, cost effective and releases fewer byproducts. In this study, an integrated technology is proposed by taking chloroform as a model THMs. This integrated technology consists of nitrogen stripping followed by anaerobic Bio-trickling Filter (BTF) or air stripping followed by aerobic BTF. Two independent and parallel aerobic and anaerobic BTFs with similar chloroform (5 ppmv) concentrations were investigated. In the anaerobic BTF, a co-metabolite and surfactant were utilized to enhance the biodegradation process. Upon the addition of Ethanol (co-metabolite) and Tomadol 25 - 7 (surfactant), the performance of the anaerobic reactor improved from an initial 49% to over 64% removal efficiency. Alternatively, the aerobic BTF was investigated under acidic environment to enhance fungi growth. The average removal efficiency under this acidic condition was 80%. The significance of the anaerobic compared to aerobic BTF was the generation of methane (0.32 mg / day) and carbon dioxide (147.4 mg/day) as the end products where as the main end product for aerobic BTF was carbon dioxide (169.7 mg/ Day). The study will further investigate the microbial diversity within the BTFs in order to get an insight of the BTFs' performances.

# REMOVAL OF VAPOUR PHASE METHANOL IN THE PRESENCE OF THIOSULPHATE USING ANAEROBIC BIOTRICKLING FILTER

***Mekonnen M. Tarekegn***
(School of post graduate studies, Department of Urban Environment and Climate Change Management, Ethiopian Civil Service University, 26148 Addis Ababa, Ethiopia)
Eldon Raj Rene Jack van de Vossenberg and Piet N. L. Lens
(Environmental Engineering and Water Technology (EEWT) Department, UNESCO-IHE Institute for Water Education, Westvest 7, 2611 AX, Delft, The Netherlands)

The feasibility of methanol removal using an anaerobic biotrickling filter (BTF) was studied under different operational conditions. Anaerobic BTF experiments were performed by changing the empty bed residence times (EBRT) and inlet methanol (gas-phase) concentrations for ~66 days.

The nutrient medium containing thiosulphate was fed to the BTF in trickling mode (pH - 8.0). The methanol removal efficiencies of the BTF at EBRTs of 5 and 3 min were 96.7% and 90.5%, respectively. Concerning concentration effects, the methanol removals at an inlet concentration of $1.4\times10^{3}$ppm$_v$ and $5.2\times10^{3}$ ppm$_v$ were 90.5% and 80.2%, respectively. A maximum methanol elimination capacity (EC) of $1.5\times10^{2}$gMeOH/m$^3$.h was achieved at an inlet loading rate (ILR) of 1.7×102 gMeOH/m$^3$.h. The average growth rate of biomass observed in the BTF reactor under steady state operation during phase V (EBRT 5 min and inlet MeOH concentration $1.0\times10^{3}$ppm$_v$), phase VI (EBRT 3 min and inlet MeOH concentration $1.41\times10^{3}$ppm$_v$) and phase VII (EBRT 3 min and inlet MeOH concentration $5.2\times10^{3}$ppm$_v$) were 0.13, 0.20 and 0.10 d$^{-1}$, respectively.

The average biodegradation rate of methanol in the BTF was 0.63 d$^{-1}$ at a maximum methanol biodegradation rate of 2.0 d$^{-1}$. While the average biodegradation rate of thiosulphate was 0.34 d$^{-1}$. The average biomass yield coefficient observed in the BTF reactor at steady state condition was 0.32 gbiomass.(gMeOH)$^{-1}$. The mass balance calaculation ascertained that ~27% of the incoming carbon and ~53% of sulphur was retained within the BTF reactor. Methanol is a hydrophilic VOC and thus it is easily scrubbed by the trickling liquid or it can be stripped by the incoming gas containing methanol. This reduced the total amount of carbon stored in the BTF reactor. It was ascertained that anaerobic biodegradation of gas phase methanol and liquid phase thiosulphate can be achieved within one bioreactor configuration which is promising for practical applications.

## INTERFACE EFFECTS FOR $CO_2$ HYDROGENATION ON $Pt_4/\gamma\text{-}Al_2O_3$.

***Yulu Liu***, Wanglai Cen and Jianjun Li
(Sichuan University, Chengdu, Sichuan, China)

Anthropogenic greenhouse gases which mainly consist of $CO_2$ have brought about climate change and are considered as a major threat faced by mankind. However, $CO_2$, a very natural C1-resource, is being paid close attention as a substitute of toxic CO in C1-chemistry hopefully to keep carbon resources recycling. And $CO_2$ recycle to useful chemicals or liquid fuel will also help to remit the greenhouse effects. Selective hydrogenation of $CO_2$ to methanol or methane has been widely studied. In this work, density functional theory (DFT) was applied to investigating the hydrogenation of $CO_2$ over $Pt_4/\gamma\text{-}Al_2O_3(110)$ catalyst. HCOO and CO are the important intermediates in $CO_2$ hydrogenation. From our results, CO formation is kinetically and thermodynamically more favorable than HCOO formation, indicating that Pt-based catalysts benefit the formation of methane. Furthermore, based on the analysis of $CO_2$ hydrogenation on Pt(111) surface, we found that the supported catalysts has a lower activation barrier (0.34 eV) than that of metal catalyst (0.59 eV) in the first step, i.e. the process of trans-COOH species formation, showing that the interfacial sites between $Pt_4$ cluster and $\gamma\text{-}Al_2O_3(110)$ slab can promote $CO_2$ hydrogenation. However, the cis-COOH species of the supported catalyst has a larger activation barrier (0.91 eV) than that of metal catalyst (0.56 eV), which means that dissociation of the former one is more difficult. Finally, further study is required to find an appropriate interface on the supported catalyst so that the compromise between activation barriers of $CO_2$ hydrogenation and COOH species dissociation could be achieved.

# A NOVEL $SnO_2$-$CoO_x$ CATALYST FOR NO OXIDATION WITH $H_2O$

***Huazhen Chang*** and Mingguan Li (Renmin University of China, Beijing, China)
Junhua Li (Tsinghua University, Beijing, China)

Noble metal catalysts have been comprehensively studied for NO oxidation which shows high catalytic activity. However, high cost limits their large-scale industrial application. In recent years, it was widely reported that cobalt-based catalysts were potential candidate for NO oxidation. Tin dioxide has been widely used as an oxidation catalyst as it can reversibly switched between $Sn^{4+}$ and $Sn^{2+}$ at a relatively low temperature. In this study, the Sn–Co–O mixed oxides catalysts were prepared by the coprecipient method and the obtained samples were studied by means of XRD, BET, $H_2$-TPR, XPS, in situ DRIFTS and NO + $O_2$ model reaction. The influences of $H_2O$ were also investigated.

A series of $SnO_2$-$CoO_x$ catalysts prepared by a co-precipitation method were investigated for NO oxidation reaction. The catalyst with a molar ratio of Sn:Co=3:1 (denoted as Sn(0.75)-Co-O) exhibited excellent catalytic activity in a temperature range of 250-350 °C. In the presence of $H_2O$, much better NO oxidation performance was also obtained over the same catalyst. NO-TPD showed that Sn(0.75)-Co-O the amounts of adsorbed NO were larger comparing with $CoO_x$ in the absence /presence of $H_2O$. It indicated that the NO adsorption capacity was improved by addition of Sn. More active sites are available for NO adsorption even in the presence of $H_2O$. XPS spectra showed that the ratio of chemisorbed oxygen increased, which contributed to enhance the redox property of Sn(0.75)-Co-O catalyst. Moreover, BET results revealed that the specific BET surface area increased over Sn(0.75)-Co-O catalyst, which are favourable for the adsorption and oxidation of NO.

# ENVIRONMENT-FRIENDLY AND HIGHLY EFFICIENT RARE EARTH BASED CATALYST FOR SELECTIVE CATALYTIC REDUCTION OF $NO_x$

Shemin Zhu, Yuesong Shen and Zhiwei Xue
(Jiangsu Collaborative Innovation Center for Advanced Inorganic Function Composites, Jiangsu National Synergetic Innovation Center for Advanced Materials (SICAM), College of Materials Science and Engineering, Nanjing Tech University, Nanjing, Jiangsu, China)
Zhimin Wang
(Shandong Gemsky Environmental Technology Co. LTD, Zibo, Shandong, China)

Nitrogen oxides ($NO_x$ for short), as one of the main causes of fog and haze and acid rain, has become the emphasis of current air pollution control. At present, selective catalytic reduction (SCR for short) of $NO_x$ has become the mainstream de$NO_x$ technology, whose technological core is the de$NO_x$ catalyst, and the major component system of the commercial de$NO_x$ catalysts is $V_2O_5(WO_3)/TiO_2$. Because the $V_2O_5$ is a toxic and water-soluble substance, once the $V_2O_5$ invasion into the human body, the health of the human body will be seriously damaged. Consequently, not only the catalyst production process will cause pollution, waste catalyst will once again threaten the safety of the environment, thereby research and development of an environment-friendly and highly efficient de$NO_x$ catalyst technology has become the great demand of the world environmental protection. Through more than ten years of scientific research, by using collaborative innovation of chemical catalysis and materials science, our R & D team successfully invented an environment-friendly and highly efficient rare earth based de$NO_x$ catalyst technology, which fills the gaps in the domestic and foreign industrial flue gas non-toxic de$NO_x$ catalyst technologies. So far, in research and development of the rare earth based de$NO_x$ catalyst, our R & D team already applied for more than 30 invention patents, published more than 60 academic papers, and obtained the original innovation intellectual property rights of China de$NO_x$ catalyst technology. The rare earth based de$NO_x$ catalyst technology successfully achieved industrialization in Shandong Gemsky Environmental Technology Co. LTD, which built the world's largest de$NO_x$ catalyst production line with a productivity of 5 million $m^3$ per year, and become the world's first and only enterprise that is able to produce environment-friendly and highly efficient de$NO_x$ catalyst. The new rare earth based de$NO_x$ catalyst exhibit better comprehensive performances and lower cost than those of the commercial vanadium-titanium system de$NO_x$ catalyst, with merits of environment-friendly chemical component, high de$NO_x$ efficiency (>90%), broad active temperature window (250°C–450°C), low conversion rate from $SO_2$ to $SO_3$ (<0.4%), high side crush strength (3MPa–15MPa), strong anti-humidity and water-resistance, strong anti-poisoning ability, reproducible for recycling use after deactivation, and waste catalyst without secondary pollution. Until now, the rare earth based de$NO_x$ catalyst has been successfully applied in over 60 actual de$NO_x$ projects in power, chemical, glass and cement industries, and take the lead in setting up the $NO_x$ "near zero emissions" project in coal-fired power plant in China. The new rare earth based de$NO_x$ catalyst technology was selected into "the directory that the State encourages the development of major environmental protection technology and equipment" of China in 2014, possessing broad prospects for international de$NO_x$ market.

# INFLUENCE OF FE LOADINGS ON DESULFURIZATION PERFORMANCE OF ACTIVATED CARBON TREATED BY NITRIC ACID

***Jiaxiu Guo***, Song Shu and Xiaoli Liu
(Sichuan University, Chengdu, Sichuan, China)

$SO_2$ is one of the major pollutants in China because of coal burning and causes several environmental problems, such as acid rain and abrasion of buildings. Therefore, removal of $SO_2$ from emission sources is a great interesting and becomes a more urgent task. Activated carbon (AC) is a major porous material. The porous structure controls the availability of surface active sites and the degree of catalyst dispersion. Fe is easy to obtain and lose electron, promoting the redox cycle and showing a good oxidation performance. In the catalytic process, this obtainment and loss of electron transfer is the key to form the active intermediate. A series of Fe supported on activated carbon treated by nitric acid are prepared by incipient-wetness impregnation with ultrasonic assistance and characterized by $N_2$ adsorption-desorption, X-ray diffraction (XRD), Fourier transform infrared spectrum (FTIR) and X-ray photoelectron spectroscopy (XPS). It has shown that Fe loadings significantly influence the desulfurization activity. Fe/NAC5 exhibits an excellent removal ability of $SO_2$, corresponding to breakthrough sulfur capacity of 323 mg/g and breakthrough time of 545 min. With the increase of Fe loadings, the generated $Fe_3O_4$ and $Fe_2SiO_4$ increase, but some generated $H_2SO_4$ can react with $Fe_2SiO_4$ and $Fe_3O_4$ to produce $Fe_2(SO_4)_3$ when $Fe_2SiO_4$ and $Fe_3O_4$ are exposed to $H_2SO_4$ for a long time. Fe/NAC1 has BET surface area of 925 $m^2/g$ with micropore surface area of 843 $m^2/g$ and total pore volume of 0.562 $cm^3/g$ including a micropore volume of 0.300 $cm^3/g$. With the increase of Fe loadings, BET surface area and micropore volume decrease. When the Fe loadings increase to 10%, BET surface area and micropore volume decrease to 706 $m^2/g$ and 0.249 $cm^3/g$. The Fe loadings influence the pore size distribution, and $SO_2$ adsorption mainly reacts in micropores at about 0.70 nm.

# MINIMIZING ENVIRONMENTAL POLLUTION AND THE EFFECTS IN NIGERIA THROUGH GREEN DESIGN AND GREEN BUILDINGS

***Iwuagwu, Ben Ugochukwu***. Dept of Architecture, Abia State Polytechnic, Nigeria
Ikechukwu onyegiri. Dept of Architecture, Imo State University, Nigeria
Iwuagwu, Ben Chioma M. Dept of Statistics, Abia State Polytechnic, Nigeria

**ABSTRACT:** The environmental impacts of buildings are enormous. Conventional buildings use large amounts of energy, land, water, and raw materials for their construction and operation. They are responsible for large amount of greenhouse gas (GHG) emissions as well as emissions of other harmful air pollutants. They also generate large amounts of construction and demolition (C&D) waste and have serious impacts on plants and wildlife. Considering the statistics, reducing the amount of natural resources buildings consume and the amount of pollution given off is seen as crucial for future sustainability. Over 40 percent of the energy used in Nigeria goes to heating, cooling, and power our buildings. And because much of this energy comes from dirty and dangerous sources like coal, oil, natural gas, and nuclear power, this accounts for nearly half of global warming pollution in the country. This high level of energy consumption pumps billions of tons of global warming pollution into the atmosphere causing ill health and depletion of the ozone layer. Green building is a high-performing, thoughtfully designed building that is very energy and water efficient, reduces its footprint on the planet, and provides a safer, healthier environment for the builders and future occupants throughout a building's life-cycle: from sitting to design, construction, operation, maintenance, renovation, and deconstruction. There are a number of motives to building green, including environmental, economic, and social benefits. However, this paper calls for an integrated and synergistic green design (GD) to both new construction and in the retrofitting of existing structures using renewable resources like sunlight through passive solar, active solar and photovoltaic techniques, using plants and trees through green roofs, rain gardens for reduction of rainwater run-off to minimize environmental pollution in Nigeria

## INTRODUCTION

Environmental pollution is one of the most pressing environmental problems and is among the critical challenges facing modern societies today. However, human activities directly or indirectly affect the environment adversely. By the activities of a stone crusher, he adds a lot of suspended particulate matter and noise into the atmosphere. Automobiles emit from their tail pipes oxides of nitrogen, sulphur dioxide, carbon dioxide, carbon monoxide and a complex mixture of unborn hydrocarbons and black soot which pollute the atmosphere. Domestic sewage and run off from agricultural fields, laden with pesticides and fertilizers, pollute water bodies. Buildings, however, contribute to environmental pollution and their environmental impacts are enormous. Conventional buildings use large amounts of energy, land, water, and raw materials for their construction and operation. They are responsible for large greenhouse gas (GHG) emissions as well as emissions of other harmful air pollutants. They also generate large amounts of construction and demolition (C&D) waste and have serious impacts on plants and wildlife. The International Energy Agency (2008) estimated that existing buildings are responsible for more than 40 percent of the world's total primary energy consumption and for 24 percent of global $CO_2$ emissions. This implies that the building sector uses more energy than other sectors like the industrial and transportation sectors. Due to increase in urban population and attendant residential pressures, energy consumption is projected to rise in buildings especially in fast growing countries (Odebiyi, Subramanian and Braimoh; 2010). This case of rise in energy consumption in buildings as a result of increase in urban population and residential pressure is also applicable in Nigeria. With the increasing building impact on energy consumption and pollution in Nigeria, it is necessary, according to Iwuagwu and Azubuine (2015) to address these challenges of unsustainable building practices by adopting more sustainable building approach such as 'green architecture to reduce the amount of pollution generated by buildings and its adverse effects on our environment.

Buildings are the dominant energy consumers in modern cities but their consumption according to IEA, (2004) can be largely cut back through improving efficiency, which is an effective means to lessen greenhouse gas emissions and slow down depletion of non- renewable energy sources. The designs of the architects and building construction methods are very important in reducing the problem of climate change. Reducing the amount of natural resources buildings consume and the amount of pollution generated and given off is seen as crucial for future sustainability. Careful selection of environmentally sustainable building materials is the easiest way for Architects to begin incorporating sustainable design principles in buildings. Design decisions regarding the selection of less environmental impact building components according to Kim and Rigdon (1998) need careful consideration during the building design process. Careful building design and materials selection can substantially reduce environmental impacts of buildings. Despite the major efforts that have been made over recent years to clean up the environment, pollution remains a major problem and poses continuing risks to health and the environment. This paper highlights the effects of pollution and climate change, stressing the significant roles buildings, designs, construction methods and material specification play in creating and reducing environmental pollution and climate change. However, this paper calls for an integrated and synergistic design to both new construction and in the retrofitting of existing structures. It concludes by recommending the use of renewable resources like sunlight through passive solar, active solar and photovoltaic techniques, using plants and trees through green roofs, rain gardens for reduction of rainwater run-off to minimize environmental pollution in Nigeria.

## EFFECTS OF ENVIRONMENTAL POLLUTION

Rise in fossil fuel burning, construction and demolition waste, transportation, general land use changes etcetera have emitted and continuing to emit increasing quantities of greenhouse gases into the Earth's atmosphere and pollute our environment. The greenhouse gases include: carbon dioxide (CO2), methane (CH4) and nitrogen dioxide (N2O) etcetera. A rise in these gases and other environmental pollutants has increased the amount of heat from the sun withheld in the Earth's atmosphere, heat that would have been normally radiated back into space. This increase in heat has led to the greenhouse effect, resulting to climate change. The climate change as a result of increase in greenhouse gas emission has led to a lot of anomaly in our environment. In 2012, about 70 percent of major global droughts occurred in Africa. Kenya, Somalia, Sudan, Malawi, Angola, Chad and Ethiopia were particularly hit hard, and more than 16 million people in those countries were affected. In the same year, floods claimed 363 lives in Nigeria and 65 lives in Niger. In 2013, heavy rains continued with major flooding in Sudan, South Sudan, Mali, South Africa, Zimbabwe, Botswana and Mozambique. Despite the major efforts that have been made over recent years to clean up the environment, pollution remains a major problem and poses continuing risks to health. Even in developed countries, however, environmental pollution persists, most especially amongst poorer sectors of society (Samet, Dearry and Eggleston,2011; Sexton and Adgate, 2001).

## BUILDINGS AS A CONTRIBUTOR TO ENVIRONMENTAL POLLUTION

Buildings consume a significant amount of our natural resources and have a wide range of environmental impacts. Considering what buildings are made of – steel, concrete, glass, and other energy-intensive materials – buildings have a high level of embodied energy. Producing and using these materials for construction depletes non-renewable resources and has environmental effects, these effects intensify the more frequently buildings are constructed, demolished and replaced. Buildings contribute to environmental pollution through:

***Energy Use.*** Conventional building materials have high embodied energy. This includes the energy of the fuel used to power the harvesting or mining equipment, the processing equipment, and the transportation devices that move raw material to a processing facility. This energy typically comes from the burning of fossil fuels, which are a limited and a non-renewable resource. The greater a material's embodied energy, the greater the amount of energy required to produce it, implying more severe ecological consequences.

***Construction Materials.*** Building construction requires the constant production and harvesting of millions of tons of a variety of raw materials especially non-renewable, to meet world-wide demand. The amount of raw materials used in buildings construction is very large. Analysis of building products, from the gathering of raw materials to their ultimate disposal, provides a better understanding of the long-term costs of materials. These costs are paid not only by the client, but also by the occupants, and the environment. Careful selection of environmentally sustainable building materials, like timber, bamboo, earth and other low embodied energy building materials, is the easiest way for architects to begin incorporating sustainable design principles in buildings. Jong-Jin (1998) asserts that, price has been the foremost consideration when comparing similar materials or materials designated for the same function. However, the price of a building component represents only the manufacturing and transportation costs, but not social or environmental costs.

***Improper Land Use.*** Millions of acres of land in Nigeria have buildings constructed on them. Buildings use large amounts of land; although the problem is the poor siting of buildings which leads to large amounts of land use. For instance, buildings built in the outskirt of existing residential or commercial areas require the construction of new roads, drainage, utility poles, and other infrastructure, which lead to, habitat destruction, land disturbance and erosion, among other things.

***Construction, Operation and Demolition Waste.*** Tons of waste is typically deposited into a landfill during the construction of new homes. Construction waste consists primarily of timber and manufactured wood products, drywall, and masonry materials. Others are a mix of roofing materials, metals, plaster, plastics, foam, insulation, textiles, glass, and packaging. Even as these materials are recyclable, in Nigeria most of it is deposited into landfills causing environmental pollution.

## OVERVIEW OF GREEN BUILDINGS.

Nowadays, human beings pay great attention to environmental protection, thus develop a new trend — the Green Buildings. It is not about the color green, but has something to do with a new architectural concept. While the definition of what constitutes a green building is constantly evolving, the Office of the Federal Environmental Executive offers a useful working definition. This agency defines this green building as: the practice of increasing the efficiency with which buildings and their sites use energy, water, and materials, and reducing building impacts on human health and the environment, through better siting, design, construction, operation, maintenance, and removal—the complete building life cycle. Environmental Protection Agency (EPA) also defines green building as the practice of creating structures and using processes that are environmentally responsible and resource-efficient throughout a building's life-cycle from siting to design, construction, operation, maintenance, renovation and deconstruction. This practice expands and complements the classical building design concerns of economy, utility, durability, and comfort. Green building is also known as a sustainable or 'high performance' building. According to Iwuagwu and Azubuine (2015), green building is an outcome of a design, which focuses on increasing the efficiency of resource use-. energy, water, and materials-while reducing building impacts on human health and the environment during the building life-cycle, through better siting, design, construction, operation, maintenance, and removal.

Buildings can incorporate many green features, but if they do not use energy efficiently, it is difficult to demonstrate that they are truly green. "Green building" is a concept that includes the design of the structure, site selection, building construction methods, building materials and landscaping practices. Green building is also referred to as "sustainable design" or "high-performance building." Green buildings offer an array of social benefits such as improved health, comfort, and productivity of occupants (Kats, Alevantis, Berman, Mills and Perlman, 2003). Economic benefits include reduction of building materials and operations costs. Social benefits include improvements in quality of life and equitable access to infrastructure services such as transportation, healthy indoor environments. Green buildings conserve vital natural resources and help to preserve fragile ecosystems, by reducing pollution and waste generation. In contrast to conventional buildings, green buildings seek to use land and energy efficiently, conserve water and other resources, improve indoor and outdoor air quality, and increase the use of recycled and renewable materials.

## HOW ENVIRONMENTAL POLLUTION IN NIGERIA COULD BE MINIMIZED USING GREEN BUILDINGS.

***Through Construction Materials.*** Building construction requires the constant production and harvesting of millions of tons of raw materials to meet demand. The amount of raw materials used in buildings is very much and has been estimated that the construction industry consumes half of all products produced by volume (Hansen, 2004). A crucial part of green buildings is the material that is used in their construction. Green building materials are generally composed of renewable rather than non-renewable resources and are environmentally responsible because their impacts are considered over the life of the product. Green building materials can be selected by evaluating characteristics such as reused and recycled content, low off-gassing of harmful air emissions, low toxicity, sustainably and rapidly renewable materials, high recyclability, durability, longevity, and local production. In Nigeria today there are a lot of green building materials available but have not been effectively utilized. These materials are abundant in nature and include, earth, thatch, bamboo, stone etcetera; they are all sustainable and have characteristics that do not pollute our environment. Using earth as a case, according to iwuagwu and eme-anele (2012), characteristics of earth include: Durability, Energy Efficiency, reusability, among others.

***Through Architectural Design.*** A major element of green building design is increased energy efficiency through incorporation of passive design, efficient lighting, renewable energy technologies, and improved thermal performance of building shells. The goal is optimization of energy performance and integration of renewable energy options. Green buildings have large glazing for proper ventilation and lighting, reducing dependency on artificial ventilation and lighting reduces energy use and invariable reduces environmental pollution. Glazing allows increased daylight (without excess heat), reducing electrical lighting need. Increasing the thermal envelope allows for as much solar heat gain during cold weather, and shading during hot weather, while minimizing heat loss. Green buildings are designed to use low embodied energy materials and renewable building materials, these renewable materials are sustainable and reduces environmental impact.

***Through Sustainable Siting.*** A sustainable site approach optimizes land use and development density using site compatibility, protection of wetlands and natural habitats, transportation access and proximity to amenities in the decision-making process. In combination, these factors can significantly reduce adverse development impacts and minimize the building's ecological footprint. Landscape elements should include native plants as alternatives to conventional grass lawns, which often depend on irrigation and pesticides. Native plants are adapted to the natural hydrology, climate, and geography of the region and have evolved in relation to other local plants. This allows native plants to provide habitat for local species. Also, native plants normally require less watering, fertilizers.

***Construction, Operation, and Demolition Waste.*** Green buildings generally seek to minimize the amount of Construction & Demolition waste they generate by recycling or reusing the C&D waste. For sites that include the demolition of existing structures, plans can be developed early in the design process to manage and reuse as much material as possible through the deconstruction, demolition, and construction processes. Demolition generates large amounts of materials that can be reused or recycled—principally wood, concrete and other types of masonry, and drywall. Rather than demolishing an entire building, all or part of a building can be deconstructed. In contrast to building demolition, deconstruction involves taking apart portions of buildings or removing their contents with the primary goal being reuse. This reduces more material production and their use, reducing the energy use in their extraction, production and transportation to the site. It also reduces destruction of natural habitats and land disturbances which encourages environmental pollution.

Minimal construction waste during installation reduces the need for landfill space and also provides cost savings. According to Jong-Jin (1998) Designing floor intervals to coincide with the standard lengths of lumber or steel framing members also reduces waste. Taking advantage of the standard sizes of building materials in the design phase reduces waste produced by trimming materials to fit, as well as the labor cost for installation.

***Local Materials.*** Using locally produced building materials shortens transport distances, thus reducing air pollution produced by vehicles. Often, local materials are better suited to climatic conditions, and these purchases support area economies. It is not always possible to use locally available materials, but if materials must be imported they should be used in a small volume.

## BENEFITS OF GREEN DESIGN AND GREEN BUILDINGS

Environmental benefit from greening buildings that can be fairly easily estimated is lower air pollutant and CO2 emissions. Emissions are reduced by decreasing energy use through energy-efficient design, use of renewable energy. Green building seek to rectify impacts of Energy Use in Buildings, Greenhouse Gas Emissions and Indoor Air Pollution, Building Water Use, Land Use and Consumption, Construction Materials and Construction, Operation, and Demolition Waste in conventional buildings. Reducing fuel and electricity consumption also lowers CO2 emissions, a greenhouse gas that is linked to climate change. Decreased use of natural gas should also reduce methane emissions to the atmosphere (methane is another greenhouse gas). The effects of the build-up of greenhouse gases in the atmosphere may include sea level rise, weather changes (e.g., increase in violent weather patterns), and impacts on agriculture. Although climate change is likely to occur gradually over a long time period, energy-efficiency measures implemented now will slow the pace of the greenhouse gas build-up and its potential effects. The major benefits include:

***Sustainable Siting.*** A sustainable site approach optimizes land use and development density using site compatibility, protection of wetlands and natural habitats, transportation access and proximity to amenities in the decision-making process. In combination, these factors can significantly reduce adverse development impacts and minimize the building's ecological footprint.

***Water Efficiency.*** Green building designs emphasize water efficiency through landscaping, and reduced wastewater generation. Some examples include integration of rainwater catchments, gray water recycling and waste water treatment systems. These sustainable technologies significantly decrease the fresh water demands on local aquifers and at the same time reduce generation of waste water compared to conventional building designs.

***Energy Efficiency.*** A major element of green building design is increased energy efficiency through incorporation of passive design, efficient lighting, renewable energy technologies, and improved thermal performance of building shells. The goal is optimization of energy performance and integration of renewable energy options.

***Building Materials.*** Green buildings use sustainable construction materials and resources with low environmental impact minimizing consumption and depletion of material resources. This includes reduction, re-use and recycling of construction materials and generated waste, rehabilitation of existing structures, and explicit attention to building durability, adaptability and disassembly. The result is a reduction of extraction, processing, transportation, and solid waste.

***Healthy Indoor Environmental Quality.*** Green buildings utilize efficient heating, ventilation and cooling systems (HVAC), low emissions paints and materials, and maximize use of natural day lighting to enhance indoor environmental quality. The result is improved health and comfort.

## BARRIERS TO GREEN BUILDINGS.

While the benefits of green buildings and sustainable development continue to show promising results, there are a number of barriers affecting realization of their full environmental, economic and social potential in Nigeria. Cost, industry awareness, and availability of green design technical capacities are the most significant operational barriers to green building design and construction. Knowledge and familiarity about green building practices has spread relatively slowly across the building industry.

## RECOMMENDATIONS

Having critically analysed the effects of environmental pollution, advantages of green buildings and the barriers to green buildings in Nigeria, the paper recommends the following, in the part of building professionals and government, to help facilitate awareness and construction of green buildings in Nigeria. Architects designing Buildings should design to eliminate waste by eliminate unnecessary finishes and make choices that use standard-sized or modular materials. Designers should consider product durability, availability, reusability, recyclability and biodegradability in the design process. Locally produced building materials are highly encouraged. Government should create more awareness on green buildings by making it a condition in all government projects. Government should also create an enabling ground for training and retraining of architects and other professional in the built industry to embrace green designs and green buildings. Government should also make it a condition for approval in the planning authorities all over Nigeria to help save our environment.

## CONCLUSION

The rate with wish we pollute our environment through our practices; if efforts are not quickly made we will suffer more severe weather as a result of our activities. Even as reducing environmental pollution as a result of construction will not correct the anomaly in our climate, it will go a long way in relieving our climate. If recommendations of this paper are followed, we achieve a better climatic condition.

## REFERENCES

Hansen, O. 2004. ''Green Buildings'' Small Business Innovation Research Special Solicitation Workshop, EPA

IEA. 2004. ''Oil Crises and Climate Challenges: 30 Years of Energy Use in IEA Countries.'' International Energy Agency, Paris, France.

International energy agency. 2008. ''Promoting Energy Efficiency Investments: Case Studies in the Residential Sector.'' International Energy Agency, Paris, France.

Iwuagwu, B.U & N.C. Eme-Anele. 2012. ''Earth Construction Technology and Design: A Positive Solution to Mass Housing In Africa''. International Journal of scientific innovations and sustainable development. Vol. 2, Number 2, 2012.

Iwuagwu, B.U & C.E. Azubuine. 2015. ''Global Warming Versus Green Architecture: African Experience''. Proceedings of International Conference on IT, Architecture and Mechanical Engineering (ICITAME'2015) May 22-23, 2015 Dubai (UAE).

Jong-Jin, K. (1998). ''Qualities, Use, and Examples of Sustainable Building Materials''. National Pollution Prevention Center for Higher Education, Ann Arbor, Michigan

Kats, G., L. Alevantis, A. Berman, E. Mills and J. Perlman. 2003. ''The Costs and Financial Benefits of Green Buildings''. A Report to California's Sustainable Building Task Force.

Kim, J. and B. Rigdon. 1998. ''Sustainable Architecture Module: Qualities, Use, and Examples of Sustainable Building Materials''. Sustainable Architecture Compendium, National Pollution Prevention Center, University of Michigan.

Odebiyi S O, S. Subramanian and A. K. Braimoh. 2010. ''Green Architecture: Merits for Africa: Nigerian Case Study''. Journal of Alternative Perspective s in the Social Sciences. Vol. 2 (2) 746 -767.

Samet J. M, A. Dearry, P. A. Eggleston. 2001. ''Urban air pollution and health inequities''. A workshop report. Environ Health Perspect 109 (3): 357–74

Sexton K, and J. L. Adgate. 2000. ''Looking at environmental justice from an environmental health perspective''. J Expos Anal Environ Epidemiol. 9: 3–8

# ENHANCING ALGAL BIOHYDROGEN PRODUCTION USING FLUE GAS DERIVED BICARBONATE AND NUTRIENT LIMITATION

***Sai Sameer Pusapaty*** and K. Sata Sathasivan (The University of Texas at Austin, TX, USA)

**ABSTRACT:** Biohydrogen generated from algal cultures can be an effective solution as cleaner alternative fuel to reducing greenhouse gas emissions. Most of the related research is focused on limited strains and little has been known so far about improving hydrogen production. Flue gas derived bicarbonate generated from the Carbon Mineralization technique can be instrumental in boosting algal growth. If it can also enhance the hydrogen production capabilities, the algal system integration can lead to sustainable biohydrogen production. In this study, the hydrogen production capabilities of 5 macro algae that were collected using standard field techniques and 6 microalgae strains obtained from the UTEX culture collection have been studied. The experiment was conducted under anaerobic conditions/dark incubation after inducing nutrient limitation through dilution. Algal growth, cell count, starch amount and hydrogen concentrations were measured before and after the dark incubation period of 7 days per batch. Results obtained from batches of algal strains grown without bicarbonate and grown with 0.1M and 0.2M flue gas derived bicarbonate over 14 days were compared. Preliminary results indicate that in 8 out of 11 algal strains bicarbonate resulted in an increased hydrogen production of up to approximately 406 ml of gas per liter of algae.

## INTRODUCTION

Extreme climate changes that are being witnessed today are the very proofs of the global warming. A major cause of the greenhouse effect is due to anthropological activities. The significant response to this complex problem is to reduce emissions. Carbon capture and storage (CCS) is a major strategy to reduce GHG emissions. Novel Carbon mineralization technology offers many advantages over the traditional CCS methods. Integrating it with algal system by using the flue gas derived bicarbonate solution as the algal feedstock can stimulate the algal growth and rapidly absorb $CO_2$, lower the overall cost and encourage widespread adaption.

Based on the existing literature, some algal strains have the ability to photosynthetically produce $H_2$ under anaerobic conditions. Hydrogen gas is seen as a future energy carrier by virtue of the fact that it is renewable, does not evolve the $CO_2$ in combustion, liberates large amounts of energy per unit weight in combustion and easily converted to electricity by fuel cells.

Algal Biohydrogen production from the integrated carbon mineralization will enable us to create a sustainable, carbon negative system. Current research is to study the effects of flue gas derived bicarbonate and nutrient depletion methods on the selected algal strains to increase the algal hydrogen production rates.

## MATERIALS AND METHODS

**Algal Strains and Cultivating Conditions.** 5 field macro algae samples and 6 micro algae strains are applied in this study. Field algae strains (TABLE 1) are collected from various water bodies around the town and the microalgae strains (TABLE 2) are obtained from the UTEX Culture Collection at the University of Texas. All the 11 strains of algae are maintained in the growth chamber with 12 hours of light and 12 hours dark cycles and kept under constant air bubbling at 20°c.

Experiments were conducted for the batches of 500 mL algal cultures (controls), 350mL algal cultures with 150mL of 0.1M flue gas derived sodium bicarbonate, 350mL algal cultures with 150mL of 0.2M flue gas derived sodium bicarbonate.

Algal cells were separated using filtering and added to DI water to induce nutrient depletion and were bubbled with Nitrogen to create anaerobic atmosphere and maintained in a dark chamber for 7 days to trigger Hydrogen production. Algal growth and starch quantification studies were conducted before and after dark incubation.

TABLE 1. List of Field Macro Algae strains collected for experimentation

| ID | DATE | LOCATION GPS COORDINATES | WATER TEMPERATURE $^{0}C$ | COLLECTION TIME | WATER PH |
|---|---|---|---|---|---|
| 1 | 8/20/15 | Latitude : 30.513036 I Longitude : -97.689046 | 26.1 | 10:18 AM | 7.71 |
| 4 | 8/20/15 | Latitude: 30.52023581558653 I Longitude: -97.73501336574554 | 26.8 | 11:14 AM | 7.31 |
| 5 | 8/20/15 | Latitude: 30.52023581558653 I Longitude: -97.73501336574554 | 26.8 | 11:30 AM | 7.32 |
| 6 | 8/20/15 | Latitude : 30.49648006306532 I Longitude : -97.76916861534119 | 27.1 | 12:20 PM | 7.71 |
| 7 | 8/20/15 | Latitude: 30.49812559017617 I Longitude: -97.76886820793152 | 27.2 | 12:45 PM | 7.12 |

TABLE 2. List of microalgae cultures obtained from UTEX Culture Collection of Algae

| ID | UTEX ID | ALGAE STRAIN |
|---|---|---|
| 1 | 2168 | *Chlorella vulgaris* |
| 2 | 1785 | *Chlorococcum rugosum* |
| 3 | 90 | *Chlamydomonas reinhardtii* |
| 4 | 2016 | *Scenedesmus obliquus* |
| 5 | 755 | *Porphyridium aerugineum* |
| 6 | LB 1355 | *Kirchneriella cornuta* |

**Bicarbonate for Photosynthesis.** Algae's efficiency at pulling inorganic carbon out of the environment is dependent on growth condition which implies the presence of an inducible carbon dioxide-concentrating mechanism in algal cells. NaHCO3 is a salt which consists of the ions Na+ and the bicarbonate anion, HCO3-. In aqueous solution, these ions are separated. Most algae can import both $CO_2$ and $HCO_3^-$ through the cell membrane. Once imported into the cell, $CO_2$ or $HCO_3^-$ accumulate mainly as $HCO_3^-$. In the carboxysome, where carbonic anhydrase (CA) exists, rapid equilibration between $CO_2$ and $HCO_3^-$ occurs. Corresponding with the accumulated $HCO_3^-$, this rapid equilibration leads to a high concentration of $CO_2$ around ribulose-1,5-bisphosphate carboxylase oxygenase (Rubisco) in carboxysomes. Carbonic anhydrase is an enzyme that interconverts carbon dioxide and hydrogencarbonate which supply Rubisco with carbon dioxide from the pool of $HCO_3^-$. In the presence of this high substrate concentration, Rubisco fixes $CO_2$ in the Calvin cycle, converting it into organic carbon (Chi, Zhanyou et al. 2011). The major antenna pigments in algae include chlorophylls, phycobiliproteins and carotenoids and the variation in the composition and relative abundance of these pigments give algae their distinctive color. Antenna complexes are proteins with many bound antenna pigments which are important in absorbing light energy. If the algal photosynthesis would increase more carbon dioxide would be removed from the environment.

**Growth quantification.** Algae cultures were maintained for 4 weeks in the growth chamber with 12 hour light and 12 hour dark cycles at 20°c. Separated test batches were moved to continuous light chamber and kept for 2 weeks prior to filtration and re-suspending into DI water and analysis via spectrophotometry, cultures were evenly mixed using sterilized stir bars. Spectrophotometry readings were taken at 680nm wavelength.

**Starch Quantification.** Starch within the algal cells is measured before and after dark incubation. To measure the starch, 30 mL of algae samples are collected and centrifuged at the highest setting (8850 Xg) for 15 min (Z, Xioa et al. 2006). Liquid is drained out and samples are stored in the freezer. Once frozen the samples are crushed with mortar and pestle to break down the cells. Using 80% ethanol the crushed cells are transferred to a tube and are boiled till the ethanol is gone removing any impurities. To 2.5mL of the algal extract, 2.5mL of amylase solution is added and kept in the incubator at 37 degrees Celsius for 30 min. 5 mL of 1M Acetic Acid is mixed and the solution is transferred to 250mL flask. 200mL of DI water is added to the flask and 5mL of iodine reagent (0.2% iodine and 2% potassium iodide) is added. Absorbance values are measured at 580nm using Spectrophotometer to estimate the starch present in the sample.

**Algal Hydrogen Production.** Algal samples are sealed in Erlenmeyer flasks using rubber corks with 1 hole and 5mm flexible airline tubing. Pinch clamps were used to seal the rubber tubes. Nitrogen is bubbled through each sample before sealing it tightly. Samples are placed in a dark incubator without any apertures for light to pass through to prevent photosynthesis. Anaerobic conditions and Sulphur deprivation eventually trigger hydrogenase and Hydrogen production (B, Tamburic et al. 2011).

**Hydrogen Quantification.** Algal samples keep using up any available oxygen in the dark anaerobic atmosphere using respiration and depending upon the algal strain at some point of time hydrogenase enzyme gets triggered and that leads to hydrogen production. Eventually algal strain adapts and repairs the PSII system to resume the photosynthesis halting the Hydrogen production. So the produced gas contains oxygen, nitrogen and possibly hydrogen and carbon dioxide in the vent of delays during sealing the rubber cork.About 3mL gas samples are collected using Gas syringe and analyzed using Shimadzu Gas Chromatograph. Percentage values of the gases are derived and approximate volumes are estimated.

To quantify the chromatogram peak areas as moles, a calibration was created. 5 ml of pure hydrogen gas was created. Pressure was also recorded using a pressure sensor. The sample was then injected into the GC and the area for the hydrogen peak was measured. Using the measured values, the volume of gas was calculated using Van der Waal Equation with the average attraction constant (a) and mole volume constant (b) were taken from literature..

$$\left(P + \frac{n^2 a}{V^2}\right)(V - nb) = nRT \tag{1}$$

## RESULTS AND DISCUSSION

**Algal Growth with Bicarbonate.** Analysis of algal chlorophyll (FIGURE 1) and starch growth (FIGURE 2) with bicarbonate using spectrophotometer absorbency values at 680nm and 580nm respectively revealed that the responses are species specific.

**Hydrogen Production.** Gas Chromatography analysis revealed the composition and the concentrations of different elements & compounds present in the injected 3mL gas samples. Hydrogen peaks in the chromatograms are easy early indicators. In figure 4, the increase in the number and length of the green bars (Batch 3 with 0.2M $NaHCO_3$) clearly indicate the positive response of algal strains with higher concentration of bicarbonate in producing biohydrogen. (0.2M) Bicarbonate triggered hydrogen production or an increase in the output in 8 out of 11 selected algal strains.

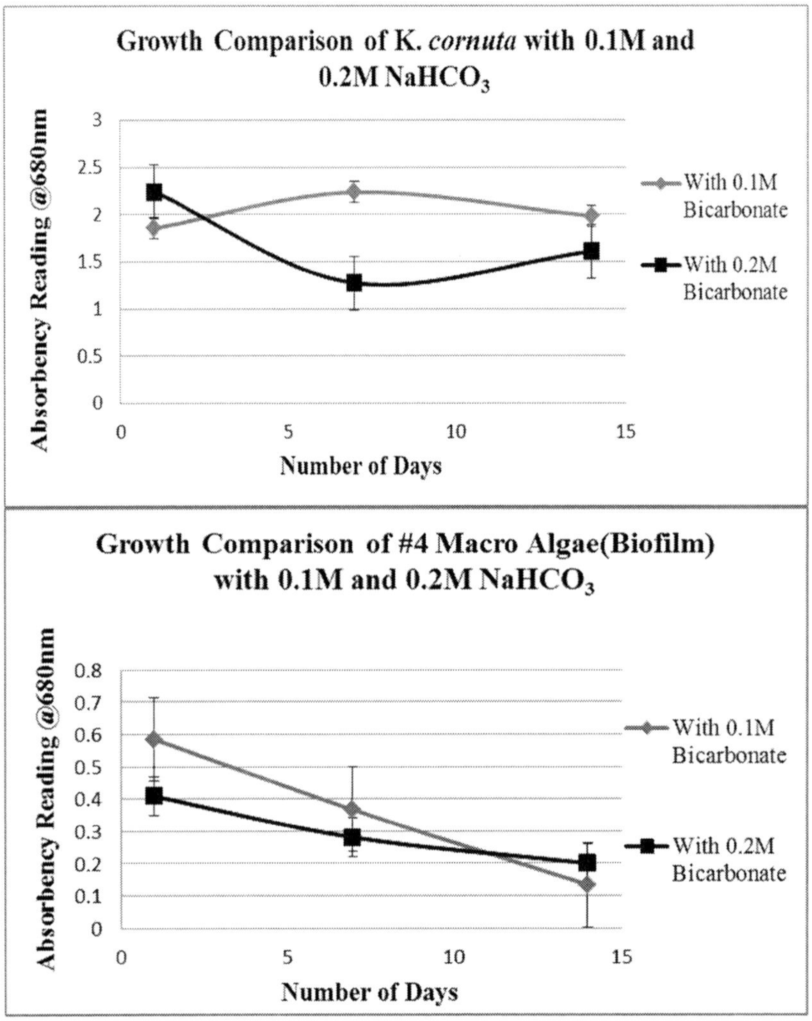


FIGURE 1. Absorbency values comparison with bicarbonate for *K.cornuta* and #4 (Biofilm)

*K cornuta, C vulgaris, C reinhardtii* and the macro strains #1, #4 (biofilm), #5 showed significant increase in hydrogen production compared to other strains. Increased hydrogen yield observed for #1, #4 (biofilm) & #7 with 0.1M bicarbonate compared to higher concentration indicates that macro algae may have a threshold to the bicarbonate tolerance. It is also interesting to see that #6 macro algae produced significant hydrogen without bicarbonate and there was no hydrogen yield when grown with bicarbonate.

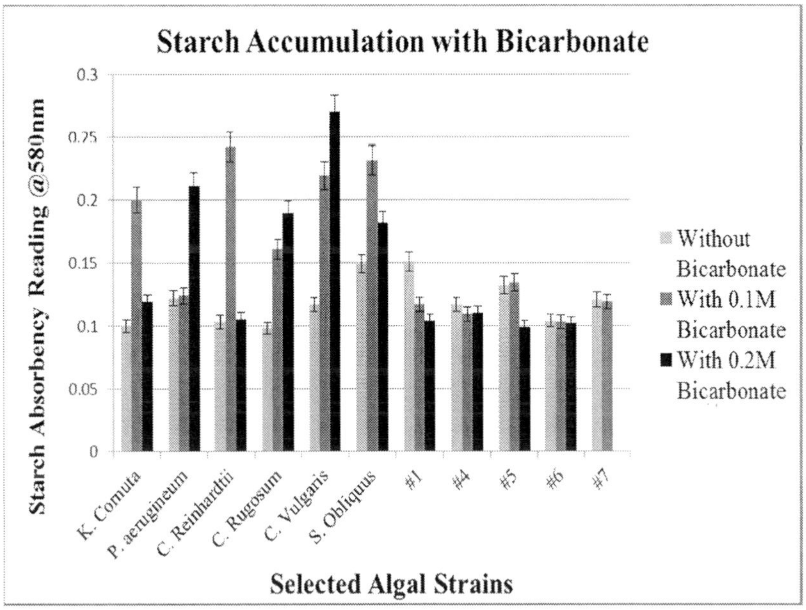


FIGURE 2. Comparison of starch absorbency values with Bicarbonate

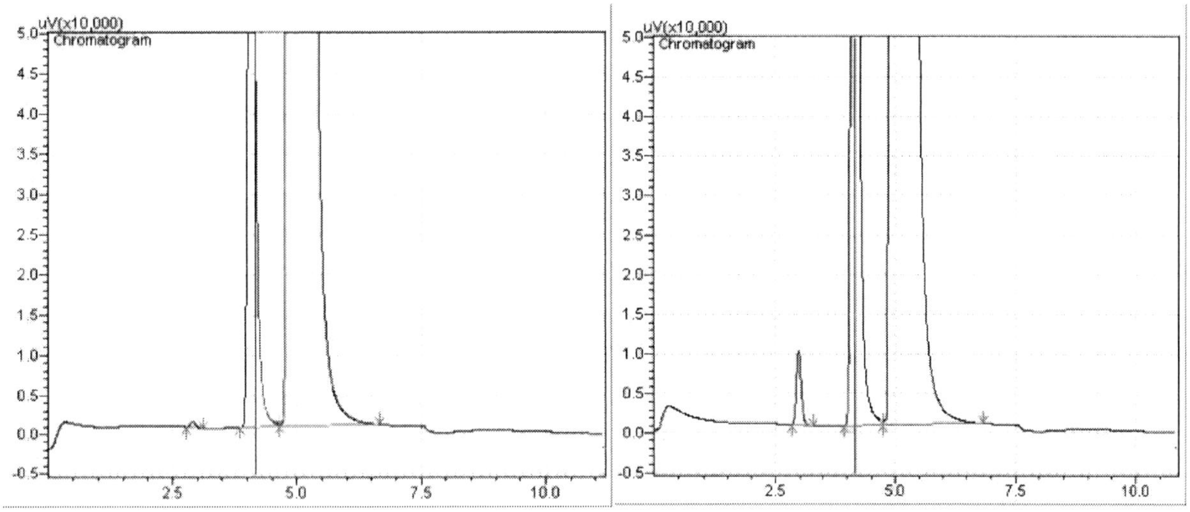


FIGURE 3. Chromatograms of *K. cornuta* before and after the dark incubation

**Algal Growth and Starch Reduction during Anaerobic Dark Incubation.** The following Scatterplots depict the correlation between the cell growth, starch reduction and Hydrogen production. Overall trend shows that an increased bicarbonate concentration led to an increased starch reduction and an increased hydrogen production.

So far most of the research was focused *on C.renhardtii*. This study indicates that other strains like *C. vulgaris, K. cornuta* and certain macro algae strains also have hydrogen production capabilities under dark anaerobic conditions and this is further enhanced by flue gas derived bicarbonate.

These promising results achieved at the expense of minimum energy and with the utilization of atmospheric $CO_2$ can lead to more fascinating findings and eventually leading to a successful carbon negative system.

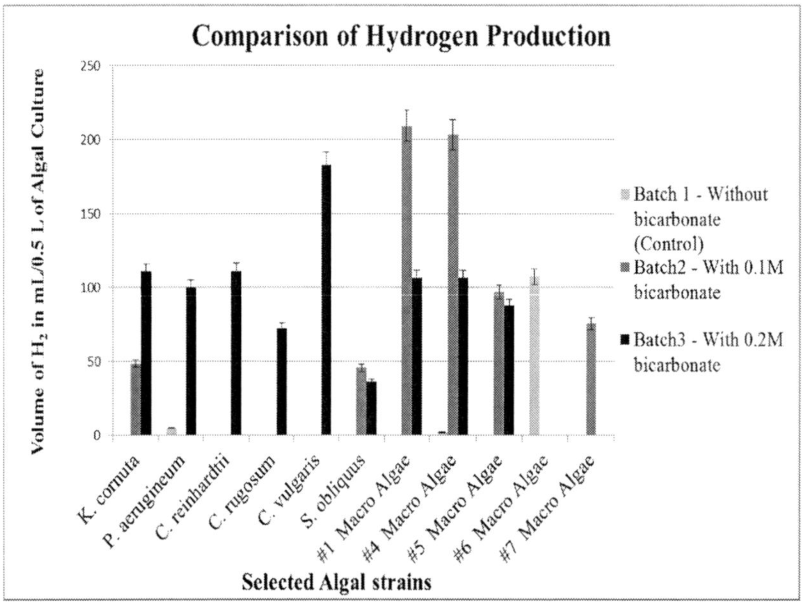


FIGURE 4. Comparison of Algal biohydrogen production in selected strains.

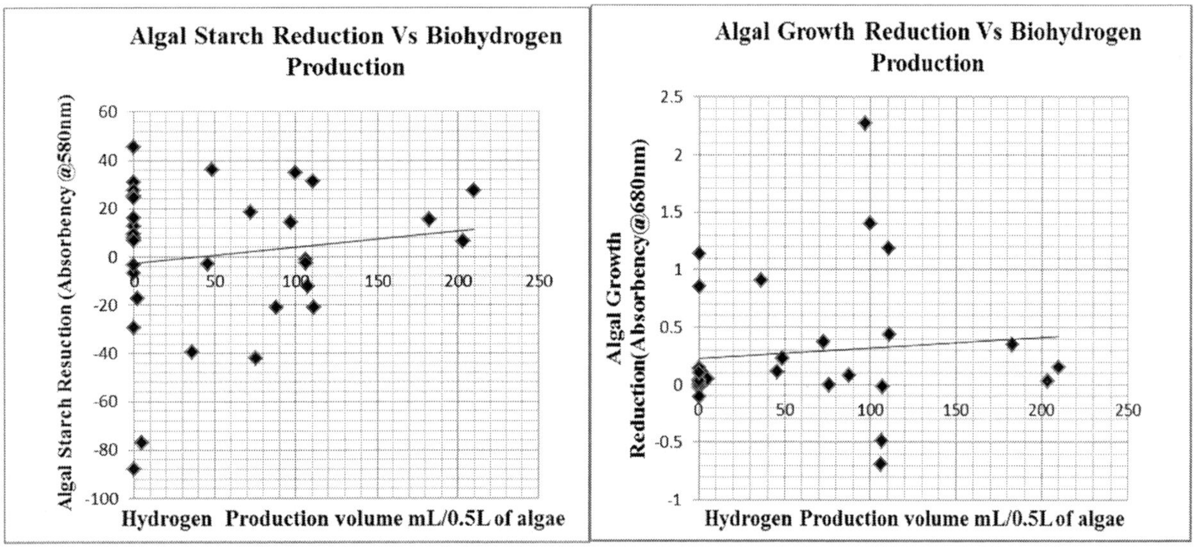


FIGURE 5. Effect of Starch & Growth Reduction on Hydrogen production

## CONCLUSIONS

This study demonstrates that bicarbonate addition can significantly affect the algal hydrogen production capability micro and macro algae, although the responses are species specific when compared under similar conditions.

Algal system integrated with Carbon Mineralization CCS Technology can utilize bicarbonate solution as algae feedstock, generate biohydrogen and other valuable byproducts from the harvested algal mass, and to recirculate the regenerated carbonate for carbon capture without spending energy. Cultivating strains of algae with higher bicarbonate absorption and growth rates have the potential to make the CCS process effective and more adaptable. Biohydrogen with 3 times the energy density as fossil fuels and water vapor as the only emission can aid in effectively remediating air pollution from flue gas emissions.

## REFERENCES


Carrieri, D., Wawrousek, K., Eckert, C., Yu, J., & Maness, P. C. 2011. "The role of the bidirectional hydrogenase in cyanobacteria." *Bioresource Technology*, *102*(18), 8368–8377.

Chi, Z., O'Fallon, J. V., & Chen, S. 2011. "Bicarbonate produced from carbon capture for algae culture." *Trends in Biotechnology*, *29*(11), 537–541.

Dragone, G., Fernandes, B. D., Abreu, A. P., Vicente, A. a., & Teixeira, J. a. 2011. "Nutrient limitation as a strategy for increasing starch accumulation in microalgae." *Applied Energy*, *88*(10), 3331–3335. http://doi.org/10.1016/j.apenergy.2011.03.012

Fernandes, B., Dragone, G., Abreu, A. P., Geada, P., Teixeira, J., & Vicente, A. 2012. "Starch determination in Chlorella vulgaris-a comparison between acid and enzymatic methods." *Journal of Applied Phycology*, *24*(5), 1203–1208. http://doi.org/10.1007/s10811-011-9761-5

GEA Mechanical Equipment. (n.d.). Separation technology for Algae Production, 8.

Hallenbeck, P. C., Abo-Hashesh, M., & Ghosh, D. 2012. "Strategies for improving biological hydrogen production." *Bioresource Technology*, *110*, 1–9. http://doi.org/10.1016/j.biortech.2012.01.103

Hallenbeck, P. C., & Benemann, J. R. 2002. "Biological hydrogen production; Fundamentals and limiting processes." *International Journal of Hydrogen Energy*, *27*(11-12), 1185–1193. http://doi.org/10.1016/S0360-3199(02)00131-3

Hemschemeier, A. C. 2005. "The anaerobic life of the photosynthetic alga *Chlamydomonas reinhardtii*," (August), 141 pp.

Hunt, P. P. 1960. "Gas Chromatography of Hydrogen-Deuterium Mixtures."

Kapdan, I. K., & Kargi, F. 2006. "Bio-hydrogen production from waste materials." *Enzyme and Microbial Technology*, *38*(5), 569–582. http://doi.org/10.1016/j.enzmictec.2005.09.015

Kim, D.-H., & Kim, M.-S. 2011. "Hydrogenases for biological hydrogen production." *Bioresource Technology*, *102*(18), 8423–8431. http://doi.org/10.1016/j.biortech.2011.02.113

Kim, S.-K., & Lee, C.-G. 2015. *Marine Bioenergy Trends and Development*. CRC Press.

Marx, F., Training, O., Training, P., Darin, C., Training, R. O., Kimberly, M., … Co-investigator, N. (2014). No Title No Title. *Igarss 2014*, (1), 1–5. http://doi.org/10.1007/s13398-014-0173-7.2

Melis, A., & Happe, T. 2001. "Hydrogen Production ." Green Algae as a Source of Energy. *Plant Physiology*, *127*, 740–748. http://doi.org/10.1104/pp.010498.740

Mertens, R., & Liese, A. 2004. "Biotechnological applications of hydrogenases." *Current Opinion in Biotechnology*, *15*(4), 343–348. http://doi.org/10.1016/j.copbio.2004.06.010

Rumpel, S., Siebel, J., Farès, C., Duan, J., Reijerse, E., Happe, T., … Winkler, M. 2014. "Enhancing Hydrogen Production of Microalgae by Redirecting Electrons from Photosystem I to Hydrogenase." *Energy & Environmental Science*, 1–10. http://doi.org/10.1039/C4EE01444H

Scoma, A., Giannelli, L., Faraloni, C., & Torzillo, G. 2012. "Outdoor H 2 production in a 50-L tubular photobioreactor by means of a sulfur-deprived culture of the microalga Chlamydomonas reinhardtii." *Journal of Biotechnology*, *157*(4), 620–627. http://doi.org/10.1016/j.jbiotec.2011.06.040

Skjånes, K., Rebours, C., & Lindblad, P. 2012. "Potential for green microalgae to produce hydrogen, pharmaceuticals and other high value products in a combined process." *Critical Reviews in Biotechnology*, *33*(January 2012), 1–44. http://doi.org/10.3109/07388551.2012.681625

Srirangan, K., Pyne, M. E., & Perry Chou, C. 2011. "Biochemical and genetic engineering strategies to enhance hydrogen production in photosynthetic algae and cyanobacteria." *Bioresource Technology*, *102*(18), 8589–8604. http://doi.org/10.1016/j.biortech.2011.03.087

Tamburic, B., Zemichael, F. W., Crudge, P., Maitland, G. C., & Hellgardt, K. 2011. "Design of a novel flat-plate photobioreactor system for green algal hydrogen production." *International Journal of Hydrogen Energy*, *36*(11), 6578–6591. http://doi.org/10.1016/j.ijhydene.2011.02.091

Tamburic, B., Zemichael, F. W., Maitland, G. C., & Hellgardt, K. 2011. "Parameters affecting the growth and hydrogen production of the green alga Chlamydomonas reinhardtii." *International Journal of Hydrogen Energy*, *36*(13), 7872–7876. http://doi.org/10.1016/j.ijhydene.2010.11.074

Tamburic, B., Zemichael, F. W., Maitland, G. C., & Hellgardt, K. 2012. "A novel nutrient control method to deprive green algae of sulphur and initiate spontaneous hydrogen production." *International Journal of Hydrogen Energy*, *37*(11), 8988–9001. http://doi.org/10.1016/j.ijhydene.2012.02.043

Vítová, M., Bišová, K., Kawano, S., & Zachleder, V. 2015. "Accumulation of energy reserves in algae: From cell cycles to biotechnological applications." *Biotechnology Advances*, *33*(6), 1204–1218. http://doi.org/10.1016/j.biotechadv.2015.04.012

Xiao, Z., Storms, R., & Tsang, A. 2006. "Corrigendum to A quantitative starch – iodine method for measuring alpha-amylase and glucoamylase activities." *Analytical Biochemistry*, *362*(MAY 2006), 146–148. http://doi.org/10.1016/j.ab.2006.01.036

# HOW DO PLANT LEAVES REMOVE DUST AND ITS ATTACHED METALS ON THE ROADSIDE OF BEIJING?

***Yang Zheng***, Liu Yanju
(Beijing Milu Ecological Research Center, Beijing, China)

Air pollution is a global urban environmental problem. As the capital of China, Beijing is keeping facing the disturbance of high concentration of air dust, thus is necessary to find improve method. Taking advantage of phytoremediation, an ecological and friendly way to improve air quality, this study aims to investigate how the urban green plants remove air dust and its attached metals by their leaves. In harvest season, leaves samples of 32 plant species were collected from heavy traffic roadsides, then their dust and dust attached metals were analysed to select those plant species which could well accumulate or remove dust and metals. The result shows that per square meter plant leaves could remove 0.52 to 23 g dust with the average of 7.50 g, follows the sequence from high to low of *Koelreuteria paniculata, Ulmus pumila, Syringa oblate, Malus micromalu, Weigela florida cv. Red Prince, Ailanthus altissima, Salix babylonica, Robinia pseudoacacia, Ligustrum × vicaryi, Euonymus japonicus* etc.. The comparison of Cd, Cr, Cu, Fe, Mn, Ni, Pb, Zn content in diswashed leaves among plant species indicates that different species could best accumulate the certain metals. *Chaenomeles speciosa* accumulated the highest Cd of 9.48 μg per gram leaf. The highest Cr value of 19.80 μg per gram leaf was observed in *Sorbaria kirilowii.* The highest Cu content of 34.12 and Fe of 867.96 μg per gram leaf both was in *Sophora japonica*. Highest Mn content of 168.77 μg per gram leaf was in *Rosa chinensis* , and highest Ni of 18.72 μg per gram leaf was in *Prunus cerasifera f. atropurpurea. Populus trichocarpa* accumulated the most Pb of 6.57 μg per gram leaf, and *Populus tomentos*a accumulated the most Zn of 141.95 μg per gram leaf. By comparing the metal contents in same species leaves diswashed and washed, some species was found to remove a certain metal more efficiently than the others. The difference between Cd in diswashed and washed leaves of *Jasminum nudiflorum* was the highest of 0.80 μg per gram leaf. Cr of 3.65 was in *Ulmus pumila*,Cu of 13.31 in *Sophora japonica*, Fe of 407.73 in *Ulmus pumila*, Mn of 20.97 in *Rosa chinensis*, Ni of 4.89 in *Prunus cerasifera* f. *atropurpurea*, Pb of 4.46 in *populus trichocarp*, and Zn of 80.23 in *Amygdalus triloba*.The difference between metals in diswashed and washed leaf was divided by washed leaf metal for the same species to get a relative metal removable ability, and the result shows that *Jasminum nudiflorum* could remove 150.5% of its total accumulated Cd, *Ulmus pumila* do 66.89% of Cr, *Sophora japonica* do 63.96% of Cu, *Weigela florida* cv.Red Prince do 220.7% of Fe, *Amygdalus triloba* do 81.8% of Mn, *Koelreuteria paniculata* do 90.5% of Ni, *Parthenocissus tricuspidata* do 2023.4% of Pb, and *Amygdalus triloba* do 245.10% of Zn.

# FUTURE PERSPECTIVE AND MITIGATION OPTIONS FOR ATMOSPHERIC MERCURY (Hg) EMISSIONS IN CHINA

***Qingru Wu*** and Shuxiao Wang
(Tsinghua University, Beijing, China)

We have presented atmospheric Hg emissions in China for the year 2020 and 2030 in this study. We develop two socioeconomic scenarios to project future energy consumption and industrial products production, including a business as usual scenario (BAU) and an alternative policy scenario (PC). For each scenario, we develop two end-of-pipe pollution control measures, including progressive (PGC) and stringent control case (SCC). We find that atmospheric Hg emissions are projected to be similar as the emissions in 2010 (approximately 539 t) under the BAU-PGC scenario. This indicates that the Hg removal by the progressive end-of-pipe pollution control measures will be offset by the Hg input from increased energy consumption and products production. In other scenarios, atmospheric Hg emissions are all less than the emissions in 2010. By using the stringent control technologies, atmospheric Hg emissions can be reduced by up to 76.9% in 2030. The share of elemental Hg will decline from 55% in 2010 to 39–41% in 2030 under PGC scenarios. However, it will still be the dominant species in 2030 under SCC scenarios.

Future emission scenarios imply different mitigation options for the convention-regulated sectors. Current legislations and progressive pollution control (eg., dust collection, desulfurization and denitration measures) will be the primary mitigation options for controlling emissions from coal-fired power plants, coal-fired industrial boilers, and nonferrous metal smelting, which will reduce 87%, 55% and 74% of current emissions (compare to 2010), respectively. Large-scale deployment of dedicated Hg removal technologies (eg., activated carbon injection and Boliden-Norzink technology) will aid to additional 2%, 35% and 15% of Hg reductions in these three sectors, respectively. For cement production, unless the stringent control measures (eg., dust shuttling) are fully applied, Hg emissions will increase by up to 22% in 2030 under the PC-PGC scenario. For municipal solid wastes incineration sectors, due to the sharp increase of incinerated solid wastes, Hg emissions will increase to 5.7–11.7 t in 2030, all larger than the emissions of 0.7 t in 2010. Thus, stringent control measures should be used in this sector. Moreover, although most of the primary emission sources in 2010 will still contribute to large amount of emissions in 2030, attention should also be paid to the emissions from non-convention-regulated sources, such as iron and steel production sector. Emission from this sector will head the emission inventory in the PC-SCC scenario in 2030.

# BIO-ASSESSMENT AND TOXICOLOGY

# RESPIRATORY SYSTEM EMERGENCY SERVICE VISITS IN ADANA, TURKEY (2007-2011)

Ertan Kara [1], Hasan Goksel Ozdilek [2], Emine Erman Kara [3]
[1] Cukurova University, Faculty of Hecicine, Dept. of Public Health, Adana, Turkey
[2] Canakkale Onsekiz Mart University, Faculty of Engineering, Environmental Engineering Dept., Terzioglu Campus, 17020 Canakkale, Turkey[139]
[3] Nigde University, Faculty of Engineering, Environmental Engineering Dept., 51245 Nigde, Turkey

**ABSTRACT:** The center of the province of Adana is located on a flat area, yet being a metropolitan area it has some ambient air quality issues. In this study, a comprehensive evaluation was completed on the effects on health, exemplified by the emergency respiratory system and asthma diagnoses, which are triggered with urban air pollution chiefly caused by motor vehicle traffic nowadays. The air pollution data that exist between 2007 and 2011 and the upper respiratory and asthma diagnosed emergency room admittances to Cukurova University Balcali Hospital were evaluated. For this purpose the official records collected for four consecutive years from Adana Cukurova University Balcali Hospital and ambient air quality monitoring stations were evaluated. The relationships between asthma and air quality parameters as well as between respiratory system diagnoses and air quality parameters were statistically evaluated using trend analysis. It is found that there is a significantly decreasing trend in the number of emergency visits due to asthma problems in Adana. However, an insignificantly changing trend in respiratory system diagnoses was determined.

***Key words:*** Air pollution, asthma, $PM_{10}$, $SO_2$, Adana-Turkey.

## INTRODUCTİON

Urbanization plays a key role in human health. Turkey has transferred itself into an urban society over the last 30 years, and it has taken big steps towards becoming one of the world's fastest growing economies. However, at the same time it has experienced air pollution related health problems. The air quality monitoring system has been automated since the late 1990s. A majority of the automated air quality monitoring stations report only particulate matter ($PM_{10}$) and sulphur dioxide ($SO_2$) in the country. Prior to this automation only some health institutions and concerned scientists collected data on air quality in Turkey. Although serious episodes of air pollution have been reported in such places as London in 1952 (Bell et al., 2004), Los Angeles in 1965 and 1984, and in New York City in 1953 (Nazaroff and Alvarez-Cohen 2001; Masters and Ela 2008), the link between air pollution and health has not been strongly established in Turkey, despite the fact that many urban centres experience air pollution related health problems (Özdilek, 2006). Özdilek (2006) compared urban centres that use natural gas and those that use mostly coal for heating purposes during winter months. He reported that pneumonia is strongly linked to urban air quality impairment (Özdilek 2006). Tomac et al. (2005) linked asthma and air quality in Zonguldak, the country's only hardcoal mining area, where the Catalagzi hard coal fired power plant has been into operation for a very long time. Berktas and Bircan (2003) examined atmospheric sulphur dioxide and particulate matter and their effects on asthma symptoms recorded at emergency rooms in Ankara, the capital of Turkey. Over the past few decades, heating systems in many Turkish cities have shifted from using conventional heating resources (primarily coal and wood) to imported natural gas. However, a dramatic increase in the number of motor vehicles, the addition of new fossil-fuel power plants, and changes in local climatic conditions (such as the urban heat island effect and the construction of new dam lakes) have had an effect. Recently, a ban on the use of sulphur containing diesel for diesel type motor vehicles and leaded gasoline for regular motor vehicles was implemented in Turkey in order to improve ambient air quality.

$PM_{10}$, defined as particulate matter with a mass median aerodynamic diameter not greater than 10 μm, is responsible for health effects diagnosed in the upper respiratory tract (Chereminisinoff, 2002) upper airways. As Nawrot et al. (2007) indicated fine particulate matter is responsible for deaths reported in summer than in winter. Sulphurous air pollution affects public health as a result of coal burning and the operation of diesel engines. Asthma and bronchitis are the main consequences of sulphur dioxide type air pollution (Stephens and Stair, 2007). These pollutants are common in the urban environment because of meteorological conditions, the intensity of traffic, fossil fuel use, and buildings that hinder air circulation. Carlslaw et al. (2007) modeled nitrogen oxide concentrations on a busy road in central London and found a decreasing nitrogen oxide concentration due to a decrease in overall traffic and meteorological circumstances not from nitrogen oxide emissions.

The metropolitan area of Adana has a population of 1,637,053 (based on 2010 data) and its annual population growth rate is 1.41% (Turkish Statistical Association, 2011). The city is approximately 215 km away from Turkey's largest coal-fired energy production plant (Afsin-Elbistan, in the province of Kahramanmaras) (Say, 2006). Adana has another fossil fuel power plant, the Sugozu coal-fired energy plant, which is 50 km away from the city of and is situated by the Mediterranean Sea. The Sugozu Power Plant utilizes imported low-level sulphur-type coal. Although Adana is located on a flat plain, the city's traffic load is thought to emit a significant amount of particulate matter and other traffic-related air pollutants into the atmosphere. The area of Adana city encompasses an east-west distance of 16 km and a north-south distance of 7 km. The urbanized area is nearly 88 $km^2$. The population density is 17,746 people per $km^2$ in the city. As of 2011, the metropolitan area has five counties: Seyhan (population 723,277), Karaisalı (population 22,981), Yüreğir (population 422,529), Sarıçam (population 120,325), and Cukurova (347,941) (TUIK, 2011).

This study examines ambient air pollution and respiratory system disease cases in Adana. We studied the temporal change in ambient air quality in Adana. The purpose of this study is to evaluate the relationship between air pollution and the number of asthmatic patients seen in the center of Adana between 2007 and 2011.

## MATERIALS AND METHODS

Meteorological data was taken from the Turkish Meteorological Institution. Air pollution data was collected from the Turkish Meteorological Institution's automated system. The data includes sulfur dioxide and particulate matter above 10 micrometer in diameter. In addition to ambient air quality, vehicular traffic-related carbon monoxide ($CO_2$), sulfur dioxide ($SO_2$), and nitrous oxide ($NO_x$) emissions between 2004 and 2010 (on an annual basis) were estimated using data about the number of motor vehicles and specifications within Adana. Both monthly and quarterly means were computed. A parametric statistical analysis was completed.

From traffic related air contaminants' emissions, the base concentrations of $SO_2$ and $NO_x$ were computed for dates when/during times when inversion occurred. Traffic-related emissions were computed from the counted number of motor vehicles and their types (using diesel or regular fuel) and their estimated years of production. While newer motor vehicles emit approximately 110 g $CO_2$ per km, older motor vehicles in Turkey emit more than 200 g $CO_2$ per km. Similarly, $NO_x$ emissions were computed. Stochiometric computations were done using such figures. The city centre was assumed to be elliptical on the ground and a dome on vertical axis. For the thermal inversion layer, the following formula was utilized:

$$\forall = \pi ab[\left(\frac{2c}{3}\right) - x + \left(\frac{x^3}{3c^2}\right)] \quad (1)$$

where,

∀: the volume of air above the city under conditions where atmospheric inversion exists ($m^3$)

a : half-length of the city from the city centre to the y axis (m)

b : half-width of the city from the city centre to the x axis (m)

c : length of the top point from the ground to the vertical (z) axis (assumed to be 100 m)

$x$ : length of the inversion layer from the ground to the vertical (z) axis (assumed to be 80 m)

The nonparametric correlation coefficient (Spearman) was computed based on seasonal median $PM_{10}$ and emergency visits diagnosed as respiratory system-related and seasonal $PM_{10}$ and emergency visits diagnosed as asthma. Finally, the Mann-Kendall trend test was applied for the same aforementioned parameters to determine whether $PM_{10}$, respiratory system diagnoses and recorded asthma cases were increasing or decreasing at the emergency room at Çukurova University Faculty of Medicine.

The variance of the Mann-Kendall trend test was computed by using the following formula:

$$\mathrm{Var}(S) = 1/18\,[(n(n-1)(2n+5))-(\sum t_p(t_p-1)(2t_p+5))] \quad (2)$$

where,

n: the number of data points (n=16 between Spring 2007 and Winter 2010)

$t_p$: the number of data points in the $p^{th}$ group.

Then Z was computed by using the following formulas:

$$Z = (S-1)/[(\mathrm{Var}(S))^{0.5}] \text{ if S is greater than 1} \quad (3)$$

$$\text{or } Z = (S+1)/[(\mathrm{Var}(S))^{0.5}] \text{ if S is a negative number} \quad (4)$$

## RESULTS AND DISCUSSION

According to the 2007-2011 data, examined on a daily basis, air pollution on more than half of the days each year exceeded the limit established for particulate matter ($PM_{10}$), which is set at 50 μg $m^{-3}$ by the European Union. For example, in 2009, 64 days exceeded the limit value for $PM_{10}$ at the Çatalan air quality monitoring station, which is approximately 20 km away from Adana city center, and 247 days exceeded the limit value for $PM_{10}$ at the Adana Governorate Building site that is situated in the city center of Adana. For 2007 a total of 161 days were above the 50 μg $m^{-3}$ limit value. For 2008, this total was 145 days. For 2009, a total of 172 days, for 2010, a total of 187 days, and for 2011, a total of 208 days were above the 50 μg $m^{-3}$ limit value.

Table 1. Respiratory system and asthma morbidities (2007-2011) in Adana metropolitan area

| Season | Respiratory system morbidity (emergency room admittance per million) | | Asthma morbidity (emergency room admittance per million) | |
|---|---|---|---|---|
| | Average | Standard deviation | Average | Standard deviation |
| Spring | 1449 | 520 | 71 | 21 |
| Summer | 852 | 300 | 45 | 10 |
| Autumn | 1402 | 529 | 62 | 17 |
| Winter | 1873 | 230 | 86 | 18 |

An aging population an increase in motor vehicles, urban sprawl, and traffic congestion are all important factors influencing health effects related to air quality. Depending on the size of the urban area, commuting distances tend to be lengthier in larger cities. In many metropolitan areas it might be hard to wean residents off their dependency on automobiles in the future as cities grow. This is expected to affect the seriousness of motor vehicle-related air pollution in such areas.

In Adana, the highest increase in patients suffering from upper respiratory system diseases occurred between September and October of 2007. Although particulate matter is the highest in August, our study shows that particulates showed no effect on the number of patients that suffered from respiratory system problems (Figure 1). Sulphur dioxide emissions at this time, however, were some of the highest for the days that data was available, and the monthly mean was 2.5 μg $m^{-3}$ in August 2007. For March, April, October, November and December, there were more than 300 patients diagnosed each month with upper respiratory system failure at Çukurova University Faculty of Medicine (Balcali Hospital). The median number of upper respiratory system failure patients treated at Çukurova University Faculty of Medicine between March and December 2007 on a seasonal basis was computed to be over 900 (Figure 1). The number of patients with upper respiratory system failure, on a monthly basis, in March and April, as well as from October to the end of the year, was above the median value. As related to the number of days on which 24-hour limit values for $PM_{10}$ were exceeded at the Adana Meteorological Station's air quality monitoring station (shown in Figure 2), it seems after September the number of patients increases remarkably, although the $PM_{10}$ monthly statistical mean is not correlated with the number of patients diagnosed with upper respiratory system failures in Adana in 2007. Not surprisingly, the highest total respiratory system emergency room visits occurred in the winter of 2011, when $SO_2$ concentration was 7 $\mu g\ m^{-3}$.

The decrease in the number of patients diagnosed with upper respiratory system-related diseases started in Mach and ended in August. In September, there is a remarkable increase in the number of patients diagnosed with upper respiratory system failures. The highest increase was recorded from September to October, and close to a 50% increase in the number of patients suffering from upper respiratory disease was recorded during that time period. The increase in patients diagnosed with upper respiratory diseases diagnosed ad Çukurova University Faculty of Medicine's Balcalı Hospital from October to November and from November to December is almost the same. This could be explained by the decreased strength of the immune system as the temperature changes in winter. Because local people are accustomed to hot weather and a relatively low level of air pollution, people tend not to become sick easily. However, sudden changes in temperature might weaken the immune system. It is found that there was no trend in the ambient seasonal $PM_{10}$ mean between spring 2007 and winter 2010.

According to the seasonal Mann-Kendall trend test for 19 consecutive seasons from spring 2007 (starting in March 2007) to winter 2011 (ending in November 2011), a significant decreasing trend was found for asthma in terms of emergency room admittances (S=-57, Var (S)=817, $Z=-1.959 < Z_{critical}= -1,645$). It was found that the level of $PM_{10}$ is closely linked to asthma (Boubel et al. 1994) despite the fact that emergency asthma cases and ambient $PM_{10}$ are not linked as being statistically significant in this study.

Table 2. Pearson correlation among variables examined in Adana between 2007 and 2011 (based on seasonal median values)

| | RSM | AM | $PM_{10}$ | $SO_2$ | WS | WD | AT | RH | AP |
|---|---|---|---|---|---|---|---|---|---|
| RSM | 1 | | | | | | | | |
| AM | **0.661** | 1 | | | | | | | |
| $PM_{10}$ | -0.170 | -0.052 | 1 | | | | | | |
| $SO_2$ | 0.192 | 0.147 | -0.053 | 1 | | | | | |
| WS | 0.058 | 0.184 | -0.057 | 0.073 | 1 | | | | |
| WD | 0.304 | 0.133 | 0.228 | 0.165 | 0.141 | 1 | | | |
| AT | -0.330 | **-0.577** | 0.188 | -0.324 | 0.038 | 0.176 | 1 | | |
| RH | -0.255 | **-0.445** | 0.056 | -0.022 | 0.345 | 0.197 | **0.447** | 1 | |
| AP | -0.172 | -0.214 | -0.175 | -0.172 | -0.240 | -0.132 | 0.153 | 0.310 | 1 |

RSM: respiratory system morbidity
AM: Asthma morbidity
$PM_{10}$: particulate matter ($\mu g\ m^{-3}$)
$SO_2$: sulphur dioxide ($\mu g\ m^{-3}$)
WS: wind speed ($m\ s^{-1}$)
WD: wind direction (degrees)
AT: air temperature (°C)
RH: relative humidity (percent)
AP: air pressure (mb)

Based on the Mann-Kendall trend test, respiratory system morbidities recorded as emergency cases experienced a statistically insignificant change between spring 2007 and autumn 2011. The test result was found as $Z=0.419 < Z_{critical}=1.645$ (S=13, Var (S)=817), which indicates a non-significant change in emergent respiratory system admittances to Çukurova University Faculty of Medicine Balcalı Hospital.

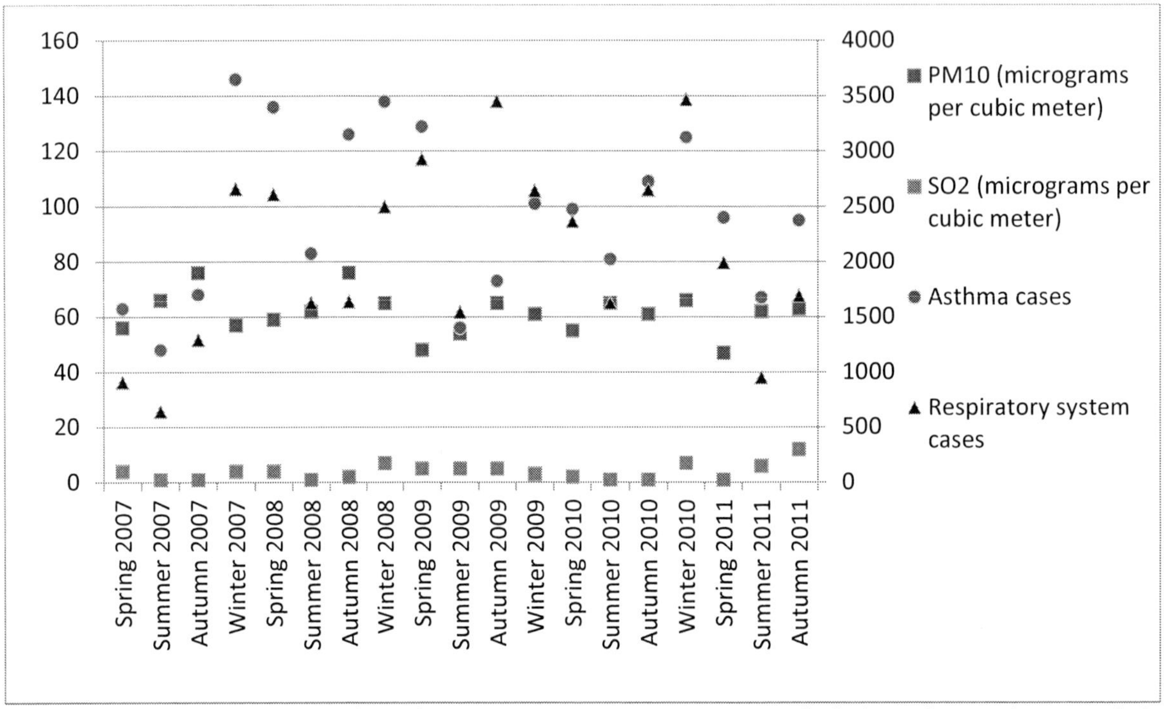


Figure 1. $PM_{10}$, $SO_2$, asthma and respiratory system emergency room admissions time series recorded at Adana Cukurova University Balcali Hospital (only respiratory system emergency cases are on secondary y axis)

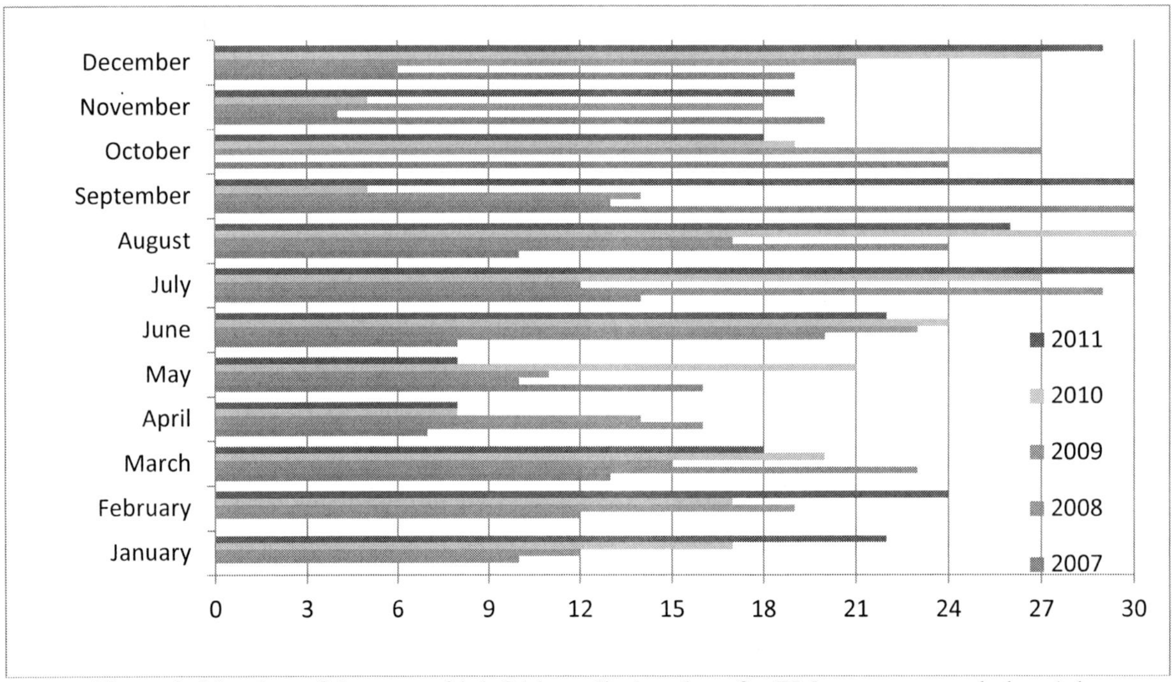


Figure 2. Number of days on which 24 hour limit values for $PM_{10}$ were exceeded at Adana Meteorological Station's air quality monitoring station. For 2007 total 161 days, for 2008 total 145 days, for 2009 total 172 days, for 2010 total 187 days, and for 2011 total 208 days are above 50 $\mu g/m^3$ limit value.

In our study it was found that ambient air quality has not improved over the period of the analysis. Ambient air quality improvement might be seen some time in the future, despite the fact that, in the future, traffic related emissions in both $CO_2$ and NOx is expected to be higher mainly due to a potential increase in private cars and traffic congestion in Adana. Adana in the 1990s underwent a transformation from an industrial to a post-industrial local economy, although in fact the air quality in Adana is somewhat less susceptible to industrial emissions than in other cities, due to the local environmental conditions. Agardy, for example, underlines that stochiometrically computed air emissions for the mass load discharge from industrial stacks are mostly lower than the actual measured emissions (Agardy, 2005).

It is known that particles in ambient air can be problematic for the human respiratory system. The European Union recommends that the number of days on which one-day limit value for $PM_{10}$ (50 $\mu g\ m^{-3}$) is met should not exceed 35 days each year (SYKE, 2008). However, problems associated with the effect of ambient air quality on respiratory health arise due to factors such as the ever-increasing traffic, meteorological variables (such as the transport of desert sand from northern Africa), construction activities, and other factors affecting air quality.

## REFERENCES


Agardy, F. J. (2005). Environmental Solutions. Agardy, F. J. and Nemerow, N.L. (editors). Elsevier Academic Press: Burlington, MA.

Bell, M.L., Davis, D.L., Fletcher, T. (2004). "A Retrospective Assessment of Mortality from the London Smog Episode of 1952: The Role of Influenza and Pollution". *Environ Health Perspect* 112 (1): 6–8.

Berktas, M. and Bircan, A. (2003). Effects of Atmospheric Sulphur dioxide and particulate matter concentrations on emergency room admissions due to asthma in Ankara. *Tüberküloz and Toraks Dergisi*. 51: 231-238.

Boubel, R. W., Fox, D. R., Turner, B., Stern, A. C. (1994). Fundamentals of Air Pollution (3rd ed.). Academic Press: San Diego, CA.

Carslaw, D. C., Beevers, S. D., Tate, J. E. (2007). Modelling and Assessing Trends in Traffic-Related Emissions Using a Generalised Additive Modelling Approach. *Atmospheric Environment*, 41: 5289-5299.

Cheremisinoff, N.P. (2002). Handbook of Air Pollution Prevention and Control. Butterworth-Heinemann, Amsterdam, the Netherlands.

Masters, G.M. and Ela, W.P. (2008) Introduction to Environmental Engineering and Science (3rd Edition) Pearson International Edition , Prentice Hall: Upper Saddle River, NJ.

Nawrot, T.S., Torfs, R., Fierens, F., De Henauw, S., Hoet, P.H., Van Kersschaever, G., De Backer, G. Nemery, B. (2007). Stronger associations between daily mortality and fine particulate air pollution in summer than in winter: evidence from a heavily polluted region in western Europe. *J. Epidemiol Community Health*. 61 (2): 166-149.

Nazaroff, W.W. and Alvarez-Cohen, L. (2001). Environmental Engineering Science. John Wiley and Sons, Inc: New York, NY.

Özdilek, H.G., (2006). "An Analogy on Assessment of Urban Air Pollution in Turkey over the turn of the new Millennium (1992-2001)" *Environmental Monitoring and Assessment* **123/1-3:** 95-112.

Say, N.P. (2006). "Lignite fired thermal power plants and $SO_2$ pollution in Turkey" *Energy Policy* **34**: 2690-2701.

Stephens, C. and Stair, P. (2007). Charting a New Course for Urban Public Health. In: State of the World: Our Urban Future. Starke L. (editor). WW Norton and Company: New York, NY.

SYKE (Finnish Environment Institute SYKE) (2008). Finland – State of the Environment. Putkuri, E., Lindholm, M. And Lyytimäki, J. (editors) 20 pp. Helsinki, Finland.

Tomac N, Demirel F, Acun C, Ayoglu F. (2005). Prevalence and risk factors for childhood asthma in Zonguldak, Turkey. *Allergy Asthma Proc*. 26: 392-402.

TUIK (2011). Address based population data (in Turkish). Accessed from the internet on January 25, 2012 from http://rapor.tuik.gov.tr.

# CARCINOGENICITY AND MUTAGENICITY ASSESSMENTS OF DIETARY EXPOSURE TO PAHs IN IMPORTED FISH PRODUCTS IN NIGERIA

***Nsikak U. Benson***, Akan B. Williams, Winifred U. Anake, Kelechi P. Eke, Adebusayo E. Adedapo (Covenant University, Ogun State, Nigeria)
Abaas A. Olajire (Ladoke Akintola University of Technology, Ogbomoso, Nigeria)

Polycyclic aromatic hydrocarbons (PAHs) occurrence and assessment of dietary exposure from imported canned sardines (*Sardinops sagax*) commercially marketed in local stores and supermarkets in Nigeria were evaluated for the first time. PAHs determinations were performed using high performance liquid chromatography (HPLC) (Agilent 1290 model) equipped with UV-VIS diodes array detector (DAD) at $\lambda =$ 210 nm and 214 nm. The percentages recoveries were higher than 96%. The degree of contamination expressed as total concentrations of PAH (TPAHs) congeners ranged between 2.53 and 35.55 $\mu g\ kg^{-1}$ dry weight (d.w.) at $\lambda = 210$ nm, and 1.30 and 27.93 $\mu g\ kg^{-1}$ (d.w.) at $\lambda = 214$ nm. The carcinogenic (BaP-TEQ) and mutagenic toxicities (BaP-MEQ) of eight priority PAHs were evaluated. Benzo(a)pyrene, benzo(b)fluoranthene, benzo(k)fluoranthene and indeno(1,2,3-d) pyrene contributed significantly to the total carcinogenic equivalents of PAHs. The mutagenic equivalents were largely influenced by equivalence factors of benzo(a)anthracene, and benzo(a)pyrene and indeno(1,2,3-d) pyrene. The estimated daily intake (EDI) and excess cancer rate (ECR) were evaluated for adults and children exposure related risks.

# DOSE RESPONSE CURVES DERIVED FROM CLINICAL OZONE EXPOSURES CAN INFORM PUBLIC POLICY

**Sabine S. Lange*** and Michael E. Honeycutt (Texas Commission on Environmental Quality, TX, USA)
Ge Tao, Lorenz R. Rhomberg, and Julie E. Goodman (Gradient, MA, USA)
Michael L. Dourson (University of Cincinnati, OH, USA)

Ozone is one of six criteria air pollutants for which regulations are set by the US Environmental Protection Agency using the National Ambient Air Quality Standards (NAAQS). The ozone NAAQS level was recently changed to 70 parts per billion (ppb) from 75 ppb, but debate is ongoing about the public health protection afforded by this change. Our study used data from human clinical studies to inform policy decisions about a protective ozone level. We plotted mean forced expiratory volume ($FEV_1$) response versus total inhaled ozone dose (calculated from ozone concentration, duration of exposure and ventilation rate) from clinical studies of 1-8 hour durations. There were clear thresholds of effect, which were consistent with the known ozone mode of action. The initial plot used data from healthy young adults; additional analyses incorporated data from children and asthmatics, but results did not differ. We estimated expected ozone doses that would be experienced by the population based on known exercise ventilation rates and durations, at ambient ozone concentrations of 75, 70 or 65 ppb (8-hour maximum average). Doses were similar at these three concentrations, and most doses were below those associated with an adverse $FEV_1$ decrement, even when different exposure times and ventilation rates were assessed. Ozone doses were estimated in Texas in 2013 at monitors that would have attained a 75 ppb or 70 ppb standard, and again found that the doses were very similar. This type of analysis can determine thresholds of ozone toxicity and realistic exposure doses, which can be crucial for choosing an ambient ozone concentration that is protective of human health.

# MITOCHONDRIAL INJURY-REGULATED JOINT HEPATOTOXICITY INDUCED BY COMBINED EXPOSURE OF PCB77 AND Cd

***CUI Jiansheng GAO Yu WANG Lixin***

(Institute of Environmental Science and Engineering, Hebei University of Science and Technology; Hebei Key Laboratory of Molecular Chemistry for Drug, Shijiazhuang, 050018, China)

Extensive research indicated that both Polychlorinated biphenyls (PCBs) and heavy metal Cadmium (Cd) could induce multiple toxicities including hepatotoxicity. However, the joint toxic effects and mechanisms caused by the combined exposure of PCBs and Cd have been rarely investigated. We here addressed their separate and joint hepatotoxicity with assessments *in vitro* in a representative human hepatic cell line SMMC-7721. Obvious joint toxicity of these two kinds of pollutants was observed through cell viability assay after 48 h combined exposure with 25 $\mu mol \cdot L^{-1}$ PCB77 and 20 $\mu mol \cdot L^{-1}$ $Cd^{2+}$. Without impact on the integrity of cell membrane, combined exposure significantly decreased mitochondrial transmembrane potential and increased mitochondrial membrane permeability followed by induced cell apoptosis. Our results also showed the increased cellular ROS level upon combined exposure. In short, we demonstrated that the combined exposure of PCB77 and Cd can incur joint hepatotoxicity, and the toxicity mechanism is that the change of the mitochondrial transmembrane potential induces cell apoptosis.

---

# PROTECTIVE ROLE OF CICHORIUM INTYBUS AGAINST MANGANESE TOXICITY IN LIVER AND KIDNEY

***Ram Prakash***
(Department of Zoology, Dr. A.G.D Bendale Mahila Mahavidyalaya, Jalgaon-425001, India)

Manganese is an essential trace element and involves widely in biological reactions. It activates numerous enzymes such asKinases,esterases, mutases, peptidases, dehydrases ect. However higher intake of manganese was recorded hazardous to cell/tissues. It was recorded in our laboratory that manganese cause hepatic cell necrosis, fibro-proliferation, bi-nucleated cells , pycnosis of nuclei , glomerulo-nephritis, glomerular damage and disturbances in plasma membrane of PCT and DCT. Manganese also decreased activity of some key enzymes such as alkaline and acid phosphatase, cholinesterase and glucose-6-phosphatase. Recently we designed experiments on rats as first were exposed with manganous chloride for thirty days and than post exposed with the extract of *Cichorium intybus* for fifteen more days . Observations were made on liver and kidney for histological and histochemical parameters. It was recorded that most lesions were reversed towards their normal conditions. Therefore it was concluded that *Cichorium intybus* has therapeutic values against manganese toxicity in liver and kidney . The observations on other parameters also support present data.

# URINARY METABOLOMICS REVEALED ARSENIC EXPOSURE-RELATED METABOLIC ALTERATION: A PROOF-OF-CONCEPT STUDY IN A CHINESE MALE COHORT

***Jie Zhang***, Weipan Xu, Heqing Shen
(Institute of Urban Environment, CAS, Xiamen, Fujian, China)

The general population may be exposed to arsenic through various sources. Although targeted urinary biomonitoring of arsenic species provides the most accurate assessment of arsenic exposure, the arsenic-related metabolic perturbations are required to be evaluated. The objective of this proof-of-concept study is to investigate arsenic-induced phenotypic metabolome changes. Urinary arsenic species such as inorganic arsenic, methylarsonic acid, dimethylarsinic acid and arsenobetaine were quantified in a Chinese adult male cohort using high performance liquid chromatography (HPLC)-inductively coupled plasma-mass spectrometry. Urinary metabolomes were analyzed using HPLC-quadrupole time of flight mass spectrometry (qTOF-MS) and the arsenic-related metabolic changes were investigated by using a partial least squares discriminant analysis model. After adjustment for age, BMI, smoking and drinking, five potential biomarkers of testosterone, hippurate, acetyl-N-formyl-5-methoxykynurenamine (AFMK), serine and guanine were identified from 61 candidate metabolites, which indicated the exposure to arsenic induced endocrine disruption and oxidative stress in humans. Some dose-dependent trends have been also observed. The combined pattern of testosterone, hippurate, serine and guanine gave an AUC of 0.91 with sensitivity and specificity = 88% and 80%, respectively. We have demonstrated that metabolomics can be used to identify arsenic-related changes at the biomolecular level. The nature of the arsenic-related metabolic changes and trends may be used to further refine population based risk analysis of arsenic exposure.

# ASSESSING OCCUPATIONAL EXPOSURE IN BLACK HAIR SALONS

***Sarah Lemelman*, *Raymond Trott,*** Anna Abrams, Allison Marill, Alexis J. Cooper, Ramya Ramakrishna, Jay Feinstein, Teleah Slater, Annie Fortnow, Cassidy Tatun, Omkar Kulkarni, Sandra Watemberg (Student Group, Environmental Studies Program, Brandeis University, Waltham, MA, United States); Faulty and instructors: Laura J. Goldin, Theodore A. Myatt, Joseph G. Allen, James H. Stewart, Matthew A. Fragala (Faulty and instructor Group, Environmental Studies Program, Brandeis University Waltham, MA, United States)

This purpose of this study was to investigate air quality in Black hair salons. Black hair salons are generally small, enclosed spaces that use different products marketed for particular types of hair styling. These products may not receive same amount of research or environmental review that other more widely marketed products would encounter and may result in exposures of consumers and salon workers to potentially harmful chemicals. We collected short-term air samples of VOCs, aldehydes, and ketones in ten salons in the greater Boston area, along with fine particulate matter ($PM_{2.5}$) and carbon dioxide. Seven out of the ten salons had fine particulate matter levels well above the National Ambient Air Quality Standards outdoor limit. A follow-up time-series investigation into sources of $PM_{2.5}$ in salons identified $PM_{2.5}$ spikes during the use of irons and hair dryers. Regarding chemical exposures, five salons had chloroform concentrations above EPA's screening level carcinogenic target risk value, and all ten salons had benzene concentrations above EPA's carcinogenic target risk level. Elevated levels of carbon dioxide in all salons also suggested inadequate ventilation. This study reveals higher than expected levels of $PM_{2.5}$ inside salons, and two carcinogens/chemicals above EPA screening levels. Further evaluation is recommended to: determine if these findings are consistent in a wider range of the salons, assess the potential health implications of these exposures, and identify which products and services are contributing to the exposures and identify controls to reduce exposures.

# DIOXIN-LIKE RATHER THAN NON DIOXIN-LIKE PCBs PROMOTE THE DEVELOPMENT OF ENDOMETRIOSIS THROUGH THE STIMULATION OF IMMUNE-ENDOCRINE INTERACTION

***Qiansheng Huang***, Yajie Chen, Sijun Dong
(Key Lab of Urban Environment and Health, Institute of Urban Environment, Chinese Academy of Sciences, Xiamen, China)

Endometriosis is an estrogen-dependent and inflammation-dependent disease with high prevalence and severe health outcome. Epidemiological studies showed the association between polychlorinated biphenyls (PCBs) and endometriosis. In contrast, little evidence is available in terms of molecular toxicology. In this study, Dioxin-like CB126 and non-dioxin-like CB153 within human exposure regimen were used to treat primary cultured eutopic endometrial stromal cells (ESCs) and EM mice model. CB126 rather than CB153 significantly enhanced the biosynthesis of 17β-estradiol (E2) secretion from ESCs in a dose-dependent way. 17β-hydroxysteroid dehydrogenase 7 (HSD17B7) was the responding gene to PCBs exposure among genes related to estrogen metabolism. CB126 decreased the methylation of HSD17B7 promoter. The elevated levels of HSD17B7 expression was also observed in the eutopic endometrium of EM patients. CB126 rather than CB153 triggered the inflammatory response through directly stimulating the secretion of inflammatory factors and indirectly inhibiting the level of lipoxin A4. Furthermore, the increasing level of inflammatory factors positively enhanced the expression of HSD17B7. Antagonism of aryl hydrocarbon receptor (AhR) diminished the endocrine and inflammatory effects by CB126. *In vivo* EM mice model confirmed the promotion of CB126 on the endometriotic lesion. The peritoneal fluid levels of E2 and inflammatory factors both increased in the EM mice treated with CB126. Taken together, we demonstrated that

CB126 instead of CB153 triggered the EM development through stimulating the estrogen biosynthesis, inflammation and their positively interaction by AhR receptor.

# BACTERICIDAL AND ANTIOXIDANT PROPERTIES OF ESSENTIAL OILS OF THE FRUITS *Dennettia tripetala* G. BAKER

**Sunday O. Okoh,** B.C. Iweriegbor, O.O. Okoh and U.U. Nwodo and A. I. Okoh
(University of Fort Hare, Alice, South Africa)

Bactericidal and antioxidant properties of unripe and ripe fruits of *Dennettia tripetala* and their potentials for the management of infectious and oxidative-stress diseases were investigated *in-vitro* in the research. Essential oil obtained from the fruit using Clevenger modified apparatus, was characterized by high resolution GC-MS, while antioxidant and bactericidal properties were investigated by spectrophotometric and agar diffusion methods respectively. The essential oils demonstrated strong antibacterial property when subjected to *Enterococcus faecium*, *Escherichia coli*, and *Staphylococcus aureus*. Unripe fruit oil (UFO) was more active than the ripe fruit oil (RFO) against test bacteria with minimum inhibition concentration (MIC) ranging between 0.05-0.15 mg/mL and that of ripe fruit oil (RFO) ranging between 0.10–0.20 mg/mL. The $IC_{50}$ for RFO (0.42 mg/mL) showed that it has higher antioxidant strength than UFO and vitamin C (0.74 and 3.38 mg/mL) and comparable to β-carotene (0.37 mg/ mL) in inhibiting 2, $2^1$-dipphenyl-1-picrylhydrazyl radicals ($DPPH^•$). Antiradical property of both oils against $DPPH^•$, 2, $2^1$-azinobis-(3-ethylbenzothiazolin - 6-sulfonic acid) diammonium salt ($ABTS^{•+}$), lipid peroxide ($LP^•$) and nitric oxide ($NO^•$) radicals were in concentration-dependant manner. Findings from this research indicates that in addition to the local uses of the UFO and RFO, the essential oil has strong bioactive compounds and have good antibacterial and antioxidant properties and may be alternative to synthetic antibiotics on future investigations.

# AFLATOXIN CONTAMINATION OF SOME EDIBLE GRAINS FROM LAGOS AND OTA MARKETS, NIGERIA

Mary Oloyede, Akan Williams and Nsikak Benson
(Department of Chemistry, Covenant University, Ota, Ogun State, Nigeria)

**ABSTRACT:** Levels of aflatoxin in maize, sorghum, millet, wheat and rice obtained from Lagos and Ota markets in Nigeria are reported. Aflatoxin determination was carried out using high performance liquid chromatograph coupled with diode array detector (HPLC-DAD). The concentrations of aflatoxins in the grains obtained from Mile 12 market, Lagos ranged between 34.3 µg/kg and 300.1 µg/kg while the concentrations in the grains obtained from Dada market, Ota ranged between 42.3 µg/kg and 1245 µg/kg. The levels of total aflatoxins were generally higher in the grains sampled from Lagos market. These values exceed the maximum limits of 10 µg/kg set by regulatory bodies in Nigeria, thus posing a great health risk for consumers. Recovery studies indicated that the method was efficient as recovery of the residue ranged between 82.2% and 96.2%. There is need for the creation of awareness on aflatoxin in order to sensitize people on health hazards associated with its contamination. In addition, there is need to promote proper practices of grain production and storage to prevent its contamination and reduce exposure.

***Keywords:*** Aflatoxin, Grains, Contamination, Nigeria.

## INTRODUCTION

Aflatoxins are by-products of fungal metabolism mainly synthesized by fungus of the genus *Aspergillus*. In Nigeria, the risk of chronic exposure to aflatoxins from susceptible foods, especially maize, a staple food, is of great concern owing to the health hazards associated with it. Dietary exposure to low doses of aflatoxins is a known risk cause of liver cancer and may also affect protein metabolism and immunity thereby leading to low immunity and malnutrition (Williams *et al.*, 2004). Consumption of highly contaminated foods results in severe, acute hepatitis also known as aflatoxicosis. Symptoms of this disease include vomiting, jaundice, abdominal pain, liver failure and sometimes death. Four different aflatoxins, $B_1$, $B_2$, $G_1$ and $G_2$ have been identified with $B_1$ being the most toxic. Human exposure to aflatoxins can result directly from ingestion of contaminated foods or indirectly from consumption of foods from animals previously exposed to aflatoxin in feeds (Odoemelam and Osu, 2009).

Grains are small, hard, dry seeds with or without attached hulls or fruit layers harvested for human or animal consumption. After being harvested, dry grains are more durable than other staple foods and this durability has made grains well suited to industrial agriculture, since they can be mechanically harvested, transported by rail or ship, stored for long periods in silos, and milled for flour or pressed for oil. Sorghum, millet, maize, wheat and rice are the most important cereals in Nigeria (Omotosho and Muhammadu-Lawal, 2008; Obatolu, 2002). Millet is the cereal of the dryer regions with little rainfall. Sorghum crops are the cereals of the savannah and stretch into the sub-humid environments. Maize and rice are crops of the humid tropic- areas with high rainfall, swamps and areas with or without water for irrigation. Rice can grow in upland areas, such as those suitable for maize. Other types of rice grow in swamps, requiring lots of water. While maize is found across Nigeria's upland and irrigated lands, rice also grows well wherever there is adequate water/irrigation (Ekwuruke, 2005).

In Nigeria, the use of cereals range from preparation of "Akamu" and "Agidi" to preparation of "Nrioka", "Tuwoshinkafa" and "Tuwomasara". Roasted corn, boiled corn or popcorn is also consumed as snacks. Cereals are also used in preparation of animal feed and fodder. They are used in the production of alcoholic beverages and fermentation products (Ekwuruke, 2005).

Storage is an important activity, which enhances marketing efficiency by providing utility. Storage is particularly important in agriculture because agricultural production is seasonal while the demands for agricultural commodities are evenly spread throughout the year. There is need to meet average demand of agricultural products by storing excess supply during the harvesting season for gradual release to the market during off-season periods thereby stabilizing food prices. Postharvest facilities or appropriate storage technology has been the major problem of Nigerian agriculture for a long time (Adejumo *et al.*, 2007). Aflatoxins are found in many countries of the world, especially in tropical and subtropical region where the warm and humid weather provides optimal conditions for the growth of aflatoxigenic molds. The optimum temperature for the growth of the moulds is 24 – 35°C, thus crops grown in tropical and subtropical conditions are more prone to aflatoxin contamination than those in temperate regions. Being a tropical country, Nigeria's conditions of temperature and humidity favour pre-harvest and post-harvest contamination of stored grains by the fungi (Odoemelam and Osu, 2009).

## EXPERIMENTAL

**Sample Collection.** This study examined two popular grocery markets namely Ota market and Mile 12 market in Lagos. Both markets are located in South West, Nigeria. Stratified random sampling was carried out to obtain 5 kg from the bulk of stored grains (rice, wheat, sorghum, millet and corn) in stores and open markets. The samples obtained were put in paper bags and refrigerated in the laboratory at $4^0$C. A laboratory size of 500g was obtained from the grounded sample and analysed.

**Apparatus and Reagents.** All glasswares used were thoroughly washed with distilled water, rinsed with acetone and dried in an oven prior to use. The apparatus used included High Performance Liquid Chromatography (HPLC) from Agilent technologies, top loading balance, orbital shaker, hot plate, hot oven, Erlenmeyer flasks, filter paper, pipette, test tubes, funnel, beaker, micro syringe, measuring cylinder and vials with TFE caps. Thestandards were obtained from Sigma –Aldrich. All reagents used were of HPLC grade. Acetonitrile was obtained from Lichrosolv, Germany and methanol was obtained from Tedia Company, USA.

**Materials and Methods.** Determination of Aflatoxins in the grains was carried out in accordance with the standard methods outlined in AOAC Official Method 994.08 with some modifications. It involved solid phase extraction with HPLC analysis.25 g of each ground sample was weighed into a 250 mL Erlenmeyer flask and 100 mL of a mixture of acetonitrile and water (84:16 v/v) was added. The mixture was shaken for one hour on an orbital shaker and then filtered through a fast qualitative filter paper.4 mL of the filtrate was passed through a SupelcoAflazea SPE cartridge. The toxin was eluted with 2 mL of HPLC grade methanol. The eluate was evaporated to dryness and reconstituted with deionized water. The mixture was constituted using a vortex mixer and 20 µL was injected into the HPLC. The HPLC was set for analysis of total aflatoxins at ambient temperature, with a flow rate of 0.8 mL/min, injection volume of 20 µL and run time of 13 minutes. The mobile phase used was an isocratic mixture of water: methanol: acetonitrile (50:40:10); analytical column: ZORBARX Eclipse XDB C18 Column, 4.6 mm × 250 mm. The diode array detector was set at a wavelength of 365 nm.

## RESULTS AND DISCUSSION

The mean levels of aflatoxins in grains from Ota and Lagos markets are shown in Tables 1 and 2 while recovery studies of some selected samples are shown in Table 3.Total aflatoxins in grains from Ota market was highest in maize with a concentration of 300.1 µg/kg while the level in sorghum from Mile 12 market was the highest. Even though $AFB_1$ was not detected in millet and sorghum samples from Ota market and in maize, sorghum and rice from Mile 12 market Lagos, it is worrisome that the level of $AFB_1$ was beyond the permissive limit of 5 ppb in all the other grains sampled (Commission Regulation, 2006).

This is because $AFB_1$ is carcinogenic and the most potent of the aflatoxins. The levels of total aflatoxins were generally higher in the grains sampled from Lagos market. This might be due to poor storage conditions and a higher level of exposure to contaminants. From the study, it is observed that sorghum has high concentrations in both markets (220.5 μg/kg and 1245.1 μg/kg). These values indicate that sorghum is highly susceptible to aflatoxin infestation. Also, it is observed that the concentration of aflatoxin in millet obtained from Ota is relatively low with a mean concentration of 34.3 μg/kg compared with 120.3 μg/kg obtained from Lagos. Generally, the absence of $AFB_1$ and $AFB_2$ in samples of sorghum obtained from both markets is remarkable. Recovery Studies of some selected samples ranged from 82.2% to 96.2% indicating that the method used was satisfactory.

**Table 1. Mean Levels (mg/kg) of Aflatoxins in Grains from Ota, Ogun State**

| | $AFB_1$ | $AFB_2$ | $AFG_1$ | $AFG_2$ | Total AFs |
|---|---|---|---|---|---|
| Maize | 10.4 | 143.9 | ND | 145.7 | 300.1 |
| Millet | ND | 11.8 | ND | 22.5 | 34.3 |
| Sorghum | ND | ND | 130.3 | 90.1 | 220.5 |
| Wheat | 13.0 | 23.5 | ND | 86.5 | 123.1 |
| Rice | 11.5 | 18.6 | 8.2 | 47.2 | 85.7 |

**Table 2. Mean Levels (mg/kg) of Aflatoxins in Grains from Mile 12 Market, Lagos State.**

| | $AFB_1$ | $AFB_2$ | $AFG_1$ | $AFG_2$ | Total AFs |
|---|---|---|---|---|---|
| Maize | ND | 567.6 | 13.6 | 21.3 | 602.6 |
| Millet | 20.3 | 12.9 | ND | 87.2 | 120.5 |
| Sorghum | ND | ND | 1203.8 | 41.3 | 1245.1 |
| Wheat | 9.1 | 5.9 | ND | 27.2 | 42.3 |
| Rice | ND | ND | 60.2 | 13.4 | 73.6 |

ND- Not Detected

**Table 3. Recovery Studies of some samples**

| Sample | Recovery (%) |
|---|---|
| Corn | 82.2 |
| Rice | 96.2 |

This study reports levels of total aflatoxins that are not within the acceptable limits of 10 ppb (EMAN, 2006). The levels of aflatoxins obtained in this study was in consonance with earlier reports and studies conducted (Bankole and Adebanjo, 2003; Gwary*et al.*, 2012; Aroworа*et al.*, 2012). The aflatoxin levels in grains obtained from open markets are, therefore, unsafe for human consumption and could pose a health risk to consumers.

## CONCLUSION

Results of this study showed that there is presence of high levels of the toxin in grains obtained from the selected markets, which exceeds the permissible limits set by regulatory bodies for consumers. The toxin was detected in all the samples with maize having the highest concentrations.Therefore, there is need to prevent aflatoxin contamination in the grains by adoption of moisture-control measures that promote improved storageand educating farmers on the importance of Good Agricultural Practices during planting, harvesting and postharvest handling of crops.

## REFERENCES

1. Adejumo, T.O.; Hettwer, U. and Karlovsky, P. (2007). Survey of maize from south-western Nigeria for zearalenone, α- and β-zearalenols, fumonisin B1 and enniatins produced by Fusarium species. *Food Additives* 24: 993-1000.
2. Arowora, K.A.; Abiodun, A.A.; Adetunji, C.O.; Sanu, F.T.; Afolayan, S.S, and Ogundele, B.A. (2012). Levels of Aflatoxins in Some Agricultural Commodities sold at Baboko Market in Ilorin, Nigeria.*Global Journal of Science Frontier Research Agriculture and Veterinary Sciences* 12.
3. Bankole, S.A. and Adebanjo, A. (2003). Mycotoxins in food in West Africa: current situation and possibilities of controlling it. *African Journal of Biotechnology* 2: 254-263.
4. Ekwuruke, H. (2011). Cereals: Grains that feed the World. *Panorama*
5. EphremGuch (2015). Implication of Aflatoxin Contamination in Agricultural Products. *American Journal of Food and Nutrition* 3(1), 12-20.
6. European Mycotoxin Awareness Network (2006). Annual Report retrieved from www.mycotoxin.org
7. Ezekiel, C.N. and Sombie, J.I. (2014) Survey of aflatoxins and fungi in some commercial breakfast cereals and pastas retailed in Ogun State, Nigeria. *Nature and Science,* 12(6); 27-32.
8. Fapohunda, S.O.; Soler C.; Ezekiel, C.N.; Manes, J.; Kayode F. and Rubert J. (2013). A survey of mycotoxins in random street-vended snacks from Lagos, Nigeria, using QuEChERS-HPLC-MS/MS. *European Mycotoxins Awareness Network* 32: 673-677.
9. Gwary, O.M.; Hati, S.S.; Dimari, G.A. and Ameh, J. A. (2012).Assessment of Mycotoxins (Total Aflatoxins and Ochratoxin-A) Contamination of Staple Cereals. *International Journal of Chemical and Biochemical Sciences* 2: 1-6.
10. IARC (2002). International Agency for Research on Cancer Monographs on the Evaluation of carcinogenic risks to humans, 80
11. Obatolu, V.A. (2002). Nutrient and sensory qualities of extruded malted or unmalted millet/soybean mixture. *Food Chem*76: 129-133.
12. Odoemelam, S.A. and Osu, C.I (2009). Aflatoxin $B_1$ Contamination of Some Edible Grains Marketed in Nigeria. *E- Journal of Chemistry* 6(3): 308-314.
13. Omotosho, O. A. and Muhammad-Lawal, A. (2010).Optimal food plan for rural households' food security in Kwara State, *Nigeria: The goal programming approach.*
14. RASFF (2011). Retrieved from http://ec.europa.eu/food/safety/rasff/index_en.html
15. Whitlow, L. W. and Hagler, W. M. (2008). Mould and Mycotoxin Issues in Dairy Cattle: Effects, Prevention and Treatment. *WCDS Advances in Dairy Technology.* 20: 195-209.
16. Williams, J. H.; Phillips, T. D.; Jolly, P. E.; Stiles, J. K.; Jolly, C. M. and Aggarwal, D. (2004). Human aflatoxicosis in developing countries: a review of toxicology, exposure, potential health consequences, and interventions. *The American journal of clinical nutrition* 80: 1106–1122.

# GENOTOXIC EFFECTS OF COAL FLY ASH IN *TRIGONELLA FOENUM GRAECUM (L.)*

***Rucha U. Raval,*** Kailash P. Patel and Mariya J. Jairajpuri
**(Department of Biosciences, Veer Narmad South Gujarat University, Surat, Gujarat, India)**
**E-mail: kailash.patel@yahoo.com**

**ABSTRACT :** The aim of this research is to study the effects of different concentration (20%, 50%, 80%, and 100%) of Fly Ash with Vermicompost and Pressmud on the root meristems of *Trigonella foenum graecum*. 10 seeds were grown in pot at 20%, 50%, 80% and 100% concentration of FlyAsh. It was determined that FlyAsh increased the rate of the mitotic division at different concentration with the increase in mitotic anomalies. Very rare anomalies were encountered in the control group and at the concentration of 20% but at concentration of 50%, 80% and 100% of FlyAsh chromosomal aberration such as sticky metaphase, laggard chromosome, bridge chromosome, vagrant chromosome etc. were observed. Mitotic index increases from control to lower concentration of Fly Ash, while at higher concentration mitotic index gets lower and chromosomal aberrations get higher. At higher concentration frequency of chromosomal bridge were very high while at lower concentration sticky metaphase normally appeared.

***Keyword:*** FlyAsh, Vermicompost, Abnormalities, Mitotic index, Chromosomal aberration.

## INTRODUCTION

Energy requirements for the developing countries are met from coal-based thermal power plants. The disposal of the increasing amounts of solid waste from coal-fired thermal power plants is becoming a serious concern. Coal ash is mostly made up of fine particles that is known as FlyAsh. In India, nearly 90 Mt of FlyAsh is generated per annum and is largely responsible for environmental pollution.

Soil application of fly ash waste has both beneficial and adverse effect on crop yield. The adverse effect is common at high rates of fly ash due to increased salinity and accumulation of toxic levels of elements. To overcome these adverse effects of fly ash, it can be supplemented by Vermicompost and additional source of organic matter like Pressmud(PM). Pressmud is a by product of sugar industries, that generates intense heat, foul odour and takes long time for decomposition. Vermicomposting is a biological technique of composting wide ranges of organic wastes with the help of the gut micro organisms of surface living earthworms. The final product of Vermicompost is a very good organic fertilizer, it reduces levels of contaminations and tends to hold more nutrients over a longer period, without causing any adverse effect on environment. India is the largest producer of fenugreek in the world(Zohary, D. and M. Hopf, 2000).

The aim of this study was to investigate the chromosomal abnormalities induced by fly ash on the root tips of *Trigonella foenum graecum* L. and also to determine the relation between mitotic index and chromosome abnormalities.

*Trigonella foenum-graecum* L (fenugreek) is cultivated as annual, winter legumes in India. The species can be grown easily in low-input, minimal environment and are good sources of calories, seed proteins, B-vitamins and minerals.

Present study was particularly carried out to evaluate whether fly ash, along with Vermicompost and pressmud can be effective for betterment of the plants.

## MATERIALS AND METHODS

10 seeds of *Trigonella foenum-graecum* were germinated in pots. 50 ml of distilled water was supplied every alternate day. The total mixture of Fly Ash, Vermicompost and Pressmud was 50gm and from that different concentrations were made as Table 1. Photomicrographs were taken on Axioscope.

**Table 1.** Parameters in the Experiments

| % Concentration | Fly Ash(gm) | Vermicompost (gm) | Pressmud (gm) |
|---|---|---|---|
| Control | 0 | 25 | 25 |
| 20% | 10 | 20 | 20 |
| 50% | 25 | 12.5 | 12.5 |
| 80% | 40 | 5 | 5 |
| 100% | 50 | 0 | 0 |

## RESULTS

The staining procedure followed was that of (Fiskesjo G 1988). Mitotic index and percentage of aberrant cells were calculated using following formula.

$$\text{Mitotic Index (\%)} = \frac{\text{No. of divided cells}}{\text{Total no. of cells studied}} \times 100$$

$$\text{Aberrant cells (\%)} = \frac{\text{No. of aberrant cells}}{\text{No. of dividing cells}} \times 100$$

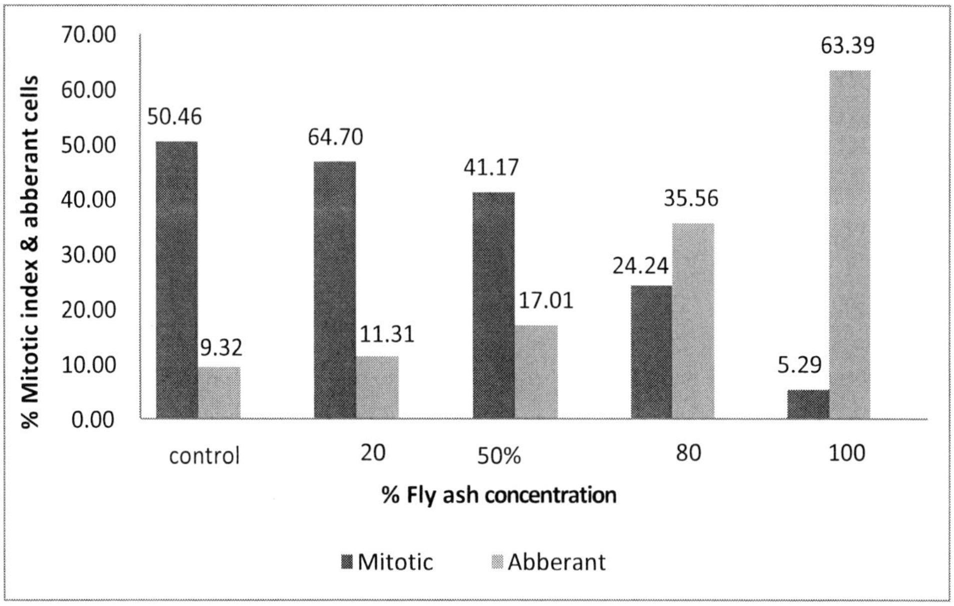


**FIGURE 1**. Mitotic Index

**Table 2.** Effect of Thermal Effluent on Root Tip Cells of *Trigonella foenum graecum.*

| Concentration (%) | Mitotic Index (%) | Aberrant cells (%) |
|---|---|---|
| Control | 50.46 | 4.34 |
| 20 | 64.70 | 2.31 |
| 50 | 41.17 | 17.01 |
| 80 | 24.24 | 35.56 |
| 100 | 5.29 | 63.39 |

The Mitotic Index is summarized in Figure 1 and Table 2. Normal chromosomes and Chromosomal abnormality caused by Fly ash are shown in Figure 2.

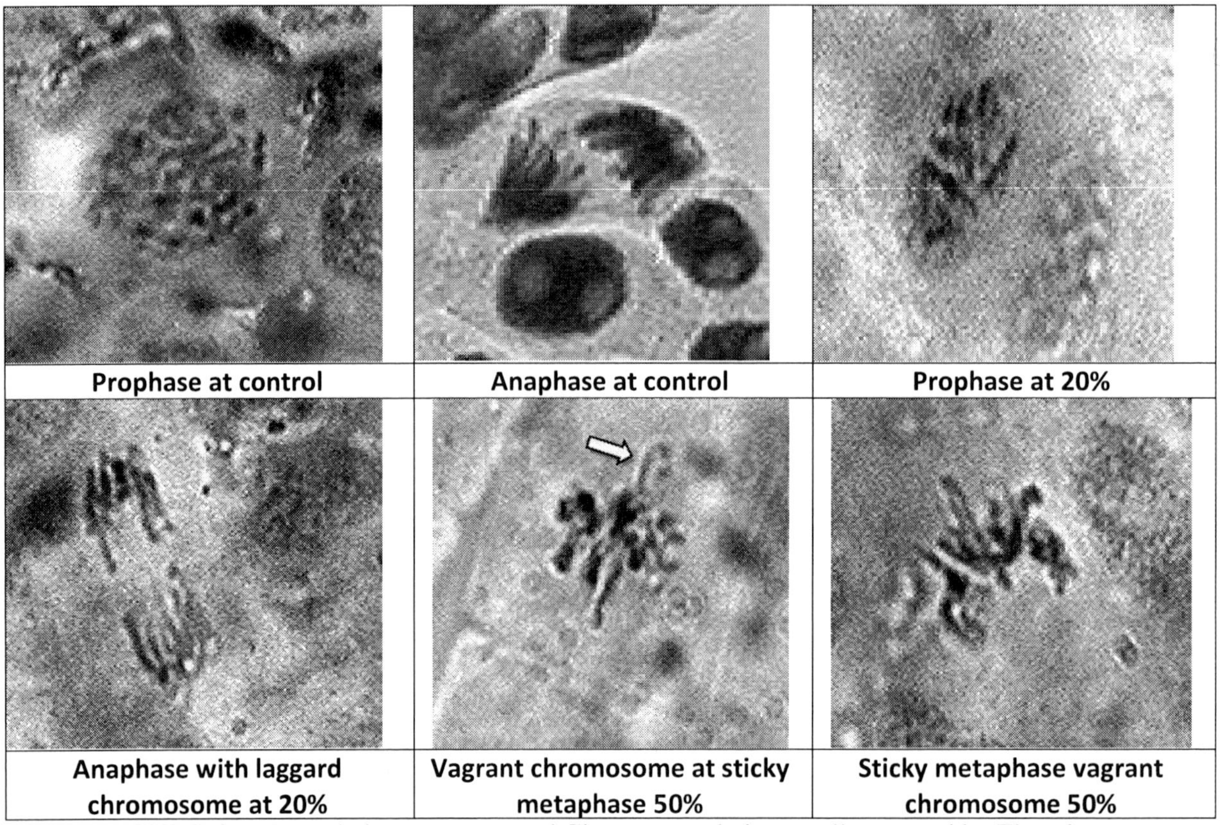


**FIGURE 2a.** Normal chromosomes and Chromosomal abnormality caused by Fly ash.

## DISCUSSION

The research revealed that fly ash increased the mitotic division creating very high chromosomal abnormalities in the root meristem of *Trigonella foenum graecum* at higher concentration. This suggests the presence of certain cytotoxic/genotoxic substances in the fly ash. At lower concentration no chromosomal abnormality is observed but at higher concentration chromosome bridges were frequently observed. Other aberration included vagrant chromosome, laggard chromosome, disturbed anaphase, sticky metaphase, micronuclei and nuclear degeneration.

Mitotic bridge at higher concentration is commonly observed abnormality, most probably formed by the breakage and fusion of chromosomes (Egypt J Bot, 1990 and Gomurgen AN, 2000). Presence of chromosome bridges may be due to stickiness or formation of dicentric chromosomes caused by breakage and reunion (Bempong AM, 1973).

Laggard chromosome resulted due to failure of the chromosomes to get attached to the spindle fibre and to move to either of the two poles (Namita Khanna*and Sonia Sharma).

Disturbed metaphase and anaphase are another most frequently observed abnormality observed in treated cells. It may be caused by the loss of activity of microtubules in spindle fibres leading to complete inhibition of spindle formation (El-Khodary S, Habib A, Haliem A, 1990).

Sticky chromosomes indicate a highly toxic, irreversible effect probably leading to cell death (Fiskesjo G 1985 and 1988). There is a hypothesis that stickiness of chromosomes may cause incomplete separation of daughter chromosomes as a result of cross-linkage chromo proteins (Kong MS, Ma TH 1999).

Stickiness in chromosomes is a result of partial dissociation and altered pattern of organization of nucleoprotein (Evans HJ, 1962). The stickiness of chromosomes makes their separation and free movement incomplete and thus they remain connected by bridges (Kabarity A, El-Bayourni AA & Habib AA 1974). This may also be due to the defective formation of the spindle apparatus (Badr A 1986 and Abraham S & Rajalakshmy BN 1989).

Also nuclear degeneration was observed with "cavity like" space, which is a starting point of apoptosis resulting in cell death due to the accumulation of heavy metals in the plants. Chromatid breaks, chromosome breaks, asymmetrical exchanges and incomplete symmetrical exchanges would give rise to acentric fragments at mitosis and at the later stage these fragments appear as micronuclei.( Kailash P. Patel *et al,* March 2013)

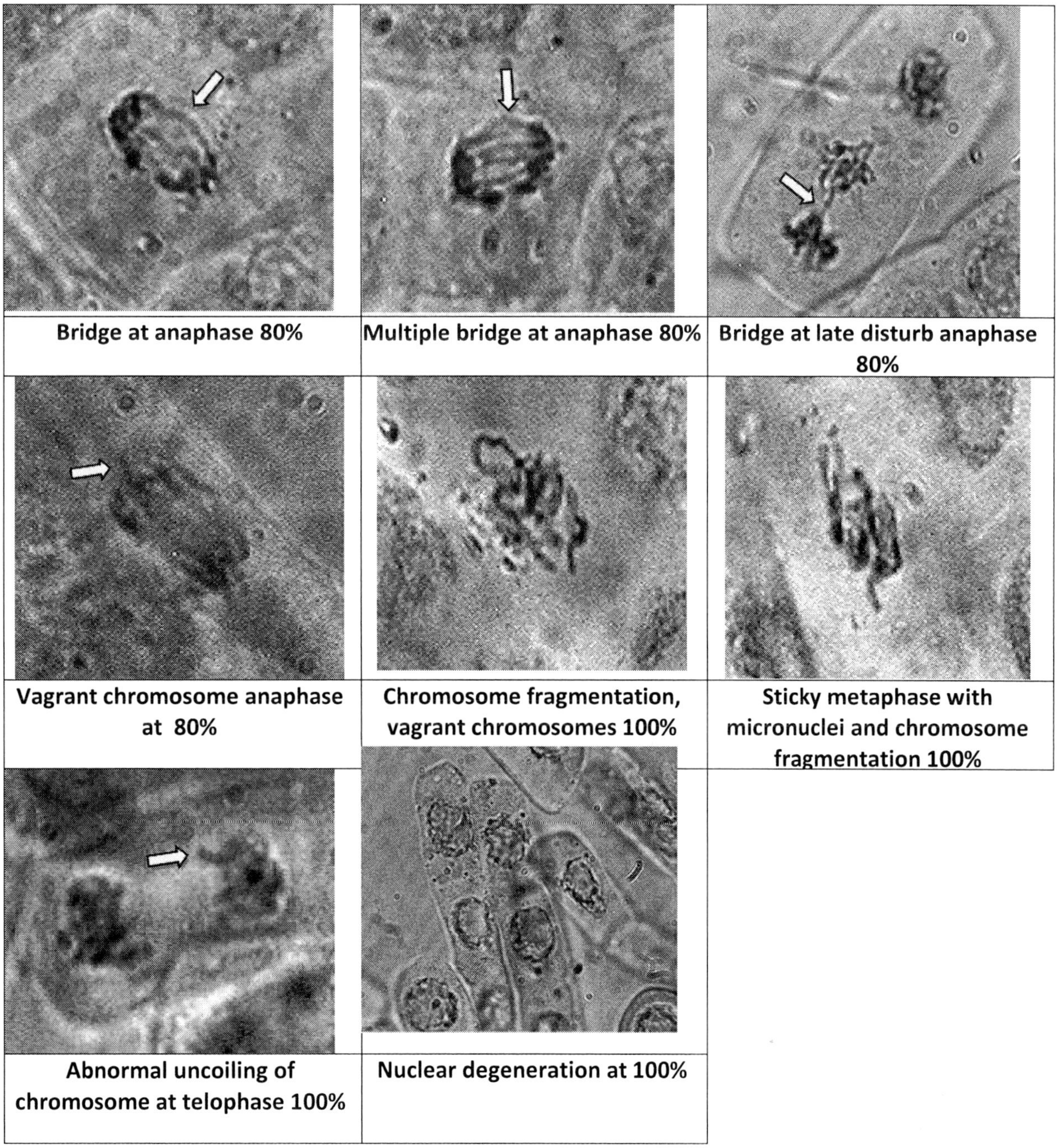


**FIGURE 2b.** Normal chromosomes and Chromosomal abnormality caused by Fly ash.

## CONCLUSION

We can conclude that the lower concentration of fly ash mixed with vermicompost and pressmud increases the mitotic index and do not cause toxic effects, whereas higher concentration of fly ash works as a toxic material for the plant.

## REFERENCES


Abraham S & Rajalakshmy BN (1989). Production of mitotic abnormalities by magnesium sulphate in Vicia faba L. Cytologia 54: 559-563.

Anis M, Aijas AW. Caffeine induced morpho-cytological variability in fenugreek, Trigonella foenum-graecum L. Cytologia 1997; 62: 343-349.

Badr A (1986). Effects of the S-Triazine herbicide tubutryn on mitosis chromosomes and nucleic acids in root tips of Vicia faba.Cytologia 51: 571-577.

Bempong AM (1973). Cytological effects of nogalmycin in Tradescantia paludosa microsporocytes. Mutation Res 21: 323-326.

Darlingtonand, C.D., L.F. La Cour, The Handling of Chromosomes, 6th Edn,George Allen and Unwin Ltd, 1976.

El-Khodary S, Habib A, Haliem A. Effect of herbicide Tribunil on root mitosis of Allium cepa. Cytologia 1990; 55: 209-215.

Evans HJ (1962). Chromosomal aberrations induced by ionizing radiations. Int Rev Cytol 13: 221-321.

Fiskesjo G (1985). The Allium test as a standard in environmental monitoring. Hereditas, 102: 99 – 112.

Fiskesjo G (1988). The Allium test – an alternative in environmental studies: The relative toxicity of metal ions. Mutat. Res. 197: 243 – 260.

Gomurgen AN. Cytological effect of the herbicide 2, 4-D Isooctylester 48% on root mitosis of Allium cepa. Cytologia 2000; 165: 383-388.

Haliem AS. Cytological effects of the herbicide sencor on mitosis of Allium cepa. Egypt J Bot 1990; 33: 93-104.

Kailash P. Patel and Kalpesh M. Patel. Cytological Changes in Trigonella Foenum-graecum (l.) under the Cadmium Stress. Journal of Life Sciences and Technologies Vol. 1, No. 1, March 2013.

Kabarity A, El-Bayourni AA & Habib AA (1974). Effect of morphine sulphate on mitosis of Allium cepa L. Root Tips. Biol Plant 16: 275- 282.

Kong MS, Ma TH (1999). Genotoxicity of contaminated soil and shallow well water detected by plant bioassays. Mutat. Res., 426(2): 221.

Manas Ranjan Senapati, Fly Ash from Thermal Power Plants – Waste Management and Overview

O. Timothy ∗, M. Idu, D.I. Olorunfemi, O. Ovuakporie-Uvo] Cytotoxic and genotoxic properties of leaf extract of Icacina trichantha Oliv South African Journal of Botany 91 (2014) 71–74.

Rahat Niyazi and Sadhana Chaurasia Vermistabilization of Fly Ash Amended With Pressmud By Employing *Eisenia Foetida*

S. Shylaja, —Effect of certain physical and chemical mutagens on plant chromosomes,‖ Ph.D. Thesis, South Gujarat University, Surat. Gujarat, India, 1994.

Solomon M, Belenghi B, Delledonne M, Menachem E, Levine A 1999. The involvement of cysteine proteases and protease inhibitor genes in the regulation of programmed cell death in plants. Plant Cell 11: 431-443.

Sugimura T S Kondo & H Takebe.1982. Environmental mutagens and carcinogens. University of Tokyo, Press and Alan R Liss Inc.New York.

# HISTOLOGICAL EFFECTS OF BISPHENOL A ON GILLS, DIGESTIVE GLANDS AND ADDUCTOR MUSCLES OF LABORATORY-REARED *Corbicula fluminea* Mull.

***Kimberly B. Benjamin***, Jessmine L. Competente and Dyan Gabrielle H. de Guzman
(University of the Philippines Manila, Ermita, Manila, Philippines)

**ABSTRACT:** Bisphenol A (BPA) is a commonly occurring industrial chemical that is present in polycarbonate plastics and epoxy resins. In this study, the toxicity of BPA was analyzed through its histological effects on the gills, digestive glands and adductor muscles of *Corbicula fluminea,* a freshwater bivalve. Forty laboratory-reared *C. fluminea* were exposed to set-ups with 1 μg/L, 2 μg/L and 3 μg/L of BPA for twenty-one days. Histological alterations such as vacuolations, necrosis, lamellar deformation, hyperplasia, loss of epithelium, necrosis, tubular alteration, neoplasia, hemocyte infiltration, hypertrophy and pyknosis were observed. These characteristics were then scored using a modified histological index. Post-hoc analysis revealed that there was a significant difference in the histological alterations observed between the tissues of exposed and unexposed clams. Moreover, it has been shown that the degree of histological damage among the tissues is as follows: digestive gland > gills > adductor muscles.

## INTRODUCTION

Bisphenol A (BPA) is a chemical commonly used in the production of polycarbonate plastics and epoxy resins- both of which are utilized in applications that makes our lives easier and convenient (Flint et al.,2014). Polycarbonate plastics are used in compact discs, medical devices as well as food and beverage containers. Epoxy resins on the other hand, are used to coat metal products such as food cans and pipes of water supply (Stahlhut et al., 2009). BPA enters freshwater ecosystems through effluents from sewage treatment plants, plastic factories and landfill leachate (Oelmann et al., 2000). According to Crain et al., (2007), BPA concentration in freshwater ecosystems may vary and can reach up to 21ug/l with the sediments containing an even greater concentration compared to the water column. BPA quickly disperse and disappear from natural water systems and this is highly attributed to adsorption by suspended materials and sediments as well as absorption of aquatic animals and plants ( Oelmann et al., 2009; Flint et al., 2014). BPA is a toxic compound and its effects on vertebrates and invertebrates have been thoroughly described (Ike et al., 2000; Kang, 2007; Oelmann et al., 2009; Flint et al., 2014 ). Mechanistically, it has been shown to affect lipid peroxidation, endocrine function and production of reactive oxygen species in vertebrate and invertebrate organisms (Lehmann, 2007; Flint et al., 2014). However, bibliographic data report limited information about the histological effects of BPA on mollusk in particular the bivalves, which are considered as suitable organisms for biomonitoring purposes due to their filter-feeding behavior, sessile status and ability to concentrate pollutants to several orders of magnitude above ambient levels in aquatic environmental (Oliveira et al., 2014). Moreover, histological effects of this compound is only limited to the gonads of invertebrates and not on gills, digestive glands and adductor muscles which are also essential in their survival (Santos and Martinez, 2014). The digestive gland being involved in pollutant detoxification, homeostasis and bioaccumulation (Mantecca et al., 2006); the gills are used for respiration and filtration (Beltran and Pocsidio, 2010); and, the adductor muscles are involved in regulating the opening of the valves of clams (Beltran and Pocsidio, 2010).

The Asiatic clam or *Corbicula fluminea* was utilized in this study, because of their wide distribution, high density as well sensitivity to pollutants such as BPA (Colombo et al., 1995, Labrot et al., 1999; Fournier et al., 2005; Lehmann, 2007). These characteristics qualify them as a good sentinel organism for monitoring the status of aquatic environment. The exposure of the clams to environmentally-relevant concentrations of 1 μg/L, 2 μg/L and 3 μg/L BPA (Hatel et al., 2012; Yang et al., 2014) under controlled laboratory conditions aims to give data on the actual histological effects of this compound on the adductor muscles, digestive gland and gills of the test organism .

## MATERIALS AND METHODS

**Experimental Design.** *C. fluminea* clam with a shell length of 20-30 mm, were obtained from a local un-impacted watershed (La Mesa, Watershed, Quezon City, Philippines). The clams were acclimatized in the laboratory for period of two weeks prior to toxicity test. During acclimatization, the clams were maintained under a 12h:12h photoperiod in a glass aquarium containing 1000L of clean aerated and dechlorinated water with sand.Further, filtration system for waste removal, constant zooplankton supply and replacement as well as quality testing of water was done every seven days.

After acclimatization, ten clams were randomly placed in a 5-L glass container labeled as control, *S1*, *S2*, and *S3*- the *control* set-up containing only depurated water and sand and the experimental set-ups containing depurated water, sand and 1.0 μg/L, 2.0 μg/L, and 3.0 μg/L of BPA, respectively. Exposure period lasted for twenty-one (21) days along with proper maintenance measures being observed.

**Histological Analysis.** After 21 days, five *C. fluminea* from each of the 4 set-ups were randomly selected. The specimen were prepared for histological analysis by prying open the shells and inserting a toothpick between the valves. The clams were euthanized and their adductor muscles, digestive glands and gills were excised and were separately placed in properly labeled vials containing Davidson's fixative for 24 hours. After the 24-hour period, the tissues were immediately transferred to 70% ethanol-containing vials until histological processing.

The tissues were processed, embedded in paraffin, sectioned at 5 microns, stained with Hematoxylin-Eosin at the National Kidney and transplant Institute and examined via and Olympus light microscope. A modified index adapted from Costa et al., (2013) was used to evaluate the intensity of cellular and morphological alterations among the fields of view. The categories for the gills and the adductor muscles were cellular and morphological changes and neoplasia while that of the digestive glands were tubular alterations, intertubular changes and neoplasia. The scores range from 0 to 3, 0 indicating 0-20%; 1 indicating 21-40%; 2 indicating 41-60% and 3 indicating 61-100% tissue aberration.

**Statistical Analysis**The Kruskal-Wallis test for non-parametric data utilizing a Mann Whitney post hoc analysis was performed to analyze the scores. Statistical comparisons were considered significant at 5% level ($p<0.05$).

## RESULTS

No mortality was observed both in the control and experimental groups during the 21-day exposure.

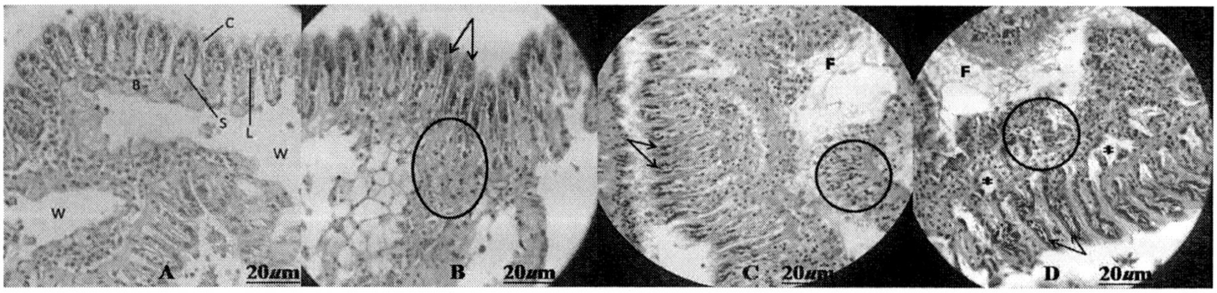


FIGURE 1. Photomicrographs of gills in control and BPA-exposed *C. fluminea* clams after 21 days. (A) Unexposed control clam showing regular arrangement of the gill filaments with a single layer of epithelial cells *(C)* supported by skeletal rods *(S)* forming a well- defined lamellae *(L)*. (B) Clams exposed to 1 ug/L BPA; (C) Clams exposed to 2 ug/L; and, (D) Clams exposed to 3 ug/L BPA. Circle=hypertrophy; F=fibrosis; arrow=hyperplasia; asterisk=necrosis.

**Gill Histopathology.** Figure 1A shows the optimal histological characteristics of the clam gills. This is characterized by a well-defined lamellae made up of skeletal rods which supports a single layer of ciliated epithelial cells. Figure 1B-C on the other hand, show the alterations brought about by varying concentrations of BPA on the tissues of the gills of *C. fluminea*. Common histological aberrations that were seen in the exposed groups include vacuolation, hypertrophy, hyperplasia, lamellar deformations and necrosis. Occurrences of other alterations such as loss of epithelium, lipofuscin aggregations, fibrosis and hemocyte infiltration were also observed as concentration increases. Statistical analysis of the median score for cellular and morphological characteristics in gills revealed that clams from all groups are statistically different from each other as shown by differences in number (Table 2). Moreover, mean ranks showed that the highest intensity of damage was seen in 3ug/L while the least was seen in the control group (Table 1). No significant neoplasia was seen in all groups.

TABLE 1. Medians and mean ranks of the cellular, morphological and neoplastic changes in the tissues of the gills of *C. fluminea* across 1, 2 and 3 μg/L of BPA. . Differences in number of median denotes significant difference using Mann Whitney Test at $p<0.05$ with n=150.

| Treatment | Cellular and Morphological Changes | | Neoplasia | |
|---|---|---|---|---|
| | **Median** | **Mean Rank** | **Median** | **Mean Rank** |
| Control | 0 | 4 | 0 | 4 |
| 1 μg/L | 1 | 3 | 0 | 3 |
| 2 μg/L | 2 | 2 | 0 | 2 |
| 3 μg/L | 3 | 1 | 0 | 1 |

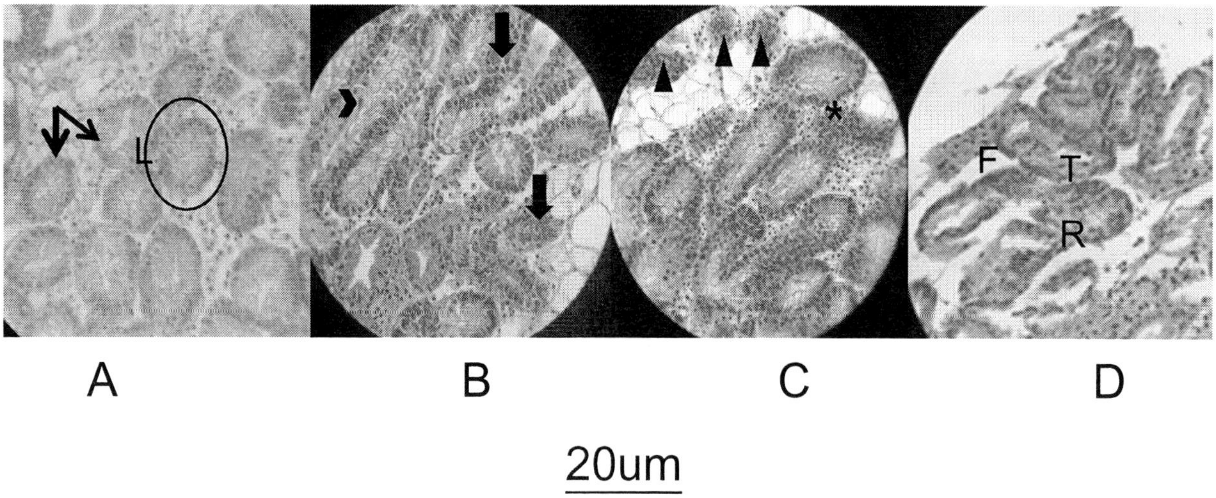


FIGURE 2. Photomicrographs of digestive gland in control and BPA-exposed *C. fluminea* clams after 21 days. (A) Unexposed control clam showing intact digestive tubules (encircled), surrounded by regular epithelium (thin arrow); (B): Clams exposed to 1 ug/L BPA; (C) Clams exposed to 2 ug/L; and, (D) Clams exposed to 3 ug/L BPA. Arrowhead=hypertrophy; thick arrow= hyperplasia; triangle= granulocytoma; F=fibrosis; asterisk= necrosis; TR=tubular regression.

**Digestive Gland Histopathology.** Figure 2A shows the normal digestive gland of clams from the control group. In its normal state, the digestive gland shows intact digestive tubules with regular epithelium surrounded by a well-defined basement membrane. Moreover, intertubular spaces consist of connective tissues without sign of neoplasia and fibroma. Figure 2B-D illustrates the effects of a 21-day exposure of clams to different concentrations of BPA. The digestive glands of the treated groups exhibited high instances of hyperplasia, hypertrophy, necrosis and granulocytoma. Moreover, few incidences of fibrosis, neoplasia and lipofuchsin aggregation were also observed. Statistical analysis revealed that in terms of tubular alterations, clams exposed to 2ug/L and 3ug/L BPA rendered similar histological effects, while the clams in the control and those exposed to 1ug/L BPA are significant. Analysis on the intertubular and neoplastic changes in digestive gland however revealed that only the control group is statistically significant (Table 2).

**Adductor Muscles Histopathology.** Figure 3A shows the adductor muscles of clams from the control group. They displayed muscle fibers that are uniformly organized and adhering with each other. The cells are fusiform in shape with a central nucleus. Figures3B-D illustrates the effect of varying concentration of BPA on the adductor muscles. Histological examination of the tissues revealed the occurrence of necrosis, loss of surface adherence and loss of muscle organization among the exposed group. Statistical analysis of the median value of cellular and morphological damage revealed that only the control group is different. Moreover, in terms of neoplasia, no significant difference was seen across all groups (Table 3).

TABLE 2. Medians and mean ranks of the tubular alterations, intertubular and neoplastic changes in the tissues of digestive gland of *C. fluminea* across 1, 2 and 3 μg/L of BPA. Differences in number denotes significant difference using Mann Whitney Test at $p<0.05$ with $n=150$.

| Treatment | Tubular Alterations | | Intertubular Changes | | Neoplasia | |
|---|---|---|---|---|---|---|
| | **Median** | **Mean Rank** | **Median** | **Mean Rank** | **Median** | **Mean Rank** |
| Control | 0 | 4 | 0 | 4 | 0 | 4 |
| 1 μg/L | 2 | 3 | 1 | 2 | 1 | 3 |
| 2 μg/L | 3 | 2 | 1 | 3 | 1 | 1 |
| 3 μg/L | 3 | 1 | 1 | 1 | 1 | 2 |

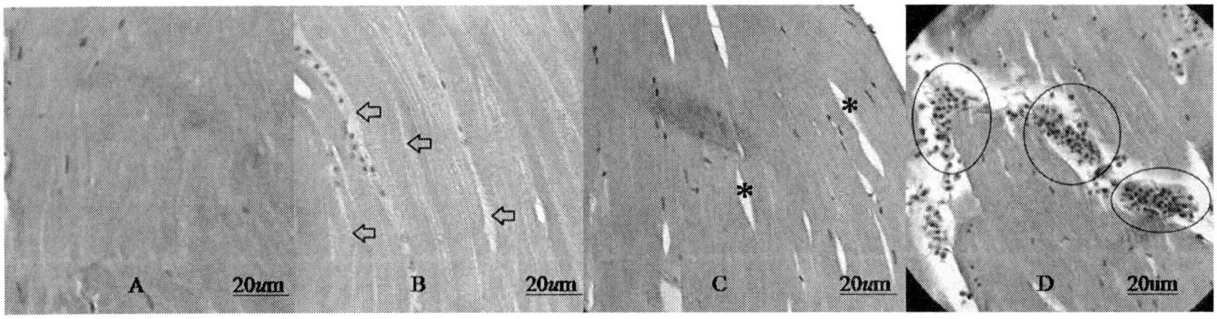


FIGURE 3. Photomicrograph of adductor muscles in control and BPA-exposed *C. fluminea* clams after 21 days. (A) Unexposed control clam showing organized muscle fibers; (B): Clams exposed to 1 ug/L BPA; (C) Clams exposed to 2 ug/L; and, (D) Clams exposed to 3 ug/L BPA. Asterisk=loss of surface adherence ; thick arrow= necrosis; solid ring= haemocyte infiltration

TABLE 3. Medians and mean ranks of the cellular, morphological and neoplastic changes in the tissues of the gills of *C. fluminea* across 1, 2 and 3 μg/L of BPA Differences in number denotes significant difference using Mann Whitney Test at $p<0.05$ with n=150.

| Treatment | Cellular and Morphological Changes | | Neoplasia | |
|---|---|---|---|---|
| | **Median** | **Mean Rank** | **Median** | **Mean Rank** |
| Control | 1 | 4 | 0 | 4 |
| 1 μg/L | 2 | 2 | 0 | 3 |
| 2 μg/L | 2 | 3 | 0 | 2 |
| 3 μg/L | 2 | 1 | 0 | 1 |

## DISCUSSION

Different organs have varying responses to different types of toxic compounds. In this study, the toxic effects of environmentally relevant concentrations of BPA on the histology of gills, digestive gland and adductor muscles of laboratory-reared *C. fluminea* were established. Based on the results, the gills, digestive glands and adductor muscles showed differential responses to the presence of BPA. Among the three organs, it was found out that the digestive gland was the most affected organ histologically and statistically due to significant tubular and intertubular alterations as well as incidences of neoplasia even at the lowest concentration. According to Hayashi et al. (2008), BPA are often directed particularly in the lumen of the tubules where they are metabolized substantiating the greater histological damage seen in the tubules. Its high sensitivity is directly attributed to its role in homeostasis, contaminant uptake, digestion, metabolism as well as detoxification process (Kanapala & Arasada, 2013; Costa et al., 2012). Next to the digestive gland, the gills of exposed clams were significantly affected as well. The gills showed a concentration- dependent response towards BPA as seen by an increasing median score in its cellular and morphological attributes. Further, the statistically significant difference in histological scores seen among concentrations would show that gills have specific responses to specific amount of BPA.

The increase in vacuolation and hypertrophy in clams exposed to 1ug/L is a physiological adaptation to limit damage (Flint, 2014). Vacuolation and hypertrophy are essential in confinement, isolation and sequestration of materials that could pose harm to the organism (Costa et al., 2012; Flint et al., 2014). At higher concentrations, vacuolations and hypertrophy prove to be inefficient as a response to the slowed metabolism of the organism. According to Leonard et al., (2014) a concentration of 2ug/L of BPA slows the metabolic processes of the cells causing them to die, thus hyperplasia or increase in cellular number take place as an adaptation to an increasing functional demand. Further, fibrosis was also common in gill tissues exposed to 2ug/L. According to Payan et al., (2014), it is a degenerative response to exposure to harmful chemicals characterized by deposition of connective tissue that can obliterate the architecture of the tissues thereby leading to loss of physiological function. At the highest concentration of 3ug/L, epithelial detachment, lamellar deformation and loss of epithelia and necrosis were seen. These mechanisms serve as barrier to prevent further entry of contaminants to damage other organs. Similar mechanism was also observed by Kumar et al., (2012) in gills of *Lamellidens marginalis*, a freshwater bivalve. Such stringent response has to be employed by the gills since it plays significant role in respiration, acid-base balance, ionic and osmotic regulation, food capture and maintenance of water current which greatly affects the physiological well-being of the organism. (El-Shenawy et al., 2007, 2009; Beltran and Pocsidio, 2010).

The least affected among the organs is the adductor muscles with alterations limited to loss of surface adherence, haemocyte infiltration, loss of muscle organization and necrosis. Statistical analysis showed that all the exposed groups had a similar median score corresponding to only 41-60% damage. The high endurance of the adductor muscles even at high concentrations of BPA is due to its function of regulating the closing and opening of the shell (Beltran and Pocsidio, 2010). In addition, Hayashi et al. (2008) noted that the adductor muscles have a very low bioaccumulation rate to BPA and its metabolites.

## CONCLUSIONS

It is concluded that digestive gland was the most affected tissue followed by the gills then adductor muscles. The differential histological responses of the tissues of *C. fluminea* in different concentrations of BPA proves that they are good indicators of environmental stressors such as BPA. Future studies on the effects of lower concentrations (<1 ug/L) on digestive gland and higher concentrations (>3ug/L) on the adductor muscles.

## REFERENCES

Beltran, K.S. and Pocsidio, G.N. 2010. Acetylcholinesterase activity in *Corbicula fluminea* Mull., as a biomarker of organophosphate pesticide pollution in Pinacanauan River, Philippines. *Environ Monit Assess* 165:331-340

Colombo, J. C., Bilos, C., Campanaho, M., Presa, M. J. R., and Catoggio, J. A.1995. Bioaccumulation of polychlorinated-biphenyls and chlorinatedpesticides by the Asiatic Clam *Corbicula fluminea*—its use as sentinel organism in the Rio-De-La-Plata Estuary, Argentina.Environ Sci Technol 29 : 914–27

Costa, P. M., Carreira, S., Costa, M. H. and Caeiro, S. 2012. Development of histopathological indices in a commercial marine bivalve (Ruditapes decussatus)to determine environmental quality. *Aquat Toxicol* 126: 442-454.

Crain, D. A., Eriksen M., Iguch,i T., Jobling, S., Laufer, H., LeBlanc, G. A. and Guillette, L. J. 2007.An ecological assessment of bisphenol-A: Evidence from comparative biology. *Reprod Toxicol* 24, 225-239.

Flint, S., Markle, T., Thompsin, S., and Wallace, E.2012. "Bisphenol A exposure, effects and policy: A wildlife perspective- a Review" *J. of Environ Mngt* 104:19-34.

Fournier, E., Adam, C., Massabuau, J. C., and Garnier-Laplace, J. 2005. Bioaccumulation of waterborne selenium in the Asiatic clam Corbicula fluminea: influence of feeding-induced ventilatory activity and selenium species.Aquat Toxicol (Amsterdam,Netherlands)72:251-260.

Hatel, A., Alavi, S.M.H., Abdulfatah, A., Fontaine, P., Rodina, M., Linhart, O. 2012. Adverse effects of Bisphenol A on reproductive physiology in male goldfish at environmentally relevant concentrations. *Ecotoxicology and Environmental Safety* 76:56-99

Hayashi, O., Kameshiro, M., Masuda, M., and Satoh, K. 2008. Bioaccumulation and metabolism of [$^{14}$C]bisphenol a in the brackish water bivalve *Corbicula japonica. Biosci. Biotechnol. Biochem., 72*(12): 3219-3224.

Ike, M., Jin, C.S., Fujita, M., 2000. Biodegradation of bisphenol A in aquatic environment. *Water Sci. Technol* 42: 31-38.

Kanapala, V. & Arasada, S. 2013. Histopathological effect of paraquat (gramoxene) on the digestive gland of freshwater snail *Lymnaea luteola* (Lamarck: 1799) (mollusca: gastropoda). *International Journal of Scientific Research in Environmental Sciences, 1*(9): 224-230.

Kang, J. H., Aasi, D. and Katayama Y. 2007. Bisphenol a in the aquatic environment and its endocrine-disruptive effects on aquatic organisms. *Crit Rev Toxicol* 37: 607-25.

Klecka, G.M., Staples, C.A., Clark, K.E., van der Hoeven, N., Thomas, D.E. and Hentges, S.G.2009. Exposure Analysis of Bisphenol A in Surface Water Systems in North America and Europe. *Environ Sci Technol* 43(16):6145-6150

Kumar, S., Pandey, R. K., Das, V.K. 2012. Dimethoate alters respiratory rate and gill histopathology in freshwater mussel *Lamellidens marginalis* (Lamarck). Journal of Applied Bioscience 38 (2): 154-158.

Labrot, F., Narbonne, J. F., Ville, P., Saint Denis, M., and Ribera, D. (1999). Acute toxicity, toxicokinetics, and tissue target of lead and uranium in the clam Corbicula fluminea and the worm Eisenia fetida: comparisonwith the fish *Brachydanio rerio.. Arch Environ Contamin Toxicol* 36:167-178.

Lehmann, D.W., Levine, J.F. and Law, J.M. 2007. Plochlorinated bisphenyl exposure causes gonadal atrophy and oxidative stress in *Corbicula fluminea* Clams. Toxic Path 35:356-365.

Leonard, J.A., Cope, W.G., Barnhart, M.C. and Bringolf, R.B. 2014. Metabolomic, behavioral, and reproductive effects of the synthetic estrogen 17 α-ethinylestradiol on the unionid mussel *Lampsilis fasciola. Aquatic toxicology, 150*, 103-116

Mantecca, P., Vailati, G. Bacchetta, R.2006. Histological changes and Micronucleus induction in the Zebra mussel *Dreissena polymorpha* after Paraquat exposure. *Histol Histopathol* 21: 829-840.

Oehlmann J., Schulte-Oehlmann U., Tillmann M. and Markert B. 2000. Effects of endocrine disruptors on prosobranch snails (Mollusca: Gastropoda) in the laboratory. Part I: Bisphenol A and octylphenol as xeno-estrogens. *Ecotoxicology* 9: 383-397.

Oehlmann, J., Schulte-Oehlmann, U., Kloas, W., Jagnytsch, O., Lutz, I., Kusk, K.O., Wollenberger, L., Santos, E.M., Paull, G.C., Van Look, K.J., Tyler, C.R., 2009.A critical analysis of the biological impacts of plasticizers on wildlife. *Phil. Trans.R. Soc. B* 364: 2047e-2062.

Oliveira, L.F., Silva, S.M.C.P. and Martinez, C. 2014. Assessment of domestic landfill leachate toxicity to the Asian clam *Corbicula fluminea* via biomarkers. *Ecotoxicol. Environ Saf* 103:17-23.

Pavan, P. G., Stecco, A., Stern, R. & Stecco, C. (2014). Painful connections: Densification versus fibrosis of fascia. *Curr Pain Headache Rep 18*(441), 1-8.

Stahlhut, R.W., Welshons, W.V., Swan, S.H. 2009. Bisphenol-A data in NHANES suggest longer than expected half-life, substantial nonfood exposure, or both. *Environ. Health Perspect* 117: 784e-789.

Santos, K.C. and Martinez, C.B.R. 2014. Genotoxic and biochemical effects of atrazineand Roundups, alone and in combination, on the Asian clamCorbiculafluminea. *Ecotoxicol. Environ. Saf.* 100: 7–14

Yang, Y., Kim, S., Hong, Y., Ahn, J., Park, M. 2014. Environmentally relevant levels of Bisphenol A may accelerate the development of Diabete mellitus in adolescent Otsuka Long Evans Tokushima fatty rats. *Toxicol. Environ. Health. Sci.* 61(1): 41-47.

# BIO-ASSESSMENT AND TOXICOLOGICAL EFFECTS OF BIOLARVICIDES ON TARGET AND NON-TARGET ORGANISMS IN RIVERS STATE, NIGERIA

**Wachukwu, C.K.**, Ollor, O.A., Ganabel, C.B, and Azike, C. A.
(Department of Medical Laboratory Science, Rivers State University of Science and Technology, Nkpolu-Oroworukwo, Port Harcourt, Nigeria)

Two biolarvicides (Bactivec and Griselesf) were used to control mosquito larvae in Rivers State, Nigeria. The aim of this study was to assess the health and environmental risks associated with the application of biolarvicides on target and non-target organisms. A total of ten (10) target organisms (mosquito larvae) and non-target organisms (tadpoles) were each exposed to different concentrations of biolarvicides (*Bacillus thuringensis* var israelensis (Bti)(Bactivec) and *Bacillus sphaericus* (Bs)( Griselesf) respectively. Ten-fold serial dilutions of the biolarvicides were made up to $10^3$. Thereafter, ten (10) target and non-target organisms each were added into the diluted samples and allowed to remain at both room temperature (20-28ºC) and atmospheric temperature (30 – 35ºC) for 14 days. They were also exposed to the undiluted biolarvicides and given the same treatment like the diluted ones. The viability or duration of the biolarvicides effect were also tested by exposing them to the environment for 7 days. Thereafter, the organisms were added to determine the effect of biolarvicides on them. Five milliliter (5ml) each of the undiluted biolarvicides (Bactivec and Griselesf) were mixed together and the organisms were added to it to check the synergistic effect on the organism. The pH and temperature were determined using the digital pH meter (model PHS-3C, UK) and the mercury thermometer. Two hundred and sixty (260) blood samples were collected and analyzed from the sprayed and control zones for malaria parasite and haematological parameters. The result showed that the undiluted biolarvicides killed the entire target and non-target organisms at ten (10) minutes of exposure, while the combined biolarvicides killed the non-target organism within 5 minutes of exposure and the target organisms were killed after 10 minutes of exposures to the biolarvicide. One in ten ($10^1$) dilution killed over 50% of the same organisms after 60 minutes exposure. At $10^2$ dilutions, the organisms were all killed after 48 hours exposure, while $10^3$ dilutions had no effect on the organisms. The results also showed that the undiluted and $10^1$ dilution can persist in the environment for long period of over seven days. The pH of the biolarvicides ranged from 6.0 and 6.9. The results of the haematological and malaria parasite showed that 162 (81%) had malaria parasites, while 38(19%) had no malaria parasites. There were no changes in the haematological parameters such as Hb, WBC and differential cell count. Therefore, it can be inferred that appropriate concentration of a given biolarvicide be used to avoid causing potential harm to other members of the ecosystem.

# ENANTIOSELECTIVE PHYTOTOXICITY OF γ-HEXABROMOCYCLODODECANE ENANTIOMERS TO MAIZE

***CUI Jiansheng, LIU Ying, WU Tong***

(School of Environmental Science and Engineering, Hebei University of Science and Technology; Hebei Key Laboratory of Molecular Chemistry for Drug, Shijiazhuang, 050018, China)

In the present study, two enantiomers of γ-hexabromocyclododecane (γ-HBCD) were prepared by preparative liquid chromatography. To investigate the selective oxidative damage of HBCD to maize, young seedlings were exposed to solutions of HBCD enantiomers and racemate with different concentrations. The root biomass, shoot biomass, and root and shoot elongation of maize were inhibited after exposure with the inhibition rates following the order (+)γ-HBCD > (rac)γ-HBCD > (-)γ-HBCD. The antioxidant enzyme activities of superoxide dismutase (SOD) and peroxidase (POD) both in roots and shoots were significantly upregulated by exposure to lower concentrations of (+) and (rac)γ-HBCD than (-)γ-HBCD. In contrast to (-)γ-HBCD, the activities of SOD and POD were all decreased by higher exposure concentrations of (+) and (rac)γ-HBCD. The intensities of histone H2AX phosphorylation were increased at first with an increasing exposure concentration of all γ-HBCD enantiomers and racemate and then decreased, indicating the enantio-specific induction of DNA damage in maize. All of these results provide solicited evidence of the significant enantioselectivity phytotoxicity of γ-HBCD to maize with a higher toxicity of (+)γ-HBCD than (rac)γ-HBCD and (-)γ-HBCD.

# EFFECTS ON LENGTH, WEIGHT AND ORIENTATION RESPONSES OF EARTHWORMS EXPOSED TO MAN-MADE ELECTROMAGNETIC NOISE

**Şükran Yalçın Özdilek**, Sevil Yalçın and Rukiye Altaş
(Çanakkale Onsekiz Mart University, Çanakkale, Turkey)

Earth has a natural magnetic field about 60 μT at the poles and 30 μT at the equator and many animals use this natural phenomenon for orientation and navigation. With the development of technology, the natural systems have to expose high electromagnetism because of heavy usages of electric and electrical devices. This man-made electromagnetic noise has increased dramatically above the natural magnetic fields in human occupied areas in the last two decades. In general, many scientists consider possible negative and positive effects of electromagnetism on human life and living system. The experimental studies indicate that magnetic and electromagnetic fields cause same changes on the living activities, such as, oxidative, genotoxic effects, survival, metabolic rates. The studies about the effects of man-made electromagnetic noise also focus on the behavioural patterns of organisms such as deviation on the migration route. The present study aimed to understand the possible effects of man-made electromagnetic noise as physiological and behavioural response on the earthworm as a model organism in laboratory conditions. The triplet experimental groups (each were composed of 20 earthworms) were exposed to 1.5 mT to 8 mT electromagnetic field using 1.5V current in a solenoid coil during one hour in a vivarium. The experimental and control designs were kept in a similar conditions and the length and weight of earthworms were measured before and after seven-day experimental design. The positions of earthworms were recorded by a photograph machine in every five minutes. The angles in terms of mean vector of each position of each earthworm beginning from the centre of vivarium were recorded by Adobe Photoshop CS6 program. The mean vectors and angles of different experimental designs and controls were compared by circular statistics. The results indicate that while the mean length of experimental group was decreased ($F=16.4$; $df= 118$; $p<0.001$) the mean weight of the earthworms had not changed ($F=2.01$; $df=118$; $p>0.05$). The orientation of the earthworms in control and experimental groups were statistically different ($p<0.005$).

# ASSESSMENT OF POLLUTION BIOMARKER AND STABLE ISOTOPE DATA IN *Mytilus galloprovincialis* TISSUES

**Şükran Yalçin Özdilek*** and Neslihan Demir
(Çanakkale Onsekiz Mart University, Faculty of Science and Arts, Department of Biology, Çanakkale, Turkey)

Mussels, *Mytilus galloprovincialis* give a rapid response to pollutants as an enzymatic antioxidant defenses with their oxidative potential. Superoxide dismutase (SOD) is one of the antioxidant defenses enzymes in mussels and converts the superoxide anion into hydrogen peroxide and this enzyme is used as biomarkers of oxidative damage. As well as many topics in ecology, stable isotopes are also signature for organic and heavy metal pollution in aquatic ecosystems. Particularly, the carbon isotopic signature is suggested as a proper sewage indicator for mussels, which filter particular organic matter from the water for feeding. Both method use different indicators for chronic response of pollution. This study aims to compare the stable carbon and nitrogen values of different mussel tissues and the changes on the SOD values of the same tissues in order to understand the relationship between two mechanisms of bio indicator processes of physiological response of mussel to pollution. The change on the level of superoxide dismutase (SOD) in the gill, hepatopancreas and mantle tissues of *M. galloprovincialis* with $\delta^{13}C$ and $\delta^{15}N$ isotopes were assessed in different locations (Kepez and Güzelyalı) in Çanakkale. The SOD values of mussel samples were found as Gill > hepatopancreas > mantle collected from Kepez and gill > hepatopancreas collected from Güzelyalı. There were significant differences among the mean SOD values of different tissues both in Kepez and Güzelyalı samples ($p<0.05$). However, there were not any significant differences among the mean SOD values of different tissues of control group. There were enrichment both in nitrogen and carbon isotope values of hepatopancreas tissues both in Kepez and Güzelyalı samples. The mean $\delta^{13}C$ value of hepatopancreas tissue and the mean $\delta^{15}N$ value of gill tissue were a distinct group for Kepez samples and the mean $\delta^{13}C$ and $\delta^{15}N$ values of hepatopancreas and gill tissues were different for Güzelyalı samples ($p<0.001$). There were a negative correlation between $\delta^{13}C$ values and SOD values in Kepez samples ($p<0.001$).

---

# GRAPHENE INDUCED EPITHELIAL-MESENCHYMAL TRANSITION IN A549 CELLS *In Vitro*

***Yanyan Liao***, Ziyan Zhang
(Institute of Urban Environment, Chinese Academy of Sciences, Xiamen, Fujian Province 3361021, China)

With the deep research of the carbon nano-materials, especially the graphene was firstly discovered since 2004， it has gradually become a "superstar material" in the region of materials science. Nano-graphene are two-dimensional structures of carbon. Due to their unique physical and chemical properties, they have been widely used in biomedical, electronic devices and composite materials research fields. They also have a very bright application prospects, such as biosensors， biological imaging, gene delivery, photothermal therapy of tumor and cancer, hydrophobic anticancer drugs, and antibacterial materials and organization of material and other fields. However, the toxicity of graphene on health remains unknown and is of great concern.

One of the most important impacts on physiological pathology is the transfer of tumor invasion. This process is called Epithelial-Mesenchymal transition (EMT). The process mainly occurred in the polarity of epithelial cells, which will be changed into mesenchymal cells and they have invasiveness and migration capability. It exists in multiple physiological and pathological processes in human body. Many studies have shown that EMT and tumor cell invasion and metastasis are closely related. Thus, this experiment was designed to observe graphene stimulate EMT mechanisms to promote the metastasis of human lung adenocarcinoma epithelial cell line (A549).

A549 cells were plated in the 6-well plates and incubated for 48 h and 72 h. Graphene samples were introduced to the cells with final concentrations of 20ug/ml, 10ug/ml, 5ug/ml and 1ug/mll. Cells without graphene exposure were taken as the control. Wound healing assay showed that cell migration rate was obviously higher in graphene treatment groups.Transwell assay demonstrated that the invasion rate was increased in graphene-treated A549 cells. The expression levels of E-cadherin, β-catenin, TIMP-1, Wnt5a, ICAM-1, vimentin and Twist1 in A549 cells were examined by Western blotting,and reverse transcription(RT)-PCR. The results showed that the expression level of epithelial markers (E-cadherin, β-catenin, TIMP-1 and Wnt5a) was reduced after exposure to graphene for 48h and 72h, while the expression of mesenchymal markers (ICAM-1, vimentin and Twist1) were significantly increased after exposure to graphene for 48h and 72h. Our data suggest that graphene may promotes A549 cells invasion, migration and tumorigenesis ability. EMT is involved in graphene treated A549 cells.

# DEVELOPING AN ASSOCIATION NETWORK FROM PROTEOME CHANGES TO ROOT PHENOTYPIC PROPERTIES FOR ALUMINUM TOLERANCE IN SWITCHGRASS

***Mahesh Rangu***, Zhujia Ye, Sarabjit Bhatti, Suping Zhou
(Tennessee State University, Nashville, TN, 37209, USA)
Theodore W. Thannhauser
(Robert W. Holley Center for Agriculture and Health,USDA-ARS, Cornell University, Ithaca, New York)

Aluminum (Al) stress is a major problem causing reduction in plant growth in acidic soil. The most noticeable symptom of Al toxicity is the development of stunted root systems on sensitive plants. Plants were grown hydroponically in Al-treated and control cultures and a proteomic analysis of plants root tissue (Switchgrass) was carried out to document differential protein expression as a function of the Al status. During short-term treatment (<1 month), plant root length was not affected, even though there was some reduction in photosynthetic activity, leaf number, plant height and biomass yield. 7614 proteins in root tip protein extract and 7097 proteins in elongation protein extract were identified. Out of these proteins 1274(17%) root tip proteins and 891 (13%) elongation proteins were showed significant differences (t-test, p-value <0.05) with low fold change. Quantitative proteomics analysis of the root-tip and the root elongation zone have identified proteins encoded by Al tolerance genes, such as homologs of AtSTAR1, proteins for heavy metal transport/detoxification, vacuolar sorting receptor 3, and NRAMP metal ion transporter 6, which were induced in both root tips and/or elongation zones. The phosphate transporter was greatly induced, but the K transporter was repressed. Various transcription factors were identified and found to be both repressed or induced depending upon their functional context. Several proteins related to root hair initiation and development were induced which may affect the total surface area for nutrient uptake including Al ions. Proteins for suberin biosynthesis and Casperian strip formation on endodermis were induced in the elongation zone, which suggests a mechanism for tighter control of mineral flux of into the stele. Proteins affecting chromatin status, gene transcription, and protein translation activity were also affected by the Al treatment. Long term Al treatments (> 2 months) also carried out with longer exposure whereby the root structure started to show the typical Al-stress phenotype. Mass spectrometry, employing tandem mass tag (TMT) labeling used to identify protein changes in plant tissues due to Al stress.

---

# SITE-SPECIFIC SOIL ARSENIC BIOAVAILABILITY: THE 21ST CENTURY NEXUS BETWEEN SITE CHARACTERIZATION AND RISK ASSESSMENT.

***N.T. Basta****, S.D. Whitacre (The Ohio State University, Columbus, OH, USA)
V. M. Hanley and P. Myers (California Department of Toxic Substances Control, Sacramento, CA, USA)
A.L. Foster (USGS, Menlo Park, CA, USA)
S.W. Casteel (University of Missouri, Columbia, MO, USA)

Risk assessment methodologies provide conservative approaches to estimate the potential for adverse effects on human health from exposure to contaminants in soil. Risk assessment often establishes reference doses and cancer slope factors based on highly bioavailable forms of chemicals. Soil often interact with chemicals reducing their potential solubility in the environment and bioavailability for absorption by humans. Often the most important risk pathway associated with human exposure to arsenic-contaminated soils is incidental soil ingestion. We studied 18 soils collected from arsenic-contaminated gold mining sites throughout California. Soil As ranged from 200 to 12,000 mg/kg and relative bioavailability, measured by soil dosing trials using Juvenile Swine, ranged from 1% to 40% which were below the USEPA default value of 60%. In vitro Bioaccessibility (IVBA) As was measured by three in vitro gastrointestinal (IVG) methods. Arsenic RBA and IVBA were greatly reduced by adsorption to reactive Fe minerals including goethite and ferrihydrite. In vitro–in vivo correlation analysis found one IVG method conservatively predicted As RBA in all study soils and provided a good estimate (within 90% CI) of RBA As (RBA As = 0.8 IVG + 4.39, $r^2$=0.82). Use of RBA As was able to identify high risk soils that require remedial action. Bioavailability and/or bioaccessibility methods can be used to measure site-specific arsenic bioavailability characterize risk at thousands of abandoned gold mining sites with elevated levels of soil arsenic throughout the state of California and elsewhere where the primary risk pathway of concern is incidental soil ingestion.

# COMPREHENSIVE EVALUATION OF IN VITRO BIOACCESSIBILITY METHODS TO PREDICT BIOAVAILABILITY OF ARSENIC IN CONTAMINATED SOILS.

***Brooke N Stevens****, Nicholas T Basta, Shane D Whitacre
(The Ohio State University, Columbus, OH, USA)
Aaron R Betts, Kirk G Scheckel, (US EPA, Cincinnati, OH, USA)
Karen D Bradham, (USEPA, Triangle Park, NC, USA)

Soil arsenic (As) is a primary human health risk driver at many DoD, DOE, and USEPA sites. Often the calculated risk can be reduced when the bioavailability of arsenic in the soil is included in the risk calculations. In vitro bioaccessibility (IVBA) methods can be used to predict in vivo relative bioavailability (RBA). In order to determine the ability of an in vitro method to predict RBA, a correlation is done producing an in vitro-in vivo correlation (IVIVC). Due to the high cost of in vivo studies the importance of in vitro method's ability to predict RBA is growing. We report the most comprehensive IVIVC study for twenty nine soils with a range in soil properties, contamination sources, and solid phase As speciation.

Arsenic sources in study soils included orchards, glass works, copper mining, smelter, pesticide, and gold mining. The ability of five international published in vitro methods to predict animal RBA As (adult mouse and juvenile swine models) was determined. IVBA As vs. RBA As was fitted using linear regression and evaluated for goodness of fit, slope, y-intercept, prediction error, and compared against proposed acceptance criteria. Soil properties relevant to As solubility and bioaccessibility were determined; including total As, reactive Fe and Al, pH, organic carbon content and clay content. Solid phase speciation was determined using Extended X-Ray Absorption Fine Structure spectroscopy and linear combination fitting (LCF). Eighteen different As mineral phases were identified by LCF analyses, the most dominant phases were As (V) adsorbed to soil minerals such as ferrihydrite and goethite. Arsenic bioavailability and bioaccessibility ranged from 4 to 80 % and < 5 to > 90% respectively. Arsenic bioaccessibility was strongly correlated ($r^2$ >0.7) with RBA As, however the methods that most resemble human physiology were better at predicting RBA As according the proposed criteria. Results from this study will help support the adoption of methods to measure bioaccessible As that predict RBA As.

# MODELLING OF ARSENIC REMOVAL FROM GROUNDWATER BY ELECTRODIALYSIS

***Rose Marie O. Mendoza*** (Far Eastern University-NRMF, Institute of Medicine, Quezon City, Philippines)
Meng-Wei Wan and Chi-Chuan Kan (Chia Nan University of Pharmacy and Science, Department of Environmental Engineering and Science Tainan Taiwan, ROC)
Maria Lourdes P. Dalida (National Graduate School of Engineering, Department of Chemical Engineering, University of the Philippines, Diliman, Quezon City)

**ABSTRACT:** A model for Arsenic removal from ground water was evaluated. At varying experimental conditions, the second order homogeneous ordinary differential equation model proposed by Yu and Admassu on ion removal (from pulp and paper mill process stream) was able to predict the concentration, current utilization and removal of As ions from ground water samples efficiently. Two model parameters specifically for arsenic systems was derived: A is a parameter used to describe the characteristics of the solution, while $\lambda^2/B$ is a parameter that is related to the stack electrical field strength. Post hoc analysis of the model and $R^2$ values of 0.9899 and above was obtained for all parameters considered which indicate that the model can adequately describe the removal of arsenic from contaminated waters. The model was used to determine the effects of experimental parameters on Arsenic removal from ground water samples in terms of applied potential, feed flow velocity and initial feed concentration. A very satisfactory agreement between model predicted and experimental values of concentration, current utilization and removal efficiencies were obtained.

## INTRODUCTION

Arsenic-related incident covers almost majority of Asia. Bangladesh, Cambodia, China, India, Myanmar, Nepal, Pakistan, Taiwan and Vietnam are countries recently has been identified to use arsenic (As) contaminated ground water for irrigation. This poses serious health hazard to people eating food from the crops irrigated with arsenic contaminated water, in addition to the fact As accumulating in irrigated soils poses a serious threat to sustainable agriculture in affected areas (Brammer and Ravensoft, 2009). Bangladesh's As-contaminated groundwater has been extensively used to irrigate paddy rice, during dry season of 75% of the total area given to rice cultivation with arsenic levels of 4-8 μg/g and reaches to 83 μg/g (Rahman and Hasegawa, 2011). Because of the similar agro-ecological and hydrogeological conditions of these South East Asian regions the threat of arsenic contamination can easily reach neighboring countries. So far, arsenic contamination is widely detected to the alluvial plains of South and South East Asia where rice (paddy) is the main crop irrigated with ground water. According to Ravensoft (2007a,b), the GIS-based-geoogical-geochemical-hydrological models predict widespread pollution of ground waters in Indonesia, Malaysia and the Philippines – where water supplies were not yet tested for Arsenic.

Mendoza et al. (2014) in their study mentioned the viability of Electrodialysis (ED) as a process to remove inorganic As from contaminated waters. The study also stressed that experiments on arsenic systems should be carefully modelled and optimized to yield higher removal rates, while generating lesser waste by-products which can be more toxic than its source. Galvanin et al. (2015) opined that model-based design of experiment is a satisfactory approach to minimize the number of trials required for actual experimental runs.

To date, there is very less attention focused on model parameters for natural aqueous systems, particularly on arsenic removal and separation using electrodialysis (ED). The Nernst-Plank equation is one of the most utilized and widely-used model for a time-dependent transport and conversion on mass through the motion of charged chemical species in fluid medium (Kirby et al., 2010). Galvanin et al. (2015) utilized the Nernst-Plank approach to model the electrodialytic recovery of multi-valent electrolytes to accomplish desalination using NaCl while Zourmand et al. (2015) combined the Nernst-Plank model with that of Navier-Stokes model to develop a mass transfer model for the prediction of ion transport through electrodialysis (ED) cell using also NaCl solution. Mohammadi et al. (2005) as well as Sadrzadeh and Mohammadi (2007) provided a statistical model of metal ion removal from textile and semiconductor effluents using electrodialysis and electrodyliasis reversal by a Taguchi L9 OA design. It was only Mendoza et al. (2014) who used the electrodialysis (ED) process in removing arsenic from as-spiked water and come up with an optimum experimental conditions based on maximized As removal or separation. Thus, the burden of treating arsenic contaminated waters that will produce streams of higher arsenic concentration is one of the drawback of the study. The need to initially model arsenic systems to predict its removal from contaminated water will be of great contribution to the environmental sustainability of the said process. Through a model that will mimic the removal of arsenic from aqueous systems and with several related studies, lesser concentrated wastes will be generated which can gear the ED process towards zero-waste technology.

This paper investigates the removal of As from actual groundwater samples through the determination of the concentration and limiting current profile using the Yu and Admassu model (2000) for metal ions removal (from pulp and paper industry) and to come up with an model-generated predictions on arsenic removal at varied experimental conditions. The study also includes the model determination of the effect of experimental parameters such applied potential, feed flow velocity, initial feed concentration and operating time which is then later compared to the actual experimental data.

## MATERIALS AND METHODS

**Materials.** Groundwater samples used in this experiment were obtained from raw water source, well no. 9 and 10 of Taiwan Water Corporation's Water Treatment Plant in Beigang Township, Yunlin, Taiwan. Characteristics of the groundwater used for this study was presented in Table 1. The samples were filtered immediately through a 0.45μm pore size cellulose acetate syringe filter to minimize arsenic III oxidation with air. To further preserve arsenic species in the filtered sample, ethylenediaminetetraacetic acid (EDTA) was added immediately to the filtrate. This preserves the distribution of arsenic species by chelation to metal ions, prevents drastic changes in sample pH and minimizes microbial activity. Filtered samples were stored in PET containers at reduced head space and refrigerated at 4°C (Bednar et al., 2004).

**TABLE 1. Characteristics of the ground water used in this study**

| Cations | Concentration mg/L | Anions | Concentration mg/L |
|---|---|---|---|
| $Na^+$ | 145.50 | $Cl^-$ | 130.559 |
| $K^+$ | 34.515 | $NO_3^-$ | <0.1 |
| $Mg^{2+}$ | 18.97 | $SO_4^{2-}$ | 2.285 |
| $Ca^{2+}$ | 20.75 | $HCO_3^-$ | 119.70 |
| $Mn^{2+}$ | 0.093 | $PO_4^{3-}$ | 6.299 |
| $Fe^{tot}$ | <0.01 | TOC | 1.19 mg. C |
| $As^{tot}$ | 0.702 | | |

*Others:* $E_h = -162.2\ mV$; $pH = 7.68$; $T = 23°C$; $EC = 1388\ \mu S/cm$

*(Obtained from: Taiwan Water Corporation's Water Treatment Plan, Well no. 9 and no. 10, Beigang County, Yunlin, Taiwan)*

The ORP and pH of the solutions was monitored using Suntex pH-ORP meter model SP-300. Current and Voltage measurements were performed using a Hao Ling HL 833 digital ammeter and voltmeter, respectively. The total As, As III and As V concentrations and concentrations of cations (Na, Mg, Ca, K, Mn and Fe) of the feed, diluate stream and concentrate stream was determined using a Perkin Elmer 2000 Optima DV Inductively Coupled Plasma - Optical Emission spectrophotometer equipped with a hydride generator (HG-ICP-OES). Anion concentration in the form of nitrates, nitrites, phosphates, sulphates and chlorides, were analyzed using a Dionex DX-120 ion chromatograph (IC) equipped with a Reagent Free Controller (RFC), carbonates were obtained by titrimetric methods and TOC by direct method for drinking water and wastewaters (Method 10129) using Hach DR 500 UV-VIS Spectrometer. Conductivities of the solution were determined using Suntex EC-410 digital conductivity meter attached to the control panel.

**Methods.** The tailor-made electrodialysis (ED) provided by Shell Kwong Sir Enterprise Co., was used to conduct As removal experiments in the recirculating batch mode. The ED stack is composed of a single hydraulic and single electrical stack with 20 cell pairs situated 0.9 mm apart. Operating temperature for the whole system was set at 25°C ± 1°C. Maximum parameters for the ED stack as recommended by the manufacturer were indicated in the study of Mendoza et al. (2014). Optimum experimental settings such as applied voltage of 17V, initial feed concentration (Co) of 700 ppb, feed flow rate of 0.033 l/s and a minimum operating time of 92 minutes used by Mendoza et al. (2014) in their study on As-spiked water was considered for model experimental runs. Groundwater samples were pre-treated in the pre-oxidation tank to transform $As^{3+}$ to $As^{5+}$, then fed into the feed tank. From the feed tank, contaminated groundwater sample was introduced in the ED stack at an optimum feed flow velocity, and the sample was desalted in 92 minutes using 17V as the applied voltage. As the feed was introduced, the tap water as the electrode flush solution was circulated into the anodic and cathodic compartments. The concentrate stream was dumped into the electrode flush tank and eventually became part of the electrode flush solution.

The samples from the diluate stream were immediately digested with 5% potassium iodide (KI), 5% vitamin C and concentrated HCl. The mixture was set aside for 25 minutes for stabilization and diluted to its marked volume using Ultrapure 18.2 mΩ deionized water. A solid phase extraction membrane, Supelco LC-SAX strong anion resin was used to immediately separate As III from As V at an elution time of 1 minute (Jang et al., 2011). The total As, As III and As V of the sample was analyzed using a Perkin Elmer Inductively Coupled Plasma-Optical Emission Spectrophotometer (HG-ICP-OES) Optima DV2000 at 228.812 nm, using Hydride Generation. A reducing agent in the form of 5% sodium borohydride ($NaBH_4$) in 0.5% sodium hydroxide (NaOH) was used counter current with the sample to prevent further sample oxidation or metal complexation phenomena. Calibration standards of 1, 5, 10, 50, 100 and 700 ppb are freshly prepared during sample analysis. Two 2 sample replicates, check samples and standards are included to countercheck the accuracy and precision of the determination of the concentrations using HG-ICP-OES. $R^2$ for each analysis was established at 0.999 or higher. Cations and anion content as well as TOC and bicarbonate were also evaluated.

## RESULTS AND DISCUSSION

In a published paper in 2000 by Z.Yu and W. Admassu, a model for the concentration and current profile was evaluated and used in removing metal ion from pulp and paper mill process stream. The model describes the transport of metal ions in the boundary layers at membrane surfaces and the limiting current density as a function of low velocity and removal efficiency. The general transport equation for the molar flux (J) driven by diffusion, migration and convection in the ED channels is governed by the Nernst-Planck equation (Yu and Admassu, 2000):

$$J_{\pm} = -z_{\pm}v_{\pm}FC_{\pm}\nabla\emptyset - D\nabla C_{\pm} + C_{\pm}u \tag{1}$$

The model was anchored on the following assumptions: (1) the potential is a function of location due to a constant current supply, and the external applied potential is considered constant thus potential gradient will also be considered constant. Thus, equation (1) becomes:

$$\frac{\partial C_\pm}{\partial t} + u\nabla C_\pm = z_\pm v_\pm F\nabla \cdot (C_\pm + \nabla\emptyset) + D_\pm \nabla^2 C_\pm \quad (2)$$

The motion of the charged particles will bring current into the system. The current intensity for the solution is defined as i=FΣzJ which gives the equation for the aqueous solution as:

$$\mathrm{i}_\pm = -F^2\nabla\emptyset\Sigma z_\pm^2 v_\pm C_\pm - F\Sigma z_\pm \mathrm{D}_\pm \nabla C_\pm + Fu\Sigma z_\pm C_\pm \quad (3)$$

To simplify the model equations the following assumptions were made:(1) for a quasi-steady state the time variation of concentration and the diffusion term along the flow direction is negligible compared to convention; and (2) neutrality of charge is satisfied in the whole channel which makes $z_+C_+ + z_-C_- = 0$ and (3) the influence of the flow profile of fluid is negligible which allowed the system to be considered as a plug flow fluid. The dilute nature of the solution makes its viscosity as that of water which will now flow the schematic flow in Figure 1 (Yu and Admassu, 2000). This follow that:

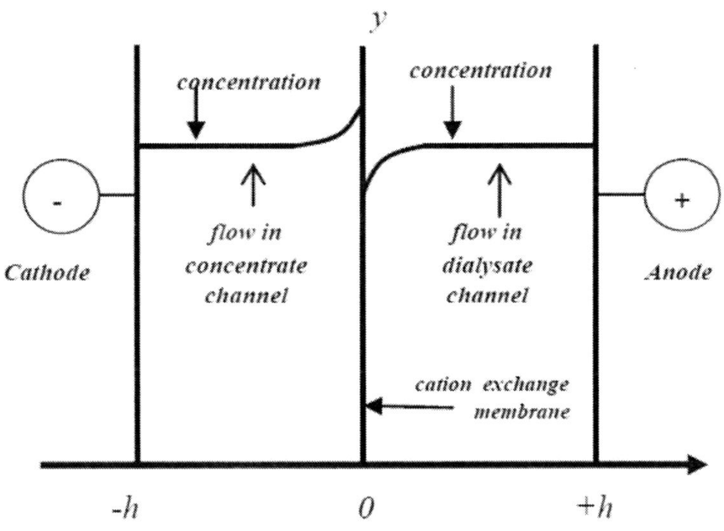


**FIGURE 1. Flow scheme of the solution in the ED stack**

Simplifying equation (3) and inserting it to equation (2) results in:

$$u\frac{\partial C_+}{\partial y} = D_+\frac{\partial^2 C_+}{\partial x^2} + z_+ v_+ F\frac{d\emptyset}{d\emptyset}\frac{\partial C_+}{\partial x} \quad (4)$$

For the current density and simplification of equation (3) is based on the assumption previously expressed considering a constant potential gradient within the stack. This makes the effect of convection in equation $uC_\pm$ negligible. The current in the anionic membrane $i_- = 0$ since the current density in the cation exchange membrane is the only one being considered in the model. Thus, the current in the cationic membrane is:

$$i_+ = -D_+ F z_+ \frac{dC_+}{dx} - \frac{F^2 z_+^2 D_+ C_+}{RT} \frac{d\emptyset}{dx} \quad (5)$$

and by principle of electroneutrality and non-ideality on membranes,

$$\frac{dC_+}{dx} = \frac{F z_+ C_+}{RT} \frac{d\emptyset}{dx} \quad (6)$$

Equation (2) is simplified by setting A = $u/D_+$ and another parameter as

$$B = {}^{z_+ v_+ F}/_{D_+} \cdot {}^{d\emptyset}/_{dx} \quad (7)$$

from a deduced dimensionless reduced concentration $C^* = C_w/C_0$, equation (2) becomes

$$A\frac{\partial c^*}{\partial y} = \frac{\partial^2 c^*}{\partial x^2} + B\frac{\partial c^*}{\partial x} \quad (8)$$

where $C_0$ is the initial concentration and $C_w$ is defned as the wall concentration in the boundary layers near the interface as shown in Figure 3 (Yu and Admassu, 2000). Analyzing Figure 2, the boundary conditions for equation (16) are set as: $x = 0$, $C^* = C_s$ ($C^*|_{x=0}$), x → $h$ ($x>x_a$), $C^* = C_h(C^*|_{x=h})$ and $y = 0$, $C^* = 1$ for $x > x_a$.

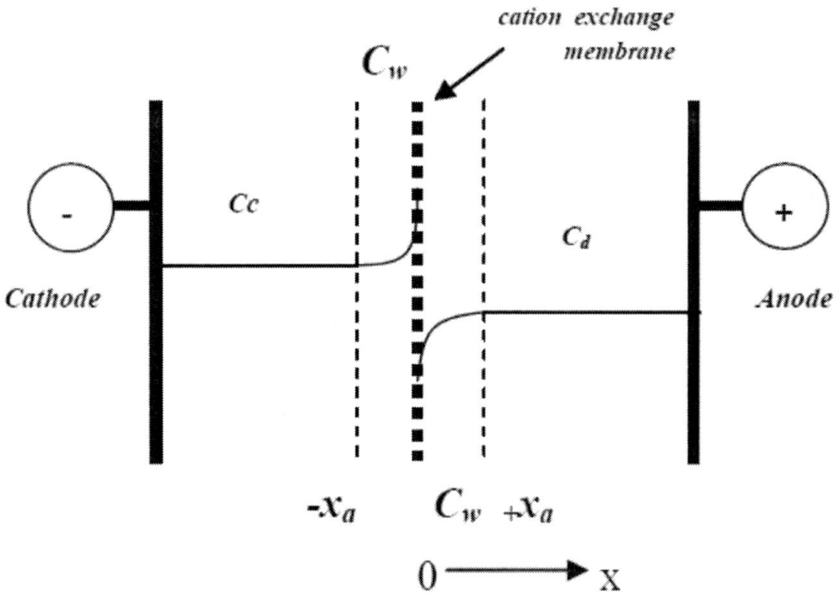


**FIGURE 2. Boundary layer used in this study for the ED process**

The transport of ions in this model assumes that it is occurs and only significant along the boundary layers, which is in the region where x is greater than 0 but less than $x_a$. From Figure 3 we can say that $x_a \ll h$ when $x > x_a$ and x → h, then x → ∞. This analysis of the boundary layers generates a general solution for equation (8) as:

$$C^*(x,y) = f(y)g(x) \quad (9)$$

which can be written as:

$$\frac{A}{f}\frac{df}{dy} = \frac{1}{g}\left(\frac{d^2g}{dx^2} + B\frac{dg}{dx}\right) \tag{10}$$

In the dialysate channel, the cation transport decreases along the flow direction (y). Introducing a constant $-\lambda^2$ to represent separation constant in the dialysate channel based on the boundary conditions set, equation (10) becomes:

$$\frac{df}{dy} + \frac{\lambda^2}{A}f = 0 \tag{11}$$

which is a second-order homogeneous ordinary differential equation with the form;

$$\frac{d^2g}{dx^2} + B\frac{dg}{dx} + \lambda^2 g = 0 \tag{12}$$

And an auxillary equation (Kreyzig, 2006):

$$K^2 + BK + \lambda^2 = 0 \tag{13}$$

Equation (13) has two roots:

$$K_{1,2} = -\frac{B}{2}\left\{1 \mp (1 - \frac{4\lambda^2}{B^2})^{1/2}\right\} \tag{14}$$

The general solution for Equation (12) (Kreyzig, 2006) is:

$$g(x) = N_1 exp(K_1 x) + N_2 exp(K_2 x) \tag{15}$$

Combining equation (11) and equation (13), the reduced concentration in the dialysate channel is:

$$C^*(x, y) = f(y)g(x) = \{m_1 \exp(K_1 x) + m_2 \exp(K_2 x)\}\exp(-\frac{\lambda^2}{A}y) \tag{16}$$

The integration constants $m_1$ and $m_2$ was determined using boundary conditions. For the purpose of simplification of equation (16) the following assumptions were made: $4\lambda^2/B^2 \ll 1$ and $\lambda^2/B \ll B$ so that $K_2 \approx -B$; at $y = 0$, $C^* = 1$ for $x > x_a$, x becomes large, exp (-Bx) → 0 faster than exp $(-\lambda^2/Bx)$ because $\lambda^2/B \ll B$. This yields,

$$m_1 = \exp(\frac{\lambda^2}{B}h) \tag{17}$$

and

$$m_2 = \frac{\exp\left(\frac{\lambda^2}{B}h\right)\left\{\frac{\left(\frac{d\emptyset}{dx}\right)_{m,d}}{\frac{d\emptyset}{dx}} + \left(\frac{RT}{z+F}\right)\left(\frac{\lambda^2}{B}\right)/\left(\frac{d\emptyset}{dx}\right)\right\}}{1+\left(\frac{d\emptyset}{dx}\right)_{m,d}/\left(\frac{d\emptyset}{dx}\right)} \tag{18}$$

and gives the concentration profile at any location in the ED dialysate and concentrate channel, respectively as :

$$\frac{C_W}{C_0} = \left[1 - \frac{\left(\frac{d\emptyset}{dx}\right)_{m,d} / \frac{d\emptyset}{dx}}{1+\left(\frac{d\emptyset}{dx}\right)_{m,d} / \frac{d\emptyset}{dx}} \exp(-Bx)\right] \exp\left(-\frac{\lambda^2}{A} y\right) \tag{19}$$

For the current density profile equations (5), (6) and (19) yields:

$$i_{w.d} = -2D_+ F z_+ C_0 \left[B \frac{\left(\frac{d\emptyset}{dx}\right)_{m,d} / \frac{d\emptyset}{dx}}{1+\left(\frac{d\emptyset}{dx}\right)_{m,.d} / \frac{d\emptyset}{dx}}\right] \exp(-Bx) \exp\left(-\frac{\lambda^2}{A} y\right) \tag{20}$$

Equation (19) and (20) is then simplified by integrating both sides with the boundary condition $C_s = C_w|_{x=0}$ which results to:

$$C_w - C_s = -\frac{i_+}{2D_+ F z_+} x_a \tag{21}$$

where $i_+$ is the current flowing in the opposite direction of $x$, F is the Faraday's constant (96,480 C/mol), $D_+$ diffusion coefficient ($cm^2/s$) of cation in the solution, z is the charge of ion and $x_a$ (mm) average thickness of the boundary layer. $C_w - C_s$ will be greater than 0 which is considered as the concentration of in the diluate compartment when boundary conditions $C_s = C_w|_{x=0}$ is considered. When $C_s \rightarrow 0$, $i_+ = i_{lim}$ and :

$$i_{lim} = \frac{2D_+ F z_+}{x_a} C_w|_{x\rightarrow 0} \tag{22}$$

From Kocherginsky (2010), equation (22) has a form:

$$i_{lim} = C_o e^{\frac{-u^*}{u}} \tag{23}$$

where $u^* = D_+ \lambda^2 y$ . Equation (23) indicates the limiting current density is a function of both initial concentration $C_o$ and flow velocity $u$. The magnitude of the parameter $\lambda^2$ is a function of initial concentration ($C_o$) removal efficiency (R%) and physical ED stack characteristics.

For equations (19), in the boundary layers, when $x \approx 0$, the concentration becomes independent of x as $x \rightarrow h$. This means that the significant removal or separation of ions mainly occurs at the membrane walls/surfaces. As a function of y, the decrease in the concentrate channel results to an increase in concentration of the concentrate channel. Due to the transport of ions in the boundary layer, the system will tend to polarize when $y = L$ as $x \rightarrow 0$. This makes the field strengths at membrane surface different in the cationic and anionic membrane surfaces. In this model, the field strengths are estimated using the continuity condition of electrical current at specified locations.

From equation (20), the value of current at entrance (y=0) is at maximum and decreases continuously as it approaches $y = L$. At the membrane surface (x=0), the current density also reaches the maximum value. In the boundary layer, the current density decreases as $x \rightarrow x_a$, beyond the boundary layer, $x \rightarrow h$ and $i \rightarrow 0$. This strongly suggests that concentration or concentration gradient is one of the factors that can influence current and current density.

From Ohms Law, the intensity of current is directly proportional to the electric field. As such there is a direct variation that can be observed between current density and the surface electrical field $(d\emptyset/dx)_{m,d}$ which in turn show a direct relationship between ion transport. This resulted in the direct proportionally between current density and initial concentration.

For parameter estimation of the model equations for concentration profile and current density at any location in the ED channel, R% is defined based on as a function of $\lambda$ is obtained as:

$$S\% = R\% = \frac{C_o - C}{C_o}\ x\ 100 = \left(1 - exp\left(-\frac{\lambda^2}{A}L\right)\right)x\ 100 \tag{24}$$

**TABLE 2. Typical values used for model calibration and model parameters for the ED process**

| Geometric Sizes | Symbol | Value | Source |
|---|---|---|---|
| Flow distance | L (mm) | 360 mm | This study |
| Flow width | W(mm) | 170 mm | This study |
| Half-ch,annel width | h (mm) | 5 mm | This study |
| Average film thickness | xa | 0.45 mm | This study |
| Number of cell pairs | | 20 | This study |
| *Characteristics of Feed Solution* | | | |
| Acerage flow speed | *u* (c,m/s) | 0.0033 | This study |
| Valence (as H2AsO4) | z | 1 | Clifford and Ghurye, 2005 |
| Initial Concentration | Co (ppb) | 700 | This study |
| Diffusion Coefficient | D (cm2/s) | $0.809x10^{-5}$ | Leaist (2007) |
| Density | ρ (g/cm3) | 1 | |
| Viscosity of water | μ( kg/ m s) | $900\ x\ 10^{-6}$ | Leaist (2007) |
| *Applied Conditions* | | | |
| Temperature | T (K) | 298 | This study |
| Constant electrical field | dθ/dx (vol/m) | 100 $kg\text{-}m/s^2C$ | Yu and Admassu (2000) |
| Removal Effieiciensy | E% | 99.56 | This study |
| *Physical Constant* | | | |
| Faraday Constant | F (C/mol) | 96480 | |
| Gas Constant | R (J/K mol) | 8.413 | |

Equation (20) is true only when 0 < (R%) < 100. An evaluation of the concentrate channel, though not so significant in this study must also be performed to facilitate transport equation facilitation. For the concentrate channel, the bulk concentration increases along the flow direction (Yu and Admassu, 2000). To compare the separation of concentrate to that of the dialysate channel, another separation constant in the form of ξ is defined related to the gain efficiency in the concentrate channel which leads to:

$$R\% = \frac{C_L - C_o}{C_o}\ x\ 100 = \left[\exp\left(\frac{\xi^2}{A}y\right) - 1\right]x\ 100 \tag{25}$$

and yields,

$$\xi^2 = \frac{A}{L}\ ln\left(1 + \frac{R'\%}{100}\right) \tag{26}$$

Non-ideality of membrane assumption ensures that there is no retention of cations at the membrane surfaces and in the membrane and indicate that the amount cations removed from the dialysate channel (R%) is equal to the amount of ions gained in the concentrate channel (R'%) which makes the magnitude of $\lambda^2 > 0$.

To estimate the magnitude of $\lambda^2/B$, model parameters based on typical values in ED process for KCl is used by the paper (Yu and Admassu, 2000), which gives an initial value for for Co and an average R% of 50 for batch ED process (Mc Neil and McCoy, 1989). This follows that:

$$\frac{\lambda^2}{B} = \frac{1}{L}\left|ln\left(1 - \frac{R\%}{100}\right)\right| \frac{uRT}{z_+ D_+ F(d\emptyset/dx)} = 0.018/\text{mm}. \tag{27}$$

$$B = \frac{z_+ v_+ F}{D_+}\frac{d\emptyset}{dx} = \frac{z_+ F}{RT}\frac{d\emptyset}{dx} = 3.823/\text{mm} \tag{28}$$

$$\frac{\xi^2}{B} = 0.011/\text{mm}. \tag{29}$$

For the model calibration, Table 2 presents the nown physical parameters of the ED stack as well as the physical and chemical properties of the solution and the membranes while to facilitate model parameter estimation, typical values for ED process in the table below is used.

**TABLE 3. Parameters obtained for Yu and Admassu Model**

| Varied Parameters | Correlation Coefficient, $R^2$ | $\lambda^2/B$ $mm^{-1}$ | A $cm^{-1}$ |
|---|---|---|---|
| **Initial Feed Concentration**[a] | | | |
| 200 ppb | 0.993 | 0.63690 | 42.0272 |
| 500 ppb | 0.966 | 0.67654 | 42. 0272 |
| 700 ppb | 0.981 | 0.73093 | 42.0272 |
| **Feed Flow Velocity**[b] | | | |
| 0.0025 m³/s | 0.994 | 1.30843 | 30.9024 |
| 0.0033 m³/s | 0.999 | 1.30843 | 42.0272 |
| 0.0042 m³/s | 0.999 | 1.30843 | 51.9160 |
| **Applied Potential**[c] | | | |
| 5V | 0.980 | 0.03133 | 42.0272 |
| 10V | 0.993 | 0.00084 | 42.0272 |
| 17V | 0.993 | 0.00004 | 42.0272 |

[a] *Apllied Potential (E)=3-20V; feed flow rate = 0.0033 m³/s; Time = 92 min.*
[b] *Apllied Potential (E)=17V; Initial feed concentration = 700 ppb; Time = 92 min.*
[c] *Feed flow rate = 0.0033 m³/s; Initial feed concentration = 700 ppb; Time = 92 min.*

The very good agreement between the experimental and model predicted values was dictated by the $R^2$ values indicated in Table 3 which was found to be greater than 0.98 on all parameters set for the study. Two model parameters were considered for this study; A is a parameter used to describe the characteristics of the solution, while $\lambda^2/B$ is a parameter that is related to the stack electrical field strength. The increasing values of $\lambda^2/B$ with respect to increase in initial feed concentration indicate efficient current utilization in the stack and efficient collision of ions during desalination within the stack channel while a constant value of A as the initial concentration increases indicates that ion transport is not limited to solution characteristics, rather it is more dictated by current and potential. Figure 3a shows that the removal of As ions from the feed for both experimental and predicted values are more efficient at applied potentials above 10V.

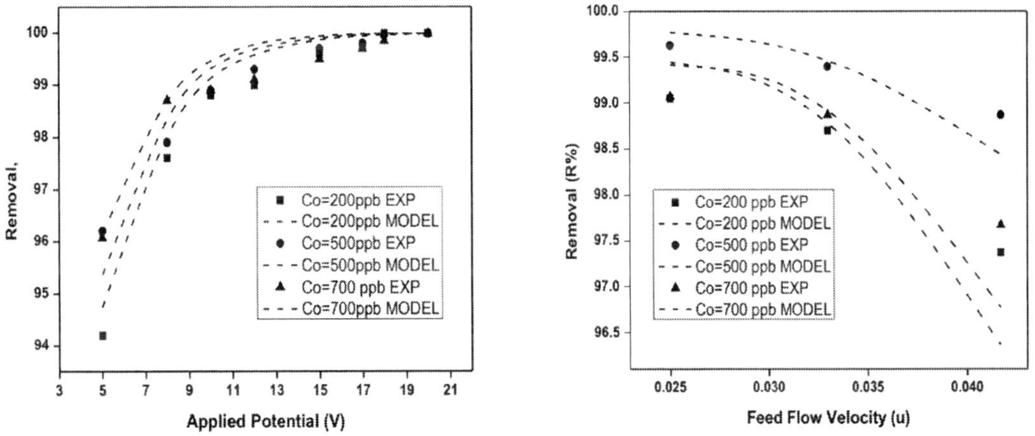


(a) (b)

**FIGURE 3. (a) Experimental and model predicted values on the effect of applied potential on As removal efficiency at variable concentrations; (b). The comparison between experimental values on the effect of feed flow velocity (u) to removal efficiency at varied initial feed concentration and constant applied potential of 15V and feed flow velocity of 0.033 l/s (2.0 l/min) and retention time of 92 minutes.**

**Effect of experimental parameters on Arsenic removal from groundwater using ED.** Model predicted values was compared with that of the experimental values in evaluating the effect of experimental parameters on arsenic removal. From Figure 3a model predicted values agrees satisfactorily with that of experimental values, confirming a direct relationship between applied potential %removal at variable concentration. The %removal was found to increase as the applied potential is increased at high or low initial feed concentrations. The same satisfactory agreement was found between % removal and feed flow velocity at high and low concentrations. However, an inverse relationship was found between %removal and feed flow velocity at varied concentrations as illustrated in Figure 3b. This can be due to the fact that the initial feed flow velocity was complemented by the applied potential resulting in a very fast transit of ions to the boundary layers, which indicates that in an ED system, very low or very high feed flow velocity are both non ideal conditions that may lead to neither increased energy consumption or polarization. Higher removal rate was found to at feed flow velocity of 0.033 l/s.

As observed from Figure 4, an increase in removal rate corresponds to decreases in current utilization. An increase in the electrical field strength parameter as applied potential is increased is a result of stack current utilization. The current in the stack is utilized to transport ions and co-ions across the membrane surfaces. As the ions are transported across, the ions in the diluate stream also decrease which results in the lesser current requirement. As shown in Figure 4, as the removal rates approaches the 90% and above region as current utilization approaches to minimum, and as expected electrical field strength is reduced. The reduction in the electrical field strength in this manner does not necessarily indicate polarization since it can also mark as the end of the desalination process. Both model generated vales and experiment a values accounts for such trend.

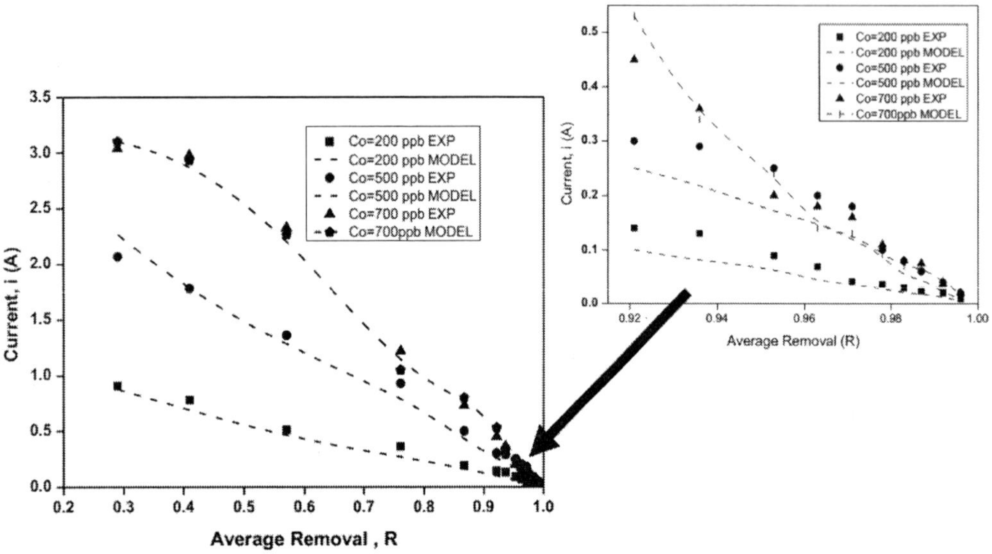


**FIGURE 4. The experimental and model predicted values of current utilization and removal efficiency at a constant applied potential of 17V and feed flow velocity of 0.033 l/s (2.0 l/min) at variable concentrations.**

## CONCLUSIONS

The Yu and Admassu model which was initially used to model the concentration and current density profile for the removal of metal ion from pulp and paper mill process stream was evaluated and found to satisfactorily predict removal of arsenic from contaminated water streams and determine the effect of experimental parameters on arsenic removal. Two model parameters was derived: A is a parameter used to describe the characteristics of the solution, while $\lambda^2/B$ is a parameter that is related to the stack electrical field strength. The increasing values of $\lambda^2/B$ with respect to increase in initial feed concentration indicate efficient current utilization in the stack and efficient collision of ions during desalination within the stack channel while a constant value of A as the initial concentration increases indicates that ion transport is not limited to solution characteristics, rather it is more dictated by current and potential. Results of this study has been found to support several literature on ED process indicating that removal rate increases as applied voltage and current density increases. In terms of initial feed concentration, the remarkably high removal rates from 500 – 700 ppb is brought about by the increase in ionic particles that intensifies the current in the stack. Medium feed flow velocity was know to be ideal for arsenic systems. The experimental data generated yielded 2 parameters than can be utilized for arsenic systems.

## ACKNOWLEDGEMENT

This paper acknowledges the Taiwan National Science Council for financially supporting this research under Contract No. NSC 101-2221-E-041-010-MY3 in collaboration with the Commission on Higher Education-Higher Education Development Program (CHED-HEDP), Republic of the Philippines.

## REFERENCES

Bednar, A.J., Gabarino, J.R.,Burkhadt, M.R., Ranville, J.F., Wildeman, T.R. (2004). "Field and laboratory arsenic speciation methods and their application to natural-water analysis". *Water Research* **38** (2). 355-364.

Brammer, H. Ravenscroft, P. (2009). "Arsenic in groundwater: A threat to sustainable agriculture in South and South East-Asia." *Environmental International*.**35**, 647-654.

Chekioua, A., Delimi, R. (2015). "Purification of $H_2SO_4$ by Pickling Bath by Fe (II) ions using Electrodialysis process." *Energy Procedia*. **74**, 1418-1433.

Galvanin, F., Marchesini, R., Barolo, M., Bezzo, F., Fidaleo, M. (2015). "Optimal design of experiments for parameter identification in electrodialysis models." *Chem. Eng. Res. Design.* **In press, accepted manuscript**, Available on line 12, November 2015.

Ghurye, G., Clifford, D.,(2004). "As(III) Oxidation using chemical and solid-phase oxidants." *J.AWWA*. **96**(1)84-96.

Kirby, B.J. (2010). "Micro- and Nanoscale Fluid Mechanics: Transport in Microfluidic Devices: Chapter 11:Species and Charge Transport" www.Kirbyresearch.com/index.cfm/wrap/textbook/ microfluidicsnanofludicsch11.html retrieved on line December 22, 2015.

Mendoza, RM. O., Kan, C-C., Dalida, ML. P., Wan, M-W. (2014). "Feasibility studies on arsenic removal from aqueous solutions by Electrodialysis." *J. Environmental Science and Health, Part A*. **49**, 545-554.

Mohammadi T., Moheb A., Sadrzadeh M., Rasmi A. (2005)."Modelling of metal ion removal from waste water by electrodialysis." *Separation Purification Technology,* **41**, 73-82.

Rahman, M.A., Hasegawa, H. (2011). "High levels of inorganic arsenic in rice in areas where arsenic-contaminated water is used for irrigation and cooking." *Science of Total Environment,* **409** (22), 4645-4655.

Sadrzadeh, M., Rami A., Mohammadi, T., (2007). "Separation of different ions from Waste water at various operating conditions using electrodialysis." *Desalination*. **54**,147-156.

Zourmand, Z., Faridirad, F., Kasiri, N., Mohammadi, T. (2015). "Mass transfer modelling of desalination through and electrodiaysis cell." *Desalination.* **359**, 41-51.

# ANAEROBIC BIODEGRADATION OF PARA-TOLUENE SULFONIC ACID, SULFANILIC ACID AND THIOPHENE-2-ACETIC ACID BY THE SULFATE REDUCING BACTERIUM *DESULFOVIBRIO PSYCHROTOLERANS*, STRAIN JS1$^T$ IN LIQUID CULTURES, SOIL AND SLUDGE MICROCOSMS

[1]Sasi Jyothsna, T.S. [1]Sasikala, Ch.* and [2]Ramana, Ch.V.

[1]Bacterial Discovery laboratory, Centre for Environment, Institute for Science & Technology, Jawaharlal Nehru Technological University Hyderabad, Hyderabad-500086, India
[2]Department of Plant Sciences, School of Life Science, University of Hyderabad, Hyderabad-500046, India
*E-mail: sasi449@yahoo.ie; sasikala.ch@gmail.com

**ABSTRACT:** A sulphate reducing bacterium *Desulfovibrio psychrotolerans,* strain JS1$^T$ could biodegrade three aromatic sulphur compounds namely p-toulene sulfonic acid (PTSA), Sulfanilic acid (SFA) and Thiophene – 2 – acetic acid (TPA) when provided as sole source of carbon under strict anaerobic conditions. The biodegradation of these compounds was studied using spectrophotometer and HPLC. The strain JS1$^T$ could grow in 25 mM and could tolerate (with feeble growth) up to 50 mM of all the three test compounds supplemented as sole carbon source however, optimum growth was at 3 or 4 mM concentration. Though none of the tested compounds were completely metabolised within the experimental period of three months, PTSA was degraded up to 82%, SFA up to 65.5% and TPA was degraded up to 72% in liquid culture. The soil and sludge microcosm studies conducted with a) the pure culture of strain JS1$^T$ alone and b) Strain JS1$^T$ along with native microbiota (as consortium) revealed that strain JS1$^T$ could degrade the test compounds more efficiently as pure culture when compared to that with its consortium. The role of native microorganisms in degradation of these compounds was insignificant. The biodegradation of PTSA and TPA was significantly reduced in sludge microcosm when compared to that in soil while, SFA degradation was similar in both soil and sludge microcosms the reasons for which are unknown. A 66%, 38% degradation of PTSA, 65%, 64% degradation of SFA and 72%, 31% degradation of TPA was noted in soil microcosm and in sludge microcosm respectively by the pure culture of JS1$^T$ . The present work in laboratory scale is a preliminary study conducted at ambient, natural conditions of soils and sludge which thus indicated the potential of *D. psychrotolerans*, strain JS1$^T$ for biodegradation of contaminated soils and sludge, which could be potentially useful in bioremediation of contaminated soils.

***Key words:*** Anaerobic biodegradation, Aromatic sulphur compounds, Microcosm studies, Sulfate reducing bacteria, *Desulfovibrio psychrotolerans*, para- toluene sulfonic acid, Sulfanilic acid, thiophene-2-acetic acid.
***Abbreviations:***SRB = Sulfate Reducing Bacteria, PTSA = para-Toluene Sulfonic Acid, SFA = Sulfanilic Acid, TPA = Thiophene-2-acetic acid.

## INTRODUCTION

Aromatic compounds form the second largest group of organic compounds in nature after carbohydrates (Shinoda *et. al.*, 2005). Hazardous aromatic compounds get into the environment in the form of diverse detergents, with oil spills, sewage from petroleum refineries and chemical plants, and with municipal waste waters. Many environments are anoxic or rapidly become anoxic due to contamination with carbon rich compounds like wastes from industrial effluents, gasoline, crude oil etc. Removal of such recalcitrant compounds becomes important as these potentially hazardous molecules may enter into drinking water supplies (El-Ashtoukhy *et al.* 2013).

As a major part of natural environment has little or no access to atmospheric oxygen, anaerobic microbes hold a major role in the processing of the nutrient cycles in nature and also in waste treatment plants where the aerobic processes may not completely remove aromatic compounds, turning researchers' interest to the study of the anaerobic metabolism of these compounds (Harwood *et. al.*, 1999). In this regard, sulfate reducing bacteria (SRB) have been extensively recognized and studied due to their ubiquitous distribution and capability in anaerobic biodegradation and biotransformation of a number of environmental pollutants under sulfate reducing conditions (Ensle and Suflita, 1995).

The aim of the present investigation was to study the ability of a sulfate reducing bacterium, *Desulfovibrio psychrotolerans*, strain JS1$^T$ (Jyothsna et al., 2008) to degrade three aromatic sulfur compounds namely para-toluene sulfonic acid (PTSA), sulfanilic acid (SFA) and thiophene-2-acetic acid (TPA) under strict anaerobic conditions. These pollutants enter into environment through effluents from textile, dye and chemical industries and petroleum products. The aerobic degradation pathway of toluene was demonstrated in *Pseudomonas testosterone* (Locher *et. al.*, 1989). A 40% degradation of sulfanilic acid under aerobic conditions by fungal strains *Phanerocheate chrysosporium* (Paszezynski *et. al.*, 1992) and *Aspergillus niger* RH19 (Faryal, 2006) were demonstrated. *Sphingomonas subartica* strain was reported to utilize sulfanilic acid as sole carbon, nitrogen and sulfur source indicating its degradation by the strain (Perei *et. al.*, 2001). Aerobic microbiological conversion of thiophenes has been studied extensively (Kobayashi *et. al.*, 2001; Hirasawa *et. al.*, 2001). However, very little information is available concerning the anaerobic conversion of these aromatic sulfur compounds.

Microcosm studies were conducted in sludge and soil microcosms to understand the degradation of test compounds under the influence of environmental and other biological interferences.. The present study demonstrating the degradation of aromatic sulfur compounds by *D. psychrotolerans*, JS1$^T$ in liquid culture and in microcosms is the first such study of degradation, by any pure culture of SRB under obligately anaerobic conditions.

## MATERIALS AND METHODS

**Growth Medium for Degradation Studies.** The pure culture of *D. psychrotolerans* strain JS1$^T$ was grown in Postgate's B medium (PBM) (Postgate, 1984) consisting of (gL$^{-1}$) $KH_2PO_4.7H_2O$, 0.5; $NH_4Cl$, 1.0; $Na_2SO_4$, 4.5; $CaCl_2.2H_2O$, 0.06, $MgSO_4.7H_2O$, 2.0, Yeast extract, 1.0, $FeSO_4.7H_2O$, 0.004, Sodium citrate, 0.3, Sodium Lactate, 3.5, Sodium ascorbate solution (1M), 1 ml and $Na_2S.9H2O$ solution (1M), 1ml. A one percent (v/v) inoculum of pure culture of strain JS1$^T$ washed twice in sterile saline, centrifuged and the culture pellet was inoculated into Postgate's B medium with the aromatic sulfur test compounds namely, para-toluene sulfonic acid [PTSA], Sulfanilic acid [SFA] and Thiophene 2-Acetic acid [TPA] supplemented as either sole source of carbon (3mM) replacing lactic acid or electron acceptor (0.5mM) replacing $FeSO_4$ respectively in 7.5 ml screw capped tubes incubated at 30$\pm$2°C. The culture was frequently sub cultured in the same medium and used for the biodegradation studies.

Turbidometry was used to monitor the growth of *D. psychrotolerans*, strain JS1$^T$. Increase in optical density (OD) (turbidity) was measured for every 24 h starting from time 0 h, up to 96 h. Optical density of the bacterial suspension was directly measured in a Systronics make (model 112) colorimeter at 540 nm (filter 7) against un-inoculated medium blank.

**Maximum Biodegradability and Biosorption Test.** Different concentrations of the test compounds (i.e. 0, 1, 2, 3, 4, 5, 10, 25 and 50 mM) were supplemented as sole carbon source replacing lactic acid. Growth as increase in optical density was measured colorimetrically at 24 h interval starting from time 0 h (i.e., 0, 24,48,72,96 and 120 h). For reading the degradation of the test compounds, two ml of the liquid culture was taken in 2.0 ml eppendorf tube, centrifuged at 8000 rpm for 15 minutes, 0.5 ml of the culture supernatant was diluted 10 times with deionized water. The diluted sample was measured for absorbance at the λ maxima of 229nm for PTSA, 280nm for SFA and TPA on a Spectron Genesys 2 Spectrophotometer.

Cells were simultaneously observed under phase contrast light microscope (Olympus BH-2) for viability from each test sample.

In order to see whether the aromatic sulfur compounds PTSA, SFA and TPA under test were being adsorbed or truly utilized by the strain $JS1^T$, live and dead (heat killed) culture/biomass of strain $JS1^T$ were separately inoculated into Postgate's B medium supplemented with 3mM of the test compound as sole carbon source and incubated at 30±2°C. Growth in terms of increase in OD was measured colorimetrically and concentrations of each of the test compounds were estimated by U.V. absorption at their respective absorption maxima of the test compounds.

**HPLC Analysis for Biodegradation of Test Compound.** The HPLC analysis of aromatic sulfur test compounds, PTSA, SFA and TPA in culture supernatants was performed at room temperature using a Shimadzu SPD-10AVP isocratic system. Luna 5μ $C_{18}$ (2) 100A column (250 x 4.6 mm) was used for the detection of metabolites in a UV-VIS detector. SFA and PTSA were detected in a solvent system containing methanol:Potassium Phosphate buffer 0.05M at pH 6.5 (40:60) at 1.0 $ml.min^{-1}$ flow rate with the detection done at 280nm, PTSA was detected in a solvent system containing methanol:water (40:60) at 1.0 $ml.min^{-1}$ flow rate with the detection done at 229nm. The retention times ($T_R$ in minutes) of SFA, PTSA and TPA were 2.6, 3.2 and 3.4 respectively in their specific solvent systems and flow rates as mentioned above.

The degradation/disappearance of three aromatic sulfur compounds PTSA, SFA and TPA by the strain $JS1^T$ was determined though HPLC at regular intervals from 0 hr, up to 15 days of incubation. Three months (90 days) old inoculated sample was also analyzed for each test compound.

**Soil and Sludge Microcosm Studies for Biodegradation.** The microcosm studies were carried out in soil collected from JNTU (Jawaharlal Nehru Technological University Hyderabad) campus and anaerobic sludge collected from the sedimentation tank of JETL (Jeedimetla Effluent Treatment Limited), Hyderabad. The sludge used in this experiment had a pH of 7.58, EC (Electrical conductivity) of 16.42 μSeimens, alkalinity of 440 mg/l, COD of 4200 mg/l (for 10 times diluted sample), TDS (Total Dissolved Solids) of 4308 mg/l. These parameters were analyzed according to the methods suggested in the standard methods for determination of water and waste water. The soil used had a pH of 7.2. The carbon, nitrogen, hydrogen and sulfur content in the soil were analzed by Elementar make CHNS analyzer comprising Vario Micro software. The soil contained 1.8% organic carbon, 0.2% hydrogen, 0.4% nitrogen and 0.02% sulfur and had a pH of 8.2.

The strain $JS1^T$ grown in PBM supplemented with 3 mM of the test compound as sole source of carbon, was centrifuged and the cell pellet was suspended in the PBM without any carbon source. This was used as inoculum in the soil and sludge microcosms spiked with the test compound. Microcosms were carried out in test tubes (25 X 150 mm) with 20g soil and/or 20 ml sludge spiked with one aromatic sulfur test compound in each at 3 mM concentration supplemented with 2 ml PBM without any carbon source in two sets of tubes. One set was autoclaved which indicated the degradation of test compound by the inoculated strain $JS1^T$ alone and another set was left unautoclaved to assess the degradation of the test compound by the native microorganisms if any. Within the autoclaved and unautoclaved sets of tubes, one set was inoculated with 2 ml of the strain $JS1^T$ and another was left uninoculated. The culture was thoroughly mixed with the soil and sludge and closed with rubber seal and flushed with Argon gas to maintain anaerobic conditions and incubated at 28 ± $2^0$C for 20 days. Soil and sludge samples (1g and/or 1ml) were drawn and analyzed for concentrations of each test compound at 5 days intervals.

For extraction of PTSA, SFA and TPA from soil and for sludge, 1g of soil or sludge was suspended in 10ml of deionized water and vortexed thoroughly followed by centrifugation at 10000 rpm for 10 min. The supernatant was extracted and diluted 10 times (10 μl supernatant in 90 μl milli Q water) and was analyzed by HPLC as mentioned in the previous segment.

## RESULTS

**Growth of $JS1^T$ on Aromatic Sulphur Compounds.** The strain $JS1^T$ grew in all media with the tested aromatic sulphur compounds with a gradual increase in growth within the tested period of 10 days.. Growth in terms of OD at 540 nm was maximum (1.98) in positive control where PBM (with 3 mM lactate as carbon source and 0.5 mM $FeSO_4$ as electron acceptor) was used and the growth was minimum (0.1 & 0.3) where PBM without any carbon or electron acceptor source was used respectively. The three test compounds were better utilized as carbon source (CS) rather than electron acceptor (EA) by the strain $JS1^T$ as evident from the result measured in terms of OD at 540 nm after 10 days of incubation as PTSA-CS (1.64) > TPA-CS (1.54) >SFA-CS (1.38) >PTSA-EA (0.97) >SFA-EA (0.90) >TPA-EA (0.63) (Fig. 1). Further experiments were carried out by supplementing these test compounds as sole carbon source.

The effect of increasing concentrations (from 0 mM to 50 mM) of each of the three aromatic sulfur test compounds, i.e., PTSA, SFA and TPA when given as sole carbon source on the growth of strain $JS1^T$ within a time course of 5 days showed that the strain $JS1^T$ grew optimally in all three test compounds at 3 or 4 mM concentration. Growth was feeble at and above 25 mM concentration of PTSA, while the strain could tolerate up to 50 mM concentration of SFA and TPA (Fig. 2).

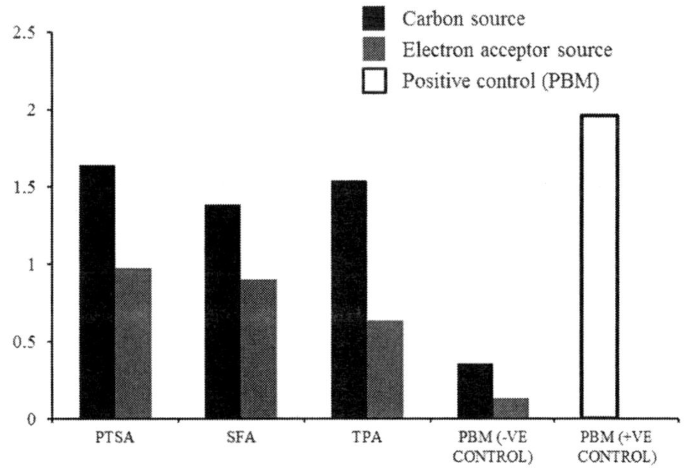


**Figure 1**. Growth of *Desulfovibrio psychrotolerans* strain $JS1^T$ on the three aromatic sulfur compounds supplemented as either sole carbon source or electron acceptor source with reference to the positive and negative controls

**Degradation of Aromatic Sulfur Compounds by Live and Dead Biomass of Strain $JS1^T$**

In the experiment conducted with live and dead biomass of $JS1^T$ to understand the loss of each test compound in the course of degradation, due to passive adsorption, if any, it was observed that the compound adsorbed was ignorable in all the three cases. There was no decrease in concentration of the test compounds inoculated with dead biomass of $JS1^T$ until 5 days of observation. Among the three test compounds, 42% of PTSA (Fig. 3-a), 47% of SFA (Fig. 3-b) and 49% of TPA (Fig. 3-c) were degraded after 5 days of incubation by the strain. None of the tested compounds were completely degraded by strain $JS1^T$ within 5 days of incubation. The compounds PTSA and SFA were gradually degraded up to day 5 while the degradation of TPA was rapid up to day 2 after inoculation and then the degradation slowed down gradually till the end of day 5 (Fig. 3).

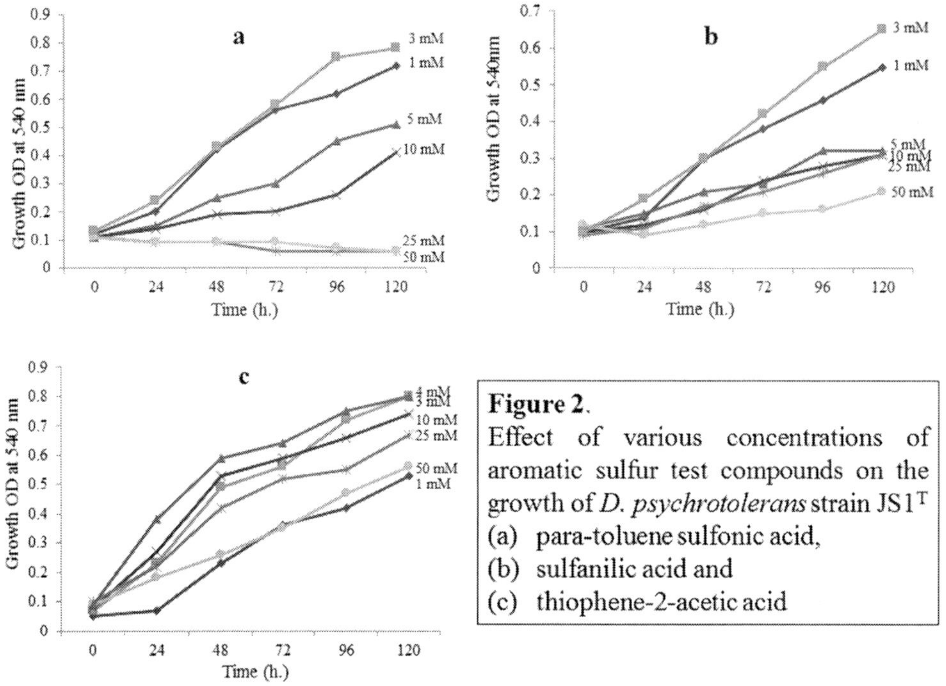


**Figure 2.**
Effect of various concentrations of aromatic sulfur test compounds on the growth of *D. psychrotolerans* strain JS1[T]
(a) para-toluene sulfonic acid,
(b) sulfanilic acid and
(c) thiophene-2-acetic acid

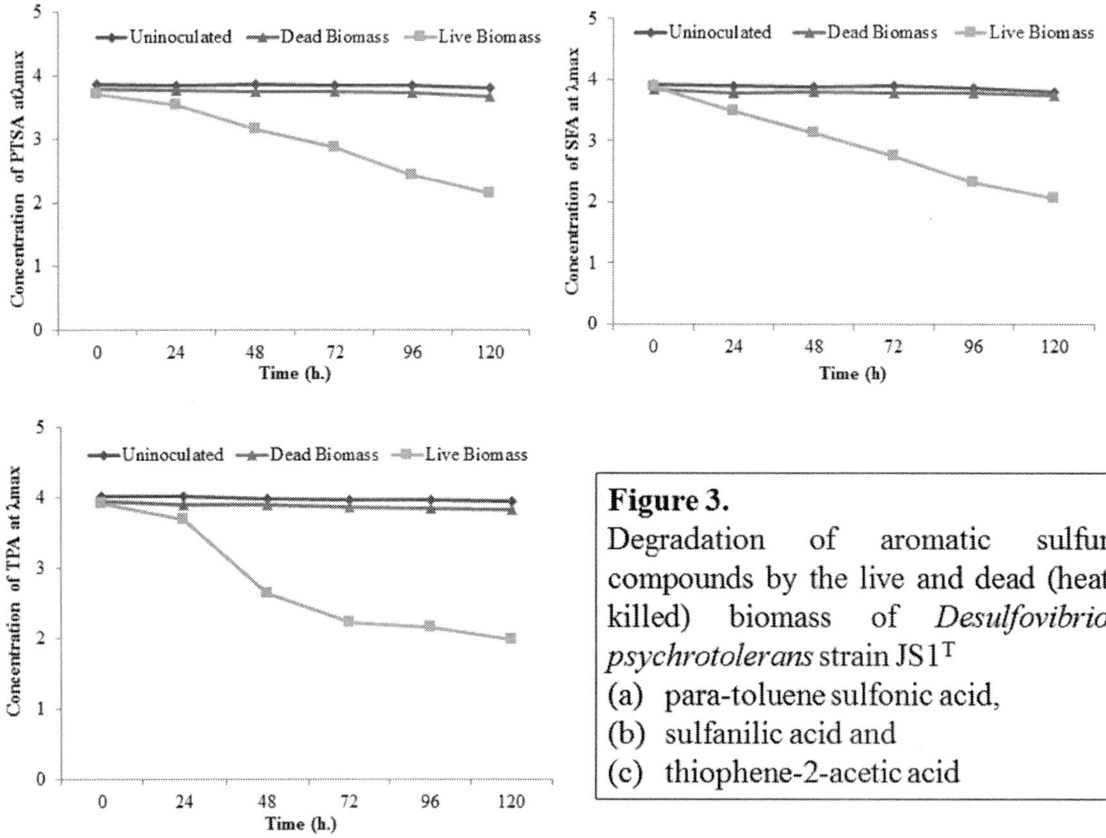


**Figure 3.**
Degradation of aromatic sulfur compounds by the live and dead (heat killed) biomass of *Desulfovibrio psychrotolerans* strain JS1[T]
(a) para-toluene sulfonic acid,
(b) sulfanilic acid and
(c) thiophene-2-acetic acid

**HPLC Determination of Degradation of Aromatic Sulfur Compounds.** The HPLC analysis of degradation of the aromatic sulfur test compounds by the strain JS1[T] showed a decrease of PTSA by 13%,

34% and 60% after 2, 4 and 15 days of incubation respectively and a maximum decrease of 83% after 90 days of incubation. Neither additional peaks nor peak shift was observed up to 15 minutes of run of the test compound in HPLC (Fig. 4-a). A decrease in concentration of SFA by 10%, 20% and 33% after 2, 4 and 15 days of incubation respectively and a maximum decrease of 75% were observed after 90 days of incubation. The decrease in the SFA peak ($T_R$ 2.6) was associated with an increase in an unknown peak ($T_R$ 3.0). But the increase in the unknown peak was not proportional to the decreasing SFA peak (Fig. 4-b). A decrease in concentration of TPA was observed by 11%, 23% and 47% after 2, 4 and 15 days of incubation respectively and a maximum decrease of 68% was observed after 90 days of incubation. The decrease in the TPA peak ($T_R$ 3.4) was associated with an increase in an unknown peak ($T_R$ 2.6). But the increase in the unknown peak was not proportional to the decrease in TPA peak (Fig.4-c). None of the test compounds were degraded completely within the tested period of three months.

**Sludge and Soil Microcosm Studies on Degradation of The Selected Aromatic Sulfur Compounds by *Desulfovibrio psychrotolerans*, JS1$^T$.** The degradation of aromatic sulfur compounds in sludge microcosms due to the individual and combined activities of *D. psychrotolerans*, JS1$^T$ and the native microorganisms (if any, present in the sludge sample) analyzed through HPLC observed after 10 days of incubation revealed a degradation of PTSA up to 38% by strain JS1$^T$ alone (Fig. 5-1a) and up to 47% by the combined activity of JS1$^T$ and native microbes of the sludge (Fig. 5-1b). SFA was degraded up to 64% by strain JS1$^T$ alone (Fig. 5-2a) and 58% by the combined activity of JS1$^T$ and native microbes of the sludge (Fig. 5-2b). TPA was degraded up to 31% by strain JS1$^T$ alone (Fig. 5-3a) and 48% degradation was observed by the combined activity of JS1$^T$ and native microbes of the sludge (Fig. 5-3b). However, no significant decrease in concentrations of PTSA, SFA and TPA was observed in uninoculated sludge samples either autoclaved or unautoclaved (Fig.5).

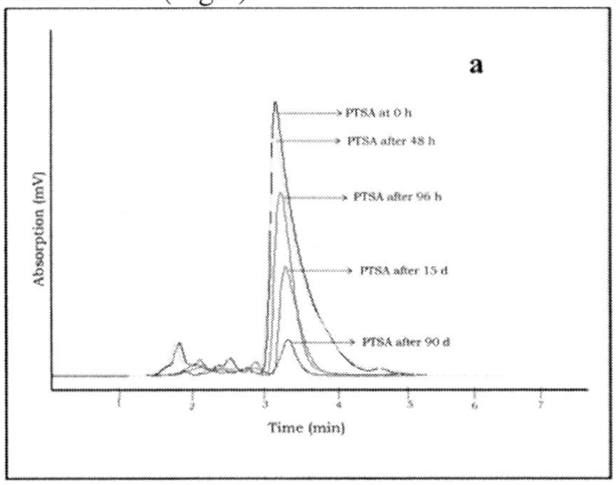


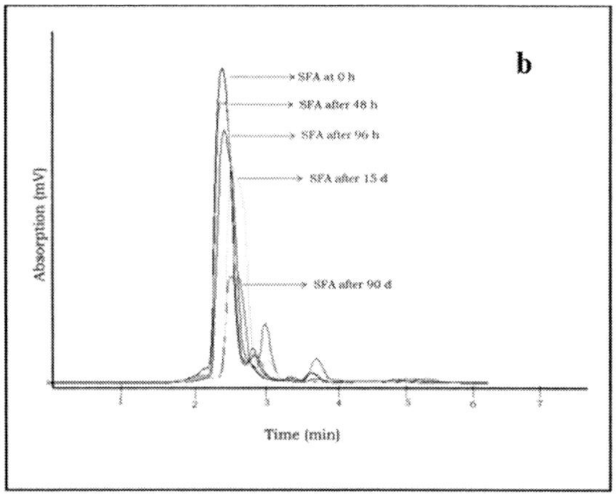


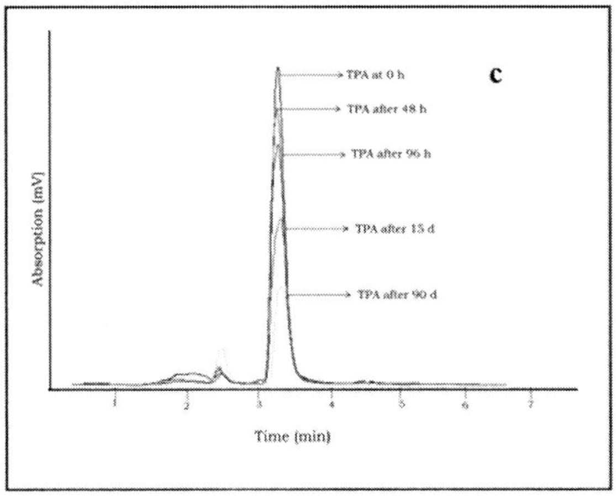


**Figure 4.**
Overlay of HPLC chromatograms showing the degradation of three aromatic sulfur compounds by *D. psychrotolerans* strain JS1$^T$
(a) para-toluene sulfonic acid,
(b) sulfanilic acid and
(c) thiophene-2-acetic acid

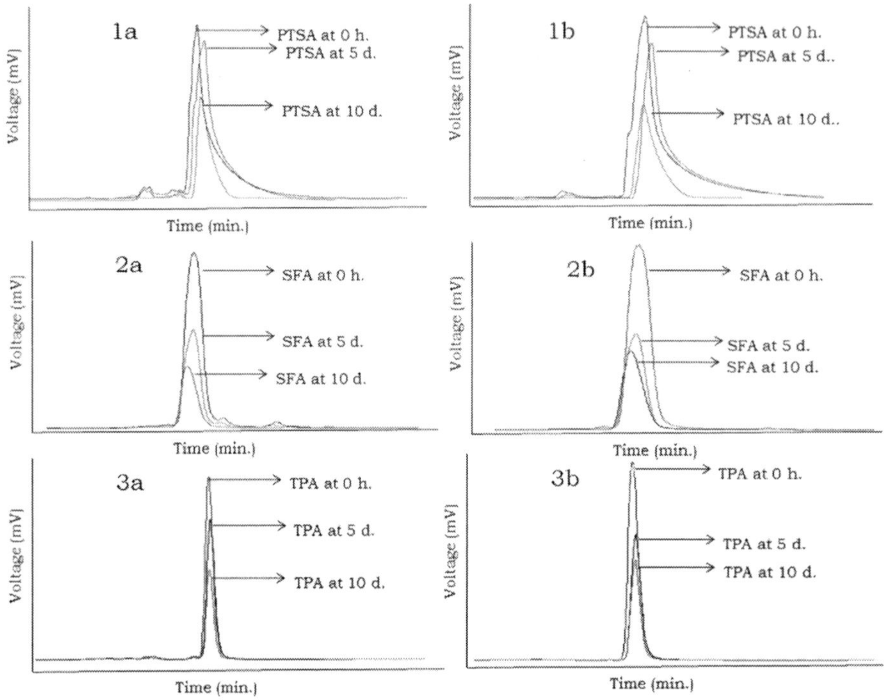


**Figure 5. Overlay of HPLC chromatograms showing the degradation of the three aromatic sulfur test compounds by *Desulfovibrio psychrotolerans* strain $JS1^T$ in sludge microcosms.**
(a) = autoclaved sludge inoculated with 2 ml log phase culture of $JS1^T$
(b) = un-autoclaved sludge inoculated with 2 ml log phase culture of $JS1^T$

Similarly, the degradation of aromatic sulfur compounds in soil microcosms revealed a PTSA degradation of up to 66% by strain $JS1^T$ alone (Fig. 6-1a) and 63% degradation by the consortium of $JS1^T$ and native microbes of the soil (Fig. 6-1b). SFA was degraded up to 65% by strain $JS1^T$ alone (Fig. 6-2a) and 40% degradation was observed by the consortium of $JS1^T$ and native microbes of the soil (Fig. 6-2b). TPA was degraded up to 72% by strain $JS1^T$ alone (Fig. 6-3a) while only 30% degradation was observed by the consortium of $JS1^T$ and native microbes of the soil (Fig. 6-3b). However, no significant decrease in concentrations of PTSA, SFA and TPA was observed in uninoculated soil samples either autoclaved or unautoclaved (Fig. 6).

## DISCUSSION

Sulfate reducing bacteria are phylogenetically and physiologically diverse group of bacteria, characterized by their versatile metabolic capabilities to use various electron acceptors and donors (Widdel, 1998). They are generally considered as the terminal oxidizers in the natural recycling of organic compounds to $CO_2$ in anoxic environments. Due to these exceptional capabilities, SRB are studied and applied in various bio-degradative tasks under anoxic regions (Widdel, 2007). In the present study, three test compounds were selected for biodegradation studies, namely para-toluene sulfonic acid (PTSA), sulfanilic acid (SFA) and thiophene-2-acetic acid (TPA). These pollutants enter into environment through effluents from textile, dye and chemical industries and petroleum products. These compounds were used as sole sulfur source or as sole carbon/electron donor source in many previous reports (Locher *et al.,* 1993; Yemashova, *et al.*, 2004 and Yamada *et al.*, 2001).

In the present study, the biodegradability of these compounds by strain $JS1^T$ was tested by supplementing them as either sole carbon source and/or sole elctron acceptor. Shcherbakova *et al.,* (2003) reported that a sulfate reducing strain *Desulfovibrio* sp. SR1 utilized PTSA as an electron acceptor. PTSA and SFA have reducible sulfur moiety in the chemical structure, and hence can also serve as electron acceptor in addition to being utilized as carbon source. TPA on the other hand was reported to undergo degradation in soil under oxygen limiting conditions when no other external electron acceptor was added (Anjana and Pranav, 2006) and dibenzothiophene served as an electron acceptor in anaerobic respiration in a report by Annweiler *et. al.,* (2001). Hence the degradation of TPA was also tested in the absence of any electron acceptor.

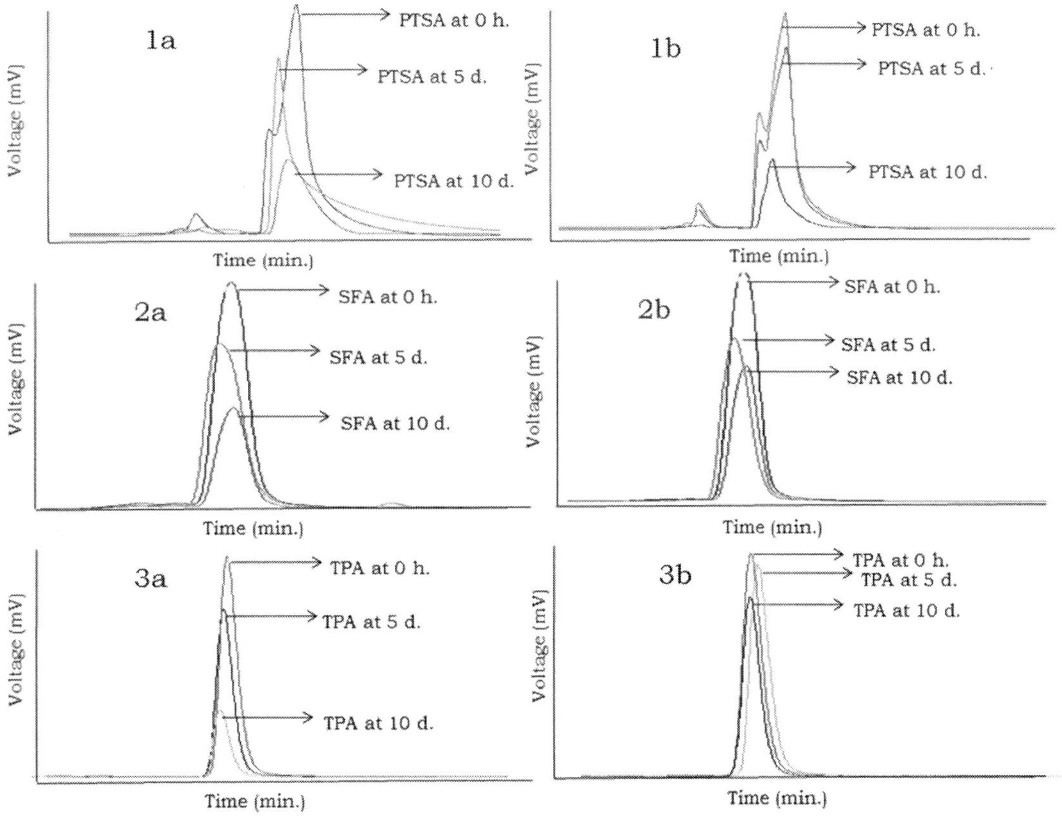


**Figure 6. Overlay of HPLC chromatograms showing the degradation of the three aromatic sulfur test compounds by *Desulfovibrio psychrotolerans* strain JS1$^T$ in soil microcosms**

(a)= autoclaved soil inoculated with 2 ml log phase culture of JS1$^T$

(b) = un-autoclaved soil inoculated with 2 ml log phase culture of JS1$^T$

*Desulfovibrio psychrotolerans*, JS1$^T$ was selected as the test organism for studying the degradation of aromatic sulfur compounds, as this was the only bacterium among 5 cultures tested, that could grow in a medium supplemented with the test compounds. Our efforts to enrich SRB degrading these compounds from 20 different environmental samples also were not successful (data not shown). The *D. psychrotolerans*, JS1$^T$ could grow in medium with the aromatic sulfur compounds supplemented as either sole carbon source or as sole electron acceptor. But as the strain JS1$^T$ could grow well in medium with these compounds supplemented as sole carbon source further degradation studies were conducted with the test compounds as carbon supplements. This result also supports the earlier report that aromatic compounds are metabolized by SRB much efficiently when they are supplemented as sole carbon source (Harms, 1993).

Hulshoff Pol, (1998) has reported that many SRB are quite resistant to overloads of certain organic compounds and to toxic upsets from aromatic compounds like alkanes, ethylbenzene, toluene, chloroform and other long chain fatty acids and can dominate in growth by competing with other anaerobic bacteria. A similar observation was made with respect to the tolerance of the test compounds by strain JS1$^T$. The *D. psychrotolerans*, JS1$^T$ was not sensitive to even very high concentrations of the test compounds. PTSA was tolerated up to a concentration of 25 mM while SFA and TPA were tolerated even up to the highest concentration of 50 mM tested compound. However, optimum growth was observed when the aromatic sulfur compounds were supplemented as sole carbon sources within a concentration of 3 and 4 mM. The pattern of degradation of the test compounds by *D. psychrotolerans*, JS1$^T$ was similar to that of its growth on these compounds.

A few SRB have been reported to adsorb certain metals and pollutants especially in activated sludge. (Byrns, 2001 and Machemer, 1992). Hence, it was essential to understand whether disappearence of the compound observed in the presence of *D. psychrotolerans*, JS1$^T$ was due to the biodegradation or simple passive adsorption. It was concluded that passive adsorption did not play a significant role since there was no significant loss (<5% loss) in the concentration of the compounds when inoculated with dead biomass (heat killed), while there was rapid decrease in their concentrtation when inoculated with live biomass within 5 days of incubation.

Though there are no reports of degradation of aromatic sulfur compounds by any SRB, *Desulfobacula toluolica* and *Desulfobacula phenolica* (Rabus *et. al.,* 1993), *Desulfosarcina cetonica* (Harms *et. al.,* 1999) have been demonstrated to degrade toluene completely to $CO_2$. The mixed cultures of SRB and other methanogenic bacteria were reported to degrade toluene. Degradation of PTSA under anaerobic conditions by mixed cultures was reported by Shcherbakova *et. al.,* (2003). Also, anaerobic degradation of dibenzothiophene (Yamada *et. al.,* 2001) and anaerobic desulfurization of benzothiophene and dibenzothiophene (Marcelis, 2003) have been reported in mixed cultures of SRB earlier, no reports exist on the anaerobic degradation of either PTSA, SFA or TPA by pure cultures of SRB. Hence, the present study on degradation of aromatic sulfur compounds by *D. psychrotolerans*, JS1$^T$ is the first such study of degradation by any pure culture of SRB. However, none of the aromatic sulfur compounds were completely degraded within the experimental period of three months (PTSA was degraded by 82%, SFA by 65.5% and TPA by 72%).

In general, laboratory studies are not accurate predictors of field degradation rates (Di et al., 1998). Bioremediation via environmental introduction of degradative microorganisms requires that microbes survive in substantial numbers and effect an increase in the rate and extent of pollutant removal in these natural habitats (Krumme *et. al.,* 1994). Measuring biodegradative activity and efficiency of the selected microbes in natural habitats is difficult due to limited accessibility of samples as well as sorption and abiotic transformation of contaminants (Margesin and Schinner, 1997). Due to these reasons, microcosm studies are generally used to understand these important parameters before switching on to on-site studies.

Microcosm studies are also known as the biofeasibility studies. Microcosms are artificial, simplified ecosystems that are used to simulate and predict the behaviour of natural ecosystems under controlled conditions (Kassen, *et al.,* 2000). These studies are useful in understanding the effect caused due to a pollutant or to determine the role of microbes in eliminating or reducing the effect of these pollutants in natural systems. The microcosm studies of degradation of pollutants by SRB conducted till now have mainly concentrated on degradation by mixed populations. An example is the degradation of toluene by SRB in oil contaminated soils (Noh *et. al.,* 2003).

In the present study the degradation of aromatic sulfur compounds and their subsequent degradation by pure culture of *D. psychrotolerans*, JS1$^T$ was carried out in both soil and sludge microcosms that mimic the naturally existing environments of contaminants. *D. psychrotolerans* JS1$^T$ was employed independently (in sterile soil and/or sludge) and also in consortium with the native microbiota (unsterile soil and/or sludge) inhabiting the soil and sludge samples. The viability of spiked pure culture of strain JS1$^T$ in soil and sludge microcosm could be checked by streaking the spiked samples on to PBM at regular time intervals during the experiment and observing for the pure colonies of the strain JS1$^T$.

The role of native microbiota alone in the degradation of the test compounds was insignificant as observed in the unsterile and uninoculated microcosms of soil and sludge spiked with the aromatic sulfur compounds. PTSA was degraded by strain JS1$^T$ much efficiently in soil microcosm (66%) than in sludge microcosm (38%). Degradation of PTSA by indigenous soil and sludge microbiota in consortium with JS1$^T$

was not very significant, but a slight increase of degradation to 47% (from 38%) in sludge and a slight decrease of degradation to 63% (from 66%) in soil were observed. The degradation of SFA by pure culture of JS1[T] was almost similar in sludge and soil microcosms (64 and 65% respectively) but the degradation in consortium with soil and sludge microflora was lowered to 58% (from 64%) in sludge and 40% (from 65%) in soil microcosms respectively. The degradation of TPA was the least in sludge microcosm (31%) while the highest in soil microcosm (72%). There was a significant decrease in degradation to 30% (from 72%) due to contribution of soil consortia and an increase to 48% (from 30%) due to contribution of sludge consortia.

The present work in lab scale is a preliminary study conducted at ambient conditions (pH, temperature, nutritional conditions etc.) of naturally occurring soils and sludge and thus indicates the potential of *D. psychrotolerans*, JS1[T] for bioremediation of contaminated soils and sludge.

## SUPPLEMENTARY

The cultures were grown in Postgate's B medium (PBM) supplemented with the aromatic sulfur compound para-toluene sulfonic acid (PTSA) or Sulfanilic acid (SFA) or Thiophene-2-acetic acid (TPA) as sole source of carbon (*CS) at 3mM concentration or electron acceptor ($^+$EA) at 0.5 mM concentration. $NH_4Cl$ (0.1% w/v) was used as nitrogen source and yeast extract (0.05% w/v) as a source of growth factor(s).

| S. No | Sulfate reducing bacteria used | Growth with PTSA | | Growth with SFA | | Growth with TPA | |
|---|---|---|---|---|---|---|---|
| | | *CS | $^+$EA | CS | EA | CS | EA |
| 1 | *Desulfovibrio psychrotolerans* JS1[T] | + | + | + | + | + | + |
| 2 | *Desulfotomaculum ruminis* JS3 | - | - | - | - | - | - |
| 3 | *Desulfovibrio dechloracetivorans* JS4 | - | - | - | - | - | - |
| 4 | *Desulfovibrio marinisediminis* JS8 | - | - | - | - | - | - |
| 5 | *Desulfovibrio carbinoliphilus* JS11 | - | - | - | - | - | - |

**Supplementary Table 1:** Utilization of aromatic sulfur compounds by Sulfate reducing bacteria

## REFERENCES

**Anjana, D. and Pranav, V. (2006).** Petroleum and hydrocarbon microbiology. In Applied Microbiology. 12-22.

**Annweiler, E., Michaelis, W. and Meckenstock, R.U. (2001).** Anaerobic cometabolic conversion of benzothiophene by a sulfate reducing enrichment culture and in a tar-oil contaminated aquifer. Appl. Environ. Microbiol. 67: 5077-5083.

**Byrns, G. (2001).** The fate of xenobiotic organic compounds in wastewater treatment plants. Water Res. 35: 2523–2533.

**Di, H.J., Aylmore, L.A.G., and Kookana, R.S. (1998).** Degradation rates of eight pesticides in surface and subsurface soils under laboratory and field conditions, Soil Science, 163: 404-411.

**El-Ashtoukhy SZ, El-Taweel YA, Abdelwahab O, Nassef EM** (2013) Treatment of petrochemical wastewater containing phenolic compounds by electrocoagulation using a fixed bed electrochemical reactor. Int J Electrochem Sci 8:1534–1550

**Ensley, B.D. and Suflita, J.M. (1995).** Metabolism of contaminants by mixed and pure cultures of sulfate-reducing bacteria. In. Larry, L. Barton., Tony Atkinson and Roger, F. Sherwood (eds.) Biotechnology Handbooks-8. Sulfate Reducing Bacteria. Plenum Press. New York. Pp. 294.

**Faryal, R., Ahmed, S and Hameed, A. (2006).** Biodegradation of 4-aminobenzene sulphonic acid by a local textile mill *Aspergillus niger* isolate Pak. J. Bot., 38: 1333-1340.

**Harms, G. (1999).** Anaerobic oxidation of *o*-xylene, *m*-xylene, and homologous alkylbenzenes by new types of sulfate-reducing bacteria. Appl. Environ. Microbiol. 65: 999–1004.

**Harwood, C.S., Burchhardt, G., Herrmann, H., and Fuchs, G. (1999).** Anaerobic Metabolism of Aromatic Compounds via the Benzoyl-CoA Pathway, FEMS Microbiol. Rev. 22: 439–458.

**Hirasawa, K., Ishii, Y., Kobayashi, M., Koizumi, K. and Maruhashi, K. (2001).** Improvement of desulphurisation activity in *Rhodococcus erythropolis* KA2-5-1 by genetic engineering. Biosci. Biotechnol. Biochem. 65: 239–246.

Hulshoff Pol, L.W., Lens, P.N.L., Stams, A.J.M., and Lettinga, G. (1998). Anaerobic treatment of sulphate-rich wastewaters. Biodegradation 9: 213-224.

**Jyothsna, T.S.S., Sasikala, Ch. and Ramana, Ch.V. (2008).** Desulfovibrio psychrotolerans sp. nov., a psychrotolerant and moderately alkaliphilic sulfate-reducing deltaproteobacterium from the Himalayas. Int. J. Syst. Evol. Microbiol. 58: 821-825.

Kassen, R., Buckling, A., Bell, G., Rainey and. Paul, B. (2000). Diversity peaks at intermediate productivity in a laboratory microcosm. Nature 406: 508–512.

**Kobayashi, M., Horiuchi, K., Yoshikawa, O., Hirasawa, K., Ishii, Y., Fujino, K., Sugiyama, H. and Maruhashi, K. (2001).** Kinetic analysis of microbial desulphurisation of model and light gas oil containing multiple alkyl dibenzothiophenes. Biosci. Biotechnol. Biochem. 65: 298–304.

**Krumme, M.L., Smith, R.L., Egestorff, J., Thiem, S.M., Tiedje, J.M., Timmis, K.N. and Dwyer, D.F. (1994).** Behaviour of pollutant-degrading microorganisms in aquifers: Predictions for genetically engineered organisms. Environ. Sci. Technolo. 28: 1134-1138.

**Locher, H.H., Leisinger, T. and Cook, A.M. (1989).** Degradation of p-Toluenesulphonic Acid via Sidechain Oxidation, Desulphonation and *meta* Ring Cleavage in *Pseudomonas (Comamonas) testosteroni* T-2. J. Gen. Microbiol. 135: 1969-1978.

**Locher, H.H., Poolman, B., Cook, A.M. and Konings, W.N. (1993).** Uptake of 4-toluene sulfonate by *Comamonas testosteroni* T-2. J. Bact. 175: 1075-1080.

**Machemer, S.D. and Wildeman, T.R. (1992).** Adsorption compared with sulfide precipitation as metal removal processes from acid mine drainage in a constructed wetland. J. Contaminant Hydrol. 9: 115-131.

**Margesin, R. and Schinner, F. (1997).** Efficiency of indegenous and inoculated cold- adapted soil microorganisms for biodegradation of diesel oil in Alpine soils. Appl. Environ. Micrbiol. 63: 2660-2664.

Noh, S.L., Choi, J.M., An, Y.J., Park, S.S. **and** Cho, K.S. **(2003).** Anaerobic biodegradation of toluene coupled to sulfate reduction in oil-contaminated soils: optimum environmental conditions for field applications. J. Environ. Sci. Health, 38: 1087-97.

**Paszezynski, A., Pasti-Grigsby, M.B., Goszezynski, S., Crawford, R.L. and Crawford, D.L. (1992).** Mineralization of sulfonated azo dyes and sulfanic acid by *Phanerochaete chrysosporium* and *Streptomyces chromofuscus.* Appl. Environ. Microbiol. 58: 3598-3604.

**Perei, K., Rákhely, G., Kiss, I., Polyák, B. and Kovács, K.L. (2001).** Biodegradation of sulfanilic acid by Pseudomonas paucimobilis. Appl Microbiol Biotechnol 55: 101-107.

**Postgate, J.R. (1984).** The sulphate-reducing bacteria, 2nd ed. Cambridge University Press, Cambridge, England.

**Rabus, R., Nordhaus, R., Ludwig, W. and Widdel, F. (1993).** Complete oxidation of toluene under strictly anoxic conditions by a new sulfate-reducing bacterium. Appl. Environ. Microbiol. 59: 1444-1451.

**Shcherbakova, V.A., Chuvil'skaya, N.A., Golovchenko, N.P. Suzina, N.E., Lysenko, A.M. Laurinavichyus, K.S. and Akimenko, V.K. (2003).** Analysis of the Anaerobic Microbial Community Capable of Degrading *p*-Toluene sulfonate. Microbiology. 72: 666-671.

**Shinoda, Y., Akagi, J., Uchihashi, Y., Hiraishi, A., Yukawa, H., Yurimoto, H., Sakai, Y. and Kato, N. (2005).** Anaerobic degradation of aromatic compounds by Magnetospirillum strains, Isolation and degradation genes. Biosci. Biotechnol. Biochem. 69: 1483-1491.

**Widdel, F. (1998).** Microbiology and ecology of sulfate – and sulfur-reducing bacteria. Biology of Anaerobic microorganisms. A.J.B. Zehnder (ed.). John Wiley and Sons, Inc. New York. 469-585.

**Yamada, K.O., Morimoto, M. and Tani, Y. (2001).** Degradation of dibenzothiophene by sulfate reducing bacteria cultured in presence of only Nitrogen gas. J. Biosci. Bioengineer. 91: 91-93.

**Yamashova, N., Telegina, A., Kotova, I., Netrusov, A. and Kalyuzhnyi, S. (2004).** Decolorization and partial degradation of selected azo dyes by methanogenic sludge. Appl. Biochem. Biotechnol. 119: 31-40.

# RELATTION AMONG TRIPHENYLTIN RECOGNITION, DEGRADATION, ION METABOLISM AND EFFECTIVE PROTEINS OF *BACILLUS THURINGIENSIS*

***Jinshao Ye*** and Linlin Wang
(School of Environment, Jinan University, Guangzhou 510632, Guangdong, China; Joint Genome Institute, Lawrence Berkeley National Laboratory, Walnut Creek 94598, CA, USA)

**ABSTRACT:** Triphenyltin (TPT) is one of the organometallic endocrine disruptors that pollutes the natural environment with its worldwide use. Thus far, information on its recognition, combination and transformation by effective proteins is limited and needs to be revealed. To this end, TPT degradation, cellular ion metabolism and proteomics of *Bacillus thuringiensis* were investigated. TPT was successively transformed to diphenyltin, monophenyltin and tin. During this process, $Na^+$, $NH_4^+$, $Mg^{2+}$ and $NO_3^-$ were released. The results of two-dimensional gel electrophoresis verified that proteins with primary and secondary structures in IEF IPG-strip gels could recognize TPT and form complexes. The up-synthesis of ribosomal proteins associated with protein translation, acyl-CoA dehydrogenase related to fatty acid metabolism, and succinyl-CoA synthetase subunit beta, succinyl-CoA synthetase subsunit alpha, ATP F0F1 synthase subunit alpha, glucose-6-phosphate isomerase and ferredoxin-NADP reductase involved in ATP synthesis were consistent with the results of the interaction between TPT and proteins in electrophoresis. This confirmed that effective proteins related to TPT transport, degradation and energy generation were also effective for TPT recognition even the active structures of these proteins were destroyed. The current findings developed a new approach for investigating interaction between proteins and target compounds, and presented a new insight into protein recognition of target compounds that the recognition by some effective proteins did not rely on their active quaternary structures.

## INTRODUCTION

Triphenyltin (TPT) is one of the endocrine disruptors that pollutes the natural environment with its worldwide use. Owing to its thermal stability, biodegradation is a major pathway accelerating the removal of TPT in the natural environments. Metabolite analysis has revealed its degradation pathway that TPT was initially triggered by benzene ring cleavage, which occurred individually and synchronously, producing diphenyltin (DPT), monophenyltin (MPT) and tin (Tang et al., 2016). Nevertheless, information concerning TPT binding and transformation by the effective proteins has remained scarce, which hinders the metabolic mediation of TPT biodegradation. To regulate TPT biodegradation and clarify the mechanism of TPT biosorption, transport and transformation, it is of critical importance to determine the relation among TPT recognition, TPT transformation and effective proteins.

As for protein identification, labeling-based mass spectrometry quantification technologies, including isotope-coded affinity tags, stable isotope labeling by amino acids in cell culture, and isobaric tag for relative and absolute quantification (iTRAQ), have been used to quantitatively analyze the proteomics in different organisms (Aebersold and Mann, 2003). To reveal the interaction between TPT and proteins at primary, secondary, tertiary and quaternary structures, both two-dimensional gel electrophoresis and iTRAQ technology were used to investigate the protein expression and identify the effective proteins involved in TPT recognition, transport and degradation in the current study.

Regarding the effective microbes for TPT degradation, *Bacillus thuringiensis* is a widespread species forming spore, which ensures the resistance of this species to a variety of stresses, including pollutant toxicity. The efficient degradation/transformation of some toxic xenobiotics, such as dimethyl phthalate (Brar et al., 2009), fipronil (Mandal et al., 2013) and uranium(VI) (Pan et al., 2015), by a portion of *B. thuringiensis* strains has been proved. A strain of this species isolated from organotins contaminated sediment samples has also been confirmed that could effectively cleave the carbon-tin covalent bonds of organotins in our previous studies (Tang et al., 2016). Therefore, *B. thuringiensis* was selected in the current study to reveal the relation about the proteomics and TPT degradation.

## MATERIALS AND METHODS

**Materials.** *B. thuringiensis* GIMCC 1.817 was an effective strain for TPT degradation and was stored at the Microbiology Culture Centre of Guangdong Province. TPT was purchased from Sigma–Aldrich. The concentrations of $KH_2PO_4$, NaCl, $NH_4Cl$, and $MgSO_4$ in the mineral salt medium (MSM) for TPT degradation were 30, 20, 30 and 10 mg $L^{-1}$, respectively.

**TPT biotransformation.** Twenty microliters of MSM containing 1 mg $L^{-1}$ of TPT and 1 g $L^{-1}$ of cells were cultured in the dark on a rotary shaker at 130 r $min^{-1}$ for 24 h. After degradation, the cells were separated and washed using pure water for protein extraction.

**Protein preparation.** The cells were suspended in 1 mL lysis buffer added with 0.2 g $L^{-1}$ phenylmethylsulfonyl fluoride, 2% v/v IPG buffer, 0.6 g $L^{-1}$ DTT, and vibrated for 30 s. The samples were frozen in liquid nitrogen thrice for 15 min each time, and subsequently treated by ultrasonication for 20 min. Nucleasae was added to the lysate at a final concentration of 1%. After incubation at 4 °C for 30 min, the cell debris was removed at 4 °C by centrifugation at 13500 r $min^{-1}$.

**Protein digestion, iTRAQ labeling and analysis.** Briefly, proteins from each sample were reduced with 10 mM DTT for 1 h at 37 °C. The cysteines were blocked with 1 μl blocking reagent for 10 min at room temperature. The protein samples were added in 10 KD Amicon Ultra-0.5 centrifugal filter devices, followed by centrifugation at 12000 r $min^{-1}$ for 20 min. After removal the liquid, the samples in filter devices were digested by 50 μl trypsin (Promega, V5280, USA) at 4% w/w overnight at 37°C. Subsequently, the samples were centrifuged, and 1 μg trypsin was added to each filter for 2 h. After centrifugation, liquid in the collection tube was collected.

The tryptic peptides were labeled with iTRAQ reagent multiplex kit (Sigma, PN 4352135, USA) according to the manufacturer's instructions. Subsequently, the labeled peptides were dried in a vacuum concentrator. The samples were then resolved with solution containing 2% v/v CAN and 0.1% v/v formicacid, centrifuged at 12000 r $min^{-1}$ for 20 min, and detected by an AB SciexTripleTOF 5600 mass spectrometer (AB Sciex, Framingham, MA, USA) equipped with a Nanospray III source.

**Two-dimensional gel electrophoresis.** Two-dimensional electrophoresis was performed to investigate the interaction between TPT and proteins with primary and secondary structures. After TPT degradation, the protein samples were diluted with hydration solution to the concentration of 400 g $L^{-1}$ added with 0.3 g $L^{-1}$ DTT, 0.5% v/v IPG buffer and 0.02‰ w/v bromophenol blue. The first dimension electrophoresis was conducted in isoelectric focusing (IEF) IPG-strip gels (13 cm, pH 4–7, GE Healthcare). Samples were first loaded by passive rehydration for 12 h. The IEF was run at 30 V for 12 h, 100 V for 1 h, 200 V for 1 h, 500 V for 1 h, 1000 V for 1 h, followed by a linear increase to and held at 8000 V until 32000 Vhr was obtained. Subsequently, the equilibrated strips were inserted on the top of 12.5% sodium dodecyl sulfate (SDS) separating polyacrylamide gel and covered with 0.5% w/v low melting point agarose. The gels were run in an electrophoresis system (Hoefer SE600, GE Healthcare) at a constant current of 5 mA $gel^{-1}$ for 30 min, and then at 15 mA $gel^{-1}$ until the marker dye reached the bottom of the gels. After in-gel digestion, the protein spots were identified.

## RESULTS AND DISCUSSION

**TPT Degradation**. To determine the dephenylation pathway, TPT, DPT and MPT at 1 mg $L^{-1}$ were degraded, respectively. Fig. 1a confirms that TPT was transformed to DPT and MPT, among which DPT reached its maximal concentration of 233 μg $L^{-1}$ at 0.5 d, followed by an almost linear decline curve. This finding illustrates that the generated DPT was further degraded effectively. However, a little of MPT was produced at 0.5 d, which indicates that MPT transformation was a rapid process, or the degradation of TPT and DPT to MPT was a slow period. The PTs curve shows that 107 μg $L^{-1}$ of TPT was transformed to tin at this interval. These results confirm that MPT dephenylation was not a rate-limiting step in TPT degradation. The decreasing curve of DPT concentration over time accompanied with MPT accumulation at 2–4 d illustrates that TPT dearylation occurred successively (Fig. 1a). The concentration of DPT (Fig. 1b) was higher than that of MPT (Fig. 1c) in the entire degradation process.

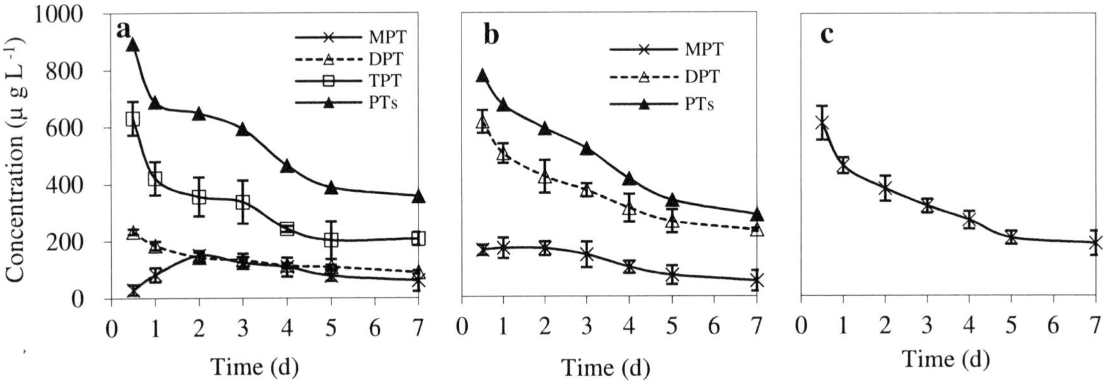


**Fig. 1** PT degradation by 0.3 g $L^{-1}$ *B. thuringiensis* for 0.5–7 d. (a) TPT degradation and its metabolites, (b) DPT degradation and its metabolite, and (c) MPT degradation.

**Ion metabolism.** The negative and positive values of ion concentration shown in Fig. 2 represent ion assimilation and release, respectively. $Na^+$, $NH_4^+$, $Mg^{2+}$, and $NO_3^-$ were released, whereas $K^+$ and $PO_4^{3-}$ were absorbed, but $Cl^-$ varied between uptake and release during the TPT degradation process. TPT biodegradation is a metabolic process that is certainly related to ion transport. Ion metabolism is primarily attributed to ion channel transport, membrane alteration, ion exchange and cell metabolism. TPT can alter the membrane structure due to the interaction between TPT and membrane components, triggering cell permeability. $Na^+$ is an essential cation for energy metabolism and ion homeostasis. Its flux is usually considered as a symbol of change in membrane permeability (Hite et al., 2015). Similar to ion exchange occurred during adsorbate binding (Ye et al. 2013), the release of $Mg^{2+}$ and $Na^+$ could balance the biosorption of PTs.

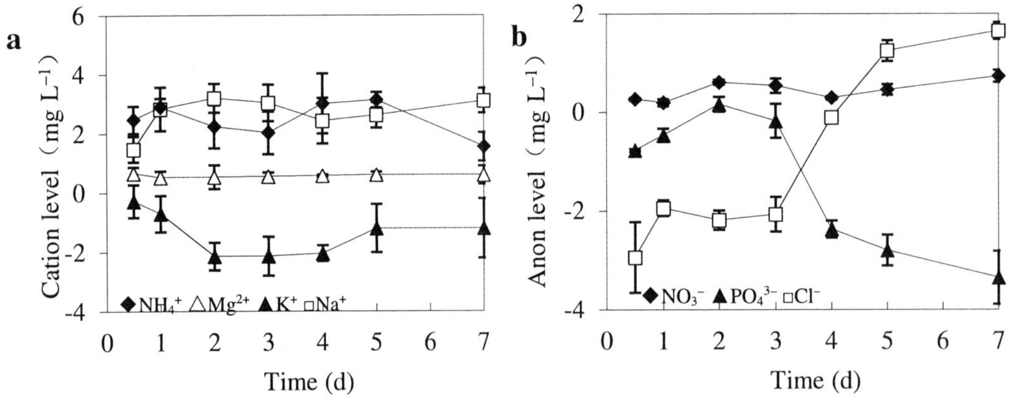


**Fig. 2** Change in (a) cation and (b) anion concentration in solution during degradation of 1 mg $L^{-1}$ TPT by 0.3 g $L^{-1}$ *B. thuringiensis*.

As a protein inhibitor, TPT depresses protein expression and triggers inactive protein degradation (D'Onofrio et al. 2013). $NH_4^+$ is a common intracellular ion and a component of proteins. Its release is related to both rose membrane permeability and protein degradation. $NH_4^+$ generated via protein and phospholipid photocatalysis was also gradually released by yeast (Thabet et al., 2013).

The accumulation of $K^+$ and $PO_4^{3-}$ might be related to their cellular functions (Fig. 2b). $K^+$ is involved in various metabolic activities, and is transported by $Na^+/K^+$–ATPase in the opposite direction as $Na^+$. $PO_4^{3-}$, an important anion for metabolism in all forms of life, takes part in various metabolic processes. Therefore, high–energy phosphate conversion could account for $PO_4^{3-}$ transport in the present study. The same direction of $K^+$ and $PO_4^{3-}$ flux during $PO_4^{3-}$ uptake by *Leishmania infantum* is consistent with the current findings (Russo–Abrahão et al. 2013). These findings obviously confirmed that the exposure to TPT is the responsible cause contributing to the increased release of ions.

**Interaction between TPT and Proteins with Primary and Secondary Structures.** To determine whether TPT recognition depended on the active structures of proteins, TPT at 1 mg $L^{-1}$ was added to the low melting point agarose. The immobilized nonlinear pH gradient strip separated proteins in the first dimension electrophoresis was located in front of this agarose. In contrast to proteins, TPT with small size in the migration process could move along the gel faster than the forward proteins. If a protein destroyed by the denaturing SDS-PAGE in the second dimension could recognize TPT, the generated complex would locate at different position in gel.

a

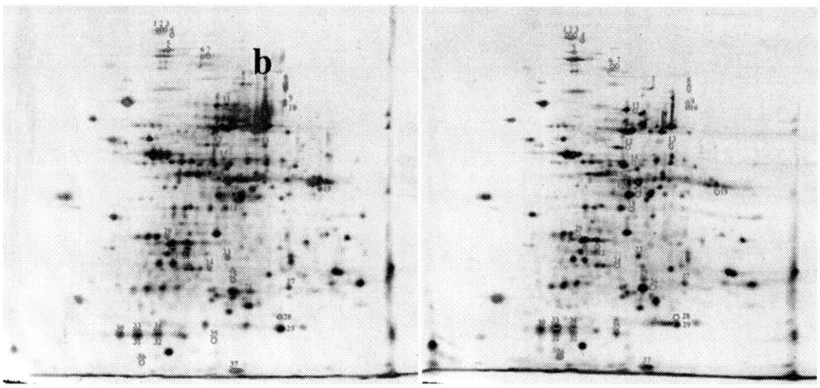


**Fig. 3** Interaction between TPT and inactive proteins during the vertical electrophoresis process. (a) control sample, and (b) experimental sample with TPT in a low melting point agarose gel.

Figure 3 exhibits that although the proteins in IEF IPG-strip gels were at inactive conditions, some of them in the 2-D electrophoresis process could recognize and combine with TPT, which revealed that pollutant recognition by some effective proteins did not rely on the active quaternary structures of proteins. The finding presents a new insight into protein recognition of target compounds. Protein identification and function category indicates that those proteins were primarily related to antioxidant defense (superoxide dismutase) and endogenous toxin transformation (glyoxalase), fatty acid metabolism and oxidation (propionyl-CoA carboxylase, enoyl-CoA hydratase, electron transfer flavoprotein subunit alpha), protein synthesis (amidotransferase, elongation factor Tu) and hydrolysis (ATP-dependent Clp protease), purine, nucleotide and fructose biosynthesis (adenylosuccinate synthase, ribose-phosphate pyrophosphokinase 1, fructose-bisphosphate aldolase). These results also infers that TPT degradation process tended to trigger the different expression of proteins associated with protein synthesis, pollutant degradation, membrane change and ATP synthesis.

**Interaction between TPT and active proteins.** The iTRAQ results show that 32 and 21 proteins were up- and down-regulated expression after TPT degradation. Among the up-regulation proteins, 14 of them enriched in a network with various functions (Fig. 4a), including biosynthetic process (GO:0009058), generation of precursor metabolites and energy (GO:0006091), nitrogen compound metabolic process (GO:0006807), phosphate-containing compound metabolic process (GO:0006796), primary metabolic process (GO:0044238) and cellular process (GO:0009987).

The coexistence of ribosomal proteins RpsB, RpsH, RplC and RPlS associated with protein translation (Fig. 4b); and the up-synthesis of acyl-CoA dehydrogenase related to fatty acid metabolism; and the up-expression of succinyl-CoA synthetase subunit beta, succinyl-CoA synthetase subsunit alpha, ATP F0F1 synthase subunit alpha, glucose-6-phosphate isomerase and ferredoxin-NADP reductase involved in ATP synthesis were consistent with the results of the interaction between TPT and proteins in two-dimensional gel electrophoresis, which confirmed that effective proteins related to TPT transport, degradation and energy generation were also effective for TPT recognition even the active structures of these proteins were destroyed by sodium dodecyl sulfate along with other reagents. Regarding the down-regulated synthesis proteins, none of them enriched in interaction network indicated that TPT degradation did not trigger significant inhibitory effects on cellular activities.

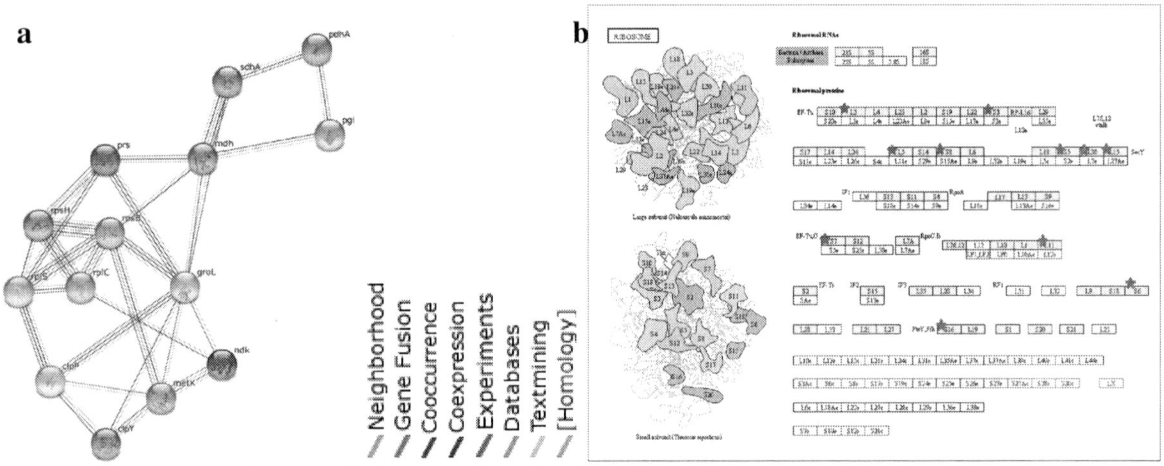


**Fig. 4** (**a**) Network of up-regulation proteins after TPT degradation, and (**b**) ribosomal proteins mapped onto KEGG pathway.

## CONCLUSIONS

TPT degradation process tended to trigger different expression of proteins associated with protein synthesis, pollutant degradation, membrane change and ATP synthesis. Effective proteins related to TPT transport, degradation and energy generation could recognize TPT, even their active structures were destroyed. The current findings developed a new approach for investigating the interaction between proteins and target compounds, and presented a new insight into protein recognition of target compounds that the recognition by some effective proteins did not rely on their active structures.

## ACKNOWLEDGEMENTS

This research was supported by the National Natural Science Foundation of China (Nos. 21377047, 21577049), and the Science and Technology Project of Guangdong Province (No. 2014A020216013).

## REFERENCES

Aebersold, R., and M. Mann. 2003. "Mass spectrometry-based proteomics." Nature 422:198-207.

Brar, S.K., M. Verma. R.D. Tyagi. J.R. Valéro, and R.Y. Surampalli. 2009. "Concurrent degradation of dimethyl phthalate (DMP) during production of *Bacillus thuringiensis* based biopesticides." J. Hazard. Mater. 171, 1016–1023.

Hite, R.K., P. Yuan, Z.L. Li, Y.C. Hsuing, T. Walz, and R. MacKinnon. 2015. "Cryo-electron microscopy structure of the Slo2.2 $Na^+$-activated $K^+$ channel." Nature 527: 198–203.

Mandal, K., B. Singh, M. Jariyal, and V.K. Gupta. 2013. "Microbial degradation of fipronil by *Bacillus thuringiensis*." Ecotox. Environ. Safe. 93: 87–92.

Pan, X.H., Z. Chen, F.B. Chen, Y.J. Cheng, Z. Lin, and X. Guan. 2015. "The mechanism of uranium transformation from U(VI) into nano-uramphite by two indigenous *Bacillus thuringiensis* strains.".J. Hazard. Mater. 297: 313-319.

Russo-Abrahão, T., M. Alves-Bezerra, D. Majerowicz, A.L. Freitas-Mesquita, C.F. Dick, K.C. Gondim, J.R. Meyer-Fernandes. 2013. "Transport of inorganic phosphate in Leishmania infantum and compensatory regulation at low inorganic phosphate concentration." BBA-Gen. Subjects 1830, 2683–2689.

Tang, L.T., L.L. Wang, H.S. Ou, Q.S. Li, J.S. Ye, and H. Yin. 2016. "Correlation among phenyltins molecular properties, degradation and cellular influences on *Bacillus thuringiensis* in the presence of biosurfactant." Biochem. Eng. J. 105: 71–79.

Thabet, S., M. Weiss-Gayet, F. Dappozze, P. Cotton, and C. Guillard. 2013. "Photocatalysis on yeast cells: toward targets and mechanisms." Appl. Catal. B–Environ. 140–141: 169–178.

Ye, J.S., H. Yin, D.P. Xie, H. Peng, J. Huang, and W.Y. Liang. 2013. "Copper biosorption and ions release by *Stenotrophomonas maltophilia* in the presence of benzo[a]pyrene." Chem. Eng. J. 219: 1–9.

# BACTERIAL COMMUNITY STRUCTURE CORRESPONDS TO PERFORMANCE IN A MICROAEROPHILIC SULFATE AND NITRATE CO-REDUCTION SYSTEM

Xi-Jun Xu, ***Chuan Chen***, Ai-Jie Wang, Duu-Jong Lee, and Nan-Qi Ren
(Harbin Institute of Technology, Harbin, Heilongjiang, China)

The succession of complex internal sulfur cycles and sulfide-oxidizing bacteria community has been observed during wastewater treatment under microaerophilic conditions or denitrifying conditions. However, current knowledge of the microbial ecology and physiology of sulfur cycles under microaerophilic denitrifying conditions is scarce, and as a result more research dedicated to understand the microbial community is required. Here we characterized the microbiology of bacterial communities stimulated by microaerophilic conditions. The results of our study indicated the superior performance of the microaerophilic reactor in terms of enhanced sulfur production and COD removal rates. We hypothesized that phylogenetic or structural features of the microbial communities could explain the enhanced performance of the microaerophilic reactor. We used Illumina sequencing with 16S rRNA gene to characterize the bacterial communities and identified *Bacteroidetes*, *Firmicutes*, *Proteobacteria*, *Spirochaetae* and *Synergistetes* members that were dominant in microbial communities. Specifically, our study revealed that it was the structural component, in terms of bacterial richness and evenness, rather than the phylogenetic affiliation of dominant bacteria, that best corresponds to reactor performance.

# A NEW AHL MOLECULE GENERATED BY NITRITE-OXIDIZING BACTERIA.

***Jie Gao*** and Guoqiang Zhuang
(Research Center for Eco-Environmental Sciences, Chinese Academy of Sciences, Beijing, China)

Revealing regulatory mechanism of nitrification is a crucial step to understand nitrogen conversion in agricultural systems and wastewater treatment. In this study, the gene *nwiI* of *Nitrobacter* was confirmed to be a homoserine lactone systhase by heterologous expression in *Escherichia coli*, which synthesized several acyl-homoserine lactone signals with 7 to 11 carbon acyl groups. A novel signal 7, 8-*trans-N*-(decanoyl) homoserine lactone (C10:1-HSL) was identified both in *Nitrobacter* and the recombined *E. coli*. Furthermore, this novel signal also triggered variance in nitrification rate and transcriptional level of genes involved in nitrification process. These results indicate that quorum sensing may play a potential role in nitrogen metabolism regulation.

# EVIDENCE OF QUORUM QUENCHING AND INHIBITION OF BIOFILM FORMATION IN *Sphingomonas Spp*

***Parul Gulati*** and Moushumi Ghosh
(Department of Biotechnology, Thapar University, Patiala-147004, Punjab, India)

*Sphingomonas* spp. is ubiquitous in the environment and can be frequently found in drinking water. Their amazing metabolic diversity, ability to survive in low nutrient conditions and oligotrophic niches qualifies its prevalence and biofilm formation. Recent reports indicate the biofilm forming capability of *Sphingomonas* in water distribution systems and therefore a hazard for human health necessitating intervention/controls. The purpose of this work was to obtain a polymeric material as a potential quorum quencher as a new strategy to control biofilm formation, through interference with the communication system. In the present study, we isolated 140 isolates of *Sphingomonas* spp. from twenty diverse municipal drinking water sites and characterized those capable of biofilm formation. Screenings of isolates were carried out by examining them on *Sphingomonas*- specific growth medium containing 100 and 50- μg/ml streptomycin and piperacillin, respectively. In addition, two sets of PCR primers targeting serine palmitoyl transferase gene (spt) were used (Yim *et al*, 2010). Growth and biofilm profiles of the positive isolate along with standard *Sphingomonas* strains were determined. N-acyl homoserine lactone (AHL) biosensors *Chromobacterium violaceum* CV026 and *Agrobacterium tumefaciens* A136 were used for detection of quorum sensing in *Sphingomonas* sp. For inhibition of quorum sensing, sublethal concentrations of quarternized polymer and standard 4-Nitropyridine -N-oxide were determined and used subsequently for quorum quenching studies. Results of our study suggest a potential of biofilm blocking by using the novel polymeric material and a possibility to further this approach for controlling *Sphingomonas* biofilms in drinking water systems.

# IDENTIFICATION OF FUNGAL ENDOPHYTES FROM MAIZE SHOOTS WITH BIOTECHNOLOGICAL POTENTIAL IN AGRICULTURE

Mashiane R.A and ***Adeleke R.A***
(Agricultural Research Council, Pretoria, Gauteng, SA)
Bezuidenhout C.C
(North-West University-Potchefstroom campus, Potchefstroom, North-West, SA)

Plant-microbe interaction plays an important role in plant health and development. Studies have indicated that almost all plants host endosymbionts. Therefore, a better understanding of endophytic fungi in maize may help to elucidate their function and potential roles more effectively in developing sustainable systems of crop production.

Our study investigated fungal endophytes associated with transgenic and non-transgenic maize shoots. Fungi were isolated from stem, leaves, tassels and seeds of both transgenic MON810 (expressing cry1Ab proteins) and non-transgenic isogenic maize lines. The fungal isolates were characterised based on sequence analysis of the internally transcribed spacer (ITS) region. This survey further allowed us to identify candidate endophytes for introduction into commercial maize cultivars for biomass enhancement.

A total of 21 and 27 isolates were obtained from transgenic Bt maize and non-transgenic isogenic maize, respectively. The fungal diversity of Bt maize did not differ significantly from that of non-transgenic maize. Taxonomically, species of the genera Epicoccum, Fusarium, Phoma, Alternaria and Nigrospora were common to both maize lines. Species of the genera Ampelomyces, Cladosporium, Pithomyces, Mucor and Acremonium were uniquely associated with Bt maize while Chaetomium, Peyronellaea and Trichoderma with non-Bt maize. The most frequently isolated group were Fusarium from the tassels, Epicoccum and Alternaria from the leaves.

The abundance of taxa in the shoots were assigned to the order Hypocreales to which mutualistic clavicipitaceous endophytes of grasses belong, suggesting great potential for utilising these endophytes for enhancing biomass production of this important agricultural crop.

# PECTINASE FROM BACILLUS SUBTILIS STRAIN BTK-27: OPTIMIZATION OF CULTURAL CONDITIONS AND APPLICATION IN COFFEE PROCESSING

***Oliyad Jeilu Oumer*** (Ambo University, Ambo, Ethiopia)
Dawit Abate (Addis Ababa University, Addis Ababa, Ethiopia)

The demand for enzymes in the global market is projected to rise at a fast pace in recent years. There has been a great increase in industrial applications of pectinase owing to their significant biotechnological uses. For this reason, this study was undertaken with aims of; isolating and screening microorganisms from coffee pulp, molecular identification of the potential pectinolytic isolates and optimizing pectinase production both on submerged fermentation and solid state fermentation. Moreover, characterizing the enzyme and proving its potential application in demucilisation of coffee. In the present investigation 95 isolates were isolated and screened for their pectinolytic activity from coffee pulp samples. The isolates with high potential pectinase activity were identified molecularly by sequencing 16s rDNA region of the isolates. The bacterium, Bacillus subtilis strain Btk27 was selected for production optimization of the enzyme based on its outstanding features. The production of pectinase was enhanced by 656% (in YEP) and 159% (in wheat bran) times on submerged fermentation and solid state fermentation, respectively. In addition, optimization steps were carried out to make the production of pectinase enzyme cost effective and commercially viable. The maximum pectinase production was observed using YEP (submerged fermentation) and Wheat bran (solid state fermentation) at initial pH of 6.5, at 37$^0$C and by supplementing the medium with 3mM MgSO4.7H2O. The maximum Pectinase activity was achieved at pH 7.5 and 50$^0$C. Also, the enzyme activity was found stimulated with Mg2+ and Ca2+ metal ions. Moreover, it was stable on EDTA, Trixton-100, Tween 80 and Tween 20. The enzyme, Km and Vmax values were identified as 1.879mg/ml and 149.6 U respectively. The potential application of the enzyme for coffee processing was studied, and it is found that complete removal of mucilage from coffee beans within 24 hours of treatment indicating the potential application in coffee processing.

# AUTHOR INDEX

# AUTHOR INDEX

# AUTHOR INDEX

# AUTHOR INDEX

# AUTHOR INDEX

**American Academy of Sciences**
9720 Town Park Drive, Suite 18
Houston TX 77036
USA

ISBN 978-1-5108-3972-4

CURRAN ASSOCIATES INC.
proceedings
.com

9 781510 839724

# Eighth International Conference on Environmental Science and Technology (ICEST 2016)

Houston, Texas, USA
6-10 June 2016

Volume 2 of 2

**Editors:**

**George A. Sorial** **Jihua Hong**

ISBN: 978-1-5108-3972-4

**Printed from e-media with permission by:**

Curran Associates, Inc.
57 Morehouse Lane
Red Hook, NY 12571

**Some format issues inherent in the e-media version may also appear in this print version.**



Printed by Curran Associates, Inc. (2017)



American Academy of Sciences
9720 Town Park Drive, Suite 18
Houston TX 77036
USA

Phone: 504-486-6332

info@AASci.org

**Additional copies of this publication are available from:**

Curran Associates, Inc.
57 Morehouse Lane
Red Hook, NY 12571 USA
Phone: 845-758-0400
Fax: 845-758-2633
Email: curran@proceedings.com
Web: www.proceedings.com

# TABLE OF CONTENTS

***On-site / Off-site Remediation***



***Landfill / Pemeable Reactive Barriers/***



***Waste Fuel Site Remediation / Waste Recycling***



***Radioactive Waste and Land Pollution***

### *Global Warming and Its Impacts*



### *Carbon Discharge Reduction*



## METALS

### *Metal Distribution*



### *Metal Removal and Remediation*



### *Speciation, Biovailability and Accumulation/ Phytoremediation*

## CHLORINATED AND OTHER PERSISTENT ORGANIC COMPOUNDS

### *Characterization of Organic Pollutants*



### *Degradation of Persistent Organic Pollutants*



## MODELING

### *Environmental Simulation*



### *Water Quality Modeling*

## GIS, DATABASE, AND REMOTE SENSING

### *GIS for Environmental Assessment / Data Management and Statistics / Environmental Remote Sensing Applications*



## ENVIRONMENTAL ANALYSIS AND MEASUREMENTS

### *Environmental Analysis*



### *Field Measurment Technologies*



### *New Method Applications*



### *Environmental Monitoring*

## SOCIETY AND THE ENVIRONMENT

### *Society and the Environment*



### *Environmental Ethics and Laws*



### *Environmental Education*



## ENVIRONMENTAL PLANNING AND MANAGEMENT

### *Environmental Quality and Planning*



### *Energy-related Environmental Problems*

**ADDITIONAL PAPER**

# LAND (SOIL, SOLID WASTE) POLLUTION AND REMEDIATION

# IMPROVING SOIL CHEMICAL PROPERTIES AND MAIZE GROWTH IN SOUTHEASTERN NIGERIAN ULTISOL USING SPENT GRAIN ASH

***Ogbonna Emmanuel Ahamefula*** (DOTECON Nigeria Limited, Lagos, Nigeria)[1]
OSODEKE, Victor Emmanuel (Michael Okpara University of Agriculture, Umudike, Abia, Nigeria)

**ABSTRACT:** A pot experiment was conducted at the Michael Okpara University of Agriculture, Umudike, Abia State, Southeastern Nigeria to determine the effect of spent grain ash on chemical properties of an acid Ultisol; its effect on the growth and nutrients uptake of maize. Six treatment levels of spent grain ash: 0 (control), 1, 2, 3, 4, and 5 ton/ha were applied and replicated three times in Completely Randomized Design (CRD) in a total of 18 pots. The effects of the treatments on soil pH, exchange acidity, exchangeable bases, available phosphorus, total nitrogen and organic matter were evaluated using standard laboratory procedures. Maize parameters measured were plant height, number of leaves, leaf area, number of roots, root lengths, stem diameter, moisture content and dry matter yield. The results showed that all the treatments significantly ($P<0.05$) increased soil pH and reduced exchange acidity, with 5ton/ ha of spent grain ash being more effective than the other treatments. For the plant parameters, 5 ton/ha of spent grain ash gave the highest significant ($P<0.05$) values except for leaf area. For soil parameters after planting, 5 ton/ha of spent grain ash significantly ($P<0.05$) increased soil pH and reduced exchange acidity. There was generally significant positive correlation between soil pH and plant parameters except for root length and dry matter weight. The study also showed that 5 ton/ha of spent grain ash was more effective in amelioration of soil acidity and enhancing the growth of maize, confirming its liming potential in an Ultisol of Umudike area of Southeastern Nigeria.

## INTRODUCTION

Soil acidity has been identified as a major problem to world food production due to the drastic reduction in crop yield as a result of depletion of basic cations (Tang *et al.*, 2003) and increase in hydrogen and aluminum ions (Mullins, 2005). Soils of Southeastern Nigeria are formed predominantly from the coastal plain sand parent materials, which are mainly Ultisols (Onwuka and Ogbonna, 2009). These soils are highly weathered and acidic in nature with a low pH, generally less than 5.5 (Osodeke, 2000). Consequently, soil acidity inhibits root growth, with the root apex being the primary affected part. It also plays a central role in aluminum and manganese toxicity and tolerance (Matsumoto, 2000).

In order to ameliorate these negative effects of acidity and ensure the sustainability of food basket in affected regions, liming practice is often employed to neutralize soil acidity and raise soil pH. Presently, attention is being shifted to the use of non-conventional liming materials and this is because it is cheaper and it is the best way of recycling such materials (Owolabi *et al.*, 2003).

The objectives of the study were to determine the effect of spent grain ash on some soil properties of an acid Ultisol and to determine the effect of spent grain ash on the growth of maize and nutrient uptake.

## MATERIALS AND METHODS

**Experimental site.** The experiment was conducted in Michael Okpara University of Agriculture, Umudike located between ($05^0\ 29^1$N; $07^0\ 33^1$E) with an altitude of 122m above the sea level (NRCRI Met. Station, 2012). The area is characterized by a humid tropical rainforest and climate with marked wet and dry season and an average rainfall of 2135.9mm per annum distributed over eight months period, ranging from March to November with bimodal peaks in July and September. The soil is deep, porous and dark in colour and derived from coastal plain sand thus classified as Typic kandiudult. Table 1 shows some initial soil properties. It was sandy loam texture with low CEC and pH.

**Field Method**. Augered surface (0-15 cm) samples were taken randomly from the study site and bulked together to form a composite. The composite soil sample was air dried at room temperature and passed through sieved with 2mm sieve then 50g were weighed into 12 liter capacity buckets each perforated at the bottom and properly labeled.

**TABLE 1. PHYSICAL AND CHEMICAL PROPERTIES OF THE SOIL USED FOR THE EXPERIMENT BEFORE THE APPLICATION OF LIME.**

| Physiochemical Properties | Values |
|---|---|
| Sand (g/kg) | 845.3 |
| Silt (g/kg) | 62.7 |
| Clay (g/kg) | 92.0 |
| Textural Class | Sandy loam |
| pH ($H_2O$) | 4.43 |
| pH ($CaCl_2$) | 3.72 |
| Organic Carbon (g/kg) | 18.0 |
| Organic Matter (g/kg) | 31.0 |
| Total Nitrogen (g/kg) | 1.50 |
| Aluminum Saturation (%) | 26.57 |
| Calcium Saturation (%) | 42.14 |
| Available Phosphorus (mg $kg^{-1}$) | 10.24 |
| Exchangeable Calcium (cmol $kg^{1}$) | 2.35 |
| Exchangeable Magnesium (cmol $kg^{-1}$) | 0.78 |
| Exchangeable Potassium (cmol $kg^{1}$) | 0.13 |
| Exchangeable Sodium (cmol $kg^{-1}$) | 0.10 |
| Exchange acidity (cmol $kg^{-1}$) | 2.00 |
| Exchangeable Aluminum (cmol $kg^{-1}$) | 1.50 |
| Exchangeable Hydrogen (cmol $kg^{-1}$) | 0.50 |
| ECEC (cmol $kg^{-1}$) | 5.36 |
| Base Saturation (%) | 62.69 |
| Ca:Al Ratio | 1.57 |

**Laboratory Methods.** The sieved samples were analyzed before and after the pot experiment. The particle size analysis was determined by the Bouyoucos (1962) hydrometer method using sodium hexametaphosphate. Soil pH was determined using Eel glass electrode pH meter (Mclean, 1982) while organic carbon was determined using dichromate wet oxidation method (Nelson and Sommer, 1982). Exchange acidity was determined by Kamprath (1976) method and total nitrogen was assessed by micro-kjeldahl digestion method as modified by Jackson (1996). Basic cations were determined by extracting 5g of soil with 30ml IN $NH_4OAC$. Potassium and Sodium were determined using flame photometer while calcium and magnesium were determined by titrating with EDTA (Tel and Hagarty, 1984). Effective cation exchange capacity was measured through the summation of exchangeable bases and acidity. The treatment, spent grain ash (by-product of brewery subjected to combustion) was analyzed before it application so as to ascertain its chemical properties, Table 2.

**TABLE 2. CHEMICAL PROPERTIES OF THE SPENT GRAIN ASH USED FOR THE EXPERIMENT**

| Properties Of The Spent Grain Ash | Values |
|---|---|
| pH ($H_2O$) | 9.21 |
| pH (CaCl) | 8.21 |
| % P | 0.27 |
| % OM | 3.64 |
| Ca (%) | 0.92 |
| Mg (%) | 0.24 |
| K (%) | 0.35 |
| Na (%) | 0.02 |
| N (%) | 0.32 |

## RESULTS AND DISCUSSION

Table 3 shows the effects of spent grain ash on some soil properties after harvesting the maize plant. The results showed that spent grain ash significantly ($P<0.05$) increased the soil pH in water when compared to the control. The results indicated that spent grain ash significant ($P<0.05$) reduced exchange acidity while five tonnes per hectare ash gave the highest reduction value of 0.90. For the exchangeable aluminum and exchangeable hydrogen, there were significant ($P<0.05$) differences among the treatments and five tonnes per hectare gave the lowest significant ($P<0.05$) values of 0.67 and 0.23 respectively. This result is an agreement with Onwuka *et al.* (2007) and Onwuka and Ogbonna (2009) who stated that the application of ash increased soil pH and reduced exchange acidity, exchangeable aluminum and exchangeable hydrogen. There was significantly ($P<0.05$) increased of soil organic carbon when compared to the control while 5 ton/ha gave the highest value of 5.32% organic carbon. The increase in the soil organic carbon could be attributed to the liming effect of spent grain ash, resulting in favoured environment for decomposing microbes (Badolucco *et al.* 1992). But, this result was in contrast to the findings of Voundi Nkama *et al.* (1998) who reported that lime application did not significantly increase soil organic.

Significant total nitrogen values above the critical level of 0.15% (Adeoye and Agboola, 1984) were obtained. This could be attributed to the increase in soil pH as a result of the treatments which in turn increase microbial activities and rates (Dee *et al*, 2003), especially the nitrifiers, which are involved in the transformation of organic nitrogen thus increasing the amount of nitrogen. Five ton/ha gave the highest significant value of 1.27 mg $kg^{-1}$ which could be ascribed to the large quantity of the treatments. This result was in agreement with Adu-Dupaal *et al.* (1994) finding, who reported that increasing the application rate of ash increases soil potassium.

**TABLE 3. EFFECT OF SPENT GRAIN ASH ON SOME SOIL CHEMICAL PROPERTIES AFTER MAIZE PLANT HARVEST IN THE POT EXPERIMENT**

| SPENT GRAIN ASH (TON/HA) | PH $H_2O$ | EX. ACIDITY | EXCH. AL | EXCH. H | OC ← | OM (%) | N → | P ← | K ($CmolKg^{-1}$) | NA → |
|---|---|---|---|---|---|---|---|---|---|---|
| 0 | 5.89 | 1.90 | 1.44 | 0.46 | 3.47 | 5.99 | 0.30 | 12.07 | 0.13 | 9.80 |
| 1 | 6.13 | 1.41 | 0.98 | 0.43 | 4.74 | 8.16 | 0.41 | 15.27 | 0.50 | 10.30 |
| 2 | 6.16 | 1.32 | 0.91 | 0.41 | 4.71 | 8.13 | 0.41 | 13.87 | 0.43 | 9.23 |
| 3 | 6.29 | 1.20 | 0.86 | 0.34 | 4.68 | 8.07 | 0.40 | 16.01 | 0.83 | 9.37 |
| 4 | 6.35 | 1.10 | 0.79 | 0.31 | 4.47 | 7.71 | 0.39 | 14.43 | 1.17 | 9.70 |
| 5 | 6.40 | 0.90 | 0.67 | 0.23 | 5.32 | 9.17 | 0.46 | 15.17 | 1.27 | 10.07 |

* Ex = Exchange; Exch. = Exchangeable; OC = Organic carbon; OM = Organic matter

The effects of spent grain ash on maize plant height in the pot experiment are shown in Table 4. The result indicated that the treatments significantly ($P<0.05$) increased the maize plant height when compared with the control. Five tonnes per hectare of spent grain ash application gave the highest significant ($P<0.05$) values of 40, 60.1 and 96.8 cm at 2, 4, and 6 WAP respectively. The highest significant ($P<0.05$) value of 110.9 cm was obtained from the treatment means at 8 WAP. This increase in the maize plant height could be attributed to the reduction of the soil exchange acidity by the spent grain ash. This finding was in agreement with what was earlier reported by Njoku *et al.* (1986), who indicated that liming increased the plant height in Umudike area of South East Nigeria.

The effect of spent grain ash on the number of leaves in the pot experiment is shown in Table 5. The results stated that at 4, 6 and 8 WAP, the values from all the treatments significantly ($P<0.05$) increased the number of leaves when compared to the control. Five tonnes per hectare of spent grain ash gave the highest values (16 leaves) across the treatments. This significant increase in the number of leaves of the maize plant could be attributed to the high content of nitrogen, above the critical level of 0.15% as indicated in the soil analysis after planting. This positive effect of nitrogen in the number of leaves could be ascribed to the increase in soil pH as a result of the application of various treatments, which in turn increase of microbial activities and rates (Dee *et al*, 2003).

Figure 1 shows the effects of spent grain ash on maize leaf area in the pot experiment. From the figure, there was no significant difference among the values obtained except at 8 WAP where all the treatments significantly ($P<0.05$) increased the leaf area when compared to the control. The control and 1 ton/ha of spent grain ash significantly (P<0.05) gave the lowest and highest significant values respectively and both treatments maintained this trend throughout the experimental period.

**TABLE 4. EFFECTS OF SPENT GRAIN ASH ON MAIZE PLANT HEIGHT IN THE POT EXPERIMENT AT 2, 4, 6 AND 8 WEEKS AFTER PLANTING (WAP).**

| SPENT GRAIN ASH (TON/HA) | PLANT HEIGHT (CM) 2 WAP | 4 WAP | 6 WAP | 8 WAP |
|---|---|---|---|---|
| 0 | 17.2 | 30.3 | 57.3 | 60.9 |
| 1 | 36.1 | 52.3 | 84.1 | 108.0 |
| 2 | 26.9 | 41.5 | 65.9 | 87.5 |
| 3 | 35.4 | 52.3 | 83.1 | 105.0 |
| 4 | 28.4 | 45.0 | 70.1 | 94.4 |
| 5 | 40.0 | 60.1 | 96.8 | 110.9 |
| LSD (0.05) | 11.85 | 16.38 | 16.63 | 19.18 |

**TABLE 5. EFFECTS OF SPENT GRAIN ASH ON THE NUMBER OF LEAVES IN THE POT EXPERIMENT AT 2, 4, 6 AND 8 WEEKS AFTER PLANTING (WAP).**

| SPENT GRAIN ASH (TON/HA) | NUMBER OF LEAVES 2 WAP | 4 WAP | 6 WAP | 8 WAP |
|---|---|---|---|---|
| 0 | 4 | 6 | 9 | 12 |
| 1 | 4 | 9 | 11 | 15 |
| 2 | 4 | 8 | 10 | 13 |
| 3 | 4 | 8 | 10 | 14 |
| 4 | 4 | 7 | 10 | 13 |
| 5 | 5 | 9 | 13 | 16 |
| LSD (0.05) | NS | NS | NS | NS |

*NS = No significant difference.

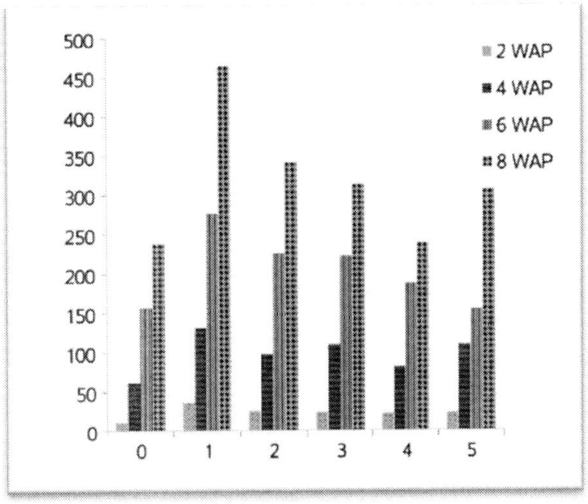


**FIGURE 1. EFFECTS OF DIFFERENT RATES OF SPENT GRAIN ASH ON MAIZE LEAF AREA IN THE POT EXPERIMENT**

Table 6 shows the effects of spent grain ash on plant growth and nutrients uptake in maize plant after harvest. The result indicated that spent grain ash significantly (P<0.05) increased the maize plant girth (cm) when compared with the control. This increase could be attributed to the potassium content of the spent grain ash, which according to Chude, *et al* (2004), is responsible for the development of the stalks.

For maize root number, spent grain ash significantly (P<0.05) increased the number of maize root when compared to the control. The result showed that both 1 and 2 ton/ha of spent grain ash gave the same number of maize root while the highest number of maize root was recorded in the 5 ton/ha treatment application. This could be attributed to the fact that the calcium in the spent grain ash suppressed the toxicity of aluminum in the soil (Franco and Munns, 1982), thus enhancing the activities of the root by creating a better environment for phosphorus release.

From Table 6, the result indicated that there were no significant differences among the values of maize root lengths obtained from the different treatments. Table 6 also indicated that the treatments significantly (P<0.05) increased both fresh weight yield (FWY) and dry matter yield when compared to the control. Three tonnes per hectare of spent grain ash gave the highest significant (P<0.05) values for both fresh weight yield and dry matter yield. The treatments significantly (P<0.05) increased the moisture content when compared to the control and 3 ton/ha of spent grain ash significantly (P<0.0) gave the highest value of 2.31% when compared to other treatments.

**TABLE 6. EFFECTS OF SPENT GRAIN ASH ON PLANT GROWTH AND NUTRIENT UPTAKE**

| Spent Grain Ash (ton/ha) | Girth (cm) | MRN | RL (cm) | FWY (g/pot) | DMY (g/pot) | MC (%) | N (%) | P (mg/kg) | K (mg/kg) |
|---|---|---|---|---|---|---|---|---|---|
| 0 | 1.38 | 16 | 18.96 | 64 | 7.2 | 0.57 | 27.94 | 94 | 31 |
| 1 | 1.9 | 27 | 17.6 | 232 | 17.8 | 2.14 | 111.07 | 422 | 106 |
| 2 | 1.35 | 27 | 16.68 | 168 | 13.7 | 1.54 | 75.90 | 311 | 104 |
| 3 | 1.66 | 23 | 17.29 | 253 | 22.9 | 2.31 | 92.52 | 508 | 168 |
| 4 | 1.48 | 22 | 18.18 | 204 | 19.2 | 1.85 | 103.10 | 496 | 147 |
| 5 | 1.95 | 29 | 19.35 | 236 | 21.5 | 2.15 | 73.96 | 745 | 179 |
| LSD (0.05) | 0.44 | 6.8 | 5.51 | 103.9 | 10.1 | 94.2 | 1.43 | 80.5 | 62.3 |

* Root length (RL); Fresh weight yield (FWY); Dry matter yield (DMY) Moisture content (MC); Root length (RL); Maize root number (MRN)

Table 6 showed the effects of spent grain ash on plant growth and nutrients (N, P and K) uptake in maize plant after harvest in the pot experiment. The results indicated that all the treatments significantly (P<0.05) increased the nitrogen uptake when compared to the control, 5 ton/ha of spent grain ash significantly (P<0.05) reduced the amount of nitrogen in the maize plant. This reduction of nitrogen in the maize plant could be as a result of the effective utilization of nitrogen in the pots by the maize plant thus giving rise to high vegetative growth of the maize plant. But this significant (P<0.05) increase in nitrogen could be attributed to the increase in soil pH as a result of the treatments which in turn increase the activities of microorganisms and bacterial rates (Dee *et al*, 2003), especially the nitrifiers, which are involved in the transformation of nitrogen thus increasing the amount of nitrogen.

Table 6 showed the effects of spent grain ash on plant growth and nutrients (N, P and K) uptake in maize plant after harvest in the pot experiment. The results indicated that all the treatments significantly (P<0.05) increased the nitrogen uptake when compared to the control, 5 ton/ha of spent grain ash significantly (P<0.05) reduced the amount of nitrogen in the maize plant. This reduction of nitrogen in the maize plant could be as a result of the effective utilization of nitrogen in the pots by the maize plant thus giving rise to high vegetative growth of the maize plant. But this significant (P<0.05) increase in nitrogen could be attributed to the increase in soil pH as a result of the treatments which in turn increase the activities of microorganisms and bacterial rates (Dee *et al*, 2003), especially the nitrifiers, which are involved in the transformation of nitrogen thus increasing the amount of nitrogen.

From Table 6, all the treatments significantly (P<0.05) increased the plant phosphorus uptake. Five tonnes per hectare of spent grain ash significantly (P<0.05) gave the highest value of phosphorus uptake. It is obvious that there is high phosphorus fixation at low pH by aluminum (Osodeke, 2000), resulting in the formation of aluminum-phosphate complex that is insoluble and unavailable to plants. This increased in phosphorus could be attributed to the application of liming material which increased the soil pH, thus increasing phosphorus availability.

For the potassium, spent grain ash significantly (P<0.05) increased the soil potassium uptake when compared to the control. Five tonnes per hectare of spent grain ash significantly (P<0.05) gave the highest value of 179 mg $kg^{-1}$ of the potassium uptake. This increase in potassium uptake could be as a result of liming effect of the spent grain ash which was in agreement with the findings of Comin *et al.* (2006).

$$2Al^{3+} + 2H_2PO_4^{2-} + 2H_2O \longrightarrow 2Al(OH)_2H_2PO_4 + 2H^+$$

The correlation matrix showing the relationship between soil chemical properties and maize growth parameters is shown in Table 7. The result showed that soil pH correlated positively and significantly with almost all the maize growth parameters except root length, fresh weight yield (FWY) and dry matter weight. Table 7 also indicated highly significant and positive correlation between soil pH, plant height and number of leaves. There was no correlation between soil acidity indices (exchange acidity, exchangeable aluminum and exchangeable hydrogen) and the maize growth parameters. Total nitrogen gave a positive and significant correlation with some maize growth parameters except for plant girth and root length. For the available phosphorus, there was no significant correlation with the maize growth parameters except with the number of leaves (where the correlation was highly significant (P<0.001)), plant height and plant girth. Potassium gave a positive and significant correlation with plant height and Dry matter yield and only exchangeable sodium was positively and significantly correlated with root length (P<0.001). From the table below, there was significant and positive correlation between soil organic matter and maize parameters except for plant girth and root length.

**TABLE 7. CORRELATION MATRIX SHOWING THE RELATIONSHIPS BETWEEN SOIL PROPERTIES AND SOME MAIZE GROWTH PARAMETERS**

| | MAIZE GROWTH PARAMETERS | | | | | | |
|---|---|---|---|---|---|---|---|
| SOIL PROPERTIES | PLANT HEIGHT | NUMBER OF LEAVES | PLANT GIRTH | ROOT NUMBER | ROOT LENGTH | FWY | DRY WEIGHT MATTER |
| PH ($H_2O$) | 0.624** | 0.641** | 0.472* | 0.512* | -0.168$^{NS}$ | 0.433$^{NS}$ | 0.318$^{NS}$ |
| EX. ACIDITY | 0.141$^{NS}$ | 0.133$^{NS}$ | 0.048$^{NS}$ | 0.382$^{NS}$ | 0.248$^{NS}$ | 0.157$^{NS}$ | 0.090$^{NS}$ |
| EX. AL | 0.066$^{NS}$ | 0.167$^{NS}$ | 0.110$^{NS}$ | 0.325$^{NS}$ | 0.200$^{NS}$ | 0.093$^{NS}$ | 0.004$^{NS}$ |
| EX. H | 0.165$^{NS}$ | -0.006$^{NS}$ | -0.078$^{NS}$ | 0.222$^{NS}$ | 0.161$^{NS}$ | 0.154$^{NS}$ | 0.165$^{NS}$ |
| N | 0.586* | 0.559* | 0.284$^{NS}$ | 0.544* | -0.198$^{NS}$ | 0.647** | 0.541* |
| AV. P | 0.555* | 0.642** | 0.514* | 0.446$^{NS}$ | -0.260$^{NS}$ | 0.373$^{NS}$ | 0.319$^{NS}$ |
| K | 0.547* | 0.434$^{NS}$ | 0.319$^{NS}$ | 0.379$^{NS}$ | 0.211$^{NS}$ | 0.455$^{NS}$ | 0.504* |
| NA | 0.182$^{NS}$ | 0.216$^{NS}$ | 0.279$^{NS}$ | 0.087$^{NS}$ | 0.594** | -0.233$^{NS}$ | -0.217$^{NS}$ |
| OM (%) | 0.574* | 0.552* | 0.271$^{NS}$ | 0.530* | -0.220$^{NS}$ | 0.640** | 0.539* |

** = Significant at P<0.001; * = Significant at P<0.05; NS = Not significant

## CONCLUSION AND RECOMMENDATION

The goal of the pot experiment was to know the effects of spent grain ash on some soil chemical properties and growth of maize (*Zea mayz*) in an Ultisol of Southeastern Nigeria. The result showed that all the treatments significantly (P<0.05) increased soil pH and reduced exchange acidity while five tonnes per hectare of spent grain ash gave the best positive effects on soil chemical properties and also revealed the potential of spent grain ash on the average improved plant parameters. Five tonnes per hectare gave the overall best results in terms of amelioration of soil acidity and the growth of maize. It is therefore recommended that 5 ton/ha should be used for the reduction of soil acidity and improvement of maize performance in Umudike area of southern Nigeria.

## ACKNOWLEDGEMENT

This research was supported by the Almighty God for His inspiration, guidance and protection in the course of the entire study. I specially appreciate the assistance made by my mentor, Dr. C. F. Isinguzo financially and my supervisor Prof. V. E. Osodeke. I equally appreciate the contributions of my lecturers, Dr. M. I. Onwuka and Dr. O. I. Nwachukwu for their support and care. I am obliged to extend my gratefulness to my beloved family, mostly my mother, Mrs. M. U. Ogbonna; sister (Chichi), brother (Chibueze) and my dearest father, Mr. L. E. Ogbonna of blessed memory.

## REFERENCES

Adeoye, G. O. and Agboola, A. A. 1984. "*Critical Level for Soil pH, Available P.K. Zn and Mn and Maize Content of P; Cu; Mn in the Sedimentary Soils of Southwestern Nigeria*". *Fertilizer Research*. 6:65-71.

Adu-Dupaal, H. K; Cobbina, J. and Asare, E. O. 1994. "*Effect of Cocoa Pod Ash on the Growth of Maize*". *Journal of Agricultural Science*. 122: 13-33.

Bouyoucos, G. A. 1962. "*Hydrometer Method Improved for making Particle Size Analysis of Soils*". *Agronomy Journal*. 54: 464 -465.

Chude, V. O; Melgori, W.B. Amapu I.Y and Ano, O. A. 2004. "*Manual on Soil Fertility Assessment*". Published by Federal Fertilizer Department in Collaboration with National Special Programme for Food Security Abuja-Nigeria.

Comin, J. J; Barloy, J; Hallaire, V; Zanette, F. and Miller, P. R. M. 2006. "*Effects of Aluminum on the Adventitious Root System, Aerial Biomass and Grain Yield of Maize Growth in the Field and in a Rhizotron*". *Experimental Agriculture*. 42(3): 351-366.

Dee, B. M; Haynes, R. J. and Graham, M. H. 2003. "*Change in Soil Acidity and the Size and Activity of the Microbial Biomass in Response to the Addition of Sugar Mill Waste*". *Biology Fertility Soils*. 32:47-54.

Franco, A. A. and Munns, D. N. 1982. "*Acidity and Aluminun Restraints on Nodulation Nitrogen Fixation and Growth of Phaseoln Vulgaria in Nutrient Solution*". *Soil Science Society of American Journal*. 46: 296-301.

Jackson, M.L. 1996. *Soil Chemistry Analysis - Advanced Course*. Published Department of Soil Science, University of Wisconsin, Madison, Wisconsin.

Kamprath, E. J. 1976. "*New Buffer pH Method for Rapid Estimation of Exchange Acidity and Lime Requirement of Soils*". *Communications in Soil Science and Plant Analysi*. 7: 637 - 652.

Matsumoto H. 2000. "*Cell Biology of Tolerance and Toxicity in Higher Plants*". *International Review of Cytology*. 200: 1 - 46.

Mclean, E.O. 1982. "*Soil pH and Lime Requirement*". *In: Pager AL, Editor. "Methods of soil Analysis, Part 2. 2nd Ed". Agronomy. Monog*. 9:199-224.

Mullins, G. L. 2005. *Sources of Lime for Acid Soil in Virdinia*, Department of Crop and Soil Environmental Sciences, Virginia.

National Root Crop Research Institute, Umudike. 2012. Meteorological Station. Nigeria.

Njoku, B. O., Enwezor, W. O. and Onyeanakwe, B. I. 1986. "*Chemical Factors Affecting Maize Growth in Acid Ultisols of Eastern Nigeria*". *Proceedings of the 14 Annual Conference of Soil Science Society of Nigeria held in Markurdi. 19th-22nd Oct. Edited by V. O. Chude*. Pp. 119-125.

Onwuka, M. I. and Ogbonna, E. A. 2009. "*Effect of Different Kitchen Residue Ash Sizes on Soil Acidity Amelioration in an Ultisol*". *Proceedings of the 33rd Annual Conference of the Soil Science Society of Nigeria held at Ekiti State*. Pp 184 - 186.

Onwuka, M. I., Osodeke, V. E. and Okolo, N. 2007. "*Amelioration of Soil Acidity Using Cocoa Husk Ash for Maize Production in Umudike Area of Southeastern Nigeria*". *Tropical and Subtropical Agroecosystem*. 7: 41-45.

Osodeke, V. E. 2000. "*Potentials of Biofertilizers for Soil Fertility Management in Southeastern Nigeria*". *Food and Fiber Production in Nigeria, Umuahia*, Pp. 277-282.

Owolabi, O., Adeleye, A., Oladejo; B. T. and Ojeniyi, S. O. 2003. "*Effect of Wood Ash on Soil Fertility and Crop Yield in South West Nigeria*". *Nigeria Journal of Soil Science*. 13:55-60.

Tang, C; Asseng, S; Diatloff, E. and Rengel, Z. 2003. "*Modeling Yield Losses of Aluminum Resistant and Aliminum-Sensitive Wheat due to Subsurface Soil Acidity Effects of Rainfall, Liming and Nitrogen Application*". *Plant and Soil*. 254:349-360.

Tel, D.A. and Hargarty, M. 1984. *Soil and Plant Analysis*, IITA, University of Ghelph.

Voundi Nkama, J. C. V; Demeyer, A. and Verloo, M. G. 1988. "*Chemical Effects on Wood Ash on Plant Growth in Tropics Acid Soils*". *Bioresource Technology*. 63: 251-260.

# VARIABLE BEHAVIOUR OF MASS TRANSFER COEFFICIENT WITH VELOCITY, LENGTH AND DISPERSION OF REACTIVE SOLUTE

***Deepak Swami***, Abhimanyu Sharma
(IIT Mandi, HP, INDIA)

This work represents the experimental and numerical investigation performed on the transport of reactive solute through stratified porous media. Numbers of experiments are performed using reactive solute (solution of NaF) transport through experimental model of stratified porous media of different lengths. The motive of the study is to correlate the effect of scale of hydraulically connected mobile-immobile region and velocity on the estimation of first order mass transfer coefficient for reactive solute transport. During the experimentation variable velocity provided for different lengths of stratification for understanding the behavior of mass transfer coefficient, under variable head gradient. The estimated values of mass transfer coefficient are analyzed for interdependency with length, velocity and dispersion coefficient. The contour plots between velocity, length of mobile-immobile interface, mass transfer coefficient and dispersion coefficient interestingly revealed the correlation in between earlier presumed independent parameters. It is observed that mass transfer coefficient has an inverse relationship with dispersion coefficient. Also first order mass transfer coefficient exhibited varying behavior with length of mobile-immobile region and velocity. Finally, variable behavior of mass transfer coefficient is obtained when the transport media is under the influence of physical non-equilibrium. The study concludes a very important relation of mass transfer between, scales of heterogeneous porous media, velocity and dispersion coefficient. It is also observed that constant first order mass transfer input in higher modelling approach does not comply with the scale effect, and need to be modified.

# PRELIMINARY STUDY OF HEAVY METAL POLLUTION IN SOIL OF INDUSTRIAL ZONE OF SURAT – INDIA

***Tank Shantilal K***
(Department of Biosciences, Veer Narmad South Gujarat University, Surat - 395007, Gujarat, India)
Mehra Bhavna K.
(Detox Corporation Pvt. Ltd., Surat – 395001, Gujarat, India)

Aggregate chemical characteristics and metal concentration in the sub surface soil ( 12 inch) of two industrial zones of Surat city, known as a part of large industrial belt of Gujarat and as the Textile hub of India , were investigated for evaluating the suitability of soil for plantation. According to WHO- SEARO (South East Asia Regional Office) the urban population of Asia is expected to double between 2000 & 2030. Soil being one of the repositories for anthropogenic and industrial wastes needs to be prevented from large quantities of pollutants being continuously introduced. Hence the present study was focused to carry out a preliminary analysis of the chemical profile and heavy metal contamination of soil in two industrial zones of Surat city. The study was carried out in the Pre- Monsoon period.

Results of soil investigation of the present study are compared to the normal content interval and maximum allowable limits (MAL) established by Kloke (1980) to calculate compliance Index ( Ic) for Plantation. Compliance index $I_C$ is the ratio between the observed concentration of metal in soil samples and the maximum accepted limits. Though the concentration of metals like Pb,Cd, Co,Zn, Cu, & Cr is found to be within the Maximum Allowable limits , the degree of soil pollution established based on the Compliance Index calculated with reference to the Maximum acceptable limits ( Kloke,1980) , shows that Cd, Co, Zn & Cu show a Significant Pollution Level 1( PS1) ( $1.0 < Ic < 5.0$) , Pb shows Potentially Significant Pollution (PPS) ($0.7 < Ic < 1.0$) and Cr shows Insignificant Pollution (PN) ($Ic < 0.7$ ) at all the target locations viz, L1,L2,L3 & L4, L5, L6 in the two separate Industrial Zones of Surat City.

# TRANSPORT AND ADSORPTION OF COLLOIDS IN SOILS: A MRI APPROACH

***Alizée P. Lehoux***, Eric Michel
(INRA, Avignon, France)
Paméla Faure, Denis Courtier-Murias, Stéphane Rodts, Philippe Coussot
(Laboratoire Navier, Champs sur Marne, France)

The ability to predict transport and retention of colloidal particles is a major environmental concern as such particles can carry adsorbed pollutants towards the groundwater or be pollutants themselves. The models currently used to predict the fate of colloids in soils are based on mechanisms inferred from breakthrough curves (evolution of concentration as a function of pore volume or time) after injection of particles into a column of porous media. In this study we aim to complement this knowledge with internal measurements by Magnetic Resonance Imaging (MRI).

MRI provides spatial distribution of colloidal particles and water content in time along the sample axis during transport experiment through a porous medium of 20 cm height. We injected several pulses of superparamagnetic nanoparticles in columns of porous media of increasing complexity: glass beads, sand, soil aggregates, and undisturbed soil.

We were able to follow suspended and adsorbed particles in glass beads and sand with MRI. From these experiments we show that dispersion of suspended particles is less important than predicted from standard measurements (breakthrough curves). This is due to entrance effects inducing an initial radial flow heterogeneity which appears as a significant dispersion in the breakthrough curve. With charged particles able to adsorb we were able to follow the adsorption dynamics as the solution advances through the porous medium.

Experiments in soil aggregates showed a strong adsorption but also a constant release, which indicates that more complex mechanisms are occurring in soil aggregates, due to heterogeneities of surface charges and complex porosity. We propose a model to explain this particular behaviour. In the last step of our study, we follow water content and particle concentration during a rain simulation under undisturbed soil.

# VERTICAL DISTRIBUTIONS OF ELEMENTS IN THE SOIL (ALBIC LUVISOL) OF SANJIANG PLAIN IN CHINA

Wei Ouyang*, Hongguang Cheng, Chunye Lin
(State Key Joint Laboratory of Environmental Simulation and Pollution Control, School of Environment, Beijing Normal University, Beijing, China)

The vertical distributions of elements in soils are critical to understand their origins and transports in soil forming processes. The objectives of this study were to investigate the vertical distribution of elements in the Albic Luvisol of Sanjiang Plain to identify the origin of toxic elements in the soil. Soil cores (0 to 50 cm depth) were collected at 33 sites in the northeastern agricultural field of Sanjiang Plain and then sliced into 10 cm slices. The concentrations of SOM and 19 elements were determined. The results showed that the concentrations of SOM, P, Ca, Cd, Hg, and Mn in the soil generally decreased with depth, while Al, Fe, Mg, K, Ti, Cr, Cu, Ni, Sc, and V concentrations roughly increased with depth. The concentrations of As, Co, Pb, and Na were higher in the middle layers than bottom and top layers. Correlation analysis showed that As and Co were significantly correlated to the mineral matrix elements Mn and Fe. Cd was significantly correlated to SOM and the mineral matrix P and Ca. Hg was significantly correlated to SOM. Pb was significantly correlated to the mineral matrix elements Mn, Al, and Fe. Cr and Ni were significantly correlated to the mineral matrix elements Al and Fe. Cu was significantly correlated to SOM and the mineral matrix elements Al and Fe. Whereas the maximal concentrations of toxic trace elements As, Cd, Cr, Cu, Hg, Ni, and Pb in the soil of Sanjiang Plain were lower than the environmental quality standards for the agricultural soils in China, Cd and Hg in the soil partially originated from anthropogenic sources. In details, about 37% of Cd and 31% of Hg in the 0 to 10 cm depth originated from anthropogenic sources. Major anthropogenic source of Cd might be phosphate fertilizer, while the anthropogenic source of Hg might be from atmospheric Hg deposition. Therefore, the long-term accumulation of Cd and Hg in the soil may be a concern in future and should be monitored regularly.

# MODELING OF PHOSPHORUS SORPTION AND BUFFERING MECHANISMS IN SELECTED SOILS OF FIJI

***Muni Sangeeta Goundar****, Satyanarayan Shashtri (Fiji National University, Lautoka, Fiji)
and R. John Morrison (University of Wollongong, NSW 2522, Australia)
*Email:sangeeta1g@yahoo.com

**ABSTRACT:** Mechanisms by which P is retained in the soil include precipitation and ligand exchange, lattice diffusion and anion exchange, which have a strong effect on the bioavailability and the transport potential of P. The main objective of this research was to use phosphorus buffer index (PBI) as an indicator of the P status of selected sugarcane growing soils in Fiji to reduce the risk of excessive fertilizer applications and leaching of P into the waterways, thus minimizing contamination of the environment. Soils used in this study were Inceptisols, Mollisols, Alfisols, Oxisols and Ultisols. PBI and phosphorus isotherm measurements were performed on selected soils. Pearson's correlation, performed using Statistix (Version 10) showed that PBI has a significant negative correlation with soil pH ($R = -0.67$, $p = 0.017$). The P sorption maxima obtained using the Langmuir equation ranged from 480 to 1314 mg/kg and Pmax had a moderate positive correlation with PBI ($R = 0.70$, $p = 0.01$). The isotherm data had a significant fit to the Freundlich sorption model, with R values ranging from 0.82 to 0.97.

***Keywords:*** Phosphorus, phosphorus buffer index, soils, fertilizers, phosphorus sorption mechanisms

## INTRODUCTION

The phosphorus status of Fijian soils along with other soil properties is declining due to the intensive cultivation of sugarcane (Morrison et al, 2005). The Fiji sugar industry has been long dependent on the use of fertilizers to improve crop production, mainly due to the low concentrations of available P and total P in most soils. However, improper long- term application of fertilizers could lead to P accumulation in agricultural soils to levels greater than those required for healthy plant growth, thus leading to the elevated soil potential for P loss to water bodies (McLaughlin et al, 2011). Fine soil particles with sorbed P may be transported via erosion and surface runoff to waterways promoting eutrophication. Furthermore, excessive applications of P fertilizer are a waste of limited resources.

Previous research on phosphate adsorption on soils of Fiji has shown that significant quantities of added P are strongly adsorbed by soils, and only 20 to 30 percent of applied P fertilizer are utilized by the plants (Chee et al, 1978; Dandy and Morrison, 1980). Adsorption reactions are considered as key processes for governing P fertilizer efficiency but then precipitation processes become important at higher level of P (McLaughlin et al, 2011). The mechanisms by which P is retained in the soil include precipitation and ligand exchange (adsorption), lattice diffusion and anion exchange. The fixation/sorption of P in soil is mechanistically complex and sorption is a term used when the mechanism of P retention is undefined (Sims and Pierzynski, 2005). Adsorption can be reversible and irreversible, fast or slow, conditional to buffering capacity of the soil and the contact time with the solution (Bhadha et al, 2012),. The concentration of P in the soil solution is controlled by the adsorption/desorption and precipitation/dissolution equilibria which are affected by the pH, anions competing with phosphate for ligand exchange sites and the levels of Fe, Al and Ca that form precipitates with phosphate ions.

Numerous studies have investigated the P sorption in the soils for environmental or agronomic purposes (e.g., Barrow, 1974; Pant et al, 2002). One issue encountered with isotherms is that the soil phosphate solution interaction is not a true equilibrium, as the soil continues to take phosphate slowly after an initial rapid pseudo-equilibrium is reached and both true chemical and surface adsorption are involved. However, the results obtained from the initial rapid uptake of phosphate from the solution can be treated using the mathematical expressions which are derived assuming that a true equilibrium is reached. The two commonly used sorption models for P sorption isotherm data are the Langmuir and Freundlich equations, which provide information such as binding energy, and sorption capacity and these equations have been developed from simple dynamic models of gas-solid interface processes (Sims and Pierzynski, 2005).

The P sorption assessment methods include: P sorption isotherms, P buffering capacity (PBC) measures (slope between equilibrium P concentration of 0.25 and 0.35 mg P/L), the single-point P-buffering indices (PBI), with methods either adjusted or unadjusted for current P fertility (Burkitt et al, 2008). It was found that the PBI values, which were calculated without the extractable P adjustments, were very reliable (Burkitt et al, 2002). In this study, a simpler method is used to determine the buffer index capacity of some selected Fijian soils.

## MATERIALS AND METHODS

**Soil Sampling Methods.** Approximately 1 kg of composite soil samples were collected from three sugarcane farms in each sampling site on Viti Levu (Rarawai Mill-R, Lautoka Mill- Lt, Penang Mill- P) and Vanua Levu (Labasa Mill-La). The R, Lt, P and La sampling sites were selected on the basis of obtaining soil samples from the different soil orders based on the USDA Soil Taxonomy (Soil Survey Staff, 1999). The soil samples selected belonged to the Inceptisols, Mollisols, Alfisols, Oxisols and Ultisols. Three farms were selected from each mill area based on the soil orders that were found to be the most extensive in that area. Inceptisols were found to be the most extensive (40%) group of soils in Fiji. Mollisols were found to be the second most extensive (24%) followed by Alfisols (10%) and Oxisols (9%).

**Soil Sample Preparation.** The field samples were air dried in trays at <40°C after removing roots, stones, living organisms, and homogenized and subsampled by coning and quartering techniques. Representative sub-samples were ground to <2 mm size for most analyses, and <250 μm sieve for carbon and nitrogen analyses. The analytical data reported represents the average of duplicate analyses of each composite.

**Methods for Soil Analysis.** Soil samples were analysed using standardised methods for pH (1:5 & 1:2.5 soil/water suspension), organic carbon (Walkley Black method), total P (strong acid digest), particle size distribution by the pipette method, oxalate-extractable aluminium (Al) and iron (Fe), exchangeable bases by ammonium acetate method, extractable P by a modified Troug method, and P sorption isotherms, as summarized in Goundar (2013). The equilibrium P concentration (EPC) for each soil type was obtained from the line of best fit of sorption curves by putting $y = 0$. The Pmax (phosphorus sorption maximum) values were assessed using the Langmuir equation and the data were also fitted to the Freundlich sorption model using Microsoft Excel.

## RESULTS AND DISCUSSION

**Chemical Parameters And Pbi For Sampled Soils.** Table 1 presents the chemical parameters measured for the sampled soils. The PBI results obtained for the topsoils tested are given in Figure 1. Overall results of the PBI of the soils studied showed low to moderate rates of PBI. The PBI values of the soils studied decreases in the following order; Oxisols > Mollisols > Ultisols > Alfisols > Inceptisols. Inceptisols showed a low PBI possibly tended to leach a higher P to the waterways (Burkitt et al, 2008).

The PBI analyses indicate that different soils have different P sorption capacities, so a common fertilizer application will not meet the P requirements for different soils. These results showed that Oxisols had the highest mean PBI of 152 ± 25 implying that P will be fixed strongly in Oxisols rendering it unavailable for plant uptake as found by Syers et al (1971) for P fixation in highly weathered soils such as Oxisols and Ultisols. Different Oxisols have a range of PBI values due to the other factors, such as, soil pH, soil texture, fertilizer application rates, the content of other elements present, and degree of weathering and vegetation as these affect the P sorption capacities in soils.

**Relationship of Pbi With Other Soil Properties for Topsoils.** There is a positive correlation between the % clay content of soil with PBI ($R = 0.76$, $p = 0.005$) as found by other researchers (e.g., Burkitt et al, 2002). Inceptisols had the lowest clay content and the lowest PBI while the commonly clay rich Oxisols have high PBIs. Pearson's correlation showed that PBI has a moderate negative significant correlation with pH ($R = -0.67$, $p = 0.017$) while soil pH had a moderate positive correlation with the available P ($R = 0.63$, $p = 0.026$). This indicates that the availability of P in soil is pH-dependent and the reactions that occur in acidic and basic soils produce insoluble P compounds (precipitates). Inceptisols (pH 6.8) had the lowest P fixation as expected according to the reviewed literature and P fixation is minimal at pH 6.5. The PBI is inversely proportional to the available P.

Table 1: Chemical parameters and PBI of 12 Fiji soils (0-29 cm). Note: Where the soils are Alf- Alfisols, Oxi- Oxisols, Mol- Mollisols, Inc- Inceptisols and Ult- Ultisols.

| Mill Area | Soil Type | pH $H_2O$ (1:5) | Oxal-Fe (%) | Oxal-Al (%) | Avail P (mg/kg) | PBI | Total P % | Ex-Ca cmol/kg | TEB cmol/kg | CEC cmol/kg |
|---|---|---|---|---|---|---|---|---|---|---|
| Lt Mill | Alf | 5.9 | 0.78 | 0.25 | 44.4 | 64 | 0.166 | 11.86 | 19 | 23 |
| Lt Mill | Oxi | 5.2 | 0.14 | 0.33 | 7.2 | 170 | 0.091 | 4.86 | 8 | 14 |
| Lt Mill | Mol | 4.9 | 1.76 | 0.30 | 51.5 | 90 | 0.138 | 14.51 | 19 | 28 |
| R Mill | Mol | 5.4 | 1.81 | 0.38 | 116.2 | 72 | 0.227 | 15.21 | 20 | 27 |
| R Mill | Inc | 5.4 | 1.20 | 0.27 | 76.0 | 54 | 0.201 | 10.68 | 15 | 21 |
| R Mill | Mol | 4.9 | 1.92 | 0.43 | 51.8 | 106 | 0.181 | 13.11 | 18 | 29 |
| P Mill | Mol | 5.1 | 1.16 | 0.32 | 7.8 | 82 | 0.083 | 9.28 | 15 | 23 |
| P Mill | Alf | 6.2 | 1.22 | 0.46 | 16.8 | 62 | 0.073 | 15.76 | 22 | 27 |
| P Mill | Inc | 6.8 | 1.05 | 0.48 | 242.9 | 33 | 0.145 | 20.80 | 27 | 31 |
| La Mill | Ult | 5.1 | 1.12 | 0.43 | 54.2 | 70 | 0.134 | 16.70 | 22 | 32 |
| La Mill | Oxi | 4.7 | 0.34 | 0.35 | 11.2 | 134 | 0.055 | 2.00 | 4 | 14 |
| La Mill | Mol | 5.9 | 0.91 | 0.43 | 59.5 | 63 | 0.088 | 19.02 | 25 | 31 |

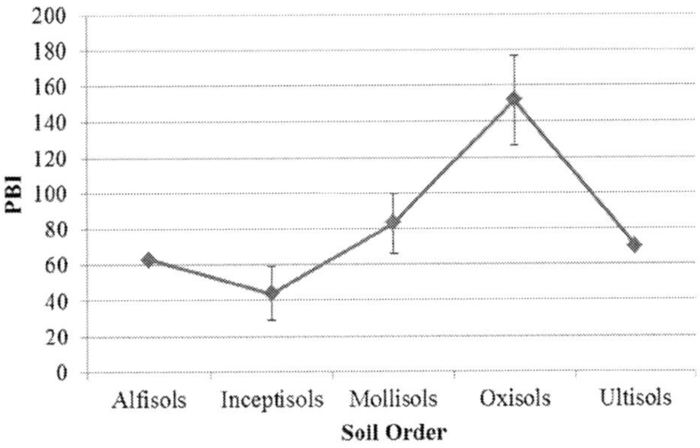


Figure 1. PBI values for different soil orders in Fiji sugarcane soils

PBI has a relatively good correlation with $Fe_{ox}$ (R = 0.70, p = 0.02) and a moderate correlation (R = 0.65, p = 0.04) with $Al_{ox} + Fe_{ox}$ when correlation is performed after the exclusion of the high values for Oxisols. The relationships of $Al_{ox}$ and $Fe_{ox}$ with P soption capacity are well documented in the literature (e.g., Pant et al., 2002). According to Weng et al (2012), the factors which control the fixation of phosphate to Fe oxyhydroxides under normal soil conditions are pH, organic matter and calcium concentration. PBI displayed a relatively strong negative correlation with exchangeable Ca (R = -0.79, p = 0.002) and the PBI also displayed a relatively strong negative relationship with TEB (R =-0.76, p = 0.0039). PBI also displayed a weak negative relationship with the cation exchange capacity (CEC).

**P Sorption Isotherms for Topsoils.** The sorption curves (Figure 2) obtained after various additions of P showed the soils with the highest sorption of P were Oxisols and the lowest sorbing P soils were Inceptisols. The effect of the available P in different soil types is understandable in the isotherm curves as soils with the high available P most probably due to the fertilizer applications resulting in less amount of P adsorbed - produces a lower curve. The PBI data compare well with the generated isotherm curves for the different soil types.

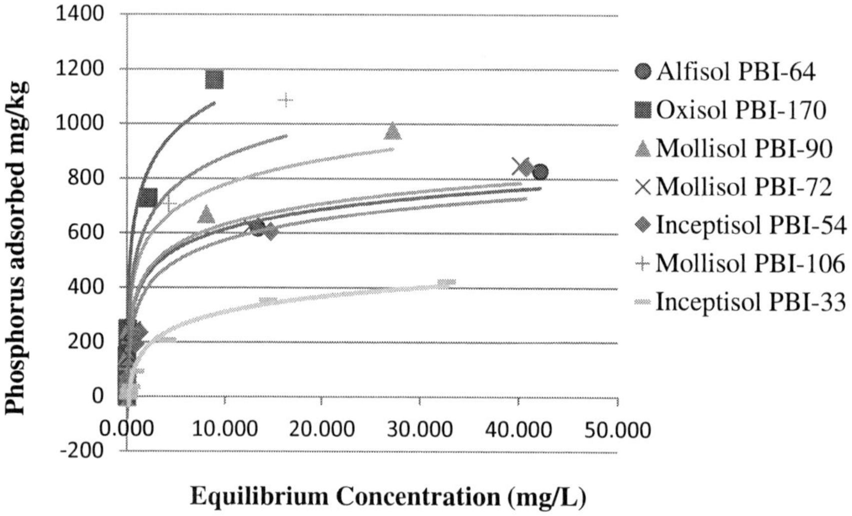


Figure 2. Sorption curves generated on different soils from soil isotherm experiments.

**Equilibrium P Concentration (Epc) for Topsoils.** The EPC values obtained in this study ranged from 0.006 to 0.224 mg/L (Table 2). Most soil types had very low values except for the few soil types such as Inceptisols from Penang Mill, and Inceptisols and Mollisols from Rarawai Mill which had high values due to the high available P. EPC values had very strong positive correlation with the available P (R= 0.92, p = <0.01) whereas a negative relationship is observed with PBI.

**Langmuir Model for Topsoils.** The Langmuir sorption model showed a highly significant fit to the isotherm data, with $R^2$ values of 0.77 or higher. Most soils fitted well to the Langmuir model, particularly the Inceptisols of Penang mill which had an R of 0.996. The P sorption maxima obtained using the Langmuir equation ranged from 480 to 1314 mg/kg (Table 2) and Pmax had a moderate positive correlation with PBI (R= 0.70 & p = 0.01) indicating that soils with a high sorption capacity will require greater amount of P fertilizer compared to soils with low sorption capacity. The Pmax obtained from the Langmuir isotherms showed a systematic decrease for soils in the following order; Oxisols, Mollisols, Alfisols, Inceptisols and Ultisols.

**Freundlich Model for Topsoils.** All soils fitted well to the Freundlich model except for the Mollisols which showed low R values compared with the other soil types. EPC and intercept of Freundlich model showed a strong negative correlation with R = -0.79 and p = 0.002. The y intercept of the Freundlich model correlated well to the PBI data. The soils with the lowest PBI such as Inceptisols from Penang had the lowest energy of sorption while Oxisols had the highest y intercept indicating the highest energy of sorption. It was also proven in this study that these soils had a high PBI.

Table 2. Data obtained from P isotherm curves and two models for twelve different topsoils of Fiji.

| Mill Area | Soil Type | EPC (mg/L) | P max (mg/kg) (Langmuir) | Slope of Freundlich equation | Intercept of Freundlich equation |
|---|---|---|---|---|---|
| Lt Mill | Alf | 0.032 | 850 | 0.350 | 2.400 |
| Lt Mill | Oxi | 0.019 | 1314 | 0.457 | 2.703 |
| Lt Mill | Mol | 0.026 | 1002 | 0.380 | 2.486 |
| R Mill | Mol | 0.033 | 861 | 0.441 | 2.322 |
| R Mill | Inc | 0.058 | 878 | 0.494 | 2.217 |
| R Mill | Mol | 0.034 | 1227 | 0.586 | 2.476 |
| P Mill | Mol | 0.003 | 809 | 0.502 | 2.949 |
| P Mill | Alf | 0.017 | 657 | 0.431 | 2.403 |
| P Mill | Inc | 0.224 | 480 | 0.631 | 1.804 |
| La Mill | Ult | 0.034 | 597 | 0.699 | 2.469 |
| La Mill | Oxi | 0.006 | 711 | 0.768 | 3.099 |
| La Mill | Mol | 0.014 | 539 | 0.617 | 2.610 |

## CONCLUSIONS

Sorption curves and the PBI data were comparable so the PBI could be considered as a simple and efficient test variable to predict the P sorption of the different soils. The P isotherm data fitted very well to both Langmuir and Freundlich models. The Inceptisol (P Mill) had the lowest P maximum (480 mg/kg) and the lowest y intercept (1.804) indicating that this soil is the lowest P sorber. Furthermore, the isotherm curve of Inceptisols (P Mill) occupied the lowest position when all the isotherm curves of the different soils were plotted in a single graph. In contrast, the Oxisols had the highest PBI with the highest P maxima and high y intercepts (1314 mg/kg & 2.703 respectively). There is a good link between the two extensively used sorption models (Langmuir and Freundlich) with the PBI results. The Inceptisols had the PBI values as low as 33 indicating that these soils have the greatest potential of P leaching to the environment. Assessments of the long-term accumulation of P in soils should be carried out at representative sites by examining the available P, PBI, total P along with an accurate data on P fertilizer or other material additions.

## ACKNOWLEDGEMENTS

The University of the South Pacific Research Office and the Faculty of Science, Technology and Environment are gratefully acknowledged for providing the financial support for this research. The staff of Sugar Research Institute of Fiji and Fiji Sugar Corporation is thanked for technical assistance. The Fiji Ministry of Primary Industries) provided particle size analyses and The University of the South Pacific, Institute of Applied Science (IAS) assisted with total nitrogen and aluminium analyses.

## REFERENCES

Barrow, N. J. 1974. The effect of previous additions of phosphate on phosphate adsorption by soils. *Soil Sciences* **118**, 82-89.

Bhadha, J. H., Daroub, S. H. and Lang, T. A. (2012). Effect of kinetic control, soil:solution ratio, electrolyte cation, and others, on equilibrium phosphorus concentration, Geodema 173-174, 209-214.

Burkitt, L. L., Moody, P. W., Gourley, C. J. P., Hannah, M. C. 2002. A simple phosphorus buffering index for Australian soils. *Australian Journal of Soil Research* **40**, 497-513.

Burkitt, L. L., Sale, P. W. G., Gourley, C. J. 2008. Soil phosphorus buffering measures should not be adjusted for current phosphorus fertility. *Australian Journal of Soil Research* **46**, 676-685.

Chee, M.T., Fergusson, J.E., Campbell, A.S. and Richmond, T.R. 1978. Phosphate sorption properties of some Fiji soils. *N.Z. Journal of Science* **21**, 157-166..

Dandy, A. J. and Morrison, R. J. 1980. Phosphate sorption isotherms of some South Pacific soils. *N.Z. Journal of Science* **23**, 399-406.

Goundar, M. S. 2013. *Phosphorus Status of Different Soils in the Sugarcane Belt of Fiji.* MSc Thesis, The University of the South Pacific, Suva, Fiji.

McLaughlin, M. J., Mcbeath, T. M., Smernik, R., Stacey, S. P., Ajiboye, B., & Guppy, C. 2011. The chemical nature of P accumulation in agricultural soils--implications for fertiliser management and design: An Australian perspective. *Plant and Soil* **349**, 69-87.

Morrison, R. J., Gawander. J. S. and Ram. A. N. 2005. Changes in the properties of a Fijian Oxisol over 25 years of sugarcane cultivation. In: *D.M. Hogarth (Editor) Proceedings International Society of Sugar Cane Technologies XXV Congress*, Guatemala **2**, 139-146.

Pant, H. K., Nair, V. D., Reddy, K. R., Graetz, D. A., Villapando, R. A. 2002. Influence of flooding on phosphorus mobility in manure-impacted soils. *Journal of Environmental Quality* **31**, 1399-1405.

Sims, J. T and Pierzynski, G. M. (2005). Chemistry of Phosphorus in Soils in Chemical Processes in Soils, Dick, W. A, Tabatabai, M. A, and Sparks, D. L eds., Soil Science Society of America Inc., Wisconsin, USA.

Syers, J. K., Evans, T. D., Williams, D. H., Murdock, J. T. 1971. Phosphate Sorption parameters of Representative Soils from Rio Grande do sul, Brazil, *Soil Science* **112**, 267-275.

Soil Survey Staff. 1999. *Soil Taxonomy: A basic system of soil classification for making and interpreting soil surveys.* Second Edition. USDA Handbook No. 436. Washington, DC.

Weng, L., Van Riemsdijk, W. H., & Hiemstra, T. 2012. Factors controlling phosphate interaction with iron oxides. *Journal of Environmental Quality* **41** (3), 628-635.

# INNOVATIVE IN-SITU OXIDATION TECHNOLOGIES FOR TREATING GROUNDWATER CONTAMINATED BY CHLORINATED SOLVENTS AND 1,4-DIOXANE

***Hua Zhong***, Mark Brusseau, Ni Yan (The University of Arizona, Tucson, AZ, USA)
Lina Zhang, Peng Cui (Hunan University, Changsha, Hunan, China)

Contamination of groundwater by chlorinated-solvent constituents, 1,4-dioxane, and related compounds is ubiquitous, and remains a significant human-health and water-resource sustainability issue for many industrialized regions. For example, in Arizona, chlorinated-solvent constituents are the primary contaminants of concern for 44 of the 50 state and federal Superfund sites. Extensive dissolved-phase groundwater contaminant plumes typically form at sites contaminated by these compounds. Clean-up of these sites will require many decades or longer under current methods. Hence, alternatives are needed for cost-effective, long-term management of sources and plumes at these sites. We are developing two approaches based on the application of in-situ chemical oxidation (ISCO) for direct treatment. Method 1 employs a permeable reactive barrier supplemented with persulfate oxidant. The barrier would be placed downgradient of a control plane (source zone or plume edge), and is for sites with shallower groundwater. Method 2 is for deeper groundwater, and comprises creation of a treatment zone wherein the natural attenuation capacity of the sediment is enhanced through injection of oxidant solutions and activators. Bench-scale batch and column transport experiments were conducted to evaluate technology potential. A variety of Fe-based media, including iron filings, $Fe_3O_4$ solids, and several Fe-containing geomedia, were tested for in situ activation of persulfate. The use of a persulfate-$H_2O_2$ binary oxidant system was examined for contaminant degradation enhancement. Emplacement of $Fe_3O_4$ and zero-valent iron nanoparticles in porous media for activation of persulfate for deep groundwater treatment was also studied. The results showed that all these methods, though to different extents, caused in situ activation of persulfate and successfully enhanced degradation of chlorinated-solvent and 1,4-dioxane contaminants.

# UTILIZING PLANTS TO LIMIT THE MIGRATION OF CONTAMINANTS FROM FRACKING-WATER SPILL SITES

***Amanda Shores***, Brittany Hethcock, Melinda Laituri
(Colorado State University, Fort Collins, CO 80525, United State of America)

Plants provide an economical and in situ remediation method for removal of organic contaminants from soils, thereby limiting their spread to groundwater. Contaminants of particular recent interest are ones found in the produced water from hydraulic fracturing methods (fracking). Produced-water spills occur on a daily basis in Colorado and release a variety of organic contaminants in high concentrations into the environment, such as benzene, toluene, ethyl benzene, *m*-xylene (BTEX) and naphthalene, all known toxins. These contaminants are mobile in soil and have the potential to reach groundwater used for drinking. Inorganic salts in produced water are found in extremely high concentrations also, often above levels in sea water. At a produced-water spill site, vegetation dies and reestablishes only after some time, possibly after the salts have been leached out by rain. I propose the use of salt-resistant local grass species to vegetate fracking sites and facilitate the degradation of the abovementioned aromatic organic compounds. To develop this technology I will conduct two greenhouse studies. The first will test the ability of seven grasses native to Colorado to thrive on local soil spiked with salt levels commonly found in water produced from fracking in Colorado. The second will test the ability of the selected salt-resistant grasses to remediate BTEX and naphthalene from CO soil spiked with these compounds at environmentally relevant levels. Soil, root and shoot tissue concentrations of BTEX and naphthalene will be assessed using gas chromatography mass spectrometry (GC-MS) to determine if the selected plant species takes up the contaminant, where the contaminant is sequestered and which species is most effectively dissipating the contaminants. I hypothesize the plant uptake of BTEX and naphthalene will be sequestered in mainly root tissue and that the presence of these selected plants will reduce the overall contaminants in the soil. This research takes steps towards limiting toxins at a produced-water spill site from migrating to groundwater, reducing human and environmental health risks associated with such spills.

---

# BIO-DRYING OF GREEN WASTES

***Ertan Durmusoglu,*** Mutala Mohammed and Ismail Ozbay
(University of Kocaeli, Izmit, Kocaeli, Turkey)

**ABSTRACT:** In this study, the effect of bio-drying process on green waste was investigated at a laboratory scale. Four different sets of bio-drying processes were conducted and the impact of bulking agent was evaluated. The laboratory scale experiment consisted of a 0.8 m3 bio-drying reactor equipped with a matrix temperature measuring and air supply system. A 50 kg of green waste was placed in the reactor with an uninterrupted airflow rate of 1.4 m3/h. The results obtained indicated that the bio-drying reduced the water content of waste with bulking agent (BA) by 27% more than green waste alone after 7 days, allowing the production of bio-dried waste with a net calorific value of 6,270 kJ/kg wet weight, 55% higher than its initial value and 38% decrease in weight in the original mass of the waste. Although on the average, bio-drying of green waste alone showed significant reduction in volume (87%), weight (55%) and increase in calorific value (150%), bio-dried matrix with BA was the most preferred option since the process produced no leachate and generated the highest calorific value.

## INTRODUCTION

The World Bank estimates that Municipal Solid Waste (MSW) generation is expected to increase to 4.3 billion tonnes by the year 2025 from the current 3 billion tonnes per year, based on population and economic rates. As developing countries continue their rapid pace of development and urbanization, these figures are expected to increase considerably. Food waste is the third largest component of generated waste and second largest component of discarded waste. Unfortunately, food waste contains high water content (Cheng et al., 2007; Zhang et al., 2008) and the indiscriminate disposal of such biodegradable waste leads to serious environmental consequences as well as significant damage to water resources when disposed in landfills. In curbing this menace, a number of innovative technologies for waste treatment such as composting, landfilling, anaerobic digestion, biodrying and thermal methods have been developed with landfilling as the most preferred option. However, as a result of the several drawbacks presented by waste landfilling and the introduction of the European Union Landfill Directive (1999/3/EC), this has led to significant reduction in the quantity of biodegradable waste disposed at landfills (European Commission, 1999; Torretta et al., 2014; Zhao et al., 2014). Henceforth, the generation of fuel from MSW by biodrying concept has recently found a new interest among researchers with the possibility to produce solid recovered fuel (SRF) (Ragazzi and Rada, 2012; Velis et al., 2013; Zawadzka et al., 2010).

Biodrying, a concept similar to composting, aims at removing or reducing water from biodegradable waste with high water content and increasing the treatability and subsequent utilization value of the Solid Recovered Fuel (SRF). The removal or reduction of moisture contents in bio-drying process involves evaporation of liquid water through aerobic decomposition of the organic material or reduction of water vapour by air flow (Cai et al., 2013; Huet et al., 2012; Velis et al., 2010). This mechanism is accomplished by relying on microorganisms, both bacteria and fungi to biologically degrade the organic component in order to reduce the moisture content while maintaining the biomass energy (Sadaka et al., 2011). Compared to traditional composting process, the essential distinguish feature of biodrying is the application of a higher ventilation rate to reduce moisture content by using the heat generated during the aerobic degradation process as well as forced aeration (Tambone et al., 2011). Also, the output from composting is stabilized organic material whereas that of biodrying is partially stabilized. Biodrying also has added advantage of pre-treating the waste at the lowest possible retention time to produce a high quality SRF. Besides these benefits, the biodrying process also renders the output material more suitable for short-term storage and lessens the transportation cost by reducing its weight through from moisture loss and partially biostabilizing it.

Initial moisture content, air-flow rate, porosity of the waste, permeability, retention time, mixing, the presence of BA and temperature have been identified by researchers as the main factors that affect biodrying process (Adani et al., 2002; Frei et al., 2004; Navaee-Ardeh et al., 2006). Among these, air-flow rate and initial moisture content are paramount with extensive research and information available on these two parameters currently for biodrying. Air-flow rate has a direct influence on the matrix temperature and drying efficiency. Adani et al. (2002) and Roy (2005) established that high air-flow rate contributes to effective and fast drying, and high calorific value. Huilinir and Villegas (2015), studied the simultaneous effect of initial moisture content and airflow rate on biodrying of sewage sludge, and found that initial moisture content has a stronger effect on biodrying, affecting the temperature and improving the water removal. Additionally, the use of BA plays a crucial role in biodrying process. The use of BA adjusts the initial moisture content and facilitates air movement due to the increase in voids ratio. It effects on biodrying has been demonstrated by some researchers including Frei et al., 2004; Zhao et al., 2011; Yang et al., 2014 and Zhao et al., 2012).

Currently, most studies on biodrying have been mainly focused on MSW, agricultural and sewage sludge. However, to the best of our knowledge, no studies of biodrying has been reported on neither green nor food waste. The present work was conducted to assess the biodrying of green waste of very high moisture content at constant air-flow rate and, the effect of BA on biodrying and leachate generation.

## MATERIALS AND METHODS

**Bio-Drying Materials and Analytical Analysis.** In this study, green waste (lettuce) collected from a canteen of University of Kocaeli was used as the raw material for biodrying. Pruning waste of 15 mm in diameter was used as BA. Four different sets of biodrying processes were performed in this study. In the first (T1) and second (T2) trial, the objective was to identify the difference in moisture and weight reduction between shredded (15-20 mm) and non-shredded (80-90 mm) green waste. These investigations were undertaken by filling the reactor with 50 kg of the shredded and non-shredded green waste. For the third (T3) and fourth (T4) trail, the reactor was filled with the same amount of green waste (shredded and non-shredded) and 10% pruning waste as BA (initial moisture content of 10.20%) was added to evaluate the effect of BA in terms of weight loss and moisture reduction on biodrying of green waste. The ratio of green waste/pruning waste was 4:1 and the total waste of feedstock was 50 kg. All trials were performed at a constant and uninterrupted air-flow rate (1.4 $m^3/h$) for seven days. The characteristics of the raw materials used in this study are presented in Table 1. The moisture content of the feedstock were analyzed following the ISO 5068-1 standard using moisture analyser (Precisa, XM 50), whereas the heat value of the biodried material was determined using IKA C-7000 model calorimeter (IKA Laboratory Equipment, Werke Staufen, Germany), in accordance with ISO 1928 standard. Elemental analysis was analyzed using Thermo Scientific Flash 2000 Elemental Analyzer (Thermo Fisher Scientific Inc., Bremen, Germany).

**FIGURE 1. Bio-reactor used for the Experiment**

**Experiment Procedure.** The biodrying process was conducted using an adiabatic box of 0.8 $m^3$ made from a stainless steel with a leachate collection system at the bottom. The detail design of the bioreactor used in this study is presented in Fig. 1. The reactor was placed on an electronic balance to monitor the waste loss during the biodrying process. The matrix temperature was monitored with a thermometer sensor inserted in the reactor (5 cm above the bottom of the perforated plate). For aeration purposes, a whirlpool pump and an air-flow meter located in the horizontal duct from the air blower to the air chamber were used. The air compressor or blower was used to supply aeration through the matrix at a constant rate while the air-flow meter monitored the air-flow rate. At the end of each trial, the final height and weight were determined.

**TABLE 1. Characteristic of the Raw Materials**

| Parameter | LW[a] | PW[b] |
|---|---|---|
| Moisture content (%) | 92.00 | 10.20 |
| Bulk density (kg/m$^3$) | 68.59 | 13.72 |
| Calorific value (kJ/kg) | 1,180 | 14,440 |

[a] Lettuce waste; [b] Pruning waste

## RESULTS AND DISCUSSION

**Moisture Content and Volume Reduction of Waste Matrix.** Table 2 summarizes the results obtained from the biodrying process of green waste operated for seven days. The moisture content of the biodried products were 82.5%, 71.5%, 45.0% and 57.0% for T1, T2, T3 and T4 process respectively. After the biodrying, the T3 process had the highest extent of moisture reduction (48.9%), followed by T4 (29.6%), T2 (13.9%) and T1 (10.3%). Commercial biodrying processes have been reported to be completed within 7-15 days with moisture losses around 25-30% (w/w) (Velis et al., 2009). For trial T1 and T2, moisture removal in the first few hours was principally achieved through leachate formation. As clearly indicated, moisture loss during biodrying process is by exothermic microbial reaction and airflow. Because biodrying is an aerobic process, the cell structure of the shredded waste matrix was destroyed, decreasing the water holding capacity which resulted in some amount of water been leaked in trial T1 and T2. However, the present of BA increased the porosity and facilitated the aerobic fermentation of the waste matrix with minimal biodegradation.

**TABLE 2. Summary of the Results obtained from Biodrying Process**

| Experiment | MC (%) | | Weight (kg) | | CV[e] (kJ/kg) | |
|---|---|---|---|---|---|---|
| | Initial | Final | Initial | Final | Initial | Final |
| T1[a] | 92.0 | 82.5 | 50 | 21.6 | 1,050 | 2,290 |
| T2[b] | 83.0 | 71.5 | 50 | 23.4 | 1,300 | 3,650 |
| T3[c] | 88.0 | 45.0 | 50 | 26.5 | 3,770 | 6,860 |
| T4[d] | 81.0 | 57.0 | 50 | 28.0 | 4,360 | 5,670 |

[a] shredded waste; [b] non-shredded waste; [c] shredded waste + BA; [d] non-shredded waste + BA; [e] Calorific Value

Clearly, the lower the moisture content, the higher the calorific or energy content of waste. The increase in calorific value of the biodried material in the present study reached 181% for T1, which corresponds to 2,290 kJ/kg. T3 had 6,680 kJ/kg, nevertheless the increase was smaller, 82% (Table 2), in agreement with previous study by Stasta et al. (2006). A mixture of green waste and BA had higher initial average calorific value, 2 times more than green waste alone. This was attributed to the higher calorific value of the BA (pruning waste) as well as the lower initial moisture content in T3 and T4. Even though the percentage increase in calorific value of the waste matrix were in the order of T2 > T1 > T3 > T4, nonetheless T2 attained the second lowest initial calorific value.

**Weight Loss Profile.** Final weight loss rates after 168 hours of biodrying were 56.8%, 53.2%, 39.2% and 36.7% for T1, T2, T3 and T4 trials respectively, with an average highest weight reduction of 4.06 kg/day for T1 and 3.14 kg/day for T4 as the lowest. In contrast, the final volume loss rates were 82.9%, 84.9%, 65.0% and 77.1% for T1, T2, T3 and T4 respectively. Leachate was generated only in T1, T2 and T3. The highest weight loss was recorded in T3 (22.92 kg) as a result of leachate formation in T1 (31.69%) and T2 (26.32%) as percentage weight loss by leachate to the total weight loss, in agreement with report previously by Robles Martinez et al., (2012). This was because the mechanical shredding of the waste matrix disintegrated the structural components such as lignins and cells present in the waste and subsequently released the bound water, thus releasing large amount of leachate (Binod et al., 2012). Additionally, during biodrying process, weight loss occurred mainly through water and solid loss from leaching, moisture from evaporation and oxidation of organic carbon or solid vitalization (Robles-Martinez et al., 2012). The results in this study established positive correlation between moisture and weight reduction. Weight loss of waste matrix for all four trials conducted in this study followed the order of T3 > T4 > T2 > T1, however, the waste matrix with BA recorded the highest moisture content reduction and weight loss, implying that waste matrix with different sizes and bulking agent could have influence on biodrying efficiency. Figure 2 shows weight loss percentages against time during seven days of biodrying. It was observed that the highest weight lossese were achieved during the first two days with T1, T2 and T3 following a similar pattern.

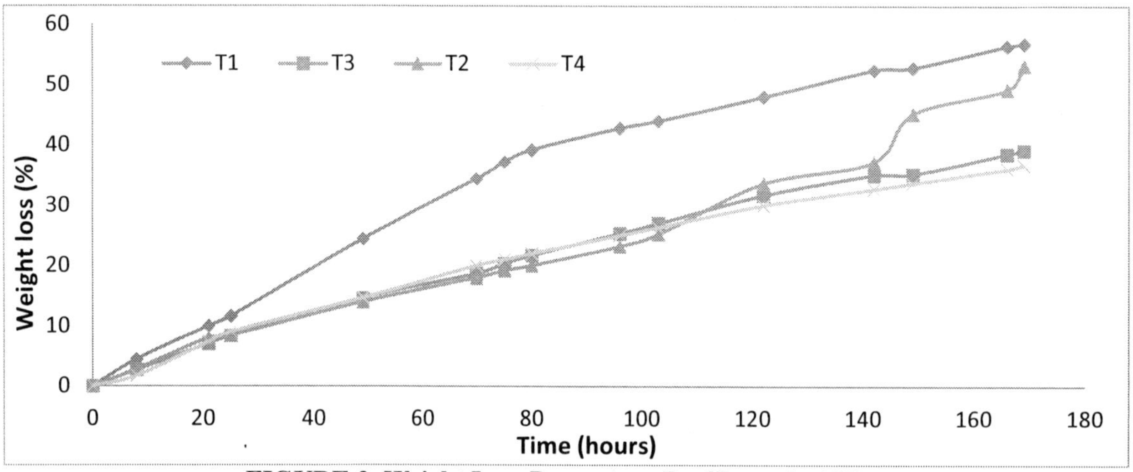


**FIGURE 2. Weight Loss Percentage Profile against Time**

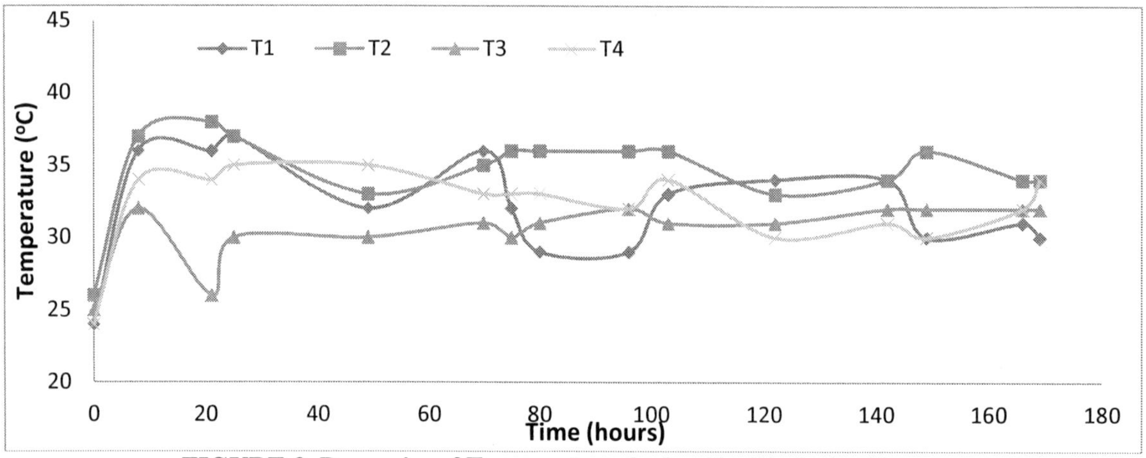


**FIGURE 3. Dynamics of Temperature during the Biodrying Process**

**Matrix Temperature Evolution.** The temperature profile of the matrix is presented in Fig. 3. The highest temperature recorded during the biodrying process was 38 °C for T2 on the 20th hour while T1 and T4 recorded the lowest as the initial temperatures, similar to that reported by Sen and Annachhatre (2015). Also, the character of changes of temperature for T1, T2 and T4 followed a similar pattern, which clearly indicated three different cycles of temperature rise and fall during the entire biodrying process. At the end of the trial processes, T1 and T2 saw a decreasing trend in temperature, with T3 maintaining a linear pattern and an increasing trend in temperature for T4, 4 °C and 2 °C more than T1 and T3 respectively. The low temperatures obtained as well as the observed increase in temperature could be attributed to lack of turning or mixing, different structural composition of the matrix and inadequate air-flow. It also evident from the temperature profile (Fig. 3) that maximum self heating of the matrix occurred within the first 24 hours (i.e. highest temperatures) showing an intense microbial activity, which is an indication that the initial period of the biodrying process is crucial.

## CONCLUSIONS

This study showed that biodrying is an effective approach for moisture and weight reduction of green waste, preserving most of its energy contents. Biodried matrix with BA (T3 and T4), recorded a higher decrease in moisture, 27% more than the matrix without BA. The use of BA had a positive impact on biodrying of green waste, both in terms of moisture reduction and calorific value increase. Calorific value of biodried matrix with BA increased, even though the results in this study revealed significant reduction of weight and volume. This implies lower cost for waste transportation and handling, nonetheless, the biodried material could not be suitable for energy generation since the calorific values, which is one of the parameters considered for SRF classification, obtained were below the required limit for solid-recovered fuel (SRF) generation. However, the effect of BA on both raw and biodried material on calorific value showed promising results. A mixed green waste with BA was found to be preferable since it provided real porosity which was crucial for moisture reduction with no leachate generation. Hence, further research or innovation is required to maximize the full utilization of bulking agent in order to preserve or increase the energy value of the biodried material. This could be achieved by varying the amount of BA.

## ACKNOWLEDGEMENTS

The authors are grateful to the Research Unit of University of Kocaeli for funding this research.

## REFERENCES

Adani, F., D. Baido, E. Calaterra and P. Genevini. 2002. "The Influence of Biomass Temperature on Biostabilization – Biodrying of Municipal Solid Waste." Bioresource Technol. 83:173-179.

Binod, P., M. Kuttiraja, M. Archana, K. U. Janu, R. Sindhu, R. K. Sukumaran and A. Pandey. 2012. "High Temperature Pretreatment and Hydrolysis of Cotton Stalk for Producing Sugars for Bioethanol Production." Fuel 92:340-345.

Cai, L., T. B. Chen, D. Gao, G. D. Zheng, H. T. Liu and T. H. Pan. 2013. "Influence of Forced Air Volume on Water Evaporation during Sewage Sludge Bio-Drying." Water Res. 47(13):4767-4773.

Cheng, H.F., Y. G. Zhang, A. H. Meng and Q. H. Li. 2007. "Municipal Solid Waste Fueled Power Generation in China: A Case Study of Waste-to-Energy in Changchun City." Environ. Sci. Technol. 41:7509-7515.

European Commission. 1999. "Council Directive, 1999/317 EC of 26 April 1999 on the Landfill of Waste. Official Journal of the European Communities, L 182 16/07/1999.

Frei, K. M., D. Cameron and P. R. Stuart. 2004. "Novel Drying Process Using Forced Aeration Through a Porous Biomass Matrix." Dry Technol. 22(5):1191-1215.

Huet, J., C. Druilhe, A. Trémier, J. C. Benoist and G. Debenest. 2012. "The Impact of Compaction, Moisture Content, Particle Size and Type of Bulking Agent on Initial Physical Properties of Sludge-Bulking Agent Mixtures before Composting." Bioresource Technol. 114:428-436.

Huilinir, C., and M. Villegas. 2015. "Simultaneous Effect of Initial Moisture Content and Airflow Rate on Biodrying of Sewage Sludge." Water Res. 82:118-128.

Mohajer, A., A. Trémier, S. Barrington, J. Martinez, C. Teglia and M. Carone. 2009. "Microbial Oxygen Uptake in Sludge as Influenced by Compost Physical Parameters." Waste Manage. 29:2257-2264.

Navaee-Ardeh, S., F. Bertrand and P. R. Stuart. 2006. "Emerging Biodrying Technology for the Drying of Pulp and Paper Mixed Sludges." Drying Technol. 24:863-878.

Ragazzi, M., and E. C. Rada. 2012. "RDF/SRF evolution and MSW bio-drying." WIT Trans. Ecol. Environ. 163:199-208.

Robles-Martinez, F., E. M. Silva-Rodriguez, T. Espinosa-Solares, B. Pina-Guzaan, C. Calixto-Mosquoeda, F. J. Colomer-Mendoza and E. Duran-Paramo. 2012. "Biodrying under Greenhouse Conditions as Pretreatment for Horticultural Waste." J. Environ. Protection 3:298-303.

Roy, G. 2005. "Modélisation Technique et Economique d'un Réacteur de Bioséchage Discontinue (In French). Technical and Economic Modelling of a Biodrying Batch Reactor." MSc Thesis, Ecole Polytechnique de Montréal, Montréal, Canada.

Sadaka, S., K. VanDevender, T. Costello and M. Sharara. 2011. "Partial Composting for Biodrying Organic Materials." Agricultural and Natural Resources, University of Arkansas, FSA1055. http://www.uaex.edu.

Sen, R., and A. P. Annachhatre. 2015. "Effect of Air Flow Rate and Residence Time on Biodrying of Cassava Peel Waste. Int. J. Environmental Technol. and Manage. 18(1):9-29.

Stasta, P., J. Boran, L. Bebar, P. Stehlik and J. Oral. 2006. "Thermal Processing of Sewage Sludge." Appl. Therm. Eng. 26(13):1420-1426.

Tambone, F., B. Scaglia, S. Scotti and F. Adani. 2011. "Effects of Biodrying Process on Municipal Solid Waste Properties." Bioresource Technol. 102(16):7443-7450.

Torretta, V., G. Ionescu, M. Raboni and G. Merler. 2014. The Mass and Energy Balance of an Integrated Solution for Municipal Solid Waste Treatment." WIT Trans. Ecol. Environ. Res. 180:151-161.

Velis, C. A., P. J. Longhurst, G. H. Drew, R. Smith and S. J. T. Pollard. 2010. "Production and Quality Assurance of Solid Recovered Fuels Using Mechanical-Biological Treatment (MBT) of Waste: A Comprehensive Assessment." Crit. Rev. Env. Sci. Tec. 40(12):979-1105.

Velis, C. A., P. J. Longhurst, G. H. Drew, R. Smith and S. J. T. Pollard. 2009. "Biodrying for Mechanical-Biological Treatment of Wastes: A Review of Process Science and Engineering. Bioresource Technol. 100(11):2747-2761.

Velis, C. A., S. Wagland, P. J. Longhurst, B. Robson, K. Sinfield, S. Wise and S. J. T. Pollard. 2013. "Solid Recovered Fuel: Materials Flow Analysis and Fuel Property Development during the Mechanical Processing of Biodried Waste." Environ. Sci. Technol. 47(6):2957-2965.

Yang, B., L. Zhang and D. Jahng. 2014. Importance of Initial Moisture Content and Bulking Agent for Biodrying Sewage Sludge. Drying Technol. 32:135-144.

Zawadzka, A., L. Krzystek, P. Stolarek and S. Ledakowicz. 2010. "Biodrying of Organic Fraction of Municipal Solid Wastes." Drying Technol. 28:1220-1226.

Zhang, D. Q., P. J. He, L. M. Shao, T. F. Jin and J. Y. Han. 2008. "Biodrying of Municipal Solid Waste with High Water Content by Combined Hydrolytic-Aerobic Technology." J. Environ. Sci. 20(12):1534-1540.

Zhao, L., W. Gu, L. M. Shao and P.. J. He. 2012. "Sludge Bio-Drying Process at Low Ambient Temperature: Effect of Bulking Agent Particle Size and Controlled Temperature. Drying Technol. 30:1037-1044.

Zhao, L., W. M. Gu, P. J. He and L. M. Shao. 2011. "Biodegradation Potential of Bulking Agents Used in Sludge Bio-Drying and their Contribution to Bio-Generated Heat." Water Res. 45(6):2322-2330.

Zhao, P., Y. Shen, S. Ge, and K. Yoshikawa. 2014. "Energy Recycling from Sewage Sludge by Producing Solid Biofuel with Hydrothermal Carbonization." Energy Convers. Manage. 78:815-821.

# A NEW CONCEPT FOR UTILIZATION OF INDUSTRIAL SOLID WASTES IN BUILDING STRUCTURES

*Arunima Shukla* and *Ashok N. Bhaskarwar*
(Department of Chemical Engineering, Indian Institute of Technology, Delhi, Hauz Khas, New Delhi 110 016, INDIA)

**ABSTRACT:** The solid-waste management is one of the important issues of present times due to rapid industrialization. Research studies reveal that globally, a total estimated annual industrial solid-wastes generation was about 11 billion tons for 2002, which is expected to increase in the coming years (Chavan and Patil, 2013).

We have developed a new technology for making light-weight aerated concrete blocks, which allows the integration of industrial solid wastes. In the present study, solid waste of autoclaved aerated concrete (AAC) industries and a by-product of coal-fired power plants known as fly ash have been successfully integrated into colloidal-gas aphrons (CGA) based aerated slurry. Experimental results have shown that it is possible to utilize flyash up to 75 % and AAC solid waste up to 25 % with respect to total mass of solids in aerated concrete.

## INTRODUCTION

Nowadays, industrial solid waste is one of the major environmental issues. The construction materials can be established as secondary industry in the area of solid waste and recycling (Chavan and Patil, 2013). There i

s plenty of literature available on the generation and utilization of flyash. In USA, the annual production of flyash is 71 metric tons, of which only 31% is being utilized (Singh and Gupta, M Singh, 2013). In India, the percentage utilization of flyash is around 10 to 15%, where more than 88 million tons of fly ash is getting generated per year (Siddique, 2004). Due to high demand of energy, the quantity of fly ash, as a solid waste, is also increasing (Singh and Gupta, M Singh, 2013). In the autoclaved aerated concrete (AAC) manufacturing industries, a considerable amount of solid-waste, to the tune of at least about 10 to 12 % of total production, is generated because of breakage due to thermal stresses arising during autoclaving, and during handling and transportation.

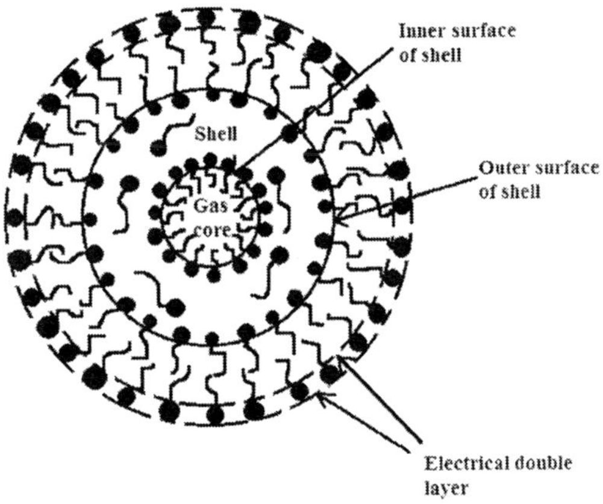


Figure 1. Conceptual diagram of CGA, as explained by Sebba

Considering the above factors, we have developed a new technology which will simultaneously manufacture light weight aerated concrete blocks, and utilize considerable amount of solid wastes generated by various industries. Aerated concrete is known for its insulating properties, as well as for its structural properties (Narayanan and Ramamurthy, 2000). Its properties mainly depend on the method of aeration, as void in the matrix is the key ingredient of light weight aerated concrete. We are working with colloidal gas aphrons (CGAs) as an aerating medium for creating the voids in the cement matrix. CGAs are micro- bubbles, typically in the range of 10 to 100 μm diameter, having high interfacial area, flow properties similar to water, and high stability (Jauregi, Mitchell, and Varley, 2000; Sebba, 1987). The conceptual diagram of CGAs is shown in Figure 1. Sebba had proposed that CGAs have inner gas phase, surrounded bi-layer of surfactant and an electrical double layer.

## MATERIALS AND METHODS

**Raw Material.** Sodium lauryl sulfate (Fisher scientific), ordinary portland cement (43 grade), flyash, autoclaved aerated concrete waste (Biltech Ltd.) of two sizes.

**Equipment.** Accelerated curing tank (Ferrotek equipments); Stirrer (Remi elecktrotechnik Ltd., 220/230 V, 1 phase, 50 Hz); Hot air oven (Macro scientific works pvt. Ltd.); Optical microscope (Motic microscope) SEM (Hitachi, TM 3000, tabletop microscope)

**Experimental Set-Up.** The experimental set-up for CGA formation consists of a spinning disc CGA generator which is an open cylindrical vessel, attached with inverted L- shaped baffles. Stirring is done by a circular disc mounted on stirrer.

## METHODOLOGY

Sodium lauryl sulfate was mixed with fixed volume of ordinary tap water for 15 minutes at 1500 rpm, in order to form a homogenous solution. This is followed by increasing the rpm of stirrer to 5500 rpm, which causes generation of waves at the surface of surfactant solution. The generated waves create thin film of gas, which being unstable, break up into small sized droplets, known as CGAs. Continuous recirculation of surfactant solution led to increase in the air hold up (up to 0.70 to 0.80) in CGAs. The CGAs generated were mixed with solid waste and cement. The samples were allowed to set for 24 hours, followed by de-moulding and steam curing at 65°C for 4 to 5 hours. The samples were then dried in oven for 24 hours, to measure oven dried weight at 110±10°C.

## RESULTS AND DISCUSSION

**Colloidal Gas Aphrons.** The CGAs were prepared by using anionic surfactant, sodium lauryl sulfate ($NaC_{12}H_{25}SO_4$). Photomicrographs of CGAs were taken using optical microscope, as shown in Figure 2. Motic software was used to determine the sizes of the micro-bubbles.

**SOLID WASTE.** Fly ash was obtained from two different resources in order to study its effect on the properties of aerated concrete. The SEM images of fly ash are shown in Figure 3. It was observed that the type 1 flyash was finer as compared to type 2. The solid waste from AAC was utilized in the form of pebbles and granules, in order to obtain re-integrated blocks of aerated concrete. Higher densities of aerated concrete were targeted to get the intact blocks in case of AAC solid waste utilization. The SEM images of different AAC waste were observed (Figure 4) to study the surface morphology of the waste. The needle-like and plate-like structures were observed in pebbles, which indicate that ettringite and calcium silicate hydrate were already developed (Stutzman, 2001). However, in case of granules, such kind of structures could not be observed.

**EFFECT OF FLY ASH ON THE PROPERTIES OF AERATED CONCRETE.** The aerated slurry has variations in the CGA distribution, which results in difference in the properties of concrete blocks. Thus, the samples, in each set of experiments, were characte--ized based on pouring of samples, the 1st poured was named as 1st cut, similarly, the 2nd and 3rd poured were named as 2nd and 3rd cut respectively. The weight percentages of cement, flyash and water were kept constant, within the range of ±0.15%. It was observed that the 1st cut was lighter, as compared to 2nd and 3rd cut, because of more air hold-up. Thus with increase in air weight % in slurry, dry density of the sample decreases, as shown in Figure 5. It was also noted that with the increase in air weight % in slurry, the compressive strength of the samples initially decreases till 0.1 air weight % w/w, and then increases slightly, as shown in Figure 6. The increase is more significant in type 2 flyash samples. This is due to the fact that porosity directly influences compressive strength (Erniati et al., 2015; Zhao et al., 2014) and the colloidal gas aphrons provide a well distributed pore size distribution in the cement matrix.

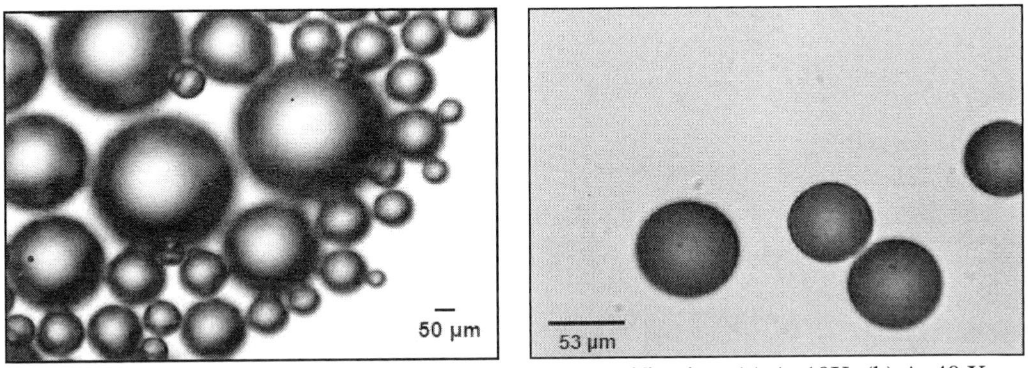


Figure 2. Photomicrographs of CGAs at different magnification: (a) At 10X; (b) At 40 X.

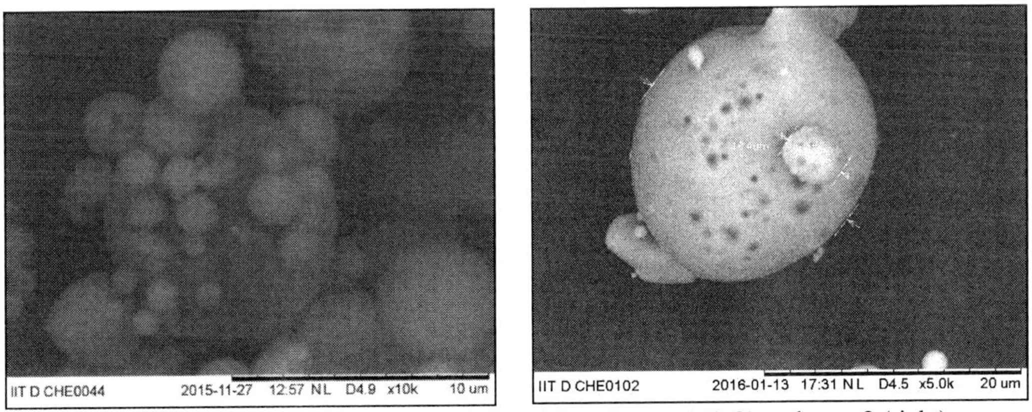


Figure 3. Scanning electron micrographs of fly ash type 1 (left) and type 2 (right).

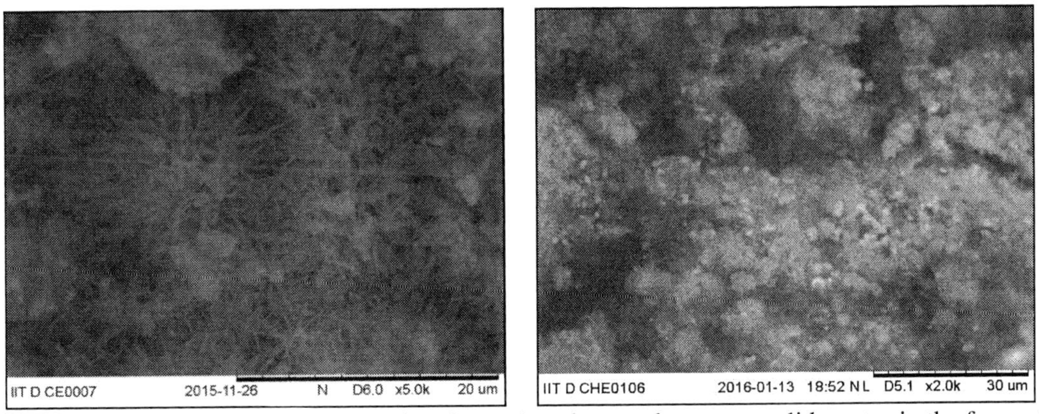


Figure 4. Scanning electron micrographs of autoclaved aerated concrete solid wastes in the form of pebbles (left) and granules (right).

All the data points were plotted (Figure 7), based on cuts, in order to check the correlation between density and compressive strength, as it was reported that compressive strength increases linearly with density (Narayanan and Ramamurthy, 2000).The range of densities from 712 to 1112 kg/m$^3$ and compressive strengths 1.01 MPa to 3.49 MPa were observed for the fly ash type 1, and densities range of 392 to 1072 kg/m$^3$ and compressive strengths 0.52 MPa to 4.6 MPa were observed for fly ash type 2, respectively. Type 2 flyash shown more linear trend as compared to type 1.

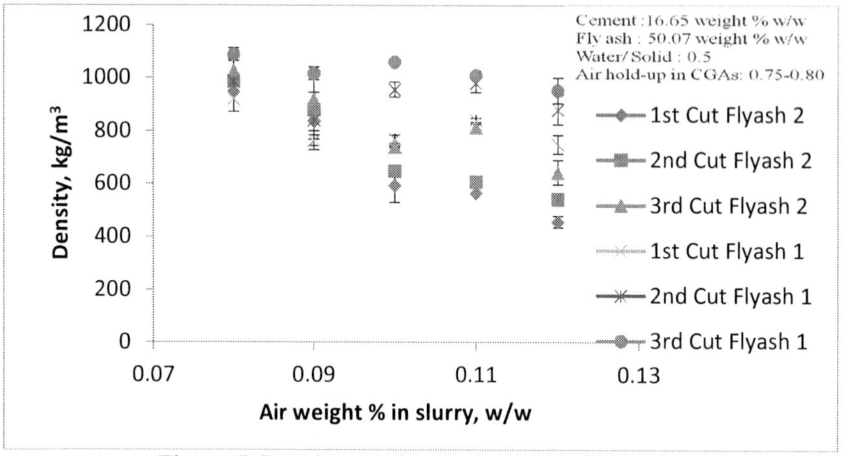


Figure 5. Density vs. air weight % w/w in slurry.

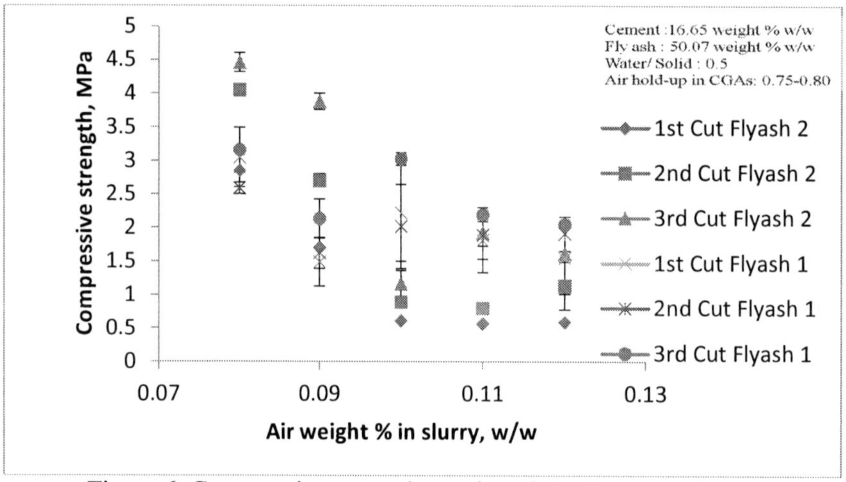


Figure 6. Compressive strength vs. air weight % w/w in slurry.

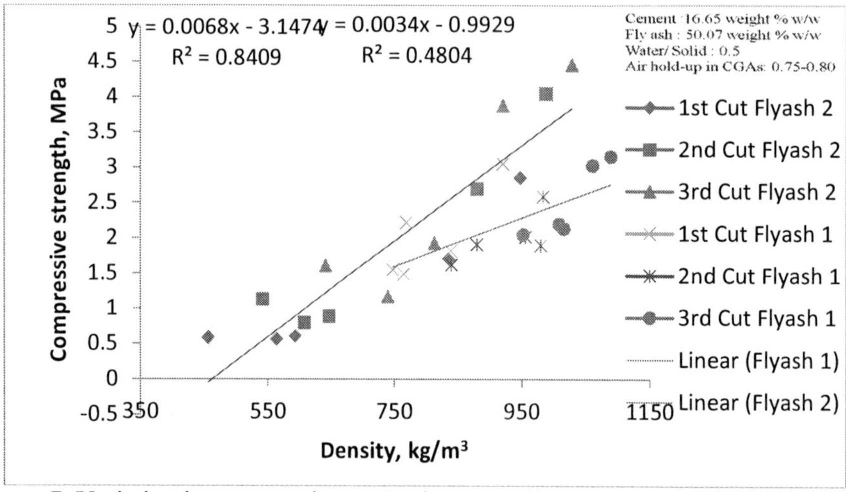


Figure 7. Variation in compressive strength vs. density based on pouring of samples.

**Autoclaved Aerated Concrete Solid Waste.** The solid waste from autoclaved aerated concrete industries was incorporated into the CGAs based slurry, varying from 5 to 25 weight % of total solid mass. The water to solid ratio was fixed as 0.5. The variation in the composition is given in Table 1.

Table 1. Varying percentages of composition with solid AAC waste

| S.No. | AAC waste (% of total solid mass) | Cement (weight %) | Flyash (weight %) | Water (weight %) | Air (weight %) |
|---|---|---|---|---|---|
| 1 | 5 | 21.09 | 42.19 | 33.31 | 0.06 |
| 2 | 10 | 19.98 | 39.98 | 33.31 | 0.06 |
| 3 | 15 | 18.88 | 37.76 | 33.32 | 0.05 |
| 4 | 20 | 17.77 | 35.54 | 33.32 | 0.05 |
| 5 | 25 | 16.66 | 33.32 | 33.32 | 0.04 |

All the samples were intact without any deformation. A good bonding between solid waste and aerated slurry was observed without any crack formation, as shown in Figure 8. The color difference between AAC waste and slurry was observed, which was marked on the sample using a marker, in order to take the SEM image of interface.

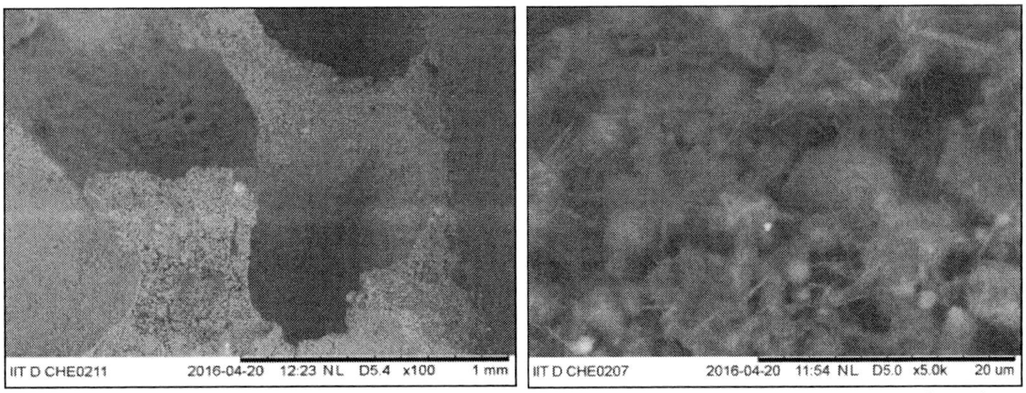


Figure 8. Scanning electron micrographs of bonding between AAC solid waste and CGAs based aerated slurry, at lower magnification (left) and at higher magnification (right) respectively.

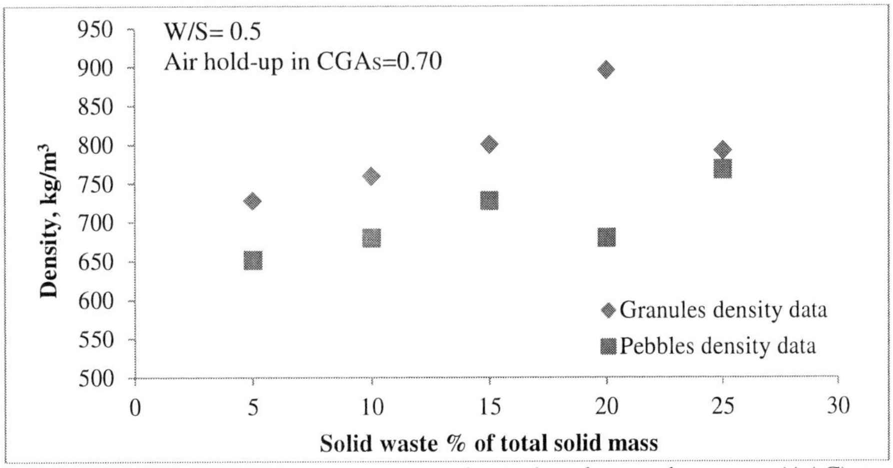


Figure 9. Density vs. solid waste percentages of autoclaved aerated concrete (AAC) sample.

The densities and compressive strengths of the samples were also measured and plotted in Figure 9 and Figure 10, respectively. The ranges of densities were from 652 to 896 kg/m$^3$ and compressive strengths 0.26 MPa to 0.75 MPa.

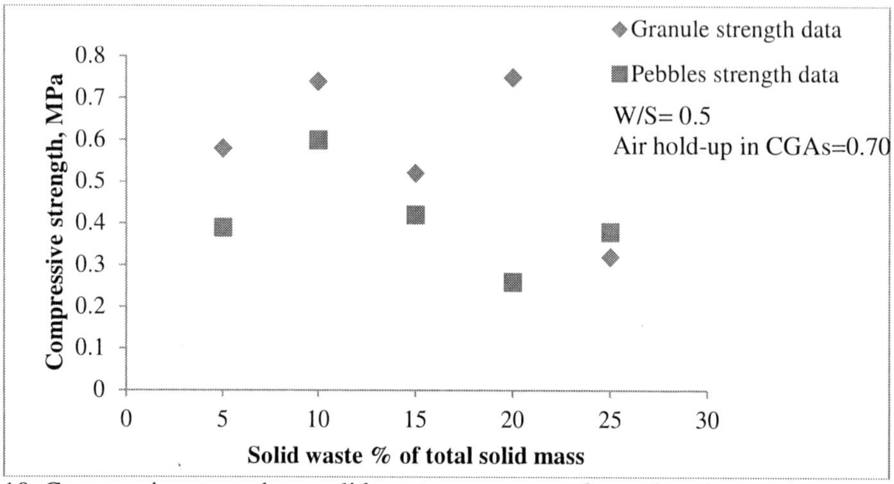


Figure 10. Compressive strength vs. solid waste percentage of autoclaved aerated concrete (AAC) samples.

## CONCLUSIONS

The results revealed that the industrial solid wastes such as fly ash, and aerated concrete, in the form of granules and pebbles, can be utilized using CGAs based technology. A wide range of strengths and densities were observed using varying composition and densities. It is, thereby, possible to utilize up to 25% of AAC solid waste with respect to total mass of solid. Moreover, there is a lot of scope for the improvement of compressive strength of the AAC solid waste sample in future, as well as utilization of other industrial solid wastes.

## ACKNOWLEDGEMENT

The authors would like to thank and acknowledge foundation for innovation and technology transfer (FITT), India for their support in patenting the work and Biltech Building Elements Limited, Palwal, India for their help in getting solid waste.

## REFERENCES

Chavan, P G, and S B Patil. 2013. "Substantial Use of Solid Waste." *International Journal of Advanced Technology in Civil Engineering* 2(1): 61–63.

Erniati, M. Wihadi Tjaronge, Zulharnah, and Ulva Ria Irfan. 2015. "Porosity, Pore Size and Compressive Strength of Self Compacting Concrete Using Sea Water."

Jauregi, Paula, Geoffrey R. Mitchell, and Julie Varley. 2000. "Colloidal Gas Aphrons (CGA): Dispersion and Structural Features." *AIChE Journal* 46(1): 24–36.

Narayanan, N., and K. Ramamurthy. 2000. "Structure and Properties of Aerated Concrete: A Review." *Cement and Concrete Composites* 22(5): 321–29.

Sebba, F. 1987. *Foams and Biliquid Foams-Aphrons*. Wiley, Chichester, Ž. England.

Siddique, Rafat. 2004. "Performance Characteristics of High-Volume Class F Fly Ash Concrete." *Cement and Concrete Research* 34(3): 487–93.

Singh, S P, and S P Gupta, M Singh. 2013. "Fly Ash Production and Its Utilization in Different Countries." *Ultra Chemistry* 9(1): 156–60.

Stutzman, Paul E. 2001. "Scanning Electron Microscopy in Concrete Petrography." *Materials Science of Concrete Special Volume*: 59–72.

Zhao, Haitao, Qi Xiao, Donghui Huang, and Shiping Zhang. 2014. "Influence of Pore Structure on Compressive Strength of Cement Mortar." *TheScientificWorldJournal* 2014: 1–12.

# RECOVERY OF COPPER FROM WASTE PRINTED CIRCUIT BOARDS (PCBS) USING ELECTRO-WINNING PROCESS

***Pradeep Kumar***, Shashi Bhushan Kumar, Poornima Pandey, Roli Saini
(Department of Chemical Engineering & Technology, Indian Institute of Technology, Banaras Hindu University, Varanasi, Uttar Pradesh, India – 221005)

**ABSTRACT:** Solid waste management is seemly a great significance topic. The constituent component part of Electronic waste (E-waste) is increasing the solid waste. The newer technology has to be developed to deal with e-waste separately. It contains hazardous substances which precipitate in to environment, if they are not handled properly. Cadmium, lead, arsenic, and chromium are most toxic metals present in e-waste. It contains some valuable materials (base metals such as iron, copper, aluminium etc.) which should be recovered. Hydrometallurgical method is suitable technique for copper recovery of metal from used PC or mobile phone. It consists of two steps first leaching and second metal recovery from the leaching solution. The mechanical processing (size reduction and magnetic separation) is an efficient preliminary operation to obtain a concentrated (Fe as the magnetic fraction and Cu as the conductive fraction in primarily phase) fraction of a specific metal from E-waste.

In present study, hydrometallurgical technique was used for recovery of copper from printed circuit boards (PCBs). Leaching was carried out in $HNO_3$ which showed a great possibility for recovery of copper. The effect of temperature, residence time, normality of leaching agent, size of PCBs, and finally solid-liquid ratio was studied in leaching operation. Electro-winning process was performed to separate copper from leaching solution. Current supply in electro-winning process was delivered by DC source across 0.37 V. The maximum copper recovery from the leachate solution in the electro-winning was obtained as 73.14 %. Overall copper recovery from PCBs to the final stage was obtained 8.641 gm/Kg of E-waste (PCBs).

***Keywords:*** E-waste, printed circuit boards, electro-winning process, copper

## INTRODUCTION

Printed circuit boards are basic component of each electronic goods, which made from polymers, ceramics and metals, mainly Cu (Calgaro et al., 2015). With cutting age technology, demand of new electronic goods increases the higher amount (approx. 20-25 million per year) of E-waste generation, which require recycling (Robinson, 2009, Orlins and Guan, 2016). Because it contained various hazardous and non-hazardous metals, these released to atmosphere and caused harmful effects on living beings (Sthiannopkao and Wong, 2013). There are several methodology available for recovery of metal from E-waste same as from ores such as pyrolysis, hydrometallurgical leaching and bioleaching in initial step. Then the finally metal from leachate will be processed by solvent extraction, cementation, activated carbon, bio-sorption and ion exchange post treatments (Cui and Zhang, 2008, Pant et al., 2012). Before recovery of metal, some basic pre-treatment steps (hammer mill, grinder mill, knife cutter, magnetic separation) are used for size reduction, separation of ferrous and non-ferrous metals (Bizzo et al., 2014). In present article, hydrometallurgical technique was used for recovery of copper in solid phase from printed circuit boards (PCBs), followed by Electro-winning process.

## MATERIALS AND METHODS

**Electronic Waste (E-waste).** Printed circuit boards from used and discarded electronic goods were segregated and collected as E-waste from "Department of Chemical Engineering, IIT (BHU) Varanasi.

**Materials.** The chemicals (AR Grade) involved during leaching of copper were nitric acid ($HNO_3$, purity 69-70%) procured from Fischer Scientific Limited as leaching agent and NaOH granules from SD Fine

Chemicals Ltd. as flow of charge in Electro-winning process. Distilled water was prepared in the laboratory usage for dilution of leaching agent.

**Experiment method.** First of all PCBs (about 1 kg) having metal strip had reduced into small pieces 10 mm × 10 mm or smaller using grinder mill and knife cutter. These contained both magnetic and non-magnetic materials, which were manually separated through magnet bar. In this article, copper was main concern metal which obtained in non-magnetic materials. This was leached out by dissolving non-magnetic materials in concentric nitric acid ($HNO_3$) as leaching agent. After filtration process, the leachate solution was treated by Electro-winning process. For this process, the cell was prepared by consisting of a copper cathode (10 mm × 20 mm) and a stainless steel plate as anode as shown in Figure 1. It was connected through wire to the DC electric source. Characterization of leachate was done by Inductive couple plasma (ICP) for determination of trace elements in a sample.

**Figure 1.** Set up of Electro-winning process

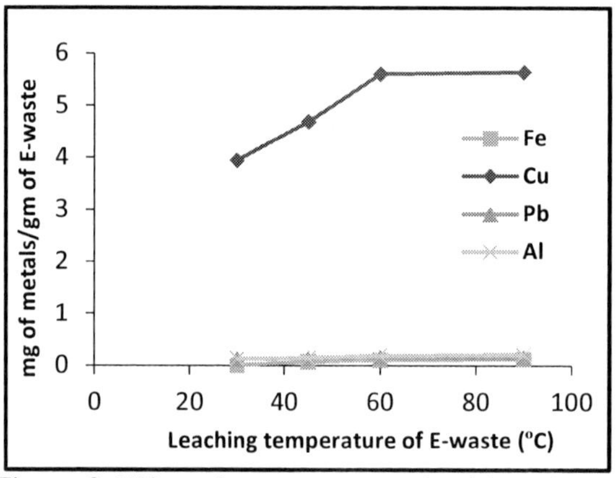


**Figure 2.** Effect of temperature on leaching of metals, Experimental condition: Sample (E-waste) = 12 gm, Solid liquid ratio = 1:3, Residence time = 180 min, Normality of the solvent ($HNO_3$) = 15.66 N

## RESULTS AND DISCUSSION

All experiments were conducted using waste printed circuit boards (PCBs) as electronic wastes (E-wastes). These were also done by variation of different parameters like temperatures, solid liquid ratios, size of PCBs, residence time, and acid concentrations on copper leaching from PCBs specimen. The obtained results were discussed below.

**Effect of Temperature.** The effect of temperature on the leaching of PCBs was studied at different temperatures of 30, 45, 60 and 90 °C or other parameters such as solid liquid ratio 1:3 (25.5 ml of $HNO_3$), residence time 180 min and normal size of PCBs (over size particle of mesh size 6) were used.12 gm of four samples were taken for leaching experiments. The results are shown in the Figure 2 that rising of the temperature has positive effect on leaching of metal in leaching solution. Leaching rate increases as temperature increases, but becomes constant at 60 to 90 °C. Initially more concentrated acid penetrate the solid material easily, but above 60 °C temperature increases viscosity of the solvent decreases rather than leaching rate becomes slow. It becomes almost constant as leaching is exothermic reaction, the heat sensitivity of solute, volatility of the solvent and possible thermal effects on the inert or residue solid limit the temperature of the reaction. At elevated temperatures, the flow of leaching agent in to pore becomes slow it cannot penetrate the solid pore.

**Effect of Solid Liquid Ratio.** The effect of solid-liquid ratio on leaching of PCBs was studied by taking four samples of equal weight (10 gm of each). Samples were maintained under solid liquid ratio 1:2, 1:3, 1:5, 1:10 by changing volume of $HNO_3$, but amount of samples were remain same. For all the four samples, other parameters were kept constant such as temperature 30 °C, leaching time 180 min and normality of solvent ($HNO_3$) 15.66 N. From Figure 3 shows that leaching rate increases on increasing solid liquid ratio. However, after certain limit it decreases because it became saturated, as further liquid molecules cannot penetrate the solid material.

**Effect of Particle Size of (PCB).** The effect of particles of crushed size of PCBs on leaching rate was studied by using particle of size greater than 3.35 mm, equal to 3.08 mm and less than 2.81 mm. 30 gm of crushed PCBs were taken and classify by different particle size with help of two sieves of BS5 and BS4. After shaking, three different particle sizes of E-waste samples were obtained. One was upper portion of the first sieve, another was in between the two sieves and finally last one was on the pan. These three samples were taken separately and weighted. It was found 12 gm upper side, 10 gm in between the sieves and remaining in the pan. Temperature of experiment was maintained constant 30 °C as well as solid liquid ratio 1:3 and residence time 180 min.

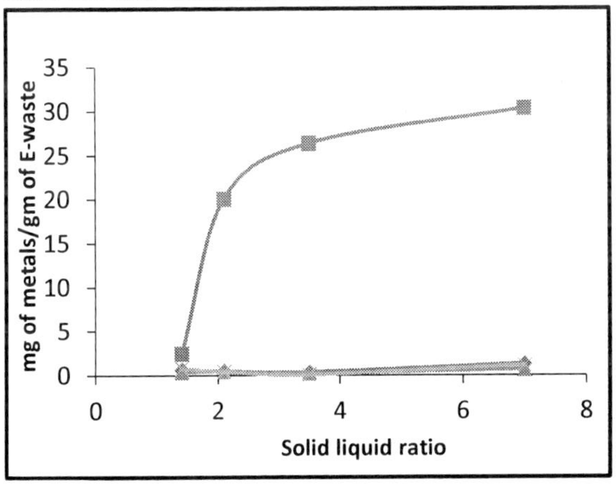


**Figure 3.** Effect of solid to liquid ratio (w/v) on the recovery of metals during leaching. Experimental condition: Residence time = 180 min, Temperature = 30°C, Normality of the solvent ($HNO_3$) = 15.66 N.

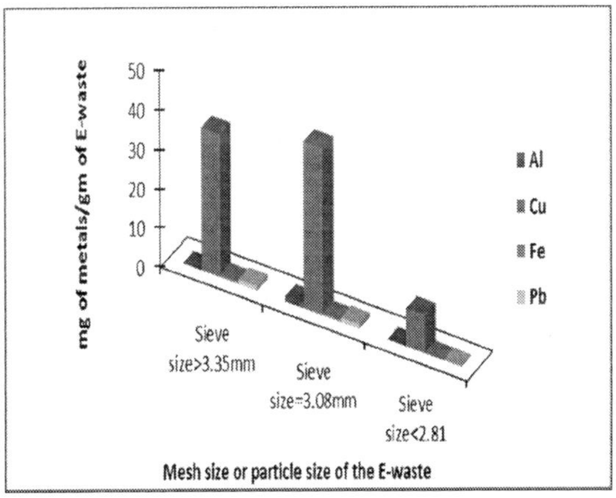


**Figure 4.** Effect of particle size of E-waste on the recovery of metals during leaching. Experimental condition: Residence time = 180 min, Temperature = 30°C, Solid liquid ratio = 1:3 and Normality of the solvent ($HNO_3$) = 15.66 N.

Figure 4 shows that as sizes of PCBs decreases extent of leaching decreases. But in real case as the size of the material decreases recovery increases because pores and surface area both increases. The leaching is directly proportional to pores and surface area. But this experiment result was not obtained as real. The reason behind that the grinder machine which was not working so effective. It crushes PCBs present but the metal strip remains greater size. When sample was separated by sieve maximum metal part remains on the upper and the middle of the sieve. On leaching sample, more metals were obtained in bigger size particle, not in smaller particle size.

**Effect of Residence Time.** Each sample (average particle of 10 mm × 10 mm) having weight of 10 gm were poured in to a beaker separately. Solid liquid ratio was 1:3 (470 gm/L) as constant for entire experiment. All four samples were filtered after time duration of 2, 4, 6, 24 hr. From Figure 5 it is observed at the beginning leaching rate increased with residence time very fast. After 4 hr leaching rate of PCBs in leaching solvent becomes constant as because the competition between different metals ions in solution. Initially there was large number of pores and surface area available, as well as no resisting material so it can penetrate solid easily, therefore leaching rate decreases after 4 hr.

**Effect of Normality of Leaching Agent.** Four samples of each 10 gm were taken for analysis of effect of normality of leaching agent during leaching. Other parameters were particle size (10mm × 10mm), temperature 30 °C, solid liquid ratio 1:3 (470 gm/L) and residence time 2 hr. Leaching rate of copper is found to increase with increase acid concentration from Figure 6, but latter it becomes constant due to the competition caused by leaching of copper at high acidity leading to increase in the viscosity of the solution that consequently inhibits the leaching process.

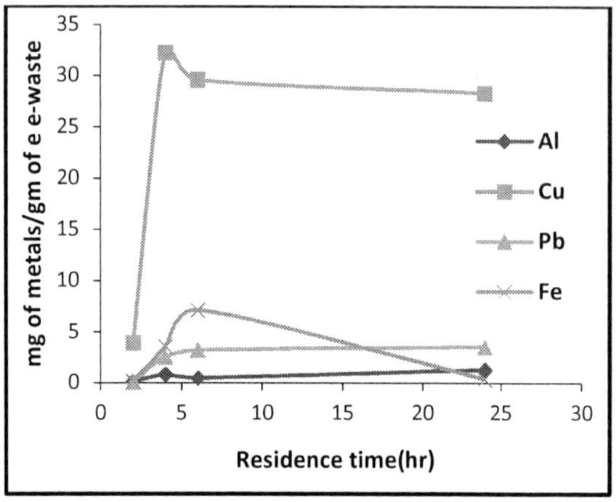


**Figure 5.** Effect of residence time on recovery of metals during leaching. Experimental Condition: Temperature = 30 °C, Solid liquid ratio = 1:3 and Normality of the solvent ($HNO_3$) = 15.66 N.

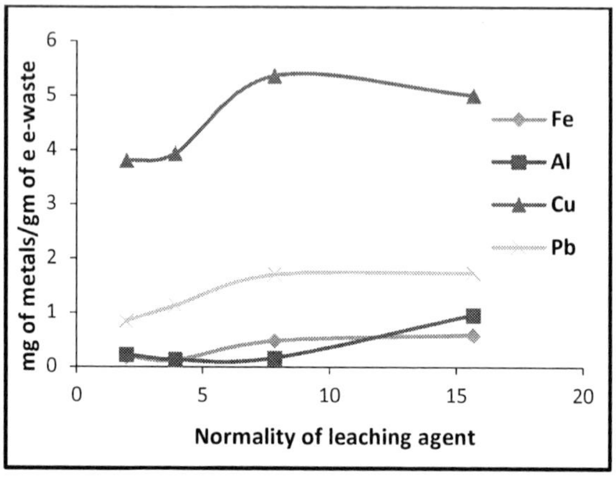


**Figure 6.** Effect of normality of leaching agent on recovery of metals during leaching. Experimental Condition: Residence time = 180 min, Temperature = 30 °C and Solid liquid ratio = 1:3.

**Recovery of copper from the leaching solution.** It was final stage of copper recovery from E-waste leachate. Electro-winning process was performed to obtained copper from leaching solution to solid form. The copper recovery in the overall process is 8.641gm/Kg of e-waste (73.14 %).

## CONCLUSION

The article had studied on the recovery of copper from the waste printed circuit boards (PCBs) as E-waste using leaching followed by electro-winning process. Analysis by inductively coupled plasma (ICP) had characterized the presence of Cu, Pb, Fe, Zn, Fe, Mg, Al, and Ca in the leachate sample of PCBs in different proportions. Leaching of PCBs was carried out in $HNO_3$ as leaching agent which showed a great possibility for recovery of copper. The variation of different parameters like temperature, residence time, normality of leaching agent, size of PCBs, and finally solid-liquid ratio was studied in leaching operation. It further revealed that best operating condition for the leaching of copper from PCBs at temperature 60°C, normality of $HNO_3$ acid 10 N, residence time 4 hr and size of PCBs about 3.08 mm. After that electro-winning process was performed in order to separate copper from leaching solution. Using DC source across 0.37 V, current supply was delivered in electro-winning process. The maximum copper recovery from the leachate solution in the electro-winning was obtained as 73.14 %. Overall copper recovery from PCBs to the final stage was obtained 8.641 gm/Kg of E-waste (PCBs). The article delivered the idea for formal recycling of devastating E-waste which could be adopted by developing countries to reduce exploitation of resources.

## ACKNOWLEDGMENTS

The authors would like to gratefully acknowledge to the Department of Chemical Engineering and Technology, I.I.T. (B.H.U.), Varanasi for offering facilities for characterization and MHRD which funded the research.

## REFERENCES

B. H. Robinson. 2009. "E-waste: An assessment of global production and environmental impacts". *Science of the Total Environment*, 408: 183–191.

C.O. Calgaro, D.F. Schlemmer, M.D.C.R. da Silva, E.V. Maziero, E.H. Tanabe and D.A. Bertuol. 2015. "Fast copper extraction from printed circuit boards using supercritical carbon dioxide". *Waste Management*, 45: 289-297.

D. Pant, D. Joshi, M. K. Upreti, R. K. Kotnala. 2012. "Chemical and biological extraction of metals present in E waste: A hybrid technology". *Waste Management*, 32: 979–990.

J. Cui, L. Zhang. 2008. "Metallurgical recovery of metals from electronic waste: A review". *Journal of Hazardous Materials*, 158: 228–256.

S. Orlins and D. Guan. 2016. "China's toxic informal e-waste recycling: local approaches to a global environmental problem". *Journal of Cleaner Production*, 114: 71-80.

S. Sthiannopkao and M. H. Wong. 2013. "Handling e-waste in developed and developing countries: Initiatives, practices, and consequences". *Science of the Total Environment*, 463–464: 1147–1153.

W. A. Bizzo 1, R. A. Figueiredo and V. F. de Andrade. 2014. "Characterization of Printed Circuit Boards for Metal and Energy Recovery after Milling and Mechanical Separation". *Materials,* 7: 4555-4566.

# WASTE: A RAW MATERIAL FOR SOIL AMENDMENT AND AN EMPLOYER OF LABOUR

*Ogbonna, Emmanuel Ahamefula** and Isinguzo, Nnadozie Sunday
(DOTECON Nigeria Limited, Lagos, Nigeria)

**ABSTRACT:** A demonstration waste control pilot project was carried out at the Ado Ekiti, Ekiti State, Southwest Nigeria, to justify the concept of waste management and environmental sustainability through youth engagement for poverty reduction in the area. The project involved a standard waste handling method which ensures a cleaner and greener environment thus reducing the harmful effects of pollution; and also, the transfer of wastes into soil amendment thereby generating cash for the sustenance of the youth through their engagement in the project. Three designated colour coded wheeler bins (green, blue and brown) of 250 each were used for collecting organic, recyclables and toxic wastes respectively through waste segregation from source of generation thus ensuring the project concept of **"3R":** reduction, recycle and re-use of wastes. A total of 750 wheeler bins were distributed to different household and designated places. Polythene bags were assigned to the green coded wheeler bins in order to forestall the spilling of organic waste. Wastes inventory, sensitization campaign of waste management, construction of a compost plant – greenhouse, transfer of composting technology or skill acquisition program, employment of young people for waste management and production of compost for agricultural purposes were employed. The recyclables wastes were sold for re-use or recycling while the toxic wastes were disposed by the government waste agent. The organic wastes were transformed into compost and sold at an affordable rate which made it readily available to the farmers for their small holder vegetable farm which had little or no previous soil amendment/conservation practices; thus justifying the need for amendment with compost. A total of 2tons of compost was turned over within 30 days interval using modified thermo aerobic biodegradation technology at the compost farm consisting of 12 cells measuring 60 square meters. A demonstration farm with *Amanrantus spp* as the test crop was used to show the effectiveness of the compost. A public training was conducted for general awareness and on the handling of waste management and compost production, during which about 500 persons received the training and a total of 20 waste handlers was employed in the pilot project.

## INTRODUCTION

The need for soil amendment and productivity, sustainable environment, organic farming and waste management for pollution reduction/eradication has called for effective and efficient disposal and conversion (composting) of wastes into useful products thus engaging manpower for a better living. The initiation of this project was aimed at showcasing the possibility of conversion of wastes to raw materials and jobs concept through segregation of wastes at source and composting. Historically, composting as an ancient biotechnology (Birendra, and Lourduraj, 2007) is a recognized practice dates to at least the early Roman Empire since Pliny the Elder (AD 23-79) and its modernization has dated since the year 1920s in Europe as a tool for organic farming (Heckman, 2006).

## PROJECT CONCEPT

The concept of this project is based on the following:

- Reduction of waste generation and segregation of wastes at source
- Zero waste discharge or disposal through the 3Rs concept therefore achieving cleaner environment
- Conversion of wastes to valuable product (compost) therefore, revenue creation
- Encouragement and improvement of organic farming
- Poverty alleviation (providing food and revenue)
- Creating job opportunities for the youths thus reducing/eradicating youth restiveness

## MATERIALS AND METHODS

**Project Area Description.** The project was executed at the Ado Ekiti GRA (within the coordinates, N07$^0$36'29" E005$^0$16'28" and N07$^0$37'54" E005$^0$16'29 ) and Erifun Compost Center (approximately N07036', Ado Ekiti, Ekiti State, Southwest Nigeria with land area of two square kilometers and a near population of 50,000 people and over 10,000 residential homes, industrial building, business area such as market and shops. Generally, Ekiti State is known for agricultural activities such as hunting and farming of much, arable (vegetable) farming.

**Field Method**. Augured surface (0-30 cm) samples were taken randomly from the study site and bulked together to form a composite. The composite soil sample was air dried at room temperature and passed through sieved with 2mm sieve and properly labeled. Borehole water and soil samples were analyzed for both physicochemical and microbiological properties of the study area so as to know the baseline of the environment and specific nutritional needs of the areas concerned.

**Project Resources:** Both skilled and unskilled manpower (youth) resources were employed. The unskilled resources involved in waste collection, segregation at various receptacles including the composting stage and transfer of wastes to labour whereas the skilled human resources were employed for both awareness and skill acquisition training in the course of this project. Materials used for the project were bill boards, PPEs, hand bills, colored coded wheeler bins, degradable or compostable wastes, composting materials among others.

**Enlightenment, Awareness and Training:** The enlightenment involved development of training strategy, procurement of sensitization materials such as posters, bill boards and media advertisement. Awareness on the need/importance of the project as well as the benefits was disclosed professionally. The training was conducted for the public for general awareness and for the waste handlers for specific skill acquisition in both wastes segregation, reduction, reuse and recycle as well as compost production. Aggressive house to house sensitization was done during bin distribution. A large number (over 100) of trainees were trained during the public training section. A total of 30 trainees were employed with an in-house training on waste management system and composting, equipped with personal protective equipment (PPEs) as well as training manual for the operations. Hand bills, material translation, coloured coded bins, bill boards, PPEs, training package and materials etc. were used to facilitate the training, awareness and enlightenment in the project geographical areas as shown Figure 1.

**FIGURE 1. ENLIGHTENMENT, AWARENESS AND TRAINING SECTIONS**

**Bin Distribution:** Demographic census and survey were carried out in the study area as shown in Figure 2. A total of 750 homes were identified and a total of 750 colored coded wheeler bins were supplied. The bins were categorized into three using coloration such as green, blue and brown of 250 each for collection of organic, recyclables and toxic wastes respectively. A set of each colored coded wheeler bin was distributed to each home so as to achieve wastes segregation at source. Branded polythene bags to be placed inside the green bins were distributed to forestall the spilling of the organic wastes.

FIGURE 2. DEMOGRAPHIC CENSUS, SURVEY AND BIN DISTRIBUTION

**Project Infrastructure/Strategy:** The infrastructures for the project were compost plant (greenhouse of a total capacity and turnover of 2.5 tons of compost per week at optimal capacity production), size reduction diesel machine (so as to increase the surface area of raw material (wastes) and speed degradation rate), borehole water, office, test farm, farm implements, waste collection and transportation system among others. The compost plant was designed to allow for adequate aeration, minimal heat loss and optimized microbial activities therefore facilitating the rate of decomposition (Figure 3). The waste collection and transportation system involved the use of transportation and waste handlers to collect the segregated wastes from homes and transport the organic wastes (green bins) to the compost plant for composting. At the point of waste collection the polyethene bags were replaced by the waste handlers. Two levels of transportation of segregated wastes were employed from the "source segregation system". Segregated wastes were transported from the "source segregated point" to the transfer center by the residents using waste silo trucks or garbage vans and finally to the waste segregation/inventory site thereafter to the recycling/composting plants. The organic wastes were weighed on delivery while the recyclables were sold for reuse or as raw material to other areas of production therefore creating avenue for income to the youth. The transportation system involved the use of compactors, carts or wheel barrows, pipes for liquid wastes (which were re-introduced back to the composting system), drainages and waste bags.

FIGURE 3. PROJECT INFRASTRUCTURE

**Project Technology:** The bio-technology was employed for the composting. This involved thermo-aeration biodegradation process using organic additives and natural catalyst by enzymatic actions of aerobes (microorganism) so as to ascertain an environmental friendly condition. A semi-intensive system of mechanization was employed due to the limited power supply thus sustainability of the project.

## RESULTS AND DISCUSSION

**Physicochemical and Microbiological Properties of the Study Area:** An initial analysis of the physicochemical and microbiological properties of the site was carried out and the result is presented in the Table 1, 2, 3 and 4. This analysis was important to know the specific nutritional needs of the areas concerned and also because the composting area is representative of a typical agricultural land area in Ekiti State; this fact is evident in the frequency at which neighboring land areas are cultivated.

The result indicated that the average concentration of Nitrogen in the soil is quite low. This important nutritional element must be adequately provided for any arable crop to do well in this area and by extension Ekiti State. Armed with this knowledge the Brown (Carbonaceous) and the Green (Nitrogenous) materials were collected, sorted, mixed and layered.

With the exception of chloride and sulfate values most parameters were low. The physical environment was predominantly clay in texture, hence further demonstrating the relatively higher levels of chloride and the rocky nature that was experienced during the borehole drilling.

**TABLE 1. MEAN VALUES OF CATIONS AND ANIONS (mg/L) IN SOIL SAMPLES**

| Sample | pH | Mg | K | Na | Cl | $SO_4$ | $PO_4$ | $NO_3$ |
|---|---|---|---|---|---|---|---|---|
| Composite sample | 5.7 | 0.040 | 0.020 | 0.011 | 19.52 | 2.97 | 0.05 | 0.12 |

**TABLE 2. AVERAGE MICROBIAL POPULATION/DOMINANT SPECIES OF SOILS FROM COMPOST PLANT SITE**

| SAMPLES | | SPECIES AVERAGE MICROBIAL COUNT (cfu/gm) | |
|---|---|---|---|
| | | ***Bacillus spp*** | ***Fungal spp*** |
| SAMPLE 1 | Control | $7.5 \times 10^6$ | $2.0 \times 10^3$ |
| SAMPLE 2 | Composite (compost site) | $10.0 \times 10^6$ | $4.8 \times 10^4$ |

There were many bacterial and fungal populations in the environment but the most outstanding and readily isolated were the Bacillus species. The control site was taken from a farm land which was about 100 meters away from the seasonal swamp.

Physical and chemical analysis of the compost site borehole water is showed in Table 3. However all the levels are within higher levels of regulatory target limits. Water sample analysis revealed absence of coliforms and presence of heterotrophic bacteria and fungi (Table 4).

**TABLE 3. PHYSICOCHEMICAL PROPERTIES OF COMPOST SITE BOREHOLE WATER**

| Sample | pH | Mg | K | Na | Cl | $SO_4$ | $PO_4$ | $NO_3$ |
|---|---|---|---|---|---|---|---|---|
| | | ← | | | Mg/l | | | → |
| Composite sample | 6.70 | 3.495 | 0.089 | ND | 80.60 | 44.80 | 28.50 | 20.10 |

**TABLE 4. RESULT OF MICROBIOLOGICAL CHARACTERISTICS OF COMPOST SITE BOREHOLE WATER SAMPLE**

| S/NO | GROUPS OF MICROOGANISMS ISOLATED | READINDS Cfu/ml |
|---|---|---|
| 1 | Total heterotrophic bacteria | $1.40 \times 10^2$ |
| 2 | Total heterotrophic fungi | $1.00 \times 10^1$ |
| 3 | Total faecal coliforms | 0.00 |

**Composting.** The urgency for environmental sustainability, youth engagement and plant nutritional need called for the compost production. The composting involved collection of segregated wastes, layering, watering, turning, temperature monitoring, air drying, sieving, weighing, bagging, sealing and marketing for agricultural purposes (Figure 4). The high demand for compost due to its environmental friendly nature led to rapid turnover per week. This aforementioned informed our decision to lay wastes based on our knowledge of their degradation rate, introduce certain already degrading wastes and additives to speed up the development of the microbes as well as the decomposition process, scavenge for more wastes and also to cover stacked-degrading-wastes' surface with dark colored tarpaulin which traps heat within the degrading wastes thereby encouraging microbial activity and warding of excessive odor and flies in the composting area. The degradation processes were observed and tested so as to ascertain different stages of the composting and daily temperature reading were also done. At maturity stage of the compost, percentage composition of the nutrient elements in the produced compost was recorded as showed in Table 5. Their impacts in a test crop (*Amarantus spp)* farm showed over 100% increase in vegetable yield when applied (Figure 4.).

**Production and Market Potentials of Compost.** There is enough market for the up-to-date composted residues from kitchens, farms, gardens and markets in Ado-Ekiti. The market potential shows at the minimum double the size than the maximum compost production potential. The agricultural sector alone would be large enough in nearly all to take up the entire compost production. Even the non-agricultural sector (e.g. landscaping, hobby gardens) shows already a sufficient market potential. Market problems in some places were caused mainly by low compost qualities and the lack of experience and knowledge about compost and the potential customers. End-of-waste standards can act in this respect beneficially by raising the awareness of the importance of the compost quality in the waste sector as a precondition for successful application and marketing.

**FIGURE 4. COMPOST PROCESSES, EFFECTS ON TEST CROP AND INAUGURATION**

**TABLE 5: BASIC PROPERTIES OF THE MANUFACTURED COMPOST**

| ITEM | ELEMENTS | PERCENTAGE COMPOSITION |
|---|---|---|
| 1 | Nitrogen | 32.2% |
| 2 | Phosphorus | 44.2% |
| 3 | Potassium | 23.6% |

**Project Impact on Youths.** Generally, a positive impact of the project was recorded on the youths. Apart from those already trained and employed thus ensuring positive living standard, the rate of youth restiveness and crime rate of the environment were reduced to the minimal level. More interest has been shown by other youths for training which now converted the project as a training or skill acquisition centre. Consequently, there comes the need for expansion or replication of the project across the nation and world at large. Some trained youths have now used their garden for composting to improve their farming system as well as getting cash into their pockets through sales of the compost.

## CONCLUSION

Conclusively, although several methods of composting exist; but with due consideration to cost, time, quality and quantity, the Indore method (thermo-aerobic technology) that we have employed here would most likely be the best. Composting materials such as organic wastes from homes and markets, poultry droppings, cow dung, pig dung, saw dust and so on are readily available without cost in Ekiti State and Nigeria in general, the market for compost also abound both in Ekiti State and outside the state. The technology we have applied is simple and can be easily replicated with little cost, all these we have passed on to the trainees given to us by Ekiti State. In addition, the compost when sold yielded revenue for the waste trainees and the State Government. These crops of young people will no doubt run with this vision of turning waste to cash while making the environment cleaner.

In addition, it is very obvious that the availability and cost of buying inorganic fertilizer is readily unavailable and high respectively thus creating difficulty to the farmers consequently very low yield being harvested. Fortunately, compost (organic fertilizer) is readily available, cheap and even more environmentally friendly when compared with the inorganic fertilizer.

This shows that all organic wastes from our domestic sources can be converted to wealth. We have proven that a zero organic waste disposal can be achieved while creating employment opportunities for the youth and facilitating a cleaner and greener environment in the Ekiti GRA environment. Last and not the least was a record of over 100% improvement in the vegetable harvest from the in situ trials.

## ACKNOWLEDGEMENT

This research was supported by DOTECON Nigeria Limited. I thank the Almighty God for His inspiration, guidance and protection in the course of the entire study. I specially appreciate the efforts and contributions of Dr. N. S. Isinguzo in the achievement of this study. I specially appreciate the assistance made by my mentor, Dr. C. F. Isinguzo financially. I am grateful to my lecturers, Dr. O. I. Nwachukwu and Dr. M. I. Onwuka for their support and care. I am obliged to extend my gratefulness to my beloved family, mostly my mother, Mrs. M. U. Ogbonna; brother (Chibueze) and my dearest father, Mr. L. E. Ogbonna of blessed memory.

## REFERENCES

Birendra, K.Y and Lourduraji, A. C. 2007. *In Organic Farming and Mycorrhizae in Agriculture.* I.K International Publishing House Pvt. Ltd, New Delhi, pp 37-58.

Heckman, J. 2006. "A History of Organic Farming: Transitors from Sir Albert Howards War in the Soil to USDA National Organic Program". *Renew Agric. Food Syst.* 21:143-150.

# THE EFFECTS OF DIFFERENT CONDENSATION TEMPERATURES ON WASTE TIRE PYROLYSIS

***Neslihan Doğan-Sağlamtimur***, Ahmet Bilgil, Tuğçe Aykanat, Öznur Çalışkan, Funda Yolcu, Elif Dilan Yıldırım, Türkan Vural, Halime Ötgün and Büşra Arıcan
(Department of Environmental Engineering, Engineering Faculty, Nigde University, Nigde, 51240, Turkey)

Despite the efforts made for the solutions, the disposal of waste tires (WTs) representing over 2% of total solid waste still remains as an important health and environmental problem. WTs have high volatile and fixed carbon contents with heating values greater than that of coal and biomass making them an ideal raw material for energy valorization process. Pyrolysis, an endothermic process, could be a good solution to WT problem because of obtained valuable solid and liquid products as well as generated low emissions.

The raw materials to be used for the pyrolysis experiments are samples of WTs rubber shreds, which will be supplied by collecting firms from Niğde, Turkey. In this study, it is aimed to obtain two phases that are liquid (pyrolytic oil) and solid (carbon black). The non-volatile carbon black remains as a solid residue, and therefore it can be recycled in worthwhile applications. The derived oils can be used as fuels or a source of refined chemicals.

The heating and cooling temperatures of WTs are important parameters that affect the reaction time, product yield and product quality. As the temperature changes, the phase of the product also changes. Lower pyrolysis temperatures usually produce more liquid products, while higher temperatures favour the production of gases. In this study, the pyrolysis reactor was designed, including modified oven for heating and its condenser units for cooling. The pyrolysis has been planned to perform five different decomposition (200, 300, 400, 500 and 600 °C) and condensation (-20, -10, 0, 10 and 20 °C) temperatures, respectively. An optimum design will be achieved for these temperature combinations. According to the results, ±50 °C intermediate values will also be examined. Expecting the values of the carbon black to remain stable, in this study, the effect of condensation temperatures for obtaining more pyrolytic oil and less gas release will be found.

# THE IMPACT OF SAW DUST BIOCHAR ON SOIL PH AND CROP YIELDS FOR CLIMATE CHANGE MITIGATION

Lukong Pius Nyuykonge and ***George Elambo Nkeng***
(Ecole Nationale Supérieure des Travaux Public (ENSTP) Yaounde, Cameroon),
Lendzemo Wirnkar Venasius
(Institute of Agricultural Research for Development, Bamenda, Cameroon)

Farmers in some regions of Cameroon are faced with the problem of high and unstable soil acidity (pH<5 in general) leading to low yields of agricultural products. The main objective of this work is to test the effect of biochar on soil pH in the central region of Cameroon (3° 51' 2" N, 11° 7' 30" E,) as a mechanism attracting carbon storage. Biochar obtained from the pyrolysis of saw dust from Iroko timber (*Milicia excels*) at a temperature of 500 °C after analysis was found to contain 44 %C, 4.6 %Ca, 1.1 %N, 0.2 %P, 1.9 %K, with a pH of 9.0, an acid neutralising capacity of 15 %$CaCO_3$, CEC of 49 cmol(+)/kg and other exchangeable cations (Ca: 8.3, K: 36, Mg: 2.8) Ca/Mg ratio 2.9. The biochar was applied in a field experiment at the rate of (5t/hectare without fertilizer, 5t/hectare with fertilizer, and the case of fertilizer alone used as the control). The biochar treatments were found to increase the final biomass, root biomass, plant height and number of leaves of Maize plant (*Zea mays*). The soil pH increased from 5.0 to 6.1 in the case of biochar without fertilizer, from 5.0 to 5.6 in the case of biochar with fertilizer and a slight decrease in pH from 5.0 to 4.9 was observed in case of fertilizer alone, in the first harvesting season (12 weeks after planting). In the next harvesting season, the treatment of biochar without fertilizer continued from 6.1 to 6.8 while that with fertilizer increased from 5.6 to 7.1 with the case of fertilizer alone increasing from 4.9 to 5.2. The greatest biomass increase due to biochar addition was found in the soils with biochar with fertilizer (23N + 10P + 5K + 3S + 2MgO + 0.3Zn) which stood at 150 % rather than biochar alone (95 %) in the first season. Over the next season, the impact increased in the case of biochar alone (from 95 to 202 %) and the treatment of biochar with fertilizer remained almost stable. The results of this study showed that biochar can be used as a soil buffer reducing the cost of expensive limestone often used thereby attracting farmers in storing carbon, a mechanism which can help mitigate against the effects of climate change.

# IMPACTS OF WASTE DISPOSAL METHODS ON HOUSEHOLD HEALTH CONDITIONS IN KENYA

Stephen Onakuse, ***Emmanuel Donkor***, Joe Bogue
(Centre for Sustainable Livelihoods, and Department of Food Business and Development, University College Cork, Cork, Ireland)

Waste management has attracted the attention of most researchers in developing countries due to the health implications attached to improper waste disposal. This paper therefore examines the impacts of waste disposal methods on health conditions of 370 households in Kenya. The Probit regression model was employed to relate waste disposal methods used by households to their health conditions. The survey reveals that majority of the households in Kenya dispose their refuse in rivers/drainage, common garbage heap, latrine whilst others use waste collectors. The incidence of water borne diseases such as cholera, malaria and typhoid is prevalent in the Kenya. Households that dispose refuse into river/drainage, common garbage heap and latrine have higher probability to contract water borne diseases as compared to those that pay waste collectors to manage their waste. Presence of health care providers in community was also found to reduce contraction of water borne diseases. The study recommends that sensitization of the health implications of improper waste disposal should be created among households so as to reduce water borne incidences.

# THE CHALLENGES OF SUSTAINABLE MANAGEMENT OF ELECTRONICS WASTE AND THE RISK ASSESSMENT OF IMPROPER MANAGEMENT PRACTICES

**E. Sahle-Demessie[a], Teri' Richardson[a], Joshua Dietrich[b]**

([a] Office of Research and Development, NRMRL, US Environmental Protection Agency, Cincinnati, OH, USA; [b]Chemical Engineering Program, Department of Biomedical, Chemical, and Environmental Engineering, University of Cincinnati, Cincinnati, OH, USA)

Electronic waste (e-waste) is currently the fastest growing waste stream, and it continues to be an important global environmental and health issue. The average useful life of electronic products has continued to decline and obsolete products are being stored or discarded with an increasing frequency. As a result, new challenges related to the management of electronic waste (e-waste) have become apparent. E-waste is hazardous, complex and expensive to treat in an environmentally sound manner. Of the more than 400 million electronic items per year that discarded, less than 20 percent is recycled in the U.S. And although, e-waste represents only two percent of America's trash in landfills, it contributes a disproportionate 70 percent of all toxic waste.

The growth in the electronics industry and the increasing trans-boundary movement of E-waste requires collaborative endeavors to ensure sustainable management. Currently, the stewardship of electronics products across their lifecycle is not sustainable, and only a fraction of E-waste is refurbished, reused or recycled. Most electronic products contain a combination of hazardous materials, toxic materials, and valuable elements such as precious metals and rare earth elements. There are risks to human health associated with the disposal of E-waste in landfills, or treatment by incineration. Innovative solutions that integrate electronics manufacturing with recycling are needed to sustainably manage the electronics currently in use, while continuing to promote the future development of innovative technologies to meet market challenges. Hence, there is increasing challenge of reducing the use of virgin materials, recovering useful elements from the waste and protecting human health and the environment from the harmful effects associated with the unsafe handling and disposal of these products (*National Strategy for Electronic Stewardship, NSES).*

EPA/ORD has current research tracking the flow e-waste materials at different stages of their life cycle, to better determine their environmental and economic impact. The collected information and data will later be used to develop a web-based information to enable domestic movement and material flow analysis of e-waste. We are also studying the health and environmental risks of the current improper managment of e-waste such as open burning and land filing.

# ON-SITU REMEDIATION OF NITRO COMPOUND CONTAMINATED SOIL BY BIOLOGICAL-ECOLOGICAL METHOD

***Quanlin Zhao***, Zhengfang Ye, and Zengxiong Li (Peking University, Beijing, China)
Zhongyou Wang (Beijing Institute of Cooperation and Innovation, Beijing, China)

Biological-ecological method was conducted to on-situ remediate nitro compound contaminated soil. High performance microorganism coupled with water and nutrients was added to the contaminated soil to biodegrade the nitro compounds. At the same time, alfalfa and reed were used for phytoremediation. The results showed that biological-ecological method can efficiently degrade nitro compound in the contaminated soil. For the contaminated soil, the average concentration of nitro compound was 877.88 mg/kg. After 12 months' remediation, the average concentration of nitro compound in alfalfa and reed remediation site decreased to 13.09 and 400.03 mg/kg, respectively. The removal efficiency of nitro compound in alfalfa remediation site was higher than that in reed remediation site, which may be caused by its developed root system.

# FENTON OXIDATION OF POLYCYCLIC AROMATIC HYDROCARBONS IN THE CONCENTRATES OBTAINED FROM THE DECONTAMINATED SOIL

***Malika Bendouz,*** Lan Huong Tran, Lucie Coudert, Guy Mercier and Jean-François Blais
(INRS-ETE, Quebec University, Quebec, QC, Canada)[18]

The decontamination of soils contaminated by Polycyclic Aromatic Hydrocarbons (PAHs) compounds, using attrition process generates concentrates containing high concentrations of PAHs. These concentrates are considered as hazardous waste and the management costs would be high.

The degradation of PAHs is a difficult task due to their low solubility, and their refractory character. However, the degradation of PAHs can be achieved through several treatment methods, such as advanced oxidation processes (AOPs). Among the AOPs, the Fenton oxidation process represents the most common, and involves the generation of hydroxyl radicals (OH°), the strongest oxidant with a redox of 2.8V.

The main objective of this study focuses to degradation of 27 PAHs and to reduce the overall toxicity in these concentrates. Different parameters have been tested to determine the operating conditions for optimal destruction of PAHs.

The results allow operating the process on the large-scale and include the study of the technical economic feasibility of the process in comparison to other concentrates management options of PAHs.

# DREDGED MATERIAL AS RESTORATION SUBSTRATE IN LANDFILL SURFACE SEALING SYSTEMS

**Morscheck, Gert[1], Astrid Lemke[1], Michael Henneberg[2]**

[1](University of Rostock, Department of Waste Management and Material Flow, 18051 Rostock, Germany. gert.morscheck@uni- rostock.de

[2](Steinbeis Transferzentrum Applied Landscape Planning, 18051 Rostock, Germany) michael.henneberg@uni-rostock.de

**ABSTRACT:** The German Landfill Ordinance requires proof of a sufficiently large field capacity for recultivation layers. In Mecklenburg-Vorpommern humus-rich and fine-grained substrate is prepared from the dredging of shipping lanes and has been used for about 15 years as landfill recultivation substrate. On four landfill sites the production and effectiveness of recultivation layers made out of matured dredged material (DM) were examined. For this purpose the used dredged material was analysed. The current functionality of the recultivation layers was examined by soil chemical and soil mechanical investigations.

***Keywords:*** dredged material, recultivation, restoration, landfill, sealing system, field capacity, HELP

## INTRODUCTION

Landfills must be secured after operation. This closure, "encapsulation" from the environment, is performed using a surface sealing system. The upper layer of this system is composed of a recultivation substrate.

The Landfill Ordinance (Ordinance on Landfill Sites and Long-Term Storage Facilities - Landfill Ordinance) (DepV 2012) sets out requirements for the recultivation layer (function and substrate).

Additional requirements on the recultivation layer are described in the recommendations of the German Ge- otechnical Society (DGGT - Deutsche Gesellschaft für Geotechnik e.V.; recommendations E 2-31 and E 2-32) (GDA 2006; GDA 2010) and Nationwide Quality Standards (BQS 7-1 2011; BQS 7-2 2011).

Currently, the feasibility of soil physical demands arising from the Landfill Ordinance is discussed extensively. It is checked what challenges arise for the life praxis of landfill construction thereof. In Mecklenburg-Vorpommern more than ten land- fills have already been restored with matured dredged material from coastal waters since the late 90s.

Due to the new requirements for substrates for recultivation layers by the current landfill regulations, already sealed landfills with matured dredged material were examined. It was examined, whether the already existing recultivation layers meet these requirements.

Can matured dredged material be installed in recultivation layers in the future?

## ORIGIN AND QUALITY OF THE DREDGED MATERIAL

The dredged material comes from waterways and harbours. Only the dumping of sand and marl in the Baltic Sea is allowed. Dredged Material with rich content of humus, silt and clay must be disposed of on land in mono disposal sites (including dewatering/flushing fields).

In Rostock this dredged material is classified, de- watered and processed in specially-developed inland disposal sites. The conditioned dredged material can be used for different purposes. The material is similar to an organic-rich, clayey-silty soil or loam.

Since the late 70s of last century, research is being conducted in Rostock for an environmentally responsible use of dredged material. The inland disposal sites of the Hanseatic City of Rostock were converted mid-90's to an industrialised dewatering and treatment system. Thus, environmentally sound and efficient management of dredged material is possible. Law requirements are met. Since 2000, Rostock utilized his dredged material (rich in organic matter and silt/clay) to 100% onshore. The recovery takes place mainly in landscaping and landfill restoration (Fig. 1).

**Particle Size Distribution (Soil Texture).** The dredged sediments are flushed into confined disposal facilities (CDF) via pipeline. The dredged material has a water content of approximately 95%. A con- fined disposal facilities is a diked area on land, used to contain dredged material. The CDF are used inter-changeably.

The particles classify in the CDF along the water stream. Stones, gravel and sand are not carried away by the water very far, at the end of the flowing water the finer and lighter components are deposited (silt, clay and organic matter) (Table 1).

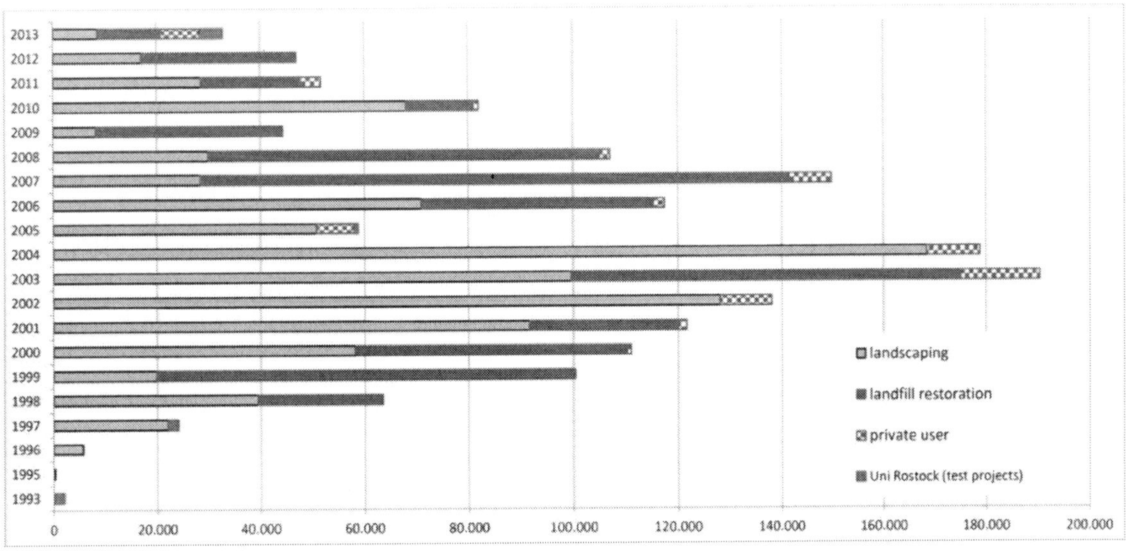


**Fig. 1.** Development of the utilisation of the Rostock dredged material (tons per year)

**Table 1.** Particle size distribution (%) of the matured dredged material (DIN ISO 11777) (arithmetic mean value = average)

| | Coarse sand | Medium sand | Fine sand | Coarse silt | Medium silt | Fine silt | Clay |
|---|---|---|---|---|---|---|---|
| average silt | 1,1 | 8,7 | 29,9 | 10,8 | 15,5 | 11,5 | 23,6 |
| average mixed soil | 1,4 | 16,3 | 56,4 | 8,5 | 57,0 | 3,0 | 8,3 |
| maximum silt | 3,0 | 22,0 | 49,0 | 18,0 | 27,0 | 19,0 | 46,0 |
| maximum mixed soil | 3.6 | 28.9 | 70.9 | 14.2 | 103.0 | 4.6 | 15.1 |
| minimum silt | 0,0 | 1,0 | 7,0 | 7,0 | 8,0 | 6,0 | 13,0 |
| minimum mixed soil | 0,4 | 7,7 | 45,7 | 3,5 | 26,0 | 1,5 | 3,0 |

The sands and gravels can be used in the construction industry. Far along in the water stream a mix of different particle sizes will settled down (mixed soils = sand-silt mixtures) and farthest from the inlet the fines particles settle down (clay-silt mixtures with organic matter). The classification along the current leads to two suitable fractions for landfill restoration (**mixed soils** and clay-silt mixtures with organic matter, called **silt**).

**Chemical Characterization of Wet Dredged Material.** The origin of the dredged material from the coastal area leads to naturally elevated salinities (SK) (Table 2).

The dredged material in Rostock is mainly influenced by the river Warnow. The Warnow flows into the Baltic Sea near Rostock. The river Warnow has a rural drainage area.

The wastewater of the entire drainage area is cleaned in sewage treatment plants. Therefore, the pollutant content in the dredged material is relatively low. After classification of the dredged material in the up- land CDF the silt is more polluted than the mixed soil (Table 3; Table 4). Compliance with the specific utilization limit and guideline values allows a wide range of applications such as the use in landfill restoration.

**Table 2.** Nutrient and salt content of silt and mixed soil

| | DM | pH | SK | $Cl^-$ | $Na^+$ | $SO_4$ | OM | $CaCO_3$ | anorg. N | P | K | Mg | total N | exchange capacity |
|---|---|---|---|---|---|---|---|---|---|---|---|---|---|---|
| | % | | % | mg/100 g soil | | | | % | | mg/100 g | | | % | mval/ 100g soil |
| | | | DM | DM | DM | DM | DM | DM | DM | DM | DM | DM | DM | |
| average silt | 58,2 | 7,3 | 1,9 | 429 | 346,1 | 514,8 | 11,1 | 8 | 2,5 | 2 | 31,4 | 127,8 | 0,5 | 26,1 |
| average MS | 71,3 | 7,4 | 1,2 | 140,9 | 114,5 | 264,2 | 5,3 | 5,6 | 0,9 | 1,8 | 12,8 | 67,7 | 0,2 | 9,2 |
| maximum silt | 71 | 8 | 4 | 1340 | 905 | 1428 | 18 | 16 | 9 | 5 | 56 | 226 | 1 | 36 |
| Maximum MS | 87,5 | 7,9 | 2,5 | 678,8 | 422 | 541,9 | 11 | 9,8 | 2 | 4,4 | 19,5 | 125 | 0,5 | 14,6 |
| minimum silt | 47 | 7 | 1 | 69 | 104 | 163 | 4 | 2 | 1 | 0,4 | 9 | 72 | 0 | 16 |
| minimum mixed soil | 60,1 | 7,0 | 0,4 | 11,5 | 28,0 | 82,0 | 0,9 | 2,5 | 0,5 | 0,4 | 7 | 14,8 | 0,1 | 4,4 |

OM – organic matter; DM – dry matter; SK – salt, MS – mixed soil

**Table 3.** Pollutant content of silts and mixed soils; and limit values

| | Pb | Cd | Cr | Cu | Ni | Hg | Zn | As | EOX | IR-H* | PAH | PCB |
|---|---|---|---|---|---|---|---|---|---|---|---|---|
| | mg/kg DM | | | | | | | | | | | |
| average silt | 21,2 | 0,4 | 35,2 | 22,4 | 14,5 | 0,3 | 100,5 | 9,6 | 0,7 | 225,3 | 0,9 | 0,033 |
| average mixed soil | 9,5 | 0,2 | 15,8 | 12,0 | 6,7 | 0,2 | 44,2 | 5,8 | 0,7 | 108,7 | 0,6 | 0,028 |
| maximum silt | 35,0 | 0,8 | 63,5 | 37,0 | 18,0 | 0,9 | 177,0 | 25,0 | 1,0 | 715,0 | 2,18 | 0,074 |
| maximum mixed soil | 23,8 | 0,3 | 32,8 | 28,5 | 10,0 | 0,4 | 94,0 | 12,5 | 1,3 | 281,0 | 3,3 | 0,066 |
| minimum silt | 10,0 | 0,0 | 12,0 | 14,0 | 11,0 | 0,0 | 36,0 | 5,0 | 0,0 | 28,0 | 0,0 | 0,009 |
| minimum mixed soil | 4,0 | 0,1 | 44,0 | 4,6 | 3,0 | 0,0 | 16,0 | 3,6 | 0,4 | 19,4 | 0,0 | 0,009 |
| BBodSchV VSW sand | 40,0 | 0,4 | 30,0 | 20,0 | 15,0 | 0,1 | 60,0 | | | | 3,0 | 0,050 |
| 70% | 28,0 | 0,3 | 21,0 | 14,0 | 10,5 | 0,07 | 42,0 | | | | 2,1 | 0,035 |
| BBodSchV VSW loam/silt | 70,0 | 1,0 | 60,0 | 40,0 | 50,0 | 0,5 | 150,0 | | | | 3,0 | 0,050 |
| 70% | 49,0 | 0,7 | 42,0 | 28,0 | 35,0 | 0,35 | 105,0 | | | | 2,1 | 0,035 |
| DepV (recultivation layer) | 140 | 1 | 120 | 80 | 100 | 1 | 300 | | | | 5 | 0,1 |

DM – Dry Matter; BBodSchV - Federal Soil Protection and Contaminated Sites Ordinance; DepV – Landfill Ordinance; VSW - precautionary value; IR-H* - IR-hydrocarbons

## USE OF MATURED DREDGED MATERIAL IN LANDFILL RESTO- RATION

Since the late 90's treated dredged material (mixed soil and silt) is used in landfill restoration. Dredged material (mainly silt) is even used in gardening, landscaping and agriculture. Especially silt has very good characteristics (nutrient holding capacity, erosion stability).

The use of the dredged material in landfill cover systems was approved individually by the authorities. Exceptions are necessary since the dredged material contains great amounts of salt and organic matter. The recommended values of the Landfill Ordinance are significantly exceeded.

**Table 4.** Pollutant content (eluate) of the silts and mixed soils and limit values according landfill ordinance

| | pH | Lf µS/cm | Chloride (mg/l) | Sulphate (mg/l) | Pb (µg/l) | Cd | Cr | Cu | Ni | Hg | Zn | As |
|---|---|---|---|---|---|---|---|---|---|---|---|---|
| average silt | 7 | 4044 | 542 | 2022 | 9,1 | 0,6 | 2,3 | 28,8 | 10,6 | n.n. | 75,8 | 3,4 |
| average mixed soil | 7.1 | 2975 | 259 | 1596 | 4,6 | 0,2 | | 11,3 | 6 | n.n. | 21,7 | 2,2 |
| DepV (recultivation layer) | 6.5 - 9 | 500 | 10 | 50 | 40 | 2 | 30 | 50 | 50 | 0,2 | 100 | 10 |

Lf – conductivity; DepV – Landfill Ordinance; VSW - precautionary value; n.n. - undetectable

The dredged material is suitable for both, forming the landfill, as well as, for building a restoration and water management layer. In Mecklenburg-Vorpommern soil material is needed for the closure of landfills and as cover of contaminated sites.

The Landfill Ordinance defines properties, which the recultivation layer has to fulfil. The recultivation layer should offer a high water storage and good plant available water capacity, helping so to reduce leachate formation. Even a temporary cover has to minimize leachate formation. The recultivation substrate also needs to ensure an adequate plant growth, a field capacity of at least 140 mm, based on the total thickness of the recultivation layer, is required (Landfill Ordinance, Appendix 1, section 2.3.1). When establishing a water regime layer, the Land- fill Ordinance calls for a field capacity of at least 220 mm, based on the total thickness of the water regime layer. In addition, the seepage may not exceed 10 per- cent of the long-term average of precipitation (usually 30 years), not more than 60 mm per year (Landfill Ordinance, Appendix 1, section 2.3.1.1). Matured dredged material also has to fulfil the requirements for landfill replacement construction materials (DepV, Annex 3, Table 2, item 9 of Table 1).

To continue the use of mature dredged material in the landfill restoration, it has to be tested on already reclaimed landfills if the new requirements of the Land- fill Ordinance (for example available water content) are complied. Observations on reclaimed landfills and experience in other fields of application of dredged material give reason to believe that the material is able to comply with the new Landfill Ordinance.

In a lysimeter test the leachate quantity was reduced up to 25% compared to loamy sand by using dredged material (silt) incorporated in the soil.

## RESULTS – TEST OF RESTORATION LAYERS

For the investigations four disposal sites covered with dredged material were selected. Following parameters at the four sites were tested:

- Current situation (leachate quantity, special problems)
- Visual assessment of the cover layer (plant growth, erosion, settlement, landslide ...)
- soil sampling (disturbed and undisturbed samples) to evaluate the residual shrinkage of the dredged material and to determine the root depth and intensity
- Geotechnical investigations under Landfill Ordinance or German Geotechnical Society recommendation (water contents, densities, shear strength on site, current field capacity (field capacity and available water content), coefficient of permeability (in-situ and laboratory), pH, organic matter, lime content
- Assessment of pollutants in the dredged material
- Investigation of the material displacement, especially salts transfer in the profile
- Examination of humus forms in the dredged material and its stability (AT4 respirometric activity)

– Modelling the water balance with the Hydrologic Evaluation of Landfill Performance Model (HELP version 3.90D)

The age of the recultivation layers was different. The landfills have been covered two to 16 years ago. The matured dredged material used for the restorations varies in its chemical and physical properties, because it comes from different dredging. None of the test fields had the same dredged material; all studies were carried out on already secured landfills.

**Current Situation.** During and after the restoration the landfills were visited often. Among others the maintenance state after restoration was recorded. If the landfills were mowed annually, grasses form the dominant plant species. If there is no maintenance (no mowing) for years, herbs and brushes grow, and first trees grow. At all landfills dense vegetation exists (high soil fertility of the substrate dredged material despite the increased salinity).

None of the landfills showed mechanical changes on the surface (erosion, slides or subsidence). The very intense rainfall of 2011 could not harm the recultivation layers. Leachate in the drainage layer of the sealing system could be observed only in one landfill. In excavation tests, cracks in the restoration layer could be observed. This was especially the case when the water content was too high during installation of the dredged material. The cracks reach several decimetres in the restoration layer. However, the shrinkage cracks are filled with soil particles. The recultivation layers are very well rooted and are tapped for water supply of the plants and evaporation loss.

**Evaluation of Chemical and Biological Characteristics of the Recultivation Layers.** Each of the four investigated landfills was sampled at two points in the entire restoration layer (about one meter deep). The measurements and the sampling were performed in three depths.

The water movements in the restoration layer led to material relocation. Especially salt and magnesium ions usually show an increase in their content with the depth. The contents of total organic carbon and lime show similar high levels as during the installation of the dredged material. The high lime content receives long- term neutral pH values (high buffering capacity). The high level of total organic substance shows the very good stability of the organic matter. The respirometric activity (AT4) is well below the specification of the landfill ordinance (5 mg $O2$/g dry matter). The relevance of origin-related high salt content in the dredged material has already been highlighted.

But individual decisions enable the reasonable use of the dredged material. The relatively low levels of available nitrogen (compared to the large supply of nitrogen in the soil) in the upper mineral recultivation layer are sufficient to enable a strong plant growth. The dredged material is low in available mineral nitrogen (0.5 to 2 mg $N_{min}$/100 g of soil; about 20 - 80 kg N/ha in the upper restoration layer of 0 - 30 cm). The levels of plant-available phosphorus are low, as known from the fresh dredged material. To satisfy the plant's needs, the content is sufficient. The levels of plant-available potassium are adequate. There is no additional fertilization required. The levels of plant- available magnesium are high to very high (35-100 mg/100 g soil). Freshly processed dredged material contains even more than 150 mg /100 g soil) (Table 5). The investigation of soil chemical parameters shows that the dredged material has a very good soil fertility determining properties for plant growth. (Table 5).

Due to the good binding capacity (high contents of total organic content and clay) as well as the excellent water holding capacity no increased discharge of salt ions and nutrients is expected.

**Evaluation of Physical Properties in the Recultivation Layers.** The Landfill Ordinance and the Nationwide Quality Standards BQS 7-1 "recultivation layers" and BQS 7-2 "water regime layers" (BQS 7-1 2011; BQS 7-2 2011) require compliance with soil physical parameters (see Chap. 3). These parameters were tested in order to prove whether the further use of the treated and matured dredged material (mixed soil and silt) for landfill closure can be allowed.

***Shear Strength.*** The in-situ shear strengths (investigated with Field Vane Shear Apparatus) vary considerably depending on the installation of the dredged material in the restoration layer. On landfills where the dredged material was slightly compacted during installation or where marl was installed as the top layer of restoration layer, high shear strengths of 70 to 80 kPa and even 100 to 120 kPa were measured, also in the top layer. The measurements on other landfill recultivation layers have a clear depth-related graduation. In the upper, loosely built layer, the determined shear strengths were only about 20 to 33 kPa.

This looseness is particularly due to the high content of organic matter and the intensive rooting. In the underlying layers shear strengths were measured by more than 100 kPa (Table 6).

**Table 5.** Chemical characteristics of recultivation layers

| landfill | depth | dry matter | pH | salt | $Cl^-$ | $Na^+$ | $SO_4^{2-}$ | organic matter | $AT_4$ | $CaCO_3$ | $N_{an}$ | P | K | Mg |
|---|---|---|---|---|---|---|---|---|---|---|---|---|---|---|
| | mm | % | | grav. % | | dry matter mg/100 g soil | | % | mg $O_2$/kg DM | % | | dry matter mg/100 g dry matter | | |
| M 1 | 100 | 78,7 | 7,2 | 0,19 | 1,9 | 1,2 | 138,2 | 4,8 | 1,0 | 3,6 | 1,3 | 4,5 | 17 | 37 |
| | 350 | 78,3 | 7,4 | 1,03 | 2,0 | 1,3 | 174,3 | 4,8 | 0,7 | 3,2 | 0,8 | 1,7 | 10 | 34 |
| | 700 | 75,2 | 7,5 | 1,13 | 0,7 | 3,8 | 187,1 | 6,9 | 1,0 | 3,8 | 0,9 | 1,9 | 9 | 74 |
| M 2 | 150 | 71,1 | 7,3 | 0,8 | 0,5 | 1,1 | 66,8 | 7,6 | 1,0 | 3,1 | 0,7 | 5,1 | 16 | 44 |
| | 400 | 73,2 | 7,4 | 1,05 | 2,0 | 2,0 | 179,4 | 5,6 | 0,9 | 3,3 | 0,3 | 2,2 | 10 | 37 |
| | 750 | 67,8 | 7,5 | 1,13 | 1,4 | 3,9 | 207,5 | 7,8 | 1,1 | 4,7 | 0,4 | 1,5 | 9 | 72 |
| N 1 | 150 | 61,8 | 7,5 | 1,1 | 4,4 | 7,4 | 284,8 | 8,2 | 1,5 | 3,4 | 0,8 | 1,2 | 17 | 37 |
| | 450 | 66,8 | 7,7 | 1,4 | 59,4 | 90,0 | 271,7 | 7,4 | 0,6 | 3,4 | 1,6 | 0,8 | 13 | 79 |
| | 850 | 62,9 | 7,6 | 1,6 | 152,5 | 149,0 | 307,8 | 7,0 | 1,1 | 2,4 | 2,2 | 0,7 | 14 | 76 |
| N 2 | 150 | 76,9 | 7,6 | 1,1 | 11,5 | 11,0 | 205,5 | 7,4 | 0,8 | 3,8 | 1,2 | 1,0 | 22 | 40 |
| | 400 | 72,1 | 7,6 | 1,6 | 103,1 | 128,0 | 274,9 | 8,7 | 1,3 | 2,9 | 1,8 | 0,7 | 12 | 92 |
| | 850 | 69,9 | 7,6 | 2,0 | 329,1 | 218,8 | 286,1 | 9,8 | 0,8 | 2,7 | 1,7 | 1,3 | 17 | 102 |
| | 1200 | 66,5 | 7,7 | 1,7 | 210,3 | 159,1 | 283,3 | 8,3 | | 3,0 | 1,5 | 0,5 | 20 | 95 |
| T 1 | 400 - 500 | 75,5 | 7,6 | 1,1 | 8,2 | 13,7 | 227,5 | 5,3 | 0,4 | 5,1 | 0,9 | 0,7 | 9 | 54 |
| | 800 - 850 | 70,2 | 7,6 | 1,3 | 35,1 | 55,1 | 262,4 | 6,5 | 0,6 | 4,2 | 1,1 | 0,6 | 13 | 77 |
| T 2 | 350 - 450 | 63,1 | 7,6 | 1,19 | 11,7 | 27,3 | 287,5 | 9 | 0,8 | 4,8 | 1,3 | 0,7 | 11 | 82 |
| | 800 - 850 | 72,2 | 7,6 | 1,21 | 17,4 | 34,3 | 249,9 | 6,1 | 0,6 | 4,6 | 1,0 | 0,6 | 11 | 71 |
| G 1 | 100 - 200 | 74,4 | 7,6 | 1,12 | 5,0 | 8,8 | 219,4 | 9,7 | 1,8 | 2,7 | 0,9 | 0,9 | 16 | 58 |
| | 400 - 500 | 66,5 | 7,6 | 1,26 | 14,2 | 29,5 | 244,8 | 8,9 | 0,9 | 2,0 | 0,9 | 0,6 | 13 | 95 |
| | 800 | 62,2 | 7,6 | 1,54 | 102,8 | 110,3 | 298,4 | 9,8 | 2,6 | 2,4 | 1,0 | 0,4 | 20 | 100 |
| | 950 | 64,2 | 7,6 | 1,51 | 95,0 | 103,9 | 307,5 | 8,6 | | 1,8 | 0,9 | 0,4 | 20 | 97 |
| G 2 | 200 - 250 | 74,4 | 7,6 | 1,11 | 7,3 | 9,4 | 232,3 | 11,3 | 1,9 | 2,6 | 1,3 | 0,6 | 14 | 48 |
| | 550 | 58,7 | 7,6 | 1,53 | 79,4 | 116,2 | 331,5 | 12,8 | 2,0 | 4,8 | 1,5 | 0,3 | 23 | 110 |
| | 800 | 57,1 | 7,6 | 1,62 | 130,1 | 148,1 | 349,2 | 11,8 | 1,9 | 4,3 | 1,4 | 0,3 | 25 | 105 |

**Table 6.** In-situ shear strength of the restoration layers (field vane test)

| Landfill | Depth | Dry Matter | In-Situ Shear Strength | Bulk Density |
|---|---|---|---|---|
| | mm | % | kPa | g/cm³ |
| M 1 | 150 | 78,7 | 26,5 | 1,11 |
| | 700 | 75,2 | 156,8 | 1,14 |
| M 2 | 100 | 71,1 | 18,4 | 1 |
| | 350 | 73,2 | 21,3 | 0,98 |
| | 650 | 67,8 | 116,6 | 1,05 |
| N 1 | 450 | 66,8 | 30,3 | 0,9 |
| | 1000 | 62,9 | 35,0 | 0,9 |
| N 2 | 150 | 76,9 | 32,7 | 0,9 |
| | 400 | 72,1 | 116,7 | 0,9 |
| | 850 | 69,9 | 109,1 | 0,9 |
| T 1 | 400 - 500 | 75,5 | 116,7 | 1,1 |
| | 800 - 850 | 70,2 | 109,1 | 1,0 |
| G 1 | 100 - 200 | 74,4 | 122,0 | 0,9 |
| | 400 - 500 | 66,5 | 136,0 | 0,9 |
| | 800 | 62,2 | 124,0 | 0,9 |
| G 2 | 200 - 250 | 74,4 | 103,5 | 0,82 |
| | 550 | 58,7 | 92,8 | 0,77 |
| | 800 | 57,1 | 97,1 | 0,83 |

***Hydraulic conductivity.*** The determination was carried out in the field with a Double ring infiltrometer (unsaturated conductivity) and in the laboratory (saturated conductivity) with sampling rings (500 cm³ or 250 cm³).

The hydraulic conductivity of the recultivation layers is very low. There was no targeted consolidation of the built-in layers. The installation was done with bulldozer or excavator. The hydraulic conductivity varies as a function of the analytical method between $8.1 \times 10^{-6}$ and $1.2 \times 10^{-7}$ m/s (laboratory method) or $2.18 \times 10^{-6}$ and $1.98 \times 10^{-8}$ m/s (field method) and de- creases from top to bottom.

***Pore volume and soil bulk density.*** The total pore volume (total porosity) of the restoration layers (matured dredged material) is even after several years (including losses due to possible minor settling after installation) usually about 50%, often over 60%. This large pore volume is also shown in the very low bulk densities of mostly 0.8 to 1.1 g/cm³ (Table 7). Both are independent of the depth. After installation no significant settlements of the recultivation layers have occurred.

***Field capacity, available water capacity and air capacity.*** The high total pore volume also enables high field capacity and available water capacity. Because of the high quantity of medium and fine pores (silty-clayey dredged material; silt), the majority of the total pore volume can hold water against gravity (very high field capacity, often more than 50%).

Often, more than half, but at least more than 35% of the pore volume is available for plants. The available field capacity is almost always more than 20% by volume, in some cases even more than 30% by volume. The requirements of the Landfill Ordinance by 140 mm (recultivation layer) or 220 mm (water regime layer) field capacity are always satisfactory, even years after the installation of the recultivation layer. Water regime layers can also be produced from dredged material (Table 7).

The low bulk density also results from the large air pore volume. In the top soil layer (0-30 cm) the air pore volume (air capacity) is partly greater than 10%. In combination with the very good water supply (very high available field capacity) and adequate available nutrients excellent conditions for rapid plant growth are given.

Field capacity and available field capacity fulfil the requirements of the Landfill Ordinance (DepV 2012), the recommendations E 2-31 and E 2-32 of the German Geotechnical Society (GDA 2006; GDA 2010) and the uniform nationwide Quality Standards BQS 7-1 and BQS 7-2 (BQS 7-1 2011; BQS 7-2 2011). The field capacity reaches up to 500 to 610 mm and the available field capacity reaches up to 230 to 290 mm by using matured silt in the entire restoration layer. These values were also measured in ten years old recultivation layers!

**Table 7.** Soil physical properties of the restoration layers

| land-fill | depth | dry matter | organic matter | Ca- CO3 | bulk density | density | PV | WV | AV | FC | aFC | aFC/FC |
|---|---|---|---|---|---|---|---|---|---|---|---|---|
| mm | | % | % | % | g/cm3 | g/cm3 | | | Vol % | | | % |
| M 1 | 100 | 78,7 | 4,8 | 3,6 | 1,11 | 2,52 | 56,1 | 37,9 | 6,1 | 50,0 | 33,7 | 67,4 |
| | 350 | 78,3 | 4,8 | 3,2 | 1,10 | 2,52 | 56,2 | 43,6 | 8,8 | 47,4 | 29,5 | 62,3 |
| | 700 | 75,2 | 6,9 | 3,8 | 1,14 | 2,47 | 53,8 | 49,1 | 0,8 | 53,0 | 23,4 | 44,1 |
| M 2 | 150 | 71,1 | 7,6 | 3,1 | 1,00 | 2,49 | 59,7 | 37,4 | 4,8 | 54,9 | 37,2 | 67,9 |
| | 400 | 73,2 | 5,6 | 3,3 | 0,98 | 2,51 | 61,2 | 41,0 | 11,8 | 49,4 | 29,1 | 59,0 |
| | 750 | 67,8 | 7,8 | 4,7 | 1,05 | 2,47 | 57,3 | 45,4 | 4,2 | 53,2 | 20,5 | 38,6 |
| N 1 | 150 | 61,8 | 8,2 | 3,4 | 0,88 | 2,42 | 63,8 | 49,3 | 12,5 | 51,3 | 25,9 | 50,5 |
| | 450 | 66,8 | 7,4 | 3,4 | 1,03 | 2,45 | 58,1 | 46,7 | 5,6 | 52,4 | 24,1 | 45,9 |
| | 850 | 62,9 | 7,0 | 2,4 | 0,89 | 2,45 | 63,8 | 55,8 | 7,9 | 55,9 | 35,1 | 62,8 |
| N 2 | 150 | 76,9 | 7,4 | 3,8 | 0,87 | 2,48 | 64,7 | 32,0 | 25,6 | 39,1 | 21,1 | 54,1 |
| | 400 | 72,1 | 8,7 | 2,9 | 0,90 | 2,43 | 62,8 | 43,3 | 9,7 | 53,1 | 30,2 | 56,8 |
| | 850 | 69,9 | 9,8 | 2,7 | 0,88 | 2,42 | 63,6 | 42,3 | 8,9 | 54,7 | 28,7 | 52,5 |
| T 1 | 400 - 500 | 75,5 | 5,3 | 5,1 | 1,12 | 2,54 | 51,4 | 49,7 | 6,1 | 45,3 | 20,2 | 44,5 |
| | 800 - 850 | 70,2 | 6,5 | 4,2 | 1,01 | 2,48 | 59,5 | 51,3 | 4,2 | 55,2 | 25,0 | 45,3 |
| T 2 | 350 - 450 | 63,1 | 9,0 | 4,8 | 0,99 | 2,42 | 59,3 | 60,1 | -3,0 | 62,4 | 24,6 | 39,4 |
| | 800 - 850 | 72,2 | 6,1 | 4,6 | 1,14 | 2,51 | 51,6 | 48,4 | 1,6 | 50,1 | 20,4 | 40,1 |
| G 1 | 100 - 200 | 74,4 | 9,7 | 2,7 | 0,88 | 2,40 | 63,3 | 44,0 | 10,9 | 52,4 | 23,0 | 43,9 |
| | 400 - 500 | 66,5 | 8,9 | 2,0 | 0,89 | 2,42 | 63,4 | 55,3 | 3,3 | 60,1 | 28,7 | 47,7 |
| | 800 | 62,2 | 9,8 | 2,4 | 0,88 | 2,43 | 63,8 | 55,2 | 4,2 | 59,6 | 24,5 | 41,1 |
| G 2 | 200 - 250 | 74,4 | 11,3 | 2,6 | 0,82 | 2,35 | 65,0 | 41,2 | 7,2 | 57,8 | 28,1 | 48,7 |
| | 550 | 58,7 | 12,8 | 4,8 | 0,77 | 2,32 | 66,8 | 59,9 | 0,8 | 66,0 | 24,2 | 36,6 |
| | 800 | 57,1 | 11,8 | 4,3 | 0,83 | 2,45 | 66,1 | 56,9 | 4,8 | 65,2 | 22,9 | 35,0 |

PV- pore volume; WV - water volume; AV - air volume; FC - field capacity; aFC - available field capacity

***Simulation of the water balance with the Land-fill Model HELP.*** The Landfill Ordinance requires proofing the percolation rates. Such a simulation of the water balance can be carried out with the model HELP (Hydrologic Evaluation of Landfill Performance).

This simulation was also done at the investigated recultivation layers (Berger 2012). The simulations were performed with the meteorological data of the standard period 1961 to 2010 and also carried out with the data from 2000 to 2010. The decade 2000 to 2010 shows significant changes with respect to the factors that are particularly challenging to the effectiveness of a restoration layer. The first decade of this century shows significantly higher annual rain- fall amounts (about 10%) and significantly higher mean annual temperatures (about 1°C) in all meteorological stations.

The calculated (decade 2000 to 2010) average annual seepage from the recultivation layer is located at the landfill site **N** at around 60 mm. The landfills **M** and **G** achieve average seepage of around 120 to 130 mm per year. The highest infiltration rates occur in landfill with around 150 to 165 mm / a (Table 8).These simulation results are very surprising. Field capacity and available field capacity are very high and on the landfills no seepage water processes from the recultivation layers were observed.

The matured dredged material with its high organic matter content seems not to behave like "ordinary" mineral soil. But the modelling approaches of HELP are based on "ordinary" mineral soils. "Whether the water holding capacity of the dredged material is higher than the capacity modelled in HELP and thus the infiltration of HELP is slightly overrated, remains a speculation" (Berger 2012, p 8). Nevertheless the results clearly show the positive characteristics of the dredged material compared to marl.

However, the question remains how ordinary substrates (soil types) should fulfil the requirements of the Landfill Ordinance. If we believe in the results of the HELP calculations even a substrate with extremely high field capacity and available field capacity seems not to be able to comply with the Landfill Ordinance.

In summary: matured dredged material (mixed soil and silt) meets in practice the soil physical requirements of the Landfill Ordinance.

**Table 8.** Mean annual water balance of the restoration layers (years 2000 - 2010) (Berger 2012)

| parameters [mm] | T 1 | T 2 | M 1 | M 2 | G 1 | G 2 | N 1 | N2 | N |
|---|---|---|---|---|---|---|---|---|---|
| precipitation | 653,1 | | 620,5 | | 662,4 | | 631,3 | | |
| surface runoff | 7,8 | 8,1 | 3,1 | 3,3 | 11,7 | 14,5 | 2,7 | 2.5 | 2 |
| PET | 665,6 | | 654,0 | | 678,6 | | 753,3 | | |
| AET | 478,2 | 493,9 | 491,6 | 486,0 | 519,6 | 510,2 | 558,3 | 558.7 | 5 |
| seepage | 165,3 | 148,9 | 122,2 | 127,8 | 122,4 | 130,7 | 61,8 | 61.2 | 6 |
| ΔWg | 1,8 | 2,2 | 3,5 | 3,3 | 8,8 | 7,0 | 8,4 | 8.9 | 8 |
| field capacity | 431 | 464 | 504 | 526 | 576 | 617 | 535 | 495 | 4 |
| available field capacity | 169 | 170 | 283 | 282 | 253 | 233 | 290 | 268 | 2 |

PET - Potential evapotranspiration; AET - Actual evapotranspiration; ΔWg - Change of soil water retention

## SUMMARY

In Mecklenburg-Vorpommern landfill sites are restored with matured dredged material since 15 years. This substrate comes from waterways and harbours. The used and tested dredged material originates from the port of Rostock, from the dredging of the waterways. This dredged material is classified, de-watered and processed in specially-developed inland disposal sites. Coarse components (gravel and sand) are used in the construction industry. The mixed soils are used in landscaping.

The organic-rich, silty and clayey substrates are used after dehydration and aging in recultivation layers of landfill sealing systems. The Rostock dredged material is not contaminated with pollutants.

The German Landfill Ordinance requires the proof of the water storage capacity (field capacity and available field capacity) for landfill recultivation layers. Whether the dredged material fulfils the demands of the Ordinance was examined on landfills already sealed with matured dredged material.

On four landfills, 2 to 15 years after their restoration, the chemical and physical status was determined (hydraulic conductivity, pore-size distribution, mechanical properties and chemical composition). It could be demonstrated that the requirements of the Landfill Ordinance regarding field capacity and available field capacity are reliably observed even many years after the restoration. The use of matured dredged material (mixed soil and silt) in recultivation layers should be continued.

**REFERENCES**

Berger, K. (2012) Simulation des Wasserhaushalts mit HELP, 3.90 D für vier Deponiestandorte in Mecklenburg- Vorpommern. Berechnungen an der Universität Ham- burg; Gutachten vom 13.04.2012, unveröffentlicht.

BQS 7-1 (2011) Bundeseinheitlicher Qualitätsstandard 7-1 "Rekultivierungsschichten in Deponieoberflächenabdich- tungssystemen" vom 23.05.2011.

BQS 7-2 (2011) Bundeseinheitlicher Qualitätsstandard 7-2 "Wasserhaushaltsschichten in Deponieoberflächenab- dichtungen" vom 20.10.2011.

GDA (2006) GDA-Empfehlung E2-31 Rekultivierungs- schichten (2006) Deutsche Gesellschaft für Geotechnik http://www.gdaonline.de/pdf/E2-31.pdf, 22:07:2012.

GDA (2010) GDA-Empfehlung E2-32 Gestaltung des Be- wuchses auf Abfalldeponien (2010) Deutsche Gesell- schaft für Geotechnik http://www.gdaonline.de/pdf/E2- 32.pdf, 22:07:2012.

DepV (2012) Landfill Ordinance - Verordnung über Depo- nien und Langzeitlager (Deponieverordnung - DepV) vom 27.04.2009, BGBl. I S. 900), die zuletzt durch Arti- kel 5 Absatz 28 des Gesetzes vom 24. Februar 2012.

# TREATMENT OF SANITARY LANDFILL LEACHATE BY MEMBRANE DISTILLATION

Gizem Sahin and Berna Kiril Mert (Sakarya University, Sakarya, TURKEY)
***Yasemin Melek Tilki****and Coskun Aydıner (Gebze Technical University, Gebze, TURKEY)
Esra Can Dogan (Kocaeli University, Kocaeli, TURKEY)[20]

As one of the most difficult wastewaters to be treated, sanitary landfill leachate has complex pollutant contents of high concentrations, and thus its treatment by novel processes has quite significance for environmental protection, economical preferences and time saving. In this regard, proactive researches on increasing of technical and economic achievements in the leachate management are encouragingly able to be pursued by promising technologies. In this work, treating the leachate from sanitary municipal solid waste landfill was aimed to achieve the appropriate effluent quality standards for direct discharge to natural receiving waters using membrane distillation (MD) in direct contact mode lying between innovative and developing membrane technologies. The experiments were carried out for conventional treatment effluent using lab–scale MD system operated at intermittent (a) and continuous (b) operating modes to determine the best conditions for 6h duration time and long–term efficiency for 6h per day at a time period of 10 days, respectively. In experiments (a), the influences of membrane type, membrane pore size, cross–flow rate, distillate temperature, trans–temperature difference, flow hydrodynamics (pulse and continuous) on the MD performance were examined. The best operating conditions were established as PTFE membrane, 0.45 μm, 200 L/h, 40 °C, 20 °C and pulse flow regime for the mentioned order of process parameters, respectively, in which rejection efficiencies of 99.96% TDS, 99.26% COD, 99.80% TOC and 99.50% $NH_4^+$ were observed as well as distillate flux of 29.8 L/m$^2$h. In experiment (b), the leachate was continuously filtrated under the best conditions determined in the (a), in which membrane cleaning was also applied at the end of filtration cycle in each day using 1% HCl, 1% NaOH and distillate for a full time of 60 min. Long–term MD operation was consistently maintained along 10 days with the performances of 24.9 L/m$^2$h flux and 99.97%, 99.60%, 99.60% and 98.85% rejections in the abovementioned order. The produced water quality was appropriate to meet the discharge criteria according to the receiving water standards described in Turkish Water Pollution Control Regulations. Using the long–term experimental data, a simulation study on full–scale application of the leachate treatment by MD process was carried out for a real–scale sanitary landfill facility having a leachate flow of 500m$^3$/d by using process design software with cost estimation module. In simulations, it was assumed that the heat demand required for heating raw leachate stream as MD feed could be met from methane gas being produced in the landfill area. Accordingly, it was calculated that the feed stream would be able to be theoretically heated up to 142 $^0$C using all biogas in the facility and hence 20% of the produced biogas could be met the heat requirement. The process simulation indicated that for directly discharging into receiving waters, the conventional effluent would be effectively treated with a unit treatment cost of 0.30\$/m$^3$ in MD that is implementable in single step for waste concentrate minimization. This study finally proved that MD–included novel treatment application seemed to be a favorable technological solution for both enviable management of leachate discharges and sustainable protection of environment.

---

# ENVIRONMENTAL MONITORING OF LANDFILLS BY A DOWNSCALING APPROACH

Mei Alessandro, Manzo Ciro, Paciucci Lucia, Allegrini, Alessia, Petracchini Francesco, Romagnoli Paola, Bassani Cristiana (Tecnoligist of Institute of Atmospheric Pollution Research - CNR (National Research Council of Italy), Research Area of Rome 1 Via Salaria Km 29.300, Monterotondo (00016), Rome, Italy)

Landfills are an important part of urban society and an accurate analysis of these area are fundamental to improve decision making strategies for planning anthropic activities. The optimization of new techniques for the monitoring of landfills represents a crucial issue since usual investigation methods are expensive and time-consuming. The application of remote sensing improves the monitoring of different environmental matrices and the assessment of gas and leachate migration. Reflectance spectroscopy approach can be adopted to investigate the stressed vegetation or soil pollution due to waste deposits developing site-depending spectral indices which are linked to bio-chemicals factors .

This paper describes a multi-source approach based on the use of remote sensing data and spatial processing techniques to monitor the environmental condition around landfill. The study area is located in the landfill of San Giovanni in Fiore Municipalty (CS) in the Southern Italy (Calabria District). Multi-source data was adopted for the application of a downscaling and multi-temporal analysis. In particular the dataset was composed by orthophotos (2000, 2006 and 2012), Worldview 2 (2012) and Pléiades (2014 and 2015) satellite imagery and aerial thermal data from Multispectral Infrared and Visible Imaging Spectrometer (MIVIS) (2011). Moreover, a field survey was performed collecting overnight thermal images and photo in photogrammetric configuration.

The orthophoto dataset supported the multi-temporal analysis not covered by spectral imagery showing a general increase of land consumption and highlights area with no-/senescent vegetation cover. The visible and near-infrared spectral regions allowed to adopt indices descriptive of vegetation and to retrieve anomalies on the land cover. The analysis of vegetation changes is performed applying an image differencing of the Normalized Difference Vegetation Index (NDVI) which analyses the spectral regions sensitive to chlorophyll absorption of red light and reflection of NIR energy scattered by cell structure. This index was calculated in both satellite and aerial imagery. Moreover Thermal InfraRed (TIR) spectral domain of MIVIS was adopted for detection of possible leakage migration and presence of biogas into the ground.

The validation of evidences of ground anomalies was accomplished by field surveys and nocturne thermal acquisitions. In addition, a photogrammetric survey was performed by pole in order to improve the analysis of both vegetation and thermal anomalies that have been detected. The integration of the visible, infrared and thermal optical sensors with field nocturnal thermal analysis and photogrammetric surveys for 3D models generation is performed. The 3-D model of some sections of the landfill supported the comprehension of the site topography and anomalies.

This study has allowed the implementation of a landfill investigation procedure integrating geomatic tools of remote and *in situ* data. The outcomes of the proposed multi-source analysis highlight warning areas inside and around the studied area. The presented approach can be considered an innovative and a performing methodology of investigation than the current classical methods, which require a great deal of work at local scale.

# PERFORMANCE OF VEGETATIVE BIORETENTION SYSTEM FOR GREYWATER REUSE IN ARID CLIMATES

*Rezaul Chowdhury*, Taoufik Ksiksi, Mohamad M. A. Mohamed and Jameelu Abaya
(United Arab Emirates University, Al Ain, United Arab Emirates)

**ABSTRACT**: Water Sensitive Urban Design (WSUD) technologies are acknowledged to play a significant role in improving the quality of stormwater in the urban landscape. However, findings have shown that greywater, the residential wastewater except that originates from toilet and kitchen, could be used to operate the vegetative bioretention system in the arid climates. A positive signal was received when investigated in the laboratory settings, however, a better picture and a more representative result can be obtained if the system is studied in a field scale. The study intended to explore further, the effectiveness of using the vegetative bioretention to treat urban greywater. This WSUD system can play a dual role in improving greywater quality and landscape aesthetics. Selection of appropriate plants for them is also an important research question. Twelve different species of native and exotic plants in triplet planted in 36 prototype bioretention columns in the laboratory. The columns were irrigated by synthetic greywater and the removal efficiency of heavy metals and other common water quality parameters were monitored on a regular basis. The retention of heavy metals in the water, plants (leaves, stem and roots) and soil were also investigated. It was found that the *Pennisetum seateceum* provides more consistency and efficiency in removing heavy metals from greywater.

## INTRODUCTION

The debate and the need for action on climate change will continue over time. Although a milestone was reached by adopting the Paris agreement in December 2015, the challenges, particularly concerning water in arid regions of the world will remain and can become worse if proper water management practices were not implemented. The United Arab Emirates (UAE) is one of the most water-scarce countries, but their water consumption rate is significantly high. Water consumption in the Abu Dhabi Emirate ranges between 170 and 200 Lpcd in flats and between 270 and 1,760 Lpcd in the villas (Environment Agency – Abu Dhabi, 2009). In a recent study, Chowdhury and Rajput (2015) observed that the water consumption in traditional villas in Abu Dhabi is more than 2500 Lpcd. Out of this, more than 80% is used for outdoor activities (gardening, car washing, for example). Because of rapid population growth in its major cities, desert greening and agricultural expansion policies and for the anticipated climate change scenarios, it is considered that water resources in the UAE will be significantly stressed in the future. Moreover, since rainfall and urban runoff are not abundant sources of water in the UAE, recycling and reuse of wastewater or greywater is considered as a potential alternative. The key technical challenges of their harvesting and reuse are appropriate treatment, storage, supply to end users at cost effective manner and their risk assessment.

Previous studies showed that reuse of alternative water resources is an economically viable solution for water demand reduction in water scarce countries (Chowdhury et al., 2015, McIntosh et al., 2013a, 2013b). Greywater (residential wastewater excluding toilet and kitchen wastewater) is becoming popular as an alternative water source in many arid and water scarce regions. Greywater generation in cities across the world is in the range of 40-80% (Allen et al. 2010; Jamrah et al. 2006; Chowdhury et al., 2015) of the indoor water consumption. Investigation on indoor average water consumption in the Al Ain city was found to be 280 Lpcd and average greywater generation estimated at 192 Lpcd in the villas. Higher water consumptions are observed in the shower (33%), dish wash (16%), toilet flush (14%) and in ablution (12%).

Further finding show that about 30% of potable water can be saved by the reuse of greywater for toilet flushing and irrigation (Chowdhury et al., 2015).

The Water Sensitive Urban Design (WSUD) systems involve several cost-effective, natural and low energy required treatment technologies for source control of stormwater runoff. Most popular WSUD systems are swales, infiltration trench, bioretention basin, permeable pavement and constructed wetlands. WSUD provides two benefits – source control of urban runoff and landscape architectural benefits to the society (Beecham and Chowdhury, 2012). Since rainfall and urban runoff are not abundant in the arid regions and that greywater is technically abundant and contains less organic loads, it is paramount to explore the applicability of its usage through WSUD.

The WSUD systems used in these studies are Permeable pavement and vegetative bioretention system. Permeable pavement is an alternative paving surface generally design to filter stormwater runoff through voids into an underlying stone reservoir where it is temporarily stored for reuse or infiltrated into ground water (Myers et al., 2009). While bioretention systems are low depth landscaped areas, which are normally underdrained and rely on engineered soils, vegetation and filtration to improve water quality and reduce runoff downstream. They are generally aimed at sustainably managing and treating runoffs from rainfall events.

Earlier studies conducted at laboratory scale by (Chowdhury et al., 2015) show that two WSUD treatment technologies, namely the vegetative bioretention basin and permeable pavement with underlying reservoir could improve most of the greywater quality parameters fit for irrigation end use. However, application-test-bed studies are essentially required for investigation that is more rigorous and which represent in a better way the real conditions of operation (Hatt et al., 2008). Another aspect is the selection of appropriate plants, which is critical to the long-term landscape amenity, functional performance, and structural integrity of WSUD systems (Australian Northern Territory Government, 2009). Plants have a functional role in treatment and for erosion protection of WSUD systems. Vegetation planted in bioretention systems need to tolerate condition where water ponds temporarily, but will drain relatively quickly.

The pilot study involving permeable pavement and bioretention system will investigate the potential soil and groundwater pollution from greywater-based irrigation. Meanwhile, analysis within the replicas with respect to plants and within the plants with respect to the water quality parameters of the column experiment will aid in selecting the most effective species from the set of plants studied. Overall, the outcomes of these studies will help to implement the multi-functional WSUD systems in the arid landscape in recycling and reusing of greywater for potable water savings.

## MATERIALS AND METHODS

Thirty-six number prototype bioretention columns were constructed and 12 different species of selected plants in triplet were planted. Firstly, the system was nursed for two months irrigating using tap water to allow the plants adapt to their new environment. Then, irrigated uniformly using measured quantities of greywater over a period. The heavy metals in the water, plants (leaves, stem and roots) and the soil were analyzed using the method of atomic absorption spectrometry. Sampling of the water effluent is twice in every month and where necessary, preserved with concentrated nitric acid and kept refrigerated before the lab analysis. Minimum of 50g dried soil sample, 0.3g dried stem and leaf samples taken at the end of every month. The sampling of the roots is only twice, in the beginning and at the end of the experimental period.

The length of the column is 60cm and the diameter is 15.8cm. The effective length is 54cm, while the 6cm (free board) left would to contain the irrigation water. At the bottom is the drainage layer, 5.3 cm thick and filled with 2-5 mm aggregate. The 8 cm thick transition layer have soil with grain size in the range of 0.50 - 02 mm. Filter media, 40 cm thick contains well graded soil 0.075 - 4.75 mm, mixed on-site in the following percentage: sand, 25%; black soil, 20%; 0.5-2mm, 35% and 2-5mm, 20%. The columns were constructed using the PVC pipes and water drained through the tap at the bottom of the system. The system was built in multiple layers in order to avoid intoxication of plants by hydrogen sulfide and other substances released in the anaerobic breakdown of organic matter typical of a single layer bioretention system (Trowsdale and Simcock, 2011).

**FIGURE 1.** Photos of thirty-six prototype bioretention columns located at the UAE University

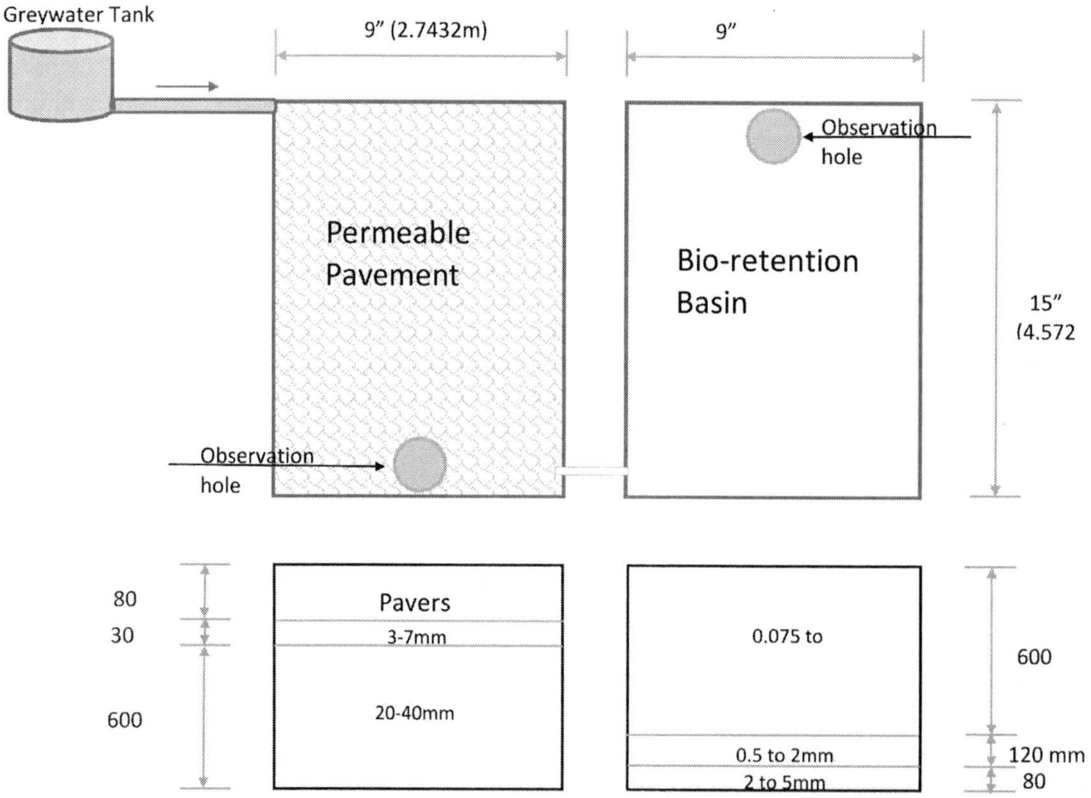


**FIGURE 2.** Layout of permeable pavement and bioretention basin constructed on the Falaz Hazza campus of the UAE University

There are 12 selected species of the plants, namely: 1. *Dodonaea viscose* (Dod) 2. *Vitex agnus* (Vit) 3. *Ficus nitida* (Fic) 4. *Jasmine sambac* (Jas) 5. *Pennisetum seateceum* (Pen) 6. *Ixora coccinaea* (Ixo) 7. *Hibiscus rosa sinensis* (Hib) 8. *Vinca rosea* (Vin) 9. *Lantana camara* (Lan) 10. *Canna indica* (Can) 11. *Alternanthera ficondea* (Alt) and 12. *Rhoeo discolor* (Rho). Irrigated with tap water, the plants had 55 days (11-09-2015 to 05-11-2015) after plantation to normalize with their new environment. Subsequently, and depending on the weather a measured quantity of greywater at a uniform rate used for the irrigation. Within the period of growth, the following amounts of fertilizers were applied three times on each plant: NPK, 1.53 g/ml; $MgSO_4$, 0.83g/ml and $CaNO_3$, 0.55g/ml.

Figures 2 and 3 shows the specification and a photo, respectively, of the constructed bioretention system in the field. The bioretention system has three layers. First, the filter layer consists of 60 cm of well-graded (0.075 to 4.75 mm) sand. Then a 12 cm transition layer of coarse sand and a drainage layer of 8cm at the bottom (2 to 5 mm gravel). The topmost layer of the system contains a wide range of soil grading. The well-graded soil is a blend of different classes of soil grades in the following ratio: normal sand (45%); 0.2 -2mm sand (35%) and 2-5mm coarse sand/crushed stone (20%). After the excavation, plastic lining was placed to prevent the seepage of water out of the system. Then, different strata in the system placed layer after layer.

On-site monitoring of some water quality parameters using a multi-parameter water quality meter would help in defining the effectiveness of the system in improving the water quality. The parameters are pH, Oxidation Reduction Potential (ORP), Turbidity, Salinity, Conductivity, Dissolved Oxygen (DO) and Total Dissolved Solids (TDS). Furthermore, additional test for Total Organic carbon (TOC), Total Nitrogen (TN), Total Phosphorous (TP) and heavy metals (Zinc, cadmium and chromium) are carried out concurrently. This applied to both permeable pavement and bioretention systems.

**FIGURE 3.** The constructed field scale vegetative bioretention systems

## RESULT AND DISCUSSION

Some plants like *Dodonaea viscose*, *Vitex agnus* and *Alternanthera ficondea* has shown some comfort since planting and after the experiment began. However, some other plants such as *Ficus nitida*, *Canna indica* and *Hibiscus Rosa sinensis* shows a little stress after plantation, they then revived during the normalization period. The stress was identified as a sign of nutrient deficiency, which was corrected using fertilizer. The water needs of the plants vary. While *Rhoeo discolour* is comfortable with small amount of irrigation water, other plants such as *Canna indica* needs a little bit more to stay healthy. Interestingly, plants like *Vitex agnus* and *Dodonaea viscose* seems to have much tolerance to both drought and wet conditions preferred by *Rhoeo* and *Canna,* respectively.

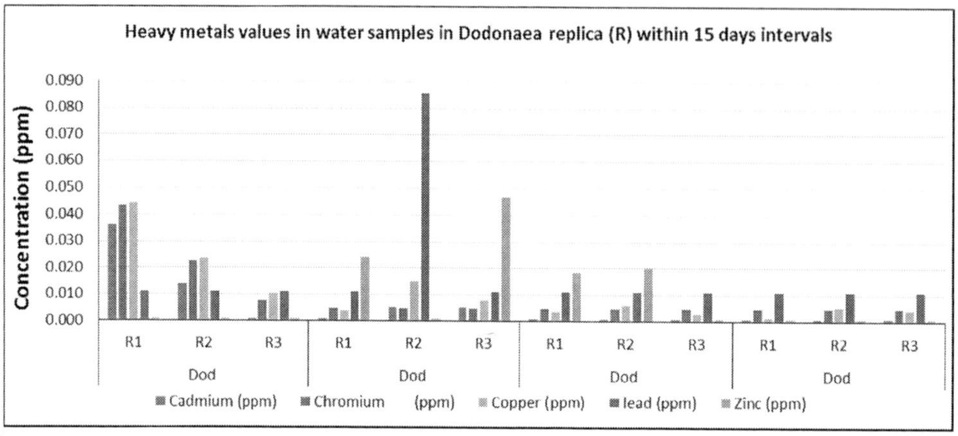


**FIGURE 4.** Heavy metals values in water samples in columns having the *Dodonaea viscose* (Dod)

The study is still under investigation. The first compartment in **Error! Reference source not found.**4 and **Error! Reference source not found.**5 is the baseline readings (raw greywater) and followed by the subsequent readings in serial order. The chromium content in the effluent seems always near the same as that of the raw greywater (about 0.0050 ppm). This implies that it is neither absorbed nor retained in the system. In addition, it is unlikely that the water dissolved the chromium in the soil. The cadmium concentration in effluents follows the similar trend of chromium. In most of the cases, the cadmium concentrations remains 0.0010 ppm for both the raw greywater and the effluent. In a few occasions, it gets down to 0.0007 ppm. For Zinc and Copper, it appears that about 99.5% of their concentration sunk into the system in all plants. Observation shows that *Pennisetum seateceum* provides more consistency and efficiency of removal compared to other plants.

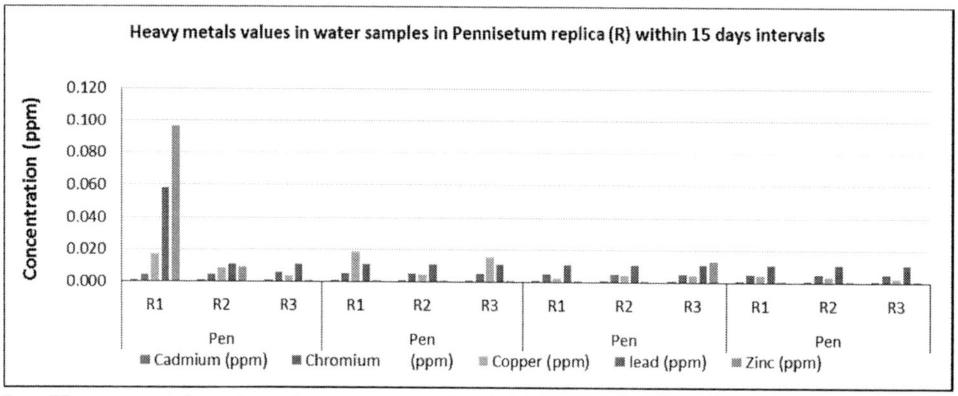


**FIGURE 5.** Heavy metals values in water samples in columns having the *Pennisetum seateceum* (Pen)

## CONCLUSION

The two experiments are to run for a year; as such, it is now too early to conclude how different plants behave in different columns of bioretention prototypes. However, preliminary results from the column experiment hinted that the some plants have a better removal capacity and consistency. Unlike chromium and cadmium, a greater percentage of zinc and copper do not drain out of the bioretention system unhindered. Furthermore, the field tests of both systems will be useful to compare the results obtained in the laboratory and in the field conditions.

## ACKNOWLEDGEMENT

The study is supported by the UAEU National Water Center interdisciplinary research grant (31R045-Research Center- NWC -2-2014) on "Greywater Reuse in Arid Climate through the Water Sensitive Urban Design Systems".

**REFERENCES**

Australian Northern Territory Government. 2009. *Water Sensitive Urban Design Vegetation Selection Guide*. Northern Territory Department of Planning and Infrastructure, Darwin, Australia.

Beecham, S. and Chowdhury, R. 2012. *"Effects of Changing Rainfall Patterns on WSUD in Australia."* Water Management 165(5):285-298.

Beecham, S. and Myers, B. 2007. *"Structural and Design Aspects of Porous and Permeable Block Pavements."* JAustCeramSoc. 43(2):74-81.

Chowdhury, R. K. and Rajput, M. A. 2015. *"Water Consumption Patterns in the Traditional Villas of Abu Dhabi."* 21st International Congress on Modelling and Simulation, Gold Coast, Australia, 29 Nov to 4 Dec 2015. www.mssanz.org.au/modsim2015.

Chowdhury, R.K., Beecham, S. and Sharvelle, S. 2015. *"Greywater Quality Changes in a Permeable Pavement Reservoir."* Water Management, DOI: 10.1680/wama.14.00107.

Chowdhury, R.K., El-Shorbagy, W., Ghanma, M. and Al-Ashkar, A. 2015. *"Quantitative Assessment of Residential Water End Uses and Greywater Generation in Al Ain City."* Water Science and Technology: Water Supply 15(1):114-123.

Hatt, B. E., Fletcher, T. D., & Deletic, A. 2009. *"Hydrologic and Pollutant Removal Performance of Stormwater Biofiltration Systems at the Field Scale."* Journal of Hydrology 365(3):310-321.

Jamrah, A., Al-Omari, A., Al-Qasem, L. and Abdel Ghani, N. 2006. *"Assessment of Availability and Characteristics of Greywater in Amman."* Water International 31(2):210-220.

McIntosh, B.S., Aryal, S., Ashbolt, S., Sheldon, F., Maheepala, S., Gardner, T., Chowdhury, R., Gardiner, R., Hartcher, M., Pagendam, D., Hodgson, G., Hodgen, M. and Pelzer, L. 2013a. *Urbanisation and Stormwater Management in South East Queensland – Synthesis and Recommendations*. Technical Report No. 106, Urban Water Security Research Alliance, CSIRO, Australia.

McIntosh, B.S., Hodgen, M., Aryal, S., Laredo, L., Wolf, L., Gardner, T., Chowdhury, R. and Maheepala, S. 2013b. *Ripley valley – An Application of GIS based Runoff Modelling to Strategic Stormwater Harvesting Assessment*. Technical Report No. 109, Urban Water Security Research Alliance, CSIRO, Australia.

Myers, B., Beecham, S., van Leeuwen, J. and Keegan, A. 2009. *"Depletion of E. coli in Permeable Pavement Mineral Aggregate Storage and Reuse Systems."* Water Science and Technology 60(12):3091-3099.

Trowsdale, S. A. and Simcock, R. 2011. *"Urban Stormwater Treatment Using Bioretention."* Journal of Hydrology 397(3):167-174.

# BIO-WASTE RECYCLING IN GERMANY

M. Nelles[1,2*], A. Schüch[1,2], A. Lemke[1], G. Morscheck[1]

[1] University of Rostock, Department of Waste Management and Material Flow, Germany
E-mail: michael.nelles@uni-rostock.de

[2] German Biomass Research Centre GmbH (DBFZ)

**ABSTRACT:** German biodegradable waste is collected separately, recycled and ecologically and economically used. Compost and digestate are used as organic fertilizer or replace peat in potting soil and plant substrates. The bio-waste recycling may also directly contribute to climate protection if the methane produced during the fermentation is used for energy production.

Around the world biodegradable waste in landfills is the main factor for the generation of the greenhouse gas methane. This environmental impact can be significantly reduced by the separate collection and recycling/use of organic waste. The separate collection of bio-waste is also a precondition for reutilizing of organic matter and nutrients. Only from separately collected bio-waste it is possible to produce high-quality compost and digestate, which are suitable for agricultural or horticultural use.

The separate collection of bio-waste from households affects the amount and composition of the residual waste. By separating bio-waste the remaining amount of waste is reduced up to a third. At the same time the residual waste contains less wet ingredients, which facilitate the waste sorting and makes the treatment in waste incineration plants more effective or even possible.

Both, the recycling of compost and digestate on soils, as well as the energy recovery of bio-waste, contribute to climate protection and resource conservation. The operation of the treatment plants determines how much of the greenhouse gases methane, nitrous oxide and ammonia is released during the process. Some plant operators need to be awakened and their awareness of climate-relevant emissions from their bio-waste treatment plants has to be raised. Also the amount of collected organic waste should be further reinforced. In 2015 the separate collection of bio-waste has to improve!

***Keywords:*** Bio-waste, Compost, Digestion, Recycling, Germany

## INTRODUCTION

Germany has implemented a separate collection for waste, household and other kinds of waste, more than twenty years ago. Diverse environmental damages, lack of landfill space and the use of finite resources led in the early 90s to a rethink in waste management. Today, climate change and energy demand are important arguments for the separate collection and utilisation of all kinds of organic wastes.

In response to the EU's waste framework directive, the Waste Management Act of 2012 (KrWG) in § 11 paragraph 1 obligates waste producers and mandated waste management authorities to collect bio-waste separately at the latest as of January 1st 2015. The term "bio- waste" in § 3 KrWG comprises yard, park, and landscape management waste as well as food and kitchen waste. The requirement in the Waste Management Act (KrWG) to collect bio- waste separately (§ 11/2 KrWG) is concretised in the Bio-waste Ordinance (BioAbfV).

This article tries to discuss all kind of organic waste but is focused on the bio-waste from private households that is collected by using bio-waste bins. Wastes collected through bio-waste bins, especially food, kitchen and yard waste, are monitored according to the European Waste Catalogue (The European Commission 2014.) laid down in the Commission Decision 94/3/EC and the German waste index regulation

(AVV) under the waste classification key 200308 (biodegradable kitchen and canteen waste, with animal residues) (Deutsche Bundesregierung 2001).

The German Bio-waste Ordinance (Ordinance on the Recovery of Bio-waste on Land used for Agricultural, Silvicultural and Horticultural Purposes) (Deutsche Bundesregierung 2013) de- fined bio-waste in general: *Waste of animal or plant origin or from fungal materials for recovery purposes, which can be degraded by microorganisms, soil-borne organisms or enzymes, including wastes for recovery purposes with high organic content of animal or plant origin or fungal materials (details in Annex 1 Bio-waste Ordinance)*

This article will use:

- Bio-waste (collected in bins): separately collected (in bio-waste bins or bags) food and kitchen waste, and yard waste from private households.
- Green waste: Separately collected yard waste from private households, and waste generated in municipal parks and in landscape management (Krause et al. 2014).
- "Organic waste" is used for all kinds of biodegradable wastes.

For the treatment of separate collected bio-waste (composting and digestion) the important laws are given with (BGK 2014):

***Bio-waste Ordinance*** (BioAbfV 1998, revised in 2012): The amended Bio-waste Ordinance (BioAbfV) of 2012 covers the application of treated and untreated bio-wastes and mixtures on land which is used for agricultural, silvicultural and horticultural purposes. It also covers suit- able raw materials, quality and hygiene requirements, and treatment and investigations of such bio-wastes and mixtures. The Bio-waste Ordinance regulates – from a precautionary perspective – the waste side (e.g. heavy metals) of the application, whereas the fertiliser law regulates the nutrient part.

***Fertiliser Law*** (DüV 2007): Gives the frame for the good code of practice of fertilising and shows special requirements for organic fertilisers. It includes the restrictions for the application of fertilisers with essential nitrogen contents in winter periods.

***Fertiliser Ordinance*** (DüMV 2012): Compost from biodegradable waste is subject to the fertiliser ordinance as a secondary raw material fertiliser (or seldom as soil improving agent). A declaration of the fertiliser type, raw material, nutrients and other product properties is obligatory. Threshold values for contaminants like PFT, PCCD or dl-PCB, included in the Fertiliser Ordinance are obligatory for compost and digestate, too.

***Federal Soil Protection Law*** (BBodSchG 1998/BBodSchV 1999): Ensures the soil function and gives among others precautionary requirements for the contamination of soils. The soil protection law is relevant for the application of compost and digestate for landscaping and recultivation.

## WHAT TO DO WITH ORGANIC WASTE?

Organic materials which the holder discards or intends or is required to discard become organ- ic waste. The organic waste can be disposed of or used in different ways. The type of disposal determines the possibilities of utilization.

## BIO-WASTE RECYCLING IS CLIMATE PROTECTION

The federal government targets to reduce the German emissions of greenhouse gases by 40 percent compared to 1990 levels until 2020. This objective can only be achieved by a sustainable energy economy - saving energy, renewable energy generation and efficient use of energy.

What could the waste industry additionally do, to achieve the resource, energy and climate objectives in the future? In particular, the use of organic waste can contribute to achieving the objectives. Today waste management saves annually about 56 million tons of carbon dioxide equivalent compared to 1990 and contributes significantly to the achievement of the climate protection goals. This has been achieved through the waste separation in the households, which is established in Germany since more than two decades. The separate collection of bio-waste and green waste takes a leading position in comparison across Europe.

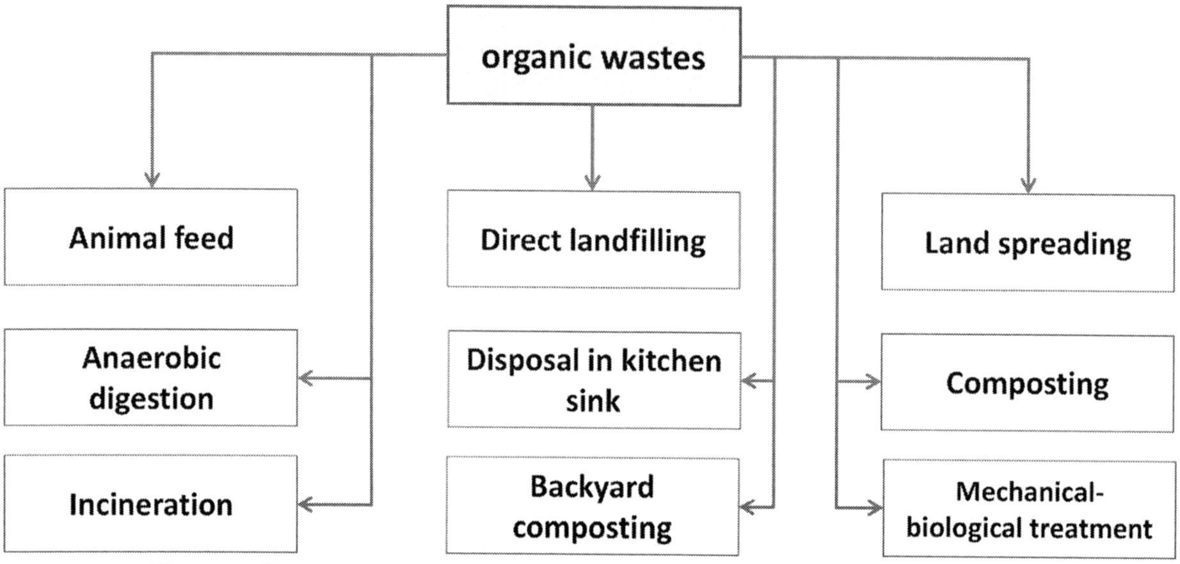


Figure 1. "Treatment" options of organic wastes (not all practised in Germany)

The share of renewable energy (such as by energy crops) in energy production should be in- creased further. The cultivation of energy crops is however in competition with food and feed production. For this reason the combined energy and material use of bio-waste and green waste is of particular interest.

A sustainable management of biogenic material flow combined material and energy recovery paths (nutrient and carbon recycling, energy supply, reducing carbon dioxide emissions by replacing fossil fuels, reducing the peat demand and lower treatment costs with extended regional added value).

Important questions are:

- what can be done to optimize collection and recycling of bio-waste
- which additionally exploitable potential exists
- how much work is required
- which benefits in relation to expenses is achieved.

The potential benefits of all biodegradable wastes should be utilized as much as possible and for this purpose the optimal combination of treatment methods has to be used.

Annually around 100 million tons of biodegradable wastes from, for example, forestry, agriculture or wastewater and waste management, arise nationwide. Of these, approximately 65 percent are technically and ecologically sensible usable, a potential of four to five percent of primary energy demand in Germany. Realising this residue potential is a significant part of the municipality's responsibility.

## ECOLOGICAL ASPECTS OF COMPOSTING AND ANAEROBIC DIGESTION OF ORGANIC WASTES

The utilization of organic waste reduces the release of greenhouse emissions compared to landfilling or incineration of wet organic waste in waste incineration plants.

The potential environmental benefits of a consistent recovery of bio-waste in Europe could achieve CO2-Eq savings of up to 50 million tons. The AD of organic waste would reach up to 7% of the overall EU target for renewable energy in 2020. Up to 42% of the bio-fuel target could be attained by the produced biogas. The AD of biodegradable waste is not only successful climate protection policy and independent energy production but also a relevant contribution to the humus formation of agriculturally used soils in

Europe. Limited resources, such as raw phosphate, can be protected by the recycling of the plant nutrients in the compost.

Not more than 11.1% of the primary energy consumption comes from renewable energy sources; less than 10% of the renewable energy comes from wastes. In the electricity sector the share of renewable energy climbed to 25.3% of the total (gross) electricity production.

Utilisation means composting and anaerobic digestion (AD) of bio-waste or incineration of woody dry green waste. But even these uses emit greenhouse gases. Emissions can be reduced by a good process control and plant construction.

The formation of greenhouse gases (GHG) as methane (CH4), nitrous oxide (N2O), ammonia (NH3) and non-methane volatile organic compounds (NMVOC) in composting and digestion processes of organic waste depends primarily on the carbon and nitrogen content in the raw material, the structure as well as the process conditions. "It depends less on the technical standard and procedural features of the treatment plant itself. The anaerobic metabolism product methane occurs also in the aerobic composting processes; the generated amounts depending on the released anaerobic activity and oxygen supply" (Cuhls et al. 2015).

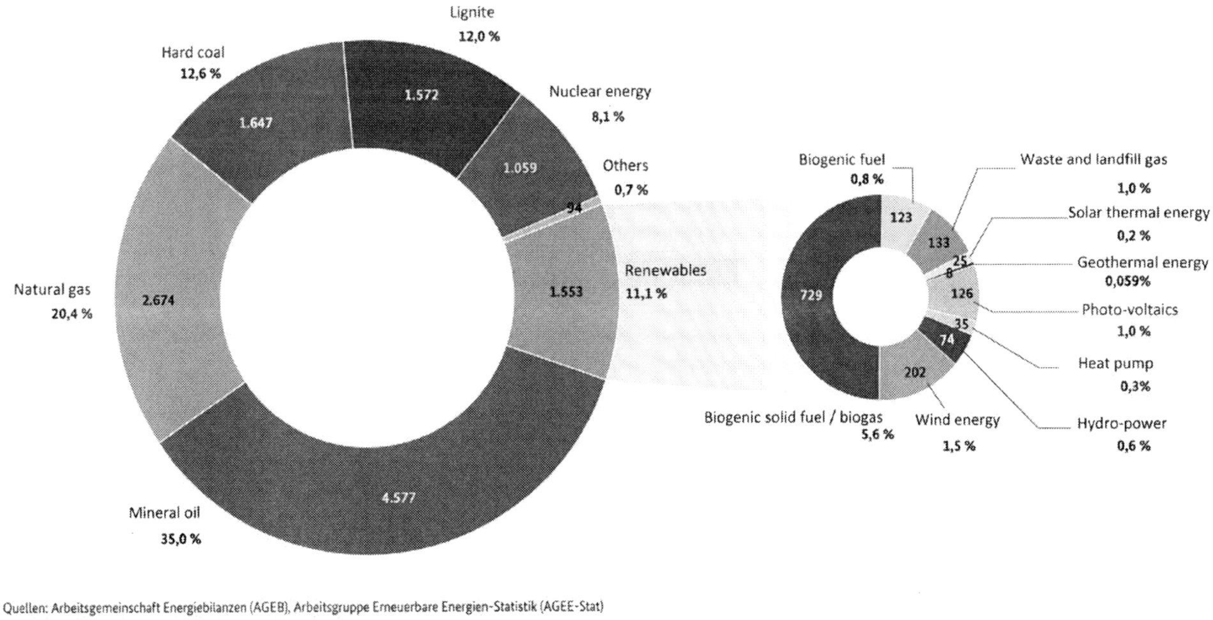


Figure 2. Share of renewable energy of primary energy consumption in Germany 2014 (AGEB 2015)

It is important that digestion plants use the total potential of the biogas. Tanks and basins with anaerobic activity, e.g. liquid fermentation residue (digestate), should be encapsulated and connected to the biogas network. Slightly with methane contaminated air flows should be used in the combined heat and power unit as combustion air or re-used as process air within the fermentation plant. In order to safely dispose of methane, exhaust air flows with high loads of methane should be combusted. On site existing biomass power stations or other combustion plants contribute to the emission control synergy (Cuhls et al. 2015).

High emissions to the atmosphere result mostly from mistakes in operation. For the biogas process reasons might be:

- inappropriate feeding system (processes in uncovered storage)
- leakage (biogas fermenter, biogas storage, pipes, valves)
- high biogas potential of the digestate (insufficient biodegradation).

In the composting or post-composting (after anaerobic digestion) process, reasons for emissions include:

- unfavourable dimensions of windrows
- insufficient use of strengthening material
- inadequate turning intensity
- high water content
- insufficient air porosity.

In the composting process all of these conditions result in a lack of air (oxygen) supply.

A closed composting process with active ventilation and exhaust gas cleaning enables a pro- cess control to reduce emissions. In practice the existing modes of operation often do not aim to reduce emissions of GHG methane and nitrous oxide (Cuhls et al. 2015).

Methane is not or only slightly decomposed in biofilters. A further development of biological processes to remove methane (e.g. in downstream methane oxidation filters) is needed.

Nitrous oxide is formed within the biofilter by NH3 conversion and its secondary products (nitrification). Therefore a removal of NH3 before the ammonia-rich gas reaches the biofilter is necessary. The solution could be acid scrubbers. The product is a dilution of ammonium sulphate, which can be concentrated and recycled as fertilizer. This applies especially to ammonia emissions from AD plants. The process conditions are crucial for the GHG emissions. With regard to the reduction of GHG emissions open windrows as well as closed composting plants (with or without AD) can be improved.

The formation of nitrous oxide can only be suppressed preventive in the biological treatment process (in situ). A subsequent reduction is not possible in any exhaust gas purification (end- of-pipe). Currently there is a lack of technical and operational requirements for a low-emission operation, in particular in order to reduce methane. Treatment of organic wastes and bio-waste management reduces the emissions of GHG, but produces also a share of the total GHG emissions.

Important are the total greenhouse gas emissions from the treatment process as well as the storage and spreading of compost and digestate in the recycling of bio-waste and green waste.

The extrapolation of the greenhouse gas emissions of CH4 and N2O and the effective indirect greenhouse gas NH3 from the operated bio-waste treatment plants (Input about 9 million tons of organic waste per year), including the emissions from the storage and spreading of compost and fermentation products results in a percentage of Germanys total greenhouse gas emissions for methane of 0.591%, nitrous oxide 0.293% and by ammonia of 0.33%. The contribution of the biological treatment and recycling of bio-waste and green waste based on the sum of total greenhouse gas emissions (CO2-equivalent) in Germany amounts to 0.066% (Kehres 2015).

## SEPARATE COLLECTION OF BIO-WASTE IN BINS, BAGS OR CONTAINERS

The waste management system in Germany is carried out in a so-called "dual system". Residual waste and organic waste are in the responsibility of the mandated waste management authorities (public administration, örE) other recyclable household waste fractions are disposed of by private waste management companies.

In 2012, the amount of bio-waste separately collected from private households and landscape management of public land in Germany amounted to 9.1 million Mg. A part of 4.3 million Mg of this amount were collected using bio-waste bins and 4.8 million Mg were green waste, collected by the mandated waste management authorities (örE).

The population specific average amount of bio-waste collected (in bins or bags) Germany- wide reaches 54 kg/capita/year and green waste 59 kg/capita/year. The collected amounts vary significantly across the German federal states (Krause et al. 2014, see Figure 3).

The mandated waste management authority can collect bio-waste in a pick-up or drop-off method. The authority has the choice to use bio-waste bins or bags as a collection container. Still not every household (waste-producer) in Germany is able to participate in separate bio-waste collection!

In 2010, private households in only 287 districts had access to a comprehensive separate collection system that uses bio-waste bins and is operated by mandated waste management authorities.

While 39 districts offered separate collection in some parts of the waste management area, 76 districts offered no bio-waste bin service at all (Oetjen-Dehne et al. 2014).

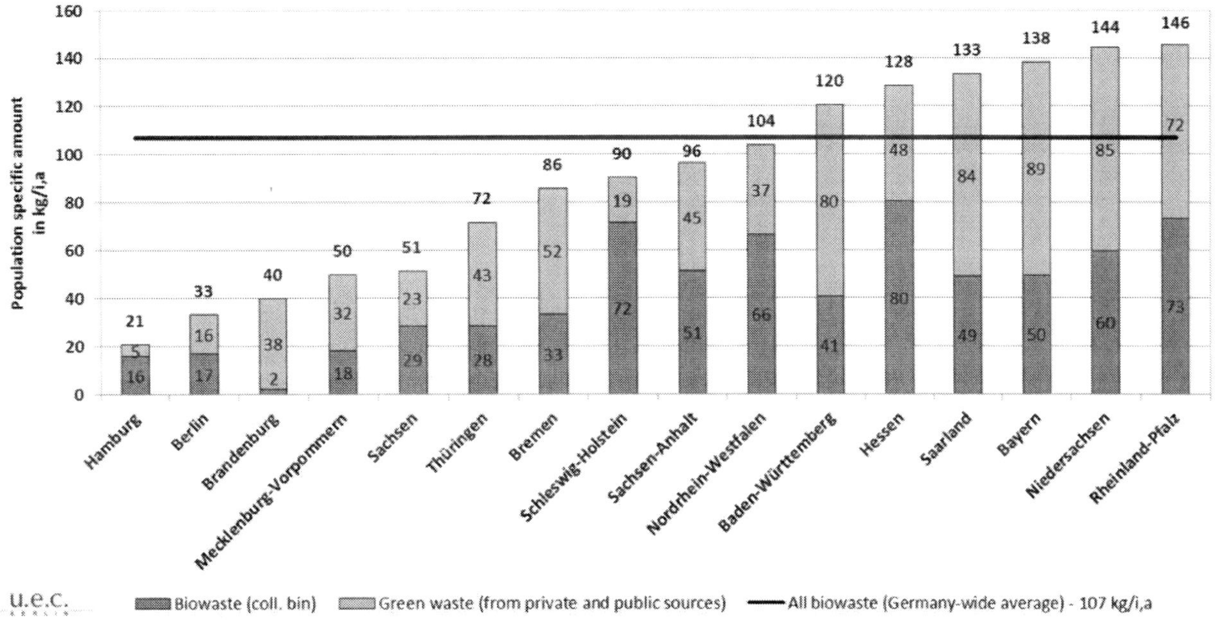


Figure 3. Separately collected amounts of bio-waste in German states in 2010 (Krause et al. 2014).

Separate collection systems for private yard waste, however, are in place in most districts. Only seven districts do not have the possibility to dispose of green waste separately through man- dated waste management authorities. In addition to various "bring systems" (green waste col- lection, green waste containers, mobile collection sites, etc.) and "pick-up systems" (kerbside collection system) for selected green waste (street collection, collection of bags of leaves, col- lection of Christmas trees, bundles collection etc.) are offered.

A survey among public waste management authorities in 2012 (Krause et al. 2014) revealed that the actual rate of access of private households to separate bio-waste collection using bio- waste bins amounts to roughly 52% Germany-wide. This number increases to 65% in areas of comprehensive separate collection systems. Nearly 40 million people in Germany do not use the bio-waste bin!

As Figure **4** shows the total amount amounts bio-waste and green waste collected in the differ- ent districts ranges from 5 to more than 200 kg/capita.

In 2015 57 to 69 administrative districts will offer no compost bin! The collection of organic waste must be further strengthened. The resource Organic waste must be used more extensive.

Only 35% (green waste and bio-waste, 3.4 and 3.9 million tons) of the theoretical bio-waste potential (21.1 million tons) has been separate collected by the public waste management au- thorities in bins, bags or containers in 2010, while roughly 23% (4.8 million tons) were collected with the residual waste. Another significant amount was disposed of in private back- yards composting and in non-recorded private business bio-waste treatment facilities.

As residual waste analyses confirm, normally most organic material contained in residual waste consists of kitchen and food waste. Further separation efforts of residual waste should therefore focus on the separation of kitchen waste from households and gastronomy. The comparison of residual waste composition in areas with and without bio-waste bins shows that already established bio-waste bin collection reduces the amount of residual waste by 15 to 20 kg of organic matter per person and year.

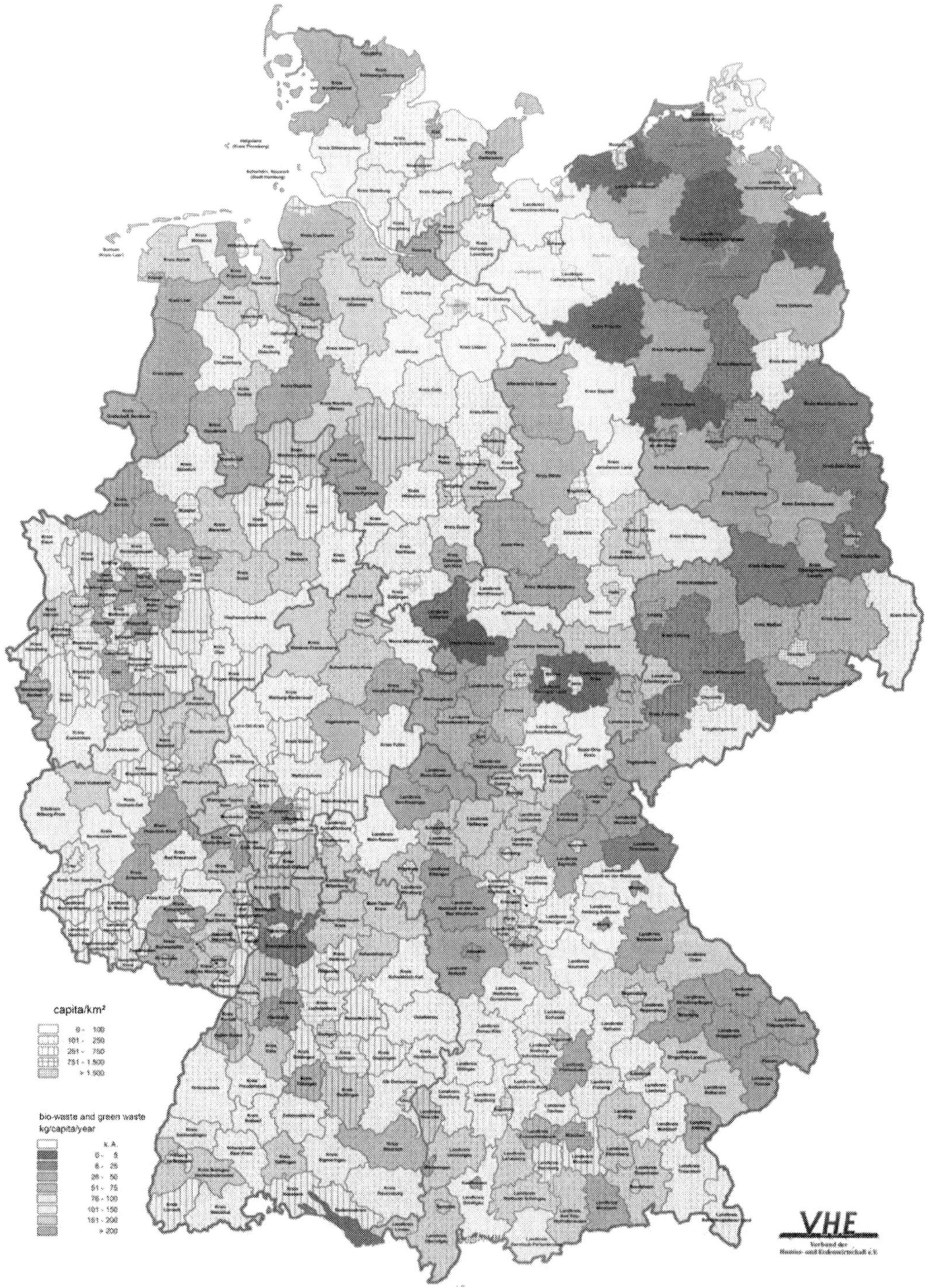


Figure 4. Population specific amounts of collected in German districts in kg/capita/year (2008 - 2011, according availability) (VHE 2012).

Additionally, yard waste is getting into bio-waste bins that would otherwise be individually composted, be illegally disposed of, or get burned. Broadly speaking, 1 kg of organic matter that is getting separated from residual waste will lead waste management authorities to gain 2 kg in yard bio-waste.

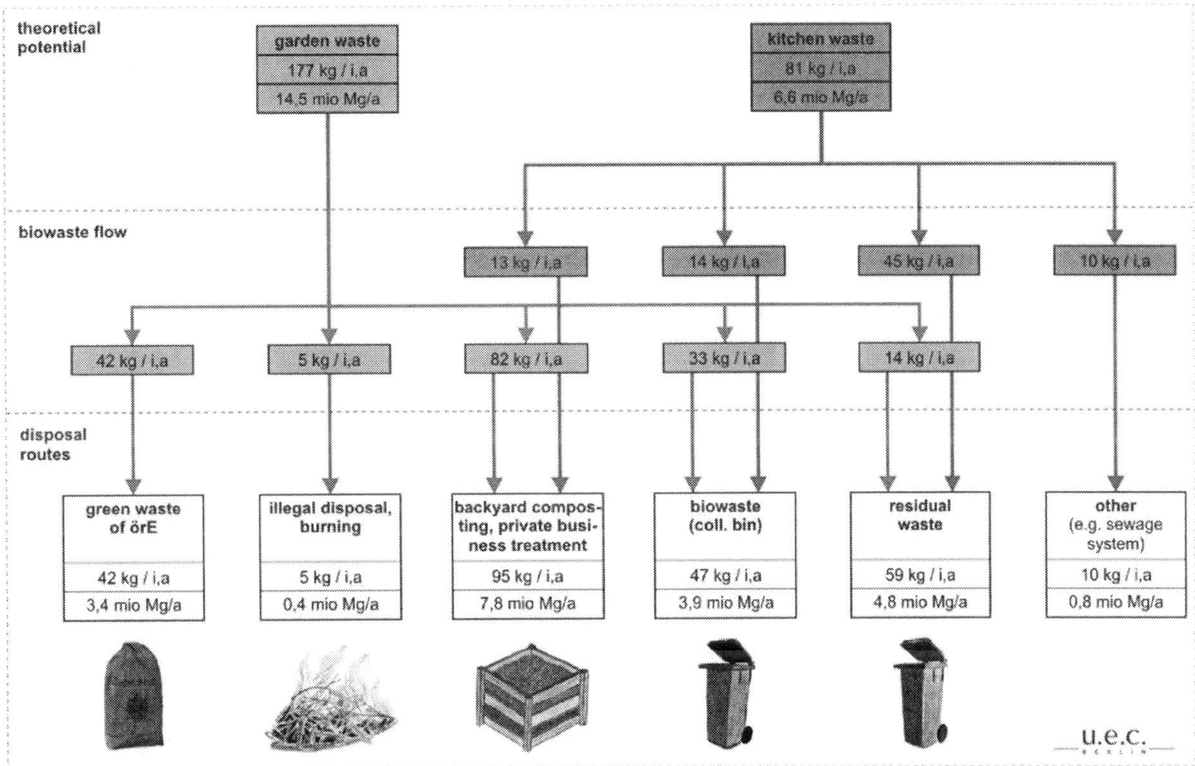


Figure 5. Bio-waste disposal routes of kitchen and yard waste in Germany in 2010 (Krause et al. 2014).

Oetjen-Dehne et al. (2014) and Krause et al. (2014) describe the bio-waste disposal situation in 2010 by using a bio-waste flow model (Figure 5).

Nevertheless, even in optimal separate collection systems and in the case of efficient use of the bio-waste bin, yearly 15 to 20 kg/capita of organic matter will usually go into residual waste, but today 59 kg/capita. In total 28.4 % (per year: 6 million tons; 74 kg/capita) of the organic waste are not collected separately!

Positive is that per year already 15.1 million tons of organic waste (184 kg/capita) are (collect- ed), treated and used (42 kg/capita green waste; 95 kg/capita back yard composting and private business treatment; 47 kg/capita separate bio bin collection). Which is more than the by the European Union estimated collectable emergence of 150 kg/capita. EU-wide previously only 50 kg/capita organic wastes are collected per year.

The collection and recycling of bio-waste can also be profitable. Electricity from bio and green waste can be reimbursed according to §27a EEG up to 50 - 75 € per ton. But the substitution effects by material recycling of bio-waste are not promoted. Currently only 5 - 15% of the organic waste is used for the production of biogas. The total potential of fermentable organic waste amounts to 11.7 million tons (Struwe 2010).

## WAYS TO INCREASE THE COLLECTION OF BIO-WASTE

In Germany the biological waste treatment looks back to a successful tradition and has become firmly established on an advanced level with about 9 million tons of separately collected or- ganic waste as part of the municipal solid waste (bio-waste in the bio bin and green waste). However, in principle an increase in collection of 2 to 4 million tons is possible.

Basically are two **collecting systems** are available.

***Kerbside collection system*** - householders are encouraged to separate targeted recyclables from their general waste and deposit this at the kerbside for regular collection.

***Bring systems*** - are also known as fixed point systems or drop-off centres, and they comprise large recycling containers in easily accessible places (household waste recycling centres or supermarkets).

The bio bin is part of a kerbside collection system, green waste, especially woody green waste is collected by a bring system.

In response to the EU's waste framework directive, the Waste Management Act of 2012 (KrWG) in § 11 paragraph 1 obligates waste producers and mandated waste management au- thorities to collect bio-waste separately at the latest as of January 1st 2015.

As outlined in the preceding presentation, improved organic waste collection is necessary. The introduction or expansion of separate collection of bio-waste in regions without or with only a small level of connection to the organic waste collection is very important.

In areas with separate collection of bio-waste could be offered measures to increase coverage rates and quality of the material flows. Objectives must be:

- the widespread removal of bio-waste and green waste from the residual waste
- optimizing of material and energy use by separation of organic waste streams for the most appropriate recovery operation.

Among other things the success depends on the following conditions:

- compulsory connection and usage of bio-waste collection systems
- regional structures of the collection area
- waste-fee system (users and pay in proportion of waste amount) for residual waste and bio-waste
- increasing the collection volumes of green waste, better collection system, banning the burning of (woody) green waste
- Better public relations work.

Expansion of the separated collection system "Bio bin" only makes sense with high quality recycling of the organic waste (high-quality recycling by material and energetic utilization).

The best recovery can be realized by fermentation (AD) followed by composting of solid fermentation residues (cascade use). Environmentally sustainable fermentation:

- Except the woody ingredients all the bio-waste is fermented
- Low-emission fermentation (including closed halls, gas-tight Storage of liquid digestate, optimized gas usage)
- Use of heat from the conversion of biogas to electricity!

Although in Germany the bio bin is already successfully introduced, there are not yet connected to the bio-waste collection authorities and people concerns about the bio-waste collection.

The resistances are against the compulsory introduction of bio-waste collection since January 2015. Separate collection results in higher costs. These costs are acceptable and must to be defrayed by the waste producer. The backyard composting captures and recycles not enough organic waste. Too much bio- waste is located in the residual waste. More than 100 kg bio-waste is collectable in bio bins even in areas with low population.

In the worldwide comparison the use of bio-waste in Germany is of a good level. The tasks for the coming years can be summarized as follows:

- All counties and cities (mandated waste management authorities - örE) have to intro- duce or improve the separate bio-waste collection (bio bin and green waste);
- Possibly higher waste fees are economically viable;
- The share of fermentation (AD) of organic wastes needs to be increased;
- The management of digestate must be improved to reduce the release of greenhouse gases;
- The use of heat from biogas electricity production needs to be increased.

## REFERENCES


1. AGEB - Arbeitsgemeinschaft Energiebilanzen. 2015. http://www.ag-energiebilanzen.de/
2. BGK – German Compost Quality Assurance Organisation 2014. Presentation of the Bundesgütegemeinschaft Kompost e.V. http://www.kompost.de/
3. Cuhls, C., Mähl, B., Clemens, J. 2015. Ermittlung der Emissionssituation bei der Verwertung von Bioabfällen. UBA (Germany's main environmental protection agency). Text 39/2015
4. Deutsche Bundesregierung (editor). 2001. Abfallverzeichnis-Verordnung vom 10. Dezember 2001 (BGBl. I S. 3379), die zuletzt durch Artikel 5 Absatz 22 des Gesetzes vom 24. Februar 2012 (BGBl. I S. 212) geändert worden ist. Berlin
5. Deutsche Bundesregierung (editor). 2013. Bioabfallverordnung in der Fassung der Bekanntmachung vom 4. April 2013 (BGBl. I S. 658), die zuletzt durch Artikel 5 der Verordnung vom 5. Dezember 2013 (BGBl. I S. 4043) geändert worden ist. Berlin
6. Kehres, B. 2015. Emissionssituation bei der Bioabfallverwertung. Bundesgütegemeinschaft Kompost e.V. www.kompost.de
7. Krause, P., Oetjen-Dehne, R., Dehne, I., Dehnen, D. & Erchinger, H. 2014. Compulsory implementation of separate collection of bio- waste Federal Environment Agency (Germany), TEXTE 84/2014. Berlin
8. Oetjen-Dehne, R., Krause, P., Dehnen, D. & Erchinger, H. 2014. Obligatory introduction of separate collection of bio-waste. Müll und Abfall. 6. 2014. 309 – 316
9. Struwe; J. 2010. Gesamtökologischer Vergleich von stofflicher und energetischer Verwertung. Abfalltag Baden-Württemberg 2010. http://www.prognos.com
10. The European Commission (editor). 2014. COMMISSION DECISION of 18 December 2014 (2014/955/EU) amending Decision 2000/532/EC on the list of waste pursuant to Directive 2008/98/EC of the European Parliament and of the Council
11. United States Environmental Protection Agency. 2014. EPA CFR40/503 Biosolids PART 503—STANDARDS FOR THE USE OR DISPOSAL OF SEWAGE SLUDGE Subpart B—Land Application
12. VHE. 2012. Bio- und Grünguterfassung in Deutschland. http://www.vhe.de/fileadmin/vhe/pdfs/ Publikationen. 17.05.2015

# UTILIZATION OF WOODWASTE AS CONTAINER MEDIA BY ACCELERATED COMPOSTING

Hoda Bakhshizadeh[1], **H. Borazjani**[2], R. C. Sloan[3], and S. S. Worthey[3]

**1** Former graduate student, Mississippi State University, Department of Sustainable Bioproducts. Current address: Soybeen Plant Pathology Lab, Bayer CropScience, Memphis, TN 38120, USA

**2** Mississippi State University, Department of Sustainable Bioproducts, Box 9820, Mississippi State, MS 39762, USA

**3** Mississippi State University, Northeast Research & Extension Center, Box 9389, Mississippi State, MS 39762, USA

The objectives of this study were: To determine the suitability of quick composted forest products woodwastes as ornamental plant media. In a 3-month accelerated composting study, woodwastes were amended with (20% & 40%) poultry litter, (1% & 2%) ammonium nitrate, or non-amended. Composting was run in fifteen 130- liter containers outdoor and samples were collected at day 90 for compost maturity (toxicity), carbon-to-nitrogen ratio and green house evaluations. The amendments containing 40% poultry litter showed significantly higher mass weight and vigor grading in transplanted zinnia and gardenia than other treatments and was comparable to commercial media. Overall, results indicated that the amendment of woodwastes with poultry litter could produce a comparable product to currently used commercial container media.

# NOVEL COUPLED SURFACE MODIFICATION AND FROTH FLOTATION SEPARATION OF HALOGENATED PLASTICS FROM HAZARDOUS WASTE (ASR/ESR) PLASTICS WITH NANOSCALE METALLIC CALCIUM COMPOSITE

***Srinivasa Reddy Mallampati,*** Je Haeng Heo and Min Hee Park
(Department of Civil and Environmental Engineering, University of Ulsan, Daehak-ro 93, Nam-gu, Ulsan 680-749, Republic of Korea)

ELVs (End of Life Vehicles) and WEEE/E-waste (Waste Electric and Electronic Equipment), are increasingly important secondary sources of ferrous metals, nonferrous metals and plastics. About 1 million ELVs are generated every year in Korea and the generation of WEEE has rapidly increased in recent years. The most problematic fractions of WEEE and ELV recycling treatments that have been mainly landfilled/incinerated in the past are shredding residues (SRs): automobile shredder residue (ASR) from ELVs and electronic waste shredder residue (ESR) from WEEE, composed of plastics, small metals, wires, rubber, textiles, etc.. Plastics are the most important portion of ASR and ESR; around 30% of WEEE plastics contain fire retardants, which are mostly based on polybrominated aromatic compounds. Approximately 50% of waste WEEE plastics are high-impact polystyrenes, with the next largest fraction being acrylonitrile-butadiene-styrene (ABS) copolymer; poly (vinyl chloride) PVC also comprise around 6–8 wt% of the total mixed plastics in ASR/ESR. The disposition and management of ASR and ESR is an emerging environmental issue of concern to the solid waste community in Korea and around the world. A Korean directive regarding resource recycling of electrical and electronic equipment and vehicles set a recycling target, including thermal recycling, at 85% by 2006 and 95% by 2015. When halogenated plastics such PVC and ABS are subjected to incineration or uncontrolled burning processes for disposal, their high bromine/chlorine content can contribute to the formation of highly toxic and persistent chlorinated/brominated dioxins. Furthermore, some of its chemical additives may leach, during landfilling. Therefore, any end-of-life product or waste stream involving products containing hazardous halogenated plastics must be separated/recovered in a way that minimizes the potential impact on human health and the environment.

Selective surface hydrophilization of PVC by a nanometallic Ca/CaO composite was successfully demonstrated in this study as a means to allow separation of PVC from other E-waste plastics. Nanometallic Ca/CaO treatment enhanced the wettability of the PVC surface; the treatment made all plastics tested more hydrophilic, but was quite selective for PVC, decreasing its water contact angle by 18°. SEM images of the plastics showed that the treatment considerably changed the surface morphology and roughness of PVC compared to other plastics. SEM-EDS elemental analysis showed that the amount of Cl detectable on PVC plastic surface significantly decreased after nanometallic Ca/CaO treatment. XPS analysis confirmed an increase of the hydrophilic functional groups C–O, C=O, and (C=O)–O on the PVC plastic surface after nanometallic Ca/CaO treatment. The treatment allowed PVC to be separated from other E-waste plastics by means of froth flotation; the plastics were separated into a PVC fraction and a PVC-free fraction, even though the plastic particles were of nonuniform size and shape. Under the condition of 100 rpm mixing, 100% of the PVC settled, and the settled fraction was 96.4% PVC; thus, the floating fraction was 100% PVC-free. The total recovery of non-PVC plastics was nearly 100%, which is greatly improved from the 20.5 wt% of light plastics that are recovered by conventional wet gravity separation. Therefore, the hybrid nanometallic Ca/CaO treatment and froth flotation separation developed herein is a simple and effective method to separate hazardous halogenated plastics from ASR/ESR. This convenient method will facilitate the industrial application of plastics flotation, and the present work provides technical insights into the process of hazardous halogenated plastics recycling. *This work was supported by National Research Foundation of Korea (NRF), funded by the Ministry of Education (2014R1A1A2055487).*

## CHARACTERIZATION OF CARBON-BASED BIOCHAR FROM GRAPESEED PYROLYSIS

***Nathaniel F. Adegboyega*** & William C. Hockaday (Baylor University, Waco, TX, USA)
Chris Cunningham & Mark Kelm (Constellation Brands Inc., Madera, CA, USA)[26]

We report on the morphological, chemical, and spectroscopic characteristics of carbon-based biochar manufactured by the pyrolysis of grape seed wastes (from the winemaking industry). The morphology of grape seeds is such that the embryo and endosperm, which comprise starches, lipids, and protein are encased in a ligno-cellulosic seed coat. The spatially-segregated chemistry of the seed make for an interesting precursor for multi-functional carbon-based sorbents. In this study we use combination of microscopy and spectroscopy to characterize the pyrolytic transformation of grape seeds to biochar carbons. Scanning electron microscope (SEM) images revealed that the general seed morphology remains intact after pyrolysis. The cell wall structure integrity is maintained while the development of an extensive ``pores`` network is restricted to the endosperm.

Pyrolysis at 350 °C resulted in an increase in elemental C concentration from 54.04 wt % to about 71.74 wt %. An extended drying (pretreatment) of grape seed sample before pyrolysis did not significantly influence the C content after pyrolysis. The solid state $^{13}$C NMR of biomass before pyrolysis showed abundant liquid-like resonances in the range 14 to 34 ppm and 128 – 129 ppm attributable to unsaturated fatty acids in grape seed oils. Pyrolysis at 350 °C caused an increase in the concentration of both aliphatic and aromatic carbons at the expense of oxygen-substituted carbons. Aromatic carbon concentrations increased further at higher pyrolysis temperatures and longer duration of heating. Biochars formed at both 350 °C and 500 °C exhibited extreme hydrophobicity (poor wettability), as revealed by a simple water droplet test.

# EFFECT OF TIRE RUBBER ASH ON BITUMINOUS MIXES USED FOR ROADWAY PAVEMENT

***Tapash Kumar Roy***
(Indian Institute of Engineering Science and Technology Shibpur Howrah-711103, W.B., India)

**ABSTRACT:** Scrap tires are abundant and alarming waste throughout the globe. Massive stockpiles of such tires occupied valuable land and also pollutes environment. Further steady increase in traffic intensity, overloading of commercial vehicle and the significant variation in daily and seasonal temperature demands the improved quality of the road for overcoming the early development of distresses in the flexible pavement constructed by conventional bituminous mix. Modification of bituminous mix by utilizing of such waste tires for construction of flexible pavement is the prime objective of this investigation. Outcome of the experimental study indicated that waste tire modified bituminous mix shows the lesser consumption of bitumen with better stability value of amounting more than 142% compared to that of conventional mix with much reduction in flow characteristics.

## INTRODUCTION

The road needs to be updated to meet the quality and quantity of performance to sustain repeated vehicle loadings, climatic changes and riding quality. In India, it is estimated that over 42 lakh kilometers of road exists which carries close to 90% of passenger traffic and 70% of freight transport. The major part of such roads are flexible in nature, deteriorated rapidly during their design life in some parts of India due to different causes e.g. high traffic intensity, overloading of commercial vehicle, improper drainage system etc. resulting early development of distress condition in the form of cracks, rutting, raveling etc. So modification in the bituminous mixtures is needed to sustain the stresses of vehicles with other changes in properties. On the other hand, due to the worldwide growth of the automobile industry and the increasing use of cars as the main means of transport have tremendously boosted tire production in recent decades. This has generated massive stockpiles of used tires. Extensive research projects were carried out on how to use used tires in different applications. The benefits of stabilizing the stone mastic asphalt (SMA) mixture in flexible pavement with shredded waste plastic was developed by Bindu & Beena (2010). Performance of rubber as an asphalt concrete modifier was evaluated by performing experiments on the shear flow properties and creep rapture behavior (Zaman et al 1995). So the reutilization of such wastes could be a suitable way of solving environmental concerns, offering low-cost recycled resources of polymer. This investigation deals with the utilization of crumb tire rubber for producing better bituminous mix by enhancing their properties.

## MATERIALS USED AND THEIR PROPERTIES

Present study involves the use of materials like aggregate, bitumen, and waste rubber for producing bituminous mixes.

**Aggregate.** The mineral aggregate used in this study were obtained from Pakur (Jharkhand) quarry. Representative samples have been tested in the laboratory and results are presented in Table-1.

**Bitumen.** In this study, 60/70 pen grade bitumen has been selected. The physical properties of the same are given in table 2 along with their permissible limits as per IS specifications.

**Waste Rubber.** Sources of waste rubber are the various commercial vehicles like trucks, busses, and trailers. In this study, waste tire of buses was used as rubber waste. Collected rubber waste has been powdered by the grinding machine and subjected to mechanical sieve analysis to get the desired particle size passes from 150 mm sieves and retained in 75 mm size. Composition of such waste rubber has been summarized in the Table 3.

**TABLE 1.** Physical properties of aggregate

| Test | Result |
|---|---|
| Impact value | 21.81% |
| Flakiness Index | 15.9% |
| Elongation Index | 14.3% |
| Abrasion Value | 23.24% |
| Specific gravity | 2.78 |
| Water absorption | 0.14 |

**TABLE 2.** Physical properties of Bitumen

| Type of test | Test Value | Permissible Limit |
|---|---|---|
| Penetration Value at $25^0$c (1/10 mm) | 68 | **60-70** |
| Ductility Value at $25^0$c (cm) | >100 | **>75** |
| Softening Point ( $^0$C ) | 54 | **40-55** |
| Specific gravity (kg/m$^3$) | 1.04 | **>0.99** |

**TABLE 3.** Composition of bus tire

| Materials | Composition in% |
|---|---|
| Rubber/Elastomer | 45 |
| Carbon black | 22 |
| Metal | 25 |
| Zinc oxide | 2 |
| Sulphar | 1 |
| Additives | 5 |

## SAMPLE PREPARATION

The mix design for Dense Bituminous Macadam (DBM) with Grading-I have been selected for this investigation by using the Marshall method. Primarily conventional bitumen was weighted and mixed with various percentages of waste tires (0%, 6%, 8%, and 10%) by weight of bitumen. Then the wastes added bitumen was kept in the mechanical stirrer for 1 hr. at 3500-4000 rpm at 180ºC to 200ºC temperature to convert the same into homogeneous modified bitumen. Next they were mixed with the preheated aggregate and filler for preparing the samples. Then Marshall Test was performed for evaluating optimum bitumen content (OBC) and other Marshall Parameters.

## RESULTS AND DISCUSSION

Specimens were prepared by using cement as filler @ 2% by wt. of aggregate with varying bitumen contents of 4.5%, 4.75%, 5.00% and 5.25% and tested and OBC was determined. Evaluated results are given in Figure 1 and Table 4.

Experimental results indicated that optimum bitumen conten of DMB mix is 5.08%. Other Marshall Properties like stability and flow value are 21.35 kN and 2.97 mm respectively which satisfied the standard specifications recommended by MORT&H.

Further Marshall Properties along with Optimum bitumen content were determined by mixing various % of waste Rubber ranging as 0%, 6%, 8%, and 10% with the bitumen by following the Marshall method and tested results have been furnished in the table 5.

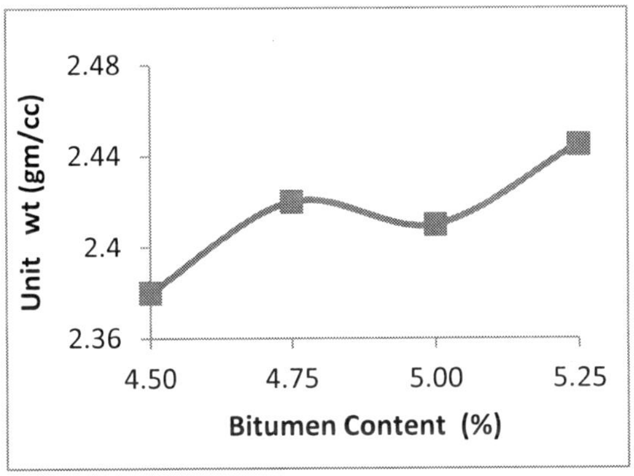


(a) Bitumen Content Vs Unit Wt

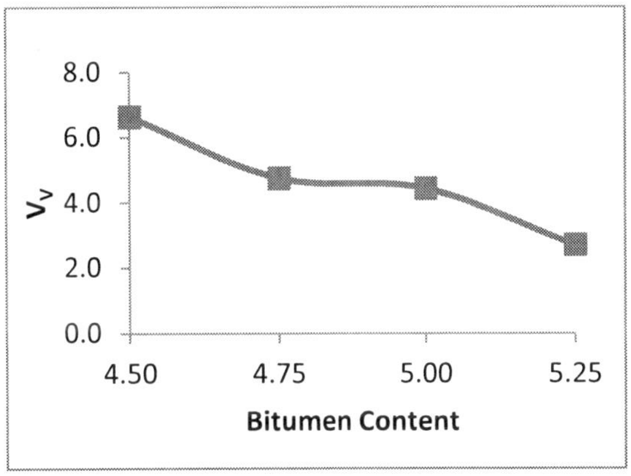


(b) Bitumen Content Vs $V_v$

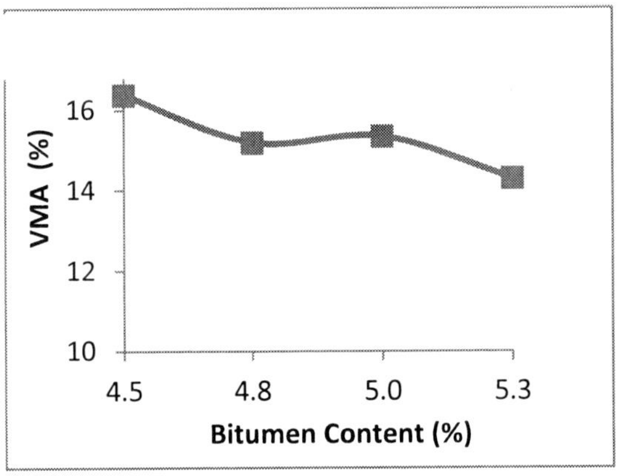


(c) Bitumen Content Vs VMA

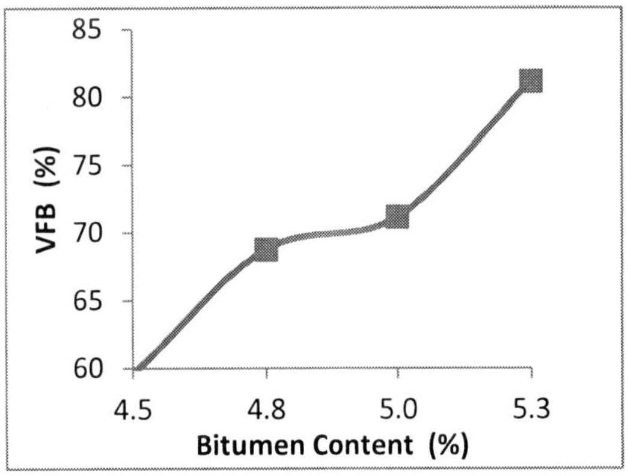


(d) Bitumen Content Vs VFB

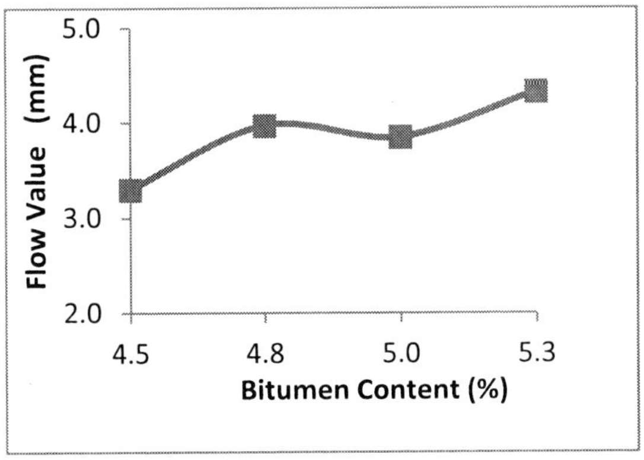


(e) Bitumen Content Vs Flow Value

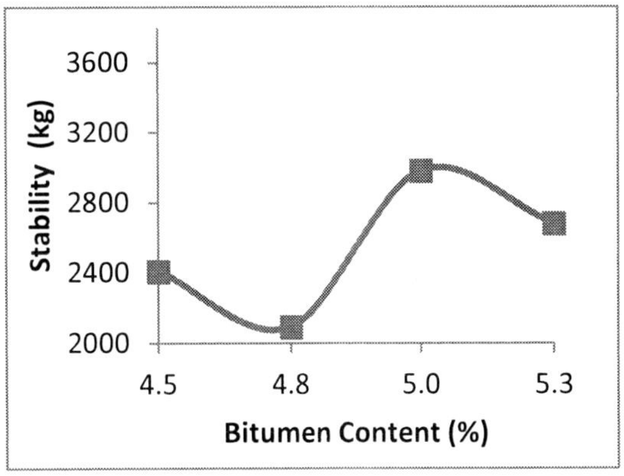


(f) Bitumen Content Vs Stability

FIGURE 1. Comparison between Different Marshall Parameter Vs Bitumen Content

Marshall Properties as evaluated in table 5 indicates that maximum stability value of amounting 51.8 kg is obtained for adding 8 % of waste rubber ash as modifier by keeping the flow value as 3.8 mm. Where optimum bitumen content remained as 4.78% for DBM mix by using said waste tire rubber ash as modifier. In comparison with the conventional mix of DBM, it is observed that waste tire modified bituminous mix shows the lesser consumption of bitumen with better stability value of amounting more than 142%. All other evaluated parameters also satisfy the standard specification of MORTH.

TABLE 4. Evaluated Marshall Properties of DBM against OBC

| Filler | Optimum Bitumen Content (%) | Vv(%) | VMA(%) | VFB (%) | Stability (KN) | Flow (mm) |
|---|---|---|---|---|---|---|
| Cement | 5.08 | 4.15 | 15.36 | 73 | 21.35 | 2.97 |

TABLE 5. Evaluated Marshall Properties by using Rubber ash as modifier

| Modifier | Quantity (%) | OBC (%) | Marshall Parameters | | | | | |
|---|---|---|---|---|---|---|---|---|
| | | | Stability (kg) | Flow (mm) | VMA (%) | VFB (%) | $V_v$ | Unit Wt (gm/cc) |
| Rubber Content | 6 | 4.95 | 47.0 | 3.75 | 14.0 | 73 | 3.7 | 2.41 |
| | 8 | 4.78 | 51.8 | 3.8 | 15.1 | 75 | 3.2 | 2.41 |
| | 10 | 4.82 | 48.7 | 3.91 | 18.3 | 72 | 5.1 | 2.37 |

## CONCLUSION

From the above investigation it may be concluded that the scrap tires discarded in the environment, including tire composition, adverse environmental effects, threats to public health and safety, and solid waste management. However, addition of such waste in construction of dense bituminous macadam reduces the optimum bitumen content compared to the conventional mix. Significant observation is made for utilizing tire rubber of 8% in the DBM mix, where stability values increased more than 142% by reducing the flow value along with the reduction in optimum bitumen content. So use of 8% tire rubber in construction of dense bituminous macadam in roadway pavement may be encouraged as a cost effective mix proportion by achieving better performance along with environmental benefits.

## REFERENCES


Bindu, C.S., K.S., Beena, (2010) "Waste plastic as a stabilizing additives in stone mastic asphalt", *International Journal of Engineering and Technology*, Vol.2(6), pp.379-387.

IRC:37-2001, Guidelines for the Design of flexible pavements, *Indian Roads Congress.*

IS: 1202-1978, "Methods for Testing Tar and Bituminous Materials; Determination of Specific Gravity", *Bureau of Indian Standards*, India.

IS: 1203-1978, "Methods for Testing Tar and Bituminous Materials; Determination of Penetration", *Bureau of Indian Standards*, India.

IS: 1205-1978, "Methods for Testing Tar and Bituminous. Materials; Determination of Softening Point", *Bureau of Indian Standards*, India.

IS: 1208-1978, "Methods for Testing Tar and Bituminous Materials; Determination of Ductility", *Bureau of Indian Standards*, India.

Zaman, A.A., A.L., Fricke, and C.L., Beatty, (1995), "Rheological properties of rubber modified asphalt", *J.Transp.* En.121, pp.461-467.

# STRUVITE RECOVERY FROM WASTEWATER AND REUSE AS AMENDMENT FOR HEAVY METALS IMMOBILIZATION IN SOIL

***Hao Wang***, Xuejiang Wang, Jing Zhang, Jingke Song
(Tongji University, Shanghai, P.R.China)

Ammonia nitrogen (AN) and phosphorus (P) were recovered from rare-earth wastewater by using the formation of struvite, which was used as the amendment with plant ash (PA) for immobilization of copper (Cu), lead (Pb) and chromium (Cr) in contaminated soil. AN removal efficiency and residual P reached to 95.32±0.73 % and 6.14±1.72 mg $L^{-1}$ under optimal conditions: pH=9.0, n (Mg):n(N):n(P) = 1.2:1:1.1, which were obtained from response surface methodology (RSM). The effects of mass ratio of precipitates and plant ash, immobilization time, soil pH, humic acid (HA) and fulvic acid (FA) on available concentration of Cu, Pb and Cr were investigated. It was shown that immobilization was most effective with the mass ratio of PP/PA at 1:3, while the PA addition further leaded to pH increase, which was beneficial for heavy metal immobilization in the bulk soil. The minimum available concentration of Cu, Pb and Cr separately reduced to 320.82 mg $kg^{-1}$, 190.77 mg $kg^{-1}$ and 121.46 mg $kg^{-1}$ with the increasing of immobilization time. HA and FA were beneficial to Cu immobilization, but both of which showed no effect or even a negative effect on the immobilization of Pb and Cr.

# HYDROTHERMAL SYNTHESIS OF ZEOLITE A FROM BOTTOM ASH OF SUGAR INDUSTRY ITS CRYSTALLIZATION

***Noor-ul-Amin*** and Yousaf Hameed
(Department of Chemistry, Abdul Wali Khan University, Mardan,23200 Pakistan)[29]

Zeolite A is indirectly synthesized from bottom ash of sugar industry using hydrothermal conversion method. Active $SiO_2$ and $Al_2O_3$ were obtained from bottom ash and aluminium sulphate respectively. The synthesis involve the dissolution of amorphous aluminosilicate releasing $[SiO_2\,(OH)_2]_2$ and $Al(OH)_4$, formation of sodium aluminosilicate gel and crystallization of zeolite A. The obtained zeolite was characterized using X-ray diffraction (XRD), X-ray fluorescence (XRF), Scanning electron microscope (SEM) and Fourier Transform Infrared spectroscopy (FTIR). The results showed that the amorphous silica obtained from ash was dissolved completely during the first hour of the crystallization stage. With the increase in crystallization time up to 3 h, a rapid growth of pure zeolite crystals reached its maximum limit. An excess time resulted in further dissolution of zeolite A and replacement by hydroxyl sodalite.

---

# RECYCLING SPENT LITHIUM-ION BATTERY FOR FUNCTION MATERIAL DEVELOPMENT

***Fu-Shen Zhang***, Meng-Meng Wang
(Research Center for Eco-environmental Sciences, Chinese Academy of Sciences, 18 Shuangqing Road, Beijing 100085, China)

In the current study, an environmental benign process namely mechanochemical approach was developed for cobalt and lithium recovery from spent LIBs. The main merit of the process is that neither corrosive acid nor strong oxidant was used. In the proposed process, lithium cobalt oxide (obtained from spent LIBs) was co-grinded with various additives in a hermetic ball milling system, followed by a water leaching procedure. Experiment results indicated that EDTA was the most suitable co-grinding reagent , and 98% of Co and 99% of Li were respectively recovered under optimum conditions: $LiCoO_2$ to EDTA mass ratio 1:4, milling time 4 h, rotary speed 600 r/min and ball-to-powder ratio 80:1, respectively. Mechanisms study implied that lone pair electrons provided by two nitrogen atoms and four hydroxyl oxygen atoms of EDTA could enter the empty orbit of Co and Li by solid-solid reaction, thus forming stable and water-soluble metal complexes Li-EDTA and Co-EDTA. Moreover, the separation of Co and Li could be achieved through a chemical precipitation approach. This study provides a high efficiency and environmentally friendly process for Co and Li recovery from spent LIBs.

# REUSE OF GREEN WALNUT SHELL TO PRODUCE DYE

Ruhsar Arabacıoğlu, ***Neslihan Doğan-Sağlamtimur*** and Ersen Turaç
(Nigde University, Niğde, Turkey)

In terms of environmental technologies, in the pyramid of waste management hierarchy, conversion of waste to another product is named as reuse. Each year very large quantities of walnut shells are released as herbal origined waste after selling walnut seeds. Therefore, development of new using areas for these wastes is required under the concept of industrial ecology. In this study, for obtaining dye from green walnut shells, following experimental steps have been planned; (a) chopping small pieces from the green wallnut shells, (b) eliminating of humidity from the chopped shells after dried in oven, (c) grinding of these shells by using grinding mill to increase of contact surface and decrease of piece size, (d) extraction processing with methods of Soxhlet extraction by using of ethanol solution, (e) evaporating of ethanol solution in controlled way. New experimental designs can be arranged in accordance to obtained results from these experiments.

The aim of this study is to obtain of natural dye from wasted green wallnut shells for dye industry. It is believed that it should be a positive step to meet the need of dye in our country. In terms of new and advanced technology, conversion of wasted green walnut shells to dye has acquired an important place in recent years.

# RECYCLING MERCURY-IMPACTED SCRAP METAL: KEY ISSUES AND RESEARCH NEEDS

***Molly E. Finster*** and Corrie E. Clark
(Argonne National Laboratory, Lemont, IL 60439, USA)

The recycling of mercury-impacted scrap metal can emit measurable amounts of mercury; however, existing characterization data are insufficient to fully understand the origin, key sources, and concentrations of mercury within scrap metal and the recycling process. Currently, industry-specific mercury emissions guidance values exist for many known anthropogenic mercury sources (e.g., coal-fired utility plants and waste incinerators), but are largely nonexistent for scrap metal processing and recycling facilities. Given the lack of significant guidance for recycling mercury-impacted scrap metal, these other values can provide a useful framework to potentially guide the development of mercury acceptance and release criteria/limits for recycling facilities. To better assess measures to protect human and environmental health from potential hazards that might be posed by the processing and melting of mercury-impacted metal, information on the origin, source, nature, and extent of mercury-impacted metal in scrap is also important for assessing measures to protect scrap metal recycling workers from potential health and safety hazards that might be posed by the processing and melting of mercury-impacted metal.

# THE POTENTIAL OF EFFECTIVE MICROORGANISM (EM) TO PROMOTE THE PHYTOREMEDIATION OF URANIUM POLLUTED WATER

Jing Zhu, Ke Chen, Renhua Huang, Gangxue Luo
(State Defense Key Laboratory of Fundamental Science on Nuclear Wastes and Environment, Southwest University of Science and Technology; School of Life Science and Engineering, Southwest University of Science and Technology, Mianyang, Sichuan, China)

The influence of effective microorganism (EM) on uranium accumulation and endurance in *Eichhornia crassipe*, *Pistia stratiotes* L. *Lemna minor* L. were investigated in uranium polluted solutions. With applied EM, the dismutase (SOD), catalase (CAT) and peroxidase (POD) activites of *E. crassipe* and *P. stratiotes* L. increased more significantly than those of the control without EM applied. In addition, the uranium accumulation amount in root, accumulation factor and transfer factor of *P. stratiotes* L. increased with 22.6%, 74.0% and 160.0% respectively, as well as the maximum uranium accumulation factor of *E. crassipes* increased with 172.7% compare to the control without EM applied. The results indicated that the EM applied are positive to the resistance, enrichment and transfer ability of *E. crassipes* and *P. stratiotes* to the uranium stress. However, the applied EM appeared to be negative to the uranium resistance and accumulation of *L. minor* L. with the activites of CAT, POD and SOD decreased, contents of malondialdehyde and proline increased and especially the chlorophyll contents and biomass decreased drastically. Therefore, our results suggested the aquatic plants *E. crassipes* and *P. stratiotes* L. combined with EM might effectively promote the potential of phytoremediation of uranium polluted water.

# GRASS AND MICROBIAL COMBINED REMEDIATION TO URANIUM-CONTAMINATED SOIL

Xiaoming Chen and ***Xuegang Luo***
(State Defense Key Laboratory of Fundamental Science on Nuclear Wastes and Environment, Southwest University of Science and Technology, Mianyang, Sichuan, China)
Xichao Hao and Ke Chen
(Southwest University of Science and Technology, College of Life Science and Engineering, Mianyang, Sichuan, China)

In order to build a grass and microbial combined remediation system for uranium-contaminated soil, the grass seeds germination, the growth and uranium enrichment ability of the grass plant under uranium stress pot soil, screening of microbe combination, and the effect of grass and microbial combined remediation were investigated respectively. The results showed seven species of grass seeds from 24 species had a relatively high germination rates at uranium stress condition. There was a promotion ability for seed germination rate at the uranium concentration less than 50 $mg \cdot L^{-1}$, while the uranium concentration more than 100 $mg \cdot L^{-1}$, had an obvious inhibition effect. The different species of grass plant had a significant growth difference under different uranium concentration. *Lolium multiflorum* had a high pot accumulation rate (10.36 $mg \cdot pot^{-1}$ uranium) at uranium concentration of 150 $mg \cdot kg^{-1}$, while *Lolium perenne* had the highest pot accumulation (3.58 mg•pot-1 uranium) at uranium concentration of 50 $mg \cdot kg^{-1}$. And *L. multiflorum* had the highest transfer factor (0.87) at uranium concentration of 100 $mg \cdot kg^{-1}$. The microbe combination including 20% *Bacillus mucitaginosus and* 10% *Bacillus subtilis* was screened out, which had a high yield of IAA, ACC deaminase and siderophore. *Lolium multiflorum* Lamk connected with this microbe combination could promote the uranium accumulation rate up to 50% than control. In conclusion, the connection of grass and microbes have a effective remediation ability to uranium-contaminated soil.

# POTENTIALLY USEFUL PLANTS FOR PHYTOREMEDIATION OF HYDROCARBONS IN SUBTROPICAL ZONE: CAMEROON CASE

***Matsodoum N. P.*** , Wanko N. A.
(University of Strasbourg, Strasbourg, France)
Ndemba Etim S. G., Djumyom Wafo G. V., Djocgoue P. F., Kengne Noumsi I. M.
(University of Yaounde I, Yaounde, Cameroon)

Identification of plants is prerequisite in the development of phytoremediation process. 4 cities in Cameroon have been investigated for the circumstance. The methodology consisted of studying vegetation of 26 sites of 36 $m^2$ each (polluted and unpolluted) in each city through floristic survey. 106 species belonging to 76 genera and 30 families have been identified in polluted sites, unlike the control sites where floristic diversity was much higher with a total of 166 species contained in 125 genera and 50 families. Among the most represented families in both sites, Cyperaceae (10), Asteraceae (8) and Amaranthaceae (5) have higher taxonomic richness in polluted sites compared to the control. Species richness and Shannon diversity index in hydrocarbon-polluted sites (8 to 18 species and 1.6 to 2.7 bits/ind.) are significantly lower than the control sites (24 to 34 species and 2.7 to 3.2 bit/ind.). Based on their relative frequency (Fri> 10%) and their relative abundance (A> 10% and 10%> A> 7%) on hydrocarbon-polluted sites, polluo-tolerant or plants potentially useful for phytoremediation of hydrocarbons are dominated by Poaceae (*Eleusine indica* (L.) Gaertn., *Cynodon dactylon* (L.) Pers., *Acroceras zizanioides* (Kunth) Dandy, *Panicum maximum* Jacq. and *Axonopus compressus* (SW) P.Beauv.) ; followed by Amaranthaceae (*Amaranthus spinosus* (L.), *Alternanthera sessilis* (L.) R. Br. ex DC†); A□teraceae (*Vernonia cinerea* (L.), *Synedrella nodiflora* (L.) Gaertn.) ; Euphorbiaceae (*Euphorbia hirta* L., *Phyllantus amarus* Schum. & Thonn.) ; Commelinaceae (*Commelina benghalensis* L.) ; Cyperaceae (*Cyperus esculentus* L.) ; Capparaceae (*Cleome ciliata* Schum. & Thonn.) and Acanthaceae (*Asystasia gangetica* (L.) T. Anderson). These plants could be tested in preliminary trials of phytoremediation to cleanup hydrocarbon-polluted soils in subtropics.

# EX-SITU PHYTOREMEDIATION OF HYDROCARBON CONTAMINATED SOIL WITH CROTALARIA PALLIDA AIT.

***Partha Pratim Baruah*** and Plabita Baruah (Gauhati University, Guwahati, Assam, INDIA)
Suresh Deka (Institute of Advanced Study in Science and Technology, Guwahati, Assam, INDIA)

The total petroleum hydrocarbon (TPH) contamination has been recognised as a serious threat to natural ecosystems. The affect of which has been frequently felt in the oil exploration, drilling and production areas as well as in the spillages, leakages and accident sites during transportation. To overcome the constraints of other competing physical and chemical alternatives, application of phytoremediation of TPH is being explored. As the selection of suitable plant species is one of the main criteria for success in phytoremediation and for development of viable remediation technology, the present endeavour explored for selecting a suitable candidate from natural habitat and identified *Crotolaria pallida* Ait. as one of the potential plant species for the purpose. Result revealed that maximum dissipation of TPH was upto 87.05 % from the soil in a progressive way in relation to growth of the plant and time which was evidenced by concomitant reduction of pH and total organic carbon and subsequent increase of water holding capacity.

# ASSESSMENT FOR REMOVING MULTIPLE POLLUTANTS BY PLANTS IN BIORETENTION SYSTEMS BASED ON ISPA MODEL

***Xiaohua Yang***, Yi Ye, Meishui Li
(School of Environment, Beijing Normal University, Beijing, China)

Bioretention system is a complex uncertain system for removing multiple pollutants by plants. To select the best plant species for removing multiple pollutants rationally in bioretention systems, an improved set pair analysis (ISPA) model, is established, in which set pair analysis theory is introduced and the weights are determined by use of the maximum entropy principle and the improved analytic hierarchy process method. And the index systems and criteria of the bioretention systems assessment are established in this study. ISPA model is used to assess five plants for removing multiple pollutants in bioretention systems. Certain and uncertain information quantity of the bioretention system is calculated by connection numbers in the ISPA model. Results show that ISPA model can fully take advantage of certain and uncertain knowledge, subjective and objective information. Finally, we make some suggestions for the bioretention system management in Beijing. This study will provide a simple and useful guide for the comprehensive assessment of multiple pollutant removal by plants in complex ecological systems.

# SYNTHESIS AND PROPERTIES OF BIODEGRADABLE POLYESTERS BASED ON BIOMASS MONOMERS

***Yang Zhang***, Jun Xu, Baohua Guo
(Tsinghua University, Beijing, China)

With increasing worldwide concern about greenhouse effects and white pollution of polymers, biobased monomers and biodegradable polymers made from renewable sources have attracted much attention recent years. Chemicals in furan family and succinic acid have been regarded as the most important building blocks derived from biomass. In our work, polyesters possessing two type molecular chain structure (random and block) which show different thermal and mechanical properties are synthesized. First, random copolymer poly(butylene succinate-*co*-butylene furandicarboxylate) (PBSF) with weight average molecular weight ranging from 55 000 to 13 000 g/mol were synthesized from 2,5-furandicarboxylic acid, succinic acid and 1,4-butanediol via polycondensation. The glass transition temperature increases continuously with increasing BF unit content from -38 °C to 37 °C. The melting temperature decreases from 113.4 °C to 69.1 °C then rise again up to 164.3 °C with increasing BF unit content from 0-100 % mol. As a result, the introducing of furan rings increases the rigidity of molecular chains and the melting temperature can be adjusted by altering the contents of furan rings in a wide temperature range through random copolymerization. Furthermore, multiblock copolymer poly (2,5-furandimethylene succinate)-*b*-poly (butylenes succinate) (PFS-PBS) were synthesized from 2,5-dihydroxymethylfuran, 1,4-butanediol and succinic acid by chain-extension reaction of dihydroxylated poly (2,5-furandimethylene succinate) and poly (butylenes succinate) using 1,6-hexamethylene diisocyanate as chain extender. The weight average molecular weight of PFS-PBS with different composition is range from 72 000 to 14 000 g/mol. The melting temperature remains relatively unchanged at 109 °C but the crystallizability decreases with increasing PFS content. Thus, the structure and properties of PFS-PBS can be modulated from a crystalline polymer with good tensile strength (13-33MPa) to an amorphous polymer of high elongation (800-1100%). So, biodegradable PFS-PBS can maintain a high end-use temperature because of unchanged molecular length of crystalline PBS block and widely tunable mechanical properties are achieved only by changing the composition of two different blocks via block copolymerization.

# ECOSYSTEM ASSESSMENT
# AND
# RESTORATION

# AN ECOLOGICAL STUDY OF REPRODUCTION IN CORINGA MANGROVE FOREST, ANDHRA PRADESH, INDIA

**Aluri Jacob Solomon Raju**
(Department of Environmental Sciences, Andhra University, Visakhapatnam 530003, India)

**ABSTRACT:** Mangroves are dynamic and unique inter-tidal ecosystems, common in tropical and subtropical coastal environments. They are among the world's most productive ecosystems and are important in protecting coasts from erosion by fierce tides, in promoting the diversity of marine organisms and fisheries by contributing a quantity of food and providing favourable habitats for animals. These economic uses of mangroves indicate that they play an important role in the lives and economies in the coastal regions of different countries. Mangrove forests are under immense threat worldwide due to their multiple economic uses and alterations of freshwater inflows by various upstream activities in catchment areas. Mangrove plants with unique adaptations play a crucial role in sustaining life in mangrove forests. Their reproductive biology is central to understanding the structural and functional components of mangrove forests. The success of sexual reproduction and subsequent population expansion in mangrove plants is linked to flowering timings, pollinators and tidal currents. Viviparous and cryptoviviparous plants are true mangroves while non-viviparous ones are mangrove associates. The dispersal propagule is seedling in viviparous and non-viviparous plants while it is seed in nonviviparous plants. In this study, viviparous and crypto-viviparous species were included for study. These species are self-compatible, self-pollinating and also cross-pollinating; such a breeding system is a requirement for the success of sexual reproduction and subsequent build up and expansion of population. They are entomophilous in the study region. The viviparous plants include *Ceriops tagal, C. decandra, Rhizophora apiculata, R. mucronata, Bruguiera gymnorrhiza* and *B. cylindrica*. The non-viviparous plants include *Avicennia alba, A. marina, A. officinalis, Aegiceras corniculatum* and *Aegialitis rotundifolia*. Sexual reproduction and regeneration events are annual in these plants and are dependent on local insects, tidal currents and nutrient content in estuarine environment. In recent times, erratic and insufficient rainfall together with industrial pollutants released into rivers is causing negative effects on the growth, development and regeneration of mangrove flora. In effect, there is a gradual decrease in mangrove cover. Added to this is continuous exploitation of mangrove plants for fuel wood, creation of shelters for cattle and changes for industrial establishments and aquaculture development in estuarine regions. As a consequence, the existing mangrove cover is struggling to survive and also not in a position to support local needs and provide livelihood opportunities through fishery resources. Further, reduced mangrove cover is showing catastrophic effects on fishing communities who live along the shore line during the period of cyclonic surges and tsunami events.

***Key words:*** Mangroves, pollination, sexual system

## INTRODUCTION

Mangrove plants are the key constituents and play a crucial role in sustaining life in mangrove forests. They display special adaptations in root system, shoot system, leaf characteristics and reproductive biology to live in the harsh environment (Tomlinson, 1986). Further, they exhibit peculiarities in seedling development and dispersal by way of vivipary. Vivipary is a functional characteristic defined as the precocious and continuous growth of the offspring when still attached to the maternal parent (Goebel, 1905); it is the norm in true mangrove plants (Tomlinson, 1986).

There are two types of vivipary, true and crypto-vivipary. True vivipary refers to a situation where embryo penetrates through the fruit pericarp and grows to a considerable size before dispersal, while crypto-vivipary is a condition in which the embryo grows continuously, but does not emerge from the fruit before dispersal. These peculiar seedling characteristics present in mangrove plants could be adaptive features to overcome the harsh tidal environment for seedling establishment (Elmqvist and Cox, 1996). Therefore, reproductive biology of mangrove plants is central to understanding the structural and functional components of mangrove forests. Further, this knowledge is essential for the restoration of degraded mangrove areas. The focus on the reproductive biology of mangrove plants has almost exclusively been on the fruit, seed or seedling dispersal stage. Surprisingly, less is known about floral biology, pollination, breeding systems and success rate of propagule production, although knowledge of the effectiveness of floral mechanics and genetic isolating mechanisms is an important prerequisite to the study of successful dispersal and establishment (Juncosa and Tomlinson, 1987; Clarke and Meyerscough, 1991; Ge et al., 2003; Chiou-Rong et al., 2005; Coupland et al., 2006).

In India, a few studies provide some preliminary accounts of floral biology and pollination in some mangrove plants (Solomon Raju, 1989; Solomon Raju et al., 1994; Subba Reddi and Solomon Raju, 1997; Pandit and Choudhury, 2001; Solomon Raju et al., 2006; Solomon Raju and Jonathan, 2008). Therefore, a more complete understanding of the reproductive biology of mangroves is useful, mainly due to the growing pressures on mangroves from coastal environment and for effective mangrove rehabilitation programmes.

The present paper is an attempt to provide information on the reproductive ecology of the genera *Ceriops, Rhizophora, Bruguiera* (viviparous), *Avicennia, Aegiceras, Aegialitis* and *Scyphiphora* (crypto-viviparous), *Excoecaria, Lumnitzera, Sonneratia, Xylocarpus, Brownlowia, Sarcolobus* and *Suaeda* (non-viviparous) occurring in Godavari and Krishna mangrove forests in the state of Andhra Pradesh, India. Floral biology, sexual system, breeding system, floral rewards, pollinators and their foraging behaviour have been investigated in these plant species. Further, fruit and seedling ecology, dispersal strategies and establishment have also been studied in these species. This knowledge is of high value for understanding the reproductive biology of the studied plant species and for conservation and management of these species. This knowledge is also useful for the restoration of degraded mangrove areas.

## MATERIAL AND METHODS

Flowering phenology was examined by conducting field trips at selected short intervals for study on all chosen plant species. Inflorescence flowering phenology was recorded by tagging some inflorescences that have not anthesed yet and then following them daily until they ceased anthesis permanently. Daily anthesis schedules and anther dehiscence schedules were carefully observed. Intrafloral events were recorded chronologically by marking some flowers at bud stage on different conspecific plants.

The aspects included flowering stage, flower organs wilting order, sepal/petal development, stamens and stigma stages and nectar production. Pollen output per anther/flower was determined as per the method given by Dafni et al. (2005). Ovule counting was made by crushing the pistil or splitting it gently and longitudinally with a blade on a glass slide spotted with lactophenol cotton blue and counting the ovules under a dissecting microscope. With the obtained pollen output and ovule number per flower, pollen-ovule ratio was determined. Stigma receptivity was examined visually and by H2O2 tests (Dafni et al., 2005). The amount of nectar secreted per flower was measured and expressed as µl of nectar/flower. The nectar sugar concentration was measured by using a Hand Sugar Refractometer (Erma, Japan) (Dafni et al., 2005). Nectar analysis for sugar types was done as per the paper chromatography method of Harborne (1973). Nectar analysis for amino acid types was done as per the paper chromatography method of Baker and Baker (1973).

Breeding systems were investigated by following the protocols given by Dafni et al. (2005). Different sets of inflorescences prior to their flowering were used for determining natural fruit set rates. Foraging schedule and forage type collected by flower visitors were observed following the methods of Solomon Raju (1989). Observations on foraging behaviour versus pollination by flower visitors were made visually. To judge the foraging activity of flower visitors, the total foraging visits of each species for the day on certain plant species were expressed in percentage. Based on this data, the percentage of foraging visits made by each species for the day was calculated. Field observations were also made regarding seed dispersal modes and subsequent establishment in mangrove forest.

## RESULTS

***Ceriops tagal* (Rhizophoraceae).** It is an evergreen shrub/tree that flowers during November-February. The flowers are born in condensed short-stalked cymes formed from dichotomizing panicles, which arise from the axils of leaves on the terminal nodes of new shoots. Flowers are short-stalked, small, white, cup-shaped, strongly fragrant, bisexual and zygomorphic. Sepals are five, small, yellowish green, valvate enclosing the inner parts until anthesis and not reflexed after anthesis. Petals are five, free, white, pubescent, two lobed, and alternating with the sepals. The lower margins of adjacent petals are held together by patches of tightly intertwining, helically coiled hairs. Each petal has three distinct clavate appendages on its distal margins. Stamens are ten, five of them antisepalous, five others antipetalous and all ten inserted on the rim of the calyx cup. Each petal encloses the antipetalous stamen and an adjacent antisepalous stamen; the two stamens remain in the petal under tension enclosed above by the clavate appendages even after anthesis. Style is slender and terminated into minute separate stigmatic lobes. The stigma stands at the height of the stamens. Disc within the stamen ring is well developed and anther lobes enclose the base of the thick filaments. Ovary is semi-inferior, three-carpelled and three-locular with a total of six ovules. The mature buds open between 1630-1800 h. The calyx lobes separate at anthesis and diverge to expose the petals. The petals with the stamens inside, two per petal do not unfold naturally throughout the flower life, but remain in a tension with the spring-loaded stamens hooded above by clavate appendages. Anther dehiscence occurs in the bud. Pollen grains are triangular, light yellow, exine smooth and 15 μ in diameter. A flower produces 14,681 + 25.62 pollen grains. The pollen-ovule ratio is 2,446:1. The stigma attains receptivity on the second day and remains receptive up to six days. But, peak receptivity occurs in third-fifth day. In this period, the white petals turn red gradually from the top to the base. A flower produces 5.65 + 1.0 μl of nectar. The nectar sugar concentration is 35-50% and the common sugars include fructose, sucrose and dextrose with the first relatively more dominant.

The nectar contains 12 amino acids which include tyrosine, glycine, methionine, proline, lysine, aspartic acid, glutamic acid, serine, cysteine, alanine, threonine and arginine. Of these, glycine, serine, cysteine, alanine and threonine are relatively dominant. The amino acids such as phenylalanine, valine, leucine, iso-leucine, tryptophan and histidine were not found in the nectar. The unpollinated flowers fall off on the seventh day. In pollinated flowers, the petals, stamens, the style and stigma drop off in this order in three-four weeks, while the fruit is in a growing stage. The sepals are persistent, become warty and spiny gradually and remain on the plant even after the shedding of propagules. The buds produced proceed to open without abortion. In open flowers, abortion rate is 42%. Of the twenty inflorescences bagged, eight flowers set fruits and, thus, the fruit set in bagged flowers is 3%. Of the sixty four open inflorescences tagged, only 115 flowers set fruit, constituting 16.3% natural fruit set. Fruit set per inflorescence varied from one to five but one and two-fruited inflorescences were more common. The pollinated flowers take four weeks to produce mature fruits with only one seed. Fruits are conical by the extrusion of the upper part of the ovary, with brown and roughened surface.

The seed has no dormancy and produces hypocotyls while on the mother tree in a span of about two months. The cotyledonary yellow cylindrical collar appears from the fruit about ten days prior to detachment of the hypocotyls. The hypocotyl is 26 cm long, distinctly ridged and hangs downwards. It is initially green, after the development of the collar, it shows a gradual colour change to brownish purple from hypocotyl end to plumule. The mature hypocotyls separate from the fruit, leaving the latter attached to the mother plant. The detached hypocotyls were found to settle in the vicinity of the mother plant. The flowers do not expose the stamens naturally but the latter attains tensed condition in the delicate petals for release by a delicate external touch. The foragers included honeybees, *Apis cerana indica, A. florea*, the fly, *Chrysomya megacephala* and the butterfly *Tirumala limniace*. They showed foraging activity throughout the day but with varying percentage of foraging visits and also species-wise, the percentage of foraging varied. Of these, honeybees foraged for pollen and nectar occasionally while the fly and butterfly foraged for the nectar consistently until the floral source exhausted. All the species approached the flower from above and probed for nectar, causing a sudden release of stamens from the petals. In effect, the pollen from the already dehiscent anthers was ejected forcibly and deposited on the underside of the foraging bee or fly. Body washings for pollen revealed the presence of pollen grains which varied from 231 to 413 per bee and from 79 to 147 per fly, suggesting that both bees and the flies have an important role in petal explosion and pollination. In case of bees, they also carried pollen loads in pollen baskets present on their legs. As each petal is independently enclosing two stamens, a single foraging visit of the bee or fly did not result in the explosion of all five petals.

Both the bees and the fly tended to visit more than one flower on the same inflorescence or different inflorescences on the same plant before flying away to visit the neighbouring trees and may return back again to visit the same flowers later. As they tended to move back and forth between trees, their foraging activity may result in substantial self- and cross-pollination. The honey bees were found to concentrate principally on *Bruguiera gymnorrhiza* and *Aegiceras corniculatum*, while the fly species exclusively on *Ceriops tagal*. The explosion of petals in open flowers was also triggered by the action of wind. The plant grows in the seaward zone and hence high winds are the characteristic of the site. Of the twenty four flowers observed, four flowers had shown petal explosion triggered by wind action indicating that 16.6% of flowers may achieve pollination by wind action. In such flowers also, the explosion of all five petals did not take place at one time.

It had not been possible to study whether wind could trigger petal explosion in the flowers located on the branches facing landward direction and mixed with the canopy where wind becomes relatively ineffective. Wind triggered petal explosion may result in autogamy.

***Ceriops decandra* (Rhizophoraceae).** It is an evergreen, semi-understory, and small to moderate tall tree with perfect flowers. It produces numerous branches from the main stem. A typical tree of 1.5 meters contains as many as 20-21 primary branches and each primary branch bears 11-15 secondary branches. It occurs mainly in the inside of the mangrove vegetation and certain individuals occur along the creeks/estuarine banks. The tree shows budding, flowering and fruiting continuously throughout the year but there is a burst of concentrated flowering during November. Some young trees show alternate flowering and fruiting phases. The production rate of flowers and fruits are more robust on trees growing in the interiors of the forest when compared to those on trees growing along the creeks/estuarine banks. The flowers are borne in condensed cymes formed from dichotomizing panicles, which arise from the axils of leaves.

The mechanical and biological features of this type of inflorescence provide continuous protection for the youngest units by a successive series of bracts, bracteoles and sepals. An inflorescence produces 5-31 flowers depending on the number of divisions of inflorescence axes. The cymes with more number of flowers are more common on the trees growing in the inner mangrove vegetation. The inflorescence takes sixten days to complete its flowering life. Flowers are small, white, cup-shaped, odorless, bisexual and zygomorphic. Sepals are five, free, small, light green, six mm long, valvate enclosing the inner parts until anthesis and not reflexed after anthesis. Petals are five, free, two-lobed, alternating with the sepals, five mm long, light green in mature bud stage and white at anthesis. They are inter-locked marginally by basal short hairs and this circumstance produces a short corolla tube crowned by the series of clavate filamentous appendages.

Stamens are ten, five of them antesepalous, five others antepetalous and all ten inserted on the rim of the calyx cup. Each stamen is two mm long, free and extends beyond the height of the stigma; the filaments are green, while anthers are light brown, dorsifixed and longer than filaments. Disc within the stamen ring is well developed and anther lobes enclosing the base of the stamens. The ovary is semi-inferior, three-carpelled and three-locular with a total of six ovules. Style is slender, green, one mm long and terminated into minute separate stigmatic lobes. The mature buds open during 0430-1100 h. The calyx lobes separate at anthesis and diverge to expose the petals, which adopt various configurations. Anther dehiscence occurs at anthesis. Pollen grains are ovate, triangular, light yellow and 16.6 μm in diameter. A flower produces 12,810 + 30.87 pollen grains. The pollen-ovule ratio is 2,135:1. The stigma attains receptivity about six hours prior to anthesis and remains receptive up to six days; but very active receptivity occurs in one-day and two-day old flowers. The stigma receptivity is notable even after six days but that receptivity appears to be non-functional in terms of pollen germination. Nectar is produced in trace amounts. The petals and stamens fall off on the seventh day of flower life. The sepals are persistent, become warty and spiny gradually and remain on the tree even after the shedding of propagules. The style and stigma dry up and drop off after ten days in fertilized flowers. The flowers that were not pollinated fall off on the fourth day. Hand pollination tests for breeding system indicated that *C. decandra* fruits through xenogamy only; the fruit set rate is 92%. Bud and flower abortion rate is 42% and 31% respectively. In open-pollinations, the fertilized flowers did not show a premature fruit drop. Fruit set per inflorescence varies from one-fur but not fruited inflorescences are more common. The pollinated flowers produce mature fruits in 50-55 days. Fruits are light green, 1.5 cm long, ovoid, conical and blunt apically. They are distinct with five-lobed persistent calyx and produce a single seed only. The embryo has no dormancy and penetrates through the seed coat and the fruit pericarp and grows to a considerable size into a spindle-shaped hypocotyl structure before dispersal while still attached to the maternal parent. This type of hypocotyl growth constitutes true vivipary.

The hypocotyl grows upright and takes 85-90 days before detachment from the fruit. It is slender, clearly ribbed, angular, sulcate, 15 cm long and broadened at the lower end. Usually, they are entirely green and occasionally purple on one side; rarely yellow hypocotyls are also produced. The green hypocotyls seem to have the potential to photosynthesize actively with water and necessary nutrients drawn from the parent tree. The purple and yellow hypocotyls seem to lack chlorophyll partly or wholly and hence doubtful that they will have a successful establishment when detached from the parent tree. Very rarely, a single fruit produces two hypocotyls which may have arisen from two seeds resulting from two ovules out of actual six ovules per flower. In fully grown hypocotyls, fruit is separated from collar which emerges shortly before detachment. The fruit set rate per inflorescence shows a pattern in accordance with the number of flowers produced and area where the tree grows within the mangrove forest. An inflorescence produces one-six fruits. The trees growing near creeks/estuarine banks produce one-four fruits per inflorescence, onefruited ones being 51%, two-fruited 34%, three-fruited 12% and four-fruited 3%. The trees growing in the inner mangrove areas produce one-six fruits per inflorescence, two-fruited ones are 45%, three-fruited 26%, one-fruited 22%, four-fruited 5% and five-fruited and six-fruited, 1% each. The foragers included *Nomia* sp., *Ceratina simillima* (bees), *Odynerus* sp. and *Polistes* sp. (wasps). The bees were quite common during September-October, while the wasps during November-December. The bees were found to collect nectar and pollen, while the wasps only nectar. These foragers visited the flowers during 1030-1630 h, more frequently during 1300-1400 h coinciding well with the availability of the number of flowers, because anthesis period is relatively lengthy and the new flowers with nectar and pollen accumulate by that time.

*Nomia* and *Odynerus* species made more percentage of foraging visits. Bees approached the flowers in an upright position, landed in the cup-shaped flowers and probed for nectar and pollen in the same or different foraging visits. In doing so, their ventral side contacted the stamens first and then stigma effecting pollination. During pollen collection, these bees rotated around the flower to collect pollen from the anthers, which are situated in one whorl against the sepals and petals. They took three-five seconds to collect the forage from each newly opened flowers and two-three seconds from differently aged flowers. Their body washings for pollen revealed that they carry 1262.7 + 428.7 (Range 35-1570) pollen grains and hence have an important role in pollination. In addition to this pollen, they also carried pollen loads in pollen baskets present on their legs. The bees were found to move between trees of *C. decandra* to collect both nectar and pollen and this inter-tree movement was considered to effect cross-pollination. As the number of new flowers per day is small in number, the bees were forced to collect forage from different trees of *C. decandra* and this flowering strategy is expected to promote cross-pollination. Wasps are about the size of a fly and construct their nest cells in sand heaps or in cavities of trees, lining them with agglutinated grains of sand or mud. The female wasps are known to feed on floral nectar after mating for the maturation of the eggs. The wasps observed on *C. decandra* collect nectar and such wasps were considered to be females. The study shows that *C. decandra* is a potential nectar source for female wasps. Wasps approached *C. decandra* flowers in an upright position, landed on the cymes, gradually moved to individual flowers and probed for nectar. They were found to move between individual trees in quest of more nectar. While probing the flower, they contacted the stamens and stigma with their underside and this resulted in pollination. The body washings for pollen indicated that they are pollen carriers; the pollen carry over ranges of eight to 49.

The number of bees and wasps visiting the flowers vary in different months, but both species occur continuously. Considering the changing number of bees and wasps in different months, both categories seem to be equally important as pollinators. The stigmatic pollen loads were analyzed to evaluate the rate of pollen deposition per stigma. The analysis indicated that each stigma receives 40.60 + 37.21 (Range 5-170) pollen grains in open-pollinations but how much of it is xenogamic pollen is not known. The pollen deposition rate from the flowers was also analyzed to evaluate the role of foragers (after four-five flower visits) to empty the anthers and subsequent transportation to the receptive stigmas. The analysis showed that 10,243.7 + 1735.6 (Range 7660-12,400) pollen grains were depleted against the pollen output by 12,810 per flower suggesting that both bees and wasps are capable of transferring pollen and effecting pollination very effectively.

Field studies indicated that *C. decandra* is used as firewood and for the constructions. Further, this species is important for its reddish brown coloured bark. The bark is known for its high tannin content ranging from 68 to 75% and dyeing with this tannin gives brown colour. The fishermen extract a reddish liquid from the bark and use it to protect cotton fishing nets from decay for a longer period. Some fishermen are involved in the trade of this bark and wood to make up their livelihood. About 80% of fishermen in 30 shore-based villages adjacent to the Godavari Delta mangrove forests depend on fishing activity for their livelihoods. They use mostly cotton nets for catching fishery sources in backwaters, while nylon nets for fishing activity in open sea water. The advantages of cotton nets include inexpensiveness and readily available because they are made locally. Nylon nets are not readily available and they are also expensive. The treatment of cotton nets is necessary to prevent damage to cotton threads due to soaking in saline water and to extend the length of its durability.

***Rhizophora apiculata* (Rhizophoraceae).** It is a medium to tall evergreen tree with profusely spreading branches, growing to a height of more than 10-12 m. The leaves are large and dark glossy green. It flowers throughout the year with profuse flowering during AugustSeptember. Flowers are borne in pairs on a stubby axis, borne below the leafy crown, that is, in the axil of a leaf scar. They are sessile and erect in position. Flowers are light yellow, 8-12 mm long, 10-14 mm wide, cup-shaped, odourless, bisexual and zygomorphic; they are situated below the leaf clusters. Calyx is characteristically hard, yellow to brown, six-eight mm long and three-four mm wide, basally cup-like, with four sepals pointed towards apex and persistent. Petals are four, alternating with sepals, five-seven mm long, creamy-white, odourless, lanceolate, glabrous and delicate. Stamens are 12, sessile, dull white to dull brown, free, anthers are five-six mm long, bilobed, introrse and sagittate. The ovary is inferior, globose with two carpels each with two glabrous ovules on axile placentation; style is thick and short, creamy white, with two creamy white to light pink stigmatic lobes. Themature buds open at 1000-1100 h; the sepals diverge least but expose the inconspicuous petals and sex organs. The petals remain flat and do not recurve. The anthers are multicellar, dehisce introrsely via the adaxial flap which falls against the base of the style in mature bud stage.

The glabrous petals do not accumulate pollen from the dehisced anthers. Stigmatic lobes are appressed in mature bud and diverged gradually after anthesis; the diverged state of stigmatic lobes indicates the commencement of receptivity and ceases around noon of the second day. The stigma has no special modifications to capture the wind borne pollen but it is thickly coated with pollen even in mature bud. The style and stigma fall off after fruit initiation. The stamens and petals drop off on the second day. The calyx remains attached to the growing fruit, expands and reflexes backward. The pollen output per anther is 46,527.1 ± 2,411.9 (Range 42,681-49,854) and per flower is 5.58.326. Pollen grains are tricolporate, dullwhite, powdery, ornamentation finely reticulate, the reticulum becoming progressively less distinct from pole to equator and 16.6 μm diameter. The pollen-ovule ratio is 1.39.581:1. The pollen protein content per anther is 8.6 μg and per flower is 104 μg. Nectar is secreted in trace amount around the ovary.

The results of breeding systems indicate that the flowers are selfcompatible and self-pollinating. The fruit set is 72.5% in spontaneous autogamy, 86% in handpollinated autogamy, 92% in geitonogamy, 93.3% in xenogamy and 55% in open pollination. The flowers release pollen into the air; the sepals and petals in open flowers sprinkled with pollen due to wind action. A delicate manual disturbance to the flowers resulted that caused the release of a cloud of pollen out of the flower. Bees were the exclusive foragers and they were *Nomia* sp., *Trigona iridipennis* and *Halictus ligatus*. They foraged during day time from 0800-1700 h for pollen and nectar from partially and completely opened flowers. Their foraging activity was consistent during profuse flowering period while it is sporadic at other times of the year. Of the total number of foraging visits of bees, *H. ligatus* made 44% followed by *Nomia* 41% and *Trigona* 15%. These bees first landed on the sepals, then investigated the flower for pollen and nectar; while doing so, their head and ventral side of the body touched the stamens and stigma and in the process they got coated with pollen. Further, they loaded pollen into pollen baskets. As all three bee species are pollen collecting bees, they made frequent visits to flowers situated on different plants which are closely and distantly spaced in order to collect more pollen. The fresh flowers available per day on a given tree are small in number and the bees made visits to different trees to collect more forage from as many flowers as possible. Further, the body washings of the bee species revealed the presence of pollen; the average number of pollen grains per bee for each species varied from 561.6 to 1006.9. It was found that thrips breed in buds and emerge when a flower opening occurs. They collected both pollen and nectar. During August-September, a single mature bud was usually found to contain many thrips moving out of the floral base when disturbed manually. Thrips were found to effect pollination as they moved in the entire flower touching the stigma. Pollinated and fertilized flowers initiate fruit development immediately and take 35 days to produce mature fruits.

Of the four ovules in a flower, one ovule produces seed. Fruit is one-seeded with persistent light green calyx. A fruit producing two seeds and hence two hypocotyls is a rare occurrence. The seed produces 550 mm long, cylindrical elongate, light green, hypocotyl with a sharp end in a time span of 55-60 days. A light brown collar emerges between fruit and hypocotyl about a month prior to the detachment of the latter. The persistent calyx and fruit part remain in place on the parent tree while the hypocotyl detaches at the collar for dispersal. The hypocotyls fall and anchor vertically in the substratum at low tide when the forest floor gets exposed; they float in water and disperse by tidal currents at high tide until settled in the mud. The radicle side of hypocotyl penetrates the soil and produces root system while plumule side produces new leaves and subsequent aerial system.

***Rhizophora mucronata* (Rhizophoraceae).** It is a medium to tall evergreen tree with profusely and horizontally spreading branches, growing to a height of more than 15 m. It flowers during June-November. The inflorescence is a four-flowered pedunculate dichotomized cyme borne in leaf axils; the flowers are pedicellate, pendulous and hanging downwards. Flowers are creamy white, 15 mm long, 10 mm wide, cup-shaped, odourless, bisexual and zygomorphic; they are situated below the leaf clusters. Calyx is characteristically hard, creamy white, 12 mm long and seven mm wide, basally cup-like, with four sepals pointed towards apex and persistent. Petals are four, alternating with sepals, ten mm long, white, hairy, odourless, lanceolate, and delicate. Stamens are eight, sessile, four opposite to petals, another four opposite to sepals, dull white to dull brown, free, anthers are seven mm long, bilobed, introrse and sagittate. Ovary is semi-inferior, globose with two carpels each with two glabrous anatropous ovules on axile placentation; style is thick and short, creamy white, with two creamy white to light pink stigmatic lobes. The mature buds open at 1000-1600 h; the sepals diverge gradually exposing the petals and sex organs. The petals also diverge and slightly reflex backwards. The anthers are multilocellar, dehisce introrsely via the adaxial flap which falls against the base of the style in mature bud stage.

The hairy petals accumulate pollen from the dehisced anthers. Stigmatic lobes are appressed in mature bud and diverged gradually after anthesis; the diverged state of stigmatic lobes indicates the commencement of receptivity and ceases around noon of the second day. The stigma has no special modifications to capture the wind borne pollen, but it is thickly coated with pollen even in mature bud. The style and stigma fall off after fruit initiation. The stamens and petals drop off on the second day. The calyx remains attached to the growing fruit, expands and reflex backwards. The pollen output per anther is 34,986.6 ± 527.34 (Range 34,007-35,729) and per flower is 2,79,893. Pollen grains are tricolporate, dull-white, powdery, and 24.9 μm long and 12.2 μm in diameter; their surface sculpture is rather smooth, with numerous small and shallow depressions. The pollen-ovule ratio is 69,973.2:1. The pollen protein content per anther is 10.2 μg and per flower is 82 μg. Nectar volume per flower is 1.03 ± 0.26 μl (Range 0.8-1.5) and accumulated around the ovary.

The results of breeding systems indicate that the flowers are self-compatible and self-pollinating. The fruit set is 52.5% in spontaneous autogamy, 63.3% in hand-pollinated autogamy, 71.4% in geitonogamy, 85% in xenogamy and 53.3% in open pollination. The flowers release pollen into the air; the sepals and petals in open flowers were found to be sprinkled with pollen due to wind action. A delicate manual disturbance to the flowers resulted caused the release of a cloud of pollen out of the flower. The flowers were foraged consistently during day time from 0800-1700 h exclusively by bees. The bees included *Nomia* sp., *Trigona iridipennis, Halictus ligatus*, *Ceratina simillima* and *Xylocopa pubescens.* An unidentified moth was also found to collect nectar occasionally. Bees collected both pollen and nectar from partially and completely opened flowers. Each bee species contributed 11-24% of total foraging visits. These bees first landed on the sepals, then entered the flower to collect pollen and nectar; while doing so, their head and ventral side of the body touched the stamens and stigma and in the process they got coated with pollen. Further, they loaded pollen into pollen baskets.

All the bee species except *Xylocopa,* are pollen collecting bees, they made frequent visits to flowers situated on different plants which are closely and distantly spaced in order to collect more pollen. *Xylocopa* bees are fast fliers and made inter-tree flight in quest of more nectar from as many flowers as possible. The fresh flowers available per day on a given tree are small in number and for this reason, the bees were compelled to make visits to different trees for more forage. Further, the body washings of the bee species revealed the presence of pollen; the average number of pollen grains per bee for each species varied from 189.3 to 951.6. Thrips breed in buds and emerge when flower opening occurs. They collected both pollen and nectar. A mature bud contained numerous thrips moving out of the floral base when disturbed manually. Thrips were found to carry pollen on their bodies and moving in the entire flower including the stigma during which they transferred pollen onto the stigma. The flowers with huge pollen production together with pollen characteristics described above were considered to be anemophilous. Further, the pollen being powdery was found to be dispersed easily due to wind action. A simple manual disturbance to flowers made the latter to release pollen into the air. Pollinated and fertilized flowers initiate fruit development immediately and take 40 days to produce mature fruits.

In fertilized flowers, only one ovule out of four ovules produced seeds. Fruit is one-seeded with woody, persistent light green and reflexed calyx. Seed produces 680 mm long, cylindrical elongate, light green, warty hanging hypocotyl in a time span of 60- 65 days. A light yellow tubular collar (20 mm long) emerges between fruit and hypocotyl about a month prior to detachment of the latter. The hypocotyl detaches at the collar for dispersal. Later, the entire fruit part also falls off. The hypocotyls settle in the substratum immediately at low tide when the forest floor gets exposed; they float in water and disperse by tidal currents at high tide until settled in the soil. The radicle side of hypocotyl penetrates the soil and produces root system, while plumule side produces new leaves and subsequent aerial system. The fruit also falls off eventually from the maternal parent.

***Bruguiera gymnorrhiza* (Rhizophoraceae).** It is a medium to tall evergreen tree with much diffused spreading branches, growing to a height of more than ten m. It flowers throughout the year with profuse flowering during April-June. Flowers are solitary, single flower in each peduncle, located in the leaf axils, usually positioned at the first (or rarely second) node below the apical shoot. Flowers are pedicellate, typically recurved, pointing away from the terminal vegetative bud, pendulous, 30-35 mm long, 19-20 mm wide, pinkish to reddish white, tubular, odourless, bisexual and zygomorphic. Calyx is characteristically hard, smooth or with grooves above lobe junctures, pinkish to reddish white, basally cup-like, rarely ribbed, with 11-14 lobes, acutely pointed, narrow and persistent. Petals are 14, bilobed, 15 mm long and four mm wide, creamy-white in mature bud, orange-brown on maturation, delicate with marginal interlocking hairs, the tips of lobes acute commonly with three filamentous appendages distally usually with a rigid straight four mm long bristle between them. The base of each petal has a cluster of smooth silky hairs. Stamens are 28 enclosed in petal pouches, two stamens in each petal, filament creamy white, ten mm long and two mm wide, anthers are creamy white initially and turn to golden brown at maturity; bilobed and basifixed. Ovary is inferior, cup-shaped with six light brown, smooth ovules; style is slender, white to light brown, filiform with three or four, 15 mm long, whitish-yellow stigmatic lobes containing small papilla and secreting mucilage; the stigma remains attached to fruit at maturity. The mature buds open at 0700-0900 h; the sepals diverge gradually presenting the closed, erect petals in the cocked position. The petals conceal the stamens in a tensed state due to the pressing of the latter against the interlocked margins of the petals. The petals bend back and remain in an erect position by the adherent ventral margins. Marginal hairs appear to be important in holding the petals in the folded position. The stamens dehisce in mature bud stage by longitudinal slits. The petal margins unzip instantaneously when triggered by external touch; then they fly apart releasing the stamens which catapult the loose pollen as a visible cloud toward the centre of the flower. If the external touch is caused by a forager, then much of the pollen would be projected onto the head and the body of the forager. Each petal explodes independently and hence multiple visits are required to trip all the petals of a flower.

Individual flowers with combinations of closed and tripped petals were found. After the petal tripping, the petals lie back against the calyx lobes with empty stamens twisted and disorganized. Petals and stamens persist for up to seven days and eventually fall off as threefour units. Untripped petals retain their tension for up to eight days, and they eventually fall off without ever opening. Stigmatic lobes are receptive from second to fourth day and show signs of withering after the petals and stamens have fallen. The pollen output per anther is 9,005.4 ± 834.95 (Range 8,004-10,469) and per flower is 2,51,856 pollen grains. Pollen grains are tricolporate, pale yellow, elliptic in equatorial view, circular in polar view, exine smooth with numerous small and shallow depressions, 29.6 μm long and 16.6 μm diameter. The pollenovule ratio is 41,976:1. The pollen protein content per anther is 1.77 μg and per flower is 49.6 μg. Copious nectar accumulates in the deep calyx cup and is retained by the petal base and its associated hairs. In newly open flowers, five-ten μl of nectar is secreted and it is continuously produced even after the abscission of petals and stamens in some flowers while it is not produced continuously in some other flowers. The flowers with the former situation were designated as un-pollinated ones while those with the latter situation were designated as pollinated flowers. A flower produces 10.49 ± 5.2 (Range 5.2 - 24) μl of nectar with a sugar concentration of 23.48 ± 4.80% (Range 18-31); glucose and fructose were present. In unvisited flowers, it overflows onto the petals where it becomes viscous by evaporation and so inhibits the explosive mechanism. The total sugar content in the nectar of a flower is 2.58 ± 1.20 (Range 1.06-5.39) mg. The nectar protein content per flower is 78.68 ± 48.56 μg (Range 44.8- 147). The nectar contains eight amino acids which include alanine, aspartic acid, glutamic acid, arginine, histidine, lysine, glycine and serine. All these amino acids are abundant except lysine. The results of breeding systems indicate that the flowers are self-compatible and selfpollinating. The fruit set is 40% in manipulated autogamy, 60% in geitonogamy, 90% in xenogamy and 87% in open pollination. The flowers are characterized by explosive pollination mechanism. They were foraged consistently during day time from 0700-1700 h exclusively by bees.

The bees included *Apis dorsata, A. cerana, A. florea, Nomia* sp. and *Halictus ligatus*. Bees collected both pollen and nectar from partially and completely opened flowers. Of the total number of foraging visits of bees, *A. florea* contributed 25%, *A. cerana* and *Nomia* each 20%, *A. dorsata* 18% and *H. ligatus* 17%. All the bees turned their heads upward to collect nectar located in the calyx cup; while doing so, they tripped the tensed petals to release a cloud of pollen explosively without being disturbed by the explosion. A single foraging visit did not result in tripping all the petals at a time. Petal explosion was found to be effected by touching sensitive basal hairs, especially those which project in the centre of the flower over the entrance to the calyx cup. The tip of the petal and its apical appendages were not sensitive but petal tripping was stimulated by fairly vigorous probing into the calyx cup.

The bees took different positions while probing and collecting nectar in relation to stamens and style. Each foraging visit invariably contacted the style and stigma and resulted in the deposition of pollen on the foraging bee. The bees were also found to collect nectar from empty flowers which are devoid of petals and stamens. They were found to move frequently between plants seeking more nectar and/or pollen as few rewarding flowers are available daily per tree. Such a foraging behaviour could make them as effective pollen dispersers causing both self- and crosspollination. Further, the body washings of all foraging bee species revealed the presence of pollen; the average number of pollen grains per bee for each species varied from 198 to 1,389. Pollinated and fertilized flowers initiate fruit development immediately and take 30-35 days to produce mature fruits. Fruit is one-seeded fleshy berry with persistent reddish calyx. Seed produces 141 ± 9 mm long, cylindrical elongate, stocky, dark green, coriaceous hanging hypocotyl with blunt apex in a time span of 45-50 days. In a sample of 200 hypocotyls, 5% were found to be damaged especially at the terminal part by the Rose-ringed Parakeet, *Psittacula krameri;* the damaged part was fleshy and hence eaten by the bird. The persistent calyx remains attached even after mature hypocotyl falls from the mother tree.

The hypocotyls settle in the substratum immediately at low tide when the forest floor is exposed; they float and disperse by tidal currents at high tide until settled in the soil. The radicle side of the hypocotyl penetrates the soil and produces root system while plumule side produces new leaves and aerial system.

***Bruguiera cylindrica* (Rhizophoraceae).** It is a medium to tall evergreen tree with diffused spreading branches, growing to a height of more than 12 m. The flowering occurs during September-March. The inflorescence is a simple pedunculate dichasium cyme with pedicellate erect flowers; each cyme is three-flowered and borne in leaf axils. Flowers are greenish-white, 10-12 mm long, short-tubed, odourless, bisexual and zygomorphic. Calyx is tubular, funnel-like (five-seven mm long and two mm wide), smooth, light green, with seveneight pointed lobes, characteristically hard and persistent. Petals are seven-eight, initially creamy white, later turning to light brown from the tip towards the base, alternating with sepals, shortly bilobed, each lobe three mm long and one mm wide, delicate with marginal interlocking minute hairs, the tips of lobes acute commonly with three filamentous appendages distally usually with a rigid straight bristle between them. The base of each petal has a cluster of smooth silky hairs. Each petal encloses a pair of stamens, filament creamy white, two mm long and one mm wide, anthers are creamy white initially and turn to light brown at maturity; bilobed and basifixed. Ovary is inferior, cup-shaped, with four light brown, smooth ovules; style is slender, white to light brown, filiform with two creamy white to yellow stigmatic lobes, two mm, situated below the height of stamens, stiff and remains attached to fruit at maturity.

The mature buds open at 0700-0900 h; the sepals diverge gradually presenting the closed, erect petals in the cocked position. The petals conceal the stamens in tensed state due to the pressing of the latter against the interlocked margins of the petals. The petals are retained in an erect position by the adherent ventral margins. Marginal hairs appear to be important in holding the petals in the folded position. The stamens dehisce in mature bud stage by longitudinal slits. The petal margins unzip instantaneously when triggered by a delicate touch; then they fly apart releasing the stamens which catapult the loose pollen toward the centre of the flower. If the external touch is caused by a forager, then much of the pollen would be projected onto the head and body of the forager. Each petal explodes non-violently and independently. In an observed set of flowers, most flowers had all petals tripped. After the petal tripping, the petals lie back against the calyx lobes with the empty stamens twisted and disorganized. Petals and stamens persist for up to three days and eventually fall as three-four units. Untripped petals retain their tension for up to six days, and they eventually fall off without ever opening. Stigmatic lobes are receptive on the second and third day and show signs of withering after the petals and stamens have fallen. The pollen output per anther is 1,289.7

± 299.29 (Range 1,021-1,828) and per flower is 20,635 pollen grains. Pollen grains are tricolporate, light yellow, powdery, ornamentation finely reticulate, the reticulum becoming progressively less distinct from pole to equator and 16.6 μm long. The pollen-ovule ratio is 5,158:1. The pollen protein content per anther is one μg and per flower is 16 μg. Nectar accumulates in the deep calyx cup and is retained by the petal base and its associated hairs. A flower produces 1.43 ± 0.31 μl (Range 0.9-2) of nectar with a sugar concentration of 15.2 ± 1.57% (Range 10-21); glucose, fructose, sucrose and maltose were present but the first sugar is dominant. The total sugar content in the nectar of a flower is 0.11 ± 0.03 (Range 0.04-0.17) mg. The results of breeding systems indicate that the flowers are self-compatible and selfpollinating. The fruit set is 40% in unmanipulated autogamy, 56% in manipulated autogamy, 63.3% in geitonogamy, 80% in xenogamy and 64.4% in open pollination. The flowers are characterized by explosive pollination mechanism. They were foraged consistently during 0700 to 1700 h by *Nomia* bee and the wasps, *Odynerus* sp. and *Polistes humilis*. The bee collected both pollen and nectar while the wasps collected only nectar. Of the number of foraging visits of bees, *Nomia* contributed 38% while the remaining percentage by the wasps. These foragers approached the flowers in upright position and probed the flowers for nectar and/or pollen; they tripped the tensed petals without being disturbed by the explosion.

One or two visits of the foragers resulted in the explosion of all flower petals. Petal explosion was found to be effected by touching sensitive basal hairs, especially those which project into the centre of flower over the entrance to the calyx cup. The tip of the petal and its apical appendages were not sensitive but petal tripping was stimulated by vigorous probing into the calyx cup. Each foraging visit invariably contacted the style and stigma and resulted in the deposition of pollen on the foraging bee. They were found to move frequently between plants seeking more nectar and/or pollen as few rewarding flowers are available daily per tree. Such a foraging behaviour could make them as effective pollen dispersers causing both self and cross pollination. Further, the body washings of all foragers revealed the presence of pollen; the average number of pollen grains per forager for each species varied from 262 to 448.2. Pollinated and fertilized flowers initiate fruit development immediately and take 18-20 days to produce mature fruits. Of the four ovules, only one ovule produces seed in fertilized flowers. Rarely, two ovules produce seeds; in such fruits, twin hypocotyls are produced per fruit. Fruit is one-seeded, creamy white, 20 mm long; persistent creamy white calyx lobes stick out at right angles to the fruit. Seed produces 136 ± 16 mm long, cylindrical elongate, slightly curved, green to purple, pendulous hypocotyl with blunt apex in a time span of 30-35 days. The persistent calyx remains attached after mature hypocotyl falls from the parent tree.

The hypocotyls settle in the mud at low tide during which the forest floor gets exposed; they float in water and disperse by tidal currents at high tide. The radicle side of the hypocotyl produces root system while plumule side produces new leaves and subsequent aerial system.

***Avicennia alba* (Avicenniaceae).** Small evergreen tree with irregular branches, growing to a height of more than 10-12 m. Following monsoon showers in June, it initiates flowering and continues flowering until the end of August. Individual trees flower for 35 ± 4 (Range 32- 48) days. Inflorescence is a terminal and axillary spicate raceme. An inflorescence produces, on average, 52.34 ± 26.96 flowers (Range 15-123) anthesing from the base to top over a period of 25 days (Range 24-28). The flowers are sessile, small (four mm long; three mm diameter), orange yellow, fragrant, actinomorphic and bisexual. Calyx is short, elliptic and has four ovate, green, pubescent, two mm long, one mm wide, sepals with hairs on the outer surface; it is persistent. Corolla has four thick, orange yellow ovate, four mm long and two mm wide petals forming a short tube at the base. Stamens are fur, epipetalous, one mm long, occur at the throat of the corolla. The anthers are dorsifixed, introrse and arranged alternate to petals. The ovary is very small (two mm long), flask-shaped, conspicuously hairy but lower part includes glandular hairs, bicarpellary syncarpous with four imperfect locules and each locule contains one pendulous ovule.

It is terminated with a one mm long glabrous style tapered to the bifid hairy stigma. The light yellow style and stigma arise from the centre of the flower and stand erect throughout the flower life. The mature buds open throughout the day but most buds opening during 0900-1200 h. The petals slowly open and take three-four hours for complete opening to expose the stamens and stigma. The petals emit fragrance at anthesis. The stamens bend inward overarching the stigma at anthesis and dehisce ½ hour after anthesis. All the four anthers dehisce at the same time by longitudinal slits. The stigma is well seated in the centre of the flower. After anthesis, the stigma grows gradually and becomes bifid on the morning of the second day. The bifid condition of stigma is an indication of beginning of stigma receptivity and it remains receptive for two days. The stigmatic lobes recurve completely. A flower lasts for six days. The petals, stamens and stigma drop off while the calyx is persistent. The pollen production per anther is 1,967 ± 31.824.3 (Range 1,929-2,010) and per flower is 7,868. The pollen grains are light yellow, granular, tricolporate, reticulate, muri broad, flat, thick; lumina small irregularly shaped, colpi deeply intruding and 24.9 μm in size. The pollen-ovule ratio is 1,967:1. The flowers begin nectar secretion along with anther dehiscence. The nectar secretion occurs in minute amount which is accumulated at the ovary base and on the yellow part of petals; the nectar glitters against sunlight. A flower produces 0.5 ± 0.1 μl (Range 0.4-0.7) of nectar with a sugar concentration of 40%.

The sugar types included glucose and fructose and sucrose with the first as dominant. The results of breeding systems indicate that the flowers are selfcompatible and self-pollinating. The fruit set is 17.5% in spontaneous autogamy, 40% in handpollinated autogamy, 62.5% in geitonogamy, 64.28% in xenogamy and 42% in open pollination. The flowers were foraged consistently during day time from 0700-1700 h by insects. The insects included *Apis dorsata, A. florea, Nomia* sp., *Chrysomya megacephala*, an unidentified fly, *Danaus chrysippus* and *Everes lacturnus*. The fly species visited the flowers in groups. The pollen was collected by bees only; they also collected nectar throughout the flower life depending on the availability. All other insects collected nectar only. Each forager species made 8.5-21% of foraging visits. All the insects probed the flowers in upright position to collect the forage. Butterflies landed on the petals, stretched their proboscis to collect nectar aliquots on the petals and at the flower base. In this process, all the insects invariably touched the anthers and the stigma; the ventral side of all insects was found with pollen. Further, the body washings of the all insect species revealed the presence of pollen; the average number of pollen grains per insect for each species varied from 67.6 to 336.2. As the nectar is secreted in minute amount, the insects made multiple visits to most of the flowers on a tree and moved frequently between trees to collect nectar. Such a foraging behaviour was considered to effect self- and cross-pollination. Pollinated and fertilized flowers initiate fruit development immediately and take about five-six weeks to produce mature fruits. In fertilized flowers, only ovule produces seed. Fruit is a one-seeded leathery pale green capsule with persistent reddish brown calyx; 40 mm long, 15 mm wide, abruptly narrowed to a short beak and hairy throughout.

Seed produces light green, hypocotyl which completely occupies the fruit cavity. An inflorescence produces three ± one fruits (Range 1-5); one-fruited ones constituted 54% followed by two-fruited ones (23%) and three-fruited (10%) and four-fruited (5%) and five-fruited (2%). Of these, 6% were damaged by the Rose-ringed Parakeet, *Psittacula krameri*; it was found to feed on the concealed hypocotyl in fruits and these fruits were subsequently empty. The fruit together with hypocotyl falls off the mother plant; it was found to settle in the substratum immediately at low tide period when the forest floor is exposed; it floats in water and disperses by tidal currents at high tide period until settled somewhere in the soil. The radicle side of hypocotyl penetrates the soil and produces root system while plumule side produces new leaves and subsequent aerial system. The fruit pericarp detaches and disintegrates when plumular leaves are produced.

***Avicennia marina* (Avicenniaceae).** It is a small evergreen tree with irregular spreading branches, growing to a height of more than three-eight m. Following monsoon showers in June, it initiates flowering and continues flowering until the end of August. Individual trees flower in 32-35 days. Inflorescence is a compound axillary or terminal cyme. An inflorescence shoot produces, on average, 47 ± 13.97 flowers (Range 26-76) anthesing from the base to top over a period of 22 days (Range 15-22). The flowers are sessile, small (six mm long; five mm diameter), orange yellow, sweet scented, actinomorphic and bisexual. Calyx is short, elliptic and has four ovate light green, two mm long sepals with hairs on the outer surface; it is persistent. Corolla has four thick, orange yellow ovate four mm long and two mm wide petals forming a short tube at the base.

The petals are glabrous inside and hairy outside. Stamens are four, epipetalous, two mm long, occur in the throat of the corolla. The anthers are basifixed, exserted, introrse and arranged alternate to petals. The ovary is very small (two mm long), conspicuously hairy, bicarpellary syncarpous with four imperfect locules and each locule contains one pendulous ovule. It is terminated with a one mm long glabrous style tapered to the bifid hairy stigma. The light yellow style and stigma arise from the center of the flower and stand erect throughout the flower life. The mature buds open throughout the day but most buds open during 1000-1300 h. The petals slowly open and take two-three hours for complete opening to expose the stamens and stigma. The petals emit a sweet fragrance at anthesis. The stamens bend inward overarching the stigma at anthesis and dehisce half an hour after anthesis. All the four anthers dehisce at the same time by longitudinal slits. The stigma is well seated in the center of the flower. After anthesis, the stigma grows gradually and becomes bifid on the morning of the second day. The bifid condition of stigma is an indication of beginning of stigma receptivity and it remains receptive for two days. The stigmatic lobes recurve completely. A flower lasts for five days. The petals, stamens and stigma drop off while the calyx is persistent. The pollen production per anther is 1643.2 ± 31.8 (Range 1600-1690) and per flower is 6572.8. The pollen grains are light yellow, granular, tricolporate, reticulate, muri broad, flat, thick; lumina small irregularly shaped, colpi deeply intruding, and 33.2 μm in size. The pollen-ovule ratio is 1,643.2:1.

The flowers begin nectar secretion along with anther dehiscence. The nectar secretion occurs in minute amount which has accumulated at the ovary base and on the yellow part of petals and it glitters against sunlight. It is quite prominent during the entire period of stigma receptivity. A flower produces 0.4 ± 0.08 μl (Range 0.3-0.5) of nectar with a sugar concentration of 38%. The sugar types included glucose and fructose and sucrose with the first as dominant. The results of breeding systems indicate that the flowers are self-compatible and self-pollinating. The fruit set is 12% in spontaneous autogamy, 33.33% in hand-pollinated autogamy, 40% in geitonogamy, 68% in xenogamy and 55% in open pollination. The flowers were foraged consistently during day time from 0700-1700 h by insects. The insects included *Halictus* sp., *Chrysomya megacephala, Eristalinus arvorum, Rhyncomya* sp., an unidentified fly, *Polistes humilis* and *Catopsilia pyranthe*. Individuals of each fly species were numerous on each tree. Bees were the exclusive pollen feeders. They also collected nectar throughout the flower life depending on the availability. All other insects collected nectar only. Each forager species made 11.5 to 17.5% of foraging visits. Percentage of foraging activity of insect organisms on *Avicennia marina* species. All the insects probed the flowers in upright position to collect the forage. *C. pyranthe* landed on the petals, stretched its proboscis to collect nectar aliquots on the petals and at the flower base. In this process, all the insects invariably touched the anthers and the stigma; the ventral side of all insects was found with pollen. Further, the body washings of the all insect species revealed the presence of pollen; the average number of pollen grains per insect for each species varied from 63.1 to 227.4. As the nectar is secreted in minute amount, the insects made multiple visits to most of the flowers on a tree and moved frequently between trees to collect nectar. Such a foraging behaviour was considered to effect self- and crosspollination. Pollinated and fertilized flowers initiate fruit development immediately and take about a month to produce mature fruits.

In each fertilized fruit, only one ovule produces seed. Fruit is a one-seeded leathery grayish green capsule with persistent reddish brown calyx; 30-35 mm long, 25 mm wide, abruptly narrowed to a short beak and hairy throughout. Seed produces light green, hypocotyls, which completely occupies the fruit cavity. An inflorescence produces 6.88 ± 2.96 fruits (Range 3-20). The Rose-ringed Parakeet, *Psittacula krameri* was found to feed on the hypocotyl by damaging the fruit pericarp; the percentage of damaged fruits is 4%. The fruit together with hypocotyl falls off the maternal parent, it settles in the substratum immediately at low tide period, it floats in water and disperses by tidal currents at high tide period until settled somewhere in the soil. The radicle side of hypocotyl penetrates the soil and produces root system while plumule side produces new leaves and subsequent aerial system. The fruit pericarp detaches and disintegrates when plumular leaves are produced.

***Avicennia officinalis* (Avicenniaceae).** It is a tall evergreen tree with irregular spreading branches giving crowned globose appearance. It grows to a height of more than 15- 20 m. The summer showers or early monsoon rains trigger flowering response. The rains vary in their occurrence and intensity even over a short distance; consequently the flowering period varies from place to place. The flowering season extends until August. The flowering density is almost uniform throughout the flowering season. Inflorescence is a terminal or axillary trichotomous panicle; each panicle produces 32 ± 11 flowers which anthese over a period of 16-25 days. Flowers are small, orange-yellow, ten mm long, ten mm wide, cup-shaped, have foetid smell, bisexual, slightly zygomorphic and oriented erect or partly horizontal. Calyx is shorter and has four ovate light green sepals with hairs on the outer surface; it is persistent. The corolla has four thick, light to orange yellow glabrous petals forming a short tube at the base; the petal margins are dull white. The adaxial petal is the broadest and shallowly bilobed. Stamens are four, epipetalous, three mm long and inserted basally on corolla.

The anthers are basifixed, exserted, introrse and arranged alternate to petals. The ovary is very small (seven mm long), conspicuously hairy, bicarpellary syncarpous with four imperfect locules and each locule contains one pendulous ovule. It is terminated with a two mm long glabrous style tapered to the unequal bifid hairy stigma; both the style and stigma are light yellow throughout the flower life. The entire female structure is over-arched by stamens above. The style is bent, situated below the adaxial corolla lobe but not in the centre of the flower. The mature buds open throughout the day with most mature buds opening during 0800- 1100 h. The petals slowly open and take two-three hours for complete opening to expose the stamens and stigma. The petals emit foetid smell at anthesis. The stamens bend inward overarching the stigma at anthesis and dehisce 0.5 h after anthesis. All the four anthers dehisce at the same time by longitudinal slits. Gradually, they become erect and bend backwards; to achieve this, they take three days. Then, the anthers became dark brown and petals turn light orange. Gradually, the bent stigma grows, becomes erect and is bifid on the morning of the third day. The bifid condition of stigma is an indication of beginning of stigma receptivity and it remains receptive until the fifth day. A flower lasts for seven days. The petals, stamens and stigma drop off while the calyx is persistent. The pollen production per anther is 2,444 ± 202.4 (Range 2078-2604) and per flower is 8,837. The pollen grains are light yellow, granular, tricolporate, reticulate, muri broad, flat, thick; lumina small irregularly shaped, colpi deeply intruding, and 33.2 μm in size. The pollen-ovule ratio is 2,209.3:1. The flowers begin nectar secretion along with anther dehiscence. The nectar secretion occurs in minute amount of the yellow part of petals and it glitters against sunlight. It is quite prominent during the entire period of stigma receptivity. A flower produces 0.65 ± 0.09 (Range 0.5-0.8) μl of nectar with a sugar concentration of 39.75 ± 1.89% (Range 36-43%).

The sugar types included sucrose, glucose and fructose with the first as dominant. The nectar amino acids included aspartic acid, cysteine, alanine, arginine, serine, cystine, proline, lysine, glycine, glutamic acid, threonine and histidine. All of these are prominent except lysine, glycine and proline. The results of breeding systems indicate that the flowers are self-compatible and self-pollinating. The fruit set is 21.42% in spontaneous autogamy, 42.85% in hand-pollinated autogamy, 63.33% in geitonogamy, 67.85% in xenogamy and 58.13% in open pollination. The flowers were foraged consistently during day time from 0700-1700 h by insects. The insects included *Apis dorsata, Xylocopa pubescens, Xylocopa* sp., *Eristalinus arvorum, Chrysomya megacephala, Sarcophaga* sp., *Euploea core, Danaus chrysippus, D. genutia, Junonia lemonias, J. hierta*, a fly and a wasp (unidentified). Individuals of each fly species were numerous at the flowers. *A. dorsata* was the only pollen feeder. It also collected nectar throughout the flower life depending on the availability. All other insects collected nectar only. Each forager species made 5.5 to 12.5% of foraging visits. All the insects probed the flowers in upright position to collect the forage. In case of Xylocopa bees, they made audible buzzes while collecting nectar aliquots from the petals. The butterflies landed on the petals, stretched their proboscis to collect nectar aliquots on the petals. In this process, all the insects invariably touched the anthers and the stigma; the ventral side of all insects was found with pollen. Further, the body washings of the all insect species revealed the presence of pollen; the average number of pollen grains per insect for each species varied from 73 to 550.2. As the nectar is secreted in minute amount, the insects made multiple visits to most of the flowers on a tree and moved frequently between trees to collect nectar. Such a foraging behaviour was considered to effect self- and cross-pollination. Pollinated and fertilized flowers initiate fruit development immediately and take about a month to produce mature fruits.

In fertilized flowers, only one ovule produces seed. Fruit is a one-seeded leathery capsule with persistent reddish brown calyx; 30 mm long, 25 mm wide, abruptly narrowed to a short beak and hairy throughout. Seed produces light green, hypocotyl which completely occupies the fruit cavity. The fruit pericarp dehisces when the hypocotyl is ready for dispersal and when the latter is still on the mother tree. The entire dehisced fruit with hypocotyl inside falls off from the mother tree. The fruit pericarp together with hypocotyl settles in the substratum immediately at low tide period; it floats in water and disperses by tidal current at high tide period until settled in the soil. The radicle side of hypocotyl penetrates the soil and produces root system while plumule side produces new leaves and subsequent aerial system. The fruit pericarp detaches and disintegrates when plumular leaves are produced.

***Aegiceras corniculatum* (Myrsinaceae).** It is a small evergreen shrubby tree with spreading diffused branches, growing to a height of four-five m. It flowers mainly from second week of February to second week of April; sporadic flowering occurs outside this period, especially during rainy season. The plant species such as *Bruguiera gymnorrhiza* (year-long bloomer), *B. cylindrica* (seasonal bloomer, September-March), *Ceriops decandra* (year-long bloomer) and *Excoecaria agallocha* (seasonal bloomer, July-August) occur in association with *A. corniculatum*. In *A. corniculatum*, an individual tree flowers for 22-25 days. Inflorescence is a simple umbel which arises in the shoot apex or in auxillary branches. An inflorescence produces 21.53 ± 4.42 flowers (Range 14-36). There is considerable synchrony of flower development in one umbel and in this state, the flowers appear in ball-like clusters. The umbels open in acropetal succession. Flowers are pedicellate, pointed in bud condition, 17 mm long, seven mm wide, white, cup-shaped, fragrant, bisexual and zygomorphic. Sepals are five, small, five mm long, creamy white, twisted, free, coriaceous, round apex and persistent. Petals are five, free, five mm long, white, twisted to the left, slightly fused at base to form short tube of five-six mm length, mouth with a dense weft of hairs and capitate hairs at the base.

Stamens are five, opposite to petals, epipetalous, adnate to the base of the corolla tube, filament creamy white, five mm long, base united to form a tube with a ring of internal and external hairs at the level of the mouth of the corolla tube. The anthers are bilobed, two mm long, inserted, sagittate, introrse and medifixed. Ovary is superior, conical, eight mm long, single-loculed with 35 ovules arranged on free central placentation in one chamber. Style is terminal, elongated, creamy white with dark brown dots, three mm long and extends beyond the mouth of the corolla tube. It has pedicellate glands at the base of the style and a nectariferous area at the base of the ovary, glabrous, soft, gradually tapering; stigma is apiculate. The mature buds open at 0600 h by slightly unfolding the petals. Gradually, the erect petals take horizontal position and finally reflex backwards exposing stamens and the single style; this entire process takes place in a time span of three hours. The anthers enclose the style and stigma and are bend towards it; both parts attain almost the same height. Anther dehiscence occurs one hour after anthesis by longitudinal slits. The pollen output per anther is 15,221 ± 1528.09 (Range 12,864-18,336) and per flower is 76,105. Pollen grains are spheroidal, tricolporate with distinct annulus, dull white, exine smooth, thick and 33.2 μm in size. The pollen-ovule ratio is 2,174:1. The pollen protein content per anther is 5.6 μg and per flower is 28 μg. Pollen grains are viable for 3.5 days. The stigma attains receptivity almost at the time of anther dehiscence and continues up to the evening of the third day. A flower produces 4.08 ± 0.6 μl (Range 2.2-5) of nectar from the ring of tissue at the base of the ovary.

The nectar sugar concentration is 36.46 ± 3.7% (Range 32-40%) and the common sugars include fructose and glucose with the former relatively more dominant. There is no significant correlation between nectar volume and sugar concentration ($r = 0.462$ at $p > 0.05$ significance level). The total sugar content in the nectar of a flower is 1.73 ± 0.39 mg (Range 0.76-2.63). The nectar contains 13 amino acids which include tyrosine, glycine, proline, lysine, aspartic acid, glutamic acid, serine, cysteine, cystine, alanine, threonine, arginine and histidine. The flowers not pollinated fall off on the fifth day. In pollinated flowers, the petals and stamens drop off on the fourth day while the fruit is in a growing stage. The sepals and style are persistent, and remain on the plant even after the shedding of propagule. Floral bud abortion is 2%. The results of breeding systems indicate that the flowers are self-compatible and selfpollinating. The fruit set is 7% in wind-pollinated flowers, 15.8% in spontaneous autogamy, 40.0% in insect-assisted pollination and 54.6% in open pollination. Fruit set per inflorescence in open pollination is 13.96 ± 5.05 (Range 5-24). The flowers are unspecialized and the stamens and style become exposed when the petals reflex backward. They were foraged during day time from 0600-1700 h. The foragers included bees (*Apis dorsata, A. cerana, A. florea, Amegilla* sp., *Nomia* sp., *Megachile* sp., *Xylocopa pubescens, X. latipes, Xylocopa* sp.), wasps (*Delta campaniforme, Polistes humilis* and *Odynerus* sp.), a fly species, *Chrysomya megacephala* and butterflies (*Catopsilia pomona, Euploea core* and a Hesperiid). All these started their forage collection from 0600 h and continued until 1700 h with varying number of foraging visits at each hour. Bees collected both pollen and nectar throughout the day and accordingly the number of foraging visits were found to be nearly consistent.

In the afternoon hours, bees were found to collect primarily pollen. Wasps, the fly and butterflies collected only nectar. They made a number of foraging visits during forenoon hours and gradually their visits were reduced towards the end of the day. Of the total number of foraging visits of insects, bees made 71%, wasps 17.1%, the fly 5.5% and butterflies 6.4%. Further, *A. dorsata* and *Xylocopa* species individually made higher percentage of foraging visits. All these insects while probing the flower for nectar and/or pollen invariably contacted the style, stigma and stamens.

These insects while probing for nectar contacted the sex organs with their ventral and/or dorsal side. In case of butterflies, the proboscis usually contacted the sex organs; the head and wings rarely contacted the floral sex organs. Bees while collecting pollen always made contact with their ventral side. Body washings of insects revealed the presence of pollen grains; the mean number varied from 51.9 to 1552. Bees were found to carry more number of pollen grains than all other categories of insects; the number of pollen grains found appeared to be related to the size of the bee species. Wasps with their smooth bodies were found to carry a small number of pollen grains. The fly was found to carry relatively more number of pollen grains than the wasps. Butterflies were found to transfer pollen grains principally through their proboscis; the proboscides were found with an average number of 93.3-117.3 pollen grains. Bees were also found to forage for pollen and nectar of *Bruguiera gymnorrhiza* occasionally. *Nomia* bees and *Odynerus* wasps also collected forage from *Ceriops decandra* occasionally. High winds prevailing in the plant area enabled the medifixed anthers with free movement to release light, dry and powdery pollen grains into the air easily. Pollinated and fertilized flowers initiate fruit development immediately and take 30-35 days to produce mature fruits. Of the 35 ovules, only one ovule enlarges and the others remain undeveloped.

The developing ovule displaces the placenta laterally while the young seed elongates. Fruit is an elongated, one-seeded capsule, light green to pink, 65 ± 5 mm long (Range 50-80), completely curved with pointed apex, persistent calyx and filled at maturity by the embryo with extended radicle and attached laterally by a long funicle-like structure. Seed coat is hard, brown, hairy with placental remains attached to mature seed coat and at the hypocotyl tip. Hypocotyl comes out of the seed coat but it does not pierce the pericarp. Mature fruit with well developed hypocotyl inside hangs downwards. They are water-buoyant and dispersed by tidal waters. After detachment from the mother plant, fruit pericarp dehisces longitudinally by absorbing water. The hypocotyl is not water-buoyant without fruit pericarp. In dehisced fruit, the green hypocotyl is exposed and produces a new plant when settled in a suitable substratum. The radicle side of hypocotyl penetrates the soil and produces root system while plumule side produces new leaves and subsequent aerial system.

***Aegialitis rotundifolia* (Plumbaginaceae).** It is a soft-wooded evergreen shrub species, growing to three m height with a basally swollen fluted axis. It occurs in association with *Ceriops decandra, C. tagal, Bruguiera gymnorrhiza* and *Excoecaria agallocha*. Its leaves excrete salt which in turn gets crystallized on the leaf surface on sunny days. The flowering occurs from second week of February to third week of April. An individual tree flowers for 20-22 days. The flowers are produced in terminal, irregular one-sided cymes with pairs of opposite linear bracteoles. An inflorescence produces 6.2 ± 1.6 flowers (Range 4-10) over a period of six or seven days. Flowers are pedicellate, small, 18 mm long, 11 mm wide, white, cup-shaped, odourless, bisexual and zygomorphic. Sepals are five, small, nine mm long, green, united basally and free apically, glabrous, coriaceous, valvate and persistent. Petals are five, alternate to sepals, 12 mm long, white, herbaceous, free above with bluntly rounded lobes, fused basally to form a corolla tube of three mm long. Stamens are five, 12 mm long, free, inserted on the corolla tube alternately with the petals. The filaments are seven mm long, whitish, glabrous, slender, forming short hollow tube with a ring of internal and external hairs at the level of the mouth of the corolla tube. The anthers are bilobed, two mm long, inserted, sagittate, introrse and basifixed.

The ovary is superior, oblong, onechambered with a single basally attached anatropous ovule. It has five grooves each extending into a lobe which in turn extending into a free style. The styles are five, free, white, each style ten mm long with an extended oblique peltate stigma initially facing inward and later facing outward. The mature buds open at 0700-0930 h by slightly unfolding a single petal first followed by the second petal and other petals within two hours. Petals reflex backward partially exposing the stamens and the styles. The styles and stigmas stand slightly below the height of the anthers, face inward at anthesis and diverge gradually moving away from the anthers but reaching closer to the petals. The anthers dehisce by longitudinal slits along with anthesis. The pollen output per anther is 288.6 ± 40.27 (Range 240-372) and per flower is 1443. Pollen grains are large, spheroidal, tricolporate characterized by prominent central wart-like sculptures, light yellow, exine rough, thick and 119.52 ± 10.49 μm in size. The pollen-ovule ratio is 1443:1. The pollen protein content per anther is 6.08 μg and per flower is 30.04 μg. Pollen grains are viable for 2.5 days. The stigma attains receptivity two hours after anthesis and continues up to the evening of the third day.

A flower produces 6.50 ± 0.8 (Range 5-7.8) μl of nectar at the flower base by the time of anthesis. The nectar sugar concentration is 46.2 ± 5.4% (Range 36-53%) and the common sugars include fructose and glucose which occur in almost equal amounts. There is no significant correlation between nectar volume and sugar concentration ($r = -0.386$ at $p > 0.05$ significance level). The total sugar content in the nectar of a flower is 3.60 ± 0.49 (Range 3.07-4.64) mg. The nectar contains 16 amino acids which include tyrosine, glycine, lysine, aspartic acid, glutamic acid, serine, cysteine, cystine, alanine, threonine, arginine, phenylalanine, proline, tryptophan, valine and histidine. The corolla gradually turns from white to dark red from day one to day four. The dark red corolla together with stamens and styles remain in place for two to three weeks during which the calyx bulges due to growing fruit inside. The calyx is persistent and inseparable from fruit. The flowers not pollinated fall off on the fourth day. The results of breeding systems indicate that the flowers are self-compatible and self-pollinating. The fruit set is 9% in wind-pollinated flowers, 25% in spontaneous autogamy, 47% in insect-assisted pollination and 60% in open pollination. Fruit set per inflorescence in open pollination is 2.62 ± 1.41 (Range 1-7). The flowers are unspecialized; the stamens and styles to become exposed when the petals reflex backward partially. They were foraged during day time from 0700-1600 h. The foragers were exclusively bees which included honey bees, *Apis dorsata, A. cerana, A. florea* and the Stingless bee, *Ceratina simillima*. All the bees collected pollen and nectar in the same and/or in a different foraging visit. Their foraging activity was primarily concentrated in the forenoon period and gradually decreased towards the end of the day. Of the total number of foraging visits of insects, *A. dorsata* and *A. florea* together made 64.5% while the other two bee species made the remaining percentage of visits. All the bees while probing the flower for nectar and/or pollen invariably contacted the styles and stamens. The *Apis* bees were also found to forage for pollen and nectar of *Bruguiera gymnorrhiza* while *C. simillima* collected pollen and nectar of *Ceriops decandra*. High winds prevailing in the plant area enabled the anthers to release light, dry and granular pollen grains into the air easily. Pollinated and fertilized flowers initiate fruit development immediately and take 30-45 days to produce mature fruits. Fruit is an elongated, bluntly pointed one-seeded capsule, light green to brown, 72 ± 4 mm (Range 65-83) long, enveloped basally by persistent calyx, funicular tube attached to seed and enlarging hypocotyl which protrudes from the seed coat but not the pericarp. Seed coat is hard, brown, embryo elongated with an extended hypocotyl. Mature fruit with well developed hypocotyl stands upright. Fruit pericarp is thin but thickened somewhat distally. It is water-buoyant and dispersed by tidal waters. It detaches along with the quadrangular calyx.

The hypocotyls settle at the mother plant if the site is not inundated due to tidal water and float in tidal waters, especially during high tide periods. In fruits that float in tidal waters, the pericarp becomes soft and ruptures longitudinally to expose the hypocotyl to saline water. The hypocotyls devoid of fruit pericarp did not float while those with it floated. Such hypocotyls float until they find suitable substratum which is sticky and silty mud. The radicle side of hypocotyl penetrates the soil and produces root system while plumule side produces new leaves and subsequent aerial system. In the study area, very few propagules were found to settle, establish and produce new plants.

## DISCUSSION

***Ceriops tagal* and *Ceriops decandra*.** It is strictly a winter bloomer whereas its closely related species, *C. decandra* is a year-long bloomer with alternate flowering and fruiting phases (Solomon Raju et al., 2006). Aksornkoae et al. (1992) reported that the occurrence of the two species at the same site is rare; we also found similar situation at the Krishna mangrove forests. The distribution of *C. tagal* in this forest indicates that it has a distinct seaward zonation and prefers well drained high saline soils, suggesting that the species is a salt tolerant mangrove with the competitive ability to grow in highly saline and partly inundated locations (Aziz and Khan, 2001). The site is flooded with water only during high tides and is well drained during low tides indicating that *C. tagal* is a higher inter-tidal mangrove specialist and the plants occurring in such sites are inundated about twenty times a month (Duke et al., 1998). Further, in such sites, rainfalls make no differences and hence, the salt content of the soil remains high and approximately uniform throughout the year. Duke et al. (1998) reported that *Excoecaria agallocha* becomes more common in the absence of *C. tagal* in such sites. At the study site, a few trees of *C. tagal* occur with some naked habitat and E. agallocha grows here and there in its association. On the contrary, *C. decandra* is not a strict seaward mangrove plant and it occurs commonly even in areas of tidal zone far away from sea shore (Tomlinson, 1986). Field studies here showed that *C. tagal* is absent in Godavari mangrove forest, while a few trees still survive in Krishna mangrove forests. Therefore, *C. tagal* being a seaward mangrove, it is highly unsuccessful to establish a good population size, while its sister species, *C. decandra*, with flexibility to survive in tidal zones even far away from the seashore, is highly successful to build up its populations to the extent of becoming a common constituent of mangrove forests.

In *C. tagal*, the floral characteristics such as white flowers, strong fragrance, complex petal-stamen configuration and production of moderate amount of nectar suggest an elaborate and specialized floral mechanism. The petals require an external delicate touch for the explosive release of stamens. The helically coiled hairs at the lower margins of the petals help to propagate explosive pollen release effectively (Juncosa and Tomlinson, 1987). The petal clavate appendages of petals in *C. tagal* lack hydathodes and abundant xylem which are characteristically present and have a role in flower function under extreme water pressure deficits during the day in *C. decandra* (Juncosa and Tomlinson, 1987). Such a state may make appendages very light and provide necessary trigger for petal explosion by delicate touch by the forager in *C. tagal*. Explosive pollination mechanism has also been reported in *Bruguiera* species for which the flower tripping agents are birds and butterflies (Tomlinson et al., 1979; Ge et al., 2003). Tomlinson (1986) reported night-flying insects, especially moths, as probable pollinators. Meeuse and Morris (1984) described the characteristics of moth flowers which include flower opening in the evening, display of overwhelming fragrance at that time, light flower colour, absence of landing platform, fringed petals for guidance, visual and olfactory nectar-guides, long and narrow corolla tube, abundance of nectar and short-tongued visitors. Baker and Baker (1983) reported that hawk moth flowers produce sucrose-rich or dominant nectar with low sugar concentration. Cruden et al. (1983) reported that small moth flowers produce relatively small volume of nectar with small amount of sugar. They also stated that moth flowers initiate nectar secretion one-three hours or even ten or more hours prior to the activity period of moths. In the *C. tagal*, nectar guides and tubular corolla are lacking, the nectar secretion begins an hour after anthesis; it is in moderate volume, hexose-rich with high sugar concentration. These characters together with the shallow nature of flowers are suitable for foraging by short-tongued bees and flies (Baker and Baker, 1983; Cruden et al., 1983).

In this study, there was no foraging activity of hawk moths or settling moths at the flowers after anthesis til late evening (up to 2200 h) and the absence of moths could be due to nonavailability of nectar at anthesis and reduced opportunities for food and breeding opportunities in harsh mangrove habitats. *C. tagal* with a few trees and a few numbers of flowers per unit of time per tree does not constitute a potential nectar station for moths. Further, adult moths do not survive for longer periods and in particular, hawk moths may survive for a period exceeding a month (Opler, 1983). Within that life span, the availability of nectar in the habitat is crucial and since *C. tagal* is unable to attract and supply its nectar requirements, the moths might have disappeared or migrated to other reliable food-rich habitats. *Apis* bees and *Chrysomya* flies make up day-time foragers for *C. tagal*. With a small number of trees and again each tree with a small number of flowers per day, *C. tagal* is not a potential pollen and nectar source for honey bees. Yao et al. (2006) also reported that this plant species is a minor pollen and nectar source for honey bees. In the study site, *Aegiceras* and *Bruguiera* flowers attract honey bees and the latter were found concentrating on these species. *Chrysomya* flies frequent the flowers of *C. tagal* daily effecting pollination but they have limited pollen transport capacity; this however, is compensated by their numbers and could bring about substantial geitonogamy and xenogamy (Faegri and van der Pijl, 1979). The petal colour change may act as a nectar guide for the flies to visit the flowers for several days. The close proximity between trees of *C. tagal* at the study site also facilitates xenogamy. The fly is present throughout the year unlike periodic bees and moths; but its presence depends on local opportunities for breeding sites (Faegri and van der Pijl, 1979). The butterfly is an occasional forager and has no role in pollination. Our observations suggest that *Chrysomya is* the primary and consistent pollinator, while honey bees are secondary and occasional pollinators. Petal explosion also occurs sporadically in nature due to wind action and this is evident in bagging experiment in which there is a negligible fruit set and also in natural conditions to some extent. The role of wind in tripping explosive pollination has also been reported in *Hyptis suaveolens* (Lamiaceae) (Solomon Raju, 1989) and *Shorea robusta* (Dipterocarpaceae) (Aluri et al., 2004). Therefore, petal explosion and subsequent pollination events are primarily vector-dependent and inadequate numbers or non-availability of pollinators are bound to result in reduced or no fruit set in *C. tagal*. Less is known about the importance of amino acids in floral nectars to foragers. Amino acids are the second most abundant class of the compounds after sugars to be found in nectar (Gardener and Gillman, 2002). Their concentrations in nectars are considerably lower than sugar concentrations. But even the slightest concentrations are important nutritionally and also contribute to the "taste" of the nectar (Baker and Baker, 1983). Honey bees respond to differences in amino acid concentration and detect amino acids. They prefer certain amino acids and their presence makes nectar more attractive (Dress et al., 1997). Tyrosine is not an essential amino acid, but may be important in the formation of sclerotin (Gardener and Gillman, 2002). Phenylalanine is a precursor of specific honey aroma component, phenylethanol (Thawley, 1969), iso-leucine is required for rapid breeding (Slansky and Feeny, 1977). Flies also prefer amino acids in nectar. In the flies, Boettcherisca peregrine and Phormia regina, proline stimulates salt receptor cells, methionine and valine stimulate sugar receptors, methionine also elicits a feeding response from flies and glycine and serine invoke an extension of the proboscis (Shiraishi and Kuwabara, 1970; Goldrich, 1973). *C. tagal* flowers with a mix of floral characteristics of moth, bee and fly flowers contain conventional protein building amino acids such as tyrosine, glycine, methionine, proline, lysine, aspartic acid, glutamic acid, serine, cysteine, alanine, threonine and arginine but are devoid of other proteinbuilding amino acids such as phenylalanine, valine, leucine, iso-leucine, tryptophan and histidine. The nectar provides an instant supply of methionine, lysine and arginine for honey bees and flies; moths if occur in the habitat also make use of this nectar for protein building. The presence of several amino acids in this nectar source stimulates feeding and may also be an important source for flower foragers. *C. decandra* as a common species in most of the regions and its distribution range may have achieved distinct out-crossing rates and accordingly, it has been reported to be an outcrosser (Solomon Raju et al., 2006). On the contrary, *C. tagal* with a small number of individuals at the study site can produce offspring with mixed mating system only if it fails to attract potential and adequate pollinators.

In line with this, we have found that it possesses such a system to produce fruit set even through autogamy. But, autogamy is negligible as realized in bagging experiment. In *Kandelia candel* and also in a crypto-viviparous species, *Avicennia marina*, negligible self-pollination has been reported in bagging experiments (Sun et al., 1998; Clarke and Myerscough, 1991). The fruit set in *C. tagal* is through geitonogamy and xenogamy. The long flower life and stigma receptivity and high pollen-ovule ratio (Cruden, 1977) indicate that the plant is primarily out-crossing. However, the close proximity of the existing trees at the study site in the course of time could bring about genetic uniformity and if this happens, then the survival of this species becomes doubtful. In *C. decandra*, the floral characteristics such as small white flowers lacking fragrance, simple stamen-petal configuration, short, thick filaments and production of trace amount of nectar suggest a simple floral mechanism. Juncosa and Tomlinson (1987) stated that the short basal hairs of petal edges have no evident function in *C. decandra*, but these hairs are well developed and help to propagate explosive pollen release effectively in fragrant flowers of *C. tagal*, in which the petal-stamen configuration is elaborate and specialized. Further, these authors reported that petal clavate appendages in *C. decandra* have abundant xylem elements with a significant reservoir of water and hydathodes at or near the termini; they have significance in flower function under extreme water pressure deficits during the day in mangrove swamps. Abundant xylem is absent in petal appendages in *C. tagal* and *K. candel*, which are pollinated at night or early in the morning (Juncosa and Tomlinson, 1987). Therefore, the abundant xylem and hydathodes and their function in petal appendages of *C. decandra* suggest that the latter is adapted for pollination during the day. Tomlinson (1986) suggested that wasps and flies are suitable for pollination in *C. decandra*. Juncosa and Tomlinson (1987) noted that trigonid bees and small insects visit *C. decandra* flowers.

In the present study, bees while collecting pollen and nectar and wasps while collecting nectar pollinate *C. decandra* flowers consistently. Female bees of *Nomia* and worker bees of *Trigona* collect pollen voraciously for brood provisioning. Female wasps collect nectar after mating for the maturation of the eggs. The plant produces a small number of flowers daily and accordingly, the pollen and nectar available is also in small quantities. Further, the anthesis process is gradual and the accumulation of new flowers with pollen and nectar begins to appear from late morning. As a consequence, bees and wasps also delay their foraging activity and appear from late morning onwards and stop foraging in the early evening. As flower number and floral rewards are small at the plant level, both bees and wasps fly between individuals of *C. decandra* in quest of more forage and this foraging behaviour brings about cross-pollination. Hand-pollination tests showed that it is an obligate outcrosser. The protandry, long period of stigma receptivity and long flower life substantiate this; pollination by bees and wasps favours outcrossing (Tomlinson, 1986). The pollen recovered from the body washings of these foragers suggests that they effect pollination. Earlier reports on bees, flies, wasps or even other insects as pollinators (Tomlinson, 1986; Juncosa and Tomlinson, 1987) and the present study on bees and wasps as pollinators suggest that *C. decandra* is strictly entomophilous and utilizes different locally available insects as pollinators. The commonness of *C. decandra* at Coringa mangrove forest provides ample opportunities for effective outcrossing and the genetic variation thus achieved would permit the species to survive well in the harsh mangrove environment. In *C. tagal*, the number of fruits per inflorescence varies from one to five, but one- and two-fruited inflorescences are most common.

The fruited flowers produce only one seed against the actual number of six ovules as in *C. decandra* and all other viviparous species of Rhizophoraceae. This characteristic may permit these plants to save resources and use them to produce one-seeded viable fruits. Despite this effort by *C. tagal*, a few propagules lack green pigment and becoming entirely yellowish or yellowish on one side and purplish on another side. Such hypocotyls have been referred to as "albino" forms which also occur in *C. decandra* and *Bruguiera gymnorrhiza* (Solomon Raju et al., 2006; Allen and Duke, 2006). These propagules are non-viable, cannot photosynthesize and die after depleting reserves if settled in the habitat. The propagules that are green first and brownish purple later are healthy and grow to their actual size. The length of hypocotyls in *C. tagal* is almost double the length of hypocotyls in *C. decandra*. Both the species of *Ceriops* show a short period of attachment to the maternal plant (Solomon Raju et al., 2006) and this characteristic is not in agreement with the report of Bhosale and Mulik (1991) that the hypocotyls of true viviparous mangrove species remain attached to the mother plant for a full year. The hypocotyl grows upward in *C. decandra* in which flowers are sessile (Solomon Raju et al., 2006), while in *C. tagal*, it grows downward which seems to be because of stalked flowers and more weight of the hypocotyls. This is an important field characteristic feature to distinguish *C. tagal* from *C. decandra*. The downwardly hanging hypocotyl is also a characteristic of Bruguiera, Rhizophora and Kandelia species. In *C. tagal*, the cotyledonary yellow cylindrical collar emerges from the fruit about ten days prior to the detachment of the hypocotyl while this structure is entirely absent in *C. decandra* (Solomon Raju et al., 2006). The cotyledonary collar is a characteristic also in *Rhizophora* in which it is reddish brown and in *Kandelia* in which it is yellow and the hypocotyl is about double the length of *C. tagal* (Aksornkoae et al., 1992). Fruit in *C. tagal* grows continuously and the seed also has no dormancy like in other mangrove species of Rhizophoraceae (Farnsworth and Farrant, 1998). This form of fruit growth and seed germination leading to the formation of hypocotyl while still attached to the mother plant represents "vivipary", the opposite of "ovipary" in which seed dormancy is the rule. The viviparous condition has been considered as an evolutionary loss of seed dormancy (Farnsworth and Farrant, 1997), however, it is an adaptive feature for the plant to overcome the harsh tidal environment for seedling establishment in the parental sites but it is not considered adaptive for dispersal either in time or space (Sun et al., 1998). The other adaptive values of vivipary include facilitation of early rooting (MacNae, 1968), buoyancy during sea dispersal (Rabinowitz, 1978), transfer of maternal nutrients to the hypocotyls (Pannier and Pannier, 1975), maintenance of embryonic osmotic equilibrium and establishment in coarse grained environments. On the other hand, vivipary incurs maternal costs to supply water and necessary nutrients. Numerous attached seedlings may constitute a substantial carbon sink to the maternal plant, a concentrated apparent resource for herbivores (Farnsworth and Farrant, 1998). *C. tagal* at the study site was found to produce 20 to 60 hypocotyls per tree and it is not known whether this small number could attract herbivores. Kairo et al. (2001) reported that viviparous mangrove species use self-planting or stranding strategy for establishment depending on forest conditions tide and stability of the soils. The self-planting strategy dominates in undisturbed mangrove forests but stranding strategy is dominant in exploited and open or naked forests (Dahdouh-Guebas et al., 1998). *C. tagal* with epigeal seed germination, elongated and pointed hypocotyls with straight curvature (Clarke et al., 2001) fall freely from the mother plant and plant themselves into the mud at the same site during low tide period.

The hypocotyls if fallen during high tides float to another site for settlement. But, our field studies do not show settlement of hypocotyls away from the mother plants suggesting that *C. tagal* uses self-planting strategy only. This is further substantiated by McGuinness (1997), who also reported that the hypocotyls of *C. tagal* in Northern Australia dispersed to very short distances; only 9% moved more than three m from the parent tree. He also mentioned that within that short distances, a high percentage of them were either damaged or eaten by animals. *C. tagal* at the study site may also be experiencing the damage or consumption by animals, especially crabs as the latter have been reported to show high predation on hypocotyls in high inter-tidal areas (Duke et al., 1998). Some of the fallen hypocotyls settled at the mother plants showed signs of weathering. Therefore, the study suggests that *C. tagal* though occurring in undisturbed and human-free site, is almost unable to add new plants and the presence of only a few individuals at the site attests this contention.

The work reported in this paper is important for initiating studies on the genetic structure of *C. tagal* population. The genetic marker analysis helps to understand the variability within and between populations. Introduction of *C. tagal* from the mangrove forests of the Sundarbans, Andaman and Nicobar islands to this site would help to enhance gene flow in order to maintain the gene diversity and expansion of population size of *C. tagal* in Krishna mangrove forests. The pollinated flowers initiate fruit growth and development immediately and produce mature fruits in about two months. Fruits are light green, ovoid, conical and blunt apically. They are distinct with five-lobed persistent calyx and produce only a single seed. The seed has no dormancy and it immediately produces spindle-shaped hypocotyls within three months, while still attached to the maternal parent. Then, it detaches from the residual fruit. The hypocotyl is slender, green, clearly ribbed, angular, sulcate, 15 cm long and broadened at the lower end. The short period of hypocotyl attachment to the maternal parent is a characteristic of cryptoviviparous species (Bhosale and Mulik, 1991). However, *C. tagal* shows this characteristic, being a true viviparous species (Selvam and Karunagaran, 2004). Further, the hypocotyl in *C. decandra* grows upright and is an important characteristic to distinguish it from *C. tagal* in which the hypocotyl grows downward (Kathiresan and Rajendran, 2003). In *C. decandra*, the hypocotyl is characteristically green and seems to have the potential to photosynthesize actively with water and necessary nutrients drawn from the parent tree. Viviparous reproduction allows hypocotyls to develop some salinity tolerance before being released from the parent tree. It provides a store of nutrients before the hypocotyls fall-off from the plant and helps in quick rooting in the muddy environment. The hypocotyl characteristics also help to develop buoyancy for distribution of the seedlings and structural stability to protect seedling from damage (Kathiresan, 2003). Therefore, vivipary could be an adaptive feature of the plant to overcome the harsh tidal environment for seedling establishment, especially in the parental sites. The small upright hypocotyls, when detached from the mother plant, float in water and settle at different places depending on the direction and extent of movement of tidal water. Such dispersal characterizes "Stranding Strategy". Pollinator species of *C. tagal* and *C. decandra* need a special mentioning here. The fly pollinators have rich sources of breeding and feeding materials in the forests, but the breeding materials are subject to flooding during high tide periods and during rainy season. Bees and wasps use above ground plant materials such as live/dead branches/wood. But, these materials are usually collected by local people for use as fuel wood. The forage collection from plants occurring in windy areas is a difficult task for them and hence these foragers tend to collect the forage from the plants occurring landward. These foragers have no difficulty to collect forage from *C. decandra* as the latter is distributed principally landward. In case of *C. tagal*, it occurs seaward where wind blows at high speed which usually prevents or minimizes the foraging activity of honeybees. In consequence, the bees tend to collect the forage from the flowers covered by the branches and foliage where there is reduced wind speed.

The importance of bark of *C. decandra* in dyeing the cotton and fishing nets here is driving the fishermen to cut down trees indiscriminately and the tree cutting rate is further driven by the trade concept in some fishermen. Most of the areas where *C. decandra* occurs have been partly cleared for its bark and wood. The cleared areas show the stumps or the basal part of the cut trees and these areas are gradually being invaded by the oviparous weed species, Excoecaria agallocha. Further, some such areas are being used as cattle shelters by some villagers. Therefore, such areas cleared of *C. decandra* trees and the still existing stretches of *C. decandra* are an important consideration for the concerned forest authorities with regard to conservation, management and artificial regeneration of this species in naked and semi-naked mangrove habitats in order to provide a continuous supply of bark tannin for fishermen for treating their cotton fishing nets.

***Rhizophora apiculata* and *Rhizophora mucronata*.** These species are polyhaline, evergreen true viviparous tree species. The former species blooms throughout the year, but shows profuse flowering for two months during August-September while the latter species flowers for about six months with profuse to sparse flowering during August-September. Mulik and Bhosale (1989) reported that *R. mucronata* flowers throughout the year with intense flowering during September-November. The flowers are sessile and borne in pairs below the leafy cluster in *R. apiculata*, while they are pedicellate and borne in four-flowered cymes within the leafy cluster in *R. mucronata*. With foliage background, the flowers of *R. mucronata* are also quite prominent and may be more attractive to flower foragers when compared to those of *R. apiculata*. In the latter species, bee foraging activity has been found to be consistent during profuse flowering period than at other times of the year. This suggests that greater floral displays during profuse flowering period have an important role in attracting bee pollinators. The year-long flowering allows the plant to set fruit continuously and this may be a fail-safe strategy against pollination limitation and propagule predation. In both the species of *Rhizophora*, the flowers exhibit certain adaptations for anemophily. The flowers are pendulous, point downward at maturity and situated below or within the leafy crown; this floral orientation is important to minimize interference of foliage for effective pollen dispersal by wind. The sepals diverge least, while the petals are glabrous, do not recurve or retain pollen grains in *R. apiculata*. But, in *R. mucronata* and in all other Rhizophora species, the petals are equipped with marginal hairs which have been shown to be promoters of anemophily by their hygroscopic movements (Tomlinson, 1986). Late morning anthesis, high pollen/ovule ratio, light powdery pollen, absence of an attractive colour and odour, absence of abundant pollinators and the presence of traces or minute quantity of nectar are some important characteristics for anemophily in the genus *Rhizophora* in general (Tomlinson, 1986) and in the studied species of *Rhizophora* in particular. Further, anther dehiscence and pollen release occur in mature bud in both Rhizophora species. In both, the late morning anthesis is another important characteristic for the effective dispersal of dry powdery pollen grains from the already dehiscent anthers due to moderate levels of temperature and humidity present at that time. The anthesis period noted in this study for *R. mucronata* does not agree with the report by Kondo et al. (1987) that the anthesis occurs in the afternoon from 1500 to 1700 h. These two plant species occur as pure stands mostly and are located in a windy environment along the creeks. This form of distribution may facilitate effective dispersal of pollen between individual trees and receptive sites of flowers receive wind-borne pollen. Kondo et al. (1987) reported that the pollen grains also have evolved characteristics for anemophily. Tomlinson et al. (1979) also experimentally proved that wind-borne pollen is abundant and hence anemophily is most efficient in such pure stands of other Rhizophora species.

Although both *R. apiculata* and *R. mucronata* have several adaptations for anemophily, their stigma is not elaborated in the manner usual for wind-pollinated species to capture wind-borne pollen. Similar structure of the stigma has been reported in all other *Rhizophora* species (Tomlinson, 1986). The absence of marginal hairs on petals, lack of elaboration of stigma, absence of odour and presence of traces or minute quantity of nectar seem to be vestigial characteristics of entomophily. Kress (1974) also related these characteristics to entomophily in *R. mangle* and *R. stylosa*. The flowers of *R. apiculata* attract the bees such as *Nomia, Trigona* and *Halictus*, while those of *R. mucronata* attract the bees such as *Nomia, Trigona, Halictus, Ceratina* and *Xylocopa*, and a moth species. Kondo et al. (1987) reported that *R. mucronata* is anemophilous, but is also pollinated by small insects like *Camponotus* sp., *Onychostylus pallidiolus* and a Collembola. In both the plant species, pollen is the principal reward since it is produced in huge amount. Further, the flowers are nectariferous but the nectar is secreted in traces or in minute quantity and hence has little importance for the foragers. The pollen and/or nectar feeding behaviour of these bees and the moth would contribute to pollination. But, their pollen feeding activity may considerably reduce the availability of airborne pollen and hence may affect the efficiency of anemophily. The copious production of pollen at flower level may compensate to some extent the pollen loss caused by pollen collecting bees. Further, thrips have been found in the floral buds of both *R. apiculata* and *R. mucronata*. Their presence in the buds indicates that egg deposition by female thrips takes place prior to bud formation; the eggs hatch, produce larvae or adult thrips by the time the buds mature. These thrips represent the suborder Terebrantia of the order Thysanoptera. The females of this group of thrips have an ovipositor with which they cut slits into plant tissue in order to insert their eggs, one egg per slit. With this ability, these thrips raise their offspring in floral buds. They are short-distance flyers, but windy areas extend their travel distance to a great extent. They feed on both nectar and pollen and in so doing contribute to both self- and cross-pollination, but the latter mode is more effective in areas where pure stands of *Rhizophora* species occur. The study suggests that the pollination mechanism in the studied *Rhizophora* species is originally entomophilous and is now being transformed to anemophily. This transitional stage of pollination mechanism is advantageous for these plant species to utilize both wind and insects as pollinating agents for the success of sexual reproduction. Tomlinson (1986) stated that animal pollination predominates in mangrove communities and in such a situation anemophily enables these species to escape the competition for insect pollinators and to set fruit in the total absence of insect pollinators. Tomlinson (1986) reported that *Rhizophora* exhibits weak protandry and is selfcompatible. He found that the isolated greenhouse plants set fruit. He has not mentioned what species of this genus are weakly protandrous and self-compatible. Coupland et al. (2006) reported that Rhizophora species show low rates of fruit set due to lack of autogamy. Ghosh et al. (2008) mentioned that a distinct trend for self-incompatibility exists in *Rhizophora*. In this study, *R. apiculata* and *R. mucronata* show anther dehiscence in mature bud and commencement of stigma receptivity soon after anthesis, suggesting that they are weakly protandrous. Tomlinson et al. (1979) reported that in these species, the petals and stamens fall off the flower on the day of anthesis leaving the central stigma alone in the empty flower. These authors also stated the stigma lacks any secretion or divergence of its two lobes. Kondo et al. (1987) reported that the stigma is two-lobed; each tip contains small papilla and secretes mucilage.

In this study, it is found that the stigmatic lobes diverge soon after anthesis and since then it is in receptive stage to receive pollen. The stigmatic lobes are mucilaginous and·have minute papilla, which may aid in capturing and retaining pollen readily. Tomlinson et al. (1979) reported that in Rhizophora stylosa, an individual flower is generally incapable of self-pollination. Hand-pollination results suggest that *R. apiculata* and *R. mucronata* are capable of self-pollination either by gravitational fall of pollen or by wind driven pollen fall on the stigma. Further, these plants set fruit through allogamy, autogamy and geitonogamy (Kondo et al., 1987). The significantly low fruit set evidenced in open-pollinations is attributable to the intense pollen feeding activity of bees and also to the wastage of pollen due to wind activity. Coupland et al. (2006) also reported low fruit set to the extent of 13% in open-pollinations.

The ability to set fruit through self- and cross-pollination is the characteristic of pioneering species in mangrove communities. *R. apiculata* and *R. mucronata* with this ability may become established as isolated individuals in new environments remote from parental source. Like in *Bruguiera cylindrica*, the flowers are four-ovuled in both *Rhizophora* species, but only one ovule develops into mature seed in fertilized and fruited flowers. The production of one-seeded fruits may be due to maternal resource constraint or maternal regulation of seed set. Fruits mature within four to six weeks and are not enclosed by the persistent and expanded calyx at any stage. The calyx therefore seems to have no role in protecting the fruit. As the fruit is thick with hard pericarp, it does not require protection from the calyx. In both *R. apiculata* and *R. mucronata*, the single seed formed in the fruit is not dormant and germinates immediately to produce a cylindrical hypocotyls or seedling, while still on the maternal parent. The hypocotyl emerges out of seed and fruit pericarp and remains naked until it is detached from the maternal parent. This is a characteristic of "true viviparous" species (Tomlinson, 1986). While still attached to the maternal parent, the seedling develops chlorophyll and actively photosynthesizes; the parent tree supplies the water and necessary nutrients (Selvam and Karunagaran, 2004). The seedling hangs downward and detaches from the residual fruit at the collar end, leaving behind its cotyledons, and falls from the maternal parent. Christensen and WiumAndersen (1977) speculated that in *R. apiculata*, the development from visible flower buds to mature propagules lasts about two years while Muniyandi (1986) reported that the propagule takes eight months to grow to full length after fertilization. But, in this study, it is found that this species produces mature seedlings from flower buds in a time span of about four months only. Muniyandi (1986) say that the seedlings of *R. mucronata* take 16 months to grow to full length after fertilization in *R. mucronata*. But, in this study, it is found that this species produces mature seedlings from flower buds in a time span of about four months only. The fallen seedlings of *R. apiculata* and *R. mucronata* plant themselves into the mud in the vicinity of the maternal parent or are stranded and planted away from maternal parent. In the latter case, the seedlings sink after some period of dispersal and exhibit growth under water (Rabinowitz, 1978). The study shows that the seedlings disperse through self-planting and stranding strategies.

Although dispersal distances have not been measured in this study, the available information shows that most viviparous seedlings do not disperse more than one km from their point of origin (Clarke, 1993), and in certain instances, as little as 65 m (Chan and Husin, 1985) or even three m (McGuinness, 1997). Davis (1940) found that *Rhizophora mangle* propagules carried by ocean currents could disperse at least 100 km. Smith (1987) reported seedling predation prior to and after detachment from the maternal parent in *R. apiculata* and *R. mucronata*. Monkeys and insects attack seedlings prior to their detachment, while crabs attack seedlings when they later fall in the mud or tidal water. Monkeys, insects and crabs collectively contribute to more than 83% of seedling predation. Bosire et al. (2005) reported that seedling predation is a mechanism to regulate individual species colonization, but in case of *R. mucronata*, the seedling predation is the least since they are very large and are not preferred by crabs. In this study, there is no seedling predation prior to detachment from the maternal parent in both the species of *Rhizophora*. Seedling predation by crabs may be there since different species of crabs have been found. The locals use crabs as a source of food and livelihood.

***Bruguiera gymnorrhiza* and *Bruguiera cylindrical*.** Both the species of *Bruguiera* are polyhaline, evergreen true viviparous tree species. *B. gymnorrhiza* is a year-long bloomer with concentrated flowering during April-June, while *B. cylindrica* shows flowering during September-March during which flowering level is almost uniform. On *B. gymnorrhiza*, pollinator activity is consistently intense during concentrated flowering period, while it is sporadic at other times of the year. On the contrary, pollinator activity is not consistent and also not intense at any point of time on *B. cylindrica*. This finding stated that pollinators preferentially visit plants with greater floral display. Therefore, fruit set is largely a function of pollinator activity during concentrated flowering period in *B. gymnorrhiza*. However, the characteristic of year-long or extended flowering in these two tree species is an important adaptation for additional fruit set; this characteristic would enable them to compensate the low fruit set rate which may result from pollinator limitation and also to compensate the loss of seedlings due to predation. Tomlinson (1986) reported that *B. gymnorrhiza* represents the large, solitary-flowered group, while *B. cylindrica* represents the small, many-flowered group. The present study shows that it is true in case of *B. gymnorrhiza* but it is not so in *B. cylindrica* in which the flowers are small, but borne in three-flowered cymes. The flowers are pendulous in *B. gymnorrhiza*, while they are either horizontal or downward or slightly erect and located outward in the crown of leaves in *B. cylindrica*; these different orientations enable the pollinators to collect the forage comfortably. In both species, the anthesis period is the same, is confined to morning hours and quite appropriate for pollination by day-active pollinators.

The field studies also indicate that their flowers receive visits exclusively from day-active pollinators. The flowers are red in *B. gymnorrhiza* while they are greenish-white in *B. cylindrica*. In both the species, the floral characteristics such as large or small flowers with concealed nectar, elaborate complex petal-stamen configuration, presence of basal clumps of hairs and marginal hairs suggest specialized explosive floral mechanism which is functional only when pollinating agents are involved. The same elaborate specialized mechanism is present in all other species of *Bruguiera* (Tomlinson et al., 1979). In *B. gymnorrhiza*, the petals are 14 and each petal encloses two stamens; this petal-stamen configuration requires multiple visits of pollinators for the explosion of all petals. In *B. cylindrica*, the flowers have only seven or eight petals with 14 or 16 stamens and closely spaced; this arrangement with reduced number of petals in relation to the small flower size is an important adaptation for the explosion of all petals of the flower and subsequent pollination in a single or two visits of the pollinator insects. Petal explosion is characteristically violent in *B. gymnorrhiza*, while it is non-violent in *B. cylindrica*. Further, similar explosive mechanism is present also in another mangrove Rhizophoraceae member, *Ceriops tagal*. Therefore, the floral mechanism in all the species of *Bruguiera* and in *C. tagal* is highly specialized and reflects an advanced state when compared to the simple floral mechanism that is present in Ceriops decandra, *Rhizophora* species and *Kandelia candel* - all belonging to mangrove Rhizophoraceae (Tomlinson, 1986; Juncosa and Tomlinson, 1987). Further, explosive pollination mechanism has been reported in a number of non-mangrove families such as Lamiaceae, Fabaceae, Onagraceae, Loranthaceae, Marantaceae, Urticaceae, Ericaceae, Fumariaceae, Musaceae, Cornaceae, Acanthaceae and Orchidaceae (Solomon Raju and Subba Reddi, 1995; 1996). In *Bruguiera,* the explosive floral mechanism involves different functional aspects. At the time the flower opens, each petal pouch enclosing a pair of anthers includes loose pollen due to anther dehiscence during mature bud stage and stays in cocked position due to floral expansion and pressing of stamens against the interlocked margins of the petals. Marginal hairs appear to be important in holding the petals in the folded position. The petal explosion occurs due to release of tension by a slight touch at its hairy base by the probing pollinator during its vigorous search for nectar in the floral cup. During explosion, the petal unzips instantly, scattering a cloud of pollen, most or part of which falls on the pollinator. In each flower, individual petals work independently and this requires more than one visit of the same pollinator species or more than one pollinator species. This observation is substantiated by the occurrence of closed and open petals in each flower. The flowers with unexploded petals due to non-receipt of foraging visit(s) fall off without pollination or without ever-opening. In *B. gymnorrhiza*, the following characters suggest adaptations for bird pollination: the flowers are recurved and typically point backwards into the crown of the tree; this facilitates an approach by a perching bird. Nectar is produced in abundance and held in the deep floral cup. The calyx is red, a colour attractive to birds. Large flowers with a heavy construction are suitable for a powerful pollinator like a bird (Faegri and van der Pijl, 1979). Azuma et al. (2002) examined *B. gymnorrhiza* flowers for scent characteristics and reported that the floral scent is lacking and the floral characteristics are indicative of bird pollination. Ghosh et al. (2008) also reported that the flowers are adapted to a range of flower visitors such as birds for pollination. Kondo et al. (1987) reported that this species is pollinated by honeyeaters, white eye and insects. Ge et al. (2003) mentioned that the flowers are pollinated by birds or butterflies. Solomon Raju (1989) reported that the flowers are exclusively pollinated by three species of passerine birds, Nectarinia asiatica, *N. zeylonica* and *Zosterops palpebrosus* at the Godavari mangrove site; then this site was ecologically healthy and there was little human interference.

The present study finds that this site is now ecologically degraded and fragmented due to land use changes. With this present situation, B. gymnorrhiza is now pollinated exclusively by pollen and nectar collecting bees consisting of *Apis, Nomia* and *Halictus* genera. These bees trip the tensed petals to release the stamens which in turn eject a cloud of pollen from the already dehiscent anthers. During this process, the bees get a pollen shower all over their body, especially on their dorsal side. Further, the bees fly between individual trees in quest of more forage; this foraging activity is important to bring about cross-pollination. The production of a small number of fresh flowers daily at tree level may also compel the pollinator bees to make inter-tree flights and effect cross-pollination. Their body washings indicated that they carry pollen and transfer the same to other flowers they visit. The study suggests that *B. gymnorrhiza* is strictly melittophilous and this mode of pollination is as efficient as ornithophily; but the pollen feeding activity of bees may affect the pollen availability rate for the receptive stigmas at the population level which in turn may affect the natural fruit set rate. However, the bees are reliable pollinators when compared to bird pollinators. The ability of the plant to utilize birds and bees is certainly adaptive and is also essentially required for the survival, colonization and expansion of its geographical range (Tomlinson, 1986). Tomlinson et al. (1979) stated that small-flowered *Bruguiera* species including *B. cylindrica* is pollinated by butterflies. These authors also mentioned the following characteristics as adaptations for butterfly pollination: small flowers in nearly erect state and displayed to the outside of the tree crown; thin branchlets which are insufficient for bird perching; greenish-yellow petals; flat calyx cup with a small quantity of nectar; and pollen release by delicate, distal stimulation of petals. In this study, the flowers have been found to have three different orientations while all other characteristics remain the same. A careful examination of these characteristics as Faegri and van der Pijl (1979) does not conform to the characteristics of butterfly-flowers. Tomlinson (1986) reported that this species is pollinated by small insects including butterflies. In this study, *B. cylindrica* has been found to be pollinated by bee and wasp species only. *Nomia* bees, *Odynerus* and *Polistes* wasps during nectar collection and the first species also during pollen collection trip all petals of a flower mostly in a single visit; then these insects get a pollen drizzle all over their body, especially on their dorsal side. Further, they fly between individual trees to collect more forage and in so doing they carry pollen on their bodies; this foraging activity is important to bring about selfand cross-pollination. Therefore, *B. cylindrica* is bee- and wasp-pollinated but not butterflypollinated. Tomlinson (1986) mentioned that bees and wasps represent a group of pollinators that nest in mangroves, and some populations are completely dependent on mangal for their existence. Ghosh et al. (2008) reported that some wasps and flies are highly dependent on mangroves for nesting. In the light of these reports, it is not unreasonable to suggest that bees and wasps are reliable pollinators since they nest in mangroves and collect forage from the same plants for their nutrition. In the present study, both the species of Bruguiera with specialized floral mechanism offer pollen and nectar as rewards to their pollinators. Pollen grains are very dry even at the time of flower-opening. Their surface sculpture is finely reticulate; they are tricolporate with numerous small and shallow depressions, and easily adhere to the body of insect pollinator so that the latter can easily transfer pollen (Kondo et al., 1987). They contain some protein content and are important in the nutrition of bee pollinators. In *B. gymnorrhiza*, nectar production is continuous in some flowers, while it is not so in some other flowers. Such nectar production pattern indicates that nectar secretion ceases in pollinated flowers, while its secretion is continuous until the stigma loses receptivity or until the flower is pollinated. The production of copious amount of nectar with dilute sugar concentration is the characteristic of bird-pollinated flowers (Baker and Baker, 1983). But, the nectar of *B. gymnorrhiza* is copious with moderate sugar concentration; it is suitable for bee pollinators as the latter tend to prefer sugar concentrations of 30 to 50% (Waller, 1972). The sugars present in the nectar include only hexoses. Roubik (1995) stated that nectar with only hexose sugars is rarely reported. Baker and Baker (1983) segregated bee flowers into those adapted to "short-tongued" bees (with less than six mm in length) and those adapted to "long-tongued" bees. These authors also reported that nectars of flowers pollinated by short-tongued bees are usually hexose-rich, while those pollinated by long-tongued bees are usually sucrose-rich. In line with this, the bees observed have tongues less than five mm length, are short-tongued and utilize the hexose nectar of *B. gymnorrhiza*. De Groot (1953) reported that insects in general and bees in particular require ten essential amino acids - threonine, valine, methionine, leucine, isoleucine, phenylalanine, lysine, histidine, arginine and tryptophan. The nectar of *B. gymnorrhiza* has three of these essential amino acids, lysine, histidine and arginine. The presence of these amino acids and also other non-essential amino acids, alanine, aspartic acid, glutamic acid, glycine and serine may have a role in giving the "taste" to the nectar (Baker and Baker, 1982). The nectar also has some protein content. Therefore, the nectar of B. gymnorrhiza with these amino acids and protein content has high nutritional value and bee pollinators are attracted to this floral source, especially during concentrated flowering period.

In B. cylindrica, the nectar is produced in small quantity, hexose-dominant and the sugar concentration is low. Bees and wasps utilize this nectar until exhausted. This observation is in partial agreement with the generalizations made by Baker and Baker (1982; 1983) who stated that the production of a small quantity of nectar with high sugar concentration is the characteristic of bee-flowers, and that the nectars of flowers adapted for pollination by shorttongued bees are hexose-rich, while those adapted for pollination by wasps are sucrose-rich. Tomlinson (1986) mentioned that pollination may favour outcrossing in *B. gymnorrhiza*. Kondo et al. (1987) reported that this species produces fruit through allogamy, autogamy and geitonogamy and hence, it might be outcrossing and inbreeding. Ge et al. (2003) stated that this species has mixed mating system with outcrossing as a main system. In this study also, *B. gymnorrhiza* has been found to set fruit through allogamy, geitonogamy and autogamy, but all modes are functional only when petal explosion is manipulated; this suggests that pollination is essentially vector-dependent and fruit set is completely a consequence of the foraging activity of bee pollinators. The fallen flowers with some or a few unexploded petals evidenced in this study indicate that wind is not an agent of pollination. Fruit set through vector-mediated autogamy or geitonogamy indicates that *B. gymnorrhiza* is self-compatible and self-pollinating. But, protandrous condition and stigma showing receptivity commencing from day two and extending its receptivity until day four indicate that it is primarily adapted for cross-pollination. The small papilla and mucilage secreted by stigmatic lobes are special adaptations to retain pollen readily; these are especially important for capturing cross-pollen (Kondo et al., 1987). Therefore, *B. gymnorrhiza* with mixed mating system is adapted for pollination by biotic agents, the classes of which may change from time to time in the same habitat depending on the local land use changes. Similar vector-dependent mixed mating system and stigma function exists in *B. cylindrica*. Such a mating system facilitates fruit set in the presence of pollinators even in isolated trees of both the species of *Bruguiera*. With the ability to set fruit through self-pollination, these species can colonize new areas and expand their distribution range. *B. cylindrica* has scattered distribution in the study sites and also here and there it has established small patches representing pure stands. This finding is in agreement with Tomlinson (1986) who stated that it may form pure stands. In mangrove Rhizophoraceae, the flowers have been reported to contain six or four ovules (Tomlinson, 1986). The flowers are six-ovuled in *B. gymnorrhiza* and four-ovuled in *B. cylindrica*; but only one ovule develops into mature seed in fertilized and fruited flowers in both the species. The production of one-seeded fruits may be due to maternal resource constraint or maternal regulation of seed set. Fruits mature within a month. The persistent and expanded calyx gives protection to the fruit; the mature fruit is well seated within the calyx and hence is not directly exposed to sunlight. Therefore, the calyx has an important role in protecting the fruit from desiccation.

In both the species of Bruguiera, the single seed formed in the fruit is not dormant and germinates immediately to produce a cylindrical hypocotyl or seedling, while still on the maternal parent. The hypocotyl emerges out of seed and fruit pericarp and remains naked until it is detached from the parent plant. This is a characteristic of "true viviparous" species (Tomlinson, 1986). While still attached to the maternal parent, the seedling develops chlorophyll and actively photosynthesizes; the parent tree supplies the water and necessary nutrients (Selvam and Karunagaran, 2004). The seedling hangs downward and detaches from the residual fruit, leaving behind its cotyledons, and falls from the maternal parent. Viviparous reproduction allows seedlings to develop some salinity tolerance before being released from the parent tree. This reproduction provides a store of nutrients before the seedlings fall off from the maternal parent and may help in quick rooting in the muddy environment. The seedling characteristics also help to develop buoyancy for dispersal and structural stability for protection against damage (Kathiresan, 2003). Therefore, vivipary is an adaptive feature to overcome the harsh tidal environment for seedling establishment. The study shows that seedlings of *Bruguiera* fall off the maternal parent freely and plant themselves into the mud or stranded and planted away from parent tree. La Rue and Muzik (1954), Rabinowitz (1978) and Van Speybroeck (1992) reported similarly for this species. Kairo et al. (2001) reported that the mode of seedling dispersal depends on the forest conditions, tides, as well as the stability of the soils. Further, Van Speybroeck (1992) reported that the seedlings can plant themselves into the mud if they are dropped from maternal parent at low water or low tide. He termed it as self-planting strategy. He also reported that the seedlings float to another site to settle and develop if they fall in the water at high tide and he termed it as stranding strategy. The study shows that the seedlings of *Bruguiera* disperse through self-planting and stranding strategies; the former strategy is functional at low tide while the latter strategy is functional at high tide. Van Speybroeck (1992) reported that the self-planting strategy dominates in undisturbed mangrove forest whereas the stranding strategy is dominant in an exploited and open forest. The stranding strategy is especially important for regeneration and colonization of naked or semi-naked habitats and also for expanding the distribution range of the species. The study found that the Roseringed Parakeet feeds on the softest part of the seedlings of B. gymnorrhiza prior to their detachment from the parent tree.

The seedlings attacked by this parakeet have not established new plants; their percentage however did not exceed 5%. The feeding on these seedlings by the parakeet may be partly attributable to the scarcity of food. There is no seedling predation prior to detachment from the maternal parent in *B. cylindrica*. Sousa and Mitchell (1999) reported that after detachment from the maternal parent, seedlings experience mortality due to crab predation. Crabs prefer to feed on small propagules as the latter facilitate easy burial in burrows, have high nutritive value and low concentration of inhibiting chemicals such as tannins. Since the seedlings of both the species of *Bruguiera* are long, crabs may not utilize them as food source but further study is required to confirm this.

***Avicennia alba, Avicennia marina* and *Avicennia officinalis*.** All the three *Avicennia* species studied are principally polyhaline evergreen tree species. *A. alba* and *A. marina* are small trees while *A. officinalis* is a tall tree. These tree species show flowering response to monsoon showers in June; the first monsoon showers seem to provide the necessary stimulus for flowering. Opler et al. (1976) and Ewusie (1980) have reported such a flowering response to light rains in summer season in a number of plants occurring in coastal environments. The flowering period extends until August in all the three species of *Avicennia* at the study sites, indicating that the flowering season is only for three months in a year. On the contrary, WiumAndersen and Christensen (1978) reported that in *A. marina*, flowering occurs during AprilMay. Further, Mulik and Bhosale (1989) noted that the flowering in this species is from April to September. These authors also mentioned that the flowering occurs during March-July in *A. officinalis*. The variation in the schedule and length of flowering season in these species may be a response to local environmental conditions and to avoid competition for the available pollinators depending on the flowering seasons and population size of the constituent plant species which vary with each mangrove forest. In all the three species, the flowers are borne either in terminal or axillary inflorescences. But, the average number of flowers per inflorescence varies with each species; it is the highest in *A. alba*, moderate in *A. marina* and the least in *A. officinalis*. This flower production rate at inflorescence level may serve as an important taxonomic characteristic for the identification of these three species.

In all, the flowers are strongly protandrous and the stamens with dehisced anthers over-arch the stigma. The stigma shows post-anthesis growth. It is erect and seated in the centre of the flower in *A. alba* and *A. marina* while it is bent and situated below the adaxial corolla lobe in *A. officinalis*. The erect stigma does not change its orientation throughout the flower life in *A. alba* and *A. marina* while the bent stigma becomes erect on day three. The stigma is bifid and appressed on the day of anthesis in all the three species; it remains in the same state also on day two in A. officinalis. The stigma commences receptivity by diverging in dorsi-ventral plane; it is receptive on day two and three in *A. alba* and *A. marina*, and on day three, four and five in *A. officinalis*. The timing of commencement of stigma receptivity in *A. officinalis* strongly contradicts with an earlier report by Subba Reddi et al. (1995) that the stigma attains receptivity three hours after anthesis with the bent stigma becoming erect. In *A. officinalis*, stigma behaviour is more advanced towards achieving cross-pollination. In all the three species, self-pollination of individual flowers is unlikely on the day of anthesis due to protandry but the stamens with dehisced anthers over-arching the stigma may facilitate the fall of pollen on the receptive stigma when the latter attains receptivity. In effect, self-pollination may occur and the same is evidenced through fruit set in bagged flowers without manual self-pollination. Further, the sequence and synchrony of flowering, and pollinator behaviour at tree level contribute to geitonogamy (Clarke and Meyerscough, 1991). Hand-pollination results indicate that it is self-compatible and fruit set occurs through autogamy, geitonogamy and allogamy. The hermaphroditic flowers with strong protandry and long period of flower life in these species suggest that they are primarily adapted for cross-pollination. Clarke and Meyerscough (1991) also reported that *A. marina* is protandrous, self-compatible and self-pollinating but the fruits resulting from spontaneous self-pollination showed a higher rate of maternal abortion reflecting an inbreeding depression. Coupland et al. (2006) reported that in *A. marina*, autogamy is most unlikely and emphasized the importance of pollen vectors to the reproductive success.

This report is not in agreement with the results obtained in handpollination experiments on *A. marina*. Primack et al. (1981) suggest that protandry promotes out-crossing in mangroves, and that insect pollination facilitates it. They also suggested that geitonogamy in coastal colonizing plants would allow some fruit set in isolated colonizing plants, and thereafter the proportion of such pollinations would decline as pollen is transferred between plants. Pollen transfer between plants in such situations would still result in sibling mating. However, this is counteracted by dispersal of propagules, canopy suppression of seedlings and irregular yearly flowering among trees in close proximity. Clarke and Meyerscough (1991) reported that in A. marina, some trees flower and fruit every year while some others do not flower every year. Those with complete canopy crops did not produce another large crop the following year. A similar pattern observed within a tree where fruit is produced on one branch and in the following year heavy flowering shifts to another branch. In the present study, all the three species of Avicennia flowered annually and the flowering is uniform on all branches within a tree. The study suggests that annual mass flowering, protandry, self-compatibility and self-pollination ability are important adaptations for *Avicennia* species to successfully colonize new areas and expand their distribution range as pioneer mangroves. All the three species of *Avicennia* are hermaphroditic and have similar floral architecture. In *A. officinalis*, the flowers are foetid and slightly zygomorphic while in the other two species, they are scented and actinomorphic. In all, the flowers are of open type and shallow with small aliquots of nectar which is exposed to rapid evaporation resulting in increased nectar sugar concentration. Corbet (1978) considered these characteristics as adaptations for fly pollination. Hexose-rich nectar is present in *A. alba* and *A. marina* while sucrose-rich nectar in *A. officinalis*. Hexose-rich nectar is the characteristic of fly- and shorttongued bee-flowers while sucrose-rich nectar is the characteristic of wasps and butterflies (Baker and Baker, 1982, 1983). The nectar sugar concentration is high and ranged from 38 to 40% in all the three *Avicennia* species.

Cruden et al. (1983) reported that high nectar sugar concentration is the characteristic of bee-flowers while low nectar sugar concentration is the characteristic of butterfly-flowers. Baker and Baker (1982) reported that the floral nectar is an important source of amino acids for insects. Dadd (1973) stated that insects require ten essential amino acids of which arginine, lysine, threonine and histidine are present in the nectar of *A. officinalis*. He also reported that proline and glycine are essential amino acids for some insects; these two amino acids are also present in the nectar of *A. officinalis*. He further stated that other amino acids such as alanine, aspartic acid, glutamic acid, glycine and serine while not essential do increase insect growth. All these amino acids are also present in the nectar of *A. officinalis*. Shiraishi and Kuwabara (1970) reported that proline stimulates salt receptor cells in flies. Goldrich (1973) reported that histidine elicits a feeding response while glycine and serine invoke an extension of the proboscis. The nectars of *A. alba* and *A. marina* have not been analyzed for amino acids and hence this aspect has not been discussed. The flowers of all the three species of *Avicennia* with differences in their structural and functional characteristics as stated above have been able to attract different classes of insects - bees, wasps, flies and butterflies. Of these, bees while collecting pollen and nectar while all others collecting nectar effected pollination and their ability to carry pollen has been evidenced in their body washings. Flies are known as short distance fliers and such behaviour largely results in autogamy or geitonogamy. Since these flies visit the flowers as large groups, there is automatically a competition for the available nectar which is secreted in small aliquots on the petals of all the three *Avicennia* species. In consequence, they shift from tree to tree in search of nectar forage and in the process they contribute to both self- and cross-pollination. All other insects are habitual long-distance fliers and affect both self- and cross-pollination. An earlier report by Subba Reddi et al. (1995) showed that only bees and flies are the pollinators of *A. officinalis* at the study sites. Tomlinson (1986) mentioned that *Avicennia* flowers are beepollinated. In Australia, *A. marina* is pollinated by ants, wasps, bugs, flies, bee-flies, cantharid beetles and moths (Clarke and Meyerscough, 1991). It is surprising to note that thrips are absent both in bud and flower stage in all the three species of *Avicennia.*

A study on this aspect is needed to understand why thrips avoided these flowers for breeding or for forage collection. Tomlinson (1986) documented that *A. alba, A. marina* and *A. officinalis* have very similar flowers and hence may well be served by the same class, if not by the same species of pollinators; when these species grow together, there is evidence of non-synchrony in flowering times, which might minimize the competition for pollinators (probably bees) and at the same time spread the availability of nectar over a more extended period. In the present study, these plant species grow together, flower synchronously, but served by the same classes of insects. There is no competition for pollen among different classes of insects since only bees collect pollen while all other classes of insects collect only nectar. Fly pollinators with their swarming behaviour at the flowers may enable the plant species to set fruit to the extent possible. Flies and bees are usually consistent and reliable when compared to wasps and butterflies. Therefore, the study shows flies and bees play an important role in the success of sexual reproduction in all the three species of *Avicennia.* Despite being pollinated by different classes of insect pollinators and having the ability to self-pollinate even in the absence of insect activity as evidenced in bagged flowers, the natural fruit set stands at 42-58% in these plant species. This low fruit set could be due to maternal abortion of self-pollinated fruits as reported by Clarke and Meyerscough (1991), non-availability of sufficient pollen to receptive stigmas due to pollen feeding activity of bees and the nutritional resource constraint to the maternal parent. Coupland et al. (2006), while reporting on fruit set aspects of *A. marina* in Australia mentioned that fruit set is not pollinator limited but resource limited. In Avicenniaceae, the flowers have been reported to contain four ovules (Tomlinson, 1986). In the present study, all the three species of Avicennia are four-ovuled but only one ovule develops into mature seed in fertilized and fruited flowers as in Rhizophoraceae. The production of one-seeded fruits may be due to maternal resource constraint or maternal regulation of seed set. Fruits grow and mature within five-six weeks in *A. alba* and within four weeks in the other two *Avicennia* species. The duration of fruit maturation is not in agreement with the report of Wium-Andersen and Christensen (1978) who stated that the development from flower bud to mature fruit takes a few months. The calyx is persistent in all the three species but it does not expand to enclose the growing fruit. Therefore, the calyx has no role in sheltering or protecting the fruit. As the fruit is a leathery capsule, it does not require any protection from the calyx. The single seed formed in the fruit is not dormant and germinates immediately to produce chlorophyllous seedling, which remains within the fruit, while still on the maternal parent. This is a characteristic of "crypto-viviparous" species; a similar situation exists in other genera such as *Aegiceras, Aegialitis, Nypa* and *Pelliciera* (Tomlinson, 1986). In all these species, fruit is the propagule; the seedling occupies the fruit cavity. The chlorophyllous seedling actively photosynthesizes while the maternal parent supplies the water and necessary nutrients (Selvam and Karunagaran, 2004). In *Avicennia* species, the propagules are small, light and the entire embryo is buoyant after detachment from the maternal parent.

Gradually, the fruit pericarp is lost exposing the leathery succulent cotyledons to tidal water. Rabinowitz (1978) reported that *A. marina* has an absolute requirement for a stranding period in order to establish since its propagules always float in tidal water. He also felt that the propagules must have freedom from tidal disturbance in order to take hold in the soil. In consequence, this species is restricted to the higher ground portions of the swamp where the tidal inundation is less frequent. In the present study, *Avicennia* species exhibit self-planting strategy at low tide and stranding strategy at high tide. However, their seedlings disperse widely in tidal water but establishment is mainly stationed in the polyhaline zone. Duke et al. (1998) reported that *Avicennia* seedlings disperse widely and are genetically uniform throughout their range. In the study areas, genetic studies are required to know whether all the three species studied are genetically uniform. When the seedlings settle, radicle penetrates the sediment before the cotyledons unfold. The first formal leaves appear one month after germination and the second pair one to two months (Wium-Andersen and Christensen, 1978). Coupland et al. (2006) reported that *Avicennia* propagules are a rich source of nutrients and attract a diverse range of insect predators which in turn influence the rate of seedling maturation. Resource constraints and insect predation on developing fruit and seedling may both act to reduce fruit set. In *A. marina* and *A. germinans*, the seedlings tend to be high in nutritive value and have relatively few chemical defences (Smith, 1987; McKee, 1995). These species tend to exhibit a pattern of very rapid initial predation (Allen et al., 2003). In the present study, seedling predation has been evidenced in *A. alba* and *A. marina* only; in both the species, the Rose-ringed Parakeet, *Psittacula krameri* attacks propagules prior to their detachment from the maternal parent. Seedling predation by crabs after detachment from the maternal parent may be expected since different species of crabs have been found in the study areas. Therefore, seedling predation may reduce the success of seedling establishment in all the three species of *Avicennia*.

***Aegiceras corniculatum.*** The plant is a mesohaline evergreen species. The flowering occurs during the dry season; sporadic flowering also occurs at other times of the year but it is especially significant during the rainy season. The floral characteristics such as morning anthesis, scent production, zygomorphic symmetry, short-tubed corolla with sexual organs exposed, pollen structural features and nectar production are adaptations for pollination by any class of animals (Baker and Baker, 1983; Cruden et al., 1983). Different workers reported different insects as pollinators of A. corniculatum - bees without mentioning the species by Tomlinson (1986); the bees Trigona iridipennis and Pseudapis oxybeloides at Coringa mangrove forest by Solomon Raju (1989); butterflies, bees, wasps and flies, and also birds in the Orissa mangrove forest by Pandit and Choudhury (2001). The present study shows that bees, wasps, flies and butterflies are the pollinators; *Xylocopa* bees are the most efficient pollinators due to their ability to collect forage quickly from each umbel and to their quick mobility between plants for want of more forage. The bees are pollen and nectar collectors while all others are nectar feeders. Pollen is a source of protein for bees and the pollen of *A. corniculatum* with some protein content is important for them. The small volume of nectar with high sugar concentration consisting of only two hexose sugars, fructose and glucose in *A. corniculatum* is stated to be a requirement for bees, flies and butterflies while sucrose-rich nectar is the requirement for wasps (Baker and Baker, 1983).

This study shows that wasps also utilize short-tubed flowers with nectar containing only hexose sugars. Further, the nectar contains four of the ten essential amino acids required by insects (De Groot, 1953). They are arginine, lysine, threonine and histidine. It also has some non-essential amino acids. These essential and non-essential amino acids in the nectar of *A. corniculatum* serve as important nutrient source for all the insects. Near synchronous anthesis in one umbel and acropetal succession of flowers resulting in the extension of flowering period are energetically beneficial and provide forage continuously until the cessation of flowering. All the insects carry pollen and pollinate the flowers without fail. *A. corniculatum* with its unspecialized flowers is capable of utilizing the locally available insect species for fruit set; the insects are especially important for cross-pollination. The pollen grains being light and dry, and small in size facilitated by the medifixed versatile anthers are carried away by wind action and this wind-driven pollen movement also contributes to a small percent of fruit set. Therefore, both insects and wind are the pollinators of this plant. In *A. corniculatum*, the anthers and stigma of a flower mature simultaneously; anthers dehisce an hour after anthesis while stigma remains receptive for the next two days. Pandit and Choudhury (2001) mentioned that the flowers of this plant would be able to self-pollinate with the simultaneous anther dehiscence and stigma receptivity, and with the stigma position at the level of the anthers. The plant is self-compatible and capable of autogamy, but suggested that insect pollinators are required for a higher level of fruit set. They also substantiated their suggestion by stating that pollen-ovule ratio falls within the range of facultative xenogamy according to Cruden (1977). Further, these authors observed low levels of genetic variation in this plant occurring in China; this may be an indication that fruit set in this plant is largely a function of self-pollination. Solomon Raju (1989) also reported that autogamy is a mode of self-pollination for fruit set in this plant. In the present study, hand-pollination results indicate that fruit set occurs through autogamy, wind and insects, but fruit set is highest only in the flowers pollinated by insects. All of this indicates that the plant though self-compatible and capable of setting fruit through autogamy is largely dependent on insects for maximizing fruit set rate.

The pollen viability and stigma receptivity for more than three days indicate are additional adaptations to achieve cross-pollination through foraging activity of insects. In this context, Primack and Tomlinson (1980) argued that if the plant is primarily a colonizing species, it would retain the need for self-fertility if it is to establish populations in isolated localities. Therefore, *A. corniculatum* with self-compatible option would be able to establish populations even in isolated localities and utilize locally available flower visiting insects for both self- and cross-pollination. In *A. corniculatum*, flower bud abortion is negligible and its occurrence may be due to defective origin. Numerous flattened ovules are embedded in the rounded, somewhat fleshy, and short-stalked free central placenta; it may cause an underestimation of ovule number. The present study shows that the ovules in an ovary are only thirty five but only one ovule produces seed in each fruit. Fruit maturation takes a month or slightly more than a month's time. Seed is not dormant and produces a hypocotyl within the fruit pericarp. The fruits hang downwards in this plant whereas the fruits stand upwards in *Aegialitis rotundifolia*. The entire fruit of *A. corniculatum* falls off when due for dispersal. Hypocotyls float only if they are with the fruit pericarp. Self-planting and stranding strategies are effective for the dispersal and establishment of hypocotyls. Therefore, the plant displays scattered occurrence and also forms pure stands in certain mesohaline areas of the mangrove forest. Bosire et al. (2005) reported higher rate of crab predation in this plant. In the present study, this aspect was not examined but the hypocotyls being small in size may contain high nutrient content and low fiber content may attract crab predators prior to their establishment.

***Aegialitis rotundifolia.*** The genus *Aegialitis* represents only two shrub species, *A. annulata* and *A. rotundifolia*. It is recently segregated as the family Aegialitidaceae because of some distinctive features from other genera of Plumbaginaceae. Some features include anomalous secondary thickening, abundant sclereids, incipiently viviparous seeds, monomorphic pollen and homostylous flowers (Weber-El Ghobary, 1984; Tomlinson, 1986). The two species do not occur together in the same forest and have distribution in different parts of the world. *A. annulata* is distributed in Australia and eastern Malaysia (Tomlinson, 1986) while *A. rotundifolia* in South Africa and South-East Asia (Kathiresan and Bingham, 2001). *A. rotundifolia* has been reported to occur in Burma, Bengal and the Andamans by Tomlinson (1986). Later, Naskar and Mandal (1999) reported this species as occurring in the Sundarbans, Andaman and Nicobar islands and Mahanadi Delta of Orissa only. Ramasubramanian et al. (2003) have not mentioned about the occurrence of *A. rotundifolia* in their published book on the mangrove flora of Krishna and Godavari deltas of Andhra Pradesh.

The present study revealed the presence of *A. rotundifolia* in Nachugunta Reserve Forest of Krishna Mangroves in Andhra Pradesh and hence it is the first record of A. rotundifolia from Andhra Pradesh. Tomlinson (1986) stated the fact that Aegialitis species prefer or require exposed sites and withstand waves and tidal action. Further, he also mentioned the fact that *A. rotundifolia* does not occur within closed mangrove communities but it may occur as back mangal if soil is highly saline. Aksornkoae et al. (1992) also reported similarly about the habitat requirements of *A. rotundifolia*. At the study site also, this species occurs in seaward, euhaline and exposed sites. As the habitat of *A. rotundifolia* is highly saline, salt in high concentrations in plant tissues is toxic and hence, must be excluded by some mechanism. The absorbed salt is excreted metabolically via salt glands present on the leaf blade (Scholander, 1968). The salt evaporates or crystallizes in a conspicuous manner on the surface of leathery leaf blade of *A. rotundifolia*. Later, the crystallized salt blows away or washes off during cool periods by the absorption of atmospheric moisture and by rain. Therefore, *A. rotundifolia* with salt excretion mechanism is highly specialized to withstand high saline soils.

The plant is a dry season bloomer but it completes flowering prior to the onset of extreme dry conditions in the month of May. During this period, fluvial discharge from rivers to the sea is almost negligible and this would result in increased salinity of seawater. A steep increase in salinity levels can be expected at the site of the plant which is characteristically seaward in occurrence. Increased salinity of seawater reportedly prevents fruiting and causes senescence of immature flowers and buds (Qureshi, 1993). This may be an important factor for the plant to cease flowering by the mid-April and supply the available resources to the growing fruits to realize maximum fruit set. *A. rotundifolia* species with anthesis during morning hours and odourless flowers indicates that it is adapted for pollination during daytime. The floral characteristics of this plant species such as short-tubed corolla with anthers at the flower entrance, the styles and stigmas situated slightly below the anthers, production of slightly moderate volume of nectar with high sugar concentration and the presence of only hexose sugars in nectar are adaptations for bee-pollination (Baker and Baker, 1983; Opler, 1983). Further, the nectar has the essential amino acids such as arginine, lysine, phenylalanine, threonine, tryptophan, valine and histidine; and also some non-essential amino acids. De Groot (1953) showed that insects in general and honeybees in particular require ten essential amino acids and seven of them are present in the nectar of this plant. The pollen also has some protein content. The flower visitors recorded are exclusively honeybees and stingless bees; the latter is also honey producers. These bees collect both pollen and nectar from the flowers. They carry pollen on their bodies and pollinate the flowers while probing for the forage. As they require more quantity of forage for honey production and brood rearing, they collect forage from as many flowers as available on *A. rotundifolia* and hence, contribute to both self- and cross-pollination. Naskar and Mandal (1999) mentioned that this plant is pollinated by the honey bee, *Apis dorsata* in the Sundarban mangroves. Bhattacharya et al. (2006) also noted that the pollen of this plant is dominant in honey collected from the Sundarbans region. Therefore, *A. rotundifolia* is primarily melittophilous. *A. rotundifolia* species flowers are weakly protandrous, self-compatible and selfpollinating. The protandry does not contribute to autogamy in the first two hours period of flower life as the stigmas lack receptivity during that period. Gradually, the stigmas diverge and stand away from the anthers while they attain receptivity to pollen. Individual flowers with this situation produce fruit through autogamy with the aid of wind or honeybees.

The pollen grains are very large and fall down on the stigmas gravitationally due to wind action, the result of which is autogamy; if there are flowers of the same age side by side in the same or adjacent inflorescences, geitonogamy may occur. However, the flower function with reference to protandry, movement of stigmas and duration of stigma receptivity suggests that the plant is primarily evolved for outcrossing. Hand-pollination results also indicate the same and the fruit set rate is highest in open-pollinations which are largely a function of foraging activity of bees. Despite high fruit set rate in this plant, its population size is small. Pollination among the individuals of this small population may lead to a reduction in genetic diversity and molecular studies on its genetic structure would enable to know the existing level of genetic variation. Transplanting the propagules from other mangrove forests like mangrove of the Mahanadi Delta of Orissa and the Sundarbans to the study site would help to enhance gene flow in order to enable the plant to build up a stable and sustainable population size. The flowers produce singled-ovuled ovary and the ovule invariably produces a single seed in fertilized flowers. This character is advantageous for the plant to save and use the resources for higher fruit set rate. Seed is not dormant and produces hypocotyl within the fruit pericarp while still on the parent plant; it is a characteristic of crypto-viviparous species (Carey, 1934; Das and Ghose, 2003). The fruit with hypocotyl inside grows upward like the upwardly growing naked hypocotyl of *C. decandra*. Since the hypocotyl is concealed, the entire capsule-like fruit falls off when due for dispersal. Fruit pericarp is essential for the hypocotyl to float until it is settled. Self-planting and stranding strategies are functional for the dispersal of hypocotyls. But, field studies indicate that only a few of those hypocotyls which have fallen at the parental sites settled well and showed further growth. Clarke and Kerrigan (2002) reported that the small-sized hypocotyls contain high nutrient content and low fiber content. The crabs prefer to consume such hypocotyls and feed on them prior to establishment. They also reported that 80% of the propagules failed to establish due to predation by crabs in *Aegialitis annulata*. In the study site of *A. rotundifolia* also, crabs are common and they may be consuming most of the hypocotyls prior to their establishment and hence, affecting the recruitment process of the plant.

## CONCLUSIONS

Both viviparous and crypto-viviparous species exhibit mixed mating system and adaptations for entomophily. In case of *B. gymnorrhiza*, the floral features suggest ornithophily but locally the plant is found to be melittophilous. The mixed mating system coupled with entomophily appears to be adaptive for the success of sexual reproduction in harsh environment which is characteristic of mangrove forest. The seed is usually the stage of the life cycle at which dispersal and the colonization of new areas occurs. It contains a reserve of food, providing the embryo with a temporary continuation of maternal support. In contrast, in both true and crypto-viviparous tree species, the seed lacks dormancy and does not perform any of these roles. The unit of dispersal in these species is not the seed but the young seedling (hypocotyl); it is naked in true viviparous species while it remained within the fruit pericarp in crypto-viviparous species suggesting that the zygote is not dependent on stored nutritional support from the endosperm or carpal tissues, but instead may be nourished directly from the maternal plant. In all the plant species, the dispersal of propagules takes place via self-planting and stranding strategies. The self-planting strategy is important in undisturbed sites while the stranding strategy is effective in exploited and open forest sites of mangroves. Therefore, the detailed information included in this paper is useful for framing effective measures for conservation and management of the studied mangrove plants as these are the characteristic species of mangrove forest. The study further provide basis for taking up extensive studies on mangrove plants for the sustainability of mangrove forest.

## REFERENCES

Aksornkoae S., Maxwell G. S., Havanond S. and Panichsuko S., 1992 – Plants in Mangroves, Chalongrat Co., Ltd., Thailand, 120.

Allen J. A. and Duke N. C., 2006 – *Bruguiera gymnorrhiza* (large-leafed mangrove): Species Profiles for Pacific Island Agroforestry, www.traditionaltree.org

Allen J. A., Krauss K. W. and Hauff R. D., 2003 – Factors limiting the intertidal distribution of the mangrove species *Xylocarpus granatum*, Oecologia, 135, 110-121.

Aluri J. B., Venkata Ramana S. P. and Subba Reddi C., 2004 – Explosive pollen release, windpollination and mixed mating in the tropical tree *Shorea robusta* Gaertn. F. (Dipterocarpaceae), Current Science, 86, 1416-1419.

Aziz I. and Khan M. A., 2001 – Experimental assessment of salinity tolerance of Ceriops tagal seedlings and saplings from the Indus Delta, Pakistan, Aquatic Botany, 70, 259-268.

Azuma H., Toyota M., Asakawa Y., Takaso T. and Tobe H., 2002 – Floral scent chemistry of mangrove plants, Journal of Plant Research, 115, 47-53.

Baker H. G. and Baker I., 1973 – Some anthecological aspects of evolution of nectar-producing flowers, particularly amino acid production in nectar, in Taxonomy and Ecology, Heywood V. H. (ed.), Academic Press, London, 243-264.

Baker H. G. and Baker I., 1982 – Chemical constituents of nectar in relation to pollination mechanisms and phylogeny, in Biochemical aspects of evolutionary biology, Nitecki H. M. (ed.), University of Chicago Press, Chicago, 131-171.

Baker H. G. and Baker I., 1983 – A brief historical review of the chemistry of floral nectar, in The Biology of Nectaries, Bentley B. and Elias T. (eds), Columbia University Press, New York, 126-152.

Bhattacharya K., Majumdar M. R. and Bhattacharya S. G., 2006 – A Textbook of Palynology (Basic and Applied), New Central Book Agency (P) Ltd., Kolkata, 126-152.

Bhosale L. J. and Mulik N. G., 1991 – Proceedings of the International Seed Symposium, David N. S. and Mohammad S. (eds), Jodhpur, 201-205.

Bosire J. O., Kairo J. G., Kazungu J., Koedam N. and Dahdouh Geubas F., 2005 – Predation on propagules regulates regeneration in a high-density reforested mangrove plantation, Marine Ecology Progress Series, 299, 149-155.

Carey G., 1934 – Further investigations on the embryology of viviparous seeds, Proceedings of Linnean Society of New South Wales, 59, 392-410.

Chan H. T. and Husin N., 1985 – Propagule dispersal, establishment, and survival of Rhizophora mucronata, Malaysian Forester, 48, 324-329.

Chiou-Rong S., Yong J. W. H. and Yang Y. P., 2005 – The *Brugueira* (Rhizophoraceae) species in the mangroves of Singapore, Especially on the new record and the rediscovery, Taiwania, 50, 251-260.

Christensen B. and Wium-Andersen S., 1977 – Seasonal growth of mangrove trees in southern Thailand, I, The phenology of *Rhizophora apiculata* Bl., Aquatic Botany, 3, 281-286.

Clarke P. J., 1993 – Dispersal of gray mangrove (*Avicennia marina*) propagules in southeastern Australia, Aquatic Botany, 45, 195-204.

Clarke P. J. and Meyerscough P. J., 1991 – Floral biology and reproductive phenology of *Avicennia* marina in south eastern Australia, Australian Journal of Botany, 39, 283-293.

Clarke P. J. and Kerrigan R. A., 2002 – The effects of seed predators on the recruitment of mangroves, The Journal of Ecology, 90, 726-736.

Clarke P. J., Kerrigan R. A. and Westpal C. J., 2001 – Dispersal potential and early growth in 14 tropical mangroves: do early life history traits correlate with patterns of adult distribution?, Journal of Ecology, 89, 648-659.

Corbet S. A., 1978 – Nectar, insect visits, and the flowers of *Echium vulgare*, in The Pollination of Flowers by Insects, Richards A. J. (ed.), Academic Press, London, 21-30.

Coupland G. T., Paling Eric I. and McGuinness Keith A., 2006 – Floral abortion and pollination in four species of tropical mangroves from northern Australia, Aquatic Botany, 84, 151-157.

Cruden R. W., 1977 – Pollen-ovule ratios: a conservative indicator of breeding systems in flowering plants, Evolution, 31, 32-46.

Cruden R. W., Hermann H. M. and Peterson S., 1983 – Patterns of nectar production and plantpollinator coevolution, in The Biology of Nectaries, Bentley B. and Elias T. (eds), Columbia University Press, New York, 80-125.

Dadd R. H., 1973 – Insect nutrition: current developments and metabolic implications, Annual Review of Entomology, 18, 881-420.

Dafni A., Kevan P. G. and Husband B. C., 2005 – Practical Pollination Biology, Enviroquest Ltd., Canada, 590.

Dahdouh-Guebas F., Verneirt M., Tack J. F., Van Speybroeck D. and Koedam N., 1998 – Propagule predators in Kenyan mangroves and their possible effect on germination, Manual of Freshwater Research, 49, 345-350.

Das S. and Ghose M., 2003 – Seed structure and germination pattern of some Indian mangroves with taxonomic relevance, Taiwania, 48, 287-298.

Davis J. H., 1940 – The ecology and geologic role of mangroves in Florida, Papers from the Tortugas Laboratory of the Carnegie Institution of Washington, 32, 303-412.

De Groot A. P., 1953 – Protein and amino acid requirements of the honey bee (*Apis mellifera* L.), Physiologia Comparata et Oecologia, 3, 197-285.

Dress W. J., Newell S. J., Nastase A. J. and Ford J. C., 1997 – Analysis of amino acids in nectar from pitchers of *Sarracenia purpurea* (Sarraceniaceae), American Journal of Botany, 84, 1701-1706.

Duke N. C., John A. H. B., Goodall J. A. and Ballment E. R., 1998 – A genetic structure and evolution of species in the mangrove genus *Avicennia* (Avicenniaceae) in the Indo-west pacific, Evolution, 52, 1612-1626.

Elmqvist T. and Cox P. A., 1996 – The evolution of vivipary in flowering plants, Oikos, 77, 3-9.

Ewusie J. Y., 1980 – Tropical Ecology, Heinemann Educational Books Ltd., London, 243.

Faegri K. and van der Pijl L., 1979 – The Principles of Pollination Ecology, Pergamon Press, New York, 243.

Farnsworth E. J. and Ellison A. M., 1997 – Global pattern of pre-dispersal propagule predation in mangrove forests, Biotropica, 29, 316-330.

Farnsworth E. J. and Farrant J. M., 1998 – Reductions in abscisic acid are linked with viviparous reproduction in mangroves, American Journal of Botany, 85, 760-769.

Gardener M. C. and Gillman M. P., 2002 – The taste of nectar - a neglected area of pollination ecology, Oikos, 98, 552-557.

Ge J., Cai B. and Lin P., 2003 – Mating system and outcrossing rates of four Bruguiera gymnorrhiza populations of mangrove in China, Nature and Science, 1, 42-48.

Ghosh A., Gupta S., Maity S. and Das S., 2008 – Study of floral morphology of some Indian mangroves in relation to pollination, Research Journal of Botany, 3, 9-16.

Goebel K. E., 1905 – Organography of Plants, Hafner, New York.

Goldrich N. R., 1973 – Behavioural responses of *Pharmia regina* (Meigen) to labellar stimulation with amino acids, Journal of General Physiology, 61, 74-88.

Harborne J. B., 1973 – Phytochemical Methods. Chapman and Hall, London.

Juncosa A. M. and Tomlinson P. B., 1987 – Floral development in mangrove Rhizophoraceae, American Journal of Botany, 74, 1263-1279.

Kairo J. G., Dahdouh-Guebas F., Bosire J. and Koedam N., 2001 – Restoration and management of mangrove systems - a lesson for and from the East African regions, South African Journal of Botany, 67, 383-389.

Kathiresan K., 2003 – Biology of Mangroves in Biodiversity in Mangrove Ecosystems, Kathiresan K. and Subramanian A. N. (ed.), UNU-UNESCO International Training Course Manual, Annamalai University, Parangipettai, 74-90.

Kathiresan K. and Bingham B. L., 2001 – Biology of mangroves and mangrove ecosystems, Advances in Marine Biology, 40, 81-251.

Kathiresan K. and Rajendran N., 2003 – Mangroves, in UNU-UNESCO International Training Course on Biodiversity in Mangrove Ecosystems, Course Manual Kathiresan K. and Subramanian A. N., (eds), Annamalai University, 138-147.

Kondo K., Nakamurat T., Tsuruda K., Saito N. and Yaguchi Y., 1987 – Pollination in *Bruguiera gymnorrhiza* and *Rhizophora mucronata* (Rhizophoraceae) in Ishigaki Island, The Ryukyu Islands, Japan, Biotropica, 19, 377-380.

Kress W. J., 1974 – The floral biology of Rhizophora mangle in south Florida, Undergraduate Honors Thesis. Biology Department, Harvard University.

LaRue C. D. and Muzik T. J., 1954 – Does mangrove really plant its seedling?, Nature, 114, 661-662. 52. MacNae W., 1968 – A general account of the flora and fauna of mangrove swamps and forests in the Indo-West pacific regions, Advances in Marine Biology, 6, 73-270.

McGuinness K. A., 1997 – Dispersal, establishment and survival of *Ceriops tagal* propagules in the north Australian mangrove forest, Oecologia, 109, 80-87.

McKee K. L., 1995 – Mangrove species distribution and propagule predation in Belize: An exception to the dominance-predation hypothesis, Biotropica, 27, 334-345.

Meeuse B. and Morris S., 1984 – The Sex Life of Flowers, Facts on File, New York, 152.

Mitchell R. J., Karron J. D., Holmquist K. G. and Bell J. M., 2004 – The influence of *Mimulus ringens* floral display size on pollinator visitation patterns, Functional Ecology, 18, 116-124.

Mulik N. G. and Bhosale L. J., 1989 – Flowering phenology of the mangroves from the west coast of Maharashtra, Journal of Bombay Natural History Society, 86, 355-359.

Muniyandi K., 1986 – Studies on mangroves of Pitchavaram (South East Coast of India), Ph.D. Thesis, Annamalai University, Parangipettai, India.

Naskar K. and Mandal R., 1999 – Ecology and biodiversity of Indian Mangroves Part - I Global Status, Daya Publishing House, New Delhi, 361.

Opler P. A., 1983 – Nectar production in a tropical ecosystem, in Bentley B. and Elias T. (eds), The Biology of Nectaries, Columbia University Press, New York, 30-79.

Opler P. A., Frankie G. W. and Baker H. G., 1976 – Rainfall as a factor in the release, timing and synchronization of anthesis by tropical trees and shrubs, Journal of Biogeography, 3, 231- 236.

Pandit S. and Choudhury B. C., 2001 – Factors effecting pollinator visitation and reproductive success in *Sonneratia caseolaris* and *Aegiceras corniculatum* in the mangrove forest in India, Journal of Tropical Ecology, 17, 431-447.

Pannier F. and Pannier R. F., 1975 – Physiology of vivipary in Rhizophora mangle L. Proceedings of International Symposium on Biology and Management of Mangroves, 2, 632- 639.

Primack R. B. and Tomlinson P. B., 1980 – Variation in tropical forest breeding systems, Biotropica, 12, 229-231.

Primack R. B., Duke N. C. and Tomlinson P. B., 1981 – Floral morphology in relation to pollination ecology in five Queensland coastal plants, Austrobaileya, 4, 346-355.

Qureshi M. T., 1993 – Rehabilitation and management of mangrove forests of Pakistan, in Towards the Rational Use of High Salinity Tolerant Plants, 1, Leith H. and Al Masoom A., (eds), Kluwer Academic Publishers, The Netherlands, 89-95.

Rabinowitz D., 1978 – Mortality and initial propagules size in mangrove seedlings in Panama, Journal of Ecology, 66, 45-51.

Ramasubramanian R., Ravishankar T. and Sridhar D., 2003 – Mangroves of Andhra Pradesh, Identification and Conservation Manual, M. S. Swaminathan Research Foundation, Chennai.

Roubik D. W., 1995 – Pollination of Cultivated Plants in the Tropics, FAO Agricultural Services Bulletin, 118.

Scholander P. F., 1968 – How mangroves desalinate water, Physiologia Plantarum, 21, 251- 261.

Selvam V. and Karunagaran V. M., 2004 – Coastal Wetlands: Mangrove Conservation and Management. Orientation Guide 1. Ecology and Biology of Mangroves, M. S. Swaminathan Research Foundation, Chennai.

Shiraishi A. and Kuwabara M., 1970 – The effects of amino acids on the labellar hair chemosensory cells of the fly, Journal of General Physiology, 56, 768-782.

Slansky F. and Feeny P., 1977 – Stabilization of the rate of nitrogen accumulation by larvae of the cabbage butterfly on wild and cultivated food plants, Ecological Monographs, 47, 209-228.

Smith T. J., 1987 – Seed predation in relation to tree dominance and distribution in mangrove forests, Ecology, 68, 266-273.

Solomon Raju A. J., 1989 – Reproductive ecology of *Ocimum americanum* L. and *O. basilicum* L. (Lamiaceae) in India, Plant Species Biology, 4, 107-116.

Solomon Raju A. J. and Subba Reddi C., 1995 – Explosive pollen release and pollination in flowering plants, Proceedings of Indian National Science Academy, B61, 323-332.

Solomon Raju A. J. and Subba Reddi C., 1996 – The explosive floral-mechanism and pollination in the genus *Hyptis* (Lamiaceae), Proceedings of Indian National Science Academy, B62, 117-124.

Solomon Raju A. J. and Jonathan K. H., 2008 – Reproductive ecology of mangrove trees *Ceriops decandra* (Griff.) Ding Hou and *Ceriops tagal* (Perr.) C. B. Robinson (Rhizophoraceae), Acta Botanica Croatica, 67, 201-208.

Solomon Raju A. J., Jonathan K. H. and Lakshmi A. V., 2006 – Pollination biology of *Ceriops decandra* (Griff.) Ding Hou (Rhizophoraceae), an important true viviparous mangrove tree species, Current Science, 91, 1235-1238.

Sousa W. P. and Mitchell B. J., 1999 – The effect of seed predators on plant distributions: is there a general pattern in mangroves?, Oikos, 86, 55-66.

Subba Reddi C. and Solomon Raju A. J., 1997 – Reproductive biology of three mangrove plant species, Indian Journal of Forestry, 20, 153-157.

Subba Reddi C., Solomon Raju A. J. and Reddy S. N., 1995 – Pollination ecology of *Avicennia officinalis* L. (Avicenniaceae), Journal of Palynology, 31, 253-260.

Sun M., Wong K. C. and Lee J. S. Y., 1998 – Reproductive biology and population genetic structure of *Kandelia candel* (Rhizophoraceae), a viviparous mangrove species, American Journal of Botany, 85, 1631-1637.

Thawley A. R., 1969 – The components of honey and their effects on its properties: A review, Bee World, 50, 51-60.

Tomlinson P. B., 1986 – The Botany of Mangroves, Cambridge University Press, New York.

Tomlinson P. B., Primack R. B. and Bunt J. S., 1979 – Preliminary observations on floral biology in mangrove Rhizophoraceae, Biotropica, 11, 256-277.

Van Speybroeck D., 1992 – Regeneration strategy of mangrove along the Kenyan coast, Hydrobiologia, 247, 243-251.

Waller G. D., 1972 – Evaluating responses of honeybees to sugar solution using an artificial flower feeder, Annals of Entomological Society of America, 6, 857-862.

Weber-El Ghobary M. O., 1984 – The systematic relationships of *Aegialitis* (Plumbaginaceae) as revealed by pollen morphology, Plant Systematics and Evolution, 144, 53-58.

Wium Andersen S. and Christensen B., 1978 – Seasonal growth of mangrove trees in southern Thailand, II, Phenology of *Bruguiera cylindrica, Ceriops tagal, Lumnitzera littorea* and *Avicennia marina*, Aquatic Botany, 5, 383-390.

Yao Y., Bera S., Wand Y. and Li C. S., 2006 – Nectar and pollen sources for honeybee (*Apis cerana cerana* Fabr.) in Qingian Mangrove Area, Hainan Island China, Journal of Integrative Plant Biology, 48, 1266-1273.

# SOME RISK ASSESSMENTS AT NUCLEAR POWER PLANTS (NPP)

Alexander Valyaev (Nuclear Safety Institute of Russian Academy of Sciences, Moscow, Russia)
Gurgen Aleksanyan (Yerevan State University, Yerevan, Armenia)
Alexey Valyaev (University of Sydney, Australia,)
Oleg Arkhipkin (Corporative University "Samruk-Kazyna" Almaty, Kazakhstan)

**ABSTRACT:** Today NPP energy production is intensive increased in the world with the growth of different threats under its exploitation. This growth is caused by different natural and manmade factors, including possible directed terrorist attacks. It is necessary to perform correct assessments of corresponding risks levels for the individual NPP. Early we predicted the irradiation doses and corresponded risks for population under implementation of Russian Federal Program:" Development of Russian atomic energy industrial complex on 2007-2020 years at 10 homeland NPP that operated in normal non-disasters regimes during two last decades. But such data are absent for NPP, that have been or will be under non prognostic emergencies. It is connected with the following facts. The part or total of needed information may be obtained only after NPP disasters. Some NPP are located in the dangerous regions and exposed to intense negative natural responses (earthquakes, tsunami, etc.) and manmade ones, when NPP are located in some dangerous conflicts zones with high level of possible terrorism threats. Here the using of classic methods of expertise risk NPP assessments are not correct and often impossible at all. Some needed thematic data may be obtained from primary virtual computer tests of individual NPP with imitation of possible disasters. It allows to plan the actions for NPP operators and special services under serious NPP disasters or may be to prevent them at all.

## INTRODUCTION

Every NPP is presented the unique very complex object for energy production. Its normal exploitation is possible under the observation of many thematic laws and rules. NPP disasters may create the great different damages with the exclusive negative and often non reversible consequences for population and environment. Usually NPPs are located in density populated industrial regions, where there is the great need in energy consumption. Besides the NPP specific technological cycle demands a lot of water and all NPP are located near the large natural or artificial water objects (WO), that greatly increase possible risks and damages, caused by NPP disasters. The constant WO management is needed especially in urgent risk situations [1-5].

Early we used our method for prediction of irradiation doses for population under implementation of Russian Federal Program:" Development of Russian atomic energy industrial complex on 2007-2020 years" [6, 7]. Only the normal non disasters regimes of 10 Russian NPP exploitation have been under the detail consideration. The representative statistic data, obtained for 2 last decades of its exploitation, were sufficient for such consistent prediction.

Here we demonstrate and analyze some possibilities of using of our method for assessments of radiation dose for population under some possible NPP disasters. The following NPP Japan Fukusima, Russian Seversk in Tomsk region, Armenian and the future NPP in Kazakhstan are under consideration.

## RESULTS AND DISCUSSION

The total vector of limited damages under NPP exploitation has been used for the fixed time interval under the following assumptions: (1) at initial state the object is in normal (non accidents) exploitation; (2) the different kinds of accidents may be occurred as noticed i =2, 3, …, m, where m is the total number of possible accidents (m=1 is corresponded to the normal regime); (3) every accident may create the different kinds of damages: (4) realization of i accident creates the damage of j kind with $P_{ij}$ probability,

$$\vec{a}_{\lim} = P(1)\vec{a}_{1n} + \sum_{i=2}^{m} \hat{P}_{ij}\vec{a}_j \qquad (1)$$

Here j is the kind of damage with $a_j$ value. Then j = 1,2, …n, where n is the total number of possible kinds of damage; where P(1) is the probability of loss formation under normal exploitation; $\vec{a}_{1n}$ is the vector of limited damage under regular exploitation. $P_{ij}a_j$ coordinate vector value in sum is equal the damage value of j kind under realization of i kind accident. Thus the matrix of loss probabilities is determined. Under absent of accidents (normal regime of exploitation) $\vec{a}_{\lim}$ is determined only the first part of (1) formula.

The main problem is $P_{ij}$ values assessment. Usually the representative statistic data for its assessment are present only for long duration of NPP normal exploitation period. Such data are absent for NPP non prognostic emergencies, when some information may be obtained only after disasters.

Some NPP are located in regions with negative natural responses such as earthquakes, tsunami, etc.) in Armenian, Japan, California NPP and the others. The serious negative manmade factors with the global wide scale treats exist for some NPP in the dangerous conflicts zones with high probability of possible terrorist attacks, including using of explosives. Using of classic methods of expertise assessments for NPP disaster are not correct (often impossible) in these cases. Some needed data may be obtained from primary virtual computer tests of concrete NPP with imitation of possible disasters. It allows to assess risk values and also to plan the actions for NPP operators and special emergency services under serious NPP disasters or may be to prevent it at all. Effectively of liquidation of any disaster, including NPP one, is determined by qualification and professional actions of special save emergency services. Especially, when the determined disaster is in the stage of its developing. In some cases developing of the happened disaster may be stopped at all. But in any case it will help to minimize and soft of its negative consequences. Also it is often necessary to transform these consequences from its possible irreversible state in reversible ones.

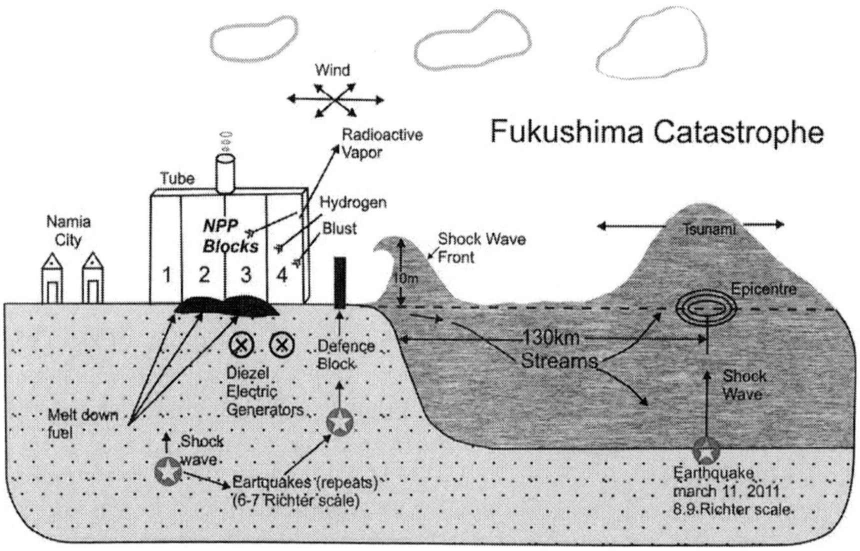


**Figure 1.** Scheme of tsunami response on FNPP.

Risks and possible ecological and economic damages from large-scale natural and manmade catastrophes in ecology-hazard regions of Central Asia and the Caucasus are presented in [8-10] with some problems of nuclear safety for U - tailing storages.

Now we pay some attention to Fukushima NPP (FNPP) catastrophe. Today FNPP presents the wide scale suitable polygon for complex detail researching of all possible NPP disasters, because it includes many events, close connected natural and manmade reasons from the stage of the catastrophe's appearance and its developing. The scheme of tsunami response on FNPP with its consequences after the earthquake on March 11, 2011 is presented in Fig.1.The main reason FNPP catastrophe was the first shock huge water wave (HWW) resulted after the earthquake, located down ocean bottom. HWW front height at the beach was about 10-15 m with its 965 km/h speed. The most dramatic scenario has taken place, when NPP external its diesel underground electric generators have been totally over dammed. The outside electric supplement from external sources, such as high voltage electric lines, was destroyed and the inside energy supplement from generators were stopped at all. The total all emergency systems (hydrogen defense, emergency cooling water injection pumps of NPP blocks, the systems of fire-protection, industrial seismic-protection, the signal system about temperature's access have been broken and switched off. Hydrogen generation inside of the reactor created its extreme explosive mixes, caused the blasts and fires, destroyed in addition all defense constructions with radioactive leakages in atmosphere and WO. The repeat shocks from next beach earthquakes with not high intensity as the first one in the Ocean didn't allow to construct additional defense from all possible pollution's penetration in environment. The direct discharges of radioactive vapor propagated in atmosphere. The defense capsule's destroyed in the first reactor's counter and melt-down nuclear fuel penetrated into the soil (heavy NPP disaster). May be some underground water channels are located here and the part of such pollution penetrated in the Pacific Ocean by this way. The noticed dramatic scenario have taken place at 3 from 8 FNPP reactors. The nearest suitable electric power source, needed for emergency equipment, is located in 6 km from FNPP. The Japan emergency specialists began to construct electric wires on the utility poles. But we think that the best solution was to put electric cable directly on the oil surface. It allowed to save substantial time.

---

According to the Russian Medicine Norma the limited irradiation human's organism doses (IHD) levels on every separated radionuclide must not more than 10 MicroSievert in year. At FNPP the irradiation level was up to 10 000 MicroSievert in hour. In the case of NPP exploitation in normal or emergency regimes IHD value is the following function:

$$IHD = F\ (RGAD + LRD) \qquad (2)$$

where RCAD is the value of radio nuclide's gas-aerosol discharges in atmosphere; LRD is the value of liquid radioactive discharges into different water objects. But our method for IHD assessment is available here. Most high irradiation risk levels are especially negative for children and embryos as it creates gene mutation at cells level under speed organism's growth. And today FNPP is constant source of shot and long -lived isotopes, that keep its activity now and will be keep during unknown period.

The main problem is the correct assessment RGAD and LRD values and its three –dimension (3D) time variations. The scientists of our Nuclear Safety Institute of Russian Academy of Sciences (IBRAE RAS) have researched some details of FNPP catastrophe in the 10 thematic articles [11] To make the correct categorization of Aqueous Media and Water Bodies by Contamination Radioactive Levels is very important [14].

Japan Government Nuclear Agency classified the F NPP catastrophe as 7-th maximum Global level. Till it was only one time for Chernobyl NPP catastrophe in 1986. The special two step test procedure has been developed in Japan for NPP test on its stability under the intense responses of earthquakes and tsunami. Only after its complete successful passing any NPP may be put into exploitation again.

It is noticed that apart from the technical mistakes in any NPP projecting and during its exploitation, there were some great political mistakes in different countries, that created negative base in future NPP accidents. After atomic industrial energy creation in USSR, at first time our Government considered that all NPPs are presented by itself the same energy objects as hydro or coal thermal stations for energy production. And it was not necessary to pay the special attention and relation to its exploitation with additional funding and to create new separate thematic Minister Department. The result was the global Chernobyl catastrophe. But under the Chernobyl only the single huge radioactive blast has been happened, that was not located near the Ocean. Today at the FNPP catastrophe we have deal with the constant intense source of environmental wide scale radioactive pollution. It is unknown the temporal final of its activity and it is impossible to assess the total risks and the connected damages. In India when the information about the serious future earthquakes have gotten all NPP reactors have been stopped. Although the similar stopping causes the RGAD and LRD increasing, but it will be not called as any disaster because there is constant monitoring over this technological process and its management.

It is important to notice that the luck of culture in nuclear safety exists at population of most countries. Although Japan population has the high level of culture in respect of different kinds of disasters (for example, to earthquakes), but its population has the great radiophobia for nuclear disasters (result of Hiroshima and Nagasaki atomic bombing in 1945). The information about the future huge earthquake and tsunami in March11, 2011 has been ignored by the official state organizations and the private energy companies, such as Tepko, that was responsible for normal FNPP exploitation Today Japan System for Prediction of Environment Emergency Dose Information(SPEEDI) has been modified to WSPEEDI.

Some China specialists consider, that the earthquake on March 11, 2011 has been stimulated by the Japan testing of underground atomic bomb's blast at the Pacific Ocean bottom. We have not information about the methods, how it is possible to differ atomic blast from earthquake at Ocean large depth.

China Mass Media informed, that the huge crater has been created at this place and in result the china ship with 100 men at its board went down. Also they consider that under surface FNPP the secrete plant for production of weapon U-235 and plutonium is located. Now agrees with the following fact that FNPP is kept as the constant intensive irradiation source.

Some problems, connected with another Armenian (Metsamor) NPP (ANPP) it is closely located near the main epicenters of last earthquakes with 7-9 balls on Richter scale. For the long-term period of APP exploitation from 1976, after Spitak earthquake in 1989 (6.8 Richter scale) APP has been conserved. In in the whole country has been appeared. Then in 1995 after the total energy absence ANPP has been put into exploitation again. It contains two VVER-400 and V230 (376 MW) nuclear reactors, that produce about 40% of the total country's electricity After the recent ANPP its active life has been prolonged on 12 years till 2026 year. But the Armenian ecological situation is constantly worse with the huge amounts of highly active, medium-active and low-active wastes has been accumulated and kept in the corresponding storages without the needed keepconditions, located near Araks river, that flows into Kura river and then in Caspian Sea. factor, that at its seismic dangerous territory are located the huge storages of the radioactive wastes That is why some US specialists call that some Armenian regions are te "black swamp" or "black quicksand."

We should like to do some remarks to the future building of one from the three NPP in Kazakhstan Republic (KR), named as Kurchatov (NPP KNPP) is planed to construct at Eastern Kazakhstan (EK) region near Kurchatov sity, located at Nuclear Semipalatinsk Test Polygon (STP) territory (Fig.2). Today KR occupies the first place in the world on U – extractions. The thematic EK thematic consideration as the the territory of global ecological risk is presented in [8, 9]. Early EK region was not seismic dangerous one. But after about 500 nuclear blasts, produced in 1949-1990 years, it became the global risk zone with the appearance of natural earthquakes in addition the negative great manmade radioactive pollution as the result of STP intense exploitation. Now STP has been closed for nuclear weapon tests.

The large hydro electric stations (HES) have been built on EK Irtysh river, the largest river of the country, with its sizeable artificial reservoirs appeared after the HES building (Fig.2). The huge water masses, accumulated in the new sizeable artificial reservoirs began to press on the surface of the ground, where many Altai high mountains massifs are located. This pressure were not desirable for our Earth at all. The Earth expresses its protest to the people for their manmade constructions in the form of stimulated new intense earthquakes. In result early non seismic region has been transformed in seismic dangerous one.

At EK territory the numerous multiple mines for extracting metals and minerals, as well as a number of industrial enterprises and their tailing dumps, including uranium, are located in the cities and their suburbs along the Irtysh River. It originates in China, further flows via the lands of Kazakhstan and Russia, including such large cities as East-Kazakhstan capital Ust-Kamenogorsk (UK), Semipalatinsk, Pavlodar, Omsk, and Tobolsk, and after its confluence with the Ob River it flows into the Arctic Ocean (Fig.2) The largest Bukhtarma HES with the 100 m height of dam has the most sizeable reservoir, the basin length is above 300 km and the depth - up to 100 m. The other UK HES with the 42 m height of dam with its huge reservoir is located in 15 km from the city.

The next Shulba HES (1) is built in 180 km from UK. Errors in the design of these HESs have caused a number of different accidents. UK (330,000 inhabitants) is one of the most contaminated towns in the world/ It locates in the narrow Irtysh mountain valley, that oversaturated by different enterprises. Ulba Metallurgical Plant produces enriched U for NPP, Be, Ta and their products. Operating UMP wastes storage, located in city center, has accumulated ~100 thousand tons of wastes, contained U, Th and their decay products. Its size is 400* 220 $m^2$ with the depth of contamination > 5 m. The level of gamma–radiation at its surface reaches 360 μR/h and increases with depth up to 1000 -6000 μR/h.

Many other operating UK large plants, such as Lead –Zinc, Titan-Magnezium, Ceramic, worked on Be base, plants, power capacitors plant, nonmetalliferous group of enterprises and Silk Cloth enterprise, use in their technologies the different poisonous and toxins, while their wastes are also located in city boundaries. In Irtysh river basin, where > 40% of HES energy in Kazakhstan is produced, the large active non-ferrous pits, precious and rare-earths metals pits with their dump nations are also located.

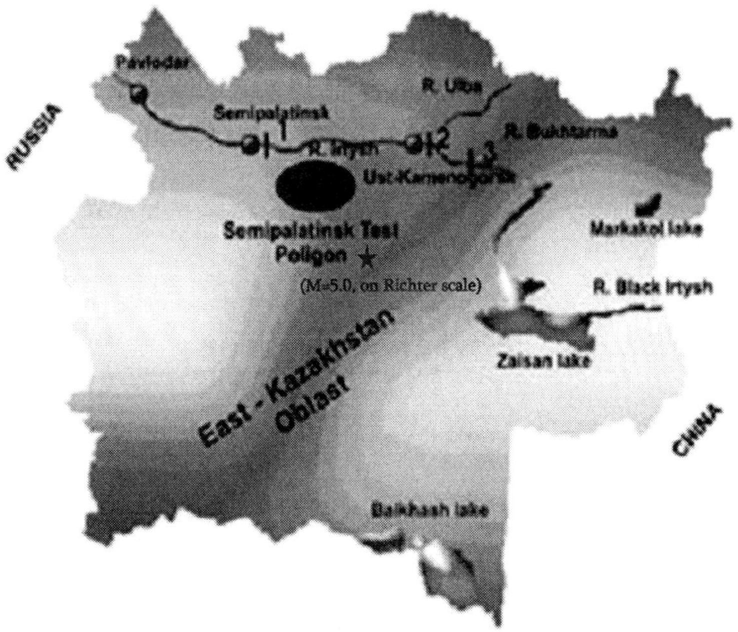


**Figure 2.** The scheme of Irtysh river flow on the territory of Kazakhstan Republic.1 – Shulba HES, 2 - Ust-Kamenogorsk HES, 3 - Bukhtarma HES. Future Kurchatov NPP will be located at Semipalatinsk Test Plolygon. ★ - Epicenter of the last earthquake on March 16, 2016 (M= 5.0 on Richter Scale)

Risks of the possible ecologic catastrophes are increased, because 2 HES are placed at upriver. We consider that the huge water masses in the manmade reservoirs press strongly on the bottom of mountain surfaces, disturb and deform their initial natural states. These factors resulted to the increasing of the frequency and intensity of the strong earthquakes, included catastrophic ones, that already happened not only near the city ( in 1990), but also in Altai mountains in Russia ( in 2003, 2005 years). The last earthquake (M= 5.0 on Richter Scale) was on March 16, 2016 near STP. Such earthquakes may cause the damages of HES dams, where in addition part of them are in non-satisfactory states, especially UK HES dam, operated >50 years. Any HES with its huge water reservoir as the any NPP are very attractive for the possible controlled terrorist acts, including with using of explosive. According to some primary assessments, in result the huge break-through damming wave with its front height ~30 m will destroy the city and its environs. All enterprises, their products and hazard impurities in the storages then will be carried out down Irtysh to many cities and after Irtysh - Ob rivers junction spread over the large territories, including Arctic Ocean through Kara Sea. It results to abrupt worsening of water quality in all cities: UK, Semipalatinsk, Pavlodar, Omsk, Tobolsk and in many inhabited localities. Irtysh-Karaganda manmade channel supplies water to Kazakhstan central regions, such as Karaganda, Kazakhstan capital Astana city and its regions.

The available downstream STP, where are buried more than 18,000 tons of radioactive waste of the total activity of 1.300.000 Ci, significantly aggravates the situation. Radionuclides of the total activity over 10.000.000 Ci were accumulated during nuclear tests in the underground wells, located within the 60 km far from Irtysh. U-235 and Eu-152 are registered as available on the territory of above 300 sq. km.

EK HES are very attractive for realization of directed terrorist acts with using of explosives. It is sufficient to destroy the Bukhtarma HES gate of upper floodgate from 3 ones, that used for passing of ships. And then the huge water mass will realize the global catastrophe with a lot of victims and great pollution including the Arctic and World Oceans. The Earth population prefer to live in regions with comfortable conditions for living in spite of high risks level. For example US California state region with intensive destroyed earthquakes and 8 US NPP constantly operate.

We shortly notice the thematic situation with the building of new NPP in Siberia near Seversk sity in Tomsk region. Here the unique Siberian NPP (SNPP) may be called as the second NPP in USSR at present the enterprise is diversifying its economic activities, providing foreign companies broad opportunities for collaboration. SNPP in its initial form was established in 1949 as a unified complex with nuclear cycle, whose main task was production of nuclear weapon components from fissile materials – namely, highly enriched uranium and plutonium. First batch of enriched uranium-235 was produced in 1953. Since the 1960's SNPP has been gradually converting to civil production and serving as an example of how such plant can successfully transition from former defense enterprises to exclusively peaceful manufacturing. Nearly all technologies of Russian nuclear industry are applied at SNPP. Such a diversification is one of the reasons that the company was able to successfully overcome periods of economic instability in Russia and develop new products and services for the international market in the early 1990's. The main little local plants, that entered in SNPP, use the unique closed nuclear cycle, from nuclear energy utilization for power generation to reprocessing and storage of nuclear wastes. After the signing of the 1991 US-Russian Agreement on Utilization of Military Uranium, the Combine became actively engaged in realization of the HEU-LEU Program ("Megatons to Megawatts", conversion of highly enriched uranium to lowly enriched form to be then used as fuel in civil nuclear industry). Apart from nuclear industry related activities, SNPP conducts continuous product diversification and introduction of new up-to-date productions. In 2005 SNPP jointly with several other Tomsk organizations established the Scientific and Industrial Center for Research and Utilization of Nuclear Energy. US companies actively take part in investment and technical reconstruction processes conducted on SNPP. Under the intergovernmental agreement between the US and Russia dated 1997, oriented to decommissioning of plutonium producing reactors, the US Department of Energy initiated the wide-scale project of SNPP heating and power station reconstruction as an alternative to heat and power produced by the reactors of SNPP Reactor Plant, which should be decommissioned by 2008. Utilization of mixed oxides of uranium and plutonium (MOX fuel) in specially design reactors is considered one of the promising directions for further development of nuclear industry, let alone the challenge of peaceful utilization of the stock of military plutonium.

Utilization of mixed oxides of uranium and plutonium (MOX fuel) in specially design reactors is considered one of the promising directions for further development of nuclear industry. Today it is under the building on the base of Brest - 300 reactor. Its main design features are the following: It is an evolutionary design with high level of reference; Consideration of safety requirements, including EUR, IAEA (INSAG inclusive); Broad application of passive systems; Using of advanced active safety systems; Low sensitivity to human errors; Complies with modern customer requirements: 60-year service life of the reactor plant equipment;

## CONCLUSION

Today it is clear that risk assessments at any NPP and another objects of atomic energy production are presented the great complex problem that has a wide spectrum of objective and subjective reasons. Only under exploitations these objects in regular non disaster regimes it is possible to assess some possible risks and damages. This situation is constantly worsened. It is connected with the intensive growth of natural and manmade catastrophes and increasing of its power, damages and consequences. Also international terrorism attacks are thickened and have more hard consequences with mass victims. We have to develop and apply high technologies for the successful struggle against terrorism [13], including the modern space techniques [14, 15]. Till today the possible nuclear terrorist acts with great global wide scale non reversible damages don't show itself. But world civilization have to be face with new serious treats, that to save and keep itself.

One of the authors of this article Alexander Valyaev was the single participant from Russia at the NATO Advance Training Course Integrated Emergency Management for Mass Casualty Emergencies", that has taken place in Florence, Italia at October 25-30, 2011. The participants from Israel and UK indicated that their countries consider the case, happened in London on 23 November 2006, the poisoning of the UK citizen A. Litvinenko by Russian A. Logovoi with using of Po -210 as the first act of radiological (the branch of nuclear one) terrorism/ But we should like to notice, that this information has not any satisfactory arguments.

The recent clearly expressed fact has confirmed that the Islam State tries to use it its criminal purposes the extreme dangerous nuclear terrorism The security officer at the Belgian NPP has been found murdered with his work pass stolen in the result of recent terror attacks in Brussels.

## REFERENCES


1. A.N.Valyaev, G.M. Aleksanyan, et. al, Managing risks to water resources in mountain regions from natural and man-made disasters" 2011, in Proc. of NATO Advanced Research Workshop (ARW):"Stimulus for Human and Societal Dynamics in the Prevention of Catastrophes: NATO Science for Pearce and Security Series. E: Human and Societal Dynamics"–vol. 80, pp. 172-188,IOS Press –Amsterdam – Berlin – Tokyo –Washington, D.C.
2. A.N.Valyaev, G.M.Aleksanyan., et.al,"Risks to aquatic ecosystems in mountains regions and its possible managment" Ibidem, pp. 191-206.
3. R.S. Minasyan, G.M. Aleksanyan., A.N. Valyaev, et. al, Large Artificial Water Reservoirs and Dams as Critical Risk Objects in Armenia, 2011, in Proc. of the NATO Advanced Research Workshop: "Stimulus for Human and Societal Dynamics in the Prevention of Catastrophes: NATO Science for Pearce and Security Series. E: Human and Societal Dynamics" –vol. 80, pp.131-138, IOS Press – Amsterdam – Berlin – Tokyo –Washington, D.C.
4. .G.M. Aleksanyan, A.N .Valyaev, K I. Pyuskyulyan, Several approaches to the solution of water contamination problems in transboundary rivers, crossing the territory of Armenia, 2006, in NATO Science Series: Proc. of NATO ARW: "Nuclear Risk in Central Asia", Kazakhstan, Almaty, June 20-22, Publ. House: Springer Science +Business Media B.V. 2008, Netherlands, pp. 201-211.
5. A.N. Valyaev, G.M. Aleksanyan, et.al, Prediction, Prevention and Management of Natural and Manmade Catastrophes at Mountain Risk Water Objects, 2013, in Materials of Intern. Conf. "MOUNTAINHAZARDS 2013"Natural Hazards, Climate Change and Water in Mountain Areas" Bishkek, Kyrgyzstan, 16-18 September, pp.112-118.

6. A.N. Valyaev, A. L. Krylov, V.N. Semenov, G.M. Aleksanyan, A.A. Valyaev, Irradiation doses at nuclear power plants at normal and emergency situations", 2012 in Proc. of NATO ARW: "Correlation Between Human Factors and the Prevention of Disasters" pp. 40-57..IOS Press –Amsterdam – Berlin – Tokyo –Washington, D.C.
7. A.N. Valyaev, A. L Krylov., V.N.Semenov, D.V Nikolisky 2011 "Prediction of irradiation doses for population under implementation of Russian Federal Program:" Development of Russian atomic energy industrial complex on 2007-2020 years". in Proc. of NATO ARW: "Stimulus for Human and Societal Dynamics in the Prevention of Catastrophes: NATO Science for Pearce and Security Series. E: Human and Societal Dynamics" –vol. 80, pp.172-188, IOS Press –Amsterdam – Berlin – Tokyo –Washington, D.C.
8. A.N. Valyaev, S.V. Kazakov, A.A. H. D. Passell et. al. 2007 "Assessments of Risks and Possible Ecological and Economic Damages from Large-Scale Natural and Man-Induced Catastrophes in Ecology-Hazard Regions of Central Asia and the Caucasus." pp. 281-299in NATO Science for Peace and Security Series -C: Environmental Security, Proc. of NATO ARW: "Prevention, Detection and Response to Nuclear and Radiological Threat", Yerevan, Armenia.
9. Valyaev A.N., Passell.H. D., Solodukhin V.P., Alexanyan G.M., et.al.,2011, "Geo- chemical and Radiological Risks in dangerous regions of Central Asia and Caucasus:, in Proceeding of NATO ARW: "Stimulus for Human and Societal Dynamics in the Prevention of Catastrophes: NATO Science for Pearce and Security Series. E: Human and Societal Dynamics" –vol. 80, pp.194-203. IOS Press – Amsterdam – Berlin – Tokyo –Washington, D.C., Edited by A. Avagyan, D. L. Barry, W. G. Goldewey, D.W.G. Reimer.
10. A.N. Valyaev, G.M Aleksanyan, et.al. 2014 "Integrated Emergency Management and Risks for Mass Casualty Emergencies". Proc. of the 7th Chaotic Modeling and Simulation Intern. Conf. Lisbon, Portugal, 7 - 10 June. pp. 507-522.
11. "Proc. of Nuclear Safety Institute of Russian Academy of Sciences, issue 13: Accident at "Fukushima -1 NPP: Response Experience and Lessons. Moscow: Nauka 2013, 246 p. ISBN 978-5-02-038468 (bound) (in Russian),
12. Kazakov S.V., Utkin S.S., Linge I.I., Valyaev A.N. 2006 "Categorization of Aqueous Media and Water Bodies by Contamination Radioactive Levels." in Proc. of Intern. Conf. "Radioactivity after Nuclear Explosions and Accidents", v.3, pp. 402-407, (in Russian) December 5-6, Moscow, Publ. House: St. Peterburg, GIDROMETIZDAT,.
13. Valyaev A.N. and Yanushkevich V.A. 2003, "Using of Acoustic Techniques for Detection of Explosives in Gas, Liquid and Solid Mediums" NATO Science Series II "Detection of Bulk Explosives: Advanced Techniques against Terrorism" (Mathematics, Physics and Chemistry), vol. 138, pp. 175 – 183, Kluwer Academic Publishers, Netherlands. Proc. of NATO ARW, June 16-21, St. Petersburg, Russia.
14. Valyaev A.N., Kazakov S.V., Aitmatov I.T., Aitmatova D.T. " Problems of Ecologic Safety under Displacement of Rock Stones, controlled from Space, at Ecological Dangerous Regions of Tien-Shyan Mountains" in Proc. the First Intern. Conf. "Earth from Space- the Most effective Solutions" (in Russian), 2003. http://www.transparentworld.ru/conference/presentations/operative.htm tyan_shyan_prsnt.zip.
15. *Valyaev A.N., Kazakov S.V. Stepanets O.V. et. al.* "Using of Space Technologies for Monitoring of Large River's Basins and Prediction of Large-Scale Natural and Man- Induced Catastrophes in Ecology-Hazard Regions of Central Asia and Caucasus" in Proc. the Second Intern. Conf. "Earth from Space- the Most Effective Solutions" Section "Space Monitoring in Problems of Management of Territories"; Moscow, 2005 http://www.transparentworld.ru/conference

# THE ELBE ESTUARY: BALANCE BETWEEN A HIGH FREQUENTED WATERWAY AND THE PROVISION OF ESS

***Carolin Schmidt-Wygasch****, Uwe Schröder and Elmar Fuchs
(Federal Institute of Hydrology, Koblenz, Germany)
Rainer Marggraf and Uta Sauer
(University of Goettingen, Goettingen, Germany)
Jan Barkmann
(Darmstadt University of Applied Sciences, Darmstadt, Germany)

The tidal River Elbe, Northern Germany, serves as an important international waterway feeding the port of Hamburg and thus indicates its function as a significant economic lifeline for that region. At the same time the estuarine riverscape holds valuable nature protected by national and European legislation. In the last centuries the estuary has been impacted by manmade alterations, e.g. by ongoing river training measures.

Several environmental directives like the Water Framework Directive (WFD European Commission 2000) and the national Federal Water Act (2009) have assigned a legal duty to the Water and Shipping Administration (WSV) not to focus only the safety and ease of navigation but also to be responsible for the enhancement of near-natural structures in and nearby the river. The new orientation towards more ecological maintenance and improvement, e.g. pristine river banks and an improved structural diversity, will lead to changes in existing ecosystem services (ESS).

Currently, the database for ESS of waterways is insufficient regarding their benefit, tradeoffs due to maintenance as well as the implementation in cost-benefit-analyses (CBA). To improve this situation our survey focuses on the development and establishment of the ecosystem service approach in water management. Key issues are the assumptions that the reconstruction of artificial bank protections leads to several benefits in terms of the ecosystem services concept and may have advantages even in monetary terms. Improving nature conservation and nature experience for the local residents of the Elbe region are two of them.

Here, we want to present and discuss our steps to assess and determine the public opinion and preferences to certain (cultural) ESS related to river banks. In a questionnaire we ask for attitudes and valuation of varying ESS provided from different bank forms (riprap vs. near natural). As long as the flood protection is ensured, three-quarters of the residents prefer near-natural river banks. An additional choice experiment (CE) determines the willingness to pay (WTP) for three main attributes, which were worked out through previous questionings. For changes within the realms of aesthetic concerns, protection of endangered species as well as natural bank development, we revealed an impressive WTP.

The results made an exemplary cost-benefit analysis (CBA) possible on a small potential bank restoration project promoting specific ESS. We could demonstrate that the primary investment costs facing a high societal benefit. For the first time important (cultural) ESS could be monetized on river banks of the tidal River Elbe. Moreover, the present results provide evidence that the evaluation of the social value of water management, or rather a river bank restoration on the basis of socio-economic methods is possible.

The research results shall find their way into the operational water management and maintenance of federal waterways. Mostly socio-economic factors are easier to communicate to decision makers than often abstract environmental issues and in consequence they could be more relevant for decision taking. The consequent implementation of the ecosystem services approach will help us to achieve a sustainable development of waterways, and on this way a higher transparency, and acceptance.

# NOVEL TAXA ARE DOMINANT IN MANGROVE SWAMP OF NIGER DELTA, NIGERIA

***Chika Christiana Nwankwo*** and Gideon Chijioke Okpokwasili
(University of Port Harcourt, Port Harcourt, Rivers State, Nigeria)

Taxonomic study has greatly progressed by the use of a high throughput technology. Metagenomics provides insights into microbial diversity which is not possible using traditional technique.Three georeferenced crude oil spilled soils and three pristine soils with no history of crude oil pollution in the mangrove swamp of Bodo West in Gokana and Bille in Degema locations of Niger Delta, Nigeria were studied. Roche 454 Metagenomic sequencer was used to produce all data sets. Bioinformatics, mathematical and statistical models were used to analyze data sets. DNA sequencing was performed by Next Generation Sequencing Technique to determine the nucleotide sequences of all microorganisms present in the soil sample using Genome sequencer 454 sequences or illumina were compared to sequences obtained from genebank by BLASTX Analysis using CLO bio software. Microbial compositions were estimated. A total of 20 phyla and 36 classes were studied in both polluted and unpolluted sites of study. It was observed that some nucleotide sequences showed no significant similarity and were classified as the potential No hits most of which belonged to the Unknown Phylum, it was dominant phylum in the polluted soils of study with abundance of 49.86% and second dominant in the unpolluted soil with abundance of 27.5% . In these study sites, the Unknown phylum had higher relative abundance when compared to *Proteobacteria* phylum. Frequency occurrence of this phylum at study sites was 91.7%. They were more abundant in the subsection of study sites for both polluted and unpolluted soils. Three non referenced bacteria isolated from mineral salt medium, showed these microbes were able to grow in the laboratory and can utilize hydrocarbons as energy source. Gram reaction revealed they are Gram positive rods. DNA extraction, PCR amplification and sequencing revealed isolates of base pair lengths, 784, 732 and 627 may be potential novel as no significant similarity was found when blast at NCBI website. Specific primers for amplification of alkM, Ppalk B and ndoB genes for hydrocarbon degradation were used for functional gene analysis on the isolates. Result showed 100% amplification of hydrocarbon degrading genes. The means value of the potential novel isolates showed that they were higher in the polluted soil when compared to the unpolluted soil. There was no significant difference between polluted and unpolluted soils as p was 0.224. Agglomerative Hierarchical Clustering method showed 70% similarity between *Gamma proteobacteria* and *No hit* class in the polluted soil , while 55% similarity was observed in the unpolluted soil. Seventy eight percent (78%) similarity was observed between the No hits and *Acidobacteria* in the unpolluted soil, other interactions exist between No hits and other classes of microorganisms. Pearsons correlation, Principal Component Analysis and Canonical Correspondence Analysis reveal the ecological interactions at polluted and unpolluted sites, variations among sites as well as influence of environmental variables on the potential novel isolates. Phylogenic trees constructed using neighborhood joining method showed evolution of novel microorganisms. Nucleotide sequences of some potential novel isolates have been submitted to the gene bank with Accession nos KT899789 to KT899790. This molecular study has revealed potential new isolates that can bioremediate crude oil polluted soil.

# PHYLOGENIC DIVERSITY OF MICROORGANISMS IN THE MANGROVE SWAMP UNDER CRUDE OIL PERTURBATION

***Nwankwo Chika Christiana*** and Okpokwasili Gideon Chijioke
(University of Port Harcourt, Port Harcourt, Rivers State, Nigeria)

Microorganisms in the environment have molecular evolution. This study is aimed at using metagenomics tool to assess DNA sequences of microbial compositions in phyla. Phylogenic study or evolution modeling has given novel insights into microbial composition and potential organisms which are important in bioremediation of crude oil polluted swamp. Soil samples were taken from Cawthornne Channel 11, Bodo West and Bille communities in Niger Delta region of Nigeria. Phylogenic analysis and estimates were analyzed using sequencer software package (version 4.J ) and compared to sequences with NCBI database using Basic Local Alignment Search Tool (BLAST). All phylogenic affiliations with any bacteria were imported and aligned. Construction of trees were done by neighborhood joining methods using Nucleotide substitution models; trees were rooted with unrooted sequences and 100 replicates were used for bootstrap analysis. Microbial diversity was characterized by 16S rRNA sequences into Operational Taxonomic Units. Specie richness and diversity index using Simpsons diversity index showed no significant difference in diversity between the polluted and unpolluted soil as p is 0.72. while specise richness was higher in polluted soil but showed no significant difference between polluted and unpolluted soils p is 0.20. Fifteen (15) phyla were studied in the unpolluted soil while 13 phyla were studied in the polluted soil, At relative abundance greater than 2%, seven phyla were studied in the polluted soils they are Unknown phylum 49.86%, Proteobacteria 21.5%, Actinobacteria 9.87%, Firmicutes 9.5%, Chloroflexi 3.56%, Ascomycota 2.0% and Bacteroidetes/Chlorobi 1.8%, while a total of six phyla were studied in the unpolluted soils:

Firmicutes>Unknown>Actinobacter>Proteobacteria>Acidobacter>Verucomicrobia. Similarly, Bodo West with 3week crude spill showed the presence of five phyla Actinobacteria >Unknown>Proteobacteria> Chloroflexi > Bacteroidetes. Bille site with 14 week spill showed the presence of seven Phyla in the order of abundance: Unknown> Proteobacteria>Firmicutes> Ascomycota> Bacteroidetes> Acidobacter. Cawthorne Channel with 50week old spill had four phyla in the order of abundance: Proeobacteria> Firmicutes>Unknown>Actinobacter. Sphingobacteria, Chlorobi and Chloroflexi showed significant difference between polluted and unpolluted soils. Variations in phylogenic diversity was observed among polluted study sites as well as between polluted and unpolluted soils. Cluster analysis using Unweighted Pair Group with Average linkages ( UPGMA) and Agglomerative Hierarchical clustering were used to construct dendograms and establish similarity between variables and observe changes in sites. Study has shown that crude oil affects phylogenic diversity and the knowledge of phylum diversity in a polluted soil is important in bioremediation processes.

# STRESS FACTORS TO FISH HABITAT IN URBAN RIVERS

***Yuta Yamauchi,*** Tetsuya Nakata, and Yutaka Sakakibara
(Waseda University, Shinjuku, Tokyo, Japan)

To assess crucial stress factors to life cycle of fish, Life Cycle Risk Assessment (LCRA) was applied to urban rivers in Tokyo. In LCRA, life cycle was divided into 3 life-stages; fry fish, adult fish, and spawning. In addition, the following eight stress factors were considered; depletion of dissolved oxygen (DO), water temperature, disappearance of habitats, shortage of depth in movement, disappearance of evacuation areas, disappearance of spawning areas, existence of obstacles, and shortage of food. In our former study, fish species in rivers in suburban and rural areas could be reasonably well predicted by the LCRA.

Comparisons of LCRA and observed results in urban rivers suggested there are additional stress factors specific in the rivers, because most discrepancies indicated that LCRA predicted the survival while no corresponding fish were observed. Therefore, main focus was put on the additional stress factors including predation by large aquatic species such as carps and red-eared slider turtles, the existence of estrogen (17alpha-ethinylestradiol; EE2), and washout by flooding. And their effects to fish habitats were studied theoretically and experimentally.

As a result, it was demonstrated that predominant feeds found in carps were waterweeds and bloodworms while those of red-eared slider turtles were crayfishes and waterweeds, indicating the predation by large aquatic species is not a crucial stress factor. Concentrations of EE2 ranged from 0.06 to 1.5 (ng/L), suggesting EE2 is also not a crucial stress factor, due to much lower concentration levels in comparison with threshold value in the literature. On the other hand, it was found that flow rates in flooding were reduced below fish swimming speeds (0.05~1.5m/s) by obstacles in rivers. Moreover, spawning periods of most fish whose survivals were not predicted by the LCRA are in May and June. This period is rainy season and large fluctuations in river water are observed. Therefore, it is considered that the additional stress factor would be attributable to the fluctuations.

## The Impact of Alternative Measures of Riverbank Fixation on Climate Change and Water Quality

***Lars Symmank**** (German Federal Institute of Hydrology, Koblenz, Germany)
Katharina Raupach (University of Goettingen, Goettingen, Germany)
Mathias Scholz and Christiane Schulz-Zunkel (Helmholtz Centre for Environmental Research-UFZ, Leipzig, Germany)

The majority of riparian zones and floodplains in Europe and North America are heavily modified due to anthropogenic impacts during the last centuries. The need for renaturation is high but rarely compatible with economic interests such as shipping and agriculture. Alternative riverbank fixations, like the replacement of riprap by reedbeds or soft wood can serve as one compromise between ecology and economy on waterways in areas with high human activities. Originally made for enhancement of biodiversity these measures involve a high potential for additional ecosystem services. We here present a first approach to estimate selected ecosystem functions in a case study of alternative riverbank fixations at the river Weser in north-west Germany. In the most cases site specific parameters needed for direct recording of ecosystem functions can only be obtained intricately. Therefore we estimate carbon fixation by biomass equations and nutrient retention by proxies derived from field investigations and published measurements. Our results show distinct additional benefits of riverbank renaturations far beyond the mere increase of biodiversity. The study illustrates the impact of even small-scale measures on climate change and water quality through measurable reductions of nutrients and carbon fixation.

# STATE TRANSITIONS AND FEEDBACK MECHANISMS DURING ECOSYSTEM DEVELOPMENT IN THE CONSTRUCTED CATCHMENT ´CHICKEN CREEK´

***Wolfgang Schaaf***, Christoph Hinz, Werner Gerwin and Markus K. Zaplata
(Brandenburg University of Technology Cottbus-Senftenberg, Cottbus, Germany)
Reinhard F. Huettl
(GFZ German Research Centre for Geosciences, Potsdam, Germany)

Landscapes and ecosystems are complex systems with many feedback mechanisms acting between the various abiotic and biotic components. The knowledge about these interacting processes is mainly derived from mature ecosystems. The initial development of ecosystem complexity may involve state transitions following catastrophic shifts, disturbances or transgression of thresholds.

To study the early development of an ecosystem and the role of feedback mechanisms at the landscape scale, we used a 10-yr time series of hydrological, meteorological, biological, geomorphological and soil data from the catchment ´Chicken Creek´ in the mining area of Lusatia, Germany (Gerwin et al. 2009). The world´s largest constructed catchment has a hillslope-shaped 6 ha size with defined boundary conditions and well-documented inner structures. The dominating substrate above an underlying 1-2 m clay layer is Pleistocene sandy material representing mainly the lower C horizon of the former landscape. The site was left to unrestricted development since 2005 and was intensively monitored (Elmer et al. 2013, Schaaf et al. 2013).

The data showed a very rapid development of the site with an increasing complexity and heterogeneity. In the first years, stochastic signals like the initial substrate conditions and external drivers like extreme weather events were the most important factors resulting in abiotic/abiotic feedback mechanisms shaping the morphology of the site and creating site diversity. Initial abiotic feedback mechanisms between water and substrate were soon followed by abiotic/biotic feedbacks between biological soil crusts, invading vegetation, geomorphology and hydrology resulting in state transitions in catchment functioning.

The scientific value of sites like Chicken Creek with known boundary conditions and structure information could help to disentangle feedback mechanisms between hydrologic, pedogenic, biological and geomorphological processes as well as a more integrative view of succession and its drivers during the transition from initial, less complex systems to more mature ecosystems. In addition, restoration in general as well as ecological engineering could benefit from this enhanced understanding.

*Environmental Science & Technology 2016 Vol. 2*

# ECOSYSTEM ASSESSMENT AND RESTORATION

Elvis Obeng Boateng
(Czech University of Life Sciences, Environmental Faculty, Sudhol, Prague, Czech Republic)

**INTRODUCTION.** Ecological restoration is the process of assisting the recovery of an ecosystem that has been degraded, damaged or destroyed. Many of the world's ecosystems have undergone significant degradation with negative impacts on biological diversity and peoples' livelihoods. Ecological restoration can be a primary component of conservation and sustainable development programmes throughout the world.

**OBJECTIVES.** Given that many people now depend on what have become degraded ecosystems to sustain their livelihoods, ecological restoration needs to address four elements. These elements are critical to successful ecosystem management. Ecological restoration should:

- Improve biodiversity conservation
- Improve human livelihoods
- Empower local people
- Improve ecosystem productivity

**METHODS.** Ecological restoration is a well-established practice in biodiversity conservation and ecosystem management. The principles, and the Attributes of Restoration Progress below, are consistent with both the scope and intent of the Convention on Biological Diversity's Principles for the Ecosystem Approach.

**HUMAN SYSTEMS**

- Ensuring all stakeholders are fully aware of the full range of possible alternatives, opportunities, costs and benefits offered by restoration.
- Empowering all stakeholders, especially disenfranchised resource users.
- Engaging all relevant sectors of society and disciplines, including the displaced and powerless, in planning, implementation and monitoring..
- Considering all forms of historical and current information, including scientific and indigenous and local knowledge, innovations and practices.
- Providing short-term benefits leading to the acceptance of longer-term objectives.
- Providing for the accrual of ecosystem goods and services

**ECOSYSTEM**

- Incorporating biological and environmental spatial variation into the design..
- Emphasizing process repair over structural replacement.
- Allowing sufficient time for self-generating processes to resume.
- Treating the causes rather than the symptoms of degradation.
- Include monitoring protocols to allow for adaptive management.

Although recent evidence indicates that restoration can be successful in increasing both biodiversity and ecosystem services, it should not be assumed that restoring biodiversity will inevitably enhance ecosystem services, or vice versa. Biodiversity and different ecosystem services might display contrasting trajectories during restoration, leading to conflicts and trade-offs. Restoration actions focusing on a particular ecosystem service could lead to negative impacts on biodiversity or provision of other services, which will need to be considered during the planning process. Resolution of conflicts in delivery of different services and biodiversity will probably require a participatory process to land-use planning.

# LAND SURFACE TEMPERATURE VARIATION AND LAND COVER CHANGES BASED ON SATELLITE IMAGERY DATA

***Anh Kim Nguyen***[1,2,3], Yuei-An Liou[3,4,5*], and Ming-Hsu Li[1]

[1]The Graduate Institute of Hydrological and Oceanic Sciences, National Central University, Taoyuan City 32001, Taiwan
[2]Institute of Geography, Vietnam Academy of Science and Technology, 18 Hoang Quoc Viet Rd., Cau Giay District, Hanoi Viet Nam
[3]Taiwan Group on Earth Observations, Hsinchu, Taiwan
[4]Center for Space and Remote Sensing Research, National Central University, Taoyuan City 32001, Taiwan
[5]Taiwan Geographic Information System Center, Taipei, Taiwan

**ABSTRACT:** Urban area is a complex eco-environment involving a variety of anthropogenic activities. Its sustainable development is affected by many factors, such as topographical appearance, hydro-meteorological environment, and social economics. Due to urban sprawl and less vegetation, urban areas often exhibit higher thermal signatures than less disturbed rural areas. Thermal signatures represent the thermal status resulting from energy balance at land-air interface. In this study, Land Surface Temperatures (LSTs) are retrieved from Landsat TM, ETM, and OLI & TIRS (Thematic Mapper, Enhanced Thematic Mapper, and Operational Land Imager & Thermal Infrared Sensor, respectively) and serve as basis to derive ecological thermal index for assessing ecological dynamics in years 1989, 2003, and 2014. The Thua Thien - Hue Province, Vietnam, is chosen as a study area because it is a coastal province and has been experiencing with natural disaster, deforestation, and rapid urbanization. Its LST is found to increase by 0.7 °C and 1.5 °C for the 1989-2003 and 2003-2014 periods, respectively. Thermal environment index maps are utilized to categorize ecological conditions into six levels (*excellent, good, normal, bad, worse,* and *worst*). To demonstrate urban development a major contributor to thermal anomaly, correlation between LST and Normalized Difference Build-up Index (NDBI) is analysed. The correlation is found to be positive as expected with coefficient values 0.87, 0.89, and 0.84 for 1989, 2003, and 2014, respectively. In contrast, LST-Normalized Difference Vegetation Index (NDVI) is found negatively correlated with corresponding coefficient values -0.81, -0.81, and -0.76, indicating that vegetation reduces thermal intensity. In addition, areas associated with *excellent, good,* and *normal* thermal environmental levels are decreased over the same period of time frames 1989-2003 and 2003-2014.

***Keywords:*** Ecological thermal indices; Eco-environment; Thua Thien-Hue Province; Landsat data

## INTRODUCTION

Eco-environment has been greatly impacted by anthropogenic processes, such as transformation of land use patterns, urban sprawl, and de-vegetation, so that urban areas often present significantly higher thermal signatures than less disturbed rural areas. Thermal environment is considered as an important aspect of the eco-environment because it represents the status of energy balance through space-time interaction and interdependence association with city micro climate, and urban ecology (Miller and Small., 2003). It has been acknowledged that urban heat island occurred mainly in Land Surface Temperature (LST), which is governed by surface heat fluxes and affected by removal of vegetation, soil, water and their replacement with common urban materials like concrete, metal and asphalt.

The Thua Thien - Hue Province has been experiencing urbanization at a rapidly rate in both population and physical size in the recent decades so that it is selected as our study area where the urban use, agricultural practice, and aquaculture activities have inevitably invaded into natural zones. These changes lead to an increasing burden on the eco-environment (Nguyen et al., 2016) by bringing water, soil, and air pollution. Unfortunately, these outcomes are enhanced by natural disasters including annual floods, landslides, and storms in the Thua Thien - Hue Province (Thanh and De Smedt, 2011; Tran et al., 2007). Thus, it is critical to monitor the land surface dynamics, in particular urban development and its effects on the urban thermal anomaly. However, uneven distribution and limitation of in situ-measurement data often cause a barrier for long–term environmental monitoring across the region. The series of Landsat satellites provide free of charge data, which stimulate various natural and man-made disaster studies and resources monitoring, and support the evaluation of magnitude, dynamics, and spatial distribution of land surface parameters and eco-environmental vulnerability using satellite-derived variables (Nguyen et al., 2014, 2016; Liou et al., 2015; Liu and Zhang, 2011; Chen et al., 2006).

This study aims to identify the trend of LST derived from Landsat TM/ETM+, and OLI thermal infrared bands and examines the influences of land cover changes patterns and its association with LST in the Thua Thien - Hue Province of different time frames from 1989 to 2003 and from 2003 to 2014. The correlation coefficients of LST and Normalize Difference Built-up Index (NDBI), and Normalize Difference Vegetation Index (NDVI) are quantified. Thermal field variance indices are computed and classified into six levels including *excellent, good, normal, bad, worse,* and *worst* as a basis for the whole framework of the current eco-environmental study.

## MATERIALS AND METHODS

**The Study Site Description.** The Thua Thien - Hue Province covers an area of 5,054 km$^2$ in the north central coastal region of Vietnam with coordinates of 15° 59'-16° 48'N and 106° 25'-107° 51'E, comprising two basins of four rivers and Hue city as the provincial capital (**Figure 1**). Due to its location in the tropical monsoon basin, the average annual temperature ranges from 21° C to 26° C. The highest temperature (approximately 41.3 ° C) normally occurs in June, July, and August, while the lowest temperature (12 ° C) occurs in November, December, and January. The average relative humidity is around 84-85%. Annual average precipitation may reach 2,500-3,500 mm in the plains and 3,000-4,500 mm in the mountains. Impressively in some years, precipitation may be much higher and reaches more than 5,000 mm in the mountains (Tong et al., 2012). In recent years, meteorological observation records, which are collected from Institute of Meteorology and Hydrology of Vietnam, also depicted that there have been significant changes in precipitation patterns in all three stations (A Luoi, Nam Dong, and Hue) of the Thua Thien - Hue Province (increase of precipitation during the rainy seasons from August to December, lessen extent from April to May, and decrease in precipitation during the drier June to July period). This trend implies that there will be very likely more floods and droughts in the future in The Thua Thien - Hue Province (UNDP, 2003; MoNRE, 2003; Tran et al., 2007).

**Data Used.** The Thua Thien - Hue Province has 3 meteorological stations, Hue, A Luoi, and Nam Dong (**Figure 1**). To evaluate land cover changes, satellite images of two seasons are collected including Landsat 5 TM images (acquired on 08 May 1989; 03 December 1990), Landsat 7 ETM+ images (acquired on 21 April 2003; 06 November 2000), and Landsat 8 OLI & TIRS images (acquired on 27 April 2014; 04 October 2014). Those images acquired in summer months are used to examine the LST and land cover changes with supporting seasonal vegetation features by using those images acquired in winter months.

**Image Processing.** To analyze the trend of LST, several steps are performed and can be described in six steps: (i) Atmospheric correction is performed for all bands by using Fast Line-of-Sight Atmospheric Analysis of Spectral Hypercubes (FLAASH) (Waner and Chen, 2001) and following the atmospheric correction, the images are geo-referenced to UTM projection, zone 48 north; (ii) Digital numbers (DNs) of band 6 are converted to spectral radiance and then brightness temperatures are computed; (iii) Land emissivity is estimated based on NDVI method (Valor and Caselles, 1996; Sobrino et al., 2004) and brightness temperatures are converted to LST (Artis and Carnahan, 1982); (iv) Visible bands of images (Landsat 5 TM image acquired on 08 May 1989; Landsat 7 ETM+ image acquired on 21 April 2003; and Landsat 8 OLI & TIRS image acquired on 27 April 2014) are selected to composite to be a single image and land cover classification was performed with Support Vector Machine method; (v) Thermal field variance index is calculated based on LST by using equation (1) (Liu and Zhang, 2011).

$$TFVI = \frac{T_s}{T_s - T_{mean}} \tag{1}$$

where TFVI is the thermal field variance index; $T_s$ is the LST of certain pixel in C degree; and $T_{mean}$ is the mean LST of the whole study area in C degree. To classify thermal field variance index into different levels, histograms are used to reveal the statistical distribution of calculated thermal field variance values from grid cells; and (vi) Calculation of NDBI and NDVI to investigate the relationship between LST and NDBI and NDVI.

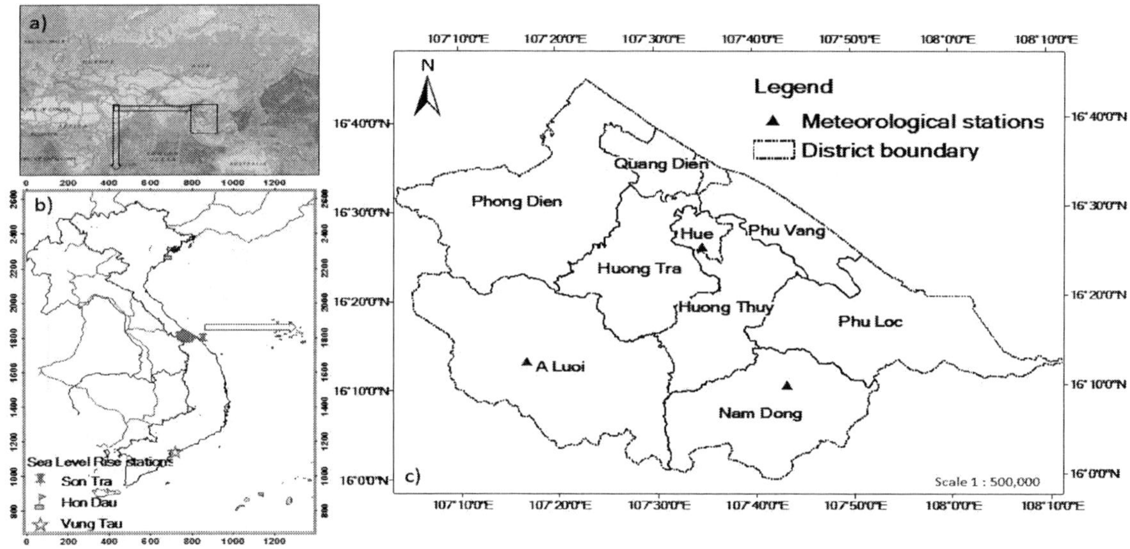


**Figure 1**. The map shows Asia (a); Vietnam and sea level rise stations (b); and Thua Thien - Hue Province (c).

## RESULTS AND DISCUSSION

**Trend of LST in the Thua Thien - Hue Province.** The LST in urban and rural areas are investigated by estimating mean and standard deviation for each district in the Thua Thien - Hue Province (**Table 1**). The LSTs in 1989, 2003, and 2014 are shown in **Figure 2**. There was an increasing trend in LST in both urban and rural areas (mean LST was increased by 0.7 $^{o}$ C and 1.5 $^{o}$ C between the 1989-2003 period and 2003-2014 period, respectively), The difference in the mean LST between urban and rural areas fluctuated from 0.9 $^{o}$C to 2.9 $^{o}$C. **Figure 2** shows the evolving patterns of LST from urban to rural areas in 1989, 2003, and 2014, and indicates each district has a different spatial-temporal pattern of LST. Annual air temperature from in situ-measurement at 3 stations show a good agreement with an increasing trend of LST derived from thermal bands of Landsat sensors (**Figure 3**). The measured air temperature increased around at 0.5 $^{o}$ C, 1.0 $^{o}$C, and 0.9 $^{o}$C at the Hue, A Luoi, and Nam Dong stations throughout the period from 1989 to 2014. However, the difference of LST from 1989 to 2014 is higher due to the time of satellite passing at around 11 AM and LST varies in different land cover types, so that the difference of LST temperature is higher as compared with that of monthly and annually measured air temperatures.

**Trends of Land Cover Changes. Figure 4** shows the increase in the physical size of the development land parallel with expansion of LST over decades (**Figure 2**). There was a dramatic increase in developed land and decrease in agricultural land and forest land. Developed land had greatest changes during different time period, from 166.12 $Km^2$ in 1989, 256 $Km^2$ in 2003, to 530.93 $Km^2$ in 2014. There was a significant decrease in agricultural land from 357.25 $Km^2$ in 1989, 228.94 $Km^2$ in 2003, to 240.88 $Km^2$ in 2014. Changes in developed land mainly occurred in urban and surrounding areas where cropland was replaced by concrete (buildings, roads, etc,) and the reduction of forest land appeared mainly in rural and mountainous areas where a lager natural forest area was gradually replaced by planted forest and cropland and building. At the regional scale, the increase of developed land, and decrease of crop land, and wood land implies that more people live in the major cities.

**Table 1.** Changes in LST of 9 districts over decades.

| | | | 08 May 1989 | | 21 April 2003 | | 27 April 2014 | |
|---|---|---|---|---|---|---|---|---|
| **No** | **District names** | **Area (Km2)** | **Mean** | **Std** | **Mean** | **Std** | **Mean** | **Std** |
| 1 | Hue (urban) | 71.9 | 25.7 | 1.4 | 26.5 | 1.6 | 27.6 | 2.1 |
| 2 | Phu Vang (urban) | 277.2 | 24.9 | 2.2 | 26.5 | 2.7 | 26.7 | 3.5 |
| 3 | Quang Dien (urban) | 162.7 | 25.1 | 2.7 | 26.1 | 3.0 | 26.1 | 3.4 |
| 4 | Phu Loc | 705.8 | 24.1 | 1.8 | 25.8 | 2.5 | 27.7 | 2.8 |
| 5 | Huong Tra | 518.2 | 24.7 | 1.3 | 24.8 | 1.4 | 26.4 | 1.9 |
| 6 | Huong Thuy | 455.9 | 24.6 | 1.7 | 25.3 | 2.0 | 27.3 | 2.0 |
| 7 | A Luoi | 1105.7 | 22.5 | 1.2 | 22.9 | 1.1 | 24.4 | 1.7 |
| 8 | Nam Dong | 643.7 | 22.8 | 1.1 | 24.0 | 1.4 | 26.6 | 2.3 |
| 9 | Phong Dien | 947.1 | 24.8 | 2.6 | 24.3 | 3.1 | 26.4 | 3.2 |

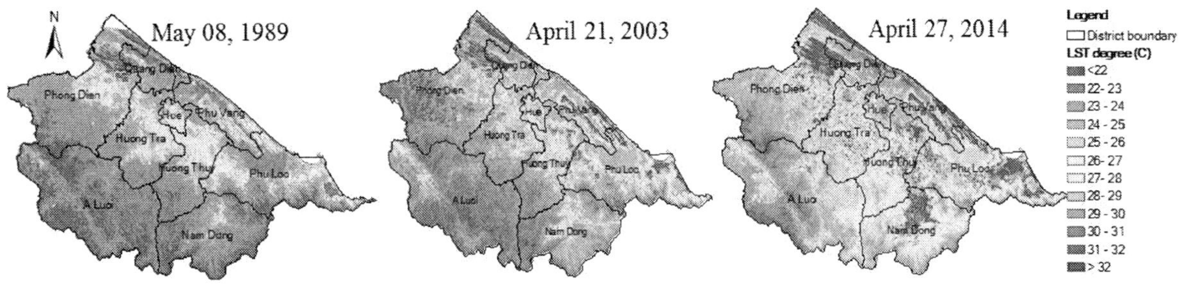


**Figure 2.** Spatial-temporal LST distribution over the Thua Thien - Hue Province derived from thermal bands Landsat sensors (Mean LSTs and Standard deviations (STDs) are 23.9 (STD 2.1); 24.6 (STD 2.4); and 26.1 (STD 2.7) for 1989, 2003, and 2014, respectively).

**Relationship between LST Variation and Land Cover Changes.** The sampling points are randomly taken from the Thua Thien - Hue Province. LST, NDBI, and NDVI values are extracted from the raster data to compute correlation coefficient among them. The LST-NDBI correlation coefficient (0.87, 0.89, and 0.84 for 1989, 2003, and 2014, respectively) indicates a positive influence of build-up land on urban temperature, implying that the build-up land can strengthen the thermal environment in the Thua Thien - Hue Province.

In contrast, negative LST-NDVI correlation (-0.81, -0.81, -0.76 in the same period) demonstrates that vegetation can weaken the thermal environment.

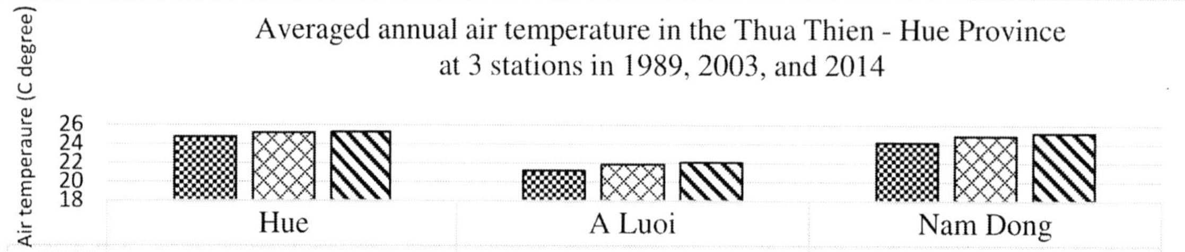


| | Hue | A Luoi | Nam Dong |
|---|---|---|---|
| 1989 | 24.8 | 21.2 | 24.2 |
| 2003 | 25.2 | 21.9 | 24.9 |
| 2014 | 25.3 | 22.1 | 25.2 |

**Figure 3.** Annual measured air temperature at 3 stations (Hue, A Luoi, and Nam Dong).

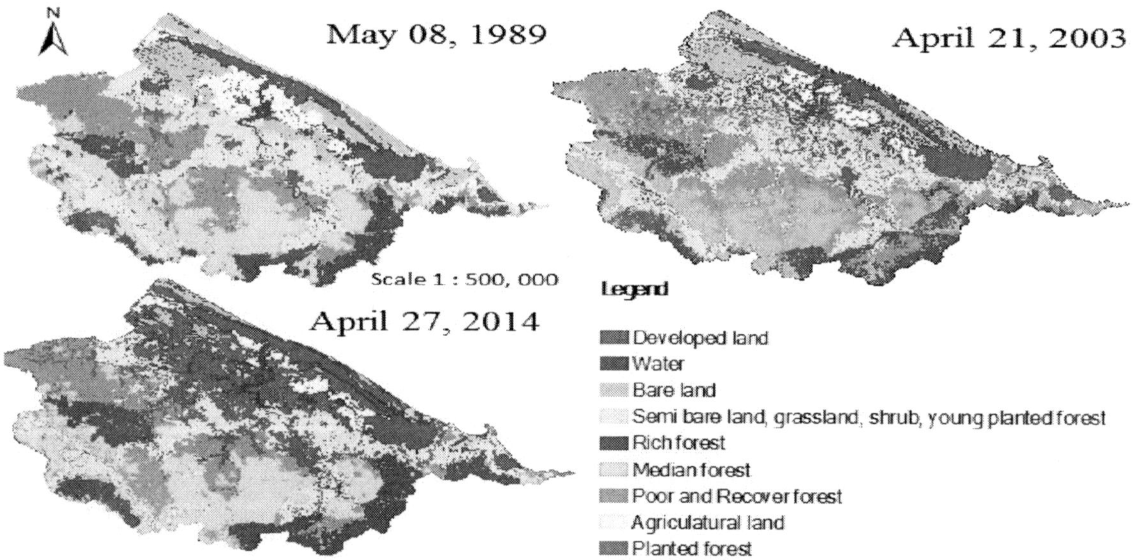


**Figure 4.** Dynamic of land cover from 1989 to 2014 derived from Landsat images (classification accuracy, Kappa coefficients are 0.84, 089, and 0.88 for 1989, 2003, and 2014, respectively).

**Thermal field variance index evaluation of the Thua Thien Hue Province.** Thermal environment index maps for 1989, 2003, and 2014 are constructed to categorize the ecological condition of the provincial area into six different levels (*excellent, good, normal, bad, worse,* and *worst*), which are shown to occupy, respectively, 77.87%, 12.68%, 1.37%, 1.30%, 2.20%, and 4.57% of the entire region of interest in 1989; 62.47%, 3.95% 8.93%, 4.49%, 4.49%, and 13.68% in 2003; and 58.27%, 3.21%, 11.09%, 5.92%, 3.75%, and 17.75% in 2014 **(Figure 5)**.

In general, from 1989 to 2003, and 2014 the areal percentage of *bad, worse,* and *worst* thermal environment levels was increased and the spatial distribution was from the low to high elevation belt parallel with time. Compared with the census data acquired in the same period of time, it is obvious that the *bad, worse,* and *worst* thermal environmental levels appeared mainly in the urban areas where social-economic activities have been intensified and the increasing trend of *bad, worse,* and *worst* thermal environment levels was in accordance with the urban population growth (about 237,800, 343,800, and 549,800 inhabitants in 1989, 2003, and 2014, respectively).

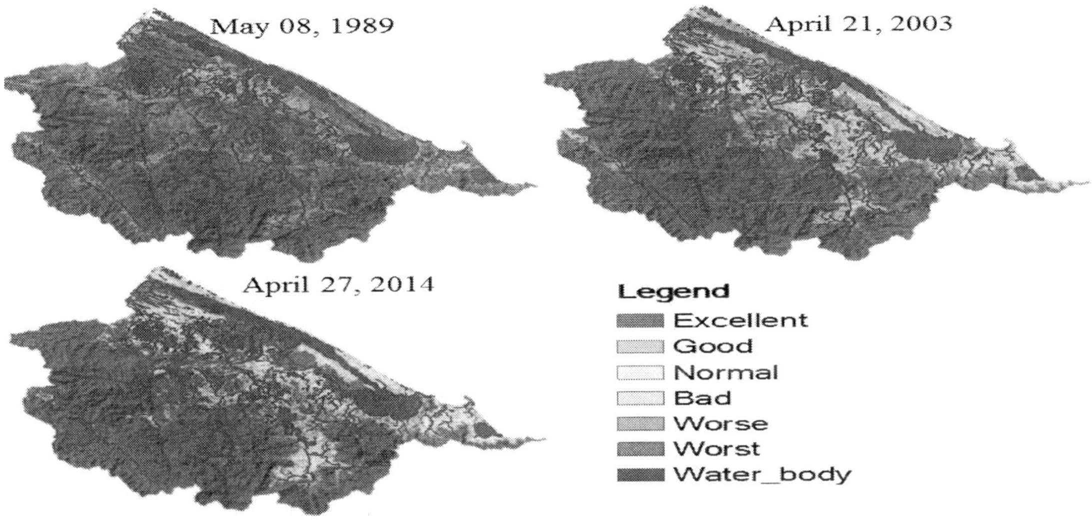


**Figure 5.** Thermal field variance index classification in the Thua Thien - Hue Province.

## CONCLUSIONS

LST is found to increase by 0.7 °C and 1.5 °C for the 1989-2003 and 2003-2014 periods, respectively. Measured air temperature data in the same period also shows an increasing trend and good agreement with LST. However, the difference of LST is higher due to the time of satellite passing at around 11 AM. LST varies in different land cover types so that LST is higher as compared with monthly and annually measured air temperatures. On the contrary, the *excellent, good,* and *normal* levels of thermal ecological environment were decreased over the same period of time. Results indicate that land use changes have great influences on urban thermal environment. The time series maps of categorization for ecological thermal indices in the Thua Thien - Hue Province can be used for future urban planning and further eco-environmental quality assessment.

## ACKNOWLEDGEMENTS

This research was financially supported by National Central University, Taiwan through the NCU International Student Scholarship. We would like to thank data support from United States Geological Survey (USGS). We also appreciate the financial support of the Ministry of Science and Technology (MOST) of Taiwan under the codes 104-2111-M-008-004 and 104-2221-E-008 -067.

## REFERENCES

Artis D.A, Carnahan WH., 1982. Survey of emissivity variability in thermography of urban areas. *Remote Sens. Environ.* 12, 313-329.

Chen, X.-L., Zhao, H.-M., Li, P.-X., Yin, Z.-Y., 2006. Remote sensing image-based analysis of the relationship between urban heat island and land use/cover changes. *Remote Sens. Environ.* 104 (2), 133–146.

Liu, L., Zhang, Y., 2011. Urban heat island analysis using the Landsat TM data and ASTER data: a case study in Hong Kong. *Remote Sens.* 3 (12), 1535–1552.

Liou, Y.A., Nguyen, A.K., Li, M.H., and Lin, C.Y. 2015. Landsat 8 operational land imager derived variables for environmental risk assessment in Taoyuan, *IGARSS*, July 26-31, Milan, Italy, DOI: 10.1109/IGARSS.2015.7325904.

Nguyen, A.K., V. Phonekeo, V. C. My, N. D, Duong, and P. T Dat. 2014. Environmental hazard mapping using GIS and AHP - a case study of Dong Trieu District in Quang Ninh Province, Vietnam. IOP Conf. Ser.: *Earth Environ. Sci.* 18 012045.

Nguyen, A.K., Y.A. Liou, M.H. Li, and T.A. Tran. 2016. Zoning eco-environmental vulnerability for environmental management and protection, *Ecol. Indic.* Vol 69, Pages 100–117. SCI. doi:10.1016/j.ecolind.2016.03.026. IF: 3.444.

Sobrino, J.A., Jimenez-Munoz, J.C., Paolini, L., 2004. Land surface temperature retrieval from Landsat TM 5. *Remote Sens. Environ.* 90, 434–440.

MoNRE, 2003. Vietnam Initial National Communication under the United Nations Framework Convention on Climate Change. *MoNRE*, Hanoi, Vietnam.

Thanh, L.N., De Smedt, F., 2011. Application of an analytical hierarchical process approach for landslide susceptibility mapping in A Luoi district, Thua Thien – Hue Province, Vietnam. *Environ. Earth Sci.* 66 (7), 1739–1752.

Tran, P., Marincioni, F., Shaw, R., Sarti, M., Van An, L., 2007. Flood risk management in Central Viet Nam: challenges and potentials. *Nat. Hazards* 46 (1), 119–138.

Tong, T.M.T., Shaw, R., Takeuchi, Y., 2012. Climate disaster resilience of the education sector in Thua Thien – Hue Province, Central Vietnam. *Nat. Hazards* 63 (2), 685–709.

UNDP, 2003. Reducing Disaster Risk, A Challenging for Development Valor, E., Caselles, V., 1996. Mapping land surface emissivity from NDVI: application to European, African, and South American areas. *Remote Sens. Environ.* 57, 167–184 (18).

Waner, T., Chen, X., 2001. Normalization of Landsat thermal imagery for the effects of solar heating and topography. *Int. J. Remote Sens.* 22, 773–788.

# CAUSES OF DAMPNESS IN RESIDENTIAL BUILDING WALLS IN JOS METROPOLIS, PLATEAU STATE, NIGERIA

***Kalada Itelima***
(Kalite Associates, Jos, Plateau State, Nigeria)

Jos Plateau is located on the North Central region of Nigeria. It is relatively undulating plain bounded by mountainous ridges. The prevailing climatic situation of the study area hinges upon alternate seasons of hot and cold weather conditions interspersed with an average period of about 6 months of rainfall, with July being the peak period. Dampness is the wetting of structural elements through moisture rise by capillary action. Dampness is one of the most serious structural defects in walls of buildings. Dampness in walls spoils paints and interior decorations, encourages the growth of bacteria and moulds, hampers aesthetics, and poses a threat to health of occupants through providing breeding conditions for mosquitoes. Dampness undermines structural integrity of wall elements, reduces thermal insulation property of building of building materials as well as affects the comfort not the occupants. Dampness causes damages to both structural and building materials and reconstruction efforts can be enormous. This study aimed at investigating the causes of dampness in the walls of some of residential buildings within Jos metropolis, Plateau State, Nigeria. The causes were diagnosed through tests, physical inspections and oral interviews. Oral interviews were held with the occupants of 500 randomly selected buildings in the study area. Fifty construction sites were also randomly selected for detailed study of the constructional practices and material adopted in building constructions within the study area. Results identified that about 80% of the houses investigated had damping defects at various degrees. The results also showed that rising damp through defective damp-proof membranes and efflorescence on walling unit serve as the major causes of dampness in residential building walls. Appropriate recommendations were made as preventive and remedial measures to the problem.

# WETLANDS

# RESTORATION OF A TROPICAL WETLAND THROUGH LOW-COST TECHNOLOGY

***Shadananan Nair***
(Nansen Environmental Research Centre – India, Kochi, Kerala, India)

Thought the wetlands play a crucial role in maintaining food and water securities and providing rural employment, environmental degradation pose a serious threat to many of them in the developing world, inviting serious shortage of reliable water and health issues. However, water quality in many of them could be improved using low-cost, environment friendly and locally available technologies, if there is proper awareness and support from local people and local administration. Present study is an example of such a success story in a tropical wetland region (eastern Kuttanad) in central Kerala in south India. Thousands of people living in the region depended on the wetland for sanitary purposes, irrigation, transportation, fishing and duck farming. Open domestic wells in the area once provided freshwater for domestic use, except during non-rainy months because of salinity intrusion from the backwater area. But, in recent decades, the region became so polluted that people have to either travel for more than a Kilometre to fetch water or to pay for it for home delivery, in a region that receives more than three times global mean rainfall. Public water supply is unreliable and the quality is poor. The wetland is highly polluted from the direct outflow of domestic waste, effluent from the local industries, outflow of contaminated water from rice fields, bathing of cattle, washing of vehicles etc. Amount of dissolved oxygen became so low that organisms such as fish and frog disappeared from most of the region. These wetlands have become inhospitable to several species of migratory birds and fish. Vectors grew and spread fast in the water, inviting serious health issues. Diseases like Chikun Guinea, Japan Encephalitis, Dengue Fever, Diarrhoea and jaundice became common. People still abstract the same water for non-cooking purpose, as there are no other alternatives. Skin diseases affect those who directly use the water for bathing. Failure in the major project in the region to control flooding and salinity intrusion in rice fields worsened the water quality, as natural flushing was obstructed. Major reason for the deterioration of water quality now is the stagnation of water. Analysis of water sample from the sixteen selected locations showed that quality is much below safety limits and serious health hazards could be expected at any time. However, experiments in the test plots in the canals joining the region with local technology using locally available material like charcoal and lime shell, and locally made cheap aerators prove that condition can be made better without much financial expenditure. Deepening and cleaning of the canals and planting of local herbs for shore protection made a lot of changes. Awareness programmes attracted local population and many of the safety information provided were new to them. This has helped to change their way of life and they show care in disposing solid material into the wetland. If the method adopted in the test plot is extended to the entire area, life in the surroundings can be brought back to earlier conditions. There should be a system for liquid waste treatment and a proper drainage system and collection facility for solid waste. Rooftop water harvesting can solve the shortage of reliable water, as the rainfall in the area is more than 300cm. This is to be done with the assistance from the local government. But, there are new challenges like industrial development associated with economic expansion, rapid rise in population, encroachment into wetlands for real estate business etc that in turn may deteriorate the wetland environment. Climate change may affect the runoff in rivers joining the wetland that may affect the water quality and water level. The proposed scheme of water diversion under national river linking project also may reduce the runoff into the wetland. Proper awareness to the public and students on the environmental issues as part of an adaptation strategy is required to overcome the crisis.

# SPATIAL DISTRIBUTION OF WETLAND VEGETATION AND ITS ECOLOGICAL FUNCTION ON RIPARIAN ZONE OF RIVERSCAPE

***Lixin Wang***
(Inner Mongolia University, Hohhot, Inner Mongolia, China)

Riparian zone is the interface or ecological ecotone between river system and adjacent terrace terrestrial ecosystems, which is an important part of the riverscape, while wetland plants and their communities on riparian zone play essential roles in maintaining ecological functions and values. The focus on riparian zone will associate Riverscape Ecology which is one of hot area of current ecological research with Wetland Ecology. Selected the riparian zones of the typical inland rivers - Wulagai River and the Xilin River in the grasslands of Inner Mongolia, China as the main object of study, with the survey of wetland plant biodiversity and plant communities, the measure of the physical and chemical characterizations of soil, the analysis of micro-topography of riparian zones, and the mapping of riparian vegetations, the formation and maintenance mechanism for the spatial distribution pattern of riparian vegetations at different scales was illustrated. In addition, based on a comparative study between the enclosed conservation plots and grazing degradation plots, the degradation characterization of wetland on riparian zone by the human disturbances was also analyzed.

On the watershed scale, the wetland vegetation on riparian zone along the longitudinal direction presents wooded swamp, shrub swamp, wet meadow, herbaceous swamp and herbaceous saline meadow, from the river source to the end of river, which is forming the spatial distribution pattern named riparian sourceward banded zonation. The Elevation, Temperature（annual mean temperature, insolation duration ）and moisture（annual precipitation）are the main environmental factors affecting the formation and maintenance of this pattern. On the river reaches scale, along the lateral direction vegetation presents aquatic plant communities, shoreline pioneer plant clusters, wet meadow communities and saline meadow communities, from river water column to terrace, which is forming the spatial distribution pattern named riparian shoreward banded zonation. The Micro-topography, spatial heterogeneity of physical and chemical characteristics of soil, the biological adaptation and inter-specific competition of wetland plants determine this pattern together.

Selectivity feed on plants and grazing trampling of livestock makes the change of dominant species in the wetland communities, the reduction of spatial heterogeneity, and also the gradually transition of the hygrophyte to the xerophytic on riparian zone. The grazing on riparian zone also leads to significantly reduction of the aboveground and belowground biomass. The obvious phenomenon of "miniaturization" occurred in the wetland plants because of grazing, that to say, the plant height, inter-node length and leaf length growing shorter. The results can provide a theoretical basis for the rational development, protection and management of wetland in the grassland of Northern China.

# IMPACT OF URBAN SPRAWLING ON EAST KOLKATA WETLAND AND ON THE ECOSYSTEM FUNCTIONING OF SUNDARBANS MANGROVE ECOSYSTEM, INDIA

***Susanta Kumar Chakraborty*** and Poulami Sanyal
(Department of Zoology, Vidyasagar University, Midnapore (West)-721102, West Bengal, India)
**Nandan Bhattacharyya**
(UGC Academic Staff College, Jadavpur University (Salt Lake Campus), Kolkata –700098)
**Ratnadip Roy and Sumana Bandhopadhayay**
(Department of Geography, University of Calcutta,35,BC. Road, Kolkata-700019)

Kolkata, the largest metropolis of India represents a vivid example of unplanned and uncontrolled type of urbanisation which is marked by severe overcrowding, illegal settlements, loss of biodiversity, destruction of wetlands, poor drainage and pollution of water regime. Eastern side of Kolkata, once dominated by marshes, tidal creeks, mangrove swamps and wetlands had been modified substantially by the colonial power and later on by the uncontrolled forces of urbanization.

East Kolkata wetland is a unique ecosystem that sustains world's largest natural water recycling model and is being considered as the oldest practice of integrated bio-resource management model. The unplanned and uncontrolled mode of urbanisation in and around Kolkata metropolitan has posed severe threats to this very sensitive, vulnerable and biologically productive wetland, the area of which has been reduced drastically(more than 60/0) during last few decades as revealed by the analysis of remote sensing satellite imageries. Such analytical tools have also been used side by side ground truth monitoring to delineate the types of wetlands along with their water quality parameters.

Sundarbans Biosphere Reserve of India, being a largest mangrove blocks of the world, is located within 30-40 km away from this wetland area and was once connected with each other through a number of rivulets, creeks and tidal channels. All these aquatic networks have been filled up and blocked because of natural siltation and other anthropogenic activities, leading to the reduction of freshwater drainage and thereby salinity invasion to Sundarbans mangrove estuarine complex.

In this context, the present paper has attemped to discuss urban sprawling and its the impact on the biodiversity and ecology of East Kolkata wetland (a Ramsar site) and thereby to Sundarbans mangrove estuarine complex (a World's Heritage site) in order to suggest the strategies for the conservation of biodiversity and mitigation on environmental perturbation.

# BEHAVIOR OF ANTIBIOTICS WITH CORRESPONDING RESISTANCE GENES IN MICROBIAL FUEL CELL WETLAND

***Shuai Zhang*** and Hailiang Song
(Southeast University, Nanjing, Jiangsu, China)

**ABSTRACT:** Pollution by antibiotics and antibiotic resistance genes (ARGs) has become a major health concern. A microbial fuel cell wetland (CW-MFC) is a new installation to simultaneously treat the wastewater and produce energy. The objective of this study was to access the removal efficiency of tetracycline (TC) and sulfamethoxazole (SMX) and development of TC and SMX resistance genes (*sulI*, *sulII*, *sulIII*, *tetA*, *tetC*, *tetO* and *tetW*) in the CW-MFCs. At the same time, electricity was produced during the co-metabolism of antibiotics and glucose. The results indicated that CW-MFCs could significantly reduce the concentration of TC and SMX in wastewater, and the removal efficiency were in the range of 92.3-94.6% and 95.0-97.5%, respectively. The removal efficiency of SMX was significantly higher than that of TC ($p < 0.05$). The satisfactory chemical oxygen demand (COD) removal rates were also obtained. The highest power density was 0.0578 W/m2 under the external resistance of 1000 Ω. The relative abundances of target ARGs showed obvious increases compared to the initial levels in the soils ($p < 0.05$).

***Key words:*** Microbial Fuel Cell Wetland; tetracycline; sulfamethoxazole; antibiotic resistance genes

## INTRODUCTION

Antibiotics are commonly served as medicine in the treatment of human and animal diseases [1]. China leads the world in antibiotic production capacity, approximately 210,000 t of antibiotics manufactured annually, of which 85% is utilized in animal agriculture and medicine[2]. However, majority of applied antibiotics taken by humans and livestock cannot be completely metabolized and are discharged directly into the environment with feces and urine[3, 4]. Excreted antibiotics can then enter river environments through a variety of pathways, including point discharges from wastewater treatment plants (WWTPs), animal feeding operations (AFOs), and fish hatcheries or via nonpoint sources [5]. It is well known that even low concentrations of antibiotics give rise to a significant risk to environmental security and human health[6]. Overuse and misuse of antibiotics pose the occurrence and spread of antibiotic resistant bacteria (ARB) and resistance genes (ARGs), which raises a major health concern. In addition, antibiotics promotes ARGs transfer between nonpathogenic and pathogenic bacteria by vertical gene transfer (VGT) and horizontal gene transfer (HGT) [3, 4, 7]. Therefore, Antibiotics, ARB and ARGs are regarded as the three worldwide public health problems by the World Health Organization[3, 8]. Tail water of WWTPs, aquaculture wastewater and agricultural non-point source pollution are the major important sources of pollution to spread ARB and ARG to environment.

A microbial fuel cell wetland (CW-MFC) system is a new device that embeds the MFC into the constructed wetland (CW) to treat the wastewater and produce bio electricity. The first CW-MFC technology was developed via placing the electrodes in the stratified aerobic and anaerobic conditions of the CW[9]. Research suggested that wetland plants can promote the cathode performance of MFCs[10]. In addition, CW-MFC was easier to build and operate than traditional MFCs[11, 12].

Fang et al. (2013) developed a CW-MFC system and obtained high decolorization rate of azo dye while producing a voltage output of 610 mV[13]. Zhao et al. (2013) achieved a power density of 9.4 mW/m$^2$ while treating wastewater by using CW-MFCs. Liu et al. (2013) developed a CW-MFC to convert solar energy into electricity by utilizing root exudates of *Ipomoea aquatica*[14]. Fang et al (2015) tested CW-MFCs for azo dye removal efficiency and power density[15]. Doherty et al. (2015) explored the effects of electrode spacing and flow direction of wastewater in CW-MFCs for investigating the potential to integrate microbial fuel cells (MFCs) into constructed wetlands (CWs) for simultaneously power generation and wastewater treatment [16].

However, these published works mostly focused on dye and nutrient in municipal and domestic wastewater, the potential and risks of this CW-MFC technology application on the removal of trace antibiotics from domestic sewage received less concern. There are several questions needed to be further clarified. For example, will the CW-MFC have high effective removal of antibiotics during the treatment process? The removal mechanism of antibiotics remains less clear. The influences of antibiotics on nutrients removal and the fate of ARGs in the CW-MFC need to be understood.

In this study, we proposed CW-MFCs to treat domestic sewage containing tetracycline (TC) and sulfamethoxazole (SMX) for eight months. LC-MS was used to determine the concentrations of TC and SMX in the water. The removal efficiencies of CW-MFCs on chemical oxygen demand (COD) were also monitored. In addition, tetracycline resistance genes (*tet* genes) and sulfonamides resistance (*sul*) genes in media and water samples were quantified by real time PCR. Five *tet* genes (*tetA, tetC, tetO, tetQ* and *tetW*) were chosen as representatives of three tetracycline resistance mechanisms because of their frequent occurrence in livestock manure and wastewater treatment lagoon[3, 4, 17-19]. Three *sul* genes (*sulI, sulII, sulIII*) were chosen as representatives of sulfonamides resistance mechanisms for their frequent report [17, 19-21]. The main objectives of this work were: (1) to study the removal efficiencies of CW-MFC on TC, SMX and COD from the domestic sewage; (2) to discuss the possible removal mechanism of TC, SMX and ARGs; (3) to assess the risks based on TC and SMX accumulation in media and its effect on electricity production and ARGs development.

## MATERIALS AND METHODS

**Experimental Design of CW-MFCs.** The CW-MFC constituted by polyacrylic plastic chamber (the internal diameter is 20 cm and the height is 55 cm) containing four layers from the top to the bottom (Fig. 1). The cathode (3 cm high) the anode (8 cm high) and of the CW-MFC were made by stainless steel mesh and activated carbon. The middle layer (28 cm high) and the bottom layer (17 cm high) were filled with soil and gravel (mass ratio of soil/gravel =1:1). Activated carbon (GAC) has been widely used as CW-MFC electrode material for its efficient electrical conductivity. Stainless steel mesh to strengthen the electron transfer was buried into the electrodes. The same *Oenanthe javanica* was transplanted into each cathode layer. The sampling ports were arranged to collect samples at an interval of 10 cm throughout the installation. The anode and cathode were externally connected by a titanium wire (diameter of 1.5 mm) of 1000 Ω resistance[15].

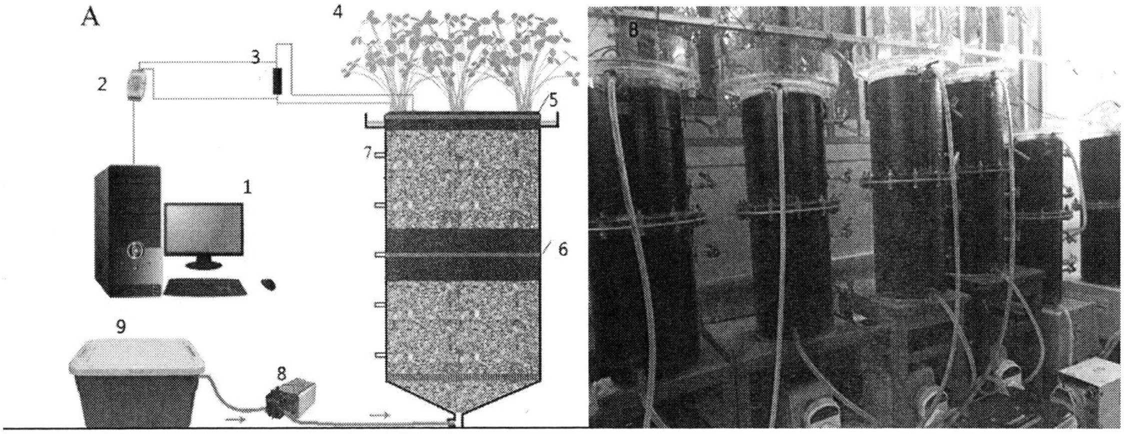


Figure 1. The structure of the CW-MFC (A: 1 computer; 2 data acquisition module; 3 resistance; 4 *Oenanthe javanica*; 5 cathode; 6 anode; 7sampling points;8 peristaltic pump;9 tank, B :The photograph of CW-MFC).

**Inoculation and Operation of CW-MFCs.** The anaerobic sludge (MLSS: 60 g/L) from the South City Municipal Wastewater Treatment Plant of Nanjing, China. The nutrient solution contained glucose (0.225 g/L), KH2PO4 (0.004 g/L), NH4Cl (0.020 g/L), NaCl (0.15g/L) and 0.20mL trace essential elements solution(contained per liter: 15 g MgSO4•7H2O, 2.2 g MnSO4•H2O, 5 mg ZnCl2, 2 g ZnSO4•7H2O, 15 g CaCl2, 1 g FeSO4 , 0.24 g CoCl2•6H2O, 10 mg FeCl3•6H2O, 2 mg NiCl2•6H2O, 0.4 mg Na2MoO4•2H2O and 1 mg CuCl2•2H2O,) [22].

After the inoculations were completed, the nutrient solution was fed continuously into the CW-MFCs using peristaltic pumps (BT100-1L, Baoding Longer Precision Pump Co., Ltd., China). The concentrations of TC and SMX in the CW-MFCs (CW-MFC1, CW-MFC2 and CW-MFC3) were 200, 500 and 800 μg/L respectively. The hydraulic retention time was 20h. The whole experiments were took place in the dark at 28±2°C. The experiment had been conducted for eight months from Apr 2015 to Nov 2015. All the cycles were repeated 3 times and error bars were used to represent standard errors of each parallel experiment.

**Quantification of Antibiotics and COD Removal.** Effluents were collected triple at three sampling periods, Jul 2015, Sep 2015. The antibiotics in the effluents and GAC were detected by LC-MS system (LCQAD-60000, Thermo Fisher Scientific, Waltham, MA, USA). Water samples were filtered through 0.45-lm fiber filters, and extracted using Oasis HLB (6 mL, Waters, USA) extraction cartridges (Huang et al., 2013). Antibiotics analysis in the samples was based on the published method [4, 5, 19, 23]. The concentrations of antibiotics were based on external calibration curves, and the correlation coefficients ($R^2$) exceeded 0.996. Concentrations of chemical oxygen demands (COD) in the effluents were measured according to American standard methods (APHA method 5220).

**DNA Extraction and ARGs Analysis.** Soil of anode, middle layer and cathode were collected triple at Jun 2015. Genomic DNA was extracted by using Power Soil DNA isolation kit (MoBio, Carlsbad, CA, USA) according to the manufacturer's instructions. DNA concentration and quality were examined by UV-9100 spectrophotometer (Lab Tech Ltd, Beijing, China) and gel electrophoresis. Eight tet genes (*tetA ,tetC, tetO, tetQ, tetW, sulII, sulII* and *sulIII*) were quantified using CFX Connect Real-Time PCR System (Bio-Rad, Shanghai, China) and SYBR Green qPCR mix (Bio-Rad, Shanghai, China). Primer sequences targeting these genes and PCR protocol were based on the previous methods [4, 17, 19, 24]. Each reaction was conducted in 25 μL on 0.1ml qPCR strip tubes kit(Thermo Fisher Scientific, Taiwan) containing 12 μL SYBR Green qPCR Mix (Bio-Rad, Shanghai, China), Primiers (forward and reverse primers, 0.6 μL each (10 μM)), 0.8 μL DNA templates and 11 μL $ddH_2O$[19]. All PCR reactions were conducted in triplicate. Plasmids carrying with target ARGs were created (Sangon Biotech, Shanghai, China) to make standard curves, and correlation coefficients ($R^2$) were higher than 0.990.

**Instrumentation and Analyses.** A data acquisition module (DAM-3057, Art Technology Co. Ltd., China) were used to monitor the voltage (V) data. The volumetric power density (Pv, W/$m^2$) is measured by the effective volume of the anode chamber ($m^2$) for the electrons used to power generation was generated in anode[15] .

The bioelectricity production of CW-MFC was estimated by the power density curve and polarization curve. A variable resister with a range of 5 Ω to 50,000 Ω was employed in this study to evaluate the power density and polarization curve. The polarization curve was judged by the current of different external circuit resistance and the electrode potential. The internal resistance (Ri) was measured by the linear region of the polarization curve.

**Statistical Analysis.** The averages and standard deviations of all data were calculated by Microsoft Excel 2010. Statistical analyses were carried out using SPSS Version 19.0. Analysis of variance and the Student–Newman–Keuls test ($p < 0.05$) were used to compare treatment means.

## RESULTS AND DISCUSSION

**Effect of Antibiotics on COD Removal.** Table. 1 shows the average COD removal efficiencies of the study period in CW-MFCs. Their cathode COD removal rates were 98.8%, 98.9±2.8% and 97.9% under the different influent conditions where antibiotics were both 400, 1000 and 1600μg/L in CW-MFCs respectively. Their anode COD removal efficiencies were 39.7%, 34.1% and 28.1 under different concentrains of TC and SMX respectively. The COD elimination rate tread for CW-MFCs were cathode > Middle layer > anode. The CW-MFCs remained stable at high level of COD removal efficiencies (97.9-98.9%) under the stress of TC and SMX in this study, for it involving adsorption, water electrolysis, microbial degradation and electro oxidation processes [25, 26]. The COD removal efficiencies were not inhibited by high concentrations of antibiotics (1600μg/ L) in CW-MFCs.

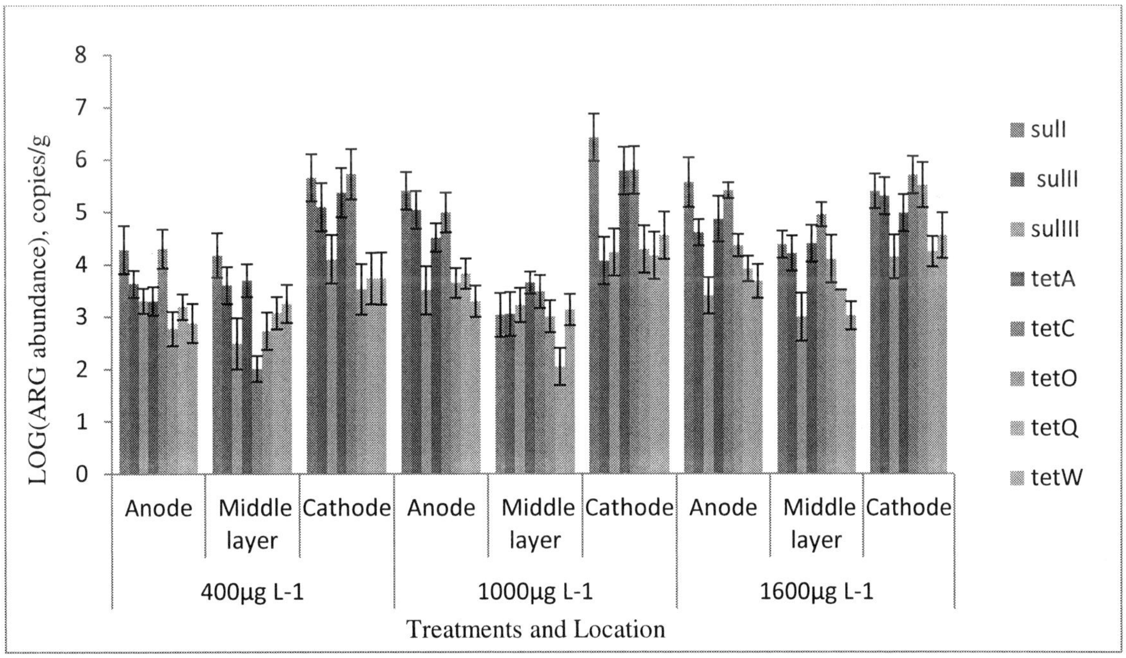


Figure 2. Log 10 the absolute abundances of *tet* and *sul* genes in CW-MFCs

Table 1. Effects of antibiotics (TC and SMX were200, 500 and 800μg/L respectively) on COD removal in CW-MFCs

| Concentrations of antibiotics ( μ g/ L) | COD removal rate (%) | | |
|---|---|---|---|
| | Anode | Middle layer | Cathode |
| 400 | 39.7±6. 3 | 78.1±6.2 | 98.8±3.2 |
| 1000 | 34.1±5.4 | 73.9±3.2 | 98.9±2.8 |
| 1600 | 28.1±5.41 | 61.7±2.6 | 97.9±4.9 |

**Removal Efficiencies of TC and** SMX **in CW-MFCs.** The target antibiotics concentrations in the influent and effluent of three CW-MFCs are shown in Table. 2. During an operation period of more than eight months, the average effluent concentrations were 14.01, 30.8 and 54.4μg/L for TC, 8.3, 15.0 and 32.4 μg/L for SMX where antibiotics were both 400, 1000 and 1600μg/L in CW-MFCs respectively. The effluent concentrations of TC were notably higher than that of SMX in all CW-MFCs. In terms of mass removal rate (Table. 2), the average mass removal efficiencies were 93.0, 93.9 and 93.2% for TC, and 95.9, 97.0

and 95.4% for SMX, respectively. Similar to the target antibiotics content in effluent, SMX showed better removal efficiency in all CW-MFCs compared to TC. However, comparisons of removal efficiencies of TC and SMX were not considered of statistically significant differences ($P < 0.05$) at different influent concentrations of antibiotics in the three CW-MFCs. CW-MFCs offered the high removal efficiencies for both TC and SMX. Among various types of CWs are most frequently used for pharmaceutical removal[4]. Previous studies demonstrated that CWs were excellent for antibiotics removal [4, 17, 27]. When using microcosm VSSF-CWs to treat pig farm wastewater containing 100 mg/L of TC, removal efficiencies of 94% was achieved [28]. So far there are few reports of CW-MFCs application on the removal of antibiotics and ARGs from wastewater. In this study, the proposed CW-MFCs obtained comparable antibiotics removal efficiencies, even though there were higher antibiotics concentrations (1600μg/L) in the influent of CW-MFC3. It is difficult to clarify which pathway contributes to TC and SMX decomposition mostly in such complicated systems. However, in this experiment, organics in the wastewater are used as a renewable resource to feed electrogenic bacteria, and electric energy was produced [15]. In addition, the instruments were in total darkness during this period. A series of biological and physicochemical processes, including substrate absorption, microbial degradation, plant uptake, hydrolysis and electric, may involve in antibiotics removal in CW-MFCs [4, 29, 30].

Table 2. Concentrations of antibiotics (TC and SMX were200, 500 and 800 μg/L respectively) in the effluent water of CW-MFCs (mean ± SD, n=3)

| Concentrations antibiotics in influent(μg/L) | Jul.2015 | | | | Sep.2015 | | | |
|---|---|---|---|---|---|---|---|---|
| | Concentrations of effluent(μg/L) | | Removal (%) | | Concentrations of effluent(μg/L) | | Removal (%) | |
| | TC | SMX | TC | SMX | TC | SMX | SMX | TC |
| 400 | 12.63±2.43 | 7.2±1.2 | 93.7 | 96.4 | 15.4±1.5 | 94±0.8 | 92.3 | 95.3 |
| 1000 | 27.1±4.81 | 12.5±2.7 | 94.6 | 97.5 | 34.5±5.5 | 17.5±2.9 | 93.1 | 96.5 |
| 1600 | 49.6±6.6 | 27.2±3.8 | 93.8 | 96.6 | 59.2±6.8 | 37.6±3.2 | 92.6 | 95.0 |

**Typical *tet* and *sul* Genes in Soil Sample.** In order to estimate the fate of ARGs in CW-MFCs, we analyzed the total absolute abundances of five *tet* and three *sul* genes in the soil of anode layer, middle layer and cathode layer on day 45 (Fig. 2). All five *tet* and three *sul* genes were detected in soli, and the absolute abundances of the genes with their log units ranged from 0 to 6.42. Previous studies have reported similar absolute abundances of *tet* and *sul* genes in CWs [4, 17]. Under the conditions of 1000 μ g/L and 1600 μ g/L antibiotics, the absolute abundances of *tet* and *sul* genes, except for *sulIII* ($p < 0.05$), in anode layer were significantly higher than those under the condition of 400 μ g/L antibiotics. It is noteworthy to agree with previous observation that the higher influent TCs concentration in CWs was selected for more copy numbers of *tet* genes in the effluent [4]. The absolute abundances of *tet* and *sul* genes in cathode layer were no significant difference compared with the different influent concentrations of TC and SMX. The absolute abundances of *tet* and *sul* genes are not influenced by different influent concentrations of antibiotic, which is accounted for the obtained comparable antibiotics removal efficiencies (Table. 2). The absolute abundances of *tet* and *sul* genes tread for CW-MFCs were cathode layer⩾anode layer > Middle layer. The high absolute abundances of ARGs in surface soils can also be attributed to its widely-distributed host bacteria[31].

**Effects of Antibiotics on Electricity Production.** To explore the performance of MFC under three different concentrations of antibiotics, testing of current densities around 400 hours were performed. Power densities of the 400 and 1000 μg/L almost constantly remained at 30 $mA/m^2$. Although the distinctions among three installations were not remarkable, the 1600 μg/L antibiotics illustrated comparative lower values, which was around 25 $mA/m^2$. With identical external resistance and electrode area, current densities substantially reflected the voltage evolutions. Voltage, determining the performance in thermodynamics, was produced by the redox reaction in air-cathode and microbial catalysis in anode [32, 33]. The increase of antibiotic from 400 to 1000 μg/L did not bring a drop of current density or voltage, which arose from a resistant of microbial to antibiotic. Comparing with lower concentration, the 1600 μg/L antibiotic in the installations impedes the biological activity by antibiotics。 Further study was conducted to identify the performance on power output.

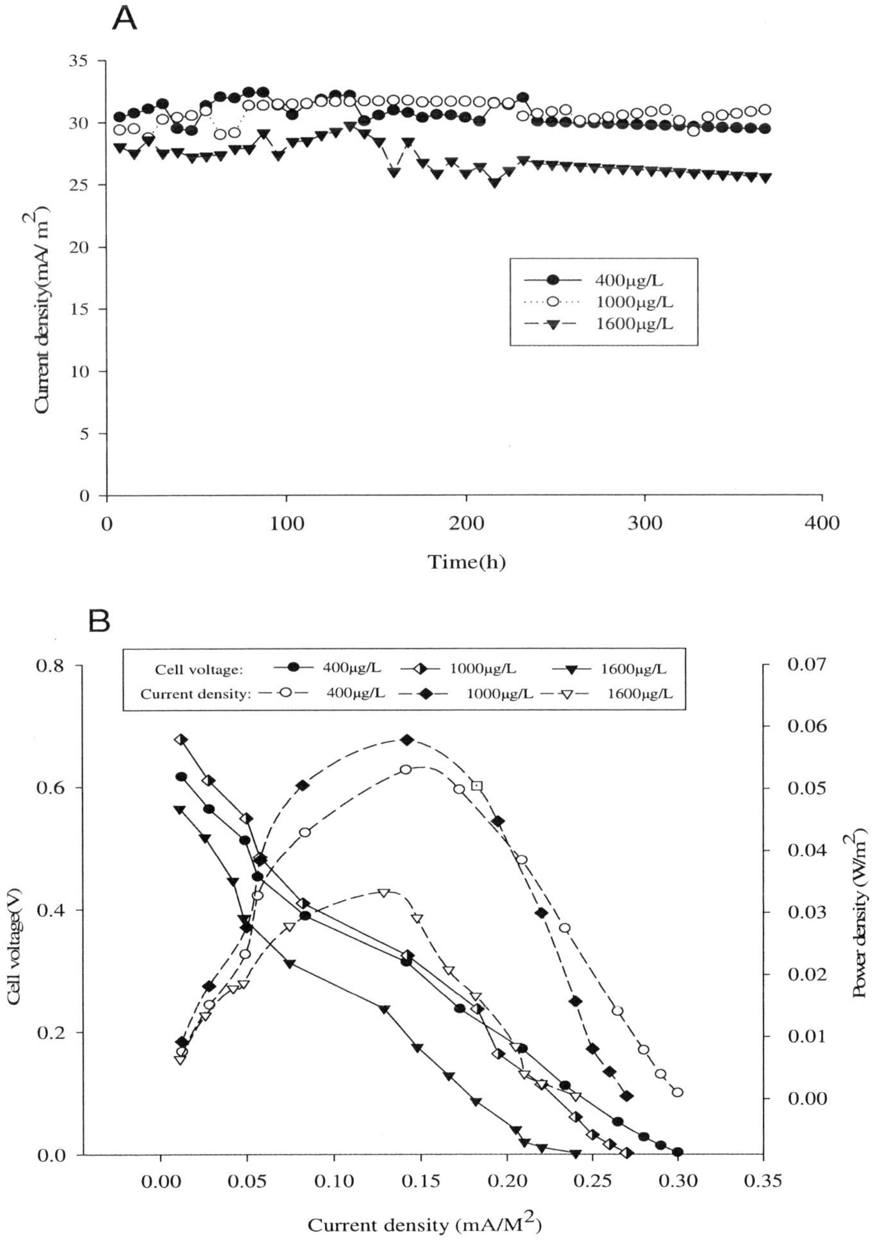


Figure 3 (A, B). Electricity research of CW-MFCs under different concentrations of antibiotics (TC and SMX were 200, 500 and 800μg/L re□pectively): (A) current den□ity; (B) polarization and power density curves

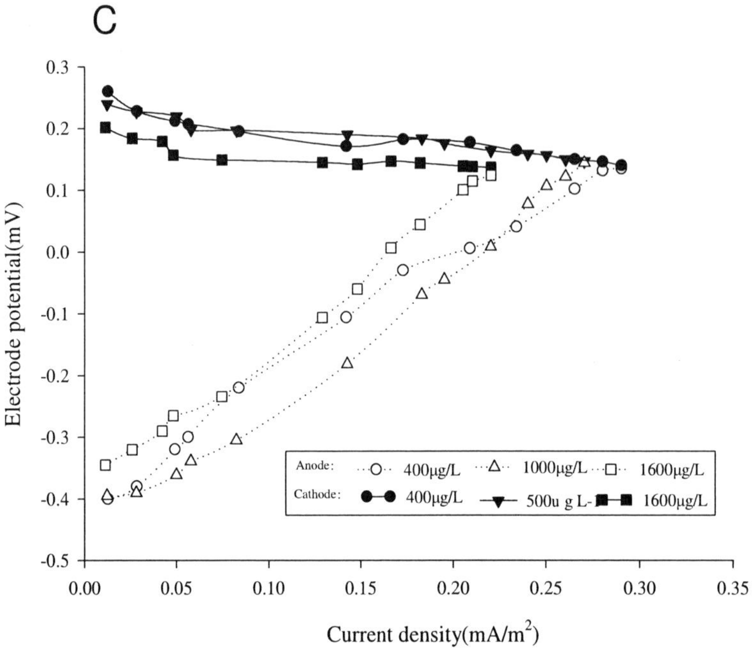


Figure 3 (C). Electricity research of CW-MFCs under different concentrations of antibiotics (TC and SMX were 200, 500 and 800μg/L respectively): (C) electrode potential

Polarization curves portrayed here visualized the variation in different external resistance. Vertex of cell voltages of the 400 and 1000 μg/L antibiotic (0.62V, 0.68V) was in 0.15 mA /m$^2$, however, the 1600 μg/L antibiotics voltage (0.41V) was in 0.13mA/m$^2$. An apparent deviation of power density between the 1600 μg/L and Lower concentrations reflected that higher concentration actually brought lower output of power. Notably, three parts namely polarization loss, ohmic loss and diffusion loss in polarization curves almost held analogous slope [32, 34], which could be the results of diffusion loss acting as predominant part in this big system [35]. Furthermore, details about electrode polarization were described below. Anode polarization profiles with identical large slopes were detected by changing their currents in large scope. However, reverse situation happened in all cathodes which maintained voltage at around 0.2V. Performance in all cathodes proved that oxygen reduction reversibility in cathodes was well enough to support big current output. Therefore, the novel structure with huge surface activated carbon and oxygen filling plants in cathodes enhanced the electrochemical reaction [36-38].

Anodes playing a role as electron output by electric activated bacteria had bad polarization in all installations maybe because of weak electron production capacitance in anaerobic conduction competing with other strains. Small system with intensive electrode potential inducing often owes better performance in anode [39, 40].

## CONCLUSIONS

This study clearly demonstrated the good performance of CW-MFCs on the wastewater containing TC and SMX based on three aspects: 97.9-98.9% for COD removal efficiencies, 93.7-97.5% for TC and SMX removal efficiencies, 0.0578 W /m$^2$ for the highest power density, respectively. Five *tet* and three *sul* genes were detected in soli, and the absolute abundances of the genes with their log units ranged from 0 to 6.42. This study employed gene quantification to study the fate of *tet* and *sul* genes in CW-MFCs. Further study to better understand the fate of potential pathogens caring *tet* and *sul* genes is warranted. It is also needed to investigate the relative abundances of *tet* and *sul* genes in CW-MFCs.

## ACKNOWLEDGMENTS

We thank the Provincial Natural Science Foundation of Jiangsu, China (SBK2014020483) for financial support.

## REFERENCES


1. Bouki, C.; Venieri, D.; Diamadopoulos, E., Detection and fate of antibiotic resistant bacteria in wastewater treatment plants: A review. *Ecotoxicology and Environmental Safety* **2013,** *91*, 1-9.
2. Luo, Y.; Xu, L.; Rysz, M.; Wang, Y.; Zhang, H.; Alvarez, P. J., Occurrence and transport of tetracycline, sulfonamide, quinolone, and macrolide antibiotics in the Haihe River Basin, China. *Environ Sci Technol* **2011,** *45*, (5), 1827-33.
3. Li, B.; Zhang, X.; Guo, F.; Wu, W.; Zhang, T., Characterization of tetracycline resistant bacterial community in saline activated sludge using batch stress incubation with high-throughput sequencing analysis. *Water Res* **2013,** *47*, (13), 4207-16.
4. Huang, X.; Liu, C.; Li, K.; Su, J.; Zhu, G.; Liu, L., Performance of vertical up-flow constructed wetlands on swine wastewater containing tetracyclines and tet genes. *Water Res* **2015,** *70*, 109-17.
5. Storteboom, H.; Arabi, M.; Davis, J. G.; Crimi, B.; Pruden, A., Identification of antibiotic-resistance-gene molecular signatures suitable as tracers of pristine river, urban, and agricultural sources. *Environ Sci Technol* **2010,** *44*, (6), 1947-53.
6. Kümmerer, K., Antibiotics in the aquatic environment – A review – Part I. *Chemosphere* **2009,** *75*, (4), 417-434.
7. Martinez, J. L., Environmental pollution by antibiotics and by antibiotic resistance determinants. *Environmental pollution* **2009,** *157*, (11), 2893-2902.
8. Davies, J.; Davies, D., Origins and evolution of antibiotic resistance. *Microbiology and molecular biology reviews : MMBR* **2010,** *74*, (3), 417-33.
9. Yadav, A. In *Design and development of novel constructed wetland cum microbial fuel cell for electricity production and wastewater treatment*, Proceedings of 12th International Conference on Wetland Systems for Water Pollution Control (IWA), 2010; 2010.
10. Villaseñor, J.; Capilla, P.; Rodrigo, M. A.; Cañizares, P.; Fernández, F. J., Operation of a horizontal subsurface flow constructed wetland – Microbial fuel cell treating wastewater under different organic loading rates. *Water Res.* **2013,** *47*, (17), 6731-6738.
11. Patil, S. A.; Surakasi, V. P.; Koul, S.; Ijmulwar, S.; Vivek, A.; Shouche, Y. S.; Kapadnis, B. P., Electricity generation using chocolate industry wastewater and its treatment in activated sludge based microbial fuel cell and analysis of developed microbial community in the anode chamber. *Bioresour Technol* **2009,** *100*, (21), 5132-9.
12. Higgins, S. R.; Lau, C.; Atanassov, P.; Minteer, S. D.; Cooney, M. J., Hybrid Biofuel Cell: Microbial Fuel Cell with an Enzymatic Air-Breathing Cathode. *ACS Catalysis* **2011,** *1*, (9), 994-997.
13. Fang, Z.; Song, H. L.; Cang, N.; Li, X. N., Performance of microbial fuel cell coupled constructed wetland system for decolorization of azo dye and bioelectricity generation. *Bioresour. Technol.* **2013,** *144*, 165-171.
14. Liu, S. T.; Song, H. L.; Wei, S. Z.; Yang, F.; Li, X. N., Bio-cathode materials evaluation and configuration optimization for power output of vertical subsurface flow constructed wetland - Microbial fuel cell systems. *Bioresour. Technol.* **2014,** *166*, 575-583.
15. Fang, Z.; Song, H. L.; Cang, N.; Li, X. N., Electricity production from Azo dye wastewater using a microbial fuel cell coupled constructed wetland operating under different operating conditions. *Biosens. Bioelectron.* **2015,** *68*, 135-141.
16. Doherty, L.; Zhao, Y. Q.; Zhao, X. H.; Wang, W. K., Nutrient and organics removal from swine slurry with simultaneous electricity generation in an alum sludge-based constructed wetland Incorporating microbial fuel cell technology. *Chem. Eng. J.* **2015,** *266*, 74-81.

17. Liu, L.; Liu, Y. H.; Wang, Z.; Liu, C. X.; Huang, X.; Zhu, G. F., Behavior of tetracycline and sulfamethazine with corresponding resistance genes from swine wastewater in pilot-scale constructed wetlands. *J Hazard Mater* **2014,** *278*, 304-10.
18. Barkovskii, A. L.; Bridges, C., Persistence and Profiles of Tetracycline Resistance Genes in Swine Farms and Impact of Operational Practices on Their Occurrence in Farms' Vicinities. *Water, Air, & Soil Pollution* **2011,** *223*, (1), 49-62.
19. Wu, D.; Huang, Z. T.; Yang, K.; Graham, D.; Xie, B., Relationships between Antibiotics and Antibiotic Resistance Gene Levels in Municipal Solid Waste Leachates in Shanghai, China. *Environ. Sci. Technol.* **2015,** *49*, (7), 4122-4128.
20. Naquin, A.; Shrestha, A.; Sherpa, M.; Nathaniel, R.; Boopathy, R., Presence of antibiotic resistance genes in a sewage treatment plant in Thibodaux, Louisiana, USA. *Bioresour. Technol.* **2015,** *188*, 79-83.
21. Dan, A.; Yang, Y.; Dai, Y. N.; Chen, C. X.; Wang, S. Y.; Tao, R., Removal and factors influencing removal of sulfonamides and trimethoprim from domestic sewage in constructed wetlands. *Bioresour Technol* **2013,** *146*, 363-70.
22. Cao, X.; Song, H. L.; Yu, C. Y.; Li, X. N., Simultaneous degradation of toxic refractory organic pesticide and bioelectricity generation using a soil microbial fuel cell. *Bioresour Technol* **2015,** *189*, 87-93.
23. Liu, L.; Liu, C.; Zheng, J.; Huang, X.; Wang, Z.; Liu, Y.; Zhu, G., Elimination of veterinary antibiotics and antibiotic resistance genes from swine wastewater in the vertical flow constructed wetlands. *Chemosphere* **2013,** *91*, (8), 1088-93.
24. Rodriguez-Mozaz, S.; Chamorro, S.; Marti, E.; Huerta, B.; Gros, M.; Sanchez-Melsio, A.; Borrego, C. M.; Barcelo, D.; Balcazar, J. L., Occurrence of antibiotics and antibiotic resistance genes in hospital and urban wastewaters and their impact on the receiving river. *Water Res.* **2015,** *69*, 234-242.
25. Ramamurthy, S. S.; Chen, Y. H.; Kamesh, K. M.; Nageswara, R. G.; Mitra, S.; Chelli, J., COLL 233-Carbon nanotube-zirconium dioxide hybrid for defluoridation of water. *Abstr. Pap. Am. Chem. Soc.* **2009,** *238*, 1.
26. Nageswara Rao, N.; Rohit, M.; Nitin, G.; Parameswaran, P. N.; Astik, J. K., Kinetics of electrooxidation of landfill leachate in a three-dimensional carbon bed electrochemical reactor. *Chemosphere* **2009,** *76*, (9), 1206-12.
27. Zhang, W.; Huang, M. H.; Qi, F. F.; Sun, P. Z.; Van Ginkel, S. W., Effect of trace tetracycline concentrations on the structure of a microbial community and the development of tetracycline resistance genes in sequencing batch reactors. *Bioresour. Technol.* **2013,** *150*, 9-14.
28. Carvalho, P. N.; Araujo, J. L.; Mucha, A. P.; Basto, M. C.; Almeida, C. M., Potential of constructed wetlands microcosms for the removal of veterinary pharmaceuticals from livestock wastewater. *Bioresour Technol* **2013,** *134*, 412-6.
29. Dordio, A. V.; Carvalho, A. J., Organic xenobiotics removal in constructed wetlands, with emphasis on the importance of the support matrix. *J Hazard Mater* **2013,** *252-253*, 272-92.
30. Hijosa-Valsero, M.; Fink, G.; Schlusener, M. P.; Sidrach-Cardona, R.; Martin-Villacorta, J.; Ternes, T.; Becares, E., Removal of antibiotics from urban wastewater by constructed wetland optimization. *Chemosphere* **2011,** *83*, (5), 713-9.
31. Chen, H.; Zhang, M., Effects of advanced treatment systems on the removal of antibiotic resistance genes in wastewater treatment plants from Hangzhou, China. *Environ Sci Technol* **2013,** *47*, (15), 8157-63.
32. Rabaey, K.; Verstraete, W., Microbial fuel cells: novel biotechnology for energy generation. *TRENDS in Biotechnology* **2005,** *23*, (6), 291-298.
33. Liu, H.; Logan, B. E., Electricity generation using an air-cathode single chamber microbial fuel cell in the presence and absence of a proton exchange membrane. *Environmental science & technology* **2004,** *38*, (14), 4040-4046.

34. Rabaey, K., *Bioelectrochemical systems: from extracellular electron transfer to biotechnological application*. IWA publishing: 2010.
35. Feng, Y.; Lee, H.; Wang, X.; Liu, Y.; He, W., Continuous electricity generation by a graphite granule baffled air–cathode microbial fuel cell. *Bioresource technology* **2010,** *101*, (2), 632-638.
36. Fang, Z.; Song, H.-L.; Cang, N.; Li, X.-N., Performance of microbial fuel cell coupled constructed wetland system for decolorization of azo dye and bioelectricity generation. *Bioresource technology* **2013,** *144*, 165-171.
37. Zhao, Y.; Collum, S.; Phelan, M.; Goodbody, T.; Doherty, L.; Hu, Y., Preliminary investigation of constructed wetland incorporating microbial fuel cell: batch and continuous flow trials. *Chemical Engineering Journal* **2013,** *229*, 364-370.
38. Villasenor, J.; Capilla, P.; Rodrigo, M.; Canizares, P.; Fernandez, F., Operation of a horizontal subsurface flow constructed wetland–microbial fuel cell treating wastewater under different organic loading rates. *Water research* **2013,** *47*, (17), 6731-6738.
39. Torres, C. I.; Kato Marcus, A.; Rittmann, B. E., Proton transport inside the biofilm limits electrical current generation by anode-respiring bacteria. *Biotechnology and Bioengineering* **2008,** *100*, (5), 872-881.
40. Torres, C. I.; Marcus, A. K.; Rittmann, B. E., Kinetics of consumption of fermentation products by anode-respiring bacteria. *Applied microbiology and biotechnology* **2007,** *77*, (3), 689-697.

# SEDIMENTS

# THE ELBE ESTUARY (GERMANY): RESISTANCE OF THE SEDIMENTS TO HYDRODYNAMIC FORCES ON MICROSCALE

Nina Stoppe and Thomas Neugebauer and Rainer Horn
(Kiel University, Kiel, Germny)
Elmar Fuchs and Uwe Schröder and ***Carolin Schmidt-Wygasch***
(Federal Institute of Hydrology, Koblenz, Germany)

The tidal River Elbe, Northern Germany, serves as an important international waterway feeding the port of Hamburg and thus indicates its function as a significant economic lifeline for that region. At the same time the estuarine riverscape holds valuable nature protected by national and European legislation. In the last centuries the estuary has been impacted by manmade alterations, e.g. by ongoing river training measures. Several national and European environmental directives like the Water Framework Directive demand for a new orientation towards renaturalisation of banks and an improved structural diversity.
However, at the moment there are still many loose ends to tie up. One of the main questions is: Which river reaches can be restored without hampering bank stability, flood protection, and navigation? To identify suitable stretches we need a sound understanding on the dynamic interaction between soils and sediments of the riverbank and vegetation development. Tidal hydrodynamics and hydrodynamic stress are of essential importance, especially to depict and estimate the soil resilience against it.

This survey focuses on soil structure and intrinsic particle-to-particle resistance capacity of the riverine sediment-soil complex. Physicochemical properties and classical soil mechanical methods were combined with rheological measurements, a methodological approach not being applied to soils usually. The latter evaluates the microstructure describing the soil´s viscoelastic deformation behavior (Integral Z). A significant result is that water content and texture are the most important factors influencing the microstructural stiffness on the particle-to-particle level. This means, that e.g., increasing soil desiccation results in a more advanced stabilization by inter-particulate menisci forces. This effect decreases with coarser grain sizes. To evaluate the interrelated effect of physicochemical properties and microstructural stiffness, statistical analyses were carried out. In this context especially soil organic matter, calcium concentration, and pedogenic iron oxides show a stabilizing response in terms of soil microstructure, and sodium vice versa. Additionally, the multivariate statistics produced so called pedotransfer functions, which allow predicting Integral Z by a specific set of easy-to-get predictor variables in non-investigated soils. This mechanism gives access to interpolate from local point data to non-examined bank areas.

Present results provide an estimation on the relative stability of the River Elbe banks mirrored by values of the rheologic target parameter Integral Z. Potential suitable stretches for bank restoration and thus a near-natural bank development can be identified. To be able to give a more precise and reliable advice to ecological engineering managers in terms of bank restructuring further site related investigations have to be carried out to precisely quantify the soil stability of pristine river banks towards hydrodynamic pressure.

# DEVELOPMENT AND EVALUATION OF NEW INDICES FOR SEDIMENT-ASSOCIATED CONTAMINANTS

***Nsikak U. Benson****, Adebusayo E. Adedapo, Omowunmi H. Fred-Ahmadu and Akan B. Williams
(Covenant University, Ota, Nigeria)
Abaas A. Olajire
(Ladoke Akintola University of Technology, Ogbomoso, Nigeria)

New indices (modified hazard quotient, *m*HQ and ecological contamination index, ECI) are developed for evaluating sediment-associated heavy metals contamination. In evaluating the proposed indices, the concentration and chemical fractionation of five heavy metals (Cd, Cr, Cu, Ni and Pb) determined in benthic sediments from five tropical ecosystems off the Bight of Bonny were used to assess the degree of contamination and estimate the extent of anthropogenic inputs from industrial activities into these ecosystems. The analysis shows that the mean concentrations (mg/kg, dw) of Cd, Cr, Cu, Ni and Pb vary from 4.33 – 5.67, 11.12 – 28.52, 30.26 – 43.72, 2.02 – 2.60 and 162.0 – 190.37, respectively. An important observation is that the mean metal levels during the wet and dry seasons did not show significant variability at all sites. The spatial distribution and severity of sediment-associated contamination by heavy metals based on the developed indices (*m*HQ and ECI) are in the descending sequence: Cd>Pb>Cu>Cr>Ni. Hence, the observed trend is in good agreement with existing pollution indices. Contamination severity index, mean hazard quotients and modified risk assessment code are also used to identify the pollution hotspots, which reflect medium risk contamination ecological systems. Aquatic pollution indicators (potential contamination index, ECI, hazard quotients, and *m*HQ) reveal significant anthropogenic contamination of Cd and Pb in the sediments, while Cr, Cu and Ni show relatively low degree of contamination. PCI generally follows the sequence Cd>Pb>Cu>Cr>Ni. Principal component analysis (PCA) and factor analysis indicate that heavy metals in the benthic sediments originate mostly from anthropogenic sources.

# GEOCHEMICAL SPECIATION AND RISK ASSESSMENT OF TRACE METALS IN SEDIMENTS FROM COASTAL ECOSYSTEMS OFF EQUATORIAL ATLANTIC OCEAN

***Nsikak. Benson***, Winifred Anake, Akan Williams, Oyeronke Akintokun, Adebusayo Adedapo
(Covenant University, Ogun State, Nigeria)
Essien D. Udosen
(University of Uyo, Akwa Ibom State, Nigeria)
Abaas A. Olajire
(Ladoke Akintola University of Technology, Ogbomoso, Nigeria)

The concentrations of Cd, Cr, Cu, Ni and Pb in estuarine benthic sediments were determined through multistep speciation scheme to evaluate their spatio-temporal distributions, selective fraction magnitude, degree of contamination and potential ecological risks. The results indicated that the metal fractionation percentages in the residual, oxidizable and reducible fractions are the most significant, while the exchangeable and carbonates bound trace metals are relatively low. High mobility and bioavailability was indicated for Cu, Cr and Ni, while Cd and Pb in sediments present low bioavailability for biota. Contamination factor ($CF_m$), degree of contamination (DC), modified degree of contamination ($mCd$), geoaccumulation index ($I_{geo}$), risk assessment code (RAC), and potential ecological risk index (PERI) were used to assess trace metals sedimentary pollution. The results indicate a prevalent moderate to high contamination of most trace metals analyzed. The contamination ranking of trace metals based on percent contribution to DC was Cd>Pb>Cu>Cr>Ni. RAC values indicate medium risk for Cd and Ni at all studied sites during the wet and dry seasons. Cd and Pb show moderate and very high individual metal potential ecological risk, respectively, while multi-elemental potential ecological risk indices ($R_I s$) indicate very high ecological risk in all the ecosystems.

# MILLING CONDITIONS EFFECT ON CONTAMINATED SEDIMENTS TREATMENT BY POZZOLANIC MATERIALS

***Rabah Hamzaoui***, Othman Bouchenafa, Abdelkrim Bennabi, Johan Colin
(Université Paris-Est, Institut de Recherche en Constructibilité, ESTP, 28 avenue du Président Wilson, 94234 Cachan, France)

Mechanosynthesis is one of the methods that allows obtaining a nanocrystalline state. It is an intensive energy process of mechanical grinding for the preparation of alloyed powders or composites in powder form. It involves a repeated fracturing and rewelding of particles, leading to size reduction and particle shape change. Planetary high-energy ball mill is machine, which contain vials mounted on a planar disc. With the rotation of the disc, the vials move in a circular and opposite direction compared to the disc rotation. The raw Materials (solid, liquid or gas with or without adjuvant) are introduced in vials with balls in order to obtain final product after milling [1-2]. In the context of sustainable development, new regulations on recycling of contaminated soil in the construction sector, leading researchers to research new techniques to value these wastes and to reuse them. The technique of solidification / stabilization (SS) can be used for treating soils or sediments contaminated, for transforming into controlled performance and quality materials suitable for further use, such as recycling in construction materials or paving materials. The objective of this research is to propose method of treatment based on the use of pozzolanic materials structurally modified by grinding (zeolites, fly ash, ....) and can be considered as an alternative to conventional methods of treatment with hydraulic binders.

# SURFACE WATER AND SEDIMENT QUALITY ASSESSMENT IN THE GBARAN-UBIE INTEGRATED OIL & GAS PROJECT, NIGER DELTA, NIGERIA

***Seiyaboh, E.I.*** and Gijo, A.H (Niger Delta University, Wilberforce Island, Bayelsa State, Nigeria)
Sikoki, F.D (University of Port Harcourt, Port Harcourt, Rivers State, Nigeria)

Surface water and sediment quality in the Gbaran – Ubie integrated Oil & Gas Project situated in the Niger Delta region of Nigeria was investigated. The physico-chemical characteristics of surface water and sediment samples collected from the project site were analyzed. Samples were collected from three (3) locations around the project site. The results obtained indicated significant differences ($P<0.05$) between the studied stations. Water samples analyzed were slightly acidic with pH values ranging from 6.63 - 6.78. Conductivity values of 86.87 - 122.60 µh mos $cm^{-1}$, TDS values of 43.20 - 60.00 mg/l, $Cl^-$ values of 28.00 - 62.60 mg/l, TH values of 23.00 – 34.67 mg/l, Ca values of 15.74 – 22.40 mg/l, Mg values of 4.29 – 6.51 mg/l, Na values of 8.64 – 9.67 mg/l, K values of 3.31 – 4.00 mg/l and Fe values of 0.170 – 0.207 mg/l were significantly higher than those recorded in other studies. Sediments in the study area were slightly acidic ranging from 6.65 - 6.85; acidic sediment can have an adverse effect on fisheries distribution and other benthic organisms. $N0_3^-$ values of 2.76 – 3.75 mg/g, $PO_4^{3-}$ values of 12.69 – 18.61 mg/g, TN values 4.53 – 5.39 mg/g, TOC values of 4.35 – 6.65 mg/g, THC values of 4.92 – 5.44 mg/g, Ca values of 5.73 – 5.77 mg/g, Mg values of 6.36 – 7.85 mg/g, Na values of 4.74 – 5.94 mg/g, K values of 4.84 – 5.54 mg/g, Zn values of 4.57 – 6.29 mg/g, Fe values of 1.66 – 1.85 mg/g, Mn values of 2.54 – 2.83 mg/g, Cd values of 1.23 – 1.55mg/g, and Pb values of 0.04 – 0.44 mg/g were significantly higher than those recorded in other studies. These results were indicative of the influence of activities within the Gbaran-Ubie integrated Oil & Gas Project on surface water and sediment quality.

# DISTRIBUTION OF ORGANOCHLORINE PESTICIDE RESIDUES IN EPIPELIC AND BENTHIC SEDIMENTS FROM LAGOS LAGOON, NIGERIA

Akan Williams[1*], Nsikak Benson[1] and Egbenya Shaibu-Imodagbe[2]

[1]Department of Chemistry, Covenant University, Canaanland, Ota, Ogun State, Nigeria.
[2]Samaru College of Agriculture, Ahmadu Bello University, Zaria, Nigeria
[*]akan.williams@covenantuniversity.edu.ng

**ABSTRACT:** Epipelic and benthic sediment samples were collected from Agboyi Creek, Oworonshoki, Ajara, Ogogoro and Tarkwa Bay in Lagos Lagoon and analyzed for organochlorine pesticide (OCP) residues using a gas chromatograph coupled with electron capture detector (GC-ECD). Sampling was conducted 4 times at the sites during the dry season months of December 2008, February 2009 and the wet season months of May, September 2009 to study the effects of seasonal variation on the samples. Sediment samples were subjected to cold extraction and clean-up. OCP residues were detected in all the samples though the levels did not show any particular pattern between the epipelic and benthic sediment during the dry and wet seasons. Detection limits of the OCPs ranged between 1.43 ng/g and 10.29 ng/g, indicating the sensitivity of the gas chromatograph at the operating conditions. The residue levels were higher in the lagoon than in the creek while a higher concentration of the residues was observed during the dry season due to reduced dilution effect. The mean recoveries of the residues ranged between 90.14 and 98.15% hence validating the methodology used in the study. Levels of OCPs residues in the sediment were within permissible residue limits.

***Keywords:*** Cold extraction, dry season, gas chromatograph, seasonal variation, wet season

## INTRODUCTION

Persistent organic pollutants (POPs) are toxic chemicals that persist in the environment and biomagnify in the food chain. Organochlorines such as chlorinated pesticides and polychlorinated biphenyls (PCBs) represent important groups of POPs, which have caused worldwide concern as toxic environmental contaminants (Covacia et al. 2005). Efforts at minimizing and eventually phasing out POPs globally gave rise to the Stockholm Convention in 2001. Residues and metabolites of many POPs are very stable, with long half-lives in the environment (UNEP 2002). The manufacture and use of organochlorine pesticides (OCPs) have been banned or restricted in developed countries. Although these bans and restrictions were enacted during the 1970s and 1980s, some developing countries are still using OCPs for agricultural and public health purposes because of their low cost and versatility in controlling various pests (Xue et al. 2006). Again, they are being used in most developing countries, including Nigeria, due to a lack of appropriate regulatory control and management of the production, trade and use of these chemicals (Darko and Acquaah 2007). There is also a paucity of data on the use of pesticides in the country, a reflection of the lack of a mechanism and planning programme in place for chemicals management as well as a low level of understanding of the environmental and public health hazards of pesticide use (Osibanjo et al. 2002).

Pesticides are chemicals used to kill or control pests. They are classified according to their chemical class or intended use. OCPs are basically insecticides composed primarily of carbon, hydrogen and chlorine and are found in the environment as a result of human activities. The use of OCPs takes many forms, ranging from pellet application in field crops to sprays for seed coating and grain storage. OCP residues enter aquatic environments through effluent release, discharges of domestic sewage and industrial wastewater, atmospheric deposition, runoff from agricultural fields, leaching, equipment washing, disposal of empty containers and direct dumping of wastes into the water systems (Yang et al. 2005). Lagoons, seas, rivers and lakes are depositories of most effluent discharges, leachates and run-offs from activities on land. The distribution of various chlorinated hydrocarbons in the marine and estuarine environment depends on residues of persistent organochlorine pesticide physicochemical properties of the eco-system as well as the partition coefficients of individual chlorinated hydrocarbons (Sarkar et al. 1997). OCPs could distribute among the components of the ecosystem, such as water and sediments. Pesticides can be bio-concentrated through biogeochemical processes and can be scavenged from the water through sorption on to suspended material before they get deposited to the bottom substrate. Different pesticides pose varying degrees and types of risk to water quality.

Sediment is one of the principal reservoirs of environmental pesticides, representing a source from which residues can be released to the atmosphere, groundwater and living organisms (Xue et al. 2006). Sediment serves as a sink from which water and biota are continuously polluted. Thus, the quality of sediment is essential in assessing the pollution status of the ecosystem (Doong et al. 2002). Persistence of these organic pollutants in sediment is possible due to their low solubility and tendency to associate with suspended particulate matter. As a result of their low water solubility, OCPs have a strong affinity for particulate matter. They are hydrophobic compounds that tend to adsorb to suspended particulate matter and benthic sediments in aquatic ecosystems. Indirect exposure to contaminated sediments takes place when fishes feed on benthic invertebrates that are ingesting particulate matter. Direct exposure through the sediment takes place by the release of contaminated particulate matter into the water column by both natural and anthropogenic processes.

The discharge of rivers into lagoons is the main transport pathway of pesticide residues. The Nigerian coastal belt has estuaries and lagoons as a transition zone between it and the numerous rivers and creeks flowing southwards into the Atlantic Ocean. Nigeria has a variety of industrial establishments producing different industrial wastes and effluents. Industrial establishments in Lagos account for over 60% of all industries in Nigeria. The proliferation of urban settlements and slums in Lagos has also led to increased human pressure and the generation of domestic effluents, which eventually find their way into the Lagos Lagoon. The lagoon, therefore, receives a complex mixture of domestic and industrial wastes and has served as the ultimate sink for the disposal of sewage. Some of the symptoms of pesticides poisoning include irritation, dizziness, tremour, tonic and chronic convulsion (Williams et al. 2013).

Organochlorine pesticide pollutant levels are usually monitored in inorganic ecosystem compartments such as water, air and sediment. Monitoring in inorganic compartments has the advantage of producing an immediate, geographically localized measure of contamination. The continuous monitoring of the pollution status of Lagos Lagoon is imperative as it is a major pelagic column in Nigeria. This study was undertaken to determine the organochlorine pesticide residues in epipelic and benthic sediments from Lagos Lagoon, Lagos, Nigeria.

## MATERIALS AND METHODS

**Study Area.** The study area for this investigation is the Lagos Lagoon, which lies between latitude 6° 24' - 6° 34' N and longitude 3° 23' - 4° 33' E in the western part of Nigeria. The map of the lagoon is shown in Figure 1. The lagoon consists of three main segments namely Lagos Harbour, Metropolitan End and Epe Division. It empties into the Atlantic Ocean via Lagos Harbour and is drained by Ogun, Agboyi, Majidun and Aye Rivers. The lagoon is shallow, with an average depth of about 1.5 m.

**Sampling Strategy.** Sampling was conducted between December 2008 and September 2009 during the dry and wet seasons to study the effects of seasonal variations on the samples. Field investigations were carried out four times at the designated sites during the dry season months of December 2008 and February 2009 and the wet season months of May and September 2009. The five sampling sites were Agboyi Creek (AGR), Oworonshoki (OWS), Ajara (AJR), Ogogoro (OGG) and Tarkwa Bay (TBY). The Agboyi Creek is one of the water bodies that drain the Lagos Lagoon and was used as the control site in this study. At Agboyi Creek, samples were collected at the zone where it empties into the lagoon while at Tarkwa Bay sampling was carried out where the lagoon discharges into the Atlantic Ocean. Sampling locations were identified with a hand-held Garmin-GPSMAP 76S-type global positioning system.

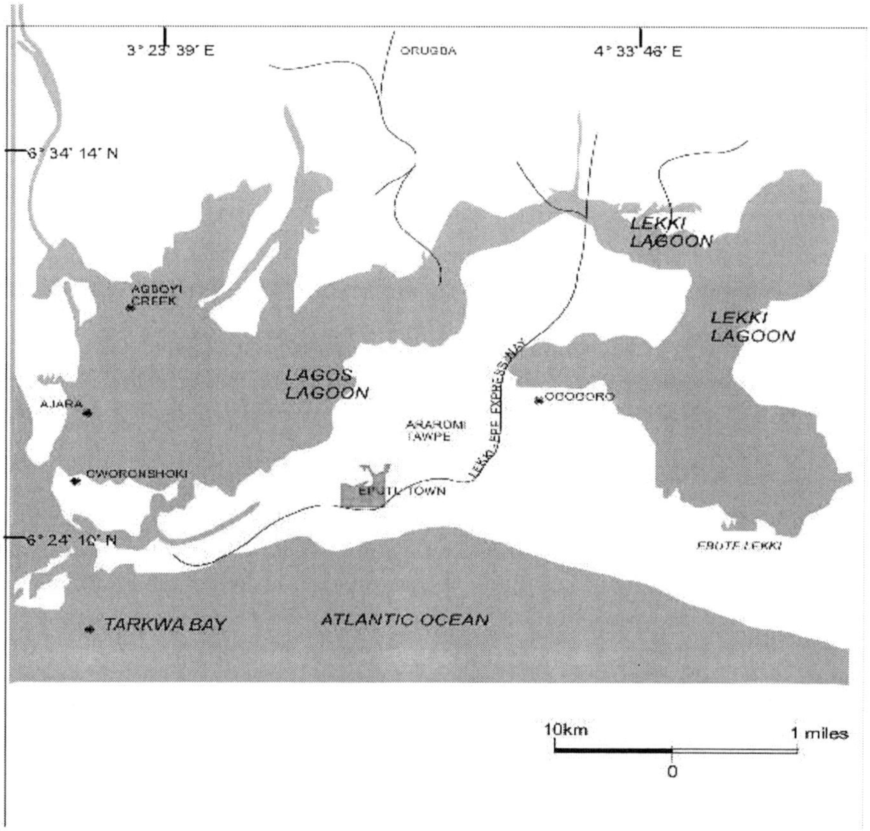


**Figure 1.** Map of Lagos Lagoon showing the sampling points
Source: Author's field survey (http://maps.google.ca/maps)

**Collection of Sediment Samples.** Epipelic (intertidal) sediments were obtained by using a short core sampler to scoop the top 1 to 5 cm of the sediments at five sites with the following coordinates: Agboyi Creek (31 N 0545069, UTM 0727598), Oworonshoki (31 N 0545159, UTM 0725638), Ajara (31 N 0545136, UTM 0727884), Ogogoro (31 N 0543288, UTM 0710820) and Tarkwa Bay (31 N 0543893, UTM 0707630). Benthic (subtidal) sediment samples were obtained at Agboyi Creek (31 N 0545254, UTM 0726974), Oworonshoki (31 N 05450361, UTM 0725474), Ajara (31 N 0545098, UTM 0727826), Ogogoro (31 N 0543330, UTM 0710845) and Tarkwa Bay (31 N 0543889, UTM 0707582) with the aid of a Shipek grab sampler. Benthic sediments were collected by lowering the grab sampler into the water bed and raising it aboard after a few minutes. Three grab sediment samples were collected at each site and mixed together to form composite samples and subsequently wrapped in labelled aluminium foil. A total of twenty epipelic and twenty benthic composite sediments were collected for OCPs determination. After collection, the sediment samples were stored in ice-packed coolers to preserve the integrity of the samples. Sediment samples were refrigerated in the laboratory at 4 °C to inactivate microbes. Pebbles, shells and vegetable matter were manually removed.

**Evaluation of Sediment Particle Sizes.** The hydrometer method was used to evaluate the particle sizes of the sediments. 50 g of the air dried sediment was ground and introduced into a plastic beaker. 400 $cm^3$ deionized water and 100 $cm^3$ sodium hexametaphosphate solution were added to the beaker. The mixture was agitated with a stir-rod and emptied into a blender where it was blended for about 30 seconds until the sediment could not be physically broken apart any further. The contents of the blender were poured into a 1000 $cm^3$ graduated cylinder and deionized water was added to the graduated cylinder up to the 1000 $cm^3$ mark. The opening of the graduated cylinder was completely covered and inverted a few times to fully suspend the sediment particles. A bouyoucos hydrometer was gently added to the sediment suspension and reading was taken 40 seconds after the cylinder inversions ended. The hydrometer was removed and cleaned. The temperature of the suspension was recorded and the hydrometer reading was corrected. The introduction of the bouyoucos hydrometer was repeated two hours later. The proportion of sand, silt and clay was determined by simple proportion from the hydrometer readings and the original mass of the sediments. The sediments were separated into the following sizes: % sand (0.20 -2.00 mm), % silt (0.02-0.20 mm) and % clay (<0.02 mm) (Brady and Weil 2007).

**Extraction of Sediment Samples for Determination of Organochlorine Pesticide Residues.** 5 g of wet sediment was homogenized with 5 g of anhydrous granulated $Na_2SO_4$ using a mortar and a pestle. Cold solvent extraction (Steinwandter 1992) was carried out using 50 $cm^3$ HPLC grade petroleum ether/acetone (1:1 v/v) mixture in a 250 $cm^3$ reagent bottle containing the homogenized sediment sample. The mixture was well shaken and the stopper was removed continuously to release the gas built up in the reagent bottle. The mixture was allowed to stand for 30 minutes and then filtered with a filter paper into a glass container (US EPA 2002).

***Pre-concentration of extracts.*** The solvent extracts were concentrated to 1 $cm^3$ using a rotary evaporator.

***Clean-up of extracts.*** Column chromatography was employed (US EPA Method 3630B) to clean-up the extracts (US EPA 1996). The glass separating column (20 cm) was packed with activated silica gel (90% < 45 μm) and washed down with n-hexane to remove any dirt. The extracts were demoisturized over 1 g of anhydrous granulated $Na_2SO_4$ and separated into two eluted fractions using mixtures of dichloromethane, hexane and acetonitrile as eluting solvents. For the first fraction, 30 $cm^3$ of a dichloromethane/hexane (20/80) mixture was used, while 30 $cm^3$ of a dichloromethane/hexane/acetonitrile (50/49.5/0.5) mixture was used for the second fraction in order to ensure that the polar acetonitrile eluted any remaining residue. The fractions were combined, concentrated to 1 $cm^3$ using a rotary evaporator and subsequently analysed for organochlorines.

**Analytical Methods.** Analysis of OCP is summarized as follows.

***Identification and determination of OCP residues by gas chromatography.*** A gas chromatograph equipped with an electron capture detector (GC-ECD) was used for the analyses of the OCPs residues. The cleaned-up extracts were dried and re-dissolved in 1.0 $cm^3$ of analar grade isooctane for injection into the gas chromatograph (Pandit et al. 2002). Blank runs were made for background correction and performance of the system. Organochlorine Pesticides II EPA Method 8081A {Mix AB # 3, cat. #32415 (ea.)} (US EPA 1996) was employed for the analyses. The detection and determination of the residues were performed by injecting 1 μL of the 1.0 $cm^3$ purified extract into the injection port of a gas chromatograph with a $^{63}Ni$ electron capture detector (GC-μECD Agilent Technology 7890A) equipped with the ChemStation software. The chromatographic conditions are given in Table 1. The purge activation time was 30 s and the run time was 17 mins. Identification of pesticide residues was accomplished using reference standards and relative retention time techniques, while the concentration of the residues was determined by comparing the peak heights of the samples with the corresponding peak heights of the reference standards of known concentrations. The stock solution of the OCP standards was purchased from Restek Corporation, USA. It contained 1000 ppm in n-hexane and was serially diluted to obtain the desired concentrations of 10 ng/mL, 20 ng/mL and 40 ng/mL. All the extracts from the sediment samples were analysed for alpha-BHC, beta-BHC, lindane, delta-BHC, heptachlor, heptachlor-epoxide (B), aldrin, dieldrin, endrin, endrin aldehyde, endrin ketone, cis-chlordane, trans-chlordane, endosulfan 1, endosulfan 11, endosulfan sulphate, methoxychlor, p,p´-DDE, p,p´-DDD and p,p´-DDT. The concentrations (ww) of the OCP residues were calculated directly by the gas chromatograph after inputing the weight of the samples and blank corrections effected.

**Table 1:** Operating conditions of the gas chromatograph

| | |
|---|---|
| Detector | Electron capture |
| Column | DB-5 fused silica capillary column (30 m length × 0.32 mm i.d. × 0.25 μm film thickness)<br>HP – 5 5% Phenyl methyl siloxan |
| Carrier gas | Helium (99.9992) flowing at 20 mL/min |
| Carrier gas pressure | 10.744 psi |
| Make-up gas | Nitrogen (99.9995%) |
| Injector temperature | 250 $^0$C |
| Injection | 1 μL Splitless |
| Detector temperature | 300 $^0$C |
| Temperature programme | Initial temp 50 $^0$C at a rate of 25 $^0$C/min to 100 $^0$C (held for 1 min), then at a rate of 5 $^0$C/min to 300 $^0$C |

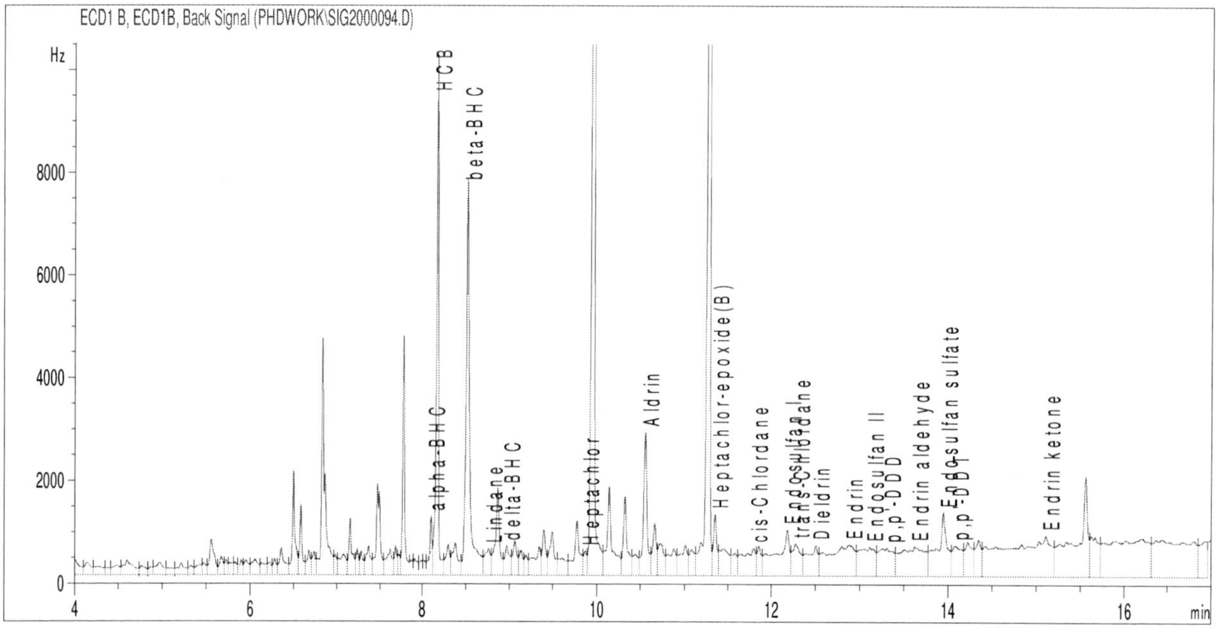


**Figure 2.** Chromatogram of organochlorine pesticide residues in epipellic sediment during the dry season in Agboyi Creek

***Quality assurance.*** Before analysis was performed, standards were run to check for the column performance, peak height, resolution and the detection limit. The correlation coefficients of calibration curves of OCPs were all higher than 0.998. The quality assurance measures included strict cleaning procedures, procedural blank, recovery of spiked standards and monitoring of detector response.

***Recovery study.*** Recovery studies were undertaken to evaluate the efficiency of the extraction procedure used. The recovery of organochlorines was carried out in replicate and was determined by spiking the previously analysed samples with the pesticide standard. The recovery values expressed in percentages were calculated from the chromatograms.

$$\% \text{ Recovery} = \frac{CS_2 - CS_1 \times 100}{CS}$$

where, $CS_1$ = concentration of pesticide residues in the sample
$CS_2$ = concentration of pesticide residues in the spiked sample
$CS$ = concentration of added pesticide

**Table 2:** Mean values of properties of epipelic sediment during the dry and wet seasons

| | Agboyi Creek | | Oworonshoki | | Ajara | | Ogogoro | | Tarkwa Bay | |
|---|---|---|---|---|---|---|---|---|---|---|
| Parameter | Dry season | Wet season | Dry season | Wet season | Dry season | Wet season | Dry season | Wet seas | Dry season | Wet season |
| Colour | Black | Grey | Black | Black | Brown | Black | Black | Brown | Black | Brown |
| % Sand | 58.00±0.12 | 58.00±0.09 | 5.00±0.06 | 5.00±0.18 | 10.00±0.11 | 5.00±0.18 | 80.00±0.14 | 5.00±0.1 | 60.00±0.10 | 60.00±0.16 |
| % Silt | 4.00±0.08 | 4.00±0.06 | 55.00±0.10 | 55.00±0.14 | 50.00±0.08 | 20.00±0.12 | 15.00±0.07 | 60.00±0. | 5.00±0.15 | 5.00±0.08 |
| % Clay | 38.00±0.09 | 38.00±0.03 | 40.00±0.06 | 40.00±0.10 | 40.00±0.12 | 75.00±0.15 | 5.00±0.10 | 35.00±0. | 35.00±0.06 | 35.00±0.13 |

**Table 3:** Mean values of properties of benthic sediment during the dry and wet seasons

| | Agboyi Creek | | Oworonshoki | | Ajara | | Ogogoro | | Tarkwa Bay | |
|---|---|---|---|---|---|---|---|---|---|---|
| Parameter | Dry season | Wet season | Dry season | Wet season | Dry season | Wet season | Dry season | Vet seaso | Dry season | Wet season |
| Colour | Black | Black | Black | Black | Brown | Black | Black | Brown | Black | Brown |
| % Sand | 3.00±0.10 | 3.00±0.02 | 5.00±0.05 | 5.00±0.11 | 8.50±0.09 | 5.00±0.03 | 5.00±0.09 | 5.00±0.0 | 60.00±0.13 | 60.00±0.08 |
| % Silt | 90.00±0.06 | 90.00±0.04 | 25.00±0.08 | 90.00±0.12 | 85.00±0.10 | 15.00±0.07 | 90.00±0.06 | 90.00±0. | 5.00±0.10 | 5.00±0.02 |
| % Clay | 7.00±0.04 | 7.00±0.06 | 70.00±0.09 | 5.00±0.14 | 6.50±0.12 | 80.00±0.07 | 5.00±0.02 | 5.00±0.0 | 35.00±0.10 | 35.00±0.12 |

***Determination of limit of detection (LOD).*** The limits of detection of the organochlorine pesticide residues were determined by replicate chromatographic runs (6 times) of the least concentration of the OCP standards (10 ng/mL) and then multiplying the standard deviation obtained by 3.

***Analyses of data.*** A total of 20 OCPs were detected and determined at various sample locations for both dry and wet seasons. Concentrations of OCP residues were calculated individually and as the sum of their isomeric forms. The mean was calculated from triplicate determinations.

## RESULTS

The results of the analyses and other related data are presented in Figure 2 and Tables 2 - 5. The mean physicochemical properties of epipelic and benthic sediments at the five sites during the dry and wet seasons are shown in Tables 2 and 3 while the mean concentrations of organochlorine pesticide residues in epipelic and benthic sediments at the sites are presented in Tables 4 and 5. The mean recoveries of the residues in sediment ranged between 90.14 and 98.15%**.** These recovery values validate the methodology used in this study. The limits of detection ranged between 1.43 ng/g and 10.29 ng/g, indicating the sensitivity of the gas chromatograph at the operating conditions. The calibration curves of the analysed pesticides present a good regression line ($r > 0.999$) in the range of explored concentrations.

## DISCUSSION

The sediments were admixtures of sand, silt and clay. An increase in the % silt in the sediments largely resulted in increasing levels of total organochlorine pesticide residues in the sediments. The persistence of OCPs in sediment is due to their tendency to associate with particulate matter.

Table 4: Mean concentrations (ng/g) of organochlorine pesticide residues in epipelic and benthic sediments during the dry season

| OCPs | Agboyi Creek | | Oworonshoki | | Ajara | | Ogogoro | | Tarkwa Bay | |
|---|---|---|---|---|---|---|---|---|---|---|
| | AGR-SS | AGR-BS | OWS-SS | OWS-BS | AJR-SS | AJR-BS | OGG-SS | OGG-BS | TBY-SS | TBY-BS |
| Alpha-BHC | 2.3±1.5 | 11.9±7.2 | 7.7±2.2 | 2.1±1.4 | 10.1±2.4 | 4.9±2.3 | 1.3±0.1 | 81.2±2.4 | 17.2±2.5 | 2.1±1.1 |
| Beta-BHC | 178.9±5.1 | 60.7±4.5 | 34.2±3.4 | 21.9±3.2 | 13.9±4.5 | 28.5±3.7 | 0.7±0.1 | 203.3±5.8 | 102.0±4.3 | 12.1±3.2 |
| Lindane | 8.2±1.6 | 24.7±2.3 | 28.4±1.8 | 15.6±6.8 | 10.1±2.1 | 11.6±2.2 | 3.6±1.6 | 56.1±8.3 | 45.7±8.2 | 4.9±2.2 |
| Delta-BHC | 9.1±3.2 | 111.2±4.1 | 37.5±8.6 | 9.5±2.3 | 16.1±3.3 | 20.3±9.4 | 0.7±0.2 | 1382.9±5.8 | 11.7±1.1 | 0.8±1.4 |
| ΣBHC | 198.5±11.4 | 208.5±18.1 | 107.8±16.0 | 49.1±13.7 | 50.2±12.3 | 65.3±17.6 | 5.3±2.0 | 1723.5±22.3 | 176.6±16.1 | 19.9±7.9 |
| Heptachlor | 5.6±2.4 | 44.9±5.6 | 33.1±5.1 | 22.8±8.2 | 9.8±4.2 | 24.9±3.3 | 92.5±5.9 | 278.3±9.9 | 119.3±5.4 | 91.8±7.2 |
| Heptachlor -epoxide (B) | 11.6±2.8 | 403.6±9.3 | 40.4±3.7 | 38.4±3.1 | 24.2±5.2 | 37.4±2.1 | 2.6±0.2 | 13681.1±4.3 | 15.4±8.3 | 2.2±2.1 |
| Aldrin | 27.3±6.3 | 14.8±4.2 | 27.9±1.6 | 4.0±1.3 | 40.0±2.3 | 36.8±6.1 | 6.5±0.4 | 210.8±5.9 | 31.6±1.2 | 12.6±2.5 |
| Dieldrin | 17.5±7.6 | 37.2±7.5 | 22.2±6.4 | 3.6±4.2 | 22.7±3.4 | 13.9±1.8 | 0.9±0.1 | 74.1±3.3 | 27.4±2.3 | 2.8±1.2 |
| Endrin | 73.8±2.1 | 139.5±2.3 | 25.3±3.2 | 15.1±3.3 | 120.8±4.6 | 16.9±4.2 | 11.8±8.6 | 170.7±2.2 | ND | 10.7±3.3 |
| Endrin aldehyde | 135.2±6.5 | 536.2±7.4 | 202.6±9.6 | 27.5±4.1 | 140.3±9.4 | 113.8±7.3 | 1.4±1.5 | 225.8±8.3 | 143.5±2.4 | 6.1±2.1 |
| Endrin ketone | 442.4±8.2 | ND | 403.6±3.8 | 21.5±8.5 | 79.4±5.6 | 451.2±2.9 | 0 | 294.6±3.5 | 238.6±8.3 | 14.5±2.2 |
| Cis-Chlordane | 18.4±3.4 | 28.4±2.5 | 35.6±2.3 | 19.4±3.3 | 23.2±3.5 | 25.7±4.1 | 0.6±0.3 | 76.5±6.4 | 60.1±5.2 | 2.8±4.2 |
| Trans-Chlordane | 16.1±8.1 | 90.8±6.2 | 29.8±2.1 | 20.1±3.7 | 31.3±5.2 | 68.3±7.8 | 0.5±0.7 | 63.1±5.5 | 45.5±3.3 | 7.7±5.3 |
| Endosulfan 1 | 30.2±2.0 | 37.7±3.6 | 26.9±5.3 | 14.3±1.2 | 13.3±9.1 | 26.7±2.7 | 0.6±0.3 | 71.8±9.2 | 62.9±4.5 | 2.0±5.2 |
| Endosulfan 11 | 30.1±4.5 | 55.3±6.3 | 39.7±3.6 | 5.6±2.1 | 44.7±3.5 | 16.4±4.4 | 0.9±0.2 | 123.3±2.8 | 260.9±5.8 | 2.2±4.19 |
| Endosulfan sulphate | 71.5±9.1 | 53.6±4.2 | 40.5±5.4 | 8.1±1.9 | 77.2±5.4 | 48.2±7.2 | 2.4±1.1 | 305.9±8.3 | 252.6±9.7 | 10.7±7.4 |
| Methoxychlor | ND | 146.7±8.2 | 22.4±8.5 | 2.8±3.8 | 47.4±8.3 | 43.6±3.6 | 0 | 58.7±2.5 | 84.7±3.4 | 7.9±8.2 |
| p,p´-DDE | ND | 43.1±5.6 | 44.2±2.5 | 5.2±5.4 | 20.9±2.6 | 39.4±8.2 | 1.2±0.3 | ND | 20.6±6.2 | ND |
| p,p´-DDD | 31.7±5.6 | 52.3±8.2 | 64.9±7.6 | 8.4±6.4 | 49.7±3.1 | 33.2±3.3 | 2.9±5.2 | 72.7±3.6 | ND | 4.6±3.1 |
| p,p´-DDT | 26.9±3.8 | 44.5±2.4 | 88.9±3.4 | 11.8±2.5 | 21.3±1.9 | 85.3±5.5 | 2.9±2.4 | 89.8±8.3 | 113.2±8.2 | 7.4±7.3 |
| ΣDDT | 58.6±9.4 | 139.9±16.2 | 198.0±13.5 | 25.4±14.3 | 91.9±7.6 | 157.9±17.0 | 7.0±7.9 | 162.5±11.9 | 133.8±14.4 | 12.0±10.4 |
| ΣOCPs | 1138.9±83.8 | 1939.1±101.6 | 1254.8±90.1 | 271.±76.7 | 811.64±89.6 | 1139±92.1 | 127±29.2 | 17521.6±64.3 | 1646±90.3 | 195±74.1 |

AGR-SS = epipellic sediment at Agboyi Creek; AGR-BS = benthic sediment at Agboyi Creek; OWS-SS = epipellic sediment at Oworonshoki; OWS-BS = benthic sediment at Oworonshoki; AJR-SS = epipellic sediment at Ajara; AJR-BS = benthic sediment at Ajara; OGG-SS = epipellic sediment at Ogogoro; OGG-BS = benthic sediment at Ogogoro; TBY-SS = epipellic sediment at Tarkwa Bay; TBY-BS = benthic sediment at Tarkwa Bay.

Sediments with higher percentages of silt accumulated higher concentrations of organochlorine pesticide residues. Methoxychlor and p,p´-DDE were not detected in the epipelic sediment of Agboyi Creek during the dry season. Endrin ketone was not also detected in its benthic sediment. p,p´-DDE was not detected at the benthic sediment of Ogogoro and Tarkwa Bay. Endrin and p,p´-DDD were not detected in epipelic sediment at Tarkwa Bay. During the wet season, p,p´-DDE, aldrin and trans-chlordane, endrin and p,p´-DDD were not detected at Oworonshoki, Ajara and Ogogoro respectively. OCP levels did not show any particular pattern between the epipelic and benthic sediment and during the dry and wet seasons. ΣBHC ranged between 5.27 ng/g in epipelic sediment of Ogogoro and 1 723.56 ng/g in its benthic sediment during the dry season. ΣDDT ranged between 4.5 ng/g in benthic sediment of Tarkwa Bay and 378.62 ng/g in epipelic sediment of Oworonshoki during the wet season. Heptachlor-epoxide (B) gave striking levels of 13 681.10 ng/g and 13 562.80 ng/g at the benthic sediment of Ogogoro and Ajara sites respectively. Total OCPs ranged between 35.97 ng/g and 17 513.80 ng/g in the benthic sediments of Tarkwa Bay and Ogogoro respectively.

There is also increasing interest in South America on organic pollutants from the coastal marine environment. Menone et al. (2001) studied the occurrence and distribution of persistent organochlorine compounds in sediments in the Mar Chiquita, Argentina coastal lagoon watershed. Heptachlor epoxide, DDT and its metabolites, and % HCH were the predominant OCPs in sediments. OCPs in sediment from the coastal marine environment of Mumbai, India have been analysed to assess their distribution in various environmental compartments (Pandit et al. 2002). HCH isomers, DDT and its metabolites were the predominantly identified compounds in all the samples. The residue levels of OCPs in this study are similar to results of earlier studies carried out on sediment at the Lagos Lagoon. Adeboyejo et al. (2011) recorded OCP levels in the sediment that ranged between 0.4 - 43.5 ng/g. In another study by Clarke et al. (2013) endosulfan I had the lowest concentration of OCP while endosulfan II had the highest mean concentration which ranged between 29.67 – 525.55 ng/g. The beta-BHC in the sediment ranged between 33.76 - 436.11 ng/g during the dry and wet seasons, respectively. The OCP levels in this study were within the permissible limits (FAO/WHO 2005; USEPA 2006).

**Table 5:** Mean concentrations (ng/g) of organochlorine pesticide residues in epipelic and benthic sediments during the wet season

| OCPs | Agboyi Creek | | Oworonshoki | | Ajara | | Ogogoro | | Tarkwa Bay | |
|---|---|---|---|---|---|---|---|---|---|---|
| | AGR-SS | AGR-BS | OWS-SS | OWS-BS | AJR-SS | AJR-BS | OGG-SS | OGG-BS | TBY-SS | TBY-BS |
| Alpha-BHC | 11.1±3.5 | 28.7±9.4 | 19.7±6.3 | 91.4±2.2 | 12.2±4.4 | 52.5±8.2 | 1.3±4.1 | 159.6±6.3 | 29.9±3.2 | 2.1±3.1 |
| Beta-BHC | 74.3±2.8 | 49.7±2.3 | 52.8±4.5 | 82.4±6.9 | 86.4±3.5 | 97.3±3.5 | 6.7±2.2 | 36.8±3.6 | 16.6±4.3 | 6.0±2.2 |
| Lindane | 25.4±8.3 | 83.3±4.4 | 51.4±2.4 | 13.2±8.5 | 26.9±8.6 | 83.1±4.4 | 3.5±8.5 | 32.3±2.5 | 26.9±2.4 | 0.5±0.8 |
| Delta-BHC | 96.3±5.2 | 20.9±2.3 | 56.5±5.2 | 11.2±2.2 | 21.1±2.3 | 44.1±6.2 | 2.2±3.2 | 9.2±7.3 | 19.4±2.2 | 0.6±0.2 |
| ΣBHC | 207.1±19.8 | 182.6±18.4 | 170.4±18.4 | 198.2±19.8 | 146.6±18.8 | 277.0±22.3 | 13.7±20.0 | 237.9±19.7 | 92.8±12.1 | 9.2±6.3 |
| Heptachlor | 49.5±4.2 | 125.6±8.5 | 99.7±9.4 | 70.3±4.3 | 60.2±5.3 | 194.6±4.5 | 580.9±9.2 | 45.1±3.2 | 556.5±5.5 | 13.2±5.2 |
| Heptachlor -epoxide (B) | 440.3±8.5 | 19.0±1.4 | 97.9±2.2 | 10.4±2.2 | 49.3±4.6 | 13562.8±2.5 | 1.0±5.5 | 40.7±8.3 | 49.0±6.3 | 0.7±1.2 |
| Aldrin | 24.2±2.2 | 43.6±2.1 | 44.7±4.4 | 26.2±7.4 | 29.5±9.2 | ND | 3.3±6.2 | 29.4±5.2 | 80.4±3.8 | 2.2±1.3 |
| Dieldrin | 33.0±8.5 | 38.9±7.5 | 64.8±7.3 | 36.7±2.3 | 37.6±4.5 | 2.0±6.1 | 0.7±2.5 | 12.5±3.1 | 48.8±2.4 | 0.6±0.2 |
| Endrin | 103.0±9.9 | 137.8±2.4 | 55.6±8.8 | 132.8±9.8 | 24.1±3.3 | 1.0±4.6 | 0.9±0.8 | ND | 49.6±8.5 | 0.8±0.2 |
| Endrin aldehyde | 334.1±4.3 | 101.8±8.5 | 226.6±4.1 | 198.8±4.7 | 419.3±8.7 | 1.1±3.8 | 1.3±0.6 | 118.2±5.6 | 375.4±5.5 | 0.6±0.1 |
| Endrin ketone | 426.8±2.8 | 691.7±5.8 | 657.8±3.8 | 148.2±8.3 | 438.9±2.2 | 0 | 0 | 332.4±8.7 | 420.6±3.6 | 0 |
| Cis-Chlordane | 61.1±3.3 | 53.7±3.2 | 75.2±2.5 | 23.9±1.1 | 52.7±2.4 | 3.0±2.3 | 0.5±0.3 | 67.3±5.3 | 76.6±6.5 | 0.6±1.2 |
| Trans-Chlordane | 34.7±7.2 | 40.8±2.3 | 141.7±8.4 | 25.0±3.2 | 89.9±4.6 | ND | 0.5±0.4 | 50.3±9.4 | 80.5±2.4 | 1.2±2.1 |
| Endosulfan 1 | 39.4±2.1 | 69.5±3.2 | 64.8±5.3 | 44.2±2.2 | 40.4±3.7 | 3.8±2.3 | 0.9±0.9 | 46.4±7.3 | 32.6±9.4 | 0.6±1.4 |
| Endosulfan 11 | 76.9±8.5 | 139.7±8.7 | 30.4±9.2 | 87.6±6.4 | 22.2±8.4 | 1.1±1.4 | 0.9±0.4 | 282.9±6.6 | 102.9±6.5 | 0.8±3.6 |
| Endosulfan sulphate | 78.2±2.4 | 114.9±5.4 | 135.4±4.9 | 151.8±2.2 | 97.6±9.8 | 6.9±1.6 | 1.5±0.4 | 105.3±3.3 | 178.6±9.6 | 2.7±7.3 |
| Methoxychlor | 20.1±5.3 | 74.2±4.2 | 84.9±7.5 | 170.2±6.6 | 61.9±2.5 | 0 | 0 | 61.3±3.2 | 141.6±5.4 | 0 |
| p,p′-DDE | 41.2±2.5 | 66.3±8.2 | 103.7±5.4 | ND | 49.5±5.4 | 1.0±1.4 | 0.7±0.2 | 54.7±9.4 | 32.2±3.6 | 0.8±3.4 |
| p,p′-DDD | 52.6±7.2 | 10.4±4.3 | 108.7±6.3 | 41.0±3.2 | 47.3±3.6 | 2.1±2.3 | 2.9±0.3 | ND | 93.1±2.3 | 2.9±4.1 |
| p,p′-DDT | 104.9±3.3 | 102.3±8.4 | 168.2±8.2 | 58.6±9.4 | 63.6±6.3 | 2.1±3.2 | 1.0±0.9 | 83.6±5.5 | 101.8±5.7 | 0.8±5.2 |
| ΣDDT | 198.7±13 | 279.0±20.9 | 380.6±19.9 | 99.6±12.6 | 160.4±15.3 | 5.2±6.9 | 4.6±1.4 | 138.3±14.9 | 227.1±11.6 | 4.5±12.7 |
| ΣOCPs | 2121±89.3 | 2102±102.5 | 2328±112.1 | 1418±93.1 | 1722±100.3 | 14054±38.3 | 608±48.6 | 1562±89.8 | 2503±111.1 | 36±42.8 |

## CONCLUSION

Results of the OCP analyses showed that a total of twenty pesticide residues were detected in the samples. The organochlorine pesticide residues were detected in all the samples though the frequency of detection of a few of the residues was less than 100%. The residue levels were higher in the Lagos Lagoon than in Agboyi Creek while a higher concentration of the residues was generally observed during the dry season.

## REFERENCES

Adeboyejo OA, Clarke EO, Olarinmoye MO. 2011. Organochlorine pesticide residues in water, sediments, fin and shell-fish samples from Lagos Lagoon complex, Nigeria. *Researcher* 3: 38-45.

Brady NC, Weil RR. 2007. *The nature and properties of soil.* Pearson International Edition.

Clarke EO, Aderinola OJ, Adeboyejo OA. 2013. Persistent organochlorine pesticides in water, sediment, fin fish (Sarotherodon galilaeus) and shell fishes, (Callinectes pallidus and Macrobrachium macrobrachium) samples from Ologe Lagoon, Lagos, Nigeria. *American Journal of Research Communication* 1:122-135.

Covacia A, Gheorgheb A, Voorspoelsa S, Maervoeta J, Redekere ES, Blustc R, Schepensa P. 2005. Polybrominated diphenyl ethers, polychlorinated biphenyls and organochlorine pesticides in sediment cores from the western Scheldt River (Belgium): analytical aspects and depth profiles. *Environment International* 31: 367-375.

Darko G, Acquaah SO. 2007. Levels of organochlorine pesticides residues in meat. *International Journal of Environmental Science and Technology* 4: 521-524.

Doong RA, Peng CK, Sun YC, Liao PL. 2002. Composition and distribution of organochlorine pesticide residues in surface sediments from Wu-Shi river estuary, Taiwan. *Marine Pollution Bulletin* 45: 246-253.

Food and Agriculture Organization (FAO) /World Health Organization (WHO). 2005. *Residues in food.* Report of joint FAO/WHO food standards programme. Rome, Italy. 2B: 61-81.

Menone ML, Aizpún de Moreno JE, Moreno VJ, Lanfranchi AL, Metcalfe TL, Metcalfe CD. 2001. Organochlorine pesticides and PCBs in a Southern Atlantic coastal lagoon watershed, Argentina. *Archives of Environmental Contamination and Toxicology* 40: 355-362.

Osibanjo O, Bashin N, Onyoyo H, Bouwman H, Yive R, Okond'Ohoka J. 2002. UNEP/GEF regionally based assessment of persistent toxic substances, Sub-Saharan Africa regional report. Available at: http://www.chem.unep.ch/pts/gr/Global-Report.pdf.

Pandit GG, Sahu SK, Sadasivan, S. 2002. Distribution of HCH and DDT in the coastal marine Environment of Mumbai, India. *Journal of Environmental Monitoring* 4: 431-434.

Sarkar A, Nagarajan R, Chaphadkar S, Pal S, Singbal SY. 1997. Contamination of organochlorine pesticides in sediments from the Arabian sea along the west coast of India. *Water Research* 31: 195-200.

Steinwandter H. 1992. Development of microextraction methods in residue analysis. In: *Emerging Strategies for Pesticide Analysis* (Eds T. Cairns, J. Sherma). 3-50. CRC. Boca Raton, FL.

United Nations Environment Programme (UNEP). 2002. Sub-Saharan Africa, regionally based assessment of persistent toxic substances. United Nations Environment Programme, Chemicals (UNEP Chemicals), Geneva, Switzerland. Available at: http://www.chem.unep.ch/Pts/regreports/ssafrica.pdf.

US Environmental Protection Agency (US EPA). 2006. Guidance for assessing chemical contaminant data for use in fish advisories. 2: Risk Assessment and fish consumption limits. Available at: http://www.epa.gov/ost/fishadvice/volum2/index.httml.

US Environmental Protection Agency (US EPA). 2002. Method 3570, Revision C. Microscale solvent extraction (MSE), Washington, DC, USA. p 9.

US Environmental Protection Agency (US EPA). 1996. Method 3630, Revision B. Silica gel Cleanup. SW-846 Manual. Washington, DC, USA.

US Environmental Protection Agency (US EPA). 1996. Method 8081A. Organochlorine pesticides by gas chromatography. Washington, DC, USA.

Williams A, Ayejuyo O, Unyimadu J. 2013. Distribution of chlorinated pesticides in shellfishes from Lagos Lagoon, Nigeria. *Journal of Marine and Biological Oceanography* 2: 1. DOI: http://dx.doi.org/10.4172/2324-8661.1000106.

Xue N, Zhang D, Xu X. 2006. Organochlorinated pesticide multiresidues in surface sediments from Beijing Guanting reservoir. *Water Research* 40: 183-194.

Yang R, Ji G, Zhoe Q, Yaun C, Shi J. 2005. Occurrence and distribution of organochlorine pesticides (HCH and DDT) in sediments collected from East China sea. *Environment International* 31: 799-804.

# DEVELOPMENT AND TESTING OF NEXT GENERATION SORBENT POLYMER EXTRACTION AND REMEDIATION FROM SEDIMENTS (SPEARS) TECHNOLOGY

***Robert DeVor**** (Vencore, Kennedy Space Center, FL, USA)
Phillip Maloney (Easi, Kennedy Space Center, FL, USA)
Jacqueline Quinn (NASA, Kennedy Space Center, FL, USA)

Ongoing research at NASA's Kennedy Space Center has shown the capability of a passive, polymer-based absorption system to be effective for the removal and treatment of sediment systems contaminated with polychlorinated biphenyls (PCBs). This technology, which has been termed **S**orbent **P**olymer **E**xtraction **a**nd **R**emedation from **S**ediments (SPEARS), utilizes a low-cost, hollow polymer filled with an environmentally friendly solvent to enhance hydrophobic, PCB molecule absorption into the SPEARS units. A recently completed, seven-month field study showed that SPEARS removed 75% of the PCBS (by mass) from contaminated sediments during this period. Longer exposer periods were not considered due to cost and time limitations, however, because the technology is driven by concentration gradients, there is strong evidence that higher removal rates could have been achieved. A next-generation design is currently being developed for use at a larger, longer-term site, with the intent of leaving the SPEARS system in place for as long as 12 to 18 months. Additional studies are currently underway to determine the effectiveness of SPEARS for the remediation of additional hydrophobic contaminants such as chlordane and pyrene (preliminary results will be presented).

# GLOBAL WARMING AND CLIMATE CHANGE

# FOREST FIRE A GLOBAL SCENARIO IDENTIFICATION AND SUSTAINABLE MANAGEMENT USING GEO INFORMATION TECHNOLOGY

JOSE T. M*, SUBIN K. JOSE*, G .MADHU**
*Christ College (autonomous), Irinjalakuda, Thrissur Kerala, jthekkan@gmail.com,josesubin@gmail.com

** School of Engineering, Cochin University of Science and Technology, Kalamaserry , Ernakulam.
Email. gmadhu@cusat.ac.in.

**ABSTRACT:** A forest fire event is influenced by ecological, human and climatic conditions. Forest fire directly cause biodiversity loss, forest degradation and climate change. It has direct ecological, economic and social impact. Fire prone area Modeling is done to have the most effective system of fire management and at the same time to reduce its deleterious effect on ecosystems, communities and landscapes. Geospatial techniques are proving to be powerful tools to assess the forest fire risks and the management. The present study deals with identification of fire prone area and there management in Shendurney wildlife sanctuary, the mega biodiversity hot spot in Western Ghats region, using GIS and Remote sensing techniques. Large extents of forests are affected by fire in every year in this region. Two types of data are used in the study i.e. spatial and non spatial data. Spatial data mainly includes Remote Sensing data, forest block, GPS field data, settlement boundaries, road/trek path network and the most important existing water bodies in that area where the non-spatial data pertains to meteorological data on temperature, rainfall . The parameters in consideration for the analysis are vegetation type, forest fragmentation, vicinity to settlement, distance from road, slope, aspect, temperature, amount of litter and rainfall. All these parameters have direct/indirect influence on the occurrence of fire and were integrated using GIS. A multi parametric weighted index model has been adopted to derive the 'fire-risk' zone/ fire prone area map. The final output shows forest fire risk area map of Shendurney wildlife Sanctuary in four categories such as very high risk, high risk, moderate risk and low risk. Based on this final output, the study attempted to give insight in the use of RS and GIS for sustainable fire management. For that, various spatial analysis have been carried out like identification of suitable sites for construction of check dams to ensure water availability in combating fires and for locating fire watch towers for better fire management. The present study effectively proved system for sustainable forest fire management.

***Keywords:*** Forest fire, GIS, Shendurney, Remote Sensing, Wildlife Sanctuary

## INTRODUCTION

Fire is a natural factor that has shaped the Earth's vegetation throughout its natural history. Fire disturbance is a major factor driving patterns of vegetation structure and composition in natural ecosystem (Whelan, 1995). Fire is clearly a global issue, affecting almost all climates and vegetation functional groups (Randerson *et al.*, 2006) Forest fire or wildland fire has become intense and more frequent in the last few decades all over the world, and is a critical issue in the biosphere–atmosphere interface (Crutzen and Andreae, 1990; Penner *et al.*, 1992; Cochrane ,2003). It has a profound effect on atmospheric chemistry, biogeochemical cycling and ecosystem structure (Schimel and Baker, 2002). Human activities can influence natural fire regimes by increasing fires in forests that would seldom burn under natural conditions (Goldammer, 2003).

In India the tropical moist and dry deciduous forests, which are frequently affected by forest fire, are extensive and account for 64% of the total forest area (Forest Survey of India, 2003). Over the past few decades, forest fires in India have received increasing attention because of the wide range of ecological, economic, social, and political impacts. Fire plays an important role in the creation and maintenance of landscape structure, composition, function, and ecological integrity (Covington and Moore, 1994; Goldammer ,1999; Morgan *et al.*, 2001) and can influence the rates and processes of ecological succession and encroachment. At the local scale, fire can stimulate soil microbial processes and combust vegetation ultimately altering the structure and composition of both soils and vegetation (Goldammer, 1999). Also, at the regional and global scales, combustion of forest and grassland vegetation releases large volumes of radiatively active gases, pyrogenic aerosols, and other chemically active species that significantly influence Earth's radiative budget and atmospheric chemistry (Andreae and Merlet, 2001), impacting air quality (Hardy *et al.*, 2001) and raise concern about risks to human health (Brauer ,1999).

Forest fire research can be considered as one of the most appropriate areas, where Geographic Information System (GIS) approach can be effectively applied. GIS can take definite advantage of the computer's capability in processing, storage and retrieval of immense data. The use of the GIS approach facilitates in integrating several variables in order to establish and focus on the problem. At the same time, it makes it possible to update or retrieve spatial information in different ways included in the database, to develop various models (Yanar and Akyurek, 2006). Geographical Information systems (GIS) are applied as tools to achieve an operational forest fire management system. With GIS models, new understanding of the fire danger situation in a management are can be gained. As an important factor of a risk analysis the damage potential that arises from fires starting at a certain point in the landscape can be estimated. (Rawat *et al.,* 1999). Earth-observing satellite sensors have been used since the 1970s to detect these changes, to quantify the burnt areas and to determine the composition and distribution of byproducts after fire (Robinson ,1991; Fuller ,2000).

The more complex and accurate fire models require spatial information that is furnished by remote sensing and Geographic Information Systems (GIS). The integration of remote sensing with GIS will be helpful for the spatial identification and management of forest fire (Bonazountas *et al.,* 2005).The integration of field data with remote sensing data is needed for the current and accurate fire risk mapping. GIS technology can serve as a vital technological core for the development of an integrated operational system in the case of forest fire crisis management. GIS with multicriteria decision making system will be helpful for the accurate prediction of different environmental phenomenon like landslide, drought and forest fire etc. (Tansey ,2004).

## MATERIAL AND METHODS

Since the study is aimed to create a GIS based methodology, both spatial and attribute data of the various thematic layers, available in the form of maps and published report were collected from the various sources. Primary data was collected using various techniques to fill any data gaps. The present study is carried out using GIS based multicriteria decision making methodology in conjunction with fuzzy logic, in a participatory decision-making framework to rank and prioritize the causative factors of fire risk in the study area. We used topographic, vegetation, climatic, social and field level parameters for evaluating the fire risk in the study area. These data were arranged in raster-based maps for further analysis. Spatial data were collected from Survey of India (SOI) Topographic map sheets of 1:25,000 and 1:50000 scale and Indian Remote sensing Satellite (IRS) satellite image. The SOI maps became the source for a number of basic thematic layers like contour, drainage, settlements, roads / trek paths and water body. IRS-1C LISS-III digital image acquired on 12 April 2009was used to prepare the land cover map. Visual interpretation of the standard False Colour Composite (FCC) with 4-3-2 RGB combination as well as digital classification of the image using ERDAS 9.1 image processing software was used for the preparation of land cover map. The classified image is field verified and found to be accurate. Details regarding trek path, settlement boundary and fire lines inside the sanctuary were collected during field visit. Data regarding fire occurrences during the last few years in study area were collected from the forest department and field visit. Temperature and humidity data was collected from the forest department meteorological stations, Indian Meteorological Department (IMD) station and also from Kerala State Electricity Board (KSEB) meteorological stations. An exhaustive field trip was carried out during the different time period of the study. Field level data collection and mapping was carried out using toposheet as reference. The real world co-ordinates were collected using GPS. The various data collected were processed and put into the Arc GIS software package.

In this study, we used Saaty's (2000) analytical hierarchy process, a MCDM (Multi Criteria Decision Making) methodology in conjunction with fuzzy logic, in a participatory decision-making framework to rank and prioritize the causative factors of fire risk in the study area. Our methodology consisted of four different components: (1) hierarchical structure development of fire risk criteria, (2) weights determination at different levels of hierarchy using linguistic variables and fuzzy sets, (3) assigning criteria weights in GIS, and (4) fire risk quantification using decision rule.

The location of existing fire watchtower was collected from the field with the help of GPS. The collected data is transferred in to GIS platform and analyse the visibility from each watch tower is analysed by using digital elevation model derived from the contour with the help of visibility analysis in Arc GIS. Based on the visibility analysis of existing watchtower suitable site for the construction of new watch tower is located. The criteria used for the identification of new watch tower is based on the invisibility from existing watchtower and the area that come under high fire risk area. Existing watchtower locations collected by the help of GPS during field study. The collected GPS location of existing watchtower location along with elevation data is plotted in Arc GIS. Visibility of existing watchtower is examined with help of spatial analyst extension in Arc GIS. For the identification of suitable check dam location flow accumulation flow direction was measured by using spatial analyst .The area with high flow accumulation and that are come under high fire risk area check dam location was identified. The base layer used for the preparation of flow direction and flow accumulation is elevation data and drainage data. The tool used for the flow accumulation analysis is the spatial analyst extension in hydrological analysis tool.

## RESULT AND DISCUSSION

The forest fire risk area map of Shendurney wildlife sanctuary is shown in Figure 1 and the fire risk area calculation is shown in Table 1. The fire risk area map is classified in to four classes, it includes low, medium, high and very high. In Shendurney wildlife sanctuary low fire risk area belongs to 41.7 Km2, medium 53.1, high 64.5 Km2, very high 11.7 Km2.

Table 1. Forest fire risk area – shendurney wildlife sanctuary

| Forest fire risk area – shendurney wildlife sanctuary | | |
|---|---|---|
| Fire risk class | Area (Km$^2$) | Area Percentage |
| Low | 41.7 | 24.38 |
| Medium | 53.1 | 31.05 |
| High | 64.5 | 37.71 |
| Very high | 11.7 | 6.86 |

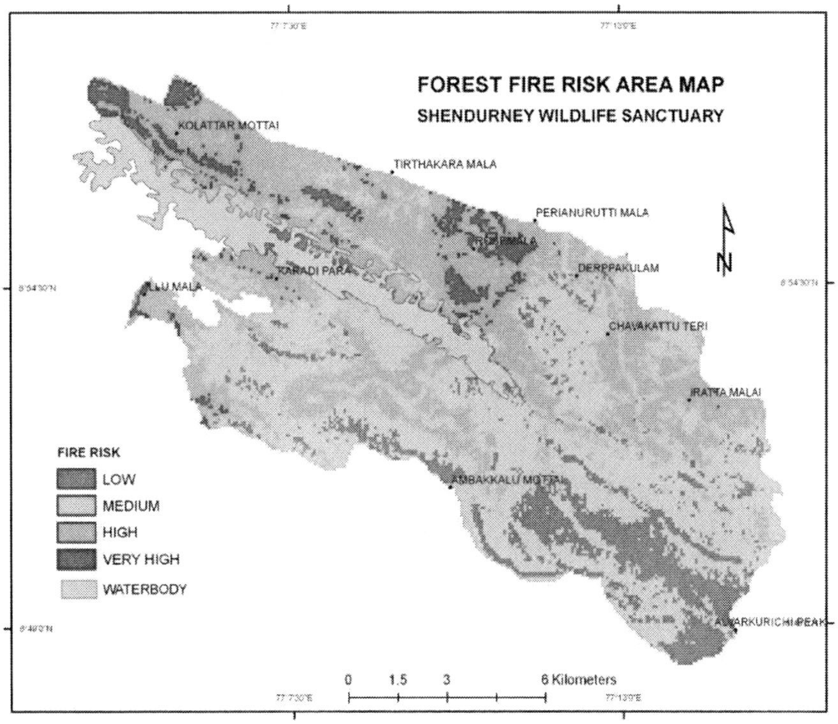


Figure 1. Forest Fire Risk Area Map of Shendurney Wildlife Sanctuary

The visibility from existing watch tower is analysed and shown in Figure 2 and based on the visibility analysis new location for watch tower is proposed. The suitable site for new fire watchtower for Shendurney is shown in Figure 3. The location for new watch tower is near to fire prone area and has high visibility. The suitable site for checkdam location of Shendurney wildlife sanctuary is shown in Figure 4. The check dam location is near to fire prone area and has high flow accumulation.

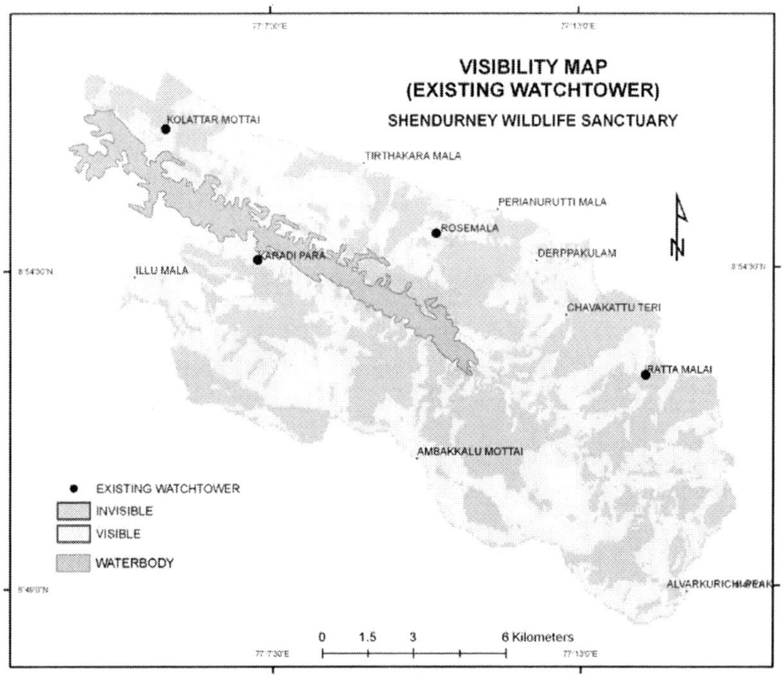


Figure 2. Visibility Map (Existing Watchtower) for Shendurney

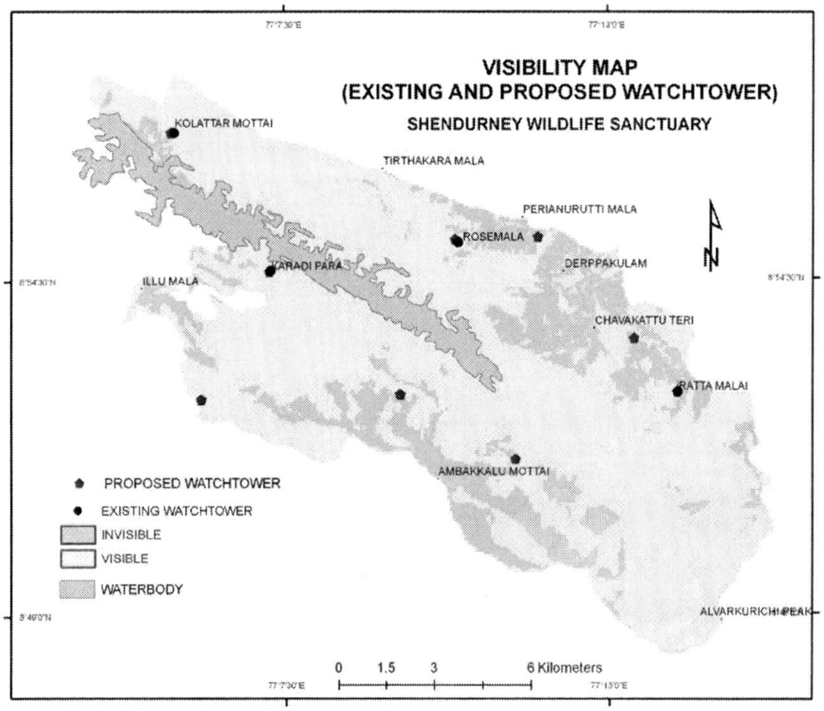


Figure 3. Visibility Map for Existing and Proposed Wachertowers

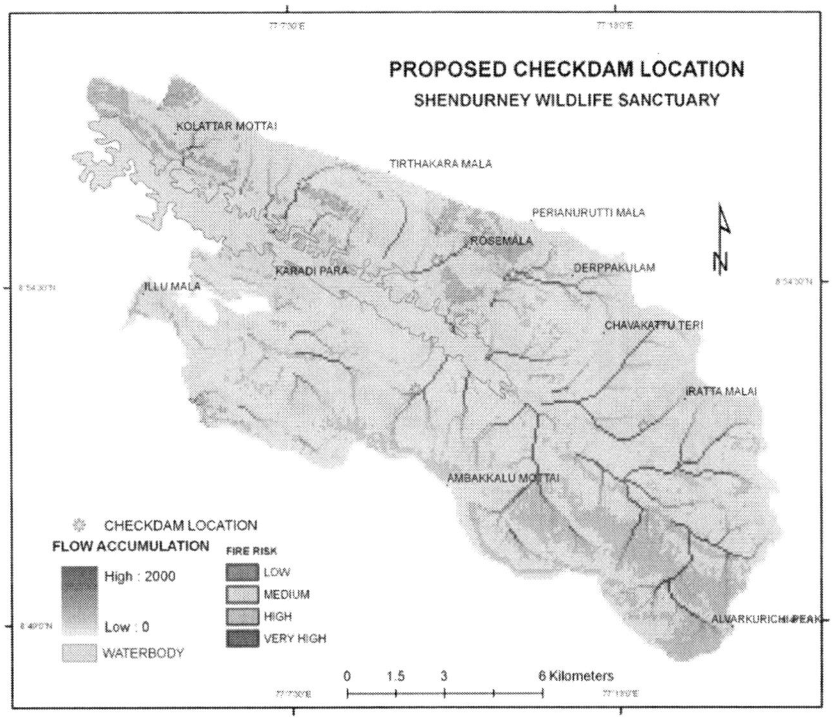


Figure 4. Proposed Checkdam Location for Shendurney

Forest fires controlled or uncontrolled have profound impacts on the physical environment including: landcover, landuse, biodiversity, climate change and forest ecosystem. They also have enormous implication on human health and on the socio-economic system of affected countries. Economic cost is hard to quantify but an estimate by the economy and environment can be provided. Most pronounced consequence of forest fires causes their potential effects on climate change. (Abhineet *et al.*, 1996). The normal fire season in India is from the month of February to May. India witnessed the most severe forest fires in the recent time during the summer.

According to the IFFN (2002) the ecological and socio-economic consequences of wild land fires in India include - Loss of timber, loss of bio-diversity, loss of wildlife habitat, global warming, soil erosion, loss of fuelwood and fodder, damage to water and other natural resources, loss of natural regeneration. Estimated average tangible annual loss due to forest fires in country is Rs.440 crore (US$ 64.70 millions approximately). The vulnerability of the Indian forests to fire varies from place to place depending upon the type of vegetation and the climate.

Lightning and rubbing of dry plant parts in wind is an important source of natural fires which have influenced dry vegetation ( Phillips ,1965; Komarek ,1968). Lightning fires have been observed and reported in the deciduous and semi-deciduous forest biomes as well as occasionally in the rain forest. Today the contribution of natural forest to the overall tropical wildland fire scene is becoming negligible. Most tropical fires are set intentionally by humans and are related to several main causative agents (Goldammer ,1996). .Deforestation activities (conversion of forest to other land uses, e.g. agricultural lands, pastures, exploitation of other natural resources); traditional, but expanding slash-and-burn agriculture; grazing land management (fires set by graziers, mainly in open forests with distinct grass strata ; use of non-wood forest products (use of fire to facilitate harvest or improve yield of plants, fruits, and other forest products, predominantly in deciduous and semi-deciduous forests); wildland/residential interface fires (fires from settlements, e.g. from cooking, torches, camp fires etc.); other traditional fire uses (in the wake of religious, ethnic and folk traditions; tribal warfare) and socio-economic and political conflicts over questions of land property and land use rights. Forest fire as an agent of regional land cover change and in modifying atmospheric chemistry.

The study attempted to give insight in the use of RS and GIS for fire management. Spatial modelling and analysis have been done in GIS environment for identification of areas prone to fire risk and subsequently response routes were suggested for extinguishing forest fires ( Porwal *et al.,* 1997). Some of the necessary components contributing to the fire behavior viz., fuel (vegetation types), topography (slope and aspect etc.) and the causes of fire (i.e., roads and settlements) have been given due weightages. The above results clearly suggest highly dynamic and spatial nature of fire events in the study area. To manage growing forest fires and associated fire hazards, as well as prioritize prescription efforts, it is essential to improve our understanding of the causative factors of fires. Earlier research relating to forest fire risk involved identifying the potentially contributing variables and integrating them into a mathematical expression, i.e., an index. In creating such an index, most of the earlier researchers focused on using meteorological data alone (Goncalves and Lourenco, 1990) or vegetation parameters (Hernandez-Leal *et al.*, 2006). Predicting the nature of "fires" may not be easy through using such indices alone. The fuzzy based GIS model attempted in our study takes a different approach, compensating MCDM through codifying the expert knowledge for forest fire risk variables and combining it with the human biophysical dimension. Most importantly, several of the causative factors of fire have inherent fuzzy characteristics.

Although humans had introduced fire in the forests of the Western Ghats at least 3500 years ago (Gadgil and Chandran, 1988), increased fragmentation of forests and continued demand on forest resources has contributed to detrimental changes in the spatial and temporal characteristics of the present fire regime (Kodandapani *et al.*, 2004). Human disturbances such as logging, grazing, and harvesting of NTFP could be responsible for altering forest fuel characteristics and maintaining these spatial patterns (Goldammer, 1988). The large number as well as the size of fires in the dry deciduous forests can be explained due to the fuel characteristics especially grass fuels with compact arrangement of fuel particles in the fuelbed (Agee, 2000), continuity in fuels (Miller and Urban, 2000), increased mean maximum daily temperature, and enhanced drying solar radiation during the fire season (Stott, 2000). Forest fires are rare in undisturbed moist deciduous forests, however past disturbance histories such as logging have now rendered the disturbed moist deciduous susceptible to fires (Woods, 1989). Once disturbed, the understorey of these forests are invaded by grass species, rendering the forests vulnerable to forest fires (Suresh et al., 1996; Freifelder *et al.*, 1998).

In addition forest fires could actually be having a net beneficial effect on the species diversity, composition, structure, and regeneration. The low fuel loads and also the low productivity (Mishra, 1983) in these ecosystems could result in patchy, low intensity fires (Oesterheld *et al.,* 1999). These fires could actually be recycling nutrients back into the soil, mitigating invasive species, reducing flammability of vegetation, and promoting regeneration of seedlings and saplings (Goldammer, 1988).

## REFERENCES

Abhineet, J., Shirish, A., Ravan, R.K., Singh, K.K., Das and Roy ,P.S. (1996). Forest fire risk modelling using Remote Sensing and Geographic Information System . *Current Science*, 70(10): 928-933.

Agee, J.K. (2000). *Wilderness fire science: a state-of-knowledge review.* In: USDA Forest Service Proceedings RMRS-P-15-VOL-5.

Andreae, M. O. and Merlet, P. (2001). Emission of trace gases and aerosols from biomass burning. *Global Biogeochemical Cycles, 15*, 955–966. doi:10.1029/ 2000GB001382.

Bonazountas, M., Kallidromitou, D., Kassomenos, P. A. and Passas, N. (2005). Fire risk analysis. *Human and Ecological Risk Assessment*, 11, 617–626. doi:10.1080/ 10807030590949717.

Brauer,M. (1999). Health impacts of biomass air pollution. In K. T. Goh, D. Schwela, J. G. Goldammar, and O. Simpson (Edt.), *Health guidelines for vegetation fire events—background papers* (pp. 189–254). Singapore: WHO.

Cochrane, M.A. (2003). Fire science for rainforests. *Nature,* 421:913–919.

Covington, W. W. and Moore, M. M. (1994). South-western ponderosa pine forest structure: Changes since Euro-American settlement. *Journal of Forestry, 92*, 39–47.

Crutzen, P.J.and Andreae, M.O.. (1990). Biomass burning in the tropics: impact on atmospheric chemistry and biogeochemical cycles. *Science,* 250:1669–1678.

Forest Survey of India. (2003). *State of forest report 2003*. Ministry of Environment and Forests, Available from http://www.fsi.nic.in/ sfr2003.html Accessed 20 Nov, 2007

Fuller, D.O. (2000). Satellite remote sensing of biomass burning with optical and thermal sensors. *Programme in physical geography* ,24:543–561

Gadgil, M. and Chandran, M.D.S. (19880. *On the history of Uttara Kannada forest.* In: Dargavel, J., Dixon, K., Semple, N. (Eds.), Changing Tropical Forests. Centre for Resource and Environmental Studies, Canberra, pp. 47–58.

Goldammer, J. (2003). Fire ecology of the recent anthropocene. *In: Proceedings of the 2nd International Wildland Fire Ecology and Fire Management Congress*, Orlando, Florida.

Goldammer, J. G. (1999). Forests on fire. *Science, 284*, 1782–1783.

Goncalves, Z. J. and Lourenco, L. (1990). Meteorological index of forest fire risk in the Portuguesemainland territory. In *Proceedings of the international conference on forest fire research* (B07, pp. 1–14). Coimbra.

Hardy, C. C., Ottmar, R. D., Peterson, J. L., Core, J. E. and Seamon, P. (Eds.) (2001). *Smoke management guide for prescribed and wild land fire: 2001 edition PMS 964 420-2.* NFES 1279.

Hernandez-Leal, P. A., Arbelo, M. and Gonzalez-Calvo, A. (2006). Fire risk assessment using satellite data. *Advances in Space Research, 37*, 741–746. doi:10.1016/ j.asr.2004.12.053.

IFFN (International Forest Fire News) ( 2002). Fire Situation In India. *International Forest Fire News*, 26: 23-27.

Kodandapani, N., Cochrane, M.A. and Sukumar, R. (2004). Conservation threat of increasing fire frequencies in the Western Ghats, India. *Conservation Biology* 18 (6), 1553–1561.

Komarek, E.V. (1968). Lightning and lightning fires as ecological forces. *International Procedings of. Annual Tall Timbers Fire Ecology conference. 8.* Tall Timbers Research Station. Tallahassee, Florida, USA.. 169-197.

Miller, C. and Urban, D.L. (2000). Connectivity of forest fuels and surface fire regimes. *Landscape Ecology* ,15, 145–154.

Mishra, R. (1983). *Indian Savannas.* In: Bourliere, F. (Ed.), Tropical Savannas. pp. 151–166.

Morgan, P., Hardy, C. C., Swetnam, T., Rollins, M. G., and Long, L. G. (2001). Mapping fire regimes across time and space: Understanding coarse and fine-scale fire patterns. *International Journal of Wildland Fire, 10*, 329–342. doi:10.1071/WF01032.

Oesterheld, M., Loreti, J., Semmartin, M., and Paruelo, J.M. (1999). *Grazing, fire, and climatic effects on primary productivity of grasslands and savannas.* In: Walker, L.R. (Ed.), Ecosystems of Disturbed Ground. Elsevier, Amsterdam, pp. 287–306.

Penner, J.E., Dickinson, R.E., O'Neill, C.A. (1992). Effects of aerosol from biomass burning on the global radiation budget. *Science,* 256:1432–1434.

Phillips, J. (1965). Fire as master and servant : its influence in the bioclimatic regions of Trans- Variations Saharat Africa. *International Procedings of. Annual Tall Timbers Fire Ecology conference..* 4. Tall Timbers Research Station. Tallahassee, Florida, USA. pp. 7-109.

Porwal, M.C., Meir, M.J.C., Hussin, Y.A. and Roy, P.S. ( 1997). *Spatial Modelling for fire risk zonation using Remote Sensing and GIS.* Paper presented in ISPRS Commission VII Working Group II Workshop on Application of Remote Sensing and GIS for Sustainable Development, Hyderabad, 1997.

Randerson, J.T., Liu, H., Flanner, M.G., Chambers, S.D., Jin, Y., Hess, P.G., Pfister, G., Mack, M.C., Treseder, K.K.and Welp, L.R.(2006). The impact of boreal forest fire on climate warming. *Science,* 314:1130–1132.

Schimel, D. and Baker, D. (2002). Carbon cycle: the wildfire factor. *Nature*, 420:29–30.

Suresh, H.S., Dattaraja, H.S. and Sukumar, R. (1996). The flora of Mudumalai wildlife sanctuary, Tamil Nadu, southern India. *Indian Forester,* 507–519.

Tansey, K., Gregoire, J. M., Binaghi, L. E.(2004). A global inventory of burned areas at 1 km resolution for the year 2000 derived from SPOT vegetation data. *Climatic Change*, 67, 345–377. doi:10.1007/ s10584-004-2800-3.

Whelan, J.(1995). *The Ecology of Fire.* Cambridge University Press.

Woods, P. (1989). Effects of logging drought and fire on structure and composition of tropical forests in Sabah, Malaysia. *Biotropica,* 21, 290–298.

Yanar, T. A. and Akyurek, A. (2006). The enhancement of the cell-based GIS analyses with fuzzy processing capabilities. *Information Sciences, 176*, 1067–1085. doi:10.1016/j.ins.2005.02.006.

# REDUCING GLOBAL WARMING IN AFRICA THROUGH TRADITIONAL AFRICAN ARCHITECTURE: CHALLENGES AND THE WAY FORWARD

***Iwuagwu, Ben Ugochukwu***. Dept of Architecture, Abia State Polytechnic, Nigeria
Ikechukwu onyegiri. Dept of Architecture, Imo State University, Nigeria
Iwuagwu, Ben Chioma M. Dept of Statistics, Abia State Polytechnic, Nigeria

**ABSTRACT:** Africa is already a continent under pressure from climate stresses and is highly vulnerable to the impacts of climate change. Many areas in Africa are recognized as having climates that are among the most variable in the world on seasonal and decadal time scales. Rising fossil fuel burning, construction, deconstruction, transportation, general land use changes etcetera have emitted, and are continuing to emit, increasing quantities of greenhouse gases into the Earth's atmosphere. These greenhouse gases include carbon dioxide (CO2), methane (CH4) and nitrogen dioxide (N2O), and a rise in these gases has caused a rise in the amount of heat from the sun withheld in the Earth's atmosphere. This increase in heat has led to the greenhouse effect, resulting in climate change. In 2012, 70 percent of major global droughts occurred in Africa. Kenya, Somalia, Sudan, Malawi, Angola, Chad and Ethiopia were particularly hit hard, and more than 16 million people in those countries were affected. In the same year, floods claimed 363 lives in Nigeria and 65 lives in Niger. In 2013, heavy rains continued with major flooding in Sudan, South Sudan, Mali, South Africa, Zimbabwe, Botswana and Mozambique. The macro- economic impact of these episodes of extreme drought and flooding is significant. It has been ascertained that construction industry alone consumes over 40 percent of energy in Africa leading to global warming and climate change, this paper therefore clamours for reduction in energy consumption in the construction industry and endorse traditional African architecture that is more sustainable and ozone friendly in reducing global warming in Africa and the world at large. This paper employs an extensive literature review and applies techniques which are common in content analysis. A thorough literature reviews was conducted to identify challenges of reducing global warming in Africa through traditional African architecture and conclude by recommending the way forward.

## INTRODUCTION

Global warming is an increase in the average temperature of the Earth's atmosphere, especially a sustained increase great enough to cause changes in the global climate. Global warming is synonymous with Enhanced green house effect, implying an increase in the amount of green house gases in the earth's atmosphere, leading to entrapment of more and more solar radiations, and thus increasing the overall temperature of the earth. Global warming is the 'talk of the town' in Africa today, especially with its detrimental effects already being brought to limelight by the recurring events of massive floods, and droughts throughout Africa. The average global temperatures are higher than they have ever been during the past millennium, and the levels of CO2 in the atmosphere have crossed all previous records. The recent trends of increasing global temperatures and incidences of extreme climate events in Africa, mainly droughts and floods, are likely to continue. According to the most recent report of the Intergovernmental Panel on Climate Change (2007), Africa is likely to warm at a faster rate than the global average during this century. According to the same report, Africa is also the continent that is least likely to be able to cope with climate change. In the East African region, average temperatures have increased by about 0.5 0C over the last century, in some other African countries, there is evidence that average temperatures have increased by as much as 1.4 0C since the 1960s (McSweeney, ''et al''. 2008). Looking further ahead, up to 4.30C change in average temperatures by the 2080s is possible according to Hepworth & Goulden (2008). A temperature rise of that magnitude would have disastrous consequences for Africa.

Climate change poses a significant and unique challenge to Africa because so much of its economy depends so much on a climate-sensitive natural resource base like rain-fed and subsistence agriculture. Dependence on such resources exposes the continent to the risks of reduction in agricultural production, municipal water supply for home use and sanitation services, industrial water use and hydroelectric power generation. While it has not yet been precisely determined how much of the recent warming was caused by human activities, the consensus among climate scientists is that most of the warming over the past 50 years was probably caused by human-produced greenhouse gases. When fossil fuels are burned to produce energy they release gases – in particular carbon dioxide ($CO^2$) – which build up in the earth's atmosphere. These gases act like a greenhouse, trapping more and more of the sun's light and heat as more fossil fuels are burned. In Africa, fossil fuels are being used at an increasing rate to generate electricity, provide fuel for transport, and heat people's homes. This means more 'greenhouse gases'. The rapidly increasing levels of greenhouse gases in the atmosphere are causing global temperatures to rise at a rate that is unprecedented in modern human history. They are also causing changes in weather patterns, altering rainfall and the timing of the seasons. At the same time as coal, oil and gas is being taken out of the ground and used for heating, electricity and transport, the world's forests are disappearing. Deforestation is also a major contributor to climate change, because trees absorb $CO^2$, the most important greenhouse gas.

Global warming has also been motivated by some other human activities like construction. Construction use more than 40 percent of world energy as a result of the high embodied energy materials use in conventional construction, especially in building construction. Using materials like steel, granite, etc that has higher embodied energy in conventional building construction contribute to global warming. It is the recommendation of this paper that in the bid to reduce global warming in Africa, that traditional building materials, like earth, wood, bamboo etcetera which have low embodied energy should be used in building construction.

## METHODOLOGY

This paper employs an extensive literature review and applies techniques which are common in content analysis. A thorough literature reviews was conducted to identify the effects of global warming in African, challenges of traditional African architecture and the way forward for traditional African architecture and their building materials in reducing global warming. The broad based, thorough and extensive reviews process aims at examining, synthesizing and recording all issues discussed by various authors in books, journals and other scholarly works. The effect of global warming in Africa and how it could be reduced using traditional African architecture becomes the focus of this paper.

## EFFECTS OF GLOBAL WARMING

Global warming has resulted in the warming of the oceans, rising of the sea levels, melting of glaciers, and diminished snow cover in the Northern Hemisphere. The recent catastrophic climatic events like the massive floods in Pakistan India and Nigeria, the Hurricane Katrina in the United States, the prolonged droughts in Australia, China, Pakistan, India, Uganda and Texas, are all the results of increased temperatures due to global warming. During the 21st century, climatic disasters occurred five times as frequently and killed or affected seventy times as many people. Between 2000 and 2004, an average of 26 climatic disasters was reported each year. Thus, the immense geological changes will continue their destruction unabated if steps to mitigate global warming are not taken.

Some of the most obvious impacts of global warming on human communities will be due to the rise in sea level that occurs mainly because ocean water expands as it is heated. Melting of ice on glaciers and polar ice caps adds to the rise. The projected total rise is estimated to be up to one metre this century and the rise will continue for many centuries. The extremely unusual high temperatures in central Europe during the summer of 2003 led to the deaths of over 20,000 people. A warmer world will lead to more evaporation of water from the surface, more water vapour in the atmosphere and more precipitation on average. On average, floods and droughts are the most damaging of the world's disasters. Between 1975 and 2002, due to flooding from rainfall over 200,000 lives were lost and 2.2 billion affected, and due to drought over half a million lives were lost and 1.3 billion affected (Jonkman, 2005). Recent estimates are that by 2050 over 10% of the world's land area will be so affected According to Houghton (2011) extreme droughts will tend to be longer, measured in years rather than months, again leading to many millions of displaced people.

Climate change poses a significant and unique challenge to Africa because so much of its economy depends so much on a climate-sensitive natural resource base like rain-fed and subsistence agriculture. In 2012, 70 percent of major global droughts occurred in Africa. Kenya, Somalia, Sudan, Malawi, Angola, Chad and Ethiopia were particularly hit hard, and more than 16 million people in those countries were affected. In the same year, floods claimed 363 lives in Nigeria and 65 lives in Niger. In 2013, heavy rains continued with major flooding in Sudan, South Sudan, Mali, South Africa, Zimbabwe, Botswana and

Mozambique. The macro- economic impact of these episodes of extreme drought and flooding is significant. Having ascertained that building industry alone consumes over 40 percent of energy in Africa leading to global warming and climate change, it has been also ascertained that global warming and its effects can be reduced especially in Africa through traditional African architecture. According to Iwuagwu and Azubuine (2015) Africa's traditional architecture made certain that its use of the resources neither diminished their availability, nor adversely affected the ecological balance upon which it relied on as an agrarian society. African traditional architecture is essentially sustainable and had evolved culturally to suit the people. Usually, earth, timber, straw, stone/rock and thatch were constructed together with the simplest of tools and methods to build simple, livable dwellings. Although globalization according to Opaluwa, ''et al''. (2012) has relegated them as being 'primitive'. But a present interpretation of sustainability has given them a new status as likely technologies for the contemporary world.

## AN XRAY OF SELECTED TRADITIONAL AFRICAN BUILDING MATERIALS

***Straw/Thatch Architecture.*** In contrast to some other materials which are not easily renewed, straw/thatch is a by-product of grown plants hence, a cultivated material. Large quantities of this material could be sourced from the immediate surrounding as the villagers cultivate much of the straw as cereals in their farms which in turn provide them a building material. Straw has in most case being used with adobe bricks or masonry walls or singularly. Even today nomad tribes within Africa still use this form of construction. Straw construction uses matted or baled straw from wheat, oats, barley, rye, rice and others as walls or covered by earthen or lime stucco. Straw bale are traditionally a waste product; it is the dry plant material or stalk left in the field after a plant has matured and harvested.

***Adobe Architecture.*** the cultural practice of the rural people indicates that adobe surely has been one of the most common and abundantly obtainable materials that influenced and sustained the rural villages in Africa. Local earth technologies of Africa have spanned form the employment of raw-earth, to refined earth brick. Generally employed was wattle-and-daub earth technology; a method that uses solid wooden post frame which is first made then filled with adobe balls to create a wall. Most often, the African builders construct the walls of their building layer by layer using the mud bricks and a slurry mixture of earth as the mortar. Once the wall dries up and binds into a monolithic structure, a dense mud plastic plaster strengthen with various additives depending on the people's culture (cow dung, goat dung, beaten straw, animal hair, animal skin fat) will be spread on by hands to smoothen the facade. The materials having being gotten from the environment certainly made it a highly sustainable practice that utilised small amount of energy and did not generate any greenhouse gas or harmful waste. As soon as it is plastered and properly covered with overhanging roofs, these earth buildings were structurally firm, environmentally sound and could exist for years as long as the day to day maintenance was adhered to.

***Bamboo.*** Bamboo is one of the oldest traditional building materials used by mankind. When you're considering potential building materials for home construction as a society we tend to focus on two or three commonly utilized and widely accepted building materials: wood, stone or concrete. Uses for bamboo can also include building construction, both in exterior and interior design elements. Unlike wood, bamboo (a member of the grass family) regenerates very quickly. It is, in-fact, one of the fastest growing plants in the world, with the fastest growth rate reaching 100cm in a 24-hr period (Farrelly, 1984). In contrast to tree harvesting, there is simply no comparison to the replenishment rate of growing bamboo. Bamboo can be harvested every three to six years for construction purposes (depending on the species); whereas trees range from 25 years (for softwoods) to 50 years (for hardwoods). Major Advantages of Bamboo include: Strength and Durability; Affordability.

## ADVANTAGES OF TRADITIONAL AFRICAN BUILDING MATERIALS

***Availability.*** Traditional African building materials are abundant in nature. These materials include earth, stone, thatch, coconut fibre etc. for example earth building technology involves the use of laterite and loamy soil that exist in abundant supply in all part of the continent. Earth has been used by our fathers and fore-fathers to erect buildings, sometimes up to two storeys high without addition of any other reinforcing materials and most of them are still standing till date.

***Affordability.*** The major reason for high cost of imported building materials in Africa includes high cost of importation and general inflation. Because of high cost of these imported building materials, the low income earners find it difficult to construct their own houses or even rent a decent house. The availability of these local building materials makes the price affordable and gives the low income earners the opportunity to construct their own houses.

***Energy Efficient.*** Environmental protective measures ensure reduction of operational energy in construction. Studies according to Iwuagwu and Azubuine (2015) reveal that the building sectors consume more than one third of the world's energy, and contribute to global warming. A typical traditional building of earth emits fewer greenhouse gases, consumes less energy, and maintains a high level of internal thermal comfort, regardless of prevailing solar radiation outside (Iwuagwu and Azubuine, 2015).

**Ozone Friendly.** The built environment contributes ultimately to global warming by its high rates of greenhouse gases emission through energy use (for cooling, heating, and lighting) and construction. Local building materials projects a possibility of total reduction to a near zero carbon emission of buildings. Local building materials are eco-friendly, climate responsive and organic protective measures to safeguard and as well minimize environmental impact. The thermal insulation, energy saving, etcetera of local building materials reduces negative environmental impact. The proximity of materials saves cost and reduces pollution by fuel burning through transportation.

***Reusability.*** Reusability is a function of the age and durability of a material. Very durable materials may have many useful years of service left when the building in which they are installed is decommissioned, and may be easily extracted and reinstalled in a new site. Windows and doors, even brick can be successfully reused. Timber from old barns has become fashionable as a reclaimed material for new construction.

***Biodegradability.*** The biodegradability of a material refers to its potential to naturally decompose when discarded. Organic materials can return to the earth rapidly, while others, like steel, take a long time. An important consideration is whether the material in question will produce hazardous materials as it decomposes, either alone or in combination with other substances. Traditional African building materials exhibit this characteristic, example include, earth, thatch, bamboo, timber etcetera.

## CHALLENGES OF TRADITIONAL AFRICAN ARCHITECTURE AND THEIR BUILDING MATERIALS

***Acceptability***. The future of any Architecture depends on the extent to which it is acceptable to the people for whom it is intended. The notion that buildings of traditional materials are substandard is the main obstacle to the development of an authentic African Architecture that is truly indigenous to the people. Perhaps issue of acceptability has brought destruction of traditional values and their replacement with alien ones.

***Durability/ Low Strength.*** The highest problem of houses built with traditional African building materials was the low strength of the houses, implying that the locally available materials have strength that is below expectation when compared with strength of the houses built with conventional materials such as cement, concrete, steel, among others. The strength of every house is important because it determines the durability and security of the house. This is consistent with the view of Venkatarama and Prasanna (2009) that one of the drawbacks for using earth alone as a material for construction is its durability which is strongly related to its compressive strength. Riza et al. (2011) further explained that most soil in their natural condition lack the strength, stability and durability required for building construction. The above indicates that the local materials for building houses lack the desired strength and improvement of their strength properties would be beneficial to the users.

***Building Tall.*** The nature and strength of these traditional building materials make it impossible for building tall. They encourage low rise buildings leading to over use of land. Millions of acres of land in Africa have buildings constructed on them. For instance, spread of buildings require the construction of new roads, drainage, utility poles, and other infrastructure, which lead to, habitat destruction, land disturbance and erosion, environmental pollution, global warming among other.

**Deforestation:** the building materials sources in Africa contribute to prevailing problem of deforestation on the continent. When timber for construction is harvested, in many cases, replacement trees are not planted, if replanted, the rate of replacement is far lower than the rate of consumption.

***Civilization.*** Local and indigenous cultures have practiced sustainable resource use due to their practical experience and human dependency on earth's life support systems. Traditional communities integrated buildings into the natural environment in terms of adequate use of cost effective and easily assessable local building material for construction and maintenance. However, importation and the use of imported building materials has put the African traditional and sustainable way of building and construction processes in danger, thus relegating our architecture

***Frequent Maintainance***. Frequent maintenance of houses built with local materials especially earth, is another challenge facing African architecture. According to Rumana (2007) there is high maintenance requirement of earthen plinth and walls which are often plastered, especially during the wet season. This is due to the low strength of the materials that make it a requirement for the frequent maintenance in order to keep the building is good condition for use. If the houses built with local materials are not maintained regularly, they will deteriorate in some few years after their construction due to their vulnerability to weather such as rain and storm.

## THE WAY FORWARD FOR TRADITIONAL AFRICAN ARCHITECTURE AND THEIR BUILDING MATERIALS

***Re-engineering of traditional African building materials.*** Traditional African building materials should be re-engineered to suit the present need of building materials. Like the compressed earth block (CEB), this is an emerging alternative to adobe blocks and wattle and daub construction. Compressed earth blocks are a creative, re-engineering of the adobe brick. Unlike the native adobe block, which is a mixture of soil, water and distinct cultural additives moulded to desired shape with the hand, the compressed earth block is supplemented in very small amounts (in most cases less than 10%) with either cement or lime component in its blending process. The blend is not worked to achieve a plastic state, but simply blended until the cement/lime and soil are thoroughly mixed. Afterwards, the mixture either machine pressed or placed in a mould and compacted with a high level of pressure applied through a hand- operated machine. After aeration, the CEBs gain a high compressive strength appropriate generally for three floors constructions but higher potentials can also be attained for up to five floors constructions [Maini, 1999]. CEB construction is more durable than wattle and daub, and is also accepted as a more refined construction method than wattle and daub.

***Combination with modern building materials.*** there should be combination of conventional building materials with traditional African building materials to add to their strength and durability. One of the disadvantages is strength, but that notwithstanding, in high rise buildings CEB could be used for the walling system not minding the height of the building while the conventional materials would be used for the foundation, reinforcement etcetera.

***Government encouragement.*** government should encourage the use of local building materials by using them in government projects all over Africa. Traditional African building materials should be included in building code of various countries in Africa, as building materials. In every construction there should be a little destruction, efforts should be made to plant more trees and make sure natural resources are not being used more than it generate. As our natural resources are quickly depleting and our homes and vehicles are damaging our planet more than ever, it's time for a change. We should detach ourselves from the notion that local building materials are low-grade material. Considering the depletion of the ozone layer, high energy consumption by buildings we will realize that local building materials are not only the past, but the future of our society and one should take into consideration the benefits that they offer.

## CONCLUSION

In general, the use of local materials needs to be supported and reinforced to produce sufficient quantities of materials of adequate quality to withstand the effects of climatic conditions which range from humid rainy seasons to extremely hot dry seasons. The submission of this paper is that with the use of traditional African building materials, we could achieve a huge reduction in global warming in Africa and save human and the environment. it is therefore necessary to promote the construction with traditional African materials and combine its potentials with modern technology so as to produce a new generation of buildings that are cheap, environmentally friendly and require no mechanical air-conditioning during hot days and nights.

## REFERENCES

Farrelly, D. 1984. *The Book of Bamboo*. Sierra Club Books.

Hepworth, N. & M. Goulden. 2008. ''Climate Change in Uganda: Understanding the Implications and Appraising the Response''. Scoping Mission for DFID Uganda.

Houghton, J. R. 2011. ''Global Warming, Climate Change and Sustainability: Challenge to Scientists, Policy Makers and Christians''. JRI Briefing Paper No 14

Intergovernmental Panel on Climate Change. 2007. ''A Report of Working Group I of the Intergovernmental Panel on Climate Change: Summary for policymakers; Royal Society. (2010).Climate change: A summary of the science. http://royalsociety.org/climate-change-summary-of-science/

Iwuagwu, B.U & C. E. Azubuine. 2015. *Global Warming Versus Green Architecture: African Experience.* Proceedings of International Conference on IT, Architecture and Mechanical Engineering (ICITAME'2015) May, 2015

Jonkman, S. N. (2005). Natural Hazards 34, 151-175

Maini S. 1999. ''Seminar on earth architecture at Alliance Françoise De Colombo''. Auroville Building Center, India

McSweeney, C., M. New and G. Lizcano. 2008. ''UNDP Climate Change Country Profiles: Uganda''. New York: UNDP.

Opaluwa E, P. Obi and O. C. Osasona. 2012. *Sustainability in traditional African architecture: a springboard for sustainable urban cities.* Proceedings of sustainable Futures: Architecture and Urbanism in the Global. South Kampala, Uganda, June 2012

Rumana, R. 2007. ''Traditional House of Bangladesh: Typology of House According to Materials and Location''. Virtual Conference on Sustainable Architectural Design and Urban Planning AsiaSustainabilityNet.upc.edu, September 2007

Riza, F. V., I. A. Rahman & A. M. A. Zaidi. 2011. ''Preliminary Study of Compressed Stabilized Earth Brick (CSEB)''. Australian Journal of Basic and Applied Sciences, 5(9) 6-12

Venkatarama, R. B. V., & K. P. Prasanna. 2009. ''Embodied energy in cement stabilised rammed earth walls''. Energy and Buildings, 42(3) 380-385.

# ENHANCEMENT OF WASTEWATER FACILITIES FOR LONG TERM GLOBAL CLIMATE CHANGES

***Amirthaganth Amithalingam*** (Parsons Brinckerhoff Inc., Seattle, Washington, USA)
D. S. Mahamah (Saint Martin's University, Lacey, Washington, USA)

Most scientists and climate science practioners agree that the global weather patterns are changing. In fact the general agreement is that the earth is warming at about 0.15°F per decade (USEPA). Although there is much disagreement on the causes, many believe gas emissions such $CO_2$ and other greenhouse gases are the culprit. The EPA indicates that the three major causes of global warming are: greenhouse effect, variation in the sun energy, and changes in reflectivity of earth surface and atmosphere. The effects of an increase in global temperature include, a rise in sea levels, changes in the amount and pattern of precipitation, expansion of subtropical deserts, threat to food security and more frequent occurrence of extreme weather events including heat waves, droughts and heavy rainfall. The operation and in particular performance of wastewater treatment plants (WWTPs) are temperature dependent. Effects of the temperature on the entire wastewater treatment process (production to treatment) can be identified as follows:

- Increased overflows- Results in more untreated sewer overflows, thus more raw sewage will be mixed with receiving bodies of water.
- Increased flooding of Treatment plants -With increased storms comes increased flooding which can be harmful to infrastructure since most WWTPs are in low, coastal areas.
- Effect on microbiology of treatment systems -Increased temperature in effluent has delicate effect on ecosystems which could affect the system negatively. In addition, biological speciation is temperature dependent and as such different bacterial populations dominate at different temperatures.

Suggested enhancement methodologies to the WWT Facilities so as to overcome impacts of long term climate change are include; utilizing Best Management Practices (BMP's), highly efficient flood management plan, bypass systems in case of emergency, backup power systems, and the continuous monitoring of influent characteristics so as to alter treatment methodologies accordingly etc. Impacts due to climate are not preventable but mitigated through BMP's and other efficient methodologies.

# AN INTEGRATED ASSESSMENT OF LOW CARBON EMISSION SCENARIOS PROPOSED IN CLIMATE POLICY

Sascha Hokamp and ***Mohammad Mohammadi Khabbazan***
(Center for Earth System Research and Sustainability, University of Hamburg, Hamburg, Germany)

In 2015 the Conference of the Parties held in Paris, France, has discussed, among other things, climate policy proposals to decarbonize the global economy before the end of the 21th century. Eventually, on this challenging task the Conference of the Parties found no consensus view but nevertheless the final agreement of Paris calls for enhanced scientific investigation of low carbon emission scenarios. We pick up this call and address how carbon emission cuts in accordance to recent climate policy proposals may allow for holding the increase of global mean temperature below 2° Celsius. To tackle this research question, we employ the socio-economic integrated assessment Model of INvestment and endogenous technological Development (MIND), which has a sophisticated energy sector module. Further, we make use of a climate module consisting of three layers: Atmosphere, Upper Ocean, and Lower Ocean; a kind of climate modelling that is frequently used in the Dynamic Integrated model of Climate and the Economy (DICE) by William D. Nordhaus. We perform, within our stylized integrated assessment model, a cost effectiveness analysis with constraints on anthropogenic carbon emissions (instead of constraints on the increase of global mean temperature). We find that carbon emission cuts from 2100 onwards to annually 0.5 Giga tons Carbon lead to a stronger increase of global mean temperature in the short-term than the business as usual scenario without any changes of climate policy. In line with the literature this contra intuitive outcome can be explained by the vanishing cooling effect of sulfur dioxide. However, considering extremely low carbon emission scenarios we find that zero emissions from 2060 become feasible if the investments for renewable energy production peak in 2020, reach a plateau from 2040 to 2050, and decline thereafter. This finding seems to be driven by the depreciation rate of installed capacity devoted to renewable energy production. Hence, decarbonizing the global economy and meeting the 2° Celsius target could be possible before the end of the 21th century; but urgent action is necessary since the time window is closing for low carbon emission scenarios recently proposed in climate policy.

# LEAPFROGGING CARBON CAPTURE SEQUESTRATION

***Adel J. Al-Khalifah***
(Saudi Electricity Company, EOA-Generation Technical Support Dept. Dammam, Saudi Arabia)

There are many barriers and knowledge gaps regarding the development of carbon dioxide capture and storage (CCS). For instances, when it comes to costs, life-cycle effects, and storage capacity. In spite of these barriers and need for further science studies, the scientists are generally very hopeful regarding CCS development. The purpose of this work is focus on two key aspects: First is the function and potential of CCS and secondly are uncertainties. The optimism among the CCS experts is tentatively explained. The explanatory flexibility of CCS is claimed to be an essential clarification for the optimism. CCS is promoted from a wide variety of perspectives, including team spirit, peace, bridge to a sustainable energy system, sustaining the modern lifestyle and compatibility with the fossil fuel lock-in. Awareness of the uncertainties and potential over-optimism is warranted within policy and decision making as they often rely on scientific forecasts and experts' judgements. Concerning economic barriers, governments should urgently consider methods to assist stakeholders to significantly drive down the cost of CCS deployment, since it is the stakeholders who will be making the majority of the financial investments. Also governments should review institutional regulatory policies to identify how the barriers to CCS deployment may be reduced. Finally, international collaboration should be a key component of any CO2 emissions reductions strategy.

# AEROBIC AND NITRITE-DEPENDENT METHANE OXIDIZERS IN THE ZOIGE WETLAND OF THE TIBETAN PLATEAU

**Anzhou Ma**, Guoqiang Zhuang, Mengmeng Cui
(Research Center for Eco-Environmental Sciences, Chinese Academy of Sciences, Beijing, China)

Methane ($CH_4$) plays an important role in global warming. However, most $CH_4$ is consumed by the methane oxidizers before its emission into the atmosphere. To investigate the impacts of $CH_4$ on global warming, it is important to understand the methane oxidizers in wetlands, the largest source habitat of atmospheric $CH_4$. In this study, we used quantitative PCR and functional gene-targeted 454 pyrosequencing to elucidate the abundance and diversity of aerobic methanotrophs and nitrite-dependent methane oxidizers (N-DAMO) in the Zoige Wetland, which is located on the Tibetan Plateau and is the largest alpine wetland in the world. Our results demonstrated type I methanotrophs were the primary aerobic methanotrophs, and *Methylobacter* comprised approximately 50% of the community. Type II methanotrophs, represented by the genus *Methylocystis*, were also detected, and the presence of N-DAMO bacteria in the Zoige Wetland was confirmed by the quantification and pyrosequencing analyses of N-DAMO 16S rRNA and *pmoA* genes. Moreover, *pmoA* ranged from $10^7$ to $10^8$ copies•$g^{-1}$ of fresh soil, and the 16S rRNA gene of the NC10 phylum ranged between $10^4$ and $10^5$ copies•$g^{-1}$ of fresh soil. The results suggest the aerobic methanotrophs and N-DAMO communities together affect $CH_4$ cycle at the Zoige Wetland.

# ENERGY CONSERVATION HELPS IN MITIGATION OF ADVERSE EFFECTS OF CLIMATE CHANGE

***Mehboob Alam Khan***
(Energy Conservation Department, K-Electric, Karachi, Pakistan)

Energy generation plays a vital role in development of any country but at the same time it also enhances the emission levels if Energy is produced through Oil, Gas or other similar resources in which carbon is emitted.

In order to reduce carbon emissions the ultimate approach is to switch towards renewable and green Energy generation resources but shifting of peak demands requires sufficient amount of time and investments, as a low hanging fruit and workable solution to get **win win** situation is to opt for Energy Conservation methodologies by which we can easy reduce the load demand as well as Generation burden hence resulting in reduced carbon emissions which is the key element of Climate change.

This Paper explains how and by what measure we can reduce the demand of Electricity hence reducing carbon emissions to achieve an overall goal of mitigating adverse effects of Climate change and Global warming.

# METALS

# SPATIAL AND TEMPORAL DISTRIBUTION OF HEAVY METALS IN COASTAL RED SEA SEDIMENTS

Bandar Almur and Andrew Quicksall (Southern Methodist University, TX, USA)
Ahmed Al-Ansari (King Abdul-Aziz University, Saudi Arabia)

**ABSTRACT:** Jeddah is the most industrialized city on the west coast of Saudi Arabia. In this study, the spatial and temporal distributions of heavy metals were obtained from near-coast Red Sea sediment cores in close proximity to Jeddah. Six elements (Cr, Cu, Mn, Zn, Pb, and Fe) from three impacted locations (Prince Naif Street, the Downtown area, and Al-Khumrah) and an upstream reference site (Salman Gulf) were analyzed via heavy acid digestion and inductively coupled plasma-mass spectrometry (ICP-MS). The average concentrations of Cr, Cu, Mn, Zn, Pb, and Fe in core sediments of all sites were 245.96, 251.82, 478.45, 623.09, 362.75, and 8506.13 mg/kg, respectively. The depth-resolved results showed that highest concentrations of Mn, Cu, and Pb were found in the top 15 cm of the core profile distributions compared to other depth sub-samples. Heavy metal concentrations in core sediments seem to be increasing toward the area of the downtown and higher in the recent years. This indicates that this area has suffered from heavy metal pollution compare to other locations in the Red Sea. Heavy metal contamination due to anthropogenic activity should be taken into account to protect the Red Sea during future growth.

## INTRODUCTION

Heavy metal contamination in sediments is a major environmental concern due to potential remobilization, bioaccumulation in fauna and flora and large mammal toxicity upon consumption of such biota (Nemati et al. 2011). Sediments are the primary sink of heavy metals in the aquatic environment due to their chemical-physical properties, such as particle characteristics, and processes including precipitation and sorption/desorption (Rath et al. 2009; Fukue et al. 2006). Determination of heavy metal concentrations in core sediments can provide important information about the current and the background levels of contamination. Furthermore, it may provide historical evidence of the anthropogenic effect in the aquatic environment (Chatterjee et al. 2007; Al-Najjar et al. 2011; Tang et al. 2010). Sediment cores can provide chronologies of contaminant concentrations and a record of the changes in concentrations of chemical indicators in the environment over time (Conrad et al. 2007; Li et al. 2012).

Here, Jeddah, Saudi Arabia is the area of study and the most industrialized city in the west coast of Saudi Arabia on the Red Sea. The coastal area of western Saudi Arabia has been subjected to pollution stress from land-based activities due to rapid increase in industrialization for several decades. The problem is greatly magnified by the production of numerous toxic chemicals that are harmful in trace concentrations (Badr et al. 2009; Usman et al. 2013; Ghandour et al. 2014). The study of depth-resolved heavy metals in coastal sediments is needed to provide useful information on human activities in the Jeddah area of the Red Sea. In the present study, four Red Sea sediment cores were collected near major industrialized areas in Jeddah, Saudi Arabia. Heavy metals were measured to investigate their spatial and temporal distribution pattern in the Red Sea. The results of this study fills the gap of research in the region and serve as a baseline for future studies in the Red Sea area. The aim of this study was to (i) assess of the spatial and temporal distribution of Cr, Cu, Mn, Zn, Pb, and Fe in the sediments of the coastal area of Jeddah; (ii) identify the origin of heavy metal pollutants in the sediment cores.

## MATERIAL AND METHODS

Four Red Sea sediment cores were collected offshore of Jeddah in January 2015. Sites were chosen to cover the coastal area, which known to be affected by land-based activities. The four different sites were selected from three impacted locations (Prince Naif Street, the Downtown area, and Al-Khumrah) and an upstream reference site (Salman Gulf). Figure 1 shows a location of the study area and sampling sites. Each sampling location was determined using a GPS and located between 21° 51' 52" N and 39° 9' 26" E. Sediment cores were collected by scuba divers, using PVC tubes of 75 cm long and 5 cm in diameter. The cores were kept in an icebox at 4°C until delivered to the laboratory for analysis. After core samples were obtained at selected locations across the study area, samples were stored frozen at -20°C for 24 hours. Sediments were recovered from the PVC tubes and sliced at 2 cm interval. Each layer was homogeneously mixed and stored in plastic vials, labeled, and stored frozen at -20°C until shipping to U.S.A.

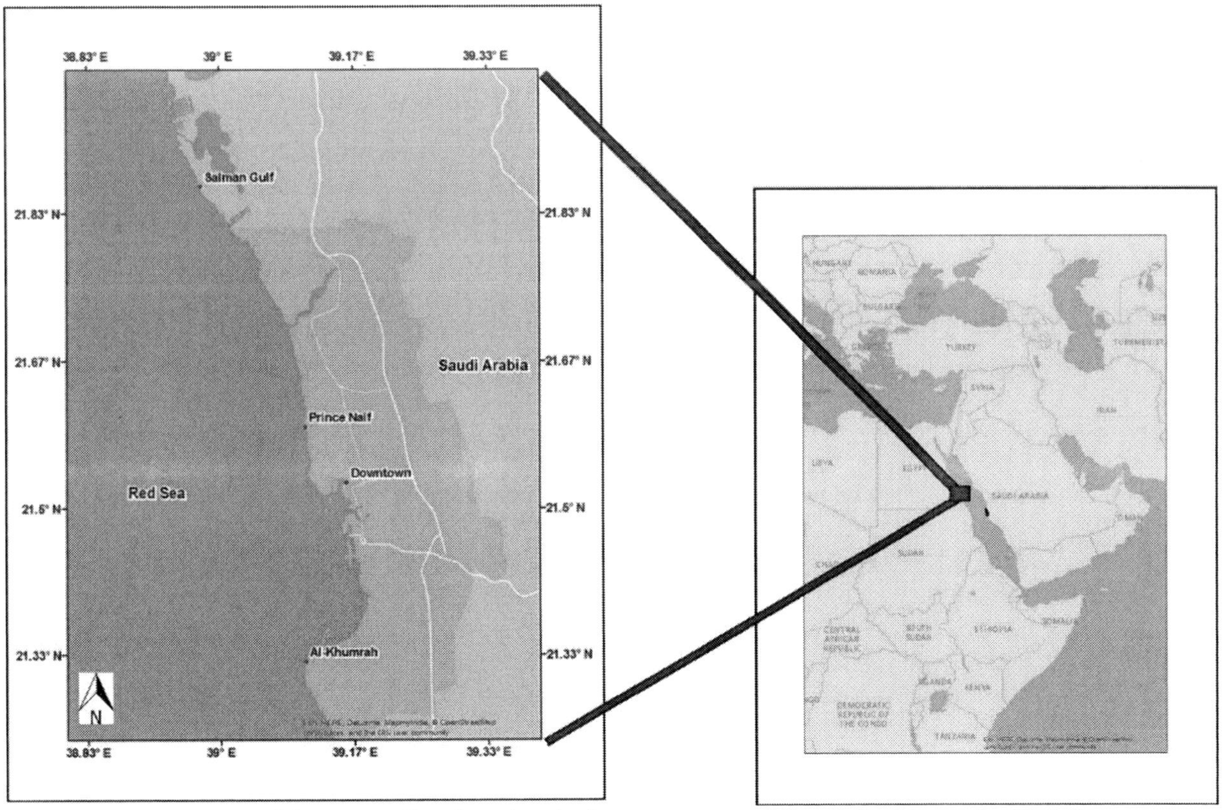


FIGURE1. Map of four sampling location in the west coast of Saudi Arabia on the Red Sea.

Sediment samples were prepared according to US EPA 3050B method to determine the total heavy metals for Cr, Cu, Mn, Zn, Pb, and Fe in core sediments. Briefly, samples were dried in an oven at 105°C and ground into a powder using a mortar and pestle. Nitric acid (70% HNO3) and hydrofluoric acid (HF) were used for the total metal analyzes. Approximately 0.20 grams of the sediment samples were digested in heavy acid with 3:1 mixture of HNO3 and HF acid. Nine ml of HNO3 and three ml of HF were added to the sediment samples inside a fume hood. Sediment samples plus HNO3 and HF acid were digested at 100°C on a hot plate until dryness. The solutions were allowed to cool to room temperature. The metal determinations were performed on ICP-MS in collision cell mode with kinetic energy dispersion (CCT-KED). For Quality Control (QC) and Quality Assurance (QA), blanks and replicates (about 10 % of the total number of samples) were analyzed to check precision under the same procedures mentioned above. A multi-element standard was used with a range from 0.5 to 75 ppb. The calibration curves were based on elements standard solution were obtained. A blank (5% HNO3) was analyzed every 10 samples. The final concentration of each metal was reported in mg/kg dry weight.

## RESULTS AND DISCUSSION

Heavy metal concentration variations at different depths for sediment cores are shown in Figure 2. Heavy metal distributions at four different locations with depths between 0 and 50 cm were evaluated throughout this study. Generally, heavy metals concentrations increase stratigraphically through the cores with a constant decrease from the top to lower level of the sediment cores. The highest concentration of heavy metals in the upper-most segments of the sediment cores compared to lower depth sub-samples for most of metals suggests increases in metals delivery to sediments in recent years (Figure 2). These high concentrations likely correspond to an increase in anthropogenic pollutant inputs from urbanization and industrialization. In this study, the results show that the average total concentrations as follows: Mn 478.45, Cr 245.96, Cu 251.82, Zn 623.09, Pb 362.75, and Fe 8506.13 mg/kg. The highest concentrations of all heavy metals were found at the downtown area.

Several authors have investigated the levels of heavy metal pollution in the marine environment of the Red Sea, especially near Jeddah (Badr et al. 2009; Al-Najjar et al. 2011; Pan et al. 2011; Usman et al. 2013; Ghandour et al. 2014). Concentrations of Mn, Zn, Cu, and Cd in sediments of Jeddah were studied by different scientist (Badr et al. 2009; Ghandour et al. 2014). In fact, most of the studies that have been carried out on the marine environment of the Red Sea have been concentrated in the central and northern parts of the coastal waters of Jeddah (Badr et al. 2009; Ghandour et al. 2014). In general, the higher concentrations of Mn in core sediments in the present study areas were similar to those found in Red Sea sediments by Hilial et al. (1988) and Ghandour et al (2014) in sediment composition of an arid coastal environment. In comparison with other studies in the Red Sea, little information on Fe concentration has been reported. In the present study, the higher concentrations of Fe were similar to Badr et al. (2009) results in Salman Gulf and Al-Khumrah areas and less than the world average concentration (48,000 mg/kg) (Venkatramanan et al. 2015).

Zn is a natural abundant element and can present in marine environment via human activates, such as food wastes, antifouling pints, and manufacturing of pesticides (Badr et al. 2009). The results of Zn were similar to Pan et al. (2011) and Badr et al. (2009) and they indicate that the high concentration of Zn might be due to recently anthropogenic activities in this area. Levels of Zn concentration in Salman Gulf, Prince Naif Street, and the Downtown areas were high from 15 cm to the surface area and significantly high in Salman Gulf in the same vertical distribution as shown in Figure 2. The higher concentrations start from the top 15 cm to the surface area indicate of being influenced from human activities from hidden non-point sources occur in the recent years or could be natural occur due to the high carbonate present in this area as $ZnCO_3$ (Othmani et al. 2015).

Linear regression analysis showed the strong positive correlation in some area between Mn and Cr and Fe (Naif Street, and the Downtown) and not correlation in other area (Salman Gulf and Al-Khumrah). The strong significant positive correlation are found in the Downtown area among Mn and Fe ($r^2 = 0.94$ $p < 0.05$), Cr and Fe ($r^2 = 0.6$ $p < 0.05$) and Cr with Mn ($r^2 = 0.54$ $p < 0.05$) (Table 1). Pattan et al. (1995) found that Cr, a redox sensitive metal, was precipitated with Mn-Fe oxide. The correlation analysis indicates that Cr, Mn, and Fe have a significant positive correlation and could indicate of a redox condition in the Downton area. Fe and Mn oxides have been recognized as playing a significant role in controlling the location of heavy metal pollution in sediments (Gasparatos 2012).

Usamn et al. (2013) studied the heavy metal contamination in mangroves from the costal of the Red Sea. Their result revealed that Cu and Cr were highly bio-accumulative in mangroves (A. *marine.*) from the costal of the Red Sea. In this study, the results of Cu indicate that Cu were following the same vertical profile of Mn and Pb, which was higher in the recent years (top 15 cm to the surface area) with a highly significant increase in the downtown area (Figure 2). The higher concentrations of Cu and organic matter in the Downtown area indicate of complexes of Cu with organic matter (Sun & Puettmann 2000; Tribovillard et al. 2006). The average concentrations of Cu were 251.82 mg/kg while sediments having a higher concentration than 60 mg/kg is classified as contaminated via the US EPA (Ingersoll and Nelson 1989). The major source of Cu in the marine environment due to antifouling paints, which is most boaters use it and content high copper concentration (CuO ~ 75%) (Srinivasan and Swain 2007). Copper also easily enters into the marine environmental through industrial discharge containing CuSO4, which is used in metal plating and fishing operations (Srinivasan and Swain 2007).

TABLE 1. Correlation coefficient matrix ($r^2$) for heavy metal concentrations in four core sediments (n=180) from coastal Red Sea (P<0.05).

Salman Gulf (Core1)

| | Cr | Mn | Fe | Cu | Zn | Pb |
|---|---|---|---|---|---|---|
| Cr | 1.00 | | | | | |
| Mn | **0.57** | 1.00 | | | | |
| Fe | 0.33 | 0.29 | 1.00 | | | |
| Cu | 0.43 | **0.69** | 0.21 | 1.00 | | |
| Zn | 0.40 | **0.67** | 0.24 | 0.41 | 1.00 | |
| Pb | 0.07 | 0.24 | 0.0 | 0.19 | 0.02 | 1.00 |

Prince Naif Street (Core2)

| | Cr | Mn | Fe | Cu | Zn | Pb |
|---|---|---|---|---|---|---|
| Cr | 1.00 | | | | | |
| Mn | **0.70** | 1.00 | | | | |
| Fe | **0.63** | **0.83** | 1.00 | | | |
| Cu | 0.49 | 0.48 | 0.27 | 1.00 | | |
| Zn | 0.00 | 0.00 | 0.00 | 0.01 | 1.00 | |
| Pb | 0.46 | 0.22 | 0.12 | **0.56** | 0.00 | 1.00 |

The Downtown (Core 3)

| | Cr | Mn | Fe | Cu | Zn | Pb |
|---|---|---|---|---|---|---|
| Cr | 1.00 | | | | | |
| Mn | **0.54** | 1.00 | | | | |
| Fe | **0.60** | **0.94** | 1.00 | | | |
| Cu | 0.19 | **0.74** | **0.69** | 1.00 | | |
| Zn | 0.29 | **0.73** | **0.69** | **0.73** | 1.00 | |
| Pb | 0.31 | **0.60** | **0.50** | 0.48 | **0.54** | 1.00 |

Al-Khumrah (Core 4)

| | Cr | Mn | Fe | Cu | Zn | Pb |
|---|---|---|---|---|---|---|
| Cr | 1.00 | | | | | |
| Mn | **0.53** | 1.00 | | | | |
| Fe | 0.21 | 0.24 | 1.00 | | | |
| Cu | 0.09 | 0.36 | 0.11 | 1.00 | | |
| Zn | 0.13 | 0.44 | 0.06 | 0.06 | 1.00 | |
| Pb | 0.66 | **0.78** | 0.23 | 0.27 | 0.33 | 1.00 |

Mean concentrations of lead in the studied sediment profiles showed high values in all cores (362.75 mg/kg). Othmani et al. (2015) reported that Pb is strongly complex with $CaCO_3$, which acts as a strong absorbent for Pb and can complex as $PbCO_3$. This is attributed to several sources such as boat exhaust systems, spilling of oil, the discharge of sewage effluents, and atmospheric input of Pb (Abo-Hilal et al. 1987). The high level of Pb in Red Sea may due to the use of gasoline as a possible reason contamination (Usman et al. 2013). Other authors assign to Pb contamination of marine sediments to boat exhaust emissions, leakage of oil from mechanized boats and sewage effluents (Badr et al. 2009).

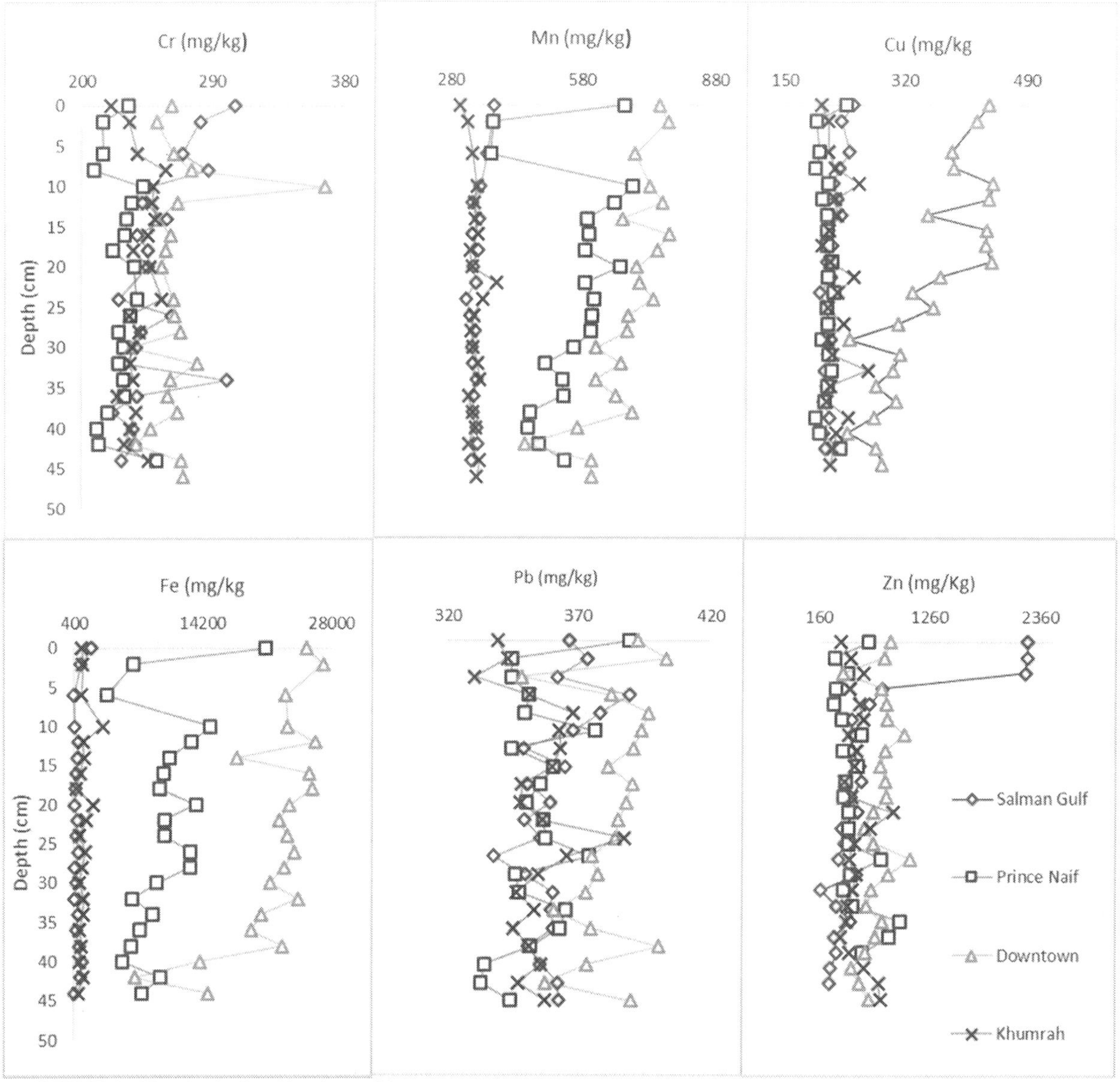


**FIGURE** 2. Vertical distribution of heavy metal concentrations in different depth of four core sediments on the Red Sea offshore Jeddah coast.

Generally, in the Red Sea, the average concentration of Mn, Cu, and Pb in all sediments followed the order: The Downtown > Prince Naif Street > Salman Gulf > Al-Khumrah. The influence of anthropogenic activities, such as fertilizers and sewage wastes are the primary metals sources on the marine environment (Qiu et al. 2011; Varol 2011). Shang etal. (2015) results indicate that the significant positive correlation among Cu, Mn, Fe, and Zn. The sources of heavy metals in the considered areas are mainly natural as well as anthropogenic through the impact of oil refinery wastes, untreated sewage effluents, and cement plants, which had been increased over the past years. The significantly positive correlations at the Downtown area of Fe ($r^2 = 0.94$ $p < 0.05$), Cu ($r^2 = 0.74$ $p < 0.05$), Zn ($r^2 = 0.73$ $p < 0.05$) and Pb ($r^2 = 0.60$ $p < 0.05$) with Mn (Table 1) showed that heavy metal pollutants were most likely derived from same source of contamination. Given the association of Mn, Cu, and Pb and copper's use as an indicator of organic complexes (Sun & Puettmann 2000; Tribovillard et al. 2006), Pb and Mn are, therefore, likely being delivered to the seafloor as organic matter complexes. Recognizing the speciation of these metals during and post delivery may play a key role in potential mitigation and/or treatment.

## CONCLUSION

Heavy metal concentrations in four sediments cores from the Red Sea near Jeddah were analyzed. The results showed high heavy metal concentrations of Mn, Cu, and Pb. Based on the heavy metal concentrations, this study indicates that the heavy metals in the core sediments were much higher in the top 15 cm. Generally, the results of heavy metal concentrations suggest that some area may be strongly impacted by anthropogenic sources with major contamination of Mn, Cu, and Pb. Future research would include isotopic data for sediment age dating by measuring the gamma emitting isotope. Sediments samples will be analyzing to measure radioactive $^{210}$Pb to assign an absolute age and sedimentation rates on Red Sea sediments.

## ACKNOWLEDGEMENTS

This research was financially supported by King Abdul-Aziz University, Jeddah, Saudi Arabia.

## REFERENCES

Al-Najjar, T., Rasheed, M., Ababneh, Z., Ababneh, A., & Al-Omarey, H. (2011). Heavy metals pollution in sediment cores from the Gulf of Aqaba, Red Sea. *Natural Science, 03*(09), 775-782.

Badr, N. B., El-Fiky, A. A., Mostafa, A. R., & Al-Mur, B. A. (2009). Metal pollution records in core sediments of some Red Sea coastal areas, Kingdom of Saudi Arabia. *Environ Monit Assess, 155*(1-4), 509-526.

Chatterjee, M., Silva Filho, E. V., Sarkar, S. K., Sella, S. M., Bhattacharya, A., Satpathy, K. K., et al. (2007). Distribution and possible source of trace elements in the sediment cores of a tropical macrotidal estuary and their ecotoxicological significance. *Environ Int, 33*(3), 346-356.

Conrad, C. F., Fugate, D., Daus, J., Chisholm-Brause, C. J., & Kuehl, S. A. (2007). Assessment of the historical trace metal contamination of sediments in the Elizabeth River, Virginia. *Mar Pollut Bull, 54*(4), 385-395.

Fukue, M., Yanai, M., Sato, Y., Fujikawa, T., Furukawa, Y., & Tani, S. (2006). Background values for evaluation of heavy metal contamination in sediments. *J Hazard Mater, 136*(1), 111-119.

Gasparatos, D. (2012). Sequestration of heavy metals from soil with Fe–Mn concretions and nodules. *Environmental Chemistry Letters, 11*(1), 1-9.

Ghandour, I. M., Basaham, A. S., Al-Washmi, H. A., & Masuda, H. (2014). Natural and anthropogenic controls on sediment composition of an arid coastal environment: Sharm Obhur, Red Sea, Saudi Arabia. *Environ Monit Assess*, 186(3), 1465-1484.sr

Ingersoll, C. G. & Nelson, M. K. (1989). Testing sediment toxicity with Hyallella azteca (Amphipoda) and Chironomus riparius (Diptera). ASTM STP 13th Symposium on Aquatic Toxicology Risk Assessment, Atlanta, Georgia, 43.

Nemati, K., Abu Bakar, N. K., Abas, M. R., & Sobhanzadeh, E. (2011). Speciation of heavy metals by modified BCR sequential extraction procedure in different depths of sediments from Sungai Buloh, Selangor, Malaysia. *J Hazard Mater, 192*(1), 402-410.

Othmani, M. A., Souissi, F., Duraes, N., Abdelkader, M., & da Silva, E. F. (2015). Assessment of metal pollution in a former mining area in the NW Tunisia: spatial distribution and fraction of Cd, Pb and Zn in soil. *Environ Monit Assess, 187*(8), 523.

Pan, K., Lee, O. O., Qian, P. Y., & Wang, W. X. (2011). Sponges and sediments as monitoring tools of metal contamination in the eastern coast of the Red Sea, Saudi Arabia. *Mar Pollut Bull, 62*(5), 1140-1146.

Pattan, J. N. Rao Ch, M. Higgs, N. C. Colley, S. & Parthiban, G. (1995). Distribution of major, trace and rare-earth elements in surface sediments of the Wharton Basin, Indian Ocean. Chemical Geology, 121, 201-215.

Qiu, Y. W., Yu, K. F., Zhang, G., & Wang, W. X. (2011). Accumulation and partitioning of seven trace metals in mangroves and sediment cores from three estuarine wetlands of Hainan Island, China. *J Hazard Mater, 190*(1-3), 631-638.

Rath, P., Panda, U. C., Bhatta, D., & Sahu, K. C. (2009). Use of sequential leaching, mineralogy, morphology and multivariate statistical technique for quantifying metal pollution in highly polluted aquatic sediments--a case study: Brahmani and Nandira Rivers, India. *J Hazard Mater, 163*(2-3), 632-644.

Srinivasan, M., & Swain., G. (2007). Managing the Use of Copper-Based Antifouling Paints. Environ Manage 39:423-441.

Sun, Y., & Puettmann, W. (2000). The role of organic matter during copper enrichment in kupferschiefer from the Sangerchausen Basin, Germany. Organic Geochemistry 31 (11), 1143-1161.

Tang, W., Shan, B., Zhang, H., & Mao, Z. (2010). Heavy metal sources and associated risk in response to agricultural intensification in the estuarine sediments of Chaohu Lake Valley, East China. *J Hazard Mater, 176*(1-3), 945-951.

Tribovillard, N., Algeo, T.J., Lyons, T., Riboulleau, A. (2006). Trace metals as paleoredox and paleoproductivity proxies; an update. Chemical Geology 232 (1-2), 12-32.

Usman, A. R., Alkredaa, R. S., & Al-Wabel, M. I. (2013). Heavy metal contamination in sediments and mangroves from the coast of Red Sea: Avicennia marina as potential metal bioaccumulator. *Ecotoxicol Environ Saf, 97*, 263-270.

U.S. Environmental Protection Agency. (1995). Test Methods for Evaluating Solid Waste.Vol. IA: Laboratory Manual Physical/Chemical Methods, SW 846 U.S. Gov. Print. Office. Washington, DC.

# VERTICAL DISTRIBUTION AND CHEMICAL BEHAVIOR OF URANIUM IN THE TAILINGS MATERIAL OF SCHNECKENSTEIN (GERMANY)

TaoufikNaamoun
(University of Sfax, Faculty of Sciences, Route de Soukra Km 3,5; P.B. 1171, Sfax, 3000, Tunisia)
Broder Merkel
(TU Bergakademie Freiberg, Gustav-Zeuner-Str. 12 . D-09596 Freiberg, Germany)

The purpose of this study was to investigate the spatial distribution of uranium and its mobility. In this work, the total concentration of uranium was determined by means of gamma and alpha spectrometry. The study of uranium mobility in selected samples from different intervals was made by the application of a seven steps sequential extraction procedure. Also hydrochemical investigation was carried out. The distribution of uranium species and saturation indices were calculated by means of the hydrochemical model PHREEQC. Results showed that uranium content ranges from 185 Bq/kg in the heap material to 6193 Bq/kg in the processed material. Also the data analysis shows the high uranium contents in most analysed water samples. Moreover, the PHREEQC model indicates the change of the tailing sediments with depth from aerobic to post aerobic or anaerobic conditions. Moreover, it illustrates the presence of uranium mostly (90%) in its high soluble form $UO_2(CO_3)_3^{4-}$. This is in agreement with the results of the selective extraction procedure which proves the association of important amounts of its contents with the carbonate phase (until 24%). Furthermore, as between 30 and 80 % of the non residual uranium is in association with the nodular hydrogenous fraction, the decrease of the Eh values in the study areas enhance its solubility.

# SYNERGETIC EFFECTS OF $Cd^{2+}$ AND $Cu^{2+}$ ON CHLOROPHYLL FLUORESCENCE OF MICROCYSTIS Aeruginosa

***Cui Jiansheng***, *Song Yanyan*
(Hebei University of Science and Technology, Shijiazhuang, Hebei, China)

*Microcystis aeruginosa* as testing plant was selected to investigated the effects of $Cd^{2+}$ $Cu^{2+}$ to its chlorophyll fluorescence property under temperature(25°C), humidity (75%RH) and photoperiod (2000-3000 lux). The separated and combined effects of different concentrations $Cd^{2+}$ and $Cu^{2+}$ on chlorophyll fluorescence intensity of *Microcystis aeruginosa* were investigated. The experiment took the $Cu^{2+}$ (0.2 mg/L) as the invariant, and $Cd^{2+}$ (0.1 mg/L, 0.4 mg/L, 0.6 mg/L, 0.2 mg/L) as the variable. The heavy metals were added to the *Microcystis aeruginosa* which was in logarithmic phase growth. Determine the chlorophyll fluorescence intensity at the same time and observe the changes, taking 96 h as a cycle. Results showed that chlorophyll fluorescence intensity were significantly upregulated by exposure to lower concentrations of $Cd^{2+}$ ( $< 0.5$ mg/L), and the changing trends is of time-dependence at higher exposure concentration. When the exposure concentration of $Cd^{2+}$ were 0.2 mg/L and 0.6 mg/L at 96 h, the inhibition ratio of chlorophyll fluorescence intensity were 0.36% and 30.48%, respectively. A similar trend of the fluorescence intensity were observed by exposure to $Cu^{2+}$ (0.2 mg/L). When its exposure time was 96 h, the inhibition ratio of chlorophyll fluorescence intensity was 25.58%. Compared to the exposure of $Cu^{2+}$, the tolerance of *Microcystis aeruginosa* was stronger when exposure to $Cd^{2+}$, demonstrating the stronger toxicity of $Cu^{2+}$ to the fluorescence intensity of *Microcystis aeruginosa*. The chlorophyll fluorescence intensity and exposure concentration was both time-dependent and concentration-dependent by exposure to $Cd^{2+}$ and $Cu^{2+}$, simultaneously. After 96 h exposure time, the inhibition ratio of chlorophyll fluorescence intensity was 48.83% when the exposure concentration of $Cd^{2+}$ and $Cu^{2+}$ were both 0.2 mg/L, the toxicity of which was much higher than that when single heavy metals was exposed. Moreover, when the exposure concentration of $Cd^{2+}$ and $Cu^{2+}$ were 0.6 mg/L and 0.2 mg/L, respectively, the inhibition ratio of chlorophyll fluorescence intensity was up to 58.14%. All these results confirm the significant synergetic effects of $Cd^{2+}$ and $Cu^{2+}$ stress on chlorophyll.

# CHARACTERISTICS OF HEAVY METAL POLLUTION FROM ANTIMONY MINING, ORE DRESSING AND SMELTING PROCESS IN CHINA

***Mengchang He***
(State Key Laboratory of Water Environmental Simulation, School of Environment, Beijing Normal University, Beijing 100875, China)

China has the most abundant antimony (Sb) resources of any country in the world, and was the highest producer of Sb, with approximately 84.0% of the world's share. Antimony mining and smelting industries produced huge amounts of wastes, and release large quantities of Sb, As and other heavy metals into the environment. In general, the Sb and As pollution from the metallurgical area is very severe, and has resulted in significant environmental problems. In recent years, more than Sb pollution accident occurred in China. The Chinese government strengthens the prevention and control of heavy metal pollution from Sb nonferrous industry, and issued the "Emission standards of pollutants for stannum, antimony and mercury industries" (GB30770-2014).

In this study, the waste waters, solid wastes and dusts from main Sb mining, ore-dressing and smelting enterprises in China were sampled. The Sb, As and other heavy metals were determined. The leaching characteristics of heavy metals in the Sb-bearing waste rock and ores were also studied. High concentrations of Sb and As were found in the mine water inrush and mineral processing wastewater. The tailings reservoir overflow also has high Sb and As concentration. The mining and smelting activities produced huge quantities of Sb-bearing waste rock and tailings. These solid wastes released high concentration of Sb and As due to the dissolution and weathering. The release percentage of Sb and As from three ores are different. The various solid wastes from smelting processes, which included water-quenched slag, arsenic-alkali residue, desulfurized slag and blast furnace dust, were enriched in a variety of heavy metals and metalloids. And more specifically the As and Sb levels in arsenic-alkali residue are up to $8.6\times10^4$ and $3.16\times10^5$ mg/kg, respectively. The smelting activities also cause serious Sb and As pollution in the soil profiles around the Sb metallurgical area. The highest degree of Sb was found in the uppermost layers of the profiles (<40 cm).

# CITIC ACID ENHANCED METALS IN REED SEEDLINGS IN ACID MINE DRAINAGE SOLUTIONS

***Lin Guo*** (Texas A&M University-Commerce, TX, USA)
Teresa J. Cutright (The University of Akron, OH, USA)

**ABSTRACT:** Acid mine drainage (AMD) with low pH and high levels of heavy metals affects many regions in the worlds. Wetland plant *Phragmites australis* (common reed) is widely used to treat AMD contaminated sites. This study investigated the effects of different levels (low 0.34 g/L, middle 17.86 g/L and high 33.62 g/L) of chelator citric acid (CA) on metal (i.e. Fe, Al and Mn) accumulation in reeds cultured in AMD contaminated solution for 12 weeks. The results indicated that CA increased Fe and Mn accumulation in both belowground and above ground tissues of reeds. The more CA was added, the more Fe and Mn were accumulated in plants. However, the Al accumulation in reeds was not affected by CA. After 12 weeks, 0.18±0.01 mg Mn/g, 56.78±7.63 mg Fe/g, and 0.60±0.01 mg Al/g were accumulated in roots; while 0.06±0.01 mg Mn/g; 6.38±0.06 mg Fe/g; and 0.19±0.01 mg Al/g were sequestered in stems of reeds treated with 33.62 g/L CA. Further research is needed to study the application of CA to enhance the phytoremediation efficiency of AMD contaminated field.

## INTRODUCTION

Acid mine drainage (AMD) is a serious environmental issue (Whitehead and Prior, 2005). AMD is produced when sulfide minerals which were initially under anoxic conditions are exposed to oxygen and water during mine activity (Akcil and Kodals, 2006). AMD can severely contaminate soils, affect water quality and pollute ecological environments due to its low pH and high concentrations of toxic elements (Peppas et al., 2000).

Wetland plants species that can survive under atrocious conditions (i.e. low pH and low levels of nutrients, high concentrations of metal concentrations), such as common reed *Phragmites australis* and cattail *Typha latifolia* are commonly used to treat AMD contaminated sites (Stoltz and Greger, 2006; Batty and Younger, 2004). However, a main disadvantage of phytoremediation is that its efficiency is affected by the mobility of metals. To overcome this limitation, chelators are applied to contaminated environment, to enhance metal bioavailability and subsequent increase accumulation by plants (Garbisu and Alkorta, 2001). The main mechanism is that chelator can react with metals to form metal-chelate complexes, thus prevent the precipitation of metals and maintain their availability for plant uptake (Marques et al., 2009). Several chelating agents, such as Ethylenediaminetetraacetic acid (EDTA) and citric acid (CA) have been studied and applied to increase the bioavailability of heavy metals and enhance the metals accumulation in plants. However, most of the research focused on heavy metals contaminated soil. Little was known whether chelator CA can be effective in increasing phytoremediation efficiency from heavy metal contaminated water. Even less is known about the effect of CA on metal accumulation in plants from AMD contaminated solutions with low pH. This study will address the knowledge gap by investigating the effects of different levels of CA on metal (i.e. Fe, Al and Mn) accumulation in reeds cultured in AMD contaminated solution for 12 weeks.

## MATERIALS AND METHODS

**Plant Sources.** Rhizomes of P. australis were randomly collected from an AMD-contaminated site located in Beaver Township, North Lima, Ohio, USA. After cleaning with distilled (DI) water, rhizomes were transferred into commercial potting soil. After 30 days, new seedlings from the wild rhizomes with similar biomass were used to initiate the seedling (i.e., young reeds) hydroponic experiments.

**Hydroponic Solutions.** The control group only contained nutrient materials, 0.250 g N ($NH_4NO_3$), 0.060 g Mg ($MgSO_4 \cdot 7H_2O$), 0.109 g P ($KH_2PO_4$), and 0.207 g K ($KH_2PO_4$ and $K_2SO_4$) per liter DI water (Guo and Cutright 2014). The treatment solution contained nutrients, synthetic AMD and different levels of CA (control 0 g/L, low 0.34 g/L, middle 17.86 g/L and high 33.62 g/L). The metal concentrations used in synthetic AMD were based on the characteristics of AMD described by Senko et al. (2008) and the metal levels of the actual AMD contaminated site (Sivaram. 2010). Synthetic AMD contained 42.94 g Fe ($FeSO_4 \cdot 7H_2O$), 0.27g Al ($Al_2(SO4)_3 \cdot 16H_2O$), 0.05 g Mn ($MnSO_4 \cdot H_2O$), 1.24 g Ca ($CaSO_4 \cdot 2H_2O$), 0.59 g Mg ($MgSO_4 \cdot 7H_2O$), and 0.28 g Na ($Na_2SO_4$) per liter of DI water. The pH was adjusted to 3.5 with $H_2SO_4$ (Senko et al. 2008). The dosages of CA were selected based on previous research (Mihalík et al. 2010) and the concentrations of metals in solution. Reeds were cultured in each hydroponic solution for 12 weeks and then harvested for analysis.

**Plant Digestion.** Upon harvesting, the roots of reeds were gently washed with DI water. Next, reeds were further separated into roots, rhizomes, stems, and leaves and air dried. One gram of dried tissue sections were digested as described in Cutright et al. (2010). The solutions were analyzed by inductively coupled plasma mass spectrometry (Perkin-Elmer Plasma 400 Spectrophotometer ICP-MS). All experiments were conducted in triplicate.

**Statistically Analysis.** The metals uptake in reeds were analyzed with one-ANOVA using the Minitab statistical package (Minitab 16). Differences between specific CA levels were identified by the Tukey test at 5 % probability

## RESULTS AND DISCUSSIONS

**Mn Accumulation.** As indicated in Figure 1, CA enhanced the Mn accumulation and translocation in all the tissues of reeds ($p<0.05$). For instance, after 12 weeks, the rhizomes in high level of citric acid (HC) contained 0.17±0.01 mg/g Mn which was also significantly higher ($p<0.05$) than that in the rhizomes grown in solutions without citric acid (NC) for the same period (0.10±0.01 mg/g Mn).

According to the results of Najeeb et al. (2009), significant enhancement in Mn accumulation and translocation in *J. effusus* plants cultured in Mn hydroponic solution was recorded under the application of CA. The elevated Mn translocation caused by chelator may related to the activation of ATPase in the plasma membrane which can produce and then increase the translocation of both essential and non-essential metals (Najeeb et al., 2009; Williams et al., 2000). It was also reported that CA can enhance metal mobility by formation of metal-chelator complex or change of binding sites (Wang et al., 2009). Furthermore, CA may alleviate the toxic effects of Mn to plants which enable to increase the accumulation of Mn in plants (Najeeb et al., 2009).

**Fe Accumulation.** As shown in Figure 2, CA also played an important role in increasing the accumulation of Fe in both underground and above ground tissues of reeds during all the growth period ($p<0.05$). After 12 weeks, the Fe levels in roots of reeds grown in solution with middle levels of CA (MC) and low levels of CA (LC) for were 33.31±3.03 mg/g Fe, 47.72±1.57 mg/g Fe, respectively. CA also increased the Fe amounts in the aboveground tissues of reeds. For instance, the leaves of reeds grown in solution HC for 12 weeks contained 3.47±0.01 mg/g Fe. It was significantly higher ($p<0.05$) than the Fe concentrations in leaves of reeds grown in solution LC for the same time period (2.05±0.01 mg/g). Citrate can maintain Fe in a soluble form by forming of stable plant-available organic-$Fe3+$complexes at low pH value (<6.8) (Jones et al., 1998), thus enhance Fe accumulation in plants.

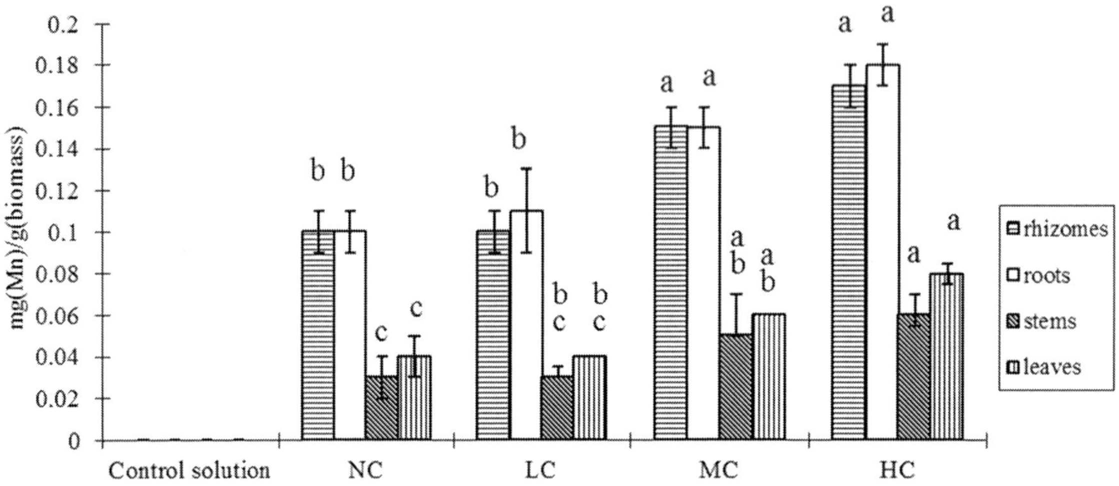


FIGURE 1. Mn concentration in the organs of reeds cultured in solution for 12 weeks. Error bars represent the standard deviation of triplicate samples. Different letters on the same plant organ indicate a significant difference at $p<0.05$.

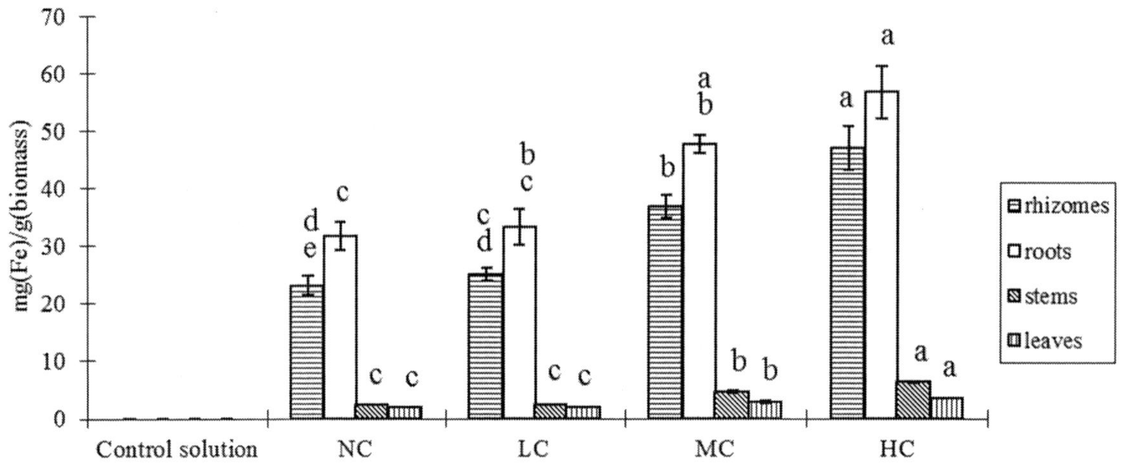


FIGURE 2 Fe concentration in the organs of reeds cultured in solution for 12 weeks. Error bars represent the standard deviation of triplicate samples. Different letters on the same plant organ indicate a significant difference at $p<0.05$.

**Al Accumulation.** However, CA also did not significantly elevated ($p>0.05$) the uptake of Al in different tissues of reeds (Figure 3). It was reported that the extent of complexation between metals and organic acid was related to the type of organic acid involved, the concentration of metals and the pH of solution, etc. (Jones, 1996). Since CA had higher affinity for Fe and Mn (Morel and Hering, 1994), and the Fe concentrations were higher than Al in solutions, most CA reacted with Fe and Mn. No citrate-Al complex were formed. That may be the main reason why CA can no improve the Al uptake in reeds.

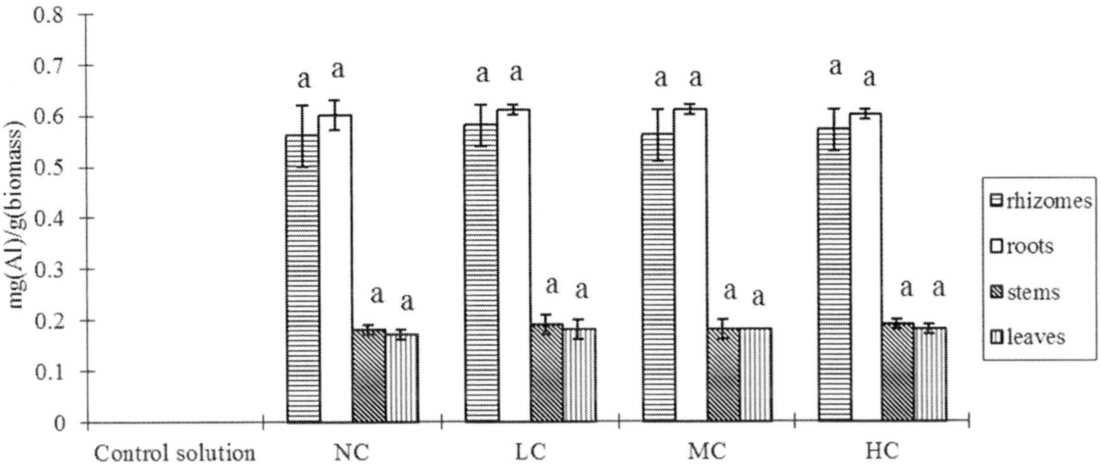


FIGURE 3 Al concentration in the organs of reeds cultured in solution for 12 weeks. Error bars represent the standard deviation of triplicate samples. Different letters on the same plant organ indicate a significant difference at $p<0.05$.

## CONCLUSIONS

The effectiveness of CA to increase the metal accumulation in was related to the concentrations of metals and CA in solution. The more CA added, the more Fe and Mn uptake in reeds. However, CA did not have important effect ($p>0.05$) on Al accumulation in reeds, since most of CA reacted with Fe and Mn which had higher affinity with CA. Thus no citrate-Al complex was formed. It is recommended that CA in conjunction with reeds to remediate AMD contaminated sites. However, a thorough site characterization should be conducted to assess the type and concentrations of metals at the site. Second, the suitable mounts of CA needed should be determined depending on the levels of metals at sites. In future, more research on the molecular genetics and physiology alternation in reeds caused by heavy metals and CA may be needed.

## ACKNOWLEDGEMENTS

The work was partially supported by the Office of Research and Sponsored Programs and Faculty Development Grants of Texas A&M University-Commerce.

## REFERENCES

Akcil, A., and Koldas, S. 2006. "Acid Mine Drainage (AMD): causes, treatment and case studies." *J. Clean. Prod.* 14 (12): 1139-1145.

Batty, L. C., and Younger, P. L. 2004. "Growth of *Phragmites australis* (Cav.) Trin ex. Steudel in mine water treatment wetlands: effects of metal and nutrient uptake." *Environ. Pollut.* 132(1): 85-93.

Cutright, T. J., Gunda, N. and Kurt, F. 2010. "Simultaneous hyperaccumulation of multiple metals by*helianthus annuus* grown in a contaminated sandy-loam soil." *Int. J. Phytoremediation.* 12(6):562-73.

Garbisu, C. and Alkorta, I. 2001. "Phytoextraction: a cost effective plant-based technology for the removal of metals from the environment." *Bioresour. Technol.* 77(3): 229-236.

Guo L., Ott D. W. and Cutright T. J. 2014. "Accumulation and histological location of heavy metals in *Phragmites australis* grown in acid mine drainage contaminated soil with or without citric acid." *Environ Exp Bot.* 105:46-54.

Jones, D. L, Darah, P. R. and Kochian, L. V. 1996. "Critical evaluation of organic acid mediated iron dissolution in the rhizosphere and its potential role in root iron uptake." *Plant Soil.* 180(1): 57-66.

Jones, D., 1998. "Organic acids in the rhizosphere -a critical review." *Plant Soil.* 205(1): 25-44.

Marques, A. P. G. C., Oliveira, R. S., Samardjieva, K. A., Pissarra, J., Rangel, A. O. S. S. and Castro, P. M. L., 2007. "EDDS and EDTA-enhanced zinc accumulation by *Solanum nigrum* inoculated with arbuscular mycorrhizal fungi grown in contaminated soil." *Chemosphere.* 70(6): 1002-1014.

Mihalík, J., Tlustoš, P.andSzaková, J. 2010. "Comparison of willow and sunflower for uranium phytoextraction induced by citric acid." *J. Radioanal. Nucl. Chem.* 285(2): 279-285.

Morel, F. M. M. and Hering, J. G., (Ed.), 1993. "Principles and applications of aquatic chemistry." John Wiley and Sons, INC.

Najeeb, U., Xu, L., Ali, S., Jilani, G., Gong, H. J., Shen, W. Q. and Zhou W. J. 2009. "Citric acid enhances the phytoextraction of manganese and plant growth by alleviating the ultrastructural damages in *Juncus effusus* L." *J. Hazard Mater.* 170(2): 1156-1163.

Peppas A., Komnitsas K. and Halikia I. 2000. "Use of organic covers for acid mine drainage control." *Miner. Eng.* 13(5): 563-574.

Wang, H., Jia, Y., Wang, S., Zhu, H. and Wu, X. 2009. "Bioavailability of cadmium adsorbed on various oxides minerals to wetland plant species *Phragmites australis.*" *J. Hazard. Mater.* 167(1-3) 641-646.

Whitehead, P. G. and Prior, H. 2005. "Bioremediation of acid mine drainage: an introduction to the Wheal Jane wetlands project." *Sci. Total Environ.* 338(1-2): 15-21.

Williams, L. E., Pittman, J. K. and Hall, J. L. 2000. "Emerging mechanisms for heavy metal transport in plants, Biochim." *Biophys. Acta.* 1465(1-2):104-126.

Senko, J. M., Wanjugi, P., Lucas, M., Bruns, M. A. and Burgos, W. D. 2008. "Characterization of Fe(II) oxidizing bacterial communities at two acidic Appalachian coal mine drainage impacted sites." *Int. Soc. Microb. Ecol.* 2(11): 1134-1145.

Sivaram, S., 2010. Assessment of bioremediation efficiency of indigenous bacteria and plants at an abandoned acid mine drainage. M.S. Thesis, The University of Akron, Akron, OH.

Stoltz, E. and Greger, M. 2005. "Effects of different wetland plant species on fresh unweathered sulphidic mine tailings." Plant Soil 276(1): 251-261.

# SILVER RECOVERY: SILVER ION TOXICITY

***Syed Sabir*** and Mansour Ibrahim Alhazza
(Chemical Engineering Department, King Saud University, P.O. Box 800, Riyadh-11421, Saudi Arabia,
*E-mail: sabirsyed2k@yahoo.com*)

**ABSTRACT:** Although we traditionally think of silver as 'second best' after gold, having high electrical, thermal conductivity, malleability and ductility are some of the characteristics that assign silver as an extremely relevant and valuable metal. Owing to these properties, silver is widely used in industrial activities including the production of mirrors, photographic films, batteries and electronic devices. Moreover, silver compounds are used as disinfectant agent on food and pharmaceutical industries due to its anti-bacteria properties. The extensive number of industrial applications increases the amount of silver concentration in spent sources, requiring a recovery before being discharged into the environment. However, there is a hot debate on the source of silver ion toxicity, because $Ag^+$ are biologically active and readily interacts with proteins, amino acid residues, free anions and receptors on mammalian, eukaryotic cell membranes, absorbed in the human body and enters the systemic circulation as a protein complex to be eliminated by the liver, kidneys degeneration and respiratory disorders and also highly toxic to bacteria at low concentrations (ng $l^{-1}$) and cause several negative health effects when disposed in large concentrations e.g. argyria or argyrosis, a disease related to skin pigmentation. Furthermore the high economic value of silver and toxicity result in the demand of recovery that both remove the metal from diverse spent sources and recover silver, in order to avoid the ecosystem contamination and bring economic advantages to the process, a critical comparison of existing technologies is analyzed for both economic viability and environmental impact was made in this presentation.

***Keywords:*** Hydrometallurgy; Silver; Recovery; Spent Sources; Ion-toxicity

## INTRODUCTION

Globally, the silver market saw rising record demand in 2015, with increasing usage of electronic equipment, photovoltaic sectors, jewelry, coin and bar posting new highs, helping to boost total silver demand to 1.17 billion ounces, instead last year and silver mine production growth slowed to 2% in 2015 and reached a record 886.7 million ounces and the overall slowdown in mine production last year is expected to continue (World Silver Survey, 2016). While we traditionally think of silver as 'second best' after gold, as distributions of medals still reflects today, pure silver holds three world records: the low contact resistance and electrical conductivity (used for audio cables, electric power switches and circuit breakers), which makes it a popular option in electronics and the highest brilliance (which lead to mirrors and optical applications) of all elements under standard conditions, and the best thermal conductivity of all metals (Syed and Mansour, 2016).

In recent year's extraction and recovery of silver was carried out either by hydrometallurgical or bio-metallurgical routes. Leaching represents the first step of the hydrometallurgical route and the recovery of silver by different techniques including cementation, chemical precipitation, adsorption, bio-sorption, electro-coagulation, electro-winning, ion-exchange and solvent-extraction etc. are very important cost-effective silver recovery and extraction process from various silver containing sources. In this book, the evaluation of lucrative and nature- friendly technologies to recover silver from primary and secondary spent sources was made (Syed, 2016).

With the growth of all these industries, the recovery and removal of silver have been intensely studied because of his present market demand poses an acute problem of recovery of silver from spent sources by new economically efficient and environmentally clean technologies (Gromov et al., 2004). Meanwhile, the phenomenal increase in the regulations against environmental pollution of silver ion toxicity for the marine, microbial, invertebrate and vertebrate community (including humans) through chains and have caused numerous diseases and disorders (Yang et al., 2012). Since it has become of great concern to recover and remove monovalent silver ions ($Ag^+$) from industrial effluents and spent sources, up to ppm level (De et al., 2010). Moreover, a great interest in silver recovery for both environmental toxicity and economic point of view (Lee et al., 2005; Park and Fray, 2009).

**TABLE 1** Silver recovery agents from leaching solutions*

| Recovery agents | Efficiency (%) |
|---|---|
| ***Cementation*** | |
| Zn from nitric acid | 100 |
| Cu from sulphuric acid | 99.2 |
| Mn from thiocyanate | 99 |
| Zn powder | 93 |
| Magnesium from cyanide solutions | 99 |
| Iron waste | 99.8 |
| | |
| ***Chemical Reduction*** | |
| Hydrogen peroxide | 95 |
| NaCl+ KOH +$H_2O_2$ | 87.30 |
| Potassium borohydride | 98 |
| Sodium dithionite | 99 |
| Cellulose (CMK) | 84.25 |
| Hydrogen reduction | 99.8 |
| Hydrazine hydrate | 87 |
| Sodium hydroxide | 95.94 |
| | |
| ***Electro-winning*** | |
| Pulsed electric field (PEF) and static cylinder electrodes (SCE) | 99 |
| Rotating cylinder electrode reactor (RCE) | 90 |
| Titanium and vitreous carbon cathodes | 95 |
| | |
| ***Ion flotation*** | |
| Dodecylamine | 100 |
| | |
| ***Solvent extraction*** | |
| Calixarene tetramide (LBC) and thio-analogue (THIO) | 99 |
| | |
| ***Ion exchange resins*** | |
| PSGA resin | 100 |
| VDP-1, VDP-2, N-0 | 100 |
| Macro-porous anion exchanger AV-17-10P | 90 |
| IRA-68 (weak base) | 98 |

*(Syed, 2016).

## RESOURCES RECYCLING PROCESS: THE DAWN

Silver containing materials arise from diverse sources, such as manganese silver ore, refractory antimony ore, silver sulfide concentrates, lead–zinc sulfide concentrates, mining wastewater, spent photography/radiography films, electronic and electrical materials, waste jewelry, silver wares, brazing alloys, catalyst, batteries, dental amalgam, orthopedic materials, spent bleach-fixing photographic/radiographic solution, electroplating solution and metallurgical processing-solution are available in many forms. Processing of such sources is, therefore complicated. However, the basic processing technology is shown in Figure 1. The pretreatment process involves washing, crushing, physical separation and incineration depending on the nature of materials and the metal obtained is of low purity. In the recovery process, such low purity metals are converted into crude silver by repeated leaching, adsorption or chemical reduction or cementation. Crude silver is subjected to the refining process to improve their purity to over 99.9% (Syed, 2016).

## SILVER RECOVERY ACCESSIBLE TECHNIQUES

Various technologies were applied for silver recovery from variety of silver containing sources, such as chemical reduction, precipitation, metallic replacement or cementation, adsorption, bio-sorbents, ion exchange, ion flotation, electro-deposition or electro-wining and solvent extraction Etc. Silver recovery agents from leachents are presented in Table 1.

## HYDRO-TECHNOLOGIES

Manuella et al., (2016) used, thermally modified bentonite clay (Verde-lodo) for batch adsorption of silver from aqueous solution. The adsorption rate was evaluated by Mass Transfer in External Film model, showed agreements between experimental and calculated data and the process is mainly controlled by external transport. Langmuir model showed best correlation coefficient and Relative Standard Deviation results and the maximum adsorption capacities for Verde-lodo clay was 61.48 mg $g^{-1}$, the result indicated that, clay is an effective adsorbent for silver recovery from aqueous sources.

Pablo et al., (2016) described two methods for extraction and concentration of silver from waste crystalline silicon photovoltaic modules. In the first method, the modules were milled, sieved and leached in 64% nitric acid solution with 99% sodium chloride; the silver concentration yield was 94%. In the second method, photovoltaic modules were milled, sieved, subjected to pyrolysis at 500 $^0$C and leached in 64% nitric acid solution with 99% sodium chloride; the silver concentration yield was 92%. The first method is preferred as it consumes less energy and presents a higher yield of silver. This study shows that the use of pyrolysis does not assist in the extraction of silver, as the yield was similar for both methods with and without pyrolysis. Figure1 shows silver recovery process diagram.

Choong, (2015) used immobilized crab shell beads to recover silver from industrial wastewater and the highest silver removal efficiency was 83%, achieved from 4.0 g of crab shell beads. The Langmuir isotherm model showed significant fit to the equilibrium adsorption data and maximum adsorption capacity of 2.951 mg/g of silver ions was achieved at the pH 6.0 and 95% of silver was recovered from adsorbed beads using 1.0 M of $HNO_3$ solution. Therefore, these crab shell beads sufficiently used to remove silver in batch and continuous process.

Sami et al., (2015) used polyamine functional resin WP-1for silver recovery from concentrated base metal chloride solutions, using batch equilibrium experiments. However, HCl was found to be the best choice for washing impurities, exclusively Cu, from the bed after the loading step. Tetrasodium salt of EDTA was found suitable especially for selective desorption of Pb from the loaded bed. Furthermore, thiosulfate and thiourea were used in silver elution from the WP-1 bed; nonetheless thiourea was preferred because of the trivial affinity of thiosulfate towards the resin. With this developed process, 72% pure silver was recovered. Subsequently, silver recovery yield depends purely on loading step length, and even 100% recovery was achieved from this process.

Meng et al. (2015) studied selective adsorption of $Ag^+$ using ion-imprinted O-carboxymethyl chitosan (ITG-OCMC) beads grafted with thiourea–glutaraldehyde from a bimetallic aqueous solution.

However, results indicated that high degree of carboxymethylation and low level of crosslinking was helped to achieve higher uptake of $Ag^+$ and the maximum uptake of $Ag^+$ was 156.32 mg $g^{-1}$ at 40 $^0C$ with an initial Ag+ concentration of 160.5 mg $L^{-1}$ and the biosorbent dosage of 1.0 g $L^{-1}$. Furthermore, Langmuir isotherm and Lagergren's pseudo-second-order kinetic was used to describe the sorption process of $Ag^+$. Consequently, FT-IR and XPS analyses confirmed that selective adsorption of $Ag^+$ was on the surfaces of ITG-OCMC beads by chelation through >C=S, amine, carboxyl and hydroxyl groups.

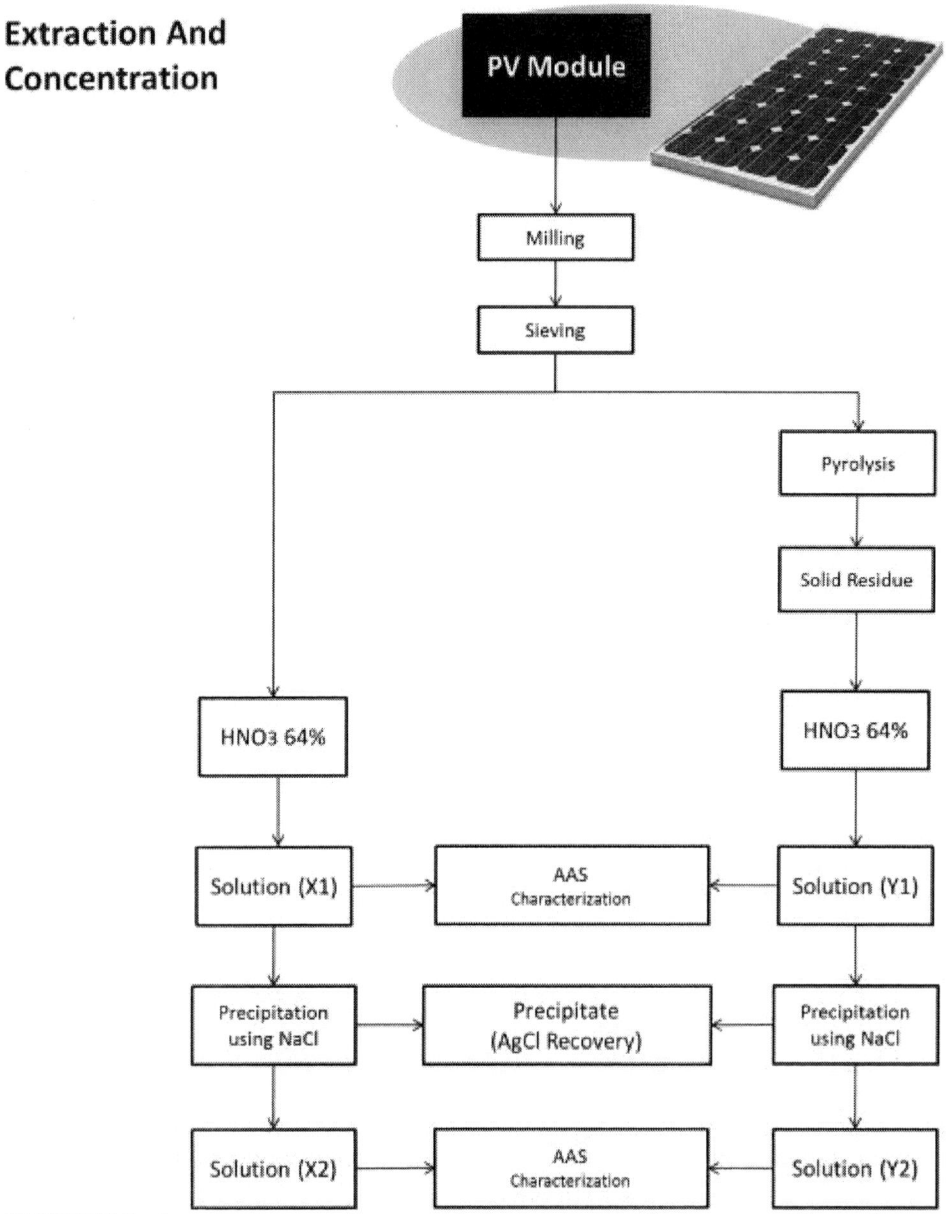


**FIGURE 1.** shows a process diagram of silver recovery from spent silicon photovoltaic modules.

Walaikorn and Thanut (2015) prepared chitosan/bamboo charcoal composite beads from blending chitosan and bamboo charcoal and used for silver ion adsorption. However, the maximum Ag+ removal was at pH 6, adsorbent dosage (0.4 g/L), interaction time (180 min), and Ag+ removal was 100%. The adsorption isotherm of chitosan/bamboo charcoal composite beads fitted well Langmuir model and the maximum adsorption was 52.91 mg/g.

Gholamreza, (2014) recovered silver ions from aqueous solution using natural Halloysite nanotubes (HNTs) and silver ions adsorption was influenced by silver ion concentration, temperature, pH, interaction time and adsorbent dose. However, adsorption increased with increasing initial silver ion concentration, pH and temperature. Pseudo-first order, pseudo-second-order and intra particle diffusion models were used to appraise the rate parameters. The adsorption followed pseudo-second-order kinetic model with correlation coefficients above 0.999. The experimental isotherm data were analyzed using the Langmuir, Freundlich and Temkin equations. The Langmuir isotherm model showed significant fit to the equilibrium adsorption data and the maximum adsorption capacity of 109.79 mg $g^{-1}$ (99.8% removal) of $Ag^+$ was achieved.

Ulisses et al., (2014) reported electrochemical recovery of silver from discarded X-ray photographic films by employing recyclable super paramagnetic carbon particles (Cmag). However, this exhibits a high affinity for $Ag^+$ ions in aqueous solution and is capable of capturing a large amount of this element because of great surface area and external magnet. Furthermore, this also provides an efficient way of transporting and concentrating the silver ions at the electrode surface. By this way electro-deposition was carried out directly from the electrode. The adsorption process at the magnetic carbon particles proceeds according to the Langmuir model, exhibiting an adsorption of 61.5 mg of silver per gram of Cmag. Consequently, complete setup for this process was shown in Figure 2.

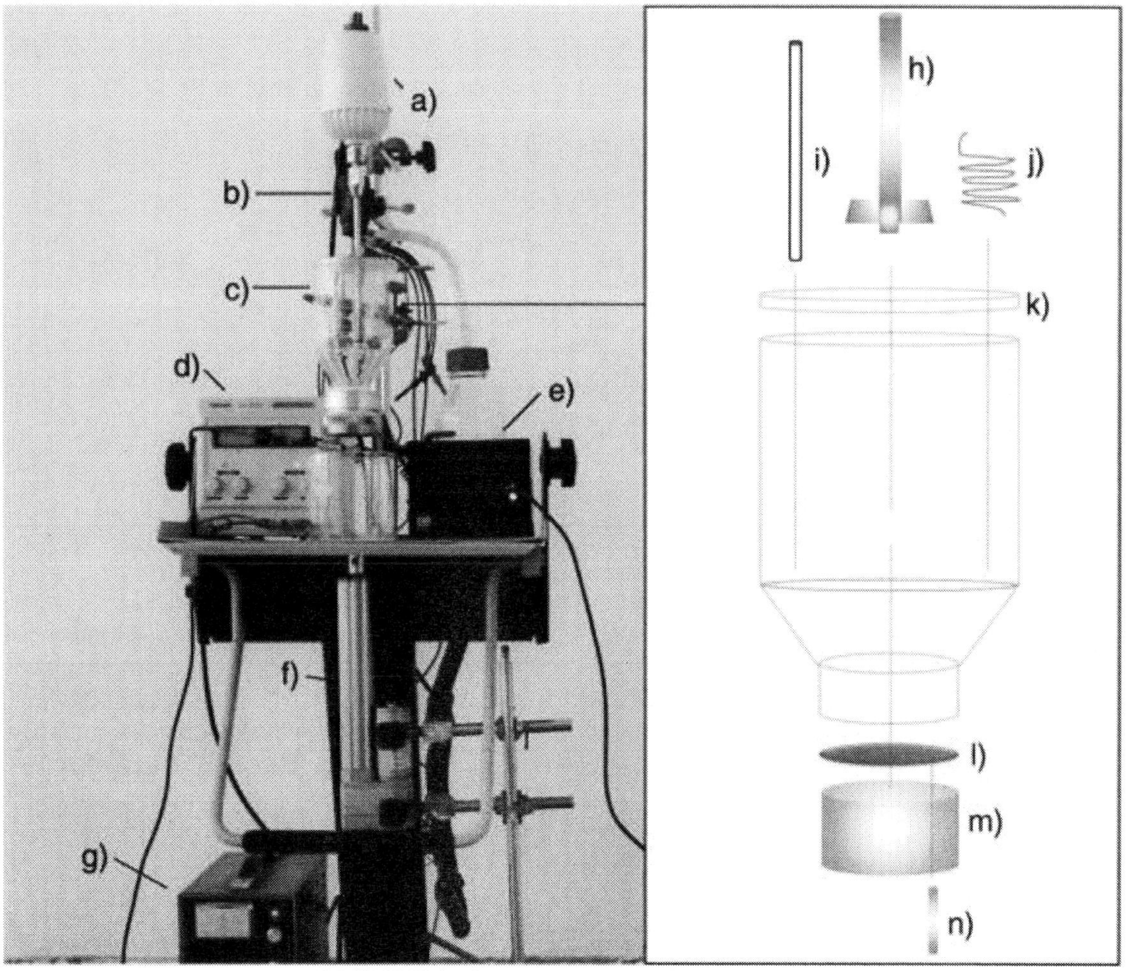


**FIGURE 2.** Complete setup of this process

## BIOSORPTION

Bio-sorption is the term given to the passive sorption and/or complexation of silver ions by biomass. The mechanisms of bio-sorption are generally based on physico-chemical interactions in silver ions and the functional groups present on the cell surface, such as electrostatic interactions. This is because of the fact that the properties of certain types of inactive or dead microbial biomass materials allow them to bind and concentrate silver ions from industrial effluents and aqueous solutions. These types of biomasses are relatively inexpensive and available in large quantities. Bio-sorption is a metabolism independent process that takes place in the cell wall, and the mechanism responsible for the metal uptake differ according to the biomass type. Bio-sorption is as promising technology for the recovery of silver from their silver containing sources (Syed, 2016). Table 2 presents various bio-sorbent for silver recovery from leaching solution.

**TABLE 2** Silver recovery bio-sorbent

| Bio-sorbents | Efficiency (%) |
|---|---|
| *Persimmon tannin gel* | 100 |
| *Cellobiose and cellulose* | Unknown |
| *Aspergillus versicolor protease* | Unknown |
| *Chitosan/amine and chitosan/ azole resins* | Unknown |
| *Valonia tannin (VT)* | 97 |
| *Aeromonas punctata* | Unknown |
| *Magnetospirillum gryphiswaldense* | Unknown |
| *Bacillus cereus biomass* | Unknown |
| *Corynebacterium glutamicum* | Unknown |
| *Pleurotus platypus* | Unknown |
| *Molecular-imprinted biosorbent* | Unknown |
| *Chemically modified chitosan resin* | Unknown |
| *Bisthiourea (BS) resin* | 95 |
| *Chitin* | Unknown |
| *Cladosporium cladosporioides* | 70 |
| *Enzyme of Bacillus subtilis ATCC 6633* | Unknown |
| *Myxococcus xanthus Biomass* | Unknown |
| *Aspergillus niger* | 100 |
| *Saccharomyces cerevisiae* | Unknown |
| *Neurospora crassa* | Unknown |
| *Alga Chlorella vulgaris (211/11b)* | 97.5 |

## BIO-HYDROTECHNOLOGIES

Yufeng et al., (2015) explicated biosorbents from brewer fermentation industry waste yeast to adsorb silver from aqueous solution. However, the FTIR result of waste yeast showed that the ion exchange, chelating and reduction were the main binding mechanisms between the silver ions and the binding sites on the surface of the biomass. Furthermore, TEM, XRD and XPS results suggested that silver nanoparticles were deposited on the surface of yeast. Subsequently, the kinetic experiments revealed that sorption equilibrium was reached within 60 min, and 93 % $Ag^+$ removal efficiency was achieved when the initial concentration of Ag (I) was below 100 mg/L. Thermodynamic parameters of the adsorption process ($\Delta G$, $\Delta H$ and $\Delta S$) identified that the adsorption was spontaneous and exothermic.

Mostafa et al., (2014) prepared magnetic cellulose xanthate sorbents and used for silver adsorption from acidic medium, the maximum adsorption capacity within 5 min was 166 mg $g^{-1}$. Moreover, the reusability of the sorbent was investigated by elution the adsorbed ions with 3.0 mL of 0.4 mol/L $Na_2S_2O_3$ solution. It was found that the recovery was (90–97%) after 3 cycles of sorption and desorption.

Manju et al., (2013) purported a new approach of silver recovery from waste mobile printed circuit boards (PCBs) by acidothiourea leaching followed by selective adsorption on low-cost biomass of agricultural waste. It was found that smaller particle size PCBs (53- 75 μm) yielded higher amounts of silver leached into the solution. Nonetheless, silver leaching was achieved in 0.5 M TU in 0.01 $MH_2SO_4$ at 60 $^0C$, under this condition, an average of 6.8 mg/g of silver was extracted from PCBs. Kinetic studies revealed that the complete leaching of silver was achieved in below 2 h. *Persimmon tannin extract* cross-linked with sulfuric acid was found to be a promising material for complete recovery of silver from the leached liquor. Furthermore, this adsorbent not lone adsorbed the dissolved silver selectively, but also reduced the adsorbed cationic species of silver to metallic silver. The proposed flow sheet is shown in Figure 3.

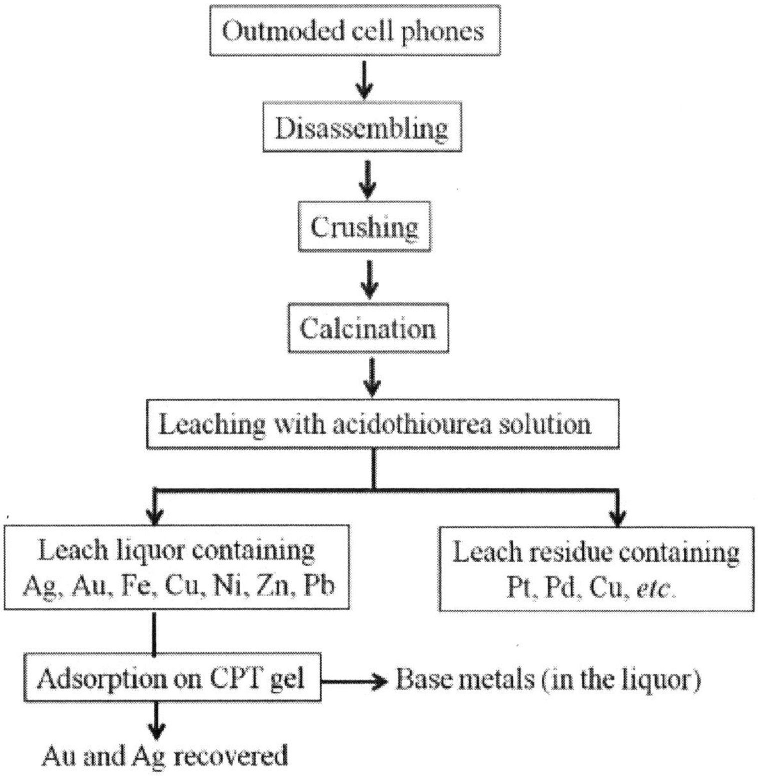


**FIGURE 3**. Flow sheet of precious metal leaching from spent mobile PCBs and recovery by adsorption (Manju et al., 2013).

## SILVER LOSSES TO THE ENVIRONMENT

In the world, maximum silver losses occur in mining industries, photographic processing and electroplating industries. Etc. But, the maximum loss occurs in mining and smelting facilities, fuel-fired power plants, manufacturing products and through waste disposal. The photographic utilize of silver was 30% this drive not account for much loss. However, direct discharge of silver containing effluent to rivers cause the silver deposition in the sediments of rivers and coastal areas of Oceans (Syed, 2005). The silver ions influence be accumulated in organisms (including human) through the food chain, and then should harm to their body. Recently, it is found that $Ag^+$ exposure caused an inhibition of active $Na^+$ and $Cl^-$ uptake, resulting in ion regulatory failure by inhibition of branchial $Na^+$-ATPase and $K^+$-ATPase in crayfish and daphnids. Henceforth, for economic as well as increasingly serious environmental problem, the active removal and recovery of $Ag^+$ has become a vigorous task (Hongyan et al., 2009).

Evaluation of silver recovery from primary and secondary spent sources and comparisons of eco-friendly and economically viable technologies are presented in Table 3.

**TABLE 3** Nature-friendly and economical viable silver recovery technologies

| Techniques | Efficiency (%) |
|---|---|
| Recovery of gold and silver from spent mobile phones by means of acidothiourea leaching followed by adsorption using biosorbent prepared from persimmon tannin | 100 |
| Preparation of a novel chelating resin containing amidoxime–guanidine group and its recovery properties for silver ions in aqueous solution | 100 |
| E-waste management through silver recovery from scrap of plasma TV monitors | 100 |
| A novel recovery technology of trace precious metals from waste water by combining agglomeration and adsorption | 100 |
| An *Aspergillus niger strain* accumulated gold and silver from solution containing high cyanide concentrations. Our results show that *A. niger* is of potential use for the removal of metals and/or the recovery of valuable metals from cyanide-containing solutions. | 100 |
| Recovery of silver from wastewater coupled with power generation using a microbial fuel cell | 99.91 |
| Preparation of fine silver powder from silver phosphate by direct electro-reduction | 99.8 |
| Silver(I) Sorption by Silica Gels Chemically Modified with Mercaptopropyl or Dipropyl Disulfide Groups | 99.9 |
| Adsorption and desorption behavior of silver ions onto valonia tannin resin | 99.6 |
| Silver recovery and chemical oxygen demand (COD) removal from waste fixer solutions | 99.55 |
| Kinetic Aspects of Gold and Silver Recovery in Cementation with Zinc Power and Electro-coagulation Iron Process | 99.5 |
| Recovery of high-purity silver directly from dilute effluents by an emulsion liquid membrane-crystallization process | 99.24 |
| An optimization study on the cementation of silver with copper in nitrate solutions by Taguchi design | 99.2 |
| A Novel Silver Recovery Method from Waste Photographic Films with NaOH Stripping | 99 |
| Study of silver precipitation in thiosulphate solutions using sodium dithionite. Application to an industrial effluent | 99 |
| A Novel Technique for Silver Extraction from Silver Sulphide Ore | 99 |
| Electrochemical recovery of silver from waste aqueous Ag(I)/Ag(II) redox mediator solution used in mediated electro oxidation process | 99 |
| Electrochemical reclamation of silver from silver-plating wastewater using static cylinder electrodes and a pulsed electric field | 99 |
| Solvent extraction of silver from nitric acid solutions by calix[4]arene amide derivatives | 99 |
| Powerful adsorptionofsilver(I)ontothiol-functionalized polysilsesquioxanemicrospheres | 99 |
| Extraction of silver from spent silver oxide-zinc button cells by using *Acidithiobacillus* ferrooxidans culture supernatant | 98 |
| A method to recover silver from waste X-ray films with spent fixing bath | 98 |
| Clean technology for the recovery of silver from processed radiographic films | 97 |
| Recovery of silver from X-ray film processing effluents by hydrogen peroxide treatment | 95 |
| Recovery of silver from silver(I)-containing solutions in bioelectrochemical reactors | 95 |

## SILVER IONS TOXICITY

There is a hot debate on the source of silver ion toxicity, because $Ag^+$ are biologically active and readily interacts with proteins, amino acid residues, free anions and receptors on mammalian, eukaryotic cell membranes, absorbed in the human body and enters the systemic circulation as a protein complex to be eliminated by the liver and kidneys and also highly toxic to bacteria at low concentrations (ng $l^{-1}$) (Lansdown, 2006; Katharina, 2011). However, silver ion absorbed in a range of 0.4–18% in mammals with a human value of 18%. Based on determination in animals, silver seems to be dispersed to all of the organs investigated, with the maximum levels being observed in the intestine and stomach. Consequently, prolonged exposure to silver may cause the silver deposition in the body, resulting in irreversible blue–grey discoloration of skin, eyes and hair caused by deposition of silver-protein complex or metabolized product (silver sulfide or silver selenide) in tissues, i.e. argyria or argyrosis (Su-juan et al., 2013). Excretion occurs via the bile and urine and following dose dependent animal toxicity findings have been described: death, weight loss, hypoactivity, altered neurotransmitter levels, altered liver enzymes, altered blood values, expanded hearts and immunological effects. Silver is precipitated in bone where $Ag^+$ displaces $Ca^{++}$ from hydroxyapatite with the potential risk of osteoporosis, but the extent of this problem is not known (Alan, 2010). Considerable proof exists suggesting that the effects induced by particulate silver are interceded via silver ions that are discharged from the particle surface (Niels and Henrik, 2014). Also, ionic silver is a toxic element of the marine microbial and invertebrate community (Ratte, 1999; Tappin et al., 2010), mainly by disabling the enzymes Na/K adenosine triphosphatase and carbonic anhydrase in animals (Morgan et al., 2004; Bielmyer et al., 2007), and interacting with thiol groups in enzymes and proteins of micro-organisms (McDonnell and Russell, 1999). Silver is strongly bio-accumulated by several marine phytoplankton, macro-algae and invertebrates (Fisher et al., 1984; Bryan, 1984), and it is known that the degree of silver accumulation by organisms is dependent on chemical speciation. The monovalent silver ion ($Ag^+$) is believed to be the utmost toxic silver species in aquatic systems and it has been shown that silver toxicity in freshwater phytoplankton is directly allied to intracellular accumulation (Campbell, 1995; Lee et al., 2005). The monovalent silver ion is the main dissolved species in fresh waters, experimental data have shown that in rivers and estuaries dissolved silver associated with colloidal macromolecular organic matter is a significant or dominant fraction of the total dissolved silver pool (Wen et al., 1997).

Ivana et al., (2014) studied the toxicity mechanisms of nano and ionic forms of silver on human hepatoblastoma (HepG2) cells. The results indicated that silver ions and citrate-coated AgNPs decreased cell viability in a dose-dependent manner. The IC50 rates of silver ions and citrate-coated AgNPs were 0.5 and 50 mg $L^{-1}$, respectively. The LDH leakage and inhibition of albumin synthesis, along with reduced ALT activity, indicated that treatment with either AgNP or $Ag^+$ ions resulted in membrane damage and decreased the cell function of human liver cells. Rating of oxidative stress markers demonstrating depletion of GSH, increased ROS production, and increased the SOD activity, indicated that oxidative stress contribute to the toxicity effects of nano and ionic forms of silver. The noticed toxic effect of AgNP on HepG2 cells was substantially weaker than that induced by ionic silver, while the uptake of nano and ionic forms of silver by HepG2 cells was same.

Fabianne et al., (2014) investigated the toxicity of silver nanoparticles (AgNP) and $AgNO_3$ on Pseudokirchneriella subcapitata, Daphnia magna and Danio rerio as model organisms to by assessing different biological endpoints and exposure periods. Organisms were periled following specific and standardized protocols for each species/endpoints, with modifications when necessary. In each test media, AgNP was characterized by Transmission Electron Microscopy (TEM) and experiments were executed by Dynamic Light Scattering to investigate the aggregation and the agglomeration behavior of AgNP under different chemical composition and test period. TEM images of AgNP in the different test media indicated dissimilar pattern of agglomeration, with some agglomerates inside an organic layer, some loosely associated particles and the presence of some individual particles. The perniciousness of both $AgNO_3$ and AgNP differ significantly based on the test species: no differences in toxicity for algae, a small difference for zebrafish and a major difference in toxicity for Daphnia magna.

Shaokun et al., (2013) appraised the cytotoxicity of Ag ions and silver chloride colloids on red blood cells and human mesenchymal stem cells (hMSCs). Subsequently, adverse effects of silver chloride on red blood cells and hMSC were viewed by SEM and LIVE/DEAD viability staining, respectively. A different tested chemical form of silver showed that, AgCl was identified to be the tiniest cytotoxic. Furthermore, a decline in the cytotoxicity of AgCl at significantly higher concentrations was observed. We attributed the reduced cytotoxicity to aggregated AgCl that limited the bioavailability of free $Ag^+$ ions; furthermore, molecular level studies are required to understand the mode of cytotoxic mechanisms.

Takamitsu et al., (2013) exposed human bronchial epithelial cells (BEAS-2B) to AgNO3 and investigated uptake and intracellular distribution of Ag, expression of metallothionein (MT), generation of reactive oxygen species (ROS), and changes in mitochondrial respiration. The culture medium concentration of Ag decreased with time and stabilized at 12 h. Furthermore, the concentration of Ag and MT in the soluble cellular fraction increased in 3 h and then decreased, indicating that cytosolic Ag relocated to the insoluble fraction of the cells. The levels of mRNAs for the major human MT isoforms MT-I and MT-II paralleled with the protein levels of Ag-MT. Subsequently, the intensity of fluorescence derived from ROS was elevated in the mitochondrial region in 24 h. Ag decreased mitochondrial oxygen consumption in a dose dependent manner and the activity of mitochondrial complex I-IV enzymes was significantly inhibited due to exposure. However, in a separate experiment, hydrogen peroxide at concentrations of 0.001% (equivalent to the concentration of $H_2O_2$ in Ag-exposed cells) removed Ag from MT. These results suggested MT was decomposed by cytosolic $H_2O_2$, and Ag released from MT relocated to insoluble cellular fractions and inhibited the electron chain transfer of mitochondrial complexes, which eventually led to cell damage.

Christina et al., (2012) studied the toxic effect of silver ions and silver nanoparticles towards bacteria and human cells and generally ascribed to a higher toxicity towards prokaryotic cells than towards mammalian cells. Furthermore, comparative studies with both silver ions (silver acetate) and polyvinylpyrrolidone (PVP)-stabilized silver nanoparticles (70 nm) showed the toxic effect of silver occurs in a similar concentration range for Escherichia coli, Staphylococcus aureus, human mesenchymal stem cells (hMSCs), and peripheral blood mononuclear cells (PBMCs), i.e. 0.5 to 5 ppm for silver ions and 12.5 to 50 ppm for silver nanoparticles. However, toxicity of silver ions and silver nanoparticles was different due to the presence of salts and biomolecules like proteins. The effective toxic concentration of silver towards bacteria and human cells is almost the same. Studied the toxic effect of silver ions and silver nanoparticles towards bacteria and human cells and generally ascribed to a higher toxicity towards prokaryotic cells than towards mammalian cells. Furthermore, comparative studies with both silver ions (silver acetate) and polyvinylpyrrolidone (PVP)-stabilized silver nanoparticles (70 nm) showed the toxic effect of silver occurs in a similar concentration range for Escherichia coli, Staphylococcus aureus, human mesenchymal stem cells (hMSCs), and peripheral blood mononuclear cells (PBMCs), i.e. 0.5 to 5 ppm for silver ions and 12.5 to 50 ppm for silver nanoparticles. However, toxicity of silver ions and silver nanoparticles was different due to the presence of salts and biomolecules like proteins. The effective toxic concentration of silver towards bacteria and human cells is almost the same.

## CONCLUSIONS

This article presents promising techniques for recovery of silver from various sources that are important subject not solitary from the point of energetic economic and high demand but also from waste treatment management and toxicity of silver ion. Subsequently, an evaluation of lucrative and nature friendly technologies to recover silver from primary and secondary spent sources and the toxicity of silver ion was made in this paper.

## ACKNOWLEDGEMENTS

Support for this research by College of Engineering, Research Center and The Deanship of Scientific Research, King Saud University, is gratefully acknowledged.

## REFERENCES

Alan D. T, Jose L. B, Charlotte B. B, Hywel E E. , Matthew D. P., Eric P. A. 2010. "Dissolved silver in European estuarine and coastal waters". *Water Research*, 44: 4204–4216.

Alan, B. G. L. 2010. "The Toxicology of Silver" Silver in Health Care: Antimicrobial Effects and Safety in Use", RSC Publication (DOI:10.1039/9781849731799)

Bryan, G.W., 1984. Pollution due to heavy metals and their compounds. In: Kinne, O. (Ed.), Marine Ecology, volume 5, part 3. John Wiley & Sons Ltd, Chichester, pp. 1289—1431.

Campbell, P.G.C., 1995. Interactions between trace metals and aquatic organisms: a critique of the free-ion activity model. In: Tessier, A., Turner, D.R. (Eds.), Metal Speciation and Bioavailability on Aquatic Systems. John Wiley & Son, Chichester, pp. 45—102.

Choong, J. 2015. "Adsorption behavior of silver ions from industrial wastewater onto immobilized crab shell beads", *Journal of Industrial and Engineering Chemistry*, 32:195–200.

Christina G., Dieter B., Alexander P., Jörg D., Bettina S., Matthias E., Manfred K. 2012. "The toxic effect of silver ions and silver nanoparticles towards bacteria and human cells occurs in the same concentration range". *RSC Advances,* 2: 6981–6987.

De, R. Q., Santoro, R., Proost, J., 2010." Kinetic transitions during Ag and Cu electrorecovery on reticulated vitreous carbon electrodes in flow-by mode". *Chemical Engineering Journal,* 162: 273–277.

Fabianne, R., Julián, A. G., Kerstin, J., Alison, C., Martin, H., Cameron, T., Amadeu, M.V.M. S., Susana, L., 2014. "Silver nanoparticles and silver nitrate induce high toxicity to Pseudokirchneriella subcapitata, Daphnia magna and Danio rerio". *Science of The Total Environment,* 466–467: 232–241.

Fisher, N.S., Bohe´, M., Teyssie´, J.L., 1984. "Accumulation and toxicity of Cd, Zn, Ag, and Hg in four marine phytoplankters". *Marine Ecology Progress Series,* 18: 201—213.

Gholamreza, K. 2014. "High removal capacity of silver ions from aqueous solution onto ch Halloysite nanotubes". *Applied Clay Science*, 90: 159–164.

Gromov, O. G., Kuz'min, A. P., Kunshina, G. B., Lokshin, E. P., Kalinnikov, V. T., 2004. "Electrochemical recovery of silver from secondary raw materials". *Russian Journal of Applied Chemistry,* 77(1): 62–66.

Hongyan, H., Haijia, S., Tianwei, T., 2009. "Adsorption of $Ag^+$ by a surface molecular-imprinted biosorbent". *Chemical Engineering Journal,* 150: 139–144.

Ivana V. V., Irena Ž., Roberta P., Ivan P., Maja D. S., Marija Ć.,Walter G. 2014. "Comparison of in vitro toxicity of silver ions and silver nanoparticles on human hepatoma cells". *Environmental Toxicology,* 31: 679 – 692

Katharina, M.F. 2011. "Give silver a shine". *Nature Chemistry*, 3: 178.

Lansdown, A.B. 2006. "Silver in health care: antimicrobial effects and safety in use". *Current Problems in Dermatology*, 33: 17–34.

Lee, D.Y., Fortin, C., Campbell, P.G.C., 2005. "Contrasting effects of chloride on the toxicity of silver to two green algae, Pseudokirchneriella subcapitata and Chlamydomonas reinhardtii". *Aquatic Toxicology,* 75:127–135.

Manju, G, Birendra, B.A, Hidetaka, K., Keisuke, O., Katsutoshi, I., Shafiq, A., 2013. "Recovery of gold and silver from spent mobile phones by means of acidothiourea leaching followed by adsorption using biosorbent prepared from persimmon tannin". Hydrometallurgy 133: 84–93.

Manuella, L. C., F. D Ambrosio, S. N .Eric and G.A.V. Melissa. 2016. "Adsorption of silver from aqueous solution onto pre-treated bentonite clay: complete batch system evaluation". *Journal of Cleaner Production,* 112: 1112–1121.

McDonnell, G., Russell, A.D., 1999. "Antiseptics and disinfectants: activity, action, and resistance". *Clinical Microbiology Reviews,* 12: 147—179.

Meng, Z., Yan Z., Robert, H. 2015. "Selective adsorption of $Ag^{+}$ by ion-imprinted O-carboxymethyl chitosan beads grafted with thiourea–glutaraldehyde". *Chemical Engineering Journal,* 264: 56–65.

Morgan, T.P., Grosell, M., Gilmour, M., Playle, R.C., Wood, C.M., 2004. "Time course analysis of the mechanism by which silver inhibits active $Na^{+}$ and $Cl^{-}$ uptake in gills of rainbow trout". *American Journal of Physiology-Regulative Integrative and Comparative Physiology*, 287: R234—R242.

Mostafa, H. B., Mehrnoosh, B., Simin, M., Farzaneh, S., Hassan, A. 2014. "Synthesis, characterization, and silver adsorption property of magnetic cellulose xanthate from acidic solution: prepared by one step and biogenic approach". *Industrial & Engineering Chemistry Research*, 53 (39): 14904–14912.

Niels H., Henrik R. L. 2014. "Oral toxicity of silver ions, silver nanoparticles and colloidal silver – A review". Regulatory Toxicology and Pharmacology, 68:1–7.

Pablo D., S. Javimczik, M. Benevit, H. Veit and A. M. Bernardes. 2016. "Recycling WEEE: Extraction and concentration of silver from waste crystalline silicon photovoltaic modules", *Waste Management* (http://dx.doi.org/10.1016/j.wasman.2016.03.016).

Park, Y.J., Fray, D.J., 2009. "Recovery of high purity precious metals from printed circuit boards". *Journal of Hazardous Materials*, 164: 1152–1158.

Ratte, H.T., 1999. "Bioaccumulation and toxicity of silver compounds: a review". *Environmental Toxicology and Chemistry*, 18: 89—108.

Sami, V., Mikko T., Mika H., Tuomo S., 2015. "Ion exchange recovery of silver from concentrated base metal chloride solutions". *Hydrometallurgy* 152: 100–106.

Shaokun, Z., Chan, D., Zaizhi, W., Xinguang, H., Kun, Z., Lihong, L., 2013. "Reduced cytotoxicity of silver ions to mammalian cells at high concentration due to the formation of silver chloride". *Toxicology in Vitro*, 27: 739–744.

Su-juan, Y., Yong-guang Y., Jing-fu L. 2013. "Silver nanoparticles in the environment". *Environmental Science: Processes & Impacts,* 15, 78–92.

Syed, S. 2016. "Silver recovery aqueous techniques from diverse sources: Hydrometallurgy in recycling". *Waste Management,* 50: 234–256.

Syed, S., M. I Alhazza .2016. "Silvery Hydrometallurgy: Recovery and Recycling", *King Saud University Press, K.S.A* (Under Publishing) Pages 103

Syed, S., 2005. "A critical review on the recovery of silver from photographic processing waste". *Journal of Saudi Chemical Society,* 9: 243–252.

Takamitsu, M., Yuta, A., Noriyuki S., Seishiro H. 2013. "Mitochondrial electron transport is inhibited by disappearance of metallothionein in human bronchial epithelial cells following exposure to silver nitrate". *Toxicology*, 305: 20– 29.

Ulisses, C., Alceu, T. S., George, W. C., Sergio, H. T., Araki, K., Henrique, E. 2014. "Silver recovery using electrochemically active magnetite coated carbon particles". *Hydrometallurgy* 147–148: 241–245.

Walaikorn, N., Thanut, J. 2015. "Removal of silver (I) from aqueous solutions by chitosan/bamboo charcoal composite beads". *Journal of Cleaner Production*, 87: 850–855.

Wen, L.-S., Santschi, P.H., Gill, G.A., Paternostro, C.L., Lehman, R.D., 1997. "Colloidal and particulate silver in river and estuarine waters of Texas". *Environmental Science and Technology* ,31: 723—731.

World Silver Survey, 2016. GFMS Limited/The Silver Institute. http://www. silverinstitute.org.

Yang, W., Xiaojie, M., Yanfeng, L., Xiaoli, L., Liuqing, Y., Lei, J., Yin, H., 2012. "Preparation of a novel chelating resin containing amidoxime–guanidine group and its recovery properties for silver ions in aqueous solution". *Chemical Engineering Journa,l* 209: 394–400.

Yufeng, Z., Dongfang, W., Hezhen X., Sung W. W., Longzhe C., Guiping W. 2015. "Adsorption of Ag (I) from aqueous solution by waste yeast: kinetic, equilibrium and mechanism studies". *Bioprocess Biosystem Engineering*, 38: 69–77.

# REMOVAL OF Hg(II) BY ENCAPSULATED THIOPHOSPHINIC EXTRACTANTS

***Janette Alba,*** Ricardo Navarro, Imelda Saucedo, (University of Guanajuato, Guanajuato, Mexico)
Eric Guibal (Ecole des Mines d'Ales, France)

The recovery of metal ions from wastewater or from effluents from metal processing has become an important challenge for the last decade due to the increasingly restricting regulations concerning water discharge to the environment (toxic elements). Mercury is an emblematic example of these metals that are concerned by the recovery regulations due to its high toxic effect (cumulative and highly toxic effect on human and animal beings, as dramatically illustrated by Minamata drama), therefore is proposed to carry out the removal of mercury by microencapsulation of extractants (EIRs), Cyanex 301 and Cyanex 302 (thiosubstituted organophosphinic acid extractants) have been successfully immobilized in alginate using matrix type encapsulation with distribution of the extractant as multi-nuclear micron-size drops. A novel process consists in the immobilization of the extractant as a single drop coated by a biopolymer layer (mononuclear type encapsulation). E-390 Encapsulator (Büchi, Switzerland) equipped with single nozzle and concentric nozzles was used for the preparation of the different materials. The objective of this study is to compare the two modes of encapsulation (i.e., matrix type versus mononuclear type) in the process of removal and recovery of Hg(II).

The study begins with the characterization of composite materials (optical microphotographs, SEM observations and SEM-EDX characterization of Hg(II) interactions with EIRs). The next step investigates the effect of HCl concentration on Hg(II) sorption for Cyanex 301 and Cyanex 302 mononuclear capsules in order to select the optimum pH for further sorption studies. Sorption isotherms and uptake kinetics are then compared for different lots of EIRs: changing the mode of immobilization, the loading of the extractant (varying the dilution of the extractant in kerosene), and the size of sorbent particles. Finally, the desorption properties (and the recycling of the material) are carried out.

Cyanex 301-based EIR is less sensitive to HCl concentration than Cyanex 302-based EIR in terms of Hg(II) sorption. Mononuclear encapsulation shows faster uptake kinetics but lower sorption capacities (in relation with lower capacity to immobilize the extractants, based on stability issues). On the other hand, matrix-type immobilization increases sorption capacity and improves the quantities of extractant that can be immobilized. The extractant loading should be optimized taking into account both the rational use of reactive groups and the accessibility to these sites: indeed, an excess of extractant may impact the accessibility and availability of reactive functions due to aggregation phenomena and viscosity effect. Metal desorption can be performed using 1 M thiourea solution (in 1 M HCl solution). The fabricated materials have good properties to carry out the removal of Hg(II).

# TECHNOLOGIES FOR REDUCING UPTAKE AND TRANSPORT OF HEAVY METALS INTO RICE GRAIN

***Yongchao Liang*** and Tingqiang Li
(College of Environmental & Resource Sciences, Zhejiang University, Hangzhou, 310058, China)
Alin Song, Zhaojun Li, and Fenliang Fan
(Institute of Agricultural Resources and Regional Planning, CAS, Beijing 100081, China)
Xionghui Ji
(Institute of Soil and Fertilizer, Hunan Academy of Agricultural Sciences, Changsha 410125, China)

Contamination of heavy metals in soil is a worldwide environmental problem. Over last decades increasing amounts of heavy metals have been released by geological activities or anthropogenic impacts and industrialization such as 1) combustion of fossil fuels, 2) mining, 3) smelting activities, 4) release of wastes and sewage water, and 5) repeated use of organic manures, chemical fertilizers (especially rock phosphate) and pesticides that contain heavy metals. Currently, approximately 20 M ha arable land is subjected to contamination of heavy metals in China, which poses high risks of food safety and human health. Specifically, uptake and transport of heavy metals into rice grain has received more and more attention recently in China. To solve this problem, some alternative *in situ* measures and/or strategies should be taken in addition of phytoremediation. Here we report some cost-effective and environmentally-friendly technologies that are proven to be highly effective in reducing accumulation of heavy metals in rice grain: 1) forming iron plaque on root surfaces via field water management and hence reducing root-to-shoot transport of heavy metals, 2) immobilizing heavy metals and hence reducing their uptake and transport via chemical amendments (fertilization or soil conditioning), and 3) reducing heavy metals uptake by using low-accumulating rice varieties. Large-scale field trials show that these "low-tech" but knowledge-intensive technologies are extremely effective in reducing accumulation of heavy metals in rice grain, and hence the environmental and food safety risks in heavy-metal-contaminated soils.

# LEVELS OF HEAVY METALS IN ORGANS OF *Oreochromis Niloticus* FROM GREAT KWA RIVER, SOUTH-EAST NIGERIA

Philomena Asuquo and Lynda-Uta Okon (University of Calabar, Calabar, Nigeria)
***Juliet Yakubu*** (Federal University Wukari, Wukari, Nigeria)

Levels of heavy metals such as Cu, Pb, Cd, Hg and Zn in the gills, offals and muscles of *Oreochromis niloticus* from Great Kwa River were observed for twelve months with the aim of evaluating the ecological risk of elevated levels of heavy metals in tissues of aquatic organisms. Results of chemical analysis with Atomic Spectrophotometry (AAS) showed that copper ranged from 0.92-1.511 µg/g (1.21±0.24 µg/g) in the muscles, 2.12-2.854 µg/g (2.42±0.27 µg/g) in offal and 1.99-3.225 µg/g (2.48±0.55 µg/g) in the gills showing the trend: gills>offal>muscle. Lead ranged from 1.88-2.25 µg/g (2.00±0.14 µg/g) in the muscles, 3.21-5.14 µg/g (3.71±0.74µg/g) in offal and 8.32-11.09 µg/g (9.49±0.98 µg/g) in the gills showing the trend: gills>offal> muscle. Cadmium ranged from 1.78-3.26µg/g (2.55±0.62 µg/g) in the muscles, 5.23-7.17µg/g (6.07±0.76µg/g) in offal and 5.63-7.66 µg/g (6.39±0.82 µg/g) in the gills showing the trend: gills>offal>muscle. Mercury ranged from 3.98-5.06µg/g with mean value of 4.36±0.42 µg/g in the muscles, 18.4-35.17 µg/g (28.72±5.67µg/g) in offal and 21.9-30.41 µg/g (25.78±2.97 µg/g) in the gills of *O. niloticus* showing the trend: offal>gills>muscle. The results showed a significant variance ($P<0.05$) between levels of heavy metals in muscles, offal and gills of *O. niloticus* due to different accumulating potentials of the organs. Based on these findings, the levels of heavy metals in organs of *O. niloticus* were within the stipulated limits. However, the present study recommends that precautionary measures need to be maintained in order to avoid possible elevated levels in the future.

# EFFECT OF SELECTED HEAVY METALS ON THE GERMINATION AND EARLY SEEDLING GROWTH OF *Jatropha curcas (Linnaeus)*

***Ogunbanjo O.R., Akintola O.O. and Awotoye O.O.**
(Federal College of Forestry, Ibadan, Nigeria)
*Email: aolunike@yahoo.com

**ABSTRACT:** Heavy metal contamination of agricultural soil has become a critical ecological concern. Over the past years there has been increasing interest in the use of live plants and other natural materials to remediate contaminated soils and water. Phytoremediation of this polluted soil with non-edible plants like Jatropha curcas offers an environment-friendly and cost-effective method for remediating polluted soil. This study thus investigated the effect of varying concentration of heavy metals on seed germination and seedling growth of *Jatropha curcas.*

Heavy metals used in this study were nitrate salts of copper (Cu), zinc Zn and lead Pb. Solutions of the salt of each of the metals were prepared using distilled water at concentrations of 0(control), 10, 20, 30 and 40mg/l. The experiment consisted of a total of 45 pots with ten seeds planted per pot and each containing 600g of sterilized and washed river sand thoroughly spiked with the prepared solutions. Seed germination and seedling growth were monitored for three weeks. Data assessed included the mean seed germination (%), mean root length (cm) and mean shoot length (cm) of the *Jatropha curcas* seedlings. All data obtained were analyzed using descriptive statistics and analysis of variance (ANOVA) using ASSITAT version 7.7 package.

The exposure of J*atropha.curcas* seeds to heavy metal salts increased the seed germination at 10mg/l but further increase in heavy metal concentration resulted in a decrease in *Jatropha curcas* seed germination. Germination index (GI) values of *Jatropha curcas* seed sown in soil spiked with zinc, lead and copper nitrates were 91-94, 88-82 and 47-87 respectively. It was observed that the GI values of *Jatropha curcas* at 20, 30 and 40 (mg/l) in Pb spiked soil were lower than 80% recommended value for phytotoxic free and mature compost. Heavy metals (Zn, Cu and Pb) concentration at 10mg/l promoted *Jatropha curcas* root lenght by 101%, 102% and 131% respectively relative to the root lenght in control experiment. It was noticed that $Cu^{2+}$ values reduced the shoot length when compared with the control at 20 and 40mg/l while $Zn^{2+}$ and $Pb^{2+}$ increased the shoot length *Jatropha curcas* even at higher concentration. Results also showed that Zn was the only metal that significantly increased the seed germination at the concentrations used in this study ($p<0.05$) while shoot length reduced significantly for seedlings in $Cu^{2+}$.spiked soils at concentration higher than 10mg/l.

Thus, the study showed that *Jatropha curcas* can germinate and grow efficiently at any zinc concentration assessed in this study; and may be grown directly in the soils that are moderately contaminated with copper and lead. Further studies need to be conducted in order to establish the maximum amount of zinc that the plants may tolerate and its ability to germinate and grow in media containing mixtures of these metals.

## INTRODUCTION

Heavy metal refers to any metallic element that has a relatively high density compared to water. These metals sometimes exist as natural components of the Earth's crust that cannot be degraded or destroyed but remain indefinitely in the environment (Raskin and Enslry, 2000). Heavy metals such as iron, copper, manganese and zinc are essential to maintain the metabolism of the human body and can serve as nutrients for plant growth while lead, mercury and cadmium have no known biological functions (Antosiewicz, 1992; Hawkes, 1997 ; Xiong, 1998; Nedelkoska and Doran, 2000 and Nicholls and Tarun, 2003).

Lead, cadmium and mercury are common contaminants in soil and at high concentrations; they may cause metabolic disorders and growth inhibition for most plant species which often lead to death (Kabata-Pendias and Pendias, 1989; Wang *et al*,2003 and Tripathi, 2007). Soils may become contaminated by the accumulation of heavy metals.

Soil contamination is a very important environmental issue which had been attracting considerable public attention over the last few decades. Heavy metal contamination of agricultural soil had become a critical ecological concern due to it's numerous adverse ecological effects. Heavy metals are released into the environment mainly by anthropogenic means such as mining and smelting, oil spills, domestic, agricultural and industrial activities (Zubillaga and Lavada, 2008).

Physico-chemicals processes such as ion-exchange, precipitation, reverse osmosis, evaporation and chemical reduction had been used to remediate heavy metal contaminated soils but had been found to be very expensive and toxic to the environment; ( Lasat, 2002 and Schnoor and McCutcheon, 2003). Over the past years there had been increasing interest in the use of live plants and other natural materials to remediate contaminated soils and water. Phytoremediation technology is aimed at providing an innovative, economical and environmentally-friendly approach to removing toxic metals from hazardous waste sites. This method makes use of metal-accumulating plants to remove heavy metals and other compounds from heavily contaminated environmental soil or water.

Plants to be used for effective phytoremediation should be non-edible, able to accumulate, degrade or volatilize the contaminants. They should also grow quickly in a range of different conditions and lend themselves to easy harvesting. Numerous plant species such as water hyacinths (*Eichhornia crassipes*), poplar trees (*Papulus spp*), forage kochia (*Kochia spp*), alfalfa (*Medicago sativa*), Kentucky bluegrass (*Poa pratensis*), coontail (*Ceratophyllum demersvm L.*) etc. had been identified and tested for their traits in the uptake and accumulation of different heavy metals.(Pandey, 2012)

Jatropha *curcas L.* commonly known as physic nut belong to the family Euphorbiaceae. It is a multipurpose small tree or large shrub and is found throughout the tropical region. Jatropha *curcas L.* is a vigorous, drought and pest-tolerant plant unpalatable to animals. It is used in Nigeria principally as a hedge plant for protecting cropland from the cattle, sheep and goats Jatropha is popularized as unique candidate among renewable energy plants due to its peculiar features like drought tolerance , rapid growth, and easy propagation, higher oil content than other oil crops , small gestation period, wide range of environ-mental adaptation,. It is regarded as a potential biofuel crop with high economic value to biodiesel and can also be used as anti-cancer medicine, pesticide, cosmetics, and for the treatment of various diseases (Jamil *et al.,* 2009; Li *et al.,* 2010a; Yadav *et al.* 2010.

Phytoremediation of polluted soil with non edible plant like *Jatropha curcas L.* offers an environment- friendly and cost effective method for remediating polluted soil. This study thus investigated the effect of varying concentration of heavy metals on seed germination and seedling growths of *Jatropha curcas L.*

## MATERIAL AND METHODS

The experiment was carried out at Federal College of Forestry, Ibadan Nigeria. The river sand used in was air dried and sieved to remove all pebbles and particles that might disturb the proper germination of the seed. The air dried river sand was analyzed in the laboratory to determine the physiochemical parameters and heavy metal contents using appropriate standard instrumentation methods. The soil used had a pH of 6.5, organic carbon -0.65%, total nitrogen- 0.65mg/kg, Zinc- 0.85mg/kg, lead- 0.35mg/kg, Copper - 0.15 mg/kg, iron -13.42 mg/kg and manganese- 8.53 mg/kg. Heavy metals salts used to spike the soil were nitrate salts of copper (Cu), zinc (Zn) and lead (Pb). Solution of salt of each of the metals were prepared using distilled water at concentrations of 0, 10, 20, 30, 40mg/l. the experiment consisted of a total of 45 pots with ten seeds planted per pot and each containing of 600g sterilized and washed river sand thoroughly spiked with the prepared solutions. Seed germination and seedling growth were properly monitored for three weeks. Data assessed include the mean seed germination (%), mean root length (cm) and mean shoot length (cm) of the *Jatropha curcas L.* seedlings. Seed germination percentage, germination index and root elongation were calculated using the formula given by Zucconi et al (1991) as follows.

$$\text{Seed Germination (\%)} = \frac{\text{No of seeds germinated in sowing medium}}{\text{No of seeds sown}} \times 100$$

$$\text{Root elongation (\%)} = \frac{\text{Mean root length in spiked medium}}{\text{Mean root length in control}} \times 100$$

$$\text{Germination Index} = \frac{\text{Seed germination}}{\text{Root elongation}} \times 100$$

All data obtained were analyzed using descriptive statistics and analysis of variance (ANOVA) using ASSITAT version 7.7 package.

## RESULTS AND DISCUSSION

**Effects of Heavy Metals on Seed Germination.** Results obtained from the experiment revealed that germination percentage of *Jatropha curcas L* was highest at 10mg/l (figure. 1). However at concentration above 10mg/l of Zn, Cu and Pb, there was a reductions in the seed germination. It can be inferred that seed germination of *Jatropha curcas L* was reduced with increasing heavy metal concentration except for zinc that did not record a significant reduction in seed germination. even at a concentration of 40mg/l ($p< 5\%$) in this study. This is in line with the findings of Claire *et al* (1991) in a study using nickel and other heavy metals on cabbage, lettuce, miller, radish, turnip and wheat and Aydinal and Marinova (2009) in the study of heavy metals on the growth of Alfalfa plant where they observed significant reduction on the effect of other heavy metals on seed germination with the exception of Zn even at higher concentration. The insignificant reduction of the seed germination to the exposure of Zn levels in the soil may be attributed to its nutritive importance to plant growth. Also the reduction in the seed germination of *Jatropha curcas L.* exposed to Cu may be attributed to the fact that Cu becomes more soluble in acidic soils. Therefore, Cu uptake by the plant would be more in acidic soil and as a result a higher reduction in the growth will occur (Marcel, 2006). As an essential micronutrient and constituent of many enzymes, at higher concentrations it creates toxicity to plants, humans and microorganisms (Perk, 2006). Lead (Pb) is one of the most common inorganic pollutants in the soils (Alkorta et al., 2004). The normal levels of Pb in soils is in the range of 0.85 to 65.8 % (Zarcinas, 2003). It is potentially toxic even at low concentrations and at concentration above 400 mg/kg, the soil is considered hazardous to human health (USEPA, 2001).

Germination index (GI) values of *Jatropha curcas L.* sown in spiked soils vary between 107-118%, 54-88% and 98-99% for zinc, lead and copper nitrates respectively (table 1). Zucconi *et al*, (1981) stated that a GI value of more than 80% is an indication of phototoxic-free and mature sowing media and this is also in line with the findings of Tiquia and Tam (1998). However, in this study only spiked soils of lead concentration higher than 10mg/l recorded GI that were lower than the suggested value which may be attributed to the release of toxic concentration of lead nitrates. This implies that Pb concentrations above 10mg/l become phototoxic to the studied plant. Generally, it was observed that, the higher the concentration of the heavy metal in sowing media, the lower the germination index value. The reduction in *Jatropha curcas L.* seed germination and increase in the GI values of Zn spiked soils may be due to it's uptake by plant as an essential micronutrient. Zn and Cu are both essential for man, plants and animals but at high concentration can be toxic. In addition, zinc plays an important role in many biochemical functions in the plants (Fox and Guerimot, 1998).

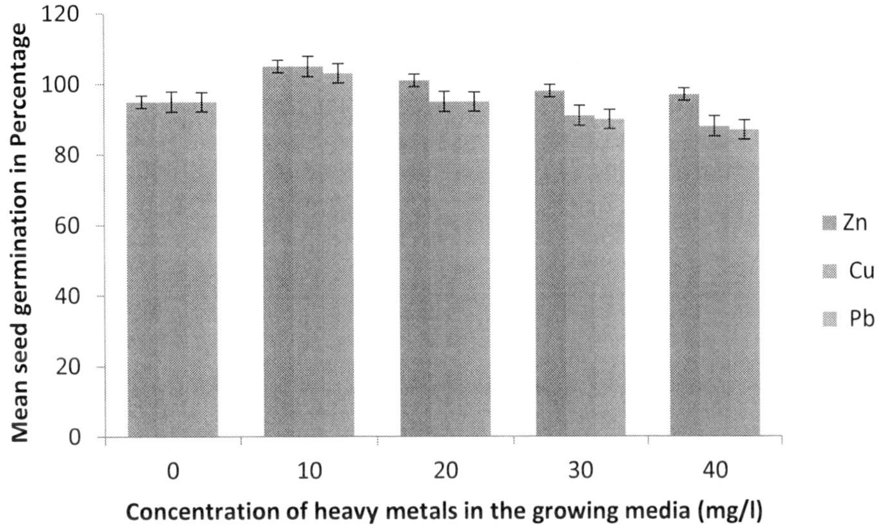


**Fig. 1. Seed Germination after Two Weeks of Exposure to Heavy Metals**

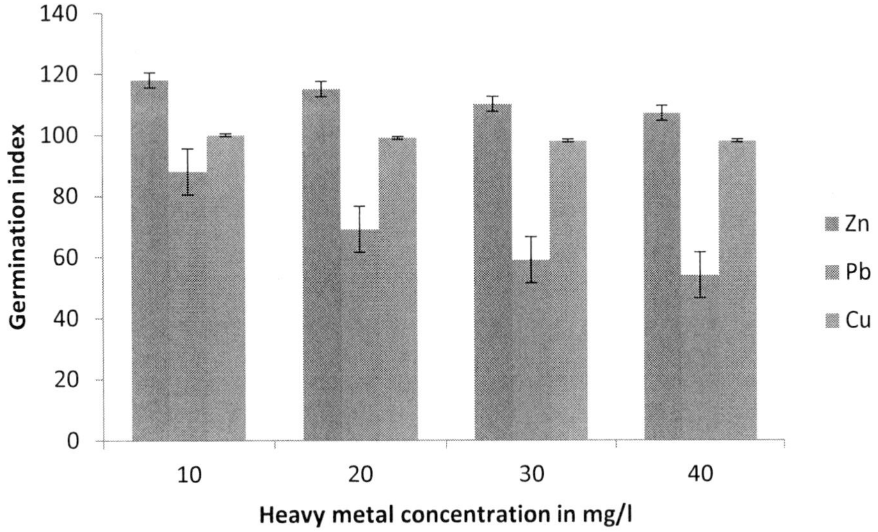


**Fig. 2. Germination Index after Weeks of Exposure to Heavy Metals**

**Effects of Heavy Metals on shoot and Root Growth.** The shoot growth of *Jatropha curcas L.* were affected by exposure to heavy metals. It was observed that the effect on the shoot growth was different when compared with the root growth (Figures 3 and 4). Copper at 20, 30 and 40mg/l reduced the shoot length of experimental plant when compared with the control while the doses of $Zn^{2+}$ and $Pb^{2+}$ increase the shoot length. However, there is significant reduction in shoot length at concentration higher than 10mg/l of $Cu^{2+}$ as shown in the control experiment ($P<5\%$) and this agreed with the findings of Once *et al* (2000) where he stated that heavy metal (Cd) reduces the chlorophyll in wheat while Chatterje and Chatterjee

(2000) found that $Cu^{2+}$ and $Cr^{2+}$ significantly decreased the water potential and $Fe^{2+}$ concentration in cauliflower.

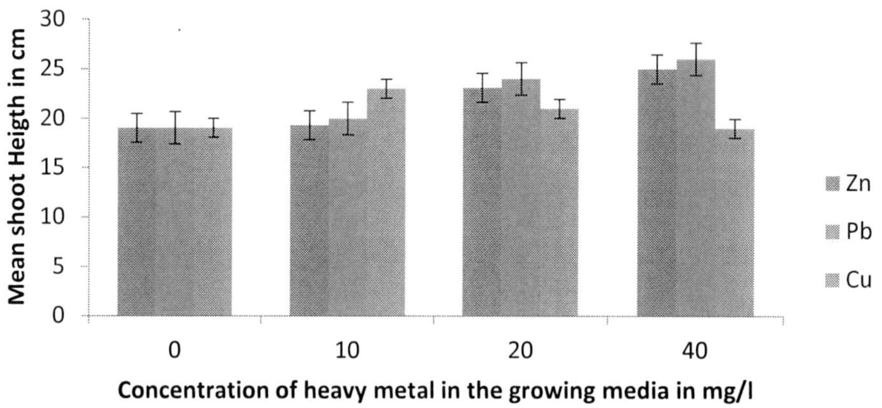


**Fig. 3. Shoot Length after Weeks of Exposure to Heavy Metal**

The root length of *Jatropha curcas L.* were affected by the exposure to heavy metals. It can be deduced from data obtained that at 10mg/l concentration of Zn and Cu salts, there was no significant effect (Fig.4). The Zn, Pb and Cu at 20, 30 and 40mg/l promote root growth giving a mean root length percentage of 101-135%, 105-137% and 100-121% respectively. This agreed with the findings of Once *et al.* (2000); and Aydina and Marinova (2009) where similar results were obtained on cadmium effect in wheat seedlings and effects of cadmium, copper, chromium, nickel and zinc on the germination and growth of Alfalfa plant respectively. The root lengths increased with increasing concentration of Pb and this may be attributed to the toxicity of Pb which promotes the root growth. The high root length values of all the metals at various concentrations when compared to control had depicted the adaptation ability of the plants to accumulate heavy metals from contaminated soil.

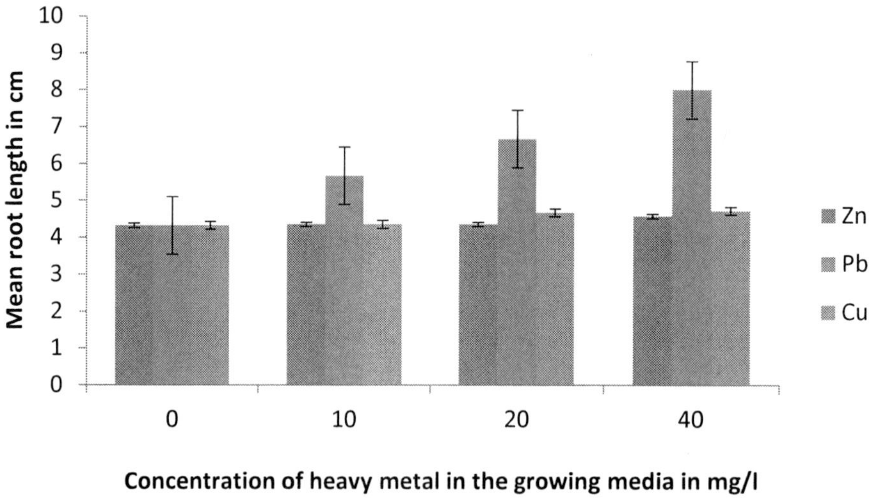


**Fig. 4. Root Length after Weeks of Exposure to Heavy Metal**

Generally, the significant effect of the heavy metals on the root and shoot growth could be attributed to the accumulation of heavy metals in plant which largely occur in the roots and to a slighter extent in leaves and stems. Plant roots promote the degradation of toxic organic material and act as a medium for soil microorganism to accelerate the biodegradation rates of organic pollutant (Palmroth *et al*, 2002 and Luhach and Chaughry, 2012). Thus, J. curcas plant has the ability to nurture well in soil with high levels of heavy metals (Macnair, 1987 and Baker, 1987).

## CONCLUSION

This study had shown that exposure of Jatropha *curcas L.* seeds to $Cu^{2+}$ and $Pb^{2+}$ concentrations have significantly affected the seed germination but slightly affected by $Zn^{2+}$. The root and shoot growth are stimulated at higher levels up to 20mg/l of $Zn^{2+}$, $Cu^{2+}$ and $Pb^{2+}$. However, *Jatropha curcas L.* was able to germinate and grow efficiently at any zinc concentration assessed in this study.

Further studies need to be done in order to establish the maximum amount of zinc that the plants may tolerate and its ability to germinate and grow in media containing mixtures of these metals.

## REFERENCES


Alkorta, I., Hernandex-Allica, J., Becerril, J.M., Amezaga, I., Aibizu, I., Onaindia, M., Garbisu, S., (2004): Chelate-enhanced phytoremediation of soils polluted with heavy metals. *Rev. Environ. Sc. Biotechnol.* 3, 55-70.

Antosiewicz, D. M. (1992). Adaptation of plants to an environment polluted with heavy metals. Acta Societatis Botanico rum Poloniae. 61:2811-919.

Aydinalp, C. and Marinova, S. (2009). The effects of heavy metals on seed germination and plant growth on alfalfa plant (*Medicago sativa*). *Bulg. Journ. Agric. Sci.,* 1547-350.

Baker, A.J.M. (1987). Metal tolerance, *New Phytol.*,106: 93-111

Chatterjee, J. and Chatterjee, C. (2000). Phytotoxicity of cobalt, chromium, and copper in cauliflower. *Environ. Pollution,* 109: 69-74.

Claire , L.C. ,Adriano, D.C., Sajwan, K.S., Abel, S.L., Thoma, D.P. and Driver, J.T. (1991). Effects of selected trace metals on germinating seeds of six plant species. *Water Air and Soil Pollution* 59: 231-240.

Fox T.L., Guerimot, M.L., (1998). Molecular biology of cationtransport in plants. *Annual Review of Plant Physiology and Plant Molecular Biology* 49, 669–696.

Hawkes J.S. (1997): Heavy Metals, *J. Chem. Educ.* 74(11): 1374.

Jamil, S., P.C. Abhilash, N. Singh and P.N. Sharma, (2009). Jatropha curcas: a potential crop for phytoremediation of coal fly ash. *J. Hazard. Mater.*, 172: 269–275

Kabata-Pendias, A. and Pendias, H., (1989). Trace elememts in the Soil and Plants. "CRC Press, Boca Raton, FL.

Lasat, M.M. (2002). Phytoextraction of toxic metals: A review of biological mechanisms. *J. Environ. Qual.*, 31:109-120

Li, X., Hou, S., Su, M., Yang, M., Shen, S., Jiang, G., Qi, D., Chen, S. and Liu, G. (2010). Major energy plants and their potential for bioenergy development in China. *Environ. Manage.*, 46: 579–589.

Luhach, J. and Chaudhry, S. (2012). Phytoremediation of heavy metals: Mechanism of tolerance in plants and advancements in techniques" *Proc: Conference on Strategies for mitigation of Environmental Degradation and Climate Change.* 83-89.

Macnair, M.R., (1987). Heavy metal tolerance in plants: A model evolutionary system, *Trends Ecol. Evolut.,* 2: 354-359.

Marcel, V.D.P (2006). Soil and water contamination from molecular to catchment scale. London Lewis publishers.

Nedelkoska, T.V. and Doran, P. M. (2000). Characteristics of heavy metal uptake by plant species with potential for phytoremediation and phytomining. *Mineral Engineering* 13:549-61.

Nicholls A, and Tarun M. (2003). Effects of lead and copper exposure on growth of an invasive weed, *Lythrum salicaria* L. (Purple Loosestrife). The Ohio J Science 103: 129-133.

Oncel, I., Y. Kele and A. S. Ustun. (2000). Interactive Effects of Temperature and Heavy Metal Stress on the Growth and Some Biochemical Compounds in Wheat Seedlings. *Environmental Pollution*. 107: 315-320.

Palmroth, M.R.T., Pitchel, J., and Puhakka, J.A. (2002). Phytoremediation of subarctic soil contaminated with diesel fuel. Bioresour. Technol. 84: 221-228.

Pandey V.C, Singh K., Singh J.Sc, Akhilesh Kumar A, Singh B., Rana P., Singh R.P. (2012): Jatropha curcas: A potential biofuel plant for sustainable environmental development. *Renewable and Sustainable Energy Reviews* 16: 2870– 2883

Perk, M.V.D (2006). Soil and water contamination from molecular to catchment scale. Tailor and Francis group. London pp 89-135

Raskin, I. and Ensley, B.D. (2000). Phytoremediation of toxic metals: Using plants to clean up the environment. John Wiley & Sons, New York.

Schnoor, J.L. and McCutcheon, S.C. (2003). Phytoremediation Transformation and Control of Contaminants,"Wiley-Interscience Inc, USA.

Tiquia S.M. and Tam N.F.Y.(2000): Co-composting of spent pig litter and sludge with forced-aeration. *Bioresource Technology* 72 1±7

Tripathi, L., Tripathi, J.N., Tushemereirwe, W.K. and Bandyopadhyay, R. (2007) Development of a semi-selective medium for isolation of Xanthomonas campestris pv. musacearum from banana plants. *Eur. J. Plant Pathol.* 117:177-186.

USEPA, (2001). Risk assessment guidance for super fund volume 111-part A, process for conducting probabilistic risk assessment. Washington DC.

Wang, Q.R., Cui, Y.S., Liu, X.M., Dong, Y.T. and Christine, P. , (2003). Soil contamination and plant uptake of heavy metals at polluted sites in China. *J. Environ. Sci. Health,* 38, 5: 823-838.

Xiong, Z.T. (1998). Lead uptake and effects on seed germination and plant growth of a Pb accumulator, *Brassica perkinensis* Rupr. *Bulg. of Environ Contamination and Toxicology* 60:285-91

Yadav, S.K., Juwarkar, A.A. , Kumar, G.P. , Thawale, P.R., Singh, S.K. and Chakrabarti, T. (2009): Bioaccumulation and phyto-translocation of arsenic, chromium and zinc by *Jatropha curcas* L.: impact of dairy sludge and biofertilizer. *Bioresour. Technol.,* 100: 4616–4622

Zarcinas B.A[1], Ishak C.F, McLaughlin M.J, Cozens G.(2004): Heavy metals in soils and crops in Southeast Asia. 1. Peninsular Malaysia. *Environ Geochem Health.*26(4):343-57.

Zubillaga, M.S. and R.S. Lavado, (2008): Accumulation and movement of four potentially toxic elements in soils throughout five years, during and after biosolid application. *Am. J. Environ. Sci.,* 4: 576-582.

Zucconi, F.; M. Forte A. M and M. de Bertoldi (1981): Biological evaluation of compost maturity. *BioCycle*, 22(2): 27–29.

# STUDIES ON EFFECT OF MIXED CROPPING TECHNIQUE ON THE PHYTOEXTRACTION POTENTIAL OF VEGETABLES OVER HYPERACCUMULATORS

Sidi.S[1], ***Tank Shantilal K***[2]

(1. Research Scholar, Dept. of Biosciences, V N South Gujarat university, Surat- 395007 India; 2. Dept. of Biosciences, VNSouth Gujarat U,niversity, Surat – 395007 India)

Accumulation of heavy metals by plants may depend on plant species and soil properties. Long-term consumption of vegetables grown with sewage water near the railway tracks can lead to the early onset of Parkinson's disease, neuron degeneration, hearing and vision impairment, and gastro-intestinal infection. The sewage water contains industrial effluents, which enter the plants through the soil. The uptake of heavy metals in cereals and vegetables is likely to be higher and accumulation of these toxic metals in human body has created growing concern in the recent days. Mixed cropping technique in the present study was tried to seek for the possibility of growing sturdy field crops like *Helianthus annus* and *Brassica juncea* along with garden vegetables like spinach and *Amaranthus viridus* to ensure decrease metal *uptake* by edible garden vegetable. The technique was proven to be unreliable as it rather triggered the efficiency of Spinacea *oleracea* and *Amarathus viridus* considerably rendering it unsuitable for mixed cropping as it would lead to unhealthy crop.

# PERSISTENT ORGANIC POLLUTANTS

# BIOLOGICAL AND INSTRUMENTAL ANALYSIS OF EMERGING CONTAMINANTS OF CONCERN: IN SINGLE AND MULTIPLE PROFILING.

***Momoh A. Yakubu***, Nina Brinkley, Syntia Kwende, Sara Munyu, Chioma Ihemadu, Fatemeh Bidabadi, Bhavin Rena, Joan Tran, Naga Naidu, Gloria Okome
(Texas Southern University, Houston, TX, USA)

Emerging Contaminants of Concern- can be described in several ways. Chemicals are being discovered in water/environments that previously had not been detected or were being detected at significantly very low levels. They were hitherto viewed as being incapable of causing harm or that they may be significantly different than expected. Advances in chemical analysis, biochemical assay development and technology have made it possible to detect chemicals that were hard to detect and evaluate for toxicity. These advances made it possible to establish the cellular and molecular mechanisms of their toxic reactions thereby raising the possibilities of injury. However, challenges remains as this information cannot be used to predict toxic effects in risk assessment evaluation of individual organism. With legislative aided effective reduction in emissions of conventional and priority pollutants from their sources in most developed countries, focus has shifted to compounds present in trace concentrations and that have only recently been thought of as pollutants. Some of these chemicals can interfere with the endocrine system of a wide range of organisms -these endocrine-disrupting chemicals (EDCs); have been the focus of much scientific attention in recent years. The interaction of multiple agents with the host's endocrine system could result in a wide spectrum of potential adverse health effects and much scientific uncertainty remains regarding the risks from these emerging pollutants. Most of the controversy surrounding the association between emerging environmental contaminants (EEC)/ EDCs and the various health endpoints is probably related to the lack of convincing evidence; the strength or weight of evidence and the association. For human health endpoints considered, concern arises from the little or lack of evidence that environmental levels of weakly active compounds elicit similar effects, either with hormonally active Pharmaceuticals or risk factors related to endogenous hormone levels. With high frequency of occurrence and/or coming from diverse sources of pollutants, they are now becoming a major concern due to their potential adverse effects on human health and environment. Despite their low concentrations within the environment and seemingly low biological activities individually – possibility exist for interactions of these diverse chemicals at low level concentrations, multiple interactions-interfacing with sensitive biochemical and hormonal pathways. These possibilities could lead to untoward and devastating consequences on human development and health - these *scenarios* are alarming. The major sources of these emerging contaminants are found in our waters -derived particularly from biopharmaceuticals, personal care products, agricultural and household used products especially multiple pesticides etc. We have over these years developed methods for biological and instrumental analysis of these contaminants as single entities or in multiple mixtures in our environmental resources, leachates into our water sources and foods as well as their effects on biological functions. Survey and analytical results (instrumental and biological) of studies of these emerging contaminants of concern will be presented with associated major health implications discussed.

---

# ANALYSIS OF NONYLPHENOL COMPOUNDS IN AQUATIC ENVIRONMENT USING GAS CHROMATOGRAPHY-MASS SPECTROMETRY

Seçil Ömeroğlu (Middle East Technical University, Tukey)
Fadime Kara Murdoch (Selçuk University, Tukey)
Faika Dilek Sanin (Middle East Technical University, Tukey)

Nonylphenol compounds are a group of organic chemicals (nonylphenol-NP, nonylphenol polyethoxylates-NPnEO and nonylphenoxy polyethoxy acetic acid-NPnEC) that are found in various industrial and commercial formulations. They are toxic and carcinogenic but beyond these, they tend to mimic hormones and interfere with the proper functioning of the endocrine system. For this reason, NP compounds are called as "endocrine disrupting chemicals" (EDCs).

Due to their surface active properties, these compounds are known to be excellent detergents, surface cleaners and lubricants that are widely used in everyday life. Although most of the countries banned their use in domestic formulations, these compounds are still present in personal care products such as shampoos, anti-wrinkle creams and deodorants. Since they have widespread use, they eventually end up in the wastewater treatment systems and environment. Most of the NP compounds are not fully degraded treatment plants, therefore, it is inevitable that they reach surface waters, sediments, etc. Many of the research showed their presence in environmental systems at a range of concentrations. Although these concentrations do not seem high, their effects are severe since NPnEOs start to degrade and form metabolites that are more toxic and persistent.

Once discharged into the receiving bodies, majority of the NP compounds remain in the water body. Unfortunately, due to their concentrations and complex matrix of the environmental systems, it is very difficult to detect these compounds in aquatic environment. Therefore, the aim of this study is to analyze the research on the measurement of NP compounds in aquatic environment and development of a rapid and reliable method to detect NP, NPnEOs and NPnECs in water systems using gas chromatography-mass spectrometry (GC-MS).

# PHOTOLYTIC AND PHOTOCATALYTIC DECOMPOSITION OF PHARMACEUTICALS IN WATER BY ULTRAVIOLET LIGHT-EMITTING DIODE

***Burhanuddin Khuzema Zaveri*** and Hyeok Choi (The University of Texas at Arlington, TX, USA)
Mohammad Reza Eskandarian and Mostafa Fazli (Semnan University, Semnan, Iran)
Mohammad Hossein Rasoulifard (University of Zanjan, Zanjan, Iran)

**ABSTRACT:** Pharmaceutical residues in the environment and their potential toxic effects have been recognized as one of the most emerging research areas in environmental chemistry. Among many treatment technologies for pharmaceuticals, ultraviolet (UV)-based oxidation has gained popularity as it has capability to decompose a broad range of refractory chemicals. As an alternative UV source to problematic conventional mercury lamps, light-emitting diode (LED), so-called UV-LED, has gained attraction. UV-LED shows many advantages such as high efficiency, low power consumption, long lifetime, no disposal problem, no warming-up time, and compactness. Although over the last decade and half several research has been done on decomposition of pharmaceutical products using heterogeneous photocatalysis and photolysis very little importance has been given to the parameters like wavelength of light, intensity of light. Analysis of SMX showed that considerable photolytic decomposition was obtained using UVC while UVA and UVC were very effective for photocatalytic decomposition process. From high intensity to low intensity brought down the degradation from 90% to 0%, thus showing its significance.

## INTRODUCTION

The occurrence of pharmaceuticals and their metabolites and transformation products in the environment is becoming a matter of concern because these compounds, which may have adverse effects on living organisms, are extensively and increasingly used in human and veterinary medicine and are released continuously into the environment (Xekoukoulotakisa.N.P.,C. Drosoua, and C.Breboua, et al,2011). Many PPCPs have been classified as emerging chemicals of concern (ECCs) and endocrine disrupting compounds, meanwhile, some industries continue to emphasize the benefits of such PPCPs in their products. Most antibiotics are often partially metabolized in the organism (i.e. less than 30%) and they are excreted as the parent substance or as metabolites into wastewaters at concentrations ranging from ng/L to g/L (Kümmerer.K, 2009). Most pharmaceuticals are known to be bio recalcitrant under aerobic conditions because their function is to remove microorganisms from body, thus escaping intact from conventional wastewater treatment plants. In response to the growing concerns about the presence of potentially toxic PPCPs and other ECCs in water resources, advanced oxidation processes (AOPs) have been widely studied and proposed to be effective for the complete degradation of such contaminants. AOPs involve the generation of transitional reactive free radicals, particularly hydroxyl radicals (HRs, $^{\bullet}OH$) and sulfate radicals (SRs, $SO_4^{\bullet-}$), which are stronger oxidizing species than conventional oxidants (Anipsitakis and Dionysiou, 2004; Son et al., 2003). Among the different advanced oxidation processes (AOP's), heterogeneous photo catalysis has emerged as a promising process for wastewater treatment, because it is a versatile and inexpensive (eliminating use of chemicals, UV radiation from sunlight, LED powered with solar radiation) technique to be applied to the treatment of different organic pollutants. When a semiconductor is illuminated with photons, whose energy is equal to or higher than the energy corresponding to the band-gap of the semiconductor, there is absorption of these photons and consequently some electrons ($e^-$) may jump from the valence to the conduction band. Simultaneously, a photo-hole is generated in the valance band ($h^+$). The electron/hole pairs migrate to the surface of the catalyst, where they can whether recombine and dissipate the input energy as heat, or react with the molecules object of degradation (Abella´n M.N., and B. Bayarri, 2007).

Ultraviolet disinfection of water has a long and well proven history. UV light has long been accepted as an effective germicidal treatment and has been installed in many major public drinking water and wastewater treatment plants. Recent developments to use UV as a means removing contaminants that cannot be removed by traditional biological process has led several research on Photocatalysis and photolysis degradation. Minimal research focuses towards the effect of UV wavelength and light intensity on overall process and problems associated with disposal of traditional UV lamps. LEDs offer potential advantages over conventional UV lamps such as a better adequacy in converting electricity into light (high quantum yields), lower power requirements, compactness and robustness, no warm-up time, no disposal problems, and potential for long lifetimes.

Our research has explored the degradation of sulfamethoxazole under UV LED. Three different UV LED (UVA, UVB and UVC) that emit light under various UV wavelengt to compare the degradation variation based on wavelength. The light source is placed at different heights and the intensity of light falling on the reactor is measured and degradation percentage of target contaminant is monitored.

## MATERIALS AND METHODS

**Chemicals.** Sulfamethoxazole (SMX) is obtained in salt form from Sigma Aldrich and are used as obtained. The chemicals used for analytical analysis acetonitrile was obtained from Sigma Aldrich and is used as received The Degussa P25 titanium dioxide is used as a catalyst for Photocatalytic experiments. 0.75 mg/L of $TiO_2$ was added to the solution which was decided based on some experiments and literature review. Most experiments revealed that an increase in catalyst loading above 1 mg/L did not show any significant increase in degradation.

**Experimental Setup.** The UVA and UVB lamps were purchased from International Light Technologies while UVC was obtained from Crystal Light. UV-LEDs and cooling system were attached using a clamp above the reactor. The solution were prepared using DI water in order to eliminate effects of any ions on the experiment. The solution was continuously agitated using a magnetic stirring bar. Current and voltage was adjusted by a DC power supply with galvanostatic operational options. Three reactor vessels with definite shape and size were considered for every wavelength experiments analysis. In order to measure different light intensities the clamp was fixed at different heights and measured using PD300RM-UV obtained from Ophir photonics.

**Analysis of Samples.** Samples were drawn at time intervals 10, 20, 30, 60, 120 and 180 minutes and filtered through 0.45 μm membranes. SMX concentration was monitored with a reversed-phase high performance liquid chromatography (1200 series, Agilent) equipped with a quaternary pump, C18 column, and ultraviolet (UV) detector. A mixture of water and acetonitrile was used as the mobile phase at water:acetonitrile ratio of 60:40% v/v with a flow rate of 1 ml/min for SMX. The UV detector was operated at a wavelength of 265 nm for SMX, which was pre-determined using a UV-visible spectrophotometer (UV 2550, Shimadzu).

## RESULTS AND DISCUSSION

The effectiveness of degradation of organic pollutants by photocatalytic process depends up on the wavelength of the emitted radiation which relates to the adsorption wavelength of target molecule, the intensity of radiation, catalyst loading, which are responsible for generation of hydroxyl radicals. The photolysis degradation depends upon the UV adsorption range of target molecule. The UV-LED has lower intensity compared to traditional mercury lamps and can be applied only if competitive degradation is observed. The figure 1 and 2 below show the degradation achieved from different wavelengths (UVA, UVB and UVC) for SMX under photolysis and photocatalysis condition respectively.

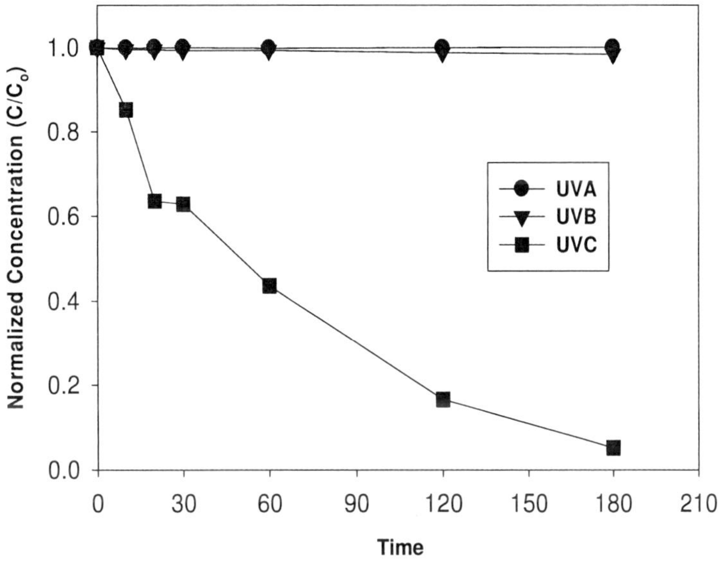


**FIGURE 1. Effect of Wavelength on Photolytic Degradation of SMX. $[SMX]_o$= 20 mg/L and $TiO_2$= 0.75 mg/L**

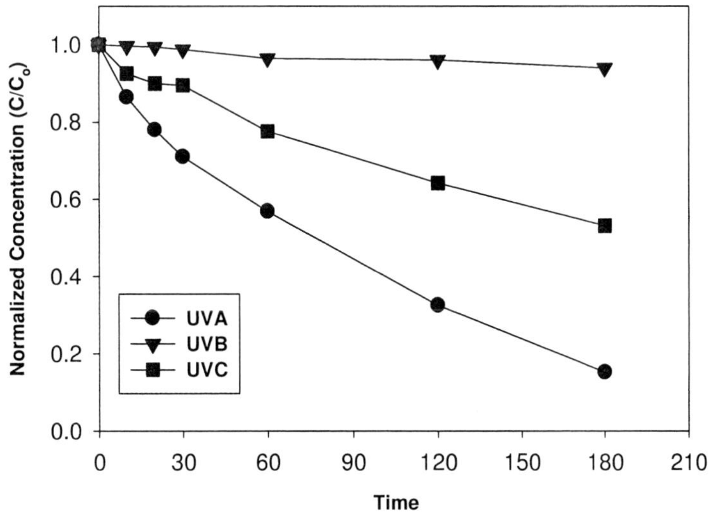


**FIGURE 2. Effect of Wavelength on Photocatalytic Degradation of SMX. $[SMX]_o$= 20 mg/L and $TiO_2$= 0.75 mg/L**

From figure 1 and 2 it can be seen that UVC performs better in both photolytic and photocatalytic process of degradation. Although UVA has its wavelength in the range of 315 nm to 400 nm which is far away from the adsorption radiation of SMX its high intensity (8.6 mW/cm$^2$) produces sufficient radicals to bring almost 90% degradation of SMX during photocatalytic process.

UVB which has its wavelength in the range of 280 nm to 315 nm does not show any sufficient degradation of SMX. The adsorption wavelength of SMX is far away from the range of UVB and also the high intensity of UVB is very low (0.1 mW/cm$^2$) and hence not a significant degradation like in case of UVA is observed in photocatalytic process.

UVC photolysis brings about 95% and 50% degradation of SMX in photolysis and photocatalysis process respectively. The peak adsorption wavelength of SMX coincides with peak wavelength of UVC lamps and hence a high degradation is observed in photolysis process in comparison to photocatalytic process where both target molecule and $TiO_2$ particles compete for the radiation thereby reducing the degradation percentage.

**TABLE 1. Effect of Intensity on Photolysis and Photocatalytic Degradation of SMX. $[SMX]_o$= 20 mg/L and $TiO_2$= 0.75 mg/L**

| SMX Degradation Observed at 180 Minutes | | | | | | |
|---|---|---|---|---|---|---|
| | UVA | | UVB | | UVC | |
| | Photolysis | Photocatalysis | Photolysis | Photocatalysis | Photolysis | Photocatalysis |
| High Intensity | 0% | 90% | 0% | 10% | 95% | 50% |
| Low Intensity | 0% | 35% | 0% | 0% | 0% | 0% |

The irradiance is a proportional to driving current and its position in a reactor. When the UV lamps are placed closed to the reactor increase of the incident photon quantity in this irradiance interval enhances the chance of photon absorption which leads to electron–hole formation and thus photo catalytic degradation. It is reported that at high irradiance values, electron–hole recombination dominates and therefore reaction rate decreases and is not anymore related to irradiance linearly but to the square root of the irradiance (Jamalia.A, and R.Vanraes, 2013).

## CONCLUSION

UVA and UVC performed better in comparison to UVB, which is most likely because of high intensity of both the lamps in comparison to UVB. A negligible amount of irradiance was measured at the bottom of the reactor which implies a non-optimal distribution of light. To improve further the performance of the reactor, it is necessary to design a structure with more uniform light distribution. Other benefits that can be expected by using UV LED, is the small bulb size enables more optimized reactor design to be developed that should reduce dose requirements further extending the opportunity that LED bulbs provide. Looking further into the future the continued development will drive down the energy associated with the use of AOP making the overall technology a more viable alternative especially in places where current systems are struggling to meet treatment needs such as the removal recalcitrant micro-pollutants such as pharmaceuticals. However, for these predictions to happen, several parameters need to improve especially the intensity in order to give UV LED/$TiO_2$ AOPs a chance of implementation at larger scale.

## ACKNOWLEDGEMENTS

Dr. Choi appreciates the financial support of the University of Texas at Arlington in form of startup fund to initiate this study.

## REFERNCES

Jamalia.A., and R.Vanraes. 2013. "A Batch LED Reactor For The Photocatalytic Degradation Of Phenol." *Chemical Engineering and Processing:Process Intensification.* 71: 43-50.

Kümmerer,K. 2009. "Antibiotics in the Aquatic Environment". *Chemosphere* 75: 417–434.

Abella´n, M.N., and B. Bayarri. 2007. "Photocatalytic Degradation Of Sulfamethoxazole In Aqueous Suspension Of $TiO_2$". *Science Direct, Applied Catalysis B: Environmental* 74: 233-241.

Xekoukoulotakisa,N.,P., C.Drosoua, C.Breboua, and E.Chatzisymeona. (2011). "Kinetics of UV-A/TiO2 Photocatalytic Degradation And Mineralization Of The Antibiotic Sulfamethoxazole In Aqueous Matrices". *Catalysis Today* 161: 163-168.

Anipsitakis, G. P., and D. D. Dionysiou. 2003. "Degradation of organic contaminants in water with sulfate radicals generated by the conjunction of peroxymonosulfate with cobalt." *Environ. Sci. Technol.* 37(20):4790-4797.

# IN-SITU DECOMPOSITION OF BIOLOGICAL TOXINS ON NITROGEN-DOPED NANOSTRUCTURED $TiO_2$ FILMS UNDER SOLAR RADIATION

***Hesam Zamankhan Malayeri*** and Hyeok Choi
(The University of Texas at Arlington, Arlington, TX, USA)

**ABSTRACT:** Contamination of water resources with natural and anthropogenic chemicals has been a huge concern worldwide, considering growing population and economy. Particularly, increasing occurrence of harmful algal blooms (HABs) alarms water and health authorities and general public. Most seriously, some algae produce and release lethal biological toxins such as microcystins (MCs). MCs are among the most powerful natural poisons. The presence of MCs in drinking water sources has raised major concern. As a result, the goal of this study is to decompose MCs, if feasible, on-site and in real-time with minimal efforts, less chemicals, and low energy inputs. To achieve the goal, we introduced $TiO_2$ photocatalysis for the generation of hydroxyl radicals which are responsible for the decomposition of MCs. High efficiency nitrogen-doped nanostructured $TiO_2$ photocatalysts, which can be activated under visible light, were developed to utilized solar radiation as a sole energy source for $TiO_2$ activation and radical generation. Considering on-site applications, $TiO_2$ was firmly immobilized onto a substrate and characterized with nanoporous, transparent, and ultrathin properties which are beneficial to photocatalytic reactivity, light utilization, and mechanical stability. Results so far showed that the $TiO_2$ film has high potential to decompose biological toxins under solar radiation within a reasonable time.

## INTRODUCTION

The increasing occurrence of harmful algal blooms (HABs) in water resources worldwide is alarming the environmental and health authorities because of their potential to release biological toxins (Vazquez et al., 2011). Of particular interest are microcystins (MCs) produced during cyanobacterial HABs (Mohamed et al., 2003). MCs are among the most powerful natural poisons. Exposure to MCs can affect the number and diversity of wild animal populations, cause bioaccumulation of toxins in the tissues of fish and shellfish, and indirectly affect other organisms through the food chain and eventually humans (Palikova, et al., 2011). Compared to particulate algae which are easily removed by conventional processes, biological toxins dissolved in water are not easy to remove and thus they are not the focus of the treatment processes. Consequently, removal of biological toxins in water resources is important, if feasible, on-site and in real-time with minimal efforts, less chemicals, and low energy inputs. The on-site treatment strategy also provides a systematic tool to protect living creatures and ecosystem in HAB sites.

Progress in chemical oxidation has resulted in advanced oxidation technologies (AOTs), one of the most powerful water treatment processes (Nfodzo, et al., 2013). AOTs are based on generation of hydroxyl radicals and sulfate radicals with high oxidation potentials. Particularly, $TiO_2$ photocatalysis has been highlighted as one of the most promising and green AOTs due to its effectiveness to generate hydroxyl radicals (Choi, et al., 2010). For the generation of hydroxyl radicals from $TiO_2$, ultraviolet (UV) radiation which provides high photon energy above the band gap of $TiO_2$ is required. This greatly inhibits the utilization of solar radiation as a sustainable energy source for the $TiO_2$ activation because only 4% of the incoming solar energy onto the earth's surface is in UV range. To utilize visible light in solar radiation for $TiO_2$ activation, dye-sensitized or metal ion-doped $TiO_2$ has been developed and the $TiO_2$ has shown promising results for the degradation of chlorinated compounds and nitrogen oxides (Bae and Choi, 2003).

Introduction of anionic dopants, especially nitrogen, to $TiO_2$ also makes it possible to reduce the $TiO_2$ band gap and thus to activate $TiO_2$ under visible light.

In addition to the intrinsic photocatalytic activity of $TiO_2$, which is directly related with its crystal properties, the structural properties of porous $TiO_2$ catalysts, such as their surface area, porosity, and pore size and distribution, are also of importance because of their potential role in enhancing the light absorbance of $TiO_2$ catalysts and the accessibility of reactants to the active catalytic sites (Zakersalehi et al., 2013). One of the methods to fabricate highly porous materials with desired pore structure and size for target-specific applications is to use amphiphilic organic molecules such as surfactants and block copolymers as pore directing agents when $TiO_2$ is formed from its molecular precursor in a sol-gel process. In particular, we used a nitrogen-containing surfactant in the sol-gel method, which is expected to produce N-doped porous $TiO_2$.

Moreover, sol stability and film homogeneity are also important. Conventional approaches immobilize already-synthesized (or commercially-available) $TiO_2$ particles onto a substrate and thus result in a non-uniform and instable thick film. In this study, $TiO_2$ precursor sol was directly deposited onto a borosilicate glass via dip-coating technique (Choi, et al., 2006). After coating, the glass substrate coated with the $TiO_2$ precursor sol was precisely heat-treated using a programmable furnace. Binding of the $TiO_2$ precursor to the glass surface at high temperatures, uniform and ultrathin nature, and nanocrystalline $TiO_2$ are believed to significantly enhance the mechanical stability of the $TiO_2$ film. Transparency is also another important factor, which is related with light penetration through the $TiO_2$/glass composite and thus overall photocatalytic reactivity. As a result, this study was focused on developing an innovative integrated material processing method to synthesize nanocrystalline N-doped $TiO_2$ with a controlled porous structure and to fabricate a robust, transparent, and ultrathin $TiO_2$ film. Preliminary results demonstrated successful decomposition of MCs on the high efficiency visible light-activated $TiO_2$ under visible light.

## MATERIALS AND METHODS

**$TiO_2$ Sol Synthesis.** Two different nitrogen containing surfactants, diethanolamine (DEA), and benzyltrimethylammonium chloride (BTAC) purchased from Aldrich were used. First each surfactant was dissolved in isopropanol (i-PrOH, Fisher) and then acidic acid (Fisher) was added to the solution for the esterification reaction with i-PrOH (Choi, et al., 2006). Finally, titanium tetraisopropoxide (TTIP, Aldrich) was added under vigorous stirring. The molar ratio of surfactant/i-PrOH/acetic acid/TTIP was 1:45:6:1. Control $TiO_2$ was also prepared with the surfactants.

**$TiO_2$ Thin Film Formation.** A borosilicate glass substrate with an effective surface area of 10 $cm^2$ was used to immobilize $TiO_2$. The substrate was dip-coated with the sol using a PTL-MM01 (MTI Corporation) dip-coating device at a withdrawal rate of 15 cm/min. After coating, the films were dried at room temperature for 1 h and then put into furnace (Paragon HT-22-D, Thermcraft). The heat-treatment temperature was increased at a ramp rate of 60 °C/h to 100 °C and maintained at this temperature for 1 h. Then the temperature was increased again to 500 °C, maintained at this temperature for 2 h, and then cooled down naturally. The dip-coating and calcination process was repeated five times.

**Photocatalytic Activity.** In order to quickly check whether the film works under visible light, first we tested decomposition of methylene blue (MB) as a model organic dye. Two 15W fluorescent lamps (Philips) were used at the intensity of 2.1 mW/$cm^2$ and mounted with a UV block filter (UV420, Bower) to cut the spectral range below 420nm. The initial concentration of MB was 5 mg/l and the volume of the solution was 10 ml. In order to measure dark adsorption, the solution was kept in dark condition for 30 min and then irradiation began and continued for 4 hours. After that, we switched target chemical to decompose from MB to MC-LR (Cayman Chemicals). The initial concentration of MC-LR was 1 mg/l.

**Analysis of Samples.** Water samples were drawn at each sampling intervals. Concentration of MB was measured using a UV–Vis spectrophotometer (UV-2550, Shimadzu) at adsorption peak of 664 nm. Concentration of MC-LR was monitored with a reversed-phase high performance liquid chromatography (1200 series, Agilent) equipped with a quaternary pump, C18 column, and ultraviolet (UV) detector. A mixture of trifluoroacetic acid (TFA, Sigma-Aldrich) and acetonitrile was used as a mobile phase at TFA:acetonitrile ratio of 80:20% v/v with a flow rate of 1 ml/min. The UV detector was operated at a wavelength of 238 nm.

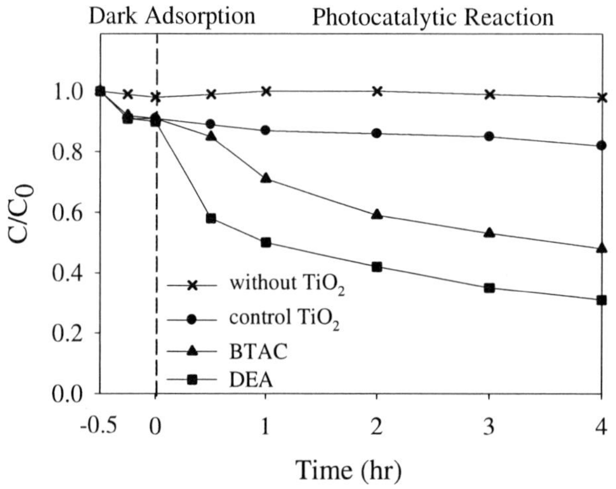


**FIGURE 1. Photocatalytic decomposition of MB by different $TiO_2$ thin films under visible light of > 420 nm**

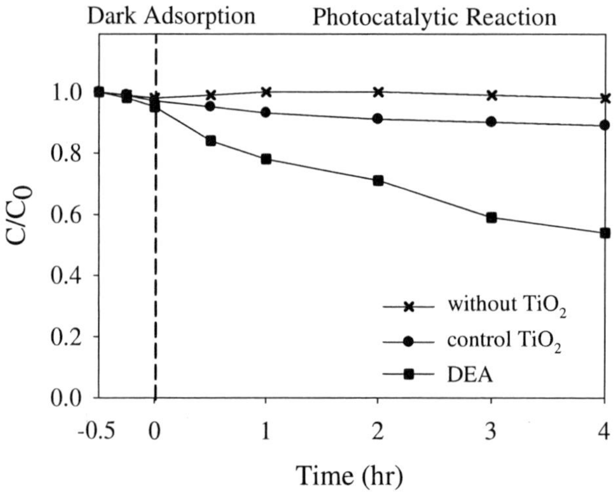


**FIGURE 2. Photocatalytic decomposition of MC-LR by $TiO_2$ thin films under visible light of > 420 nm**

## RESULTS AND DISCUSSION

Photocatalytic activity of different $TiO_2$ thin films was measured in terms of MB decomposition, as shown in Figure 1. As expected, there was no removal (adsorption) in the dark condition in the absence of $TiO_2$. All three $TiO_2$ materials showed around 10% adsorption of MB in the aqueous solution. Under visible light, there was no photolysis of MB in the absence of $TiO_2$. Control $TiO_2$ which was made without the surfactants showed negligible photocatalytic decomposition of MB under visible light. However, N-doped $TiO_2$ thin films made of BTAC and DEA showed significant decomposition. In particular $TiO_2$ made of DEA achieved around 70% removal of MB after 4 hours. The result implied the surfactant BTAC and DEA are effective to improve the physicochemical properties of $TiO_2$ for visible light activation (Vaiano et al. 2015).

Since $TiO_2$ made of DEA showed the highest reactivity with MB under visible light, we tested the $TiO_2$ for the decomposition of MC-LR, as shown in Figure 2. As expected, there was no photolysis of MC-LR by visible light because MC-LR is a strong chemical and control $TiO_2$ made without the surfactants did not show any reactivity under visible light. $TiO_2$ made with DEA showed successful decomposition of MC-LR at around 40% after 4 hours.

## CONCLUSION

This study was focused on developing an innovative integrated material processing method to synthesize nanocrystalline N-doped $TiO_2$ with a controlled porous structure and to fabricate a robust, transparent, and ultrathin $TiO_2$ film. Preliminary results demonstrated successful decomposition of MC-LR on the high efficiency visible light-activated $TiO_2$. Our research effort was to develop a new sustainable approach to decompose MCs, if feasible, on-site and in real-time with minimal efforts, less chemicals, and low energy inputs, is under progress. The effort includes characterization of the materials for explaining the high reactivity under visible light, control of calcination temperature for manipulating porous structure and nitrogen content, utilization of many other N-containing surfactants, and test of the materials under solar radiation. This study implies high potential for utilizing solar radiation as a sole energy source for the on-site treatment of MCs by using the $TiO_2$ photocatalytic system.

## ACKNOWLEDGEMENTS

This research was supported by the Texas Higher Education Coordinating Board through the Norman Hackerman Advanced Research Program Fund (THECB13311).

## REFERNCES

Bae, E., and W. Choi. 2003. "Highly enhanced photoreductive degradation of perchlorinated compounds on dye- sensitized metal/$TiO_2$ under visible light." *Environ. Sci. Technol.*, 37(1), 147-152.

Choi, H., S. R. Al-Abed, D. D. Dionysiou, E. Stathatos, and P. Lianos. 2010. "TiO2-based advanced oxidation nanotechnologies for water purification and reuse" in: *Sustainability Science and Engineering, Volume 2: Sustainable Water for the Future*(Eds. Isabel I. Escobar and Andrea I. Schafer, ISSN 1871-2711), pp. 229–254, Elsevier Science, Netherlands.

Choi, H., E. Stathatos, and D. D. Dionysiou. 2006. "Synthesis of Nanocrystalline Photocatalytic $TiO_2$ Thin Films and Par- ticles Using Sol-Gel Method Modified with Nonionic Surfactants." *Thin Solid Films*, 510(1-2):107-114.

Mohamed, Z. A., W. W. Carmichael, and A. A. Hussein. 2003. "Estimation of microcystins in the freshwater fish Oreochromis niloticus in an Egyptian fish farm containing a Microcystis bloom." *Environ. Toxicol.* 18(2):137-141.

Nfodzo, P., D. D. Dionysiou, and H. Choi. 2013. "Water Supply and Treatment." in: *Encyclopedia of Environmetrics Second Edition* (Eds. Abdel H. El-Shaarawi and Walter W. Piegorsch), pp. 1712-1726, John Wiley & Sons, Chichester, UK.

Palikova, M., J. Mares, R. Kopp, J. Hlavkova, S. Navratil, O. Adamovsky, L. Chmelar, and L. Blaha. 2011. "Accumulation of Microcystins in Nile Tilapia, Oreochromis niloticus L. and Effects of a Complex Cyanobacterial Bloom on the Dietetic Quality of Muscles." *Bulletin Environ. Contam. Toxicol.* 87: 26-30.

Vaiano, V., O.Sacco, D. Sannino, and P. Ciambelli. 2015. "Nanostructured N-doped $TiO_2$coated on glass spheres for thephotocatalytic removal of organic dyes under UV or visible light irradiation." *Applied Catalysis B: Environmental,* 170:153-161.

Vazquez, E. J. N., I. G. Lizarrage, C. J. B. Schmit, A. C. Tapia, D. J. L. Cortes, F. E. H. Sandoval, A. H. Tapia, and J. J. B. Guzman. 2011. "Impact of harmful algal blooms on wild and cultured animals in the Gulf of California." *J. Environ. Biol.* 32: 413-423.

Zakersalehi, A., M. Nadagouda, and H. Choi. 2013. "Suppressing NOM access to controlled porous $TiO_2$ particles enhances the decomposition of target water contaminants." *Catalysis Communications*, 41:79-82.

# DECHLORINATION OF DDT AND ITS PRODUCTS USING PALLADIZED BACTERIAL CELLULOSE IN A REACTOR

Vyjayanthi Jeevan Prakash and ***Sumathi Suresh****[*] (Centre for Environmental Science and Engineering, Indian Institute of Technology Bombay, Powai, Mumbai, India, e-mail: [*]sumathis@iitb.ac.in)

**ABSTRACT:** A rotating catalytic contact reactor comprising of zerovalent palladium immobilized on bacterial cellulose was used in the present investigation in both batch and continuous modes for the dechlorination of DDT, DDD and DDE. Treatment of aqueous solution of 72 mg $L^{-1}$, DDT, 99 mg $L^{-1}$ DDD and 100 mg $L^{-1}$ DDE solubilised in 0.05% (v/v) biosurfactant resulted in greater than 99% removal by dechlorination within 30 minutes of reaction when the reactor was operated in batch mode using molecular hydrogen as the reductant.. Approximately 10 L of 72 mg $L^{-1}$ DDT solution could be treated with 90% efficiency when the reactor was operated in continuous flow mode with a hydraulic retention time (HRT) of 1 hour in the presence of hydrogen. GC-MS analysis confirmed the formation of completely dechlorinated hydrocarbon skeletons of DDT, DDD and DDE, namely, diphenylethane and diphenylethylene, as the end products, thereby implying the removal of all alkyl and aryl chlorine atoms during the course of reductive reaction. The foregoing results thus provide a successful demonstration of the dechlorination efficiency of the catalytic system towards the target compounds studied.

## INTRODUCTION

DDT (1,1,1-trichloro-2,2-bis (4-chlorophenyl) ethane), DDD (1-chloro-4-[2,2-dichloro-1-(4-chlorophenyl)ethyl]benzene), DDE (1,1-*bis*-(4-chlorophenyl)-2,2-dichloroethene) are toxic and highly stable hydrophobic compounds that are known to bioaccumulate and biomagnify along the food chains. The US Environmental Protection Agency as well as the Stockholm Convention on persistent organic pollutants classified DDT, DDD and DDE as priority pollutants (Gryglewicz and Piechocki, 2010). Although use of DDT was banned worldwide in 1972, it continues to persist in soil, sediments and ground water bodies. Thus development of effective treatment technologies for the removal of DDT and its products is critical to the restoration of environment.

It is acknowledged that DDT, DDD and DDE are susceptible to reductive dehalogenation reactions that lead to detoxified products that are vulnerable to biodegradative processes. Based on this premise reductive degradation technologies based on the application of monometallic (such as $Fe^0$, $Mg^0$ and $Pd^0$) and bimetallic systems (examples include $Fe^0/Pd^0$, $Mg^0/Pd^0$) have been developed for the degradation of recalcitrant chlorinated compounds. Anodic corrosion of the base metals such as $Fe^0$, $Mg^0$ can supply electrons for the direct dehalogenation of target compounds. However such reactions are slow and often result in the accumulation of partially dechlorinated products. Therefore combination of the base metal with an efficient catalytic metal such as $Pd^0$ is desirable for complete dehalogenation of pollutants. Palladium-based catalysts in association with molecular hydrogen ($H_2$) gas have been also applied for aqueous-phase destruction of halogenated organic compounds in groundwater at ambient pressure and temperature (Mekhaev et al., 2011; Jujjuri and Keane, 2010). These studies cite several key advantages of Pd which are: (i) effective for a wide range of contaminants (ii) presents high rates of reaction and efficiency (iii) requires mild reaction conditions and iv) generation of fewer toxic end products that require minimal follow-up treatment. Molecular hydrogen has been reported to dissociate to form atomic hydrogen (called hydride) at room temperature at the surface of the hydrogenation catalyst Pd to form highly reactive Pd hydride which has a low reduction potential (-2.2V). However, commercial applications of palladium have been impeded due to the fact that it is an expensive catalyst and its recovery after the reduction reaction is very poor. This problem can be overcome by immobilizing the catalyst onto suitable support materials which enhances its reusability which is the goal of the present investigations. The overall goal of our research project was to explore the possibility of using metallic palladium that has been immobilized irreversibly on bacterial cellulose (palladized bacterial cellulose) as a versatile and novel bio-inorganic catalyst to mediate the reductive dechlorination of 3 persistent organochlorinated pesticides, namely, DDT, DDD and DDE, in a reactor. The specific objectives of our research project were to (1) design and develop a rotating catalytic contact reactor, which uses palladium immobilized on bacterial cellulose for the treatment of water spiked with DDT, DDD and DDE; (2) study the kinetics of removal of DDT, DDD and DDE in the reactor operated in batch mode; (3) to study the efficiency of removal of DDT in reactor operated in continuous flow mode; and 4) identify the end products formed as a result of reductive removal of the target compounds. Extracellular cellulose synthesized by bacteria offers the following attractive features as a biogenic support matrix: a) ability to precipitate noble metals such as palladium, gold and silver and immobilize the metal particles within its cellulose network (Evans *et al.,* 2003) b) can be grown to any desirable size and shape and applied in reactors c) Can be prepared in situ from a wide range of biodegradable substrates and d) easily biodegradable.

## MATERIALS AND METHODS

**Design and Operation of Rotating Catalytic Contact Reactor (RCCR)**: The design characteristics used for development of the reactor used for the present study were adapted from Patel and Suresh (2008). The schematic diagram of the RCCR is shown in Figure 1.

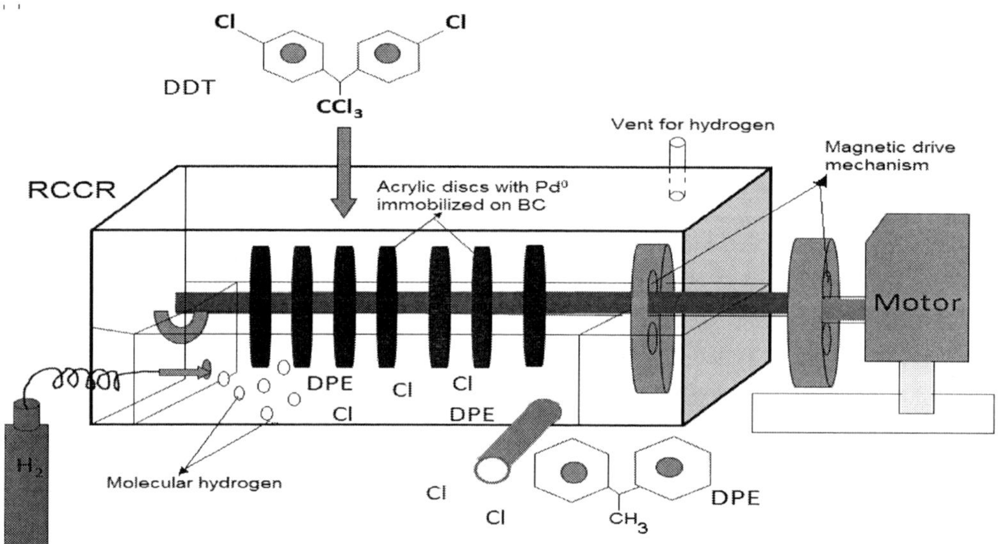


FIGURE 1. Schematic diagram of the rotating catalytic contact reactor (RCCR).

The rectangular reactor consisted of 7 circular discs of 100 mm diameter providing a total surface area of 1142 $cm^2$. The discs were mounted on a horizontal shaft, which was rotated using a magnetic drive mechanism. Effective liquid volume of the reactor was ~650 ml while the speed of rotation was 8 RPM for all experiments. The top of the reactor was covered with a tight lid (although no precaution was taken to keep the reactor free of oxygen). Molecular hydrogen was bubbled into the liquid phase of the reactor at a flow rate of 40 ml $min^{-1}$ and 0.1 kg $cm^{-2}$ pressure. Continuous purging of hydrogen was required as hydrogen is a reactant and is required for the reductive dehalogenation reactions. Rotation of discs in RCCR facilitated exposure of immobilized palladium alternately to the liquid phase containing the substrate solution and the gaseous phase containing molecular hydrogen to perform effective reduction. The reactor was provided with a vent for hydrogen to prevent explosive conditions and maintained under atmospheric pressure and room temperature (28°C - 30°C).

**Production of Bacterial Cellulose in the Laboratory and Its Palladization**: Microbial culture capable of producing cellulose was prepared in the laboratory as per the conventional method followed in Philippines (Thampan, 1981) with modifications as suggested by Patel and Suresh (2008). Three cups of orange fibre pulp was mixed with 6 cups of water and 3 cups of sugar and left undisturbed in a wide-mouthed plastic container covered by a thin cloth. A thin layer of soft, white, jelly like film (bacterial cellulose) of ~2 mm thickness was observed after 12-15 days. Cellulose-producing bacteria have been reported to adhere to cellulose deposits and grow actively at the cellulose-air interface (Serafica *et al*., 2002). Hence cellulose films were scraped and used as the source of bacterial inoculum for production of cellulose in RCCR.

**Development of Bacterial Cellulose Film and Immobilization of Metallic Palladium:** The medium for growth of bacterial cellulose-producing culture consisted of 65 g $L^{-1}$ of sucrose and 25 g $L^{-1}$ of acetic acid added to 650 ml of filtered mature coconut water. The resulting medium was autoclaved and then introduced into the reactor and inoculated with bacterial cellulose deposits. The discs were rotated at slow speed for even deposition of bacterial cellulose until a film thickness of ~2 mm was attained. Pre-treatment and palladization of BC was conducted according to the protocol described by Evans *et al*., (2003). As a first step, discs were immersed for 24 hours in 1% NaOH to kill bacterial cells adhering onto the cellulose following which they were washed thrice with water and then equilibrated in 0.5 M sodium acetate solution (pH of 5) for 2 h. Finally bacterial cellulose immobilized on the rotating discs was incubated in 650 ml of 5 mM $K_2PdCl_6$ solution prepared in 50 mM sodium acetate solution (at 90°C, pH 4.5) until all the discs turned black due to the deposition of zerovalent palladium (3 hours). The palladized discs were subsequently washed in water to remove any residual salt of palladium or unbound $Pd^0$.

**Kinetics of Dechlorination of DDT, DDD and DDE Using RCCR in Batch Mode:** Kinetic studies were conducted in batch mode by introducing 650 ml of DDT (72 mg $L^{-1}$) or DDD (99 mg $L^{-1}$) or DDE (100 mg $L^{-1}$) solubilised in 0.05 % of JBR biosurfactant into the reactor. The order of reaction and rate constant values were computed. Removal could be influenced by both non-reactive sorption (adsorption of target molecules onto BC matrix) and reactive sorption followed by dechlorination. Hence, the effects of both these factors on the extent of removal were determined. Aliquots (3 ml) were collected at chosen time points, extracted thrice with cyclohexane and 0.2 µl of the pooled extracts were analyzed by GC-ECD. Dechlorination was also performed using Tween 80, a commercially available chemical, as the solubilising agent for DDT.

**Dechlorination of DDT Using RCCR in Continuous Flow Mode:** DDT (72 mg $L^{-1}$) was passed through the reactor in the presence of hydrogen. Constant inflow and outflow rates were maintained using a dual channel "Masterflex" (Cole-Parmer) peristaltic pump. The effluent was pooled continuously and analyzed for residual DDT concentrations using GC-ECD. Because complete mixing of the liquid phase occurred in the reactor, the relationship between the concentration of target substrate, DDT in the reactor (or the effluent) to that in the reactor inlet (under steady state conditions) can be given by equation 1 as given below.

$$C_e = \frac{C_0}{1 + k\theta} \qquad \textbf{(1)}$$

where, $C_e$ and $C_0$ are the concentrations of DDT (mg $L^{-1}$) in the reactor/effluent and inlet to the reactor, respectively, $\theta$ = hydraulic retention time (min) and k = first order reaction rate constant ($min^{-1}$).

**Gas Chromatography with Electron Capture Detector (GC-ECD) and Gas Chromatography-mass Spectroscopic (GC-MS) Analyses**: Extracted samples were analyzed using an Agilent model 6890 gas chromatographic instrument equipped with $Ni^{63}$ electron capture detector (ECD). HP-5 capillary column of 0.32 mm ID, 0.25 μm film thickness and 30 m length was used. Injection was made in splitless mode using nitrogen as the carrier gas. The following temperature programme was used: the initial oven temperature of 150°C with hold time of 4 min, ramped at 6°C $min^{-1}$ up to final temperature of 290°C held for 4 min. Detector temperature was set at 300°C. Residual concentrations of DDT, DDD and DDE were quantified from peak areas obtained through automated integration. GC-MS analyses of extracted samples were performed using Perkin Elmer GC system (model Clarus 500) interfaced with a mass detect or Agilent Technologies 7890A GC system coupled with 5975C mass detector. An Elite 5MS (methyl polysiloxane) column of 0.25 mm ID, 0.25 μm thickness and 30 m length was used with helium as the carrier gas (flow rate 1.0 ml $min^{-1}$). The column temperature was ramped as follows: initial temperature of 100°C with hold time of 1 min, ramped at 8°C $min^{-1}$ to final temperature of 290°C held for 10 min. The ion trap was operated at 70 eV with a scan range of m/z from 30 to 500. A sample of 2.0 μl was injected in splitless mode. Intermediates and end products were identified by comparing the mass spectral data with the existing Wiley library data base.

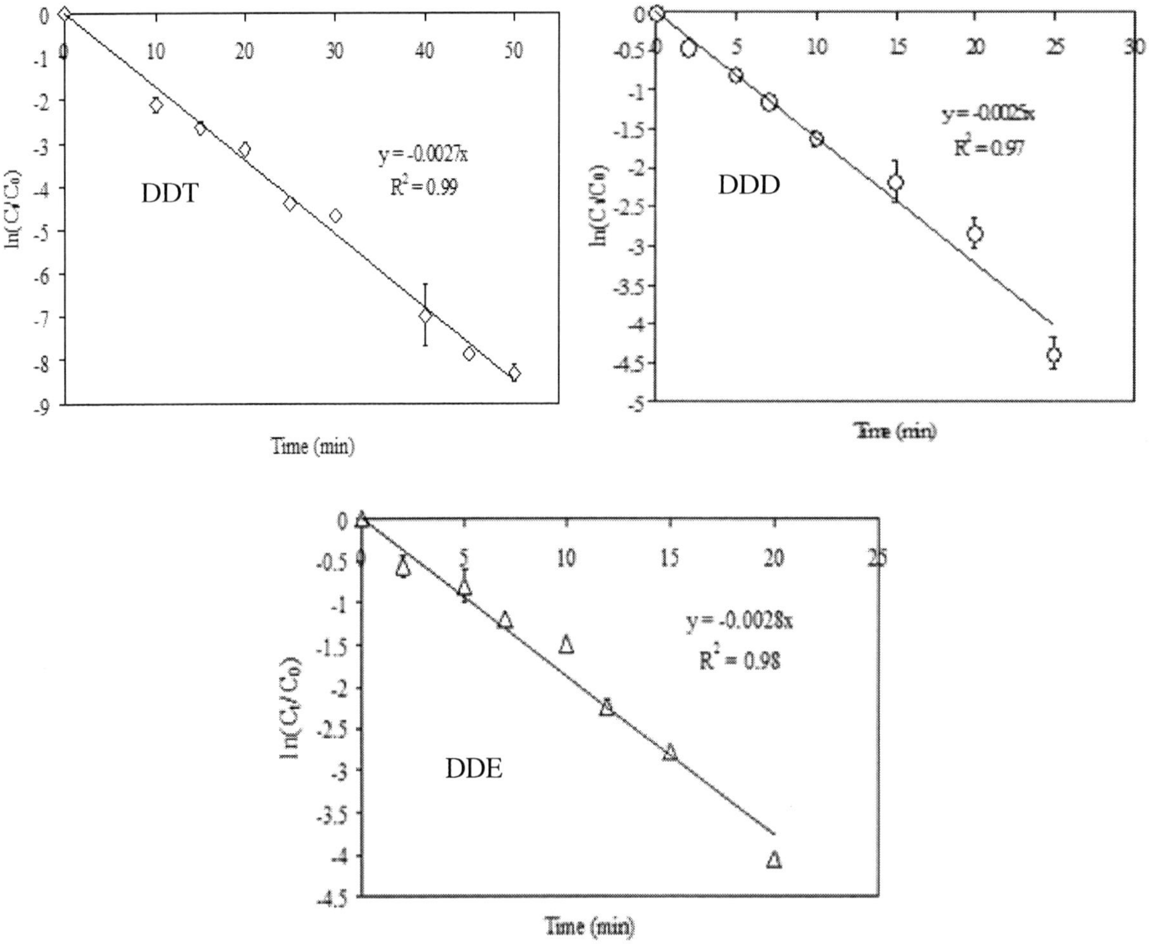


**FIGURE 2. First order kinetic plots for the removal of DDT, DDD and DDE in RCCR operated in batch mode using 0.05% (v/v) biosurfactant and $H_2$ as the reductant.**

## RESULTS AND DISCUSSION

**Removal of DDT, DDD and DDE Using RCCR in Batch Mode:** Figure 2 compares first order kinetic plots for the removal of DDT, DDD and DDE in RCCR. It may be noted that sorption contributed to approximately 35%, 28% and 21% removal (data not shown) while reaction in the presence of hydrogen caused >99% dechlorination of all three compounds. The observed first order rate constants were computed to be 0.0027, 0.0025 and 0.0028 $min^{-1}$ $mg^{-1}$ $Pd^0$ for DDT, DDD and DDE respectively. pH of the treated solutions turned acidic probably due to the release of chloride ions. Partially dechlorinated intermediates such as DDD, DDE and DDMS could not be detected by GC-ECD probably due to high rate of dechlorination of sorbed target molecules and their partially dechlorinated intermediates. Removal of DDT solubilized in 0.05% Tween 80 also followed pseudo first order kinetics with comparable first order rate constant of 0.0029 $min^{-1}$ $mg^{-1}$ $Pd^0$ (plot not shown).

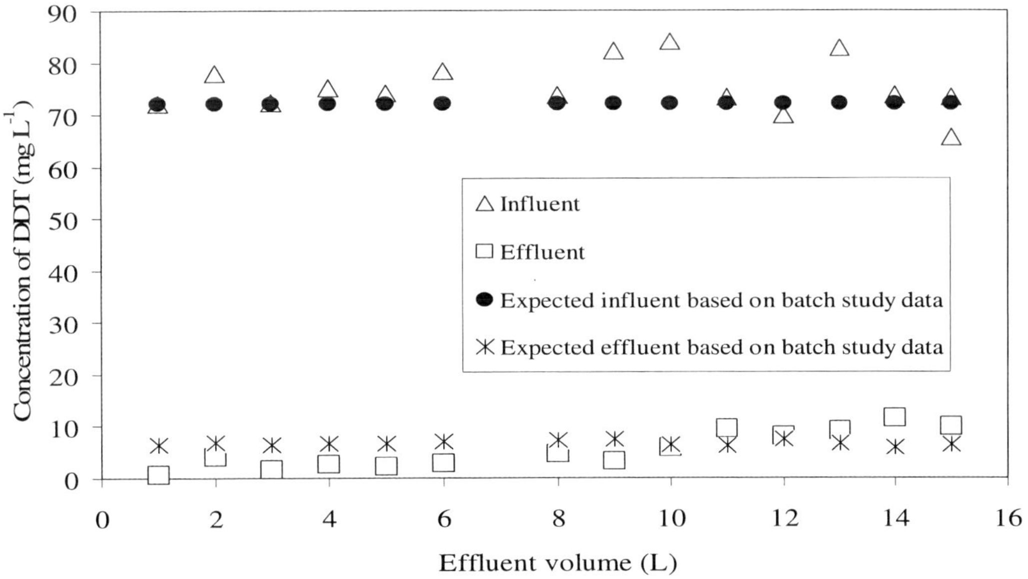


**FIGURE3. Removal of DDT in RCCR operated under continuous flow mode using $H_2$ as the reductant.** Experimental condition□: bio□urfactant = 0.05% (v/v); RPM = 8; $Pd^0$ content = 63 mg $g^{-1}$ dry wei□ht of BC; hydraulic retention time = 1 h; flow rate = 10.83 ml $min^{-1}$.

**Removal of DDT Using RCCR in Continuous Flow Mode**: By substituting the first order rate constant value of 0.17 $min^{-1}$ (0.0027 $min^{-1}$ $mg^{-1}$ $Pd^0$) and setting θ at 1 h, the expected outlet DDT concentration was calculated to be 6.5 mg $L^{-1}$ giving a predicted efficiency of approximately 94% for the removal of 72 mg $L^{-1}$ DDT solution. It can be observed from Figure 3 that the concentrations of DDT in the effluent (obtained experimentally) closely matched with that of computed value (until the pooled volume of effluent reached 10 L). Intermediates or end-products of reaction in the reactor might be responsible for the deterioration of catalytic efficiency above this

**Identification of End Products of DDT, DDD and DDE Dechlorination in RCCR Treated Water by GC-MS Analyses:** The identity of intermediates and end products formed following 50, 20 and 20 minutes of reaction of DDT, DDD and DDE in RCCR, respectively, were determined by GC-MS. The total ion chromatogram and the corresponding fragmentation pattern after reaction of 72 mg $L^{-1}$ DDT are shown in Figures 4 (a) and 4 (b) respectively. Weiss and LaPierre (1980) reported that DDT and DDD yield a base peak of 235 while DDE gives a base peak of 246 in the corresponding mass spectra. Figure 4 (a) reveals the presence of a dominant ion eluting at 8.6 min. Based on the molecular ion fragmentation of this peak as seen in Figure 4 (b), the product was identified as 1,1-diphenylethane (DPE), characterized by m/e of 182. A base peak of m/e 167 is expected to be formed due to the elimination of $-CH_3$ from DPE and presence of the above peak as seen in Figure 4 (b), provided further confirmatory evidence for the formation of this product. Formation of DPE is expected to occur by the successive or sequential removal of 3 alkyl and 2 aryl chlorine atoms from DDT. Based on the GC-MS analyses the product obtained after reductive dechlorination of DDD and DDE in RCCR were determined to be 1,1-diphenylethylene, which is the expected chlorine-free carbon backbone of DDD of DDE (MS data not shown). Complete dechlorination of DDT to form DPE and DDD and DDE to form diphenylethylene by zerovalent palladium immobilized on various supports using hydrogen as the reducing agent have been previously reported (Hashimoto *et al.*, 2010).

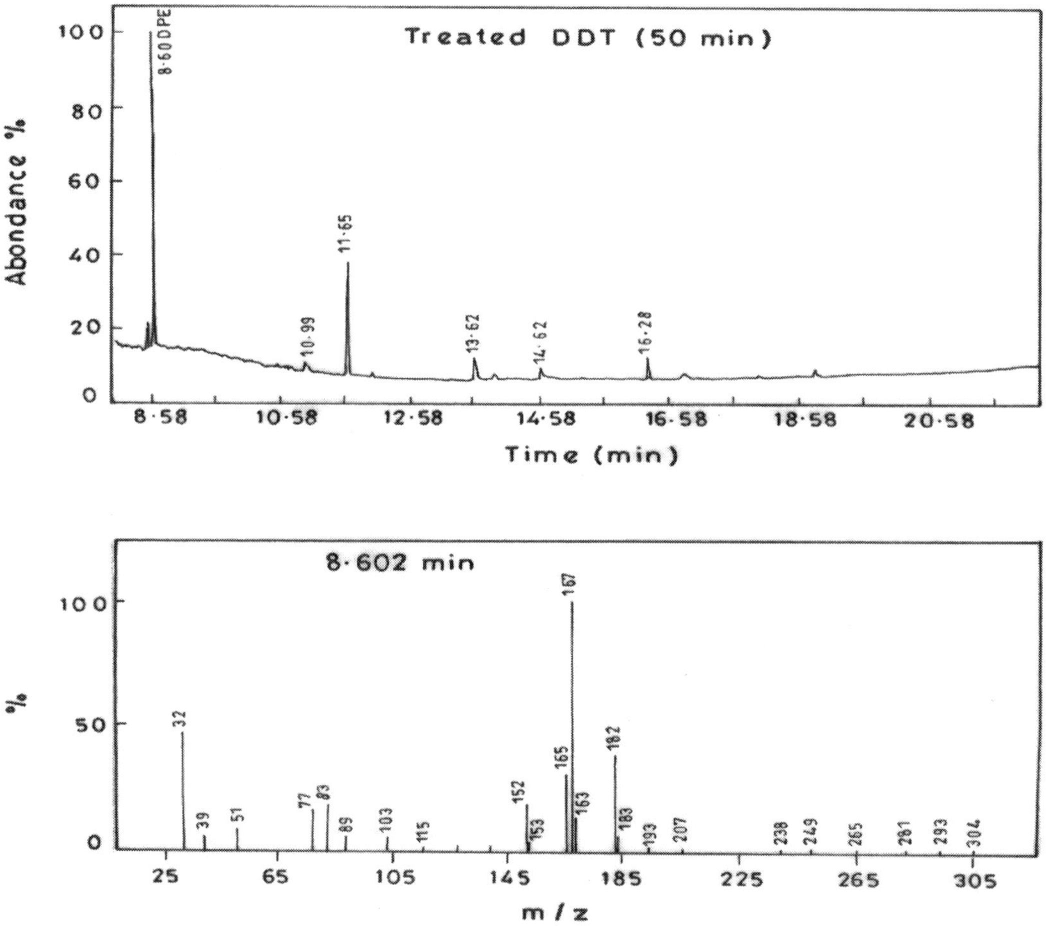


**FIGURE 4. (a) Total ion chromatogram of DDT (72 mg $L^{-1}$) following 50 min of reaction with palladized bacterial cellulose and hydrogen in RCCR (b) fragmentation pattern of the product ion eluting at 8.6 min.**

## CONCLUSIONS

Results obtained through this investigation demonstrate the catalytic efficiency of zerovalent palladium immobilized on the biogenic support matrix, bacterial cellulose. Complete dechlorination of

DDT, DDD and DDE could be achieved in RCCR under batch and continuous flow modes. Reusability of the expensive catalyst, $Pd^0$ immobilized on bacterial cellulose is expected to lower the cost of the technology. Due to the extremely small particle size, large surface area and high *in situ* reactivity, RCCR may be applied for remediation of ground and surface water bodies.

## ACKNOWLEDGEMENTS

The authors thank DST and CSIR, Government of India, for providing financial support. The authors acknowledge IIT Bombay for providing infrastructural facilities to execute the project and services provided by Sophisticated Analytical Instrument Facilities, IIT-Bombay for GC-MS analyses.

## REFERENCES

Gryglewicz, S. and W. Piechocki. 2010. "Conversion Pathways of DDT and its Derivatives During Catalytic Hydrodechlorination". *Pol. J. Environ. Stud.*, 19(4): 715–721.

Mekhaev, A.V., M. G. Pervova1, O. P. Taran, I. L. Simakova, V. N. Parmon, M. A. Samorukova, V. P. Boyarskiy, T. E. Zhesko, V. I. Saloutin, and Y. G. Yatluk. 2011. "Liquid-Phase Dechlorination of Toxic Man-made Products Using Nanodispersed Palladium Catalysts Sibunit". *Chemistry for Sustainable Development.* 19: 173–180.

Hashimoto, N., T. Hara, S. Shimazu, Y. Takahashi, T. Mitsudome, T. Mizugaki, K. Jitsukawa, and K. Kaneda. 2010. "Complete Hydrodechlorination of DDT and its Derivatives Using Hydroxyapatite-Supported Pd Nanoparticles Catalyst". *Chem. Lett.* 39: 49–51.

Patel, U.D. and S. Suresh. 2008. "Complete Dechlorination of Pentachlorophenol Using Palladized Bacterial Cellulose in a Rotating Catalyst Contact Reactor". *J. Colloid Interf. Sci.*, 319: 462–469.

Serafica, G., R. Mormino and H. Bungay. 2002. "Inclusion of Solid Particles in Bacterial Cellulose". *Appl. Microbiol. Biotechnol.*, 58(6): 756–760.

Jujjuri, S. and M. A. Keane. 2010. "Catalytic Hydrodechlorination at Low Hydrogen Partial Pressures: Activity and Selectivity Response". *Chem. Eng. J.* 157: 121–130

Evans, B. R., H. M. O'Neill, V. P. Malyvanh, I. Lee and J. Woodward. 2003. "Palladium-Bacterial Cellulose Membranes for Fuel Cells". *Biosens. Bioelectron.* 18(7): 917–923.

Weiss, A. H., and R. B. Lapierre. 1980. "Gas Chromatographic/Mass Spectroscopic Analysis of Partially Dechlorinated 1,1-Diphenylethane and Ethylene". *Environ. Int.* 3: 353–357.

Thampan, P. K. 1981. *Handbook on Coconut Palm.* Oxford & IBH Publishing, New Delhi, India

# THE POTENTIAL OF SPORE FORMING BACTERIA IN HEAVY OIL SPILL CLEAN UP

**Biji Shibulal**, Saif Al-Bahry, Yahya Al-Wahaibi, Abdulkadir Elshafie, Ali Al-Bemani, Sanket J
(Sultan Qaboos University, Muscat, Sultanate of Oman)

The Sultanate of Oman, located in the Southeast of Arabian Peninsula is seriously threatened by oil pollution as Oman is one of the major oil producing countries in the Middle East and about 30-40% of the oil produced is passed along the Oman Coast. Oil spills may occur due to drilling rigs, routine shipping, run-offs and dumping or human negligence and have severe ecological and economic consequences. Clean-up and remediation of heavy crude oil is a challenge. Bacteria isolated from heavy oil contaminated environment are accustomed to heavy oil and could be much effective in its bioremediation. The objectives of the study were to isolate heavy crude oil degrading spore forming bacteria and to study their responses to heavy oil spills. A total of 42 different spore forming bacteria were isolated and confirmed as heavy oil biodegrading ones. The bacterial isolates were primarily representatives of genera *Bacillus* and *Paenibacillus*. Two different isolates which showed maximum growth at high concentrations of heavy crude oil (7% w/v, API $<6^0$) in Bushnell Haas media (BHM) were selected for the study. The isolates were identified by 16S rDNA sequencing as *Bacillus licheniformis* strain BG2 (KP119098) and *Bacillus subtilis* strain BS2 (KP119102) and their nucleotide sequences were submitted to NCBI GenBank, USA. Degradation efficiency was measured using Gas Chromatography–Mass Spectrometry (GC–MS). The heavy crude oil, degraded for a period of one week, was compared with those of a naturally biodegraded oil and the effect of these bacteria on biodegradation of heavy crude oil was studied. The GC-MS analysis showed a significant effect of these organisms in the biodegradation on heavy crude oil on the ninth day of incubation (Tstat≥Tcritical, $P \leq 0.05$). The results showed that the overall reduction of total heavy crude oil 58% and 62% for BG2 and BS2 respectively compared to that of the control. It can be concluded these strains can be used as an effective eco-friendly candidate for heavy oil spill clean-up can be introduced to a spill in order to hasten biodegradation.

# SEPARATION OF POLYBROMINATED DIPHENYL ETHERS IN FISH FOR COMPOUND-SPECIFIC STABLE CARBON ISOTOPE ANALYSIS

**Yanhong Zeng***, Xiaojun Luo, and Bixian Mai
(Guangzhou Institute of Geochemistry, Chinese Academy of Sciences, Guangzhou, 510640, China)

Polybrominated diphenyl ethers (PBDEs) are a group of brominated flame retardants, which have been added to a large variety of commercial products. Due to their widespread industrial use, PBDEs have been detected as a ubiquitous pollutants in the environment. Moreover, studies reported that PBDEs can bioaccumulate and undergo biotransformation in wildlife and humans. Debromination occurring in biota is a confounding factor to evaluate bioaccumulation behaviour of individual PBDE congener. It is very important to understand whether a specific PBDE congener in organism is an accumulated contaminant or the debromination product of the higher brominated congeners. Compound specific isotope analysis (CSIA) using gas chromatography in combination with isotope ratio mass spectrometry (GC-IRMS) is a suitable tool to study such issues.

Purification and isotope analysis method were developed to accurately measure the stable carbon isotope ratio of PBDEs with three to six substituted bromine atoms in fish samples. Pooled fish samples were treated with concentrated sulfuric acid, separated using a complex silica gel column, and finally purified with an alumina/silica gel column using a proper elution solvent. The purities of extracts were verified by gas chromatography and mass spectrometry in the full scan mode. The average recoveries of all compounds across the purification method were between 60% and 110%, with the exception of BDE-154. The stable carbon isotopic compositions of PBDEs can be measured with a standard deviation of less than 0.5‰. No significant isotopic fraction was found during the purification of the main PBDE congeners. A significant change in the stable carbon isotope ratio of BDE-47 was observed in fish carcasses compared to the original isotopic signatures, implying that PBDE stable carbon isotopic compositions can be used to trace the biotransformation of PBDEs in biota.

# MODELING

# DESIGN OPTIMIZATION OF GAS-TO-LIQUID BIOSLUDGE MANAGEMENT SYSTEMS

***Rana Abbass***, Emad Imam (The American University in Cairo, Cairo, Egypt) and Udeogu Ongwusogh (Qatar Shell Research and Technology Center, Doha, Qatar)

**ABSTRACT:** The optimum management system of biosludge varies according to the type of the biosludge, applicable environmental legislations as well as site and climate conditions. Biosludge generated from plants treating wastewater of Gas-To-Liquid (GTL) plants was used for experimental investigation of its treatability characteristics. Lab results characterized the biosludge to have an organic content of more than 80% of its dry solids content. Aerobic digestion (AD) of the GTL biosludge achieved solid content reduction values of as high as 70% in 9 weeks with corresponding decay constant of $0.0145 day^{-1}$. An optimization design model was developed for the management of GTL biosludge. The optimum design obtained by the model excluded the use of the Aerobic Digestor (AD) and comprised of the centrifuge, drying beds (DB) and landfill disposal resulting in a unit cost of $7.26/ton. To meet the EPA requirement, the design required 21 days in the AD and 8 days in the DB with a unit cost of $9.97/ton. Applying the design model to domestic and refinery biosludge that have higher decay constants has resulted in management costs of $6.01/ton and $4.55/ton, respectively.

## INTRODUCTION

Biosludge is the end product of biological, physical or chemical wastewater treatment plants (WWTPs) (Guyer, 2011). As environmental awareness spreads across industries and cultures, environmental regulations have constrained how WWTPs deal with biosludge waste. Moreover, biosludge treatment demands a considerable percentage of any WWTP's costs. Such regulations require the proper design and installation of a functional and feasible biosludge treatment system as part of any WWTPs while taking into consideration environmental and operational constraints. It is also imperative to treat the produced biosludge to achieve volume reduction and stabilization to be suitable for safe disposal or reuse in accordance with environmental legislation.

The U.S. Environmental Protection Agency (EPA) has proposed certain techniques to achieve volume reduction and stabilization known as Processes to Significantly Reduce Pathogens (PSRPs) (U.S. EPA, 2003). Such processes include aerobic digestion (AD) where biosludge is agitated with oxygen for a certain residence time at the prevailing temperature to achieve stabilization of the organic content (U.S. EPA, 2003). Other methods include air-drying, composting and lime stabilization (U.S. EPA, 2003). As for anaerobic digestion, biosludge is treated in the absence of air at high temperatures to produce stabilized biosludge and biogas (U.S. EPA, 2003).

Once sludge is left to separate, it is mostly highly putrescible organic matter hence; a stabilization treatment technique is inevitable to produce inert biosludge fit for reuse or disposal (Appels et al., 2008). Volume reduction follows digestion and includes dewatering using centrifuge units and drying beds (DB) (Appels et al., 2008).

Biosludge composition differs widely depending on the type of wastewater and the treatment train dictating different treatment processes. The optimum biosludge management system should also account for the applicable environmental legislations as well as site and climate conditions. Therefore, there is no standard solution to treat biosludge. Engineers usually follow certain technical design codes to decide on and size the suitable treatment units to treat the biosludge at hand. These codes disregard the effect of external factors like for example cost and availability of land, variations in ambient temperature from one season to another and other site-specific conditions. Design codes are resorted to due to a lack of models that assist the designer to decide on the optimal line-up for the treatment of a certain biosludge composition.

Literature on optimizing biosludge management systems in general is limited. However optimization studies were carried out. Tang et al. developed a model to prioritize sludge treatment alternatives to select the optimum process for a situation (Tang et al., 1997). The model focused on developing countries where the trade-off between cost and effectiveness must consider socio-cultural and environmental conditions (Tang et al., 1997). Linear programming models have been developed to optimize the design of sludge treatment processes at the lowest cost but were soon found to be inadequate for optimization of these processes as they have a more dynamic nature (Tang et al., 1997). An Analytic Hierarchy Process (AHP) was used owing to its ability to take into account environmental and social aspects (Tang et al., 1997). Parameters considered included land size, location and cost, economic factors, climatic conditions, sludge nature and age, local construction skills and community support (Tang et al., 1997). However, the AHP is not quantitative enough to obtain the optimum design of the sludge management system. On the other hand, Bertanza et al. emphasized the importance of optimizing the dewatering stage in the biosludge management system (Bertanza et al., 2014). They carried out an experimental methodology to evaluate the technical and financial performance of dewatering devices that can be used as a Decision Support System (DSS) (Bertanza et al., 2014). The limitation of this study lies in its focus on only one treatment stage however; this choice was made due to the high volume reduction potential achieved through dewatering coupled with sludge's poor dewatering qualities.

This study focuses on the design of an optimum management system for biosludge generated from plants treating wastewater of Gas-To-Liquid (GTL) plants, which is important in many areas, such as the Gulf region, with high land cost, warm arid climate and a lack of sludge re-use markets. Therefore, a tailored management system is needed for such conditions, which is likely to be different from the conventional sludge management systems. This research hence, consists of two parts: (i) experimental investigation of the characteristics of GTL biosludge and its aerobic treatability, and (ii) developing a design model to obtain the optimum design of a GTL biosludge management system.

## MATERIALS AND METHODS

**Experimental Work.** The treatment of GTL biosludge using aerobic digestion was simulated in a lab experiment to derive the GTL decay constant. Samples were collected from the GTL WWTP on a weekly basis for lab analysis. The lab set-up was an aerobic digester batch set-up, which is also referred to as a die-out design where aerobic bacteria are put in a non-limiting environment given that the supply of oxygen is in excess allowing for maximum oxidation rate of Volatile Suspended Solids (VSS) (Bhargava & Datar, 1988). Furthermore, other performance critical parameters like pH and temperature are kept within optimal range to eliminate their effect. The used lab scale reactors were equipped with an aerator, a mixer, pH, temperature and Dissolved Oxygen (DO) probes. Nine batch experiments were run over the course of 12 weeks in three phases. Both influent biosludge and final biosludge samples were characterized and analyzed. Samples were extracted from the reactor at predefined Sludge Retention Times (SRTs), on a weekly basis. To study the performance of a batch reactor, chosen response variables were tested and monitored including the digestion rate of aerobic bacteria which is tested through monitoring the reduction rate in solid content especially VSS through monitoring Volatile Solids (VS) destruction and Chemical Oxygen Demand (COD) concentrations (Rao & Baral, 2011).

Aerobic digestion causes stabilization and volume reduction owing to the process of endogenous metabolism being its core activity (Wang & Shammas, 2009). Endogenous metabolism is a function of the active fraction of the bacterial population in biosludge (Wang & Shammas, 2009). Hence a drop in solids during aerobic digestion will only be due to a decrease in the active biomass or in other words the organic content in the digester. GTL biosludge was analyzed for solid content distribution showing that the VSS content is more than 80% of the Total Suspended Solid (TSS). This proves that aerobic digestion is an optimal treatment technique for GTL biosludge stabilization and volume reduction, since it acts on the VSS content.

The main aim of running those experiments is to derive the GTL biosludge specific decay constant to be fed into the optimization model. To derive $k_d$, the change in solids content can be represented by a first order biochemical reaction represented by equation 1 (Bhargava & Datar, 1988):

$$\frac{dX}{dt} = -k_d X \qquad (1)$$

Where;

dX/dt = rate of change of solids per unit of time, mg/L/d; $k_d$= reaction rate constant, $d^{-1}$;

X= concentration of solids content remaining at time t in the aerobic digester, mg/L; and

t= time, d

Rearranging equation 1 and integrating both sides results in equation 2 (Park et al., 1997), which is used to estimate $k_d$.

$$-\ln\left(\frac{X_t}{X_o}\right) = k_d t \qquad (2)$$

Where;

$X_t$= cell mass at time t (mgVSS/L); and $X_o$= initial cell mass (mgVSS/L).

The resulting GTL biosludge plot has produced a $k_d$ value of 0.0145 at 20°C. The higher the value of $k_d$, the faster the decay rate indicating a higher organic constant. This is evident by the high value of decay constants for shrimp processing, dairy products and municipal waste biosludge, which range between 0.02 and 0.5 $day^{-1}$ (Mardani et al., 2011). While on the other hand, industrial, tannery and textile biosludge samples yielded considerably lower decay constant values in the range of 0.012 $day^{-1}$ indicating lower organic content (Ramdani et al., 2012).

**The Optimization Model.** The model aims at simulating a biosludge management system through tracking the changes in biosludge content as it flows from one treatment unit to the other. The model consists of three sub-models; first, the simulation sub-model, which is used to track the daily water and solids content of the inflow biosludge through the AD, centrifuge and DB. It also predicts the daily final sludge amounts collected from the DB and transported to the landfill (LF) for disposal.

The AD stage is responsible for the reduction in solid content although the flow rate $Q_l$ entering and leaving the reactor stays the same, as shown in Figure 1. The percentage drop in solids (D) is calculated as shown in equation 3 where $S_1$ and $S_2$ are solid concentrations in and out of the digester while $k_{d(T)}$ is the decay constant at a specified temperature (T) and (t) is the residence time measured in days.

$$D\ (\%) = \frac{S_2 - S_1}{S_1} \times 100 \qquad (3)$$

$$where\ S_2 = S_1 e^{-k_{d(T)}t}$$

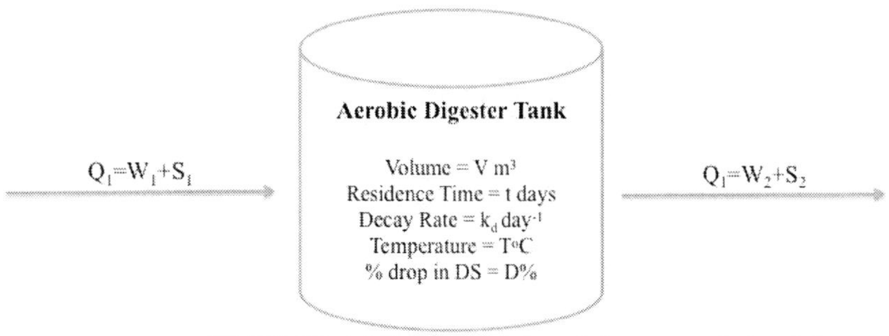


**FIGURE 1. Flow and Mass Diagram of the AD**

The centrifuge is then modeled as it receives an inflow of biosludge $Q_1$ to produce two streams, a water stream known as centrate $W_R$ and a biosludge stream $Q_2$ known as sludge cake. The sludge cake is made-up of water content $W_3$ and solid content $S_2$, which is equal to the influent solids concentration since the centrifuge does not affect the solid content. The centrifuge accepts a range of influent solid concentrations chosen based on the range of concentrations produced by the AD. It then produces an effluent with a predefined solid content percentage $P_c$ chosen as a specification of the procured dewatering unit. $P_c$ and $S_2$ are used to calculate $Q_2$ and consequently $W_3$ and $W_R$ are found. Equations 4, 5 and 6 quantify the centrifuge operation.

$$Q_2 = \frac{S_2 \times 100}{P_c\%} \tag{4}$$

$$W_3 = Q_2 - S_2 \tag{5}$$

$$W_R = W_2 - W_3 \tag{6}$$

Drying beds (DB) receive biosludge with a certain water content $W_3$ and a solid content of $S_2$ where they do not affect the solid content. Volume reduction achieved in the drying beds is done through the loss of moisture content, which is owed to two processes, water drainage $W_s$ through seepage into the leaching system and water evaporation $W_E$. Drying beds operations is illustrated in figure 2.

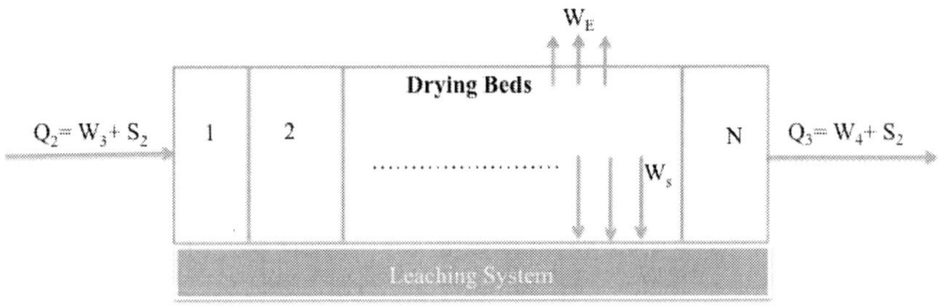


**FIGURE 2. Flow and Mass Balance of Drying Beds**

The second sub-model is the design sub-model, which is used to size the AD, the centrifuge, and DB based on selected values for sludge residence times. The AD is designed based on the design volume flow rate of the inflow biosludge $Q_1$ and its Dry Solids (DS) contents. According to the design; $Q_1$ in $m^3/d$, the volume (V) of the AD in $m^3$, should ensure a design residence time (t) in days, as per equation 7. AD should be equipped with an efficient aeration system where the DO concentration in excess of 2mg/L must be maintained (U.S. EPA, 2003). The aeration system of the AD is comprised of an air blower, a piping system and air diffusing equipment, which need to be appropriately sized to ensure proper aeration. Equation 8 empirically relates the blower power $P_w$ in horsepower (hp) to the airflow rate in standard cubic feet per minute (SCFM). To calculate the number of diffusers needed to uniformly spread the airflow from the blower across the volume of the tank, the blower airflow rate is divided by the airflow rate per diffuser disc, which typically ranges between 0.5 and 4.5 SCFM and is chosen as 3 SCFM for this model (Xylem, Inc., 2012).

$$Volume\ V\ (m^3) = Flow\ rate\ \ Q_1\ (m^3/d) \times Residence\ Time\ \ t\ (d) \tag{7}$$

$$Blower\ Power\ P_w\ (hp) = 0.0517 \times Air\ Flow\ Rate\ (SCFM) + 0.8849 \tag{8}$$

The sizing of the centrifuge or its power rating depends on the volumetric flow rate of biosludge into the centrifuge unit. Furthermore, the centrifuge is set to achieve a certain degree of dryness given a certain range of DS content of influent sludge.

Drying beds are paved shallow basins, equipped with leaching systems and divided into equally sized beds where the daily batch of biosludge is moved from one bed to the other to allow for proper aeration, increase evaporation and prevent the release of odors. The length of each cell can then be estimated using equation 9 (Guyer, 2011). The number of beds is equal to the number of days the biosludge will stay in the DB. This is a decision made by the designer based on the percent dryness that needs to be achieved to meet landfill requirements, environmental standards and economic constraints.

$$Area\ of\ one\ bed = \frac{Volume\ of\ a\ daily\ batch\ of\ biosludge}{Sludge\ layer\ thickness} = Length\ of\ bed \times Width\ of\ bed \qquad (9)$$

This design sub-model generates alternatives where each alternative is a combination of different values of the design variables. Two main design variables are varied to identify the optimum design: residence time in the AD and time spent in the DB. The residence time in the AD sets tank volume yielding a different digestion rate and solids concentration of the AD effluent.

The third sub-model is the cost and optimization sub-model, which estimates the costs for each design alternative. The minimum cost alternative is chosen as the optimal treatment line-up recommendation. The minimum cost is a trade-off between the cost of biosludge being aerated for a longer time and the drying and disposal cost savings achieved by smaller volumes of produced biosludge.

To quantify the optimization process, it is imperative to realize the investments made for each scenario. Hence, cost functions that vary according to the treatment units sizing options were developed to derive the unit cost of handling one ton of biosludge over an annual average performance. Cost functions were divided into capital expenses (CAPEX) and operational expenses (OPEX). The Cost Recovery Factor (CRF) formula was used to find the annual CAPEX for each cost item where CRF is multiplied by the CAPEX to give the ánnual cost which is then used to find the cost per day. CAPEX items included the cost of land, material and construction costs of the AD tank, the cost of the blower and its diffusers and the cost of a centrifuge. The OPEX costs included operation and maintenance cost of digestion, dewatering and handling the sludge at the drying beds as well as its transportation and disposal in the landfill (LF). The CAPEX and OPEX per ton figures are summed to give a numeric value that reflects the performance of each alternative.

The sludge flow rates and composition and their treatment are simulated on a daily basis over a typical year to account for the effect of changes in the sludge and the variable climatic conditions since simulating the effects of external variations are expected to affect performance and costs. Also, running the model over one year allows the designer to monitor the variations in performance and cost and vary the design accordingly, which allows flexibility in design choices that result in the optimal system design.

## RESULTS AND DISCUSSION

Varying the residence time in the AD tank, keeping other variables constant yields treatment alternatives with different unit costs. The chosen point is preferably the minimal cost point unless other environmental legislations dictate the validity of one alternative over the other. Increasing the residence time of the AD stage affects the volume of dry solid content going through the centrifuge and the drying beds and ultimately disposed of in the landfill. An increase in residence time reflects as a reduction in cost of dewatering, drying and disposal. However, it also translates into an increase in costs due to the need for larger AD tank and aeration equipment. After choosing the optimal AD residence time, the designer keeps it constant and starts varying the time biosludge is left in the DBs. Increasing the time in the DB reduces the final volume of the disposed biosludge, which in turn reduces the costs of trucking and landfill occupation. However, the longer the time a biosludge batch is left in the DB, the higher the capital costs. Consequently, the minimum cost point is chosen such that the maximum dryness is achieved with minimal increase in cost. Finding the minimum cost point achieved by each variable suggests the optimal design configuration for the treatment line-up.

The optimization model was tested using different applications to obtain the optimal treatment line-up. The main application used for this study is GTL biosludge.

The first two applications test the optimum management system of GTL biosludge under different constraints. In the first case, no constraints were imposed on the management system. The optimum system suggested the exclusion of the AD from the line-up and retaining the centrifuge, and the DBs for duration of 8 days. The unit cost for sludge management was $7.26/ton including transportation and disposal to LF of 22.18 tons of biosludge. The second case, the sludge management was to meet the EPA requirement of limiting the final sludge to 20 tons. The design retained the AD with a residence time of 21 days plus 8 days in the DBs with a unit cost of $9.97/ton.

Applying the design model to domestic and refinery biosludge that have higher decay constants has resulted in a drop in the amounts and costs of biosludge produced to 16.8 tons/d of sewage sludge costing $6.01/ton and 10.21 tons/d of refinery sludge costing $4.55/ton.

## CONCLUSIONS

Experimental investigations show that the biosludge generated from plants treating wastewater produced from GTL plants has an organic content of more than 80% of its dry solids, which suggests the feasibility of its treatment through digestion techniques whether aerobically or anaerobically. The optimization model, the design sub-model can be used for initial sizing and conceptual design. On the other hand, the simulation sub-model can track the solids content, sludge volumetric flow rates, volumes, amount of polymers in the centrifuge, airflow and power requirements for the aeration system as well as drainage water and evaporated water from the drying beds. The estimates can be a good base to predict the capital and operation cost of the various treatment units.

## ACKNOWLEDGEMENTS

Support for this research by the American University in Cairo Environmental Engineering Department and the Qatar Shell Research and Technology Center is gratefully acknowledged.

## REFERENCES

Appels, Lise, Jan Baeyens, Jan Degreeve, and Raf Dewil. 2008. "Principles and Potential of the Anaerobic Digestion of Waste-Activated Sludge." *Progress in Energy and Combustion Science* 34 (6): 755–81.

Bertanza, Giorgio, Matteo Papa, Matteo Canato, Maria Cristina Collivignarelli, and Roberta Pedrazzani. 2014. "How Can Sludge Dewatering Devices Be Assessed? Development of a New DSS and Its Application to Real Case Studies." *Journal of Environmental Management* 137 (May): 86–92.

Bhargava, Devendra, and Madhav Datar. 1988. "Progress and Kinetics of Aerobic Digestion of Secondary Sludges." *Water Research* 22 (1): 37–47.

Guyer, J. Paul. 2011. *Introduction to Sludge Handling, Treatment and Disposal*. NY, USA: CED Engineering.com.

Mardani, Sh., A. Mirbagheri, M.M. Amin, and M. Ghasemian. 2011. "Determination of Biokinetic Coefficients for Activated Biosludge Processes on Municipal Wastewater." *Iran Journal of Environmental Health Science* 8 (1): 25–34.

Park, Jae Kwang, Jenchie Wang, and Gerald Novotny. 1997. "Wastewater Characterization for Evaluation of Biological Phosphorus Removal- Biological Kinetic Parameter Estimation." *Ecology and Natural Resources Collection*, 14–23.

Ramdani, Abdellah, Peter Dold, Alain Gadbois, Stéphane Déléris, Dwight Houweling, and Yves Comeau. 2012. "Biodegradation of the Endogenous Residue of Activated Sludge in a Membrane Bioreactor with

Continuous or on-off Aeration." *Water Research* 46 (9): 2837–50.

Rao, P. Venkateswara, and Saroj S. Baral. 2011. "Experimental Design of Mixture for the Anaerobic Co-Digestion of Sewage Biosludge." *Chemical Engineering Journal* 172: 977–86.

Tang, S. L., C. L. Wong, M Phil, and K. V. Ellis. 1997. "An Optimization Model for the Selection of Wastewater and SIudge Treatment Alternatives." *Water and Environment Journal* 11 (1): 11–23.

U.S. EPA. 2003b. "Control of Pathogens and Vector Attraction in Sewage Sludge: Under 40 CFR Part 503." Guidance. Environmental Regulations and Technology. Cincinnati, OH 45268: U.S. Environmental Protection Agency.

Wang, Lawrence K., and Nazih K. Shammas. 2009. "Aerobic Digestion." In *Handbook of Environmental Engineering*, 8:635–69. Totowa, NJ, USA: Springer.

Xylem, Inc. 2012. "Aeration Products for Energy-Efficient Biological Treatment." Product Report. Charlotte: Sanitaire.

# PROBING THE WATER QUALITY TRANSFORMATION OF UNDERGROUND CAVERNS IN A TROPICAL REGION USING ELCOM-CAEDYM MODEL

Ming Chen, Jian Li, and ***Xiaosheng Qin*** (School of Civil and Environmental Engineering, Nanyang Technological University, Singapore)

**ABSTRACT:** This study aims to preliminarily model the water quality variations in an underground cavern environment using the coupled hydrodynamic and ecological model ELCOM-CAEDYM. The water quality transformation in terms of nitrogen and chlorophyll-*a* under both light and dark environment was analyzed. The ecological part of the model (i.e. CAEDYM) was calibrated and verified by experimental data. It was found that the chlorophyll-*a* concentration (phytoplankton biomass) decreased under the dark and light conditions with the decreasing rate being faster in a dark environment. The ammonia concentrations under both dark and light conditions increased rapidly first and then declined to an equilibrium state. The trend for the nitrite concentrations differed under these two conditions. Nitrite concentrations under the light condition only slightly fluctuated during the entire experiment, while they varied significantly under a dark environment. Nitrate concentrations generally increased with time except for the initial stage under both conditions. The modeling results demonstrated that ELCOM-CAEDYM could reasonably reproduce the experimental observations. Different scenarios for hypothetical design were also run for ascertaining the critical factors controlling the water quality in an underground cavern.

## INTRODUCTION

Water shortage is a global problem today due to uneven distribution of water resources and fast economic and population growth. In 2006, it was reported that about one third of the world population (about two billions) was suffering from water scarcity (Oki and Kanae 2006). Unfortunately, the situation for water shortage is getting worse due to climate change and human-induced activities. In the past decades, Singapore has been developing at an accelerating rate, leading to an increasing water demand. Although Singapore receives considerable rainfall throughout the year, she has limited land to collect and store rainwater. The use of underground caverns has proved to be an effective technology to mitigate water resources shortage (Broch 2007; Zhao and Bergh-Christensen 1996). Due to enclosed and dark environment in underground caverns, water quality may deteriorate, rendering the water unsuitable for human usage. Thus, research is needed to investigate and predict how water quality changes during the storage in underground caverns.

Light is an important factor associated with many processes in an aquatic ecosystem. There is evidence that light controls the concentrations of dissolved inorganic phosphorus and dissolved total phosphorus at the water-sediment interface in Taihu, China (Jiang et al. 2008). It was reported that light intensity was related to nitrogen cycling rates in lake Maracaibo, Venezuela (Gardner et al. 1998). Many other researchers also investigated the light effects on various aspects of aquatic ecologic systems (Spears et al. 2008; Dalai et al. 2014; Bhuvaneshwari et al. 2015). Generally, an open environment allows oxygen to enter the water, while enclosed environment limits the normal air exchange between water and air interface. Although the underground reservoir may still be subject to intermittent episodes of inflows with certain nutrient and pathogenic loads, the absence of solar radiation would lead to severe limitations on algal dynamics, nutrient uptake and cycling. Previously, few efforts have been made in exploring water quality variations in underground caverns (Trolle et al. 2012). In this study, a preliminary modeling study is to be carried out for exploring the water quality changes in an underground cavern environment, with the aid of experimental test.

## MATERIALS AND METHODS

**Bench-scale Experiment.** Singapore is a tropical country, with a total land area around 700 km$^2$ and population about 5 million. The annual rainfall is around 2,400 mm and mean daily temperature varies generally from 23-32°C (Gin et al. 2011). To better understand the mechanisms of the ecological processes in an underground environment in such a region and also to support verification of numerical modelling, a bench-scale experimental test was conducted. A local lake near Nanyang Technological University (NTU), Singapore, was selected for water sampling. The nutrification level in this lake is found higher than the general Singapore raw water and is useful for our experimental purpose. Two stainless-steel water tanks with one having light and the other without light were prepared to store the water samples. FIGURE 1a shows the dimension (40cm×30cm×20cm) and configuration of the water tanks. The tank with light input has its top open to sunlight to ensure phytoplankton photosynthesis, and the one without light was covered to mimic the underground environment. Both tanks have connection pipes to atmosphere and are not completely sealed, which is similar to real cavern condition. The experimental test started from September 9$^{th}$, 2015 to October 27$^{th}$, 2015. The analyzed parameters include dissolve oxygen (DO), pH, temperature, total carbon (TC), inorganic carbon (IC), ammonia, nitrate, nitrite and Chl-*a*. The experimental instruments used for the measurement of these parameters are: DO: YSI 5000 DO meter; pH: pH meter (Mettler toledo)**;** Temperature: YSI 5000 DO meter; TC and IC: TOC Analyzer (Shimadzu); Ammonia: Ammonia TNTplus, ULR; Nitrate: Nitrate TNTplus vial test, LR; Nitrite: Nitrite TNTplus, LR; Chl-*a*: UV-vis Spectrophotometer (UVmini-1240, Shimadzu).

**Ecological Modeling.** The ELCOM-CAEDYM model was used to simulate the water quality deterioration in an underground cavern environment. The CAEDYM module was customized to reproduce the water quality changes in the experimental water tanks. The configuration and preprocessing of the model mainly include water-tank meshing, bathymetry processing, boundary/initial condition setup, and parameter estimation. The metrological factors were considered major driving forces influencing the water quality in the water tank as there was no inflow into the water tank and the water body was always static. The solar radiation and air temperature used in the modelling process was assumed following the Sine function as they varied diurnally. The air temperature was set to 29 °C considering no major changes for the studied period. The model used the actual size of the water tanks to study water quality changes with and without light input (see FIGURE 1a). The settling of the particle nutrients and phytoplankton cells may cause limited vertical distribution of concentration; in areal direction, it is assumed spatially uniform due to the small size of the system. The mesh size used in the ecological modeling was 0.02 m with grid numbers 20×15×10, which was deemed sufficiently dense for capturing the chemical/biological reactions in the tank.

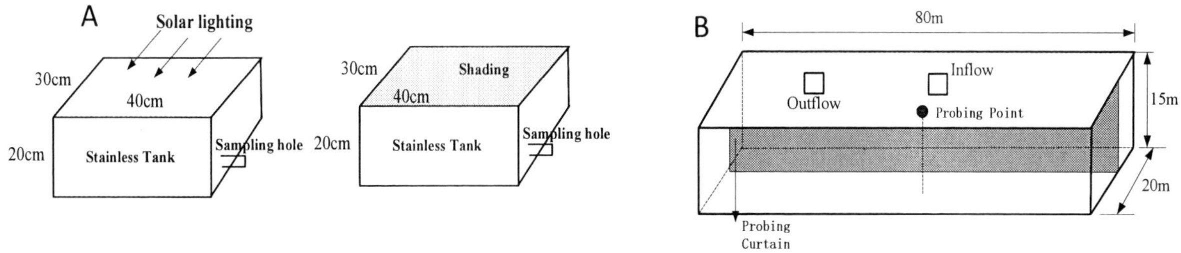


**FIGURE 1.** (a) Water tank configuration and (b) hypothetical underground water cavern.

A number of water quality variables including ammonia ($NH_4$), nitrate ($NO_3$), dissolved oxygen (DO), water temperature, pH, TC, and Chl-*a* concentration were measured. The phosphate ($PO_4$)

concentration was below the detection limit of our equipment and its initial concentration in the model was provided from the related literature on Singapore reservoir water quality. The initial concentrations of $NH_4$, $NO_3$, DO, water temperature, pH, TC, and Chl-*a* were set based on the laboratory measurement data listed in TABLE 1. The initial concentrations of nutrients and Chl-*a* were higher in the light condition than those without light as some reactions may already have started before the first measurement on September 9th, 2015.

**Modelling a Hypothetical Underground Water Cavern.** The modelling of a hypothetical underground water cavern was carried out assuming frequent inflows into the cavern. This was to simulate the water input from surface catchment under high storm events. Also, the hydrodynamic cycling would affect the internal water quality changing process. As shown in FIGURE 1b, the modeling case was designed as a rectangular tank with dimensions of 80 m long, 20 m wide and 15 m high. The total volume of the modelling case is 24,000 $m^3$. One inflow point and one outflow point were setup in the middle of the left and right sections, respectively. The wall was assumed to be a concrete wall which means no light could penetrate into the tank. The size of the case was referred to two real projects including one pilot underground oil cavern built in Bukit Timah granite bedrock in Singapore (Zhao and Bergh-Christensen 1996) and one underground water storage cavern built in Trondheim, Norway (Broch 2007).

**TABLE 1.** The initial condition of water quality variables in modeling

| Variables | Without Light | With Light | Unit | Value obtained from |
|---|---|---|---|---|
| DO | 5.20 | 5.80 | mg/L | Measured |
| Water temperature | 28.0 | 28.0 | °C | Measured |
| pH | 8.03 | 8.06 | - | Measured |
| $NH_4$ | 0.037 | 0.041 | mg/L | Measured |
| $NO_3$ | 0.095 | 0.151 | mg/L | Measured |
| PO4 | 0.01 | 0.01 | mg/L | Reference (Low et al. 2010) |
| ON | 1.10 | 1.10 | mg/L | Reference (Low et al. 2010) |
| OP | 0.05 | 0.05 | mg/L | Reference (Low et al. 2010) |
| Chl-*a* | 54.91 | 115.77 | μg/L | Measured |

The factors influencing the water quality changing process in the studied underground water storage tank include the inflow/outflow rates, the discharging frequency and duration, and the pollutant loads. The nutrients and Chl-*a* concentration of the inflow water will be studied through scenario analysis. But some other influencing factors such as the chemicals from the concrete wall, the suspended particulate matter and microorganisms were not considered in this study.

The initial conditions of DO, water temperature, nutrients and Chl-*a* concentrations were the same in all the scenarios of modeling studies, which aims to compare the effect of different boundary conditions on the underground water quality changing process. The total Chl-*a* concentration was the sum of the phytoplankton groups. Three kinds of algae including Cyanobacteria, Nodularia and Freshwater Diatoms were chosen according to the literature related to the reservoirs in Singapore (Low et al. 2010), and the ratio of the above algae was set as 2:2:1. The simulation period was set to 60 days for all scenarios.

In total, 4 cases were modelled to examine the effects of different boundary conditions on the water quality changing process in the hypothetical underground water storage cavern. Case-1 was used as a standard benchmark case and other cases will be used for comparison. All the boundary conditions used in the scenarios are listed in TABLE 2. The nutrient concentrations including both nitrogen and phosphorous increased in Case-2. The inflow Chl-*a* concentration was 10 times higher in Case-3, and the inflow frequency and duration all decreased by half in Case-4. It should be noted that the general Chl-*a* concentration in Singapore raw water is usually lower than 100 μg/L; we set it at a higher value for theoretical analysis only.

## RESULTS AND DISCUSSION

**Changes in Water Quality.** Water sampling was performed continuously from these two water tanks for water quality analyses over about two months. It was found that DO concentrations exhibited a similar trend under the two types of conditions (FIGURE 2). Their mean DO concentrations were at similar levels (5.76 mg/L for no light and 5.78 mg/L for light conditions, respectively). We did not observe significantly different patterns for TC, IC and pH concentrations between water samples under light and dark conditions. However, the curves for Chl-*a* concentrations were notably different under these two conditions, although all of them showed a generally decreasing tendency. The declining trend in Chl-*a* concentration under the dark condition was faster than that under the light one. Temporal plots of ammonia, nitrate and nitrite were shown in FIGURE 2. The mean ammonia, nitrate and nitrite concentrations were 0.053, 0.540 and 0.036 mg/L for the dark condition, respectively, and 0.045, 0.515 and 0.022 mg/L for the light condition, respectively. There was a similar trend for the ammonia concentrations under the dark and light conditions. Both observed ammonia concentrations increased rapidly within the first week, and then quickly declined. After about 10 days, they would reach an almost equilibrium state. It is noted that, the trend for the nitrite concentrations was largely inconsistent under these two conditions. Nitrite concentrations under the light condition only slightly fluctuated during the entire experiment, while they varied significantly under the dark condition. Nitrate concentrations generally increased with time except for the initial stage under both conditions. But the increasing rate was obviously higher for the tank with dark condition.

**TABLE 2.** Scenarios modeling boundary conditions

| Variables | Case-1 | Case-2 | Case-3 | Case-4 | Unit |
|---|---|---|---|---|---|
| Discharge | 5.0 | 5.0 | 5.0 | 5.0 | $m^3/s$ |
| Frequency | 1 | 1 | 1 | **2** | /10days |
| Duration | 6 | 6 | 6 | **3** | hours |
| Temperature | 29.0 | 29.0 | 29.0 | 29.0 | °C |
| DO | 8.0 | 8.0 | 8.0 | 8.0 | mg/L |
| $NH_4$ | 0.06 | **0.20** | 0.06 | 0.06 | mg/L |
| $NO_3$ | 0.15 | **1.50** | 0.15 | 0.15 | mg/L |
| $PO_4$ | 0.01 | **0.02** | 0.01 | 0.01 | mg/L |
| ON | 0.60 | **3.0** | 0.60 | 0.60 | mg/L |
| OP | 0.02 | 0.02 | 0.02 | 0.02 | mg/L |
| Chl-*a* | 91.5 | 91.5 | **915** | 91.5 | μg/L |

**Modeling Results for Lab Water Tanks.** CAEDYM model was used to simulate the changing process of water quality variables to repeat the lab test results. The variation of nitrogen nutrient components has been given a priority for analysis. The phosphate ($PO_4$) concentration was very low and limited the phytoplankton growth in the reservoir of Singapore (Gin et al. 2011), so it was not analyzed. The concentrations of ammonia would decrease gradually in the simulated period (30 days) because ammonia would transform into nitrate through the nitrification process and also the phytoplankton cells adsorption. From FIGURE 2a, the simulated ammonia concentration changed following a similar trend with observed one regardless of with light or without light. From FIGURE 2b, the nitrate concentration would consistently increase from 0.1 to about 1.0 mg/L until the end of simulation period. The Chl-*a* concentration dropped from an initial high concentration to nearly zero due to the death of phytoplankton cells (FIGURE 2c).

Nevertheless, the total nitrogen in the water tank shows only a slight decrease which means the total mass of nitrogen was kept in equilibrium. The simulated nitrogen nutrients and Chl-*a* concentrations are in a good agreement with the observed data. It seems that the relatively weak solar lighting in the laboratory would not significantly affect the chemical reaction and phytoplankton dynamics; the adapted

CAEDYM could well reproduce the ecological process in the dark environment. Generally, both the simulated results and measured data indicate that (i) the phytoplankton would die because of the limiting factors from phosphorous nutrients and organic carbon, (ii) the nitrate concentration, that are transformed from the initial organic nitrogen, would rise and keep at a high level, and (iii) the transformation rates of ammonia and nitrate under the dark environment are somewhat higher than those under the light one, but the decay rate of Chl-*a* is somewhat lower under the light condition than under the dark one. The observed DO concentration decreased before the 10th day and then increased due to re-aeration from the atmosphere (FIGURE 2d). Interestingly, the trends for DO concentrations under light and dark conditions were very close, as shown in FIGURE 2d. The simulated DO concentration agrees with the observed data well to a large extent.

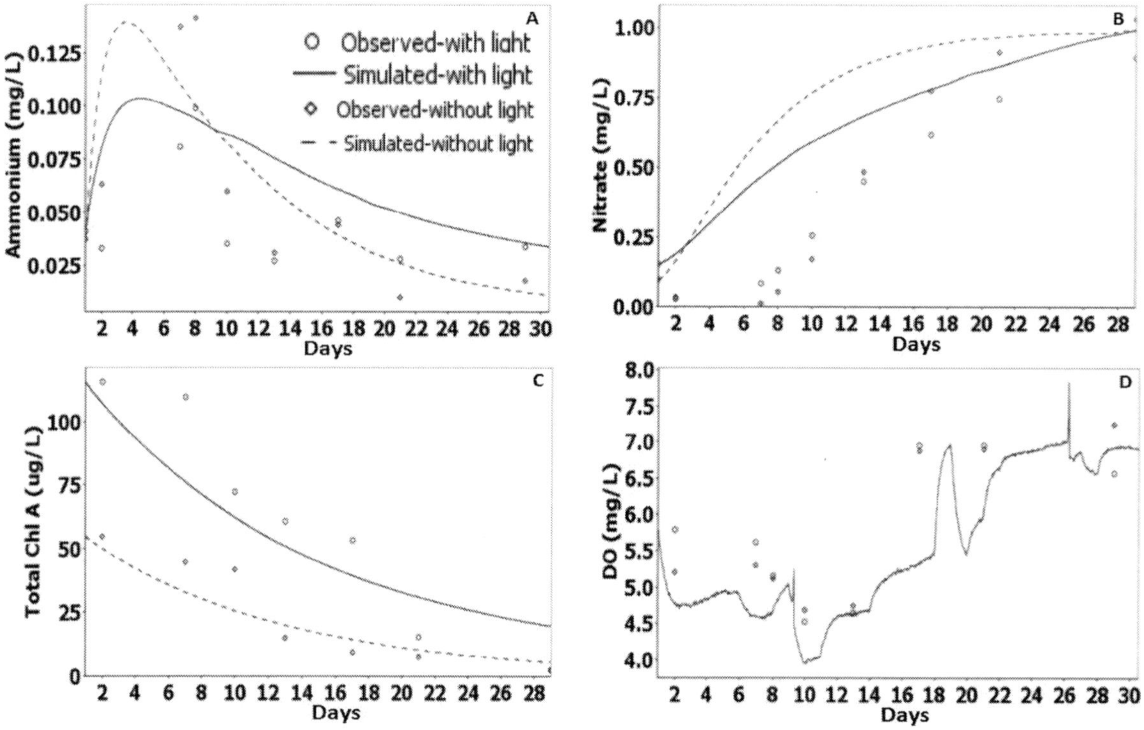


**FIGURE 2.** Dynamic trend of the measured variables (a: Ammonia, b: Nitrate, c: Chl-a, and d: DO).

**Modelling Results for the Hypothetical Underground Cavern.** Based on the parameter setting for the lab model under no light condition, we conducted a modeling study for the hypothetical underground cavern. It should be noted that, there are multiple possibilities of results under different scenarios (e.g., DO concentration, inflow patterns, pollutant loadings, location of probing profile). Herein, only one scenario was presented (as shown in FIGURE 3). The $NH_4$ concentration variations were affected by the boundary conditions, as demonstrated from FIGURE 3a where the $NH_4$ concentrations of Cases 2, 3 and 4 were different from those of benchmark Case 1. The increase of $NH_4$ concentrations in the inflow water had a significant impact on those in the underground cavern. From FIGURE 3b, with a high Chl-*a* concentration, the frequency and duration of inflow had a minor effect on the nitrate concentrations. However, these nitrate concentrations would largely increase when the nitrate concentrations in the inflow water increase as reflected by Case 2. The Chl-*a* concentration would decrease gradually in all the simulation periods when no inflow was presented (FIGURE 3c). The presence of inflow would increase the Chl-*a* concentrations. The TN concentrations were higher in Case-2 and Case-3 with higher inflow nutrients and Chl-*a* concentrations than in Case-1 and Case-4, where the inflow boundary conditions related to N and Chl-*a* would affect the TN concentration more obviously than other factors (FIGURE 3d). This study only gives a projection of possible water quality changes under a hypothetical cavern case, with the main purpose of demonstrating the capability of ELCOM-CAEDY. The results may vary significantly under different settings of operating modes or physical characteristics of the cavern.

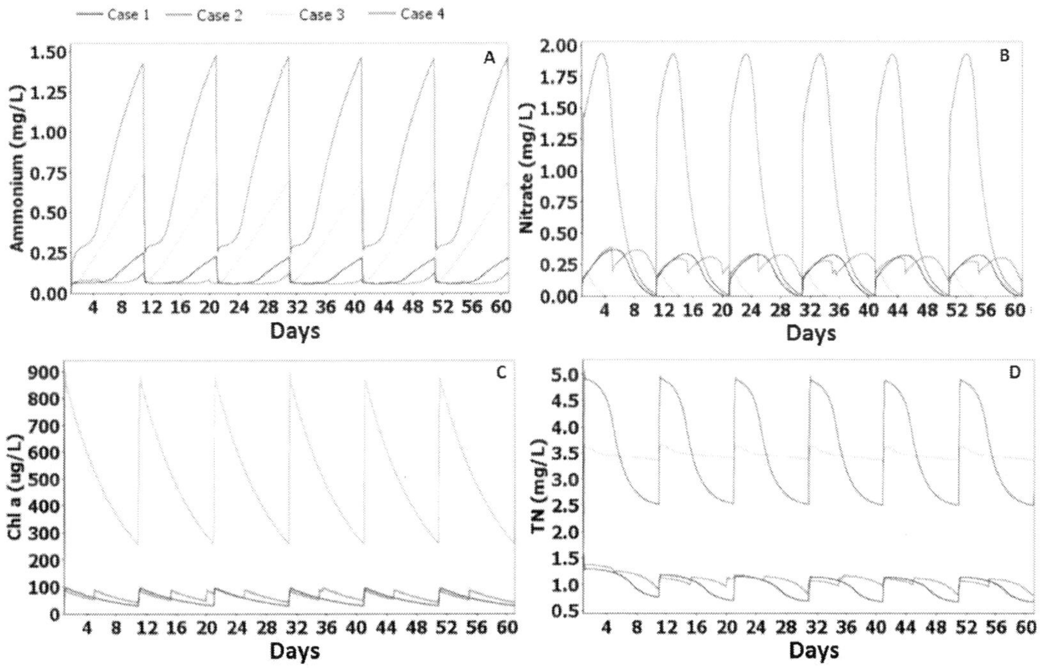


**FIGURE 3.** Variations of water quality variables at the probing point.

## CONCLUSIONS

This study conducted a preliminary modeling study on the water quality variations in dark and light environment using the coupled hydrodynamic and ecological model (i.e. ELCOM-CAEDYM). The experimental data was used for model validation. Different varying trends for Chl-*a*, ammonia, nitrite, and nitrate concentrations were found in the experiment and they could be well reproduced by the applied model. Some scenarios for a larger scale system were run for ascertaining the critical factors controlling the water quality in an underground cavern environment.


## ACKNOWLEDGEMENTS

This material is based on research/work supported by the Land and Liveability National Innovation Challenge under L2 NIC Award No. L2NICCFP1-2013-3 (WBS no.: M4061545.D63).


## REFERENCES


Bhuvaneshwari, M., et al. 2015. "Cytotoxicity of ZnO NPs towards fresh water algae Scenedesmus obliquus at low exposure …, visible and dark conditions." *Aquat Toxicol*, 162: 29-38.

Broch, E. 2007. "Use of the Underground in the City of Trondheim, Norway." 11th ACUUS Conference: Underground Space: Expanding the Frontiers, Sep. 10-13 2007, Athens - Greece.

Dalai, S., et al. 2014. "Studies on interfacial interactions of TiO2 nanoparticles with bacterial cells under light and dark conditions." *B Mater Sci*, 37(3): 371-381.

Gardner, W. S., et al. 1998. "Nitrogen cycling rates and light effects in tropical Lake Maracaibo, Venezuela." *Limnol Oceanogr*, 43(8): 1814-1825.

Gin, K. Y.-H., et al. 2011. "Comparison of nutrient limitation in freshwater and estuarine reservoirs in tropical urban Singapore." *J Environ Eng*, 137(10): 913-919.

Jiang, X., et al. 2008. "Effects of biological activity, light, temperature and oxygen on phosphorus release processes … of Taihu Lake, China." *Water Res*, 42(8): 2251-2259.

Low, E., et al. 2010. "Top-down control of phytoplankton by zooplankton in tropical reservoirs in Singapore." *Raffles B Zool*, 58(2): 311-322.

Oki, T., and Kanae, S. 2006. "Global hydrological cycles and world water resources." *Science*, 313(5790): 1068-1072.

Sanudin, N., et al. 2014. "Feeding Activity and Growth Performance of Shrimp Post Larvae Litopenaeus vannamei Under Light and Dark Condition." *J Agr Sci-Cambridge*, 6(11): 103-109.

Spears, B. M., et al. 2008. "Effects of light on sediment nutrient flux and water column nutrient stoichiometry in a shallow lake." *Water Res*, 42(4): 977-986.

Trolle, D., et al. 2012. "A community-based framework for aquatic ecosystem models." *Hydrobiologia*, 683(1): 25-34.

Zhao, J., and Bergh-Christensen, J. 1996. "Construction and utilization of rock caverns in Singapore Part D: Two proposed cavern schemes." *Tunn Undergr Sp Tech*, 11(1): 85-91.

# DEPTH FILTRATION THROUGH POROUS MEDIA

**Hooman Fallah***
(Department of Petroleum Engineering, Islamic Azad University, Firoozabad Branch, Firoozabad, Iran)
*Email address: hooman.fallah@gmail.com[88]

**ABSTRACT:** Porous medium occurs in numerous processes of environmental, chemical, petroleum and civil engineering. Several investigations have been done in order to understand the mechanisms of the transport of particulate suspension flow through porous medium. Transport of particulate suspensions and colloids in porous media is accompanied by particle capture and consequent permeability decline. In general, deep bed filtration studies have been conducted to analyse the mechanism involved in the processes of capturing and retaining particles occurs throughout the entire depth of the filter and not just on the filter surface. In this work, the steady-state transport equation is presented and the solution to the complete advective-dispersion equation for particulate suspension flow has been derived for the case of a constant filter coefficient. This model includes transport parameters which are particle advective velocity and particle longitudinal dispersion coefficient. This theoretical investigation of the transport of particles flowing in porous media is limited to flows with low Reynolds number (linear and laminar flow) and high Peclet number.

***Keywords:*** Porous Media, Particle Advective Velocity, Longitudinal Dispersion Coefficient, Filter Coefficient

## INTRODUCTION

Transport through porous medium occurs in numerous processes of environmental, chemical, petroleum and civil engineering. Transport of particulate suspensions and colloids in porous media is accompanied by particle capture and consequent permeability decline. It occurs in oil reservoirs during sea or produced water injection, drilling fluid invasion causing formation damage, filtration of completion fluid, fines migration during production of heavy oils in low consolidated reservoirs resulting in productivity. Prediction of particle propagation and retention by mathematical modeling is an essential stage during planning and design of above-mentioned industrial processes. It helps in design of injected water treatment, in particle sizing of drilling and completion fluids, in sizing the gravel pack and designing the sand screens. In general, deep bed filtration studies have been conducted to analyse the mechanism involved in the processes of capturing and retaining particles occurs throughout the entire depth of the filter and not just on the filter surface [1].

In this work, the steady-state transport equation is presented and the solution to the complete advective-dispersion equation for particulate suspension flow has been derived for the case of a constant filter coefficient. This model includes transport parameters which are particle advective velocity, particle longitudinal dispersion coefficient and filter coefficient. The numerical model needs to be developed for general case of a transition filter coefficient. The individual parameters are analysed according to dimensional analysis argument. Here, a particle mass transport equation is developed which includes the transport parameters, advective velocity and longitudinal dispersion coefficient. A solution to the complete advective-dispersion equation for particle transport was derived for the case of constant filter coefficient. The individual parameters are analysed according to dimensional analysis argument.

In order to develop the theories understanding of the basics of flow in porous media is needed advective fluid velocity and longitudinal dispersion coefficient. This theoretical investigation of the transport parameters of particles flowing in porous media is limited to flows with low Reynolds number (linear and laminar flow) and high Peclet number (advection dominates diffusion). The theoretical development used dimensionless numbers to define the transport parameters and incorporated them into an advective-dispersion equation describing particle transport.

## ADVECTIVE VELOCITY

The averaged particle velocity in the porous media, has been found to be either the same or slightly higher than that of the carrier fluid **[2]**. This deviation is caused by the particle`s size. The expected difference can be determined by analyzing the velocity profiles of both the fluid and the particles in a pore **[3]**. The model has been formulated for a capillary tube which has a constant rate with the following assumption: No interactions between the particles and the wall, suspension is well-mixed with a constant concentration across the cross section .There is no transverse flow.

$$U_S=U_O\left\{1-\left(\frac{r}{r_o}\right)^2\right\} \quad (1)$$

$$(0 < r < r_o)$$

$$U_P = U_O\left\{1-\left\{\frac{r}{r_o}\right\}^2-\gamma\left\{\frac{a_p}{r_o}\right\}^2\right\}(0 < r < r_o - a_p)$$

As the particle travels through a tube, Brownian motion and shear action will cause the particle to travel across the entire cross-section of the tube except that the center-line of the particle will be excluded from the immediate region of the wall due to its radial dimension.

After the particle has traveled far enough longitudinally through the tube, the particle will have spent equal amounts of time in all radial position across the capillary tube [4].

Integration of the velocity profile over the range of possible radii shown in Eq. (1) for both the particle and fluid yields the higher average velocity for the particle than that for the carrier fluid.

The average fluid velocity, $V_S$, in a capillary tube is:

$$V_S = \frac{U_O}{2} \quad (2)$$

The average velocity of a particle, $V_P$, in a capillary tube is:

$$V_P = U_O\left(1-\frac{1}{2}\left\{1-\frac{a_p}{r_o}\right\}^2-\gamma\left\{\frac{a_p}{r_o}\right\}^2\right) \quad (3)$$

By inspecting Eq. (2) and (3), the particles are expected to have a larger average velocity than the carrier fluid velocity. This enhanced velocity of the particle can be expressed as a fractional difference between the two average velocities:

$$\Delta V = \frac{V_P-V_S}{V_S} = 2\frac{a_p}{r_o} - \frac{7}{3}\left(\frac{a_p}{r_o}\right)^2 \quad (4)$$

This equation shows that as the radius of the particle increase, the difference between the average particle velocity and the average fluid interstitial velocity also increase. This increase is not unbounded but reaches a maximum $\Delta V$ value for $\frac{a_p}{r_o} = \frac{3}{7}$; as $\frac{a_p}{r_o} > \frac{3}{7}$ the velocity difference decrease.

In physical sense, the pore radius can be estimated to approximately equal to one-fifth of media grain diameter ($r_o = \frac{1}{5}d_g$); therefore the lar□e□t po□□ible particle to be able to fit throu□h the porou□bed has a radius to this pore radius ($a_p = r_o$).

For a particle with $a_p = r_o$, the particles have been shown to collect on the bed surface in a cake [4].

These references show that the onset of deep bed filtration occurs for a particle radius $a_p$ less than one-twentieth of the media grain diameter ($a_p < \frac{1}{20}d_g$). Particles with the radii larger than this will not transport into the bed, but will collect on the surface. By letting $d_g = 5r_o$, the largest particle which will transport has a radius equal to one-fourth of the pore radius ($a_p \leq \frac{1}{4}r_o$).

## LONGITUDINAL DISPERSION COEFFICIENT

An important element of any dispersion model is the representation of the geometry of the porous medium. Houseworth [5] has thoroughly reviewed such longitudinal dispersion model for solute tracers.

Instead of modeling the internal structure of a porous medium, dimensional analysis is used to analyze the problem. In this study, the effect of mechanisms is expected to scale with the pertinent transport variables.

The pertinent variables for solute dispersion are:

$D_L = longitudinal\ dispersion\ coefficient$ ;

$D = free\ fluid\ molecular\ diffusion\ coefficient\ of\ solute$ ;

$V_S = fluid\ interstitial\ velocity$ ;

$and\ d_g = media\ grain\ diameter\ (L)$.

From the Buckingham pi theorem, the following pairs of groups are formed:

$$\frac{D_L}{V_S.d_g} = F\left[\frac{V_S.d_g}{D}\right], or\ alternatively,$$

$$2.\frac{D_L}{D} = F\left[\frac{V_S.d_g}{D}\right] \quad (5)$$

$$where: Pe = Peclet\ number = \frac{V_S.d_g}{D}$$

Experimental data for solute longitudinal dispersion in uniform media show good correlation with these dimensionless groups [5], [6].

When the Peclet number is grater than 1, the two groups can be reduced to one:

$$\frac{D_L}{V_S.d_g} \approx Constant \quad (6)$$

Where: $Pe_D = dynamic\ Peclet\ number = \frac{V_S.d_g}{D_L}$.

An order to magnitude approximation for the longitudinal dispersion coefficient for solutes can be made with:

$$D_L = V_S.d_g \quad (7)$$

Particle longitudinal dispersion is expected to be similar to that of solutes.

Currently, no particle breakthroughs have been performed by others form which particle longitudinal dispersion coefficient be determined.

## TRANSPORT OF FLUID THROUGH POROUS MEDIUM

In this part, the steady-state transport equation is presented and the solution to the complete advective-dispersion equation for particulate suspension flow has been derived for the case of a constant filter coefficient. This model includes transport parameters which are particle advestive velocity and particle longitudinal dispersion coefficient [7].

These parameters have been defined by dimensional analysis using the pertinent variables of the porous media system.

**Particle Advective Velocity.** The result of the size exclusion for particles flowing in capillary tube can be written as equation (3) where $\boldsymbol{\gamma} = \frac{2}{3}$. By using the equations (2) and (4) we will have:

$$V^* = 1 + \Delta V \tag{8}$$

$$where\colon V^* = \frac{V_P}{V_S} \tag{9}$$

As the particle size increase, the difference between particle velocity and fluid velocity increase.

**Particle Longitudinal Dispersion Coefficient.** The previous stated method uses a single velocity and length scale and does not model the internal structure of porous medium. In modeling particle dispersion, the following variable substitutions are used:

$$\boldsymbol{D \rightarrow D_P}$$

$$\boldsymbol{V \rightarrow V_P}$$

$$\boldsymbol{D_L \rightarrow D_{LP}}$$

$$where\colon D_{LP} = particle\ dispersion\ coefficient\ ;$$

$$D_P = particle\ molecular\ diffusion\ coefficient\ ;$$

$$V_P = particle\ velocity\ ;$$

$$d_g = media\ grain\ diameter\ (L);\ and$$

$$d_p = particle\ diameter\ (L).$$

Particle size variable can be removed by using the particle properties as shown, provided $\boldsymbol{\frac{d_p}{d_g} << 1}$. also the effect of particle size is included in the enhanced advective velocity for the particles.

This analysis shows that particle and solute longitudinal dispersion are similar. When the particle Peclet number ($\boldsymbol{Pe_p} = \frac{V_P d_g}{D_P}$) is grater than 10, the two groups can be reduced to one :

$$\boldsymbol{\frac{D_{LP}}{V_P d_g} = constant} \tag{10}$$

$$where\colon Pe_p = \frac{V_P d_g}{D_P} = particle\ dynamic\ Peclet\ number.$$

An order of magnitude approximation for the longitudinal dispersion coefficient for particles can be made with:

$$\boldsymbol{D_{LP} = V_P d_g} \tag{11}$$

As mentioned before, the dimensional argument for defining the longitudinal dispersion coefficient is only valid when $\mathrm{Pe_p} \gg 1$. For uniform media, this restriction is seen to be $\mathrm{Pe_p} > 10$ [5].

Flow conditions are simultaneously limited to the linear, laminar regime for which the Reynolds number must be less than 10.

**Steady-State Transport in Porous Medium Equation and Solution.** Particle removal or filtration occurs as a particle suspension flows through a porous medium due to the interaction of the advecting particles and grains of the medium. Iwasaki [8] is credited with being the first to express filtration a first-order decay of particle concentration with distance:

$$\frac{\partial C}{\partial x} = -\lambda C \tag{12}$$

$$1. C(x = 0) = C_0 \tag{13}$$

And a solution in dimensional terms:

$$C = C_0 exp[-\lambda x] \tag{14}$$

Or, in dimensionless terms:

$$C^* = exp[-\lambda^* x^*] \tag{15}$$

$$where{:}\ C^* = \frac{C}{C_0};$$

$$C_0 = C(x = 0);$$

$$\lambda^* = \lambda d_g;$$

$$\mathbf{x}^* = \frac{\mathbf{x}}{\mathbf{d_g}};$$

A complete equation of steady-state filtration can be formulated by using the general steady-state advection-dispersion equation of transport for particle concentration with a sink term to describe particle removal due to filtration:

$$0 = D_{LP}\frac{\partial^2 C}{\partial x^2} - V_p\frac{\partial C}{\partial x} - V_p\lambda C \tag{16}$$

The following semi-infinite medium boundary conditions are:

$$C(x = 0) = C_0;$$

$$\lim_{x\to\infty} C(x) = 0. \tag{17}$$

The solution which is shown in dimensional terms is derived in appendix:

$$C(x) = C_0 exp[\alpha x] \tag{18}$$

$$where{:}\ \alpha = \frac{1}{2}\frac{V_P}{D_{LP}}\left(1 - \sqrt{1 + 4\lambda\frac{D_{LP}}{V_P}}\right)$$

In dimensionless terms the transport equation becomes:

$$0 = \frac{\partial^2 C^*}{\partial x^{*2}} - Pe_{Dp}\frac{\partial C^*}{\partial x^*} - Pe_{Dp}\lambda^* C^* \tag{19}$$

With the same boundary conditions:

$$C^*(x^* = 0) = 1(30)2. \lim_{x^*\to\infty} C^*(x^*) = 0 \tag{20}$$

And a solution in dimensionless terms:

$$C^* = exp[\alpha^* x^*] \qquad (21)$$

$$\alpha^* = \alpha d_g = \frac{1}{2} Pe_{DP}\left(1 - \sqrt{1 + \frac{4\lambda^*}{Pe_{DP}}}\right) \qquad (22)$$

## CONCLUSION

In this work, a mass balance particle transport equation which includes filtration has been developed. This model includes transport parameters which are particle advective velocity, particle longitudinal dispersion coefficient and filter coefficient. The steady-state transport equation is presented and the solution to the complete advective-dispersion equation for particulate suspension flow has been derived for the case of a constant filter coefficient.

This work recommends to be investigated by particle longitudinal dispersion calculation from experimental data. Besides, the numerical model needs to be developed for general case of a transition filter coefficient.

## *APPENDIX*

### *A.1 Solution Derivation*

Consider the one-dimensional steady-state particle advective-dispersion equation which includes the removal term to account for filtration effects:

$$0 = D_{LP}\frac{\partial^2 C}{\partial x^2} - V_P\frac{\partial C}{\partial x} - \lambda V_P C \qquad (A.1)$$

$where: C = concentration$;
$x = longitudinal\ distance\ (L)$;
$D_{LP} = longitudinal\ dispersion\ coefficient\ for\ particles$;
$V_P = average\ particle\ interstitial\ velocity$ ; and
$\lambda = filter\ coefficient\ (L^{-1})$.

With the following boundary conditions:

$1. C(x = 0) = C_0$; and

$$2. \lim_{x\to\infty} C(x) = 0. \qquad (A.1.1)$$

For convenience, the x-variable is allowed to range from negative to positive infinity ($-\infty < x < \infty$), although the equations are only applied for $x> 0$. this avoids difficulty at $x=0$, because small dispersion is allowed. In dimensionless form, the transport equation becomes:

$$0 = D_{LP}\frac{\partial^2 C^*}{\partial x^{*2}} - Pe_{Dp}\frac{\partial C^*}{\partial x^*} - \lambda^* Pe_{Dp} C^* \qquad (A.2)$$

$where: C^* = \frac{C}{C_0}$;
$C_0 = C(X = 0)$;
$d_g = media\ grain\ diameter$;
$x^* = \frac{x}{d_g}$;
$Pe_{Dp} = \frac{V_p d_g}{D_{LP}}$; and $\lambda^* = \lambda d_g$.

With the same boundary conditions:

$1. C^*(x^* = 0) = 1$; and

$$2. \lim_{x^*\to\infty} C^*(x^*) = 0. \qquad (A.2.1)$$

In order to derive a solution, try the following as a solution:

$$C^*(x^*) = exp[\alpha^* x^*] \qquad (A.3)$$

Check the equation (A.3) by substituting into Eq. (A.2), this results in second-degree polynomial in term of $\alpha^*$, and two roots of this polynomial are:

$$\alpha_1^* = \frac{1}{2}Pe_{DP}\left(1 - \sqrt{1 + \frac{4\lambda^*}{Pe_{DP}}}\right);\ \alpha_1^* = \frac{1}{2}Pe_{DP}\left(1 + \sqrt{1 + \frac{4\lambda^*}{Pe_{DP}}}\right)$$

Using these two roots, Eq. (A.3) becomes:

$$C^*(x^*) = Aexp[\alpha_1^* x^*] + Bexp[\alpha_2^* x^*] \qquad (A.4)$$

The constants of this equation can be determined by applying the boundary conditions:

$$1.\ C^*(x^* = 0) = 1 = A = B; and$$
$$2.\ C^*(x^* \to \infty) = 0 = (A)(0) + (B)(+\infty)$$
$$\therefore A = 1\ and\ B = 0.$$

By substituting these constants into equation (A.4), the solution to equation (A.2) becomes:

$$C^*(x^*) = exp\left[\frac{1}{2}Pe_{DP}\left(1 - \sqrt{1 + \frac{4\lambda^*}{Pe_{DP}}}\right)x^*\right] \qquad (A.5)$$

### *A.2 Solution Approximation*

Consider an approximation for Eq. (A.5) by simplifying $\alpha^*$ :

$$\alpha_1^* = \frac{1}{2}Pe_{DP}\left(1 - \sqrt{1 + \frac{4\lambda^*}{Pe_{DP}}}\right) \qquad (A.6)$$

Perform a Taylor series expansion of the radical portion of $\alpha^*$ of Eq. (A.6) by considering a function f(x):

$$f(x) = \sqrt{1 + x}$$
$$where: x = \frac{4\lambda^*}{Pe_{DP}}.$$

The Taylor series approximation is:

$$\sqrt{1 + x} = 1 + \frac{1}{2}x - \frac{1}{8}x^2 + \frac{1}{16}x^3 - \frac{5}{128}x^4 \ldots$$

Since this series is an alternating series, the truncation error must be less than the first truncated term (absolute values). Substituting the first two terms of the series into Eq. (A.6) for the radical portion of $x^*$ yields:

$$x^* \cong \frac{1}{2}Pe_{DP}\left(1 - \left\{1 + \frac{2\lambda^*}{Pe_{DP}}\right\}\right) = -\lambda^* \qquad (A.7)$$

Substituting the approximation for $\alpha^*$, Eq. (A.7), into the exact steady-state equation:

$$C^* = exp[-\lambda^* x^*] \qquad (A.8)$$

Using Eq. (A.8), $\lambda^*$ can be directly calculated from concentration and position measurements. Also, Eq. (A.8) is the same result that would have been derived by ignoring the dispersion term originally.

### *A.3 Approximate Versus Exact Solution*

The error using the approximate solution can be determined from the relative error of the two solutions, Eq. (A.3) and (A.8):

$$\Delta C^* = \frac{C_a^* - C_e^*}{C_e^*} = \frac{exp[-\lambda^* x^*] - exp[\alpha^* x^*]}{exp[\alpha^* x^*]} \qquad (A.9)$$

$$where: C_a^* = approximate\ solution;\ and\ \ C_e^* = exact\ solution.$$

Eq. (A.9) can be simplified as follows:

$$\Delta C^* = exp[-x^*(\lambda^* + \alpha^*)] - 1 \qquad (A.10)$$

The Taylor series expansion for Eq. (A.10) is:

$$exp[-x^*(\lambda^* + \alpha^*)] = 1 - x^*(\lambda^* + \alpha^*) + \cdots \qquad (A.11)$$

So, the approximate value for $\alpha^*$ can be determined using the Taylor series for $\alpha^*$ found in section A.2:

$$\alpha_1^* = \frac{1}{2}Pe_{DP}\left(1 - \sqrt{1+x}\right) = -\frac{1}{2}Pe_{DP}\left(\frac{1}{2}x - \frac{1}{8}x^2 \ldots\right) \leq -\frac{1}{4}Pe_{DP}x + \frac{1}{16}Pe_{DP}x^2$$

Substitute for x:

$$\alpha^* \leq -\lambda^* + \frac{\lambda^{*2}}{Pe_{DP}} \qquad (A.12)$$

Add $\lambda^*$ to Eq. (A.12) in order to determine the argument of the exponential term of Eq. (A.10) so that Eq. (A.11) can be evaluated:

$$\alpha^* + \lambda^* \leq \frac{\lambda^{*2}}{Pe_{DP}} \qquad (A.13)$$

Substitute this approximation into the series expansion expression, Eq. (A.11), of the exponential term in Eq. (A.10):

$$|\Delta C^*| \leq |1 - x^*(\lambda^* + \alpha^*) - 1| \leq \frac{\lambda^{*2}x^*}{Pe_{DP}} \qquad (A.14)$$

The absolute value of the error is the absolute value of $\Delta C^*$ of Eq. (A.14):

$$\therefore |Error| = |\Delta C^*| \leq \frac{\lambda^{*2}x^*}{Pe_{DP}} \qquad (A.15)$$

## Nomenclature

*dg*= media grain diameter (L)
*dp*= particle diameter (L)
*c* = suspended particle concentration in carrier fluid
*σ*= particle retained concentration
*kdet*= detachment rate coefficient
*U* = flow velocity
*US* = fluid velocity
*UP* = particle velocity
*UO* = fluid centerline velocity
*r* = radial distance
*ro* = capillary radius
*ap* = particle radius
*p* = dynamic pressure
*x* = longitudinal distance
*DL* = longitudinal dispersion coefficient (L·T–1)
*D* = free fluid molecular dispersion coefficient of solute (L2·T–1)
*VS* = fluid interstitial velocity (L·T–1)
*Pe* = Peclet number = *SgV.dDL*
*PeD* = dynamic Pec let number = *SgV.dDL*
*C* = particle concentration (M·L–3)
*X* = longitudinal position (L)
*λ* = filter coefficient (L–1)
*WS* = particle settling velocity

*VS* = fluid interstitial velocity
*Ppf* = densities of particle and fluid, respectively
*g* = gravitational acceleration
*H* = Hamakar constant (ergs)
*NG* = gravitional group = $\eta G$
*DLP* = particle longitudinal dispersion coefficient (L2·T–1)
*DP* = particle molecular diffusion coefficient in a free fluid (L2·T–1)
*VP* = particle velocity (L2·T–1)

---

## REFERENCES

R. Farajzadeh, "An Experimental Investigation into Internal Filtration and External Cake Build Up," *MSc. thesis*, Delft University of Technology (2004).

C.G. Enfield., G. Bengtsson., "Macromolecular Transport of Hydrophobic Contaminants in Aqueous Environments", Ground Water, Vol.26, No.1, 1988, pp. 64-70.

E. A. DiMarzio., C. M. Guttman, "Separation By Flow", Macromolecules, Vol. 3, No. 2, 1970, pp. 131-146.

J.P. Herzig, , D.M. Leclerc, P. Le Goff, "Flow of suspension through porous media –application to deep filtration" *J. Ind. Eng. Chem.* 65(5), 8-35 (1970).

J.E. Houseworth, "Longitudinal Dispersion in Nonuniform, Isotropic Porous Media," Ph.D. Thesis, W.M. Keck *Laboratory of Hydraulics and Water Resources, California Institute of Technology*, Report No. KH-R-45, (1984).

Foppen, J.W.A., Schijven, J.F "Evaluation of data from the literature on the transport and survival of Escherichia coil in aquifers under saturated conditions," *J.Water Res.* 40,401-426 (2006).

J.F. Schijven, S.M. Hassanizadeh, "Removal of viruses by soil passage: overview of modeling processes, and parameters," Crit. Rev. Environ. Sci. *Technol.* 30(1), 49-127 (2000).

T. Iwasaki, "Some notes on sand filtration," *Water Works Ass.* 1591-1602, 1937.

# ENVIRONMENTAL QUALITY AND FOREIGN TRADE IN TURKEY: A CO-INTEGRATION ANALYSIS WITH A STRUCTURAL BREAK

**Nilgün Çil***
(Istanbul University, Istanbul, Turkey)

**ABSTRACT:** This paper explores the existence of a long run relationship between environmental quality and foreign trade indicators for Turkey over the period from 1974 to 2011. For this purpose, we use carbon dioxide emission as an environmental quality indicator and export, foreign direct investment, GDP per capita as explanatory variables. In our empirical analysis, we use Perron (1997) unit root test and Gregory-Hansen (1996) co-integration test. Both tests allow for a structural break. The obtained results indicate that there is evidence of co-integration between carbon dioxide emission, export, foreign direct investment and GDP for Turkey.

## INTRODUCTION

Carbon dioxide emission is one of the major factors causing global warming, climate change and environmental pollution and is also an important indicator of environmental destruction. The increase in the carbon dioxide emission which is emitted into the atmosphere by the use of fossil fuels such as gasoline, coal and natural gas, increases the temperature of the earth's surface. So, it is critical to understand the relationship between carbon dioxide emission and its determiners.

In environmental economics literature, the relationship between environmental destruction and income is called the environmental Kuznets curve (EKC) hypothesis. According to this hypothesis, environmental pollution increases as a country develops, but after a turning point, begins to decrease as income increases. This hypothesis derives its name from the study of Kuznets (1955) who proposed a similar relationship between income inequality and economic growth. Since the seminal work of Grossman and Krueger (1991), several studies have examined the relationship between pollution and economic growth for different countries, using various methodologies. Some of these studies support the environmental Kuznets curve hypothesis (see Galeotti et al. (2006), Millimet et al. (2003)) and some others fail to find evidence for environmental Kuznets curve hypothesis (see Dinda et al. (2000), Perman and Stern (2003)).

In recent years, economists argue that foreign trade affects the environment, both locally and globally. Expanding trade from domestic market to international market not only increases market share of each country but also rising competition among the nations and improve efficiency of utilizing scarce resources. But on the other hand, environmental economists claim that the costs of spreading trade to international markets are depleting natural resources and rising pollution emissions that ultimately deteriorates environmental quality [Antweiler et al. (2001), Copeland and Taylor (2001), Chaudhuri and Pfaff (2002)].

Since there is not enough empirical research that investigates the relationship between environmental quality and foreign trade indicators, this paper empirically examines the long-run relationship among carbon dioxide emission (as an environmental quality indicator), export, foreign direct investment and GDP per capita for Turkey over the period 1974-2011. During this period, many internal and external developments take place in Turkish economy. In other words, the Turkish economy was influenced by both international developments (e.g. the oil crises of 1973 to 1974 and Asia Crisis of 1996) and national developments including the structural transformation and transition to a free market economy after 1980, capital account liberalization in 1989 along with the economic crises in 1994, 2000 and 2001.

Especially within the framework of the structural transformation program of 1980, the rapid industrialization breakthrough in Turkey caused an increase in the amount of greenhouse gas emitted into the atmosphere as it increased the demand for energy. In the 2007 report of the United Nations, Turkey was announced to be the country with the most dramatically increasing emission of gases that cause the greenhouse effect with a ratio of 74.4% for the period between 1990 and 2004. The reasons for the dramatic increase are listed as emissions due to the energy sector and industrial sector and use of fossil fuels. After these internal and external developments, it is very important to take into account existence of structural breaks in our empirical analysis. As argued by many authors before, conventional unit root and co-integration tests lose power when existing structural break(s) are not taken into consideration. From this point of view, we investigate the unit root properties of the mentioned series by using Perron (1997) unit root test, which takes into account one structural break. After obtaining necessary conditions, we explore long run relationship by using Gregory-Hansen (1996) co-integration test with a structural break.

The balance of the paper is organized as follows: The next section presents the materials and methods, the third section gives results and discussions. Finally, the last section concludes.

## MATERIALS AND METHODS

This paper examines the long run relationship between carbon dioxide emission, export, foreign direct investment and GDP by considering existence of structural breaks. We explore the unit root properties of the series by using Perron (1997) unit root test, which takes into account one structural break. We also use Gregory-Hansen (1996) co-integration test with a structural break for long run relationship between these series.

**Perron(1997) Unit Root Test.** Nelson and Plosser (1982) argue that almost all macroeconomic time series have a unit root. The presence or absence of unit root helps to identify some features of the underlying data generating process of a time series. The commonly used unit root method to test unit root properties of a time series is Augmented Dickey Fuller (ADF) test. But, Perron (1989) claims that standard ADF tests are biased towards the non rejection of the null hypothesis in the presence of a structural break. Following this argument, Perron (1989) develops a modified Dickey-Fuller unit root test, which includes a dummy variable for one known or exogenous structural break. This test allows for a structural break under both the null and alternative hypothesis. The test equations take into account the existence of three different kinds of structural breaks: a break in the intercept, a break in the trend and a break in both intercept and trend. The equation that takes into account an exogenous structural break in both intercept and trend can be seen below:

$$y_t = \alpha_0 + \alpha_1 DU_t + d(DTB)_t + \gamma DT_t + \beta t + \rho y_{t-1} + \sum_{i=1}^{p} \phi_i \Delta y_{t-i} + e_t \tag{1}$$

where $DU_t$ represents a change in the intercept; $DT_t$ represents a change in the trend and $TB$ is the break date.

However, this procedure is criticized on including structural break as exogeneous. Then, Perron(1997) develops one endogeneous structural break version of this unit root test. In this version, the first model allows only a change in the intercept under both the null and alternative hypotheses and the unit root test is performed using the t- statistic for testing $\alpha = 1$ in the following regression:

$$y_t = \mu + \theta DU_t + \delta(DTB)_t + \beta t + \alpha y_{t-1} + \sum_{i=1}^{k} c_i \Delta y_{t-i} + e_t \tag{2}$$

where $DU_t = 1(t > TB)$ and $(DTB)_t = 1(t = TB + 1)$ with $1(.)$ the indicator function. Under the second model, both a change in the intercept and trend are allowed at time $TB$. The test is performed using the t-statistic for the null hypothesis that $\alpha = 1$ in the regression:

$$y_t = \mu + \theta DU_t + \gamma DT_t + \delta(DTB)_t + \beta t + \alpha y_{t-1} + \sum_{i=1}^{k} c_i \Delta y_{t-i} + e_t \qquad (3)$$

with $DT_t = 1(t > TB)t$. Under the third model, a change in the trend is allowed but both segments of the trend function are joined at the time of break. The other detailed information on the test can be seen in the original paper of Perron (1997).

**Gregory-Hansen (1996) Co-integration Test.** Since the standard co-integration methods lose power in the existence of structural breaks, we apply Gregory-Hansen (1996) co-integration test in order to investigate the long run relationship between environmental quality and foreign trade indicators for Turkey by allowing structural break. The importance of Gregory-Hansen (1996) test lies in the fact that conventional co-integration tests are biased towards non-rejection of the null of no co-integration when there exists any structural break(s) in the co-integrating relationship. Gregory and Hansen (1996) suggest three models for a possible endogenous structural break in the co-integrated relation. They use modified Augmented Dickey Fuller, $Z_a$ and $Z_t$ tests in order to test the null of no co-integration against the alternative of co-integration with an unknown regime shift. The first model considers level shift (Model C), where the second one considers level shift and trend (Model C/T). The last model is the regime shift model (Model C/S), which has the following form:

$$y_{1t} = \mu_1 + \mu_2 \varphi_{1\tau} + \alpha_1^T y_{2t} + \alpha_2^T y_{2t} \varphi_{1\tau} + e_t \qquad (4)$$

In Model 4, $\mu_1$ denotes the intercept before the regime shift where $\mu_2$ represents the change in intercept at the time of the regime shift. Similarly, $\alpha_1$ shows the slope coefficients before the regime shift and $\alpha_2$ is the change in the slope coefficients after the shift. They suggest to apply Augmented Dickey Fuller (ADF) unit root test and $Z_a$ and $Z_t$ tests of Phillips (1987) to the residuals ($e_t$). Since the true dates of the breakpoints are not known, we estimate the Model 4 for all the possible breakpoints and determine the dates of the structural breaks via selecting the dates, which give the minimum test statistics. The test statistics are defined as follows:

$$ADF^- = \inf_{\tau \in T} ADF(\tau) \qquad Z_T^- = \inf_{\tau \in T} Z_t(\tau) \qquad Z_\alpha^- = \inf_{\tau \in T} Z_\alpha(\tau)$$

where $T = (0.15n, 0.85n)$ shows the sample area which we search for the breakpoints. Critical values are available in Gregory and Hansen (1996).

## RESULTS AND DISCUSSION

This paper uses annual data on the carbon dioxide emission (metric tons per capita) as an environmental quality indicator, export of goods and services (US \$), foreign direct investment net inflows (US \$) and GDP per capita (US \$) for Turkey over the 1974-2011 period. The data are obtained from World Development Indicators website. Before the analysis, we take the natural logarithms of the series. The labels for the natural logarithms of the carbon dioxide emission, the natural logarithms of the export, the natural logarithms of the foreign direct investment and the natural logarithms of the GDP per capita are LCO2,

LEX, LFDI and LGDP respectively. Before testing for cointegration, the unit root properties of the series are explored by using Perron (1997) unit root test with a structural break. The results can be seen in Table 1.

The results in the table show that we failed to reject the null hypothesis of unit root for all the series. After insuring that the series are difference stationary, it is possible to apply cointegration test. For this purpose, we apply Gregory and Hansen (1996) cointegration test with a structural break. The results for the full break model are reported in Table 2.

**TABLE 1: Perron(1997) unit root test results**

| Series | Test Statistic | Time of break |
|---|---|---|
| **LCO2** | -4.013 | 1984 |
| **LEX** | -4.378 | 1980 |
| **LFDI** | -5.055 | 2004 |
| **LGDP** | -3.524 | 1983 |

**Note:** The critical values are -6.32, -5.59 and -5.29 for 1%, 5% and 10% significance levels, respectively.

**TABLE 2: Gregory Hansen (1996) cointegration test results**

| Test Statistic | Lag length (k) | Time of break |
|---|---|---|
| $-6.188^{b}$ | 0 | 1991 |

**Note:** $^{b}$ indicates statistically significance at 5% significance level. The used critical values are -6.51 and -6.00 for 1% and 5% significance levels, respectively.

The results of Gregory and Hansen (1996) cointegration test indicate that there exists a long-run relationship with a structural break among the variables and the break in 1991 is statistically significant. The test results reveal that the structural break in 1991 can be refer to the financial liberalization period in Turkey. It means that financial liberalization affected the relationship between carbon dioxide emission, export, foreign direct investment and GDP per capita for Turkey.

## CONCLUSIONS

In this paper, an attempt has been made to investigate the long run relationship between carbon dioxide emission, export, foreign direct investment and GDP for Turkey over the period 1974-2011. For this aim, we use Perron (1997) unit root test and Gregory-Hansen(1996) co-integration test in our empirical analysis. These tests allow for existence of one structural break. According to obtained results, it is found that there is a long run relationship between carbon dioxide emission, export, foreign direct investment and GDP for Turkey with the structural break in 1991. This break can be refer to the financial liberalization period of Turkey.

## ACKNOWLEDGEMENTS

This research is supported by Scientific Research Projects Unit of Istanbul University.

## REFERENCES

Antweiler, W., B. Copeland, and M. S. Taylor. 2001. "Is Free Trade Good for the Environment?." *American Economic Review*. 91: 877–908.

Chaudhuri, S., and A. Pfaff. 2002. "Economic Growth and the Environment: What Can We Learn from Household Data?." *Columbia University Discussion Paper Series*. No: 0102-51.

Copeland, B. R., and M. S. Taylor. 2001. "International Trade and the Environment: A Framework for Analysis." *National Bureau of Economic Research Working Paper*. No: 8540.

Dinda, S., D. Coondoo, and M. Pal. 2000. "Air Quality and Economic Growth: An Empirical Study." *Ecological Economics*. 34: 409-423.

Galeotti, M., A. Lanza, and F. Pauli. 2006. "Reassessing the Environmental Kuznets Curve for CO2 Emissions: A Robustness Exercise." Ecological Economics. 57: 152-163.

Gregory, A. W., and B. E. Hansen. 1996. "Residual-Based Tests for Cointegration in Models with Regime Shifts." *Journal of Econometrics*. 70: 99-126.

Grossman, G., and A. Krueger. 1991. "Environmental Impacts of a North American Free Trade Agreement." *National Bureau of Economics Research Working Paper*. 3194, NBER, Cambridge.

Kuznets, S. 1955. "Economic Growth and Income Inequality." *American Economic Review*. 45: 1–28.

Millimet, D.L., J.A. List, and T. Stengos. 2003. "The Environmental Kuznets Curve: Real Progress or Misspecified Models?." Review of Economics and Statistics. 85: 1038-1047.

Nelson, C.R., and C.I. Plosser. 1982. "Trends and Random Walks in Macroeconomic Time Series." *Journal of Monterey Economics*. 10: 139-162.

Perman, R., and D.I. Stern. 2003. "Evidence from Panel Unit Root and Cointegration Tests that Environmental Kuznets Curve does not Exist." *The Australian Journal of Agricultural and Resource Economics*. 47: 325-347.

Perron, P. 1989. "The Great Crash, the Oil Price Shock and the Unit Root Hypothesis." *Econometrica*. 57: 1361-1401.

Perron, P. 1997. "Further Evidence on Breaking Trend Functions in Macroeconomic Variables." *Journal of Econometrics*. 80: 355-385.

Phillips, P. C. B. 1987. "Time Series Regression with a Unit Root." *Econometrica*. 55: 277–301.

# ENVIRONMENTAL KUZNETS CURVE HYPOTHESIS AND STRUCTURAL BREAKS: EVIDENCE FROM TURKEY

***Burcu Kıran***
(Istanbul University, Istanbul, Turkey)

**ABSTRACT:** This paper tests the environmental Kuznets curve hypothesis which means an inverted U-shape relationship between various indicators of environment degredation and income per capita for Turkey over the period 1960-2011. In our empirical analysis, we use Narayan and Popp(2010) unit root test which is flexible enough to allow for at most two structural breaks in the level and trend, and also Hatemi-J(2008) cointegration test which calculates three residual based test statistics for cointegration by taking into account two possible regime shifts. The results reveal that there is evidence of cointegration between carbon dioxide emission, energy consumption and income with two structural breaks. Hence, the environmental Kuznets curve hypothesis is valid for Turkey.

## INTRODUCTION

The Kuznets curve which implies an inverted U-shaped curve relationship between per capita income and income inequality is first suggested by Kuznets(1955). According to Kuznets(1955), there is a changing relationship between per capita income and income inequality. As per capita income increases, income inequality also increases at first and then starts declining after a turning point. Following Kuznets curve, the same inverted U-shaped relationship is found between environmental degradation and income per capita. This relationship developed and tested by Grossman and Krueger(1991) is known as the Environmental Kuznets Curve hypothesis. This hypothesis indicates that as per capita income grows, environmental impact increases, reaches a peak and then declines. In other words, pollution levels increase as a country develops, but begin to decrease as the rising incomes pass beyond a turning point (Odhiambo, 2012).

Grossman and Krueger(1995) point out that there are three accesses for economic growth to affect environmental degradation, namely scale effect, composition effect and technique effect. Scale effect occurs as pollution increases with the size of the economy. Panayotou(2003) states, when the production of an economy shifted mainly from agriculture to industry, pollution intensity increases. Therefore, it could be argued that scale effect will do harm to environment. However, composition and technique effects will improve environment. When the industrial structure enhances further, from energy-intensive heavy industry to service and technology-intensive industries, pollution falls as income grows. This is known as the composition effect. Actually, technique effect goes with the composition effect. Technical progress makes it possible to replace the heavily polluting technology with cleaner technology.

The aim of this paper is to test environmental Kuznets curve hypothesis by investigating the long term relationship between carbon dioxide emission (an indicator of environment degredation), energy consumption and income for Turkey over the period from 1960 to 2011. Although numerous studies exist in the literature based on this hypothesis, there are some reasons for choosing Turkey in our analysis. Lise(2006) and Lise and Montfort(2007) report these reasons in a multi-dimentional perspective. The first reason is that Turkey is a candidate for becoming an European Union (EU) member in the near future. So, Turkey has strict environmental obligations to fulfill in order to qualify for full membership. According to the Commission of the European Communities, the EU aims at reducing environmental pollutants 30% below the 1990 levels by 2020. Hereafter, it is expected that Turkey will be under strong pressure from the EU to comply with the Union's regulations on environmental policy (Halıcıoğlu, 2009). The second reason to find Turkey as an interesting country is that Turkey has a strategic position as a gas and oil transportation country. And also, Turkey is listed as an Annex 1 country and has not yet set a greenhouse gas emissions reduction target. The last reason is that the Turkish economy has a boom-bust structure in the recent past. According to the 2006 report of World Energy Council Turkish National Commitee, Turkey's contribution to carbon dioxide emissions is quite low; in 2003 per capita carbon dioxide emissions were at 2.87 tons, much lower than the OECD average of 11.08 tons and also its share in total OECD emissions was 1.59% and in world emissions was 0.81%. But, in recent years, there is a rapid increase in the carbon dioxide emissions of Turkey. In 2008, carbon dioxide emission was the leading greenhouse gas with the ratio of 81.1%. The results of greenhouse gas emissions inventory show that total greenhouse gas emissions in Turkey are 401.9 million tons on a carbon equivalent basis in 2010 and 81.23% of these emissions is carbon dioxide.

There are limited number of empirical studies which test the environmental Kuznets curve hypothesis for Turkey and their results are mixed based on different time periods, different variables and different methodologies [Lise(2006), Akbostancı et al.(2009), Halıcıoğlu(2009), Soytaş and Sarı (2009), Öztürk and Acaravcı(2010), Omay(2013), Yavuz(2014)]. The contribution of our paper to the literature is that, specifically, we test the validity of the environmental Kuznets curve hypothesis by using Narayan and Popp (2010) unit root test and Hatemi-J (2008) cointegration test in the same analysis. Both tests take into account existence of two structural breaks. Narayan and Popp (2010) unit root test is flexible enough to allow for at most two structural breaks in the level and trend. And also, Hatemi-J (2008) cointegration test calculates three residual based test statistics by taking into account two possible regime shifts. It is clear that taking into account the structural breaks is necessary because the long run relationship between carbon dioxide emission, energy consumption and income might change and the shifts can occur in the cointegrating vector. Hatemi-J (2008) mentions about many reasons to expect structural changes in the relationship between the series such as economic crises, technological shocks, changes in the economic actor' preferences and behaviour accordingly, policy and regime changes, and organizational and institutional evoluation.

The rest of the paper is constructed as follows: The next section presents the materials and methods, the third section reports results and discussions. Finally, the last section concludes the paper.

## MATERIALS AND METHODS

This paper intends to test environmental Kuznets curve hypothesis by taking into account existence of two structural breaks in the relationship between carbon dioxide emission, energy consumption and income. For unit root properties of the series, we use recently developed Narayan and Popp (2010) unit root test which can identify two structural breaks in the level and trend. For long run relationship between the series, we apply Hatemi-J (2008) cointegration test which considers two possible regime shifts.

**Narayan and Popp (2010) Unit Root Test.** Perron(1989) argues that the exclusion of structural breaks while modeling the unit root often leads to accepting the false null hypothesis. Since this argument, many authors addressed this issue by identifying structural breaks endogenously and exogenously. However, Poop(2008) noticed that spurious regression arises from different interpretations of test parameters under the null and alternative hypothesis, because of the parameters affect the selection of the break date. This problem is solved by Narayan and Popp(2010). Following Schmidt and Philips (1992), they developed an ADF type test for the case of innovational outlier (IO) where the data generating process is constructed as an unobserved component model. Narayan and Popp(2010) take into account the following data generating process of a time series $y_t$ :

$$y_t = d_t + u_t \quad (1)$$

$$u_t = \rho u_{t-1} + \varepsilon_t \quad (2)$$

$$\varepsilon_t = \psi^*(L)e_t = A^*(L)^{-1}B(L)e_t \quad ; \quad e_t \sim iid(0,\sigma_e^2). \qquad (3)$$

Here, $y_t$ has two components; $d_t$ and $u_t$. $d_t$ is deterministic component and $u_t$ is a stochastic component. $A^*(L)$ and $B(L)$ are the lag polynomials. It is assumed that the roots of these lag polynomials which are of order $p$ and $q$, respectively, lie outside the unit circle.

In the construction of this test, two different specifications both for trending data are considered: One specification allows for two breaks in the level (denoted M1) and the other specification allows for two breaks in level and trend (denoted M2). These model specifications (M1 and M2) differ in deterministic component $d_t$. For M1, $d_t$ is defined as follows:

$$d_t^{M1} = \alpha + \beta t + \psi^*(L)(\theta_1 DU'_{1,t} + \theta_2 DU'_{2,t}) \qquad (4)$$

And also $d_t$ which is defined for M2 can be seen in below:

$$d_t^{M2} = \alpha + \beta t + \psi^*(L)(\theta_1 DU'_{1,t} + \theta_2 DU'_{2,t} + \gamma_1 DT'_{1,t} + \gamma_2 DT'_{2,t}) \qquad (5)$$

In both specifications, $DU'_{i,t}$ and $DT'_{1,t}$ are defined as follows, respectively:

$$\begin{aligned} DU'_{i,t} &= 1(t > T'_{B,i}) \quad ; \ i = 1,2 \\ DT'_{i,t} &= 1(t > T'_{B,i})(t - T'_{B,i}) \ ; \ i = 1,2. \end{aligned} \qquad (6)$$

Here $T'_{B,i}$, $i = 1,2$ describe the true break dates, $\theta_i$ and $\gamma_i$ parameters indicate the magnitude of the level and trend breaks, respectively. The inclusion of $\psi^*$ in corresponding equations enables breaks to occur slowly over time. Specifically, it is assumed that the series responds to shocks to the trend function the way it reacts to shocks to the innovation process $e_t$ (Vogelsang and Perron, 1998). This process is known as IO model. In order to test unit root hypothesis for M1 and M2, the IO-type test regression can be derived by merging the structural model (1)-(5). Narayan and Popp(2010) use t-statistics in order to test the null hypothesis of a unit root ($H_0 : \rho = 1$) against the alternative hypothesis ($H_1 : \rho < 1$).

**Hatemi-J (2008) Cointegration Test.** According to Gregory and Hansen (1996), the cointegration tests which do not take into account the presence of structural changes or regimes shifts, have low power. Following this argument, they introduce a new cointegration test procedure, which allows for an endogenously determined break in the cointegrating relationship. Their test includes three alternative forms of structural break: level shift (Model C), level shift with trend (Model C/T) and regime shift (Model C/S). The model with regime shift (Model C/S) can be constructed as follows:

$$Y_t = \alpha_0 + \alpha_1 D_t + \beta_0 X_t + \beta_1 (D_t * X_t) + u_t \qquad (7)$$

where $D_t$ is a dummy variable equal to 0 if $t \le \tau$ and 1 if $t > \tau$. Here, the unknown parameter $\tau$ denotes the timing of the change, $\alpha_1$ denotes the change in the intercept coefficient at the time of the shift and $\beta_1$ represents the change in the slope of the cointegrating equation. Given that the timing of structural break is unknown a priori, Gregory and Hansen (1996) propose a suite of tests: the commonly used ADF test statistic and the extensions of the $Z_t$ and $Z_\alpha$ test statistics of Phillips (1987). By calculating these statistics, the null hypothesis of no cointegration is tested against the alternative of cointegration with structural break. The Equation (7) is generalized by Hatemi-J (2008) for two structural breaks in the cointegrating relationship. The equation which takes into account the possibility of two structural breaks can be seen in below:

$$Y_t = \alpha_0 + \alpha_1 D_{1t} + \alpha_2 D_{2t} + \beta_0 X_t + \beta_1 (D_{1t} * X_t) + \beta_2 (D_{2t} * X_t) + e_t \qquad (8)$$

where $D_{1t}$ and $D_{2t}$ are dummy variables constructed as follows:

$$D_{1t} = \begin{cases} 0 & if \quad t \leq \tau_1 \\ 1 & if \quad t > \tau_1 \end{cases} \quad \text{and} \quad D_{2t} = \begin{cases} 0 & if \quad t \leq \tau_2 \\ 1 & if \quad t > \tau_2 \end{cases}.$$

Here, $\tau_1$ signifies the period before the first break and $\tau_2$ signifies the period before the second break. In order to test the null hypothesis of no cointegration, ADF, $Z_t$ and $Z_\alpha$ test statistics are calculated. The critical values can be found in the original paper of Hatemi-J (2008).

## RESULTS AND DISCUSSION

To test the validity of the environmental Kuznets curve hypothesis for Turkey, we use the following linear logarithmic quadratic model construction:

$$\ln CO_{2t} = \beta_0 + \beta_1 \ln EC_t + \beta_2 \ln GDP_t + \beta_3 \ln GDP_t^2 + e_t$$

Here, $\ln CO_2$ refers to the natural logarithm of the carbon dioxide emission per capita (metric tons) which is an indicator of environment degredation, $\ln EC$ is the natural logarithm of the energy consumption per capita (kilograms of oil equivalent), $\ln GDP$ represents natural logarithm of the gross domestic product per capita (with constant prices), $\ln GDP^2$ is the natural logarithm of the square of gross domestic product per capita. We use annual data over the period 1960-2011 obtained from World Development Indicators database of World Bank. As a first step, we neglect existence of possible structural breaks and start our analysis by investigating unit root properties of the series with conventional Augmented Dickey Fuller (ADF) and Philips and Perron (PP) unit root tests. The test results are tabulated in Table 1.

**TABLE 1. The results of ADF and PP unit root tests**

| | ADF | | PP | |
|---|---|---|---|---|
| **Series** | **Level** | **First difference** | **Level** | **First difference** |
| $\ln CO_2$ | -2.6190 | -7.0409[a] | -2.6677 | -7.0422[a] |
| $\ln EC$ | -2.3923 | -6.8914[a] | -2.4461 | -6.8899[a] |
| $\ln GDP$ | -2.9317 | -7.1164[a] | -3.0356 | -7.1214[a] |
| $\ln GDP^2$ | -2.8572 | -7.0914[a] | -2.9908 | -7.0969[a] |

**Note:** [a] indicates rejection of the unit root null hypothesis at the 1% significance level.

According to the obtained results, ADF and PP unit root tests show that all the series in the analysis are nonstationary in level but stationary after first differencing. In other words, these results give evidence to support that the integration order of the series can be considered as one ($I(1)$) for all series. Since Perron (1989) argues that ignoring a structural break can lead to false acceptance of the unit root null hypothesis, we extend our unit root analysis with structural breaks. For this purpose, we use a new unit root test proposed by Narayan and Popp (2010). This test is flexible enough to allow for at most two structural breaks in the level and trend of the series. The results based on M1 (breaks in level only) and M2 (breaks in level and trend) are reported in Table 2.

The results in Table 2 imply that the null of unit root hypothesis cannot be rejected for all $\ln CO_2$, $\ln EC$, $\ln GDP$ and $\ln GDP^2$ series. These results are consistent with ADF and PP unit root tests. The implication of this evidence is that validity of the environmental Kuznets curve hypothesis can be tested by using cointegration methods. Conventional cointegration methods are biased towards accepting the null of no cointegration and if there is structural breaks in the relationship, these standard tests may produce spurious cointegration results. By considering that the long run relationship between mentioned $\ln CO_2$, $\ln EC$, $\ln GDP$ and $\ln GDP^2$ series might change and the shifts can occur in their cointegrating vector, instead of conventional cointegration tests, we apply Hatemi-J (2008) cointegration test in the presence of two unknown structural breaks. The results can be seen in Table 3.

**TABLE 2. The results of Narayan and Popp (2010) unit root test with two structural breaks**

| | M1 | | | M2 | | |
|---|---|---|---|---|---|---|
| **Series** | **Test statistic** | **TB1** | **TB2** | **Test statistic** | **TB1** | **TB2** |
| $\ln CO_2$ | -2.238 | 1977 | 2000 | -4.065 | 1977 | 2000 |
| $\ln EC$ | -2.488 | 1978 | 2000 | -4.156 | 1978 | 2000 |
| $\ln GDP$ | -3.839 | 1979 | 2000 | -4.622 | 1979 | 2000 |
| $\ln GDP^2$ | -3.974 | 1979 | 2000 | -4.129 | 1979 | 2000 |

**Critical values for unit root test**

**Model M1 (Breaks in level only)**

| 1% | 5% | 10% |
|---|---|---|
| -5.259 | -4.514 | -4.143 |

**Model M2 (Breaks in level and trend)**

| 1% | 5% | 10% |
|---|---|---|
| -5.949 | -5.181 | -4.789 |

**Note:** The critical values for endogenous two breaks are from Narayan and Popp (2010).

**TABLE 3. The results of Hatemi-J (2008) cointegration test with two unknown structural breaks**

| Test statistics | Statistic value | $\tau_1$ | $\tau_2$ |
|---|---|---|---|
| $ADF^*$ | -8.079[a] | 1972 | 1984 |
| $Z_t^*$ | -8.370[a] | 1979 | 1993 |
| $Z_\alpha^*$ | -57.383 | 1972 | 1984 |

**Note:** [a] indicates statistically significance at the 1% significance level. The critical values for $ADF^*$, $Z_t$ and $Z_\alpha$ test statistics are collected from Hatemi-J (2008). These are −7.833, −7.352 and −7.118 for $ADF^*$ and $Z_t$ statistics; −118.577, −104.860 and −97.749 for $Z_\alpha$ statistic at the 1%, 5% and 10% significance levels, respectively.

The results indicate that $ADF^*$ and $Z_t^*$ statistics give evidence in favour of the existence of long run relationship between carbon dioxide emission, energy consumption and GDP series with two structural breaks. The meaning of this result is that environmental Kuznets curve hypothesis is valid for Turkey. When the attention is given to the identified structural breaks, it is seen that the results are mixed. At a glance, the first break points to the 1972 while the second break points to the 1984 for $ADF^*$ and $Z_\alpha^*$ statistics. These break points refer to the oil crisis of 1973 and structural transformation reforms which opened up the Turkish economy to international forces after 1980s, respectively. According to the $Z_t^*$ statistic, we identify the first break in 1979 and the second break in 1993. These break points are consistent with the important economic events of Turkish economy. The break in 1979 refers to the period which there is an increase in energy consumption and carbon dioxide emissions due to the growth in the industrial sector and the break in 1993 provides apparent evidence for 1994 economic crisis.

## CONCLUSIONS

In this paper, the environmental Kuznets curve hypothesis is tested by investigating the long run relationship between carbon dioxide emission, energy consumption and income for Turkey over the period from 1960 to 2011. For this purpose, we employ Narayan and Popp(2010) unit root and Hatemi-J(2008) cointegration tests. Both tests take into account the existence of two structural breaks. The obtained results indicate that there is evidence of cointegration between carbon dioxide emission, energy consumption and income with two structural breaks, hence the validity of the environmental Kuznets curve hypothesis for Turkey. It is also seen that the identified structural breaks point to the important economic events of Turkish economy such as oil crisis, structural transformation reforms, carbon dioxide emission increases and economic crisis.

## ACKNOWLEDGEMENTS

This research is supported by Scientific Research Projects Unit of Istanbul University.

## REFERENCES

Akbostancı, E., S. Türüt-Aşık, S., and G. İ. Tunç. 2009. "The Relationship between Income and Environment in Turkey: Is There an Environmental Kuznets Curve?." *Energy Policy*. 37(3): 861-867.

Gregory, A. W., and B. E. Hansen. 1996. "Residual-Based Tests for Cointegration in Models with Regime Shifts." *Journal of Econometrics*. 70: 99-126.

Grossman, G., and A. Krueger. 1991. "Environmental Impacts of a North American Free Trade Agreement." *National Bureau of Economics Research Working Paper*. 3194, NBER, Cambridge.

Grossman, G., and A. Krueger. 1995. "Economic Growth and the Environment." *Quarterly Journal of Economics*. 110(2): 353-377.

Halıcıoglu, F. 2009. "An Econometric Study of CO2 Emissions, Energy Consumption, Income and Foreign Trade in Turkey." *Energy Policy*. 37(3): 1156-1164.

Hatemi-J, A. 2008. "Tests for Cointegration with Two Unknown Regime Shifts with An Application to Financial Market Integration." *Empirical Economics*. 35(3): 497-505.

Kuznets, S. 1955. "Economic Growth and Income Inequality." *American Economic Review*. 45: 1–28.

Lise, W. 2006. "Decomposition of $CO_2$ Emissions over 1980–2003." *Energy Policy*. 34: 1841–1852.

Lise, W., K. V. Monfort. 2007. "Energy Consumption and GDP in Turkey: Is There a Cointegration Relationship?." *Energy Economics*. 29: 1166–1178.

Narayan, P. K., and S. Popp. 2010. "A New Unit Root Test with Two Structural Breaks in Level and Slope at Unknown Time." *Journal of Applied Statistics*. 37: 1425-1438.

Odhiambo, N. M., 2012. "Economic Growth and Carbon Emissions in South Africa: An Empirical Investigation." *Journal of Applied Business Research*. 28(1): 37-46.

Omay, R. E. 2013. "The Relationship between Environment and Income: Regression Spline Approach." *International Journal of Energy Economics and Policy*. 3: 52-61.

Ozturk, I., and A. Acaravcı. 2010. "CO2 Emissions, Energy Consumption and Economic Growth in Turkey." *Renewable and Sustainable Energy Reviews*. 14(9): 3220-3225.

Panayotou, T. 2003. "Economic Growth and the Environment. Economic Survey of Europe." 2(2): http://www.unece.org/fileadmin/DAM/ead/pub/032/032_c2.pdf.

Perron, P. 1989. "The Great Crash, the Oil Price Shock and the Unit Root Hypothesis." *Econometrica*. 57: 1361-1401.

Phillips, P. C. B. 1987. "Time Series Regression with a Unit Root." *Econometrica*. 55: 277–301.

Popp, S. 2008. "New Innovational Outlier Unit Root Test with A Break at An Unknown Time." *Journal of Statistical Computation and Simulation*. 78: 1143-1159.

Schmidt, P., and P. C. B. Phillips. 1992. "LM Test for a Unit Root in the Presence of Deterministic Trends." *Oxford Bulletin of Economics and Statistics*. 54: 257-287.

Soytas, U., and R. Sari. 2009. "Energy Consumption, Economic Growth and Carbon Emissions: Challenges Faced by An EU Candidate Member." *Ecological Economics*. 68(6): 1667-1675.

Vogelsang, T., and P. Perron. 1998. "Additional Tests for a Unit Root Allowing for a Break in the Trend Function at an Unknown Time." *International Economic Review*. 39(4): 1073–1100.

Yavuz, N. Ç. 2014. "CO2 Emission, Energy Consumption and Economic Growth for Turkey: Evidence from a Cointegration Test with a Structural Break." *Energy Sources, Part B: Economics, Planning and Policy*. 9(3): 229-235.

# A LINEAR GENETIC PROGRAMMING APPROACH FOR DISCHARGE FORECASTING. JONATHAN BARGE AND HATIM SHARIF

Jonathan Barge and ***Hatim Sharif***
(University of Texas at San Antonio, TX, USA)

This study focuses on assessing the ability of Linear Genetic Programming (LGP) to model the rainfall-discharge relationship and forecast next-day discharge. The approach is quite different from more complex modeling environments such as physically based models, conceptual models, and more intricate data-driven techniques such as Artificial Neural Networks (ANNs). The main difference between LGP and most data-driven techniques in that it deals with multiple solutions while optimizing the functional relationships between input variables. Individual models are optimized through the application of crossover and mutation operators, which mimic Darwinian evolutionary processes. Model results indicate that a minimum number of input variables are necessary for deriving highly robust solutions, as the application of LGP as a standalone technique returned strong validation values while using only two days of observed discharge and one day of precipitation information. The potential of LGP was found to be further enhanced when embedded into hybrid model designs. Clustering the input space through Kohonen's Self Organizing Map (SOM) proved to allow LGP to create a better fit to each data cluster. Decomposing the original streamflow time-series into a number of subcomponents through the data Ensemble Empirical Mode Decomposition (EEMD) decomposition technique and utilizing that information as additional input data significantly improved results. Combining all three techniques in an EEMD-SOM-LGP design returned the best performance statistics of any model produced.

# CFD MODELLING OF CYLINDRICAL VORTEX FLOW REGULATORS PERFORMANCE - SIMULATION AND EXPERIMENTS

***Patryk Wójtowicz**** and Małgorzata Szlachta
(Wrocław University of Technology, Wrocław, Poland)

Vortex flow regulators (also known as hydrodynamic flow regulators) are used in wastewater and storm water collection systems for flow throttling and control. Vortex regulators are not only very efficient energy dissipaters but also effective atomizers, which is beneficial for sewer aeration. Maintaining a proper oxygen balance is important to preserve against putrefaction, odours and concrete sewer corrosion. In this study, we present the experimental and numerical results for the cylindrical type of vortex regulators (CVR).

Laboratory and pilot scale devices were examined in a closed-circuit test rig. For CFD modelling, the commercial solver ANSYS Fluent as well as OpenFOAM were used. We performed testing for a range of different RANS, URANS and LES (Large Eddy Simulation) turbulence models to find that unsteady incompressible two-phase URANS Reynolds Stress Model (RSM) yielded the best results and was less computationally expensive than LES. The application of seven-equation Reynolds stress turbulence model allowed us to take into account the complex hydrodynamics of vortex regulators. Specifically, this model addresses the effects of turbulence anisotropy (Reynolds stresses), streamline curvature, swirl and rotation. To minimize numerical diffusion, we used only hexahedral meshes. To study mesh dependence, we tested meshes from 200k to 1.5 million cells. In most cases, the meshes composed of 500k cells were finally used in simulations. The initial steady solution was obtained with the use of two-equation k-epsilon model. The converged k-epsilon solution was used as boundary conditions for URANS simulations. The important parameters of CFD simulation were as follows: Reynolds stress model for turbulence closure with the enhanced wall functions. Pressure-based (segregated) solver with PRESTO! scheme for pressure interpolation and SIMPLE scheme for pressure-velocity coupling. In the final solution, we used the second order upwind scheme. In order to obtain the converged solution, we applied dynamic simulation control for unsteady RSM simulation (e.g. by adjusting Courant number). Grid adaption was also employed (only with Fluent simulations) – mesh was optimized on the basis of velocity gradients. The CFD simulations were compared with experimental data (pressure and velocity profiles, inlet pressure and flow rate, photo measurements – spray angle, air core diameter, air core offset). All calculations were run in parallel on the Macintosh computer (running Linux openSUSE) with 12 GB RAM and Intel quad-core i7 3.4 GHz.

The comparison of computed pressure profiles with the experiments exhibited a high level of accuracy. The major quantities of interest, such as pressure drop, discharge coefficient and minor loss coefficient were accurately predicted (within 5% of error). The swirl flow features, such as air core diameter and (asymmetrical) shift vectors from the outlet center were well captured in the CFD simulations. The selected Reynolds-stress turbulence model was suitable for simulation of complex vortex flow in hydrodynamic flow regulators.

# VULNERABILITY ASSESSMENT OF THE CENTRAL GULF OF MEXICO COAST USING A MULTI-DIMENSIONAL INDEX APPROACH

David E. Dismukes and ***Siddhartha Narra***
(Louisiana State University, Baton Rouge, LA, USA)

The coastal communities of the central Gulf of Mexico (GOM) form a highly productive and complicated human, physical, and natural environment that interact in ways unique compared to other coastal areas of the globe. Past studies on understanding coastal resiliency and developing coastal vulnerability indices for this region have mainly focused on climate change and sea-level rise, with more recent research directed towards socio-economic and demographic factors. However, the interactions of climate change with non-climatic drivers of the coastal ecosystem such as economy and infrastructure concentration are often overlooked. To support the development of policies relating to coastal management and climate change, it is vital to integrate all the relevant parameters. This paper presents a relative vulnerability assessment of the Central GOM coast by incorporating climatic, geological, socio-economic, demographic and economic variables in a GIS system. A multi-dimensional Coastal Vulnerability Index (CVI) developed based on these parameters is used to rate coastal segments into different classes based on their relative range of vulnerability. We study the relationship between energy infrastructure and the physical and human aspect of communities to identify and prioritize communities, and the proximate infrastructure most at risk from coastal climate change. Special focus is directed towards the concentration of pipeline and transportation infrastructure in this region when developing these vulnerability indices. Spatial analysis will be a component part of the index-based approach to characterize, organize, and analyze data in order to assess coastal community vulnerability in areas supporting critical energy infrastructure.

# PREDICTING THE LONG-TERM LEACHING BEHAVIOR OF COAL ASH DEPOSITS IN MINE SITE: A COLUMN STUDY

***Mina Mohebbi****, Farshad Rajabipour, Barry E. Scheetz
(Pennsylvania State University, University Park, PA, USA)

Combustion of coal to generate electricity produced about 116 million tons of coal combustion products (CCPs) in 2013. Approximately, 56% (vol.) of CCPs consists of pulverized coal (PC) fly ash and fluidized bed combustion (FBC) ash. 43% of fly ashes are used for beneficial applications such as concrete construction, soil stabilization, and land reclamation, and the remaining 57% are deposited in landfills or surface impoundments. Reclamation of mine sites with coal ash has been shown to potentially alleviate the negative effects of mining activities on the environment. These effects include disturbance of large areas of land, risk of land subsidence, and risk of contamination of water resources due to acid mine drainage and heavy metal leaching. The alkaline pH of fly ash has been proved to neutralize the acid mine drainage, and consequently precipitate the dissolved elements out of the mine leachate, thereby reducing contamination to water resources. However, the long-term (several decades) leaching of harmful trace elements from coal ash deposits to subsurface aquifers is still not fully understood, and continues to be a major concern.

In this study, CrunchFlow, a quantitative reactive-transport model, is developed to predict the long-term (30 years) leaching risk of elements from a Pennsylvania PC fly ash in a 1D leaching column geometry considering the geochemical conditions of mine sites. This model assists in correct identification of the leaching mechanisms of major (Si, Al, Fe, Ca, K, S), minor (Na, Mg), and trace elements of interest (i.e., As, Mo, B). Simulation considers fly ash mineralogy, host phases for target elements, and environmental/mine conditions (i.e., influent pH and porosity evolution), and predicts leaching by accounting for solubility controlled phase dissolutions as well as delayed release/leaching of elements from ash phases exposed to flowing water for long durations.

The simulation results revealed that dissolution, precipitation of secondary phases, adsorption, and surface complexation are the main mechanisms controlling the leaching of trace elements in long term. The influent water pH was changed from 5.5 (pH of rain water flow) to 8 which shows the alkalinity and neutralization potential of studied fly ash. To evaluate the accuracy of the model, it is applied to leaching column test geometry, and the model results are compared with the experimental data. If successful, this model could be scaled up in future to an actual mine site, and be used to evaluate the risk of groundwater contamination. Such information are crucial to find appropriate mitigation techniques to retain the harmful elements in solid phases and prevent their leaching to underground water resources.

# INVESTIGATING, MODELING AND PREDICTION OF TRICLOSAN AND CARBAMAZEPINE IN SURFACE WATERS USING EFDC: CASE STUDY IN THE SHAHE STREAM

Xiao Yuan, Shiyu Li*, Jiatang Hu, Yuying Li, Ziyun Wang
(Guangdong Provincial Key Laboratory of Environmental Pollution Control and Remediation Technology, School of Environmental Science and Engineering, Sun Yat-sen University, Guangzhou 510275, China)

**ABSTRACT:** The aim of this study was to preliminary attempt to model pharmaceuticals and personal care products (PPCPs) in surface water bodies, and to use EFDC model to forecast. As two kinds of PPCPs, triclosan (TCS) and carbamazepine (CBZ) were generally used in disinfectant or anticonvulsants agent. Hydrolysis, biodegradation and photolysis as the major routes of TCS and CBZ in this study. A central composite design methodology has been investigated to determine the relevant natural parameters (including pH, DO, salinity, temperature, light intensity, humic acid) for their degradation as far as possible in laboratory. Several important impacts on the parameters were discussed. The Shahe Stream has been selected, of which hydrodynamic condition was extremely simple and pollution sources were very specific, in order to reduce the difficulty of simulation under natural conditions. Then, the fate and behavior of TCS and CBZ during the stream were investigated using a 2D numerical flow and transport model with EFDC. Results showed that the predicted values of the model were basically higher than the measured values, and the prediction deviation between 23%-52% of TCS, 23%-37% of CBZ. It proved that PPCPs can be simulated by using the numerical model in simple water, but must pay attention to the process of PPCPs about adsorption and desorption between water and sediment.

***Keywords:*** triclosan, carbamazepine, pharmaceuticals and personal care products (PPCPs), modeling, surface waters

## INTRODUTION

Active ingredients in pharmaceuticals, personal care products are collectively referred to as PPCPs, including not just prescription drugs and biological agents, but also commonly used diagnostic reagent, nutrition, perfumes, and skin care, shampoo hair care, oral health and so on (Daughton and Ternes, 1999). The presence of PPCPs in surface waters has been frequently reported in the literature(Yoon *et al.*, 2010; Ortiz de Garcia *et al.*, 2013; Carmona *et al.*, 2014). China is a large country with high production and consumption of PPCPs owe to its economic development and population growth, which may result in serious pollution by PPCPs(Peng *et al.*, 2014; Wu *et al.*, 2014; Wang *et al.*, 2015). The technology for prediction and monitoring of PPCPs in different areas of China is urgent.

PPCPs selected in this study are triclosan (TCS) and carbamazepine (CBZ) (Table. 1).Triclosan [5-chloro-2-(2,4-dichlorophenoxy)phenol] is abroad spectrum antibacterial agent used in cosmetics, household cleaning products, toys, medical devices, plastic materials, textiles, and kitchen utensils (Bedoux et al., 2012). TCS is used in great quantities in the daily chemical products, along with the wastewater discharge into the natural environment, the concentration of TCS about the related test was 0.01 ug/L to 10 ug/L. Xu reported that the concentration of TCS was 0.030 ug/L to 0.100 ug/L in the Victoria harbor of Hongkong, and 0.025 ug/L to 0.035 ug/L in the Pearl River (Xu *et al.*, 2007). Just one year later, another study pointed out that the concentration of TCS was reached 0.035 ug/L to 1.023 ug/L in the same area(Peng *et al.*, 2008). Furthermore, the concentrations of TCS was detected higher in the sewage treatment plant, such as Wu investigated that the average concentration of TCS were 0.1585 ug/L, 0.0225 ng/L in the primary treatment effluent and the secondary treatment effluent of the sewage treatment plant in Shenzhen city(Wu *et al.*, 2007). Another target, carbamazepine [5H-dibenzo[b,f]azepine-5-carboxamide]is an antiepileptic drug belongs to anticonvulsants that commonly used in control certain types of seizures by reducing abnormal electrical activity in the brain(Mohapatra et al., 2014). It was the most frequently detected pharmaceutical residue in water bodies(Zhang *et al.*, 2008), and related to the growth rate of microorganisms and to effluent concentrations as a possible anthropogenic marker and sewage markers in the aquatic environment(Clara *et al.*, 2004; Kuroda *et al.*, 2012). Based on the significant correlations between PPCPs, it can also as the chemical indicators for predicting the overall PPCPs contamination(Wang *et al.*, 2015). In Pearl River, the concentrations of carbamazepine in surface water were lower than those in USA and European countries(Yang *et al.*, 2013), but it was still detected with a detection frequency of 81% and median concentration of 15.6 ng/L, and the maximum concentration of carbamazepine(43.1 ng/L) was also found in the Shijing River which was tributaries of Pearl River(Zhao *et al.*, 2010).

**Table 1.** Physicochemical properties of the target compounds

| Common Name | Carbamazepine[a] | Triclosan[a] |
|---|---|---|
| Therapeutic groups | Phsychiatrics-antiepileptics | Disinfectants |
| Abbreviation | CBZ | TCS |
| Chemical Formula | $C_{15}H_{12}N_2O$ | $C_{12}H_7C_{13}O_2$ |
| Chemical Structure | 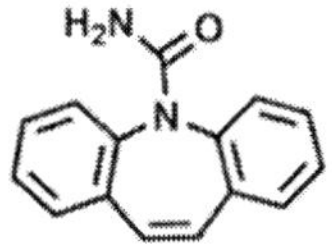 | 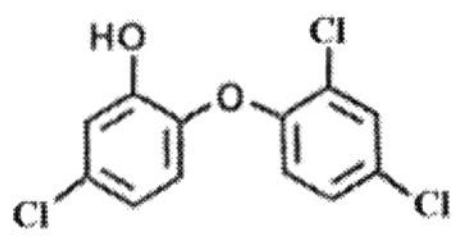 |
| CAS No.[b] | 298-46-4 | 3380-34-5 |
| Mol. Mass (g mol-1) | 236.27 | 289.54 |
| Solubility (mg/L)[c] | 17.7 | 10 |
| $pK_a$ | 7/13.9 | 4.5/8.1 |
| $LogK_{ow}$ | 2.47 | 4.80/5.34 |

[a] (Kosma *et al.*, 2014); [b] (Deo, 2014); [c] (Stackelberg *et al.*, 2007).

Although there are a lot of reports about the concentration and distribution of TCS and CBZ, and theirs degradation mechanism, processing technology in the laboratory research, but the data collected in such studies rarely provide quantitative information on the various processes that determine the fate of the compound in the natural environment. To our knowledge, almost no information on which water quality model about the fate of TCS and CBZ in surface water is available in the literature.

## MATERIALS AND METHODS

**Chemicals and Reagents.** Standards of TCS (99.0%), CBZ (99.5%), $^{13}C_3$-caffeine and simatone were purchased from Dr. Ehrenstorfer (Augsburg, Germany). Stock solutions of the standards were prepared by dissolving each compound in methanol or ultrapure water. The solvents, HPLC-grade methanol, water (Lichrosolv), HPLC-grade phosphoric acid and formic acid 98% were provided by Merck (Darmstadt, Germany). Sulfuric acid, sodium hydroxide, sodium chloride, humic acid (fulvic acid, 90%), disodium ethylenediamine tetraacetate ($Na_2EDTA$) were all purchased from Aladdin (Los Angeles, USA).

**Sampling Site and Sample Collection.** Sampling sites (S1-S6) were located in the upstream of the Shahe Steam (Fig. 1). The Shahe Stream is located in the center of Guangzhou city, which is the provincial capital of Guangdong province in China. It's discharged into the Pearl River which flows through the Pearl River Delta and finally merges into the South China Sea via the Pearl River Estuary. The Pearl River is the third largest river of China. A wastewater treatment plant (WTP) is located in the upstream of Shahe Stream, the plant facilities used half buried type structure, using biological membrane technology to deal with the sewage near the living area. Its tail water through the pipe into the Shahe Stream.

After field survey and measurement, the pollution sources is single the this area. At the same time, due to the regional water mainly comes from the WTP and the stream was artificial built and hardening. Therefore, the study can effectively reduce the difficulty of the simulation by further simplifying the hydrodynamic conditions. This choice can not only try to avoid some unpredictable influence to achieve the purpose of this study, but also provide good foundation for deeper research in the future.

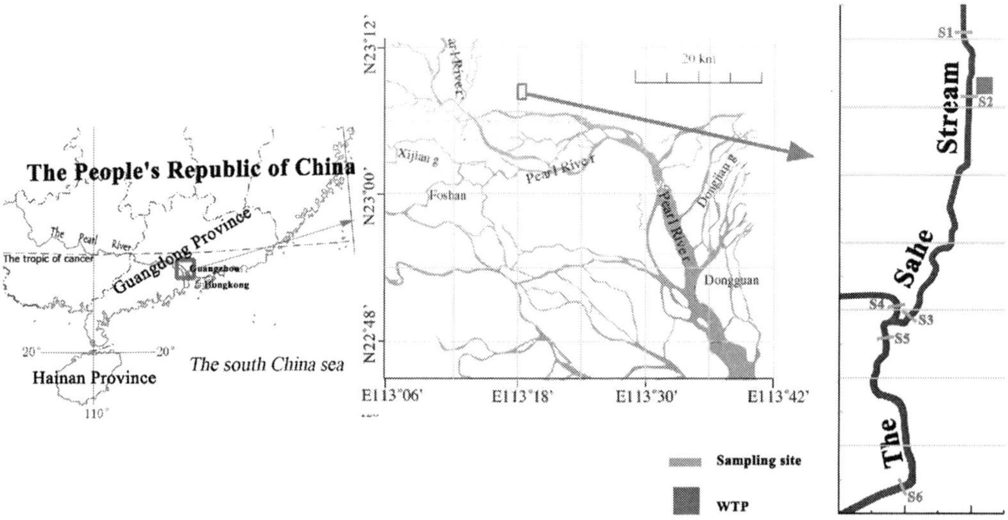


**Fig. 1.** Location of the sampling sites

Sampling was respectively performed on March, 2014. All sampling sites were sampled during both the level period (Peng *et al.*, 2008).All of the samples was taken at the middle of each cross-section, 0.5 m from the water surface.

Grab water samples (1 L) were collected in clean polypropylene bottles, from the middle of the river width at ca. 30 cm depth (Carmona *et al.*, 2014). Before sample collection, each bottle was thoroughly pre-rinsed with ultrapure water in the laboratory and then, rinsed with sample water three times prior to sample collection.

All water samples analyzed in this study were previously filtered with a 0.45 μm cellulose nitrate membranes of 47 mm by CNW (Düsseldorf, Germany) in the laboratory. Then, samples were adjusted pH to 2.5 with concentrated sulfuric acid to suppress microbial activity and kept in their original package at 4 °C until the analysis that was always performed within two days.

**Sample Preparation and Extraction.** Typical concentrations of antibiotics in the environmental water samples were less than 1 μg/L(Boyd et al., 2003; Richardson et al., 2005), so Solid-phase extraction (SPE) methods should be using to cleanup and preconcentration at the beginning of the pretreatment procedures. Procedures for preparation of water samples for instrumental analysis were described in detail previously (Yang *et al.*, 2013).

A 500 mL sample was extracted using SPE cartridges (Strata C18-E cartridges, 3mL 500mg; Phenomenex, Los Angeles, USA) on a 16-position vacuum system (Supelco). Briefly, 60 mg SPE cartridges were conditioned with 6 mL of methanol, 1 mL acidified methanol (0.1 % formic acid in methanol, v/v), followed by 2 × 6 mL of HPLC-grade water at a flow rate of 2 mL $min^{-1}$. Samples were spiked with 10 mL 5 g $L^{-1}$ $Na_2EDTA$ and 50 μL 500 μg $L^{-1}$ $^{13}C_3$-caffeine and loaded on the cartridges at 5 mL $min^{-1}$. After extraction, the cartridges rinsed with 2 × 6 mL of HPLC-grade water, dried under vacuum for 5 min, to remove excess of water, and eluted with 4 × 2 mL of acidified methanol. The extracts were then evaporated and reduced to -50 μL under a gentle nitrogen stream, reconstituted with 250 μL of methanol-water (10:90, v/v). Then, the extracts were added and mixed ten microliters of internal standards (2.5 mg $L^{-1}$ simatone). Addition of internal standards in the final extract compensates possible matrix effect, but does not compensate any of the possible procedure losses (Petrovic *et al.*, 2014). Finally, the extracts were filtered with 0.45μm syringe filters into 2 mL amber autosampler vials containing 250 μL inserts.

**The Degradation Experiments.** The migration and degradation process of PPCPs is mainly divided into the biodegradation, photolysis, hydrolysis, the adsorption and desorption in surface water (Pal *et al.*, 2010). Among them, photolysis is the main chemical way except adsorption process which PPCPs degradate in surface water, the reaction is mainly influenced by light intensity, pH, dissolved oxygen and chloride ions (Onesios *et al.*, 2009; Batchu *et al.*, 2013). Besides photolysis reactions, hydrolysis and microbial degradation is also an important way, including microbial response affected by pH, temperature, illumination, dissolved oxygen, nutrient, microbial species and quantity, the hydrolysis reaction is mainly influenced by pH, temperature, organic carbon and dissolved oxygen (Luo *et al.*, 2011). In consideration, the degradation experiment combined with three reaction process (biodegradation, photolysis, hydrolysis), and each factor in surface water bodies actual values range, by central composite experiment method. Central Composite Design (CCD) is the most frequently used five level fractional factorial designs for the construction of second-order response surface model. One of the advantages of CCD is the possibility to explore the whole of the experimental region (Daghrir *et al.*, 2014). Each reaction process set as eight factors five levels. The experiment would serve initial concentration PPCPs ($C_0$), pH, temperature (Temp), light intensity (LH), time (T), and salinity, humic acid (AH), and the concentration of dissolved oxygen (DO) as variable factors(Table. 2).

**Table 2.** Experimental Parameters.

| Name | Units | -1 | +1 | -α | +α |
|---|---|---|---|---|---|
| Initial concentration ($C_0$) | ug/L | 192.61 | 807.39 | 0.00 | 1000.00 |
| Time (T) [a] | h | 23.11 | 96.89 | 0.00 | 120.00 |
| Temperature (Temp) | °C | 10.93 | 33.07 | 4.00 | 40.00 |
| pH | | 6.39 | 7.61 | 6.00 | 8.00 |
| DO | mg/L | 1.73 | 7.27 | 0.00 | 9.00 |
| Salinity [b] | ‰ | 3.09 | 12.92 | 0.01 | 16.00 |
| Humic acid (AH) | mg/L | 5.79 | 24.23 | 0.02 | 30.00 |
| Light intensities (LH) [c] | lux | 282.43 | 817.57 | 100.00 | 1000.00 |

[a] The value of the time (T) were 18.24, 71.76, 0, 90 min in the photolysis reaction process,
[b] The value of the salinity expressed by the concentration of sodium chloride.
[c] The value of the Light intensities (LH) were zero in the biodegradation reaction process and hydrolysis reaction process.

Before the degradation experiment, the raw water which used by the degradation experiments sampled from the Shahe Stream had been filled by membrane filter (0.45 um) to remove the impurities in the water. Moreover, the raw water would be filled by membrane filter (0.22 um) to filter sterilization in photolysis and hydrolysis experiments. An rotary photochemical reactor (BL-GHX-V, Bilon, Shahai, China) (Fig. 2) and a solar-simulated light source (350 W xenon lamp with 290 nm filter, λ> 290 nm) were used to perform the degradation experiment. in a dark-room (Ma *et al.*, 2014). For controlling and maintaining the temperature, the constant temperature condensate tank drove the recirculated water to flow around the quartz tube.

**Instrumental Analysis And Quality Control.** Target chemicals were determined using a high-performance liquid chromatography system (1100 LC, Agilent, Santa Clara, USA), which consisted of two G1312A Binary HPLC pumps and a diode array detector. The gradient separation and the conditions in HPLC were done at an ZORBAX SB-C18 column (4.6×150 mm, 5μm, Agilent, Santa Clara, USA) was maintained at 30 ℃, with injection volumes of 20μL and the mobile phase flow rate was 1 mL $min^{-1}$. The mobile phase comprised the mobile phase A was ultrapure water with HPLC-grade phosphoric acid (pH = 3.0) and the B was methanol. The gradient elution was set as follows: started at 40 % B, 0–2-min linear gradient to 45 % B, 2–10-min linear gradient to 80 % B, then the eluent was brought up to 90 % B and maintained for 10 min.

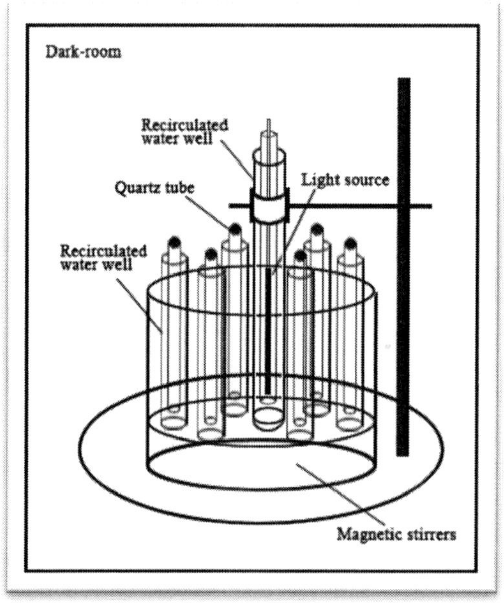


**Fig. 2.** Rotary photochemical reactor

The PPCPs with a signal-to-noise ratio more than 3 were considered as detectable ones. The average recovery ratios of CBZ, TCS, $^{13}C_3$-caffeine were 98.15 %, 85.30 %, 97.75 % with, respectively, the relative standard deviations were all less than 3 %, suggesting that the extraction procedure and instrumental analysis method for PPCPs in sampling was reliable. Furthermore, it should pointed out here that this study used $^{13}C_3$-caffeine just to assurance of the PPCPs extraction protocol, not to quantify other compounds (Chen *et al.*, 2013).

**The Degradation Modeling.** In this study, a two-dimensional numerical model, Environment Fluid Dynamics Code (EFDC) , was applied in the modeling and prediction of the TCS and CBZ in the target area. The EFDC was originally developed by Virginia Institute of Marine Science as authorized by US EPA (Jeong *et al.*, 2010). It solves finite-differenced forms of the hydrostatic Navier-Stokes equations which uses stretched or sigma vertical coordinates and Cartesian or curvilinear orthogonal horizontal coordinates to represent the physical characteristics of a water body(Sucsy and Morris, 2001).

The reach area used for simulation which summing up to a length of 3 km by an orthogonal curvilinear grid. The model grids consist of 264 × 782 cells in the horizontal direction, and the horizontal grid spacing ranged from 1.59 to 6.33 m. The water depth in each grid ranges from 0.51 m to 1.13 m. In the vertical direction, one layer was employed, i.e. horizontal two-dimensional model(Wu and Xu, 2011). The hydrodynamic and mass transportation calculation time steps were both set as 86400 seconds.

After the generalization of the area, a set of input files including initial concentrations and boundary conditions need to be configured to drive the EFDC model. In order to simplify the hydrodynamic model, The boundary conditions such as inflows and the initial concentrations of the water quality factors were setup in efdc.inp file. This study set the drainage of WTP and the only tributary to constant flow. On the premise of no precipitation, the terms of source and convergence of the flow could keep the balance by measured in this area.

## RESULT AND DISCUSSION

**Occurrence of TCS and CBZ in Effluent and Surface Waters.** The two PPCPs, TCS and CBZ were detected in most of samples from the Shahe Steam, the tributary of the Shahe Stream as well as the effluents from the WTP. The concentration of TCS and CBZ descended which was accompanied by the flow direction as a whole. Finally, the concentrations of TCS and TCC in the downstream of the Shahe Stream (S6) were below the detection limits (LOD: 50 ng/L for both compounds). Table. 3 shows concentrations of TCS and CBZ in surface water samples. Values of target compounds ranged from less than 64.38ng/L to 5.45ug/L. Maximum concentrations were measured in the tributary (S4).The WTP was another source of TCS and CBZ in Shahe Stream, the WTP couldn't dispose all of PPCPs in spite of its process had used the membrane filtration treatment technology(Yang *et al.*, 2011), on the contrary, some studies show that CBZ in all cities mainly originated from WTP effluents(Wang *et al.*, 2015). In all sampling site, the concentration of CBZ was much higher than TCS. There are maybe two reasons to explain this phenomenon: The one was that Jingxi Southern Hospital and a medical university were located in the upstream of the tributary. Medical wastewater contains many CBZ which become the main source in Shahe Stream. The other one was that the degradation rate of TCS is far higher than that of CBZ.

**Table 3.** TCS and CBZ (ug/L) in surface water samples.

| | S1 | S2 | S3 | S4 | S5 | S6 |
|---|---|---|---|---|---|---|
| Conc.$_{TCS}$ (ng/L) | 290.70 | 428.51 | 161.50 | 411.06 | 64.28 | ND [a] |
| Conc.$_{CBZ}$ (ng/L) | 1348.45 | 3862.52 | 4117.79 | 5454.64 | 9963.00 | 3312.24 |
| Temperature (℃) | 25.0 | 25.8 | 25.5 | 25.5 | 25.5 | 25.1 |
| DO | 2.78 | 2.35 | 7.37 | ND [a] | 6.34 | 8.07 |

[a] ND, not determined, below the limit of detection

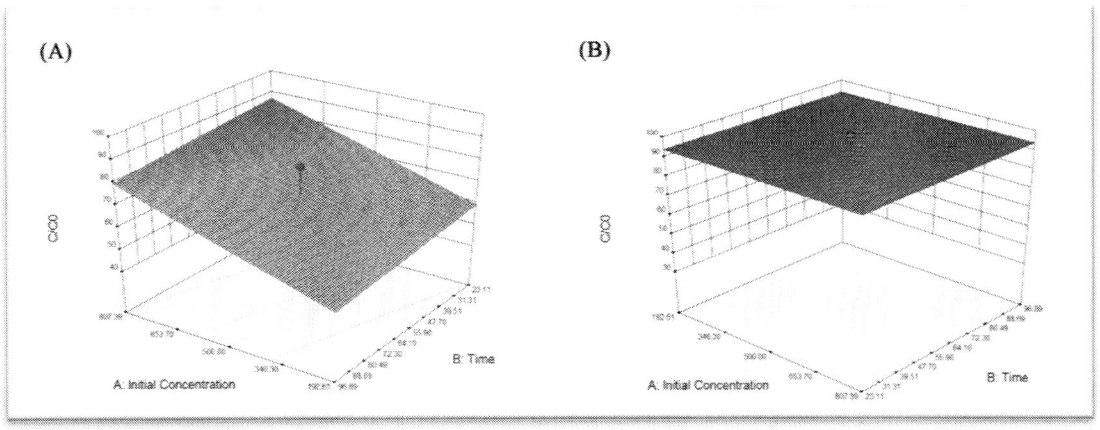


**Fig. 3.** Hydrolysis of CBZ (A) and TCS (B), as a function of time and a ratio about residual concentration (C) to initial concentration (C0). Other conditions: 22.00℃ temperature, 7.00 pH, 4.50mg/L DO, 8.01 Salinity, 15.01mg/L Humic acid.

**Hydrolysis, Biodegradation and Photolysis.** All statistical analyses were conducted by Design Expert 8 for Windows (Stat-Ease Inc., Minneapolis, USA).The software can set of quadratic polynomial curve, establishing the mathematical model, drawing the response surface.

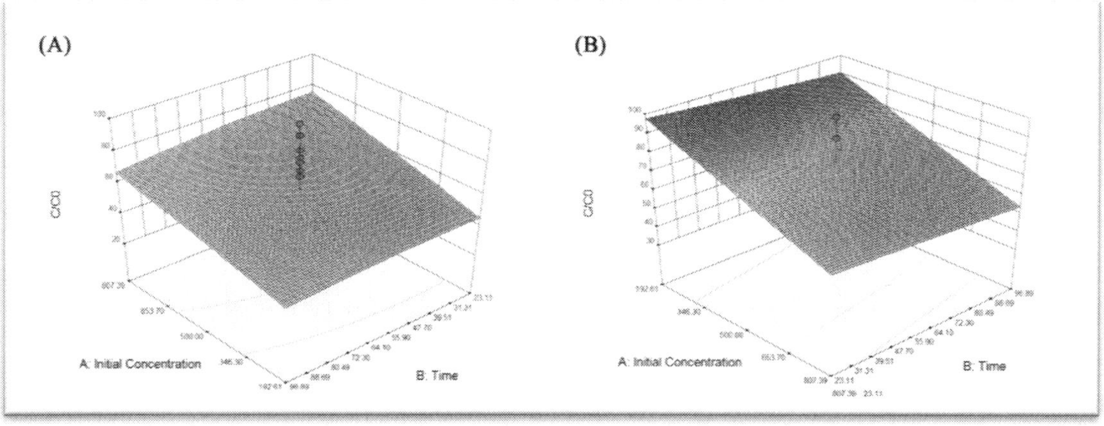


**Fig. 4.** Biodegradtion of CBZ (A) and TCS (B), as a function of time and a ratio about residual concentration (C) to initial concentration (C0). Other conditions: 22.00℃ temperature, 7.00 pH, 4.50mg/L DO, 8.01 Salinity, 15.01mg/L Humic acid.

***Hydrolysis.*** The Hydrolysis of PPCPs mainly electrophilic group of nucleophilic reagent and molecular reaction, and replace a leaving group. Usually, the hydrolysis of organic matter in water is first order reaction. In the process of hydrolysis, pH took the main effect, and temperature also had important influence to hydrolysis. Other factors, like DO, salinity, humic acid, light intensity hadn't find that obvious effects on PPCPs' hydrolysis. Moreover, from Fig. 3 we can find that the initial concentration maybe influences the hydrolysis rate of CBZ but TCS doesn't. Along with the time, the hydrolysis rate of CBZ was falling. Due to the very low Henry's law constant ($pK_a$) and the resistance against strong bases, we can expect that hydrolysis can be neglected for CBZ and TCS(Tixier *et al.*, 2003). In this study, we also confirmed this view. In the hydrolysis experiment, The hydrolysis of TCS was fewer than 10%, and the hydrolysis of CBZ was negatively correlated with initial concentration, the higher the concentration of CBZ, the less its hydrolysis. Combining with the Shahe Stream, we could also think that the hydrolysis reaction of CBZ was very small in the total degradation process.

***Biodegradation.*** Biodegradation of PPCPs has been extensively studied in wastewater treatment systems(Onesios *et al.*, 2009; Mohapatra *et al.*, 2014; Durán-Álvarez *et al.*, 2015), but the degradation of them are very difficult to determine under natural conditions. From Fig 4, we can find that the initial concentrations can influence the biodegradation rates of CBZ and TCS. During the period of experiment, the concentrations of CBZ and TCS failed to reach half of the initial concentrations, namely the half-life time of biodegradation. Although CBZ and TCS were all biodegradable during the test, but the biodegradation rate of TCS was far less than CBZ could be seen clearly.

Biodegradation conforms to the first-order degradation equation as Eq. (1):

$$C_t = C_0 \cdot e^{-k_b t} \quad (1)$$

where $C_0$ and $C_t$ are the concentrations of the target pollutants at time 0 and t, respectively; $k_b$ is the first order biodegradation rate constant, and t is the experimental time (in days for biodegradation experiments).

From Fig. 5, although DO, pH and Salinity has influenced to the CBZ biodegradation certainly, but the temperature effect on the biodegradation rate plays a main role. The most suitable temperature for biodegradation of CBZ are between 25~19℃. When the temperature is less than 19℃ or over 25℃, it doesn't favor the bacteria growth .So we used the Arrhenius function, Eq. (2):

$$k_b = k_{b20} \cdot \theta^{(T-20)} \quad (2)$$

where $k_{b20}$ is the first order biodegradation rate constant when the temperature is 20℃, T is the water temperature in the reaction (℃), and θ is the temperature correction factor. The $k_{b20}$ of CBZ is obtained by solving that is approximately equal to 1.5443E-2 $d^{-1}$, and θ is 1.0002; And the $k_{b20}$ of TCS is approximately equal to 1.5397E-2 $d^{-1}$, and θ is 0.0087.

According to related research, the biodegradation half-life time of CBZ was between 3000 ~ 5600 h in two rivers, because these rivers may contain different microbes, so there will be quite different of the biodegradation rate in different rivers(Kimura *et al.*, 2014). And half-life time was determined by using Eq. (3):

$$t_{1/2} = \frac{\ln 2}{k_b} \quad (3)$$

where $t_{1/2}$ is the half-life time for each compound under the experimental conditions (days), so the half-life times of CBZ and TCS were 43 ± 2 days for CBZ and 45 ± 1 days for TCS..

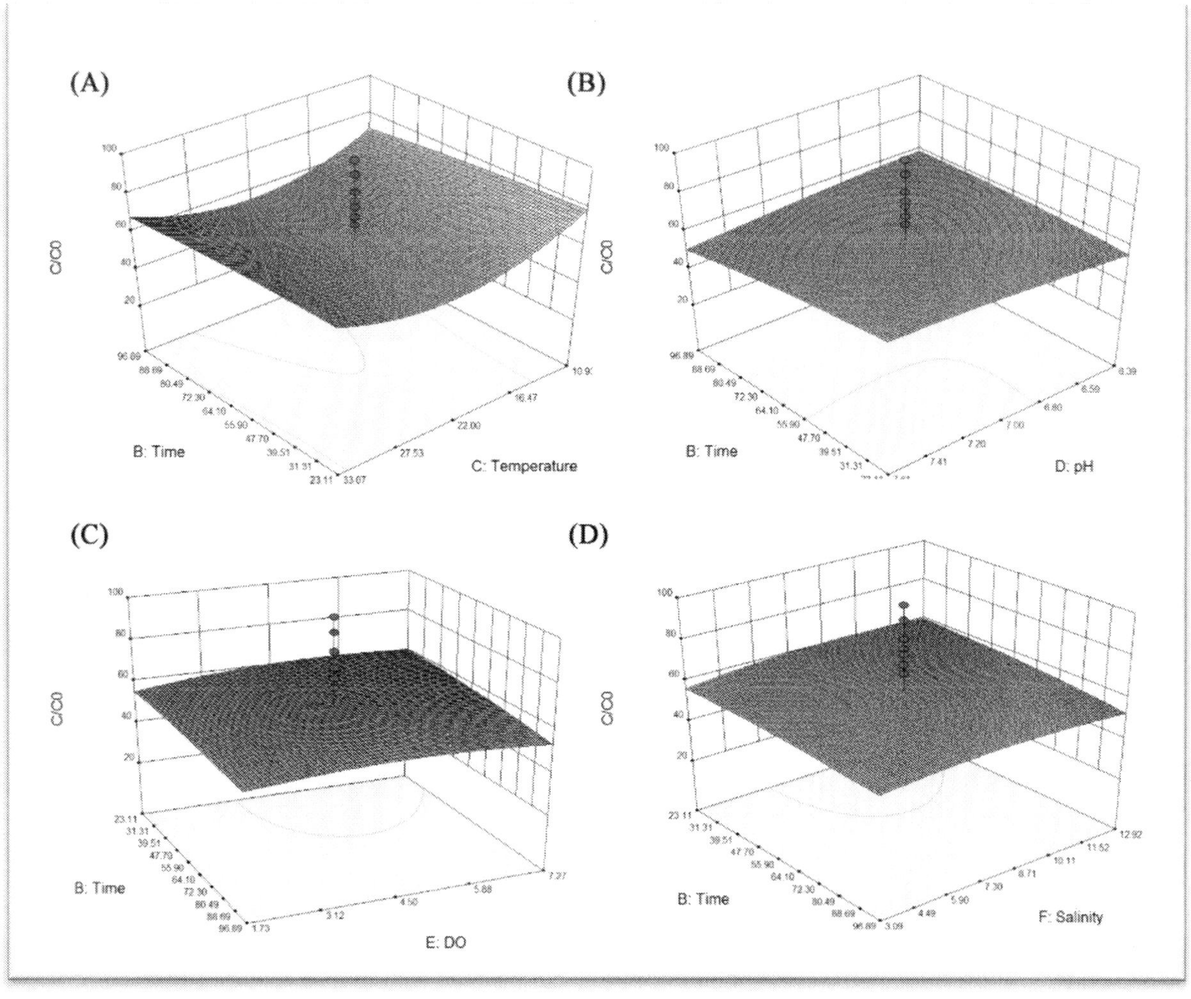


**Fig. 5.** The parameters effect on biodegradtion of CBZ, as a function of time and a ratio about residual concentration (C) to initial concentration (C0). Other conditions: 22.00℃ temperature, 7.00 pH, 4.50mg/L DO, 8.01 Salinity, 15.01mg/L Humic acid.

***Photolysis.*** Photolysis reactions can be divided into direct photolysis, indirect photolysis and photosensitivity photolysis. In the literature(Fabbri *et al.*, 2015), the important photochemical parameters are available (direct photolysis determined by the quantum yields and indirect photolysis involves reactive transients). Moreover, photolysis reactions via direct photolysis is expected to be an important degradation pathway, compared to other processes such as sorption or biological degradation, in the environmental fate of TCS(Bianco *et al.*, 2014). Direct photolysis is determined by the amount of the chemical when quantum yield, as well as photolysis absorption wavelength(Aranami and Readman, 2007; Dann and Hontela, 2011). Indirect photolysis, reactive transients such as reaction rate constants with ·OH, $CO_3^-$ and the triplet states of chromophoric dissolved organic matter, $^3$CDOM*(Bianco *et al.*, 2014). In this case, we can see the photolysis of CBZ and TCS happened quickly under the condition of laboratory. Due to the low initial concentrations of the reactants , the difference of the photolysis rates between CBZ and TCS is extremely small, so we think that both of them were mainly indirect photolysis.

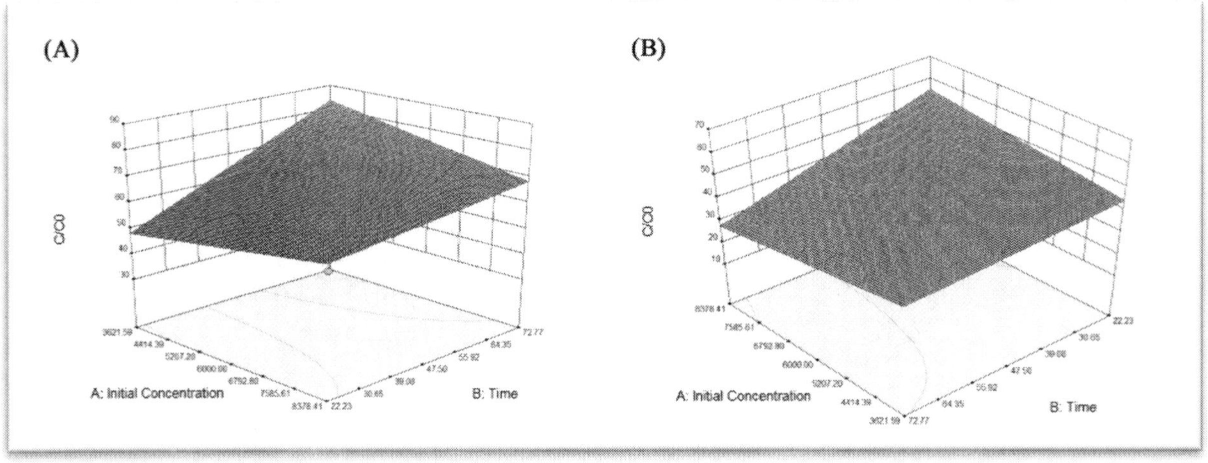


**Fig. 6.** Photolysis of CBZ (A) and TCS (B), as a function of time and a ratio about residual concentration (C) to initial concentration (C0). Other conditions: 22.00℃ temperature, 7.00 pH, 4.50mg/L DO, 8.01 Salinity, 15.01mg/L Humic acid.

From Fig. 7, we can see that all parameters had effect on photolysis of TCS. For molecules of acid and alkali solution from the group of antibiotics, pH can significantly affect the photochemical reaction(De Laurentiis *et al.*, 2014), TCS degraded more easily under the condition of alkaline because this condition promoted the reaction between humic acid and TCS. DO and humic acid provided active oxygen reactants, however, because of their photoionization, competitive relationship between them and TCS which can inhibit the photolysis of TCS. Salinity had important influence to the photolysis. When the low concentration of salinity could promote the photosensitivity of TCS. However, when the salinity was increased to a certain extent, the chloride ions released that would be inhibit the indirect photolysis of TCS. Worth noting is that temperature showed a negative correlation with the photolysis of TCS, raising the temperature is not good for photolysis of TCS in this study, and the reason is not clear now.

The photolysis also conforms to the first-order degradation equation as Eq. (1), but there are too many factors affecting photolysis rate. In order to simplify the photolysis rate of various factors of influence, we used the Design-Expert software to process the data and use the Quadratic equation for fitting to get the Eq. (4) for CBZ and Eq. (5) for TCS:

$$C_{CBZ}=-5641.03+0.439686A+0.266565B-31.1352C+1603.668D+99.64726E-6.31369F+5.263092G-6.657635312H-0.00029848AB+0.002879818AC-0.063961331AD+0.004627611AE+0.001724795AF+0.001317929AG+9.02243E-07AH+0.122107505BC-1.720424763BD+0.304967623BE+0.078051509BF-0.286997603BG+0.008100496BH+0.16302462CD+0.89250154CE+0.125101639CF+0.684069113CG-0.002023905CH-26.32559608DE+1.567497878DF-2.644672393DG+0.428268019DH+0.291545646EF+0.755724953EG+0.100098074EH+0.199106181FG+0.025159551FH+0.037131861GH-4.76803E-06A^2+0.138897604B^2+0.062492738C^2-85.53735285D^2-1.837417777E^2-1.737470199F^2-0.286250728G^2+0.002851302H^2 \quad (4)$$

$$C_{TCS}=4.38+0.65A-2.03B+3.37C+0.17D-1.65E+3.03F+2.97G-1.80H-0.84AB+2.90AC-2.27AD-3.18AE-2.95AF-0.16AG+4.28AH-9.26BC-5.95BD-0.20BE+0.34BF-2.26BG-4.17BH+3.94CD+6.55CE+6.9C2F+7.13CG-3.97CH-4.60DE+2.12DF+1.42DG+7.49DH+4.34EF+0.19EG+3.60EH+1.09FG+10.30FH+9.92GH-0.73A^2+0.17B^2-2.04C^2-0.65D^2+0.27E^2+1.20F^2-2.60G^2+11.05H^2-2.16ACE-6.62ACF-7.14ACH-6.61ADE-6.89ADF-0.48ADH-1.23AEG+1.72AFG-3.08AGH-6.54BCE-1.19BCF+3.12BCH+5.82BDE+2.64BDF-8.95BDH+5.49BEG-0.25CDE+2.15CDF-1.70CDH+6.23CEF+4.04CEG \quad (5)$$

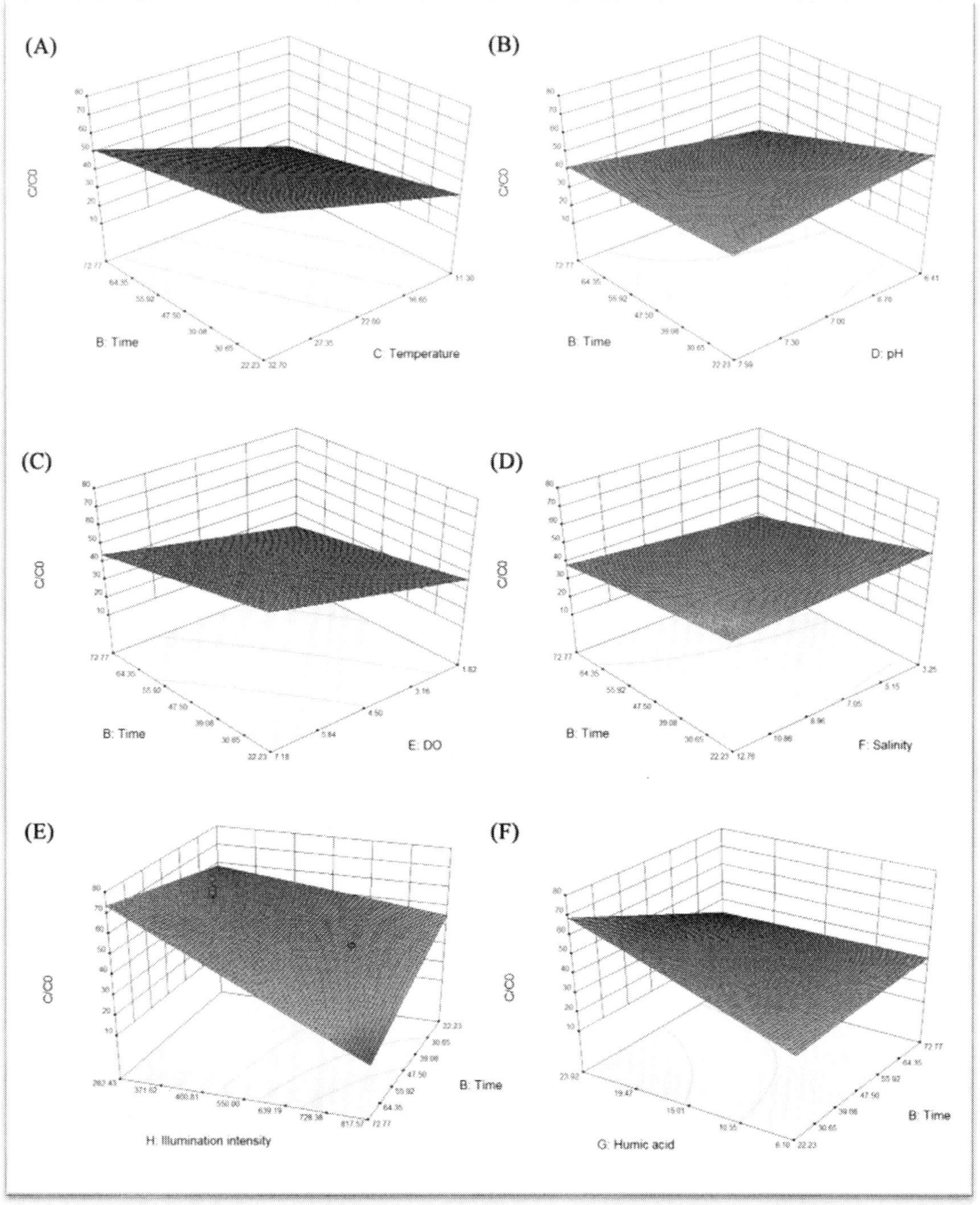


**Fig. 7.** The parameters effect on photolysis of TCS, as a function of time and a ratio about residual concentration (C) to initial concentration (C0). Other conditions: 22.00℃ temperature, 7.00 pH, 4.50mg/L DO, 8.01 Salinity, 15.01mg/L Humic acid.

A, B, C, D, E, F, G, H means the initial concentrations of the pollutions(mg/L), the reaction time (min), water temperature (℃), pH, DO (mg/L), salinity (‰), humic acid (mg/L), light intensity (Lux), respectively.

These equations of the F values were 2.99 and 4.38, the p values were 0.003 and 0.002, it shows that the model is significant. Then, we use the corresponding data with samplings in the above equations, get the corresponding $C_{CBZ}$ and $C_{TCS}$. Finally, the photolysis rate constant had been solved with Eq. (1), the $k_p$ is the first order photolysis rate constant, the $k_p$ of CBZ is 0.1081 $d^{-1}$, and the $k_p$ of TCS is 5.9322E-2 $d^{-1}$.

**Modeling the Targets in Surface Waters.** EFDC model was calibrated with sequential trials by modifying hydrodynamic and water quality parameters. Water balance showed good agreement between simulation and daily observation at the Shahe Stream. This stream has special terrain and the boundary condition which is simplified to its hydrodynamic conditions, so we set the Shahe Stream as one-dimensional unsteady flow. The highest water level and lowest water level were initialized at 5.80 m and 8.96 m and steadily fell to 5.36 m and 11.2 m.

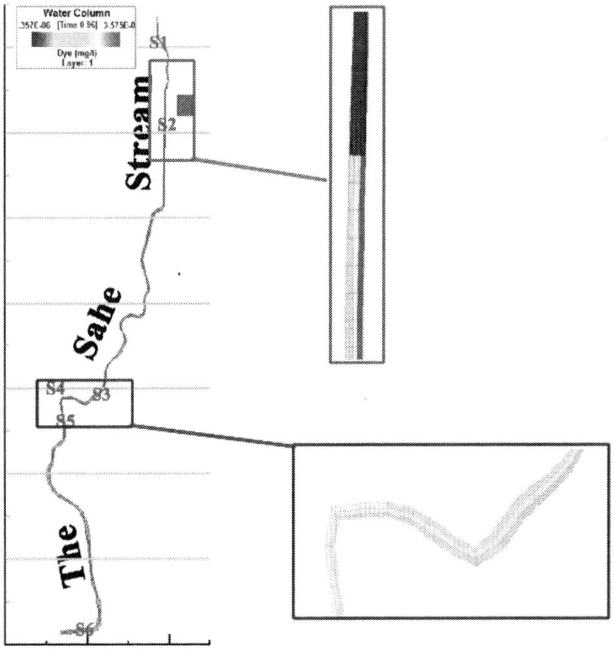


**Fig. 8.** CBZ in simulation by EFDC

Due to the bottom of the river are the rocks, the study ignored the adsorption and desorption. And the hydrolysis of CBZ and TCS were relatively weak, so the degradation of equation, Eq. (6) is:

$$C_t = C_0 \cdot e^{-kt} = C_0 \cdot e^{-k_b t} + C_0 \cdot e^{-k_p t} \quad (6)$$

The degradation rate constant k was 0.6324 $d^{-1}$ for CBZ and 0.6560 $d^{-1}$ for TCS. Using the Dye module by EFDC, CBZ and TCS were simulated respectively. The time of simulation is the same day at sampling. The relative deviation (RD) of water level simulation was calculated at 23%-52% of TCS, 23%-37% of CBZ indicating good fittings.

## CONCLUSIONS

CBZ and TCS undergo hydrolysis, biodegradation and photolysis in surface water. Under the condition of the same nature and the low concentrations of pollutants, degradation of CBZ and TCS are given priority to photolysis since hydrolysis of both types is relatively weak.

To our knowledge, this is the first report of modeling of CBZ and TCS with the degradation mechanism in the surface water. It demonstrates that PPCPs can be modeled and predicted in natural environment. However, the adsorption and desorption of PPCPs in water-sediment has not been considered because special at the bottom of the Shahe Stream, and failing to consider more water quality parameters. Moreover, the degradation rate of CBZ and TCS in EFDC model was globally set as constant, which is unreasonable as degradation rate is a dynamic coefficient varied across the Shahe Stream. So degradation rate value set as a constant maybe suitable in the Shahe Stream only. The dynamic degradation rate coefficient and the adsorption and desorption of PPCPs in water-sediment should be further investigated around the Shahe Stream in the next step and more studies should be conducted.

## ACKNOWLEDGMENTS

This study was financially supported by "Monitoring, prediction and control technology of Pharmaceutical and personal care products（PPCPs）in the surface water and drinking water system" funded by the major cultivation program of Sun Yat-sen university. Authors thank Jia Huang, Hanjie Yang and Dehong Liu for their help during sampling collection and modeling.

## REFERENCES

Aranami, K. and J.W. Readman, 2007. Photolytic degradation of triclosan in freshwater and seawater. Chemosphere, 66(6): 1052-1056. Available from http://www.ncbi.nlm.nih.gov/pubmed/16930676. DOI 10.1016/j.chemosphere.2006.07.010.

Batchu, S.R., V.R. Panditi, K.E. O'Shea and P.R. Gardinali, 2013. Photodegradation of antibiotics under simulated solar radiation: Implications for their environmental fate. Science of the Total Environment, 470-471C: 299-310. Available from http://www.ncbi.nlm.nih.gov/pubmed/24144935. DOI 10.1016/j.scitotenv.2013.09.057.

Bedoux, G., B. Roig, O. Thomas, V. Dupont and B. Le Bot, 2012. Occurrence and toxicity of antimicrobial triclosan and by-products in the environment. Environmental Science and Pollution Research, 19(4): 1044-1065. Available from http://www.ncbi.nlm.nih.gov/pubmed/22057832. DOI 10.1007/s11356-011-0632-z.

Bianco, A., D. Fabbri, M. Minella, M. Brigante, G. Mailhot, V. Maurino, C. Minero and D. Vione, 2014. New insights into the environmental photochemistry of 5-chloro-2-(2,4-dichlorophenoxy)phenol (triclosan): Reconsidering the importance of indirect photoreactions. Water research. Available from http://www.ncbi.nlm.nih.gov/pubmed/25179274. DOI 10.1016/j.watres.2014.07.036.

Boyd, G.R., H. Reemtsma, D.A. Grimm and S. Mitra, 2003. Pharmaceuticals and personal care products (ppcps) in surface and treated waters of louisiana, USA and ontario, canada. Science of the Total Environment, 311(1-3): 135-149. DOI 10.1016/s0048-9697(03)00138-4.

Carmona, E., V. Andreu and Y. Pico, 2014. Occurrence of acidic pharmaceuticals and personal care products in turia river basin: From waste to drinking water. Science of the Total Environment, 484: 53-63. Available from http://www.ncbi.nlm.nih.gov/pubmed/24686145. DOI 10.1016/j.scitotenv.2014.02.085.

Chen, Y.S., S. Yu, Y.W. Hong, Q.Y. Lin and H.B. Li, 2013. Pharmaceutical residues in tidal surface sediments of three rivers in southeastern china at detectable and measurable levels. Environmental Science and Pollution Research, 20(12): 8391-8403. Available from http://www.ncbi.nlm.nih.gov/pubmed/23764981. DOI 10.1007/s11356-013-1871-y.

Clara, M., B. Strenn and N. Kreuzinger, 2004. Carbamazepine as a possible anthropogenic marker in the aquatic environment: Investigations on the behaviour of carbamazepine in wastewater treatment and during groundwater infiltration. Water research, 38(4): 947-954. Available from http://www.ncbi.nlm.nih.gov/pubmed/14769414. DOI 10.1016/j.watres.2003.10.058.

Daghrir, R., A. Dimboukou-Mpira, B. Seyhi and P. Drogui, 2014. Photosonochemical degradation of butyl-paraben: Optimization, toxicity and kinetic studies. Science of the Total Environment, 490: 223-234.

Available from http://www.ncbi.nlm.nih.gov/pubmed/24858220. DOI 10.1016/j.scitotenv.2014.05.006.

Dann, A.B. and A. Hontela, 2011. Triclosan: Environmental exposure, toxicity and mechanisms of action. Journal of Applied Toxicology, 31(4): 285-311. Available from http://www.ncbi.nlm.nih.gov/pubmed/21462230. DOI 10.1002/jat.1660.

Daughton, C.G. and T.A. Ternes, 1999. Pharmaceuticals and personal care products in the environment: Agents of subtle change? Environmental Health Perspectives, 109(Suppl 6): 907-938.

De Laurentiis, E., C. Prasse, T.A. Ternes, M. Minella, V. Maurino, C. Minero, M. Sarakha, M. Brigante and D. Vione, 2014. Assessing the photochemical transformation pathways of acetaminophen relevant to surface waters: Transformation kinetics, intermediates, and modelling. Water research, 53: 235-248. Available from http://www.ncbi.nlm.nih.gov/pubmed/24525071. DOI 10.1016/j.watres.2014.01.016.

Deo, R.P., 2014. Pharmaceuticals in the surface water of the USA: A review. Current Environmental Health Reports, 1(2): 113-122. DOI 10.1007/s40572-014-0015-y.

Durán-Álvarez, J.C., B. Prado, D. González, Y. Sánchez and B. Jiménez-Cisneros, 2015. Environmental fate of naproxen, carbamazepine and triclosan in wastewater, surface water and wastewater irrigated soil — results of laboratory scale experiments. Science of The Total Environment, 538: 350-362. Available from http://www.sciencedirect.com/science/article/pii/S0048969715305313. DOI http://dx.doi.org/10.1016/j.scitotenv.2015.08.028.

Fabbri, D., M. Minella, V. Maurino, C. Minero and D. Vione, 2015. Photochemical transformation of phenylurea herbicides in surface waters: A model assessment of persistence, and implications for the possible generation of hazardous intermediates. Chemosphere, 119(0): 601-607. Available from http://www.sciencedirect.com/science/article/pii/S0045653514008959. DOI http://dx.doi.org/10.1016/j.chemosphere.2014.07.034.

Jeong, S., K. Yeon, Y. Hur and K. Oh, 2010. Salinity intrusion characteristics analysis using efdc model in the downstream of geum river. Journal of Environmental Sciences, 22(6): 934-939. Available from http://www.sciencedirect.com/science/article/pii/S1001074209602011. DOI http://dx.doi.org/10.1016/S1001-0742(09)60201-1.

Kimura, K., Y. Kameda, H. Yamamoto, N. Nakada, I. Tamura, M. Miyazaki and S. Masunaga, 2014. Occurrence of preservatives and antimicrobials in japanese rivers. Chemosphere, 107: 393-399. Available from http://www.ncbi.nlm.nih.gov/pubmed/24556546. DOI 10.1016/j.chemosphere.2014.01.008.

Kosma, C.I., D.A. Lambropoulou and T.A. Albanis, 2014. Investigation of ppcps in wastewater treatment plants in greece: Occurrence, removal and environmental risk assessment. Science of the Total Environment, 466-467: 421-438. Available from http://www.ncbi.nlm.nih.gov/pubmed/23933429. DOI 10.1016/j.scitotenv.2013.07.044.

Kuroda, K., M. Murakami, K. Oguma, Y. Muramatsu, H. Takada and S. Takizawa, 2012. Assessment of groundwater pollution in tokyo using ppcps as sewage markers. Environmental science & technology, 46(3): 1455-1464. Available from http://dx.doi.org/10.1021/es202059g. DOI 10.1021/es202059g.

Luo, Y., L. Xu, M. Rysz, Y. Wang, H. Zhang and P.J. Alvarez, 2011. Occurrence and transport of tetracycline, sulfonamide, quinolone, and macrolide antibiotics in the haihe river basin, china. Environmental science & technology, 45(5): 1827-1833. Available from http://www.ncbi.nlm.nih.gov/pubmed/21309601. DOI 10.1021/es104009s.

Ma, D., G. Liu, W. Lv, K. Yao, X. Zhang and H. Xiao, 2014. Photodegradation of naproxen in water under simulated solar radiation: Mechanism, kinetics, and toxicity variation. Environmental Science and Pollution Research, 21(13): 7797-7804. Available from http://www.ncbi.nlm.nih.gov/pubmed/24638836. DOI 10.1007/s11356-014-2721-2.

Mohapatra, D.P., S.K. Brar, R.D. Tyagi, P. Picard and R.Y. Surampalli, 2014. Analysis and advanced oxidation treatment of a persistent pharmaceutical compound in wastewater and wastewater sludge-carbamazepine. Science of the Total Environment, 470-471: 58-75. Available from http://www.ncbi.nlm.nih.gov/pubmed/24140682. DOI 10.1016/j.scitotenv.2013.09.034.

Onesios, K.M., J.T. Yu and E.J. Bouwer, 2009. Biodegradation and removal of pharmaceuticals and personal care products in treatment systems: A review. Biodegradation, 20(4): 441-466. Available from http://www.ncbi.nlm.nih.gov/pubmed/19112598. DOI 10.1007/s10532-008-9237-8.

Ortiz de Garcia, S., G. Pinto Pinto, P. Garcia Encina and R. Irusta Mata, 2013. Consumption and occurrence of pharmaceutical and personal care products in the aquatic environment in spain. Science of the Total Environment, 444: 451-465. Available from http://www.ncbi.nlm.nih.gov/pubmed/23287535. DOI 10.1016/j.scitotenv.2012.11.057.

Pal, A., K.Y. Gin, A.Y. Lin and M. Reinhard, 2010. Impacts of emerging organic contaminants on freshwater resources: Review of recent occurrences, sources, fate and effects. Science of the Total Environment, 408(24): 6062-6069. Available from http://www.ncbi.nlm.nih.gov/pubmed/20934204. DOI 10.1016/j.scitotenv.2010.09.026.

Peng, X., C. Wang, K. Zhang, Z. Wang, Q. Huang, Y. Yu and W. Ou, 2014. Profile and behavior of antiviral drugs in aquatic environments of the pearl river delta, china. Science of the Total Environment, 466-467: 755-761. Available from http://www.ncbi.nlm.nih.gov/pubmed/23973541. DOI 10.1016/j.scitotenv.2013.07.062.

Peng, X., Y. Yu, C. Tang, J. Tan, Q. Huang and Z. Wang, 2008. Occurrence of steroid estrogens, endocrine-disrupting phenols, and acid pharmaceutical residues in urban riverine water of the pearl river delta, south china. Science of the Total Environment, 397(1-3): 158-166. Available from http://www.ncbi.nlm.nih.gov/pubmed/18407320. DOI 10.1016/j.scitotenv.2008.02.059.

Petrovic, M., B. Skrbic, J. Zivancev, L. Ferrando-Climent and D. Barcelo, 2014. Determination of 81 pharmaceutical drugs by high performance liquid chromatography coupled to mass spectrometry with hybrid triple quadrupole-linear ion trap in different types of water in serbia. Science of the Total Environment, 468-469: 415-428. Available from http://www.ncbi.nlm.nih.gov/pubmed/24055661. DOI 10.1016/j.scitotenv.2013.08.079.

Richardson, B.J., P.K. Lam and M. Martin, 2005. Emerging chemicals of concern: Pharmaceuticals and personal care products (ppcps) in asia, with particular reference to southern china. Marine pollution bulletin, 50(9): 913-920. Available from http://www.ncbi.nlm.nih.gov/pubmed/16038943. DOI 10.1016/j.marpolbul.2005.06.034.

Stackelberg, P.E., J. Gibs, E.T. Furlong, M.T. Meyer, S.D. Zaugg and R.L. Lippincott, 2007. Efficiency of conventional drinking-water-treatment processes in removal of pharmaceuticals and other organic compounds. Science of the Total Environment, 377(2-3): 255-272. Available from http://www.ncbi.nlm.nih.gov/pubmed/17363035. DOI 10.1016/j.scitotenv.2007.01.095.

Sucsy, P.V. and F.W. Morris, 2001. Salinity intrusion in the st. Johns river, florida. ML Spaulding (ed.), ASCE, pp: 120-139.

Tixier, C., H.P. Singer, S. Oellers and S.R. Müller, 2003. Occurrence and fate of carbamazepine, clofibric acid, diclofenac, ibuprofen, ketoprofen, and naproxen in surface waters. Environmental science & technology, 37(6): 1061-1068.

Wang, Z., X.-H. Zhang, Y. Huang and H. Wang, 2015. Comprehensive evaluation of pharmaceuticals and personal care products (ppcps) in typical highly urbanized regions across china. Environmental pollution, 204: 223-232. Available from http://www.sciencedirect.com/science/article/pii/S0269749115002109. DOI http://dx.doi.org/10.1016/j.envpol.2015.04.021.

Wu, C., X. Huang, J.D. Witter, A.L. Spongberg, K. Wang, D. Wang and J. Liu, 2014. Occurrence of pharmaceuticals and personal care products and associated environmental risks in the central and lower yangtze river, china. Ecotoxicology and environmental safety, 106: 19-26. Available from http://www.ncbi.nlm.nih.gov/pubmed/24836873. DOI 10.1016/j.ecoenv.2014.04.029.

Wu, G. and Z. Xu, 2011. Prediction of algal blooming using efdc model: Case study in the daoxiang lake. Ecological Modelling, 222(6): 1245-1252. Available from http://www.sciencedirect.com/science/article/pii/S0304380011000020. DOI http://dx.doi.org/10.1016/j.ecolmodel.2010.12.021.

Wu, J.-L., N.P. Lam, D. Martens, A. Kettrup and Z. Cai, 2007. Triclosan determination in water related to wastewater treatment. Talanta, 72(5): 1650-1654. Available from http://www.sciencedirect.com/science/article/pii/S0039914007002135. DOI http://dx.doi.org/10.1016/j.talanta.2007.03.024.

Xu, W.H., G. Zhang, S.C. Zou, X.D. Li and Y.C. Liu, 2007. Determination of selected antibiotics in the victoria harbour and the pearl river, south china using high-performance liquid chromatography-electrospray ionization tandem mass spectrometry. Environmental pollution, 145(3): 672-679. Available from http://www.ncbi.nlm.nih.gov/pubmed/16996177. DOI 10.1016/j.envpol.2006.05.038.

Yang, X., F. Chen, F. Meng, Y. Xie, H. Chen, K. Young, W. Luo, T. Ye and W. Fu, 2013. Occurrence and fate of ppcps and correlations with water quality parameters in urban riverine waters of the pearl river delta, south china. Environmental Science and Pollution Research, 20(8): 5864-5875. Available from http://www.ncbi.nlm.nih.gov/pubmed/23608973. DOI 10.1007/s11356-013-1641-x.

Yang, X., R.C. Flowers, H.S. Weinberg and P.C. Singer, 2011. Occurrence and removal of pharmaceuticals and personal care products (ppcps) in an advanced wastewater reclamation plant. Water research, 45(16): 5218-5228. Available from http://www.ncbi.nlm.nih.gov/pubmed/21864879. DOI 10.1016/j.watres.2011.07.026.

Yoon, Y., J. Ryu, J. Oh, B.G. Choi and S.A. Snyder, 2010. Occurrence of endocrine disrupting compounds, pharmaceuticals, and personal care products in the han river (seoul, south korea). Science of the Total Environment, 408(3): 636-643. Available from http://www.ncbi.nlm.nih.gov/pubmed/19900699. DOI 10.1016/j.scitotenv.2009.10.049.

Zhang, Y., S.-U. Geißen and C. Gal, 2008. Carbamazepine and diclofenac: Removal in wastewater treatment plants and occurrence in water bodies. Chemosphere, 73(8): 1151-1161. Available from http://www.sciencedirect.com/science/article/pii/S004565350800996X. DOI http://dx.doi.org/10.1016/j.chemosphere.2008.07.086.

Zhao, J.L., G.G. Ying, Y.S. Liu, F. Chen, J.F. Yang, L. Wang, X.B. Yang, J.L. Stauber and M.S. Warne, 2010. Occurrence and a screening-level risk assessment of human pharmaceuticals in the pearl river system, south china. Environmental Toxicology and Chemistry, 29(6): 1377-1384. Available from http://www.ncbi.nlm.nih.gov/pubmed/20821582. DOI 10.1002/etc.161.

# SPACE/TIME GEOSTATISTICAL ESTIMATION OF CHLORIDE ALONG MARYLAND RIVERS USING A COVARIANCE MODEL WITH RIVER DISTANCES

***Prahlad Jat***, Marc L Serre
(University of North Carolina at Chapel Hill, NC, USA)

Chloride, which is widely used for road deicing, is an emerging pollutant that is causing substantial increases in the salinity of the Nation's ground and surface waters and is threatening our aquatic environment. In this work we use chloride monitoring data collected in Maryland to estimate its concentration along all river miles of three sub basins (Gunpowder-Patapsco, Severn and Patuxent) from 2005 to 2014.

A previous study on Dissolved Oxygen Carbon (DOC) using the Maryland Biological Stream Survey (MBSS) data collected in Maryland in 1996 found that DOC geostatistical estimates obtained using a covariance model based on Euclidean distances are more accurate that estimates obtained using a flow-weighted (weighted asymmetric hydrologic distance) covariance model. However, that study did not report results for a covariance model based on river distances (distances measured along river network), unlike several subsequent studies which proposed and used covariance models based on river distances to study water quality along river networks.

We queried the MBSS database for all chloride measurements from 2005 to 2014 in our study area, which resulted in a dataset of 390 space/time chloride values. Using this dataset we performed a geostatistical analysis using a covariance with an Euclidean or river metric, and we compared their accuracy based on a leave one out cross validation. We found that the $R^2$ between measured and estimated values was 0.72 when using the river distance, compared to 0.65 when using an Euclidean distance. This work is the first to demonstrate that the river distance covariance model is better than the Euclidean distance covariance model for the geostatistical estimation of chloride in stream waters of three sub basins in Maryland. This finding indicates that processes that are distributed along river networks such as highways (a known source of chloride), and vegetation buffers (a known attenuation factor), are important drivers of the distribution of chloride along rivers. The implication of this finding are that (1) the river covariance model should be used in place of the Euclidean covariance model whenever modeling chloride concentration in these sub basins, and (2) the areas of impaired water identified using the river covariance model are confined along the river branches, which are easier to target for remedial purposes than areas spreading across multiple parallel branches.

A limitation of this work is that the river distance was only used for the covariance model. Our results suggest that the use of river distances should be expanded to other aspect of space/time modeling, such as mean trend modeling or spatial regression.

# MODELING OF SPREADING OF SUBMARINE MINE TAILINGS IN A NORWEGIAN FJORD

Øyvind Leikvin[1], ***Venkat Kolluru***[2], Guttorm N. Christensen[1], Magdalena Kempa[3], Torulv Tjomsland[3]
([1]Akvaplan-niva AS, Tromsø, Norway; [2]Environmental Resources Management, Inc., Philadelphia, USA;
[3]Norwegian Institute of Water Research (NIVA), Oslo, Norway)

Four different submarine disposal sites in Repparfjorden situated by the town of Kvalsund, Norway, were assessed for the suitability as dumping ground for the disposal of fine tailings from the beneficiation of ores in Wolf Ridge and Nussir, operated by a confidential client. The area northeast of Fægfjordholmen is the only one that has a large enough volume of water with low bottom currents that was found appropriate, and thus impacts were assessed through consequence modeling for compliance with the Norwegian and OSPAR environmental regulations. Simulations of the spread of tailings were made with a 3-D hydrodynamic, water quality and sediment transport model called GEMSS. The model was calibrated to available ADCP measurements. The model was then used to evaluate the tailings depositional footprint at the bottom of the fjord and total suspended sediment concentrations (TSS) in the water column. The calibrated model was used to run many different scenarios to find the best possible location for the release of tailings through an outfall pipe. It was found that it was necessary to have a relatively deep discharge point (> 50 meters) to the north of Markoppneset, in order to get the tailings to be deposited within the prescribed fjord bed area. An area of 2.4 $km^2$ got more than 10 cm footprint thickness, while an area of 7.4 $km^2$ got more than 1 mm thickness. The flocculated tailings deposited mainly on the bottom, while only 2% were suspended in the water column. Though the near-field turbidity due to TSS was very high, the far-field region showed a small increase of turbidity in the water column that is comparable to background values measured in the fjord. The copper content in the tailings was designated with chemical status class V (very bad). The pore water concentrations in the modeled tailings footprint were similarly high in the same chemical status class, as calculated by the equilibrium constants. Ecotoxicological tests showed that the tailings had potentially toxic effects on marine organisms, but it was found a NOEC (no-observable-effect-concentration) of 32% tailings. The natural sedimentation in the fjord will eventually dilute the tailings within a few years after cessation of the deposit.

# GIS, DATABASE, AND REMOTE SENSING APPLICATIONS

# ESTIMATION OF CROP COEFFICIENTS USING REMOTE SENSING AND GIS: A CASE STUDY OF KHUSKERA-BHIWARI NEEMRANA INVESTMENT REGION, RAJASTHAN, INDIA

***Rohit Goyal***, Kuldeep Tiwari, and Priyamitra Munoth
(Malaviya National Institute Of Technology Jaipur, Jaipur, Rajasthan, India)

**ABSTRACT:** Satellite data based estimation of crop coefficient ($K_c$) is one of the widely accepted methods for irrigation water management for large and/or inaccessible areas. Similarities between the crop coefficient curve and a satellite-derived vegetation index showed potential for modelling the crop coefficient as a function of the vegetation index. Therefore, in this paper the possibility of directly estimating the crop coefficient from moderate resolution satellite reflectance of crops of Rabi (winter) season has been investigated. Two different approaches, single ($K_{c\_single}$) and dual ($K_{c\_dual}$) crop coefficients have been used and are compared with standard curves from FAO. In present work, NDVI data from MODIS (MOD13Q1) sensor on-board Aqua/Terra satellite has been used to estimate both single and dual crop coefficient at the cell level for agriculture area of the study area. This study is conducted using data of Rabi (winter) season for climate conditions in the Khuskera-Bhiwari Neemrana Investment Region, located in Rajasthan, India. Land-use in around 70% of the study area is agriculture. Major crops grown in this area for Kharif season are Beans (Gwoar) and Millet and for Rabi season Wheat and Mustard. $K_c$ values have been estimated for wheat and mustard crops at 16 days interval. Comparison of these two different approaches with standard curve shows that dual crop coefficient approach is more accurate for the estimation of $K_c$ values.

## INTRODUCTION

Knowledge of crop water requirements is an important practical consideration for designing and managing irrigation systems (Mohan and Arumugam 1994; Kashyap and Panda 2001; Dhotre *et al.,* 2015). The water requirement varies widely from crop to crop and also during the period of growth of individual crops. Thus estimation of crop water demand considering the cropping pattern is an area of research, which has attracted the attention of water resources planners and engineers. In India, agriculture accounts for about 85% of total annual groundwater draft. Also it fulfills about 60% irrigation and 80% drinking water requirements (Kushwaha and Goyal, 2016). The crop coefficient ($K_c$) represents crop specific water use and is essential for accurate estimation of irrigation requirement of different crops in the any area (Tyagi *et.al.,* 2000). Crop coefficients are properties of plants used in predicting evapotranspiration (ET). The most basic crop coefficient, Kc, is simply the ratio of ET observed for the crop studied over that observed for the well calibrated reference crop under the same conditions (Wikipedia, 2016). The crop coefficient can be calculated by using FAO-56 approach, which is based on the concepts of combining reference evapotranspiration $ET_0$ and crop coefficients ($K_c$). There are two methods used to estimate crop evapotranspiration from remote sensing data: the single and the dual crop coefficients. In the single crop coefficient ($K_{c\_single}$), the effect of crop transpiration and soil evaporation are combined into a single $K_c$ (Kamble *et al.*, 2013). The dual crop coefficient consists of two coefficients: a basal crop coefficient ($K_{cb}$) and a soil evaporation coefficient ($K_e$) (Er-Raki et al., 2007). The crop coefficient values have been calculated by both approaches in this paper. MODIS vegetation indices like Normalized Difference Vegetation Index (NDVI) and Enhanced Vegetation Index (EVI) data are widely used for vegetation and crop study (Sharma *et al.,* 2016). NDVI can be used as an indicative parameter for vegetation density (Singhal and Goyal, 2012). Several studies have been specifically dedicated for estimating $K_c$ from vegetation indices (VI). Er-Raki *et al.,* (2007) tested three methods based on the FAO-56 ''dual'' crop coefficient approach to estimate actual evapotranspiration (AET) for winter wheat under different irrigation treatments in the semi-arid region. The three methods differ in the calculation of the basal crop coefficient ($K_{cb}$) and the fraction of soil surface covered by vegetation ($f_c$). The first approach strictly follows the FAO-56 procedure, with $K_{cb}$ given in the FAO-56 tables and $f_c$ calculated from $K_{cb}$ (No- Calibration method). The second method uses local $K_{cb}$ and $f_c$ values estimated from field measurements (Local-Calibration method) and the last approach used a remotely-sensed vegetation index to estimate $K_{cb}$ and $f_c$ (NDVI-Calibration method). The results showed that there exists a good potential for modeling $K_{cb}$ from NDVI values. Rafn *et al.* (2008) estimated the evapotranspiration parameter $K_c$ or $K_{cb}$ using readily available satellite-derived NDVI values. Results from a case study in Idaho for irrigated agriculture indicated that the NDVI/$K_c$ method has significant potential to estimate $K_c$ because it is comparably fast, easy, and less costly to apply, and does not require images from the thermal band. Pakhale *et al.* (2010) calculated the irrigation water requirement of wheat crop for Rabi season from 1999 to 2003 in Karnal district of Haryana state, India. Area under wheat cultivation was determined using Landsat ETM+ image by applying Artificial Neural Network (ANN) classification technique. Potential Evapotranspiration and crop coefficient for wheat was used for estimating crop water requirement. Kamble *et al.*, (2013) developed a linear regression model between NDVI and the $K_{c\_single}$ from a moderate resolution satellite data (MODIS). They showed the correlation between the NDVI-estimated $K_{c\_single}$ and the measured $K_{c\_single}$ with an $R^2$ of 0.91 and 0.90, while the root-mean-square error (RMSE) for $K_{c\_single}$ were 0.16 and 0.19, respectively. Development of daily crop coefficients and their seasonal variability have not been studied for major crops under tropical climatic conditions. Thus, the present study was undertaken with the objective of deriving the crop coefficients for major crops using remote sensing data. The objective of this study was to investigate the applicability of time-series MODIS 250m NDVI data to the development of spatially representative $K_c$.

## STUDY AREA

The study area selected for present study is Khushkheda-Bhiwadi-Neemrana Investment Region (KBNIR), located in the Alwar district, Rajasthan state, India (Fig. 1). The study area is situated in North-East of Rajasthan between 27°54'33" N and 28°03'20" North latitudes and 76°24'06" E and 76°35'40" East longitudes covering a geographical area of about 162.4 km². The climate of the KBNIR area is semi-arid and very hot in summer and extremely cold in winter. The maximum temperature of the region has been 48°C and the minimum is up to freezing point. Average annual temperature is 26°C. The average annual rainfall is 588.7 mm. The major crops grown in this area in Rabi (winter) season are wheat and mustard and in Kharif (summer) season are barley and beans.

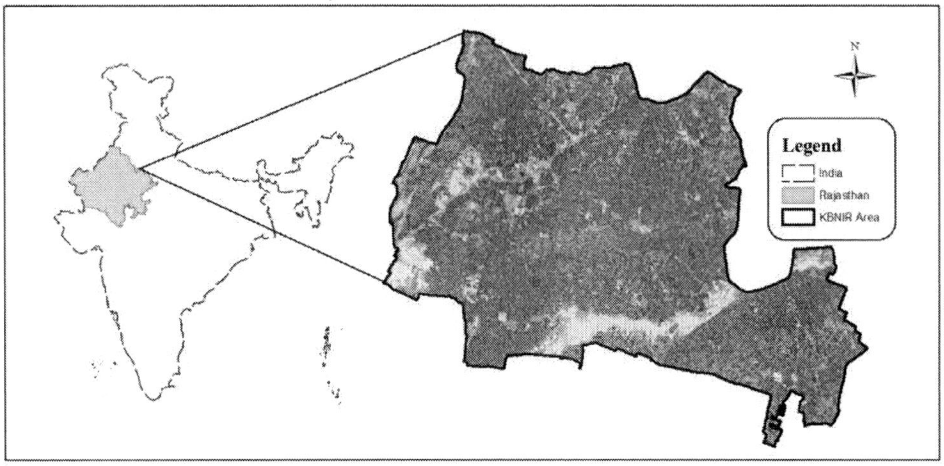


**FIGURE 1. Study area (KBNIR) map**

## METHODOLOGY

The data used for this study includes NDVI images of MOD13Q1 from MODIS sensor on-board Aqua/Terra satellite spanning from 16th Sept. 2014 to 23rd April 2015. The MODIS instrument offers possibilities for large-area crop mapping by providing a daily global coverage of high-quality, intermediate resolution (250 m) data. MOD13Q1 images were downloaded from the National Aeronautic and Space Administration (NASA) Land Process Distributed Active Archive Centre (LP DAAC, 2016) for the study area. The NDVI utilizes reflectance of the canopy in the near-infrared (NIR) and red (R) bands of the spectrum. The value of NDVI ranges theoretically from -1 to +1 but practically from approximately -0.2 to 0.7. This vegetation index is given by:

$$NDVI = \frac{NIR-R}{NIR+R} \quad (1)$$

Where NIR is the reflectance in near infrared band and R is the reflectance in the red band. In this study two modals developed by Er-Raki *et al.,* (2007), which gives the values of dual crop coefficient ($K_{c\_dual} = K_{cb} + K_e$) and Kamble *et al.,* (2013), which gives the value of single crop coefficient ($K_c$) are used. According to Er-Raki *et al.,* (2007) the main plant parameters needed to calculate the dual crop coefficient ($K_c$) is the values of basal crop coefficient ($K_{cb}$) at three crop growth stages (initial, mid-season and maturity respectively), the soil evaporation coefficient ($K_e$) and the height of vegetation ($f_c$). The detail equations of all these parameters are given below. The $K_{cb}$ can be defined as follows:

$$K_{cb} = 1.07 \times \left[1 - \left(\frac{NDVI_{max}-NDVI}{NDVI_{max}-NDVI_{min}}\right)^{\frac{0.84}{0.54}}\right] \quad (2)$$

Where $NDVI_{min}$ and $NDVI_{max}$ are the minimum and maximum values of the NDVI associated with bare soil and dense vegetation, respectively. The soil evaporation coefficient ($K_e$) and the height of vegetation ($f_c$) is given as follows:

$$K_e = 0.25 * (1 - f_c) \quad (3)$$

$$fc = 1.18 * (NDVI - NDVI_{min}) \quad (4)$$

The equation (3) shows that $K_e$ decreases with the development of crop (As ground cover fraction increase, $f_c$ will increase) and becomes negligible when the crop is well developed and completely covers the soil i.e. at ($f_c$) = 1

Kamble *et al.* (2012) developed a simple linear regression model to establish a general relationship between a NDVI from a moderate resolution satellite data (MODIS) and the single crop coefficient ($K_c$), which is given as:

$$K_{C\_single} = 1.457\, NDVI - 0.1725 \quad (5)$$

The $K_c$ values are calculated for wheat and mustard crops for Rabi (winter) season for particular date of the year based on the phenological cycle of the respective crops, when NDVI values were available.

## RESULTS AND DISCUSSIONS

MODIS data is re-projected from the sinusoidal (SIN) projection to the standard UTM projection and resampled to spatial resolution (23.5 m) with the available LULC map created by LISS-III sensors to identify agriculture and non-agriculture area. Before applying the FAO dual approach for each pixel, the images were masked to keep only wheat, mustard and other crops fields. Re-classification method was applied to the 16-days NDVI time series to produce the crop. As vegetation types can be characterized using their seasonal (or phonological) variations in the NDVI time series (Mingwei et al., 2008) so wheat and mustard crops were delineated by examining NDVI temporal profiles from the crop growing season. The crop calendar for Rabi (winter) season is given below in Fig. 2. The mustard crop is sown in September and harvested in March while the wheat crop is sown in October and harvested in April as shown in Fig. 2.

| Crop | Sep 14 | Oct 14 | Nov 14 | Dec 14 | Jan 15 | Feb 15 | Mar 15 | Apr 15 |
|---|---|---|---|---|---|---|---|---|
| Wheat | | | | | | | | |
| Mustard | | | | | | | | |

Harvest

**FIGURE 2. Crop calendar of Rabi (winter) season of India.**

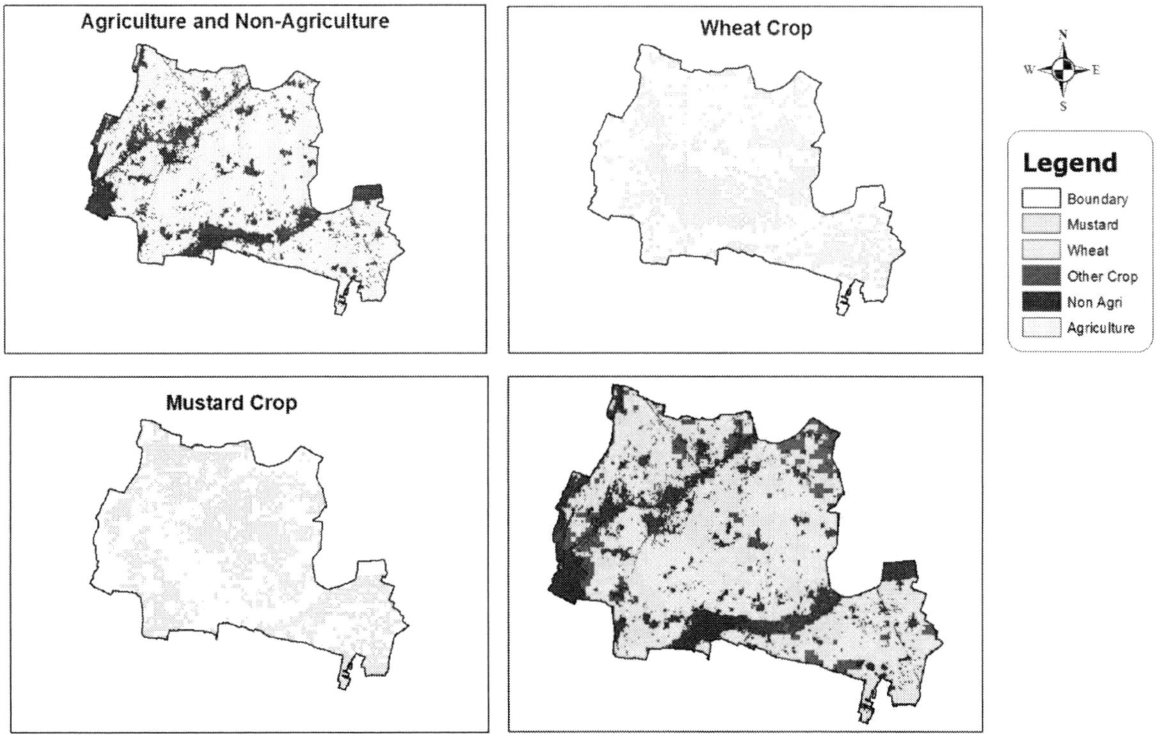


**FIGURE 3. Distribution of wheat and mustard in KBNIR area in 2014-15.**

A crop classification map was generated using GIS tools as shown in Fig. 3. The area identified for mustard crop was 55%, for wheat crop 39% and for other crops, it was 6%. There was little difference between the crop area estimated by the MODIS data and the statistics already available from Agriculture department, government of Rajasthan.

As mentioned above, both $K_c$ single and dual were derived by using equations 2 to 5. The date wise $K_c$ values of wheat and mustard are listed in table 1.

The crop coefficient ($K_c$) for a crop varies throughout the growing season, and its full value will not only depend on the crop development stage but also on the sowing date and climatic conditions. The variation in crop coefficient values for wheat crop at various growth stages estimated by different methods and standard value by FAO, are given in Fig. 4. There is a constantly increasing trend in $K_{c\_dual}$ and $K_{c\_single}$ during the crop development growth stage beginning from 16/10/14 to 18/02/15. The peak value of $K_{c\_dual}$ around 1.12 is reached in mid-February while 0.88 for $K_{c\_single}$ in the beginning of February. During the development stage good correlation were found between $K_{c\_dual}$ and $K_{c\_FAO}$ as shown in Fig. 4. The $K_{c\_dual}$ declined rapidly to 0.488 and $K_{c\_single}$ to 0.196 during the last crop growth stage covering the period from 28/02/15 to 14/04/15. It is observed that $K_{c\_dual}$ values are in general 0.35 more than $K_{c\_single}$ for wheat crop.

The estimated $K_{c\_FAO}$ values of wheat by the FAO method during initial, crop development and reproductive stages are 0.3, 1.15 and 0.3.

**TABLE 1: NDVI and $K_c$ values for wheat and mustard crops**

| | | Wheat Crop | | | Mustard Crop | | |
|---|---|---|---|---|---|---|---|
| Date | Day of Year | NDVI | $K_{c\_dual}$ | $K_{c\_single}$ | NDVI | $K_{c\_dual}$ | $K_{c\_single}$ |
| 16-10-2014 | 289 | 0.2537 | 0.6959 | 0.1971 | 0.3288 | 0.6982 | 0.3065 |
| 01-11-2014 | 305 | 0.2539 | 0.6962 | 0.1975 | 0.2443 | 0.6335 | 0.1835 |
| 17-11-2014 | 321 | 0.395 | 0.8443 | 0.403 | 0.2589 | 0.6678 | 0.2047 |
| 03-12-2014 | 337 | 0.4813 | 0.913 | 0.5287 | 0.4137 | 0.8782 | 0.4302 |
| 19-12-2014 | 353 | 0.5497 | 0.954 | 0.6284 | 0.4994 | 0.9344 | 0.5551 |
| 01-01-2015 | 1 | 0.6326 | 1.0453 | 0.7491 | 0.6027 | 1.0392 | 0.7057 |
| 17-01-2015 | 17 | 0.6516 | 1.0949 | 0.7769 | 0.6699 | 1.0869 | 0.8036 |
| 02-02-2015 | 33 | 0.726 | 1.1078 | 0.8852 | 0.7076 | 1.0933 | 0.8585 |
| 18-02-2015 | 49 | 0.6925 | 1.1258 | 0.8364 | 0.7281 | 1.0897 | 0.8884 |
| 06-03-2015 | 65 | 0.6616 | 1.0732 | 0.7915 | 0.6632 | 1.0703 | 0.7937 |
| 22-03-2015 | 81 | 0.461 | 0.7361 | 0.4992 | 0.6329 | 1.059 | 0.7497 |
| 07-04-2015 | 97 | 0.2532 | 0.4882 | 0.1964 | Crop Harvested | | |
| 23-04-2015 | 113 | 0.2128 | 0.5884 | 0.1375 | | | |

Similarly, the variation in crop coefficient values for mustard crop at various growth stages can be seen in Fig. 5. The $K_{c\_dual}$ and $K_{c\_single}$ values increase during the crop development growth stage beginning from October to starting of February. The estimated $K_{c\_FAO}$ values of mustard by the FAO method during initial, crop development and reproductive stages are 0.25, 1.28 and 0.3. The average difference in Kc_dual and Kc_single for mustard crop has also been observed to be approximately 0.35. The Kc value is modified by variation in climate, mean daily wind speeds and variable crop height.

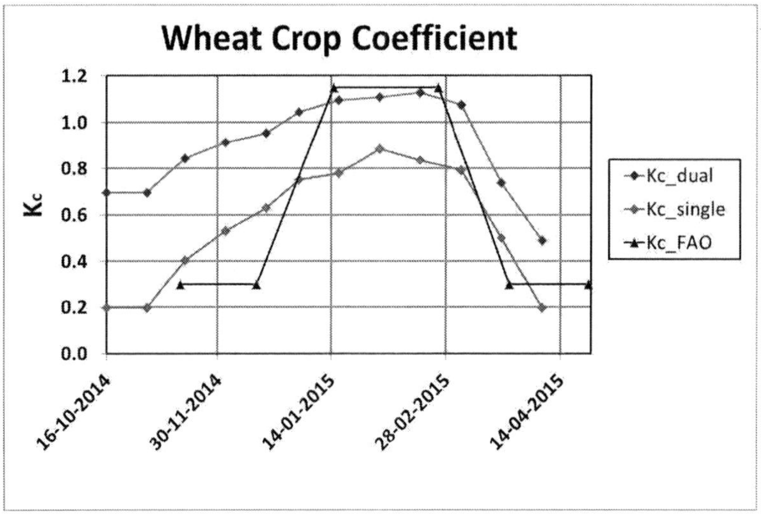


**FIGURE 4. Kc curves for wheat crop**

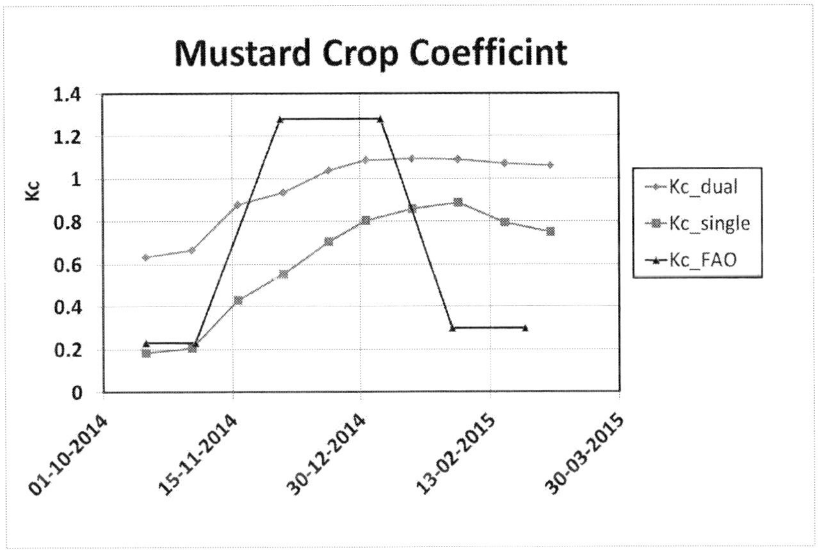


**FIGURE 5. $K_c$ curves for mustard crop**

## CONCLUSIONS

This study has demonstrated that time-series MODIS 250 m NDVI data provide a viable option for regional-scale crop mapping, however small error in estimation may be due to the mixed pixels at the 250 m spatial scale level. The present study shows that Remote Sensing and GIS integrated approach can be used for the estimation of crop coefficients for different crops. For the study area, the $K_{c\_dual}$ value varies from 0.6959 to 0.5884 for wheat crop and from 0.6982 to 1.059 for mustard crop while the $K_{c\_single}$ value varies from 0.1971 to 0.1375 for wheat crop and 0.3065 to 0.7497 for mustard crop. It is observed that in general $K_{c\_dual}$ values are 0.35 more than $K_{c\_single}$ values. There is good correlation between $K_{c\_dual}$ and $K_{c\_FAO}$ for both the crops hence it can be concluded from the study that dual crop coefficient approach is more accurate than the single crop coefficient approach for the estimation of $K_c$ values. The local climatic parameter and soil characterises also effects the variation in $K_c$ values.

## REFERENCES

Chen, J. M., Chen, X., Ju, W. and Geng, X. 2005. "Distributed hydrological model for mapping evapotranspiration using remote sensing inputs". Journal of Hydrology. 305:15–39. http://doi.org/10.1016/j.jhydrol.2004.08.029

Tyagi, N. K., Sharma D. K. and S. K. Luthra. 2000. "Evapotranspiration and Crop Coefficients of Wheat and Sorghum." Journal of Irrigation and Drainage Engineering. 126 (4): 215–22. doi:10.1061/(ASCE)0733-9437(2000)126:4(215).

Dhotre, ST, Sharma, G., Goyal, R. and Parikh, SK. 2015. "Soil Moisture Estimating Governing Models – Parameter Estimation- A Review," Int. J. of Emerging Tech. and Adv. Engg. 5(2): 165-172. www.ijetae.com

Er-Raki, S., Chehbouni, A., Guemouria, N., Duchemin, B., Ezzahar, J. and Hadria, R. 2007. "Combining FAO-56 Model and Ground-Based Remote Sensing to Estimate Water Consumptions of Wheat Crops in a Semi-Arid Region." Agricultural Water Management. 87 (1): 41–54. doi:10.1016/j.agwat.2006.02.004.

Er-Raki, S., Chehbouni, A. and Duchemin B. 2010. "Combining Satellite Remote Sensing Data with the FAO-56 Dual Approach for Water Use Mapping in Irrigated Wheat Fields of a Semi-Arid Region." Remote Sensing. 2 (1): 375–87. doi:10.3390/rs2010375.

Gontia, N. K., and Tiwari, K. N. 2010. "Estimation of crop coefficient and evapotranspiration of wheat Triticum aestivum in an irrigation command using remote sensing and GIS". Water Resources Management. 24(7): 1399–1414. http://doi.org/10.1007/s11269-009-9505-3

Kamble, B., Kilic, A., and Hubbard, K. 2013. "Estimating crop coefficients using remote sensing-based vegetation index". Remote Sensing. 5(4): 1588–1602. http://doi.org/10.3390/rs5041588

Kashyap, P. S., and Panda, R. K. 2001. "Evaluation of evapotranspiration estimation methods and development of crop-coefficients for potato crop in a sub-humid region". Agricultural Water Management. 50(1): 9–25. http://doi.org/10.1016/S0378-3774 (01 00102-0).

Kumar, R., Shankar, V., and Kumar, M. 2011. "Development of Crop Coefficients for Precise Estimation of Evapotranspiration for Mustard in Mid Hill Zone- India Abstract : Mid-season Late-season". Universal Journal of Environmental Research and Technology. 1(4): 531–538.

Kushwaha, K. and Goyal, R. 2016. "Methodology for the Estimation of Groundwater Flux across Simplified Boundary using GIS and Groundwater Levels", J. of Current Science. 110 (6): 1050-1058.

LP DAAC, USGS. 2016. "Vegetation Indices 16-Day L3 Global 250m" *URL: https://lpdaac.usgs.gov/dataset_discovery/modis/modis_products_table/mod13q1*

Mohan, S., and Arumugam N. 1994. "Crop Coefficients of Major Crops in South India." Agricultural Water Management. 26: 67–80. doi:10.1016/0378-3774(94 90025-6.

Pakhale, G., Gupta, P., and Nale, J. 2010. Crop and Irrigation Water Requirement Estimation by Remote Sensing and GIS : A Case A Case Study Of Karnal District, Haryana, India. International Journal of Engineering and Technology. 2(4): 207-211

Rafn, Eric B., Bryce Contor, and Daniel P. Ames. 2008. "Evaluation of a Method for Estimating Irrigated Crop-Evapotranspiration Coefficients from Remotely Sensed Data in Idaho." Journal of Irrigation and Drainage Engineering. 134 (6): 722–29.

Sharma G, Zaidi, S and Goyal, R. 2016. "Trend analysis of temporal variations in evi with respect to rainfall of jaipur district". International Journal of Research in Engineering and Technology, 05 01

Singhal, V and Goyal R. 2012. "A methodology based on spatial distribution of parameters for understanding affect of rainfall and vegetation density on groundwater recharge", European Journal of Sustainable Development.1 (2): 85

Wikipedia. 2016. "Crop coefficient", URL: https://en.wikipedia.org/wiki/Crop_coefficient, accessed in April 2016.

# GEOSPATIAL MAPPING OF GROUNDWATER QUALITY USING GIS

***Prachi Singh*** and ***Raj Mohan Singh***
(Department of Civil Engineering, MNNIT, Allahabad, India)

**ABSTRACT:** The groundwater is a highly useful and abundant natural resource. However, groundwater is vulnerable to pollution and depletion, and needs to be managed carefully to avoid serious economic and environmental costs. The contamination of aquifer not only threatens public health and the environment, it also involves large amounts of money in fines, lawsuits, and cleanup costs. Once groundwater is contaminated, it may be difficult and expensive to clean up. Therefore, it is imperative to know the spatial and temporal extent of the pollution so that necessary remedial steps may be taken. In this work the spatial distribution of groundwater quality has been developed for quality parameters such as pH, TDS, EC, Ca, Na and $NO_3$. These water quality parameters of 22 locations each varying in four seasons from 2005-2007 are shown in GIS framework. Different kriging interpolation techniques are used for spatiotemporal representation of water quality parameters in aerial study area.

## INTRODUCTION

Groundwater is the major source of irrigation and drinking water in both urban and rural India. Even in those areas where surface water is abundant, the groundwater is used as practical source of water for public supply, agriculture and industry due to its quality, easy accessibility, reliability and relatively low cost associated with its use. However, groundwater is vulnerable to pollution and depletion, and needs to be managed carefully to avoid serious economic and environmental costs. Contamination of groundwater poses serious threat to people.

The assessment of spatial correlation in hydrochemical variables is an important tool in the analysis of groundwater chemistry. In geostatistical methods, kriging technique is quite popular. The technique is applicable to cases such as determining groundwater level, estimation of hydrochemical distribution of soil properties, and other estimations. Kumar and Remadevi (2006) have compared various variogram models, i.e., spherical, exponential, and Gaussian semivariogram models to fit the experimental semivariograms in identifying the spatial analysis of groundwater levels in small part of Rajasthan area, India. The results indicate that the kriged groundwater levels satisfactorily matched the observed groundwater levels. Ahmadi and Sedghamiz (2007) used ordinary and universal kriging methods in identifying the spatial and temporal analysis of monthly groundwater level fluctuations in the arid region, the Darab Plain, in the south of Iran. Results revealed that the spatial and temporal variations of groundwater level and groundwater level fluctuations were underestimated by 3 and 6 %, respectively. Very low variation and acceptable errors support the unbiased hypothesis of kriging. Shamsudduha (2007) used different interpolation procedures in evaluating the most appropriate prediction method for the estimation of arsenic concentrations in the shallow aquifer in Bangladesh. In this case, ordinary kriging on original arsenic concentrations and their residual values produced better prediction models regarding the cross-validation, mean prediction error, and biasedness analyses.

The role of geographic information system (GIS) software in analyzing spatial distribution of groundwater has been investigated by many authors such as Mehrjardi et al. (2008). Pradhan (2009) studied groundwater potential zonation for basaltic watersheds. Ayazi et al. (2010) investigated disasters and risk reduction in groundwater. Manap et al. (2012) applied the probabilistic based frequency ratio in groundwater potential. El Afly (2012) and Machida et al. (2012) applied an integrated Geostatistics and GIS technique in groundwater. Neshat et al. (2013) estimated groundwater vulnerability to pollution, and Manap et al. (2013) studied prediction of groundwater potential zones.

Present work investigate the application of various spatial models to interpret the spatial distribution of the groundwater quality and to predict the general trend of the spatial distribution of groundwater in the recharge area of Muzaffarnagar district, Uttar Pradesh in Yamuna-Krishni sub-basin. In this study, kriging techniques in the framework of GIS software (ArcGIS Geostatistical Analyst) are used.

## DESCRIPTION OF THE STUDY AREA AND DATA COLLECTION

The study area falls in the District of Muzaffarnagar between Longitudes $77^0 05'$ and $77^0 27'$ E and latitude $29^0 15'$ and $29^0 41'$ N (Fig.1) and covers an area about 1100 $km^2$. It lies in the interfluves of Ganga and Yamuna rivers in the western most part of the Uttar Pradesh. The Rivers Yamuna and Krishni forms the western and eastern boundaries, respectively. Heavy withdrawal of groundwater has set a declining trend of water table over the decade in the study area.

District Muzaffarnagar is rectangular in shape and it is situated between the district of Saharanpur on the north, Meerut and Baghpat on the south, and is bounded by the Ganga on the east and the Yamuna on the west. The main rivers, the Ganga, the Kali, the Hindon and the Yamuna have played an important role in carving the topography of the district and divide it into four distinct tracts (Nevill, 1903, Varun, 1980) i.e. (i) the Ganga Khadir (ii) upland upto Kali Nadi (iii) the Kali Hindon doab and (iv) the Hindon Yamuna doab. The study area Krishni-Yamuna sub basin is a part of Hindon-Yamuna doab. Out of total water resources of the Yamuna-Krishni sub-basin, only 8.1% is derived from the surface water sources and 91.9% from the groundwater. The Eastern Yamuna Canal and its distributaries are main source of surface water irrigation (CGWB, 2008). Water quality data is available for 22 wells collected in the four different seasons. The observation data for 6 parameters are used for spatiotemporal representation using different kriging techniques.

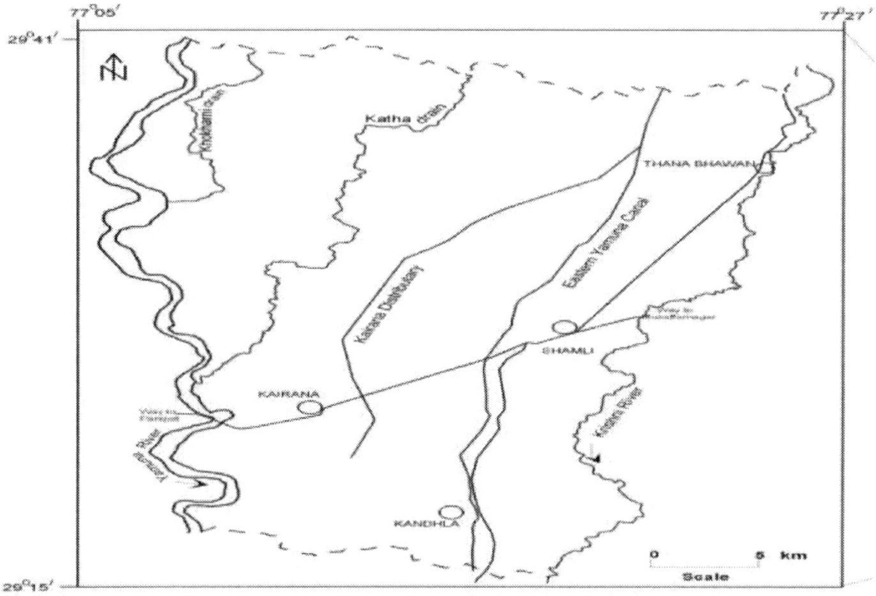


Figure **Error! No text of specified style in document.** Map of the Study Area

## METHOD OF SPATFIOTEMPORAL ANALYSIS

Spatial distribution of groundwater chemistry parameters is analyzed using GIS software. In this research, classification of concentration values to identify water quality for drinking water purposes follows the approach by McNeely et al. (1979). Spatiotemporal methodology are developed using the following steps:

1. Preparation of base map.
2. Georeferencing and geometric correction

3. Digitization
4. Assigning attributes to the study area.
5. Interpolation using different ordinary kriging.
6. Representation of different water quality parameter by contour map.

Geostatistical analysis is used to model the spatial distribution of groundwater chemistry. Geostatistics can be regarded as a collection of numerical techniques that deal with the characterization of spatial attributes employing primarily random models in a manner similar to the way in which time series analysis characterizes temporal data (Olea 1999). It deals with spatially autocorrelated data that have a basic structure or spatial patterns which can be manifested in (semi)variogram analysis. (Semi)variogram is a characterization of the spatial correlation of the variables under study. The semivariogram shows the relationship between the lag distance on the horizontal axis and the semivariogram value on the vertical axis. Lag distance is the distance between the measurements of a particular property. From the semivariogram, the spatial correlation of a spatially varying property can be described. The semivariogram value increases from low to high values indicating higher spatial autocorrelation at the small lag distance (Nayanaka et al. 2010).Theoretically, to calculate the semivariogram, the following formula is commonly used:

$$\lambda(h) = \frac{1}{2n(h)} \sum_{i=1}^{n(h)} [Z(x_i) - Z(x_i + h)]^2 \quad (1)$$

Where $\gamma(h)$ is the semivariogram value for the lag distance (h), n(h) is the total number of the variable pairs separated by a lag distance (h), and Z(x) is the value of the variable.Kriging is a technique of making optimal, unbiased estimates of regionalized variables at unsampled locations using the structural properties of the semivariogram and the initial set of data values (David 1977). The general equation of the kriging method has the following form:

$$Z^*(x_0) = \sum_{i=1}^{n} \lambda_0^i Z(x_i) \quad (2)$$

Where $\lambda_0^i$ is the weight associated with each data point i (i =1, 2, 3,…, n), $Z(x_i)$ is the observed value at point xi, $Z^*(x_0)$ is the predicted value at point x0, and n is the number of sample points. The values of the weights, $\lambda^i_0$, are estimated by minimizing the kriging (error) variance ($\sigma^2$) given by,

$$\sigma^2 = \frac{1}{n} \sum_{i=1}^{n} [Z^*(x) - Z(x)]^2 \quad (3)$$

The unbiased condition in kriging method is expressed as

$$E[Z^*(x_i) = E[Z(x_i)] \quad (4)$$

Where E[] is the expected value operator.

Eq.4 will lead to the formula of De Marsily (1984),$\sum_{j=1}^{n} \lambda_0^i = 1$ which means that the sum of the weight should be 1.The kriging system of equations that should be solved are as follows:

$$\sum_{j=1}^{n} \lambda_0^i \gamma(x_i - x_j) + \mu = \gamma(x_i - x_0) \quad (5)$$

$\sum_{j=1}^{n} \lambda_0^i = 1$, i=1,….,n, Where μ is a Lagrange multiplier, and $\gamma(x_i - x_j)$ is the semivariogram between two points $x_i$ and $x_j$.

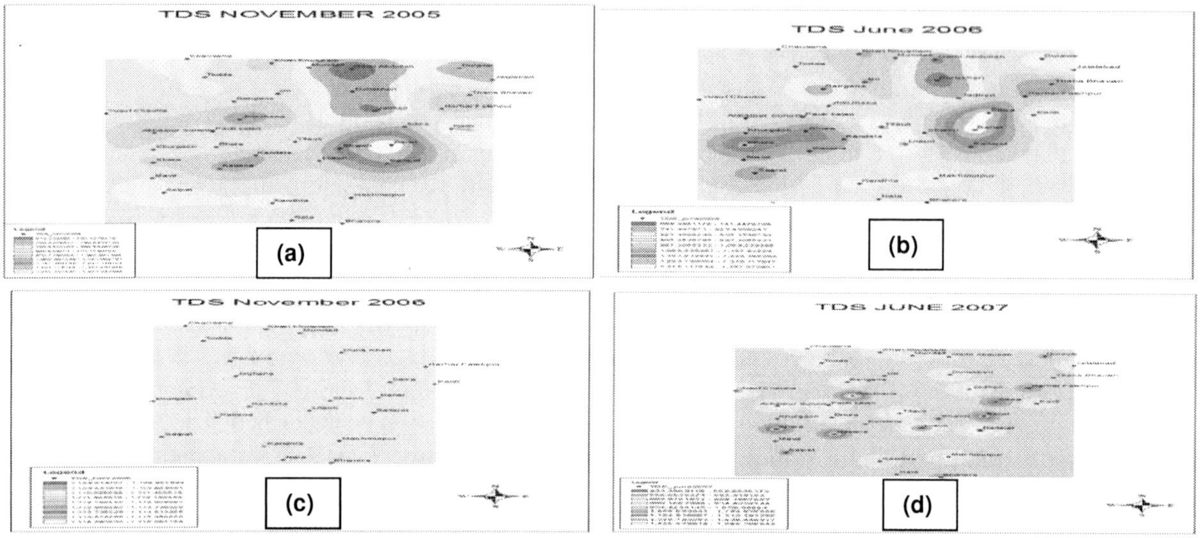


Figure 2. Interpolation of TDS Using Spherical Kriging in (A) Nov 2005, (B) June 2006, (C) Nov. 2006 & (D) June 2007

## RESULT AND DISCUSSION

The spatial distribution of TDS for four different seasons and five different kriging are shown on the maps and is presented in Figs. 2 to 7. From the various kriging techniques used in the spatial distribution, it is clear that spherical kriging gives much accurate and clear distribution pattern as compared to other kriging method.

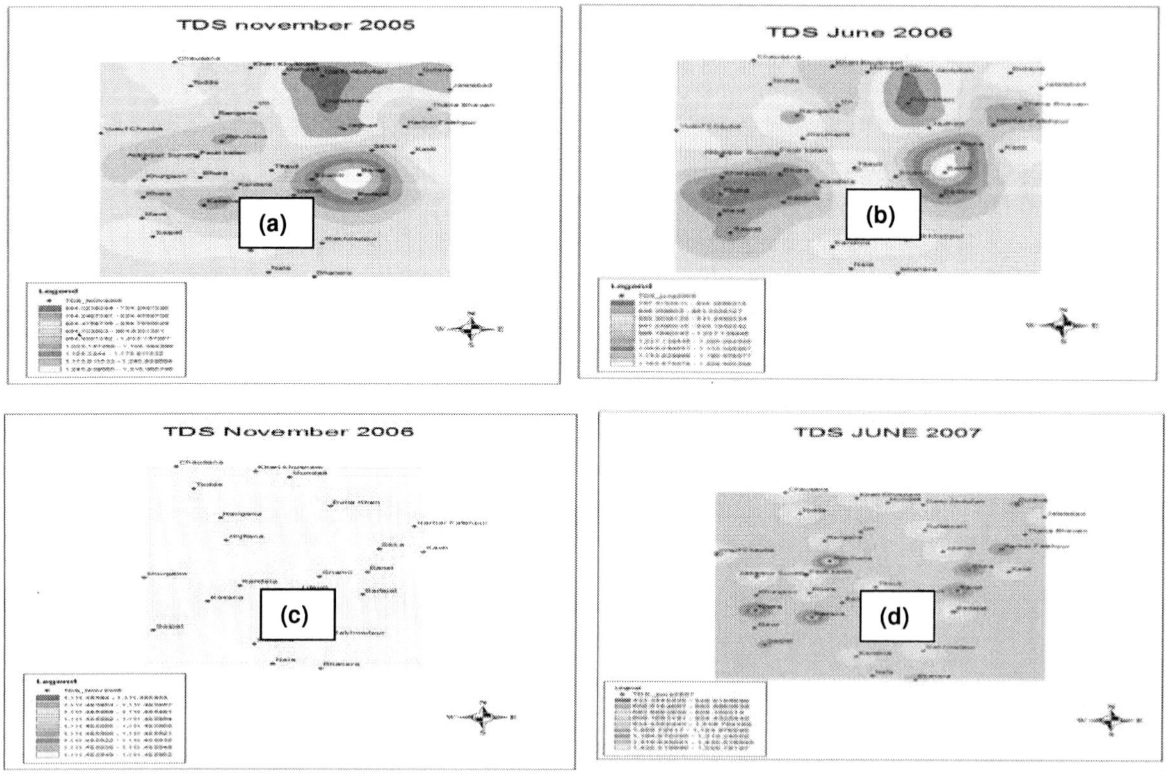


Figure 3. Interpolation of TDS Using Circular Kriging in (A) Nov 2005, (B) June 2006,

(C) Nov. 2006 & (D) June 2007

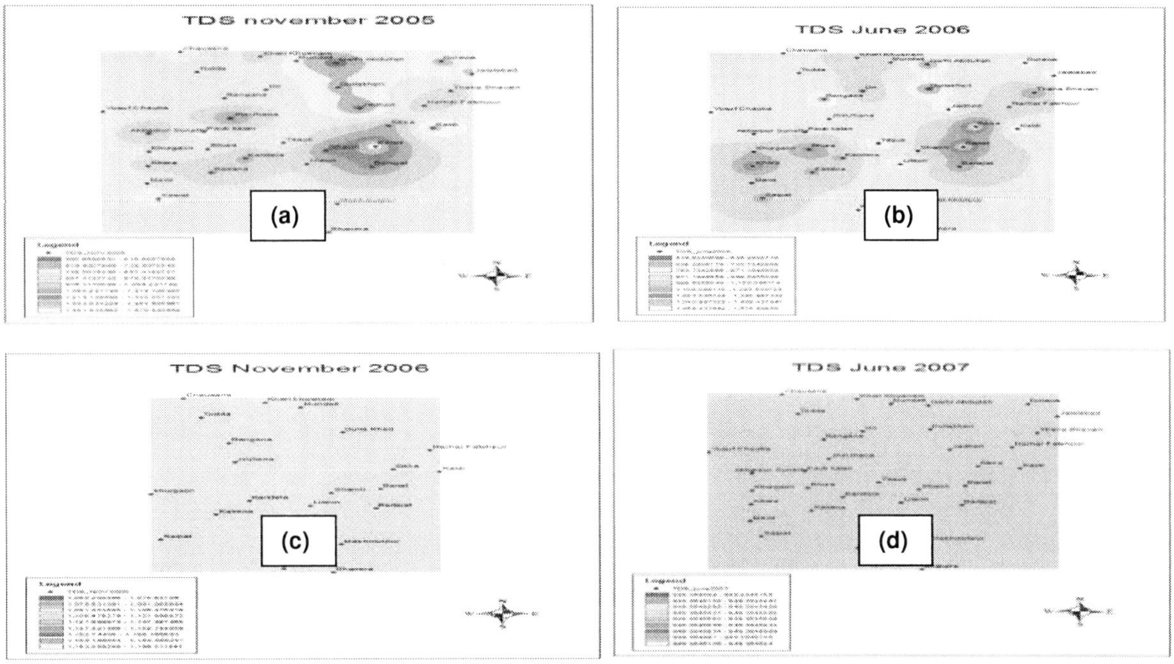


Figure 4. Interpolation of TDS Using Exponential Kriging in (A) Nov 2005, (B) June 2006, (C) Nov. 2006 & (D) June 2007

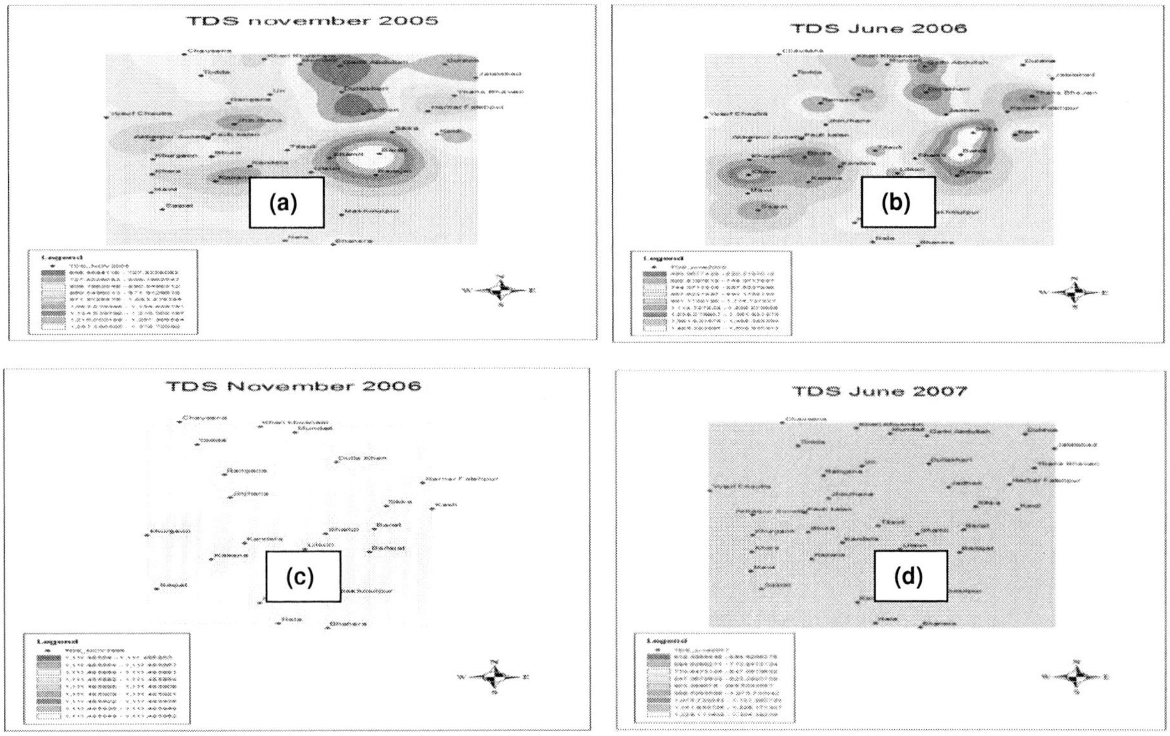


Figure 5. Interpolation of TDS Using Gaussian Kriging in (A) Nov. 2005, (B) June 2006, (C) Nov. 2006 & (D) June 2007

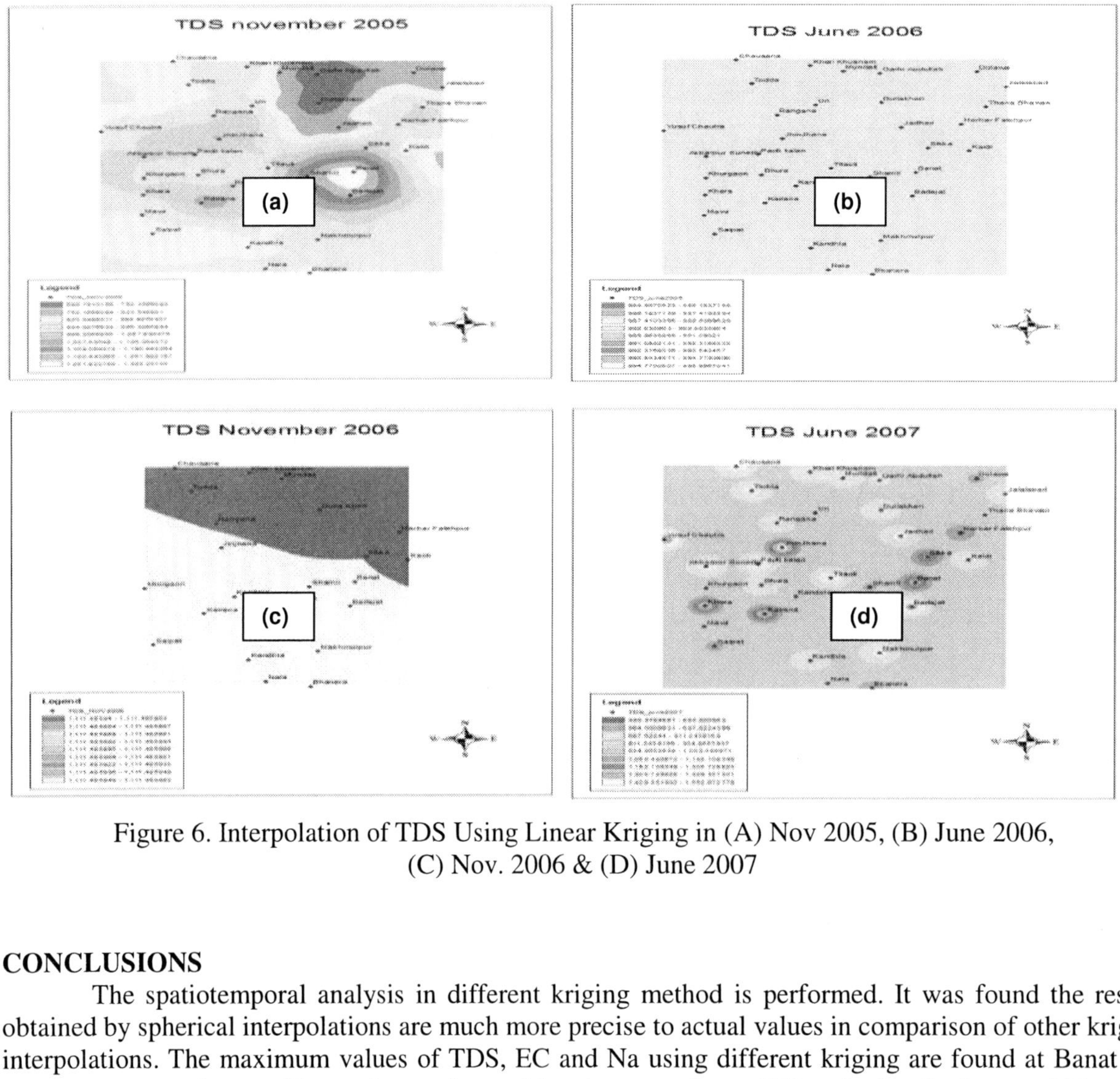


Figure 6. Interpolation of TDS Using Linear Kriging in (A) Nov 2005, (B) June 2006, (C) Nov. 2006 & (D) June 2007

## CONCLUSIONS

The spatiotemporal analysis in different kriging method is performed. It was found the results obtained by spherical interpolations are much more precise to actual values in comparison of other kriging interpolations. The maximum values of TDS, EC and Na using different kriging are found at Banat and nearby areas, maximum pH was observed at makhmulpur, maximum $NO_3$ at harhar fatehpur and khera.

## REFERENCES

Ahmadi S.H., A. Sedghamiz.2007. "Geostatistical analysis of spatial and temporal variations of groundwater level". Environ Monit Assess. 129: 277–294. doi: 10.1007/s10661-006-9361-z

Ayazi M.H., S. Pirasteh,A. K. P. Arvin, B. Pradhan, B. Nikouravan, S. Mansor. 2010. "Disasters and risk reduction in groundwater: Zagros Mountain Southwest Iran using geoinformatics techniques". Disaster Adv 3(1): 51–57

Central Ground Water Board (CGWB).2008. *Groundwater Flow Modelling and Aquifer vulnerability assessment studies in Yamuna–Krishni Sub-basin, Muzaffarnagar District*: Project no 23/36/2004.

David M. 1977. "Geostatistical ore reserve estimation". Elsevier, Amsterdam

De Marsily G. 1984. Spatial variability of properties in porous media: a stochastic approach. In: Bear J, Corapcioglu M (eds) Fundamentals of transport phenomena in porous media. Martinus Nijho, Dordrecht, pp 721–769

El Afly M. 2012. Integrated geostatistics and GIS techniques for assessing groundwater contamination in Al Arish area, Sinai, Egypt. Arab J Geosci 5(2): 197–215. doi: 10.1007/s12517-010- 0153-y

Kumar V, Remadevi. 2006. "Kriging of groundwater levels—a case study". J Spat Hydrol. 6(1): 81–92.

Manap M.A, H. Nampak, B. Pradhan, S. Lee, W. N. A. Sulaiman, M.F. Ramli. 2012. Application of probabilistic-based frequency ratio model ingroundwater potential mapping using remote sensing data and GIS. Arab J Geosci. doi: 10.1007/s12517-012-0795-z

Manap M.A., W. N. A. Sulaiman, M. F. Ramli, B. Pradhan, N. Surip. 2013. A knowledge driven GIS modelling technique for prediction of groundwater potential zones at the Upper Langat Basin, Malaysia. Arab J Geosci 6(5): pp. 1621–1637

McNeely R.N., V. P. Nelamnis, L. Dwyer.1979. Water quality source book: a guide to water quality parameter. Inland Waters Directorates, Water Quality Branch, Ottawa

Mehrjardi R.T., M. Z. Jahromi, S. Mahmodi, A. Heidari. 2008. "Spatial distribution of groundwater quality with geostatistics (case study: Yazd-Ardakan Plain)". World Appl Sci J. 4(1): pp. 9–17

Nayanaka V.G.D., W. A. U. Vitharana, R. B. Mapa .2010. "Geostatistical analysis of soil properties to support spatial sampling in a paddy growing alfisol". Trop Agric Res 22(1): 34–44. doi: 10.4038/tar.v22i1. 2668

Neshat A., B. Pradhan, S. Pirasteh, H. Z. M. Shafri .2013. Estimating groundwater vulnerability to pollution using modified DRASTIC model in the Kerman agricultural area. Iran Environ Earth Sci. doi: 10.1007/s12665-013-2690-7

Olea R.A. 1999. Geostatistics for engineers and earth scientists. Kluwer Academic, Boston

Pradhan B .2009. "Ground water potential zonation for basaltic watersheds using satellite remote sensing data and GIS techniques". Cent Eur J Geosci 1(1): pp.120–129. doi: 10.2478/v10085-009-0008-5

Shamsudduha M .2007. "Spatial variability and prediction modeling of groundwater arsenic distributions in the shallowest alluvial aquifers in Bangladesh". J Spat Hydrol. 7(2):pp. 33–46.

# DEVELOPMENT AND APPLICATION OF THE INTEGRATED WATER DISTRIBUTION MANAGEMENT SYSTEM

***Patryk Wójtowicz**** (Wrocław University of Technology, Wrocław, Poland)
Andrzej Ziółkowski (Warsaw School of Information Technology, Warsaw, Poland)
Jan Studziński (Systems Research Institute, Warsaw, Poland)

In recent years, a rapid growth of IT, automation and measurement technologies accelerated the implementation of computer aided decision systems in the water sector. We present research and innovation aimed to meet the need for a scalable, automated and robust solution for water utilities that can integrate existing (commercial, in-house and open source) software and hardware solutions into one package, benefiting in a synergetic effect.

We demonstrate the results of tests in two contrasting, both in terms of type and size, case-studies in Poland, facing different short-term and long-term challenges. First was a large-scale main water transportation system of the Upper Silesian Waterworks in Katowice serving to almost 3 million customers. In the other case-study, we implemented ICT system in a typical medium-size municipal water supply of Głubczyce (population of ca. 13 000). These two case-studies provided us with a unique opportunity to test various approaches to calibration as well as automatic hydraulic model updating and validation. The key elements of this system have been examined, namely: GIS, telemetry (AMI and SCADA), hydraulic model as well as tools for multi-criteria optimization and forecasting. We explain the details of our custom semi-automated calibration procedure, where we applied the combination of fuzzy logic and genetic algorithms.

Specific aspects of data collection, model calibration and simulation were discussed. Special focus was also given to the development of the hydraulic model using in-house MOSKAN-W modelling software - EPANET based mobile (Javascript) for two very different water distribution systems. GIS is the central element of integrated system, which was used to manage and process input and output data from the model. In particular, we demonstrate the integration between GIS system and other modules.

# ENVIRONMENTAL ANALYSIS AND MEASUREMENTS

# ENVIRONMENTAL IMPLICATIONS OF FOOD AND PLA WASTE MANAGEMENT OPTIONS

***Shakira R. Hobbs***[a], Jay P. Devkota[a], Prathap Parameswaran[b] and Amy E. Landis[a]
([a]Institute of Sustainability, Glenn Department of Civil Engineering, Clemson University Clemson, SC, 29634 USA; [b]Department of Civil Engineering, Kansas State University, Manhattan, KS 66506, USA)

**ABSTRACT:** The study assesses the environmental impacts of anaerobic digesting, composting and landfilling of food and PLA waste via the Environmental Protection Agency's Waste Reduction Model (WARM) and life cycle assessment (LCA) of anaerobic digestion model. WARM contains composting and landfilling food and polylactic acid (PLA) waste, but not anaerobic digestion. Compost, landfill and anaerobic digestion of food and PLA waste data was extracted and assessed from WARM, Ecoinvent and literature to show environmental impacts. Results show landfilling 200 tons of food and PLA waste per year produces the most $CO_2$ equivalent at 132 tons/yr. Composting resulted in negative emissions for food waste, however PLA does not degrade in industrial composting facilities and are instead sent to landfills. Anaerobically digested food and PLA waste is the best scenario, in terms of reduced carbon emissions, when biogas is used to heat the digester. In contrast, anaerobic digestion requires the most energy, even when biogas is utilized in comparison to landfilling and composting. Although WARM shows energy and carbon impacts from food and PLA waste, it does not show process breakdowns. Results from the study highlight the limitations of WARM and show processes that can be further optimized to reduce environmental impacts.

## INTRODUCTION

Sending food to landfills has serious and detrimental environmental impacts. Food waste in landfills breakdown anaerobically and produces a greenhouse gas, methane, 25 times more potent than $CO_2$ (IPCC, 2007). Although landfills can recover gas, only 20% of landfills are capable of recovering the gas (EPA, 2016a). Food waste can be redirected from landfills to compost and anaerobic digestion facilities to produce nutrient rich soil amendments and produce methane for combine-heat and power.

Recently, handling food waste has become more complicated because it is often 'contaminated' with bioplastic such as PLA. Bioplastic are often used in food packaging and increasingly used to achieve zero-waste efforts and sustainability goals. While PLAs are intended to be composted after use alongside food waste, industrial composting facilities managers report that PLAs do not fully biodegrade in their systems (Meeks et al., 2015; Siracusa et al., 2008). Many composting facilities are rejecting PLAs and sending them to landfills (Hottle et al., 2013). New sustainable solutions are needed to manage both food and PLA waste. PLA and food waste can be sent directly to anaerobic digesters without being cleaned or sorted. These waste management scenarios may be more environmentally friendly and energetically positive when compared to compost facilities and landfills.

The aim of this study was to understand the energy and carbon impacts of composting, landfilling and anaerobically digesting food and PLA waste. WARM (an EPA waste reduction model) has compost and landfill options to estimate energy use and green house gas (GHG) emissions from food and PLA waste. However, there is no model that exists that considers anaerobic digestion of food and PLA waste to estimate energy use and GHG emissions. Therefore, we developed a new anaerobic digestion model based on data from Ecoinvent and literature. This model can be used by designers and practitioners for optimizing processes with an aim for reducing environmental impacts and understanding trade-offs.

## METHODS

**Goal and Scope Definition.** The goal of this LCA was to evaluate and compare three different end-of-life waste management options: compost, landfill and anaerobic digestion in terms of environmental impacts and energy use. The functional unit for the LCA was normalized to 200 tons to reflect the amount of food and PLA waste produced per year (EPA, 2016b; NatureWorks, 2012). 200 tons was used because the capacity of anaerobic digester was 200 tons.

The analysis begins once food and PLA waste is discarded as shown in Figure 1. Other processes outside of the system boundary (dotted line) are also shown in the background and are not considered in the analyses since they do not deal with use-phase or end of life processes. In this analysis, food waste consists of: beef, poultry, grains, bread, fruits and vegetables and dairy products–this also includes uneaten and prepared food from commercial, non-commercial and industrial sources (Buzby et al., 2014). PLAs are often used for food packaging such as forks, cups, lids and thin films, which are also considered in this analysis.

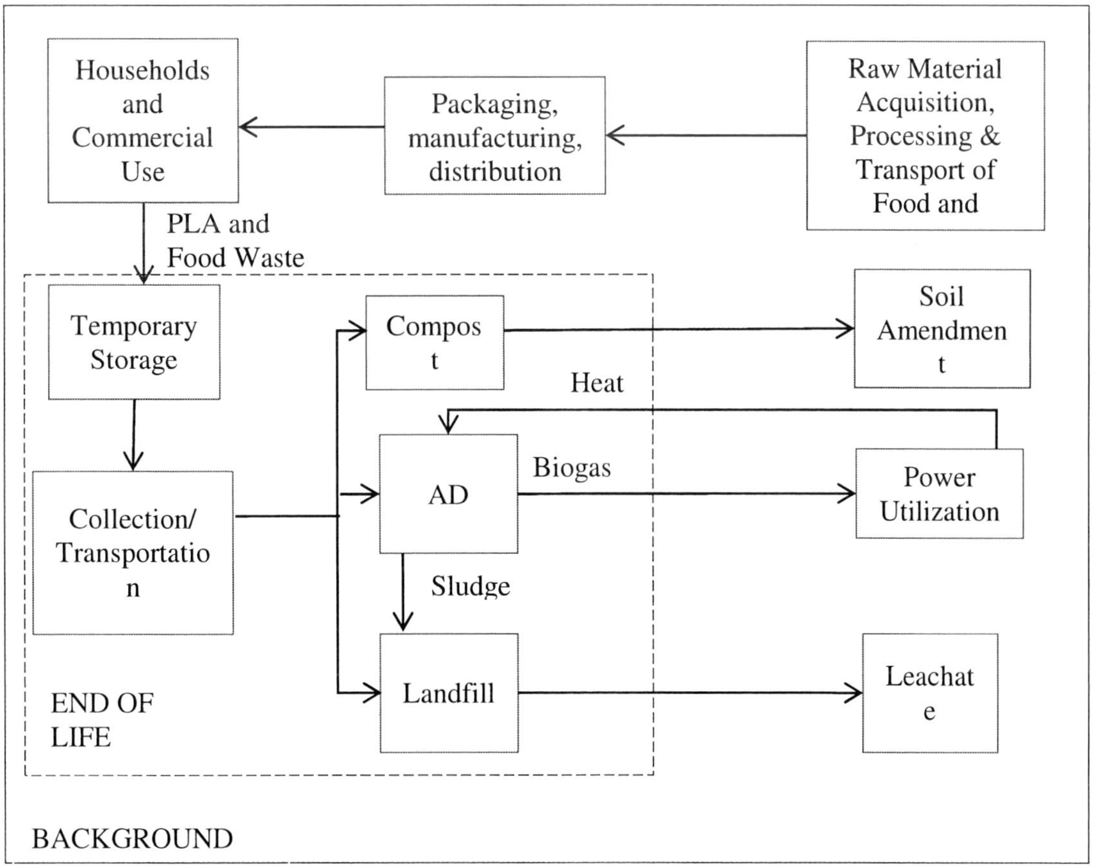


**FIGURE 1. Life cycle of food and PLA waste.**

**WARM an AD LCA Model**. Waste Reduction Model (WARM) was created by the EPA to assist waste solid managers in environmental assessment of material types and waste management practices (EPA, 2015). WARM allows users to select material inputs for solid waste management options and compare different treatment options. Results are reported as metric tons of carbon dioxide equivalent (MTCO2E), metric tons of carbon equivalent (MTCE) and units of energy (million BTU). WARM has waste management options such as: recycling, landfilling, combusting and composting, but does not have anaerobic digestion.

End of life-cycle data was used to create a model for anaerobic digestion that shows the carbon and energy impacts of digesting food and PLA waste. Similar system boundaries, inputs and outputs were used based on WARM. However, WARM gives limited life cycle inventory (LCI) data, which makes it difficult to replicate the model. Therefore, data was collected from WARM, Ecoinvent and from literature. Life Cycle Impact Assessment was completed using the TRACI 2.1 V1.01 impact characterization–Global Warming Potential (GWP) and energy use was calculated.

***Scenario 1, the landfilling scenario.*** This scenario describes the management system for handling food and PLA waste. WARM was used to assess landfilling emission factors, which are made up of:
1. Transportation $CO_2$ emissions from landfilling equipment; 2. Biogenic carbon stored in the landfill; 3. CO2 emissions avoided through landfill gas-to-energy projects" (EPA, 2015).

***Scenario 2, the composting scenario.*** WARM was used to assess this scenario and describes the management system for handling food and PLA waste. Compost emission factors are calculated based: 1. "Collecting and transporting the organic materials to the central composting site; 2. Mechanical turning of the compost pile; 3. Non-$CO_2$ GHG emissions during composing (primarily $CH_4$ and $N_2O$); 4. Storage of carbon after compost application to soils." (EPA, 2015).

***Scenario 3, the anaerobic digestion scenario.*** Anaerobic digestion LCA model was used to assess this scenario and describes the management system for handling food and PLA waste. Data was retrieved from (Ecoinvent, 2010) and (Salter, 2008). Anaerobic digestion factors are calculated based on: 1. Collecting and transporting the organic material to central AD site; 2. Operation of anaerobic digester; 3. $CO_2$ and Methane emissions avoided through AD gas to energy products or utilization of natural gas; 4. Transportation $CO_2$ emissions from landfilling equipment.

## RESULTS AND DISCUSSION

The results show significant differences in energy use and environmental impacts between the waste management techniques assessed in this study. The results from the study emphasize the importance of showing individual processes rather than impacts from food and PLA waste only.

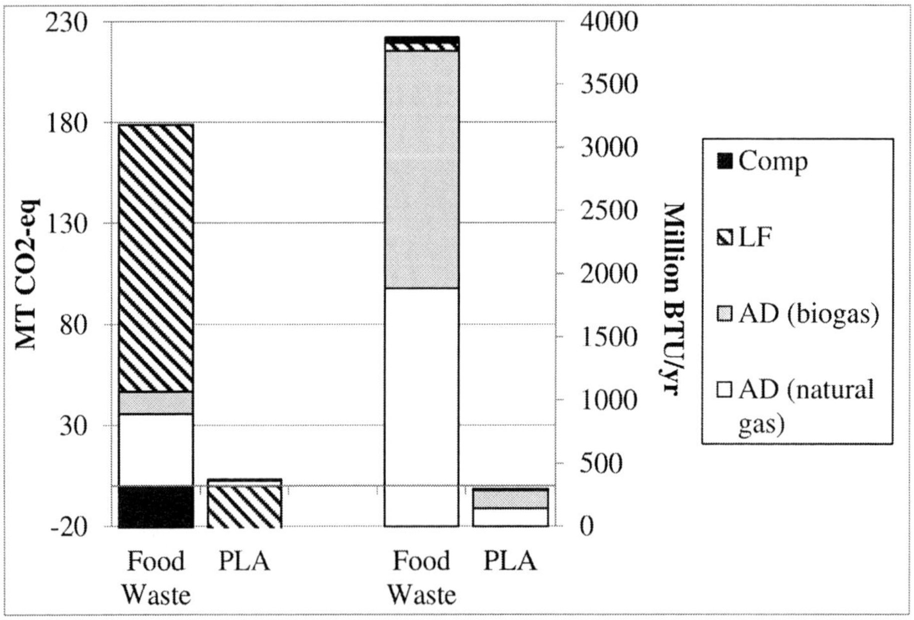


**FIGURE 2. Global warming potential and energy use of food and PLA waste compared to four waste management techniques.**

**Comparison of Four Waste Management Options Results.** Results showed anaerobic digestion, utilizing natural gas or biogas, had significantly lower $CO_2$ emissions than landfilling (Figure 2). Utilizing natural gas to operate the digester instead of biogas results in 26.6 more MT $CO_2$ eq (Figure 2). The additional surplus of biogas from the anaerobic biogas scenario can be sold or stored for future uses. Therefore, using biogas, produced from anaerobic digestion, to operate anaerobic digester facilities further reduces the global warming potential as opposed to utilizing natural gas. However, there is no significant difference in energy use between utilizing biogas versus natural gas. Furthermore, landfilling and composting food and PLA waste requires significantly less energy to dispose and treat than anaerobic digestion. Landfilling food and PLA waste resulted in significant GWP collectively, 132 MT $CO_2$ eq. compared to compost at -30 MT $CO_2$ eq. avoided emissions (Figure 2). Emissions are avoided from food and PLA when composted since WARM considers composting food and PLA waste to biogenic, not anthropogenic like landfilling food waste. However, there is some debate in the literature on whether or not PLA produces anthropogenic or biogenic waste.

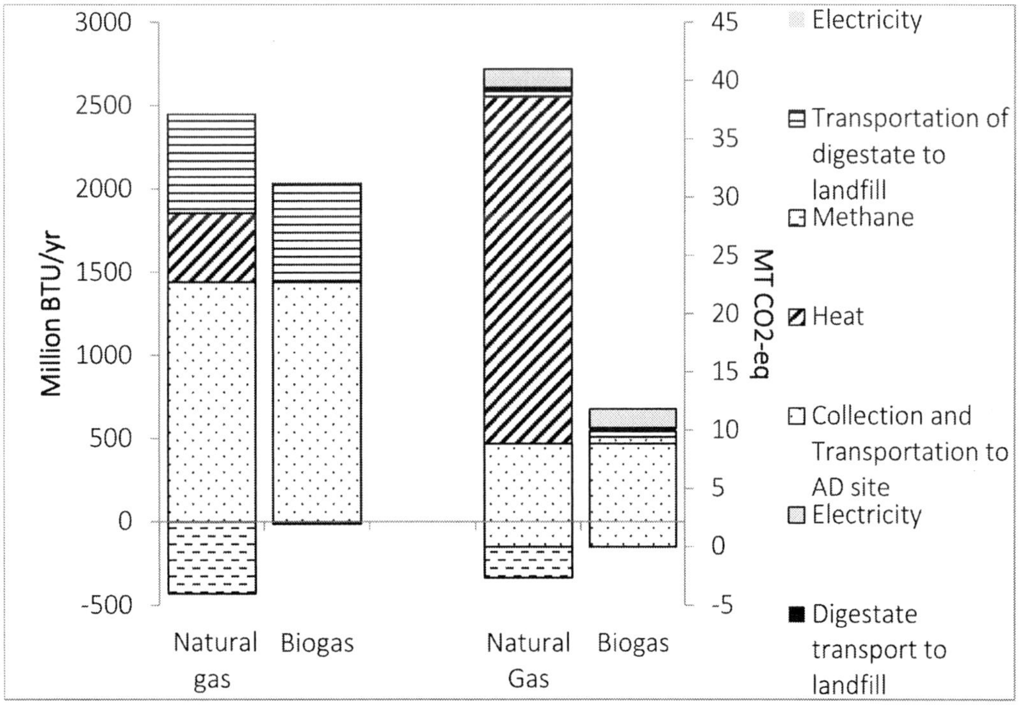


**FIGURE 3. Energy use and Global Warming Potential of individual processes of anaerobic digestion utilizing biogas and natural gas**

WARM does not distinguish between amorphous and crystalline PLA and states that PLA does not degrade in landfills, but degrade in composting facilities resulting in negative PLA emissions for the waste management option landfill and compost. There is some debate in the literature. For example, crystalline PLA, which are typically forks and cups, do not degrade in landfills (Kolstad et al., 2012), therefore emissions are avoided from PLA in landfill. Furthermore, composting facilities in the US are reporting that PLA does not degrade at the same rate as organic materials, such as food waste, and are rejected and sent to landfills (Hottle et al., 2013). However, amorphous PLA, which are typically thin films, degrade in anaerobic conditions like landfills (Kolstad et al., 2012) and produce methane and carbon dioxide–i.e. GHG. There are experiments, which also show that amorphous and crystalline PLA biodegrade under thermophilic anaerobic conditions (Krause & Townsend, 2016; Vargas et al., 2009)–same conditions as landfills.

**Anaerobic Digestion (Biogas And Natural Gas) Process Results.** Collection and transportation of food and PLA waste to anaerobic digestion site, heat, electricity, methane and transportation of digestate to landfill are ancillary processes that are used to create the model. 415 million BTU/yr of natural gas is needed to operate the anaerobic digester with about 430 million BTU/yr produced as a byproduct (Figure 3). Methane produced from the digester can be substituted for natural gas and used to power and operate the anaerobic digester leaving a byproduct of 15 million BTU/yr of methane. Utilizing natural gas as a heating source for the anaerobic digester contributed to 38.4 MT CO2-eq compared to utilizing biogas at 11.8 MT CO2 eq. Substituting natural gas for biogas drastically reduces MT CO2 eq. (Figure 3) and demonstrates the benefits of showing individual processes.

Majority of the energy used in the anaerobic digestion scenario is the collection and transport to the anaerobic digester site at 1440 million BTU/yr (Figure 3). Quantifying and showing individual processes assist in waste management operators' decision-making and provide concise guidance highlight sustainable waste handling techniques.

## CONCLUSION

The study assessed the environmental impacts of anaerobic digestion, composting and landfilling of food and bioplastic-PLA waste via WARM and anaerobic digestion LCA model. WARM does not have anaerobic digestion as a waste management option. Therefore, data was extracted and assessed from Ecoinvent, literature and WARM to create a LCA model for anaerobic digestion and to estimate the environmental impacts. Results showed that, landfilling 1 tons of food and PLA waste per year produced the most $CO_2$ at 0.55MT $CO_2$ eq. Composting resulted in negative emissions for food waste, however PLAs do not degrade in industrial composting facilities and are instead sent to landfills. Negative emissions from composting were because WARM considered PLA to be biogenic emission, despite literature stating otherwise. Among the waste reduction scenarios considered in the study, anaerobically digesting food and PLA waste was reported as the best scenario, in terms of reducing $CO_2$ emission, provided that the produced biogas was used to operate the digester. However, anaerobic digestion required the most energy (10 Million BTU/yr) to digest 1 ton of food and PLA waste, even when biogas was utilized. The anaerobic digestion scenarios highlight the importance of showing individual processes. It also assists in identifying individual process that require and emit the most energy and $CO_2$ that can be further optimized.

## ACKNOWLEDGEMENTS

This material is based upon work supported by the National Science Foundation under CBET Grant No. 1066658.

## REFERENCES

Buzby, J. C., Wells, H. F., & Hyman, J. (2014). *The Estimated Amount, Value, and Calories of Postharvest Food Losses at the Retail and Consumer Levels in the United States* Retrieved fromEcoinvent. (2010). *Ecoinvent v2.2 database*. Dübendorf, Switzerland: Swiss Centre for Life Cycle Inventories.

EPA. (2015). Waste Reduction Model (WARM). from Environmental Protection Agency (EPA)

EPA. (2016a). Energy Projects and Candidate Landfills. https://www3.epa.gov/lmop/projects-candidates/index.html

EPA. (2016b). Turning Food Waste into Energy at the East Bay Municipal Utility District (EBMUD). *Reducing Food Waste*. https://www3.epa.gov/region9/waste/features/foodtoenergy/food-waste.html

Hottle, T. A., Bilec, M. M., & Landis, A. E. (2013). Sustainability assessments of bio-based polymers. *Polymer Degradation and Stability, 98*(9), 1898-1907.

IPCC. (2007). *Climate Change 2007-:Synthesis Report*. Retrieved from Geneva, Switzerland: https://www.ipcc.ch/pdf/assessment-report/ar4/syr/ar4_syr_full_report.pdf

Kolstad, J. J., Vink, E. T. H., Wilde, B. D., & Debeer, L. (2012). Assessment of anaerobic degradation of Ingeo polylactides under accelerated landfill conditions. *Polymer Degradation and Stability*(97), 1131-1141.

Krause, M. J., & Townsend, T. G. (2016). Life-Cycle Assumptions of Landfilled Polylactic Acid Underpredict Methane Generation. *Environmental Science & Technology Letters, 3*(4), 166-169. doi:10.1021/acs.estlett.6b00068

Meeks, D., Hottle, T., Bilec, M. M., & Landis, A. E. (2015). Compostable biopolymer use in the real world: Stakeholder interviews to better understand the motivations and realities of use and disposal in the US. *Resources, Conservation and Recycling, 105, Part A*, 134-142.

NatureWorks. (2012). NatureWorks Broadens INGEO Product Portfolio with Sulzer Proprietary Production Equipment Retrieved from http://www.natureworksllc.com/News-and-Events/Press-Releases/2012/09-05-12-Sulzer-equipment-for-increased-Ingeo-production

Salter, A. (2008). *Anaerobic Digestion: Overall Energy Balances–Parasitic Inputs & Beneficial Outputs* Paper presented at the Sustainable Organic Resources Partnership - Advances in Biological Processes for Organics and Energy recycling Birmingham

Siracusa, V., Rocculi, P., Romani, S., & Rosa, M. D. (2008). Biodegradable polymers for food packaging: a review. *Trends in Food Science & Technology, 19*(12), 634-643.

Vargas, L. F., Welt, B. A., Teixeira, A., Pullammanappallil, P., Balaban, M., & Beatty, C. (2009). Biodegradation of treated polylactic acid (PLA) under anaerobic conditions. *Transactions of the ASABE, 52*(3), 1025-1030.

# DEUTERATED MONITORING COMPOUNDS FOR BETTER ACCURACY AND PRECISION MEASUREMENT OF GC/MS ENVIRONMENTAL DATA

***C. Appleby***
(U.S. Environmental Protection Agency, 1200 Pennsylvania Ave., NW, MC5203P, Washington, DC 20460, USA)

The use of surrogate compounds to measure method performance in Gas Chromatography/Mass Spectroscopy (GC/MS) methods for environmental monitoring is not a new practice. All EPA-approved methods require the use of three to six compounds; however only a few are deuterated analogs of target analytes. Deuterated analogs are more representative of target analytes, thereby providing more information regarding matrix effects while measuring the accuracy and precision. Since 2001, the EPA's Office of Superfund Remediation and Technology Innovation's Contract Laboratory Program (CLP) has required laboratories to add over a dozen deuterated monitoring compounds (DMCs) to each sample, all analogs of target analytes. Developed to improve data quality used in decision-making processes, this approach ultimately reduced the cost to the Superfund Program. This presentation will show, with thousands of data points, how incorporating more DMCs into EPA-approved GC/MS methods has improved data quality and provided cost savings to the Agency, and how it may benefit the entire analytical chemistry community.

# PRETREATMENT METHOD FOR TOTAL ORGANIC CARBON (TOC) ANALYSIS OF SUSPENDED SOLIDS-CONTAINING WATER SAMPLES

Han-Saem Lee, So-Hui Kim, and ***Hyun-Sang Shin*** (Seoul National University of Science & Technology, Seoul, Republic of Korea)
Jin Hur (Sejong University, Seoul, Republic of Korea)
Keun-Heon Lee (HUMAS Co., Ltd. Daejeon, Republic of Korea)

Total organic carbon (TOC), an index of the total amount of organic matters in water, is widely used as a tool to measure the organic pollutant loading of water samples and to manage wastewater treatment plants. However, the samples often contain wide variety of particulate matters depending on the origin, which make it complicated to analyze the accurate and precise determinations of organic carbon. The purpose of this study was to evaluate the effect of pretreatment by ultra-sonication, filtration and dilution on TOC analysis for the samples with high levels of suspended solid (SS) and organics. The SS samples used in this study included sewage sludge, algae and soil particles. Two different methods for the TOC analysis, high temperature catalytic oxidation and wet chemical oxidation with sulfate, were also compared with regard to their TOC response efficiency for the SS samples. The probe-type sonicator was more effective in reducing particle sizes of SS and producing higher TOC repsonce efficiency than bath-type one. Dilution of samples followed by sieving (200~300 um) produced higher TOC than the dilution after sieving, and the differences were increased as the dilution ratio decreased from 1/50 to 1/200. The results of this study showed that the pretreatment procedures with ultrasonic crushing, dilution and sieving can effectively be applied to analyze TOC for high concentration of suspended solids-containing samples. It also demonstrates that the two different TOC oxidation method yields comparable results at low concentration of suspended solids (< ~20 mg/L) in water.

# EXPERIMENTAL STUDY ON SURFACE FLOW MEASUREMENT USING LARGE SCALE PARTICLE IMAGE VELOCIMETRY TECHNIQUES

Aadhi Naresh[1] and M. Gopal Naik[2]
(Department of Civil Engineering, Osmania University, Hyderabad, Telangana, India)
[1]ayyaure@gmail.com; [2]mgnaikc@gmail.com

**ABSTRACT:** The rapidly increasing demands on the water resources around the globe and most of the world make it imperative that they can be determined with high degree of precision as is practicable. Stream flow determinations will require a greatly expanded network of hydrometric stations and high degree efficiency in their operation. The standard methods such as trained hydrographers and rational formulae are competent to make stream flow measurements but there are becoming increasingly scare. The latest techniques such as ADCV, UHF Doppler Radar, LIF and Large Scale Particle Image Velocimetry (LSPIV) are more intensive use of high-speed computing devices, which are made available. LSPIV technique shoots the target area using a video camera and can obtain the instantaneous flow field to overcome the limit of the measuring equipment based on existing sensors such as LDV and ADV. In this study, LSPIV techniques has been adopted to measure the velocity and discharge of surface flow performed on the tilting flume at Fluid Mechanics Laboratory, University College of Engineering, Osmania University. The LSPIV device consists of an adjustable tripod with a video camera. The digital video camera acquired sequences of images of the surface flow velocities. It is observed from the study that, the velocity of fluid particles are varies from minimum 0.20 m/s to maximum of 0.415 m/s for different discharges. Floating objects like polypropylene particles are used as floating tracers to track the flow direction and their velocities. The average velocity has been calculated with applying the multiplication factor. The simulated flow velocity and discharges using LSPIV have been compared with the actual measurement values calculated through experimental results. Two different types of bed materials such as coarse aggregates Bermuda type grass vegetation are used for flow variations. It is found from the study that, the flow parameters varies within error of ±10%. However, the results obtained from LSPIV techniques indicate that the adopted technique is more reasonable and have good agreement with volumetric method results.

***Keywords:*** Stream Flow, Large Scale particle Image Velocimetry.

## INTRODUCTION

Open channel shallow flows are defined as largely unidirectional, turbulent shear flows occurring in a confined layer (Jirka, 2001). These confined flows lead to the development of strong large scale two-dimensional turbulent structures mutually interacting with the omnipresent small three-dimensional turbulence scales. The main geometrical scale for these flows is their depth. Shallow flows are omnipresent in various areas of the surrounding landscape and situations. They are typically encountered in small streams, ephemeral rivers, ponds, and river floodplains in the natural environment. Ubiquitous shallow flows in urban landscape are the storm water runoff over impervious surfaces (e.g., pavements and compacted soils) and gutter flows. Given their reduced depth, shallow flows are often associated with low-velocities over the cross section. Documenting flow features in these measurement situations is usually problematic for both traditional methods (e.g., propeller meters) and the newer acoustic-based instruments. The mechanical-based techniques require a minimal water depth to accommodate the instrument measurement volumes (of the order of several centimeters), and they cannot accurately acquire near-zero velocities. The newer acoustic techniques (such as Acoustic Doppler Velocimeter, Acoustic Doppler

Current Profiler) are quasi-non-intrusive but they require immersion of the probe in the water body (Simpson, 2001). The minimum measurement volumes of the acoustic instruments are of typically larger than those of mechanical methods and they also encounter difficulties in capturing low-velocity flows (Church et al., 2012). More promising for measuring velocities in natural-scale shallow flows are image-based techniques, collectively labeled herein as Large-Scale Particle Image Velocimetry (LSPIV). These techniques stem from the Particle Image Velocimetry (PIV) methodology extensively used in laboratory fluid mechanics studies since the mid-1980s (Adrian, 1991). PIV is essentially a non-contact, non-intrusive velocity measurement technique that quantifies the movement of small and light particles moving within illuminated planes transecting the body of a fluid. The particles, polypropylene tracers, are expected to accurately follow the underlying flow and to uniformly "seed" the area to be measured.

Under natural light sources it would adequately illuminate the particles visualizing the flow motion. Images of the illuminated surfaces are captured with cameras using various combinations for the timing of the lighting and camera acquisition rate (Adrian, 1991). Velocities are eventually obtained using pattern-recognition algorithms based on statistical methods that allow to "guess with known confidence" where the image patterns are moving in a series of images recorded at known time intervals. Specifically, the recorded images are divided in small "interrogation areas" that are evaluated using statistical means (auto- or cross-correlations) to extract velocities. The process of interrogation is repeated for all areas of the recording leading to a whole-field velocity distribution over the imaged area. LSPIV is a newer sub-family of the PIV technique and it is applied to moving fluid bodies with the objective to quantify velocities at the free surface. The first attempts to extend the PIV methodology for quantifying the velocity at the free surface was made in Japanese rivers in early 1990s (Fujita and Komura, 1994). Given that the areas subjected to velocity measurements are considerable larger that the laboratory-size PIV measured areas this variation of the technique is labeled "Large-scale" PIV. The technique is subsequently applied in laboratory conditions for quantifying velocity at the free surface in hydraulic models for riverine environment (Muste et al. 2008). Ambient light is typically used for visualizing tracers on the free surface in field or laboratory measurements.

The main distinctions between PIV and LSPIV are the size of the measurement volume and the size and characteristics of the tracers used for visualization of the flow. Since the targeted area for LSPIV velocity measurements is order of magnitudes larger than those measured with conventional PIV, the image recording is usually made from an oblique angle. The flow tracers are also considerable larger for LSPIV compared with the micron-size particle used in the laboratory PIV. Practically, the tracers are not anymore individual particles as in PIV but clusters of particles of larger sizes that collectively form floating patterns at the free surface by the moving water body. The oblique angles used for recording inherently introduce distortion of the physical space images due to geometric perspective. The distortion is removed using few selected points contained in the images in conjunction with a geometrical transformation that relates the camera with real-world coordinates. The result of this transformation is that the LSPIV estimated velocities are provided as vectors with actual magnitudes and orientations everywhere over the imaged area. Although LSPIV has been successfully employed in a variety of field and laboratory conditions, there are substantially few reported measurements acquired with the technique in shallow water bodies. Often time these flows are also slow (at times close to zero velocities) and confined within changing and difficult to access boundaries such as in the floodplain flow situations. The non-intrusive and two-dimensional (2D) nature of the measurements coupled with its capability to instantaneously provide velocity over large areas, makes LSPIV a good candidate for the situation and at times the only one possible measurement solution (Nord et al., 2009). The goals for this study are to illustrate that: a) to measure surface flow velocity using Large Scale Particle Image Velocimetry, b) To compare the accuracy of the PIV measurements and decrease the work that has to be done manually and c) To optimize the certainty for each vector and the fluency of the mean frame, the part of the channel that is photographed should be covered with tracers, for each frame of the series

## MATERIALS AND METHODOS

In recent years, methods have been developed for using an LSPIV system to measure discharge in open channel. The USGS procedure for evaluating emerging open-channel discharge measurement technologies (Melcher et al., 2002) adopted. In general, the methods developed in this study follow the steps (Harpold A.A. et al., 2006) as shown in Figure 1 and initially, an appropriate location is selected for optimum installation of the equipment using established guidelines (ISO, 1997; Rantz et al., 1982). The images of 8fps are recorded with Camera (Canon EOS 7D) is mounted on a Vanguard tripod and adjusted to remove erroneous illumination of flow. The effect of an oblique camera angle on LSPIV accuracy is investigated using 0°, 15°, 30°, 45° and 90 °oblique angles to get best picture(Harpold A.A. et al., 2006). The channel is first seeded with tracer particles that mimic the fluid movement. Subsequently, successive images of the channel surface are recorded and saved. The image quality is enhanced to improve tracer visibility, and spatial distortion is removed prior to velocity estimation. The surface velocity field is estimated using statistical correspondence methods. A representative open channel velocity field is estimated by correcting the surface velocity to account for a non-uniform vertical velocity profile. After the velocity field is estimated, the channel dimensions and stage measurements are used to estimate discharge with area-velocity methods (e.g., midpoint method).

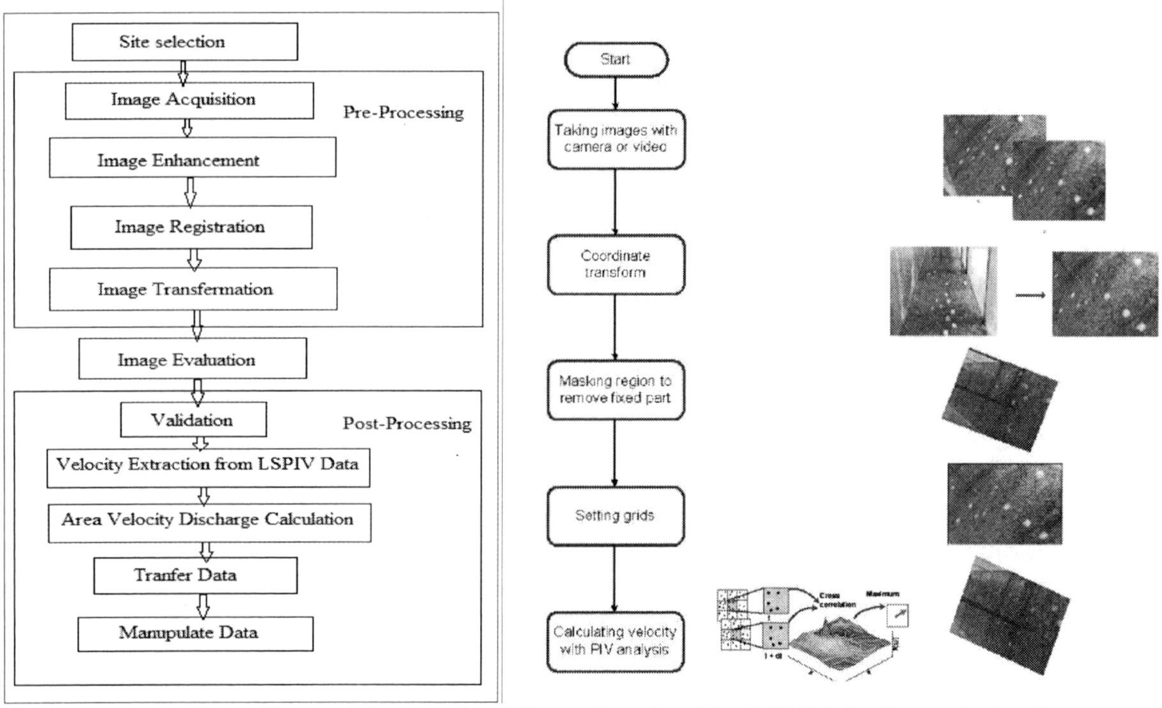


**FIGURE 1.** Flow Chart of Generalized Steps Involved in LSPIV for Low-Order Stream and LSPIV Measurement Sequence

**Pre-Processing.** Pre-Processing is must for removing the basic errors obtained in the images. Further, the images are checked with all the errors which going to cause in the post processing. The errors mainly affect the velocity vectors of seeding tracers in the water.

**Image Acquisition.** Good image acquisition is crucial for obtaining accurate measurements with LSPIV. There are three key components to image acquisition using LSPIV. First, there must be adequate features on the stream or channel surface to capture velocity measurements across the image plane. A homogeneous distribution of particles at a medium seeding density is optimum for image evaluation and the tracers must occupy at least one pixel. Second, the channel must be sufficiently illuminated to create contrast between the tracers and the water surface. Finally, the camera must be able to capture the image field at the desired frame rate, field of view, and resolution. Appropriate incorporation of all of these components is a prerequisite for the development of a functioning LSPIV system. The image acquisition equipment and procedures used in the laboratory experiments is developed after an extensive review of the literature. The camera used in the study is Canon EOS 7D DSLR camera. This camera is selected because of its high resolution 5184 x 3456, 8 fps continuous shooting speed in image mode, and robustness for outdoor and indoor use. After that a PIVlab tool is used to collect images and the high image capture rate is necessary to capture reasonable pixel displacements for high velocities. An Acer ASPIRE 5733 computer with 4GB RAM, and 2.40 GHz operating speed is used to collect and analyze the data. All forms of PIV analysis using PIVLab must start with importing a series of images.

**Image Enhancement.** Several aspects of image acquisition can introduce error, such as poor illumination, glare, and shadows. Image enhancement methods can remove some of these errors prior to image evaluation. Image enhancement is performed by altering the pixel values of the recorded image using image-processing software. Typical image enhancement methods used in LSPIV include increasing the signal-to noise ratio, attenuating background noise, and improving the contrast and brightness of the image, DPP (Digital Photo Professional) software is used for Raw Image Processing, Viewing and Editing of images.

**Registration and Transformation.** In LSPIV laboratory and field applications, the camera is usually at an angle of 90° fixed to the tripod arranged in the channel flow, which introduces spatial distortion. The distortion is corrected through image transformation, which relates the pixels to their physical locations. The transformation can be corrected through implicit parameters of the camera. LSPIV applications, the transformation is explicit, as ground reference points (GRP) are used to numerically optimize the parameters. In the laboratory experiments, registration points are fixed using gauge placed in the flume prior to data collection. Prior to the experiment, the pixel and physical locations of four reference points with respect to flume are found for all the camera angles (0°, 15°, 30°, and 45°). The average dimensions of the pre-processed images are approximately 5184 × 3456 pixels. The images are analyzed to determine tracer displacements using cross-correlation algorithms.

**Image Processing Parameters.** Several parameters must be specified prior to image evaluation, including the interrogation window size, region of interest (ROI), window-offset scheme, and grid spacing (Adrian, 1991). Optimization of the interrogation window size and location is a principal means of acquiring accurate displacement measurements. The window size must be small enough to preserve the spatial scale of interest (any scale smaller than the window size is lost) and avoid second-order effects (e.g., displacement gradients). The ROI sizes are selected to be four times the maximum displacement in order to deliver minimum displacement error. The ROI sizes varied depending on the size of the image and the average tracer displacement. The laboratory procedure with ROI window size as a first pass, and then a (128, 64) offset window as a second pass. The cross-correlation coefficients and corresponding displacement vectors are calculated at the user-specified grid spacing of 32 pixels. This produced a 12 × 20 or more vector surface velocity field. The image evaluation procedures are used in the laboratory flow analysis. Our image velocimetry algorithm uses the cross- correlation coefficient as a similarity index (Fujita et al., 1998). Cross correlation is computed between an interrogation area (IA) in the first image and interrogation areas located within a search area (SA) in the second image. The pair of particles showing the maximum cross-correlation coefficient is selected as a candidate vector. The cross-correlation coefficient, $R_{ab}$, is defined as;

$$R\left(a_{ij},b_{ij}\right)=\frac{\sum_{\substack{1\le i\le M_i\\1\le j\le M_j}}\left(A-\overline{A_{ij}}\right)\left(B_{ij}-\overline{B_{ij}}\right)}{\left[\sum_{\substack{1\le i\le M_i\\1\le j\le M_j}}\left(A-\overline{A_{ij}}\right)^2\sum_{\substack{1\le i\le M_i\\1\le j\le M_j}}\left(B_{ij}-\overline{B_{ij}}\right)^2\right]^{2}} \quad (1)$$

Where $M_i$ and $M_j$ are the respective sizes of the interrogation areas in pixels. $\overline{A_{ij}}$ and $\overline{B_{ij}}$ are the respective distributions of grey-scale intensity in the two images. $\overline{A_{ij}}$ and $\overline{B_{ij}}$ are the mean intensities in that interrogation area used to normalize the intensities. The most probable displacement of the fluid from point in a time interval is the one corresponding to the maximum correlation R (aij, bij). Our image processing algorithm is similar to the correlation imaging velocimetry of Fincham and Spedding (1997). Both algorithms use a variance normalized correlation, in which each pixel in the IA is equally weighted, such that the background is just as important as the particle images.

**Post Processing.** Post processing is the analysis window in which all the experimental parmameters can be changed to our requirements.

**Data Processing.** The PIV algorithms are some of the most robust available and well-tested in laboratory settings. After the desired images have been selected for import into PIVLab it is necessary to choose the settings with which the program will analyze the images. It is here that the user is able to setup the interrogation area and the interrogation offset that will determine how PIVLab tracks the particles. It is also possible to setup multiple passes to try to increase the accuracy of the measurements. The multiple pass options work in the exact same way as the initial interrogation window setup. Using ROI can also drastically reduce computation time by eliminating part of the image from the analysis. Image Masks work in the opposite way from ROI. Instead of choosing the area that you wish to analyze, instead you are choosing the area which you chose to ignore.

**Vector Validation.** This particular set of options allows the user to put local filters on the image to change how PIVLab analyses the data. These filters can remove vectors based on how many standard deviations they are away from the mean or local median.

**Image Evaluation Using PIV.** Image evaluation techniques use image intensity fields to measure tracer displacements and estimate a surface velocity field. Image evaluation is a critical step in LSPIV, requiring the most specialized techniques. Most LSPIV research has used correspondence techniques to determine particle displacement. Cross-correlation algorithms work well for the velocities and the low seeding densities found in open-channel applications.

**Discharge Estimation.** The LSPIV velocity measurements are used to calculate discharge using the same area-velocity procedures as other devices (e.g., current meter). In the laboratory, a surface correction factor of 0.95 is used to estimate average velocity from surface velocity. This value is determined from preliminary testing that indicated a parabolic vertical velocity profile. The Plexiglas flume provided minimal resistance, necessitating a smaller correction coefficient. However, at shallower depths, the velocity profile may have changed, which is not reflected in the surface correction coefficient.

## EXPERIMENTAL STUDY AND DATA ANALYSIS

An Experiment is conducted on a tilting flume having 12m long, 0.5m wide, and 0.6m depth and a collecting tank with attached pitot tub. The flow is allowed to release from the elevated tank directly connected through pipes and a constant head is maintained by operating the value. The flow measurements have been taken with various discharges as well as for varying slopes with constant discharges. The actual discharges are calculated through tank measurements. The Concurrent measurements are conducted using LSPIV technique and volumetric method. Measurements are acquired near the flume approximately 10 m long from the inlet of a channel Schematic depicting the experimental setup at Fluid Mechanics laboratory as shown in Figure 3. The LSPIV images are taken using a commercial HD camera shown in Figure 2 (Canon EOS 7D) and fixed to the adjustable tripod. Volumetric method acquires two transects to take readings but LSPIV covering two transects in image. The vector fields at the surface and in the verticals across the channel are acquired with relatively small amount of effort is less than hour. Continuous images are recorded for different depth of flows to allow for several analyses. Both the techniques are adopted, analyzed manually and the data acquisition software. However the technique requires the tracers to track the flow.

The measured data from a flume with vegetation, coarse aggregate is tabulated in table1 and is used to calculate various flow measurements such as velocity, discharge. The equation is used for the calculation of these values are shown below.

$$V = \frac{C_1 R^{2/3} S^{1/2}}{n} \tag{2}$$

$$Q = A \text{ x } V \tag{3}$$

**FIGURE2**. Canon Eos 7d Professional Camera

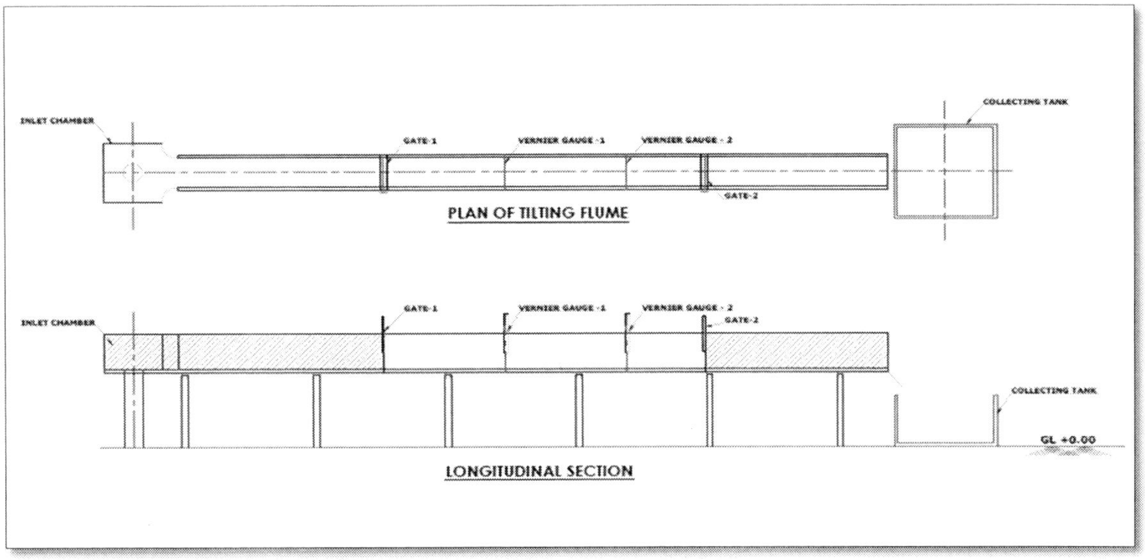


**FIGURE 3.** Schematic Depicting the Experimental Setup at Fluid Mechanics Laboratory

## RESULTS AND DISCUSSIONS

Measuring discharges with traditional methods is time consuming and other errors. Local remote sensing, using imaging systems and appropriate image analysis techniques, could provide a reliable and cost-effective technique for discharge measurement in some situations. This study examines an image-based non-contact approach in an experiment on the open channel. PIV is used to estimate surface velocities from floating particles that are moving along with the channel flow. Velocity–area methods are then used to estimate discharge at a channel cross section. In most laboratory applications of PIV, flow seeding is needed to assure that there are sufficient flow tracers to compute velocities. This experiment uses artificially introduced seeding particles. Secondly, the ground reference points are required to correlate in the PIVLab with actual measurements.

**TABLE 1.** Experimental Condition, PIV Surface Velocity and Flume Velocity under Coarse Aggregate.

| Date | Water Level(m) | Weather | Recording Local Time | Channel Slope | PIV velocity m/s | Flume Velocity m/s | PIV Discharge m3/s x 10^-3 | Actual Discharge m3/s x 10^-3 |
|---|---|---|---|---|---|---|---|---|
| | 0.053 | | 8:04PM | | 0.267 | 0.258 | 6.8 | 6.83 |
| 2/9/2014 | 0.058 | | 8:10 PM | | 0.268 | 0.259 | 7.5 | 7.51 |
| | 0.06 | Sunny | 8:14 PM | 1/1200 | 0.263 | 0.272 | 8.03 | 8.16 |
| 6/9/2014 | 0.06 | | 11:49 AM | | 0.267 | 0.278 | 8.33 | 8.34 |
| 6/9/2014 | 0.064 | | 11:53 AM | | 0.297 | 0.293 | 9.37 | 9.37 |
| | 0.045 | | 12:01 PM | | 0.297 | 0.29 | 6.25 | 6.52 |
| | 0.052 | | 12:04 PM | | 0.268 | 0.263 | 6.81 | 6.83 |
| 6/9/2014 | 0.055 | Sunny | 12:07 PM | 1/600 | 0.283 | 0.279 | 7.5 | 7.67 |
| | 0.057 | | 12:10 PM | | 0.297 | 0.296 | 8.33 | 8.43 |
| | 0.06 | | 12:18 PM | | 0.297 | 0.311 | 9.37 | 9.33 |
| | 0.047 | | 12:25 PM | | 0.291 | 0.275 | 6.42 | 6.46 |
| | 0.05 | | 12:28 PM | | 0.268 | 0.274 | 6.81 | 6.85 |
| 6/9/2014 | 0.052 | Sunny | 12:31 PM | 1/400 | 0.291 | 0.286 | 7.5 | 7.43 |
| | 0.054 | | 12:34 PM | | 0.297 | 0.308 | 8.33 | 8.31 |
| | 0.057 | | 12:37 PM | | 0.332 | 0.331 | 9.37 | 9.43 |
| | 0.045 | | 12:48 PM | | 0.267 | 0.273 | 6.08 | 6.14 |
| | 0.048 | | 12:52 PM | | 0.297 | 0.302 | 7.03 | 7.24 |
| 6/9/2014 | 0.05 | Sunny | 12:54 PM | 1/300 | 0.297 | 0.314 | 7.75 | 7.85 |
| | 0.053 | | 12:57 PM | | 0.313 | 0.317 | 8.33 | 8.4 |
| | 0.056 | | 12:59 PM | | 0.332 | 0.348 | 9.78 | 9.74 |
| | 0.042 | | 1:05 PM | | 0.297 | 0.303 | 6.25 | 6.36 |
| | 0.046 | | 1:08 PM | | 0.297 | 0.306 | 7.03 | 7.03 |
| 6/9/2014 | 0.049 | Sunny | 1:12 PM | 1/240 | 0.2697 | 0.312 | 7.5 | 7.64 |
| | 0.05 | | 1:16 PM | | 0.325 | 0.337 | 8.33 | 8.42 |
| | 0.053 | | 1:19 PM | | 0.335 | 0.356 | 9.37 | 9.43 |
| | 0.04 | | 1:28 PM | | 0.332 | 0.32 | 6.25 | 6.4 |
| | 0.044 | | 1:30 PM | | 0.297 | 0.311 | 6.81 | 6.84 |
| 6/9/2014 | 0.046 | Sunny | 1:33 PM | 1/200 | 0.332 | 0.325 | 7.5 | 7.47 |
| | 0.048 | | 1:36 PM | | 0.332 | 0.349 | 8.33 | 8.37 |
| | 0.05 | | 1:39 PM | | 0.369 | 0.377 | 9.37 | 9.42 |
| | 0.038 | | 3:38 PM | | 0.335 | 0.343 | 6.25 | 6.51 |
| | 0.041 | | 3:42 PM | | 0.335 | 0.353 | 7.25 | 7.23 |
| 6/9/2014 | 0.044 | Sunny | 3:45 PM | 1/171 | 0.384 | 0.387 | 8.03 | 8.51 |
| | 0.046 | | 3:47 PM | | 0.38 | 0.379 | 8.65 | 8.71 |
| | 0.047 | | 3:51 PM | | 0.415 | 0.403 | 9.37 | 9.47 |

**LSPIV Velocity Measurements.** The velocity vectors measured during the flow release and are performed at different depths for seven slopes. It appeared that the data values are near to the actual velocities but velocity coefficient 0.95 (Tom Vanden Berghe) is used in the laboratory studies for measuring the discharges. In case of flow under vegetation the effect is more in actual measurement. This relatively large variation of flow for the effective height of grass may be explained by the contraction of bed material because of stagnation water.

**LSPIV Parameterization and Computations.** The image sampling frequency depends on apparent water velocity and is adjusted to obtain pattern displacements of 124 pixels, which is an acceptable compromise to obtain significant displacement of well-correlated patterns. As the view points and velocity magnitudes are different for the varying flows. Image sequences have variable quality, and some of them have sun reflection and shadows. However all the image sequences are used for LSPIV analysis. The LSPIV correlation is performed on squares of gray-scale pixels which are called interrogation areas (IA). The maximum correlation search is restricted to a rectangular domain which is called search area (SA). Surface Velocities of a Channel with Coarse Aggregate and Vegetation as a Bed Material using LSPIV (Slope: 1:1200) are analyzed. Likewise for every slope discharges are measured and tabulated 1 and 2. The comparison of LSPIV surface and discharge measurements with concurrent methods gives a valuable indication of the LSPIV discharge reliability and accuracy. The deviations for both the methods of discharge values an average are -0.075% and +0.094%. The resulting discharge deviation is±0.05% for vegetation whereas -0.02% is for coarse aggregate.

**TABLE 2.** Experimental Conditions, PIV Surface Velocity and Flume Velocity under Vegetation.

| Date | Water Level(m) | Weather | Recording Local Time | Channel Slope | PIV velocity m/s | Flume Velocity m/s | PIV Discharge x 10^-3 m3/s | Actual Discharge m3/s x 10^-3 |
|---|---|---|---|---|---|---|---|---|
| 30/08/2014 | 0.051 | Sunny | 4:24PM | 1/1200 | 0.249 | 0.240 | 6.349 | 6.12 |
| | 0.054 | | 4:28 PM | | 0.283 | 0.263 | 7.641 | 7.1 |
| | 0.058 | | 4:31 PM | | 0.281 | 0.270 | 8.149 | 7.83 |
| | 0.061 | | 4:34 PM | | 0.281 | 0.280 | 8.57 | 8.54 |
| | 0.064 | | 4:37 PM | | 0.334 | 0.310 | 0.01 | 9.92 |
| 30/08/2014 | 0.047 | Sunny | 5:46 PM | 1/300 | 0.242 | 0.265 | 5.687 | 6.22 |
| | 0.051 | | 6:37 PM | | 0.222 | 0.220 | 5.661 | 5.61 |
| | 0.056 | | 6:40 PM | | 0.205 | 0.200 | 5.74 | 5.6 |
| | 0.058 | | 6:43 PM | | 0.242 | 0.240 | 7.018 | 6.96 |
| | 0.06 | | 6:46 PM | | 0.268 | 0.270 | 8.04 | 8.1 |
| 30/08/2014 | 0.048 | Sunny | 7:05 PM | 1/240 | 0.263 | 0.260 | 6.312 | 6.24 |
| 30/08/2014 | 0.052 | | 7:08 PM | | 0.242 | 0.262 | 6.292 | 6.81 |
| 2/9/2014 | 0.054 | | 12:34 PM | | 0.335 | 0.290 | 9.045 | 7.83 |
| 2/9/2014 | 0.057 | | 12:37 PM | | 0.335 | 0.300 | 9.547 | 8.55 |
| 2/9/2014 | 0.059 | | 12:40 PM | | 0.336 | 0.320 | 9.912 | 9.44 |
| 2/9/2014 | 0.061 | Cloudy | 2:38 PM | 1/600 | 0.336 | 0.300 | 9.912 | 9.15 |
| | 0.065 | | 2:42 PM | | 0.298 | 0.310 | 9.685 | 0.01 |
| | 0.067 | | 2:48 PM | | 0.297 | 0.325 | 9.949 | 0.01 |
| | 0.068 | | 2:52 PM | | 0.297 | 0.326 | 0.010 | 0.011 |
| | 0.071 | | 2:55 PM | | 0.332 | 0.328 | 0.011 | 0.011 |
| 2/9/2014 | 0.052 | Cloudy | 5:22 PM | 1/400 | 0.291 | 0.253 | 7.566 | 6.57 |
| | 0.056 | | 5:29 PM | | 0.297 | 0.255 | 8.316 | 7.14 |
| | 0.057 | | 5:32 PM | | 0.297 | 0.263 | 8.464 | 7.49 |
| | 0.059 | | 5:35 PM | | 0.299 | 0.280 | 8.82 | 8.26 |
| | 0.062 | | 5:37 PM | | 0.297 | 0.310 | 9.207 | 9.61 |
| 2/9/2014 | 0.051 | Cloudy | 12:48 PM | 1/200 | 0.268 | 0.250 | 6.834 | 6.37 |
| | 0.052 | | 12:51 PM | | 0.297 | 0.280 | 7.722 | 7.28 |
| | 0.054 | | 12:54 PM | | 0.297 | 0.308 | 8.019 | 8.31 |
| | 0.056 | | 12:57 PM | | 0.297 | 0.300 | 8.316 | 8.4 |
| | 0.058 | | 13:00 | | 0.335 | 0.326 | 9.715 | 9.45 |
| 2/9/2014 | 0.052 | Cloudy | 1:20 PM | 1/171 | 0.257 | 0.251 | 6.682 | 6.52 |
| | 0.054 | | 1:23 PM | | 0.268 | 0.265 | 7.236 | 7.15 |
| | 0.056 | | 1:26 PM | | 0.267 | 0.270 | 7.476 | 7.56 |
| | 0.059 | | 1:28 PM | | 0.268 | 0.275 | 7.906 | 8.11 |
| | 0.061 | | 2:28 PM | | 0.297 | 0.310 | 9.058 | 9.45 |

## CONCLUSIONS

A new LSPIV technique has been adopted to measure the velocity and discharge of surface flow performed on the tilting flume and it is observed from the study that, the flow velocity varies from minimum 0.20 m/s to maximum of 0.415 m/s for different discharges. It is observed from the Table 1 that, the error in velocity variation is found to be of $\pm 1\%$ when compare with both LSPIV values and the experimental results values when the coarse aggregates used as channel bed materials. It is found from the measurements that, the error in velocity variation is less than 1% when compare with both LSPIV values and the experimental values and the bed material used as Bermuda grass type vegetation. It is concluded based on the results that, the adopted new LSPIV techniques is more suitable for stream flow measurements and have good agreement with volumetric method results. This LSPIV technique is very much useful and reliable for surface flow measurements where there is a difficulty to measure directly. This technique also has advantage for measurement of flow circulation, shear rate and normal strain rate analysis.

## REFERENCES

Adrian, R.J. 1991. Particle-imaging techniques for experimental fluid mechanics. Annual Review of Fluid Mechanics 23 (1): 261–304.

Church, M., Biron, P., Roy A. 2012. Gravel-bed Rivers: Processes, Tools, Environments. John Wiley & Son.

Fincham, A.M., and Spedding, G.R. 1997. "Low cost, high resolution DPIV for measurement of turbulent fluid flow". *Experiments in Fluids* 23 (1997) 449-462.

Fujita, I., Komura, S. 1994. "Application of Video Image Analysis for Measurements of River- surface Flows". Annual *Journal of Hydraulic Engineering*, Japan Society of Civil Engineers, 38: 733-738.

Fujita, I., Muste, M., Kruger, A. 1998. "Large-Scale Particle Image Velocimetry for flow analysis in Hydraulic Engineering Applications". *Journal of Hydraulic Research* 36(3), 397-414.

Harpold, A. A., Mostaghimi, S., Vlachos, P.P., Brannan, K., Dillaha, T. (2006). "Stream Discharge Measurement Using A Large-Scale Particle Image Velocimetry (LSPIV) Prototype." *American Society of Agricultural and Biological Engineers* Vol. 49(6): 1791–1805.

ISO 1997. Optimum installation of the equipment using established guidelines.

Jirka, G.H. 2001. "Large scale flow structures and mixing processes in shallow flows". *J. Hydraul. Res,* 39, 567-573.

Melcher, N. B.,Costa, J.E., Haeni, F.P., Cheng, R.T., Thurman, E.M., Buursink, M., Spicer, K.R., Hayes, E., Plant, W.J., C. Keller, and Hayes, K.. 2002. "River discharge measurements by using helicopter-mounted radar." *Geophysical Res. Letters* 29(22): 411-414.

Muste, M., Fujita, I., Hauet, A. 2008. "Large-scale particle image velocimetry for measurements in riverine environments". *Water Resources Research* 44(4):W00D19.

Nord, G., Esteves, M., Lapetite, J.M., Hauet, A. 2009. "Effect of particle density and inflow concentration of suspended sediment on bedload transport in rill flow". Earth Surface Processes and Landforms 34(2): 253-263.

Rantz, S.E.1982. "Measurement and Computation of Streamflow". Measurement of Stream and Discharge, vol. 1, Water Supply Paper 2175. US Geological Survey, Washington.

Simpson, M.R. 2001. Discharge Measurements Using a Broad-Band Acoustic Doppler Current Profiler. Open-File Report 01-1, U.S. Geological Survey, Sacramento, CA.

Tom Vanden Berghe (2013). "Image processing for a LSPIV application on a river". Thesis Submitted to the ingenieurswetenschappen: bouwkunde.

# RADIOSONDE LAUNCH UNDER HIGH SURFACE WIND

***Boris S. Yurchak***
(EPA, Arlington, VA, USA)

**ABSTRACT**: Under emergency situation accompanied with hazardous atmospheric releases of pollutants from chemical or nuclear enterprises, the reliability and continuity of the upper-air data flow, which are primarily measured with radiosonde, are of particular importance for mapping the transport of contamination. However, the ability to launch the radiosonde can be strongly impacted by the high surface wind speeds under inclement weather. To increase the speed range of the successful launch, the cone-shaped tower is suggested to be used for launch. Use of the cone-shaped tower will allow for a greater initial height of instrumental box pre-start position as compared with the routine launch method. The tower uses the wind force jointly with the lifting force of the balloon to deliver the instrumental box up to the top of the tower. The installation of the lower part of the tower into the wind shelter secures the operator from contamination and wind impacts and provides the radiosonde launch even in case of very strong wind (i.e., gale, ~ 18–24 m/s). The shelter has a sliding roof with an opening. It is reasonable to convert some of aerological stations, located close to the potentially dangerous enterprises, into proposed secure all-weather facility.

## INTRODUCTION

Air pollution dispersion as a result of an accident in potentially dangerous enterprises (PDE) of chemical and nuclear industries is governed by the physical parameters of the Atmospheric Boundary Layer, which are primarily measured with radiosonde (upper-air) sounding. The radiosonde (RS) consists of the balloon filled with a gas lighter than air (hydrogen or helium) and connected with a harness string to the instrumental box consisting of the sensors for meteorological parameters and with the radio transmitter for wirelessly delivering the data to the ground receiver. The balloon filling and other preflight preparation and maintenance procedures are conducted in a shelter (NWS Manual, 2007). The contemporary upper-air (aerological) sounding is carried out up to the moderate surface wind (to 12 m/s) by the operator, usually from the open launching pad located at some distance from the shelter. The RS launching under conditions of the strong surface wind at the speeds of higher than 12 m/s is a complex, physically hard and sometimes even dangerous work for the operator. There is no reliable way to protect the operator from contamination deposition during the launch. The objective of the research presented is to work out a method of RS launching at the wide range of the surface wind speed, which provides safe launch of the instrumental box and ensures the protection of the operator from impacts of strong wind and contamination/debris.

## AN EQUATION OF THE SUCCESSFUL LAUNCH

The RS launching at the presence of the wind at the launching pad is provided by the use of some well-known measures briefly considered below using the criteria of the successful launch formulated in (Yurchak, 2013, 2014a):

$$Z_g(t_0) \equiv \frac{z(t_0)}{L} = y + \frac{\pi}{2}\frac{\overline{W}}{\sqrt{gL}} - 2\sin^2\frac{\theta}{2} > 0 \tag{1}$$

where $Z_g(t_0)$ is the relative height of RS payload at the moment $t_0$; $t_0 \approx (\pi/2)\sqrt{L/g}$ ($L$ is the length of the harness string, $g$ is the acceleration of gravity) is the time needed for the harness string to take the vertical position after releasing the RS; $z(t_0)$ is the height of the instrumental box above the surface at the vertical position of the harness string at the time moment $t_0$; $y = h/L$, where $h$ is the initial height of the instrumental box above the surface at the moment of releasing the RS; $\overline{W}$ is the mean speed of the balloon ascent during the time $t_0$ (equation (20) in (Yurchak, 2013)); $\theta$ is the angle of the deviation of the harness string from the vertical at the moment of releasing the RS that depends on surface wind speed (equation (2) in (Yurchak, 2013)). The above equation (inequality) (1) is the theoretical basis for the method presented below.

## A CONE TOWER AS THE BASIC ELEMENT TO IMPROVE CONDITION OF RS LAUNCH

As follows from (1), the increase in the initial height of RS release $h$ causes the increasing of the relative height $y$ and, consequently, makes the left part of the inequality greater that mitigates the satisfaction of its condition (… > 0). This is the primary idea of the proposed technique. It should be noted that known solutions of the problem are based on temporary shortening of the length of the payload harness string ($L$) with so-called temporarily harness shortening devices (THSD) which brief review has been provided in (Yurchak, 2014a). To put the idea into practice, it is proposed to launch the RS using a cone tower as that schematically shown in Figure 1.

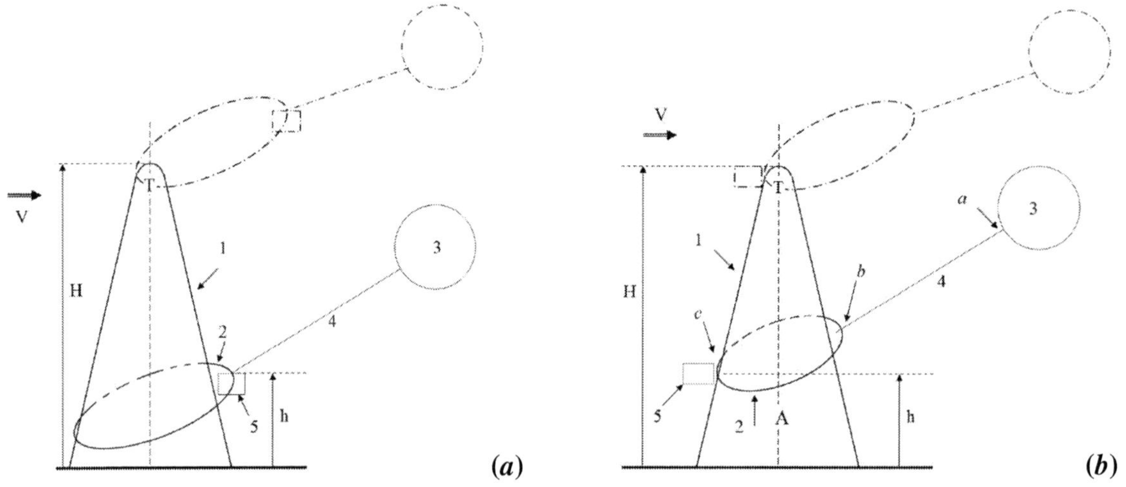


**FIGURE 1. A scheme of RS launch using the cone tower. (*a*) The launching from the leeside of the tower. (*b*) The launching from the windward of the tower using the sliding loop as a constituent of the harness string. Designations of the elements are in the text.**

An operational principle of the cone tower *A* is the following: the transportation of the sliding loop *2* (embracing a perimeter of the tower *A*) is directed along the inclined conic wall *1* with the harness string *4* stretched from the side of the balloon *3* moving under the pressure of the wind force and its own lifting force. Due to the inclination of the lateral wall *1* which lifts the sliding loop *2* and the instrumental box *5* connected with it, to the top of the tower *T*. The inclined surface of the lateral wall *1* should be smooth in order to minimize the friction with the sliding loop *2*. To provide the unimpeded separation of the sliding loop *2*, the top *T* of the tower is rounded. The launch can be carried out at the joining of the instrumental box *5* and the sliding loop *2* both from the lee and windward sides (Figures 1a and 1b, respectively). In the latter case, the sliding loop *2* becomes a constituent of the harness string 4 after the separation from the tower. Other versions of the launching pattern as well as the physical substantiation of the RS harness moving along to the top of the tower are provided in (Yurchak, 2014a, 2014b). The resume of this analysis states the relationship between wind speed and constructive features of the tower to make this moving possible as follows:

$$\cot\left[\theta(V_{\max}) - \frac{\omega(r_T)}{2}\right] \geq k_f \tag{2}$$

where $\theta(V_{\max})$ is the angle of the deviation of the harness string from the vertical at the moment of releasing the RS that depends on surface wind speed $V_{max}$ (equation (2) in (Yurchak, 2013)); $\omega(r_T) = 2\cot^{-1}(r_T/H)$ is the vertex angle of the cone tower of base radius $r_T$ and the height $H$; $k_f$ is the friction factor of the sliding loop on the surface of the lateral wall of the tower. The smaller the difference between the angle of deviation of the harness string θ and the vertex angle ω, the better is the meeting of inequality (2). Equation (2) helps to find some constructive features of the tower. For example, if one expects $V_{max}$=20 m/s then $\theta$=80$^0$ (see (2) in (Yurchak, 2013)). If $k_f \leq 0.22$ then $\theta$-$\omega$/2=77.35$^0$ and it results $\omega$=5.3$^0$. If $H$=10-12 m then $r_T \approx 0.5$ m. It should be noted that even in the case of very high wind speed, when the chord of the balloon is oriented practically horizontally (i.e., θ~ 90$^0$), the sliding loop 2 has the nonzero component of the driving force which brings the balloon to the upper point of the tower. According to its functional purpose, the cone tower is named a cone launching tower (CLT). The essential property of the CLT is the delivery of the instrumental box to the top of the tower using only the energy of the wind and balloon lift, without using any additional devices and mechanisms. To launch the RS using CLT, the use of THSD is not ruled out from the general point of view. However, a possibility of THSD usage in practice needs additional field testing. If the upper part of the CLT is made of an electromagnetically transparent material, the antennas of the radio equipment including the radar can be installed in the tower for the reception of the telemetry data (Yurchak, 2014b).

## SECURE RS LAUNCH USING THE SHELTER

The CLT only is not sufficient for the implementation of the method because it is very difficult for the operator to reach the tower from inflation room and to dock the RS to the sliding loop due to obstructing impact of the strong wind. To resolve the issue, the secure method of the RS launching using the CLT that is partially installed in the room-shelter can be suggested, Figure 2. The launching procedure is as follows. When installing the CLT *A* in the shelter *6*, both halves of the sliding roof *8* are made with half-round notches that adjoin close to the prominent part of the cone tower *A* in the closed position. This enables carrying out the prestart preparation of the RS in the absolutely wind- and precipitation / contaminant sedimentation–proof room. To provide the unhampered launching of the filled balloon *3* through the open roof *8* along the cone tower *A*, the second (upper) sliding loop *7* is used, Figure 2a. It is docked with the appendix *a* (see location of point *a* in Figure 1b) of the balloon *3*. To launch the RS, the roof *8* slides apart and the operator moves the balloon *3* through the open roof *8* gradually releasing the harness string *4*. The balloon *3* contacting the edges of the open roof *8* is prevented by the upper sliding loop *7* which does not allow the balloon *3* impacted by the wind to separate from the tower *A* at the level of the roof *8*. When reaching the tower top *T*, the upper sliding loop *7* slips off (Figure 2b) and the balloon *3* separates from the tower *A*. After that, the operator releases the instrumental box *5* docked to the lower sliding loop *2* which is transported by this loop to the tower top *T* along the lateral wall *1* under the influence of the tight harness string *4*. The contact of the string *4* with the internal half-rounded edge of the open part of the roof *8* and even its flexure do not affect at all the ability of the instrumental box *5* to reach the tower top *T* freely and the ability of the lower sliding loop *2* to slip off it subsequently for providing the free flight of the RS, Figure 2c. Other versions of the launching pattern from the shelter using sliding loops as harness string constituents are provided in (Yurchak, 2014a, 2014b). During the RS launching after the opening of the roof *8*, the operator remains protected from the direct wind impact and is subjected to the short-term tilt directed (due to the wind impact) precipitation / sedimentation only. If the room-shelter *6* is provided with proper ventilation and heat insulation from auxiliary rooms of the aerological station, the preflight maintaining of the RS can be carried out both at the open and close roof *8* of the shelter *6*. It also should be noted when the roof is opened, the difference of wind speed over the shelter and inside the shelter causes the pressure difference directed (from high to low) out of the shelter. That prevents penetration of the contamination into the shelter and thus is favorable from the point of view of the health protection.

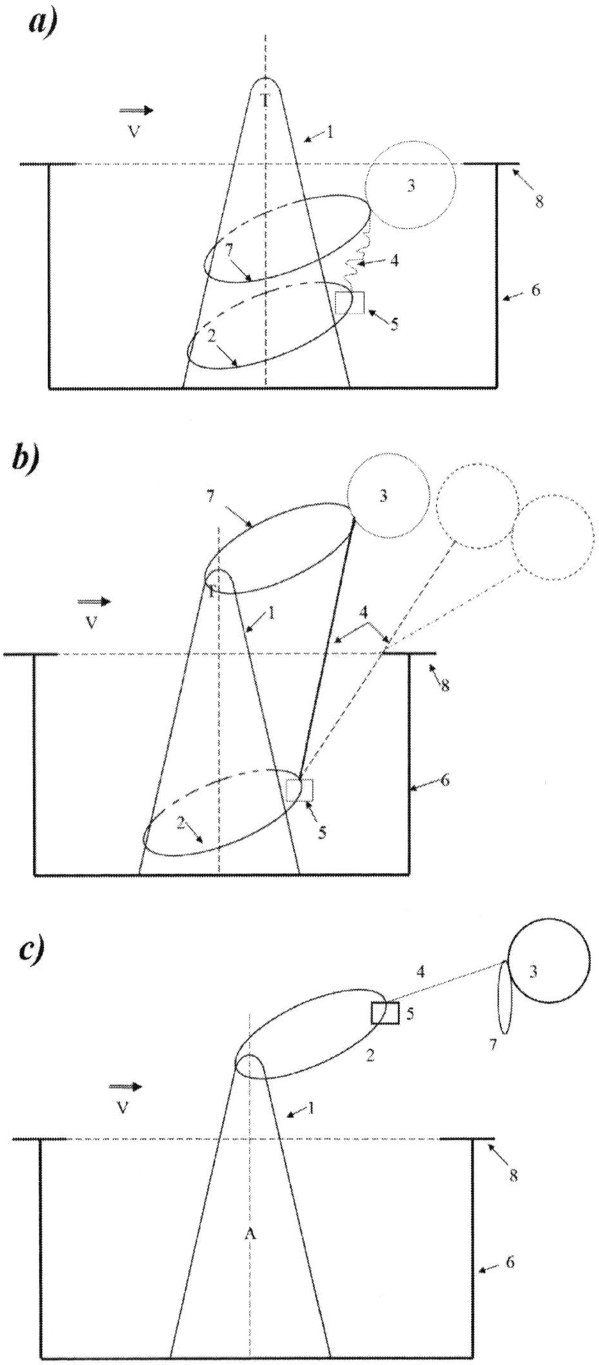


**FIGURE 2. Radiosonde launch from the shelter using the cone tower. Designations are in the text.**

## FIELD MODEL EXPERIMENT

The field model test of the RS launching from the CLT with one and two sliding loops was carried out for the experimental qualitative testing of the feasibility of the proposed technique. The S-7879O signal road cone made of polyvinyl chloride with the height of 70 cm and diameters of base and top equal to 27.5 cm and 5.6 cm, respectively, was used as a cone tower model. The latex balloon having a shape close the ellipse in the inflated state with the major and minor axes of 42 cm and 30 cm, respectively, was used. The strong wind, that causes the deviation of the harness string from the vertical of about $80^0$, was generated by the industrial vacuum cleaner blowing out. The harness string made of the nylon thread had a length equal to the cone height. The sliding loops were made of the same thread. The instrumental box was imitated by the light cylinder container with diameter of ~ 4.5 cm. As a result of the test, the major principle of the method was corroborated: the balloon moves the sliding loop with the payload fixed to it to the top of the tower using the natural wind force and its own aerostatic lift. It was revealed, additionally, that with using two sliding loops, the sharp upward jerk of the harness string that takes place at the moment of the separation of the upper loop from the tower is reduced. At the same time, the lower loop with the model of the instrumental box reaches the cone top very quickly and moves away from it. It may be assumed this occurrence is due to the disappearance of the friction of the upper loop that, in turn, causes additional acceleration of the lower loop movement along the cone wall. Therefore, this feature is an additional favorable factor for the unhampered launching of the instrumental box from the shelter.

## CONCLUSIONS

To extend the range of the surface wind speed that allows for successful radiosonde launch, it is substantiated to use the cone tower; it provides the delivery of radiosonde elements to the desired launching height using the wind energy only. The possibility of the technique is corroborated experimentally by means of the field model testing. To protect the operator against wind, precipitation and contamination from hazardous air-borne pollutants, the cone tower is partly installed in the shelter with the sliding roof. It is shown how the unhampered launching of the radiosonde from the shelter can be done. The radiosonde launch using the cone launching tower within the shelter can be carried out. Not going beyond the area of the aerological station considerably improves the working conditions of aerologists, especially under extreme weather conditions and contaminated environment. To have the reliable and uninterrupted data flow of air physical parameters and to have protected working conditions for personnel of the aerological stations, it is reasonable to convert some of aerological stations, located close to the potentially dangerous enterprises, into proposed secure all-weather facility. The novelties of the method and of the aerological station type have been confirmed by the US patent (Yurchak, 2014b).

## REFERENCES


National Weather Service. 2007. Manual 10-1401. "Operations and Services, Upper Air Program NWSPD 10-14, Radiosonde Observations." Washington, D.C.

Yurchak, B.S. 2013. "An Assessment of Radiosonde Launch Conditions Affected by the Surface Wind." *Russian Meteorology and Hydrology*, 38(3):159-167.

Yurchak, B.S. 2014a. "A Technique of Radiosonde Launch under the Surface Wind of High Speed." *Russian Meteorology and Hydrology*, 39(1):38-46.

Yurchak, B.S. 2014b. "Cone Tower Based Facility and Method for Launching an Atmospheric Sounding Device under Strong Winds." Patent No.: US8,857,759 B2. Date of Patent: Oct. 14, 2014. Untied States Patent and Trademark Office. Washington, D.C.

# THE CRITICAL PARAMETERS FOR HOW TO ADDRESS OIL/HYDROCARBON BASEDMATERIAL SPILLS

***Steven Pedigo***
(OSEI Corp, P.O. Box 515429, Dallas, Texas 75251, USA)[106]

The purpose of these actions is to insure a safe for humans, non toxic to marine species clean up that will permanently remove spills from the environmentwithout creating secondary clean up processes trade offs or adverse effects, if possible.

Three basic parameters are main considerations whendetermining how to clean up an oil/fuel hydrocarbon based material spill. (1) Is the product/process safe for humans? (2) Is the product/process safe for aquatic marine life? (3) Does the product/process permanently remove oil/fuel from the environment in accordance to the US Clean water act, without creating any secondary clean up problems/processes.

There is one Solution that meets these 3 basic criteria, that has been used on over 33,000 spills since 1989, Oil Spill Eater II (OSE II).

The process developed with OSE II emulates mother natures own process, except OSE II speeds the process of converting oil/fuels/hydrocarbon based materials to CO2 and water in a matter of days to weeks, while mother nature may take 25 to 50 years, allowing the lingering toxicity of the oil I the environment. Oil Spill Eater II the proven, safe for humans, non-toxic to aquatic species, means to permanently remove oil/fuel spills from the environment, without creating any secondary spill problems or clean up processes.

This information is presented to show how and why decisions should be based for spill response and show there is a safe effective means to meet the US CleanWater act, and requirements set forth in other countries.

# AN ALTERNATIVE METHOD TO OVERCOME INCONSISTENT CELLULOLYTIC SCREENINGS WHEN USING GRAM'S IODINE

***Joshua OHair***, Terrance Johnson, Anthony Ejiofor and Suping Zhou
(Tennessee State University, Nashville, TN, USA)

Gram's Iodine is currently used in detection of endoglucanase, a cellulase enzyme which cleaves beta 1-4 glycosidic bonds in carboxymethylcellulose (CMC). However, when a zone of degradation was noticed around non-cellulolytic *Escherichia coli* JM109, Gram's Iodine Assay was further scrutinized. It was during this investigation that a clear ring of degradation around several bacterial colonies was observed in plates containing no cellulose, and on plates containing agar only. However, when agar was removed as a solidifying agent and substituted with phytagel, zones of clearing around all cellulolytic organisms were altered. This led to the conclusion that phytagel should be used as a designated solidifying agent when using the Gram's Iodine Assay for cellulase screening. The use of the improved method can overcome the inconsistency problems for detecting cellulose activity on agar-CMC plates.

# LANDSCAPE CHANGES IN LOPÉ NATIONAL PARK, GABON: TO WHAT IS THE EXTENT OF SAVANNAH THICKENING?

***Gilbert Gauci***
(Environmental Management, University of Stirling, Stirling, Scotland, FK9 4LA, United Kingdom)[108]

Savannah thickening and forest encroachment are two phenomena occurring as parallel processes throughout Africa threatening the seasonal use of the habitat by large mammals and the existence of rare specialist and endemic species found in the savannah that are unable to grow in a forest environment. For the past 23 years Lopé National Park, Gabon has been practicing prescribed burning of the savannah in order to slow down the process. Understanding the extent and effects of this process will aid park managers to develop management strategies to tackle conservation threats. The present study aimed at quantifying the extent of savannah thickening in Lopé National Park. A habitat classification was used to classify the change in habitats between open savannah and colonising forest. Using digital photography taken in each cardinal and ordnance direction in October 2008 and February 2016, 10 sites were examined. Using the free software paint.net, a standard grid was overlaid on the photos and each square was assigned a colour according to the habitat classification. The grid technique detected thickening in 6 out of 10 sites, however, thickening was not always detected due to the size of the squares and a qualitative analysis was applied to further describe habitat change. The photo monitoring method is a low-cost method that can be easily replicated. The study concludes with recommendations on how to monitor savannah thickening by making use of citizen science and remote sensing.

# SOCIETY
# AND
# THE ENVIRONMENT

# PUBLIC PERCEPTION AND ACCEPTANCE OF FERTILIZERS FROM HUMAN URINE AMONG TURKISH CITIZENS HOLDING UNIVERSITY DEGREES

Ayse D. ALLAR EMEK, Nihan YILDIZ DOGAN, ***Bilsen BELER BAYKAL***
(Istanbul Technical University, Istanbul, Turkey)

**ABSTRACT:** Human urine is a viable source of fertilizers, through direct use as well as indirect routes. Proper use of this source will contribute to global sustainability, however such practice will not be effective without the acceptance of the final consumers in a target community. Public perception and approval for/acceptance of fertilizers from human urine was investigated in this work among Turkish citizens holding university degrees in two surveys: a generalized one and a more specific one based on the results of the former. The use of human urine as fertilizer received considerable acceptance among citizens with higher education. Indirect use received preference over direct use. Landscape application was more highly accepted in comparison to food stuff. Use in public areas and commons received higher preference as compared to self-use in households. Plants which were cooked received higher acceptance than those which are eaten raw. The level of acceptance decreased with the higher levels of possible urine contact with food stuff. An appreciable acceptance was also observed for industrial plants grown using urine/urine-based fertilizer. Installation of urine diverting toilets (UDT) received acceptance provided that they are supplied free of charge.

## INTRODUCTION

Stream segregation is a recent domestic wastewater management approach which claims that domestic wastewater is not a waste to be discarded but a source to be revaluated. Segregation has to be done at the point of generation and one of the options is Ecological Sanitation (ECOSAN) which suggests separation as yellow/brown/grey water. Of those streams, yellow water is separately collected human urine which is rich in nutrients, with over 80% of nitrogen and 50% of phosphorus in conventional domestic wastewater, but only 1% by volume (Beler Baykal, 2015). Nutrients are indeed problematic from the perspective of water pollution however; they also constitute the major ingredients of fertilizers. Urine contains macronutrients nitrogen, phosphorus and potassium in a form which is readily available to plants. The idea is to recover nutrients in urine to return them back to soil and to reuse them as fertilizer. This can be done either through direct application of urine onto soil, or through indirect routes, where not the entire urine itself but only nutrients recovered from urine using different processes, are applied onto soil in due course. Direct use of human urine lends concerns including pathogens, pharmaceuticals, hormons and salinity, however, elimination of these drawbacks can be accomplished by processes such as struvite precipitation, ion exchange/adsorption, stripping/absorption. In addition to contributing to sustainability through recovery/recycling/reuse due to this fertilization potential, source separation and ECOSAN is an alternative for conventional water/wastewater management where water pollution can be managed. Urine diverting toilets (UDTs) and urinals help to collect urine separately to make it available for further reuse. Combining the information that the global fertilizer equivalent of human excreta amounts up to about 35% of the global fertilizer consumption, and that one person's urine provides enough fertilization capacity to produce 200 kg of cereals, based on information given by Wach (2007), on an annual basis, with the fact that natural fertilizers are actually excreta of animals, the use of human urine in agriculture seems reasonable.

Several pieces of research work in the literature have provided justification that human urine is a effective source of fertilizers (for example Morgan, 2003; Beler Baykal et al., 2011a; Kocaturk and Beler

Baykal, 2012; Andersson, 2015; Ranasinghe et al., 2016). However, putting this proposal into practice on a large scale requires acceptance by the community. Table 1 is prepared to summarize surveys about the acceptance of urine as fertilizer based on those reported in the literature. A survey of those investigations on a general sense shows that typically about half of the respondents have shown acceptance towards the use of urine as fertilizer. As can be seen from the Nigerian case, when given evidence from an actual demonstration field, acceptance by the community seems to increase remarkably. The table also shows that a significant group of farmers were not aware of the fertilizing potential of urine. In a previous survey made with Turkish citizens who were residents of Istanbul, the acceptance for natural fertilizers was 100% while only about half thought synthetic fertilizers were acceptable. There was a considerably higher acceptance for indirect use of human urine as fertilizer after processing as compared to direct use. Landscape plants received a much higher acceptance as compared to food stuff. The use of human urine was more acceptable for plants eaten cooked in comparison to those eaten raw. Self-consumption of plants fertilized with human urine received a slightly lower (3-11%) acceptance than approval of the use of fertilizers (Beler Baykal et al., 2011b).

**Table 1.** Surveys about the acceptance of urine as fertilizer

| Country | Community | Focus | Opinion | Reference |
|---|---|---|---|---|
| Switzerland | Farmers | Use of fertilizer is a good idea | 57% yes | Lienert et al., 2003 |
| Switzerland | Consumers | Will eat vegetables fertilized with urine? | 72% yes | Pahl-Wosti et al., 2003 |
| | | Preference for urine over artificial fertilizer | 80% yes | |
| Indonesia | Farmers | Use urine and feces based fertilizers? | 80% yes | WSP, 2010 |
| | | Eager to inform customers about the type of fertilizer used | 40% yes | |
| South Africa | Homedwellers with UDDT* | Currently use fertilizers from toilet pits | 2% yes | Okem et al., 2013 |
| | | Use own urine as fertilizer | 53% yes | |
| | | Use other people's urine as fertilizer | 21% yes | |
| Nigeria | Local community | Use urine as fertilizer (no demonstration) | 8% yes | Sridhar et al., 2005 |
| | | Use urine as fertilizer (after showing a demonstration field) | 80% yes | |
| | | Willingness to build urine diversion toilet (after showing a demonstration field) | 95% yes | |
| Ghana | Farmers | Are not aware of human urine as fertilizer | 65% no | Cofie et al., 2011 |
| | Marketers | Accept to sell vegetable fertilized urine | 57% yes | |
| | Consumers | Accept to buy vegetable fertilized with urine | 54% yes | |
| Turkey | Residents of Istanbul | Acceptance of natural fertilizer | 100% yes | Beler Baykal et al., 2011b |
| | | Acceptance of synthetic fertilizer | 49% yes | |
| | | Direct use of urine as fertilizers | 23-55%** | |
| | | Use of fertilizers from urine after processing / indirect use | 68-91%** | |

* UDDT: Urine diversion dehydration toilet
** for different types of plants: landscape plants, cooked crops, uncooked crops

The aim of this paper is to summarize the results of two preliminary surveys in an attempt to assess the perception and level of acceptance of the use of urine and urine-based fertilizers amongst well educated Turkish citizens. Two sets of questionnaires were done in two consecutive years with respondents holding university degrees. Questionnaires were aimed at finding out public attitude in the segment with higher education towards the utilization of urine as fertilizer for agricultural and landscape purposes as well as for industrial plants, and urine diverting toilets.

## MATERIAL AND METHODS

Of the two sets of questionnaires, the first one was directed towards general acceptance of reuse of human urine while the second was focused upon more specific issues taking first one as basis. The first survey had 252 respondents while the second one had 358, a total of 610 respondents, all of which received university education (bachelor degree or higher education). In addition to personal data like age, gender, education, occupation, residence etc., the first questionnaire was directed at assessing (Allar, 2015)

- acceptance of natural and synthetic fertilizers
- approval of using urine directly as fertilizer for specified products
- willingness to self-consume specified products that were fertilized directly using urine
- approval of using processed urine as fertilizer for specified products (i.e. indirect use)
- willingness to self-consume specified products that were fertilized with processed urine (i.e. indirect use)

where the specified products were landscape plants, plants which are consumed after cooking, and plants which are consumed raw.

The second questionnaire explored further (Yildiz-Dogan et al., 2015)

- approval of the use of urine directly/indirectly for different types of plants which included cereals, nuts and fruits which grow on trees, fruits that grow on the ground, which had different chances to be in contact with the urine/urine-based fertilizer applied
- approval of the use of urine directly/indirectly for different types of green areas which included landscape areas, parks, stadiums, home gardens, picnic areas, school gardens
- approval for self-use of industrial plants which included cotton, tobacco/cigarettes, sugar/chocolate, wicker basket fertilized with urine
- motivations/obstacles for using UDTs
- willingness to install and use UDTs

The participants were Turkish citizens who were currently living in Istanbul, in the range of 18-60 years of age, the largest percentage of which was between 20-30. The participants were from different cities of the country, educational backgrounds and occupations. 55% (335 respondents) of the respondents were female, while the rest 45% (275 respondents) were male.

Most of the questions were in multiple-choice format, which requested respondents to choose the most appropriate answer. Additionally, respondents were asked to state any concerns and comments regarding the use of urine as fertilizer. No information about urine-based fertilizers and UDTs were given, so that participants could make decisions based on their existing knowledge on urine, environmental pollution and sustainability.

The surveys were conducted through both face-to-face and internet questionnaires. The analysis of the surveys was based on results as simple head counts and their respective percentages.

## RESULTS AND DISCUSSION

Almost all of the participating individuals (98% of the total of 610 respondents) indicated that they would accept the use of natural fertilizers, however, only 22% approved of the use of synthetic fertilizers, a considerably lower percentage as compared to Table 1.

Figure 1, summarizing the outcome of the first survey, shows the approval of using urine and willingness for self-consumption of products that were fertilized with urine directly or indirectly for landscape plants, cooked and raw crops. The approval of using urine and willingness for self-consumption of plants fertilized through indirect use of urine was higher as compared to direct use. This was true for all three groups of plants questioned. The gap between direct and indirect use was less pronounced for landscape use as compared to plants which are eaten either raw or cooked. The approval and willingness for self-consumption was highest for landscape plants and lowest for plants which are consumed raw. With Turkish citizens holding university degrees, the results in terms of approval and willingness for self-consumption of plants grown with urine-based fertilizers were comparable, within ±2%. The acceptance levels for both were 74% for direct use and 86% for indirect use for landscape plants, 45% for direct use and 65% for indirect use for cooked crops, and 31% for direct use and 56% for indirect use for uncooked crops. This is to be compared to 3-11% lower willingness for self-consumption as compared to acceptance observed in the previous study (Beler Baykal et al., 2011b) without any differentiation in terms of educational level. The difference was more obvious when direct and indirect use was compared as well as in terms of different type of plants produced as shown in Figure 1. Indirect use was clearly preferred over direct use and landscape plants received the highest acceptance, followed by cooked crops and finally uncooked crops.

Following the first survey, more detailed questioning was adopted in the second survey. Different types of food groups were chosen to observe how people's attitudes will change with the changing proximity of urine-based fertilizer to the food products including fruits from ground, fruits from trees, nuts, and cereals. Considering the entire set of products, the maximum level of acceptance of direct use was 11% for nuts. Acceptance for indirect use was in the order of cereals, nuts, fruits from trees, fruits from ground, cooked vegetables and non-cooked vegetables and ranged between 36-54% as shown in Figure 2. This was in line with the results of the first survey. As expected, the level of acceptance decreases with the higher levels of possible urine contact with food, i.e. the lowest acceptance was for fruits from the ground which will have the highest probability of contact. Likewise, the level of acceptance for application on uncooked vegetables was lower as compared to cooked ones due to higher possible bacteriological risks associated with the former. Of the total respondents, 32-56% said they would rather have no use of urine-based fertilizers applied to food stuff. Of the products questioned, the highest disapproval was for uncooked vegetables with 56% followed by fruits from ground with 53%. Under all circumstances, indirect use received far more acceptance as compared to direct use and the levels of acceptance were 5-8 times that of direct use.

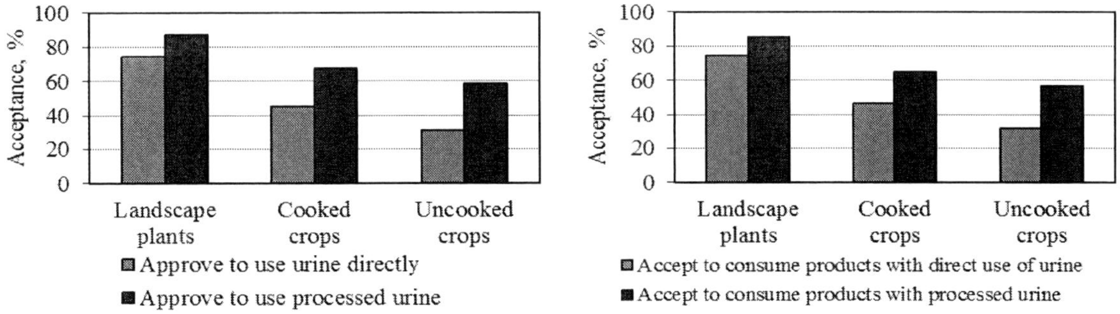


**FIGURE 1.** Approval of using urine as fertilizer directly or indirectly and willingness to consume products that were fertilized with urine directly or indirectly

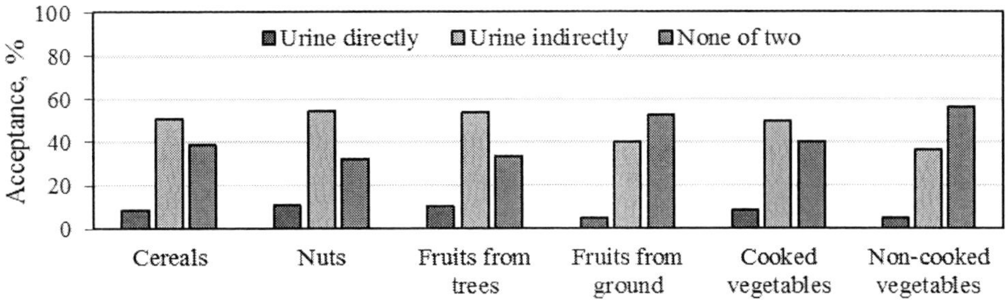


**FIGURE 2.** Application of human urine for various food types

A more detailed survey was also included in the second part regarding green areas which was detailed as landscape areas, parks, stadiums, home gardens, picnic areas and school gardens. As expected, people were observed to question the use of urine more when the point of use was in closer proximity. As shown in Figure 3, stadiums, parks and landscape areas received higher acceptance for both direct and indirect use as compared to home gardens, picnic areas, school gardens. Once again, under all circumstances, indirect use received far more acceptance as compared to direct use, as in the case of food stuff and the levels of acceptance for indirect use were 2-3 times that of direct application. 19-41% of the respondents indicated that they would prefer not to have any urine-based fertilizers applied.

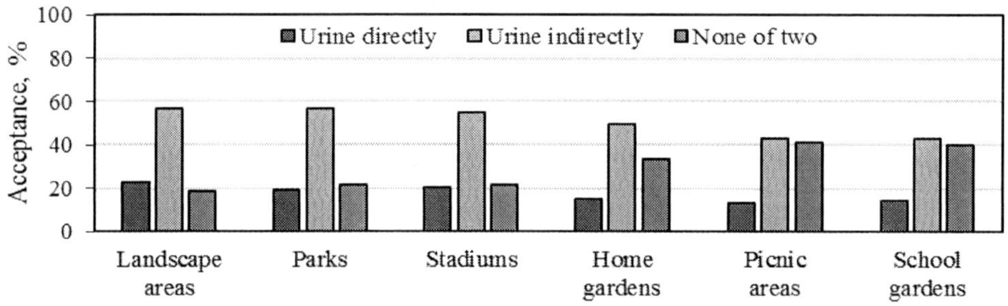


**FIGURE 3.** Application of human urine in various green areas

Another question in this study was directed towards finding out participants' attitude to wearing a cotton dress manufactured in a farm fertilized with urine/urine-based fertilizer. According to the results, 65% of the respondents answered as 'Yes, I would wear one' while, 34% of all did not find this idea appealing. This high ratio was taken as a starting point for investigating the reuse of human urine for non-edible industrial products as examplified by cotton, and motivated the investigation of the use of urine for other industrial products. As an attempt to begin this, another small survey was conducted with 107 people to see acceptance towards the use of industrial products fertilized with urine/urine-based fertilizer. The results have revealed that at least 50% of the respondents were willing to use those products including chocolate made of sugar produced with fertilizer of urine origin and cigarettes as shown in Figure 4. The highest acceptance in this group was for wicker basket with 77%. Acceptance for cotton was 62% for this part which was comparable with 65% observed in the main survey.

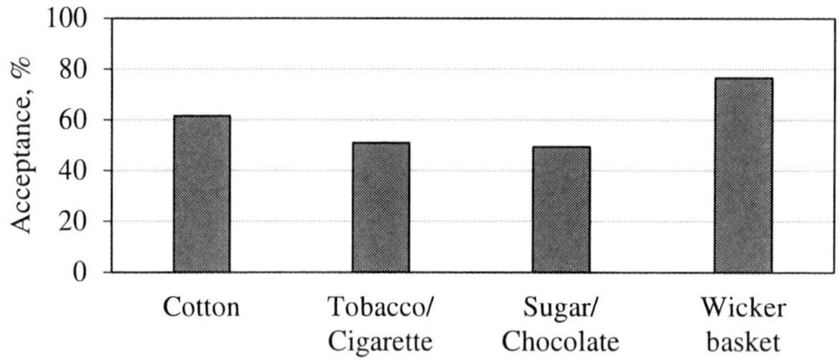


**FIGURE 4.** Acceptance of the use of human urine for industrial plants/production

Going back to the results of the second survey, 44% of the respondents said they were aware of the fertilization value of human urine. In response to the question 'Do you think urine can be collected separately to make fertilizer out of it?' 82% of the respondents answered as 'Yes' which is considered as a high ratio showing that people think it can be done. Similarly, respondents replied positively with 81% to practice it themselves.

The most interesting point regarding UDTs was the willingness to pay money for these toilets. The idea of paying extra money for the installation of a UDT was rejected by 70% of the participants. However, when the question was asked as 'Would you allow a UDT to be installed at your home for free?', acceptance raised to 77% as seen in Figure 5. These numbers illustrate how the financing/funding of this new approach is linked with the acceptance level.

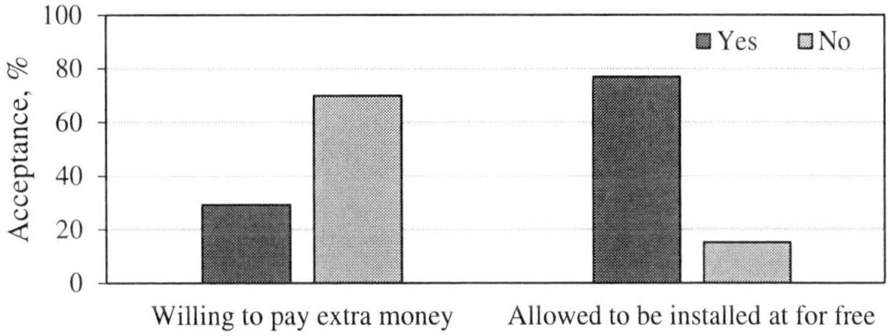


**FIGURE 5.** Willing to pay money for UDTs

Throughout both parts of the survey, the participants were asked to state their concerns and comments about the use of urine/urine-based fertilizers, too. Those who indicated concern were uncomfortable about the idea of using urine from a pshychological point of view. They also had uncertainties about the reliability of utilization including human health aspects, especially pharmaceuticals, and pathogenic safety in terms of the use of urine as fertilizer. Several had expressed other corcerns including odor and religious reasons.

## CONCLUSION

Results have indicated that as opposed to almost 100% acceptance of natural fertilizers (animal excreta), synthetic fertilizers got the approval of less than 25% of the respondents. The use of human urine as fertilizer received considerable acceptance. Indirect use received an obvious preference over direct use. Landscape application was the most highly accepted one. Use in public areas and commons received higher preference as compared to respondants' own premises. Plants which were cooked received higher acceptance than those which are eaten raw. The level of acceptance decreased with the higher levels of possible urine contact with food stuff. An appreciable acceptance was also observed for industrial plants grown using urine/urine-based fertilizers. Installation of UDTs received quite a high level of acceptance provided that they are supplied free of charge.

Source separation of urine and recovery of nutrients is a promising source of fertilizers. Creating value from this 'waste' by recovering and reusing plant nutrients in urine will lead to a sustainable future through closing the nutrient loop. There is an appreciable amount of acceptance of this approach by highly educated Turkish citizens. This will most probably be enhanced through future public education.

## REFERENCES

Allar, A. D. 2015. *Alternatives for Indirect Use of Urine as Fertilizer through Processing with Clinoptilolite (*İnsan İdrarının Gübre Olarak Dolaylı Kullanımında Klinoptilolitle Muamele ve Alternatifleri). PhD Thesis, Istanbul Technical University, Istanbul, Turkey (in Turkish with extended abstract in English).

Andersson, E. 2014. "Turning waste into value: using human urine to enrich soils for sustainable food production in Uganda." J. Clean. Prod. 96:290-298.

Beler Baykal, B. 2015. "Stream segregation in household use: A review of grey water as an alternative source of water and yellow water as an alternative source of fertilizers." Water Quality, Exposure and Health 7:27-37.

Beler Baykal, B., A.D. Allar, and S. Bayram. 2011a. "Nitrogen recovery from source-separated human urine using clinoptilolite and preliminary results of its use as fertilizer". Water Sci. Technol. 63(4):811-817.

Beler Baykal, B., A. D. Allar, and E. D. Bozkir. 2011b. "A preliminary survey of the public acceptance of the use of human urine as fertilizer in Turkey." III. International Congress Smallwat11 on Wastewater in Small Communities. Towards the Water Framework Directive (WFD) and the Millennium Development Goals (MDG), April 25-28, Seville, Spain.

Bonvin C., B. Etter, K. M. Udert, E. Frossard, S. Nanzer, F. Tamburini, and A. Oberson. 2015. "Plant uptake of phosphorus and nitrogen recycled from synthetic source-separated urine." Ambio 44(2):217-227.

Cofie, O., P. Amoah, I. Egyir, N, Adamtey and F. Tettey-lowor. 2011. "Demonstration on the use of urine in urban agriculture." SWITCH Sustainable Water Management in the city of the future, Final Report. http://www.switchurbanwater.eu/outputs/pdfs/W5-2_GEN_RPT_D5.2.4_Demonstration_on_the_use_of_urine_in_urban_ agriculture.pdf.

Kocaturk, N.P., and B. Beler Baykal. 2012. "Recovery of plant nutrients from dilute solutions of human urineand preliminary investigations of pot trials". Clean - Soil, Air, Water 40(5): 538-544.

Lienert, J., M. Haller, A. Berner, M. Stauffacher, and T. A. Larsen. 2003. "How farmers in Switzerland perceive fertilizers from recycled anthropogenic nutrients (urine)." Water Sci. Technol. 48(1):47-56.

Morgan, P.,2003. "Experiments using urine and humus derived from ecological toilets as a source of nutrients for growing crops." 3th World Water Forum, March 16-23, http://aquamor.tripod.com/KYOTO.htm.

Okem A. E., S. Xullu, E. Tilley, C. Buckley, and E. Roma. 2013. "Assessing perceptions and willingness to use urine in agriculture: A case study from rural areas of eThekwini municipality, South Africa." J. Water Sanitation and Hygiene for Development 3(4): 582-591.

Pahl-Wosti, C., A. Schönborn, N. Willi, J. Muncke, and T. A. Larsen. 2003. "Investigating consumer attitudes towards the new technology of urine separation." Water Sci. Technol. 48(1):57-65.

Ranasinghe E. S. S., C. L. S. M. Karunarathne, C. K. Beneragama, and B. G. G. Wijesooriya. 2016. "Human urine as a low cost and effective nitrogen fertilizer for bean production." Procedia Food Science 6:279-282.

Sridhar, M. K. C., A. O. Coker, G. O. Adeoye, and I. O. Akinjogbin. 2005. "Urine harvesting and utilization for cultivation of selected crops: trials from Ibadan, South West Nigeria." The 3rd International Ecological Sanitation Conference, 23-26 May, Durban, South Africa. http://conference2005.ecosan.org/papers/sridhar_et_al.pdf.

Yildiz-Dogan, F. N., M. S. Shihab, F. T. Citak, B. Beler-Baykal, M. E. Pasaoglu, and A. Ozturk, 2015. "A preliminary survey of public willingness and acceptance of segregation and use of human-urine as fertilizer in Turkey." 2nd ICSAE 2015 International Conference on Sustainable Agriculture and Environment, Abstract Book, 116, September 30-October 3, Konya, Turkey. http://www.2ndicsae.org/docs/abstrack_book.pdf.

Wach, F. G. 2007. "Sustainable water management and Ecological Sanitation." TUBITAK Marmara Research Center Chemistry and Environment Institute, Zer0-M Course on Sustainable Water Management and Technologies, April 16-17, Kocaeli, Turkey.

WSP 2010. Ecological sanitation, social factors impacting use of EcoSan in rural Indonesia (Researcher Entin Sriani Muslim). Water and Sanitation Program (World Bank). https://www.wsp.org/sites/wsp.org/files/publications/WSP_EcoSan.pdf (accessed on July 24, 2015).

## Fragmented Stakeholders: Unaccountable Representatives

***Yaw Amo Sarpong***; Nelson Owusu Ansah, ***Francisca Amoatemaa***
(Bureau of Integrated Rural Development, KNUST, Ghana)
Emmanuel Marfo
(Forestry Research Institute of Ghana, Ghana)

Democratic representation has gained increased attention globally as a mechanism of promoting good governance and management of the environment and natural resources. Stakeholders are critical players in every democratic representation process. However, little scientific work has been done to assess the nature of stakeholder groups and whether the current model of representation has indeed been an effective mechanism in ensuring grassroot participation in the natural resource sector. This study therefore examined the peculiar challenge of stakeholder organization and participation in the natural resource sector and how that may affect democratic representation. A sample of 30 stakeholder groups who participated in a multi-stakeholder dialogue was selected for the study. The survey results showed that 37% of the stakeholder groups were not organized into any formal grouping. Content analysis and observation done further indicated that the 37% constitute grassroot stakeholder groups who are rarely included in decision-making process in the management of natural resources. Using the democratic representation theory as a lens to investigate the participation approach, it was realized that, the engagement process fails to promote accountability since representatives are unable to solicit views and provide feedback to the larger constituents. It further confirms the notion that participation in the natural resource sector is superficial and a mere rhetoric. The de facto management approach alienates the grassroot from substantive participation in the decision-making process.

# SOME REMARKS TO RISK ASSESSMENTS OF NUCLEAR POWER PLANTS (NPP)

***Alexander Valyaev[1]***, Gurgen Aleksanyan[2], Alexey Valyaev[3]
([1]Nuclear Safety Institute of Russian Academy of Sciences, Moscow, Russia; [2] Yerevan State University, Yerevan, Republic of Armenia; [3]The University of Sydney, Australia, NSW 2049)

Today NPP using for energy production is constantly increasing in the world with the growth of different threats, for example, caused by directed terrorist attacks. It is necessary to make possible correct assessments of corresponding risks levels in common case and for single NPP from moment of NPP projecting, building and especially during its exploitation. It demands to provide exclusive attention and special conditions under realization of complex integrated emergency NPP management.

We use our universal formula for assessment of the total vector of limited losses under NPP exploitation for the fixed time interval under the following assumptions: (1) at initial state the object is in normal (non accidents) exploitation; (2) the different kinds of accidents may be occurred as noticed i =2, 3, …, m, where m is the total number of possible accidents (m=1 is corresponded to the normal regime); (3) every accident may create the different kinds of losses. (4) realization of i accident creates the loss of j kind with $P_{ij}$ probability,

$$\vec{a}_{\lim} = P(1)\vec{a}_{1n} + \sum_{i=2}^{m} \hat{P}_{ij}\vec{a}_j \qquad (1)$$

Here j is the kind of loss with $a_j$ value. Then j = 1,2, …n, where n is the total number of possible kinds of losses; where P(1) is the probability of loss formation under normal exploitation; $\vec{a}_{1n}$ is the vector of limited loss under regular exploitation. $P_{ij}a_j$ coordinate vector value in sum is equal the loss value of j kind under realization of i kind accident. Thus the matrix of loss probabilities is determined. Under absent of accidents $\vec{a}_{\lim}$ total vector is determined only the first part of (1) formula.

The correct assessment of loss probability matrix elements is the main global problem. If representative statistic data, obtained for long NPP exploitation period, are present, then some of its elements may be assessed by statistic methods. We have predicted the irradiation doses and corresponded risks for population under implementation of Russian Federal Program:" Development of Russian atomic energy industrial complex on 2007-2020 years at 10 homeland NPP that operated during some last decades. But early some types of NPP disasters were absent at researched NPP. Using of classic methods of expertise assessments is not correct in this case. Some needed data may be obtained from primary virtual computer tests of concrete NPP with imitation of possible disasters. It allows to assess risk values and also to plan the actions for NPP operators and special services under serious NPP disasters or may be to prevent them at all. These problems are under detail consideration in our communication.

# COST-RISK ANALYSIS OF SOLAR RADIATION MANAGEMENT

Mohammad Mohammadi Khabbazan, ***Elnaz Roshan***, and Hermann Held (Center for Earth System Research and Sustainability (CEN), University of Hamburg, Grindelberg 5, 20144 Hamburg, Germany)

Potential side-effects of solar-radiation management (SRM) should be taken into account before society decides on its implementation as a climate policy option to alleviate anthropogenic global warming. Here we evaluate SRM alongside mitigation considering precipitation change as the so far most prominent side-effect of SRM and probabilistic knowledge about climate sensitivity. To do so, we perform a cost-risk analysis, which trades off expected welfare-loss from mitigation and SRM costs, and the risk from transgressing climate targets. We maximize social welfare for the following three scenarios: temperature-risk-only, precipitation-risk-only, and both-risks. Our simulations show that in the temperature-risk-only scenario, perfect compliance with 2°-temperature target is attained for all numerically represented climate sensitivities while the 2°-compatible precipitation corridor is violated. Our analysis of the precipitation-risk-only scenario displays that precipitation and temperature remain confined to their targets for all and more than half of climate sensitivities, respectively. When both precipitation and temperature risks are activated with equal weights in both-risks scenario, almost 95% and perfect compliance can be achieved for temperature and precipitation targets, respectively. The results suggest that welfare in joint-mitigation-SRM analysis converges to welfare in no-climate-policy option. However, the average welfare-loss from mitigation cost plus climate risk is about 3.6% over the scenarios in a mitigation-only analysis compared to no-climate-policy option.

# IN LIGHT OF INFRASTRUCTURE AND CONSTRUCTION AFFORDABILITY, CAN SOUTHEAST QUEENSLAND CATER FOR CURRENT POPULATION FORECASTS?

***Kirsty Miller*** (Undergraduate Student at the Queensland University of Technology, Brisbane, Queensland, Australia)
David Hood (Queensland University of Technology, Brisbane, Queensland, Australia)

Without a significant increase in development of infrastructure and better housing affordability, the region of Southeast Queensland, Australia, will not be able to cater for current forecasts in population growth. This study investigates the discourse of population growth, the impacts that substantial growth can cause, and looks at solutions to these issues. It is proposed that when humans recognize that we live on a finite planet, with finite resources, we simply cannot provide for infinite population growth.

Most infrastructure in this densely populated region is aged, and plans for restoration do not adequately consider resilience to extreme weather events likely from global warming. Furthermore, due to increasing urbanisation over recent years, there is a high demand for dwellings in the region. This is pushing rental and property prices beyond the reach of most incoming residents. More and more are relying on housing assistance. With increasing population size, these stresses exacerbate. The growth of the region needs to be focused on consolidation with the better assistance from Government through planning and support programs. As urbanization has already resulted in highly populated suburbs it is recognized that current growth management plans often sacrifice the liveability of the region. Opportunities such as utilising biophilic urbanism to relieve the pressures of population growth are discussed. Based on the research in this paper, recommendations are proposed for further research.

# SCIENCE, POLITICAL POLARIZATION, AND HOW HYDRAULIC FRACTURING IS FRAMED AS ENVIRONMENTALLY SAFE OR RISKY

***Michelle L. Edwards***
(Texas Christian University, Fort Worth, TX, USA)

**ABSTRACT:** Environmental science issues have become increasingly politically polarized in America (McCright and Dunlap 2011). From energy production to global climate change, conflicts originating between political party elites have begun diffusing through the general public. Additionally, social scientists have found an increasing divergence in trust in the scientific community among liberals and conservatives over the past 30 years, with conservatives becoming increasingly less trusting of the scientific community (Gauchat 2012). However, scholars have recently found support for the "contextual thesis" that both liberals and conservatives express distrust in science when it challenges their ideological worldviews (Nisbet et al. 2015). Using semi-structured, in-depth interviews, this study explores societal perceptions of hydraulic fracturing to extract unconventional natural gas resources in the Dallas-Fort Worth region of Texas. Results demonstrated that key informants who have been actively involved in the debate over "frac'ing" in this region – both in support of "frac'ing" and in opposition to it – often rely on scientific arguments to frame the hydraulic fracturing process as either environmentally safe or risky. Similar to the "contextual thesis," participants often identified scientific findings, or interpreted the absence of findings, in ways that likely fit their ideological worldviews.

## INTRODUCTION

The United States continually seeks new sources of energy to support its growing demands. One of these new sources of energy is unconventional natural gas, which, due to recent technological advances, can now be extracted from the ground using hydraulic fracturing or "frac'ing" (or "fracking"). While some have praised frac'ing for its contributions to energy production, others have questioned whether its potential benefits outweigh its potential risks, particularly in terms of its environmental impacts. Some researchers have begun to explore residents' perceptions toward hydraulic fracturing. This study furthers this research by considering how individuals involved in the debate in the Dallas-Fort Worth region use scientific arguments to frame frac'ing as either environmentally safe or risky.

**Increased Use of Hydraulic Fracturing.** Though hydraulic fracturing was originally developed in the late 1940s, advances in horizontal drilling techniques and the discovery of large new gas reserves have recently driven a massive increase in this form of natural gas extraction (Ladd 2013). According to the Energy Information Administration (2016), hydraulic fracturing accounts for about two-thirds of U.S. natural gas production. One prominent site for this energy development has been the Barnett Shale field underneath the Dallas-Fort Worth region. Large-scale horizontal drilling began in this region around 2005, with higher productivity areas tapped first (Jackson et al. 2014). By 2011, median initial production rates had actually increased, in part due to companies increasing production through "tailoring their practices to local geology," as well as increasing the intensity of extraction (Jackson et al. 2014:331-332). Refracturing wells has also contributed to estimated ultimate recovery in the region.

Many individuals have questioned whether the potential economic benefits of this type of energy development outweigh its potential risks. Major concerns expressed about hydraulic fracturing often relate to its potential environmental impacts, including: increased water pollution for groundwater and surface water, decreased water availability, increased air pollution and greenhouse gas emissions, and induced seismicity associated with hydraulic fracturing and wastewater disposal (Jackson et al. 2014).

---

**Public Perceptions of Hydraulic Fracturing.** Scholars have also examined the public's perceptions of hydraulic fracturing. In the Barnett Shale region, social scientists focused originally on the perceptions of rural residents in communities experiencing unconventional energy development (Anderson and Theodori 2009; Theodori 2009; Wynveen 2011). These studies highlighted both the perceived positive economic benefits and improved services resulting from this development, as well as the potential concerns residents held about the impacts of frac'ing on public health, community safety, and the environment.

However, as development has expanded, public opinion researchers have also expanded their sampling frames beyond those directly affected by hydraulic fracturing. National surveys of American adults have demonstrated that only a minority of Americans are aware of and have formed an opinion on frac'ing (Boudet et al. 2014), with fairly similar percentages of supporters and opponents (Davis and Fisk 2014). Political views play a key role in predicting Americans' perceptions on frac'ing, with political conservatives more likely to be supporters of frac'ing (Boudet et al. 2014); and Democrats more likely to support greater regulation of frac'ing (Davis and Fisk 2014). Researchers have also found political conservatives to perceive frac'ing as less risky compared to liberals (Choma et al. 2016).

Scholars have also found a negative relationship between support for hydraulic fracturing and how much a person had heard or read about it (Boudet et al. 2014). Those with greater basic objective knowledge about frac'ing also tend to have higher perceptions of risk related to frac'ing than those with less knowledge (Choma et al. 2016). However, Choma et al. found that frac'ing knowledge tended to "strengthen conservatives' belief that fracking does *not* present risks" (2016:113). While these studies are useful for capturing the public's overall support or opposition to hydraulic fracturing, they do not capture how stakeholders with varied views on frac'ing use science to frame their position on risk.

## METHODS

**Data Collection.** In 2015 and 2016, I conducted in-depth interviews with key informants in the Dallas-Fort Worth (DFW) region to gain insight into the public's perceptions of hydraulic fracturing as environmentally risky or safe. I interviewed individuals who represent a wide variety of views including: residents, scientists, community activists, realtors, industry representatives, business owners, and government regulators familiar with frac'ing in the region. As past research has demonstrated, semi-structured in-depth interviews allow a researcher to "explore a topic more deeply" through allowing respondents to "shape the order and structure of the interview" to fit their own experiences (Esterberg 2002:87).

**Analysis.** Following transcription of the interviews, I analyzed the data using the two-stage coding techniques described by Esterberg (2002). In the first stage, I utilized open coding to analyze each line of the interview for conceptual categories. In the second stage, I sorted the earlier-found information into more clearly defined thematic groups and reanalyzed the interviews using focused coding.

## RESULTS

Participants discussed how science impacted perceptions of hydraulic fracturing as environmentally safe or risky along three major themes: 1) hydraulic fracturing as a complex, scientific process, 2) selective interpretation of science and scientists, and 3) absence of scientific information.

**Hydraulic Fracturing as a Complex, Scientific Process.** Opponents and supporters of hydraulic fracturing described this process as complex and highly technical. For some participants, their confidence in the safety of hydraulic fracturing was rooted in the technical nature of this process. For example, one participant described his experience visiting a frac site:

> *People watch the movie, for instance, "There Will Be Blood," and they think, "my God, this is just the wild West. You go out, you dig a hole, oil shoots all over the place..." The reality is, I was on a frac site about four years ago, and the thought that actually came to mind to me was that it was boring... I'm standing in the trailer or office building or whatever they call it, and there are a dozen*

*people sitting there all looking at computer screens and assessing these millions of data points that are coming up. It's a constant assessment of scientific and engineering principles just to frac a single well. The industry's predisposition is to leverage science and facts and data to show how this is safe and show why it is safe.*

However, others argued that the technical complexity of the process increases the probability that an unpredicted event could occur, with potentially negative environmental implications. For example, one respondent explained: "There are so many moving parts that I do believe in most cases the hydraulic fracturing event is probably not at fault for any problems, but just the sheer logistics of how many wells are drilled, some things failed." Several of these participants cited scientific studies which have measured error related to different parts of the process of unconventional natural gas extraction, such as cement casing failure rates.

Part of the disagreement among participants as to the amount of risk involved in hydraulic fracturing had to do with how they defined the term "fracking." Some participants expressed concern that "fracking" is a "catch-all term" used to capture the many processes included in unconventional natural gas extraction. For these participants, calling everything "fracking" is problematic for measuring the different risks involved in the various steps of unconventional natural gas extraction, as well as in identifying causes of environmental problems. For example, one participant discussed how "as scientists, we are always very careful to draw a distinction between the actual process of hydro-frac'ing fresh shale to extract the natural gas versus drilling the well and extraction activities, and then the wastewater disposal associated with it, because we need to actually physically, mathematically model each set of processes."

However, other stakeholders argued that disconnecting frac'ing from the other components of unconventional natural gas extraction is misleading. For example, one participant stated:

*They're trying to say, "Well, the fracking itself didn't cause that. That was something else." What they're not saying is that something else would never occur if there wasn't fracking. To us, it's all part and parcel of the same operation. When you start clearing a pad site, from that point on everything is what is commonly referred to as "fracking," even though, technically speaking, frac'ing is always about where you fracture the well itself...You don't have wastewater injection if you're not frac'ing to begin with. They're just splitting hairs and trying to use semantics to avoid the reality when they say frac'ing doesn't lead to earthquakes.*

By using different definitions of "fracking," participants were able to come to vastly different conclusions regarding whether hydraulic fracturing is safe or risky.

**Selective Interpretation of Science and Scientists.** In response to questions about how science impacts the public's perceptions of hydraulic fracturing, many participants also discussed how the public selectively interprets science and scientists depending on their predetermined views on this issue. Several stakeholders identified this as a potential explanation for how opponents and supporters of frac'ing could reach very different conclusions using access to the same scientific information.

For example, a number of respondents discussed the various interpretations of the Environmental Protection Agency's (EPA) 2015 report assessing the potential impacts of hydraulic fracturing on drinking water resources. Two major findings presented in the draft form of this report were discussed across a variety of media sources: 1) the EPA did not find evidence of "widespread, systemic impacts on drinking water resources" (Brady 2015), and 2) the EPA found "'specific instances' when fracking 'led to impacts on drinking water resources, including contamination of drinking water wells'" (Banerjee 2015). As one participant explained: "If you were to read Fox News and Huffington Post about that report, you'd get two drastically different takes on it. I think people's preconceived notions are what govern the way they're going to think about it." Participants described how, to really understand the report, the public would need to read past the headlines and understand the "facts [that] are buried" in the report, as well as the limitations to the report's data. From the perspective of an opponent to hydraulic fracturing:

*The EPA told several cities and towns, "Do not drink your water. Stop today. Don't shower with it. Don't put your kids anywhere near it." All that is in the EPA report, but the report was politically*

> *corrupted... so the average person in Fort Worth is going to say, "Oh, it's okay to frack. No worries anymore." But that's not necessarily the truth... Also, it's very important to know that the EPA report, even though it's done by the EPA and it took five or six years, it was still a relatively small sample and a lot of that sample was provided by the industry. Someone like me with my perspective, I know enough about the corruption in this industry that it's a tainted report, simple as that. It can't be totally scientific.*

Though both supporters and opponents of frac'ing critiqued the credibility of the EPA as the sponsor of this study, they each highlighted separate parts of this study as evidence confirming their claims that hydraulic fracturing is scientifically found to be safe or risky.

A number of participants also described their own experiences working as scientists on issues related to hydraulic fracturing as examples of how the public selectively interprets science. For example, one participant joked that their research project is like a "Rorschach test. Everyone listens to us, but what they hear us saying depends on how they felt about it coming in" and "we get accused of bias from every side." Others expressed similar experiences:

> *Let's say I went and gave a presentation. Half the room might think I'm a junk scientist, and the other half of the room might think that I'm a great scientist. That's based solely on their predetermined opinion on the issue. None of them are going to look at my CV and read any of the papers we've done. They're not going to know the quality of the work and the quality of the journals we publish in.*

Several scientists also pointed to the key role funding plays in predicting how the public is going to perceive one's work. One participant explained, "If you're not affiliated with industry, then you get accused of being biased against it. If you are affiliated with industry, you're biased for it. There are a lot of political ramifications for who gets what type of data and what their results actually mean." Even though participants in this study argued that their "source of funding was never going to change [their] results," they worried about how it would affect perceptions of their work.

Some participants proposed that finding less traditional funding sources, utilizing internal university grants, or personally funding their own research may be some of the few currently viable ways to support scientific work on a controversial topic like this one. However, researchers also argued that the need to pursue nontraditional funding sources "dissuades a lot of researchers from getting into the area." Participants with scientific and non-scientific backgrounds both argued that the contentious nature of this topic made "totally scientific" research more challenging, with several participants citing worries about academic freedom and others discussing the difficulty drawing a line between research and regulatory or court matters. Some participants even described how divisions over this topic affected data collection:

> *I can tell you that the people that we take samples from run the full gamut. Some will let you take water because they're very much afraid of what's happening to their water. Some are absolutely indifferent. Some will let you come on and take the water because they're absolutely sure that you're not going to find anything... My colleague has had people come and hug him and thank him for doing this work and he's also had guns pulled on him.*

Thus, the public's selective interpretation of scientific research and scientists begins before data is even collected and continues long after the publication of results.

**Absence of Scientific Information.** The third theme that emerged from conversations about hydraulic fracturing risk had to do with how participants discussed the absence of scientific information. Many participants argued that more data are needed to assess the impacts of hydraulic fracturing. For example, one participant discussed their initial reaction to viewing the documentary "Gasland": "In this documentary, they're making some pretty significant statements. [I thought] there must be a lot of data on this issue. We looked... and the truth of the matter was no, there wasn't." Participants posed questions about the entire gas development process such as "exact locations, timing, when things start, how long they go on, volumes of fluids being put down there as well as components of what's in them, what they do with the waste products, where that stuff goes, if there are ever any surface spills." As this participant explained, "There's a million links in the chain where something could go wrong, and those are the kinds of things we just don't have data on."

A number of participants also noted the difficult nature of assessing causality (e.g., does hydraulic fracturing cause water contamination?) when so much information is missing from scientists' analyses, including baseline data for comparisons, and exact information on chemicals used during hydraulic fracturing operations. Several participants discussed companies' nondisclosure of proprietary information, such as mixtures of chemicals or where faults are located, as contributing to this problem. However, others noted that even when scientists can access proprietary data, for places which companies have determined to be "not that economically viable," companies have "stopped spending millions of dollars collecting the data," and thus, the issue is sometimes not lack of access, but lack of collected data.

While most of the scientists interviewed specifically noted that their goals were to answer research questions and not to prevent energy development, the opponents of frac'ing more explicitly argued that until a sufficient amount of research has been completed, companies should not be allowed to move forward. For example, according to one participant:

> *We're getting more and more data. Do we need more? Yeah, but we need more data about earthquakes in general. Here's what I think about this. If there is even a sliver of doubt that this is dangerous to the public, shut it down until we know. If we shut it down, we know for certain there will be no danger, but if we keep going, we aren't so certain... I am really worried. Is Fort Worth going to become a toxic dump?... What's going to happen... not tomorrow, not even five to ten years down the line, but twenty-five years down the line?*

For this participant, as well as others that expressed the precautionary principle regarding hydraulic fracturing, the financial benefits or "mailbox money" especially did not outweigh the known and unknown *long-term* environmental risks of hydraulic fracturing. One participant said, "That's why we fought this so hard. Because once they get the wells in, it's too late. You can't go back. You can't undrill it... Those wells are going to be fracked for the next fifty years and that means all this mess is going to continue."

In contrast, several participants who viewed frac'ing as less environmentally risky claimed that a sufficient amount of study had already been done. One participant explained:

> *Typically the people who are advancing the argument... "We don't have enough information on this to move forward" are the people who don't want it to ever occur anyway... It's also, I would say, incredibly misleading to suggest that we don't have enough information on this. Frac'ing started in 1947... We know [it is safe] from over sixty years of experience, and that's not just experience, that's sixty years of data points, of understanding what works, what doesn't, trial and error, figuring these things out... Are we actually asking for more study, or are we trying to advance something else?*

From this perspective, the argument of "not enough information" is a strategic one intended to prevent hydraulic fracturing indefinitely and is typically used by individuals who are critical of oil and gas drilling. However, some supporters of hydraulic fracturing did occasionally use the argument of "not enough information" to explain their lack of confidence in studies that have demonstrated negative environmental impacts of hydraulic fracturing. For example, one participant described his critique of a past study of water contamination based on the lack of baseline data: "Ideally, you would've tested those wells twenty years ago, before the drilling started. How do you know those wells weren't [already] contaminated? You don't." Thus, the absence of scientific information is problematic for people who view hydraulic fracturing as risky, as well as those who view it as safe.

## CONCLUSIONS

A number of participants likened the debate over hydraulic fracturing in the United States to the debate over climate change, where political polarization has played a key role in explaining different levels of concern, as well as differences in trust in the scientific community (McCright and Dunlap 2011). In a study of perceptions of hydraulic fracturing in Texas, a largely conservative state, one might expect to find participants often expressing their *distrust* in the scientific community. To the contrary, most participants of this study framed their discussion of risk related to hydraulic fracturing in scientific terms demonstrating the important role science plays in how participants justify their views on hydraulic fracturing as a risky or safe process. This work finds support for the "contextual thesis" which suggests that liberals and conservatives engage in motivated reasoning, where "equally informed citizens may employ scientific evidence in very different ways depending on their ideological orientation" (Nisbet et al. 2015:38). Residents in the Dallas-Fort Worth region seemed to be concerned about scientific perspectives on frac'ing, but in ways that fit their prior viewpoints on this issue.

**ACKNOWLEDGMENTS**

This research was supported in full by grants from the Texas Christian University Research and Creative Activities Fund and the Junior Faculty Summer Research Program.

**REFERENCES**

Anderson, B.J., and G.L. Theodori. 2009. "Local Leaders' Perceptions of Energy Development in the Barnett Shale." *Southern Rural Sociology* 24:113-129.

Banerjee, N. 2015. "Fracking Has Contaminated Drinking Water, EPA Now Concludes." Retrieved 30 May 2016. (http://insideclimatenews.org/news/05062015/fracking-has-contaminated-drinking-water-epa-now-concludes).

Boudet, H., C. Clarke, D. Bugden, E. Maibach, C. Roser-Renouf, and A. Leiserowitz. 2014. "'Fracking' Controversy and Communication: Using National Survey Data to Understand Public Perceptions of Hydraulic Fracturing." *Energy Policy* 65:57-67.

Brady, J. 2015. "EPA Finds No Widespread Drinking Water Pollution From Fracking." Retrieved 30 May 2016. (http://www.npr.org/sections/thetwo-way/2015/06/04/412047602/epa-finds-no-widespread-drinking-water-pollution-from-fracking).

Choma, B.L., Y. Hanoch, and S. Currie. 2016. "Attitudes toward Hydraulic Fracturing: The Opposing Forces of Political Conservativism and Basic Knowledge about Fracking." *Global Environmental Change* 38:108-117.

Davis, C. and J.M. Fisk. 2014. "Energy Abundance or Environmental Worries? Analyzing Public Support for Fracking in the United States." *Review of Policy Research* 31:1-16.

Energy Information Administration. 2016. "Hydraulically Fractured Wells Provide Two-Thirds of U.S. Natural Gas Production." Retrieved 30 May 2016. (http://www.eia.gov/todayinenergy/detail.cfm?id=26112).

Esterberg, K.G. 2002. *Qualitative Methods in Social Research.* McGraw Hill, Boston, MA.

Gauchat, G. 2012. "Politicization of Science in the Public Sphere: A Study of Public Trust in the United states, 1974 to 2010." *American Sociological Review* 77(2):167-187.

Jackson, R.B., A. Vengosh, J.W. Carey, R.J. Davies, T.H. Darrah, F. O'Sullivan, and G. Pétron. 2014. "The Environmental Costs and Benefits of Fracking." *Annual Review of Environment and Resources* 39:327-62.

Ladd, A.E. 2013. "Stakeholder Perceptions of Socioenvironmental Impacts from Unconventional Natural Gas Development and Hydraulic Fracturing in the Haynesville Shale." *Journal of Rural Social Sciences* 28:56-89.

McCright, A.M., and R.E. Dunlap. 2011. "The Politicization of Climate Change and Polarization in the American Public's Views of Global Warming, 2001-2010." *The Sociological Quarterly* 52:155-194.

Nisbet, E.C., K.E. Cooper, and R.K. Garrett. 2015. "The Partisan Brain: How Dissonant Science Messages Lead Conservatives and Liberals to (Dis)Trust Science." *The ANNALS of the American Academy of Political and Social Science* 658:36-66.

Theodori, G.L. 2009. "Paradoxical Perceptions of Problems Associated with Unconventional Natural Gas Development." *Southern Rural Sociology* 24:97-117.

Wyneen, B.J. 2011. "A Thematic Analysis of Local Respondents' Perceptions of Barnett Shale Energy Development." *Journal of Rural Social Sciences* 26:8-31.

# IMPLEMENTATION OF ENVIRONMENTAL LAWS AND ETHICS IN INDIA

C.V. Rajeswari
(Government Degree College, Pattikonda, 518380, A.P, India)

Sustainable development can be achieved through effective implementation of Environmental Laws and Ethics. The nature of the of the individual to protect the environment is not only to punish the culprit but also to balance the eco system. Even though in India, evolution of environmental laws and ethics can be dated back to Pre-Vedic age, strict adherence to law has not been achieved till now due to lack of proper awareness among public and also due to lack of punishment for breaking the laws. In this paper environmental legislation in India and loop holes in its implementation are elaborated upon and new laws to be introduced to protect the environment are also focused.

# ENVIRONMENTAL EDUCATION AT THE MARITIME EDUCATION TRAINING

***Tanzer Satır*** (Istanbul Technical University (ITU), Tuzla-Istanbul, Turkey)
Neslihan Doğan-Sağlamtimur (Nigde University, Nigde, Turkey)

Environmental education has different definitions. This education refers to organized efforts to teach how natural environments function, particularly, how human beings can manage behavior and ecosystems to live sustainably. It is a multi-disciplinary field integrating disciplines such as biology, chemistry, physics, ecology, earth science, atmospheric science, mathematics, and geography. It is a process that allows individuals to explore environmental issues, engage in problem solving, and take action to improve the environment.

Environmental education is an important part of Maritime Education Training (MET) that provides and educates manpower for maritime industry. Marine pollution is one of the key issues at the maritime sector. Most of the maritime universities and institutes have increased the number of environmental courses. In this study, authors will analyze and compare environmental courses in the maritime universities and institutes in the world. The study's aim is to find ideal environmental education for the MET system.

# EDUCATION PROGRAM TO THE ENVIRONMENT: AN ADVOCACY OF A TRAVEL AGENCY IN AVAO CITY, PHILIPPINES

Ruben A. Neri; Recto B. Campos; ***Lilia F. Panchito***; Elisa P. Madrazo (Research and Marketing of Surelite Travel and Tours, Davao City, Philippines 8000; City Environment and Natural Resources, Davao City, Philippines)[117]

Davao City is situated in the southern most of the Philippines. With a total population of approximately 1.6 million people, the city boasts of its 4 icons; namely: Waling- waling (vanda sanderiana)-Queen of the Orchids; Mt. Apo- the tallest mountain in the Philippines; Philippine Eagle- 2nd biggest raptor in the world; and the Durian- an exotic and King of the fruits. Davao City was judged by Numbeo.com that it is the 4th safest city in the world. That is why tourists from all walks in life do come and visit our place.

Travel and tour, nowadays, is a lucrative business in the tourism industry. People got the chance to visit places and adventures, may it be foreigners or local residents as your market. For having served the industry for a decade, it is the advocacy of Surelite Travel and Tours, Davao City, Philippines to give back to Mother Nature what is due to her. Educating to the students from kinder, elementary, high school and college on what is the importance of the environment to mankind. *Environment is* the complex of physical, chemical, and biotic factors (as climate, soil, and living things) that act upon an organism or an ecological community and ultimately determine its form and survival. (Merriam Dictionary). In short, Environment is the natural world, the place we live in and its surroundings.

Our company visited schools within the city and inform administrators the advocacy of having an educational program to the environment. The administrators are so happy and willing to have this said program. In fact they will integrate it to their school curriculum for the next school year and such succeeding years. They say that some of our youngsters are so busy on their gadgets found in their houses and in school. They do not do household chores when they are hooked in the computers. Students seldom have time to explore their surroundings and be in contact with Nature. Some are alien to other things found in nature. We think it is high time that we give back to Nature what is due to them. Our company formulated different modules to different grade levels and activities done during the trip. It is a different fieldtrip to the students.

As a result, the administrators are so pleased that such an educational program does exist in our city. In fact the city government thru the City Environment and Natural Resources Office is really very much willing to help us in this advocacy. Students have the chance to plant a tree in this trip. Although, the trip is so tedious and tiresome, it is worthwhile to the students and the administrators.

It is high time now that such an advocacy exists in the tourism industry especially that we have our global warming through climate change.

## STUDENTS' ATTITUDES TOWARDS THE ENVIRONMENT

Mile Srbinovski, ***Veton Latifi*** & Zamir Dika
(South-East European University- Tetovo, The Republic of Macedonia)

The objective of this study is to determine the students' attitudes towards the environment. A 13-item, 5-point bipolar Likert – type scale instrument was used. The statements included several environmental parameters: utilization of natural resources, interconnection of components in the biosphere, consequences of the environment pollution, anthropogenic influence, the role of the environmental consciousness for the survival of the Earth, characteristics of nature, environmental balance, etc. The study included 51 students from the South East European University- Republic of Macedonia. All of them have been chosen The Environmental Protection course as free elective. This research was conducted at the beginning of this course.

Results indicate that more students display pro-ecological conceptions and they are aware of the negative impact humankind has on nature. Here we engage in a deep discussion and comparison of the obtained results in relation to the factors which influence students' environmental attitudes in the Republic of Macedonia.

# DOES VIEWING DOCUMENTARY FILMS AFFECT ENVIRONMENTAL PERCEPTIONS AND BEHAVIORS?

**HenryJanpol** (Nova Southeastern University, Ft. Lauderdale, FL., USA)
Rachel Dilts (Walden University, Minneapolis, MN., USA)

This research examined the question of whether viewing of documentary films about the natural or human built environment can exert a measurable influence on people's behaviors and perceptions toward the environment. To explore this question, two different documentary films were viewed by college student subjects. One film emphasized the natural environment, while the other focused on the built environment. Eighty-one subjects were included in the study. Each subject was randomly assigned to one of two conditions (watching a film about dolphins or a bridge building video). Immediately after viewing the film, participants played a computer generated game designed to assess their implicit attitudes about the natural and built environments. Finally, after completion of the computerize assessment instrument, participants took part in a charitable donation activity. They were given an opportunity to make a monetary donation to either one of two causes (helping dolphins or a fund for bridge rejuvenation).

Results yielded significant differences in subjects' environmental perceptions regarding the natural or human-built worlds, as measured by the computer game, depending on which video they had viewed. Significant differences were also found between the two treatment groups on the monetary donation variable. Findings are discussed as well as implications for future research in communication about environmental conservation and documentary films.

# ENVIRONMENTAL PLANNING
# AND
# MANAGEMENT

# A STUDY ON VULNERABILITY OF ALAHSA GOVERNORATE TO GENERATE URBAN HEAT ISLANDS

***Ilham S. M. Elsayed***
(King Faisal University, Hufof, Eastern Province, KSA)

**ABSTRACT:** The purpose of this study is to investigate Alahsa Governorate status and its vulnerability to generate urban heat islands. Alahsa Governorate is considered as the largest oases in the Arabic Peninsula with the largest conventional oil field in the world that accounts for more than half of the cumulative oil production of Saudi Arabia. Nevertheless, literature reviewed shows that no researches were done for this region concerning urban heat island. The data was collected from authorized bodies who are control weather station networks over Alahsa Governorate, Eastern Province, Saudi Arabia. Although, the number of weather station networks within the region is very limited and the analysis using GIS software and its techniques is difficult and limited, the data analyzed confirms an increase in temperature for more than $2^{O}C$ from year 2004 to year 2014. Such increase is very considerable whenever human health and comfort are the concern. The increase of temperature within one decade confirms the availability of urban heat islands. The study concludes that, Alahsa Governorate is vulnerable to create urban heat islands and more attention should be drawn to strategic planning for this governorate that is developing with a high pace and a considerable increasing levels of urbanization.

## INTRODUCTION

Alahsa Governorate is located in the South-eastern corner of Saudi Arabia occupying the southern part of the eastern region between 17-26 latitudes and 48-55 longitudes. Alahsa climate is hot summer, cold wet winters, and mostly cloudy atmosphere, seasonal rains in winter and autumn, and the region is exposed to sand storms from time to time.

Alahsa Governorate covers a vast area of around 530 000 $km^2$, representing 68% of the area of the eastern region and 24% of the area of Saudi Arabia. It is considered as the largest oases in the Arabic Peninsula. It is famous of palm cultivation (Mohamed, I. Sharaf. 1996). There are more than 10 thousand hectares of cultivated land for more than two million palms that produces the best and most famous types of dates in Saudi Peninsula.

According to Census 2010, the population of Alahsa province is 1, 063, 112 people. The Saudi citizens counted for 870.577 and 192.535 residents. The Governorate consists of many cities and villages in addition to the most famous empty desert that is called Empty Quarter. Hofuf and Mubarraz cities are the oldest urban centers in this region and most developed areas (Alahsa Municipality, 2015). Hofuf city is considered as the capital city because of its economic role in the region. Its leading economical role importance increased after the discovery of oil in Eastern Province of the Kingdom. Several oil centers are located in Alahsa Governorate, among them the largest oil field in the world that accounts for more than half of the cumulative oil production of Saudi Arabia, which is named Al- Ghawar. It measures 280 by 30 km.

**Historical Background.** As a result of population growth and continuing migration to both Hofuf and Mubarraz largest and most urban developed cities (Alahsa Municipality, 2015), urban development passed over four stages during the period 1957-2012.

**First stage before 1963.** This stage is considered as the city origination phase. Water is one of the most important foundations for the emergence of the region and the stability of human. Wherever springs and water wells are located, communities are formed. Hofuf, Mubarraz are famous of their permanent springs and water wells that helps in formalizing the cities and affect the level of urbanization by increasing the rate of migration. Hofuf, Mubarraz were characterized by tribal groupings communities having their own lands and natural resources.

The first phase was characterized as the phase of looking for the best position to stabilize population making use of the presence of wells and springs as a source of permanent water, enough to provide drinking water and Agriculture. The stability of the region was supported by the discovery of oil in 1939 in Eastern region that led to the evolution of national income and creation of municipal system that affects action plans for development and supervision. Figure 1. below shows an aerial view to Hofuf and Mubarraz during first stage of growth before 1963.

**Figure 1.** First growth stages for Hofuf and Mubarraz before 1963

**Second stage 1963-1973.** This stage is considered as growth and configuration phase. The region passed through many urban growth stages during this ten years. Recognized economic and social changes were recorded during this decade. Noted urban growth rates increased with parallel development especially in the development of transport networks. A five-year development programs started with the concurrent trebling increase in oil prices. That directly affected the whole country income and lead to doubling of national incomes. New neighborhoods were built at both Hofuf and Mubarraz towns. Population grown as well as commercial activities increased on main streets in the heart of the metropolis. The expansion of development at that time was in southern and Western directions of the region and its commercial centers. At that stage, major services and Government departments were centered at Hofuf town and a transportation system connected the town with Mubarraz town. Therefore at this stage Hofuf was considered as administrative capital of the surrounding territory.

**Third stage 1973-1994.** This is stage is considered as the evolution and jump development phase. . As a result of the accumulation of wealth and escalating economic development, a substantial urban growth rates were recorded in Alahsa specifically in Hofuf and Mubarraz. At this period, employment of international labor to implement development and reconstruction operations started. This phase was characterized by urban extension towards the West and South, resulting in the emergence of residential buildings on both sides of the new roads that stretch beyond the capital of Hofuf. Due to the increased migration of nomads, districts were formed as Alrogaigah District. At this stage Hofuf start to take its current form as a capital city.

**Fourth stage 1994- 2012.** This stage is the recent recorded phase (Alahsa Municipality, 2015). A considerable increase in urban growth in Alhasah Governorate took place; from 7649 to 22869 Acre of built area. While the growth in Hofuf city was from 3730 to 14877 Acre that influenced urban development to present situation. Moreover, the approved expected urban growth is 38000 Acres in 2028 by adding approximately 19000 Acres to Hofuf and Mubarraz City. Figure 2. Below shows urban growth stages and trends in Hofuf and Mubarraz.

## MATERIAL AND METHOD

Data collected from authorized bodies, who are controlling weather station network over the area under study, Alahsa Governorate, Eastern Province, Saudi Arabia. The total number of weather station networks within the region is very limited. After a comprehensive and long investigation, the number of weather station that cover Alahsa Governorate are only three weather stations:

**Alahsa Meteorological Station.** Alahsa Meteorological Station is counted as the main and oldest authorized governmental meteorological station in the region. It works under ministry of Defence and Aviation, presidency of Meteorology and Environmental Protection, National Meteorology and Environmental Center. It is located inside Alahsa Airport premises at 25 17 53 N Latitude and 49 29 11E Longitude, 178.17 m Elevation. The data collected from this station is for 2004 and 2014 over a period of three months that are counted as the hottest months of the year (Mohamed, 1999). Accordingly, the data used are for June, July and August for the two previously mentioned years.

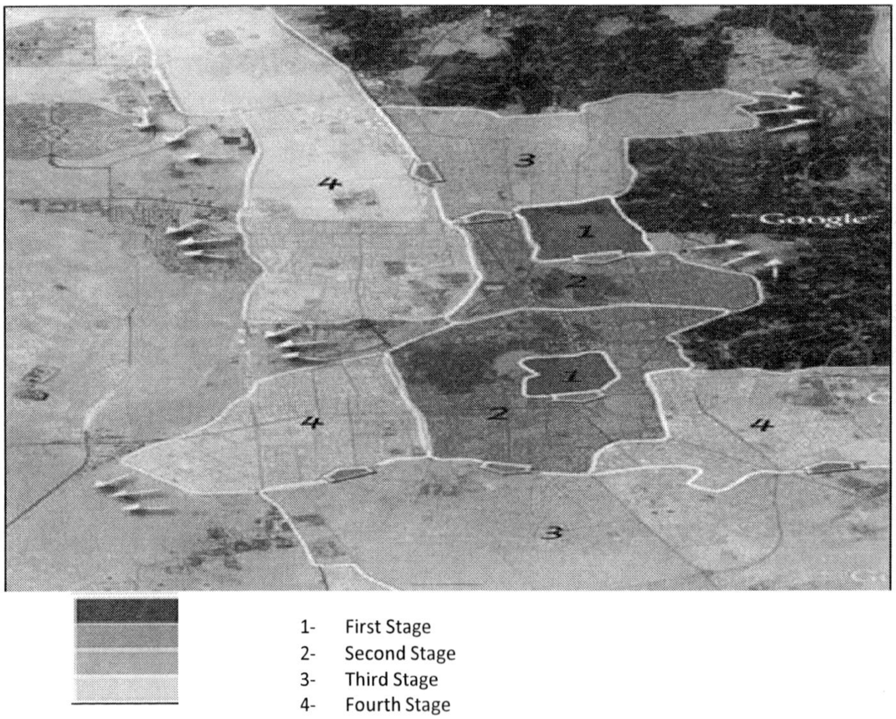


**Figure 2**. Stages and trends of urban growth in Hofuf and Mubarraz City

**Training, Agricultural & Veterinary Research Center (TAVRC) Meteorological Station.** The TAVRC Meteorological Station is located inside King Faisal University (25.3401 latitude and 49.5968 longitude) with in the premises of the Training, Agricultural & Veterinary Research Center. The data collected from this station is for less than two years. It is for whole 2013 and almost 2014, as the authorized bodies confirm that, this data is the maximum data that they can provide. According to the purpose of the study, the data used from this station is confined to year 2014 to match that data collected from Alahsa Meteorological Station. While data for 2013 is excluded from analysis.

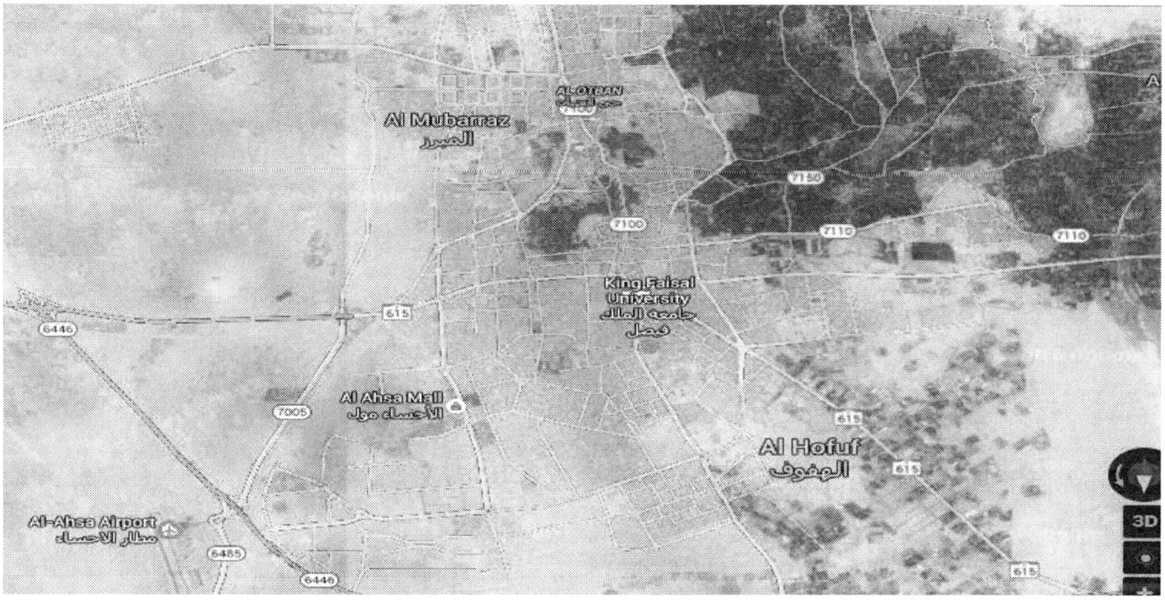


**Figure 3.** The location of Alahsa Airport and King Faisal University Courtesy: Google maps, Saturday, April 30, 2016.

**College of Engineering Meteorological Station**. The third station is located inside King Faisal University too. It is running in collaboration and control of King Abdulaziz City for Science and Technology at College of Engineering, King Faisal University premises. It is the latest station built before around three years. Although such data is very recent, it will help in comparing the data collected from other station for year 2014. Nevertheless, as authorized body mentioned, data retrieval process is on hold waiting for new password and user names after renewing contract between King Faisal University and King Abdulaziz City. Therefore, it is not included with the data used for this study.

The data collected from the above mentioned stations is used mainly to investigate the temperature pattern over the Governorate. Figure 3. below shows the location of Alahsa Airport and King Faisal university using Google maps on Saturday, April 30, 2016. TAVRC Meteorological Station is located within the most populated well developed urban area in Hofuf city. While Alahsa Meteorological Station is located at the far end of the city in a less populated and less developed area. (Streutker, 2003) defined the urban heat island as the difference in temperature between the urban and rural areas. Accordingly, the research investigates and analysis the temperature measured at TAVRC and Alahsa stations during June, July and August of 2014. Moreover, it investigates on temperature measured at Alahsa station in 2004 and 2014 for months of June, July and August.

## RESULTS AND DISCUSSION

**TAVRC and Alahsa stations during June, July and August of 2014.** Chart 1. Below shows the average maximum temperature measured in degree Celsius at TAVRC and Alahsa station during June, July and August 2014. The difference in maximum temperature between the two stations in the three previously mentioned months is more than 1.5 $^{O}$C. It reached 2.0 $^{O}$C during June, 2014. Which confirmed that the city centers are always hotter than suburban areas (Elsayed, 2012).

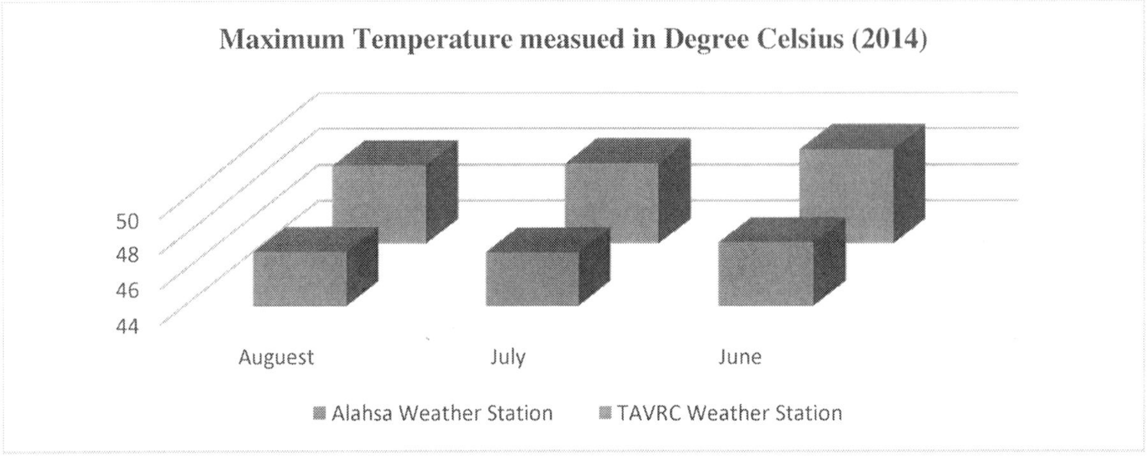


**Chart 1.** Maximum temperatures at TAVRC and Alahsa station during June, July and August 2014

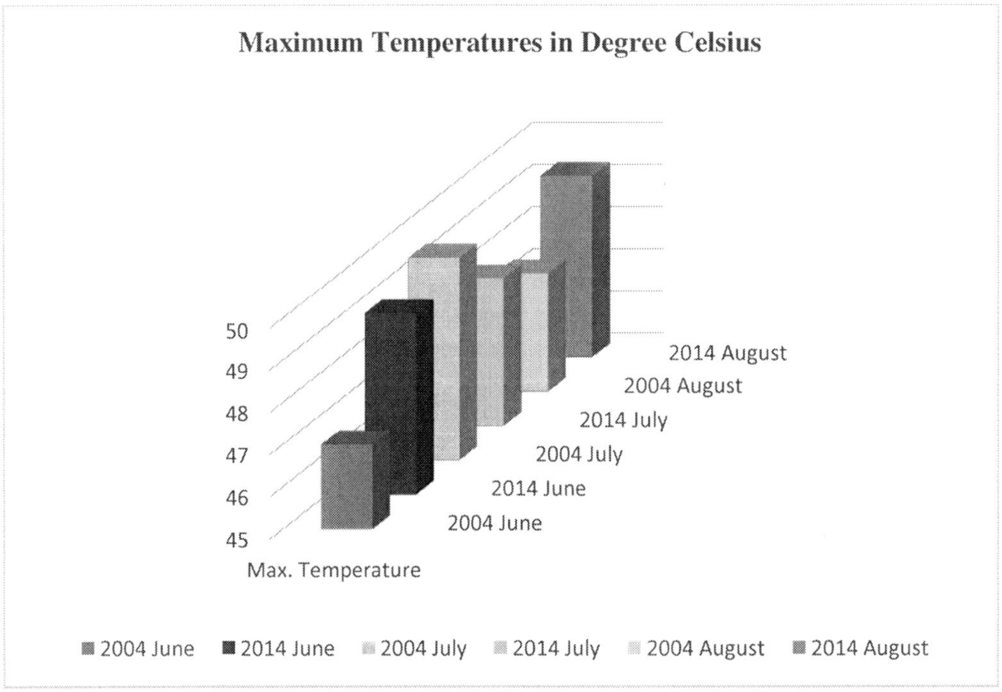


**Chart 2.** Maximum temperatures during June, July and August, 2004 and 2014

**Alahsa station in 2004 and 2014.** As mentioned previously the study focuses on the hottest months of the year June, July and August. Chart 2. Below shows the average maximum temperatures measured during June, July and August year 2004 and 2014. The difference in temperature between June 2004 and June 2014 is 2.3°C. While, it is 1.5°C for August. The most astonishing result is that collected for July 2004. It shows the heights temperature recorded. It is even higher than that recorded for July 2014. Which is somehow contradicts the research assumptions that expected increase in temperature in 10 years period.

Nevertheless, it is recognized that this high temperature is accompanied with the heights relative humidity and lowest mean wind speed and velocity during the study period. Therefore, it is excluded. On the other hand, chart 3. below shows the average maximum temperature with respect to average mean wind speed, average wind velocity and population for June 2004 and 2014. The chart confirms that there is an increase of temperature with respect to increase of wind speed, velocity and population.

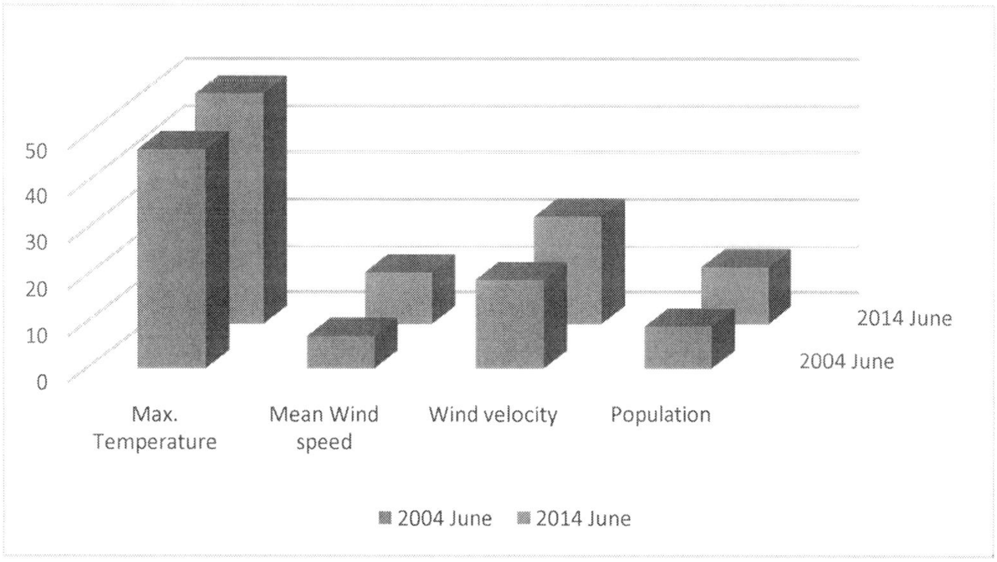


**Chart 3.** Max. temperature ($^{O}$C), mean wind speed, velocity and population, June 2004 and 2014

## CONCLUSION

The results of the study confirms previous studies (Elsayed, I. 2012a &b, Smoyer, K. E. 1997, Valazquez-Lozada, A. 2002) that show that, as the population increases the temperature increases. The study proves that, there is an increase of temperature counted for 2.3$^{O}$C for one decade period from 2004 to 2014. Moreover, it shows that there is variation in temperature intensity between the two weather stations considered for the research from June to August 2014 counted for 2.0$^{O}$C difference. The variation in temperature between Alhasah and TAVRC station confirms the availability of UHI within the area under study. Such variation on temperature confirms the need of extensive research to measure the intensity of this UHI, nucleus of UHI and exact number of HUIs available within the Governorate.

The study concludes that, Alahsa Governorate is vulnerable to generate urban heat islands and recommended further studies to locate the nucleus of the UHI. That will contribute much to support decision makers in strategic planning and future development of a governorate that is developing with a high pace and considerable increasing levels of urbanization.

## REFERENCES

Al Zawad, M. F. 2008. *Impact of Climate Change on Evaporation, Surface Runoff and Soil Moisture on a Regional Scale in Saudi Arabia*. M. Sc. Thesis, King Fahd University Of Petroleum and Minerals, Dhahran, Saudi Arabia.

Alahsa Municipality. 2015. *The Current Conditions of the Present Alahsa: Second Report*. Alahsa Municipality, Eastern Province, Saudi Arabia.

Elsayed, I. S. 2012a. "A Study on the UHI of the City of Kuala Lumpur, Malaysia." *King Abdul-Aziz University Journal, Sciences of Meteorology Environment and Arid land Agriculture*. 23 (2)

Elsayed, I. S. 2012b. "Mitigation of the Urban Heat Island of the City of Kuala Lumpur, Malaysia." *Middle-East Journal of Scientific Research, IDOSI Publications.* 11 (11): 1602-1613

Hafner, J.1996. *The Development of Urban Heat Islands in the Southeast Region of the United States in the Winter Season (Global Warming).* Ph. D. Thesis, University of Alabama, Huntsville, AL.

Ken, D. S., Evans, K., Blane, H., Michele, L., Kallur, S., and James, D. 2015. "Vulnerability to climate change in three hot spots in Africa and Asia: Key issues for policy relevant adaptation and resilience building research." *Springer.*

Mohamed, F. Atta. 1999. "Climatic Conditions in Alahsa, Saudi Arabia." *Research Consulting Service Center Journal*: 1 (1).

Mohamed, I. Sharaf. 1996. "Climatic Risk to Agriculture in Oasis of Alahsa." *Journal of Faculty of Arts, Alexandria University, Egypt.*

Smoyer, K. E. 1997. *Environmental Risk Factors in Heat Wave Mortality in St. Louis.* Ph. D. Thesis, University of Minnesota, Minnesota, MN.

Streutker, D. R. 2003. *A Study of the Urban Heat Island of Houston, Texas.* Ph. D. Thesis, Rice University, Taxes, TX.

Valazquez-Lozada, A. 2002. *Urban Heat Island Effect Analysis for San Juan, Puerto Rico.* M. Sc. Thesis, University of Puerto Rica, San Juan.

## OPTICAL-ELECTRONIC EQUIPMENTS OF ECOLOGICAL DEDICATION

**Ruben Asatryan** (National Institute of Metrology, Yerevan, Armenia)
Hamlet Karayan(Yerevan State University, Yerevan, Armenia)
Norayr Khachatryan (National Institute of Metrology, Yerevan, Armenia)

Currently sharply increased interest in environmental issues, which is primarily due to the ever-increasing contamination of the environment. According to the latest data on the study of atmospheric pollution in industrial developed countries, the main sources of pollution are industrial and energy facilities and transport, which accounted for over 80% of the total amount of pollution. The major components of air pollution are gaseous compounds of carbon, nitrogen and sulfur, as well as solid and liquid aerosol formation, which are of particular concern for the normal functioning of humans and other biological objects. Therefore, the creation of optoelectronic devices and systems with the best metrological parameters that enable the operational analysis of basic physical and environmental parameters, and distant monitoring of the atmosphere and air infrared environmental control of vast forest spaces (for detection of fires in the early stages of their development) and pipelines of natural gas is a very important task. The present work is devoted to presenting the results of research and development work on the elaboration and manufacturing of optical-electronic instruments for environmental purposes to explore the basic physical and ecological parameters of the atmosphere, as well as monitoring forest spaces and main gas pipelines.

# AN APPROACH FOR ENVIRONMENTAL RISK ASSESSMENT OF ENGINEERED NANOMATERIALS USING ANALYTICAL HIERARCHY PROCESS (AHP) AND FUZZY INFERENCE RULES

**Emel Topuz*** and C.A.M van Gestel
(VU University Amsterdam, Faculty of Earth and Life Sciences, Amsterdam, The Netherlands)

The usage of Engineered Nanoparticles (ENPs) in consumer products is relatively new and there is a need to conduct environmental risk assessment (ERA) to evaluate their impacts on the environment. However, alternative approaches is required for ERA of ENPs because of the huge gap in data and knowledge compared to conventional pollutants and their unique properties that make it difficult to apply existing approaches. This study aims to propose an ERA approach for ENPs by integrating Analytic Hierarchy Process (AHP) and fuzzy inference model which provide a systematic evaluation of risk factors and reducing uncertainty about the data and information, respectively.

Risk is assumed to be the combinantion of occurrence likelihood, exposure potential and toxic effects in the environment. A hierarchy was established to evaluate the sub factors of these components. Evaluation was made with fuzzy numbers to reduce uncertainty and incorporate the expert judgements. Overall score of each component was combined with fuzzy inference rules by using expert judgements. Proposed approach reports the risk class and its membership degree such as Minor (0.7). Therefore, results are precise and helpful to determine the risk management strategies. Moreover, priority weights calculated by comparing the risk factors based on their importance for the risk enable user to understand which factor is effective on the risk. Proposed approach was applied for Ag (two nanoparticle with different coating) and $TiO_2$ nanoparticles for different case studies. Results verified the proposed benefits of the approach.
(**Acknowledgement:** This study was funded by the Scientific and Technological Research Council of Turkey (TUBITAK) 2211 Grant Program)

# WHAT IS ETHICAL?
# AN ANALYSIS OF THE OPPOSING FORCES OF ENVIRONMENTAL SUSTAINABILITY

Terry L. Polen
(West Virginia Department of Environmental Protection, University of Maryland University College)

**ABSTRACT:** There is a duality, or triality, between the opposing forces of sustainability; the environmental, economic, and community. By using reinforcing and balancing causal loops we can show that a simple ethical question regarding values of right and wrong regarding the environment is not so simple. This systems thinking analysis will show the ethical, and practical, causes and implications of environmental policy and decision making that surround the current national and international environmental 'climate'. Additionally, an understanding of the drivers both individually and corporately is derived using the Polen Universal Field Management Theorem.

## INTRODUCTION

It seems simple enough, what is ethical? Something is ethical or it is not, correct? Well, maybe not. It may be descriptive ethics (thou shalt not cheat on your taxes) or normative ethics (you 'ought' to pay your taxes). Yet life is just not that simple is it? Paying your fair share of taxes is what you are required to do, and what you 'ought' to do, but what is your fair share? Is it really ethical to take all of those deductions knowing that there are starving children in Africa that could be fed by your contribution to the government?

What about the environmental aspects? What 'ought' government to do? What is right? For example, a coal fired electrical utility is critical to the lifestyle that many have and wish to retain. Conversely, what of the pollutants? What about the workers and their livelihood? Will they have another good paying job that they can move into, to or will they be placed out of work and into the public safety net? What of the parks and hospitals? What implication and import will an environmental policy decision really have? Is the greater good, the greater good? Who decides and how do they? These are the seemingly simple, yet oh so deep.

## DISCUSSION

> *To be, or not to be-that is the question: Whether 'tis nobler in the mind to suffer the slings and arrows of outrageous fortune or to take arms against a sea of troubles, and by opposing end them. Shakespeare, Hamlet Act III, Scene I.*

This is seemingly the most well-known of all of the Bard's phrases had nothing to do with environmental sustainability. Yet the implication is similar. What is right? What should we do? The ethical conundrum that encompasses most issues is also present in the consideration of environmental sustainability. The commonality of an 'obvious' path is quite uncommon. Environmental, like social and community issues do not lend themselves to simple trite answers:

- Every politician wants 'jobs'
- Every politician wants 'clean air'
- Every politician wants 'strong communities'

But what if these issues are not synergistic? What if these issues actually work against each other? Would the high paying jobs in a commercial sector facility degrade the air? In doing so will that same manufacturing facility create the tax structure for the park, yet increase the noise and light and pollution around said park?

What was once accepted is now rejected - at least by a segment of the populace. Then what is right? Ethicists and Philosophers and even Engineers have been struggling with this conundrum for millennia.

If we can but understand that there is 'logic' to the other side, then we have already traveled farther down that road than most societies. The implications and interactions of internal and external forces are not, by nature, linear. However, the causes and implications are and can be shown, analyzed, and therefore understood. Consider our own ethics that are described below (Figure 1). Something that is legal and ethical is easy to think of; maybe feeding the homeless. Something that is illegal and unethical is easy to think of; maybe torture. What of something that is legal and not ethical? Some would consider abortion in this bin. What of something that is illegal and ethical? The American Revolution would fill this category for Americans. Not so for the British of the time.

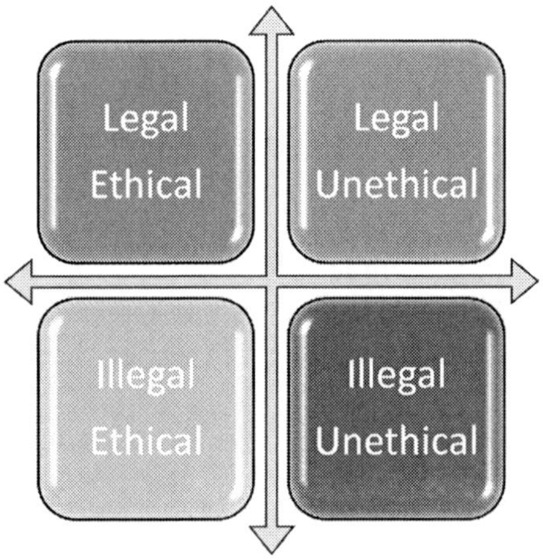


Figure 1. Legal, Illegal, Ethical and Unethical Relations

The common definition of sustainability from United Nations (U.S. EPA, 2011) is that there is a 'sweet spot' in the overlapping priorities of Economic, Environmental, and Societal needs. It is that spot that is the most sustainable and it is this to which we strive. We understand the overshadowing one priority for another can lead to difficulties. Pollution impacts in many places of the world would fit here.

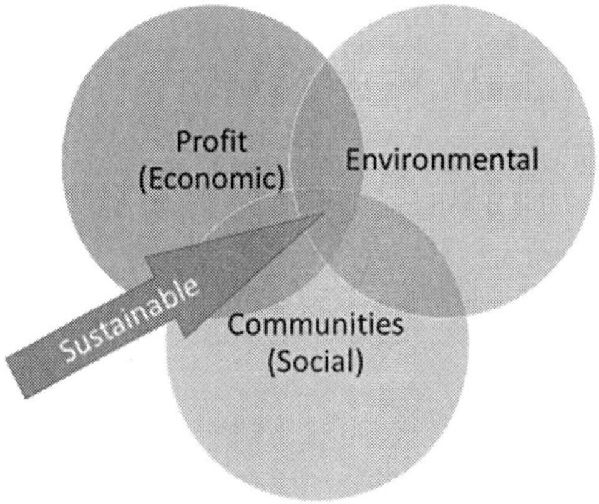


Figure 2. Impact of Economic, Environmental and Social factors on Sustainable Development

A common definition of sustainability speaks of the opposing forces of sustainability i.e. the environmental economic and community issues that interact with each other (Figure 2). However these three forces both support and oppose each other at the same time. If one of them 'tramples' over another, difficulties arise, certainly in the societal realm. It is common in the history of the world to see areas where the economic portion has overshadowed that of community and that of society. It is common in the history of the world to see we that are in environmental issues are completely overshadowed by the power of the dollar. In the United Nations interpretation of sustainability, these three opposing forces have the sweet spot, a place within the Venn type diagram whereby all three opposing forces will balance. It is in that space that you will find the greatest sustainability. The above figure shows how this is typically denoted. Yet this mortification of a Venn diagram with its sweet spot can write on his face inaccurate. These opposing forces fight against each other and for area to be sustainable there must be a balance yet that balance is different for each individual as they consider that place. Each culture and subculture will view that 'sweet spot' differently. So what is right? What is ethical?

Therefore the second diagram in this analysis shows a further continuation and possibly a more accurate description (Figure 3).

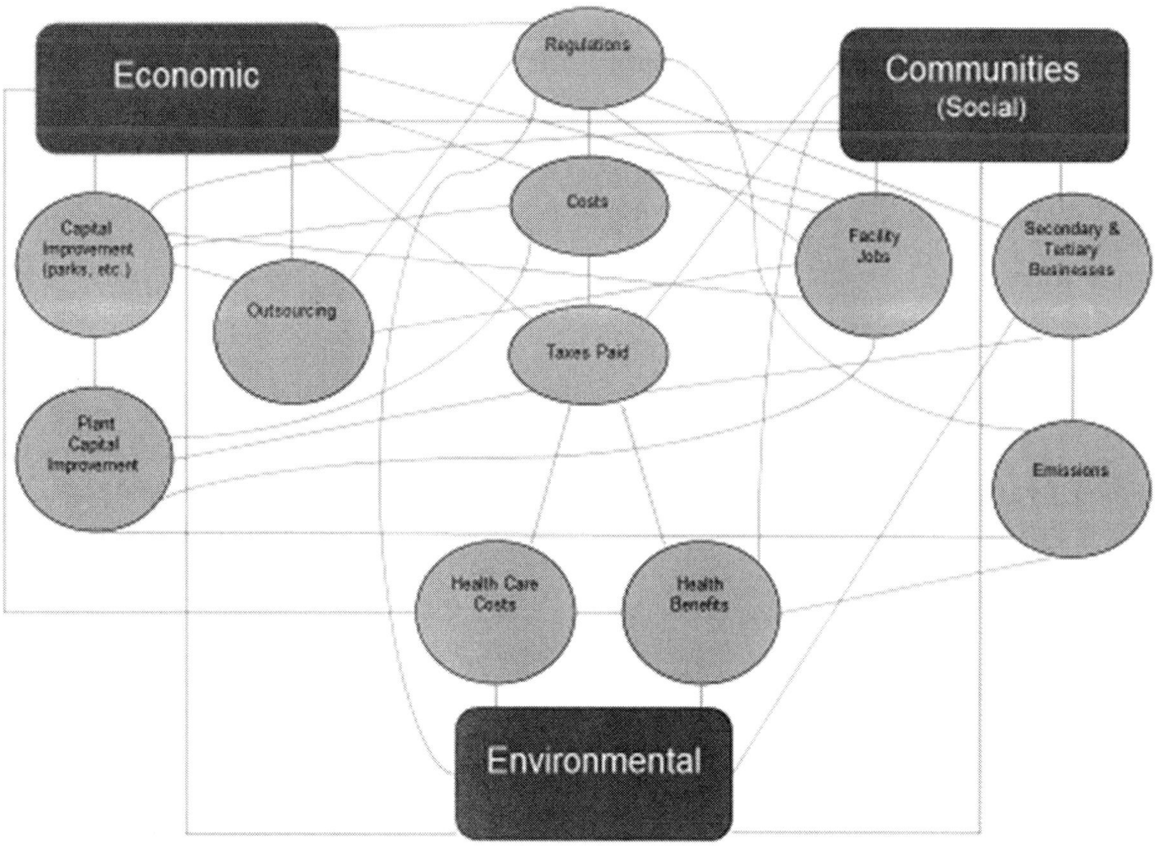


Figure 3. Interactions among Economic, Social and Environmental factors

In previous work (Polen, 2014), it was shown that the model of sustainability is incorrect in that it has many tendrils that pull and push on the environmental, social, and economic spheres. Using the causal loop analysis above, it was shown that there is a strong interrelation of the spheres. Reinforcing loops and balancing loops work like so many carnival attractions. If a new business comes to the community, the tax base will increase which will help build infrastructure (roads, bridges, parks, etc.), but at the same time, that plant brings increased truck and rail traffic, as well as increased water and air pollution. But that same plant additionally brings needed jobs to the community.

While laborers and construction workers and potential new employees are thrilled, environmentalists worry about the impact on those same individuals and the vegetative and animal species that will be effected by the new facility.

Both sides firmly believe that they are right and well within their rights to encourage, or stop, said facility. They also 'know' that the other side is 'wrong' or even 'nuts'. Their side is only common sense.

- They need jobs.
- They need clean water.
- What good are jobs without clean water?
- What good is water without someone there?
- And the debate roils.

The UN model defines the conjunction of the spheres in some great societal and cultural Venn diagram. Yet, even this view is too simplistic to truly create public policy that is common sense and that is the best for all. Policy makers must understand that yin and yang between the various forces.

While this diagram moves us forward. This too is limited by the two dimensional thinking that has created the problem initially.

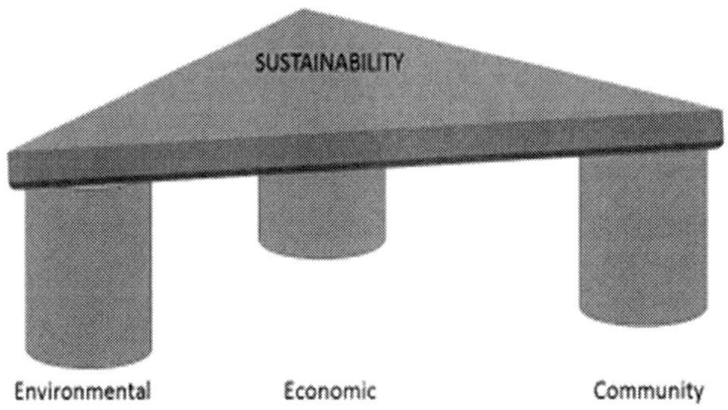


Figure 4. Construction of Solid Sustainability

When we consider 'what is ethical', the triangular table above gives the best representation of the process. Any of the legs of the stool, when limited, can cause the stool itself to be unstable. So a balance of these forces must be maintained to maintain the equilibrium that leads to long term success (Figure 4). Yet this mainlining of equilibrium brings us full circle to the original question of 'is it ethical'?

To whose ethics do we agree? To whose belief structure? Why are you right when I, with a well thought out and passionate argument, wrong?

Ethicist and philosophers from Heraclitus to Socrates to Plato to Aristotle to Bentham to Mill to us posit that we must pursue what is best for the most. Bentham described this as felicitous calculus (Mill, 1863). What makes the most people happy? This is the basis for utilitarianism. The calculation of the best for the most. Consider road construction. While eminent domain is used to remove a few house and homesteads that have been part of that family for generations, additional untold thousands and tens of thousands will be able to use that road for commutes and community. For family, faith, and a better future. The road and the transportation itself will aid in economic development that will create hope and a future for untold multitudes. Yet the road itself destroys wetlands and old growth forests. Protesters stand before bulldozers. What is the balance?

Let us first consider the individual needs and wants of 'why do they do that'?

The author, in previous works (Polen, 2015) has posited that individuals output or outcomes are derived from the following equation:

*Polen's Universal Field Management Theorem (UFMT)*
*Output=Obedience [(Pistos*Agape)+(Leadership*Motivation)]*
*Output=O(PA+LM)*

Where Pistos and Agape are terms from Kione Greek for faith and love respectfully.

In relation to sustainability, this equation can be used to describe not only 'why do they do that', but additionally, why they believe that that is ethical.

Let us simplify the sides in the road discussion to environmental groups, economic groups (including businesses, construction workers, and economic developers), and the community (including the public at large and the local politicians).

The environmentalists see a degradation in the air quality and a negative impact to the water and other environmental media. The economic developers see an opportunity for growth to their area. The community sees a resurgence of tax base that will fund parks and hospitals, with a corresponding new need for the tax base, community, and hospitals.

Using the UFMT equation, all sides have a leadership component. Someone has a strong opinion. Someone believes that they are right and the other side is wrong. These groups are well motivated to do the 'right thing'. The Agape, or love of community (or individual house if the concept is used in that fashion), work, environment, money and etc. along with the faith (Pistos) that what they are doing is not only right, but will do the positive thing is yet another driver.

The individuals involved must be obedient to their cause, or side of the discussion. Obedience is the key to the other terms in the equation. Individuals may well feel strongly, yet never act on those beliefs.

## CONCLUSION

Hence the balancing act. It is not a sweet spot that all agree upon. It is a table whose legs must not be unbalanced too far. Otherwise the people and the environment and the society itself will be hurt. So who is right in a contentious situation? Who is wrong? Who is on the side of the greater good? Often lawmakers and policy makers contend that they are completely unbiased, yet their decisions impact without the due course of thought regarding what is really, truly ethical.

The common refrain of it is only 'common sense' has become ubiquitous in most political debates, arm chair societal quarterbacks, and new environmental and economic policies. How stridently we attack renewable fuels when we know that coal miners will be on the unemployment lines is a discussion for utilitarians, and should be for policy makers.

## REFERENCES

Gottlieb, A. (2000). *The Dream of Reason.* New York: W. W. Norton & Co.
Mautner, T. (2005). *Dictionary of Philosophy* (2nd ed.). London: Penguin Books.

Mill, J. S. (1863). *Utilitarianism.* London: Parker, Son, and Bourn.

Pittenger, D. J. (2002). Deception in Research: Distinctions and Solutions from the Perspective of Utilitarianism. (G. P. Koocher, Ed.) *Ethics and Behavior, 12*(2), 117-142.

Polen, T. L. (2014). A Systems Thinking Analysis of Environmental and Economic Forces on Sustainable Community Development. *Seventh International Conference on Enviromental Science and Technology.* Houston.

Polen, T. L. (2015). *The Hidden Hand of Management.* Mustang, OK: Tate Publishing and Enterprises.

Rawls, J. (1955). Two concepts of rules. *The Philosophical Review, 64*, 3-32.

Russell, B. (1903). *The Principles of Mathematics.* New York: Cambridge University Press.

Ryckman, T. (n.d.). *Russell's Theory of Descriptions.* Retrieved from Philosophy at Lawrence University: https://www2.lawrence.edu/fast/ryckmant/Russell's%20Theory%20of%20Descriptions.htm

Shakespeare, W. (2005). *William Shakespeare, The Complete Works* (2nd ed.). (S. Wells, & G. Taylor, Eds.) Oxford: Claredon Press.

U.S. EPA. (2011). Sustainabilty and the U.S. EPA. Washington, DC: National Academies Press.

# CLIMATE VARIABILITY AND SUSTAINABLE DEVELOPMENT: THE CASE OF NIGERIA'S NIGER DELTA REGION

**Chika Ogbonna,** Eike Albrecht and*René Schönfelder
(Brandenburg University of Technology Cottbus Senftenberg, Germany)
(*University of Lübeck Institute for Software Engineering)

**ABSTRACT:** Current and future climate associated events endanger sustainable development (SD) in the Nigeria's Niger Delta region. The situation was examined by analyzing a 30-year temperature and rainfall data from 1980 to 2010 obtained from Nigerian Meteorological Agency (NIMET). It was found that temperature trend in different States of the region that made up the study area was increasing while rainfall was variable and declining. Scientific literatures that reveal varying evidence on the effects of climate variability in the region were reviewed and used to complement the results. Nonetheless, the effects of climatic change such as seasonal and erratic flooding, coastal erosion, and variation in temperature and rainfall threaten people's wellbeing and contributed further to environmental degradation. Adaptation measures that need to be addressed to reinforce sustainable development initiatives were suggested. The findings will be useful to decision-makers, non-governmental organizations, relevant institutions, and government agencies in Nigeria.

## INTRODUCTION

Climate variability and climate change have become major environmental challenges facing developing countries in the 21st century. The Second National Communication to the UNFCCC (2014) affirms that the Nigeria nation in general faces climate-related impacts such as severe floods, windstorms, heat waves, ocean surges, and a wide range of climate extremes that have negative impact on the country's socio-economic activities. Projections show that a rise of 0.5m in sea level could result in the loss of 35% coastline of the highly-productive delta (INDC 2015). The area faces extensive climate risks ranging from storm surges, increasing risks of flooding, variable but extreme rainfall and heat waves. However, besides seasonal flooding, the 2012 flooding incident in the coastal Niger Delta was devastating in the area. On the other hand, following the global climate temperature trend, observed climate shows that temperature in Nigeria have been increasing in the last five decades and it has become more significant since the 1980s and this is also the case in the Niger Delta (FME 2014). The Fifth Assessment Report of the International Panel on Climate Change (IPCC) concludes that "climate variability and change is a threat to sustainable development" (IPCC 2014a). The concept of sustainable development started in Stockholm, Sweden during the 1972 United Nations Conference on Human Environment. The World Summit on Sustainable Development held in Johannesburg, South Africa from the 2 to the 4 September 2002 reaffirms the commitment to Sustainable development. Sustainable development is a process of development that seeks to achieve, in a balanced manner, the three pillars: social development, economic development and environmental protection. However, the concept of sustainable development was popularized by the Brundtland report published by the World Commission on Environment and Development (WCED) in 1987, also known as ''Our Common Future'' (UN 2010). The report defined sustainable development as ''development which meets the needs of the present without compromising the ability of future generations to meet their own needs" (WCED 1987). While Agenda 21 adopted by more than 178 governments at the 1992 Earth Summit brought a comprehensive plan to table, the 2012 Rio+20 conference asked countries to renew their commitment to sustainable development (IPCC 2014a). The relevance of discussing the potential effects of climate variability and its connection to sustainable development in this paper is due to the obvious fact that climate change and variability is anticipated to pose major threat to future sustainable development in Nigeria and particularly in the Niger Delta. In essence, such climate impacts make it hard to achieve sustainable development. This study therefore, gives insight into how climate associated events endanger current and future sustainable development and present viable adaptation measures that need to be addressed to reinforce sustainable development initiatives.

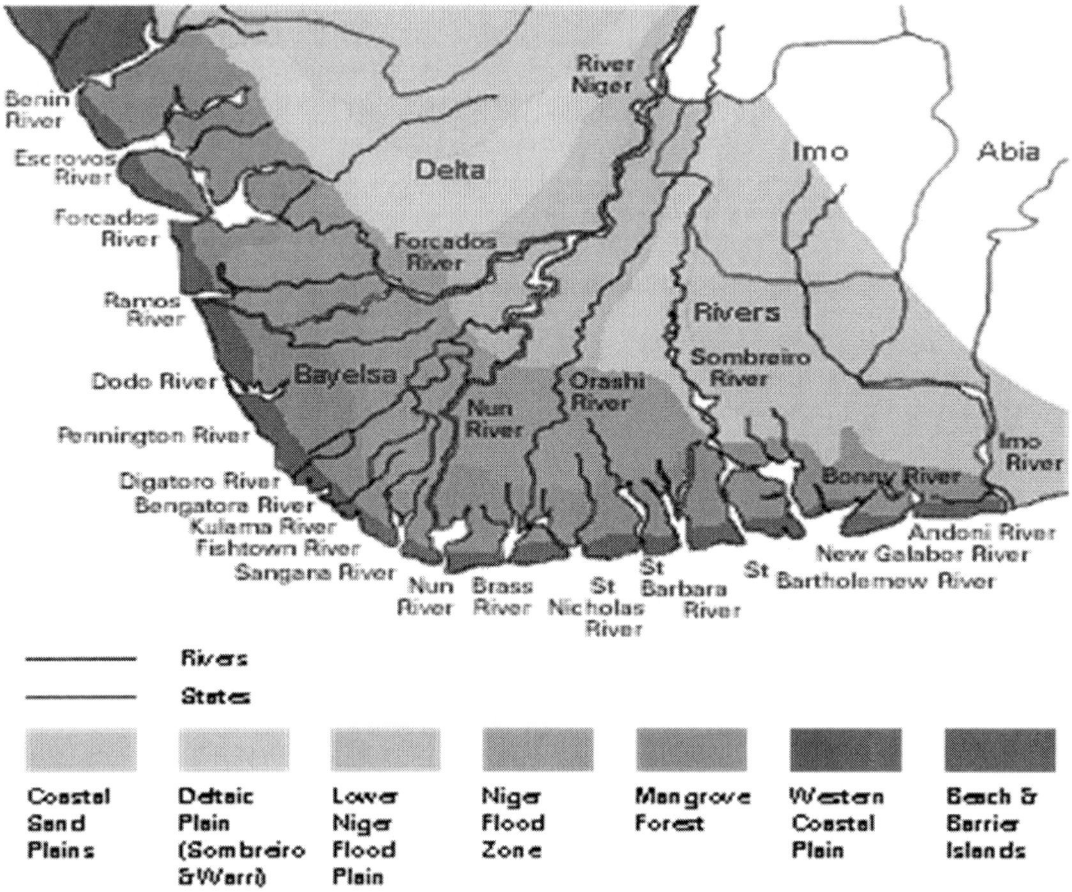


**Figure 1.** Map of Niger Delta Nigeria (study area) showing, Rivers and Vegetation. Three focal States for this study include Delta, Imo, and Rivers State. Source: UNDP Niger Delta Biodiversity Project.

The Niger Delta region is located on the Atlantic Coast of Southern Nigeria, where the river Niger splits into numerous distributaries (Figure 1). It is one of the largest delta regions in the world and the rich part of Nigeria in terms of natural resources (Ogbonna 2014). For example, it has large oil and gas deposits and biodiversity. The region covers an area of about 70,000 km$^2$, starting from a coastal zone of swamps, which extends Northwards towards the rainforest that gradually merges with woodlands and savanna grasslands in central Nigeria (UNEP 2011). The Niger Delta is located on latitudes from 4°25' N to 6°00' N and longitudes from 5° 00' E to 7°5' E.

## MATERIALS AND METHODS

The observational data relates to climate variation within a 30-year period ranging from 1980 to 2010. The minimum rainfall data and maximum temperature data used in trend analysis was collected from the Nigerian Meteorological Agency (NIMET), Lagos. The temperature and rainfall data were imported into 'R' statistical software and plotted using ggplot2. The monthly maximum temperature is averaged for an indication of heat waves and the monthly rainfall is added up to get the annual rainfall. These values were plotted in a line chart and then extended by a linear regression curve showing the trend. The study also relied on the analysis of various secondary materials and literature, relevant journals, government policy papers, research papers, and books that provides evidence of climate impacts and vulnerability in the Niger Delta region.

## RESULTS AND DISCUSSION

**Trend Analysis of Temperature Data.** As can be seen on the result (Figure 2), the average monthly maximum temperature at Warri, Owerri and Port Harcourt showed an increasing trend in temperature patterns. In Warri (Delta State), the lowest temperature of 30.3°C was recorded in 1985 while the highest temperature of 32.3°C was observed in 1994. Results also showed that from 1980 to 2010, the average temperature in Warri, Delta State increased by 0.14°C per decade, whereas the average monthly maximum temperature in Owerri, Imo State over the 30-year period indicated a gradual increase in temperature. The lowest maximum temperature in Owerri was recorded in 1985 at 31.4°C, which rose again rapidly beyond 32°C around 1987, and gradually fluctuated until the highest temperature of 33.10°C was attained by the end of 1998. The linear trend indicates a progressive increase in temperature with time recorded in years. The temperature trend analysis also showed that the average temperature over the years increased by about 0.21°C per decade in Owerri. On the other hand, the monthly maximum temperature at Port Harcourt, Rivers State varied between 1980 and 2010. The lowest temperature recorded within this period was in 1992 with a record of 30.25°C, which rose and fluctuated within 31°C and 31.75°C between 1994 and 2005. After 2005, average temperature increased again and fluctuated until the highest was reached in 2010 at 32.05°C. Overall, temperature did not increase extensively in Port Harcourt in comparison with Owerri and Warri as the trend analysis indicated a slight rise in temperature as the years progressed. Results also show that the monthly maximum temperature in Port Harcourt increased at a rate of 0.042°C per decade.

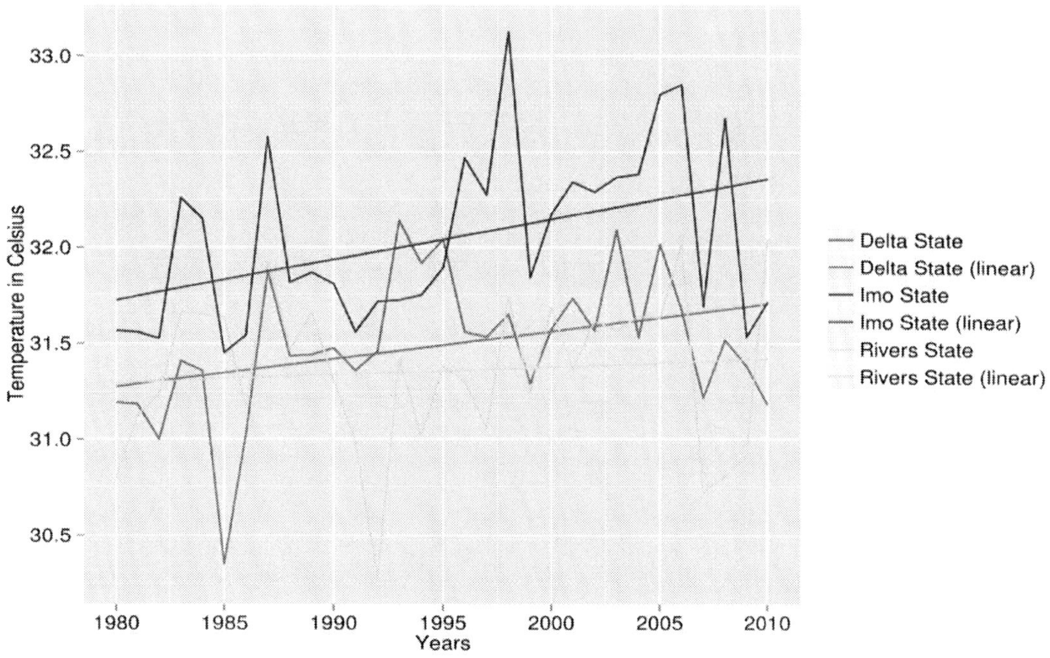


**Figure 2**: Maximum temperature and yearly average from Owerri (Imo State), Warri (Delta State) and Port Harcourt (Rivers State) synoptic stations respectively.

**Trend Analysis of Rainfall Data.** The graph (Figure 3) indicates that Delta State experienced a record high of rainfall of about 3400 mm in 1995, whereas a record low of rainfall of about 2,400mm occurred in 1981. Observed rainfall data in Port Harcourt, Rivers State shows a wide variation in rainfall. There was a sharp decline in rainfall pattern recorded in 1982 at 2,200 mm with an all-time record low of 2,000 mm in the years 2000 and 2008, while a peak in rainfall pattern of 2,800 mm was observed in 2007. In Imo State, there was an average rainfall decline to 1,550 mm and 1,650 mm in 1983 and 1998 respectively, a sharp increase to 2,300 mm in 2007 and ultimately, a decline to 2,200 mm in 2010.

The entire results correspond to the existing research on the Niger Delta region (BNRCC 2011; Ike and Ezaziye 2012, Okorie et al. 2012) and with the global predictions of the IPCC 2007 and 2014b. It also strengthens confidence that temperature anomalies in the States follow the same trends of global warming. High values in average monthly maximum temperature clearly depict the intensity of maximum temperature (heat waves) all through the years. The results also corroborated the research conducted by Okorie et al. (2012) in Owerri, Imo State which found that "an increase in surface temperature and torrential rainfall causes flash floods and flood disasters, increasing frequency and intensity of extreme weather events such as thunderstorms and unpredicted rainfall patterns, as well as other climate related disasters." However, while rainfall data show fluctuating variations, temperature data shows an upward trend. Given the fluctuating rainfall pattern in Figure 3, one may wonder why the flood event in the area is increasing besides other socio-economic factors that may contribute to seasonal flooding. In this vein, the IPCC posits that ''if an increase in high-intensity rainfall events is concurrent to the peak of the rainy season, this may result in widespread flooding'' (Sylla in IPCC 2015). This is also the case in the coastal Niger Delta communities as information derived from the relevant literature indicates that increase in flood events are recurrent during the peak rainy seasons. The situation is compounded by interactions among multiple environmental stressors.

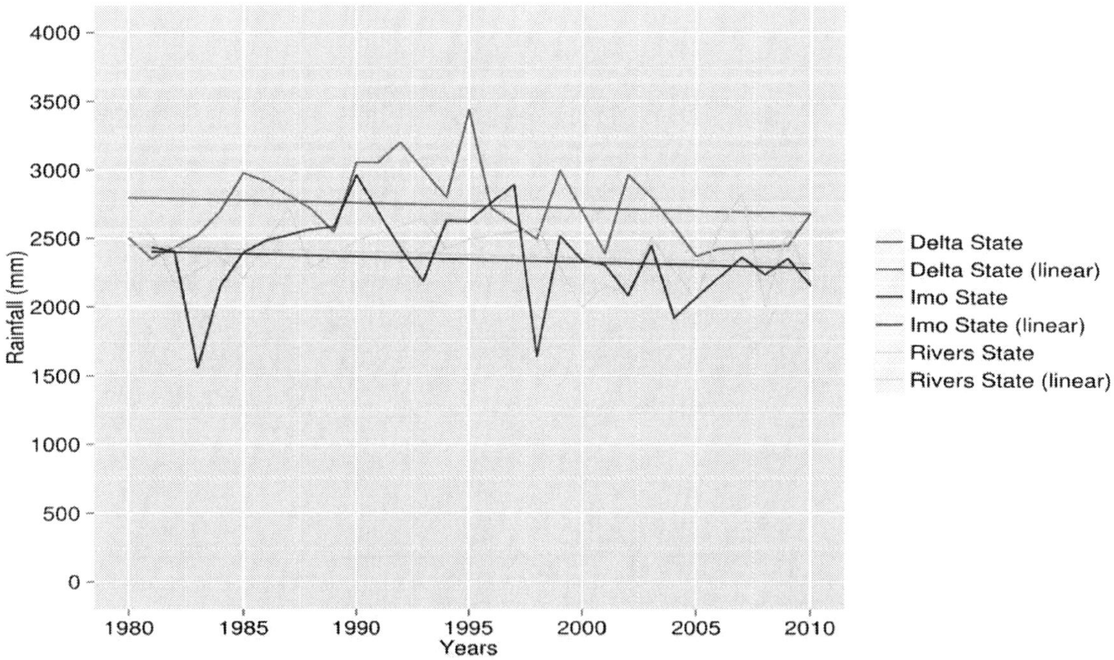


**Figure 3**. Average annual maximum rainfall for Owerri (Imo State), Port Harcourt (Rivers State) and Warri (Delta State)

**Goal 1**

- **End poverty in all its form everywhere:** The increasing climate-related events such as floods endanger the eradication of poverty in the region. For instance, the 2012 flood in the coastal Niger Delta heightened poverty as many community people lost their livelihoods and life savings, coupled with sudden increase in prices of food commodities in the communities (TACC 2013, PDNA 2013).

**Goal 2**

- **End hunger, achieve food security and improve nutrition and promote sustainable agriculture**: Effects of climate variability (e.g. high temperatures, increasing trend of heat wave intensities and decreasing rainfall pattern) influence food security, especially as farm households, depend on subsistence agriculture and small-scale fisheries. As large scale irrigation is seldom practiced due to lack of irrigation facilities, crop production is dependent on weather and climatic conditions (NEST 2011;TACC 2013;Ogbonna 2014).

**Goal 5**

- **Achieve gender equality and empower all women and girls**: Climate variability associated extreme events possess different implications to men and women in the area during extreme events, accelerating gender inequality as women folks are most vulnerable and socially disadvantaged than men (TACC 2013). For example, pregnant women experience more stress from excessive heat condition (TACC 2013); and negative impacts during the 2012 flood on female entrepreneurship was significant (PDNA 2013).

**Figure 4**. Effects of Climate related events on Sustainable Development
Source: NEST (2011); TACC (2013); PDNA (2013); Ogbonna (2014)

The variability in climate portrayed by both temperature increase and changes in rainfall seasons casts light on how climate variables could affect the human environment and sources of people's livelihoods such as ecosystem services, human health, agricultural activities, and fishing for example. Nonetheless, the study by the United Kingdom Department for International Development (DFID) predicts a rise in temperature of up to 3.2°C by 2050 under a high climate change scenario in every part of Nigeria. It also predicts a significant sea level rise from the 1990 levels to 0.3m by 2020 and 1m by 2050 along the Nigerian coasts (DFID 2009, FME 2014). However, considering the fact that the coastal Niger Delta settlement is a low-lying area, the future predictions of rise in sea levels call for action.

**Implications for Sustainable Development.** The above analyses show evidence of observed climate variability and its potential risk to people and communities in the area and draw on the Nigeria's Intended National Determined Contribution to the Paris Agreement which states that "climate change poses a significant threat to the achievement of development goals, especially those related to eliminating poverty and hunger and promoting environmental sustainability" (INDC 2015). In other words, achieving sustainable development entails dealing with climate vulnerability (IPCC 2014c). The increasing trend in temperature and irregular yet decreasing trend in rainfall pattern portends extensive negative environmental and socioeconomic implications unless developmental policies consider resilient and robust adaptation actions. For instance, the following three new sustainable development goals are potentially threatened by current and anticipated climate and weather related impacts in the region (Figure 4).

## CONCLUSION

Section 20 of the Nigerian 1999 constitution requires that ''the state shall protect and improve the environment and safeguard the water, air and land, forest, and wildlife of Nigeria." The United Nations Framework Convention on Climate Change provides the basis to adapt to the impacts of climate change; Nigeria ratified the Convention on the 29 August 1994 (Ogbonna and Albrecht 2015). Article 4.1 of the Convention is crucial for undertaking adaptation and paragraph 1(e) of the Article asked all parties to the Convention to cooperate in preparing for adaptation in the areas of coastal management and agriculture with particular reference to Africa. Drawing upon the above study analyses, climate variability combined with other environmental stressors exerts potential pressure on current and future sustainable development initiatives in the Niger Delta. Adaptation will enhance social, economic and environmentally sustainable management of the region's natural resources. In this respect adaptation measures that need to be addressed to reinforce sustainable development initiatives include the following but not limited to: provision of accurate weather and climate predictions, though not a panacea to extreme weather events, but will enhance proactive adaptation actions and improve self-protective behaviors; scaling up research that would improve traditional agricultural practices. For example, improved crop species that could be resilient to extreme temperatures and variable rainfall patterns. Such could help avert influences on hunger, food security and poverty. Application of the precautionary principle is imperative to reduce climate risks, protect coastal settlements, infrastructures, and people's livelihoods in the Niger Delta. The provision of appropriate legal and policy frameworks, which will mean bringing together all necessary regulatory tools and environmental instruments that can enhance adaptation implementation, are crucial in this regard.

## REFERENCES

Albrecht, E. and Ogbonna, C.U. 2015. Strategic Environmental Assessment as a Tool to Integrate Climate Change Adaptation: A Perspective for Nigeria. In: Leal Filho, W. (ed.), Handbook of Climate Change Adaptation, Springer-Verlag, Berlin Heidelberg, pp. 1239–1260.

Federal Ministry of Environment (FME). 2010. National Environmental, Economic and Social Development Study (NEEDS) for Climate Change in Nigeria. Federal Ministry of Environment, Abuja, Nigeria. Special Climate Change Unit, pp. 45.

Federal Ministry of Environment (FME). 2014. Second National Communication on Climate Change. Federal Ministry of Environment of the Federal Republic of Nigeria, Abuja: MEFRN, pp. 132.

INDC. 2015. Nigeria's Intended Nationally Determined Contribution, available at:http://www4.unfccc.int /submissions/indc/Submission%20Pages/submissions.aspx,accessed 05.26.2016.

IPCC. 2014a. Climate Change 2014.Impacts, Adaptation, and Vulnerability. Cambridge University Press, Cambridge University Press, Cambridge, United Kingdom.

IPCC. 2014b. Climate Change 2014. Synthesis Report Contribution of Working Groups I, II and III to the Fifth Assessment Report of the Intergovernmental Panel on Climate Change IPCC, Geneva.

IPCC. 2014c. Climate Change 2014. Mitigation of Climate Change. Contribution of Working Group III to the Fifth Assessment Report of the Intergovernmental Panel on Climate Change. Cambridge University Press, Cambridge, United Kingdom.

NEST 2011. Reports of Research Projects on Impacts and Adaptation. Building Nigeria's Response to Climate Change. Ibadan, Nigeria: Nigerian Environmental Study/Action Team (NEST), pp. 330.

Ogbonna, C.U. 2014. Adaptation to Climate Change in Developing Countries: A need in the Niger Delta region of Nigeria. In: Albrecht, E. Schmidt, M., Mißler-Behr, M., Spyra, S.P.N. (eds). Implementing Adaptation Strategies by Legal, Economic and Planning Instruments on Climate Change. Springer Berlin Heidelberg, pp. 165–187.

Okorie F. C, Okeke I., Nnaji A., Chibo, C., and Pat-Mbano, E. 2012. "Evidence of Climate Variability in Imo State of Southern Eastern Nigeria." Journal of Earth Science Engineering 2: 544–553.

Post-Disaster Needs Assessment (PDNA). 2013. Nigeria Post-Disaster Needs Assessment. A Report by the Federal Government of Nigeria, with Technical Support from the European Union, United Nation, World Bank, and other Partners, Nigeria Post-Disaster Need Assessment 2012 floods. Federal Government of Nigeria Abuja, pp. 154.

Sylla, M.B. 2015. "Regional Climate Change over West Africa: Trends, Shift of Climate Zones and Timing of the Wet Extremes." In: Stocker, T.F., D. Qin, G.-K. Platter, and M. Tignor (eds.), IPCC. 2015. Workshop Report of the Intergovernmental Panel on Climate Change Workshop on Regional Climate Projections and their Use in Impacts and Risk Analysis Studies, pp. 154–155. IPCC Working Group I Technical Support Unit, University of Bern, Bern, Switzerland.

Territorial Approach to Climate Change (TACC). 2013. Biophysical and Socio-Economic Assessment of the Nexus of Environmental Degradation and Climate Change, Delta State, Nigeria, pp. 234.

UNDP Niger Delta Biodiversity Project, available at: http://www.undp.org/content/dam/undp/documents/projects/NGA/Niger%20Delta%20Biodiversity_Prodoc.pdf, accessed 20.05.2016

United Kingdom Department for International Development (DFID). 2009. Impact of Climate Change on Nigeria's Economy, DFID Abuja, Nigeria.

United Nations (UN). 2010. Sustainable Development: From Brundtland to Rio 2012. Background Paper Prepared for Consideration by The High Level Panel On Global Sustainability at Its First Meeting, 19 September 2010. International Institute for Sustainable Development (IISD), pp. 26.

United Nations Environmental Program. 2011. Environmental Assessment of Ogoniland. United Nations Environmental Program, Nairobi, Kenya, pp. 257.

World Commission on Environment and Development: Our Common Future (WCED). 1987. Report of the World Commission on Environment and Development: Our Common Future, available at http://www.un-documents.net/our-common-future.pdf, accessed 05.15. 2016.

# URBAN ENVIRONMENTAL OVERALL PLANNING AND ITS ROLE IN URBAN SUSTAINABLE DEVELOPMENT IN CHINA

***Luyan Wang***, Yang Zhao, Lina Tang, and Jingzhu Zhao
(Institute of Urban Environment, CAS, Xiamen, Fujian Province, 361021, China)

Urbanization is facilitating economic development in China and around the world but is limited by land resources and environmental problems. Unplanned urbanization restricts sustainable development in China. However, the present urban environmental planning only emphasizes the prevention and treatment of the existing civilian and industrial wastes but lacks the efforts in urban environmental overall planning that intends to turnaround environmental degradations from the source. Urban overall planning and land-use overall planning represent technical supports for urban ecological and environmental protection from the angles of spatial pattern and land usage, and make it possible for an environment-development-construction-land integrated planning system of urban sustainable development. The overall planning of urban environment aims at (1) explicit urban resource carrying capacity and environmental capacity, and correctly regulating the rates in urban population growth and economic scale expansion; (2) defining ecological red lines, promoting ecological functions, improving environmental quality, and sustaining resource utilization; (3) comprehensive functional classification of urban environment, realizing classified management and administration of urban environment, and guiding the regulations of industrial distribution and structure; and (4) recognizing major sources of environmental risk, defining red lines of environmental risk, and developing control systems of environmental risk. To ensure the effective role of urban environmental overall planning in urban sustainable development in China, it is essential to clarify legal status of urban environmental overall planning, strengthen the concepts of ecological and environmental spatial management, and attach great importance to the ideas of regional coordination.

# PERFORMANCE ENHANCEMENT OF EVAPORATIVE COOLING TECHNOLOGY IN THE COMPANY OF LOWER ENVIRONMENTAL IMPACT

***S.P. S. Rajput*** and Amrat Kumar Dhamneya
(Maulana Azad National Institute of Technology Bhopal, India 462003)

The objective of this paper is to identify the development of different geometry and material of cooling pads used in cooling for performance enhancement of direct evaporative cooling. Nearly everyone discussed that the effectiveness of direct evaporative cooling decreases as temperature increases. Focusing on human thermal comfort at minimum cost, low maintenance, inexpensive, minimum energy consumption and to minimize the emission of greenhouse gases effect, researchers and designers are developing new methods for evaporative cooling technique in place of conventional air-conditioning. India has different weather conditions at different places throughout the year. Central India climate is best favourable for air-conditioning with evaporative cooling. To increase the effectiveness different evaporative cooling method are being used in combined operation with other cooling measures, like indirect/direct evaporative cooling (IDEC) with vapour compression refrigeration system (VCRS), indirect evaporative cooling with desiccant wheel, direct evaporative cooling with desiccant, direct evaporative cooling with ground storage water and direct evaporative cooling with VCRS. In this paper the concluding remark will be on the performance enhancement of direct evaporative cooling technique in favour of cooling effectiveness and better human thermal comfort keeping in mind the minimum consumption of electricity and lower environmental impact. Selection of different evaporative cooling technique for air-conditioning according to the weather conditions will be applied according to appropriate condition. This study will focus on heat exchanger configuration and its material composition, distribution and treatment of water flow, cost-effectiveness and minimum environmental impacts. Emphasis is laid on extracting the heat from atmospheric air by passive technique as well as minimizing the conventional load of air conditioner.

# ADAPTATION STRATEGIES TO MITIGATE CLIMATE CHANGE IMPACTS ON URBAN STORMWATER.

**Ashantha Goonetilleke** (Queensland University of Technology, Brisbane, Queensland, Australia)
An Liu (Shenzhen University, Shenzhen, China)
Erick R. Bandala (Desert Research Institute, Las Vegas, NV, USA)

The water environment is particularly vulnerable to climate change due to the predicted changes to the weather patterns. Studies have shown that even with aggressive mitigation efforts, the impacts of climate change will worsen and, appropriate adaptation efforts are needed. However, attention in relation to climate change impacts on the water environment is primarily focused on water resources with limited attention given to the impacts on stormwater. The need for concern stems from the fact that stormwater runoff exerts a significant influence on freshwater resources. Furthermore, urban stormwater is an important resource for the future as urban areas are facing increasing deficiencies to meet the needs of an ever growing population, which are being further exacerbated by climate change. The predicted climate change impacts will influence the characteristics of the primary stormwater pollutant processes, namely pollutant build-up and pollutant wash-off. Additionally, the changes in rainfall characteristics will result in changes to the stormwater runoff hydrograph. Therefore, understanding the linkages between climate change and stormwater runoff characteristics is important for developing effective adaptation strategies to mitigate the adverse consequences of climate change on the water environment. The paper provides an overview about changes to flow processes and pollutant processes and current knowledge gaps in relation to urban stormwater. Additionally, a broad discussion is provided on the options available for adaptation to cope with climate change impacts on stormwater runoff quality and quantity. It is proposed that adaptation should be viewed as an opportunity to create a sustainable development agenda. Accordingly, two key approaches are proposed to advance the sustainable development agenda: low impact development to underpin urban development; and the identification of water sensitive opportunities in the urban environment based on local needs and imperatives.

# ENVIRONMENTAL AND SOCIAL SUSTAINABILITY

Sukhmander Singh
(Santa Clara University, Santa Clara, California 95053, USA)

Sustainable development should include both environmentally and socially sustainable solutions. This is so because environment, society and technology are interlocking parameters. Society's economic system influences the production systems that are appropriate to that society. The production system in turn impacts the ecosystem. It is now well recognized that the production system with an ever increasing per capita consumption can lead to the depletion of resources as well as to the pollution of the ecosystem; and hence the importance of sustainable solutions. Since the production is often motivated by profit- which can be exploited by the rich, even in the use of a sustainable solution, there is a need for a sustainable solution to be both environmentally and socially just. A mode is presented for addressing sustainability in society. It is argued that both environmentally and socially sustainable solutions can be achieved by incorporating locally available renewable or recycling material with the use of local labor. Such an approach would help set up employment base for local people who otherwise may be denied such an opportunity should a large scale system /entrepreneurship is allowed to employ an approach of mass production versus production by masses. In some poor countries where poverty, hunger and poor sanitation exist, environmentally acceptable, economically accessible and socially sustainable solutions must be found to bring about a developed and socially just society. These solutions ought to be simple, inexpensive and environmentally safe. Examples of such solutions are presented in the paper. It is recommended that affordable technologies based on the concept of sustainability be developed to help marginal communities. The paper points out that there's a need for interference between technology and society while developing technology based sustainable solutions.

# ECOTOURISM AND SUSTAINABLE DEVELOPMENT; EVIDENCE FROM LOPE NATIONAL PARK, GABON

**Cornelius FRIMPONG**
(Department of Biological and Environmental Sciences, University of Stirling, Stirling, Stirlingshire, FK9 4LA, Scotland, UK)

Ecotourism is widely applied to the conservation of tropical forest resources as a means of raising funds for conservation and to support social and economic development of local communities. With these three objectives, ecotourism has been embraced as a sustainable management approach. On this premise, this study has produced a new schematic diagram to guide ecotourism conceptualization. Consequently, using the case of Lope National Park, Gabon, this study engaged a purposive sample of 20 conservation scientists and Park staff (Gabonese and non-Gabonese) to analyse how the three sustainability objectives are achieved and how this is influenced by respondents' nationality. Data analysis and hypotheses testing were conducted through MANOVA, Wilcoxon signed rank test and paired-sample t-test in SPSS Version 21 (IBM). The study found that ecotourism is perceived to fulfil all three objectives of sustainability. However, statistical analysis showed significant difference in the distribution of ecotourism revenue between Gabonese and non-Gabonese [$F$ (3, 16) = 5.66, $p$ = .008, partial $\eta 2$ = .515]. Follow-up ANOVA conducted revealed that, whereas Gabonese distributed revenue nearly equally across the three pillars, non-Gabonese placed overemphasis on conservation [$F$ (1, 1.21) = 17.056, $p$ = .001, partial $\eta 2$ = .487]. Consequently, ecotourism management at Lope was found to fall short of meeting international standards. The study concluded with recommendations and a suggestion for further research to broaden this scope by recruiting a larger sample of various stakeholders.

# STAKEHOLDER PARTICIPATION IN INTEGRATED WATERSHED MANAGEMENT: MODELS AND PROCESS

***Aysegul Tanik**** and Cigdem Kanber
(Istanbul Technical University, ITU, Turkey)

**ABSTRACT:** Stakeholder participation in sustainable watershed management, the related participation models and process forms the main part of this study. Especially in developing countries, the main concern is to increase public awareness and encourage public representatives to participate in management issues. In that sense, four participation models; Commentary Model, Social Learning Model, Joint Planning Model and Consent/Consensus Model are introduced. Buyukmenderes watershed of Turkey is selected as the pilot watershed. Five main provinces share this watershed making the management a crucial process. Upon investigation, education, socio-economic and development levels are found well above the country's average. The current legal and administrative structure permits the establishment of Watershed Management Committees (WMC); however, the members though limited to 25 exceed 70 in multi-provincial watersheds. A new structure is proposed in this study, namely Watershed Protection Association (WPA) to overcome this participation problem. Training of WMC and WPA members are detailed under the principle of *learning together to manage together*. At the initial phase (2-3 years) of watershed management studies, Social Learning Model is selected for the pilot watershed. After gaining experience on participatory process, Joint Planning Model is recommended. Developing countries may follow a similar pathway to introduce public in watershed studies.

## INTRODUCTION

Integrated Watershed Management aims to define the natural resources within its boundaries, and to determine any sectoral conflicts while enabling coordination among any activities and policies related to the current legal and administrative structure with the participation of stakeholders representing the main water-related actors and users in the most compatible manner (Heathcote, 1998; Randhir, 2007; US EPA, 2013a). Practices of different stakeholder participation models heading to increase the well-being of public have so far shown that applicable models may differ among watersheds and that the development level of the watershed is an important concern governing the selection of the appropriate model (US EPA, 2013b; Kanber, 2015). One of the available models may be initially selected and applied for a certain period of time may then be quitted and another model may be preferred. Four of the models are referred in this study. These are; Commentary Model, Social Learning Model, Joint Planning Model and Consent/Consensus Model as shown in Figure 1 (Url-1).

Turkey bearing 25 watersheds is at a stage of preparing the watershed management plans and trying to get acquainted with watershed management principles and policies regarding public participation which is a common case for most of the developing countries. One of the essential strategies to the achievement of the objectives of integrated watershed management relates to the promotion of community participation (ESCAP, 1997; US EPA, 2013c). Public participation is a planned effort to involve citizens in the decision-making process and to present and resolve citizen conflict through mutual two-way communications. Therefore, this research attempts to set a methodology on how public may participate and/or may be represented in watershed management studies. Various participation models will be briefly outlined; suitable ones that seem to best fit the selected watershed of Turkey, namely Buyukmenderes Watershed, will be mentioned. A watershed protection association will be proposed and its structural framework and projected responsibilities will be defined regarding the legal and administrative status of the country.

## MATERIALS AND METHODS

**Participation Models.** Four of the models of public participation are given in Figure 1 and the relationships among these models are schematically shown in Figure 2. All the referred models aim to introduce stakeholders to the participation system. The main concern is to increase public awareness and encourage public representatives to participate in management issues especially in developing countries. In a way, the question is '*how can we include different stakeholders in watershed management*'?

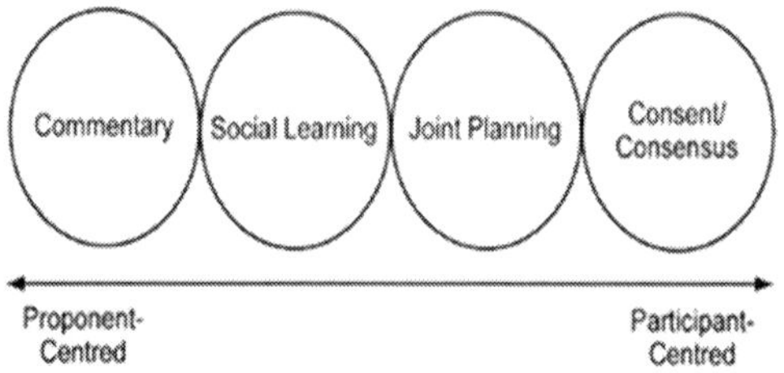


**FIGURE 1.** Four Models of Public Participation (Url-1)

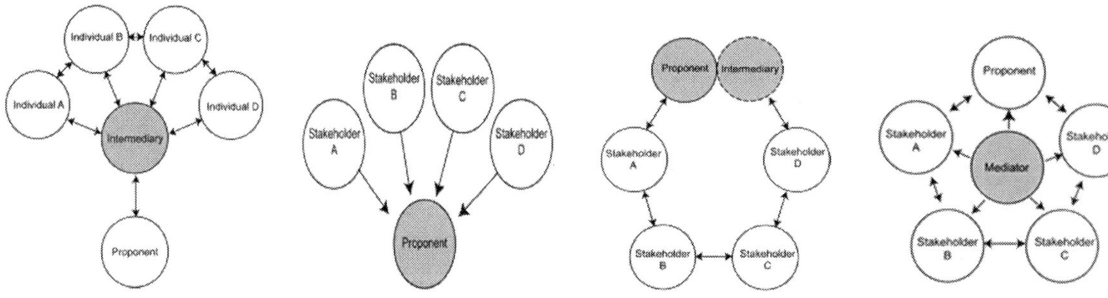


(a) (b) (c) (d)

**FIGURE 2.** Relationship among the participation models; (a) commentary model, (b) social learning, (c) joint planning, (d) consent/consensus (Url-1; Kanber, 2015)

**Brief Information on the Watersheds of Turkey.** Administratively, the country has 81 provinces with political boundaries that do not match with the hydrological boundaries of 25 watersheds. As might be imagined, the provinces neither bear similar surface areas nor the watersheds. Information defining the watershed characteristics regarding their geographical status, climatic conditions, ecological characteristics as well as the socio-economic welfare may all be gathered on provincial basis. However, the provinces share watersheds but not always fully as seen in Figure 3. Some provinces are shared by more than 2 provinces which make the situation of collecting information quite tiresome. Neighborhood provinces, even though they take place in the same watershed, bear different physical characteristics as well as different financial and social status. Under these circumstances, management becomes more complicated. However, these administrative problems are partly overcome by the publication of two legislations; one is about formation of watershed management committee, duties, working procedures and regulations initially published in 18.06.2013 with no 28681in the Official Newspaper, and revised in 20.05.2015 with no 29361, and the other is National Watershed Management Strategy (2014-2023) put into force on 04.07.2014 with no 29050.

**Physical Structure of Buyukmenderes Watershed.** 10 provinces share this coastal watershed located along the Aegean Sea. Aydin Province constituting 27.45% of the basin almost fully lie (95%) within the watershed, and thus acts as the coordinator province. Denizli Province covers 33.65% of Buyukmenderes

Basin and almost 70% of the province is within the basin. These two provinces are followed by Usak, Afyon and Mugla where 67.87%, 23.04% and 19.71% of their respective areas lie within the basin. Apart from these 5 provinces the other 5 have shares less than 5%, thus, they are not considered as important provinces (Figure 4).

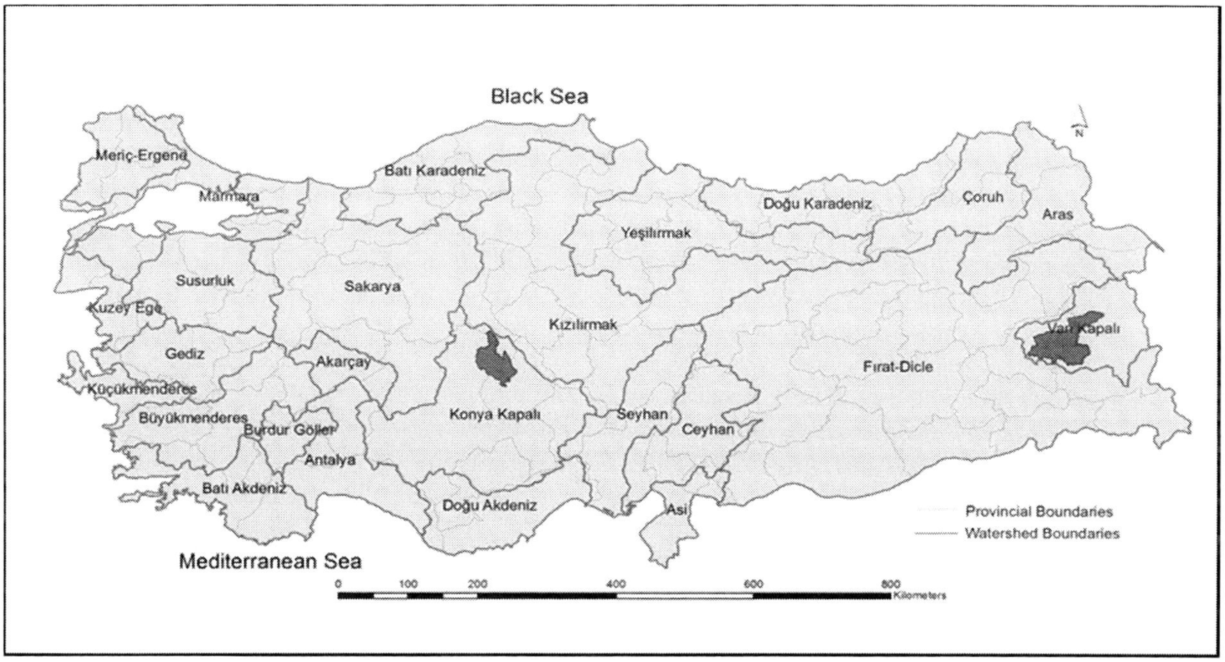


**FIGURE 3.** Watersheds of Turkey

**FIGURE 4.** Sharing provinces of Buyukmenderes Watershed

Inspecting education level of these 5 major sharing provinces, one can easily observe that the level of education is well above the country's average. In that sense, public awareness and consciousness to environmental and especially to water issues in the basin is quite high in parallel to development level. Similarly, the socio-economic status of the basin is also higher than the other neighborhood coastal basins.

Based on the 2013 values, per capita income level is rated as 'high medium income' according to World Banks criteria in the major provinces of the basin despite some differences in the socio-economical and financial status among the sharing provinces. For example, the coordinator province, Aydin, is ranked as the 3th province of the basin in terms of development level (Gul and Cevik, 2014). However, all the sharing provinces are placed in the first 44 row among the overall number of 81 provinces indicating the above average position of the basin (Kanber, 2015).

**Institutional Structure of Buyukmenderes Watershed.** According to the current legislation, Watershed Management Committee (WMC) is limited to 25 people including the governor of the coordinator province. However, for this specific basin consisting of 5 major provinces, this number is calculated as 73 (Kanber, 2015). Only the coordinator province authorities are 21 people. Thus, it is for sure that WMC will not be able to fulfil principles like equality, right of representation and transparency. Moreover, convergence of '*public-private sector-people*' cannot be maintained leading to almost no public participation in environmental issues. Therefore, this structure will not be sufficient for involving public to management. A new autonomous system or formation is necessary to work under WMC.

**WMC Training and Social Learning Process.** Social learning has a basic principle *'learning together to manage together'*. In that sense, a training program for the WMC members at the beginning is proposed. To give training to 73 people is a serious job. Thus, the following questions need to be replied;

- How, when and where these committee members will come together for training?
- Who will give this training?
- What will be the scope and framework of this program?
- Who will cover the expenses of this training program?

It will be quite healthy if this training is given at the early stages of watershed planning process. Thus, the committee members will be furnished with information on the importance of public participation in integrated watershed management studies, and these members on their way back to their sub-basins will communicate with other stakeholders and satisfy the principle of *learning together to manage together*. An establishment of a new structure is also proposed in this study; namely, Watershed Protection Association (WPA), whose structure will be briefly defined in the following part. The training center seems to be better selected by this new association by paying attention to ease of transportation for the committee members. At a start, the place preferred is Regional Directorate of State Water Works in Aydin Province. Probable trainers are recommended to be the team who has already been trained by EU presentatives within the context of 'Training the Trainees Project'. The living, accommodation and traveling costs for training the committee members must be covered by the General Directorate of Water Management under the Ministry of Forestry and Water Affairs, and they should be permitted from their work places to attend this training program. Those members who complete this process with success may be awarded by giving certificates of attendance. As such, they will be encouraged to work as volunteers in management studies. Similarly, those who fail must be discarded from the system. The training program may be considered to last 3 days. These programs may be repeated every two years. Certified members and candidate members may lose these lives, may become older, may pose unstable behavior or lose their enthusiasm during the two years' time; thus, repetition of these programs may be a must to gain new trained members.

WMC members may still lead to lack of sufficient public participation and may act as a supreme board at the watershed. Therefore, a new structure is needed to fulfill the participatory management approach.

**Formation of Watershed Protection Association (WPA).** As defined by the legislation of WMC, the members have to meet monthly. In order to fulfill the requirements of public participation, another association is proposed to be formed under WMC. This formation is named as Watershed Protection Association (WPA) which is actually necessary for a more healthy, fair and consistent public participation.

After the completion of the first run training program to WMC; some of the certified members may take part in the formation of WPA, and WPA members have to be trained by the similar training program. It is then that the successful members of WPA may participate actively in the integrated watershed management efforts and especially in the preparation of watershed management plans, and also in the application of the 'preventive programs' during the application of the management plans. Members of the WPA may consist of public and private sector representatives together with representatives of various other stakeholders like land owners, farmers, fishermen, district or regional representatives, local municipality authorities, various state officers, employers and sectoral representatives, women and youth, public service members, religious officers, universities, colleges, environmental agencies, active NGO's, soil and water protection regional representatives, irrigation associations, national and local media. The association will be a legal personality; therefore, it needs at least a general secretary, office manager and office personnel salaried in its constitution. Furthermore, it will need accounting personnel for arranging the cost of rent and collection of participation shares together with a computer expert, graphic designer and public relations expert (PR). For the field trips and surveys, a vehicle and at least a driver have to be hired. At this point, it is of utmost importance to have sufficient number of representatives of public in WPA. It may be attained through different commissions devoted to training, field applications and PR. Members of these commissions may be represented well by equal distribution of public participation. As such, these commissions will work in close contact with public and feedback gained will increase the consciousness among public towards IWM. Figure 5 shows the schematic diagram of the proposed WPA structure.

## RESULTS AND DISCUSSION

High public awareness and participation is expected in this watershed as its development and education levels are comparatively higher than the country's average together with the income rates. In that sense, effective and active participation in IWM attempts will not be surprise and public interest to environmental issues is expected to be high. A similar training program is also suggested to the members of WPA. Successful members of WMC who have passed from this training may train the members of WPA; forming an example of *learning together.*

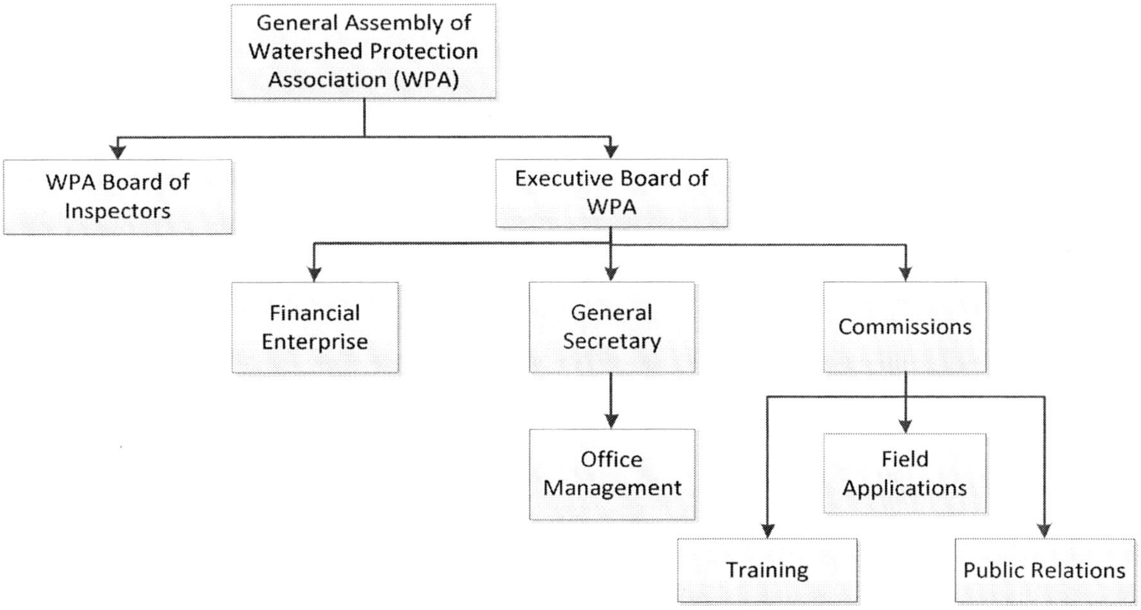


**FIGURE 5.** Proposed WPA structure

At this stage, one of the referred participation models has to be selected for this purpose. Out of the 4 models referred, it is suggested that Social Learning and Joint Planning Models may best fit to

Buyukmenderes Watershed. In the first 2-3 years of time, Social Learning Model may be practiced; however, after this initial stage as experience will be gained on public participation, Joint Planning Model may be followed considering the high interest of public on IWM.

## CONCLUSIONS

Structure of a developing country is always dynamic. Therefore, legal and administrative legislation regarding watershed management tend to change frequently. This study is prepared according to the legislation of April 2015. It is for sure that the legislation of 2013 on the formation of Watershed Management Committee has been an important threshold regarding watershed scale studies. With this legislation for the first time in the country, public participation is encouraged despite many discrepancies. Public awareness and consciousness has still not been widespread, and cooperation and collaboration among various institutions is not at a suitable level. Simply, this study proposes a new structure under the WMC that will communicate more effectively with inhabitants of the watershed. Such an establishment is expected to be successful especially in watersheds where the number of the sharing provinces are high like is the case in this selected watershed.

A fixed participation model is not recommended for all the watersheds of the country as each bears different education, development and socio-economic level. In case of lower education and income levels, proponent-centered commentary model may be preferred. In this model, final decisions are taken by the central authority. The model does not require public participation at the beginning and may be applied by ease.

## REFERENCES

EC-HarmoniCOP. 2005. *Learning Together to Manage Together - Improving Participation in Water Management*, Handbook, EC-funded output of the project Harmonizing Collaborative Planning within the thematic program "Energy, Environment and Sustainable Development" (1998-2002 /Contract No. EVK1-CT-2002-00120).

Erdogan, R. 2013. "Advances in Landscape Architecture- Chapter 6", In M. Ozyavuz (Ed.), *Stakeholder Involvement in Sustainable Watershed Management Environmental Sciences,* InTech, DOI: 100.6772/55798. http://www.intechopen.com/books/advances-in-landscape-architecture/stakeholder-involvement-in-sustainable-watershed-management

ESCAP. 1997. *Guidelines and Manual on Land-use Planning and Practices in Watershed Management and Disaster Reduction*, Economic and Social Commission for Asia and the Pacific (ESCAP), United Nations, 133 p.

Gul, E., and B. Cevik. 2014. *Searching the Development Level of the Turkish Provinces with 2010 and 2012 Data*, Economic Researches Division, February 2014 Report, Turkish IsBank, (in Turkish). http://webcache.googleusercontent.com/search?q=cache:HD4WkK9xrUEJ:https://ekonomi.isbank.com.tr/UserFiles/pdf/ar_03_2012.pdf+&cd=1&hl=tr&ct=clnk&gl=tr

Heathcote, W. I. 1998. *Integrated Watershed Management Principles and Practice*, Library of Congress Cataloging-in-Publication Data, John Wiley & Sons, Inc., NY, USA.

Kanber, C. 2015. *Stakeholder Participation in Integrated Watershed Management: Models and Process*, M.S. Thesis, Istanbul Technical University (ITU) Graduate School of Science Engineering and Technology, Istanbul (in Turkish).

O'Faircheallaigh, C. 2009. "Public Participation and Environmental Impact Assessment: Purposes, Implications, and Lessons for Public Policy Making", *Environmental Impact Assessment Review*, 30(1):19-27.

Randhir, T. O. 2007. *Watershed Management Issues and Approaches*, IWA Publishing, UK.

Url- 1: http://www.hardystevenson.com/research/articles.php. Decision making on Controversial Issues of National Scope: Models for Involving the Public.

US EPA. 2013a. *Getting in Steps: Engaging Stakeholders in Your Watershed*, 2nd Ed. EPA 841-B-11-001, US EPA Office of Water, Nonpoint Source Control Branch (4503T) Washington, DC. http://cfpub.epa.gov/npstbx/files/stakeholderguide.pdf

US EPA. 2013b. *Public Participation Guide*. http://www2.epa.gov/international-cooperation/public-participation-guide-view-and-print-versions

US EPA. 2013c. *A Quick Guide to Developing Watershed Plans to Restore and Protect Our Waters*. http://webcache.googleusercontent.com/search?q=cache:5QYcP7WNE6wJ:water.epa.gov/polwaste/nps/upload/watershed_mgmnt_quick_guide.pdf+&cd=1&hl=tr&ct=clnk&gl=tr

Webler, T., H. Kastenholz, and O. Renn. 1995. "Public Participation Impact Assessment: A Social Learning Perspective", *Environmental Impact Assessment Review*, 15(5):443-463.

# EFFECTIVENESS OF SOUTH AFRICA'S BIODIVERSITY POLICIES IN PROTECTING THE INDIGENOUS VEGETATION OF THE SOUTPANSBERG REGION

**Priscilla Ntuchu Kephe**
(University of Limpopo, Turfloop, South Africa)
Brilliant M. Petja
(Water Research Commission, Pretoria, South Africa, University of Limpopo,Turfloop, South Africa)

The management of protected areas and biosphere reserves needs to be revised in the light of growing - problems related to the declining biomass and composition of the indigenous vegetation in the Soutpansberg Mountain range. This paper analyses key policies, achievements, gaps and constraints in the implementation of South Africa's key biodiversity policies in the Soutpansberg. Issues examined include threats to vegetation such as habitat transformation (composition and biomass) and changes in the local climate of the area. It was found out that despite remarkable achievements and shifts in conservation paradigms, the development of a widely accepted national policy on biodiversity and an increase in the number of the country's protected areas, Soutpansberg continues to lose plant species and vegetation composition. In order to safeguard what is left of the indigenous vegetation of the Soutpansberg, the protection of plants species and habitats should be prioritise, more especially since the vegetation of the Soutpansberg influences the local climate of the area.

# RENEWABLE ENERGY DEVELOPMENT

# PHOTOGALVANIC CELL: AN EFFICIENT DEVICE FOR PHOTOCHEMICAL CONVERSION AND STORAGE OF SOLAR ENERGY

***Chhagan Lal*** (Department of Chemistry, Harcourt Butler Technological Institute, Kanpur, India)
E-mail: clal9940@gmail.com

**ABSTRACT:** Photogalvanic effect was studied in photogalvanic cell containing EDTA as reductant, mixed dyes (combination of two dyes) as phtotosensitizer i.e. EDTA-NMB-SFO, EDTA-NMB-FG, EDTA-BG-FG, EDTA-BG-CB, EDTA-NMB-CB. The used reductant was Ethylene diamine tetra acetic acid and photosensitizers were new methylene blue (NMB), Safranine-O-(SFO), Fast Green (FG), Brilliant Green (BG), Celestine Blue (CB), in different combinations. The photopotential, photocurrent, conversionefficiency, power of the cell and performance of the photogalvanic cells were determined. The effects of various parameters on the electrical output of the cell were observed and a mechanism has also been proposed for the generation of photocurrent in photogalvanic cells.The photogalvanic cells (system) can be used for 95.0, 90.0, 72.0, 65.0, 45.0 minutes in dark respectively.

***Keywords:*** Photogalvanic cell; Dyes; Fill factor; Conversion Efficiency

## INTRODUCTION

Although, the photogalvanic effect was first of all reported by Rideal and Williams (1925) but the iron-thionine system was systematically investigated by Rabinowitch, 1940).Later on, it was investigated by many workers from time to time Fox and Kabir-ud-din (1979), Stevenson and Errelding (1981), Ameta et al (1985). Many workers have used different photosensitizers in different photogalvanic systems and observed a reasonable electrical output as compared to iron-thionine system. Some of the photosensitizers used in solar cells are proflovine (Eisenberg and Silverman, 1961),Tolusafranine (Kaneko and Yamada, 1977), Riboflavin (Murthyetal,11980), Flavinemononucleotide (Yamase,1981), Poly(N-acylamido methylthionine (Tamilarasan and Natrajan,1981), Toluidine blue (Murthy and Reddy,1983) Brilliant Cresyl blue (Jain et al,1984) Methylene blue, (Ameta et al,1989) and Gangotri et al,1996), Safarnine (Gangotri and Regar,1997) etc. but the use of mixed dyes (Gangotri and Lal, 2005,Yadav and Lal, 2010, Lal and Gangotri 2011, Yadav and Lal, 2013 combinations of two dyes) as photosensitizer and reductant received a little attention. Therefore, present study was undertaken.

In order to increase the electrical output, conversion efficiency and storage capacity of the cells, the different combinations of two dyes (mixed dyes) system have been used in the photogalvanic cells in the present work and the role of mixed dyes has also been investigated.

## EXPERIMENTAL

EDTA-disodium salt (MERCK), Sodium hydroxide (Ranbaxy),Methylene blue (NMB), Safranine-O(SFO), Fast green (FG), Brilliant green (BG), Celestine blue (CB), were used in the present work. All the solutions were prepared in the doubly distilled water. A mixture of the solutions of the dyes, reluctant, sodium hydroxide was taken in an H-shaped glass cell. A platinum electrode ($1.0\times1.0$ $cm^{2)}$ was dipped into one limb of the cell and a saturated calomel electrode (SCE) was kept in the other. The whole system was placed in dark till a stable potential was obtained, then the platinum electrode was exposed to a 200W tungsten lamp (ECE) and the limb containing SCE was kept in dark. A water filter was placed between the exposed limb and the light source to cut off infra-red radiations.

The photochemical bleaching of mixed dyes was studied potentiometrically. The photopotential and photocurrent generated by the system-mixed dyes/ reluctant/ OH/ hv were measured by a digital pHmeter (Model HPG-2001) and a microammeter (Osaw, India) respectively. The (i-v) characteristics of the cells were studied using an external load (Log 470K) in the circuit.

## RESULTS AND DISCUSSION

**Significant Observations of the Systems.** The important observations of different systems are given in Table 1, which are reflecting the overall outcome of the present studies, and justified the significance of these cells from the solar energy conversion and storage point view.

**Table 1. Significant Observations of the Systems**

EDTA= $2.2\times10^{-3}$M; Temperature = 303K; Light intensity = 10.4mWcm$^{-2}$; pH =12.8

| **Systems→** <br> **Observations↓** | EDTA-NMB+SFO <br> NMB =$1.08\times10^{-5}$ <br> SFO =$3.6\times10^{-5}$ | EDTA-NMB+FG <br> NMB = $7.2\times10^{-5}$ <br> FG = $4.8\times10^{-5}$ | EDTA-BG+FG <br> BG = $1.32\times10^{-5}$ <br> FG = $2.80\times10^{-5}$ | EDTA-BG+CB <br> BG = $1.24\times10^{-5}$ <br> CB = $4.4\times10^{-5}$ | EDTA-NMB+CB <br> NMB = $8.0\times10^{-5}$ <br> CB = $4.00\times10^{-5}$ |
|---|---|---|---|---|---|
| (i) Open Current Voltage Voc (mV) | 1235.0 | 998.0 | 930.0 | 894.0 | 795.0 |
| (ii) Photopotential ΔV (mV) | 740.0 | 738.0 | 690.0 | 636.0 | 580.0 |
| (iii) Potential at Power Point Vpp (mV) | 640.0 | 652.0 | 580.0 | 550.0 | 480.0 |
| (iv) Equilibrium Photocurrent ieq (μA) | 108.0 | 120.0 | 112.0 | 93.0 | 88.0 |
| (v)Maximum Photocurrent Imax(μA) | 139.0 | 142.0 | 129.0 | 138.0 | 101.0 |
| (vi) Current at Power Point ipp (μA) | 70.0 | 90.0 | 75.0 | 60.0 | 65.0 |
| (vii) Short Circuit Current isc (μA) | 110.0 | 120.0 | 112.0 | 93.0 | 88.0 |
| (viii) Rate of Generation (μA) | 7.0 | 5.9 | 6.1 | 6.9 | 8.2 |
| (ix) Power at Power Point (μW) | 81.40 | 88.56 | 77.28 | 59.14 | 51.04 |
| (x) Fill Factor (η) | 0.32 | 0.48 | 0.41 | 0.39 | 0.44 |
| (xi) $t_{1/2}$ (min) | 95.0 | 90.0 | 72.0 | 65.0 | 45.0 |
| (xii) Charge time (min) | 112.0 | 129.0 | 132.0 | 117.0 | 145.0 |
| (xiii) Conversion Efficiency (%) | 0.4309 | 0.5642 | 0.4182 | 0.3173 | 0.3010 |

**Current-Voltage (i-v) Characteristics of the Cell.** The short circuit current ($i_{sc}$) and open circuit voltage ($V_{oc}$) of the photogalvanic cells were measured with the help of a multimeter (keeping the circuit closed) and with a digital pH meter (keeping the other circuit open), respectively. The current and potential values in between these extreme values were recorded with the help of a carbon pot (log 470K) connected in the circuit of multimeter, through which an external load was applied .The i-v characteristics of the photogalvanic cells containing mixed dyes – EDTA systems were studied and it was observed that in all the five cases, i-v curves deviated from their expected regular rectangular shape. The power point (a point on the curve where the product of potential and current maximum) in these i-v curves were determined and their fill factor were calculated using the formulae: The i-v characteristics of the photogalvanic cells are given in Table 2.

$$\text{Fill Factor} = V_{pp} \times I_{pp} / V_{oc} \times I_{sc} \quad (1)$$

Where, $V_{pp}$ = Potential at power point; $I_{pp}$ = current at power point; $v_{oc}$ = Open circuit voltage; $i_{sc}$ = Short circuit current

**Table 2. i-v Characteristics of the Photogalvanic Cells**
EDTA = $2.2\times10^{-3}$M; Light intensity =10.4mWcm$^{-2}$; pH =12.8

| EDTA-Dyes Systems | $V_{oc}$ (mV) | $I_{sc}$ (μA) | $V_{pp}$ (mV) | $I_{pp}$ (μA) | η |
|---|---|---|---|---|---|
| EDTA-NMB+SFO<br>NMB =$1.08\times10^{-5}$<br>SFO =$3.6\times10^{-5}$ | 1235.0 | 110.0 | 640.0 | 70.0 | 0.32 |
| EDTA-NMB+FG<br>NMB =$7.2\times10^{-5}$<br>FG =$4.8\times10^{-5}$ | 998.0 | 120.0 | 652.0 | 90.0 | 0.48 |
| EDTA-BG+FG<br>BG =$1.32\times10^{-5}$<br>FG =$2.80\times10^{-5}$ | 93.0 | 112.0 | 580.0 | 75.0 | 0.41 |
| EDTA-BG+CB<br>BG =$1.24\times10^{-5}$<br>CB =$4.4\times10^{-5}$ | 894.0 | 93.0 | 550.0 | 60.0 | 0.39 |
| EDTA-NMB+CB<br>NMB =$8.0\times10^{-5}$<br>CB = $4.00\times10^{-5}$ | 795.0 | 88.0 | 480.0 | 65.0 | 0.44 |

On the basis of fill factors calculated, the order of the efficiency of photogalvanic cells in the presence of mixed dyes is: EDTA- NMB + CB >> BG + FG > NMB + SFO > NMB + FG > BG + FG

**Conversion Efficiency of the Cells.** The conversion efficiency of the all these systems were calculated using the output at power point and the power of incident radiations. The systems at the optimum condition were also exposed to sunlight. The system of optimum conditions was also exposed to sunlight. The conversion efficiency and sunlight conversion data for these photogalvanic cells are reported in Table - 3.The conversion efficiency of these systems were calculated using the formula.

$$\text{Conversion Efficiency (CE)} = \frac{V_{pp} \times I_{pp}}{\text{Light Intensity (mW/ cm}^2\text{)}} \times 100\ \% \quad (2)$$

The observed conversion efficiencies were determined with a tungsten lamp of 200W (as light source), which has a light intensity around 15 times less than direct sunlight. Therefore, the conversion efficiencies have experimentally been observed as 6.46, 8.44, 6.27, 4.75 and 4.50.The observed low conversion efficiencies in comparison to theoretical conversion efficiency of photogalvanic cells (18%) seems due to the lower stability of dyes, back electron transfer and aggregation of dyes molecules around the electrode, etc.

**Table 3. Conversion Efficiency and Sunlight Conversion Data**

EDTA = $2.2\times10^{-3}$M; Light intensity =10.4mWcm$^{-2}$; Temperature =303K; pH=12.8

| EDTA-mixed dyes systems | Conversion Efficiency (%) | Sunlight Conversion Data | |
|---|---|---|---|
| | | Photopotential (mV) | Photocurrent (μA) |
| EDTA-NMB+SFO<br>NMB=$1.08\times10^{-5}$<br>SFO=$3.6\times10^{-5}$ | 0.4307 | 1335.0 | 130.0 |
| EDTA-NMB+FG<br>NMB=$7.2\times10^{-5}$<br>FG=$4.8\times10^{-5}$ | 0.5632 | 1360.0 | 155.0 |
| EDTA-BG+FG<br>BG=$1.32\times10^{-5}$<br>FG=$2.80\times10^{-5}$ | 0.4182 | 1305.0 | 110.0 |
| EDTA-BG+CB<br>BG=$1.24\times10^{-5}$<br>CB=$4.4\times10^{-5}$ | 0.3173 | 1280.0 | 127.0 |
| EDTA-NMB+CB<br>NMB=$8.0\times10^{-5}$<br>CB=$4.00\times10^{-5}$ | 0.300 | 1375.0 | 100.0 |

**Table 4. Performance of the Photogalvanic Cells in Dark**

EDTA = $2.2\times10^{-3}$; Light intensity =10.4 mWcm$^{-2}$; pH =12.8

| EDTA-mixed dyes systems | Power (μW) | Power at power point (μW) | $T_{1/2}$ (min) |
|---|---|---|---|
| EDTA-NMB+SFO<br>NMB =$1.08\times10^{-5}$<br>SFO =$3.6\times10^{-5}$ | 81.4 | 44.8 | 95.0 |
| EDTA-NMB+FG<br>NMB =$7.2\times10^{-5}$<br>FG = $4.8\times10^{-5}$ | 88.5 | 58.6 | 90.0 |
| EDTA-BG+FG<br>BG =$1.32\times10^{-5}$<br>FG =$2.80\times10^{-5}$ | 77.2 | 43.5 | 72.0 |
| EDTA-BG+CB<br>BG =$1.24\times10^{-5}$<br>CB = $4.4\times10^{-5}$ | 59.1 | 33.0 | 65.0 |
| EDTA-NMB+CB<br>NMB = $8.0\times10^{-5}$<br>CB = $4.00\times10^{-5}$ | 51.0 | 31.2 | 45.0 |

Based on all these observations, the fine EDTA-Mixed dyes photogalvanic system are arranged in order of their decreasing efficiencies. EDTA-NMB+FG > EDTA-NMB+SFO > EDTA-BG+FG> EDTA-BG+CB > EDTA-NMB+CB

**Performance of the Cell.** The performance of the photogalvanic cells containing EDTA mixed dyes systems was studied by applying the desired external load to have the potential and current corresponding to power point, after removing the source of illumination. It was quite interesting to observe that the performance of the cell in dark is affected appreciably in the presence of mixed dyes. The time t ½ was determined after removing the source of power. The time t ½ has been used as the measure of storage capacity of the cells. The performance of cells was studies and comparative results are summarised in the Table 4.

The order of storage capacity or utility of photogalvanic cells in terms of t ½ in the dark was:
EDTA-NMB+SFO>NMB+FG>BG+FG> BG+CB>NMB+CB

**Mechanism.** On the basis of these observations, a mechanism has also been proposed for the generation of photocurrent in photogalvanic cell as:

***Illuminated Chamber:***

$D+D^{1} \rightarrow D^{*}+D^{1*}$
$D^{*}+D^{1*}+2R \rightarrow D^{-}+D^{1-}+2R^{+}$

***At platinum electrode:***
$D^{-}+D^{1-} \rightarrow D+D^{1}+2e^{-}$

***Dark Chamber:***
At counter electrode
$D+D^{1}+2e^{-} \rightarrow D^{-}+D^{1-}$
$D^{-}+D^{1-}+2R^{+} \rightarrow D+D+2R$

## CONCLUSIONS

According to observed photogalvanic effect in all these five system containing EDTA as reductant and mixed dyes (combination of two dyes molecule) systems, the EDTA-NMB+FG system was most efficient in all the ways and there is still hope to improve the output and storage capacity if it is thoroughly exploited. Through exploration of photogalvanic effect may give a pathway in the field of solar energy conversion and storage and a breakthrough is possible at any movement it is really planned and properly intensified. Workers have developed the photogalvanic cells containing reductant and photosensitizer with reasonable electrical output, but by combining two different photosensitizers the electrical output has increased 1.5-2.0 times as the observations show. More efforts are required to increase conversion efficiency and storage capacity by selecting suitable substances to achieve the desired results for efficient and commercial viability of the photogalvanic cells.

## ACKNOWLEDGEMENT

The authors are thankful to the UGC, New Delhi for financial assistance.

## REFERENCES

1. Rideal, E.K. and Williams, D.C. 1925. "Photogalvanic effect."*J. Chem. Soc.* 127: 258-269.
2. Rabinowitch, E. 1940. "The Photogalvanic Effect I. The Photochemical Properties of the Thionine-Iron System." *J. Chem. Phys.* 8: 551-59.
3. Rabinowitch, E. 1940. "The Photogalvanic Effect II. The Photogalvanic Properties of the Thionine-Iron System" *J. Chem. Phys.* 18, 560-566.
4. Fox, M.A. and Kabir-ud-din. 1979. "A New Carbonic Photogalvanic Cell." *J. Phys. Chem.* 83: 1800-1801.
5. Stevenson, K.L. and Errelding, W.F.1981. "A Photogalvanic Cell Utilizing the Photodissociation of Iodine in Solution." *Solar Energy.* 27(2): 139-141.

6. Ameta, S.C., Jain P.K., Janu, A.K. and Ameta, R. 1985. "Studies in Photochemical Conservation Of Solar Energy: Use Of Brilliant Cresyl Blue-Mannitol System In Photogalvanic Cell." *Energy J.* 58: 8-12.
7. Eisenberg, M. and Silverman, H.P. 1961. "Photo-electrochemical cells." *Electrochim Acta.* 5(1–2): 1–12.
8. Kaneko, M. and Yamada, A. 1977. "Photopotential and Photocurrent Induced By a Tolusafranine Ethylenediaminetetraacetic Acid System." *J. Phys. Chem. 81* (12): 1213–1215.
9. Murthy,A.S.N.and Reddy, K.S. 1980. "Photogalvanic Effect in Riboflavin—Ethylenediaminetetraacetic Acid System." *Int. J. Energy Res.*, 4(4): 339–343.
10. Yamase, T. 1981. "Production of Hydrogen by a Photogalvanic Cell with a Flavine Mononucleotide-Edta System." Photochem. Photobiol. 34(1): 111–114.
11. Tamilasaram, T. and Natarajan. 1981. "Photovoltaic Conversion by Macromolecular Thionine Films." *Nature*". *292*: 224 – 225.
12. Murthy, A.S.N. and Reddy K.S. 1983. "Studies on Photogalvanic Effect in Systems Containing Toluidine Blue." *Solar Energy*. 30(1): 39-43.
13. Jain, P.K. and Ameta, S.C. 1984. "Use of Brilliant Cresyl Blue in Photochemical Conversion Of Solar Energy." *Z. Phys. Chem* (Leipzig). 265(4): 841-2.
14. Ameta, S.C., Chittora, A.K., Gangotri, K.M. and Khamesra, S. 1989. "Use of Azur C-Glucose System In Photogalvanic Cell For Solar-Energy Conversion." *Z. Phys. Chem. (Leipzig)*. 270, 607-612.
15. Gangotri, K.M., Regar, O.P., Lal, C., Kalla, P., Genwa, K.R., and Meena,R. 1996. "Use Of Tergitol-7 In Photogalvanic Cell For Solar Energy Conversion And Storage: Toluidine Blue-Glucose System." *Int. J. Energy Res.* 20(7): 581–585.
16. Gangotri, K.M. and Regar, O.P. 1997. "Use of Azine Dye as a Photosensitizer in Solar Cells: Different Reductants-Safranine Systems." *Int. J. Energy Res.* 21(14):1345–1350.
17. Gangotri, K. M. and Lal, C. 2005. "Use of Mixed Dyes in Photogalvanic Cells for Solar Energy Conversion and Storage: EDTA-Methylene Blue and Thionine System." *J. Power and Energy* (Part A). 219: 315-320.
18. Yadav, S. and Lal, C. 2010. "Efficient Solar Energy Conversion and Storage Through Photogalvanic Cell Based on EDTA: Brilliant Green and Fast Green System." *Int. J. of Green Energy*, 8(2): 265-274.
19. Lal, C.and Gangotri, K.M. 2011. "Energy Conversion and Storage Potential of Photogalvanic Cell Based on Mixed Dyes System: Ethylene Diaminetetraacetic Acid-Toluidine Blue-Thionine." *Environmental Progress & Sustainable Energy*. 30(4): 754–761.
20. Yadav, S. and Lal,C. 2013. "Optimization Of Performance Characteristics of A Mixed Dye Based Photogalvanic Cell for Efficient Solar Energy Conversion and Storage." *Energy Conversion and Management.* 66: 271-276.

# THE ENVIRONMENTAL BENEFITS OF ACHIEVING HIGH SOLAR PENETRATIONS IN THE UNITED STATES

Ryan Wiser and ***Dev Millstein*** (Lawrence Berkeley National Laboratory, Berkeley, CA, USA)
Trieu Mai, Jordan Macknick, Alberta Carpenter, Stuart Cohen, Wesley Cole, Bethany Frew, and Garvin Heath (National Renewable Energy Laboratory, Golden, CO, USA)

We assess three key potential environmental benefits of achieving high solar penetrations (14% and 27% of U.S. electricity demand by 2030 and 2050, respectively): greenhouse-gas (GHG) emissions reductions, air-pollution health and environmental impacts, and water-use reductions. To provide context for future scenario analysis, we first assess the recent environmental benefits of the 2014 U.S. solar fleet. In this case, we use the Environmental Protection Agency (EPA)'s Avoided Emissions and geneRation Tool (AVERT) to estimate the characteristics and location of fossil generation sources displaced by year-end 2014 solar capacity. Looking forward, we use a scenario-analysis approach in which the 14%-by-2030 and 27%-by-2050 "SunShot Vision" scenario is compared with a "No New Solar (NNS) baseline" scenario in which no new solar is deployed after 2014. This framework allows us to assess the potential benefits of all incremental solar deployment. We employ the National Renewable Energy Laboratory's Regional Energy Deployment System (ReEDS) electric sector capacity-expansion model to conduct our analysis. For each of the three benefit categories, we take the modeled output from AVERT or ReEDS and then apply additional tools—as necessary—to assess potential benefits in physical and, where feasible, monetary terms. We qualify the study results appropriately, highlighting areas of uncertainty.

Our analysis finds that a future U.S. electricity system in which solar plays a major role would result in enduring environmental benefits globally, nationally and locally, and that the existing fleet of solar power plants is already offering a down-payment towards those benefits. We find that achieving the SunShot Vision scenario from 2015-2050 would reduce 10% of life-cycle GHG emissions from the power sector, reduce premature mortalities from air pollution by 25,000–59,000, and reduce power-sector water withdrawal and consumption by 4% and 9%, respectively.

The total monetary value of the GHG and criteria air pollution benefits of the SunShot Vision scenario exceed $400 billion in present-value terms under our central estimates, equivalent to roughly 3.5¢/kWh-solar. Focusing on just the existing end-of-2014 fleet of solar power projects, recent annual benefits equal more than $1.5 billion under our central estimates, equivalent to 4.7¢/kWh-solar. These are central estimates, with a sizable range of uncertainty. GHG reductions are valued based on standard estimates of the global social cost of carbon. U.S. air pollution benefits are valued based on standard estimates of the value of reducing premature mortality and morbidity outcomes. The figures exclude the value of reduced water use—for which monetary quantification was not feasible.

It is important to recognize that the environmental benefits of solar are strongly dependent on not only the amount of solar deployment, but also the location of that deployment—solar that displaces higher-emitting coal generation has substantially larger environmental benefits than solar that displaces lower-emitting gas-fired generation. Although not addressed in this work, decision makers will naturally wish to compare these potential environmental benefits with, among other things, the potential costs and risks introduced by adding solar to the electric system as well as the potential impacts of solar on local ecosystems and communities.

# ENERGY CONSUMPTION PATTERN AND GREENHOUSE GASES EMISSION: A CASE STUDY ON POTENTIALITY OF BIOGAS AS A RENEWABLE ENERGY TECHNOLOGY AND ITS ROLE IN REDUCING GREENHOUSE GASES EMISSION

***Maheshwor Poudel*** (Central Department of Environmental Science, Tribhuvan University, KTM, Nepal)
Pabitra Adhikari (College of Applied Sciences-Nepal, Tribhuvan University, Kathmandu, Nepal)
Ramesh Man Singh (Centre for Energy and Environment - Nepal, Kathmandu, Nepal)

**ABSTRACT:** The study shows the current energy consumption pattern and Green House Gases emission. The study is focused to analyze the status of Biogas as a Renewable Energy Technology (RET) and its role in reducing the consumption of traditional energy resources and GHGs emission. The total energy consumption among the sampled households was found to be 7225.72 GJ per year with 14.06 GJ per capita per year. Fuelwood was found to be the dominant energy resource which comprises 93.73% of the total energy and rest 6.27% of energy was supplied by other sources like electricity, kerosene, solar and LPG. The per capita biomass fuel consumption was 13.18 GJ per year, whereas non-biomass fuel consumption was 0.868 GJ. The per capita Greenhouse Gases (GHGs) emission was found to be of 1209.11 kg $CO_2$e per year of the study area. The annual GHGs reduction was found 7.62 t$CO_2$e per biogas plant i.e. GHGs emission of 205.62 t$CO_2$e per year was reduced from the 27 biogas plants available in the sampled households and the study showed that the possibility of another 57 biogas plants of size 6 $m^3$ which can reduce another 433.20 t$CO_2$e per year of GHGs emission. The study showed that the use of biogas saved 46.67% of fuelwood and if all the 84 potential biogas plants were constructed from the dung produced annually, the amount of fuelwood saved would be 217.32 tons per year per household. The simple payback period of the biogas plant was found to be 3.06 years excluding the subsidy given by the Government of Nepal. Therefore, the RET used in the study area had either reduced or substituted fuelwood as an energy source. The reduction on GHGs emission by the installation of biogas and high potentiality of biogas on the basis of livestock clearly indicates biogas as a suitable RET in reducing GHGs emission with efficient energy use.

***Key words:*** *Energy Consumption Pattern, Fuelwood, GHGs emission, Biogas, RET*

## INTRODUCTION

Energy is an important development indicator, which provides vital inputs for survival and economic development. Human activities particularly the intense use of carbon based energy sources and unsustainable use of biomass resources is increasing greenhouse in the earth's atmosphere resulting global warming in a way that could profoundly affect the global climatic system (Parikh, 2004).

The overall energy consumption of Nepal is largely dominated by the subsistence use of traditional forms of energy such as fuelwood, agricultural residues and animal waste. The intensive use of fossil fuels and unsustainable use of biomass resources from human activities are emitting GHGs which are causing greenhouse effect in the earth's atmosphere resulting global warming process which in turn have resulted the climate change as indicated by various scientific studies (Shakya and Shrestha, 2006). Fuelwood harvesting has been identified as one of the most significant causes of forest decline in rural areas of developing countries (Bhatt and Sachan, 2004). Fuelwood is the biggest energy resources in Nepal providing about 77% of the total energy demand exerting immense pressure on the forest resources (WECS, 2010).

Biogas, solar power, wind power, biobriquettes and micro level hydropower are categorised into renewable energy resources in Nepal. Biogas and biobriquettes are examples of modern interventions, wherein traditional energy sources undergo transformations into modern types of fuels. In case of Nepal, about 1% of total energy consumption comes from the renewable sources (WECS, 2010). The RETs based on the locally available resources with increased efficiency and environment friendly alternatives seems very feasible for controlling GHGs emission and protecting environment.

The research is focused on determining the amount of energy consumed in terms of firewood and GHGs emission among the biogas user and non user households in the hilly region of Nepal. It will help to calculate the firewood consumption among biogas user and non user households and find out the potentiality of biogas as an alternative source of energy in the study area.

## METHODOLOGY

**Study Area.** The studied site was Sankhupati Chaur VDC of Kavrepalanchok (Figure 1). The area of Sankhupati Chaur VDC is 6.95 km$^2$ and is located between the latitude 27$^0$34'24.3"N to 27$^0$57'05" N and longitude 85$^0$33'26.5"E to 84$^0$56'02" E and the elevation of 1523 masl.

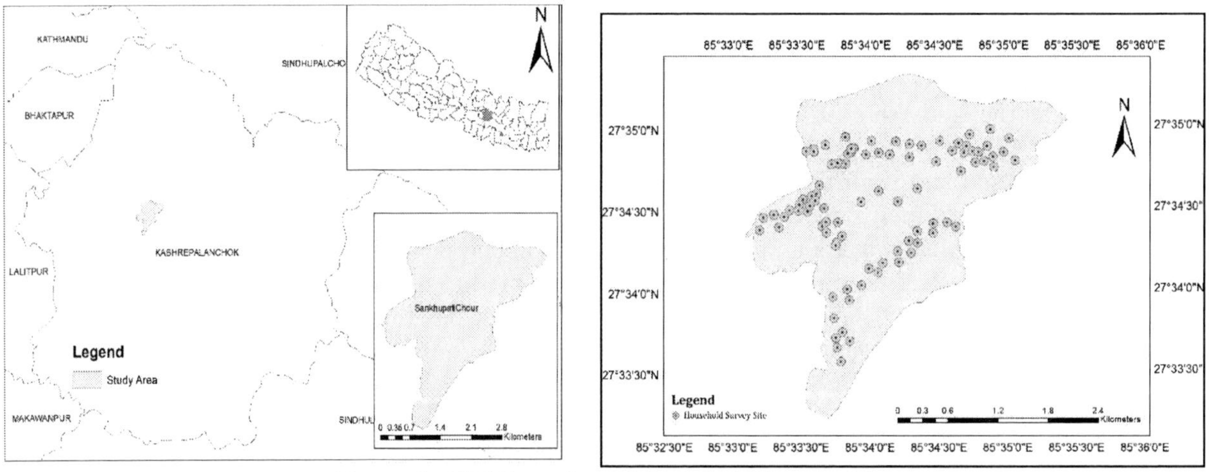


Figure 1. GPS Location of Sampled Household

The VDC is dominated by warm temperate climate with maximum temperature of 32$^0$C and minimum temperature of 0$^0$C. The average annual rainfall is 1200 mm. 41.58% of VDC is covered by forest. The major forest types include *Pinus roxburthii* forest, *Schima castanopsis* forest, *Alnus nepalensis* forest, *Pinus wallichii* forest, *Rodhodendron spp.* forest. The species are mainly fodder species like *Dudhilo, Uttis, Hattipaila, Khanayo, Mauwa, Painyu* and *Bakaino.*

**Data Collection.** Simple random purposive sampling was used to determine the sample size. Among 722 households of Sankhupati Chaur VDC, 95 households were taken as a sample household (Arkin and Colton, 1963). Out of 95 households, 15 households of marginalized people within the VDC were also selected considering they were not in random selection. It is because marginalized people are using the less efficient technology of energy consumption technology and they have less capacity to adapt with the efficient energy consumption technology.

A semi-structured questionnaire was used for data collection in household survey. Direct observation about the energy consumption practice and its environmental impacts was noted during the field visit. The weight of 1 *bhari* was determined by taking average weight of 10 different *bhari*s of fuelwood. Key informants were interviewed with structured checklists in order to verify the gathered relevant information. Focus Group Discussion was also carried out with the local stakeholders to get information about the past and present condition of forest, stoves use, training and awareness campaign on

RETs, their socio-economic status shifting etc. Check list was prepared to include RETs trend, environmental impacts and socioeconomic status.

**Data Analysis and Presentation**

***Estimation of Biogas Potential.*** The sampled households were asked about the number of cattles owned to determine the daily dung production. Considering the daily dung production of 10 kg per cattle by assuming the animals were partly stall fed and partly grazed (BSP, 2009) and depending on average number of adult and young cattle, the total amount of dung produced in the area was estimated and the gas produced from the dung was calculated considering the gas produced per kg of dung to be 0.032 $m^3$ (Werner et al.1989).

Gas produced per kg of dung = 0.0230 – 0.040 $m^3$

(Average value = 0.032 $m^3$)

***Calculation of Amount of Dung Required Generating 1m³ of Gas.***

0.032 $m^3$ of gas is produced from 1 kg of dung

To produce 1 $m^3$ of gas = 1 / 0.032 kg of dung is required

= 31.25 ~ 31 kg of dung

***Calculation of Amount of Fuelwood Saved.*** Amount of fuelwood saved was calculated on the basis of energy content on fuelwood as stated by Vaidya and Gautam (2011) which states that 1 ton of dry fuelwood contains 16.75 GJ of energy.

Table 1. Energy Consumption Pattern

| Energy Sources | Consumption per day | Consumption per month | Consumption per year | Annual Energy Consumption in GJ | Per Capita Energy Consumption in GJ |
|---|---|---|---|---|---|
| **Fuelwood** | 1123.17 kg | 33695.20 kg | 404.34 ton | 6772.86 | 13.18 |
| **Electricity** | 82 Units | 2460 kWh | 29.52 MWh | 106.27 | 0.21 |
| **Kerosene** | 1.43 L | 42.90 L | 0.51 KL | 18.49 | 0.04 |
| **LPG** | 4.70 kg | 140.86 kg | 1.69 ton | 83.22 | 0.16 |
| **Biogas** | 680.22 MJ | 20406.52 MJ | 244.88 GJ | 244.88 | 0.48 |
| **Total** | | | | 7225.72 | 14.06 |

*(Source: Field Data, 2015)*

***Calculation of Number of Trees Saved and Areas of Forest Protected.*** The equivalent forest area saved was estimated based on the study of Winrock and Eco Securities (2004) on impact of biogas on forest which states that 1 ha of forest contains 32.81 tons of fuelwood. The estimated number of trees saved through potential biogas plants was based on the estimation of an annual saving of 11.6 trees per biogas plant (Devkota, 2007).

***Estimation of GHGs Emission Reduction.*** 1 kg of fuelwood generates 1.518 kg equivalent of carbon emission (Smith et al., 2000). GHGs emission reduction in terms of carbon dioxide was obtained as follows:

GHGs Emission Reduction = Amount of fuelwood saved X GHGs emission per kg of fuelwood.

***Calculations and Analytical Process***

- Average weight of 1 *bhari* of fuelwood was determined by weighing randomly 10 *bhari* of fuelwood and found to be 30.8 kg.
- Annual income saving from reduction in fuelwood was calculated at the local rate of NRs. 140 per *bhari*.

- Per capita fuelwood consumption = (average fuelwood consumption per family per day X total number of families in the ecosystem) / total number of people in the ecosystem (Maharjan, 2008).
- The estimated number of technically potential biogas plant was based on the daily dung available from cattle as 10 kg dung/cattle (BSP, 2009).
- Emission factor for combustion of fuels were also converted to $CO_2e$ as given by IPCC (1996) and Smith et al. (2000).
- The methane leakage calculation was based on the methane leakage calculated by Devkota (2003).

## RESULT AND DISCUSSION

**Energy Consumption Pattern**. Fuelwood, electricity, kerosene, LPG and biogas were the energy sources used in Sankhupati Chaur VDC. The total annual energy consumption among the sampled household was found to be 7225.72 GJ so that the per capita energy consumption was found to be 14.06 GJ per year (Table 1). The annual per capita energy consumption from the fuelwood was 13.18 GJ, whereas electricity, kerosene, LPG and biogas share 0.21 GJ, 0.04 GJ, 0.16 GJ and 0.48 GJ respectively.

**Possibility of Biogas Plant.** The study showed that the possibility of total 84 biogas plants among the sampled households. Out of which 27 biogas plants had already been constructed so the study shows that the possibility of another 57 biogas plants of size 6 $m^3$. It was calculated by assuming the total dung produced per animal per day to be 10 kg and the amount of dung required to run one 6 $m^3$ of biogas plant is 31 kg/day (BSP, 2009) (Table 2).

Table 2. Biogas Potentiality

| Livestock | No. of Livestock | Total No. of Livestock | Dung produced per animal per day | Total dung produced day | Dung required to run one 6 $m^3$ plant | Potential of 6 $m^3$ of biogas plant |
|---|---|---|---|---|---|---|
| **Cow / Ox** | 179 | 261 | 10 | 2610 | 31 | 84.19 |
| **Buffalo** | 82 | | | | | |

*(Source: Field Data, 2015)*

**Amount of GHGs Reduction from Biogas Plant**. All together 205.62 $tCO_2e$ per year of GHGs emission was reduced from the 27 biogas plants of sampled households of Sankhupati Chaur VDC (Table 3).

Table 3. Amount of GHGs Reduction

| Biogas Size ($m^3$) | Hill ($tCO_2e$/year/plant) | No. of Plants | Total Reduction ($tCO_2e$/year) |
|---|---|---|---|
| **4** | 5.43 | 4 | 21.72 |
| **6** | 7.6 | 18 | 136.80 |
| **8** | 9.42 | 5 | 47.10 |
| **Total** | | 27 | 205.62 |

*(Source: Field Data, 2015)*

If all the potential biogas plant of size 6 $m^3$ were constructed from the dung produced annually than the total amount of GHGs reduction will be 638.82 $tCO_2e$ per year.

**Estimation of Number of Trees Saved and Areas of Forest Protected.** The study showed that the total potentiality of 84 biogas plants was found in Sankhupati Chaur VDC. If all these 84 biogas plant of size 6 $m^3$ were constructed then total number of trees saved will be 974.4 per year. Hence, the forest area protected per household per year will be 0.0789 ha (Table 4).

Table 4. Estimation of Equivalent Area of Forest Protected

| Particulars | Comparing Biogas User and Non-Biogas User Households |
|---|---|
| **Annual Fuelwood Saving (tons/year/HH)** | 2.5872 |
| **Equivalent Forest Area Protected (ha/HH/year)** | 0.0789 |

*(Source: Field Data, 2015)*

If all the 84 potential biogas plants were constructed from the dung produced annually, the amount of fuelwood saved would be 217.32 tons per year per household.

**Fuelwood Consumption among Biogas User and Non-User Households.** The comparison was done between biogas user and non-user households (20 households from each group) which showed that non-biogas user households use 5544 kg of fuelwood annually, whereas biogas users use only 2956.8 kg of fuelwood. Therefore, there was the saving of 2,587.2 kg of fuelwood which is equivalent to NRs. 11,760 per year per household. The study showed that the use of biogas saved 46.67% of fuelwood per year per household (Table 5).

Table 5. Fuelwood Consumption among Biogas User and Non-User Households

| Particulars | Biogas User HH | Non-Biogas User HH |
|---|---|---|
| **Average HH fuelwood consumption per month (kg)** | 246.4 | 462 |
| **Average HH fuelwood consumption per year (kg)** | 2956.8 | 5544 |
| **Expenditure in fuelwood per year per household (NRs.)** | 13,440 | 25,200 |
| **Fuelwood saved per year per household (kg)** | 2587.2 | - |
| **% of fuelwood saved** | 46.67% | - |
| **Expenses saved per year per household (NRs.)** | 11,760 | - |

*(Source: Field Data, 2015)*

**Estimation of GHGs Emission from Energy Sources.** The per capita GHGs emission was found to be 1209.11 kg $CO_2e$ per year. The study showed that 613791.76 kg $CO_2e$ was emitted with the consumption of 404342.40 kg fuelwood annually, whereas 5263.97 kg $CO_2e$ was emitted from with 404342.40 kg of LPG. Likewise, consumption of 514.80 L of kerosene per year emits 1278.76 kg $CO_2e$. Similarly, biogas emits 1088.86 kg $CO_2e$ from the consumption of 244879.20 MJ of energy. Only 0.53 kg of $CO_2e$ emitted with the consumption of 29520 kWh of electricity so electricity is known as a clean form of energy (Table 6).

**Economic Analysis.** From the field survey it was found that, NRs. 36,000 will cost for the construction of biogas plant of capacity of 6 $m^3$ including all operational cost, maintenance cost and labour cost, excluding the amount of subsidy given by GoN. This shows that the simple payback period in term of fuelwood saved for the construction of 6 $m^3$ biogas plant will be only 3.06 years (Table 7).

**Impact of Biogas on Environment and Health.** The main cause of pollution in the households is attributed to firewood burning. Apart from this, contamination of faecal waste produces obnoxious smell that is detrimental to human health and environment. Therefore, construction of biogas plant with the connected toilets helps in the better management of animal dung and human waste with better environment.

Table 6. Total $CO_2e$ Emission from Energy Sources

| Energy Sources | Annual Energy Consumptions | GHGs | $CO_2e$ | Annual GHGs Emission (kg $CO_2e$) |
|---|---|---|---|---|
| **Fuelwood** | 404342.40 kg | $CO_2$ | 1.406 kg/kg | 613791.76 |
| | | $CH_4$ | 0.084 kg/kg | |
| | | $N_2O$ | 0.028 kg/kg | |
| **Electricity** | 29520 kWh | | 0.0000179 (ef) | 0.53 |
| **Kerosene** | 514.80 L | $CO_2$ | 2.457 kg/L | 1278.76 |
| | | $CH_4$ | 0.007 kg/L | |
| | | $N_2O$ | 0.020 kg/L | |
| **LPG** | 1690.37 kg | $CO_2$ | 3.0689 kg/kg | 5263.97 |
| | | $N_2O$ | 0.0452 kg/kg | |
| **Biogas** | 244879.20 MJ | | | 1088.86 |
| **Total** | | | | 621423.89 |

*(Source: Field Data, 2015)*

Table 7. Payback Period of 6 $m^3$ Biogas Plant

| Plant Capacity ($m^3$) | Amount of Fuelwood saved per year per household (kg) | Amount of Expenses saved per year per household (NRs.) | Total cost of the plant excluding subsidy (NRs.) | Pay Back Period in years |
|---|---|---|---|---|
| **6** | 2587.2 | 11,760 | 36,000 | 3.06 |

*(Source: Field Data, 2015)*

Biogas is an alternative form of energy due to easy availability of raw materials and its efficient technology. Therefore, the indoor air pollution can be reduced through biogas stove and reduce the respiratory diseases and eye related problems to those working in the kitchen. On the other hand, there will be conservation of forest resources, wildlife habitat and natural balanced ecosystem.

Besides this, biogas is a clean form of energy, where Nepal is gaining huge amount of revenue from CDM project.

## ACKNOWLEDGEMENTS

The authors would like to thank Central Department of Environmental Science for providing the opportunity to carry out this research. The authors are equally thankful to Centre for Energy and Environment Nepal for all the technical support to conduct the research work. The authors are thankful to all the participants of this study for their immense support and providing required information to conduct this research.

## REFERENCES

Arkin, H., and Colton, R.R., (1963). *Tables for Statisticians*. New York: Barnes and Noble.

Aryal, S., (2007). *Adoption of Renewable Energy Technology Towards Sustainable Harvesting of Fuelwood from Community Forest* (Unpublished M.Sc. Thesis). Central Department of Environmental Science, Tribhuvan University, Kathmandu Nepal.

Bhatta, B.P., and Sachan M.S., (2004). Firewood Consumption along with altitudinal gradient in Mountain Villages of India. *Biomass and Energy.* 27(1):69-75

BSP, (2009). *Biogas as Renewable Source of Energy in Nepal Theory and Development.* Biogas Support Programme, Lalitpur, Nepal.

Devkota, G.P. (2007). *Renewable Energy Technology in Nepal.* An Overview and Assessment.

Devkota, G.P., (2003). *Final Report on Analysis of Biogas Leakages from Household Digesters.* Winrock International, Kathmandu, Nepal

IPCC, (1996). *Revised Guideline for National Greenhouse Gas Inventory.*

Maharjan, R., (2008). *Study on Energy Consumption Pattern and its impact on Environment with reference to Okharpauwa VDC, Nuwakot* (Unpublished M.Sc. Thesis), College of Applied Science, Tribhuvan University, Kathmandu, Nepal.

Parikh, J. (2004). Clean Development Mechanism and Sustainable Development. *Integrated Research and Action for Development.* New Delhi.

Shakya, S. R., and Shrestha, J. N. (2006). Contribution of renewable energy technology for greenhouse gas emission in Nepal. *Proceeding of first national conference on renewable energy technology for rural development,*12-14 October, 2006, Kathmandu, Nepal.

Smith K.R., Uma, R., Kishore, V.V.N., Zhang J., Joshi, V., and Khalil, M.A.K., (2000). Greenhouse Implications of Household Stoves: An Analysis for India. *Annual Reviews Energy Environment* 25, 741-763.

Vidhya, S., and Gautam, R. M., (2011). *Energy Outlook of Nepal.* Informal Sector Research and Study Centre. First Edition, pp.315.

Warner, U., Stohr, U., and Hees, N. (1989). *Biogas plants in Animal Husbandary.* GATE/GTZ, Gmbh.

WECS. (2010). *Energy Sector Synopsis Report.* Water and Energy Commission Secretariat, GoN. Kathamandu Nepal.

Winrock and Eco Securities, (2004). *Nepal Biogas Programme, CDM Baseline Study 2003.*

# CLONING OF GENES ENCODING FOR CELLULOLYTIC ENZYMES FROM METAGENOME IN GOAT'S RUMEN

***Santosh Thapa***, Hui li, Suping Zhou, Sarabjit Bhatti and Roger J. Sauvé
(Department of Agricultural and Environmental Sciences, College of Agriculture, Human and Natural Sciences, Tennessee State University, TN, USA)

The increasing fuel demands has caused in rise in price of crude oil and so the need for alternative source of bioenergy. Among the potential alternative bioenergy resources, cellulosic biomass is considered as a sustainable raw material for biofuel production. The cellulosic bioconversion into fermentable sugar is a complex process and requires the synergistic action of endoglucanases, exoglucanases and β-glucosidases. Those enzymes are hosted in the microbial genomes in the rumen of herbivores. To be able to tap into those resources is expected to have a significant impact on cellulosic biofuel production, but a majority of those microbial species are not accessible using laboratory culture techniques. This study was performed to develop molecular approaches for using these genome resources. Goat rumen fluid was collected immediately after goats were slaughtered. Metagenome DNA was extracted and sequenced using next-generation sequencing. Following assembly and annotation, several gene sequences encoding for cellulases and xylanases were found in the metagenome database. PCR primers were designed to amplify the gene of interests using metagenome DNA as the template. The amplified products were cloned into pET 101 vector. After confirmation of gene sequences, those putative cellulase enzymes were expressed in the BL21 Star (DE3) system. Several genes with high cellulase and hemicellase activities have been isolated on this project. This work has provided the foundation for using these microbial genomic resources in the lignocellulosic bio-ethanol industry. The project was supported by NIFA grant NIFA-2010-02417-223076.

# ANAEROBIC AND SUBSEQUENT PHOTOSYNTHETIC PROCESS FOR BIOHYDROGEN AND BIOPLASTIC (PHB) PRODUCTION

***Arun. A****, Dinesh. G. H and Satheesh Murugan. R
(Alagappa University, Karaikudi, Tamilnadu, India)
S. Karthik Sundaram
(PSG college of arts and science, Coimabtore, India)

The faster intake of these non-renewable resources resulted in both, the increase $CO_2$ concentration in the atmosphere and the rapid depletion of fossil resources. Thus research efforts have been devoted in developing new energy alternatives. Compared with fossil fuels as traditional energy sources, hydrogen ($H_2$) is a promising candidate as a clean energy carrier in the future because of its high-energy yield ($122kJ^{-1}$) and producing only water instead of greenhouse gases on burning. Increasing awareness of environmental pollution has generated a resurgence of interest in biological methods for the production of biodegradable polymers.

The present industrial interest of poly-β-hydroxybutyrate (PHB); Hydrogen production via dark fermentation has the greatest potential as a pre-treatment step that can be followed by a suitable secondary process step, such as bioconversion of volatile fatty acids to other products, such as poly hydroxyalkanoates. The present study is carried out for bio hydrogen and PHB production by anaerobic, photo bacterial and their subsequent process using rice straw and husk hydrolysates.

Optimization of substrate (rice straw and husk hydrolysates) (10-100%); pH (5.5-8.0) for anaerobic, photo fermentative and subsequent anaerobic - photo fermentative $H_2$ and PHB production was carried out using the microbial isolates. In anaerobic condition by *Bacillus subtilis* isolate the maximum volume of cumulative hydrogen production was observed in 70% (14.576 ml) rice husk; 30% (14.576 ml) rice straw hydrolysates at pH 7; while in phototrophic condition the isolate R able to produce maximum volume of cumulative hydrogen production in 70% (76.524 ml) by utilizing rice husk hydrolysate at pH 7.There was 5.25 fold increase in cumulative hydrogen production when comparing anaerobic with photo fermentation. The maximum of PHB (136 mg/l) was produced in 80% of rice husk hydrolysate by under anaerobic fermentation at 80% rice husk hydrolysate at pH 5.5 and the isolate G and R produced 28 mg/l of PHB by utilizing both substrates at various concentrations.

The present work was able to obtain concomitantly the dual benefits of producing bioplastic and alternative energy source such as $H_2$.

# ALGAL OIL PRODUCTION POTENTIAL UNDER DIFFERENT MUNICIPAL WASTEWATER DELIVERY SCENARIOS

Tyler Brown and ***Yiwen Chiu***
(California Polytechnic State University, San Luis Obispo, CA, USA)

A lot of research has focused on finding alternative biofuels that are economically and environmentally beneficial to meet the renewable fuel production goals proposed by the Energy Independence and Security Act. Among all the potential biofuel production sources, algae is a prime focus standing apart from other feedstock alternatives resulting from its per-acre production potential. Former studies stated that algae showed promising potential to produce 55 to 341 times more biofuel than other conventional energy feedstock such as corn and soybean. However, algal oil also raises controversial concerns due to negative environmental footprints. One of the significant impacts is the excessive water demand for algae cultivation and refinery maintenance, ranging from 22 to 8357 liters of water per liter of algal oil production. A proposed solution is to introduce municipal wastewater as a nutrient sources for algae cultivation, which can also contribute to the reduction of water demand.

However, a known challenge of using municipal wastewater to aid algal oil production is resulted from the geographically decoupling relation between refinery and wastewater treatment plants. Therefore, to investigate the this concept, we conducted a study aiming to determine the change of algal oil production potential in California under different spatial scenarios varying the distance between refinery sites and existing wastewater plants. We geographically integrated the distribution of suitable land for refinery development with the locations and discharge capacity of existing wastewater plants in California to construct the analytical framework. The results will elaborate upon the feasibility and opportunity of using wastewater to support algal oil production as a water and nutrient source.

# ECONOMICS OF GREEN VERSUS DRY FOREST HARVEST RESIDUE FEEDSTOCK FOR THE AVIATION FUEL SUPPLY CHAIN

Rene Zamora-Cristales and ***John Sessions***
(Oregon State University, Corvallis, Oregon, USA)

Forest harvest residues are produced as a byproduct of commercial timber harvest and represent a source of readily available material for aviation fuel production. However, these forest harvest residues, when green, contain high moisture content that can increase the costs of processing and transportation and may reduce the yield of potential extractable polysaccharides for biofuel production. Different moisture management strategies can be implemented to reduce the moisture content prior to processing but the most important question is what are the economic and environmental trade-offs between collecting processing and transporting green versus dry residue, moreover what is the opportunity cost of letting the material sit while drying. Our main objective was to estimate the implications in processing, transport and pretreatment of using Douglas-fir green versus dry forest harvest residues. Our analysis was focused on harvest residues ground at the landing and loaded and transported in chip vans. Specifically, we discuss differences in bulk density, bark and needles content, and polysaccharides yield. We processed 150 tons of green and 150 tons of dry forest harvest residues from two similar 45 year-old Douglas-fir stands. Dry residue moisture content when ground was estimated at 15% (wet basis). Green residue was collected immediately after harvest and immediately ground. The green residue had an average moisture content of 60%. Results indicate that the proportion of bark, needles and other substances in green residue is higher than in dry material (14% versus 8%). Chemical analysis from the pretreatment process revealed that polysaccharides content in dry residue are 26% higher than in green residue. Lignin content in dry residue is 17% higher compared to green residue. Implications in processing and transportation economics are discussed.

# CLONING AND CHARACTERIZATION OF GENES ENCODING CELLULOLYTIC ENZYMES SCREENED FROM GOAT RUMEN METAGENOME

***Santosh Thapa***, Hui Li, Sarabjit Bhatti, and Suping Zhou
(Tennessee State University, Nashville, TN, 37209, USA)

One of the major limitations for the efficient bio-conversion of lignocellulosic biomass to bio-fuel is the perverse nature of the plant cell wall, which is composed mostly of cellulose, hemi-cellulose and lignin. Goat rumen harbors an assemblage of microbes hypothesized to evolve a plethora of enzymatic strategies for hydrolyzing cellulose and hemi-cellulose into its constituent sugars for subsequent fermentation to bio-fuels. However, the vast majority of such microbial life remains uncatalogued as they are recalcitrant to in-vitro culture. In this study, we have leveraged a functional metagenomics screen for the identification of cellulolytic enzymes from the bacterial community in the rumen of goat. Metagenome DNA was extracted and sequenced using next-generation sequencing. Following assembly and annotation, 14 gene clones encoding for cellulases, hemicellulases and xylanases have been procured from the metagenome. PCR primers were designed against the assembled gene sequences, and PCR reactions were performed using metagenome DNA as the template. The amplified PCR products were cloned using TA cloning strategy. Gene sequences were confirmed using Sanger sequencing and submitted to NCBI database. The genomic and proteomic analyses of cellulolytic and hemi-cellulolytic bacterial species aim to identify and clone novel hydrolytic enzymes, which can be further sub-cloned and expressed to construct highly active recombinant proteins that render as a potential candidate for various biotechnological applications. The present study shows that a functional metagenomics approach can be an innovative maneuver to isolate previously uncharacterized cellulases from the rumen environment.

# PREPARATION AND PROPERTIES OF MA-PA/EG COMPOSITE PHASE CHANGE MATERIAL FOR ENERGY STORAGE

**Jianyun Wu[1], Yanping Yuan[2], Yaguang Yuan[2]**
(1.School of Environmental Science and Municipal Engineering , Lanzhou Jiaotong University , Lanzhou 730070 , China; 2. School of Mechanical Engineering, Southwest Jiaotong University ,Chengdu 611756 , China)

**ABSTRACT:** Phase change energy storage is an environmental-friendly new technology and the front research direction in energy area, which uses the latent heat that phase change material storaged and released can effectively solve the inconsistency between supply and demand of energy in time and space.

This research using the myristic acid(MA) and the palmitic acid(PA) as phase-change material monomers to obtain the MA-PA eutectic mixture, which is based on theoretical calculation and experiment. Besides, using the porous absorbing material, expanded graphite(EG),as primary structure to prepare the MA-PA/EG composite phase-change material(PCM) by porous adsorption method. The performance of phase-change material has been analyzed by the differential scanning calorimetry(DSC), thermal accelerate circulation experiment and heat storing and releasing experiment. The results show that, on the understanding that the best mass ratio of MA-PA eutectic mixture and EG is 15:1, the melting and freezing temperature of the MA-PA/EG composite PCM are 44.42℃ and 44.99℃, the latent heat of fusion and solidification are 156.9J/g and 155.3J/g. There is few difference as compared with MA-PA eutectic mixture; after conducting 1000 time's thermal accelerate circulation experiment on the composite PCM, the heat stability and chemical stability of this material is still excellent; the heat storing and releasing rate of composite PCM is improved effectively after adding the EG, also there is no liquid side leakage during the solid-liquid phase-change process. This composite PCM has extensive application in the areas of solar power, energy conservation, air condition and etc.

***Key words:*** composite phase change material; myristic acid- palmitic acid; expanded graphite; porous adsorption method; thermal properties

## INTRODUCTION

PCMs absorb and release a great amount of heat in phase change process, its heat storage density is larger than general energy storage material. Additionally, the heat storage and release processes of PCM happen in the same temperature, it also has good thermal stability. Therefore, PCM is appropriate for space heating/cooling applications, for example, it can be used in building envelope, industrial waste heat recovery, solar heat and textile industry, which has a wide application prospects [1-3].

PCMs are divided into organic and inorganic matters, or the combination of them. Among all investigated PCMs, fatty acids are the best, which has many superior properties, including proper melting temperature range, high heat capacity, congruent phase change temperature, little or no supercooling during phase transition, non- corrosive [4]. In addition, mix melting different fatty acids can obtain the eutectic mixture, which decreases the temperature of PCM, and expands the temperature range in its application field [5]. Yuan et al. [4] calculated the temperature and latent heat of phase change of PCMs mix melted with 15 kinds of fatty acids. The temperatures was between 10.2℃ and 51.5℃, while the minimum of latent heat is 138.6J/g, and the maximum is 187.5 J/g. Meng et al. [6] obtained the ternary fatty acids/EG composite PCM by canoic acid(CA), lauric acid (LA) and PA, the melting temperature is 16.8℃ and the melting enthalpy is 140.5 J/g. Fatty acids PCMs have advantages of inexpensiveness and large latent heat, but need containers which make extra thermal resistance and leakage[7]. Common methods of encapsulating PCM into containers are sol-gel method, melt infiltration method, porous adsorption method, microcapsule method and polymerization method [8]. Cai et al. [9] used $SiO_2$ nano-fiber as supporting material to obtain the CA-LA-PA/ $SiO_2$ nano-fiber composite PCM. Cao et al. [10] used carboxy methyl cellulose-1 as container. Wang et al. [11] chose polymer alloy as matrices. Hadi et al. [12] selected sodium pyruvate and sodium palmitic as supporting materials, and got two kinds of composite PCM. These shape-stabilized materials can effectively prevent the leakage of PCM, furthermore, thermal conductivity increased with the adding of these materials. Zhang et al. [7] selected EG, diatomite and other 7 kinds of high thermal conductivity materials as supporting structures of PCM, the results showed that EG has the most remarkable effect. EG not only enhances the thermal conductivity but also prevents leakage of PCM, therefore, EG has been widely applied.

Two kinds of fatty acids and EG are used to obtain the MA-PA/EG composite PCM, and the mass ratio has been calculated. The properties of PCM has been researched by DSC, the accelerated circulation test of heating and cooling, heat storage and release test and thermal conductivity test.

## EXPERIMENTAL

**Experiment Materials and Equipment.** Experiment materials: Expansible graphite (EG, 80 mesh, expansion ratio:200ml/g, carbon content:99%) was obtained from Jinrilai Electronic Materials Factory, Qingdao, China. Myristic acid (MA, 98% pure), Palmitic acid (PA, AR) were supplied by Aladin Industrial Corporation, Shanghai China.

Experimental equipment: Differential scanning calorimeter (DSC, TA Q20, America), $N_2$ atmosphere, heating rate 5℃/min, range of temperature:10℃-80℃,the melting temperature and latent heat of PCM in DSC curve are represented by $T_m$ and $H_m$. The freezing temperature and latent heat are shown as $T_f$ and $H_f$. Magnetic stirrers with thermostat of heat-collected type (DF-101S) is provided by Yuhua Equipment Limited Liability Company in Gongyi City. KBr tabletting, range of tested frequency: 400cm$^{-1}$-4000cm$^{-1}$. Metal bath (CHB-T2, BIOER ThermoQ, China). Temperature data logger (Agilent 45980A); Thermostat water bath. Copper-constantan thermocouple (type T). Thermal constant analyzer (Hot Disk 2500-OT) produced by Hot Disk Company in Sweden, the range of thermal conductivity: 0.005-500W/mK, measurement accuracy ≤±3%, type of test probe: Kapton, probe model:5501, radius:6.403mm.

### Preparation Process

***Preparation of EG.*** Put a certain amount of EG in to drying oven at 60℃ for 16h, then put it in to microwave oven, heat for 30s, which will obtain EG with good adsorption property.

***Preparation of Eutectic mixture MA-PA.*** Put defined amount of MA and PA(68/32 wt%) into beaker, then put the beaker into thermostatic waterbath at 80℃. After the specimen melted, magnetic stir it at the speed of 400r/min for 30min. Make sure that MA and PA are well-mixed. At last, nature cool the mixture, then MA-PA binary eutectic mixture is obtained.

***Preparation of MA-PA/EG composite PCM.*** Put several parts of EG into beakers, each of EG is 0.8g, next add MA-PA with different qualities. Put the beakers in to oven at 70℃ for 24h, the mixture should be stirred every 8h to ensure that EG can evenly adsorb MA-PA. After that, cooled down to room temperature, then a series of MA-PA/EG in different ratio is obtained, which named from M1 to M10.

### Confirmation of the Ratio of Composite PCM

***Confirmation of the theoretical ratio of MA-PA.*** The phase-transition temperature and quality ratio of MA and PA in MA-PA eutectic mixture can be calculated by Schroder's formula [13]:

$$T_m = \left[ \frac{1}{T_i} - \frac{R \ln X_i}{H_i} \right]^{-1} \quad \text{(i=A,B)} \qquad (1)$$

$T_m$ — the melting point of eutectic system, K;

$T_i$、$H_i$ — the melting point and latent heat of each component, K、J/g

$R$ — gas constant, 8.315J/(mol•K)

$X_i$ — mole fraction of component i in eutectic system

The latent heat of fusion of MA and PA are 182.6 J/g and 189.6 J/g, and the melting temperatures of MA and PA are 54℃ and 62.0℃[5], which is higher than the needed temperature, therefore, the PCM with lower phase-transition temperature is required. In order to obtain the PCM with the proper temperature, the formula (1) is used to calculate the melting temperature of MA-PA, and the mass ratio of MA and PA in eutectic mixture in this situation is 68:32. The melting temperature showed in DSC curves of MA-PA with this mass ratio inosculates theoretical value, which demonstrates that the theoretical ratio is 68:32.

***Confirmation of the best ratio of MA-PA/EG.*** Latent heat of phase change is an important index when PCM used in condense heat recovery system. Under the proper temperature, the larger latent heat, the better heat storage performance will be. Therefore, in MA-PA/EG, the more MA-PA it contains, the larger latent heat of PCM will be. EG serves as supporting material, which has no effect on latent heat of MA-PA, however, the content of EG influences heat-conducting property and side leakage of composite PCM. Therefore, there is a best mass ratio between MA-PA and EG when the content of MA-PA is highest without side leakage. In order to obtain the best composite PCM, MA-PA/EG with EG and MA-PA in different mass ratio (1:11, 1:12,1:13, 1:14, 1:15, 1:16, 1:17, 1:18, 1:19, 1:20) is prepared, and named them from M1 to M10. Weigh 0.200g MA-PA/EG in different mass ratio on filter paper as fig.1 shows. Put it into oven at 65℃ for 90min. The result shows that there are no obvious impressions around M1-M5, which means that there's no MA-PA leaks out under the adsorption of filter paper. While there are impressions around M6-M10, and are growing with the increased content of MA-PA. This is caused by the excessive amount of MA-PA, which is beyond the absorption capacity of EG, therefore, the extra MA-PA leaks out and form the impression. The amount of leaked MA-PA can be reflected by impression. The results show that the best mass ration between MA-PA and EG is 15:1.

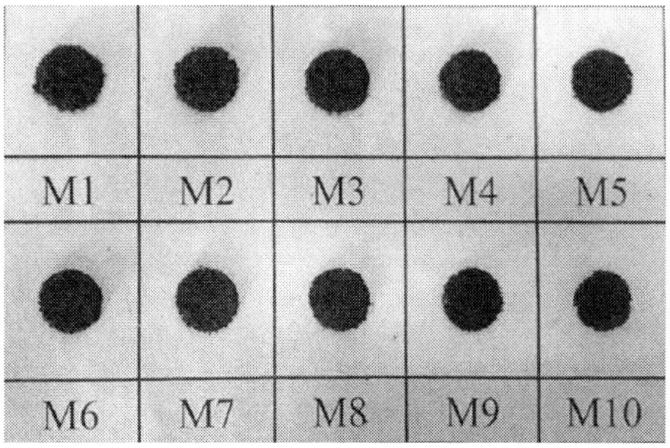


Fig. 1. MA-PA/EG in different mass ratio

## RESULTS AND DISCUSSIONS

**Phase-transition Temperature and Latent Heat of MA-PA/EG Composite PCM.** The phase- transition temperature and latent heat of eutectic mixture MA-PA and MA-PA/EG composite PCM have been tested by DSC. The DSC curves of MA-PA and MA-PA/EG are showed in Fig.2 and Fig.3. The data show in Table.2 and Table.3.

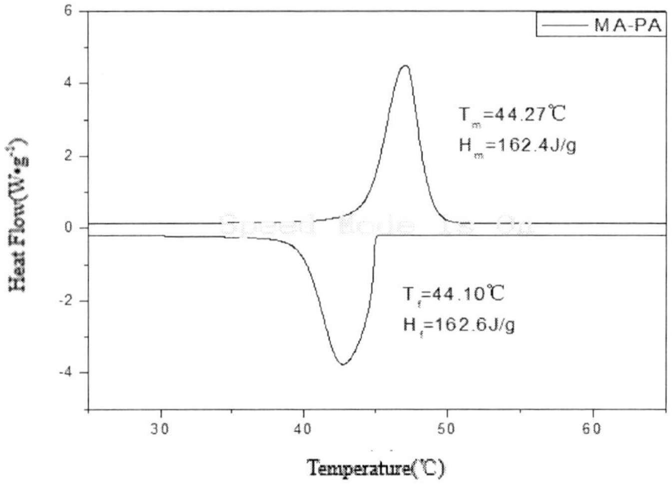


Fig. 2. DSC curves of MA-PA

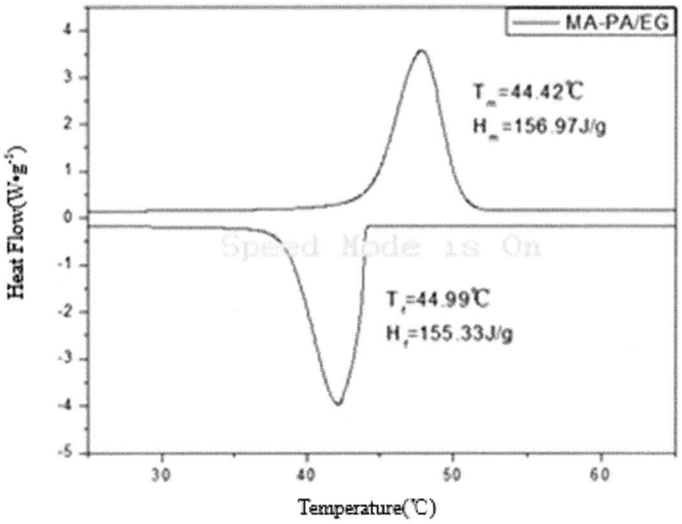


Fig. 3. DSC curves of MA-PA/EG

Table 1. DSC data of MA-PA

| Group | Meltin□ temperature ℃ | Latent heat of fu□ion J/□ | Freezin□ temperature ℃ | Latent heat of □olidification J/□ |
|---|---|---|---|---|
| 1 | 44.41 | 159.9 | 44.00 | 161.6 |
| 2 | 44.17 | 163.7 | 44.20 | 164.2 |
| 3 | 44.20 | 163.6 | 44.11 | 162.0 |
| Avera□e | 44.27 | 162.4 | 44.10 | 162.6 |

Fig.2 shows that there is only one melting peak of MA-PA on the DSC heating curve, which illustrates that MA and PA are eutectic. Additionally, in the DSC curve of MA-PA/EG, the melting and freezing temperature are presented as 44.2℃ and 44.99℃, the latent heat of fusion and solidification are showed as 156.9J/g and 155.3J/g. Contrasting these thermal properties with MA-PA can find out that the phase-transition temperatures are basically same. Besides, EG has little influence on latent heat of composited PCM, which demonstrates that EG only supports MA-PA but has no effect on the behavior of melting and freezing. The latent heat of phase change is proportional to the content of MA-PA. Since the latent heat and content of MA-PA are known, according to formula 2, the theoretical value of latent heat of composited PCM can be obtained, the results of latent heat of fusion and solidification are 152.25J/g and 152.44J/g. The theoretical value is smaller than the value of actual measurement, which may be caused by experimental error. The results show that, MA-PA/EG composite PCM has a widespread application on solar energy utilization and condense heat recovery.

$$H_{ml}=\eta H_m \tag{2}$$

$H_{ml}$—the latent heat of MA-PA/EG composited PCM, J/g

$H_m$—the latent heat of MA-PA MA-PA, J/g

$\eta$—the mass fraction of MA-PA in composited PCM, $\eta=M_m/M_{ml}$ ($M_m$ is the quality of MA-PA, g; $M_{ml}$ is the quality of MA-PA/EG, g.)

Table 2. DSC data of MA-PA/EG

| Group | Meltin□ temperature °C | Latent heat of fu□ion J/□ | Freezin□ temperature °C | Latent heat of □olidification J/□ |
|---|---|---|---|---|
| 1 | 44.43 | 154.2 | 44.98 | 151.0 |
| 2 | 44.43 | 158.2 | 45.01 | 157.4 |
| 3 | 44.41 | 158.5 | 44.98 | 157.6 |
| Avera□e | 44.42 | 156.9 | 44.99 | 155.3 |

**Heat stability of MA-PA/EG composite PCM.** After a long term and multiple times of thermal storage and release process, the composited PCM should still keep a stable thermal property and chemical property, which ensures the long useful life. An experiment of accelerated thermal cycling of MA-PA/EG for 1000 times has been done. The DSC curves of MA-PA/EG after circulation is showed in Fig.4. And the tested melting and freezing temperature of MA-PA/EG after circulation list in Tab. 3, as well as the tested latent heat of fusion and solidification.

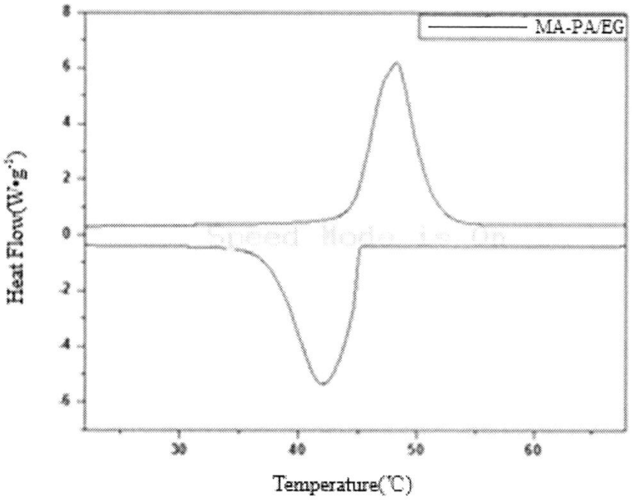


Fig. 4. DSC curves of MA-PA/EG after circulation

The data show that, after 1000 times of circulation, the melting temperature and freezing temperature have changed -0.06℃ and 0.32℃, while the latent heat of fusion and solidification have changed -3.25J/g and -3.84J/g. It can be seen that MA-PA/EG composite PCM has a good heat stability.

Table 3. DSC data of MA-PA/EG after 1000 times of circulation

| Group | Meltin□ temperature °C | Latent heat of fu□ion J/□ | Freezin□ temperature °C | Latent heat of □olidification J/□ |
|---|---|---|---|---|
| 1 | 44.43 | 151.8 | 45.44 | 150.82 |
| 2 | 44.29 | 155.5 | 45.18 | 152.1 |
| Avera□e | 44.36 | 153.65 | 45.31 | 151.46 |

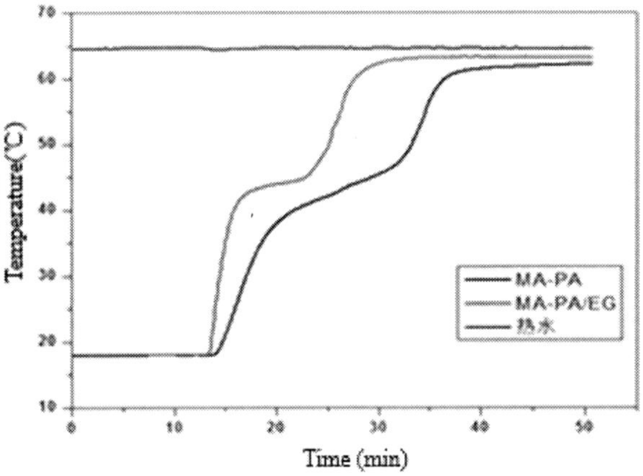


Fig. 5. The melting curves of MA-PA and MA-PA/EG

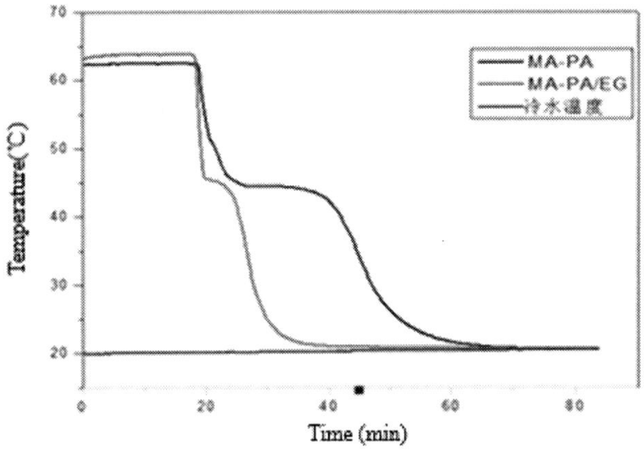


Fig. 6. The freezing curves of MA-PA and MA-PA/EG

**Thermal energy stored and discharge performance of MA-PA/EG composite PCM**. The speed of thermal energy storage and release of PCM influence its usage. Put the test tubes separately fitted with MA-PA and MA-PA/EG for 45ml into thermostatic waterbath at 65℃. Test the temperatures of MA-PA and

MA-PA/EG by temperature data logger. Record the data every 10s. After about 50min, the temperature of MA-PA and MA-PA/EG achieves a constant. Then put the test tubes separately fitted with MA-PA and MA-PA/EG into thermostatic waterbath at 20℃. Record the data every 10s. After about 70min, the temperature of MA-PA and MA-PA/EG achieves a constant. The data of thermal energy stored and discharge experiment is processed by Origin7.5. The melting and freezing curves of MA-PA and MA-PA/EG are presented in Fig.5 and Fig.6.

Fig. 5 shows that, the temperature of MA-PA/EG increases fast from 20℃ to 42℃, the process only takes 3min, while MA-PA spends 9min. In the initial stage, the temperature of MA-PA increased much slower than that of MA-PA/EG. The solid-liquid transition of MA-PA happens when the temperature is between 42℃ and 46℃. This process takes about 9min, which is slow. The time increased by 38.9% from MA-PA/EG. With the elevation of temperature, the heating rate of MA-PA is much slower than that of MA-PA/EG when the temperature ranges from 45℃ to 65℃. During the whole melting process, MA-PA/EG reaches stability first, which is 10min faster than MA-PA. Similarly, the Fig.6 shows that, the temperature of MA-PA/EG composite PCM reduces quickly when the temperature drop from 63℃ to 45℃, which only takes 1.5min, while MA-PA spends about 4.5min. The solid-liquid transition of MA-PA and MA-PA/EG happens when the temperature is between 45℃ and 43℃. This process takes MA-PA/EG 5min, while MA-PA 18min. When the temperature ranges from 43℃ to 20℃, the freezing rate of MA-PA is much slower than that of MA-PA/EG. During the whole freezing process, MA-PA/EG reaches stability first, which is 30min faster than MA-PA. The results demonstrate that EG serves as primary structure, absorbing MA-PA eutectic mixture into its micro pores, which maintains the energy storage and temperature control properties of PCM, saving much time.

**Heat-conducting property of MA-PA/EG composite PCM.** Thermal conduction of MA-PA/EG is also an index to evaluate its property. In order to improve it, EG is added into MA-PA. At first, suppress MA-PA and MA-PA/EG into columns (diameter:30mm, quality:9g) by tablet press. Then test the heat-conducting property of MA-PA and MA-PA/EG. The results show as Table 4 and Table 5.

Comparing the heat conductivity coefficients of MA-PA and MA-PA/EG, it is showed that before adding EG, the heat conductivity of MA-PA is 0.3030, after EG added, it rises to 6.462. Obviously, EG enhances the thermal conductivity substantially, which illustrates that adding of EG can effectively improve the heat-conducting property of PCM, it also provides a guarantee of heat storage and stability properties of MA-PA/EG composite PCM.

Table 4. Heat conductivity coefficient of MA-PA

| Group | Output power W | Testing time s | Temperature rise K | Characteristic time s | Heat conductivity coefficient |
|---|---|---|---|---|---|
| 1 | 0.03 | 160 | 0.577 | 0.622 | 0.3086 |
| 2 | | | 0.577 | 0.638 | 0.2973 |
| Average | | | | | 0.3030 |

Table 4 Heat conductivity coefficient of MA-PA/EG

| Group | Output power W | Testing time s | Temperature rise K | Characteristic time s | Heat conductivity coefficient |
|---|---|---|---|---|---|
| 1 | 0.5 | 5 | 0.674 | 0.476 | 6.609 |
| 2 | | | 0.681 | 0.479 | 6.479 |
| 3 | | | 0.673 | 0.488 | 6.298 |
| Average | | | | | 6.462 |

## CONCLUSION

(1) The MA-PA eutectic mixture that can be used in air-conditioning condense heat recovery system,

solar energy utilization and phase change wall has been obtained.

(2) The MA-PA/EG composite PCM with the best mass ratio (15:1) has been prepared. Its melting and freezing temperature are 44.42°C and 44.99°C. Its latent heat of fusion and solidification are 156.9J/g and 155.3J/g.

(3) EG has no effects on the phase transition temperature and latent heat, only serves as primary structure.

(4) MA-PA/EG composite PCM has a good thermostability and chemical stability.

(5) Adding of EG holds the temperature control property of PCM, which shorten the phase change time, also enhance the thermal conductivity.

## REFERENCES

[1] Zhang Y P, Hu H P, Kong X D. Phase change energy storage—theory and application[J]. Press of University of Science and Technology of China, Hefei, 1996.

[2] Li Haijian, Ji Zhijiang, Wang Jing, Xin Zhijun. Research status of phase change composite materials [J]. Materials Review,2008,22(Z2): 248-251

[3] LI Zhisheng, ZHANG Shuting. Research progress and application of phase change materials [J]. Building Energy Efficiency,2016,44:57-60

[4] Yuan Yanping, Zhang Nan, Tao Wenquan, Cao Xiaoling, He Yaling. Fatty acids as phase change materials: a review[J]. Renewable and Sustainable Energy Reviews, 2014, 29: 482-498.

[5] Yuan Yanping, Bai Li, Niu Ben, Theoretic predicton of phase change temperature and latent heat of fatty acids eutectic mixture [J]. Materials Review,2010,24(2): 111-113

[6] Meng Xin, Zhang Huazhi, Zhai Ziming, et al. Preparation, encapsulation and thermal properties of fatty acid /expanded graphite composites as shape-stabilized phase change materials. Chemical Journal of Chinese Universities. 2012, 33(03): 526-530

[7] Zhang Yinping, Ding Jianhong, Wang Xin, et al. Influence of additives on thermal conductivity of shape-stabilized phase change material[J]. Solar energy materials and solar cells, 2006, 90(11): 1692-1702.

[8] Jiang Yunyun, Zhang Yuzhong, Zheng Shuilin. Preparation of composite phase change materials and its application development[J]. China Non-metallic Minerals Industry. 2011(3),4-7

[9] Cai Yibing, Sun Guiyan, Liu Mengmeng, et al. Fabrication and characterization of capric–lauric–palmitic acid/electrospun $SiO_2$ nanofibers composite as form-stable phase change material for thermal energy storage/retrieval[J]. Solar Energy, 2015, 118: 87-95.

[10] Cao Lei, Tang Yaojie, Fang Guiyin. Preparation and properties of shape-stabilized phase change materials based on fatty acid eutectics and cellulose composites for thermal energy storage[J]. Energy, 2015, 80: 98-103.

[11] Wang Yi, Wang Shaoying, Wang Jianping, Yang Rui. Preparation, stability and mechanical property of shape-stabilized phase change materials[J]. Energy and Buildings, 2014, 77: 11-16.

[12] Fauzi H, Metselaar H S C, Mahlia T M I, et al. Thermal characteristic reliability of fatty acid binary mixtures as phase change materials (PCMs) for thermal energy storage applications[J]. Applied Thermal Engineering, 2015, 80: 127-131

[13] Lv Shilei, Zhu Neng, Feng Guohui. Euctectic mixtures of capric acid applied in building wallboards for heat energy storage [J]. Energy and Buildings, 2006,38(6):708-711

# ECO-ECONOMICAL NON-CONVENTIONAL METHOD FOR ELECTRICITY GENERATION

***Seema Vats***
(University of Delhi, New Delhi, INDIA)

Energy is the primary and most universal measure of all kinds of works by human beings and nature. In this world we see the expression of flow of energy in one form or the other. Energy in the form of electricity plays a very important role in our lives. The modern age has, therefore, been truly called the "age of electricity". So today our whole lifestyle is dependent on electricity but we know that the resources to generate electricity are limited. There is an urgent need to generate electricity through non –conventional methods which are not only non-polluting but also cost effective. Also there is dire need to install multiple non-conventional techniques at one point to extract maximum electrical energy for further use. In the light of above, we have presented the method of clubbing the energy generated from roller as well as from piezoelectric materials at the same installation cost. This will be very cost effective method in a way that it does not require separate arrangement for installing piezoelectric materials.The mechanical energy is being converted into an electrical energy. A wooden prototype model with PZT ceramic for the same has been developed and studied. Also we have tried to explore possibility of generation of electricity by using PVDF films (piezoelectric) as another substitute in combination with roller. The generated electricity can be used at toll gates for charging battery operated vehicles and street lights on highways.

# AUTHOR INDEX

# AUTHOR INDEX

# AUTHOR INDEX

# AUTHOR INDEX

**American Academy of Sciences**
9720 Town Park Drive, Suite 18
Houston TX 77036
USA

ISBN 978-1-5108-3972-4
9 781510 839724

CURRAN ASSOCIATES INC.
proceedings.com